PETER TREES - 2005

Sewers

Dedicated to my wife, Margaret, who suffered the gestation period but passed away before this volume was published.

G.R.

Sewers
Replacement and New Construction

Edited by

Geoffrey F. Read MSc, CEng, FICE, FIStructE, FCIWEM,
FIHT, MILE, FconsE, Fcmi, MAE

Consulting Civil and Structural Engineer and
Visiting Lecturer, Department of Civil and Structural Engineering, UMIST, UK

ELSEVIER
BUTTERWORTH
HEINEMANN

AMSTERDAM • BOSTON • HEIDELBERG • LONDON • NEW YORK • OXFORD
PARIS • SAN DIEGO • SAN FRANCISCO • SINGAPORE • SYDNEY • TOKYO

Elsevier Butterworth-Heinemann
Linacre House, Jordan Hill, Oxford OX2 8DP
200 Wheeler Road, Burlington MA 01803

First published 2004

Copyright © 2004, Elsevier Ltd. All rights reserved

Permissions may be sought directly from Elsevier's Science and Technology Rights
Department in Oxford, UK: phone: (+44) (0) 1865 843830; fax: (+44) (0) 1865 853333;
e-mail: permissions@elsevier.co.uk. You may also complete your request
on-line via the Elsevier homepage (http://www.elsevier.com), by selecting
'Customer Support' and then 'Obtaining Permissions'

British Library Cataloguing in Publication Data
Sewers: replacement and new construction
   1. Sewerage 2. Sewerage – Maintenance and repair
   I. Read, Geoffrey F.
   628.2

Library of Congress Cataloguing in Publication Data
Sewers: replacement and new construction/edited by Geoffrey F. Read.
   p. cm.
   Includes bibliographical references and index.
   ISBN 0 7506 5083 4
   1. Sewerage–Design and construction. I. Read, Geoffrey F.
   TD678,S48 2004
   628'.2–dc22   2004040939

ISBN 0 7506 5083 4

Typeset by Replika Press Pvt Ltd, India
Printed and bound in Great Britain

Contents

About the Editor   xi

Preface   xiii

1   Development of Sewerage Rehabilitation   1
Geoffrey F. Read

   1.1   Introduction   1
   1.2   Sewerage systems   1
   1.3   General principles of sewerage   5
   1.4   Public utilities   6
   1.5   Outline design   6
   1.6   Development of public health engineering   7
   1.7   Sewer rehabilitation   9
   1.8   Strategy   16
   1.9   Environmental impact of sewer collapses   17
   1.10 Cost of sewer collapses   18
   1.11 Funding   19
          Bibliography   19

2   New Construction   20
Geoffrey F. Read

   2.1   Rates of flow – foul sewage   21
   2.2   Rates of flow – surface water sewerage   23
   2.3   Storm sewage overflows   27
   2.4   Soakaways   29
   2.5   Inverted syphons   29
   2.6   Manholes   32
   2.7   Backdrop manholes   33
   2.8   Access to manholes   35
   2.9   Pipes   36
   2.10 Pumping stations and rising mains   41

3   Site Investigation   47
Ian Whyte

   3.1   Introduction   47
   3.2   Preliminary sources survey (the desk study)   56
   3.3   Main ground exploration   60
   3.4   Construction investigations: records and feedback   89
   3.5   Geotechnical reports   90
          References   92

4   Site Investigation and Mapping of Buried Assets   **95**
Nick Taylor

4.1   Desktop study of existing plans   95
4.2   Tracing of utilities on site   96
4.3   Mapping of findings   99
4.4   Site investigation – intrusive methods   99
4.5   Conclusion   99

5   Traffic Management and Public Relations   **100**
R.G. Daintree

5.1   Introduction   100
5.2   The New Road and Street Works Act of 1991   101
5.3   Consultation   102
5.4   Summary of steps in consultation and public relations process   105
5.5   Signing and statutory requirements   108
5.6   One-way working   111
5.7   Give and take system   111
5.8   Priority signs   111
5.9   Traffic control by Stop/Go boards   111
5.10 Traffic control by portable traffic signals   112
5.11 Signing and lane marking   112
5.12 Maintenance   113
5.13 Reinstatements   114
      Bibliography   115

6   Aspects of Sewer Design   **116**
Ian Vickridge

6.1   Introduction   116
6.2   Survey and scoping   116
6.3   Preliminary design   116
6.4   Hydraulic design   118
6.5   Determination of sewer size   119
6.6   Structural design   123
6.7   Future developments   130
      Bibliography   131

7   Open-cut and Heading Construction   **132**
Geoffrey F. Read

7.1   Open-cut construction   132
7.2   Pipe laying   144
7.3   Sewers near existing structures   145
7.4   Backfilling and reinstatement   145
7.5   Heading construction   147
      References   149

8   Tunnel Construction   **150**
Malcolm Chappell and Derek Parkin

8.1   Introduction   150
8.2   Geological/topographical aspects   152

8.3 Linings 166
8.4 Soft ground tunnelling 170
8.5 Ground treatment 172
8.6 Mechanism of the tunnelling process 175
8.7 Surveying/alignment 179
8.8 Temporary works 190
8.9 Future 192

9 On-line Sewer Replacement in Tunnel **193**
Geoffrey F. Read

9.1 Introduction 193
9.2 Procedure – open-cut 193
9.3 Heading 193
9.4 Segmental tunnelling 196
9.5 Pre-cast concrete tunnel linings 196
9.6 Sewer replacement tunnels 197
9.7 Access shafts 197
9.8 City centre access problems 201
9.9 Tunnel construction 203
9.10 Dealing with existing flow 206
Bibliography 209

10 Pipejacking **210**
Malcolm Chappell and Derek Parkin

10.1 Introduction 210
10.2 Technical aspects of pipejacking 212
10.3 Pipe design and manufacture 217
10.4 Surveying and alignment 219
10.5 Temporary works 221
10.6 The finished product 222

11 Management of Construction **224**
Malcolm Chappell and Derek Parkin

11.1 Introduction 224
11.2 Types of contract 224
11.3 CDM Regulations 225
11.4 Contract management plan 226
11.5 Site organisation 226
11.6 Training 227
11.7 Site support 229
11.8 Audits 234
11.9 Conclusion 235

12 Impact Moling **236**
Norman Howell

12.1 Introduction 236
12.2 The technique 238
12.3 New pipes 246
12.4 Ground conditions 247

12.5 Pipe installation   249
12.6 Applications   252
      References   253

13 The Pipe Ramming Technique   254
Norman Howell

13.1 Introduction   254
13.2 The type of pipes to be installed   254
13.3 Equipment   258
13.4 Site preparation works   262
13.5 Pipe installation   265
13.6 Applications   269

14 The Pipebursting Technique   272
Norman Howell

14.1 Introduction   272
14.2 Principle of the system   273
14.3 Scope of the system   273
14.4 Replacement capabilities   273
14.5 Ground conditions   274
14.6 Ground movement   276
14.7 Replacement pipe options   279
14.8 Lateral and service connections   281
14.9 Pipebursting system and equipment options   281
14.10 Site operational requirements   292
14.11 Application of the system   295

15 Horizontal Directional Drilling   297
Brian Syms

15.1 Introduction   297
15.2 Directional drilling means what?   300
15.3 Equipment and requirements   301
15.4 Planning a directional drilling project   305
15.5 Guidance systems   309
15.6 Drainage applications   311
15.7 Advantages and disadvantages   313
15.8 Case studies   320
15.9 Future development   320
      Acknowledgements   321
      Bibliography   321
      Appendix   322

16 Vacuum Sewerage   327
Geoffrey F. Read

16.1 Introduction   327
16.2 Basic considerations   328
16.3 System operation   328
16.4 Interface value monitoring system   332
16.5 Design details   332

16.6	Piping design	332
16.7	Station design	333
16.8	Recent vacuum sewerage projects in the UK	336
16.9	General observations	338

17 Social or Indirect Costs of Public Utility Works 339
Geoffrey F. Read and Ian Vickridge

17.1	Introduction	339
17.2	Social and environmental effects of public utility works	341
17.3	The environmental impact of public utility works	345
17.4	Environmental impact assessment	346
17.5	Legislative mechanisms to control social costs	347
17.6	Developing the concept of road space rental to minimise social costs	349
17.7	The New Road and Street Works Act 1991	350
17.8	Preliminary conclusions	351
17.9	The determination of road space rental charges	354
17.10	Problems of assigning monetary values to noise, vibration and air pollution	356
17.11	Overall effect of road space charging on the economy	356
17.12	Working shaft location	357
17.13	The future	357
	Bibliography	358
	Appendices	359

18 Factors Affecting Choice of Technique 367
Geoffrey F. Read and Ian Vickridge

18.1	Introduction	367
18.2	Technical feasibility	367
18.3	Cost	371
18.4	Time	374
18.5	Safety	374
18.6	Environmental impact	375

19 Civil Engineering Contract Management 378
Geoffrey F. Read and Geoffrey S. Williams

19.1	Introduction	378
19.2	Claims under the main contract	381
19.3	Subcontract claims	413
19.4	Claims by the contractor against the subcontractor	424
	Bibliography	429

20 Project Management 430
Brian Syms

20.1	Introduction	430
20.2	Project evaluation	432
20.3	Feasibility stage	442
20.4	The design process	449
20.5	Risk	451
20.6	The Construction (Design and Management) Regulations 1994	455
20.7	Procurement	462

20.8   Construction and site management   466
20.9   Other contract types   468
20.10 Summary   474
       Acknowledgements   476
       Bibliography   477

21 Sewer Safety   **478**
M.J. Ridings

21.1   Introduction   478
21.2   Current legislation   478
21.3   Types of confined spaces   482
21.4   Hazard identification   482
21.5   Personal protective equipment   483
21.6   Gases and gas detection   485
21.7   Breathing apparatus   493
21.8   Ventilation   497
21.9   Safe working procedures   498
21.10 Rescue   504
21.11 Hygiene   505
21.12 Summary   506
       Bibliography   506

22 Contract Supervision   **507**
Barry H. Lewis

22.1   Introduction   507
22.2   Contract responsibilities   508
22.3   Programming the work   516
22.4   Service diversions and accommodation works   519
22.5   Possession of the site   520
22.6   Site supervision   522
22.7   Measurement and payment   538
22.8   Completion   543
22.9   Contract strategy   544
       Bibliography   550

Index   **551**

About the Editor

Geoffrey F. Read, MSc, CEng, FICE, FIStructE, FCIWEM, FIHT, MILE, FconsE Fcmi, MAE

Geoffrey F. Read is a Consulting Civil and Structural Engineer, a Member of the Academy of Experts and a Director of Deakin Walton Limited, Consulting Civil and Structural Engineers of Sale, Cheshire, where he has particular responsibility for sewerage/drainage and flooding as well as the provision of expert witness services generally. In addition he has his own consulting engineers' practice in Alderley Edge, Cheshire.

He held a Research Fellowship for three years at the University of Manchester, Institute of Science and Technology, where he was responsible for researching the social and environmental costs associated with the various sewerage rehabilitation techniques. He was then retained as a Visiting Lecturer dealing with sewerage and environmental impact assessment – an appointment he still holds.

Geoffrey has had many years' design and construction experience including over 12 years as City Engineer, Manchester, and Engineer to Manchester Airport. It was during this period when the city centre suffered a considerable number of major sewer collapses. Not surprisingly he became widely known as one of the leading campaigners for the rehabilitation of sewerage infrastructure. He is known worldwide as a recognised authority on sewerage rehabilitation and on the social and indirect costs of public utility works in highways. He was for a number of years Advisor on Highways and Sewerage to the Association of Metropolitan Authorities, a past Chairman of the City Engineers' Group and Founder President of the Association of Metropolitan Engineers.

He previously held appointments with six other local authorities prior to the Manchester appointment and was also Engineer to the Bolton and District Joint Sewerage Board for some 14 years until it became the responsibility of North West Water.

He had responsibility for directing among other capital works a large programme of the rehabilitation of Manchester's Victorian brick and pipe sewers involving renovation and on- and off-line replacement.

To enable the general public to readily appreciate the void size following a sewer collapse, the Editor was, while holding the appointment in Manchester, responsible for introducing a comparison factor which has been adopted worldwide, namely, the *DDB factor*. That is to say the number of double decker buses which could be contained within the particular cavity.

Author of a variety of technical papers including several research papers on the social costs of public utility works Geoffrey also contributed to the Sewer and Culvert Sections of *The Maintenance of Brick and Stone Masonry Structures* for E. & F. N. Spon.

Geoffrey Read's experience covers a broad spectrum of civil and structural engineering but latterly has tended to be concentrated on sewerage. He directed a social-economic cost/benefit study for a major flood alleviation and sewerage improvement scheme (£75m) in a large industrial city as well as developing and implementing a strategy for rehabilitation of the effluent drainage system in a major industrial complex.

He regularly provides an expert witness service for solicitors and insurers over a full range of professional services.

Preface

For how unseemly it is when you are speaking about sewers to use high-sounding expressions.

Cicero, Orator, XXI

Volume I in this series concentrated on the repair and renovation of sewers and particularly renovation techniques related closely to the recommendations contained in the Water Research Council's *Sewerage Rehabilitation Manual* (SRM) and the wealth of information it contains.

There are many instances where renovation is not a practical option and replacement of the sewer becomes necessary after appropriate engineering investigation. This volume deals with the aspect of the subject and the closely related technique of completely new construction. The SRM has provided a great deal of detail for the design of renovation schemes and is an excellent common sense treatise on the subject but does not yet address the aspect of replacement. It was intended to provide a framework for decision making while still allowing the necessary flexibility for drainage engineers to apply their experience and skill to reach the most appropriate cost-effective solution. One might hope in due course that the SRM will be extended to cover conventional and new methods of reconstruction which may well materialise in the next few years but in the interim every endeavour is made in this volume to review current techniques.

As indicated in Volume I, having planned the two volumes in the series and formulated the chapter contents so as to provide a logical pattern without significant overlap, the next step was to decide which chapters I would write and which chapters would be allocated to authors who had specialised in particular aspects of the work. This would then give the reader a wide base of technical expertise and opinion. To this end I am indebted to former local authority colleagues along with colleagues in consultancy and those now working for water companies and in the contractual and academic fields. This approach has in practice lengthened the preparation period but hopefully will have resulted in overall benefit to the reader.

It has always been my aim that the two books would be of interest to middle-level technical staff in water companies and consultancies as well as for university undergraduates to assist them in bridging the gulf between theory and practice.

Volume I has proved popular in both the UK and overseas – I trust Volume II will be similarly received.

The various techniques described are not intended in any way to be exhaustive but represent perhaps the most popular current solutions. It should always be remembered that there is not a standard solution to all sewer problems. It is – as I have frequently expressed – still very much a case of 'horses for courses'.

I greatly appreciate the valuable help and assistance provided by my friends and colleagues towards the production of this volume and hope that it receives widespread acceptance as Volume 1 has done.

It is too much to hope that I have managed to eliminate all errors from the text and diagrams but I hope these will be minimal.

Geoffrey F. Read

1

Development of Sewerage Rehabilitation

Geoffrey F. Read MSc, CEng, FICE, FIStructE, FCIWEM, FIHT, MILE, FconsE, Fcmi, MAE

1.1 Introduction

Sewerage is usually defined as a network or system of sewers and associated works designed for the collection of waste water or foul sewage, conveying it via pipes, conduits and ancillary works, discharging it at a treatment works or other place of disposal prior to returning it to the environment in appropriate condition.

Discharge to natural water courses of foul sewage or of any sewage containing putrescible matter cannot be permitted unless either the quantity is so small that dilution prevents nuisance or the sewage is treated in such a manner that the 'purified' effluent is not offensive or deleterious when so discharged. The quantity of water in rivers in Great Britain is generally not sufficiently high to give the degree of dilution (i.e. not less than 500 times the dry weather flow) that is considered necessary for the permissible discharge of crude sewage and therefore the foul sewage from all British inland towns and many coastal towns has to be treated at a disposal works.

As well as domestic waste water the network takes the used water of business and industry and accommodates part or all of the surface water, namely, the flow of storm water from roofs and paved surfaces of all types.

1.2 Sewerage systems

Towns are sewered according to three basic systems:

1. The *combined* system in which the foul and surface water sewage are discharged into one system of sewers which lead to the sewage treatment works or point of outfall, this being the format existing in the older towns and cities.
2. The *separate* system in which the foul sewage is carried by an individual system of sewers to the sewage treatment works or point of outfall while surface water is carried away by a number of local systems each discharging at various points into natural watercourses or discharging into soakaways – the system generally being found in the newer towns and cities.
3. The *partially separate* system in which the greatest part of the surface water is dealt with by the surface water sewers while the surface water from the backyards and back part of roofs is passed to the foul sewers.

All these systems have their specific uses. The separate system is considered the best for most purposes, whereas combined drainage is desirable where the runoff from road or other hard surfaces is so foul that it requires some form of treatment before passing to a water course. The partially separate system is sometimes applied where the local water consumption is not sufficiently high to keep the foul sewers clean.

Because of the time required to design, finance and construct such facilities the engineer is obliged to estimate future population and industrial growth. The type of industry present or anticipated is particularly important owing to the wide range of water required by various industries – up to 2 27 000 litres of water may be required to produce a ton of steel and up to 4 55 000 litres to produce a ton of paper.

To maintain the health and general well-being of the community the efficient disposal of all types of refuse including sewage is clearly of paramount importance. In Volume I we examined in some detail the earlier forms of sewerage and sewage disposal and traced the development of public health engineering from early times to the present day. Works of sanitation are known to have existed in ancient times and historians have shown that drainage systems were used in Roman cities over 2000 years ago and by other civilisations at a much earlier period in history.

Civilisation has been described as the art of living in towns. Some people, even as long as 5000 years ago, were skilled in that art – others as recently as 500 years ago neglected it and suffered the consequences.

The first civilisations grew in Egypt, Crete, Iraq, northern India and China and they all learned at some stage to provide good water supplies and some sort of drainage system. The need to have convenient access to water has always been one of the most powerful influences on human life and settlement patterns. Everyone needs water for personal health and hygiene – indeed for life itself. As water circulates in the hydrological cycle (Fig. 1.1) with its own momentum, everyone has, in principle, the same equitable right to share in its abundance and its scarcity. The development of the ability to carry water enabled early

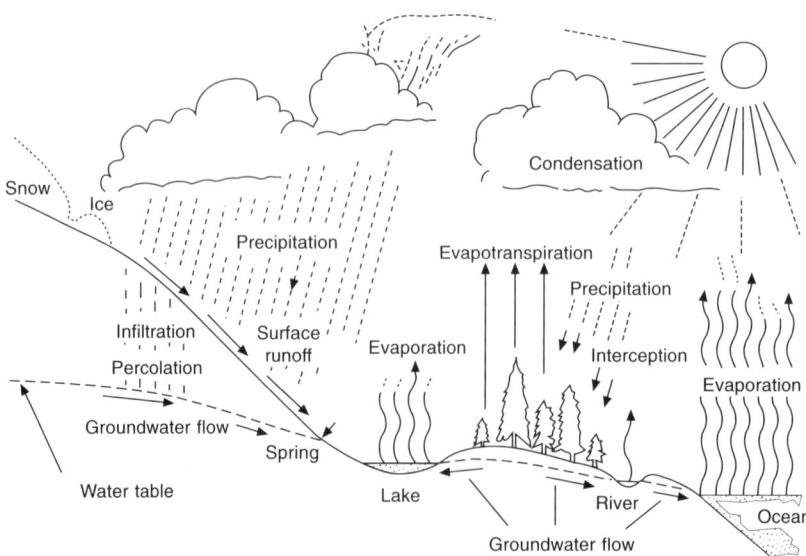

Figure 1.1   The hydrological cycle (Reproduced with permission of Longman Group (Phillips/Turton, *Environment, Man and Economic Change*)).

people to extend their hunting range and gave them the possibility of greater mobility generally, but for permanent settlement and the cultivation of crops, etc., ready access to water or a frequent and reliable rainfall is essential.

Sewers were commonplace in the Indus valley (now western Pakistan) around 2500 BC. In Mohenjo-Daro, one of the largest towns of that early civilisation, every house had a latrine, many had bathrooms and there was a large public bath.

The location of settlements was usually conditioned by the availability of a water supply. However, as a community grew it would need to supplement its water supply and for public health to organise the safe and efficient disposal of its waste products – especially if occurring in congested or confined conditions. The Greeks instinctively equated hygiene with health and, like some other ancient civilisations, had very clear ideas on how to achieve it (Figs 1.2 and 1.3). On the island of Crete, for instance, people knew how to build drainage systems long before the time we think of as 'ancient Greece' and the Minoan palace of Knossos even had clay drainage pipes taking away human waste – the pipes being apparently tapered to increase the velocity and provide a self-cleansing flow – all some 3600 years ago.

Years later, water supplies and sewers were common at the time of the Roman civilisation – the famous sewer of ancient Rome, the Cloaca Maxima ('Cloacina' was the goddess of sewers), was built about 588 BC to drain the valleys between the Esquiline, the Viminal and the Quirinal to prevent disease and to carry away the surplus water to the River Tiber. It was originally a natural water course but had to be artificially diverted where building made this necessary and in the sixth century BC the Romans also arched over – a similar treatment to that given to the water courses in the centres of our industrial towns during the early nineteenth century. The Romans' fondness for baths and fountains is well known and they certainly appreciated the importance of a proper sewerage system, epitomised by the shrine to Venus Cloacina, although the River Tiber was ill-treated – nearly all the city's waste water and sewage were eventually allowed to flow into it including the flow from the Cloacina Maxima.

Roman engineers introduced their techniques throughout their vast empire; even today, at such ancient bathhouses as that at Bath, may be seen the lead pipes for bringing water in and the drains to take away the used water.

Notwithstanding the efforts of the Romans during their occupation of this country there was really no subsequent effort to provide proper sanitation – as we know it today – in the United Kingdom and indeed in other modern countries until the early nineteenth century.

In the garrison towns and frontier outposts of northwest Europe the disposal of human waste was largely in keeping with the 23rd chapter of Deuteronomy which described withdrawal outside the camp. But in the rapidly growing cities of the high Middle Ages, the first attempts at organised waste removal were appearing. Cesspools were introduced in place of the privy but throughout the Renaissance the general practice was to dump all waste in the town's gutters hopefully to be flushed through the primitive 'surface water drains' next time it rained.

Not surprisingly, these crude sanitary arrangements contributed to the spread of disease. John Snow, a nineteenth century English physician, compiled a list of outbreaks of cholera, which he believed had moved westwards from India reaching London and Paris in 1849. He traced a London occurrence to a public well, known as the Broad Street Pump, in Golden Square, which he determined was being contaminated by nearby privy vaults. This was a noteworthy achievement, especially since it predated by several years the discovery of the role of bacteria in disease transmissions.

(a)

(b)

Figure 1.2   (a) Remains of second/third century AD toilets at KOS. (b) KOS Culvert intersection –
part of drainage from Thermopylae Roman baths between second century BC and first century AD.

Figure 1.3 Clay water pipes c. first century BC at KOS. Note: Lime deposits inside pipes and moulded pipe joints,

The sewerage systems as we now understand them have developed from experience gained virtually only over the last 200 years – sewerage has in fact only really developed since the advent of a piped water supply and the consequential need for an effective means of removal of the greater quantity of waste water.

Modern foul water sewerage is descended from the permitted discharge of domestic waste into the underground sewers laid for the purpose of draining rainwater from the streets, etc. How this came about is illustrated in the report of the London County Council, in which it is mentioned that:

A few years before the passing of this Act [*i.e. the Michael Angelo Taylor's Act, 1817*], an invention had been introduced which was to have a very important effect on sanitation. This was the water closet, which offered facilities, before unknown, for the entire removal of sewage. At first the watercloset was made to discharge not into the sewers, but into a cesspool, the ancient receptacle for offensive household refuse, the contents of which were removed from time to time. The large addition thus caused to the contents of the cesspools, however, made it necessary to introduce overflow drains running from them into the street sewers; and other reasons also gave inducement to discharge the sewage, with the aid of a sufficient water supply, direct from the water closet to the street sewers. Originally, the discharge of offensive matter into the sewers was a penal offence, and so continued up to about the year 1815. Afterwards it became permissive to drain houses into sewers, and in 1947 it was made compulsory.

These sewers were originally banked-up open watercourses, intended solely for the purpose of carrying off the surface drainage.

1.3   General principles of sewerage

Modern sewers, although they are usually constructed underground, are designed for the purpose of hydraulics, as if they were open channels. Although they are usually round pipes or culverts, the crown of the pipe or the arch of the culvert may be considered as being there only to prevent the ground above from falling in. Except in the comparatively rare conditions of surcharge, the crown of the pipes is not designed with the intention of confining the flow of sewage, this being done by the invert and sides.

Sewers are considered in practice to be discharging the maximum capacity for which they are designed when flowing just full – usually for the greater part of the time they are flowing only partly full.

1.4   Public utilities

A sewerage system is one of a number of vital public utilities upon which the modern community is so dependent. The exact value to the public of any particular service is extremely difficult to ascertain. It is not possible to say precisely the degree of expenditure that can be justified in overcoming nuisance and inconvenience – expenditure is largely regulated by practice, current opinion and government policy backed up more recently by environmental considerations generally originating from policies determined by the EC.

Unfortunately as so much of the infrastructure is not visible to the general public a lack of appreciation of its importance is commonplace until such time as it requires attention – the latter varying from unplanned emergency works through to planned maintenance and renewal; all of which have an impact on the day-to-day life of the community which normally reacts at any early stage notwithstanding the longer-term benefits.

1.5   Outline design

Foul sewerage design is based on the number and density of buildings, the number of families per building, the size of the family and the varying habits of the population in regard to the use of water.

Sewers should be laid at gradients that will produce velocities sufficient to prevent permanent deposit of solids. A velocity of 0.75 m/s occurring sufficiently frequently is usually enough to sustain self-cleansing conditions and avoid long-term deposition of solids.

For the adequate design of surface water sewers, the engineer must have knowledge of the topography, together with details of the intensities of rainfall of the particular district. There is clearly an economic level to the intensity of storm that can reasonably be catered for and in practice such sewers are usually designed for the worst storms likely to occur every year or two years.

In view of the size of the flows involved, a surface water sewerage system is usually designed if possible as a gravity system, pumping being kept to a minimum. For a foul sewerage scheme, it is often necessary to compare the alternative engineering considerations and economies of gravity sewers and a pumping scheme. Nevertheless in a flat district pumping or a vacuum system may be the only practical solution. The structural design of buried pipelines was developed in America in the early twentieth century and has been in use in the UK for many years. The importance of structural design became even more apparent with the development of flexible joints for all types of pipe and with the increasing use of plastics.

1.6   Development of public health engineering

As mentioned earlier, in Volume I we traced in some detail the development of public health engineering from as long as 5000 years ago through to the present day.

It was emphasised that public health in Britain and in fact elsewhere in Europe did not automatically improve with the passage of time. In the Roman period, for instance, going to the toilet was usually a clean, comfortable and relaxing experience. Hundreds of years later, in the Middle Ages, toilets were often quite dangerous and disgusting places to enter.

The Industrial Revolution was really the turning point in the development of public health in Britain. It was within this period of rapid change and invention that civilisation in this country entered a new era and a multiplicity of factories and industrial plants rapidly developed in the townships, especially those having abundant supplies of river water, coal and other raw materials such as existed in Manchester.

The population rapidly began to leave the countryside and the agricultural way of life, attracted by the relatively higher wages offered by the new industries and towns began to develop at an alarming rate as they entered the nineteenth century.

Previously the country had been primarily agricultural in character but with the massive population movement to the quickly developing towns and cities, the Victorians had to take some rapid action to combat the health problems that quickly developed. A scenario of masses of people living immediately adjacent to the factories where they worked using water and producing waste of all types meant both a separate supply of potable water and means of disposing of it when used before it became a health hazard was of paramount importance.

Water is often described as the foundation of life but in an industrialised community it quickly becomes a health hazard in itself unless properly treated.

Until the middle of the nineteenth century most British towns obtained their water from convenient wells, springs and rivers, which quickly became polluted from the overloaded piecemeal surface water drainage system which in the main only collected filthy water from the rapidly developing areas to discharge it into the nearest water course where it soon again became drinking water! At that time the link between polluted water and disease did not appear to have been taken seriously. The early mains did not supply water constantly. As the factories grew there was an ever-increasing demand for water power as a result of the installation of water wheels and the damming of water courses to provide the necessary operating head for machinery, in addition to the needs for processing water. This retention of flow and the pollution resulting from the many industrial processes reduced the supply for domestic purposes although we should always remember that the Victorians did not wash or bath to any extent. Nevertheless as a result of severe health problems – cholera killed many thousands of people in the first epidemic of 1831–2 – they were forced to quickly come to terms with the situation – and Britain's early combined sewers were born from a necessity to eliminate the breeding grounds of urban disease. For example, Manchester – the birthplace of the Industrial Revolution – saw its population increase from 5000 in 1750 to 1 42 000 in 1842 and by 1867 it had reached 3 63 000 and some 7 00 000 by the end of the Victorian period.

In fact Manchester experienced the most rapid growth of any city in this country and is described in *Manchester through the Ages* by David Rhodes. By 1840, Manchester had been transformed from an eighteenth century provincial market town to the foremost manufacturing and technological centre in England. It became the centre of the manufacturing industry for cotton, in addition to machinery of all types, machine tools as well as coal mining, bridge building, construction of railway locomotives, gas works, chemical plant, etc. The public health problems of town housing, already bad, soon become acute and the results were justifiably described as both *foul* and *fatal*. The life expectancy of mechanics, labourers and their families was only some 17 years in Manchester compared with a figure of 38 years in a rural community such as Rutland.

The speed of this massive population increase and the buildings that went with it resulted in many kilometres of sewer and water main being constructed over a relatively short period with the result that although much of it is still in use today parts of it have reached the end of their working life over a relatively short period. A similar pattern emerges in regard to the massive networks of private culverts. During the Industrial Revolution, extensive culverting of minor rivers and streams, often spanned by early buildings, became commonplace, and due to a lack of maintenance work their condition today is generally much worse than the adjacent public sewerage networks constructed at about the same time. The situation is reflected over many of our northern industrial towns and cities which grew up during the Industrial Revolution. Against this background one can appreciate why so many sewer collapses have materialised over a relatively short period. Where towns and cities have developed more gradually as in the south, the infrastructure has followed a similar pattern and is also likely to wear out over a longer period although other factors may come into play.

In Volume I we researched the development of the national sewerage networks using Manchester – which possesses the oldest extensive sewerage network in the country – as a typical example. It was reported that initially the two main forums of construction were the 'U'-shaped brick sewer with stone flag tops and the egg-shaped butt-jointed clayware pipe sewers – both still operational under the heart of Manchester today.

Clayware pipes had begun to appear again in this country about the middle of the nineteenth century – such pipes having been known to the Romans but the technique apparently disappeared with their civilisation. From about 1850 the standard Manchester sewer was constructed using butt-jointed ovoid clayware pipes which had been introduced by John Francis, the then Surveyor to Manchester's Paving and Soughing Committee ('sough' – an underground channel) and they were extensively used until about 1880. Apparently, the brick 'U'-shaped sewer type was still favoured rather than clayware pipes when sewers of a larger capacity were required. Generally construction was by tunnelling – round shafts (blind eyes) usually 900 mm internal diameter of double skin brickwork – were sunk 10.8 metres apart and the ground between then excavated to the exact form of the invert without the use of timber. The cost of this particular pipe was claimed to be less than the equivalent brick sewer and construction time was halved. Its additional advantage lay in its shape and the resultant hydraulic benefits obtained. Its main disadvantage, although apparently this was not realised at the time, was the fact that each section of pipe was butt jointed to the next, clay puddle being used to seal it in some instances but more commonly left open so that the sewers could also serve – it was hoped – as land drains. This type of construction with its 'joint' failure over years of continuous use is one contributory factor to today's legacy of sewer dereliction in Manchester and elsewhere.

Circular clayware pipes were introduced around 1880 and represented a considerable improvement but although these were spigot and socketed, the sockets were initially only manufactured over the lower half of the pipe. It was again apparently intended that the sewers would also act as land drains.

The dual function which these early sewers were apparently intended to provide is a recurring feature of all the early sewer types including brick sewers and suggests that the brickwork was actually laid dry so that any ground water would infiltrate.

The use of clayware pipes continued to increase in the latter part of the nineteenth century with brick construction predominating for the larger conduits. This latter form of labour intensive construction continued into the early part of the twentieth century until the introduction of concrete pipes and in due course pre-cast concrete bolted segments, although for many years the latter technique involved an inner lining of engineering brickwork in cement mortar – a practice which has now virtually ceased in view of the concrete quality now obtained.

1.7 Sewer rehabilitation

It is obvious that all forms of engineering construction deteriorate with time and require consequent maintenance and repair. Without doubt, rehabilitation in some form or other has taken place since the first appearance of sewers, namely, in Roman and Greek times or even earlier, utilising the particular techniques then available.

Life cycle management in relation to sewerage networks is of paramount importance – whether they are the responsibility of the water companies or private/industrial commercial sewers – if the maximum cost effective life span is to be obtained. Sewers are hidden from public view and consequently there has always been a tendency to neglect them – out of sight, out of mind!

If repairs and renovations are not carried out at the proper time the useful service life is shortened with the risk of a dangerous collapse situation developing, as shown in Figure 1.4.

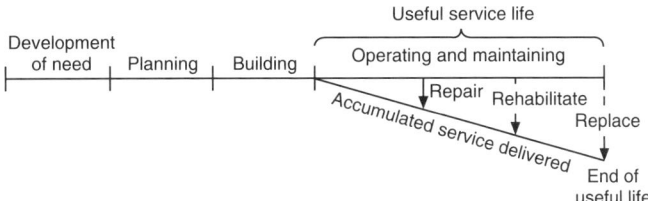

Figure 1.4

In sewerage generally, the actual conduit represents some 20% of the initial cost depending on the size of the pipe – excavation, temporary support, backfilling and reinstatement (both temporary and permanent) account for the remainder – so that the largest part of the infrastructure asset is 'the existing hole in the ground'. Figure 1.5, which comes from the former Institution of Public Health Engineers' Report on 'Trenchless Pipelaying', illustrates graphically the relative costs of the individual components involved in constructing a sewer.

If, therefore, a new pipe or the equivalent can be inserted inside the old sewer or it can be renovated without further excavations, theoretically a saving of up to 80% of the traditional cost can be made in addition to the savings on social or indirect costs which might have arisen.

Consequently in recent years sewer renovation has been increasingly advocated as a means of coming to terms with the immense problem of dealing with an often outdated

Table 1.1 Implied life of critical sewers. The table below sets out the implied life of critical sewers taking OFWAT's published information. It uses figures previously published in the CROSS Newsletter, and develops initial calculations by Edward Naylor. The figures for each sewerage company have been sent to that company's chief executive. There is not space to include their replies, but we will include some of their comments in a future edition. CROSS supporters may wish to use these figures in any activities they may undertake to argue for greater investment by the undertakers in their sewers.

Company	Total length of critical sewers*, km	Critical sewers renovated 90–6**, km	Critical sewers renovated 96–7**, km	Critical sewers renovated 97–8**, km	Critical sewers replaced 90–6**, km	Critical sewers replaced 96–7**, km	Critical sewers replaced 97–8**, km	Total critical sewers renovated or replaced 90–5, km	Total critical sewers renovated or replaced 96–8, km	Avge p/a sewers renovated/replaced 90–8, km	Avge p/a sewers renovated/replaced 96–8, km	Implied asset life on 8 year view – years	Implied asset life on 2 year view – years
Anglia	8191	27	7	29	34	2	16	61	54	14.375	27	569.8	303.37
Dwr Cymru	4321	20	12	8	64	16	8	84	44	16	22	270.06	196.41
North West	10 674	108	19	23	101	19	24	209	84	36.625	42	291.44	253.1
Northumbrian	5982	118	19	49	16	1	1	134	70	25.5	35	234.59	170.9
Severn Trent	7471	84	11	5	234	31	21	318	68	48.25	34	154.84	219.7
South West	1815	31	1	1	16	1	0	47	4	6.375	2	284.71	907.5
Southern	6460	19	2	3	15	1	0	34	6	5	3	1292	2153.3
Thames	18 936	137	58	35	67	23	14	204	130	41.75	65	453.56	291.32
Wessex	2841	51	12	6	16	1	1	67	20	10.875	10	261.24	284.1
Yorkshire	6846	19	4	17	2	9	9	21	39	7.5	19.5	912.8	351.07
										av. p/a per co	av. p/a per co	overall av.	overall av.
Total/all	73 537	614	145	176	565	104	94	1179	519	21.225	25.95	472.504	513.086
								av.p/a total 235.8 km pa	av.p/a total 259.5 km pa	av.p/a total 212.25 km pa	av.p/a total 259.5 km pa	using av. p/a total ren/reps 346.46 years	283.38 years

* reported in 1999; source OFWAT W2000 JR-T-16 included in private correspondence

** Source: Reports on the financial performance and capital investment of water companies in England and Wales, 1995–96, 1996–97 and 1997–98, OFWAT, Birmingham

Note 1: Figures for 1998/9 were published after this table was first devised, but figures show total critical sewers renovated at 184 km and 76 km replaced. The average total renovation/replacement per company was 26 km, although for some companies the figures were 0.

Note 2: In 1997, 8% of critical sewers were categorised as Grade 4 – some brick loss/badly made connections/moderate loss of level and 2% were categorised as Grade 5 – collapsed/severely deformed/missing inverts/extensive areas of missing fabric or bricks, i.e. 7350 km (4560 miles) in the two worst categories. For non-critical sewers, 9% of the total length of 2 32 264 km were categorised as in Grades 4 and 5, that is 20 900 km (12 980 miles).

Table 1.2   Implied asset life of critical sewers based on rate of renovation and replacement over the past 11 years. The following table includes data from the report for 2000–1 and shows that two water companies have now joined the '1000 year club' and now expect their critical sewers to last over a thousand years at the present rate of renewal

Company	Critical sewers – total length (km)	Total length replaced/renovated in 00/01 (km)	Total length renovated/replaced 90–01 (11 years) (km)	Implied asset life based on activity since 1990
Anglia	8191	9	150	601
Dwr Cymru	4321	2	158	301
United Utilities*	10 674	106	481	244
Northumbrian	5982	6	306	215
Severn Trent	7471	2	451	182
South West	1815	1	51	391
Southern	6460	1	43	1653
Thames	18 936	11	479	435
Wessex	2841	20	119	263
Yorkshire	6846	3	75	1004
Total	73 537	161	2313	350

*Formerly North West Water
Source: OFWAT: Financial performance and expenditure of the water companies in England and Wales.

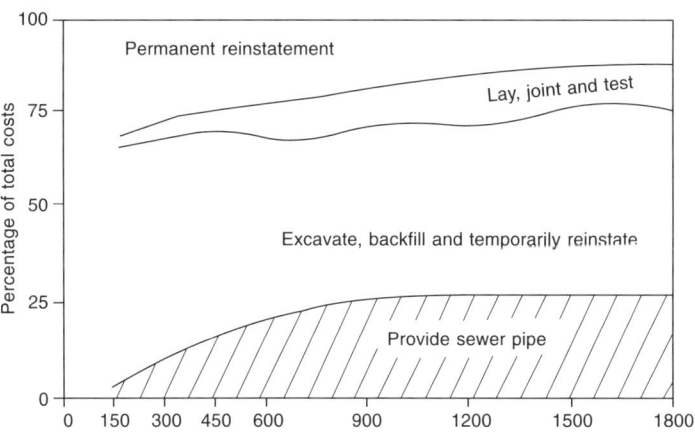

Figure 1.5   Costs of provision of pipe as a proportion of total cost.

sewerage network now faced by the water industry generally. This principle of maximising the potential of the existing 'hole in the ground' is further endorsed in Water Research Council's (WRC) *Sewerage Rehabilitation Manual*. A main factor favouring renovation of a sewer against its complete replacement is reduced initial cost thereby permitting more widespread use of limited financial resources. Sewer renovation should also considerably reduce the surface disruption (particularly if the alternative is sewer replacement in open cut) and hence reduce the 'social' costs associated with the work. While these latter costs can be considerable, representing perhaps three or even ten times the capital value of the scheme, other factors need to be fully assessed when considering renovation work. It should be remembered that renovation is only one of the options to be considered in any sewerage improvement programme. Nevertheless it should always be appreciated that renovation in itself will still involve considerable resources in time, manpower and funding. The potentially hazardous nature of some renovation works and the long-term implications of renovated sewers should never be overlooked. Notwithstanding the saving in social costs the author is conscious of the need to fully assess other long-term factors when considering the option of renovation in comparison with replacement. For instance, the difficulty in obtaining the design conditions for an old dilapidated brick sewer, probably with external cavitations, so as to ensure the designed lifespan materialises is a significant factor. A synopsis is given of the problems in Volume I, which have arisen during actual renovation schemes, leading to a prognosis of the way in which some of the problems may be overcome.

Rehabilitation has in fact been defined in the WRC's *Sewerage Rehabilitation Manual* as covering all aspects of upgrading the performance of existing sewerage systems and includes repair, renovation and renewal. Renovation is not a panacea merely an important option worthy of consideration at the outset – in fact, when the manual was launched the WRc suggested that it was not a case of renovation or renewal but rather 'has renovation been considered as an option?' The manual has provided a great deal of detail for the design of renovation schemes and is an excellent common sense treatise on the subject for which WRc are to be congratulated – although it does not offer any fundamentally dramatic new concepts it was intended to provide a framework for decision making which is consistent and technically justifiable, while still allowing the necessary flexibility for civil engineers in this field to apply personal judgement in particular situations.

The main theme of the new approach – which has been described as the new age in sewer technology – is to help engineers to maintain 'the hole in the ground' so as to prevent costly excavations and often high social costs, to optimise hydraulic performance and maximise the use of renovation.

At this stage one might hope that perhaps Volume IV of the manual might follow in due course so as to include coverage of conventional and new methods of reconstruction that may well materialise during the next few years. Rather than a scheme being 'problem orientated' the new strategy is a review of the complete drainage area followed by a fundamental reassessment of the network as a whole from which it is assumed the problems will emerge.

Overall the policy suggested concentrates pre-emptive rehabilitation on the critical sewers which generally make up about 20% – although in larger UK towns and cities the figure is much higher of the national network leaving the problems in the others to be dealt with via reactive response. Of the 20% it is hoped that 5% will enjoy a failure-free situation with the likelihood of failure in the remaining 15% being significantly reduced.

Sewer renovation has to date appeared in different parts of the world for various but completely different reasons. In North America, for instance, the prime justification has been to control infiltration while in hotter Middle East countries the need has been related to serious corrosion difficulties in relation to concrete pipes. In the UK structural deterioration, flooding and pollution are the main problems.

In comparison the main advantage of on-line sewer replacement methods are that reliance on the uncertain structural properties of existing sewers is not required and opportunities may be taken during the planning of reconstruction work to rationalise the system and generally to improve access to the network as a whole.

Cost effectiveness remains the primary design criterion and if a renovation scheme, taking into account all the safety and quality control requirements cannot be shown to be cost effective, the *WRc Manual* strategy would recommend discarding the renovation options. If the engineer decides that for a variety of possible reasons the old sewer is not in fact suitable for renovation then *replacement* should be considered. This could be replacement on-line which, although having to deal with the existing flow during the works, has an advantage so far as the connections of laterals is concerned. There may nevertheless be a case for off-line replacement perhaps retaining the old sewer in a minor capacity of making it no longer part of the primary network and thus carrying less flow.

Although it is clearly possible to replace both on- and off-line by construction in open cut this is no longer normally acceptable in more densely trafficked urban areas and construction operations normally proceed via the use of trenchless technology.

If the engineer is able to prove that the social and environmental costs which are likely to occur if open cut is used added to the construction costs of the smaller conduit, are likely to be less than the equivalent costs of the alternative larger size tunnel construction, then in such an unlikely event work would proceed via open cut.

The most common method where complete on-line replacement is necessary is via the traditional 1520 mm diameter precast reinforced concrete (usually shield driven), bolted sequential construction generally with smooth invert on the line of the existing sewer (Fig. 1.6, 1.7). Although generally oversized from a hydraulic consideration, tunnels of this size permit the existing structure to be encompassed within the face excavation and enable connections to be readily located and satisfactorily made without excavation from the surface.

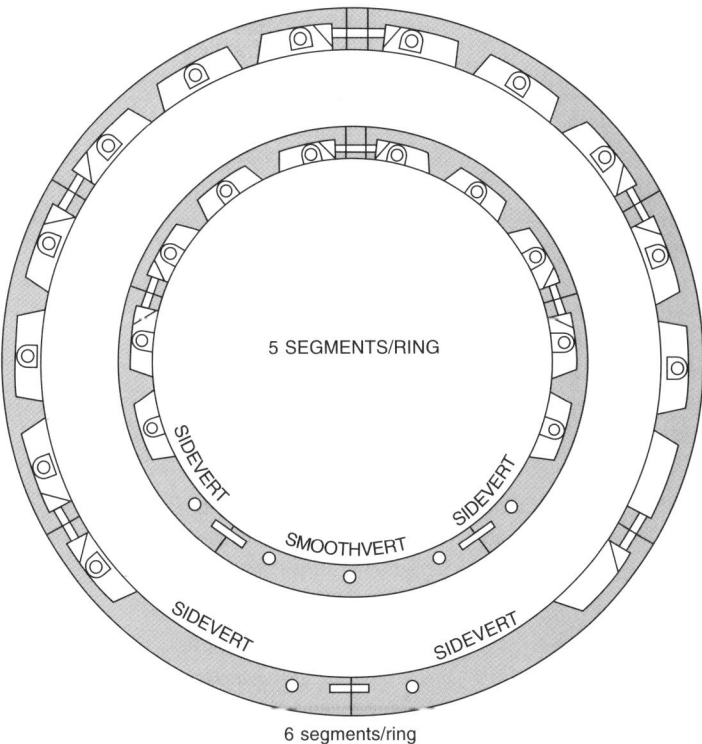

Figure 1.6   Typical 5 and 6 segment bolted tunnel lining units with smooth invert.

Figure 1.7   Existing brick sewer being replaced by bolted segmental tunnel.

Where the existing sewer wanders from the tunnel face (which is not usual – Fig. 1.8), it is a relatively easy matter to remove side segments and head out to pick up any isolated connections. On completion, full man access is possible for future maintenance, which

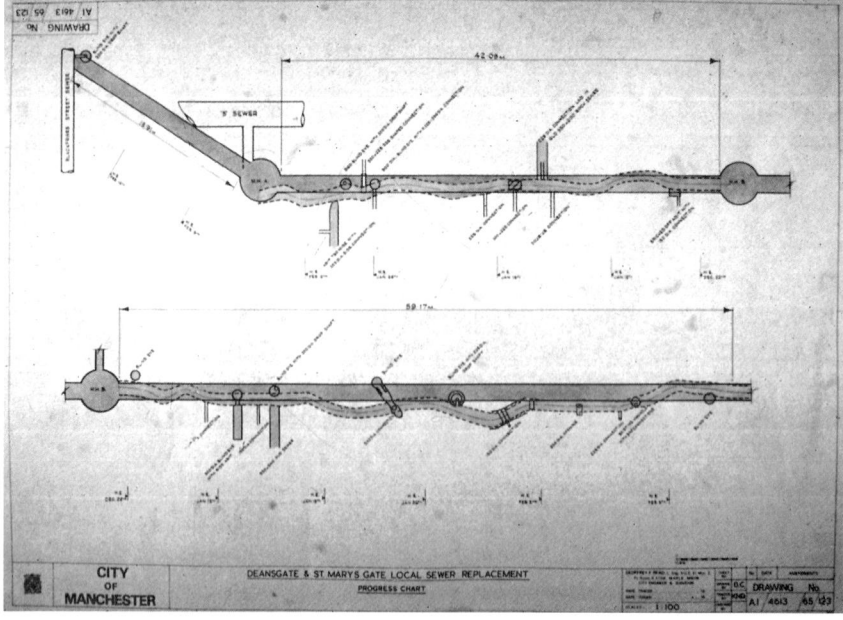

Figure 1.8   Existing sewer deviating from line of new tunnel.

should then become more cost effective. In addition, this arrangement does away with the need for manholes at junctions with minor connecting sewers.

In considering the economics of the situation it should not be forgotten that a completely reconstructed sewer has a longer life than a renovated one, especially since connections are more readily made and more effectively protected from further collapse.

It is considered that when the effective lifespan of a renovated sewer has been reached, there will normally be no alternative to replacement by a new sewer, but time will tell.

Taking into account all the factors which influence the selection of renewal methods, such as mode of sewer construction and surrounding ground, depth below ground level and the location of the sewer relative to other aspects of the environment, it is clearly not possible to arrive at a 'standard' solution applicable to all situations; the choice for each particular sewer length must be taken on its merit.

However, sufficient experience has been gained from renewal work (particularly in Manchester) for guidelines to be suggested for assessing the feasibility of renovation methods. Renovation methods within the criteria laid down in the manual are acceptable for the larger man entry and man accessible sewer but the condition and size of the earlier and smaller sewers (less than 900 mm diameter after lining) make renovation by manual labour completely unacceptable.

The prospects are better in these sewers for the use of renovation techniques relying on remote installation arrangements (e.g. Insituform, sliplining and impact moleing). For sub-man entry sewers, however, these techniques do not offer any convenient means of dealing with voids external to the sewer and this is considered a serious disadvantage when dealing with the early brick sewers and the open jointed clayware pipelines of this era which the Victorians also designed to act as land drains.

Inevitably, the risks associated with all aspects of sewer renovation are greater than those involved in replacement.

However, there are many situations where renovation is the obvious answer but notwithstanding the excellent work carried out by WRc as a result of which a 50-year design life can be anticipated, there is nevertheless a certain reluctance apparent in the use of these 'new' materials so far as sewerage is concerned, perhaps not unrelated to the problems that have arisen in regard to other 'new' techniques such as system buildings, particularly in high rise flats, high alumina cement. Nevertheless, renovation is now an option which must be carefully examined as part of the overall engineering feasibility investigation undertaken in connection with sewerage rehabilitation.

It is the degree of control that can be exercised over the execution of the renovation work that is of paramount importance to the longevity and success of the system. It must be fully understood that size limitations of the existing sewer considerably affect the quality of workmanship and supervision that can be achieved.

If the replacement is to be off-line there is still usually a case to be made for the use of a bolted (man access) tunnel which facilitates the reconnection of laterals. If the old sewer is to remain as a rider sewer dealing with the localised flow only, this criterion is no longer significant and the provision of a smaller diameter pipeline – still constructed via trenchless technology – may be the answer.

Just as it is a case of 'horses for courses' in sewerage rehabilitation generally, the same situation applies where replacement is under design consideration. East situation should be treated individually and a solution determined without predisposition towards any one technique.

When selecting appropriate rehabilitation methods (whether renovation or replacement) assessment of the adequacy of the existing sewer is required. First, its hydraulic carrying capacity for present and future needs (some renovation methods may significantly reduce sewer cross-sectional areas) must be considered. Second, to assess the condition of the existing sewer structure in the short term for carrying out inspection and investigation as well as possible renovation work from within it and its long-term viability as part of the sewerage system.

1.8 Strategy

The problem of sewer dereliction is immense. It is perhaps the greatest problem facing the water companies since the design and construction of the sewerage interceptors and treatment works, so far as Manchester is concerned, some 90 years ago. On the national scale sewer dereliction is one aspect of underground service dereliction (mainly sewers and water mains) which may be compared with the housing problem of the mid-twentieth century, which was overcome by the dramatic slum clearance programme of the post-war years. It is extremely unlikely, unless some major catastrophe occurs, that the underground dereliction problem will ever attract a similar level of public attention and resources and the national drive which was harnessed during the slum clearance era. Surface development and improvements, because the results are visible to all, are more likely to be favoured compared with underground infrastructure.

It should always be remembered that Britain's sewerage network has, in practice, been an important factor in ensuring the health and prosperity of the public for a considerable period, although some of it has now become a liability as a result of the lack of maintenance.

It is worthwhile noting in this context that one other public service industry, nemely, British Gas, because of the different characteristics of natural and town gas had little alternative but to embark on an extensive programme of renovation and/or replacement of its underground network within a relatively short period. Some 47 000 km of new mains were laid in ten years and in this case 'age' was not the major factor responsible.

Although the water industry can take heed from this achievement there are two important differences. First, the costs for replacing relatively small diameter shallow gas mains are a different order of magnitude to those for replacing larger and deeper sewers. Second, British Gas, following the revitalisation of the industry with the exploitation of natural gas, has been able to finance such a replacement programme. Traditionally the water industry has in the past made little allowance for replacement costs in setting its charges to the consumer. Although attempts are now being made to rectify this situation it is recognised that this could not be fully achieved within the present system without apparently increasing charges to unacceptable levels. Thus the problem is one of enormous magnitude tainted with the realisation that resources for its eradication are extremely limited and likely to remain so over the foreseeable future.

In the past, the policy dictated by the availability of finance has allowed events to determine the priorities for action, i.e. only nominal resources were allocated for investigations and maintenance, the incidents being dealt with as they occurred. Such works were essentially of an emergency nature in costs and manpower resources and are not easily predicted or effectively managed. The results of such an approach are that the extent of the problem is never fully recognised. For the civil engineer the basic objection to this thinking is that it contravenes the principles of planned and efficient use of resources. The continuation of such an approach led to the major crisis situation in Manchester reported in Volume I, which could have even resulted in a complete breakdown of the sewerage system.

It was interesting to note in early 1998 that the House of Commons' Environment Sub-committee included in their recommendations that water companies should have a renewal programme in place by 2002 to ensure that sewers are replaced as quickly as they deteriorate. To comply with this long overdue sewer replacement proposal water companies would have to considerably increase their level of investment. They currently replace some 1% of faulty sewers per year against a background of 8% to 10% of the network being in very poor condition. Since 1991 only around 1200 km of the total network (3 20 000 km) has been replaced.

In a paper on 'The Development, Renovation and Reconstruction of Manchester's Sewerage System', which I presented to the Manchester Literary and Philosophical Society in 1982 while holding the appointment of City Engineer, Manchester, I warned 'that a losing battle is being fought and dereliction is taking place much more rapidly than renewal'.

Sewers are hidden from public view and consequently there has also been a tendency to neglect them – in fact, some have never been repaired since construction.

Engineering works do not last forever – even Victorian sewers – and regular maintenance works and ultimate replacement are of paramount importance if collapses, with the serious associated public health and injury risks, are to be avoided.

It is generally accepted that particularly the older parts of the British sewerage network are in poor structural condition, their hydraulic performance is often inadequate and they are having to carry much greater external live loads than those envisaged at the time of construction. Consequently the need for substantial civil engineering works on a regular basis is unquestionable in order to deal with sewers having serious structural defects and where the lack of capacity is causing river or stream pollution – often with associated public health risks!

The immediate construction cost of these works is high – probably over £200 million per year – but in addition there are indirect or social costs arising out of the consequential impact on the community – perhaps varying between three and ten times the civil engineering costs in the most critical places. These social costs have in general to be met by the public at large or the highway authority without recourse to the water company. Clearly at some stage the industry will have to move towards a fully evaluated cost-benefit approach to take into account such social costs along with the anticipated benefits.

1.9   Environmental impact of sewer collapses

It is clear that serious environmental and public health consequences result from the ageing sewer network.

The personal injury risks for the public travelling over a worn-out network is also considerable and would doubtless result in greater priority being given to rectification if it were a structure above ground such as a bridge rather than an underground problem – typically out of sight, out of mind! In addition the immediate consequences of a sewer collapse may include the contamination of drinking water supply, flooding of habitable properties with crude sewage or other equally obnoxious and unacceptable environmental conditions. The results of such problems – particularly where a number of collapses occur concurrently – has an impact on the normal business and commercial life of a town or city which, in turn, can lead to a loss of trading confidence in its future well-being.

The cavity associated with a sewer collapse is usually onion shaped with an undercut format at the top adjacent to the road surface. These often spectacular collapses highlight the immense problem of sewer dereliction which now has to be faced against a background

Figure 1.9   Bolted segmental tunnel replacing 'U' shaped brick sewer with flag top.

of underfunding. The potential risk of vehicles falling into these large cavities is horrifying to consider particularly when one is immediately aware of the potential risk to life and limb. To enable the general public to readily appreciate the void size the author introduced a comparison factor which has now been adopted worldwide, namely, the DDB factor. That is to say the number of double decker buses which could be engulfed in the particular cavity – hopefully each not carrying the maximum of 92 passengers. The largest cavity in Manchester to date has been 4DDB which necessitated excavating some 21 m to the 3.66 m diameter trunk sewer.

1.10   Cost of sewer collapses

A typical sewer collapse account for a city centre location might be made up as follows:

Direct costs to water company	£
1. Repair and reinstatement of damaged sewer	45 000
2. Business loss claims under Public Health Act legislation	1 77 000
	2 22 000

Indirect cost to general public, etc	
1. Losses to bus or tram operator	10 500
2. Traffic disruption	82 815
3. Damage to other utilities plant	90 000
4. Emergency services	2 000
5. Flooding	70 000
	£2 55 315

Therefore, the likely cost to a water company and the public of a major sewer collapse is £4 77 315.

Based on the recent Manchester scenario of six major collapses per year in the central area the overall cost (direct and indirect) just for localised repair could be:

$$£477\ 315 \times 6 = £2\ 863\ 890$$

This assessment of course ignores the potential ever-present personal injury risk in such an emergency situation.

(The above figures are based on estimates produced for a socio-economic cost-benefit analysis for a large industrial city in the north of England – not Manchester – for which the author was responsible.)

1.11 Funding

When sewerage was a local authority responsibility, it had to compete at budget time with other more sociably attractive services such as housing, social services, education, etc., and did not attract adequate funding. Privatisation of the industry was intended to overcome this and enable adequate finance to be made available. However, this has not obtained, the socially attractive services have been replaced in practice by shareholders who wish the maximum return on their investment and European legislation related to river and coastal pollution. Clearly the first water regulator (OFWAT) who was appointed in 1989 following the privatisation of the water and sewerage companies has to ensure that the sewerage network is obtaining adequate funding within the overall budgets and that a comprehensive planned maintenance strategy is introduced forthwith.

Over the last few years a new organisation has developed *viz.* Campaign for the Renewal of Older Sewerage Systems (CROSS) and Figs 11.1 and 11.2 which they have published highlight the implied asset life of critical sewers based on the rate of renovation and replacement over the last 8 and 11 years which confirm the seriousness of the present situation.

Bibliography

1. The Federation of Civil Engineering Contractors, 'Public Health and the Engineers'.
2. Read, Brian *Healthy – A Study of Urban Hygiene*, Blackie, 1970.
3. House of Commons' Environment, Transport and Regional Affairs Committee. 'Sewage Treatment and Disposal', Stationery Office.
4. Proceedings of ISTT Conference 'No-Dig 90', Rotterdam.
5. Water Research Centre. '*Sewerage Rehabilitation Manual*', fourth edition.
6. Read, Geoffrey F. 'Sewer Dereliction and Renovation – an Industrial City's View'. Repair of Sewerage Systems Conference, Institution of Civil Engineers, London, 1981.
7. BS 8005 Parts 1 and 2.

2

New Construction

Geoffrey F. Read MSc, CEng, FICE, FIStructE, FCIWEM, FIHT, MILE, FconsE, Fcmi, MAE

Replacement of a sewer, whether on-line or off-line, clearly has similarities to the techniques used for the construction of new sewers whether these be in urban areas or green field sites. In the case of replacement, the need for this has no doubt emerged as a result of the development of a drainage area plan which has established the flow along with other significant factors in the various parts of the network under review. Possible rationalisation and reapportionment of flow is part of the exercise which in turn leads to confirmation of the flow requirements in the 'new' sewer to be constructed. In new construction the engineer has first to determine the flow to be dealt with which will involve the development of a drainage area plan – somewhat different in format to that referred to earlier – but developing a detailed network proposal often for the extension of an existing system of which the new sewer may only be one element.

System planning within a catchment area should consider the complete hydraulic scenario including the sewerage network, storm sewage overflows, storage facilities, pumping installations and the receiving waste water treatment works along with the effects of their discharges on the receiving water courses. A review of the hydraulic performance of the whole system to confirm that proposed additions or modifications to the network do not result in overloading or premature operations of overflows, etc. is necessary.

Structure plans and local plans can assist in the planning and design of new and extended sewerage systems although the possible impact of changing land use should not be overlooked. It will be possible to make an assessment of the extent of the impervious area in the catchment area under review leading to a calculation of the likely surface water runoff. Similarly an assessment can be made from these plans as to the forecast population changes leading to a determination of likely foul sewage flows.

To assess the likely flow in relation to foul sewerage, the water supply authority will be of assistance in making a forecast of water consumption in a particular area. Allowance should be made for distribution losses and consumption that does not result in discharge to sewers. It should not be overlooked in relation to commercial and industrial premises, supplies for which are normally metered, that some may, in addition, obtain water from rivers, canals, wells or boreholes directly but these will be normally covered by abstraction approvals from the relevant water companies from whom details should be available. In addition, details of existing or proposed trade effluent discharges will have to be determined. These discharges relate to the water used in processing, etc. and may contain a variety of matter in solutions or suspensions. They differ markedly from typical domestic foul sewerage and can raise special problems such as:

1. The effects they may have on the fabric of the sewer
2. Possible damage to sewer operatives during maintenance work
3. Effects on sewage treatment processes, etc.

The water companies are responsible for controlling the quality and quantity of discharges to the sewer and make charges depending on the strength and make-up of the discharge for treatment. Compliance with laid down conditions or in order to reduce the charges may require the industrialist to undertake pretreatment of the effluent prior to discharge.

Where there is existing similar industrial development information will be available in regard to constituents and amount of discharge. Where industrial sites are underdeveloped, planning policies may identify the nature and size of likely development which together with information on trade effluents from similar industries elsewhere may enable some approximation to be made as to details of the likely future discharge although this cannot be other than very broad based. Generally for undeveloped industrial sites one should design the sewerage network to be capable of taking a strong effluent in view of the ever present risk of accidental discharge (Fig. 2.1).

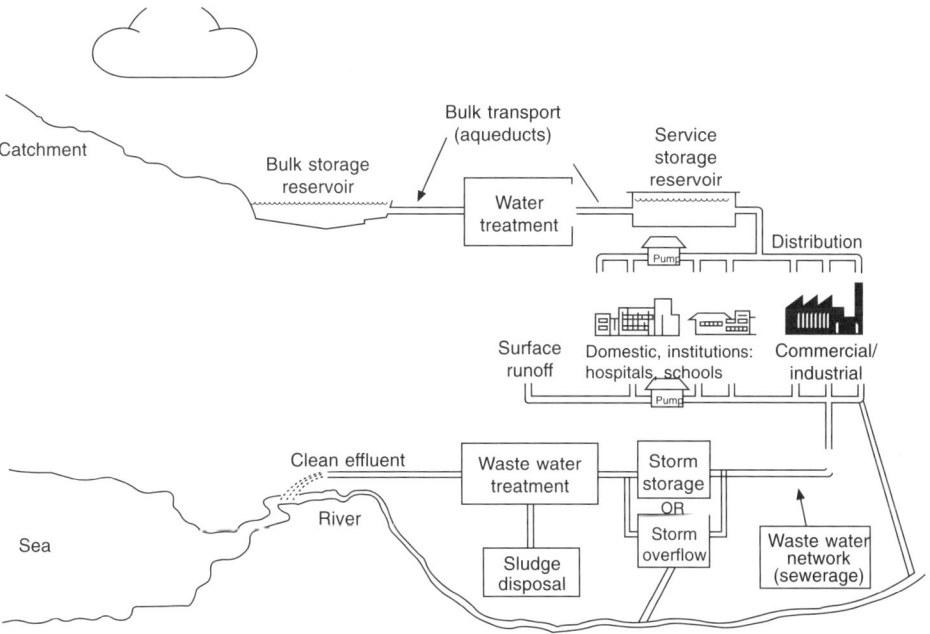

Figure 2.1 The water cycle.

2.1 Rates of flow – foul sewage

Whereas the design of a surface water sewer is dependent on rainfall intensities and the extent and permeability of the catchment area, rates of flow of foul sewage are dependent on the distribution of population and on the rate at which water is used. The average flow is called the dry weather flow (DWF) and is the average rate of flow of domestic and industrial waste water, together with an allowance for infiltration – even with modern jointing systems it is still prudent to make allowance for the latter.

The discharge of domestic waste water is very similar to that of water demand and accordingly it is generally assumed that the peak flow of domestic sewage (excluding industrial wastes) is the same as that of water demand, namely, about twice the average rate of flow (2 DWF) (Fig. 2.2). To take account of diurnal peaks and the daily and seasonal fluctuations in water consumption together with an allowance for infiltration a maximum flow of 6 DWF is used for design.

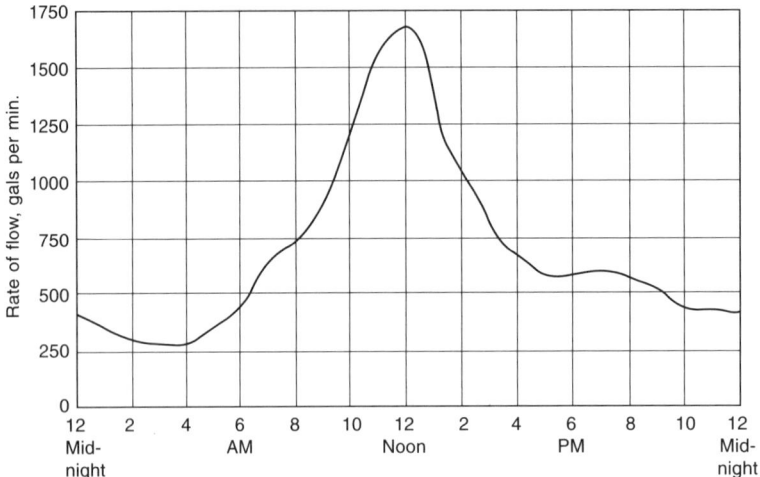

Figure 2.2   Fluctuation of flow of domestic sewage during the day.

Normally a rate of flow per head including an allowance for infiltration is 220 litres/day.

At 3.5 persons per house, the figure of 220 litres is equivalent to a domestic DWF of 770 litres per day per dwelling. On the usual design basis for a separate system this is normally rounded down for new development to 4000 litres per dwelling (approx. 6 DWF).

In terms of population densities and assuming the above figure of 220 litres per capita per day, the DWF can be calculated from the following:

$$Q = \frac{D \times A}{393} \text{ litres/second}$$

Where Q is the DWF in litres/second
D is the population density in persons/hectare
A is the catchment area in hectares

In the case of combined and partially separate sewers, the predominant flow is the storm element with the overall design figure including an allowance for foul sewage flow. The latter part of the DWF is calculated as previously and the design's capacity allowed for is 2.5 DWF or 1500 litres per day per dwelling.

In addition to the flows calculated from the normal residential population figures, further allowance must be made for the flow from schools, hospitals, industrial premises, etc. Generally one assumes a daily flow of 100 litres per capita from day schools. For boarding schools, hospitals and similar establishments, the flow can be taken on the same basis as that for normal dwellings.

For domestic waste from industrial premises it is usual to allow for at least 50% of the rate of flow from normal domestic premises, i.e. if the DWF from the latter is taken as

220 litres per head per day, the minimum allowance for domestic sewage from industrial premises should be 110 litres for each person employed.

The rate of discharge of industrial waste water will vary from factory to factory depending very much on the process involved. When planning for future industrial development where the types of industry are unknown the trade element in the DWF is usually taken as 2 litres per second per hectare for normal industries and 4 litres/second/hectare for wet industries or small sites.

2.2   Rates of flow – surface water sewers

As the name implies surface water sewers are designed to collect and convey surface water, during periods of rainfall, to a suitable water course, where such sewers form part of a separate system.

The amount of flow to be accommodated is primarily related to the rainfall intensity and the type of surface on which it falls – the latter depending on the proportion of paved and unpaved areas.

If the surface water sewers form part of a combined system they will normally ultimately discharge to a waste water treatment works. In such cases storm sewage overflows would be provided at suitable points to minimise the ongoing flow so that during times of storm excess diluted flows are diverted to neighbouring water courses or a storm relief sewer taking the flow from a number of overflows to a river. As these overflows can be sources of pollution the present policy is towards separate rather than combined sewers or the provision of storm water storage facilities from which the flow can be returned for onward treatment after the storm.

Except for very small areas, the design of surface water sewers is based on anticipated rates of rainfall. As a result of experiments carried out early in the twentieth century Lloyd-Davies was the first investigator to set out a sound theory for design.

The main principles which Lloyd-Davies proved were:

1. That the variations in the rate of rainfall during a storm and in the volume of water retained in the sewer system may be neglected.
2. That the storm water runoff from any drainage area is in direct proportion to the percentage of the surface that is impermeable to water (Table 2.1).
3. That the greatest runoff from a drainage area occurs when the duration of the storm concerned is approximately equal to the time taken for the sewage to travel from the furthest part of the system to the point at which flow is required to be known (the time of concentration).

Various formulae have been suggested for the calculation of rainfall intensities in terms of the duration of the storm. The best known are probably those which have become known as 'the Ministry of Health' formulae for storms with an expected frequency of once per year:

1. For $t = 5$ to 20 mins

$$R = \frac{750}{t + 10} \text{ mm/hour}$$

2. For $t = 20$ to 120 mins

$$R = \frac{1000}{t + 20} \text{ mm/hour}$$

Table 2.1   Typical impermeability factors

Type of surface	Factor(%)	Type of surface	Factor(%)
Urban areas, where the paved		Heavy clay soils	70
areas are considerable	100	average soils	50
Other urban areas, average	50–70	light sandy soils	40
residential	30–60	vegetation	40
industrial	50–90	steep slopes	100
playgrounds, parks, etc.	10–35	Housing development at	
General development – paved areas	100	10 houses per hectare	18–20
roofs	75–95	20 houses per hectare	25–30
lawns – depending on slope and subsoil	5–35	30 houses per hectare	33–45
		50 houses per hectare	50–70

Where R is the rate of rainfall mm/hour

t is the duration of the storm in minutes (i.e. time of concentration + time of entry)

The two formulae relating to storms of an expected frequency of *once per year* which would be appropriate for catchments with a surface gradient average greater than 1% as indicated in Table 2.2.

Table 2.2   Storm return periods for design of small schemes

Location	Return period
	Years
Sites with average surface gradient greater than 1%	1
Sites with average surface gradient 1% or less	2
Sites where consequences of flooding are severe, e.g. existing basement properties adjacent to new developments	3

A table of rates of rainfall was published in CP 2005 giving rates in mm/hour for storms likely to occur once every six months through to once in 100 years and for times of concentration from 5 to 120 minutes, this is reproduced in Table 2.3.

The decision as to which frequency of recurrence of storm to adopt is basically one of economics. Sewers designed to carry storms expected, say, once every ten years will be more expensive than those designed for one year storms, but on the other hand the risk of occasional flooding will be less. Practice in the UK is generally to design sewers on storm intensities expected once every year or two years, although the basis of once every five years has been approved in some instances. Various formulas and methods have been used in the past to determine the likly rate of runoff and either the use of the Lloyd-Davies method or the Transport and Road Research Laboratory hydrograph method is suggested.

The Lloyd Davies formula, which is applied to each section of the catchment in turn, is:

$$Q = 2.75Ap \cdot R \cdot 10^{-3}$$

Where Q = runoffs in cumecs
Ap = area in hectares
R = rate of rainfall in mm/hour

Table 2.3 Rainfall intensities

Frequency of recurrence, once in	Intensity (mm/h)	
	$t = 5$ to 20 min	$t = 20$ to 120 min
0.5 years	$\dfrac{580}{t + 10}$	$\dfrac{760}{t + 19}$
1–0 years	$\dfrac{660}{t + 8}$	$\dfrac{1000}{t + 20}$
2–0 years	$\dfrac{840}{t + 8}$	$\dfrac{1200}{t + 19}$
5–0 years	$\dfrac{1220}{t + 10}$	$\dfrac{1520}{t + 18}$
10–0 years	$\dfrac{1570}{t + 12}$	$\dfrac{2000}{t + 22}$

Extract from CP 2005 'Sewerage'.

In the past, the most commonly used design methods have been the rational (Lloyd Davies) method and the TRRL hydrograph method. Since 1981 these techniques have been updated by the methods in the Wallingford Procedure, developed by Hydraulics Research Station, with the assistance of the Institute of Hydrology and the Meteorological Office, under the supervision of the DoE/NWC Working Party.

With the older methods a design storm was selected and pipe sizes calculated to carry the resulting flows without surcharge. It was assumed that an adequate protection from surface flooding was thus provided. With the newer methods, the same approach is taken for sizing the pipes, but the design can be checked using a simulation program (WASSP-SIM) to establish the performance of the system under storms of longer return period. If an adequate protection against flooding is not found, the design should be adjusted. WASSP-SIM is principally used to check the hydraulic performance of existing systems. (See Chapter 8 of Volume I *Repair and Renovation*).

Example calculation to determine the diameter of a single branch surface water sewer according to Lloyd-Davies method.

Assume an area of 2.2 hectares with 80 houses to be built on it.

Then density $= \dfrac{80}{2.2} = 36.36$ houses per hectare

According to Table 2.1 the impermeability is 47.26%

therefore impermeable area $= \dfrac{47.26}{100} \times 2.20$

$$= 1.04 \text{ hectares}$$

Assume length of sewer is 300 metres and the fall of the site is 1.5 metres, then sewer inclination is 1 in 200.

The designer now has to make a preliminary guess as to the size of pipe required. If he chooses a 225 mm pipe at an inclination of 1 in 200 then discharge = 0.033 cumecs at a velocity of 0.86 m/second.

Sewer is 300 m long, therefore time of flow through it is $\dfrac{300}{0.86} = 349$ seconds

If time of entry is 300 seconds, then time of concentration will be 649 seconds. Then rainfall intensity according to Table 2.3 assuming a one year storm

$$= \frac{660}{t + 8} = \frac{660}{10.848} = 35 \text{ mm/hour}$$

By Lloyd-Davies, run of $2.75 \times 1.04 \times 35 \times 10^{-3}$

$$= 0.10 \text{ cumecs}$$

but the 225 mm pipe only discharges 0.033 cumecs – it is too small.

Check with the next pipe size, namely, a 300 mm pipe at 1 in 200 discharge = 0.077 cumecs at a velocity of 1.045 metres/second.

Time of flow now becomes $\frac{300}{1.045} = 288$ seconds plus time of entry 300 seconds

Therefore time of concentration = 588 seconds

$$\text{Rainfall intensity} = \frac{600}{9.8 + 8} = 37.07 \text{ mm/hour}$$

by Lloyd Davies, runoff = $2.75 \times 1.04 \times 37.07 \times 10^{-3}$

$$= 0.106 \text{ cumecs}$$

The 300 mm pipe is therefore nearly sufficient, so that the designer knows that a 375 mm diameter pipe will be adequate.

The designer now knows the necessary size of pipe for the bottom end of the 300 metre surface water sewer but at the top end there will be little flow and a much smaller diameter would suffice. This means he would in theory have to make a separate calculation for every individual length of sewer working from the top end downwards. Suppose the distance between the first manhole at the top end and the next is 100 metres, one could then assume that the impervious area draining to it is approximately 0.34 hectares. A 225 mm pipe will satisfy the requirement giving a discharge of 0.0359 cumecs at a velocity of 0.86 metres per second.

In the case of the second length the time of concentration is made up 300 seconds time of entry, 86 seconds through the 275 mm pipe plus the unknown time of flow through the length between the second and third manholes. It will be safe to guess that this will be a 300 mm pipe discharging 0.077 cumecs at a velocity of 1.045 metres per second.

The check calculations would then be:

Length of pipe 100 metres

Time of flow through 300 mm pipe $= \frac{100}{1.045} = 95.6$ seconds

Time of concentration = 300 + 86 + 95.6 = 481.6 seconds

Then rainfall intensity $= \frac{660}{8.02 + 8}$

$$= \frac{660}{16.02} = 41 \text{ mm/hour}$$

Assuming the area draining to the bottom end of the 300 mm pipe is 2×0.34 hectares then runoff $= 2.75 \times 2 \times 0.34 \times 40 \times 10^{-3} = 0.076$ cumecs and the 300 mm pipe will be adequate.

Finally, the last length, i.e. the 375 mm pipe, may be rechecked. The time of concentration then being the time of entry plus the time of flow through the 225 mm pipe, plus the time of flow through the 300 mm pipe, plus the time of flow through the 375 mm pipe, the last being 84 seconds. This equates to a value of 565.6 seconds and a runoff of 37.88 mm/hour which is well within the capacity of the 375 mm pipe.

2.3 Storm sewage overflows

These overflows are used for keeping the flows in combined sewers down to an agreed maximum and for passing the surplus into the nearest water course (provided sufficient dilution can be obtained), surface water sewer or storm tanks.

As indicated previously, these arrangements, particularly on the older parts of the system, can often result in serious pollution as a result of overloading and are nowadays kept to a minimum notwithstanding the need on economic grounds to limit the size of the combined sewer downstream. It is recommended that overflows should not normally be installed on combined sewers less than DN 500.

For new sewers, the storm sewage overflow should generally be set such that:

$$Q = DWF + 1360P + 2E$$

Where Q is the continuation flow retained in the sewer system and passed forward for treatment

DWF in litres per day

P is the population served by the sewer upstream

E is the average trade effluent flow in litres/day

An overflow may take the form of one (Fig. 2.3) or two side weirs. a stilling pond, a syphon or vortex with peripheral spill weir, covered in detail in Chapter 14 of Volume I.

The detailed hydraulic design of overflows is outside the scope of this volume but pollution reduction can be greatly improved by the provision of downstream storage for the diverted flow in the form of a storage tank.

RETENTION OR STORAGE TANKS

Both on-line and off-line are commonly used in the rehabilitation of sewerage systems for a number of reasons including the following:

1. To alleviate known problems of flooding and surcharge
2. To reduce the frequency, magnitude and duration of the 'spilled' flow from storm water overflows on combined sewers and hence to minimise the pollution and discharge to the receiving water course
3. To regulate the flow into the downstream sewer system and eventual discharge into the waste water treatment works
4. To optimise the retention of the pollutant load within the system.

The term 'storage' refers to the concept of including an additional retention capacity in the form of an oversize pipe, a tunnel or specifically designed storage tank whether these are on-line or off-line.

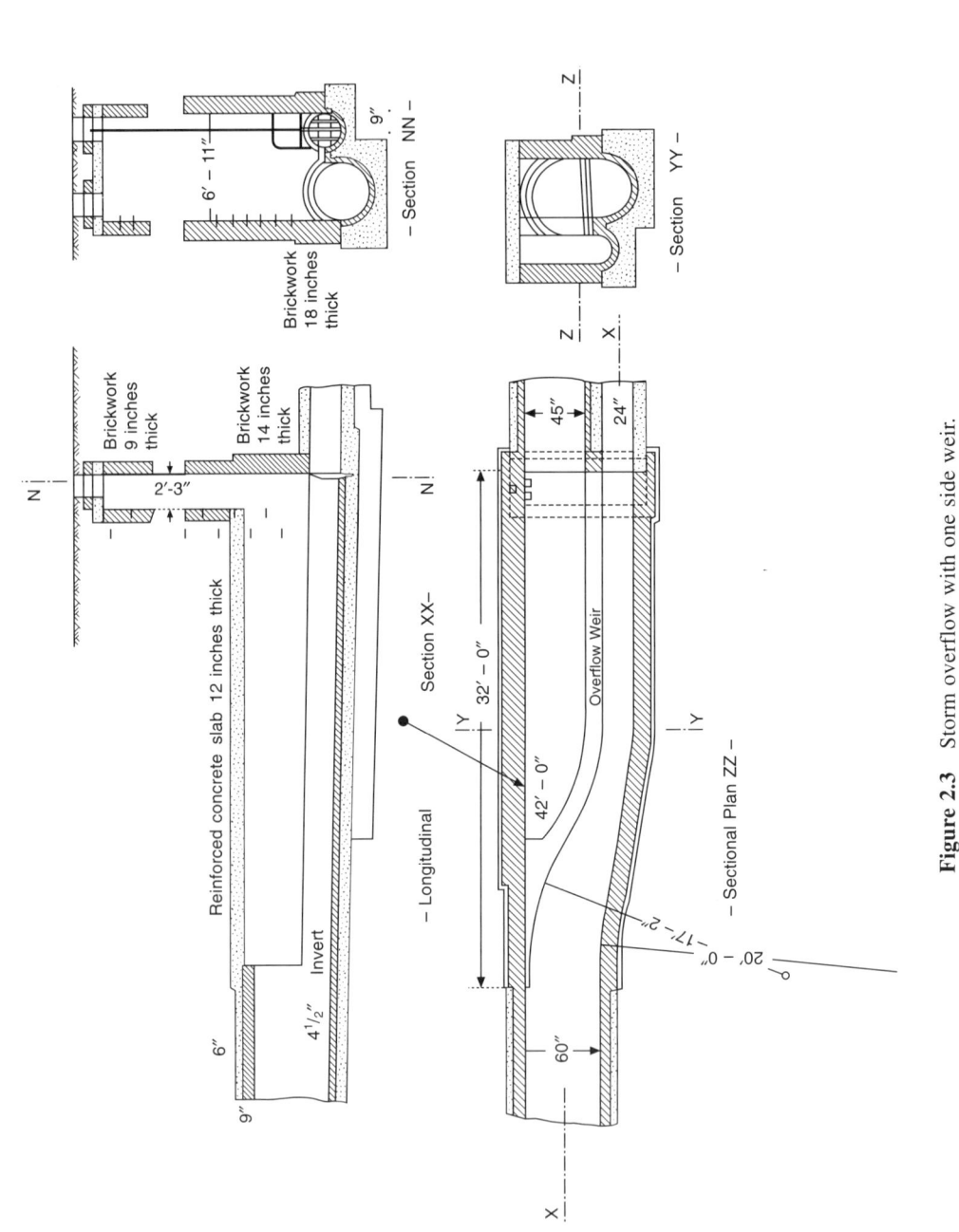

Figure 2.3 Storm overflow with one side weir.

A proposal to provide retention facilities should be based on a careful economic appraisal of the circumstance. The cost of such provision including maintenance should be compared with the costs of any improvements which would otherwise be required to existing sewers.

The optimum size of retention tanks in sewerage systems can be derived using WASSP-SIM. An appropriate outlet throttle being assumed to give the required carry-on flow. Various tank sizes can then be tested under a range of design storms. The critical storm duration will be found to be between two and four times the time of concentration of the upstream network. Storage tanks are installed more frequently to upgrade the performance of existing systems than as part of new systems.

Retention tanks should not be used on separate foul sewerage systems except:

1. At pumping stations
2. At the inlet to small waste water treatment plants where large fluctuations in flow may impair the treatment process.

Where new areas are being developed, it is often found that the existing water courses will be inadequate for the increased rate of runoff from the newly created impervious areas and new surface water sewers. In such circumstances some form of balancing or storage of part of the runoff may be preferable as an alternative to the construction of expensive capital works to increase the water course's capacity. It may be found that the provision of a balancing reservoir or water meadow, together with a small diameter outlet surface water sewer may be more economical than the construction of a sewer of larger capacity. Such an arrangement may mean that an existing surface water sewer through a densely built-up area can be retained and the cost and inconvenience of duplication avoided.

A flood plai is often then available for use as an open space when not required to retain flood water. Appropriate banking may be necessary to contain the storm flow and prevent flooding of any adjoining properties.

2.4 Soakaways

Where practical surface water drainage should be discharged into a sewerage system. Sometimes, however, where the ground is suitable it may be more economical to construct soakaways. Individual soakaways at each gulley, etc. may be the pattern or a number may be linked via a storm water sewer which itself discharges to a soakaway.

A soakaway is a pit providing soakage and storage capacity, its base and sides are open jointed to facilitate the percolation of water into the surrounding subsoil. The pit can be filled with rubble to obviate the need for a slab or a roof slab – if the former is used extra capacity should be provided to allow for this. If rubble infill is used it should be sealed at the top to prevent the ingress of soil. A capacity equal to 12 mm of rainfall over the area to be drained should be adopted, i.e. assuming 100% impermeability.

Small pre-cast concrete soakaways have capacities of up to 0.65 m^3 per metre depth while specifically manufactured concrete soakaways are obtainable with capacities up to 3.5 m^3 per metre depth. Particular attention should be paid to provision of oil spillage interception in order to protect the performance of the soakaway and to safeguard the ground water.

2.5 Inverted syphons

An inverted syphon is the term given to describe a system of pipes connecting from the lower end of one gravity sewer, dropping under a valley or other obstructions and rising again at the other side to connect into the head of another gravity sewer with a sufficient difference in the levels of the two sewers to give a suitable hydraulic gradient in the syphon (Fig. 2.4). They pose serious maintenance problems and if construction above ground level on piers is not acceptable in view of environmental consideration there is little alternative.

The principal disadvantages of inverted syphons are:

1. Because the gradient of the invert is not continuously downhill, large particles sliding down the invert will collect unless the velocity is periodically high.
2. Because the syphon is below the hydraulic gradient it is always full of sewage and therefore at times of slack flow the velocity must be extremely low with consequential silting.

The main points to consider when designing a syphon are:

1. The velocity of flow must be sufficient to carry away the solids.
2. The total length is the distance along the centre line of the pipework including bends, etc.
3. The syphon is designed as a pipe under pressure.
4. The required discharge under wet and dry weather conditions must be fully investigated.
5. Owing to the danger of silting the pipe or pipes must be designed so that they can easily be cleaned out.
6. If possible an overflow should be provided so that in the event of the syphon becoming surcharged or choked the flow may be diversified.

Calculations for the head lost in an inverted syphon should be based on the losses due to friction through the pipe or pipes which will be high owing to the roughness due to silting plus entry losses, plus losses due to the bends.

Entry losses can be calculated from:

$$H = \frac{Kv^2}{2g} \text{ m}$$

Where v is velocity in m/s

g is acceleration due to gravity = 9.806 m/s^2

K is a constant, usually taken as 0.025

Losses in each bend in the pipeline can similarly be calculated, i.e.

$$H = \frac{Kv^2}{2g} \text{ m}$$

with K at 0.5 for normal medium bends and 0.75 for short bends

The total losses due to entry and bends (excluding friction losses) in a normal inverted syphon using four medium bends would be as follows:

$$H = 5 \times \frac{0.5v^2}{2g} \quad \text{or} \quad 0.125v^2 \text{ m (approx.)}$$

This loss of head is then added to the friction losses which depend on the length of the system.

Example of a syphon design. It is required to design a syphon to take minimum and maximum flows of 1.5 cubic metres per minute and 10 cubic metres per minute. Length of syphon including the vertical legs is 210 metres and the maximum difference of level between the inlet and outlet pipes is 3 metres. There are few easy bends in the syphon. Find

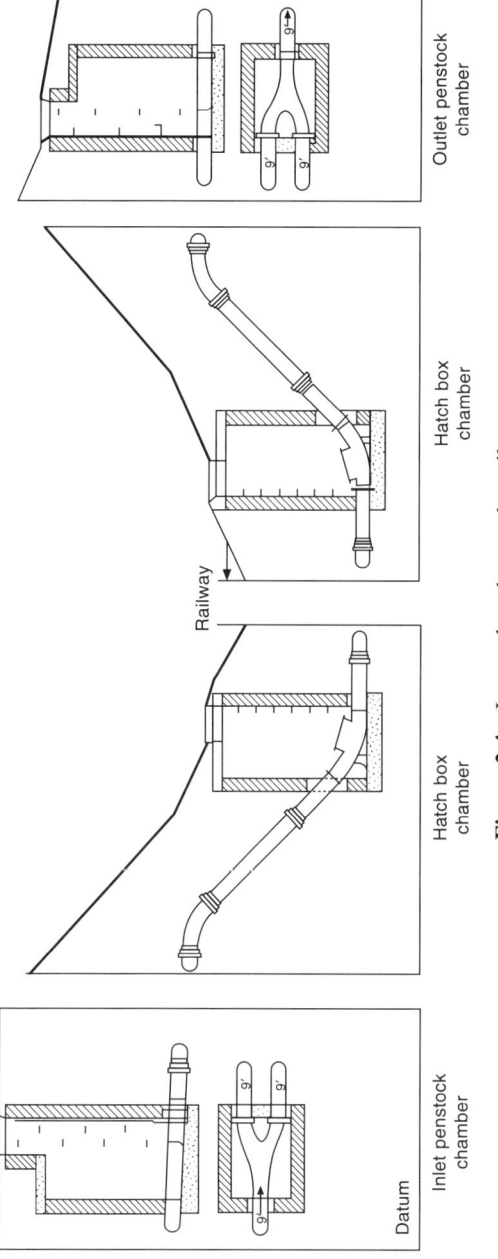

Figure 2.4 Inverted syphon under railway.

a suitable sized pipe so that the above conditions are met and so that the minimum velocity in dry weather conditions is such as to prevent excessive deposits of silt.

In a problem of this type the solution is arrived at by the trial of various pipe diameters and the choice of the one which satisfies the conditions for discharge and minimum velocity.

Assume the pipe to be 375 mm diameter. Loss of head due to friction H_F is determined as follows:

$$H_F = \frac{4fl}{d}\frac{v^2}{2g}$$

$$= \frac{4 \times 0.01066 \times 210 \times v^2}{0.375 \times 2 \times 9.806}$$

Loss of head in syphon = H_F + head loss due to bends + the head lost at entry

$$H_F = \frac{4 \times 0.01066 \times 210 \times v^2}{0.375 \times 2 \times 9.806} + \frac{5 \times 0.5v^2}{2 \times 9.806}$$

$$= 1.217 + 0.127$$

$$H_F = 1.34v^2$$

$$3 = 1.34v^2 \text{ and } v = \frac{3}{1.34} = 1.50 \text{ metres/second}$$

Discharge per minute $= 1.50 \times 60 \times A$

$$= 1.50 \times 60 \times 0.11$$

$$= 9.9 \text{ cubic metres per minute which almost satisfies the maximum}$$
$$\text{flow conditions}$$

when discharging 2 cubic metres per run the velocity $= \dfrac{2}{9.9} \times 1.50$

$$= \underline{0.30 \text{ metres per second}}$$

This velocity is on the low side and will only just prevent fine silt and light materials being deposited, but during certain periods of the day the flow will increase and help to prevent silt-up.

There is normally more grit in suspension in a combined or partially separate system than in a separate foul system – consequently the need to achieve a self-cleansing condition in the former is more essential than in the latter. Some means of flushing will often be necessary where the velocity is not self-cleansing.

Bearing in mind the danger of silting consideration should be given in the design for the duplication or multiplication of the pipeline with provision for diverting the flow from one pipe to any other.

The pipes should be laid so that they drain to some lower point, from which they may be cleaned out when the inlet penstocks are closed.

2.6 Manholes

Manholes are provided as a means of access for inspection and testing and for the clearance of obstructions in sewers. Smaller sewers, not more than 1 m in diameter, cannot be easily entered for cleaning, etc. and therefore a manhole is necessary at every change of alignment or inclination, at the head of all sewers and at every junction of two or more sewers as well as wherever there is a change in size of sewer. Manholes were previously sited no further apart than 100 m but with the arrival of high pressure jetting equipment this spacing has been increased to 150–180 m. In the case of larger sewers with a diameter greater than 1 m where a man can enter, the frequency of manholes is reduced. It is suggested that the spacing is governed by:

1. The distance which silt or other obstructions may have to be conveyed along the sewer to the nearest manhole for removal
2. The distance for which materials for repair can be conveyed through the sewer
3. Ventilation requirements for men working in the sewer.

In general a spacing of 180–200 m is suggested for straight runs although on sewers of 1.8 m diameter or over, constructed in tunnel, this could be increased to 300 m.

There are four main forms of construction for manholes:

1. Pre-cast concrete diameter tubes or rings (Fig. 2.5)
2. Brick
3. In-situ concrete
4. Shafts – usually constructed from pre-cast reinforced concrete segments (Fig. 2.6).

For other than large deep sewers the use of pre-cast concrete tubes is the most common form of construction. These tubes are specially manufactured for manholes and are complete with pre-cast inverts, taper sections, shaft sections and cover slabs. Chamber diameters will vary with sewer diameter and are available from 900 to 1800 mm diameter, with shafts standardised at 675 mm diameter. Sections are available depending on what 150 mm adjustments can be made to the depths of chambers and shafts.

Joints between chambers, taper and shaft units are usually rebated, ogee or tongued and grooved, sealed with cement mortar or mastic.

It is also possible to obtain rectangular pre-cast box culvert units, complete with step irons; where there is corrosive ground water or effluent it may be desirable to construct manholes using non-corrodible materials such as PE, unplasticized PVC or GRP.

Further information on the detailed design of manholes is contained in Chapter 7, Volume I, *Repair and Renovation*.

2.7 Backdrop manholes

In the interest of economy it is often desirable to lay a sewer at a gradient sufficient for the hydraulic requirements and then to connect to a lower sewer by means of a backdrop manhole – (Fig. 2.7). The incoming sewer being led into a vertical pipe outside the manhole which in turn terminates at its lower end in a 90° bend at or just above the invert level of the lower sewer. Such a stepped arrangement may also be necessary when designing a sewer for construction in steeply sloping ground so as to minimise the velocity and depths of construction.

Up to a difference in level of 1800 mm at 45° a ramp should be formed in the last part of the upper sewer and a rodding eye provided just as in the case of a vertical backdrop.

In very large sewers the backdrop should incorporate some form of water cushion to reduce the turbulence and velocity. A method of doing this is to discharge below the water level in a stilling bay of adequate capacity which, in turn, discharges over a weir at a higher level than the invert of the outgoing pipe.

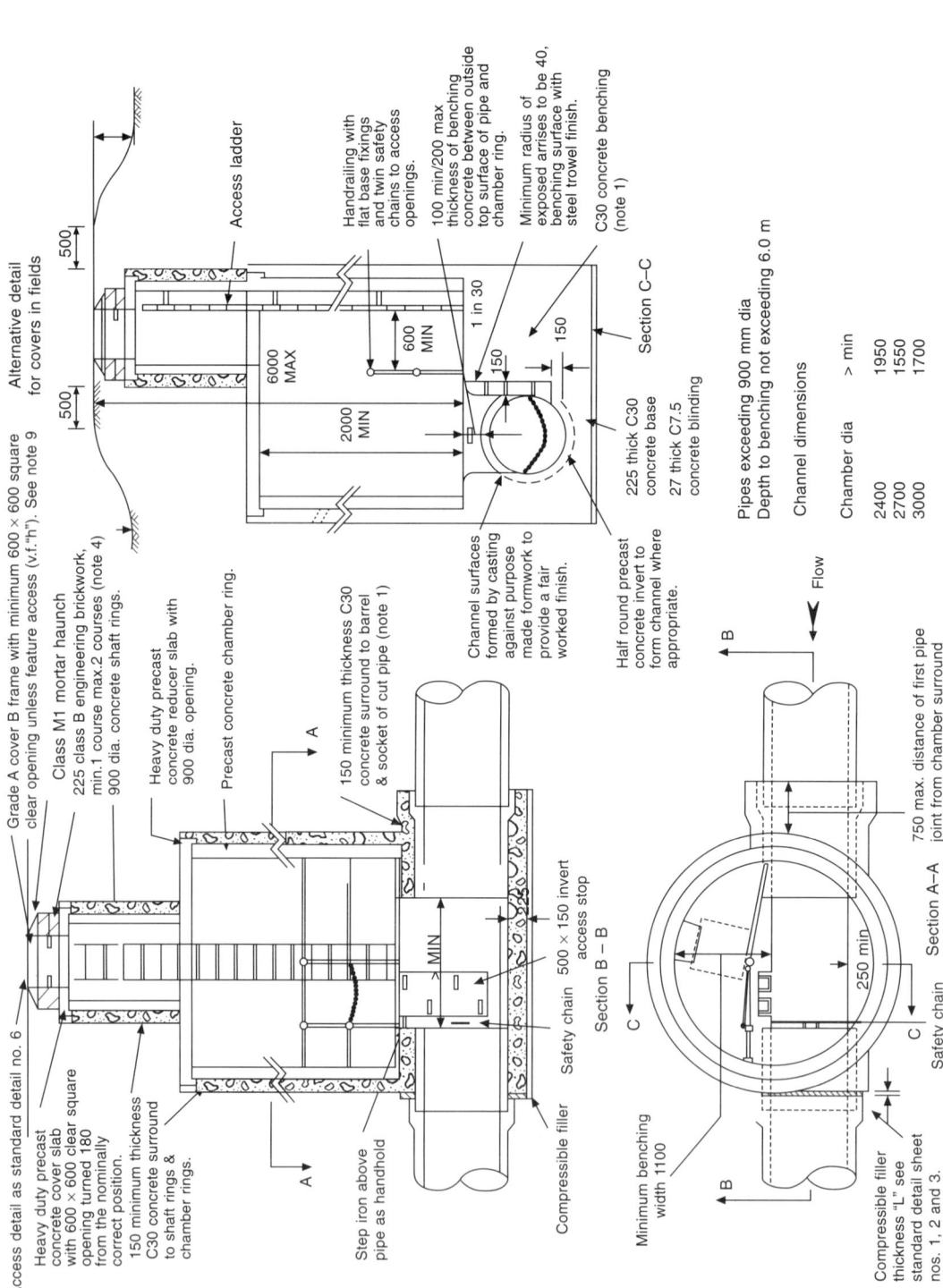

Figure 2.5 Standard detail drawing for pre-cast concrete manhole (North West Water).

Access detail as standard detail no. 6

Grade A cover B frame with minimum 600 × 600 square clear opening unless feature access (v.f."h"). See note 9

Alternative detail for covers in fields

500

500

Heavy duty precast concrete cover slab with 600 × 600 clear square opening turned 180 from the nominally correct position.

Class M1 mortar haunch

225 class B engineering brickwork, min.1 course max.2 courses (note 4)

900 dia. concrete shaft rings.

Access ladder

150 minimum thickness C30 concrete surround to shaft rings & chamber rings.

Heavy duty precast concrete reducer slab with 900 dia. opening.

Precast concrete chamber ring.

Handrailing with flat base fixings and twin safety chains to access openings.

6000 MAX

2000 MIN

600 MIN

1 in 30

100 min/200 max thickness of benching concrete between outside top surface of pipe and chamber ring.

Minimum radius of exposed arrises to be 40, benching surface with steel trowel finish.

C30 concrete benching (note 1)

A

A

150 minimum thickness C30 concrete surround to barrel & socket of cut pipe (note 1)

150

150

225 thick C30 concrete base

27 thick C7.5 concrete blinding

Section C–C

Channel surfaces formed by casting against purpose made formwork to provide a fair worked finish.

Half round precast concrete invert to form channel where appropriate.

Step iron above pipe as handhold

> MIN

500 × 150 invert access stop

Safety chain

Section B – B

Compressible filler

Pipes exceeding 900 mm dia
Depth to benching not exceeding 6.0 m

Channel dimensions

Chamber dia	> min
2400	1950
2700	1550
3000	1700

Minimum benching width 1100

B

C

C

B

Compressible filler thickness "L" see standard detail sheet nos. 1, 2 and 3.

Safety chain

Section A–A

250 min

Flow

750 max. distance of first pipe joint from chamber surround

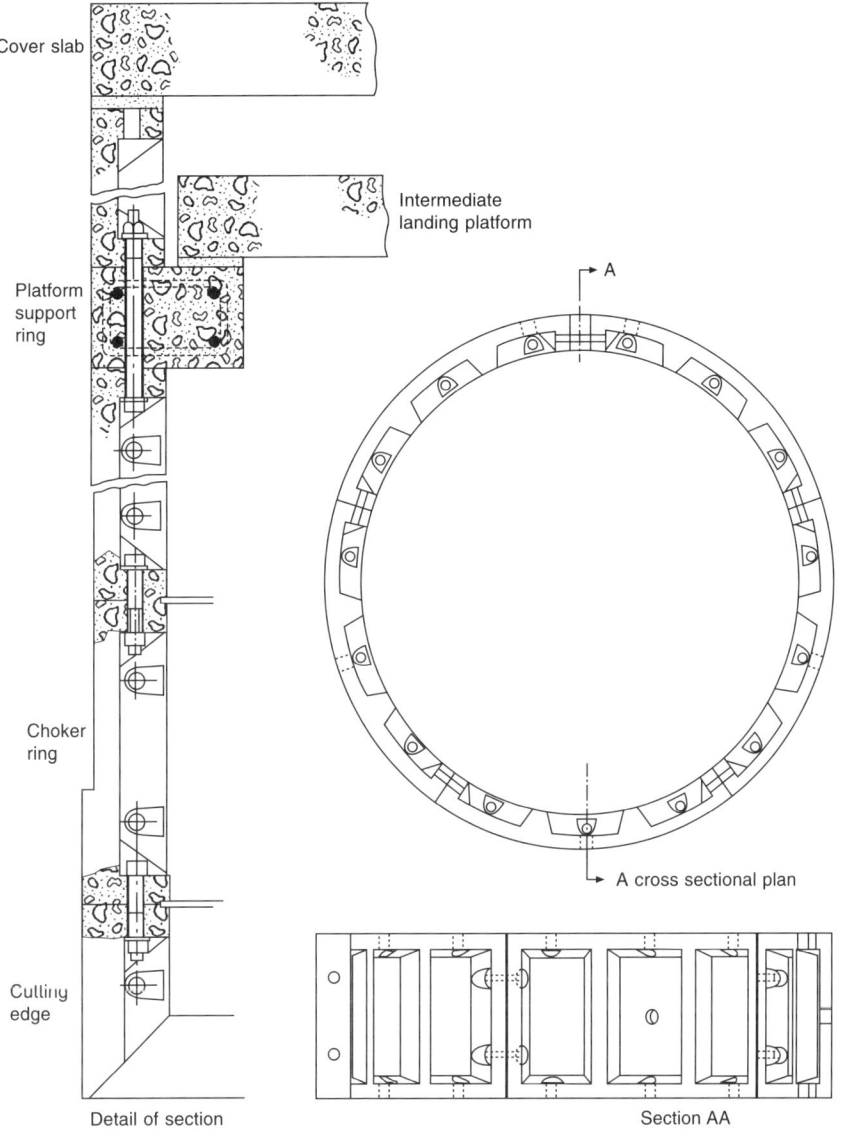

Cover slab

Intermediate
landing platform

Platform
support
ring

A

Choker
ring

A cross sectional plan

Cutting
edge

Detail of section

Section AA

Figure 2.6   Principal components of pre-cast, ribbed segmental rings (Charcon Tunnels Ltd).

Another technique used in larger sewers is to provide a cascade or steep ramp composed of steps to break up the flow with a bypass pipe for smaller flows constructed within the main flight of steps.

2.8   Access to manholes

In general, except at very deep manholes, access is normally provided by step irons. These are uniformly spaced at 300 mm vertically and staggered horizontally. For deep manholes (over about 3.5 m deep) ladders of heavy pattern galvanised steel are preferable because they are easier to use when wearing heavy boots. They are fixed to the sides using gun

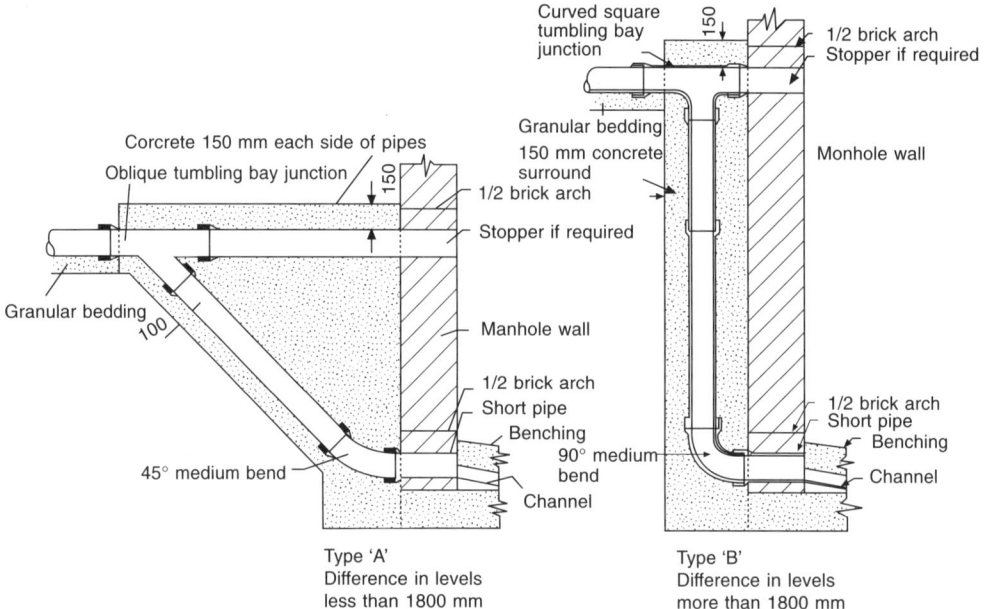

Figure 2.7   Typical backdrop manhole. Note: rigid joints may be used in lieu of flexible joints where these are surrounded with concrete.

metal bolts. Ladders should not rise more than 6 m without the provision of an intermediate platform or landing and this should be arranged so the line of the ladder is broken. It emphasises that adequate clearance should be provided around the ladder to ensure safety. Clearances at the back of any rung should not be less than 225 mm to give adequate allowance for feet, and on the user's side of the ladder this should be not less than 65 mm. The ladder stringers should be adequately supported from the manhole wall at intervals not exceeding 3 m. The top rung or step iron should be within 450 mm of the top of the frame. For shafts of 900 mm diameter or greater and where 550 mm to 600 mm clear opening covers are used they should be positioned so that the edge of the opening in the frame is vertically above the top rung of the ladder or the front edge of the step iron.

The access shaft of a manhole in a road should be brought up to a suitable level to allow a manhole cover and frame to be bedded on a maximum of three courses of regulating engineering brickwork or pre-cast concrete manhole sealing rings – this latter arrangement enabling the level of the cover to be easily adjusted as dictated by the level of the highway.

2.9   Pipes

It is current practice to use only flexibly jointed pipes for both sewers and rising mains. Rigid jointed pipes which do not deform appreciably under their design load are nevertheless still manufactured although their use is very limited – the actual joint in the latter being made by caulking a compound or working a cement mortar into the joint or by bolted flanges or by welding the pipes together. They should not be used in conjunction with granular bedding or where settlement of the ground is anticipated.

2.9.1   RIGID PIPES

Clay

Vitrified clay pipes have been manufactured since about 1845. They are now available in nominal diameter from DN 75 to DN 1000 in lengths up to 3 m and are currently not normally glazed. They are suitable for gravity flow under atmospheric pressure and should comply with BS 65.

Four classifications of pipe are available:

1. Normal
2. Surface water only
3. Perforated for land drainage
4. Extra chemically resistant.

Various forms of flexible mechanical joint are available as well as the traditional rigid joints.

Concrete

Concrete pipes are manufactured in accordance with the appropriate Part of BS 5911 – summarised below – and can be either unreinforced or reinforced.

> **BS 5911: Part 1** specifies requirements for flexibly jointed pipes in nominal diameters DN 150 to DN 3000 in standard lengths 0.45 m to 5 m (3 m for pipes DN 600), in three strength classes for sewage or surface water (and for jacking pipes).
>   **BS 5911: Part 3** specifies requirements for pipes with ogee, rebated joints, in nominal diameters DN 150 to DN 1800, in standard lengths 0.45 m to 2.5 m, in three strength classes, used principally for land drainage and surface water and for headings.
>   Precast glass composite concrete pipes, strengthened by continuous alkali-resistant glass fibre rovings, are covered by DD 76: Part 1 (see note 1). These pipes are available for sewage under gravity flow in diameters from DN 150 to DN 1800, lengths from 0.45 m to 5 m, and in three strength classes. Flexible in-wall rebated joints are used with elastomeric joint rings complying with BS 2494.
>   Precast concrete pipes strengthened by chopped zinc-coated steel fibres are covered by DD 76: Part 2. The pipes and fittings are suitable for sewage or surface water under gravity, and are available in nominal diameters from DN 375 to DN 1200, in three strength classes, in lengths of 0.45 m to 2.5 m. Joints are generally of the flexible, spigot and socket type, but in-wall design is also catered for.
>
> (The above is an extract from BS 8005: Part 1: 1987)

Various manufacturers are now producing specially designed concrete pipes for trenchless technology techniques, namely microtunnelling and pipejacking. The former involved pipes from DN 300 to DN 900 with an integral steel banded joint elastomeric seal and joint packer. Jacking pipes are available in sizes from DN 900 to DN 3000 and have either an in-wall joint with rolling elastomeric ring or a steel banded joint similar to microtunnelling joints. In both cases the external surface of the pipe without external joint sockets is smooth for easy insertion through the ground during installation.

Following the pattern initiated by the Victorian engineers who used ovoid stoneware pipes to give better velocity characteristics at low flow when compared with circular pipes, concrete ovoid flexible jointed pipes are now available. These are available in two sizes – 800 × 1200 mm and 600 × 900 mm – for use with both foul and surface water.

Asbestos – cement

These pipes are manufactured either to BS 486 as pressure pipes suitable for use in rising mains or to BS 3656 for sewerage gravity pipelines. The former covers pipe diameters from DN 50 to DN 2500 in lengths of 3 m, 4 m and 5 m.

The latter deals with pipe diameters from DN 100 to DN 2500 in pipe lengths of 3 m, 4 m and 5 m.

Cast iron and ductile iron

Iron pipes are used for rising mains and for gravity sewers where they are above ground and should comply with one of the following:

BS 437   'Cast iron Spigot and Socket Drain Pipes – for low pressure'.
This makes provision for centrifugally cast pipes with flexible joints in lengths of up to 5.5 m.

BS 4622   'Grey Iron Pipes and Fittings'.
These are manufactured in the size range DN 50 to DN 225 with flexible (5.5 m lengths) or flanged joints (4 m lengths).

Steel

These are sometimes used for inverted syphons, gravity sewers above ground and rising mains. They need protection against corrosion from sewage. Manufactured to BS 534 in diameters up to 1800 mm (bitumen lined welded pipes) and up to 300 mm (bitumen lined seamless pipes).

Flexible pipes

These deform under load – the extent of the deformation depending upon the stiffness of the pipe and the compaction of the immediate surrounding fill.

Corrugated metal

Manufactured in steel from DN 150 to DN 2600 and in lengths from 6 m to 9 m.

Unflanged corrugated liner plates

Used in the construction of pedestrian tunnels and culverts, generally supplied galvanised with longitudinal corrugations.

Glass reinforced plastic (GRP)

Suitable for gravity sewers or working under pressure. They should comply with BS 5480: Parts 1 and 2 in sizes DN 25 to DN 4000 and lengths from 3 m to 12 m.

Flexible joints and available in eight preserve and stiffness classes (Fig. 2.8).

High density polyethylene (HDPE)

These should comply with BS 6437 in sizes from DN 16 to DN 500 for pressures from 2.5 bar to 10 bar in lengths from 10 m to 12 m.

Figure 2.8   GRP pipes are lightweight yet incredibly strong.

Unplasticised PVC

For gravity flow pipelines to comply with BS 4660 for nominal sizes DN 110 and DN 160 and with BS 5481 for nominal sizes DN 200 to DN 630 (plastic pipes designated by their outside diameter) in lengths up to 6 m.

2.9.2   SEWERS LAID ABOVE GROUND

In situations where environmental considerations do not predominate it may be acceptable to construct a new gravity sewer above ground in order to carry it across a depression or valley (Fig. 2.9). Pipes raised above ground level for more than a matter of a few feet should be protected at the ends by means of collars of spiked iron railings to prevent children walking or climbing over them.

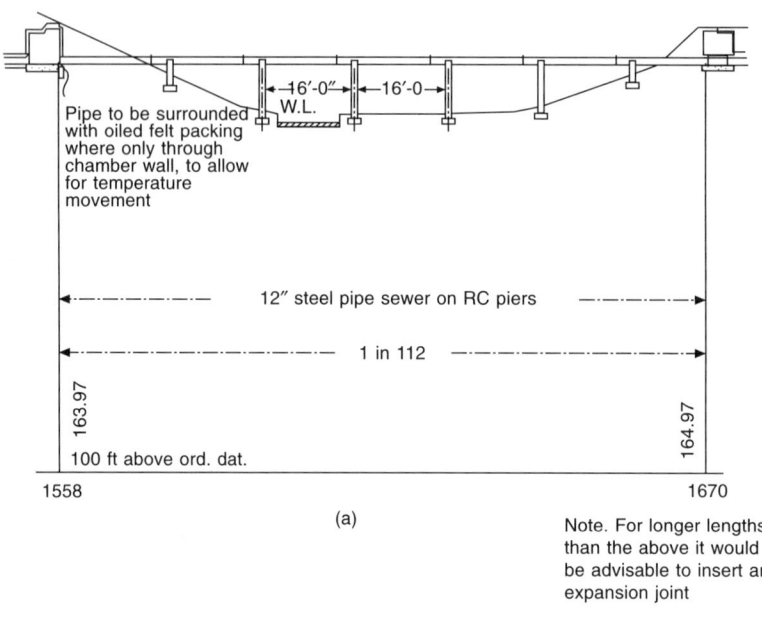

(a)

Note. For longer lengths
than the above it would
be advisable to insert an
expansion joint

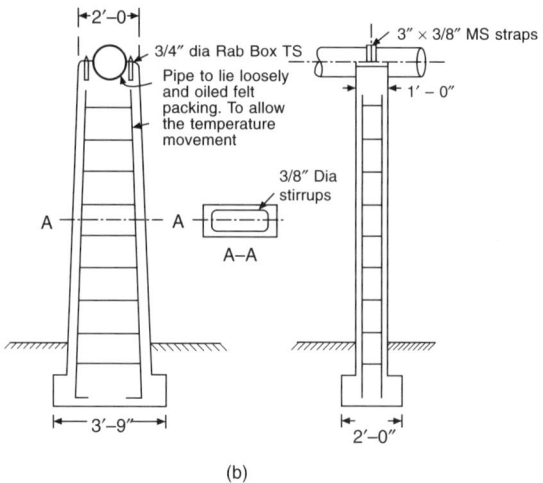

(b)

Figure 2.9   (a) Sewer carried on RC piers over stream and low ground. (b) Detail showing steel pipes
on RC concrete piers.

In exceptional circumstances such as in the spanning of rivers and railways the following
might be used although it is not intended to deal with the design in detail as it would be
more appropriate in a book on structures:

1. Steel flanged pipes supported on steel trestles
2. Steel flanged pipes supported on masonry, brick, mass or reinforced concrete piers
3. Sewer carried by plate girders, box girders, steel trusses, reinforced concrete arches or
   beams
4. Steel spigot and socket pipes supported by piers at appropriate intervals.

Example using steel flanged pipes:
Assume the required pipe is 600 mm bore and 615 mm external diameter
Flanged pipes are normally designed as continuous beams

Therefore BM is $\dfrac{Wl}{12}$

where W is total weight of full pipe and l is span in metres.
For 600 mm pipe $W = 60N$.

then BM $= \dfrac{60 \times l}{12}$

Safe working BM from Table 2.4 is 87 kNm

therefore $\dfrac{60 \times l}{12} = 87$

therefore $l = \dfrac{87 \times 12}{60} = 17.4$ m

but pipes are 5 m in length, so a safe working span of 15 m would be used as indicated in Fig. 2.9.

A fundamental requirement of flanged pipework is its ability to support an external bending moment. The magnitude of these permissible bending moments is related to the weight of the pipe and its contents for a given unsupported span. The length of the unsupported spans is limited by the need to confine stresses due to the combined effects of internal pressure, bolt tightening and bending moments within safe limits. These same limits are in turn applied to flanged pipework subjected to loads caused by thrusts due to internal pressure, e.g. at changes in direction. *The limits are such that it is recommended that flanged pipe is NOT buried.*

The safe working bending moment values are given in Table 2.4.

2.10 Pumping stations and rising mains

It is desirable in an ideal situation, for the treatment works to be at the lowest point of a sewerage system and for all sewage to flow to that point by gravity. Although pumping is reduced to a minimum because of its cost and because gravitational methods are more reliable, many sewerage networks have of necessity to involve one or more pumping stations. Often the relative economics have to be examined in detail so that the capital cost of deep sewers can be compared with the capital and annual costs of pumping.

It is suggested that the following circumstances may make the pumping of foul or surface water either necessary or advisable:

1. Avoidance of excessive depths of sewer
2. The drainage of low lying parts of an area
3. The development of an area not capable of gravitational discharge
4. Overcoming a physical obstruction, such as a hill, water course, railway, etc.
5. Rectification of the effects of running subsidence
6. Obtaining sufficient head for waste water treatment plant
7. Raising sewage to storage facilities.

Where possible, pumping sewage at the treatment works, *after* preliminary treatment, is

Table 2.4   Sale working bending moment values

Nominal size (DN)	Bending moment (kNm)
80	1.8
100	2.3
150	4.0
200	6.0
250	8.5
300	26.0
350	33.7
400	42.0
450	51.2
500	62.7
600	87.0
700	115.7
800	146.0
900	181.0
1000	221.5
1100	265.0
1200	312.5
1400	422.5
1600	547.5

These figures apply to both welded-on and integrally cast flanges.

preferable to the crude sewage, as the efficiency of the plant will be increased and the capital and maintenance costs will be less.

Generally rotodynamic or centrifugal pumps are used with foul or combined sewage and axial flow or screw pumps for dealing with surface water sewage where large volumes have to be pumped against low heads.

The former can have either vertical or horizontal drive spindles – the choice being generally based on the priming arrangements and site conditions. On a site, subject to flooding the motors and other electrical gear, above flood level, or if this is not practical the pumps can be of the submersible type. Submersible sewage pumps are now quite common, particularly for the smaller installations. These are arranged to slide down guide rails and seat automatically on the permanent discharge connection. The weight of the pump forces the mating flanges into contact thus providing a seal on the discharge side. The high standard of modern pumping plant and the development of automatic control and latterly of telemetry (via radio or telephone network) has meant that quite large pumping stations can be operated unmanned with maintenance visits only being required.

A recent development intended for isolated housing estates and similar conditions is the self-contained packaged pumping station which houses the pumps and motors (with a wet well if required) together with all valves and switchgear in one factory built unit. Such an arrangement is suitable for flows up to about 200 m^3/hour.

Clearly the pumps in a pumping station have to be capable of dealing with a wide range of flows, from low at night to peak daytime flows. Sewage cannot be stored for long periods in the pump well or septic conditions and consequential odours as a result of the generation of hydrogen sulphide will develop. It is usual therefore to install more than one duty pump, so that one will deal with low flows and further pumps will come into operation as the flow increases.

Standby pumping capacity must always be provided. A small station will usually have two duty pumps and one standby, while larger stations must have different installed capacity to cope with the flow if one of the larger pumps is out of action.

When pumps are dealing with foul sewage it is necessary to ensure that they have special 'unchokable' impellers capable of passing comparatively large solids. It is generally accepted that the pump should be capable of passing a 100 mm diameter ball. The usual unchokable impeller has only one or two blades set tangentially across the eye so there is no leading edge. It is the leading edge of a conventional impeller that tends to collect string and rags which can build up and eventually lead to a blockage.

Various forms of self-cleansing centrifugal pumps are available in which a cutting knife works in conjunction with the impeller – the combined unit then acting as both a disintegrator and a pump. This type of pump is suitable for isolated pumping stations when screening at the inlet would not be convenient.

The maximum discharge rate from a pumping station when *all* the duty pumps and pumping mains are in use should be equal to, or preferably greater than, the maximum design rate of flow to the station. The minimum pumping rate should achieve a self-cleansing rate of flow in the pumping mains.

Electricity is normally used to drive the pumps. In the UK a 415 V 3 phase supply is normal for motors up to about 150 kW to 200 kW. A standby electricity supply is frequently provided by taking a second feeder from a different substation or by switching to a standby diesel generating plant.

Usually the control of pumping plant is based on the level of liquid in the wet well – the operation of a pump starter being activated by the closing of electrical control circuits.

The common devices used are:

1. Floats
2. Electrodes
3. Air pressure discharge bubbles
4. Ultrasonic beams
5. Photoelectric light beams
6. Pressure transducers.

The conventional sewage pumping station houses vertical spindle pumping units with the pumps located in a dry well, drawing sewage from an adjacent wet well (Fig. 2.10).

The depth of the wet well is determined by the depth of the incoming sewer – to avoid silting the highest *cut-in* level of the normal duty pumps should be below the normal dry weather flow depth in the sewer -- often the cut-in level is set some 150 to 300 mm below the sewer invert. The *cut-out* or stop level is usually set at about 400 to 500 mm below the start level.

The wet well usually extends the full length of the dry well and is usually divided into two compartments for easy maintenance with a penstock in the partition wall Allowing for the 400 to 500 mm depth referred to above the width of the wet well can then be calculated.

In sewage pumping stations it is normal to provide pumping capacity for 6 DWF and to design for a maximum of 15 starts per hour. The storage capacity in a pumping well between 'start' and 'stop' levels must be equal to the output of the largest pump, i.e.:

$$Q = D/60$$

Where Q is the storage capacity in m^3
  D is the pump discharge in m^3/hour

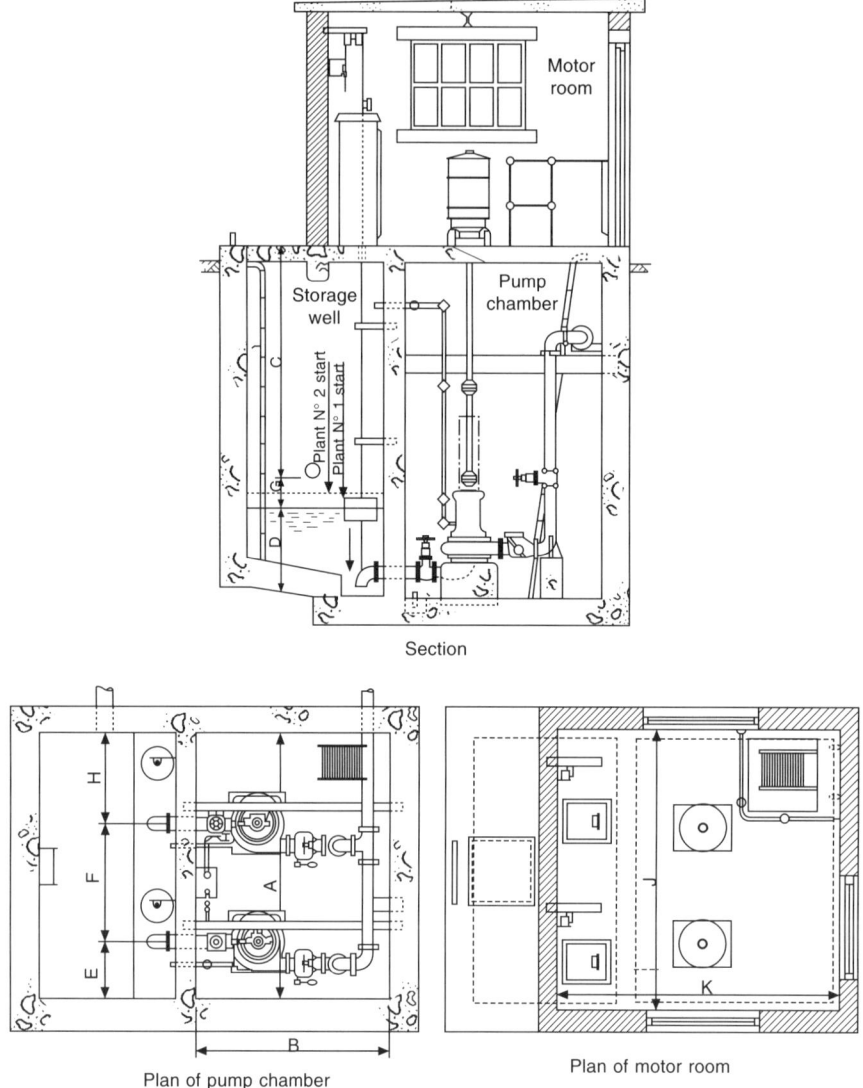

Figure 2.10 Conventional sewage pumping station.

Having established suitable pump outputs and designed the rising main for a specific total manometric head the pump power in kilowatts can be determined as follows:

$$P = \frac{Q \times H}{3.67r} \text{ kW}$$

Where Q is the output in m³/hour

H is the total manometric head in metres

r is the efficiency as a percentage

Example

Pump output 500 m³/h

Manometric head 21.4 m

Pump efficiency 40%

thus, power of required pump $P = \dfrac{500 \times 21.4}{3.67 \times 40}$ kW

$$= 73 \text{ kW}$$

Assumed motor is 85% efficient, then

motor power $= 73 \times \dfrac{100}{85}$

$$= \underline{86 \text{ kW}}$$

A rising main will normally follow the contours of the ground but the individual gradients of sections of this pipe link will not influence its design.

Rising mains are designed to withstand the total manometric head on the pumps, i.e. static head plus friction losses plus the friction through the pumps and station pipework and valves plus the entry and exit head losses as this is the head that will have to be overcome in the vicinity of the pumps. Should the pipeline cross a valley, the static head at the crossing may be greater and should be included when assessing the total head on the pipeline. The rising main must be of such a diameter that solids deposited while the pumps are stopped will be scoured out when pumping takes place. A velocity of between 0.75 and 1.20 m/second is suggested.

The design of pumping plant must be based on an accurate assessment of the total manometric head (i.e. static head plus friction losses) and one must calculate the friction head in rising mains as accurately as possible.

The Hazel–Williams formula is generally used to determine the friction head:

$$h = \frac{1128 \times 10^9}{4.87} \times \frac{(Q)}{(C)} \, 1.85$$

Where h is the friction head in metres per 1000 m

Q = flow in m³/hour

d = inside diameter of main in mm

C = friction coefficient from Table 2.5

The velocity of flow according to the same formula is:

$$v = 10.93 \times 10^{-3} \; \frac{C(d)^{0.63}}{(4)} \times 5^{0.54}$$

Where v = velocity in m/s

s = hydraulic gradient

It will usually be necessary to choose a number of possible diameters of pipeline and to calculate the friction losses for each one – the final choice being a matter of economics, i.e. the diameter selected to give the lowest overall *annual* cost.

Wash-out valves are fixed on rising mains at all low points in order that the mains may be emptied for repair purposes, etc. and connections laid so that they can discharge into a convenient foul sewer. Air valves are placed at all high points to prevent air lock.

Table 2.5 Friction coefficients for use with the Hazel–Williams formula

Pipe material	Condition	Value of 'C'	
		300 mm diameter and under	Over 300 mm diameter
Uncoated cast iron	18 years old	100	–
Uncoated cast iron	10 years old	110	–
Uncoated cast iron	5 years old	120	–
Uncoated cast iron	Smooth and new	125	130
Coated cast iron	Smooth and new	135	140
Coated steel	Smooth and new	135	–
Uncoated steel	Smooth and new	140	145
Coated asbestos cement	Clean	145	150
Uncoated asbestos cement	Clean	140	145
PVC	Clean	150	–

Rising mains for crude sewage will not be less than 100 mm diameter.

The total manometric head on the pumps (i.e. static head plus friction losses) is the head the pipes will have to withstand in the vicinity of the pumping station but additional allowance may have to be made for the effects of surge in the cases of high pumping velocity or in very long rising mains.

3

Site Investigation

Eur Ing Ian Whyte BScTech, DipASE, CEng, FICE, MIHT

3.1 Introduction

Site investigations are made to determine the ground conditions for planning, design and construction purposes, and also for reporting the safety of existing works and for investigating situations where failure has occurred. Even with a full and adequate investigation it is not possible to be certain about the nature, distribution and properties of the ground and ground water. An element of uncertainty is inherent to all construction works; the skill of the engineer is to reduce the uncertainty to acceptable limits. Unfortunately some investigations are less than adequate and projects are then exposed to ground problems which may not have been foreseen. The economic consequences of ground-related problems could be significant in terms of both cost and delay. Research has shown that the level of risk to a project reduces dramatically by properly planned and executed site investigation. Investment in resources for site investigation has a direct benefit to clients from the reduction in uncertainty and improved reliability of estimates.

The greatest uncertainty to a project is present during the early stages of development: risks have to be identified, analysed and managed. Geotechnical works are particularly high-risk activities and are most uncertain at early project stages. Methods of investigation have evolved to high levels of technical sophistication and varieties of technique. Traditional practices for ground exploration originated many years ago and can be in conflict with the geotechnical requirements. The works have to be planned and directed in a critical manner through assessment of the risks and uncertainties. Such approaches are advocated in the developing Eurocodes for design (BSI 1995).

For sewer works the investigation of ground conditions is an essential component and must be undertaken to the highest levels of good practice. Once the proposed layout of a new sewer or sewers has been determined a detailed survey of the subsoil along the route is required (including the 'desk' study). This information is necessary not only for the structural design of the sewer but also to make proper provision during the contract for ground support and possible dewatering.

Conditions of contract normally assume that the designer has provided the tenderer with information on ground conditions and the method of working. Rates in the tender will be primarily based on this information, including an assessment of the risks involved. The importance of proper site investigation cannot be overstressed.

For sewerage work and in particular where pumping stations, storm retention tanks, etc. are involved, the information obtained from boreholes should be supplemented by other methods, for example a number of trial holes. These should confirm the type of ground support that will be needed and will give a clearer picture of subsoil water conditions.

When trial pits are excavated it may be possible for these to be left open for a long period to help study subsoil water levels, for example during periods of increased runoff in nearby water courses, etc. (consideration has to be given to safety issues). The proximity of soil water to the ground surface can affect the design of sewers as in some circumstances it can be more economical to lay shallow sewers (with pumping stations) to avoid the expense of deeper construction.

In cases which involve the connection of a new system to an old system of sewers (which is not to be reconstructed) the infiltration of ground water to the old network may be of significance and should be estimated by observation of the right time flow – preferably when the ground water level is high.

It is always to be remembered that a site investigation before construction can only reveal factual data over a relatively small part of the length of a sewer. It may not identify all variations in ground conditions, i.e. the ground risks can never be eliminated with 100% certainty. The art of good site investigation is to reduce the levels of ground risk to acceptable values; this requires good judgement and experience.

3.1.1   AIMS AND STRUCTURES OF SITE INVESTIGATION

The primary aims of an investigation are:

1. To decide on the relative suitability of different sites or distinct areas of one site, likely changes in the environmental conditions of the site (and adjacent areas) due to the construction and operation of the project
2. To allow rational, safe and cost-effective designs
3. To discover and evaluate possible problems in the construction of both permanent and temporary works
4. To reduce the risks from unforeseen ground conditions, thereby decreasing the likelihood of changes in design and methods of working during construction and consequent claims
5. To assure the parties financing the project that the end product will meet the required performance criteria relating to ground/structure interaction.

The main stages of a site investigation are:

Stage I: Desk study: A preliminary appreciation of the site and ground conditions
Examination of existing information on the site history, geology, industrial archaeology, performance records (of existing tunnels for renovation) provides data to assess the feasibility of the works, information for sewer alignment, preliminary cost estimates and for the planning and design of the main ground exploration. The desk study report is also invaluable for the interpretation of the main ground exploration.

Stage II: Ground exploration before construction
Preliminary investigations may be required to confirm potential hazards suspected from the desk study and to confirm the design of the main exploration. Preliminary surveys can include selected boreholes/probes, trial pits, specialised local surveys (e.g. geophysical works) and preliminary testing.

The main ground investigation provides information for the final alignment, design and construction of the sewer. It involves borings, probes, pits, *in-situ* and laboratory tests. Other works include geophysical methods, trial holes and adits, permeability testing, instrumentation (e.g. piezometers) and building and other structure condition surveys. A major failing in site investigation is to believe that the ground exploration is the investigation. All too often the desk study phase is missed out or incomplete (Building Research Establishment, 1987; Mott MacDonald and Soil Mechanics Ltd, 1994) and subsequent investigations during construction are not properly planned and managed.

Stage III: Ground investigation during construction

Observation of site conditions during construction, and investigation where necessary, confirms and supplements the earlier stages. Such investigations provide agreed records on conditions encountered during the sewer construction and feedback to the designer to confirm the design assumptions. Should conditions vary from those predicted then further investigations during construction (e.g. extra boreholes, observation of ground movements and piezometers, grouting trials) may be required.

The ground investigation during construction provides records not just to confirm the design but also results in the as-built report essential for future maintenance and extensions to the works. For example, a cut and cover sewer under construction along a road in a town centre experienced serious difficulties from base heave of the excavation, sewer alignment, loss of ground and temporary ground water control. Talks with a local resident revealed that the same problems had been encountered some 15 years before on previous sewer works but these had not been recognised since records had not been preserved from these earlier works.

3.1.2 GROUND HAZARDS IN SEWER CONSTRUCTION

Predicting the potential hazards (knowing the enemy) provides a start for the site investigation process. The following discussion is not exhaustive and each project should be considered on an individual basis, particularly where sewers are penetrating previously developed sites and obstructions/contaminated land are real risks.

Jardine and Johnson (1994) report a framework for the assessment of ground risks and similar approaches are advocated in ENV7: Geotechnical Design (BSI 1995). The preconditions that produce a ground hazard are:

- The ground
- The content of the ground
- The actions of the ground
- The property/development of the site
- The people on site or making decisions which influence the works.

Some potentially hazardous conditions of the ground are listed in Table 3.1.

Other factors, which influence the ground uncertainty, are:

- Incompleteness: Features that are not identified but lead to behaviour different to that assumed. Ground investigation reports are frequently incomplete (Matheson and Kier, 1978).
- Uncertainty: Features detected in the site investigation may not readily be quantified or represented adequately. Common uncertainties relate to the site geology, ground profile and ground water conditions.

Table 3.1 A-Z of potential ground hazards

A:	Archeology, aquiclude, artesian water, asphyxiation
B:	Bearing failure, boulders, blow, boils, bedrock
C:	Collapse settlement, corrosion, creep, contamination, cavitation, chemicals, cracks, compressibility
D:	Differential settlement, dewatering, dissolution, driveability
E:	Erosion, earthquake, earth pressure, excavations
F:	Foundations, fissures, fill, flowslide, frost heave, fault
G:	Groundwater, gases, gouge
H:	Heave, horizontal stresses, hardrock, heavy ground
I:	Instability, inflows, inrushes, incompetent ground
J:	Joints
K:	Karst
L:	Landslip, loss of ground, leaching, liquefaction, landfill gas, lawsuits
M:	Movement, moisture change, mudflow, methane
O:	Obstructions, organics, overbreak
P:	Pollution, progressive failure, peat, piping, pore pressure
Q:	Quicksand, quagmires, quick clays
R:	Rockfall, residual soil, rupture, ravelling, running ground
S:	Subsidence, shear zone, settlement, seepage, swelling, shrinkage, solution, slickensides, squeezing ground, spalling
T:	Tension cracks, time, tremors
U:	Unsuitability for use, uplift, undrained strength, ultimate bearing capacity
V:	Vulcanicity, voids, vibration, volume change
W:	Weathering, water table, weakening, wells
X:	Xenoliths (foreign bodies)
Y:	Yielding
Z:	Zone of shearing or yield

- Communications: A failure to communicate potential hazards leads to risk. Site investigations cover projects from concept to completion and the potential exists for known information not to be transmitted to all parties involved, for example staff on site. The perception of risk can alter between planner, designer, contractor and client.

For sewers, particular hazards exist for the various forms of rehabilitation and new construction. Some examples are given below.

Replacement

Particular risks are associated with

- Stability and safety of existing sewer
- Ground conditions and profile, both longitudinally and laterally
- Access for investigation
- Obstructions
- Cavitation and voids
- Possible contamination
- Sewer condition and alignment
- Connection locations
- Old shafts (possibly now sealed at the surface).

Read (1986) discusses the options to be considered when deciding as to whether to renovate a sewer (i.e. to restore the existing sewer structure to an acceptable condition) or to replace the sewer (i.e. to construct a new sewer section along the line and level of the existing sewer).

Key factors are:

1. The dimensions and alignment of the sewer (horizontally and vertically). The major axis dimension of a sewer has to exceed about 900–1000 mm for man access. Sewer alignments can wander and gradients can be variable, and on occasions be the reverse to that designed. Thorough and comprehensive investigation is necessary to report the history and location of a sewer. The hydraulic capacity has also to be estimated.
2. The condition of the existing sewer. Brickwork can be uneven and erratic with irregular bonding and open joints. Missing bricks can frequently occur. Sewers in erodible soils (mainly silt and sand soils) can develop large external cavities and voids from either cavitation from the flowing sewer or from flowing ground into the sewer when infiltration develops from external water present permanently or at variable intervals (e.g. seasonal variations). The cavities can develop to a large scale and when collapsed can engulf cars and large vehicles, e.g. a bus.
3. Health, safety and environmental factors. Wandering alignments and varying gradients can produce deposits of debris and silt which lead to toxicity and gases. Rodents can also be a problem.

New construction: tunnels

Ground hazards relating to tunnels are discussed by McKusger (1984), see Table 3.2.

Data from 84 tunnels in America (US National Committee on Tunnelling Technology, 1984) found the frequency of problem conditions as shown in Table 3.3. The impact rating represents the incidence of claims per occurrence.

It is to be noted that ground water problems contribute to many of the conditions, water plays a significant role in tunnel construction difficulties and the problem tabulations do not give as much weight to ground water as it deserves.

The American study showed a direct correlation between the accuracy of cost estimate (in relation to final outturn costs) and the level of site investigation. It was recommended that expenditure for site investigation be increased from an average of 0.44% of estimated project cost from the case studies to at least 3%. About 60% of claims were for large amounts; analysis showed that both the level and number of claims related inversely to the amount of investigation. The more known about the ground in advance of construction reduces the level of uncertainty. This leads to better planning and forecasting and a reduction in the degree and severity of claims. Data from highway site investigations in the UK (Mott MacDonald and Soil Mechanics Ltd, 1994) produced similar evidence (Fig. 3.1). A comprehensive analysis of investigation costs for tunnels in the UK is not published. West *et al.* (1981) reported that UK case histories showed the cost of ground exploration to be generally less than 3% of the cost of the works, and may be as low as 0.5%.

Guidelines for avoiding and resolving disputes in underground construction in America were reported in 1989 and extended in 1991 to general projects through an equitable risk-sharing philosophy between client and contractor and are worthy of study for UK practices. In particular, the adoption of a Geotechnical Baseline Report for a contract has much to be commended (ASCE, 1997).

A report from the Institution of Civil Engineers (Clayton, 2001) gives guidance on managing the geotechnical risk and includes the viewpoint of the client, designer and contractor. Examples of risk registers are given and case histories reviewed.

Table 3.2   Ground hazards in tunnels

Description	Behaviour	Typical conditions
Ravelling (slow/fast)	Exposed materials drop from roof, sidewalls after exposure. Due to loosening or to 'overstress' and 'brittle' fracture (ground breaks along distinct surfaces, opposed to squeezing ground). In fast ravelling the process starts within a few minutes.	Residual or granular soils may be fast ravelling below the water table, slow ravelling above. Stiff fissured clays may be slow or fast ravelling depending on the degree of overstress.
Squeezing	Ductile plastic yield: ground intrudes plastically into the tunnel without visible brittle fracture or loss of continuity and without perceptible increase in moisture content.	Low strength soils. Soft to firm clays can squeeze at shallow to medium depths. Stiff clays may move in combination with ravelling at exposed surface and squeeze at depth behind the face.
Swelling	The ground absorbs water, increases in volume and expands slowly into the tunnel.	Overconsolidated clays with PI greater than about 30, particularly if clay mineral is expandable.
Running	Granular soils cut to slopes steeper than the angle of repose run until the slope flattens to the angle of response.	Clean dry granular soils. Moisture may allow the material to stand for a brief period before it runs – this is termed cohesive running.
Flowing	A mixture of soil and water flows into the tunnel like a viscous fluid. It can flow from the invert as well as from the face, crown and walls. If not stopped the tunnel can fill with material over a considerable distance.	Silt, sand or gravel without clay content (to give cohesion and plasticity) and below the water table. Can occur in sensitive clay if disturbed.
Bouldery	Shields or forepoles are jammed when advanced. Blasting or careful excavation ahead of shield may be necessary.	Glacial tills, some landslide deposits, some residual soils.
Ground surface subsidence	Tunnelling in weak ground can cause substantial damage to surface structures. Settlement is a function of the type of ground, size and depth of tunnel, method of tunnelling, type of lining. Settlement is produced from ground strains resulting from material losses in tunnelling.	Subsidence most severe when tunnelling granular soils below the water table (ground losses from ravelling, flowing, etc.). Movement very dependent on quality of subsidence control at the tunnel face. Subsidence least severe in rocks, very stiff/hard clays and sands above the water table.

New construction: cut and cover

Potential hazards that produce risk in cut and cover construction include:

- Excavated materials and suitability for replacement.
- Ground water flows to the cut and temporary works to control the ground water. Deficiencies in control can produce running sand conditions that rapidly lead to instability, collapse and surface settlement of adjacent ground.
- Temporary support to the sides of excavations with internal props or external anchorages. The installation of ground support systems can produce hazards; for example, driven sheet piles can declutch and split leading to instability on excavation.

Table 3.3 Problem conditions, frequency and impact rating (84 tunnels).

Condition	% Problems or occurrence	Impact rating
Blocky/slabby rock, overbreak, cave-ins	38	42
Ground water flow	33	18
Running ground	27	33
Squeezing ground	19	42
Obstructions (boulders, piles, rock)	12	92
Face instability, soil	11	45
Surface subsidence	9	22
Methane gas	7	28
Noxious fluids	6	66
Spalling, rock bursts	6	66
Hard abrasive rock, TBMs	5	40
Face instability, rock	5	20
Flowing ground	5	80
Mucking	5	40
Pressure binding, equipment	4	100
Roof slabbing	4	25
Soft zones in rock	4	50
Steering problems	4	0
Soft bottom in rock	2	100
Air slaking	1	0
Existing utilities	1	0

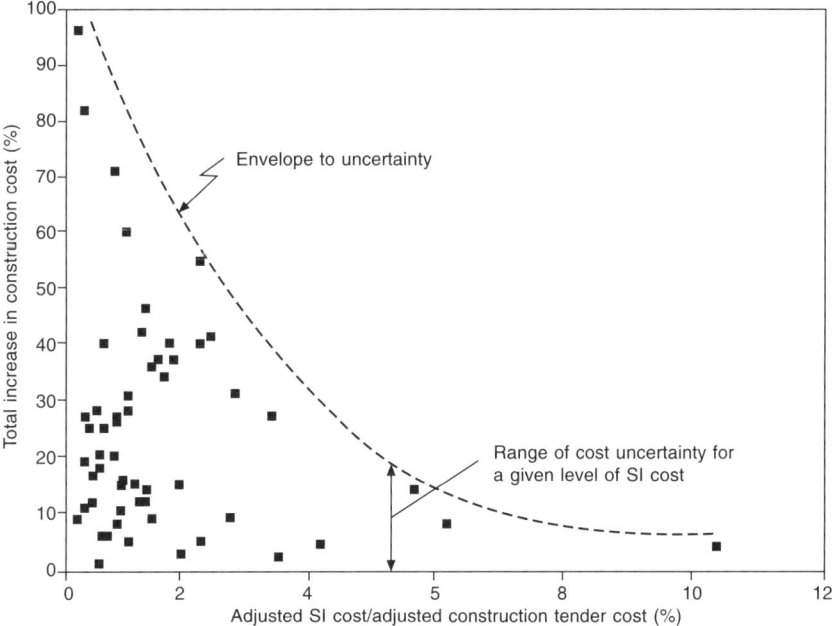

Figure 3.1 Relation of construction cost increase to SI expenditure (after Mott MacDonald and Soil Mechanics Ltd, 1994).

- Degradation of the materials at the base of the excavation when these are exposed and disturbed by construction.
- Movements of the base from hydraulic pressures in underlying soil layers. If the pressure is not relieved then blowouts can occur with catastrophic instability developing. Poorly backfilled exploration boreholes can produce serious problems from underlying water-bearing layers.
- Ground subsidences adjacent to excavations. If the ground water is controlled then settlement is generally small for dense sands, clayey sands and sandy clays. Movement increases as clay strength decreases to soft and very soft consistencies and settlements become very large if 'running' ground conditions develop from poor ground water control (Peck, 1969; Geddes, 1977).

3.1.3   MANAGING THE RISK

The perception of risk depends on one's point of view – one man's loss can be another man's gain. Engineers and professionals in construction are technologically trained and tend to see risks from that viewpoint. Clients and their financial advisers view events in terms of economic power and the market. Ground risks are a problem and an insufficient investment in time and money resources to do an investigation produces inadequate knowledge that leads to technical and financial consequences. Solutions to the lack of proper investigations have been sought through procedural change and forms of contract. Jardine and Johnson (1994) state that none of these initiatives has had an effect on the bulk of site investigation works. This is also evidenced by many piling contractors ('customers' of a site investigation) who report that ground investigation reports are no better now than in the 1980s when standards were abysmally low. A survey of the industry (Anon, 1995) clearly demonstrates the lack of satisfaction with site investigations. It is of interest to note that a report by Mott MacDonald and Soil Mechanics Ltd (1994) showed the only construction cost index in the UK to have fallen over the period 1979–91 was the Ground Investigation Price Index – this fell by about 24% as compared to the Retail Price Index which rose by about 140%.

In construction the most serious effects arising from risk events are:

- Failure to keep within the cost estimate
- Failure to achieve the completion date
- Failure to meet quality and operational requirements.

Ground conditions pose real hazards to construction, particularly in sewer rehabilitation and new construction. Control of the risk is by good management and communications: the provision of resources (time to do the work and finance to investigate and report) to identify and quantify hazards followed by the design, implementation and monitoring of control measures. Whyte (1995) discusses these issues in detail. A flowpath recommended by Whyte and Tonks (1993) is shown in Fig. 3.2. Guidance on planning, procurement and quality management is given by Uff and Clayton (1986, 1991), Site Investigation Steering Group (1993) and ASCE (1991). Dumbleton and West (1976) provide a guide to site investigation procedures for tunnels.

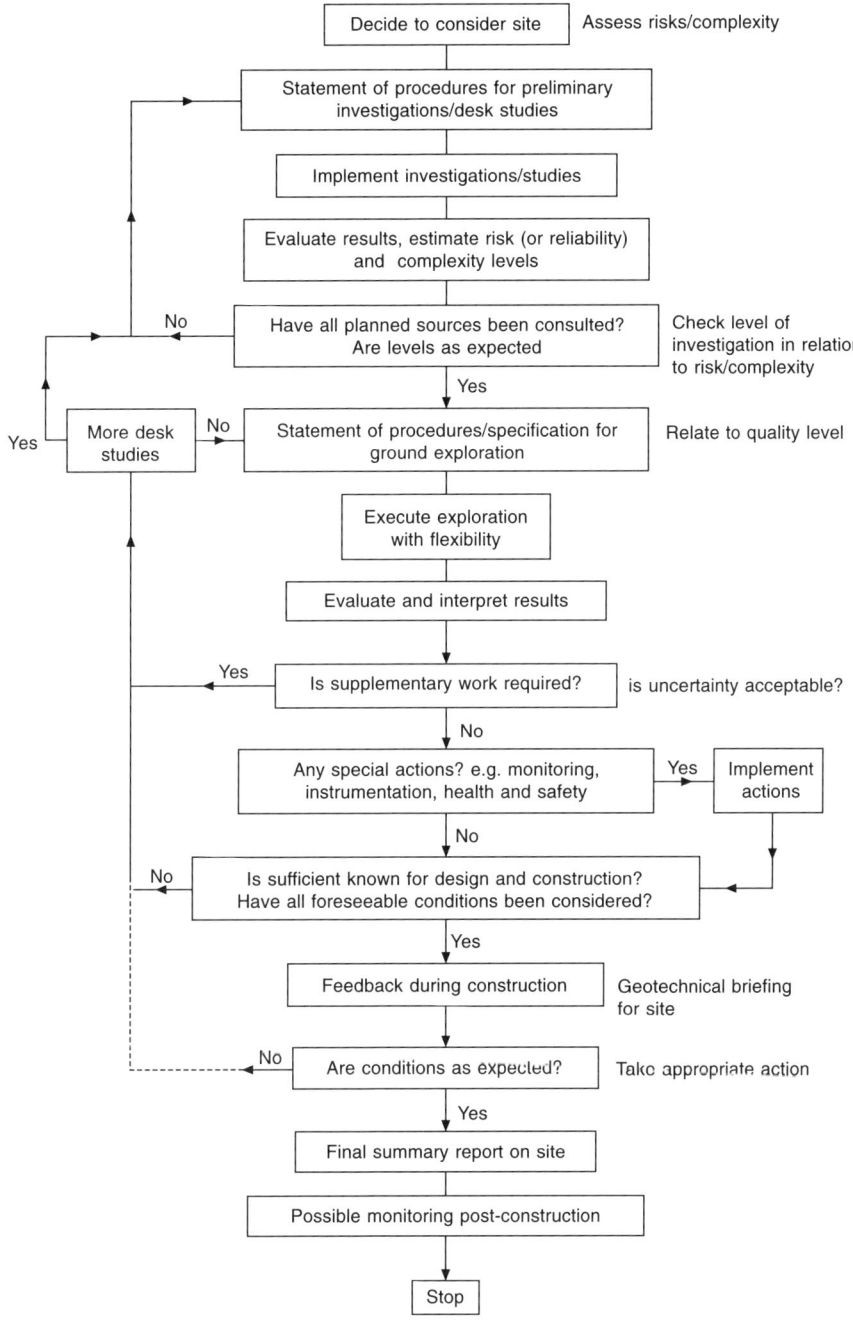

Figure 3.2 Site investigation flowchart (after Whyte and Tonks, 1993).

3.2   Preliminary sources survey (the desk study)

3.2.1   *INTRODUCTION*

The preliminary sources survey (desk study, walk-over survey and possible preliminary explorations) is essential to any successful site investigation. The objective is to determine as much information as possible on ground conditions using available sources of information, supported where necessary by some exploratory work, and site inspection. The survey should be completed before the main ground exploration as the data obtained is used to plan the borings and tests, etc., such that these are both sufficient and cost effective. Table 3.4 reports the results of a 1987 questionnaire on desk study information for site investigations (Peacock and Whyte 1992).

Table 3.4   Desk study information

Question	Client	Consultant	Site investigation firm
Who does it?	44%	33%	23%
Done before site investigation tender documents are prepared?	70%	75%	–
How much is spent as a percentage of site investigation costs?	12.5%	15%	5.5%

Site investigations form a low-cost component of construction, the preliminary sources survey is a very low-cost part of the site investigation. A full 'desk study' can be expected to be achieved for about 0.1% to 0.3% of the cost of construction (i.e. less than £5000 for a £1M contract). A large quantity of information is provided at negligible cost and forms the most cost-effective part of a site investigation. The main problem in implementing an adequate study is having sufficient time to do the work: obtaining information from sometimes obscure sources may take weeks. Effective preliminary sources surveys require good time management; the study should be started at the earliest stages of project planning. A lack of time produces ineffective work.

In a survey of routine foundation design practice carried out by Roscoe and Driscoll (1987) the use of desk study information was considered as being the minimum essential for routine practice. At 42% of sites fewer than four sources were consulted and at only 5% of the sites were more than five sources used. The Mott MacDonald and Soil Mechanics Ltd (1994) report on highways also commented on inadequate preliminary sources surveys leading to construction problems, cost escalation and delay. An example relating to a tunnel construction is reported by Parkinson (1975). In this case, a new tunnel started to remorselessly disintegrate as highly acidic ground water attacked the concrete lining. The source of the acid was a nearby steel mill which for 50 years had been disposing of sulphuric acid into a convenient hole (groundwater pH was as low as 1.9 in parts).

In another case, reported by Whyte (1995), a micro-bore sewer tunnel drive encountered ground problems including an unforeseen petrol tank at a previously dismantled petrol station site and damage to a water main. The petrol tank was removed by a heading driven from a construction shaft and the water main was repaired in open excavation. The project costs escalated by about one-third due to these and other ground conditions. Risk analysis modelling showed the sensitivity of delay and cost escalation related to ground uncertainty and that better site inspection significantly reduced the uncertainty and hence the level of financial risk to the client is reduced. Analysis showed that a delay of one week on site could invoke costs in excess of a full quality site investigation (including the desk study). It was suggested that the cost estimate for a site investigation would be better related to the cost of a week's delay to a project rather than as a percentage of an estimate. A good site inspection should pay for itself by avoiding the risk of at least one week's delay to a project.

3.2.2 SOURCES OF INFORMATION

In the UK Perry and West (1996) and Clayton. *et al.* (1995) provide detailed references to sources of information. Table 3.5 summarises the information which can be obtained from a desk study and sources which provide the data.

Perry and West (1996) give contact addresses for the organisations listed in Table 3.5. It is good practice to plan a preliminary sources survey by setting up an advance checklist of the sources of information to be consulted. This should be reviewed as the work progresses so that any additional searches indicated from the records uncovered can be identified and followed through. The main problem with these surveys is having sufficient time to do the necessary detective work: cost is usually very small and should not be a problem to an informed client.

3.2.3 SITE INSPECTION SURVEY ('WALK-OVER')

An inspection of the site and surrounding areas, including discussions with local residents, greatly increases the information available about a site. The inspection, however, has to be planned and be a follow-on to the 'desk study' since this reveals problem areas worthy of detailed inspection. If you are unaware of a problem (e.g. the potential for landslip) then evidence on site may not be recognised and the problem is overlooked. When forewarned by geological or other information that a problem can exist then the site inspection aims to positively seek evidence, and the problem is less likely to be overlooked.

The object of the site investigation is to confirm, amplify and supplement information detected from the preliminary sources. The inspection consists of two operations: the site inspection and inquiries with locals. The site inspection involves a visual examination of the site in conjunction with maps (topographic and geological), site plans and, possibly, aerial photographs. Perry and West (1996) suggest the following equipment to be useful on an inspection:

- Notebook, pencil, measuring tape, compass, camera, clinometer, binoculars
- Ordnance survey and geological maps, site plans, preliminary geotechnical map, notes and sections, block diagrams
- Air photographs, viewing board, viewing aid, pocket stereoscope, 'Chinagraph' pencil. Portable hand auger, penetrometer, geological hammer, trowel, polythene bags and labels
- Penknife, trowel, hand lens (×10), 50% hydrochloric acid (in small polythene dropping bottle with screw cap).

Appendix C of BS 5930 Code of Practice for Site Investigations (British Standards Institution 1999) provides a useful checklist for procedures and methods of carrying out an inspection. Features to note and observe are:

Table 3.5   Sources of information

Information	Source
1. Site geology	Geological maps – both published and unpublished Publications and records • Regional guides • Memoirs • Journals Groundwater records • Hydrogeological maps • Well records • Papers and reports Engineering geology maps Applied geology maps Air photographs
2. Soil survey	Soil maps Data and report archives
3. Land use: agriculture and planning	Land utilisation survey Land classification maps Planning maps
4. Topography, morphology	Ordnance survey Archive records Air photographs Tithe, enclosure, estate maps
5. Mines and minerals	Geological maps and records Record offices, minerals office Surveyor/mining firms
6. Previous land use	Old maps (topographical, geological, etc.) Old air photographs Airborne remote sensing Archaeological records Mining records
7. Local knowledge	Previous site inspection reports (from official or client records) Local clubs, societies, libraries, universities
8. Land stability (subsidence, landslides)	Mining records National landslide databank
9. Contaminated land	Government records Surveys of contaminated land Insurance plans Trade and business directories Topographical maps Commercial databases
10. Existing construction and services (gas, water, electricity, sewers)	Construction drawings Topographical maps Utility records
11. Meteorological information	Meteorological records

- Land forms:
  - General features such as slopes, changes in form, etc.
  - Glacial features
  - Evidence of mass movements such as hummocky or broken ground, structural damage, alignment of posts, trees, fences
- Geology:
  - Exposures, noting materials, discontinuities. Samples can be taken
  - Vegetation, land use. Small hand auger holes can be made
- Ground water and surface water:
  - Springs and seepages, vegetation, evidence of erosion, old wells. Location of ponds and streams, evidence of flooding, water levels, direction and rate of flow
- Site access
  - Access for ground exploration equipment and vehicles, location of overhead obstructions such as power cables, photographs of entrances and access routes
- Existing structures:
  - Inspect and record details, note cracks and damage. Photograph structures
- Industrial/archaeological survey:
  - Evidence of past land use (old quarries, tips), made ground and contaminated land, old workings, archaeological remains, ancient monuments.

Local enquiries supplement the survey with information provided from persons and organisations with local knowledge. Enquiries can be made with:

- Local engineers and surveyors:
  - Local contractors familiar with the area.
  - Local authority/agency staff responsible for the area
- Statutory undertakers and utility firms:
  - Electricity, gas, water, telecommunications, sewerage, waste water
- Archives:
  - Records in libraries, local authorities, historical/engineering societies (e.g. archaeological clubs, history societies, geological societies, etc.), schools, colleges and universities
- Inhabitants:
  - Evidence from inhabitants, workers, farmers, factory owners.

3.2.4   PRELIMINARY SOURCES SURVEY REPORT

The evidence from the desk study and site inspection should be collated into a report. The report should state the sources of information planned to be consulted at the start of the investigation and identify any which have not been consulted along with the reasons why. The report summarises the results of the investigation and by use of maps, sections and block diagrams the anticipated geotechnical conditions stated. Areas of particular concern and uncertainty are highlighted for further investigation and exploration. Potential hazards, as listed in Tables 3.1, 3.2 and 3.3, are identified. In sewer works particular attention should be given to:

- A geotechnical plan and section along the sewer line showing expected strata, potential variations
- Possible ground water conditions, both near the surface and at depth (e.g. evidence of artesian water)

- Problem areas such as made ground, faults, subsidence areas, fault breccia and gouge, dykes and sills
- Evidence on depth to bedrock and topography of the sub-drift surface
- Potential for obstructions, both natural (e.g. boulders) and man-made (e.g. tanks, quarries)
- Any old mineshafts, wells, adits, boreholes. Sites of demolished structures, possible old foundations, basements, filled ponds. Changes in river courses, old culverts.

For tunnels, information of particular importance is:

- Buried channels and valleys
- Soil/rock interface, bands of hard ground
- Rock faults
- Fissured or heavily jointed ground
- Ground water levels
- Soft soils, silt, sand, peat.

A preliminary survey report has several advantages:

1. It is obtained at low cost and at an early planning stage provided sufficient time has been given to collate the information
2. It identifies potential hazards and areas of uncertainty and allows engineering decisions to be formulated
3. The report enables the main ground exploration to be planned, designed and executed in a cost-efficient and effective way. It is helpful to the interpretation of exploration information
4. The information is of value to the designer, contractor and engineer constructing the works.

On completion, the main ground exploration is planned with budget and time estimates. Contracts for exploration works are prepared, tendered and on award the main ground exploration starts.

3.3   Main ground exploration

Prior to the main ground investigation it may be necessary to conduct preliminary explorations (borings, pits) in areas of particular hazard or uncertainty suggested by the preliminary sources survey report. A geophysical survey can also be of value in the preliminary works. These explorations are useful in planning the main works with greater efficiency.

3.3.1   PLANNING THE EXPLORATION: BOREHOLE LOCATION AND DEPTH

For tunnel construction, boreholes should be located at each shaft and portal position and at key positions along the route to allow interpretation of the geological structure and materials along the alignment. Probes, e.g. dynamic penetrometers or static penetration test, and simple rapid drilling methods, e.g. window sampling, can be used to infill detail and variations between borehole locations. Geophysical exploration can also be used between borings.

At shaft sites boreholes can usually be sited on axis but consideration has to be given to how effectively these can be sealed and backfilled to avoid water ingress from aquifers during construction (Collins, 1972). If water is likely to pose a hazard then the borings should be outside the shaft location and two, or more, may be needed to establish the geological structure at the shaft. Boreholes should not be omitted from shaft positions. Along the sewer line boreholes should be sited a few metres on either side of the proposed line (allowing for later minor alignment changes) so as not to interfere with construction of the tunnel. For cut and cover works borings may be located within the strip to be excavated but further borings off alignment should be considered to assess the three-dimensional ground profile. Inclined or horizontal boreholes (if access is available) may assist in determining the ground structures.

Key boreholes should be taken to some considerable depth below the sewer line in order to help with the geological succession and identify strata (particularly aquifers) which may pose a hazard to the construction. The minimum depth (D) for key borings for cut and cover sewers is:

$$D = 2 \times \text{depth to invert level in excavation}$$

If a confined aquifer with high piezometric level is encountered then this depth, D, becomes a minimum for all borings. For bored tunnels, the minimum depth is $2 \times$ sewer tunnel diameter below the invert. It is recommended that borings be taken to greater depths since the vertical alignment may be designed deeper at a later stage and allowance should be made for this contingency.

For sewer replacement along an existing alignment additional probes/surveys may be possible from inside the existing sewer if safe access is available. Read (1986) suggests the minimum principal dimension of an existing sewer for man entry exploration to be 1000 mm.

3.3.2 METHODS OF EXPLORATION

Trial pits

To depths of about 4 m trial pits are effective and low-cost methods of exploration can be considered for shaft positions and shallow sewers. Direct ground logging is made and large representative samples obtained. Care in location is necessary since trial pits do disturb a considerable zone of ground. In addition, excavations are hazardous for man entry and suitable support and safety measures have to be in place.

Boreholes

The principal methods of boring used in many countries are:

1. Soft ground:
   - Light cable percussion
   - Power augering
   - Washboring
2. Hard ground:
   - Rotary drilling
     - open hole
     - coring.

Kent *et al.* (1992) give details of the drilling technology.

Light cable percussion

This method is popular for soft ground drilling due to its simplicity, versatility and ability to drill through many variable materials, including obstructions. The method, shown in Fig. 3.3, is an adaptation of standard well-boring methods used for many centuries. The subsoil boring rig consists of a 1 tonne diesel engine and a tripod type derrick about 6 m high from which the legs fold down to form a simple mobile trailer for transport.

The boring tools consist of:

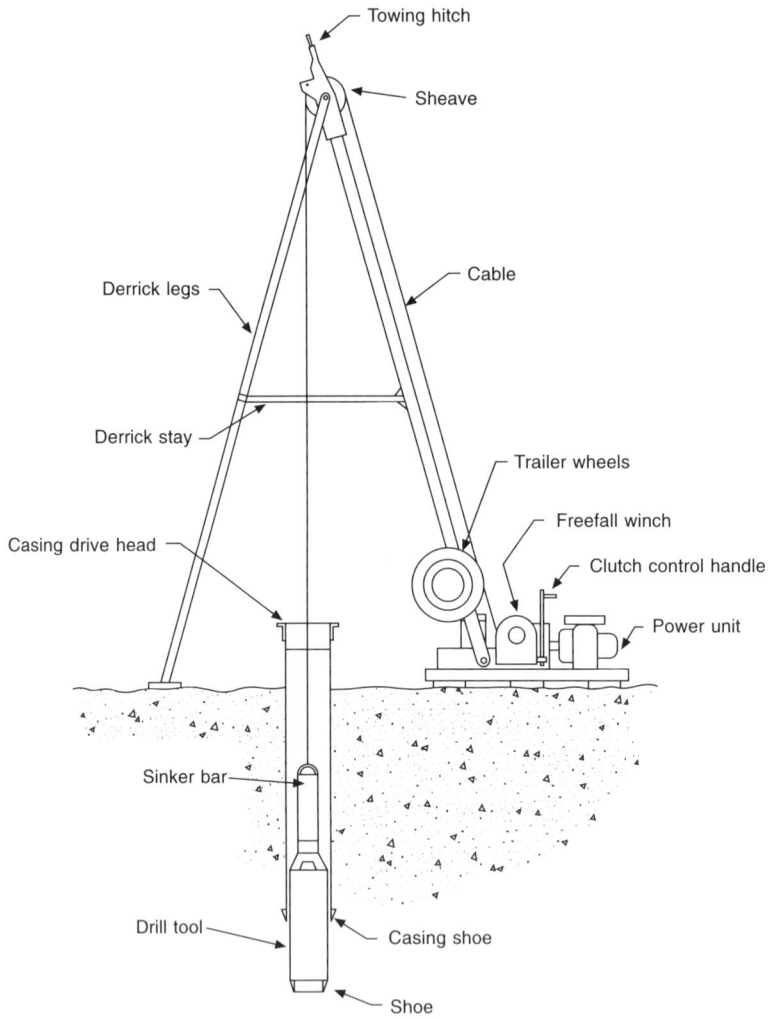

Figure 3.3 Light cable percussion drilling rig.

- The shell: A heavy cylindrical tube about 1.5 m long with a metal or leather clack valve placed between the cutting shoe and body of the shell/bailer. This is used in non-cohesive materials such as sand, gravel and silts. Casing is required to support the sides of the borehole. Sufficient water must be present (at least to cover the clack valve) for drilling to progress – thus water may have to be added to the borehole.

- The clay – cutter: A heavy metal tube about 1.5 m long with windows cut into the side to facilitate emptying of material from the tube. This is used in cohesive soils such as clay and clay mixtures. Casing may not be necessary in stable formations but is inserted in weaker ground to support the hole sides.
- The sinker bar: This is a heavy cylindrical weight added to the boring tools to aid percussion drilling, for example in water-filled holes where additional weight is required.
- The chisel: Heavy chisels are used to break up obstructions or hard materials. The debris is cleaned from the borehole by the shell.
- The casing: Casing is normally about 1.5 m long and has typical diameters of 150, 200, 250 and 300 mm. In non-cohesive soils casing is either driven or falls under its own weight to the bottom of the borehole to provide continuous support. In cohesive soil, the operator can bore ahead of the casing, usually about 1.5 m or one casing length, and the casing is then advanced by driving. It is not usual for casing to be advanced below the borehole base but this may be necessary when difficult drilling conditions develop. Any such advancement must be recorded on the drill log.

Under favourable conditions it is possible to drill to 60 m depth. The diameter of the borehole is a function of the depth, generally each casing size can be driven to a maximum depth of 20 m. Deep boreholes, deeper than about 15–20 m, have to start at large diameter and have casing reduction, i.e. a 60 m deep borehole could be as follows:

Depth (m)	Casing diameter (mm)
0–15	300
15–30	250
30–45	200
45–60	150

The logging requirements for boring are specified in BS 5930 (1981). It is important that all drilling details are recorded since the method of drilling can be in conflict with geotechnical quality requirements. Examples are listed below:

1. Boreholes without water balance: It is normal for boreholes to be drilled without maintaining a high internal water head in balance with ground water pressures. In non-cohesive soils, and in some cohesive soils such as laminated clays, ground water flows upwards through the borehole base (due to the presence of casing) and produces severe disturbance due to loosening and softening of the strata. The disturbance can extend 3–5 diameters below the borehole base and thus influence both sample quality and *in-situ* test data. Thus boring without water balance enables water entries to be noted, along with sealing of the water by casing, but at the cost of probable loss of sample quality and test reliability is improved by maintaining a water balance in the borehole to eliminate seepage through the base, (Fig. 3.4). For best practice, therefore, two borings are needed: one without water balance to investigate strata details with aquifers and aquicludes and one with water balance for sampling and *in-situ* testing.
2. Casing levels: Many drillers are paid on an incentive bonus related to quantity of drilling ('meterage'). This encourages fast drilling and there may be a reluctance to insert casing leading to loose spoil at the borehole base and cavitation/overbreak along the borehole sides. Sampling and testing may not be entirely in 'undisturbed' ground. Drilling ahead of casing also encourages loss of verticality of the hole. In all ground other than stiff homogeneous clays the casing should be kept at or within 1.5 m of the

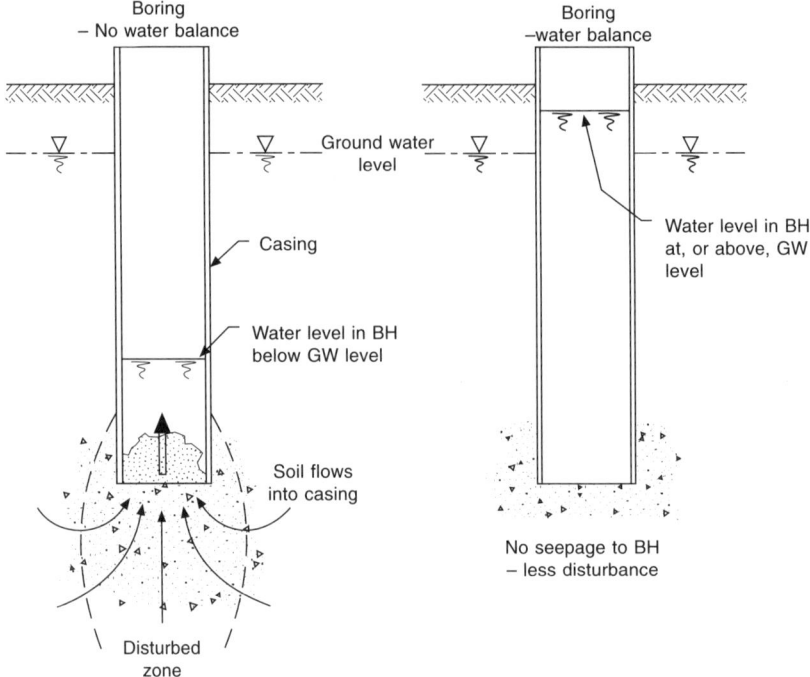

Figure 3.4   Boreholes with, and without water balance.

boring base. Casing should not be advanced beyond the bottom of the borehole except in very special circumstances.

3. Rate of drilling: Problems of softening, loosening and piping can be aggravated by the drilling action. A rapid raising of the shell/claycutter produces a pump action in the borehole with suctions and possible inflow of materials leading to disturbance below the base. The problem is overcome by use of drilling tools of a smaller diameter than normal to reduce the pump action effects.

The reliable and accurate logging of boreholes and sampling/*in-situ* testing requires careful and professional drilling. For this to be achieved the site operations have to be properly planned, specified and procured. It is not achieved through open competitive tendering to drill 'X' boreholes at a site and provide a report.

Power augering

Clayton *et al.* (1995) describe various forms of augers. For site investigation works a popular machine is the continuous flight auger with a hollow stem through which samples can be recovered and *in-situ* tests performed.

Continuous flight augers are best suited to drilling cohesive soils. Problems can occur in soils containing cobbles and boulders (depending on the diameter of the borehole) and when working below the ground water level in cohesionless soils. The machines need considerable mechanical power to operate and are relatively heavy, generally mounted on large vehicles. Access problems can occur along with trafficability on sites with weak surface soils. As with percussion drilling, augering can conflict with geotechnical requirements.

1. Thrust: Heavy downward thrust to aid penetration of the auger into the soil can cause displacement and disturbance to the soil ahead, i.e. in the sample/test zone, particularly in soft and firm soils.
2. Logging: Soil in the auger flights becomes mixed during drilling with consequent problems in strata identification and logging. Consecutive sampling can overcome the problem but with a reduction in drilling rate. Logging also then requires the samples to be extruded – this may be at some later time.
3. Sampling: The maximum sampler diameter is 100 mm due to restrictions in the diameter of the hollow stem. In some systems, the sampler design has to be modified to suit the hollow stem and base plug. Such modifications may increase aspects such as area ratio and inside clearance ratio and result in a lowering of sample quality. The sampler specification should be checked against quality requirements.

The hand auger provides a means for drilling and sampling boreholes up to 200 mm diameter and to maximum depths of about 5 m.

Washboring

Washboring uses water jets from the base of drill rods to advance the borehole. The rods are rotated and surged up and down by the drilling crew using a rope wound around a spinning cathead. The method was widely used in the USA but is now being replaced by power auger drilling.

Washbores produce small diameter borings (usually about 66–75 mm diameter) and historically are significant in that the SPT was developed with this equipment. This explains the 150 mm seating drive in the SPT test (this being at least $2 \times$ borehole diameter below the base into 'undisturbed' soil) and why SPT N-values from washbores are generally larger than from cable percussion bores. In the washbore, the SPT hammer is raised by a hemp rope wound around the spinning cathead, thus 'free' fall is impeded as the rope unwinds on release. In the UK, the hammer is released by a trip mechanism and imparts a greater energy to the sample resulting in lower blow counts. The SPT N-value is thus also dependent on the borehole diameter, for example BS 1377 (1991) restricts the maximum bore diameter to 150 mm (i.e. the test is performed in soil at a depth below $1 \times$ borehole diameter). Clayton (1995) describes these issues in more detail.

Rotary drilling

Rotary drilling is in five groups, Kent *et al.* (1992) describe the methods:

- Direct flush ('open' hole)
- Diamond core
- Masonry core
- Reverse circulation
- Auger (see above).

In site investigations rotary core drilling is used to prove geological formations and to determine characteristics such as rock classification, rock quality designation, compressive strength, permeability and other physical properties.

The recovery of a sample of rock with a rotating diamond bit is relatively new, its introduction in 1863 being attributed to a Swiss engineer, Leschot. The basic principle is to cut an annulus of rock with a diamond-tipped bit and then to retain the rock core in a hollow tube that is brought to the surface. Many refinements and variants now exist and the art of drilling is now a sophisticated business. Diamond-cored boreholes range between 20 mm and 150 mm in diameter and may reach depths of 2000 m. Larger cores are possible with specialist drilling rigs.

The prime objective is good core recovery without excessive drill-induced fractures. One hundred per cent recovery is the aim and a minimum 90% recovery should be required unless it can be shown this is impracticable. In general, the larger the core size the better the recovery. Recovery is enhanced by the core barrel selected:

- Single tube core barrels: The simplest type suitable only for massive and uniform hard rocks – rarely used in practice.
- Double tube rigid core barrels: Both outer and inner tubes are fixed to the core barrel head with the bit attached to the outer barrel. Suitable for hard and non-fractured formations only.
- Double tube swivel core barrels: A bearing permits the inner tube to remain stationary while the outer tube rotates with the cutting bit. Best suited to fractured medium and hard rocks and can be successful in soft friable formations. Special barrels, with inner split tubes, permit recovery from very weak/highly weathered formations and may even recover overburden samples
- Triple tube swivel core barrels: A modified version of the double tube system in which a split liner is placed within the inner tube. Suitable for soft friable formations.

The methods of boring are summarised in Table 3.6 and compared using the scale:

Table 3.6   Drilling methods (after Kent *et al.* 1992)

Strata description	Cable percussion boring	Continuous flight augers	Rotary drilling		
			Direct flush	Diamond core	Reverse circulation
Sands and gravels Alluvial deposits	1	1[a]	2	3[b]	1[c]
Boulders and cobbles Boulder clays	3	4	3	4	3
'Soft' sediments: clay, marl, shale	2	2	1	3	1
Consolidated sediments: mudstone, chalk, sandstone	2	3	1	3	1
'Hard' sediments: limestone, dolomite gritstone	4	4	1	1	2[d]
Metamorphic: slate, marble, schist	4	4	2	1	4
Igneous: gabbro, dolerite, basalt, granite	4	4	2	1	4

Notes: (a) Can be difficult once ground water encountered.
       (b) Adapted equipment to core sands, alluvials. Gravel difficult to drill.
       (c) Near surface strata will require protection of lining tubes.
       (d) Heavy loading required on bit. Other drilling methods may be preferable.

1 = Best
2 = Possible
3 = Difficult
4 = Not recommended

Backfilling boreholes

Poorly backfilled boreholes can cause construction problems in deep excavations and tunnels, for example air loss in compressed air workings and water inflows from deeper aquifers. The best backfill practice is to grout the borehole through a tremie pipe using a cement/bentonite mix, say about 4:1.

Pits and probes

Trial pits provide a quick and economical method of exploration to about 4 m depth and allow a direct examination of the ground. The location of pits has to be considered since they disturb a considerable volume of ground. For man access safety has to be provided through adequate support and escape means.

The disadvantages of trial pits are:

* Problems in excavating below ground water level
* Increased complexity and cost for depths below about 4 or 5 m.

A wide range of static and dynamic penetrometers is commercially available. These offer relatively low cost methods to explore strata and ground variations between boreholes, for example:

* Mackintosh probe
* Dynamic probe (light/heavy)
* Weight sounding
* Window sampling.

Dynamic penetration probes are generally available. Typically, a solid cone is driven continuously into the ground (e.g. a 50 kg weight falling 500 mm) and the blows for each 100 mm penetration are recorded. It is not possible from the results to determine the soil type being tested and correlation can be difficult. Where conditions are favourable, e.g. a soft stratum overlying a stiff layer, then probing provides a cost-effective profiling method. Tentative correlations with geotechnical test data can be made (Tonks and Whyte, 1988).

Window sampling is a lightweight, portable drilling system in which 1.5 m length tubes are driven quickly into the ground. Depths to about 7 m are possible through reduction in tube diameter from 66 mm at the start to 25 mm at the end. Slots, or windows, are cut in the tubes and thus allow a continuous log to be made.

The location of probes/window sampling has to be chosen with care since such holes can be difficult to backfill effectively. They should be located outside any underground workings unless conditions are favourable, i.e. no potential hazard to the workings from gas/water entry.

3.3.3 GEOPHYSICAL INVESTIGATIONS

Background

Geophysical methods have been used in investigation since the 1950s and in the last 10 to 15 years have progressed and developed into powerful tools due to technological advance and processing capability.

Geophysical investigations fall into two main categories:

1. Potential field methods which utilise gravitational, magnetic, electromagnetic, electrical and thermal fields within the earth. Anomalous zones are detected and then interpreted to infer geological features.
2. Induced methods in which signals are sent from seismic, electrical or electromagnetic sources and the ground response is measured.

Geophysical surveys provide data which measures lateral and vertical variations of a physical property which can then be interpreted. There are two main approaches in geophysical surveys for engineering works:

1. Surveys established on a grid basis over a site with measurements made at grid points or constant line separation. Contours can be drawn which show up anomalous zones, or 'hot spots'. Such anomalies may relate to ground features such as faults, swallow holes, buried fill materials, mineshafts, etc.
2. Surveys to establish vertical variations in a property through measurements along a horizontal profile. The ground response is analysed and the geological structure is inferred from mathematical models. For an effective survey the model has to interact with and be conditioned by factual geological information from boreholes and field mapping.

The first approach, i.e contour mapping, is much quicker and less expensive than the second approach. The modelling of geological structures in three dimensions through vertical and lateral profiling can cost about five times more than a survey to produce a simple contour map of a geophysical property.

In considering geophysics the following five factors have a major influence on the choice of method and output performance.

1. Penetration: The required depth of investigation into the ground is important. For example, some methods such as ground probing radar have limited depth of penetration and can be further reduced under certain conditions, e.g. waterlogged ground.
2. Vertical and lateral resolution: The resolution, i.e level of measurement accuracy, has to relate to the intended measurement target. Thus radar achieves excellent resolution when significant penetration of the ground is achieved. In seismic surveys the seismic reflection method may only detect vertical profiles to a resolution of tens of metres since the low frequencies fail to resolve fine details. Higher frequencies produce finer resolution but at the expense of environmental noise and problems of signal attenuation. Technical advances over the last 10 to 15 years have greatly enhanced the resolution of geophysical measurements, for example signal enhancement through microprocessor-based digital techniques to produce measurements to fine limits.
3. Contrast in physical properties: For a geological structure to be identified by geophysics then it has to exhibit a contrast to the signal being used to identify it. For example, the boundary between two geological strata will only be found through seismic refraction if seismic velocity distinctly changes at the boundary. Similarly, in mineshaft surveys it is difficult, and may be impossible, to detect a shaft through magnetic anomaly if the shaft is backfilled with material from the surrounding ground. If, however, the shaft is bricklined or filled with ferrous debris then it can be detected as a magnetic anomaly.
4. Signal to noise ratio: If the signal to noise ratio is low then the required signal, e.g seismic or electromagnetic reponse, may not be observed within the ambient noise. This

is an area of significant technological improvement over the last 10 to 15 years through the use of electronic filtering techniques and digital enhancement.
5. Factual information on the site: All data has to be interpreted and correlated. Thus factual information, e.g. boreholes at key positions, provides the evidence to refine a mathematical model for a site and improves the reliability of interpretation between the boreholes.

Geophysics has advanced significantly in recent years due to high quality equipment accompanied by hardware for data acquisition, recording and processing. It is now possible to undertake 'live' interpretation of geophysical data during a survey and thus adjust the field measurements to improve the reliability of the information. This flexibility, however, produces new difficulties in control and management communication and requires collaboration between the geophysicist, geotechnical engineer and client.

Geophysical methods are likely to be used more often for sewer investigations since they are cost effective, provide 2-D and 3-D images of ground conditions and are indirect and non-intrusive. Thus, for example, in sewer replacement surveys structural information on existing sewer linings, and possible external cavities, can be obtained with ground probing radar (Nakamura *et al.*, 1994). The methods of investigation are both complex and extensive in nature (Kearey and Brooks, 1990) and specialist advice should be obtained. A brief commentary on geophysical methods is given below:

Gravity: Changes in micro-gravity values are measured, the changes arise from vertical and lateral density variations in the ground.

Advantages: Can be carried out in areas where background 'noise' prevents use of electromagnetic and seismic surveys. The ground penetration is greater than with magnetic surveys.

Disadvantages: Data needs specialist and experienced operators, is relatively slow to obtain and has poor lateral resolution.

Magnetic: This is carried out with a portable magnetometer and sensors mounted on staffs. Measurement is taken for the intensity of the earth's total magnetic field. Can be used to locate features such as mineshafts, drums, buried services.

Advantages: A quick reconnaissance method for ferrous targets. High measurement rates enable good lateral resolutions to be obtained. When a gradient array is used (two or more sensors used simultaneously) then shallow ferrous targets are detected with good resolution.

Disadvantages: Measurements are affected by 'noise', i.e. interference from buried and overhead cables, vehicles, pipes, fences. The interpretation can be difficult in order to model depths and volumes. This is particularly so if ferrous targets are clustered at depths below about 3 m, then resolution of such targets can be poor.

Electromagnetic: A time-varying electromagnetic field is generated using handheld equipment and induces currents which produce a secondary field with a strength proportional to conductivity.

Advantages: Can detect both ferrous and non-ferrous targets relatively rapidly and can be used as a metal detector to a depth of about 3 m. Processing of the data can lead to indications of disturbed ground conditions in otherwise undisturbed areas. It may be used to interpret variations in ground water quality.

Disadvantages: Measurements are affected by 'noise', i.e. interference from buried and overhead cables, pipes, fences. Quantitative modelling requires repeated measurements at different array geometries. Measurements of ground conductivity are limited to less than about 100 ms/m.

Thermal: Thermographic surveys measure temperature differences in the ground. Surveys normally require a helicopter and a flight path along a sewer line can be a useful screening in the preliminary survey.

Infrared photography detects differences in reflected energy. It can highlight areas of vegetation distress that may indicate disturbed ground, contaminated land or gas. Vegetation distress can, however, be produced naturally due to seasonal and other effects on plant growth. Aerial surveys are normal though model aircraft have been used.

Advantages: Detection of temperature differences and reflected energy at relatively low cost and relatively rapidly.

Disadvantages: Seasonal and weather effects can influence the interpretation of survey data; shadows can be problematical.

Seismic reflection: A shock wave is induced in the ground, often by a hammer blow on a steel plate. An array of geophones detects both compression (P) and/or shear (S) waves that have been reflected on an acoustic boundary. These surveys have been carried out in practice over many years.

Advantages: For rock excavations an estimate of rippability is possible. The P and S wave velocities can be combined with density to estimate elastic moduli. Thickness and depth of lithological units may be measured. It may be possible to detect the depth of the ground water table. Vertical boundaries can be measured, e.g. old quarry walls, etc.

Disadvantages: Requires the ground strata to have distinctive seismic velocities that increase with depth. Data production is relatively slow and may be interfered with by background noise, e.g. traffic, ground vibration.

Electrical resistivity: The apparent resistivity of the ground is measured using an array of electrodes that are placed in the ground. Can be used for both traversing and vertical profiling.

Advantages: May differentiate between saturated and unsaturated soils and may provide profiles and depths of strata. Detection of cavities, faults and fissures is possible. The equipment is compact and portable.

Disadvantages: Resistivity may not be as cost effective as other methods (e.g. magnetic survey). It is affected by local noise, e.g. electric cables, railway lines. If the ground is highly heterogeneous then interpretation is difficult. In soundings, large electrode spacing may be necessary to obtain a significant depth of penetration (say 50 m or more).

Ground probing radar: A radar unit produces a pulse of electromagnetic waves in the microwave frequency and the antenna detects reflected signals.

Advantages: The results can be immediate; data acquisition is rapid. High resolution of near surface targets, ferrous and non-ferrous, is possible, e.g. pipes (plastic and metal), voids, disturbed ground.

Disadvantages: Poor penetration in high conductivity ground, e.g. saturated clay. A contrast in dielectric properties is needed to detect a target. Can be sensitive to noise, e.g. metal structures, radio transmitters, power lines can produce signals which can saturate a sensitive receiver. Expert processing of data is necessary and a licence is required for operation.

Planning a geophysical investigation

Where scrutiny of case histories shows a geophysical survey to have produced poor results (Darracott and McCann, 1986) then the cause can be attributed to:

- Inadequate and/or bad planning of the survey
- Poor specification (inappropriate science)
- Human factors: Inexperienced personnel and poor management/communications.

The choice of the best technique for a given situation requires specialist experience and advice. Geophysical surveys have to be integrated with the overall investigation and be complementary to other activities. The majority of engineers are not sufficiently knowledgeable to specify and procure the correct technique and the following procedure is recommended:

1. Appoint a geophysicist as a technical adviser to advise on:
   - The geophysical nature of the problem
   - The most appropriate technique
   - Suitability of the site for geophysics, e.g. possible background noise and interference from cables, pipes, metallic elements such as sheet piles, etc.
   - Programme and cost estimates
   - Presentation and interpretation of results.
2. Review the preliminary sources study to aid the selection of the best geophysical technique.
3. Conduct a geophysical test survey to compare methods and techniques and to optimise the detailed investigation.
4. Analyse data in a preliminary form at the site and modify investigations if required.
5. Check and calibrate data against correlation boreholes and site observations, e.g. outcrops, dips.

In common with all investigatory methods the adoption of good management, communications and partnership with specialists encourages value and success in the survey. A rigid adherence to often inappropriate specifications and inflexible contract conditions can lead to conflict and dissatisfaction with the result. More particularly, however, the survey may then be inadequate with subsequent high risk to the contract from 'unforeseen' events that could have been 'foreseeable'.

3.3.4 SAMPLING AND SAMPLE QUALITY

Purpose and types of sample

Samples are recovered from the ground for two main purposes:

- To describe the various strata such that a geological borehole log can be prepared
- To provide material for examination and testing in the laboratory.

To meet these objectives there are two basic types of sample:

- Disturbed samples taken from the drilling tools as the borehole is being advanced, or from the cutting shoe of tube samplers. Disturbed samples should allow the type of ground to be identified but often any fabric or structure is destroyed. Different drilling tools produce different degrees of disturbance.
- 'Undisturbed' samples obtained from sampling devices normally tube samplers in soils and rotary core samplers in rocks. All samplers experience disturbance to some degree and the quality is a function of the design of the sampler tube, its method of insertion into the ground and the method of drilling.

Sample quality and disturbance

In practice the specification for a site investigation will require that 'undisturbed' samples be obtained. This is never the case since no sample is ever completely undisturbed; what has to be considered is the degree of disturbance along with how representative the sample is for subsequent test purposes. Rowe (1972) modified a German classification system for sample quality that was then adopted in BS 5930 (1981, 1999). Table 3.7 outlines the quality levels for soil sampling.
   Sample disturbance decreases sample quality and is caused by:

- Borehole drilling effects
- Sampler design and method of sampling.

Borehole drilling effects have been described in section 3.3.1; ground disturbance from drilling can readily extend 1–4 borehole diameters below the base of a borehole and possibly to as far as $10 \times$ borehole diameters or more when 'blowing' conditions develop under hydraulic flow through the borehole base. Such disturbance can be minimised by good drilling practice.
   Disturbance can occur to both 'disturbed' and 'undisturbed' samples. Disturbed samples are affected through:

- The cutting action of the drill tool breaking the ground into small pieces
- Water or drilling fluid in the borehole leading to the suspension of fines and loss of small particles from the sample.

This latter form of disturbance is of particular significance since simple index tests, e.g. particle size distribution curves, are then erroneous and can lead to engineering problems relating to drainage/pumping or grout take/grout choice.
   Sampler design and method of sampling both influence the degree of disturbance. Hvorslev (1949) and Clayton, *et al.* (1995) provide detailed discussions on soil disturbance. The following points are of significance:

- Remoulding of the soil: Samplers that displace a large volume of soil relative to the

Table 3.7   Quality levels for soil sampling

Quality class	Typical sampling procedure	Purpose	Properties that can be reliably determined
1. 'Undisturbed'	Piston sampler with water balance	Laboratory data on *in-situ* soils	Classification, moisture content, density, strength, compressibility, permeability, coefficient of consolidation
2. 'Slightly disturbed'	Pressed or driven thin or thick-walled sampler with water balance	Laboratory data on *in-situ* insensitive soils	Classification, moisture content, density, strength, compressibility
3. 'Substantially disturbed'	Pressed or driven thin or thick-walled sampler, water balance in permeable soils	Laboratory data on remoulded soils, fabric examination	Classification, moisture content. Remoulded properties
4. 'Disturbed'	Bulk and jar samples: cohesive soils	Laboratory data on remoulded soils, sequence of strata	Classification. Remoulded properties
5. 'Heavily disturbed'	Bulk samples, washings: granular soils	Approximate sequence of strata	None

Notes: (a) Samples should be well sealed with 25 mm of paraffin wax and labelled immediately on recovery.
  (b) Samples are best stored in a vertical position at 80% minimum constant humidity and 10°C maximum temperature. Freezing must be avoided.
  (c) Extrusion should be in a vertical position. Extrusion and testing should be performed as quickly as possible after sampling.

Table 3.8   Routine sampling methods and quality class

Ground condition	Sample method		
	Bulk/jar samples	SPT	U100
Granular soils with gravel, cobbles and boulders	5	5	–
Sand	5	4	4
Silt	5	4	3–4
Lightly overconsolidated clays (many soft clays)	4	4	2–3
Overconsolidated clays (many stiff clays)	4	4	2–3
Clays with gravel, cobbles and boulders	4–5	5	3–4–5

Note: U100 samplers provide better quality samples when they are clean and have undamaged cutting shoes with acceptable area and inside clearance ratios.

volume of soil sampled cause remoulding. A numerical measure is provided by the area ratio, see Fig. 3.5.

$$A = \frac{D_w^2 - D_c^2}{D_c^2}$$

Where D_w = outside diameter of the cutting shoe
        D_c = inside diameter of the cutting shoe

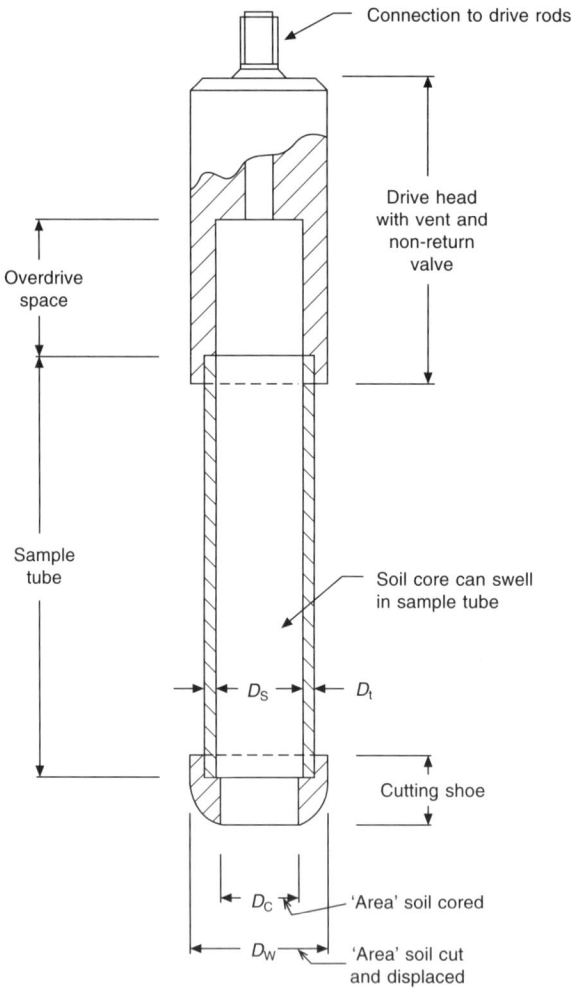

Figure 3.5   Thick-walled tube sampler.

For minimum disturbance the area ratio should be less than about 10% to 12%, values attained by thin-walled tube samplers and piston samplers. Thick-walled tube samplers, e.g. the U100, have area ratios of about 25% to 30%. The SPT sampler area ratio exceeds 100%, i.e. more soil volume is displayed during driving than the volume of core recovered.

• Volume change of the soil: Samplers that have cutting shoe diameters smaller than the tube diameter permit the soil to increase volume by expansion and stress relief. If this occurs under water then the moisture content increases (with softening of the sample) while if it occurs in air then the samples become partially saturated. Partial saturation leads to anomalies and error in many laboratory tests. A numerical measure is provided by the inside clearance ratio:

$$IC = \frac{D_t - D_c}{D_c}$$

Where D_t = inside diameter of the sample tube. For minimum volume change the inside clearance for 'short' samples (length to diameter ratio less than 5) should be in the range 0% to 0.5%, values attained by thin-walled tubes and piston samplers. The U100 sampler has typical inside clearance ratios of between 1% to 3%.

• Method of driving: Sampling drive methods can be rated as follows:

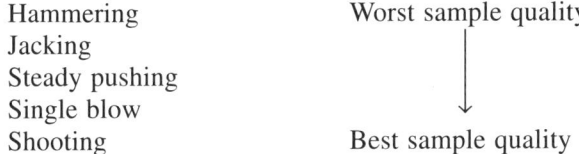

Hammering   Worst sample quality
Jacking
Steady pushing
Single blow
Shooting   Best sample quality

Routine practice with thick-walled tubes (e.g. U100) is to hammer the sampler either from the top of the hole or down the hole using a jarring link. Piston samplers are commonly jacked or pushed into the ground. Clayton *et al.* (1995) suggest a satisfactory sampling speed of 25 mm/s.

Representative samples

The size of sample governs the maximum dimensions of test specimens, such specimens should be representative of the material to be tested if the results are to be reliable. Criteria for sample size are:

• Specimens to possess representative soil fabric details for reliable permeability and consolidation behaviour
• Specimens to represent fissuring, joints, particle size distribution to give reliable measures of strength and stiffness
• Sufficient material to be provided for the range of tests required.

Rowe (1972) found the following minimum specimen sizes for various soil parameters; these are summarised in Table 3.9.

Sampling and engineering practice

From the above, the quality of soil samples is influenced by:

• Drilling technique
• Sampler design and recovery method
• Disturbance from (a) remoulding and (b) volume change
• Sample size (representative specimen).

Thus standard and routine procedures produce specimens of variable, and generally low, quality. For engineering works of low risk the results from such methods provide a cost-effective approach provided it is recognised that results will be subject to uncertainty.

Table 3.9   Minimum specimen size (mm)

Soil	Undrained strength, C_u	Angle of shearing, $\varnothing'$	Compressibility M_v	Consolidation C_v
		Parameter		
Non-fissured soil	100 to 250	100	75	250
Fissured soil				

Notes: (a) Larger specimen sizes may be necessary to represent fissure spacings and soil fabric details.
(b) Smaller sizes may be possible in non-fissured uniform fabric-free soils. These are geologically rare.
(c) 'Routine' 38 mm diameter specimens are rarely representative for measuring soil strength. 75 mm diameter specimens are not representative for measuring the coefficient of consolidation.
(d) For soil classification, etc., the minimum sample dimension should be five to ten times the maximum soil particle diameter. For particle size distribution and compaction tests, therefore, large bulk samples are required for coarse soils, see Table 3.10.

Table 3.10   Sample sizes for various tests

Soil type	Maximum soil Particle size (mm)	Particle size distribution	Compaction	Soil stabilisation
		Mass of disturbed soil sample (kg)		
Clay, silt, sand	2.0	1	25–60	100
Fine and medium gravel	20	5	25–60	130
Coarse gravel and cobbles	200	1000–2000	60(a)	160–250(a)

Note: (a) In standard tests the maximum particle size is 20 mm, larger particles being discarded. The test specimen is thus a soil of different grading to the *in-situ* soil and care is necessary in interpreting data.

Developing data for design and construction is then more of an art than a science and interpretations should be cross-correlated where possible through relationship testing, empirical data and local knowledge and experience.

For projects where risks to life and property are high, which can include sewer works, then the required quality has to be specified and implemented. This can require non-standard and non-routine methods to obtain reliable data such that any uncertainty is acceptable. The evidence from project cost modelling (Whyte and Peacock, 1988) is that the cost of such investigations is repaid many times from the reduction in capital uncertainty and the reduction in delay risk to the works.

3.3.5   IN-SITU TESTING

In-situ testing can be divided into the following three categories:

1. Tests in boreholes:
   - Standard penetration tests (SPT)
   - Permeability tests
   - Packer tests
   - Down-hole geophysical logging

- Pressuremeter test
2. Tests in trial pits:
   - Penetration tests (hand)
   - Plate bearing tests
   - Vane tests
   - Density tests
3. Tests at ground/formation level:
   - Plate bearing
   - Vane tests
   - Density
   - California bearing ratio (CBR)
   - Static and dynamic probing
   - Self-boring pressuremeter.

The most common tests are SPT and permeability tests in boreholes along with static and dynamic probing. The intention of an *in-situ* test is to provide a direct measure of the property under investigation with the test being, hopefully, on undisturbed ground.

Standard penetration test

The test records the resistance of the soil to penetration by a standard sampler (or solid 60° cone in coarse soil) when driven by blows from a standard drop hammer. Test details are given in BS 1377 (1991). The driving resistance is recorded as the number of blows (N) for the sampler to be driven the final 300 mm over a 450 mm drive length. The test is popular since it can be readily incorporated within conventional drilling techniques and it is inexpensive since it takes only about 15 minutes to execute. The SPT, however, is subject to all the problems of borehole base disturbance described for soil sampling and N-values can be subject to uncertainty. Whyte (1985) and Clayton (1995) discuss details. Clayton *et al.* (1995) state that, due to problems of 'boiling' in granular soils, the light percussion boring rig used in the UK cannot give good results for the SPT even when the highest levels of specification and supervision are supplied. Best results with the test are from small diameter (about 75 mm) rotary or washbored holes, with mud flush, where drilling tools are lifted slowly from the borehole and the casing is kept no more than 1 m above the base of the borehole.

Permeability testing

Permeability testing in boreholes is:

1. Rising or falling head tests
2. Constant head tests
3. Borehole packer, or Lugeon, tests.

RISING OR FALLING HEAD TESTS
This test is generally used in lower permeability soils – in high permeability ground the flow of water to or from the soil is too rapid for reliable measurement. In principle, the test consists of cleaning out the bottom of a cased borehole, applying a hydraulic head greater than (or less than) that in the ground and then measuring the flow of water into the soil at intervals of time, until equalisation. The test may be made more reliable in determining permeability when performed on a piezometer installation in a borehole, or by setting the

borehole casing some 1 m to 3 m above the base with medium gravel fill in the unlined section.

The test result can be influenced by water leakage on the outside of the casing and through joints in the casing. Test results become less reliable for fine soils such as silts and clays unless the test zone can be properly sealed. Piezometer tests are preferred for silts and clays for this reason. BS 5930 (1981) discusses the test and data interpretation.

CONSTANT HEAD TEST
This is generally used in ground of relatively high permeability. In the test, water is pumped into the borehole such that a constant head, above that in the ground, is maintained. The rate of flow is measured and recorded at time intervals. By adjustment of a simple bleed off valve system the rate of water flow can be varied to produce a range of constant heads above the ambient level. Soils of low permeability can be subject to constant head testing in piezometer installations, though if too high a head is applied hydraulic fracturing can occur. This fracturing develops when the water pressures exceed the *in-situ* total stresses. If hydraulic fracturing occurs then the results are totally misleading and the apparent permeability is much higher than the soil value by as much as several orders of magnitude.

PACKER TESTS
Packer tests, also known as Lugeon tests after the French inventor, are normally used in rocks to measure the rate at which the formation surrounding the borehole will accept water under pressure. The results are interpreted either in terms of permeability or in Lugeons (water flow in litres per metre of test section per minute at a pressure of 1000 kN/m^2). The packer test is similar in principle to the borehole permeability test used in soils, the main difference being that the length of borehole to be subjected to test is isolated by means of expanding packers inflated by gas pressure from a nitrogen bottle. Packer testing is particularly suited to determine whether or not cavities, fractures and fissures in rocks are open, i.e. of high water flow, or are filled with gouge, i.e. of low water flow. Testing can be by means of a single packer which isolates the base section of a borehole for test, or by a double packer mounted one above the other to seal a length of borehole. Typically, test zones are 3 m to 5 m in length and the maximum water pressure head is limited to avoid hydraulic fracture. The minimum packer inflation pressure is normally about two to three times the applied water pressure, with an upper limit of about 2000 kN/m^2. The length of packer should be at least five diameters of the borehole.

Common difficulties relate to inefficient sealing of the test section by the packer through:

- Irregular borehole walls, for example when drilling a zone of fractured rock
- Oversize borehole diameter, for example when drilling through a band of weak rock in otherwise more competent strata.

Single packer tests are preferred since in a double packer system the effectiveness of the lower packer seal can be difficult (if not impossible) to check. Single packer tests are, however, more costly since they have to be performed as the borehole is advanced. Double packer tests are usually carried out on completed boreholes. BS 5930 (1999) provides further information on the test and interpretation of data.

Static probing, cone penetration test (CPT)

Static cone testing consists of pushing a 60° cone of face area 10 cm^2 into the ground at a constant rate of penetration of 2 cm/second. The response of the ground to the cone is measured. In its simplest form, the cone end resistance and side sleeve friction are measured. In more advanced versions pore water pressure response is measured (the piezocone) and in specialist applications ground contaminants can be detected (environcone). From the data it is possible to correlate with soil type (from the ratio of side friction to cone end resistance) and geotechnical properties such as strength, relative density and stiffness (particularly for granular soils).

Details on CPT applications are given by Meigh (1987) and standards such as BS 1377: Part 9 (1991) and ASTM D3441 (1986). With advances in data logging and processing, cone tests can be analysed as they progress. This gives the engineer the opportunity to vary the test sequence and programme on site.

Cone tests should always be considered in conjunction with boreholes in order to check on empirical correlations on a site-specific basis. The depth of penetration is limited to the reaction weight (normally the vehicle) and hard strata or layers may not be penetrated. Cone tests should not be used where artesian ground water is to be encountered since sealing the water flow may not be possible. Equally, care has to be taken on contaminated land sites where unsealed probe holes may permit cross-contamination through ground water flow.

Dynamic probing (DPT)

Dynamic probing consists of driving a solid cone continuously into the ground using repeated hammer blows. The number of blows to drive the cone each 100 mm increment is recorded ($N100$), rod friction being estimated from torque measurements. Probing can be classified as 'light', 'medium' or 'heavy' dependent on the hammer weight and fall: typical values are a 50 kg weight falling 500 mm. Attempts have been made to correlate the results with SPT N-value and undrained shear strength, C_u (Tonks and Whyte 1988). The test procedure is given in BS 1377: Part 9 (1991) and German DIN 4094, Parts 1 and 2 (1980).

The DPT is ideally suited to ground profiling between boreholes where marked differences in strata hardness exist, e.g. weak ground over more competent soil, old ponds, etc. The method is rapid and portable and can provide much data at low cost. It is not possible, however, to correlate the $N100$ blowcounts with soil type and interpretation can be difficult unless layers exhibit markedly different penetration resistances.

3.3.6   GROUND WATER

Ground water can be a major factor in the design and construction of sewers in both open cut and cover and tunnelling. Determining ground water conditions with reliability and accuracy is very difficult and is likely to be more difficult than finding geotechnical properties such as strength, stiffness, etc. Engineers are well advised to consider ground water investigations more seriously within the general site investigation programme.

In the climatic conditions of the UK, most soils are virtually fully saturated with water at shallow depths beneath the surface. The water pressure depends upon the seepage conditions and physics of the soil mass. The pressure can be less than atmospheric pressure (i.e. negative in value due to suction from capillarity and osmosis) or greater than atmospheric (i.e. positive). The ground water table is defined as being the imaginary surface in the soil water system along which the pore water pressure is at atmospheric pressure (i.e. zero pressure). The water table is not analogous to the surface of a lake or stream (underground lakes and rivers do not exist except in cavernous limestone rocks), the ground can be saturated for many metres above the water table due to suction effects.

Under conditions of seepage water flows from zones of high total head (h) to low head in both the suction and positive pressure regions. Thus a correct interpretation of water pressure distributions (which includes the water table) requires knowledge of the total head and its variation throughout the soil mass. It is to be noted that flowing water, or seepage, can be expected to be the norm in nature rather than the hydrostatic condition frequently assumed by engineers.

Methods of observing ground water rely on a flow of water to a measuring system until an equilibrium state is achieved. The time taken for this to occur is known as the response time. This response time is a function of two main parameters, the volume of water flow necessary to achieve equilibrium and the ground permeability.

Ground water observations in boreholes

Ground water entries into boreholes give an indication of ground water conditions but need careful interpretation as they can misrepresent *in-situ* states. Furthermore, such inflows are likely to seriously disturb the ground around and below the borings (see section 3.3.1). Similarly, standing water levels in borings may mislead due to the long response time, for example a 'dry' boring does not necessarily indicate a low water table since insufficient time has elapsed for water to flow into the hole to achieve equilibrium. Equally, borings full of water do not necessarily reflect high water tables but may show that the borehole acted as a sump during heavy rainfall.

Interpretations of water inflows and standing water levels from open borings have to be made with caution and engineering judgement. In general, boreholes are not sufficiently reliable instruments to measure ground water conditions and piezometers are preferred. For surveys of works such as sewer projects a planned piezometer installation should be made for each scheme.

It is important that where boreholes are drilled 'dry' (i.e. no attempt to maintain a water balance) that accurate ground water levels are recorded. The following is recommended:

1. When ground water is encountered entering the hole, stop boring and observe and record the rise in water level over a 30 minute period.
2. Record water levels in the borehole at the start and end of a daily shift and immediately prior to each sampling/*in-situ* test operation. This indicates whether or not the sample or test result is likely to have been influenced by disturbance at the base due to water inflow.
3. Recover samples of ground water for analysis. Samples should be taken as soon as possible after inflow from the water entering as seepage. Clean glass or inert plastic bottles with airtight lids can be used. For bacteriological test purposes the specimen containers should also be sterilised.

Piezometers

Piezometers range in type from standpipes and standpipe piezometers through to hydraulic pneumatic and electrical piezometers and are chosen with respect to suitability and required response time (Hanna, 1973). Standpipes and standpipe piezometers are shown in Fig. 3.6 and are discussed below.

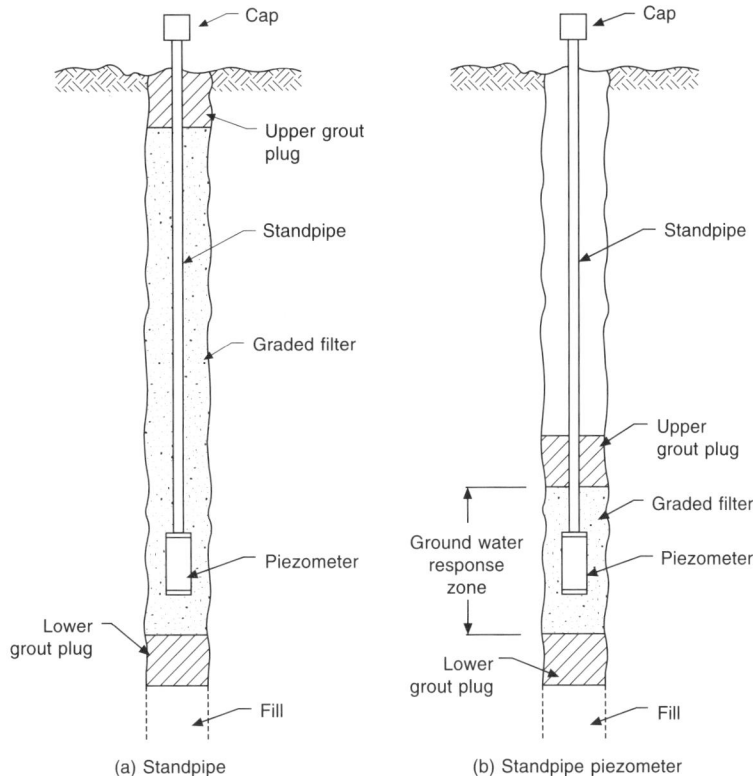

Figure 3.6 A standpipe and a standpipe piezometer.

Standpipes

A standpipe consists of an open-ended tube of about 10 mm to 20 mm diameter which is placed in the borehole with the bottom section of the pipe either perforated or connected to a 250 mm long, 40 mm OD porous filter tube. The space between the pipe and borehole wall is filled with sand or fine gravel up to near the ground surface. The top of the hole is sealed with a concrete or cement/bentonite mix to prevent surface water entering the borehole. Water levels are then recorded by dipping the standpipe at time intervals.

Standpipes are very simple to install but can lead to problems in interpretation, particularly if more than one aquifer is intersected by the borehole. The disadvantages are:

1. A standpipe does not detect seepage patterns in the ground. If soil layers act as a series of aquifers and aquicludes then the standpipe water levels can be totally misleading and meaningless. This is of particular importance in tunnel works such as sewers.
2. The response time for equalisation of water level in the standpipe with that in the ground can be long, often days and even weeks.
3. Cross contamination can occur between potentially contaminated near surface water and deeper water bearing strata.

For these reasons, standpipes in general should not be specified. Standpipe piezometers are preferred.

Standpipe piezometers

In a standpipe piezometer the plastic tube is connected to a porous tip made from low air entry ceramic. This tip is set in the centre of a response, or test, zone of the borehole. This zone is often about 750 mm long and the tip is surrounded by sand (a washed sharp sand is suitable) or fine gravel. The zone is sealed from the rest of the borehole by a cement and bentonite grout above and below the zone (1 bentonite:1 cement:8–12 water) with pre-mixed cement/bentonite balls tamped and compacted immediately above and below the sand/gravel response zone. Vaughan (1969) describes sealing practice. An important final check is to use the cable of the dipmeter to check the inside of the standpipe and tip is clear to the bottom and that no soil particles are within. If there is an obstruction then a joint has probably parted and it is then necessary to clean out and start the assembly procedure again. (This check is best made before placing the final borehole grout to the surface.) Water levels are then recorded from a dipmeter.

In very low permeability clays equalisation of water in the standpipe with ground water pressure heads may take weeks or months to achieve. This can be made quicker if the standpipe is topped up with water and equalisation obtained from a falling head rather than a rising head. The top-up level need not be to ground level.

A number of piezometers should be installed to help interpretation where seepage occurs, for example Fig. 3.7 compares a simple hydrostatic ground water regime with an under-drainage situation. Attempts have been made to place multiple level standpipes in a single borehole but these are not always successful due to practical difficulties in making effective seals to the response zone.

3.3.7   LABORATORY TESTING

Laboratory testing is an integral part of the investigation and dictates the type and frequency of the sample taken so as to produce representative data. In routine works, conventional samples may not be of the highest quality and produce specimens for test that can influence the reliability of the measured parameter. Since sample quality may be suspect then test data has to be cross-correlated with possible influences from the drilling method and other test results, possibly through empirical means.

Details of laboratory tests and techniques are found in textbooks such as Head (1980, 1982, 1986) and National Standards such as BS 1377 (1991) and ASTM standards. It is important that standards are adhered to but there is evidence (Matheson and Keir, 1978) that with ground explorations competitively procured this may not be the case in practice. Engineers should ensure that testing laboratories do comply with standards and only employ those with third party quality assurance such as NAMAS accreditation. Periodic visits to testing laboratories are recommended.

For sewer works the laboratory tests most often required consist of:

1. Classification and index tests:
   - plastic and liquid limit
   - moisture content
   - particle size distribution
   - compaction tests
   - particle density tests
2. Geotechnical parameter tests:
   - strength: total and effective stress

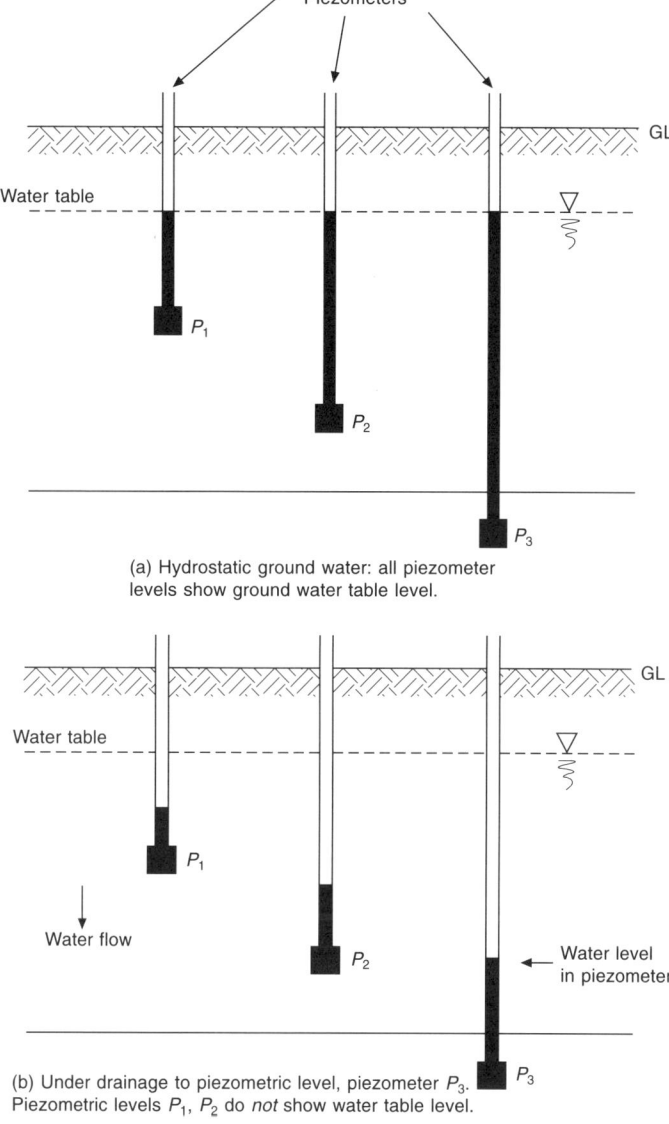

Figure 3.7   Ground water and piezometric level.

- stiffness: oedometer tests
- consolidation: oedometer
- permeability
- chemical tests.

Classification and index tests

PLASTIC AND LIQUID LIMIT (W_P, W_L)
The tests indicate limits to the plastic behaviour of soil; at the plastic limit the remoulded soil becomes sufficiently strong and stiff to be brittle; at the liquid limit the soil becomes so weak it flows like a fluid. A useful index of behaviour is the consistency index (CI):

$$CI = \frac{W_L - W}{W_L - W_P}$$

Where W is the moisture content of a soil (for fine soils) or the moisture content of the soil matrix (i.e that proportion with particles finer than 0.425 mm).

At the plastic limit, the remoulded soil strength is at least 110 kN/m^2. At the liquid limit the soil strength is about 1.6 kN/m^2. Thus the plastic range, or plasticity index (PI), is an index to the sensitivity of soil strength to small change in moisture content. The fall in strength from the plastic to liquid limit approaches a factor of 100 and is of an exponential nature: small moisture variations at or near the plastic limit produce larger order strength change than do moisture variations at or near the liquid limit (Whyte, 1982).

Figure 3.8   'Perfect site investigation?' – Iman's Palace, Sana's, Yemen. Photo Ian Whyte.

A knowledge of moisture contents and plasticity limits thus forms a useful empirical check on undrained shear strength data.

PARTICLE SIZE DISTRIBUTION
The distribution of soil particles is found by sieving and sedimentation testing of a representative specimen. The results are used to classify coarse soils and can be used as an empirical guide to soil behaviour for properties such as permeability, grout take and in ground water control. Details of correlations can be obtained from texts on soil mechanics.

Soil behaviour is governed by the finest fractions (finest 10% for permeability, finest 30% for strength). The most significant part of the grading curve is therefore the finest 30% or so. Care has to be taken to ensure that such fines are not lost in suspension in water during sampling.

Samples for grading analysis are fully disturbed and may not represent the *in-situ* geological fabric of the ground. Engineers should be mindful of a soil's geology and description when assessing results, particularly in sewer works.

Figure 3.9 Two percussion drilling rigs – general view. Photo courtesy of EDGE Consultants UK Ltd, Manchester.

Geotechnical parameter tests

STRENGTH TESTS

The strength of soils is most frequently measured by the triaxial compression test, less frequently by the shear box and laboratory vane test. The conventional procedures were developed in the post-war period and have remained essentially unchanged from that time. Particular points to note are:

1. The size of borehole (often 150 mm or 200 mm) was originally governed by the availability of percussion well-boring drilling rigs.
2. The borehole size dictated the maximum sample size, thus 100 mm nominal diameter samplers were adopted.
3. The necessity to produce failure envelopes from tests dictated three test specimens, thus small diameter (38 mm diameter by 76 mm long) specimens became the norm and dictated triaxial equipment development.

These procedures developed along principles of convenience and habit rather than from considerations of what best represents the material properties of soils. At the time of development, major research studies on sampling, etc., were not concluded and the results arrived too late to significantly influence the methods adopted. A consequence is that the conventional approach to strength testing and interpretation may not yield reliable data on material properties and results can be highly variable (by not being representative) and uncertain.

UNDRAINED TESTS

The test data is useful in sewer works for assessing stability of excavations (sides and base), tunnelling methods, pipe jacking resistance, etc.

In the undrained test, the soil strength is measured under constant volume conditions for a saturated soil and an apparent cohesion, C_u, is found. This 'cohesion' is not a measure of any natural 'cementing' between soil particles but is a result of the particular failure condition imposed by shearing at constant volume. The strength is thus a function of

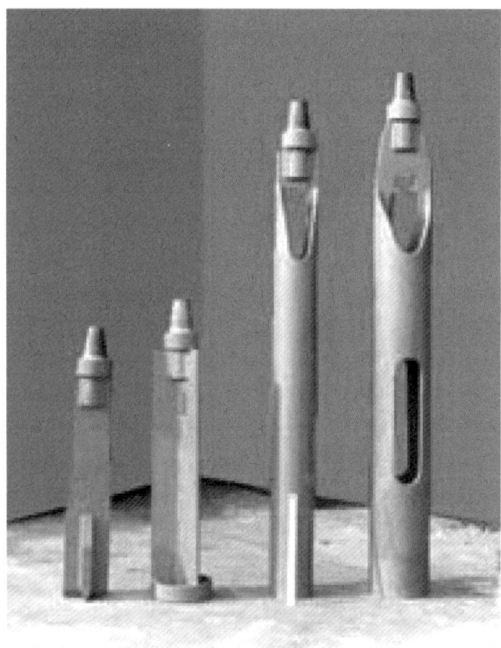

Figure 3.10   Percussion rig drilling tools – chisels, clay corer and sand bailer. Photo courtesy of Dando Drilling International Ltd, Littlehampton, West Sussex.

moisture content (see above for remoulded soils and plasticity states). The practical consequences are:

1. The measured value for C_u will only be relevant to the *in-situ* undrained strength if the moisture content of the test specimen and the soil *in-situ* are identical. Drilling and sampling methods may allow variations to occur, e.g. by swelling in the sample tube.
2. If there is a moisture content difference from sample disturbance, etc., then the measured strength will differ from the *in-situ* strength. An approximate indication of the sensitivity to change in moisture content is given by the plasticity index. The less plastic the soil, the more severe is the strength difference.
3. If the test specimen is partially saturated (possibly from sampling disturbance) then volumetric changes occur in the test as different cell pressures are applied, i.e partial saturation violates the constant volume criterion of undrained testing. The result is test data with $\emptyset_u$ values, i.e. apparent angles of shearing resistance, and interpretation is difficult. The problem is particularly developed for soils with degrees of saturation below about 80% to 85%.
4. Undrained strength is significantly influenced by soil fabric features such as fissures, pedal structures, etc. Small specimens such as 38 mm diameter samples are often not sufficiently representative of the *in-situ* material and result in widely varying test results. This leads to interpretation difficulties. A better, but often still approximate, procedure is to test 100 mm diameter specimens. The cost of testing 100 mm diameter specimens is not significantly different to the cost of testing 38 mm diameter samples.

Full records should be kept of borehole drilling, sampling conditions, descriptions of soil on the borehole log and laboratory test specimen and measurements of moisture content, density, liquid and plastic limits. These records assist the interpretation of data.

Figure 3.11   Window core sampling. Photo Ian Whyte.

EFFECTIVE STRESS TESTS

Effective stress tests yield values for the angle of shearing resistance, $\emptyset'$. Since soils, by definition, are not cemented the apparent effective cohesion, C', should be zero. Failure envelopes are known to be slightly curved and the fitting of a linear Mohr–Coulomb envelope to test data can result in small values for the cohesion intercept that may be used (with caution) for design. Values of cohesion C' greater than about 1.5 to 2.0 kN/m^2 often reflect difficulties with the test rather than a soil property and should not in general be relied upon for design. Practical points on the test are:

1. The effective stress parameters are less sensitive to soil fabric features relative to the undrained parameter C_u. 38 mm diameter specimens can be used but 100 mm test specimens are to be preferred.
2. The test has to be carried out sufficiently slowly to permit either full equilibration of pore water pressures (in an undrained test with pore water pressure measurement) or full drainage of pore water (in a drained test). Should a test rate not be sufficiently slow then the data analysis results in a serious overestimate in the cohesion, C' (i.e. values larger than about 1 to 2 kN/m^2) and a serious underestimate in the angle of shearing resistance $\emptyset'$. Such errors can produce design errors for soil stability.
3. Interpretation may need to take account of three possible states:
   - Peak strengths where the soil is rapidly dilating or generating changes in pore water pressure
   - Critical state strengths where the soil is being sheared under constant volume conditions or pore water pressures are not changing
   - Residual state strengths where the soil is failing on a pre-existing slip surface.

Current design practice is often based on critical state strengths, particularly when progressive failure conditions can develop within the designed structure.

Figure 3.12   Percussion drilling rig. Photo shows the U100 sampler being attached to the drill cable.
Photo courtesy of EDGE Consultants UK Ltd, Manchester.

OEDOMETER TESTS
Oedometer tests measure the soil stiffness (or compressibility, m_v) under one-dimensional compression and allow the coefficient of consolidation, c_v, to be determined. Thus estimates can be made for settlement and the rate at which settlement occurs. In conventional tests specimens are typically 75 mm diameter and 20 mm thick though hydraulic oedometer cells now permit tests on specimens up to 250 mm diameter (provided large diameter undisturbed samples are obtained). Practical points to note are:

1. Conventional test methods, properly conducted, produce reasonably reliable estimates for soil stiffness or compressibility for one-dimensional compression (Padfield and Sharrock, 1983).
2. Conventional tests are usually totally unreliable for estimating the coefficient of consolidation c_v. Estimates can be in error by orders of magnitude and frequently result in settlement times being predicted at too slow a rate, i.e. field settlements occur more rapidly than the prediction. For reliable estimates of c_v then geological fabric features have to be accounted for and large (250 mm diameter) undisturbed specimens may be required (Rowe, 1972). Alternatively, field construction trials on instrumented sections can be undertaken and back analysed.

PERMEABILITY TESTS
Laboratory testing of permeability obtains a relatively inexpensive, albeit potentially misleading, estimate of soil permeability. Practical points to note are:

1. The constant head permeameter is best suited to clean sands and gravels with permeabilities down to about 10^{-4} m/s.
2. The falling head permeameter is for permeabilities in the range 10^{-4} to 10^{-7} m/s, i.e. silt and fissured clay.

3. Oedometer tests can produce permeability estimates less than 10^{-7} m/s, i.e. intact clays.
4. Misleading values can result from non-representative and disturbed samples, particularly if the soil macro-fabric is disturbed or misrepresented.
5. Air in test specimens, and temperature effects, can influence permeability estimates. More reliable measurements of permeability are obtained from *in-situ* tests (e.g. piezometers) or field pumping trials.

SOIL CHEMICAL TESTS
Tests on contaminated land are not discussed here. Routine tests for soil and ground waters are pH value, sulphate content and organic content. Test details are given in BS 1377 (1991) and their effects are discussed in Tomlinson (1995) and BRE Digest 250 (1991).

3.4 Construction investigations: records and feedback

Information from the preliminary sources survey (desk study) and ground investigation will always be subject to uncertainty. The level of uncertainty, however, is reduced with increasing quantity and quality of information. Since there is a residual uncertainty then investigations should continue into the construction phase. BS 5930 (1999) lists the purposes of construction investigations to be:

1. To check the adequacy of the design
2. To check on safety of the works and the adequacy of temporary works
3. To check the findings from the ground investigation and to assess these by feedback
4. To check assumptions on ground conditions and ground water in relation to construction
5. To provide agreed information in the event of a dispute

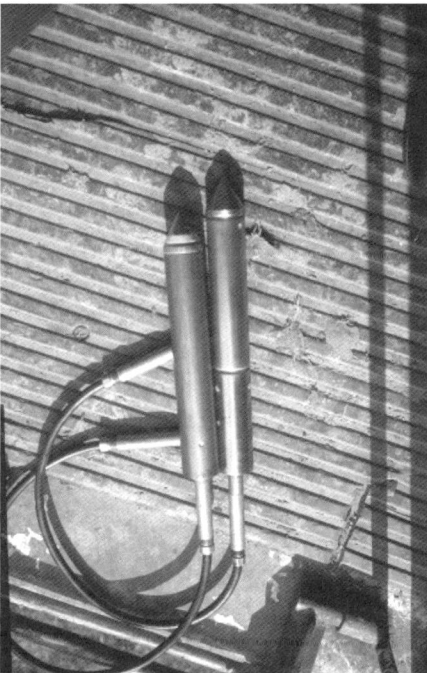

Figure 3.13 Static cone and piezocone. Photo Ian Whyte.

6. To check on instrumentation installations
7. To decide on the best use of excavated materials
8. To assess the choice of construction plant and method.

In addition:

9. To provide a record of 'as-built' conditions for future maintenance and other record purposes, e.g. health and safety.

Prior to the start of construction a briefing report on the expected ground and ground water conditions should be prepared for the site staff. The maintenance of construction investigation records and management of the feedback pays dividends when, and if, unexpected and unforeseen ground or ground water is encountered. Delays to the works and design changes are then kept to a minimum.

3.5 Geotechnical reports

An example of good reporting practice is provided by the Department of Transport (1992) based on highway construction. The model can be extended to other works, such as sewers, and amended to suit replacement/new construction situations. The sequence is as follows:

1. Procedural Statement Number One: A statement outlining how the preliminary sources study is to be conducted: information sources to be consulted, cost and time estimates.
2. Preliminary Sources Study Report: Information from the desk study and any preliminary explorations, identification of possible hazards and levels of risk.
3. Procedural Statement Number Two: A statement outlining the proposed extent and quality levels required from the main ground investigation. Identification of method of procurement and cost/time estimates.
4. Ground Investigation Report:
   • Factual report
   • Interpretative report
   Information from the main ground survey, including borehole logs, geophysical surveys, *in-situ* testing, laboratory data, geotechnical and geological plans, sections; analysis and interpretation of data; design and construction recommendations; need for additional works.
5. Site Briefing Report: Summary report for site staff on anticipated conditions, instrumentation requirements, additional surveys, feedback and monitoring requirements. Communications for feedback, response options if hazards develop or unforeseen conditions develop.
   This report can highlight areas of particular risk and uncertainty, the observations to be made and can list anticipated responses to the encountered conditions. Such responses include feedback (with contact names and organisations) and options for investigation and additional monitoring. Any requirements for advanced probing (e.g. in tunnels) and investigations during construction (e.g. further borings, geophysical surveys) can be included in the briefing report.
   Typical observations/investigations during construction include (the following list is not exhaustive and can be extended for particular works):
   • Continuously updated geological/geotechnical plans and sections as work proceeds
   • Regular samples for measuring properties and specification requirements
   • Photographic records, including video
   • Frequency and orientation of discontinuities such as joints, faults, etc.

- Ground water observations, inflows, rates, control
- Observation and geological recording of ground conditions
- Hazards encountered, e.g. old ponds, buried channels, mine shafts, contaminated land, etc.
- Records of additional probes, boreholes, geophysical surveys, *in-situ* and laboratory tests
- Data from instrumentation, e.g. piezometers, settlement gauges, etc.
- Weather conditions.

6. Post Construction Geotechnical Report: This report records as-built conditions, construction experiences, equipment and methods employed; records of hazards and difficulties during the works and measures taken to overcome these and requirements for future maintenance and monitoring of the works, health and safety implications.

3.5.1 COMMENTARY ON REPORTS

Disagreement can develop between the client and contractor as to the level of detail to be provided with contract documents, i.e. whether to include both factual and interpretative ground investigation reports. An alternative approach has been found to be successful in America and can be considered in other countries (US National Committee on Tunnelling Technology, 1984). This is the Geotechnical Baseline Report (ASCE, 1997).

In sewer works the risk of unforeseen conditions is high and changes are costly and lead to delay. The contractor has a right to know the anticipated conditions and what the designer believed these to be for the design of the works. He also needs to know about the interpretation of ground effects on the construction and temporary works. The designer is in a unique position to provide a report that develops in parallel with the design. This report, the Geotechnical Baseline Report, is developed from information contained in all earlier geotechnical reports, is written in simple language and details the ground conditions anticipated by the design. The document should form part of the contract specifications and be a contract document. It forms an agreed geotechnical baseline for the contract.

In America the use of a geotechnical baseline report has resulted in more uniform bid prices with less exposure to claims involving the interpretation of ground data. Clients initially expressed concern over the use of such a report. The key is the recognition that the document allows the client to appreciate how ground conditions influence the management of risk. Conservative baselines reduce the risk of unexpected cost increase but increase contract bid prices. The client, under advice, has control over the trade-off between risk and price. The geotechnical design report demands a thorough assessment of geotechnical design and construction, it produces better practice than when the contract emphasis is on factual data and price. Details of the system are reported by the ASCE Underground Technology Research Council (1989, 1991, 1997). Guidance on managing the geotechnical risk, including tunnels, is given by Clayton (2001).

A case history on a managed programme of ground investigation for sewer relief tunnels for the city of Cambridge is reported by Petrie and Hobden (1998). The contract was completed on time and within budget. If a managed programme of investigation had not been carried out then risks to the client and contractor would be significantly greater, for example:

- Health and safety in construction
- Ground movement and surface settlement

- Structural damage
- Delays with consequential environmental damage due to overflows into the River Cam
- Significant contractual issues.

Quantifying or valuing such risks is virtually impossible. A subjective approach demonstrates the costs of remedial actions and the provision of contingencies far exceed the cost of the managed programme of ground investigation. The most significant factor in the success of the project was the working relationship, 'partnering', between the client, designer and contractor. The integration was essential to solve difficulties during occasions where ground conditions were not as expected

References

American Society of Civil Engineers (1989) Avoiding and resolving disputes in underground construction – successful practice and guidelines. Tech. Comm. on Contracting Practices, Underground Tech. Research Council. ASCE (Construction Division), New York, 24 pp. + Appendices.

American Society of Civil Engineers (1991) Avoiding and resolving disputes during construction – successful practices and guidelines. Tech. Comm. on Contracting Practices, Underground Tech. Research Council. ASCE, New York, 82 pp.

American Society of Civil Engineers (1997) Geotechnical baseline reports for underground construction: guidelines and practices. Technical Committee on Geotechnical Reports of the Underground Technology Research Council. ASCE (Construction Division), New York, 39 pp.

American Society for Testing and Materials (1986) ASTM D 3441 – 86, standard test method for deep, quasi-static, cone and friction cone penetration tests of soil. ASTM, Philadelphia.

Anonymous (1996) SI Psyche. *Ground Engineering*, 22–27, March.

Building Research Establishment, (1987), Site Investigation for low rise building: desk studies. BRE Digest 31 BRE, Watford.

Building Research Establishment, (1991), Sulphate and acid resistance of concrete in the ground. BRE Digest 363 BRE, Watford.

British Standards Institution (1981, 1999) Code of practice for site investigations, BS 5930. BSI, London.

British Standards Institution (1991) British standard methods of test for soils for civil engineering purposes, BS 1377: Parts 1–9. BSI, London.

British Standards Institution (1991) Eurocode 7: Geotechnical Design, part 1, general rules. DD ENV 1997-1 BSI, Watford.

Clayton, C. R. I. (1996) The Standard Penetration Test (SPT): Methods and Use CIRIA Report 143. CIRIA, London.

Clayton, C. R. I. Matthews, M. C., Simons, N. E. (1995) *Site Investigation* (second edition). Blackwell Science, Oxford, 584 pp.

Clayton, C. R. I. (2001) Managing Geotechnical Risk – improving productivity in UK building and construction. The Institution of Civil Engineers, Thomas Telford, London, 80 pp.

Collins, S. P. (1972) Discussion on Smith, W. *et al.*, Municipal tunnelling from a contractor's viewpoint. *Tunnels and Tunnelling*, May 1972, 245–246.

Darracott, B. W., McCann, D. M. (1986) Planning engineering geophysical surveys. *Geological Society, Eng. Geol. Sp. Pub.* No. 2, 85–90.

Department of Transport (1992) Design Manual for Roads and Bridges: Vol. 4, Geotechnics and Drainage, Section 1, Earthworks: Part 2 HD22/92 ground investigation and earthworks procedure for geotechnical certification. HMSO August 1992.

Deutsches Institüt für Normung (1980) DIN 4094 Dynamic and standard penetrometers, Part 1: Dimensions of apparatus and method of operation; Part 2: Application and evaluation. DIN, Berlin.

Dumbleton, M. J., West, G. (1976) Preliminary sources of information for site investigations in Britain. Transport and Road Research Laboratory Report LR403 (revised edition). Crowthorne, Berks.

Geddes, J. D. (1977) (ed.) *Large Ground Movements and Structures*. Pentech Press, London.

Hanna, T. H. (1973) *Foundation Instrumentation*. Trans. Tech Publications, Cleveland, Ohio.

Head, K. H. (1980) *Manual of Soil Laboratory Testing, Volume 1: Soil Classification and Compaction Tests*. Pentech Press, London.

Head, K. H. (1982) *Manual of Soil Laboratory Testing, Volume 2*. Pentech Press, London.

Head, K. H. (1986) *Manual of Soil Laboratory Testing: Effective Stress Tests, Volume 3*. Pentech Press, London.

Hvorslev, M. J. (1949) Subsurface exploration and sampling of soils for civil engineering purposes. Waterways Experimental Station, Vicksburg VSA.

Jardine, F., Johnson, S. (1994) Risks in ground engineering: a framework for assessment. Risk, Management & Procurement in Construction, 7th Annual Conf., Centre of Construction Law & Management, King's College, London and CIRIA, London.

Kearey, P., Brooks, M. (1990) *An Introduction to Geophysical Exploration*. Blackwell Scientific Publications, second edition. Oxford, 296 pp.

Kent, T., England, B., Morris, R., Wakeling, R. (1992) *Drilling Technology, Parts 1 and 2*. British Drilling Association (Operations) Ltd, Brentwood, Essex.

McCusker, T. G. (1982) Soft ground tunnelling, in *Tunnel Engineering Handbook* (eds Bickel J. O., Knesel, T. R.). Van Nostrand, New York.

Matheson, G. D., Keir, W. G. (1978) Site investigation in Scotland. Transport and Road Research Laboratory TRRL Report 828, Livingstone, Scotland.

Meigh, A. C. (1987) Cone penetration testing: methods and interpretation, CIRIA Ground Engineering Report: In-situ testing. CIRIA, London/Butterworths, London.

Mott MacDonald and Soil Mechanics Ltd (1994), Study of the efficiency of site investigation practices. Transport Research Laboratory TRL Project Report 60, Crowthorne, Berks.

Nakamura, T., Kubota, H., Inagaki, M. (1994) Utilization of waterway tunnel survey system using ground penetrating radar. Proc. Congress Tunnelling and Ground Conditions, Cairo. Balkema 407–412.

Padfield, C. J., Sharrock, M. J. (1983) Settlement of structures on clay soils. CIRIA special Publication 27, PSA Civil Engineering Technical Guide 38. CIRIA, London.

Parkinson, J. (1975) Tunnel faces acid test – and disintegrates. *New Civil Engineer*, 30 Jan., 21/22.

Peacock, W. S., Whyte, I. L. (1988) Site investigation and risk analysis. Proceedings Institution of Civil Engineers, Civil Engineering, May, 74–82.

Peck, R. B. (1969) Deep excavations and tunnelling in soft ground. Proceedings 7th Inst. Conference on Soil Mechanics and Foundation Engineering, state-of-the-art report, Mexico, 225–290.

Perry, J., West, G. (1996) Sources of information for site investigations in Britain (revision of TRL Report LR403). Transport Research Laboratory TRL Report 192, Crowthorne, Berks, 47 pp.

Petrie, J. L., Hobden, C. J. (1998) Cambridge Riverside Tunnel Project – a case history. Proc. Seminar the Value of Geotechnics in Construction. Inst. Civil Eng., November, Construction Research Communications Ltd, London.

Read, G. F. (1986) The development and rehabilitation of Manchester's sewerage system. Proc. ICE North Western Association Centenary Conference Infrastructure Renovation and Waste Control, Manchester 121–130.

Roscoe, G. H., Driscoll, R. (1987) A review of routine foundation design practice. Building Research Establishment Report BR104, Garston.

Rowe, P. W. (1972) The relevance of soil fabric to site investigation practice: 12th Ranking Lecture. *Geotechnique* 22(2), 195–300.

Site Investigation Steering Group (1993) Planning, procuring and equality management. *Site Investigation in Construction, Volume 2*. Thomas Telford, London, 30 pp.

Tomlinson, M. J. (1995) *Foundation Design and Construction*, sixth edition. Longman Scientific and Technical, Harlow, 536 pp.

Tonks, D. M., Whyte, I. L. (1988) Dynamic soundings in site investigations: some observations and correlations. Proc. ICE Conference Penetration Testing in the UK, Thomas Telford, London, 105–113.

Uff, J. F., Clayton, C. R. I. (1986) Recommendations for the procurement of ground investigation. CIRIA Special Publication 45 CIRIA, London, 44 pp.

Uff, J. F., Clayton, C. R. I. (1991) Roles and responsibility in site investigation. CIRIA Special Publication 73. CIRIA, London, 42 pp.

US National Committee on Tunnelling Technology (1994) Geotechnical site investigations for underground projects. Volume 1: Overview of Practice and Legal Issues, Evaluation of Cases, Conclusions and Recommendations. Subcommittee on Geotechnical Site Investigations. US Nat. Comm. Tunn. Tech Commission on Engineering and Technical Systems, National Research Council. National Academy Press, Washington DC, 183 pp.

Vaughan, P. R. (1969) A note on sealing piezometers in boreholes. *Geotechnique*, 19, 405–413.

West, G., Carter, P. G., Dumbleton, M. J., Lake, L. M. (1981) Rock Mechanics Review: Site investigation for tunnels. *International Journal Rock Mechanics & Mining Science & Geomechanical Abstracts*, Vol. 18, 345–367.

Whyte, I. L. (1982) Soil plasticity and strength, a new approach using extrusion. Ground Engineering, Vol. 15, No. 1, January.

Whyte, I. L. (1985) Some anomalies in interpreting SPT tests. Proc. Int. Conf. Construction in Glacial Tills and Boulder Clays, Edinburgh, Engineering Technics Press, 121–129.

Whyte, I. L., Tonks, D. M. (1993) Project risks and site investigation strategy. Proc. Conf. Risk and Reliability in Ground Engineering. ICE, London, Thomas Telford (pub. 1994), 100–112.

Whyte, I. L. (1995) Ground uncertainty effects on project finance, in *Risk, Management and Procurement in Construction* (eds. Uff J., Odams A. M.). Centre of Construction Law and Management, King's College, London, 263–297.

Whyte, I. L. (1998) Managing site investigations: Practice and the value of change. The Value of Geotechnics in Construction. Inst. Civil Eng., November, Construction Research Communications Ltd, London, 13–25.

4

Site Investigation and Mapping of Buried Assets

Nick Taylor

Site investigation is a vital step in the design process for the replacement or new construction of sewers. As by definition most of the construction is going to take place below surface, potential problems are not immediately apparent. It is therefore essential to minimise these problems and the resultant costs. Within the budget, appropriate provision is vital for site investigation – thorough planning, appropriate to the scale of the project, will ensure that the time schedule is optimised, and thorough investigation in the course of planning will obviate most of the major difficulties that are likely to be encountered.

Ground conditions and the presence of buried utilities are the two primary considerations for investigation – whether installation is planned by open cut or trenchless technology. An understanding of the geology to be encountered is significant. When designing a gravity system, the cost of rock excavation to achieve the desired gradient can severely influence the costings. The presence of mass foundations on a brown field site would restrict the construction technique – pipejacking through concrete is hardly viable. The constraints of designing a replacement gravity system may dictate that the only feasible route requires the relocation of one or more utilities. In such a case, the knowledge of the precise position of all the utilities within the design route will facilitate planning as well as preventing unforeseen delays when excavation reveals the inevitable obstacles.

Investigation should advisedly follow four principal steps and adhere to the following order, the logic of which will become apparent, commencing with non-intrusive methods:

1. Desktop study of existing plans
2. Tracing of utilities on site by non-intrusive methods (EML and GPR)
3. Mapping of findings
4. Site investigations (boreholing and trial pits).

4.1 Desktop study of existing plans

Much can be determined from existing plans. A comprehensive examination of the evidence available will assist the designer to determine which technique or combination of techniques is most cost effective. Studies will also provide a structure for the site investigations and determine which techniques are required. Here again the acquisition of existing data follows a logical sequence.

If the planned sewer follows a green field route, the ground conditions may be deduced from geological maps and substantiated by any local borehole information such as that available from the British Geological Society. Generally, such information is diagrammatic in representation but will aid in planning trial pits or boreholes.

Access to historical maps may shed some light on previous land use – not just on the likelihood of encountering made-up ground and the obstructions of foundations but the existence of a historical site will jeopardise the project schedule while archaeological investigations take place.

Naturally if the replacement of a sewer over an existing route is being planned, or the route lies through urban land, then the ground conditions are probably known and the presence of existing utilities is of greater significance. All utilities are required to maintain and make available records of their buried plant and the designer is faced with just that – a separate plan for each utility – and the problem of combining all the data into a usable format. Historically, such information was manually appended to hard copy of the Ordnance Survey 1:2500 and where available 1:1250 scale drawings.

The advent of GIS (Geographical Information Systems) has resulted in digitally held data, often a scanned (or at best digitised) version of the former.

The danger here is that too great an emphasis is placed on the information without regard to the mode of creation. Merely because the data is presented in a pleasing computer format does not mean that it is accurate. Furthermore the facility to view the data on screen at any desired scale serves to compound the felony. Data originally drawn at a scale of 1:2500 on an Ordnance Survey background may be viewed at say 1:500 thereby increasing the error factor by 5. As a detailed design may be required at a scale of *at least* 1:500 in order to show sufficient detail, the accuracy of transposing the utility information is negated by this error factor.

The accuracy of such plans in whatever format is highly variable and reliance thereon must be governed by the scale of the project. The majority of sewers are located under the carriageway and at a depth of greater than 1 metre – a matter of benefit since the majority of services are laid in the footway. For a small scheme, inspection of the records *may* provide sufficient information, but more often than not, the record information is woefully inadequate and further investigation is highly advisable.

If any boreholing or coring and sampling is planned then the location of any utility in the vicinity of the point of investigation is crucial. The cost of replacing a damaged fibre optic cable may well exceed the entire budget of the scheme – notwithstanding the additional claims for loss of business.

4.2   Tracing of utilities on site

Buried utilities fall into two categories by nature of their conduit:

1. Metal pipes and cables
2. Non-metallic pipes and cables.

Metal pipes may vary in diameter from 50 mm to 1200 mm or even larger. Cables with metal content may vary from single core to oil-cooled EHV cable. Non-metallic pipes may be plastic, clayware, concrete or asbestos and similarly vary from 50 mm to man entry tunnels or culverts.

Two types of location technique are commonly used: electromagnetic location (EML) and ground probing radar (GPR). Other techniques such as magnetrometry, resistivity, seismic investigation and dousing are available, but fall outside the accepted mainstream methods. Due consideration must be given to the fact that neither method traces the actual utility but instead detects either the signal emanating from or induced into the pipe or cable (EML) or the reflection from the target (GPR). Both techniques have advantages and disadvantages but, as will be seen, are highly complementary.

4.2.1   ELECTROMAGNETIC LOCATION

EML traces the route of the signal of a metal conduit or the passage of a signal emitting sonde passed through the pipe or culvert. EML has the following advantages:

- Follows a continuous route
- Utility route is generally identifiable to type
- Unaffected by weather or ground conditions
- Provides plan position and depth
- Easily marked on the surface
- Readily portable and inexpensive equipment
- Tracing can be a one man operation.

EML has the following disadvantages:

- Does not detect plastic pipes unless accessible to a signal emitting sonde
- Requires an experienced operator for optimum results.

EML is a detective process, whereby the operator commences by tracing the utility most readily identifiable, progressively eliminating the services and working towards the least identifiable service. The operator either connects directly to the service or induces a signal into the pipe or cable with a transmitter and proceeds to follow the signal with a receiver. In its basic form, the signal is produced as an audible tone varying in strength. Supplemented with a visual display, either analogue or digital, the peak signal indicates a position vertically above the service. More sophisticated receivers are equipped with a means of interpreting depth. At the point of maximum signal strength the depth button is pressed and a measurement of depth is given as factor of signal strength.

EML will locate the signal from buried pipes and cables with a traceable metallic content to a depth of 2–3 metres below surface or the passage of a signal emitting sonde to a depth of 10–15 metres. The accuracy of tracing is a direct function of the actual depth – generally accepted as a factor of $\pm$ 10% of the depth in both plan and depth, i.e. at a depth of 1.60 m the accuracy may be considered as $\pm$ 0.16 m. When tracing the passage of a sonde, a further constraint must be taken into account – the sonde cannot be guaranteed to lie in the invert of the pipe. The presence of silt or rubble may cause the sonde to wander, as may a curve in the pipe. The larger the diameter of the pipe being investigated, the greater the problem.

Gradually the pattern of the utilities is revealed as the anomalies are minimised. The findings are generally marked on the surface with road marking paint showing position, type and depth. Once the EML investigations are complete unless construction is to take place immediately without further design, a record of position should be taken for reference. This could be as simple as a sketch plan with taped dimensions related to permanent detail such as kerb lines, buildings or permanent street furniture or as sophisticated as a full three-dimensionally co-ordinated computer image.

4.2.2 GROUND PROBING RADAR

The ground probing radar (GPR) method utilises UHF/VHF electromagnetic pulses to produce graphic depth sections. It can be considered as an electromagnetic analogy of marine sonar. GPR has many useful applications in the field of shallow, land-based site investigation. Examples include the detection and location of voids and buried obstructions; the mapping of geological interfaces such as soil or rock layering and depth to bedrock; identifying ancient landfill and examining archaeological sites. GPR is also used as a non-destructive way of investigating building materials to locate reinforcing, voids, delaminations and internal layering within concrete and masonry.

Various antenna units are used with the radar system, depending on the survey requirements. Smaller, higher frequency antennas are useful for low depth, high resolution work such as the non-destructive testing of concrete. Low frequency units offer greater depth penetration and are therefore mainly applicable to site investigation. The high signal losses in most ground conditions in the UK limit the maximum depth penetration to between 2 and 20 metres, depending on ground conductivity.

UHF/VHF signals are transmitted into the ground from an antenna held in direct contact with the surface. The signal is reflected by the target, received by the antenna and recorded as a factor of time against distance. The passage of the antenna over the ground surface is measured and a graphic section produced showing the targets identified. Accurate depth profiles are obtained by moving the antenna along marked survey lines, producing a two-dimensional image of reflecting boundaries below the survey line. The horizontal scale is obtained by marking regular intervals of distance along the profile, using a marker switch. The vertical time axis is calibrated in nanoseconds per centimetre and converted to depth by using standard equations. The radar record produced is therefore a graph of reflection time against distance along the survey line.

As opposed to EML, GPR identifies the target as an intercept on a vertical section. The survey is carried out by passing the antenna over the ground recording a series of parallel sections on a rectilinear grid. Each graphical section is then compared to the previous or adjacent sections. The reflected events on the record can be correlated with physical interfaces within the ground such as layering, air cavities or bedrock, by the careful application of interpretation techniques. Unless the profile is taken close to an associated physical feature, GPR does not identify the service as in the case of EML – it does, however, have the advantage of locating plastic pipes.

GPR locates planar horizons and voids enabling the location of major obstructions such as mass concrete or foundations and geological anomalies such as rock or running sand. GPR has the following advantages. It can detect:

- Voids
- Changes in ground conditions
- Plastic pipes
- Mass concrete, etc.
- Foundations

GPR has the following disadvantages:

- Does not trace linear features other than by interpretation of successive profiles
- Does not reliably trace small services or one service above another
- Influenced by ground conditions and surface standing water.

4.3   Mapping of findings

Once the non-intrusive investigations are complete, the results may be collated into a permanent record. If a topographical survey has been commissioned, the findings can be appended, and the EML routes can be surveyed with standard land survey methods and accurately presented. GPR results may be similarly surveyed – especially the start and finish point of the profile, though generally as post-field work interpretation is required, the findings are added to the drawing by a CAD operator.

Drawings may then be produced showing the accumulated subsurface data and the next stage of planning decisions facilitated.

4.4   Site investigation – intrusive methods

The non-intrusive methods create an overall picture without the need for excavation – they do not dispense with the need for boreholes or trial pits, but optimise their usage. Once the presence of obstructions and buried utilities is identified and presented on plan, the optimum position of the boreholes and trial pits can be determined. Boreholes and core sampling will confirm the ground conditions identified either from the geological plans or radar profiles and will be positioned in an area clear of buried pipes and cables – as identified by electromagnetic tracing survey. Similarly, if trial pits are needed, these may be positioned to locate services unidentifiable conclusively by the aforementioned techniques and affording direct connection to metal pipes or cables, enabling further EML survey and thus serving to increase the value of information so obtained.

As mentioned previously, the results of the various surveys can be appended to the Ordnance Survey drawing although this should only be carried out with due consideration of the accuracy factor involved.

4.5   Conclusion

Investigation pays dividends and diminishes the risk of unplanned events. Due consideration should be taken of the below-ground hazards that may be encountered and the appropriate steps taken to minimise the accompanying risk. As a minimum, electromagnetic tracing should be carried out and from this any further action necessary will be determined.

Proper planning should save the financial cost of the investigations and facilitate the smooth execution of the work.

5
Traffic Management and Public Relations

R.G. Daintree BSc, CEng, MICE

5.1 Introduction

Probably the most politically and media sensitive part of the work is the effect the works have on traffic. This is not new, for many years it has been the practice of reporters to look for any event that catches the public eye in order to sell copy and any form of roadworks will always draw the attention of the reader or listener to the latest drama on the roads. Our predecessors were very aware of this and of the predominant need to ensure safety for users of the highway. It is interesting to see the following extract from *Sewers* by Bevan and Rees published in 1937 – nearly 60 years ago:

> Here is shown a gantry for excavated material from the shaft head gear, hoardings to isolate all the Contractor's plant and operations from the road, and automatic traffic lights to control the passing traffic through the restricted roadway. In spite of the magnitude of the shaft and the adjacent plant, everything is very compact and one can say that the motorist has received every consideration.

Without wishing to be critical of that work in any way, today we need to ensure that it is not just the motorist who is considered when we are setting up our diversion and temporary accommodation works to enable the general public to pass and repass in complete safety on our highways. Foot traffic is of just as much importance and sadly has tended to be overlooked to some extent in the past when diversionary works are being considered. Nowadays the public is extremely sensitive to any disturbance to their normal pattern of life regardless of the cause.

The problem that has to be faced in dealing with traffic flows on roads in urban areas is that the roads themselves are mostly unable to cope well with the traffic that uses them because they are underdesigned for modern traffic. Prior to the 1939–45 war the use of our roads by motor traffic was low and even in the 1950s very few families owned a motor vehicle. Today many families have at least two cars, yet in our towns and cities much of the road structure remains as it was pre-war. This is not to say that road construction has been totally at a standstill in this time – there have been many improvements with ring roads and traffic diverted away from town centres by other bypass routes, yet the basic road pattern under which most of our sewer systems lie has remained largely unaltered. At peak traffic hours many roads leading into and out of main centres of business are very congested and traffic movement is generally slow. When the area of road available for traffic is further restricted to accommodate the construction of a new or replacement sewer, the traffic congestion will undoubtedly become worse and unacceptable to the road user.

It will not matter how technically intricate and skilled the works themselves are to perform, if traffic is seen to be delayed by an apparently unnecessary operation great criticism of the management of the work will undoubtedly follow. How many times have we all driven along a motorway or principal traffic route and, finding ourselves in a stream of slow moving traffic, felt it incumbent on ourselves to find some criticism of the works in progress which are causing us to be delayed? Radio and TV commentators seem to take great delight when interviewing site managers or the client's representative to criticise the traffic management arrangements besides the unfitness of carrying out works at all on a particular stretch of busy road. In order to minimise this type of criticism it is advisable to have a show of activity at all times on site and particularly at peak hours – motorists cannot see what is happening below ground and can obtain a misleading impression, for example, from safety personnel lounging on barriers at the top of a working shaft, apparently doing nothing.

It is nowadays as essential a part of the design programme to pre-plan traffic diversions as it is to design the works. Indeed it could well be the most expensive aspect of the work, particularly if public consultation is poorly done, and claims for loss of profits result from businesses adjacent to the work or on one of the diversion routes, or from haulage companies who have lost time on deliveries, or from members of the public who have lost earnnings because of late arrival at their place of employment. The more intensive the pre-consultation with such people and the more that is done to accommodate their needs the less likely are claims to be made. Indeed it is often the case that the more that people know about what is going on the more sympathetic and tolerant they are of inconveniences caused by the work.

5.2   The New Road and Street Works Act of 1991

The New Road and Street Works Act of 1991 has made considerable changes to the responsibilities of all public utilities and highway authorities, when compared with those responsibilities which existed in the Public Utilities Street Works Act of 1950. It is not the intention of this chapter to describe all of the differences and readers are recommended to read the documents for themselves to assess how they are affected in full. It is perhaps, however, relevant to mention a few of the differences which are of direct concern to the subject matter we are looking at now. The 1991 Act makes those carrying out streetworks fully responsible for their excavations and reinstatements to proper standards. In order to ensure that the Act is complied with, there is a requirement that companies carrying out work in a public highway have their staff trained and accredited in the various activities associated with work in or on a public highway. The legislation for the qualification of supervisors and operatives is contained in Statutory Instrument 1992 No. 1687 and basically states that as from 1 August 1992 supervisors must be trained within a two year period and operatives within a five year period. Undertakers and contractors must ensure that all supervisors and at least one operative on a site at all times have the prescribed qualifications for the work being undertaken. Details of the various units are contained in Schedules 1, 2, 3 and 4 of the Regulations. So far as traffic management is concerned it is of particular importance that appropriate training and accreditation is obtained in the use of the code of practice 'Safety at Street Works and Road Works' known as the *Blue Book* and of which more is said later.

Highway authorities are responsible under the Act for the co-ordination of all works in a street. In order to comply with this part of the Act they have had to compile a streetworks register which is intended to be easily accessible to all relevant parties. The register records all the significant activities in relation to each job, e.g. the dates of serving each notice, excavation commencement, reinstatement commencement and completion, dates of each inspection and any relevant comments. In order to minimise traffic disruption highway authorities are empowered to restrict access to particular highways at certain times and where possible, when more than one undertaker proposes to carry out works in a street, to arrange for them to carry out their works in a programmed manner, possibly at the same time, on a particular length of street while it is restricted to traffic usage.

If an undertaker's works are proposed to be done following substantial highway works the highway authority may, by notice, restrict the works to be carried out until after the expiry of 12 months from completion of the highway work. In order to make progress on sewer renewal work in this situation, it may be possible to do work by agreement with the highway authority using no dig methods as mentioned earlier.

In order to carry out works on a public highway which restricts traffic flow it is necessary to obtain a temporary traffic order (see the typical layout of the formal advert in Fig. 5.1), but before this can be obtained there is a considerable amount of negotiation and consultation which needs to take place.

ANYTOWN METROPOLITAN BOROUGH COUNCIL, TEMPORARY ROAD TRAFFIC REGULATION ORDER, ROAD TRAFFIC REGULATION ACT 1984 – SECTION 14(AS AMENDED) ORDER NO 25/234

Notice is hereby given that the Council intends, not less than seven days from the date of this Notice, to make the following Temporary Road Traffic Order(s):

(234) Anytown Metropolitan Borough Council (ALPHA STREET) (Temporary PROHIBITION OF TRAFFIC) Order 1997. The effect of the order shall be to PROHIBIT TRAFFIC on ALPHA STREET, from BETA STREET to GAMMA STREET.

The alternative route(s) for traffic will be via:

1. GAMMA STREET, LAMBDA STREET, BETA STREET.
2. BETA STREET, DELTA STREET, GAMMA STREET.

The council is satisfied that this prohibition is necessary in order to facilitate sewer works. It is expected that the works will be in progress between 8th March 1997 and 11th May 1997, WEEKENDS ONLY.

The Order shall come into operation on the 8th March 1997 and may remain in force for a maximum period of 18 months. If all necessary works are completed within a shorter period, the Order shall cease to have effect at the end of that shorter period.

ALFRED PROHIBITOR.
SOLICITOR TO THE M.B.C.

Town Hall
Anytown.

Dated 28th February 1997.

Figure 5.1

5.3 Consultation

For major projects the New Road and Street Works Act of 1991 requires that one month's advance notice of the works be given to the street authority. In practice it is usual for designers of a sewer reconstruction or renewal scheme, which involves excavation in or below a public highway, to involve the street authority, traffic police and other undertakers in discussions about the proposed work at a very early date in the design process. It is usual for this to be almost the first thing that is done after it has been decided that the work is necessary.

These preliminary discussions will cover the extent that the works are likely to impinge upon: the user of the affected streets; the street furniture; other undertakers' apparatus; other traffic routes; and other works which may be planned, or in progress, on or adjacent to the routes affected by the sewer works, such as new building works or refurbishment of premises.

It may well be that these discussions will put constraints on the type of design being considered which had not previously been realised. This could be, for example, by limits imposed on the time that parts or all of the streets would be available to be closed or have restricted traffic use, or by the presence of underground equipment or other physical objects which would be expensive to divert or remove.

Following the discussions with the street authority and undertakers, a survey of the route along which the work is to proceed and the proposed diversion routes may reveal existing offices and businesses which require special consideration in order to maintain access during the work and without which the client is likely to be involved in costly litigation in claims for loss of income. For example, a local fast meals shop may have a regular passing trade from drivers who habitually use the route affected by the diversion. Preventing the parking of vehicles outside the premises may result in a large loss of income to that business which will undoubtedly result in claims. It may well be possible to provide signs to indicate that access to this shop is available, say, from a rear street. This is not to say that claims will not occur if only a limited access can be maintained, the claims will, however, be minimised and early co-operation is more likely to be in the client's favour when the claims are considered in court.

Access to and emergency egress from premises in the event of fire must also be considered. Where there are existing fire escape provisions at a premises they must not be blocked or interfered with without the agreement of the chief fire officer and the proprietor of the establishment. Any alternative arrangements that are made must also be to the satisfaction of the chief fire officer. In this connection there are many locations within the public highway system where hydrants are located and to which, in the event of a fire, the local fire brigade must have quick and easy access. Once access along a particular route is restricted it may be that access to some hydrant locations is denied and before finalising any traffic management scheme the chief fire officer should be given the opportunity to assess his force's capability to fight fire with the remaining resources. Similarly access for ambulances must be considered and where appropriate alternative arrangements made with the agreement of the health authority.

During the route survey, or as soon after an initial survey of the route or routes is completed, each property having special access needs should be visited, the need for the work to be done should be explained and the alternative access possibilities explored with the manager or owner of the property. This should be done as early as possible in the design process. The access provisions that can be made should be explained and if necessary modifications to the diversion or the design or the method of construction made wherever possible to ensure least impact of the works upon the function of the premises.

Claim's for loss of profit as a result of sewerage works are covered by section 278 of the Public Health Act 1936 and particularly in the case of town centre sewerage schemes can be considerable depending on the type of business amounting to a settlement figure even approximating to 10% of the overall project cost. The planning of both the route and the timing of the scheme will obviously have an effect upon trades and businesses.

It is essential for the engineer to:

1. Take positive steps to identify trades/businesses likely to be affected by possible option for the works.
2. Consider the types of business and how their particular trading seasons will be affected by the funding of the proposed works.
3. Consider alternative routes.
4. Estimate the cost of loss of profits regarding disruption and evaluate alternative working methods.
5. Plan to site shafts and compounds in areas of least business activity in order to reduce the disruption.

Similarly access provision to domestic properties needs to be taken into account. Occupiers of property often arrange for deliveries of furniture, large electrical appliances, etc. to their house, or for the removal of their belongings to another property. It will be of help to them and to the contract if they are aware of the restricted access dates as far in advance of the works as possible, in order to avoid confliction. Circular letters to each occupier can minimise difficulties of this type.

Press releases to appropriate newspapers, describing the proposed work and the expected diversions to traffic with proposed start and finish dates, can also prove to be helpful and, if contact telephone numbers are given, persons affected may be given the opportunity to make contact with the designers of the scheme. This enables all persons to have the opportunity to be fully aware of the expected restrictions to traffic. It is usual that at this time in the consultation process the various temporary traffic orders that will be needed to enable the highways to be temporarily restricted will need to be advertised in the local press. This would be done by the traffic authority on behalf of the instigator of the work and access to the highway will not be legally possible until the orders have been advertised for the relevant prescribed periods of time and duly authorised. Currently temporary traffic orders are dealt with under the Road Traffic (Temporary Restrictions) Act 1991 which made new provisions in place of sections 14 and 15 of the Road Traffic Regulation Act 1984. These new provisions enable a traffic authority to restrict or prohibit temporarily the use of a road, or of any part of it, by vehicles or pedestrians, for various reasons, which include works that are proposed to be executed on or near the road. When making the order the traffic authority is required to have regard for the existence of suitable alternative routes which are available for traffic affected by the order. There are specified time limits on the orders of six months in relation to footpaths, bridleways, cycle tracks or by-ways open to all traffic and in any other case 18 months. The Act, however, says that the time limit of 18 months shall not apply to an order if the traffic authority is satisfied and it is stated in the order that it is satisfied that the execution of the works in question will take longer than 18 months, but in any such case the authority shall revoke the order as soon as the works are completed. Where an order has been made restricting the period to 18 months and for any reason overruns this time, application for an extension has to be made to the Secretary of State who may extend the order by up to six months.

The above types of consultation and spreading of advance information at an early date will enable designers to be aware of special access problems which could influence the location of access shafts to the work, contractors' working space, the siting of site facilities and the method of working, for example the possible use of 'no-dig' techniques over some or all of the works to minimise disruption of normal life.

It is of equal importance that fire, ambulance, local passenger transport authorities, and Taxi Drivers Associations are involved in discussions as soon as a draft diversion route has been established with the street authority. The police, fire and ambulance vehicles particularly have a legal requirement to attend a call within prescribed periods of time, and it may be possible to make special arrangements for their call-outs; indeed it is in all our interests to see that their times of travel along a particular route are interfered with as little as possible. Benefit can also be obtained from advance liaison with the Automobile Association and the Royal Automobile Club, when major traffic routes are affected.

Other considerations that need to be borne in mind when considering possible diversion routes with the various parties to the consultation process are:

- The routes used by abnormal loads – advice on this can be given by the street authority or the traffic police
- The routes used to transport prisoners to or from local courts/prisons – these are often chosen for their high security value
- Access to and from banks in relation to the transport of money
- The presence of low bridges or of bridges with either a width or load restriction on them
- The relocating of bus stops and the guidance needed to direct the public to the temporarily resited stop with safe pedestrian access (this would be done in agreement with the local transport authority)
- The effect of the chosen route(s) on adjacent streets or other main routes.

The point of this latter item is that often, drivers having found that their normal route is suffering from traffic delays, transfer themselves on to another route into their centre of business. This route may be through a residential area, which could create danger, or along streets which are inadequate to cope, either in width or junction capacity, with the added traffic load. In order to reduce this risk it may be necessary to restrict traffic usage by temporary traffic orders or the introduction of temporary traffic calming measures of a more physical nature.

During the course of the work there may be changes to the way in which the work has to be carried out which cause changes to the traffic routes or to the available carriageway or footway. In all such cases the traffic police and the highway authority must be informed and be in agreement with any changes which have to be made. Following this the emergency services and other utilities must be advised and where a transport authority is affected, on a bus route, for example, they too must be advised. It is possible that the alterations to the work may cause delays to its progress and the public should be kept informed of this as well as the amendments to the traffic management measures.

5.4 Summary of steps in consultation and public relations process

The need for the new construction work to be done having been established, the consultation and public relations process is summarised as follows:

1. Examine the site of the proposed work and assess the likely effect on traffic of the works as initially visualised and form an opinion on the expected need to close off roads and divert traffic.
2. Arrange to meet with the street authority, traffic police and other undertakers to explain your work proposals and suggestions on the way traffic is likely to be affected. At this

(a)

(b)

Figure 5.2 (a) Besides the courtesy notice relating to 'Business as Usual', the photograph shows the confined width in which the contractor is working. (b) This is a view from the opposite end of the section of the works at an off-peak traffic time, and indicates again the confined way in which the contractor has contained the working site to minimise any possible loss of trade. (c) Here the central reservation has been reduced in width to obtain an increased width of carriageway on the outer lane. Along the shop frontage parking lay-bys are being provided to enable shoppers and delivery vehicles to park when work is not in progress along that frontage of shops. (d) Work is in progress laying the new sewer along the middle lane, traffic is using the widened outer lane. Access to the shops fronting the short length of the works site is by foot from the parking lay-bys before and after the fenced-off site.

(c)

(d)

Figure 5.2 *Contd*

meeting it is likely that some ideas of the types of diversions or traffic controls will be discussed and a clearer picture of what will be needed from a traffic management point of view will emerge.

3. Carry out a more detailed survey of the route of the works and of any other route which will be affected by diverted traffic, for example, during this survey all establishments

domestic or commercial which are likely to have access difficulties should be noted for individual consultation.

4. Send a circular letter to the domestic properties whose access is going to be affected, giving details of the work, the diversion, the expected start and finish dates and a point of contact if the recipient wishes to have further information or to give you details of particular difficulties they may have.

5. Either send a similar letter to the affected commercial properties or visit them personally to discuss the likely problems that may be experienced and ways in which the problems may be overcome. In this latter case a confirmatory letter to the relevant person is always advantageous when settling disputes at a later date.

6. At this stage photographs of all properties fronting the works and the diversion routes should be taken as they may be required in connection with any subsequent claims for structural damage.

7. Meet again with street authority, traffic police, other undertakers and include this time fire brigade, ambulance, passenger transport authority or operators as appropriate, and a representative of the Taxi Drivers Association if operative in that area. At this meeting the diversion routing should be clarified and finalised, bearing in mind any special issues like bridge strengths and heights, abnormal load routes, street processions or carnivals.

8. Send formal notice of intention to do works to the street authority to enable the relevant traffic orders to be advertised in relevant newspapers in time for the expected start date.

9. Send a press release to the relevant local newspapers describing the work to be done and giving details of the traffic diversions; a small plan of the routing of the diversions is often helpful and when printed in the newspaper attracts the attention of the public to the scheme and leads to their discussion with others who may not have seen the proposals in the press.

10. Advise the local radio and television operators of the details of the scheme at least a fortnight in advance of the expected start date.

11. During the course of the work events may occur which alter the planned diversion and local radio should be kept informed of such changes as they happen to enable the details to be broadcast on the traffic bulletins. It is of course essential that such changes are done in consultation with the street authority and traffic police and the passenger transport undertakings, fire brigade and ambulance people informed.

12. Depending on the significance of any changes it may also be desirable to send a further letter of explanation to fronting properties as in (4) and (5) above

5.5   Signing and statutory requirements

The majority of today's roads are saturated with heavy fast moving traffic and drivers need to be very alert to changing road conditions, particularly those created by the need to carry out construction work on or below a road. Sometimes the very presence of the works can cause drivers' attention to be distracted while they find out what is going on at an adjacent construction site. Advance warning signs stating what the work is are not mandatory, but can help the motorist to relax and pay more attention to the diversion ahead. Such signs should be able to be read by a driver without difficulty and consequently lettering should be clear and the wording be as concise as possible, to reduce distraction time.

It is of great importance to ensure the safety of all members of the public in their passage along a highway, this includes drivers, pedestrians, people with disabilities, children, the elderly and people with prams. It is therefore essential that the works and the diversion routes are clearly signed and guarded in a safe and readily understandable manner and that the signing and guarding is maintained that way throughout the course of the work. Standardisation of signs is always of benefit and full details of the types of signs and their uses which are suitable are contained in the Road Traffic Regulation Act of 1984 and the *Traffic Signs Manual*, published by the Department of Transport.

Chapter 8 of the *Traffic Signs Manual* gives excellent advice on signing and guarding; it is not, however, a mandatory document. It is recommended for use in conjunction with 'Safety at Street Works and Road Works' – this is a code of practice issued by the Secretaries of State for Transport, Scotland and Wales under sections 65 and 124 of the New Road and Street Works Act 1991; this document is mandatory, as such it is a criminal offence to fail to comply with its directives.

The diversion chosen may be a simple case of restricting the width of a carriageway at a few locations to permit the excavation of working shafts, in which case, it remains that the safety measures referred to in section 65 of the New Road and Street Works Act are met. The code of practice for 'Safety at Street Works and Road Works', often referred to as the *Blue Book*, describes the minimum carriageway widths for single lane or passing two-way traffic; the use of Stop/Go boards or temporary traffic signals in detail, and the salient details are described later in this chapter.

If, however, the full-scale closure of a main traffic route requiring the diversion of traffic onto one or more other traffic routes is needed, an assessment of the expected traffic transfer load should be done. It is likely that the street authority will have details of the peak traffic flows along main traffic corridors from their traffic count records, together with information on the capability of existing junctions and traffic signals, carriageway widths and pavement strengths to cope with the forecasted increased flow.

The New Road and Street Works Act, section 77, defines the liability of cost of use of an alternative route on a lower category of road and requires the undertaker to indemnify the highway authority in respect of costs reasonably incurred by them in:

- Strengthening the highway, so far as that it is done with a view to and is necessary for the purposes of its use by the diverted traffic; or
- Making good damage to the highway occurring in consequence of the use by it of the diverted traffic.

It follows therefore that such costs will be in addition to the cost of carrying out the sewer reconstruction and, as said at the beginning if this chapter, this can be very high.

On a main traffic route it is likely that retiming of traffic signals will be needed, the position of traffic bollards to facilitate turning movements will need adjusting, and the level of street lighting along the whole of a route or at individual junctions may need improving. Work of these types should only be done in conjunction with advice from the local traffic authority. The retiming of traffic signals will of course need to be redone when the diversion is taken off.

Similarly it may be necessary to improve or change the location of access to premises to fulfil promises made to owners of premises at the earlier meetings, and subsequently to restore them. One of the most difficult of these can be the arrangements to provide access to petrol/diesel filling stations, where the turning loci of large vehicles can impinge greatly on the available road width on a restricted carriageway.

When considering the traffic load on a diversionary route and the effect of the additional traffic load on that route, it is often of value to consider advising traffic to take a totally different route into the centre of a town than the one they would normally travel on to reduce the traffic load on the route affected by the works. This can mean that advisory signs may be needed possibly several miles in advance of the works site.

One has to cater for pedestrians, the disabled, and the blind as well as vehicular traffic. These all have their own special criteria. The code of practice, the *Blue Book*, describes in some detail the type of signing, the toe boards for blind persons to tap their sticks against and so on; what is not mentioned, perhaps because it shouldn't need to be said because of other legislation in the Highways Act of 1980, is the need to ensure that surfaces of pavements should be kept clear of obstructions and dirt at all times, so that a safe passage can be obtained. A blind person, for example, will have enough difficulty locating the board against which to tap his or her stick but if the surface is uneven that person is likely to feel that they may have wandered on to the site of the work rather than be on a temporary footway diversion. Wheelchairs are another case where one needs to take special care to ensure that not only is there a level area for the vehicle to move along but that there is sufficient space to manoeuvre it around or over obstacles. Often one sees a disabled person in a self-propelled wheelchair unable to mount a kerb; or in a situation where, because of works being carried out, footway width has been restricted to such an extent that there remains an insufficient width for all four wheels of the wheelchair to be on the same level. Thus forcing the occupant to travel in a lopsided manner with two wheels in the channel and two wheels on the footway level and heading for a capsize situation. The code of practice states that temporary pedestrian ways should never be less than 1 metre wide and wherever possible they should be 1.5 metres or more in width. I understand from manufacturers that the maximum width of a self-propelled wheelchair is 905 mm. As the reader will appreciate there is little room to manoeuvre in either of the two dimensions for the temporary pedestrian route, and certainly no room to pass. In practice it is helpful to provide passing places of at least 2 metres width on pedestrian routes which are at or near the minimum prescribed width. These should be spaced at about 25 metres apart dependent on the pedestrian usage.

Available road width is of great importance to the safe passage of traffic, after the working space for the construction works has been established. The *Blue Book* gives advice on this:

> For two lane single carriageway roads, used by HGVs past a roadside works site, two-way traffic must have an unobstructed width of at least 6.75 metres. Where this cannot be achieved the Contractor's Supervisor should contact the Public Transport co-ordinator. A width of less than 5.5 metres is too narrow for two-way working and if this is the best width that can be provided then cones should be used to restrict the width to not more than 3.7 metres and traffic should be controlled by one of the systems described below for one-way working. At a works site which is in the middle of the road a lane width of 3.25 metres is recommended on both sides of the site, it may be possible to reduce this to 3.0 metres (absolute minimum) or to close one side completely and set up traffic control on the other side. Where traffic is expected to consist only of cars and other light vehicles the lane width may be reduced to 2.75 metres (desirable minimum) or 2.5 metres (absolute minimum). If these widths cannot be achieved the Traffic Authority should be contacted for guidance.

The code of practice permits the contractor's site supervisor to decide if, and what type of, traffic control is needed on the basis of:

- The available road width
- The amount of traffic
- Whether or not drivers can see oncoming traffic beyond the works.

5.6   One-way working

Shuttle lane is defined as an area of carriageway, where owing to a temporary restriction traffic has to flow first in one direction and then the other.

5.7   Give and take system

This should only be used when there are fewer than 20 vehicles passing the works site in both directions in any period of three minutes (400 vehicles per hour), there are fewer than 20 heavy goods vehicles passing the works in an hour and the speed limit is 30 mph or under, the length of the works site from the start of the lead-in taper to the end of the exit taper is no more than 50 metres and drivers approaching from either direction can see both ends of the site. With this system the decision to pass the site is left to the discretion of each driver on the basis that the road ahead is clear to them. The minimum available lane width is quoted at 3.25 to 3.7 metres unobstructed width.

5.8   Priority signs

The conditions for using this system are: two-way traffic flow must be fewer than 42 vehicles counted over a three minute period (850 vehicles per hour), the length of the works from the start of the lead-in taper to the end of the exit taper is no more than 80 metres and drivers can see from a point 60 metres before the coned-off area to a point 60 metres beyond the end of the coning on roads restricted to 30 mph or less. On roads with higher speed restrictions the system is permitted provided clear visibilities are obtainable of 70 metres on 40 mph roads, 80 metres on 50 mph roads and 100 metres on 60 mph roads. Priority is normally given to vehicles going up a steep gradient or vehicles in the unobstructed lane going past the works. The signs are:

1. A rectangular plate with a light blue background showing a large white arrow pointing upwards in the direction of priority traffic flow on the left side of the plate and a smaller red arrow pointing downwards on the right of the plate. This sign is placed facing traffic which has priority.
2. A circular prohibition sign with a red circular edging containing on a white background a large black arrow pointing downwards on the right of the plate and a smaller red arrow pointing upwards on the left of the plate.

5.9   Traffic control by Stop/Go boards

Traffic can be controlled manually using Stop/Go boards when the relation of the two-way traffic flow to the length of the site is no more than the following:

Site length (metres)	Maximum two-way flow	
	Vehicles per 3 minutes	Vehicles per hour
100	70	1400
200	63	1250
300	53	1050
400	47	950
500	42	850

It the shuttle lane is shorter than 20 metres in length a single board operated either at one end or in the middle will be sufficient. Where a road junction occurs within the restricted length it may not be appropriate to use Stop/Go boards; it is always advisable to get further advice from the traffic authority in this case. In all cases special care needs to be taken by trained operatives. Where there is a bend in the road or operatives are out of sight of each other it may be necessary to use two-way radio communication to help with traffic control. The operator showing 'Go' to traffic should always be the one to reverse the board first to show 'stop', and the other board should not be reversed to show 'Go' until all the traffic has moved out of the shuttle lane. For works near a level crossing Stop/Go boards are of great benefit and probably the safest system to use. Railway management should always be consulted before any works are carried out close to or affecting a level crossing.

5.10   Traffic control by portable traffic signals

Before using portable traffic signals the local traffic authority should be advised and their authority obtained in writing if the signals are proposed to be used at a site containing a road junction.

Portable traffic signals can be used at most sites which have an overall length of 300 metres or less; it is, however, important to ensure that the exits to the restricted length remain unblocked. If this does happen take further advice from the traffic authority. If the use of the signals causes traffic to block up within the vicinity of a level crossing consultation with railway management as well as the traffic authority will be necessary.

5.11   Signing and lane marking

Before any temporary traffic signs or lane markings are introduced onto any public highway it is essential to see that existing traffic is not put into any kind of danger by a partly signed diversion or restriction in carriageway width. If there are parked vehicles in the length of road being affected by such changes the owners of the vehicles should be contacted and asked if they would please move them and an explanation given. If a full lane is being taken out of use, say for the installation of temporary traffic signals for one-way traffic, the use of Stop/Go boards on a temporary basis should be considered, certainly there should never be an instance where two-way traffic is permitted to continue to use a single lane pending the operation of temporary traffic signals.

Guidance on the location of signs and cones is given in the code of practice for various traffic situations, and in addition the traffic section of the local authority will no doubt give helpful advice. It is, however, worth adding a few points that may also be helpful.

Signs stood on a footway are an obstruction to pedestrian movement and a hazard to the blind. If the works are of short duration then one might accept that such signs are probably acceptable, but if the works are likely to last longer than a week it is worthwhile considering pole mounting of the signs. This has several advantages in that the signs are:

1. Readily visible to vehicle drivers
2. Difficult to be interfered with by roisterers, hence less of a problem to keep in the correct location
3. Not an obstruction and can release either more working space or more space for traffic.

Pole mounting will require the agreement of the traffic authority to use existing sign poles or street lighting columns or additional temporary poling, and will need to be done with some sensitivity, particularly in relation to visibility sight lines. The signs should not interfere with existing sight lines to remaining traffic signs or traffic signals and if placed during the autumn remember that leaf growth on trees can obscure any poorly placed signs that need to remain into the spring. The fixing of the signs should be by means of non-ferrous straps to minimise damage to the poles by staining. Nylon webbing with self-tightening buckles has been found to be an inexpensive solution, and can be reused. The traffic authority may have their own patent system and may by arrangement erect the signs for you, alternatively the AA or RAC may carry out erection work on a rechargeable basis.

Lane markings which exist and are being temporarily removed should be completely obliterated preferably by burning off with hot air; painting over these lines is rarely satisfactory in that the old lines show through, particularly in wet weather and even in sunny weather when the sun is at a low angle. If temporary lane markings are needed they should be painted onto the carriageway using a paint acceptable to the traffic authority.

The placement of cones is a comparatively simple job, so much so that often they are carelessly placed and insufficiently weighted to withstand even the slightest winds. Most cones have their own weight ballast placed at the bottom; frequently after usage the ballast holder may have become damaged and it is good practice to inspect all cones before use to ensure that they are all of the proper type and weight and that a responsible person checks the placement of the cones for correct spacing longitudinally and transversely. It is often useful for the odd paint mark to be put on the carriageway to locate lane widths between cones in case the cones become displaced while the work is in progress.

Barriers are also prone to being moved and wherever possible they should be weighted down with sandbags or have sufficient self-weight to withstand wind or the occasional bump. The space between barriers should be from the furthest projection points of the support for the rails and not just between the rails. The code of practice shows cones used as supports but in practice there appear to be a multitude of different methods in use. Some supports are made of metal and some of wood and in both cases the bases form a cross. The author spotted one of these made from 100 mm by 50 mm timber that had a foot projection of 450 mm into the pedestrian route, no allowance had been made for this encroachment onto the minimum space of 1 metre for pedestrians and there was a major hazard situation.

Of equal importance to the provision of new signing for the diversion is the temporary removal or covering over of previously existing signs for the normal traffic flow system which are in confliction with the guidance to traffic on the diversion route. If the signs are covered over it is important to ensure that the covering material, besides being dense so that it can't be seen through, is firmly secured to prevent it being blown off in high winds or simply pulled off by the weight of water which has been retained in folds of the cover in times of rain.

5.12 Maintenance

The condition of the signing and guarding of the works should be inspected at regular intervals. Where works are regularly in progress adjacent to the signs this should present little difficulty, it is, however, all too often that people on site move barriers or cones to give themselves a little more space to carry out a particular operation and in doing so create a danger to passing traffic. It is therefore important to stress to site managers, foremen and gangers the need to ensure that if signing and guarding has to be temporarily moved for a short period it should be replaced as quickly as possible and that a person should be appointed to keep watch on the traffic while the danger exists and to take action to prevent accidents occurring.

The signing of the diversion should also be regularly inspected. In a busy city or town it is not uncommon for roistering revellers to use barriers and cones for their own unpredictable purposes after they have completed their course of imbibing. This can sometimes have strange effects on the movement of traffic which can be dangerous and the cause of considerable disruption. The displacement of barriers at the site of the works is also a source of danger particularly when interfered with outside normal working hours. If the site does not have a security patrol on duty it is often helpful to ask the local traffic police if they would advise a site manager of any problems which are reported to them outside working hours. It should also be standard practice to provide a 24 hour telephone contact number on site notice boards for any member of the public to contact in the event of any problems which they may see. On motorway works this number is usually referred to as a 'Cones Hotline', and is displayed at frequent locations along the route of the coned-off areas.

During the course of the work it may be necessary to move sections of the diversion along the course of the work. It is advisable to have a responsible person checking that all unnecessary signs and cones are removed, that all site gear has been cleared away and that surfaces of the carriageway and footways have been safely reinstated, before reopening the section to traffic. This of course applies at the end of the contract as well, but is mentioned here because all too often areas are opened to traffic in some haste to enable the next section of the work to progress, and problems, which could have been avoided, occur because of signs remaining that are no longer relevant; or of surfaces, which have been roughly patched instead of receiving a sound repair, becoming dangerous.

5.13   Reinstatements

As each part of the sewering work is completed the surface reinstatement of the carriageways and footways should be commenced and completed, as soon as is practicable, to enable them to be restored to traffic. The New Road and Street Works Act of 1991 places the responsibility for carrying out the reinstatement work on the undertaker who opened the surfaces and the local highway authority is responsible to see that the works are done in a satisfactory manner. Chapter 6 of volume I mentions that it has been normal practice to reinstate the highway following sewer or other utilities work with temporary construction and to return at a later date once the excavated area has settled and become sufficiently compacted to permit the final surface materials to be laid. The point is made in that chapter that this causes further disruption to traffic and in many heavily trafficked locations highway authorities will restrict permanent reinstatement time to an occasion when traffic is light, such as weekends in a city centre location, to prevent another encroachment onto the road users' domain when traffic usage is high. Weekends are of course a time when labour costs are at their highest and it will inevitably cost more to do this than to do an effective one-off reinstatement.

The execution of permanent reinstatement works immediately on completion of the main sewering works, known as one-pass reinstatements, is to be encouraged but a great deal of experimental work still needs to be done as it is not easy to achieve in practice. The use of traditional compacted backfill materials such as graded crushed stone, lean mix concrete, pulverised fuel ash, with a foundation of reinforced concrete surfaced with bituminous macadam or hot rolled asphalt, does not lend itself to single-pass work, because of the difficulty of obtaining 100% compaction. All too often the settlement of highway surfaces laid above a trench backfill which has not been properly compacted has resulted in surface cracks along the edges, permitting the ingress of water followed by frost damage and potholing. Some useful experimental work has been done with foamed concrete which has in many cases been successful as a backfill material; drawbacks have, however, been found when the foamed concrete has flowed into adjacent ducts or chambers. The advantage of using this material is that it sets reasonably quickly to enable surfacing with bituminous materials to be carried out in 24 hours. It is light in comparison to more traditional materials and develops friction with the sides of the trench or shaft by filling the irregularities thus reducing the amount of downward movement that would have occurred with a plug of the much heavier normal concrete. The choice of method of reinstatement for particular locations is a matter for discussion between the undertaker and the highway authority and in any event should comply with the requirements of the 'Specification for the Reinstatement of Openings in Highways', being a code of practice approved by the Secretary of State for Transport and issued for use under the terms of sections 71 and 130 of the New Road and Street Works Act 1991.

Bibliography

Sewers by Bevan and Rees.
Article from the Trenchless Technology workshop 1/7/93.
New Road and Street Works Act 1991. Published by HMSO.
Road Traffic Regulation Act 1984. Published by HMSO.
Traffic Signs Manual. Published by HMSO.
Safety at Streetworks and Roadworks: A Code of Practice. Published by HMSO.
Sewers Rehabilitation and New Construction Repair and Renovation. Published by Arnold.

Acknowledgement is due to John Kennedy (Civil Engineering Ltd) Chaddock Lane, Worsley, Manchester M28 1XW for their help during the production of this chapter, and to my friends and ex-colleagues at Manchester City Council for their helpful comments during the production of this chapter.

6

Aspects of Sewer Design

Ian Vickridge BSc(Eng), MSc, CEng, MICE, MCIWEM

6.1 Introduction

The design of a sewer system includes both hydraulic and structural considerations as well as economic ones. In this chapter only an overview of the major aspects of design will be given together with comments on design aspects that cannot easily be found elsewhere. Detailed design considerations are covered comprehensively elsewhere, and the reader is referred to the texts cited in the bibliography.

The design process includes the initial survey and scoping, the preliminary design, and the detailed design, all of which are tempered by cost considerations. As with most engineering design the process is iterative.

6.2 Survey and scoping

An assessment of the area to be served must be made initially by studying maps of the area, determining the population to be served, and making some estimates of the flows that will drain into the proposed system. Much of the early work can be done from contour maps if they are available and up to date. However, in some cases such information may not be readily available and a full survey of the area will be required.

6.3 Preliminary design

Typically a 1/2500-scale contour map is required to carry out the preliminary design, which entails determining an initial proposed layout of the sewer network. Sewerage networks are normally laid in straight lines between access chambers. In the past, the maximum distance between access chambers on straight line sewers was taken to be approximately 100 metres, but more recently the allowable distance between chambers has been increased and may be up to 300 metres, depending on the diameter of the sewer (see Chapter 2). However, additional chambers should be provided at all changes of direction and gradient. Sewer pipe gradients should be uniform between access chambers to provide continuity of flow and prevent silt and debris from collecting in low spots. In order to minimise the depth of the sewer, the pipe gradient should follow approximately the ground surface gradient, although the actual gradient of the pipe will be dictated predominantly by hydraulic considerations.

The preliminary layout design of a sewer network is a process of trial and error. The sewer network must pass close to the properties it is intended to serve and in most cases this means that the main sewers follow the street network, with subsidiary pipes (or laterals) connecting individual properties to the main sewer (see Fig. 6.1). A suitable preliminary layout for the sewer network must first be drawn on a large-scale map of the area, taking into consideration the proximity of the buildings to be served, the local topography, and the method of construction.

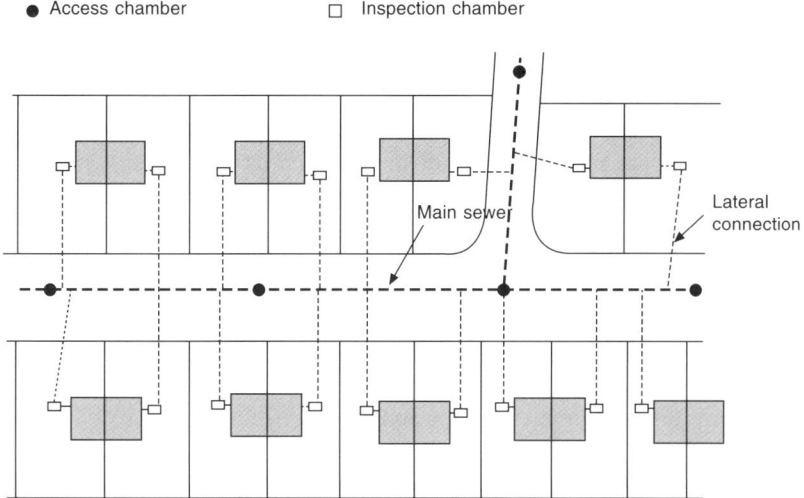

Figure 6.1   Typical sewer layout.

Although the typical arrangement shown in Fig. 6.1 is most common, alternative arrangements have been advocated whereby lateral connections are made directly into access chambers rather than into the sewer itself (see Fig. 6.2). Although this alternative arrangement results in longer pipe runs for the laterals, it has benefits where lateral connections

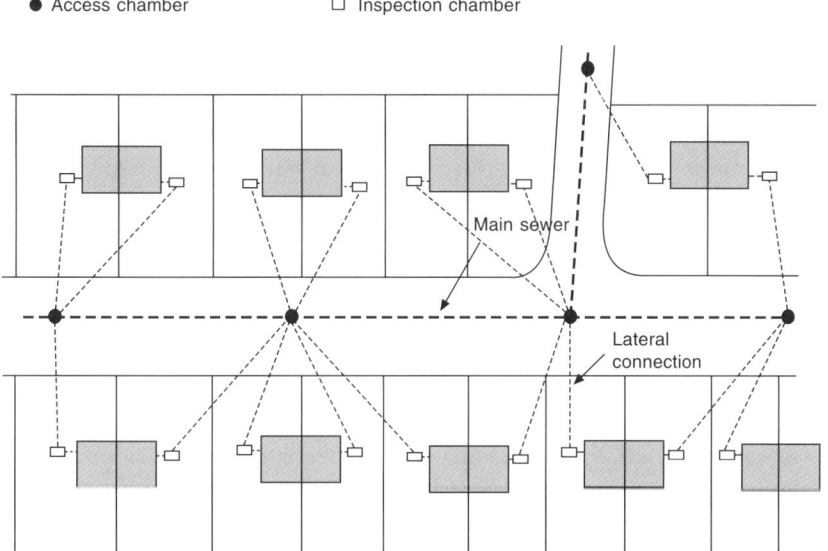

Figure 6.2   Sewer layout with laterals connected to manholes.

to the sewer are difficult to make, as may be the case for deep sewers, or where sewers are installed using microtunnelling or horizontal directional drilling techniques. There may also be long-term benefits. As sewers deteriorate with use and time, there will be a need to repair and rehabilitate them in the future. It is generally cheaper and less disruptive to reline an old sewer than to replace it, but in conventional sewer layouts this means that holes need to be cut in the lining to accommodate and reconnect the laterals. This can affect the overall integrity of the lining and will increase the costs. Connecting laterals directly into access chambers will thus reduce the long-term life cycle costs associated with the future repair and rehabilitation of the system.

The method of construction should also be considered at the preliminary layout stage, particularly if a trenchless method is to be considered. It is possible with trenchless methods to consider a sewer route passing under obstacles such as buildings, water courses, and railways. Such a route would not be possible using conventional open-trench sewer installation techniques. If a trenchless method such as microtunnelling is being considered, the location of the drive and reception shafts must be chosen carefully, as these will also form access chambers on the completed sewer. The excavations for these shafts will be open for some time during the construction process and they will therefore need to be positioned where they will cause minimum disruption to traffic flows and not, for example, at a busy road intersection.

The next step in the design is to draw a longitudinal section, or profile, of the ground surface along the proposed route of the sewer. These sections are normally drawn with a horizontal scale of 1/2500 and a vertical scale of 1/250. Suitable preliminary gradients for each sewer length can then be selected and also drawn on the section. Gradients are selected to ensure that all flow can be accommodated at an adequate velocity through the system. At the same time, the designer will attempt to limit the depth of the sewer below the ground surface, as this will have a significant impact on the cost. At this stage, the designer must use his or her past experience to arrive at a preliminary solution, which will subsequently be tested and amended by more rigorous calculation.

6.4 Hydraulic design

Once the preliminary layout and longitudinal section of the sewer network has been decided, the diameter of each pipe must be calculated and the gradient checked. To do this, it is first necessary to estimate the maximum likely flow in each pipe, and this will depend on whether or not the sewer is to carry surface water, foul water, or both. Foul water flows are estimated on the basis of the average daily per capita water consumption multiplied by the number of people served by the specific section of the network under consideration. This provides an estimate of the average flow, which is then multiplied by a factor to give an estimate of the maximum or peak flow. The calculation of surface water flows is more complex and is based on hydrological principles describing the relationship between precipitation and runoff. The peak flow depends not only on rainfall characteristics, the area and shape of the catchment contributing to flow at each section in the network, and the nature of the ground surface and the underlying soil, but also on the diameter and gradient of the sewer network itself. Further details of flow estimation are provided in Chapter 2 of this volume.

Determination of the required gradient and diameter of any pipe length in a sewer network is thus an iterative process, which starts with the designer making an informed estimate of the required diameter, and proceeds with increasingly more accurate estimates of flow rates and velocities until a suitable diameter and gradient is finally selected. Because of the iterative nature of the process and the complexity and size of most sewer networks, the hydraulic design of sewer networks is best done on a computer using one of a number of sophisticated software packages, such as Micro Drainage (see http://www.microdrainage.co.uk) that are currently available for this purpose.

Many of these programs are based on either a form of the rational method or the hydrograph method developed by the TRL (formerly the TRRL) in the UK (see White, 1978 for further details). This method uses a system of referencing pipe lengths in a sewer network that has subsequently been widely adopted. Figure 6.3 provides an example to illustrate the principle.

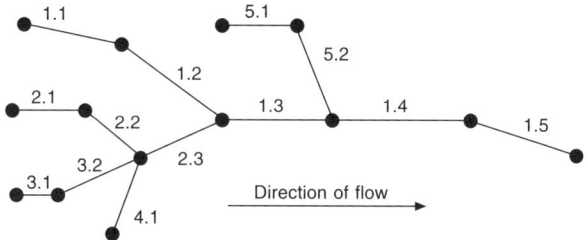

Figure 6.3   Pipe referencing system.

Referencing of all the pipes is of the form $x \cdot y$, where x refers to the branch of the network, and y refers to the number of the pipe in that branch. Reference numbers are assigned in sequence starting from the upstream end of a main branch. This branch is assigned the number 1 and the first pipe length then becomes 1.1. Other pipes in the branch then become 1.2, 1.3, and so on. When another branch meets the first branch, it is assigned the number 2 and the pipes are then numbered 2.1, 2.2, etc. from the upstream end to the point where the second branch meets the first. These rules are applied until the entire network has been referenced.

In the hydraulic design of each pipe in the system, the diameter and gradients of the upstream pipes must be determined first, as this will affect the flows in pipes further downstream. In the example shown in Fig. 6.3, pipe 1.3 cannot be considered until gradients and diameters of pipes 1.1, 1.2, and all pipes in branches 2, 3 and 4 have first been determined. The referencing system outlined above is required for computer simulation programs to perform the computations in the correct order and thus identify suitable pipe diameters for each section of pipe in the network.

6.5   Determination of sewer size

The diameter and gradient of a sewer must be chosen to meet two basic criteria. First, the pipe must be able to accommodate the calculated peak flow rate, and second, the velocity of flow must be sufficiently high to keep the sewer clean and free of grit and other solids that could settle in the invert of the pipe. This minimum (or 'self-cleansing') velocity is usually taken to be between 0.90 m/s for small diameter pipes to 0.75 m/s for larger pipes.

Mean velocity for a given diameter and gradient can be calculated using the Colebrook–White formula:

$$v = 2\sqrt{(2gdi)} \, \log \left[\frac{k}{3.7d} + \frac{2.51\mu}{d\sqrt{(2gdi)}} \right]$$ (6.1)

Where v = velocity
d = diameter
g = gravitational acceleration
i = hydraulic gradient
k = roughness
μ = kinematic viscosity (1.141×10^{-6} m^2/s for water at 15°C)

The *Sewer Rehabilitation Manual* provides the following suggested values for pipe roughness for sewers, k:

For $v > 1.5$ m/s, $k = 0.3$ mm
For 1.5 m/s $> v > 1.0$ m/s, $k = 0.6$ mm
For 1.0 m/s $> v > 0.75$ m/s, $k = 1.5$ mm

So, for a given gradient and diameter of pipe, the velocity of flow can be calculated and hence the flow rate or discharge, Q, can be calculated from the continuity equation:

$$Q = Av$$ (6.2)

Tables and charts, based on the Colebrook–White formula, are available to facilitate the calculation of flow and velocity for a variety of pipe sizes and gradients (see, for example, HMSO, 1998 and CPDA, 1999). The flows and velocities provided normally correspond with the pipe flowing full.

However, because the flow, depth, and velocity in a sewer will vary considerably, depending on the time of day and the amount of rainfall entering the system, it is necessary to check that the velocity requirements are met under these varying flow conditions. To illustrate this, an earlier and less complex formula for velocity, the Manning equation, is given below:

$$v = \frac{1}{n} m^{2/3} i^{1/2}$$ (6.3)

Where v = velocity of flow (m/s)
n = Manning roughness coefficient
m = hydraulic radius (m)
= A/p
A = cross-sectional area of flow (m^2)
p = wetted perimeter (m)
i = hydraulic gradient

The flow rate, Q, can then be calculated from equation (6.2) as before.

It is clear that the hydraulic radius, and hence the velocity and the flow rate, will vary with the depth of flow, y. The parameters A, d, y, and p are illustrated in Fig. 6.4.

Using the Manning equation, the actual velocity (v) and discharge (Q) at any depth of flow (y) can be computed and compared to the 'full flow' discharge (Q_0) and 'full flow' velocity (v_0). Table 6.1 shows the relationships.

It should be noted that the velocity, when the depth of flow is half the pipe diameter, is exactly the same as the velocity at full flow and that, when the depth of flow is between a half and full flow, the velocity is actually greater than for full flow conditions. Thus, in

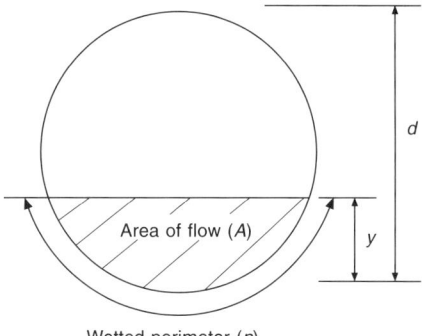

Area of flow (*A*)

Wetted perimeter (*p*)

Figure 6.4 Partial depth flow.

Table 6.1 Proportional depth relationships

Proportional depth (y/d)	Proportional velocity (v/v_0)	Proportional discharge (Q/Q_0)
0	0	0
0.02	0.128	0.001
0.04	0.213	0.003
0.06	0.283	0.007
0.08	0.345	0.013
0.10	0.400	0.021
0.12	0.450	0.031
0.14	0.496	0.042
0.16	0.539	0.056
0.18	0.580	0.071
0.20	0.618	0.088
0.22	0.654	0.107
0.24	0.688	0.127
0.26	0.720	0.149
0.28	0.750	0.172
0.30	0.779	0.197
0.35	0.846	0.264
0.40	0.904	0.338
0.45	0.955	0.417
0.50	**1.000**	**0.500**
0.55	1.038	0.585
0.60	1.071	0.671
0.65	1.097	0.755
0.70	1.117	0.835
0.75	1.130	0.909
0.80	1.136	0.974
0.85	1.134	1.027
0.90	1.121	1.063
0.95	1.092	1.072
1.00	1.000	1.000

many cases, the velocity of flow will not fall below the required self-cleansing velocity over a wide variation in flow rate. For example, imagine a sewer designed to carry the peak flow rate when flowing full at a velocity of 1 m/s. From Table 6.1, it can be seen that the velocity will still be 0.75 m/s when the proportional depth is 0.28, at which point the discharge will be 0.172 × peak flow, or approximately one-sixth of the peak flow. This clearly allows for a suitably wide and acceptable variation in discharge without compromising the minimum velocity requirement.

However, problems may arise when engineers attempt to design sewers for fairly flat areas and, for cost reasons and to minimise pumping requirements, they are tempted to use slack gradients for the sewers. To illustrate the problem, assume that the peak flow rate of 4 × DWF for a particular length of sewer has been calculated as 16 l/s, and that the topography suggests a maximum gradient for the sewer of 0.003. A self-cleansing velocity of 0.7 m/s is required.

A check is done on a suggested pipe diameter of 225 mm. This would normally be done using hydraulic tables or a computer program, but for the purposes of the example, the Manning equation will be used, with a Manning coefficient (n) of 0.012.

$$v = \frac{1}{n} m^{2/3} i^{1/2}$$

$$m = A/p = Ad^2/4Ad = d/4$$

$$v = (1/0.012)(0.225/4)^{2/3}(0.003)^{1/2}$$

$$= 0.67 \text{ m/s}$$

Discharge $Q = Av$

$$Q = Ad^2v/4$$

$$= A(0.225)^2 \times 0.67/4 = 0.0265 \text{ m}^3/\text{s} = 26.5 \text{ l/s}$$

The 225 mm diameter pipe is thus capable of carrying the peak flow of 16 l/s, but the velocity does not quite meet the minimum requirements for self-cleansing. At this point the designer may be tempted to choose a larger pipe, as this will give a higher velocity at full flow.

The full flow velocity in a 300 mm pipe at a gradient of 0.003 is calculated as follows:

$$v = (1/0.012)(0.300/4)^{2/3}(0.003)^{1/2}$$

$$= 0.81 \text{ m/s}$$

and $Q = A(0.300)^2 \times 0.81/4 = 0.0573 \text{ m}^3/\text{s} = 57.3 \text{ l/s}$

This gives the appearance of a more satisfactory solution as both the flow and velocity requirements seem to be satisfied. However, a 300 mm pipe would flow at less than half its full capacity for the calculated peak flow (16 l/s), and it is worth investigating the actual velocities and depths of flow in the two situations.

In the 225 mm diameter pipe, the proportional flow at peak flow is 16/26.5 = 0.6. From Table 6.1 it can be seen that the proportional depth would be approximately 0.56 and that the proportional velocity would be approximately 1.04, thus giving a flow depth of 0.6 × 225 = 135 mm, and a velocity of 1.04 × 0.67 = 0.7 m/s. Thus, the actual velocity at peak flow does just meet the minimum velocity requirement.

The 300 mm diameter pipe will have proportional flow of 16/57.3 = 0.3, giving a proportional depth of 0.37 and a proportional velocity of 0.87. The depth of flow is thus 111 mm and the velocity is 0.70 m/s. It can be seen that the velocity of flow at peak flow is the same for a 300 mm pipe as for a 225 mm diameter pipe.

However, if we consider the situation in both pipes at the DWF of 4 l/s, the depth of flow in the 225 mm diameter pipe can be calculated to be 59 mm with a velocity of 0.48 m/s, whereas in the 300 mm pipe the depth is 16 mm and the velocity is just 0.23 m/s. This is less than half the velocity that would be attained at this flow rate in the 225 mm diameter pipe.

Although this is just one specific example, it is important to note that simply increasing the diameter of a pipe will not necessarily increase the velocity of flow. Many designers would say that for a given gradient, a smaller pipe will keep cleaner than a larger one.

6.6 Structural design

A sewer, like any other pipeline, is a complex structure whose structural integrity relies on the properties of the pipeline material, the diameter and wall thickness of the pipe, and the vertical and horizontal support provided by the surrounding soil and other supporting structures. The loading on the sewer is difficult to accurately determine as it depends on internal pressures within the pipe, the depth of the pipe, the width of the trench in which the pipe is laid, the varying depth of the water table, the nature of the soil above the pipe, and loads from traffic and structures superimposed at the surface. This complexity of soil conditions and superimposed loads, coupled with a lack of precise, easy to use analytical tools for modelling the interaction between the pipe, the soil and the superimposed loads, means that a pragmatic and empirical approach to the structural design of sewers must be adopted.

There are four main aspects of the structural design of pipelines to consider, three of which relate to the operation of the pipeline and the fourth concerns the transportation and installation of the pipe.

As far as the operation of a pipe is concerned, the three factors to consider are:

- The capacity of the pipe to resist internal pressure
- The capacity to resist external forces
- The capacity to resist stresses generated from movement and/or dimensional changes of the pipeline.

The last of these does not normally need to be considered by sewer designers, apart from taking measures to ensure that pipes are installed according to commonly accepted good practice. This will include ensuring that trenches are prepared properly with no hard spots which could induce point loads on the pipe, providing adequate support for the pipe, and ensuring that flexible joints are provided to allow for any lateral and vertical movements that may occur.

In addition to these operational considerations, the pipe must also be able to resist all forces imposed during transportation to the site, storage on site, and installation.

6.6.1 INTERNAL PRESSURE

Although internal pressure does not normally need to be considered in gravity sewers, it may be a major consideration in pumping mains. Resistance to internal pressure is determined by the wall thickness and the strength of the pipe material and, for thin-walled pipes, can be analysed using simple ring design procedures.

Figure 6.5 shows a free body diagram for half a circular pipe under an internal pressure

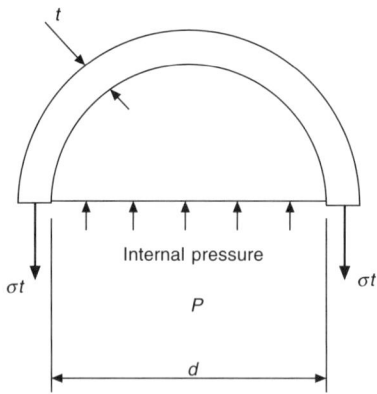

Figure 6.5 Simple ring design.

of P. For a unit length of pipe, the vertical upward force is given by Pd, and this is balanced by the vertical downward force of $2\sigma t$.

Thus, $P = 2\sigma t/d$ (6.4)

or $t = Pd/2\sigma$ (6.5)

where σ = hoop stress in pipe wall
P = internal pressure
d = internal diameter of pipe
t = pipe wall thickness

The analysis can thus be used to determine minimum wall thickness for a particular design pressure, or to determine maximum pressure for a given wall thickness. It is normal to apply a safety factor of 2.

6.6.2 *EXTERNAL FORCES UNDER OPERATIONAL CONDITIONS*

External forces arise from the load applied by the soil above the pipe, any external water pressure, and loads on the surface from traffic and structures. In general, the soil load increases with the depth of the pipe, while the traffic load decreases with depth. Shallow pipes thus need special protection against the high traffic loading and very deep pipes need to be protected against the high soil loads. Near the surface, traffic loading predominates and, as depth of cover increases, the combined traffic and soil load actually decreases to a minimum value and then increases as soil loading starts to predominate.

It is important to note that a buried pipe is a composite structure consisting of both the pipe itself and the surrounding soil. Both the soil loading on a pipe and the support given to a pipe by the surrounding soil depend not only on the depth, but also on the pipe diameter, the material properties of the pipe, the nature of the soil, and the trench width.

The earliest attempts to develop a theory for the soil load on a pipe were those by Marston as early as 1913. This work was advanced by Marston and his colleagues Schlick and Spangler at Iowa State University and, although various subsequent adaptations to this work have been suggested, the equation given below still encompasses the basic relationships they derived. Their work still forms the basis for numerous standards and codes of practice throughout the world, including the European Standard for the structural design of buried pipelines, BS EN 1295, which was published in 1998.

Figure 6.6 shows a pipe of external diameter, D, in a trench of width, B, with a cover depth of H. The fill load, W, on the pipe is given by:

$$W = C_d \gamma B^2 \tag{6.6}$$

Where W = load on the pipe
$\quad\quad C_d$ = load coefficient
$\quad\quad B$ = trench width
$\quad\quad \gamma$ = unit weight of backfill material

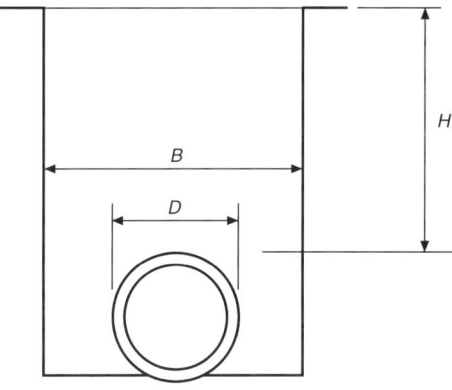

Figure 6.6

C_d is a function of the depth of cover (H), the width of the trench (B), the angle of internal friction of the backfill, and the angle of friction between the backfill and the trench sides.

For very shallow or wide trenches ($H/B < 1$), C_d is approximately H/B and the equation becomes:

$$W = \gamma H B \tag{6.7}$$

It is clear that the width of the trench has a significant influence on the load experienced by the pipe, and equation (6.7) would seem to indicate that the load on a rigid pipe will increase indefinitely with increase in trench width. However, this is not the case and, for wide trenches, a different theory based on the pipe being buried beneath an embankment can be used.

Rigid (concrete, clayware), semi-rigid (ductile iron) and flexible (PVC, PE, GRP, thin-wall steel) pipes are used for sewers and, as these various types of pipe behave quite differently, they each require a different form of analysis. Essentially rigid pipes fail due to excessive shear or bending stress at relatively low deflections, while flexible pipes fail through excessive deformation or buckling. Rigid pipes thus rely mainly on the material strength of the pipe material coupled with some support from the surrounding soil; flexible pipes have lower inherent strength but deform to take advantage of passive support from the surrounding soil. Figure 6.7 illustrates the difference.

Research has shown that, for flexible pipes in well-compacted backfill, the load depends more on the pipe diameter rather than simply on the trench width, and the loading equation becomes:

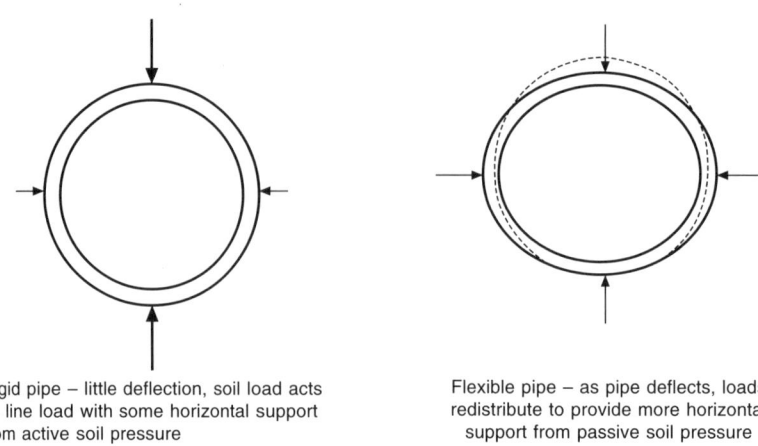

Rigid pipe – little deflection, soil load acts
as line load with some horizontal support
from active soil pressure

Flexible pipe – as pipe deflects, loads
redistribute to provide more horizontal
support from passive soil pressure

Figure 6.7   Rigid and flexible pipes loading and support.

$$W = C_d \, \gamma BD \tag{6.8}$$

Again, as with equation 6.7, for shallow cover depths, C_d is approximately H/B, and the load is given by:

$$W = \gamma HD \tag{6.9}$$

For design purposes, the loads imposed must be resisted by the pipe strength and any enhancement provided by the surrounding soil. For rigid pipes, the loads imposed must not exceed the pipe strength, but for flexible pipes the maximum load will be controlled by limiting pipe deflection to an allowable limit.

The strength of a rigid pipe is usually expressed in terms of a crushing load, measured in a laboratory by a three edge bearing test. The actual load that can be carried will generally be greater than the crushing load due to the enhancement provided by the surrounding bedding material, and in general terms this can be expressed as:

$$\text{Design load} = \frac{(\text{Crushing load}) \times (\text{Load factor})}{(\text{Factor of safety})}$$

The load factor expresses the degree of support provided by the bedding material.

The maximum load for a flexible pipe is determined by deflection criteria and in general terms this can be expressed as:

$$\text{Deflection} = \frac{(\text{Vertical load})}{(\text{Pipe stiffness component}) + (\text{Embedment stiffness component})}$$

The embedment stiffness is a function of the properties of the bedding material and surrounding soil, and thus quantifies the support provided. The embedment stiffness component of untamped backfill may be only about 10% of the embedment stiffness provided by a very well-compacted backfill. This illustrates the importance of backfill compaction to the performance of flexible pipes.

The situation is thus extremely complex and for a full exposition of the structural behaviour of buried pipelines, the reader should refer to the work by Watkins and Andersen (2000) cited in the bibliography to this chapter. The soil not only exerts a vertical load on the pipe but it also provides horizontal support which, as shown above, depends on the nature of the soil surrounding the pipe and the material properties of the pipe itself. Although the early research work on the application of soil mechanics to pipeline design by Marston, Schlick and Spangler at Iowa State University is still relevant, much subsequent work has been published on the subject.

6.6.3   RIGID PIPE DESIGN

The bedding around the pipe has a significant impact on the overall strength of the pipeline, and this led to the development of the concept of bedding factors and to their inclusion in design codes. These bedding factors present a way of including the enhancement provided by embedment in the design of buried rigid pipelines.

Typical bedding types together with bedding factors for vitrified clay pipes are shown in Fig. 6.8. It should be noted that the bedding factors given are for clay pipes in wide trenches and that lower values apply to other pipes.

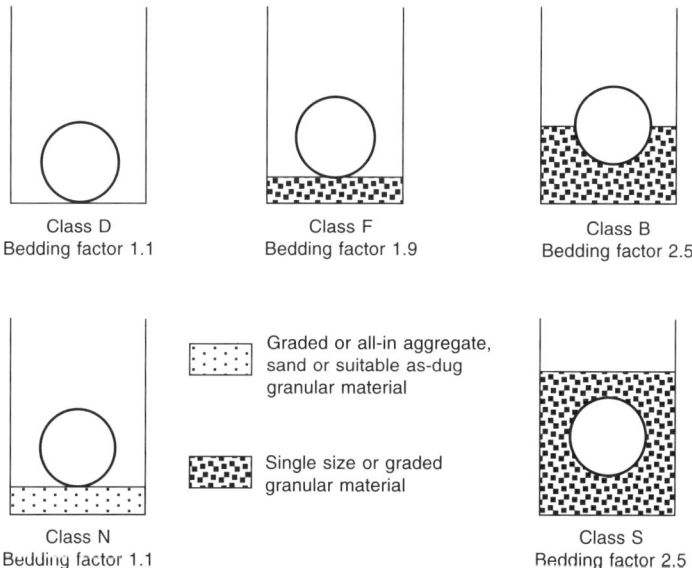

Figure 6.8   Pipe bedding classes and factors for glazed vitrified clay pipes

The basic design equation is:

$$\text{Applied load} = \frac{(\text{Bedding factor}) \times (\text{Crushing strength})}{\text{Factor of safety}}$$

The structural design procedure starts with an estimation of the applied loads from both soil and traffic, and then a suitable combination of pipe strength and bedding is selected to resist the loads. The factor of safety is taken to be 1.25.

In practice tables and charts are available to determine the loading. For example, the Clay Pipe Development Association (CPDA) publish tabulated values of soil (or fill) loading for different trench widths and cover depths. A sample is shown in Table 6.2.

Similar tables are available for the wide trench condition.

Traffic loads have also been tabulated for a variety of conditions and a sample of these,

Table 6.2 Fill loads for narrow trenches

Cover (m)	Width of trench (m)				
	0.7	0.8	0.9	1.0	1.1
	Load due to fill (kN/m)				
1.0	11.47	13.40	15.33	17.27	19.21
1.2	13.29	15.59	17.90	20.22	22.54
1.4	14.99	17.65	20.33	23.02	25.72
1.6	16.56	19.58	22.62	25.67	28.74
1.8	18.02	21.38	24.78	28.19	31.63
2.0	19.38	23.08	26.82	30.59	34.38

as published by the CPDA, is given in Table 6.3 for a loading condition of two wheels of 30 kN with an impact factor of 2.

Table 6.3 Traffic loads – two wheels 30 kN, impact factor 2

Cover (m)	Pipe diameter (mm)				
	150	225	375	450	600
	Load due to traffic (kN/m)				
1.0	6.81	9.96	16.16	19.05	24.66
1.2	5.26	7.71	12.46	14.73	19.02
1.4	4.17	6.12	9.93	11.77	15.30
1.6	3.38	4.96	8.07	9.58	12.51
1.8	2.78	4.09	6.66	7.92	10.37
2.0	2.32	3.42	5.58	6.64	8.72

Similar tables are available for other traffic loading conditions.

Once the loading has been determined, a suitable combination of pipe strength and bedding can be selected to resist the imposed load.

The crushing strength of a rigid pipe is a function of pipe diameter, wall thickness, and the pipe material properties, and suppliers of pipes can provide a range of strengths to suit different situations. As previously noted, the bedding factors for different forms of pipe support shown in Fig. 6.8 are for vitrified clay pipes in wide trenches. Other bedding factors will apply for other pipe materials.

It should be noted that the fill load should be determined for both the narrow trench and the wide trench conditions and the lower of the two values is taken for design purposes. The design load is then the sum of the fill load and the traffic load. In practice, design would normally take the following steps:

- Determine the applied loads from both fill and traffic
- Multiply the applied load by the factor of safety
- Choose a pipe with an appropriate crushing strength
- Determine the required bedding factor
- If the calculated bedding factor is too high, choose a pipe with a higher crushing strength
- Determine new bedding factor.

In summary, the structural analysis of buried pipelines is extremely complex and still not fully understood. As a consequence the practical design of sewers is normally carried out using loading conditions derived from appropriate tables or charts and matching these to pipe strengths, provided by pipe manufacturers and suppliers, and bedding conditions, detailed in relevant standards and codes.

6.6.4 DESIGN OF SEMI-RIGID AND FLEXIBLE PIPES

BS EN 1295-1: 1998 outlines distinct procedures for the design of semi-rigid and flexible pipes. The first step in the procedure for the design of semi-rigid pipes is to select the type of pipe embedment and compaction. The standard provides various classes from which to choose. Classes *B*, *D*, and *S* are similar to those shown for rigid pipes in Fig. 6.8, although subclassifications for *D* and *S* are given as detailed in Table 6.4. The embedment classes are applicable to both semi-rigid and flexible pipes.

Table 6.4   Embedment classes for semi-rigid and flexible pipes

Embedment class	Description	Notes
S1	Single size gravel	Normally processed
S2	Graded gravel	materials
S3	Sand and coarse grained soil with more than 12% fines	
S4	Coarse grained soil with more than 12% fines OR Fine grained soil, liquid limit <50%, medium to no plasticity, more than 25% grained material	'As dug' soils
S5	Fine grained soil, liquid limit < 50%, medium to no plasticity, less than 25% grained material	Only recommended for semi-rigid pipes
B1	Upper surround as for S3 or S4, lower surround as for S1 or S2	Not recommended for pipes of less than 10 kN/m² stiffness
B2	Upper surround as for S5, lower surround as for S1 or S2	

Procedures and data are given in the standard to allow the pipe/soil stiffness factor to be calculated. This is then used to determine the soil pressure, which in turn is used to determine pipe deflection and pipe wall bending stress, both of which must lie within acceptable limits for satisfactory design.

The procedure for flexible pipe design is similar to that for semi-rigid pipe design, apart from requiring a check on buckling capacity, which must exceed the design loading by a factor of safety of 2.

6.6.5 OPERATIONAL LOADS FOR PIPES INSTALLED USING TRENCHLESS METHODS

The discussion above relates to the most common method of installing buried pipelines, which is by excavating a trench, placing the pipe within it and then backfilling the trench. The soil loading equations are based on the 'prism' theory, which essentially states that the load on the pipe is equivalent to the weight of the prism of backfill material above the pipe. Of course, this is modified somewhat, for certain conditions, to allow for frictional resistance between the backfill and the original trench and to allow for some arching action within the soil itself. The determination of the support provided to the pipe is also based on backfill conditions, notably the degree of compaction that can be achieved.

The theories and equations developed for analysis and design of buried pipelines do not therefore necessarily apply to pipes installed using trenchless techniques such as pipebursting, microtunnelling, and directional drilling. Trenchless methods, as the name implies, do not require a trench to be excavated, and therefore the soil above and around the installed pipe is not disturbed in the same way it is when conventional open trench methods are used. In some cases, such as where holes are drilled or tunnelled through stiff cohesive self-supporting soils, it is possible that there will be little or no soil loading on the pipe at all. In any case the undisturbed soil surrounding the pipe will generally be much denser than artificially compacted backfill and will thus provide much greater horizontal support to the pipe. A pipe installed by trenchless methods should therefore be more resistant to loads, less likely to suffer from settlement and consequent cracks, and provide a longer asset life.

6.6.6   INSTALLATION AND TRANSPORTATION FORCES

Pipes may well experience higher loading conditions during transportation and installation than during their normal operation after installation. This may be due to impact loading during transit, excessive loading through high stacking during storage, bending during lifting operations, compression during pipejacking, tension when pulling pipes in a directional drilling operation, or external pressure from drilling mud during directional drilling.

Although it is rare for stacks of pipes to be damaged under normal static loading, it is certainly possible for dynamic or impact loads to cause cracking or yielding, particularly in brittle pipes. Impact loading can arise when individual pipes or bundles of pipes are dropped, or when insecurely fastened pipes bounce up and down on trucks during transit. There may be particular problems when pipes are handled several times in transit, such as off-loading from a truck to a ship and then reloading to another truck on arrival at the destination port. Brittle pipes, such as fibre reinforced cement pipes, may suffer from micro-cracking under such loads, a condition almost impossible to spot but one which may well come to light once the installed pipe is put under pressure.

Whereas pipes used for pipejacking or microtunnelling must be specially designed to withstand the high compressive forces exerted by the jacking equipment during installation, pipes installed by directional drilling will be subject to tensile forces when being pulled into the bore. Care must be taken to ensure that an adequate arrangement for clasping the pipe is used to avoid high local stresses, particularly when plastic pipes are being used. Unexpectedly high tensile forces can be experienced when pulling a pipe into a directionally drilled bore if there are excessive deviations in the bore profile, or if hydra-lock occurs and exerts pressure on the front face of the towing head and pipe. Hydra-lock may also cause high drilling fluid pressures around the pipe and this may lead to buckling or flattening of the pipe. Care must therefore be taken to ensure that drilling fluid flows and pressures, and the pull back forces are regularly monitored during the installation of the pipe.

6.7   Future developments

Most engineers are still unaware of the potential for installing sewers using trenchless techniques, such as microtunnelling, pipebursting, and directional drilling. These techniques not only significantly reduce traffic disruption, they also offer the opportunity for a completely different approach to the preliminary layout and design of sewerage networks. When open trench methods are used, the depth of the pipe is a major cost factor. However, if trenchless methods are used, increased depth does not significantly affect the cost and this means that designers can be offered the option of installing sewers at greater depths without a cost penalty. Also, with trenchless methods it is possible to route a pipe under obstacles that it would be inconceivable to do using open trench methods. These factors mean that a completely different approach to routing and preliminary design of sewer networks can be adopted, as long as this is taken into consideration at an early stage in the design process.

Sophisticated computer-based methods for the hydraulic design of sewers are now available and in common use. This has facilitated a much more effective and efficient use of sewer pipes, not only for conveying waste water, but also for storing it during peak flow conditions and thus minimising pollution of water courses from storm water overflows.

Water authorities and companies are now approaching the problem of water and waste water engineering from the perspective of *demand* management rather than *supply* management – in other words, attempting to manage the demand for water rather than continually increasing supply to match an increasing demand. This raises issues of not only reducing domestic waste water, but also reducing surface water runoff into sewerage systems. Approaches such as recycling domestic grey water and harvesting rainwater from roof catchments will have a considerable influence on the flows expected in sewerage systems. Another approach is that of porous pavements, designed to allow surface water to infiltrate into the soil rather than being channelled into sewerage systems. This will lead to a much more integrated approach to the design of urban developments.

The structural behaviour of the soil/pipe composite is still not fully understood. In particular, the implications of soil loading and support on pipes installed using trenchless methods needs to be more fully researched. A greater understanding of this soil/pipe behaviour could lead to better and more economic use of pipes and may identify further economic benefits, such as improved life expectancy, of trenchless methods.

Bibliography

1. White, J. B. *Wastewater Engineering*. Edward Arnold, 1978.
2. Hydraulics Research. *Tables for the hydraulic design of pipes: 7th edition*. HMSO, 1998.
3. Clay Pipe Development Association. *The specification, design and construction of drainage and sewerage systems using vitrified clay pipes*. CPDA Ltd, 1999.
4. BS EN 1295-1: 1998. *Structural design of buried pipelines under various conditions of loading*. British Standards Institution.
5. Watkins, Reynold King and Anderson, Loren Runar. *Structural Mechanics of Buried Pipes*. CRC Press, 2000.
6. Young, O. C. and Trott, J. J. *Buried Rigid Pipes – Structural Design of Pipelines*. Elsevier, 1984.
7. WRc/WAA *Sewerage Rehabilitation Manual* (editions in 1984, 1986, and 1994).
8. Butler, David and Davies, John W. *Urban Drainage*. E & F N Spon, 2000.

7

Open-Cut and Heading Construction

Geoffrey F. Read MSc, CEng, FICE, FIStructE, FCIWEM, FIHT, MILE, FconsE, Fcmi, MAE

7.1 Open-cut construction

Whether we are dealing with sewer replacement or new construction in open-cut as distinct from trenchless construction it will usually be necessary to provide some form of temporary support to the adjoining ground and in certain circumstances to any adjoining structure, except of course where unsupported sloping sides are acceptable. Generally very little construction can take place without recourse to excavation.

The choice between excavating a trench with supported vertical sides or with unsupported sloping sides is generally a matter of economics where sufficient width is available. However, there are circumstances where timbered trenches are unavoidable such as in a restricted working area which is generally the situation where space is not available for a wide trench, or in water-bearing sands which would result in the unsupported ground slumping to a very flat angle of repose. The unsupported trench has the advantage of clear working space, unhampered by struts, but a deep trench would require the slopes to be cut well back to ensure the safety of workmen, and the cost of the extra excavation together with the additional cost of carefully replacing and ramming the backfilled soil might well outweigh the cost of supporting a vertical sided trench to the same depth.

It is generally advisable to allow the contractor to choose his own method of excavation provided that it complies with the specifications and the drawings. The pipeline will, however, have been designed for a specific maximum trench width and it must be clearly set out in the contract that any proposals by the contractor to excavate beyond the design width must be approved by the engineer. The cost of any additional work which may then be required to maintain the strength of the pipeline must be borne by the contractor.

Trenches should normally be dug to a minimum width of 300 mm plus the diameter of the pipeline where this is 150 mm or more. Additional width should be included for any timbering or sheeting. The Clay Pipe Development Association suggest a trench width of $(1.33d + 200)$ mm with a minimum width of 500 mm. In made ground, peat and coarse sand where these are waterlogged, trenching may prove impossible and trenchless techniques would be used.

Some excavations can be very deep while others may be superficial. In between, probably the most prolific form of excavation, outside bulk digging, is in trenches and this is true in the case of sewerage.

It is a regrettable fact of life that the collapse of excavations is a major source of accidents in the construction industry, many of which are *fatal*.

An analysis of *fatal* accidents in trenches during the period 1973–80 showed that they have related to:

Unsupported excavation	(63% of cases)
Working ahead of support	(20% of cases)
Inadequate support	(14% of cases)
Unstable slopes of open-cut	(3% of cases)

It is not surprising therefore that excavations attract a good deal of safety legislation in relation to those who 'manage' such work and the operatives who do the work.

There are no codes or standards which specifically deal with the support of excavations. BS 8004: 1986 'Code of Practice for Foundations' and BS 6031: 1981 'Code of Practice for Earthworks' make only limited reference to the support of excavations. There are nevertheless a significant number of publications which provide authoritative guidance on the safe support of excavations. Two of these in particular will be specifically referred to within this chapter.

Trench excavations is defined in '*Timber in Excavations*' as 'an excavation whose length greatly exceeds its width'. It may have vertical sides, which usually require strutting from side to side or battered sides not requiring any support. A trench generally being accepted as not exceeding 5 m in width.

The object of sheeting and strutting excavations is to protect workmen within them and to limit ground movement which may result in damage to adjacent property and any nearby public utilities equipment. To achieve these objectives *first-class* workmanship is essential at every stage, i.e. in *installation*, in *use* and during *dismantling*.

Double sides support is the most common technique utilised and Fig. 7.1 illustrates the key features. The temporary support structure maintains the 'status quo' between the exposed excavated faces. The sheeting on both sides collects the earth pressure on the face

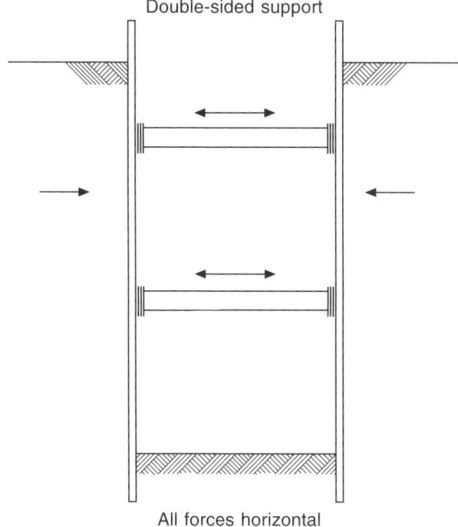

Figure 7.1   Double-sided support to trenches. All forces horizontal.

and transfers it via the walings to the struts, which provide equal and opposite force at each end and so maintain the equilibrium of the system, all forces involved being horizontal.

The support of excavations can be divided into methods needing either standard solutions or methods calling for purpose design by experienced civil engineers.

Examination of the three basic types of support confirms that only double-sided support is suitable for the use of standard solutions. All loads are horizontal and a status quo situation then exists. With raking supports, (Fig. 7.2) vertical loads are involved and detailed knowledge of ground conditions and soil strength is essential. In the case of ground anchors, (Fig. 7.3) their installation and design is very much a specialist technique requiring

Single-sided – with raking support

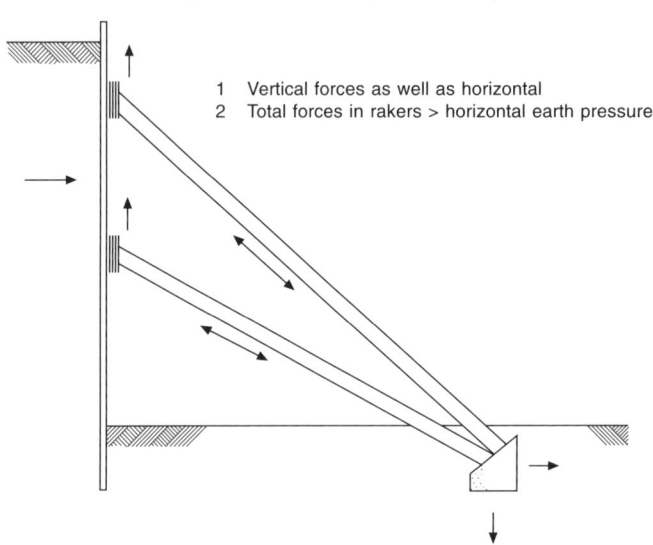

1   Vertical forces as well as horizontal
2   Total forces in rakers > horizontal earth pressures

Figure 7.2   Raking shore support internally. Forces induced. Note uplift present.

Single-sided support

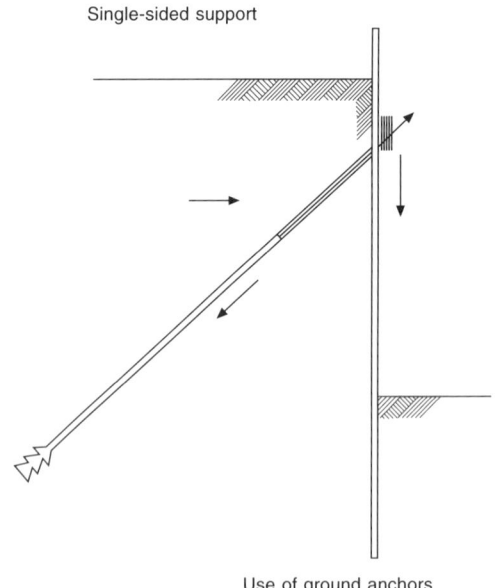

Use of ground anchors

Figure 7.3   Ground anchors externally. Forces induced. Note downward force on system.

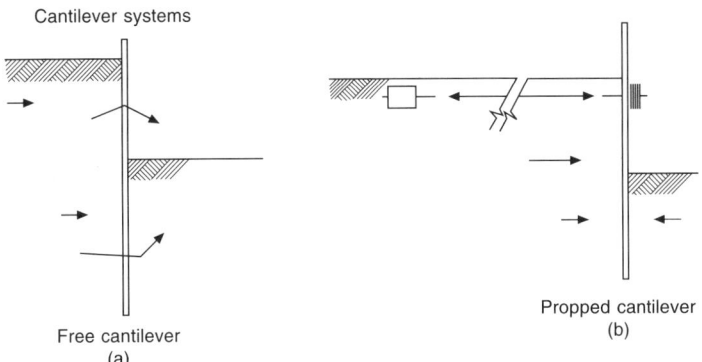

Figure 7.4 (a) Essential features of cantilever support. (b) Propped cantilevers.

appropriate equipment and knowledge – the same sort of specialist knowledge is also needed where cantilevers are involved. (Fig. 7.4)

There are a number of standard solutions for double-sided support in trench excavations but if accidents are to be avoided certain criteria must be applied particularly in relation to the depth to which they should be used. The recommended limitations on the use of standard solutions whatever their source are given below where the depth limitation is kept to 6 m.

7.1.1 LIMITATIONS OF STANDARD SOLUTIONS IN THE SUPPORT OF EXCAVATIONS

There is general agreement that the following limitations on the use of standard solutions should apply in the support of excavations.

Criteria

Standard solutions should only be used for the support of excavations:

1. Where open-cut not exceeding 6 m in depth is feasible
2. With double-sided support, in non-water-bearing ground, to an excavation not exceeding 6 m deep
3. In shallow pits not exceeding 6 m deep
4. In water-bearing ground, where water problems have been eliminated by other means (e.g. well pointing), within the limitations of (2) and (3).

Procedures

In the above situations the procedures below should be followed:

1. When deciding the batter of an open-cut excavation, proper account must be taken of the material and its characteristics and the safe slopes recommended in either trenching practice or timber in excavations.
2. Do not assume that excavation in rock is necessarily stable. Look for sloping strata, fissures and loose material after blasting. Support unless the material is stable.
3. Supervision should make sure that persons erecting or removing supports have been adequately instructed on the methods to be used.

4. Where proprietary methods are being used, the procedures must be strictly in accordance with the manufacturers' instruction.
5. All relevant legislation must be complied with.

7.1.2   *EXCAVATIONS UP TO 6 M IN DEPTH*

It is unnecessary and unrealistic to calculate lateral pressures on supports to excavations shallower than 6 m unless hydrostatic pressure is taken into account. Lateral pressures at shallow depth can be highly variable in any given soil type. For example, a clay will shrink away from behind the timbering in dry weather, sometimes forming a gap down which elements of dry clay may fall. The onset of wet weather then causing the clay to swell and if the gap has become filled with clay debris, the swelling forces on the timbering may be high enough to cause crushing or backfilling of the struts. In stiff or compact soil higher stresses than from earth pressure can be caused in struts and walings by hard driving of a folding wedge at the ends of the struts. Again, if dry timber is used to support an excavation in water-bearing ground the seepage through the runner onto the walings and struts and the effects of rain falling on the latter can cause heavy stresses in the timbers, due to swelling of the dry timber. Expansion of the soil at times of hard frost can also result in heavy loads on timbering.

Accordingly, it can be argued that the usual method of designing timbering by rule of thumb for shallow excavations is fully justified. The size of timbers given by such methods has been based on centuries of experience and takes into account the desirability of reusing the timber as many times as possible, the requirement of withstanding stresses due to the swelling of dry timber and the necessity or otherwise of driving wedges lightly at the ends of the struts (or of jacking the ends of the struts) to prevent yielding of the sides of the excavation. They in fact bear little or no relationship to the stresses arising from earth pressure. The reader may in fact have noticed that the size of timbers used in the shallower excavations is more or less the same on any job no matter what type of soil is involved or even the depth of excavations (Fig. 7.5)

Sizes of timber struts for excavations to a depth of about 6 m are only nominal and as mentioned above are calculated on a 'rule of thumb' basis.

One such rule applicable to square struts for excavation from 1.2 to 3 m width is:

Strut thickness (mm) = 80 × Trench width (m)

It is convenient practice generally to use struts of the same depth as the walings and often of the same scantling in order to avoid too many timber sizes. Thus 225 mm × 75 mm struts would be used with 225 mm × 75 mm walings, 300 mm × 150 mm struts with 300 mm × 150 mm walings and so on.

The spacing of the struts is governed by the length of the walings. Thus 225 mm × 75 mm walings are usually supplied in 3000–4800 mm lengths. Three struts are a convenient number for a pair of 3600 mm walings giving openings of 3400 mm for the three struts, with the pair at the abutting end of the walings at about 200 mm centres.

It is possible to obtain longer lengths of waling but it is generally more convenient to work with the shorter length because of the difficulty in threading long walings for lower settings through the previously set frames.

Shorter walings are also preferred for convenience in striking timbering when backfilling. The two publications referred to earlier provide valuable data for standard solution assessment: *Trenching Practice* specifically relates to trenching support and covers tables for the support

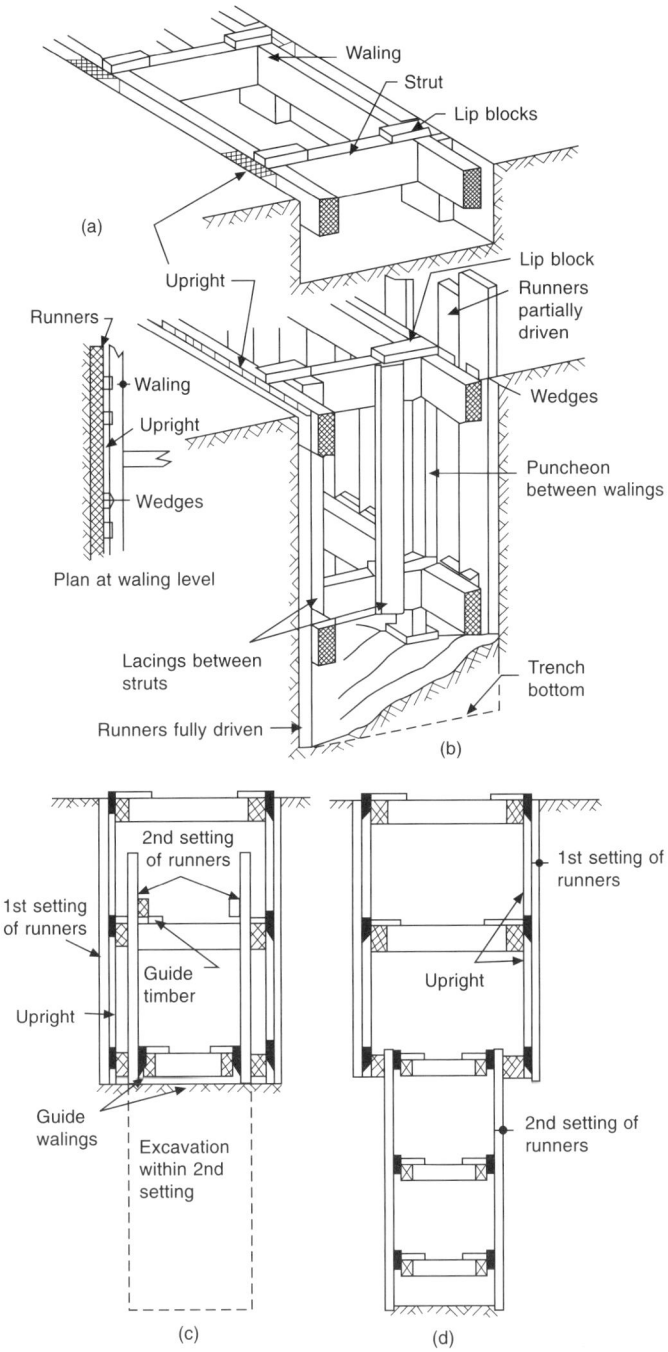

Figure 7.5 Support of excavations in loose soil with runners. (a) First-stage excavation. (b) Second-stage excavation. (c) Pitching second set of runners. (d) Completed excavation.

needed to various types of ground, the material need, good practice in installation and backfilling and removal of supporting material; in *Timber in Excavations* standard solutions are covered in much the same way. While it has already been emphasised that wide single-sided excavations need specialist input for the design of the required support, the worked examples in these publications will allow the method planner to reach a close assessment of the likely requirements at tender stage.

7.1.3   PROPRIETARY SUPPORT SYSTEMS FOR TRENCH EXCAVATION

Proprietary excavation support systems for trench excavations are now widely used because of the rapid installation and removal facility provided, the avoidance of risk to the operatives and the lack of skilled experienced timbermen.

The principal types listed in CIRA Report 97 are:

1. Hydraulic frames (waling frames)
2. Shores
3. Boxes
4. Sliding panels and
5. Shields (drag boxes)

All those systems are suitable for use in stable ground which can stand unsupported for a sufficient length of time to lower the members into the trench. Operatives are not then required to work within unsupported excavation. They are suitable for working in stiff clay, damp sands and gravels and weak weathered rocks. They can be used in water-bearing gravelly sands, sands and sandy silt if the ground water has been previously removed via a well point or bored well system.

1. Hydraulic frames: Trench sheets (runners) are set in the excavation at a spacing depending on the stability of the ground (Fig. 7.6); the sheets being pushed into the soil to a depth below the trench bottom so that they are temporarily self-supporting. The lower frame, consisting of a pair of walings connected by hydraulically operated struts, is then lowered into the trench with the pressure hoses attached to a pump on the ground surface – the latter is then operated to expand the struts and force the sheets against the trench sides. The upper frame follows with the same procedure.
2. Shores: This arrangement is similar to (1) except that the walings are omitted and each pair of trench sheets is forced against the trench wall by a single hydraulic sheet. It is used where the ground requires less support than (1).
3. Boxes: These are assembled at ground level and lowered into a pre-dug trench or lowered into a partly dug trench and then sunk as a caisson to the final level. They can be adjusted in width by selecting a suitable strut length and can be used in deep trenches by stacking one box above another (Fig. 7.7).
4. Sliding panels: These are assembled by pushing two pairs of slide rails (soldiers) into a partly dug trench (or from the surface in unstable ground). The steel sliding panels are then set between the rails and are pushed down progressively, with the rails, as the excavation is taken deeper. In deep trenches the panels can be stacked one above another (Fig. 7.8).
5. Shields (drag boxes): Strictly speaking these are not support methods at all. They are designed to be pulled along a trench, once excavated, by the digger to protect the workmen in the trench from falls of earth. As such they do not push out to hold the sides of the excavation.

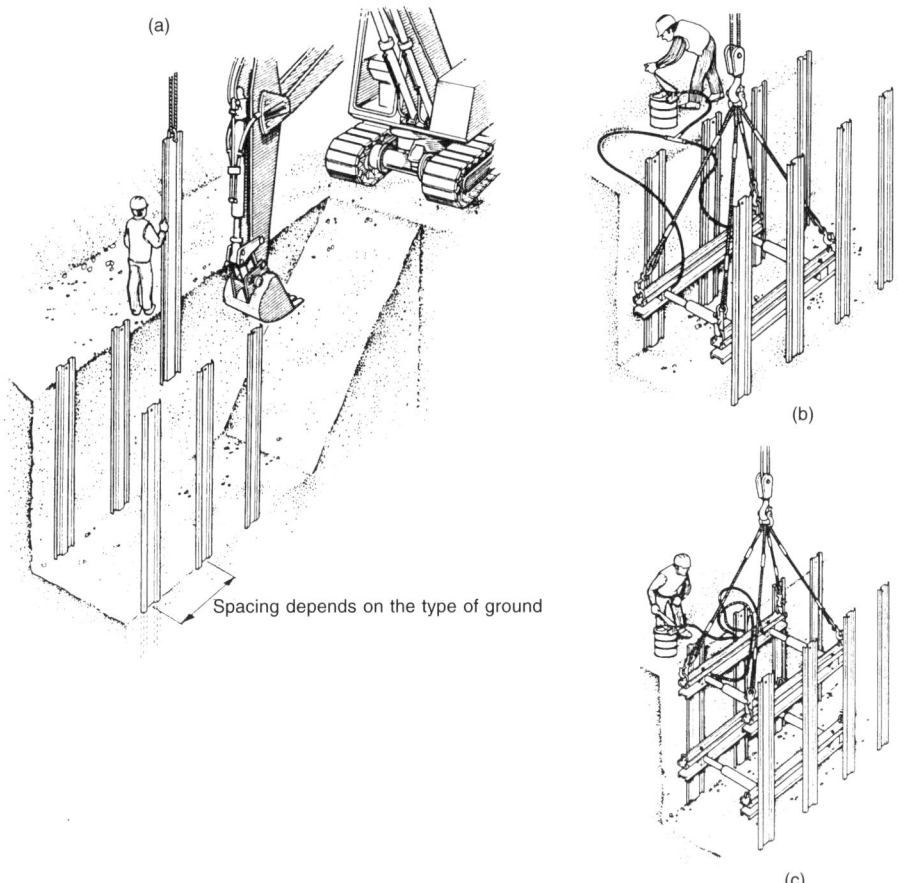

Spacing depends on the type of ground

Figure 7.6 Excavate trench to full depth over short length. (a) Place trench sheets and toe them in. Connect hydraulic struts to pump. Lower complete frame into trench. (b) Inflate struts and remove hydraulic hoses (from outside the trench if system allows) (c) Install further frames as required.

The box formation is rigid and narrower than the trench. It is clearly important that the excavator in use has sufficient power to pull the box along. The forward movement being assisted by a cutting edge on the front (Fig. 7.9).

Systems using boxes, sliding panels or shields cannot be used economically if service pipes or cable ducts cross the line of the excavation at close intervals.

7.1.4 *STEEL SHEET PILING*

When working in extremely bad ground, the use of interlocking steel sheet piling in the form of an extended cofferdam many be the only solution available. Steel sheet piling's main advantage is its ability to retain ground with water-free conditions or keep out water while counteracting outfalls into rivers or the sea. Such piling is designed with interlocking sides but to ensure complete water tightness, expensive caulking or other measures may be necessary.

Figure 7.10 illustrates the type of interlock used and also shows the way in which steel

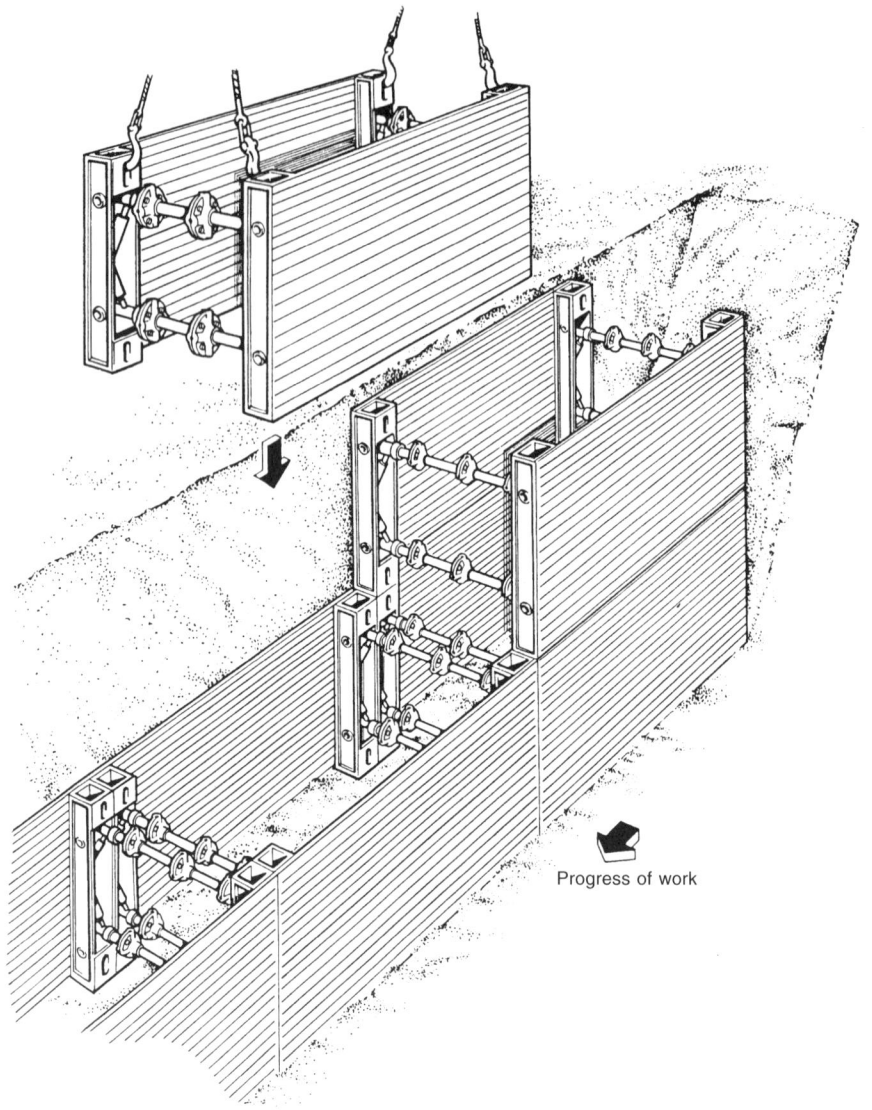

Progress of work

Figure 7.7   Boxes

sheet piling provides a good section modulus against bending – in fact it is the only really satisfactory material for use in cantileveral support methods.

Sheet piling also has an unimportant role to play where a water bearing strata overlays an impervious one. By driving the piles into the impervious strata water can be cut off from entering the excavations.

7.1.5   *NARROW TRENCHING AND MINIMAL EXCAVATION*

Trenching machines (ladder or wheel type) were popular some year ago but most contractors now prefer to use backacters or drag shovels. Trenches can work to depths of about 4 m and are still useful for very narrow trenches in open ground, being particularly useful for

Figure 7.8 Sliding panel.

the laying of small diameter water mains. Backacters can work to depths of about 4 m while drag shovels can operate to depths of 11 m, being particularly useful in open country where the excavation involves sloping sides.

Excavations should commence at the lower end of a pipeline and proceed upstream and will allow any subsoil water to drain away from the working area. If necessary temporary subdrains should be laid at the bottom of the excavation, leading to a sump at the lower end – the extra depth being filled with compacted granular material.

7.1.6 DEWATERING

Where the proposd excavation foundation level is below the ground water table and the grading of the surrounding ground is suitable, it may be practical to temporarily lower the ground water table locally by pumping during the period of the excavation.

The quantity of water to be removed from trenches and other excavations to enable construction to take place must be estimated in advance by reference to the site investigation

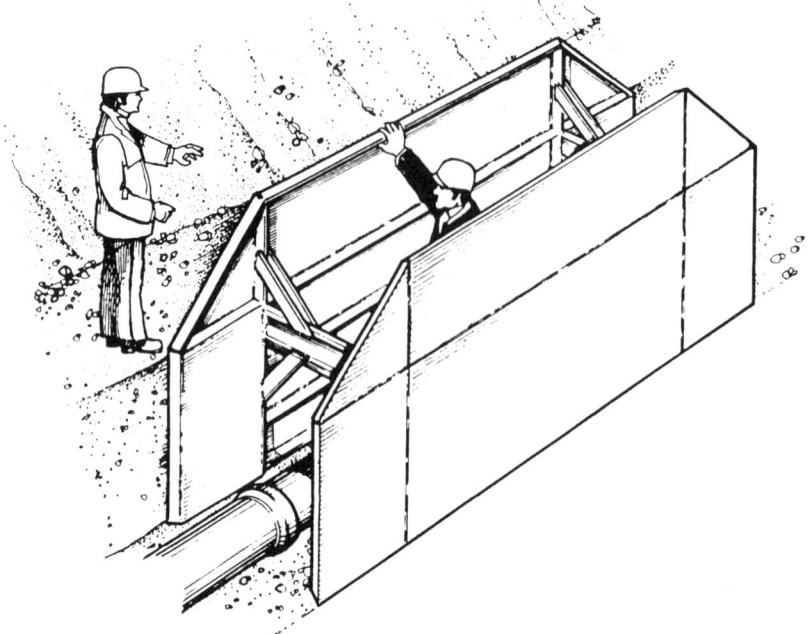

Figure 7.9 Shield.

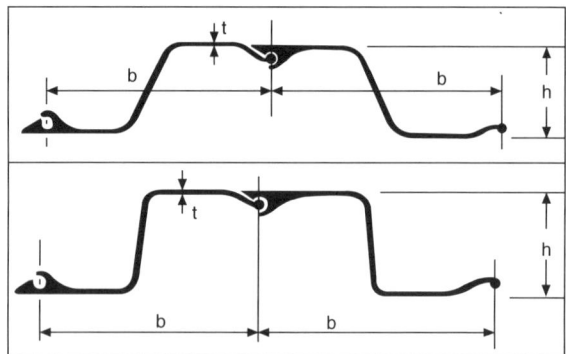

Figure 7.10 Sheet piling.

details so that suitable techniques can be adopted. If necessary the drainage of trenches by subdrains to one or more sumps may be adequate – the water level in the sump being controlled by pumping more or less continually using appropriate pumps. In such instances the subdrains are constructed below formation level and covered with gravel or stone up to formation level. Subdrains can be laid at one or both sides of the trench discharging to pump sumps clear of the trench line. Where possible sub drains are grouted up after pipe laying is complete. (Fig. 7.11).

In suitable soil conditions when the inflow of water is too great for normal pumping, well-point dewatering can be adopted. This method of lowering the ground water cannot be applied to soils containing more than 10% by weight of grains less than 0.03 mm diameter.

The well-point system consists of a series of vertical 'well-point' pipes and risers (approx. 40–50 mm diameter) sunk into the water bearing stratum on one or both sides of

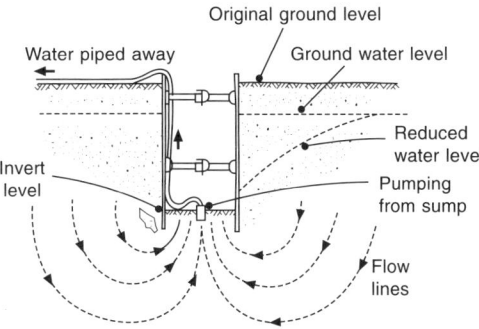

Figure 7.11 Sump pumping.

the trench. These are connected through short horizontal pipes and non-return valves to a horizontal leader pipe (150 mm diameter) which in turn is connected to a vacuum pump. The well points are sunk into the ground by 'jetting', i.e. by forcing water through them to scour away the ground beneath the well point. Well pointing has the advantage of drawing water away from the trench and in suitable conditions is effective in lowering the water by 4 to 6 m. It will also reduce the hydrostatic head on the trench support system. Its greatest practical use is in sands.

Well points are usually installed at 0.6 to 2 m centres. In certain grounds it may be necessary to pre-bore the holes. The efficiency of the well points is then increased by filling the annulus with sharp sand.

Table 7.1 is an approximate guide to the spacing and drawdowns of well points. A range of trench widths from 0.6 to 2.0 m and a depth limit of 6 m have been assumed.

Table 7.1   Spacing and drawdown of well points

Ground description	$D_{10}^{(f)}$ (mm)	*In-situ* permeability 'k' (m/s)[d]	Well-point spacing (m)[g]	Time (days) to effective draw-down Single side	Double side	Maximum drawdown at trench in m[e] Single side	Double side
Coarse sand possibly with gravel[a]	1.0 to 0.6	10^{-2} to 10^{-3}	0.6 to 1.0	<1	<1	5	6
Medium sand possibly with coarse sand and gravel[b]	0.6 to 0.2	10^{-3} to 10^{-4}	1.0 to 1.2	1 to 3	1 to 2	5	6
Fine sand possibly with medium and coarse sand and gravel and some silt[c]	0.2 to 0.06	10^{-4} to 10^{-5}	1.0 to 1.5	3 to 10	1 to 5	4.5	5.5
Silty sand possibly with gravel[d]	0.06 to 0.02	10^{-5} to 10^{-6}	1.0 to 2.0	10 to 20	5 to 10	4	5

Notes: (a) Probably requires sumps in trench.
    (b) Allow for trench base drain and sumps as well.
    (c) May require special well points with improved vacuum. Even so this may not work.
    (d) Confirm *in-situ* permeability by prior pumping tests if possible.
    (e) Well-point depth must be at least 1 m greater than required drawdown at trench.
    (f) Maximum grain size of the smallest 10% by weight of the grading.
    (g) Connections on header pipes generally at 0.75 and 1.0 metre centres.

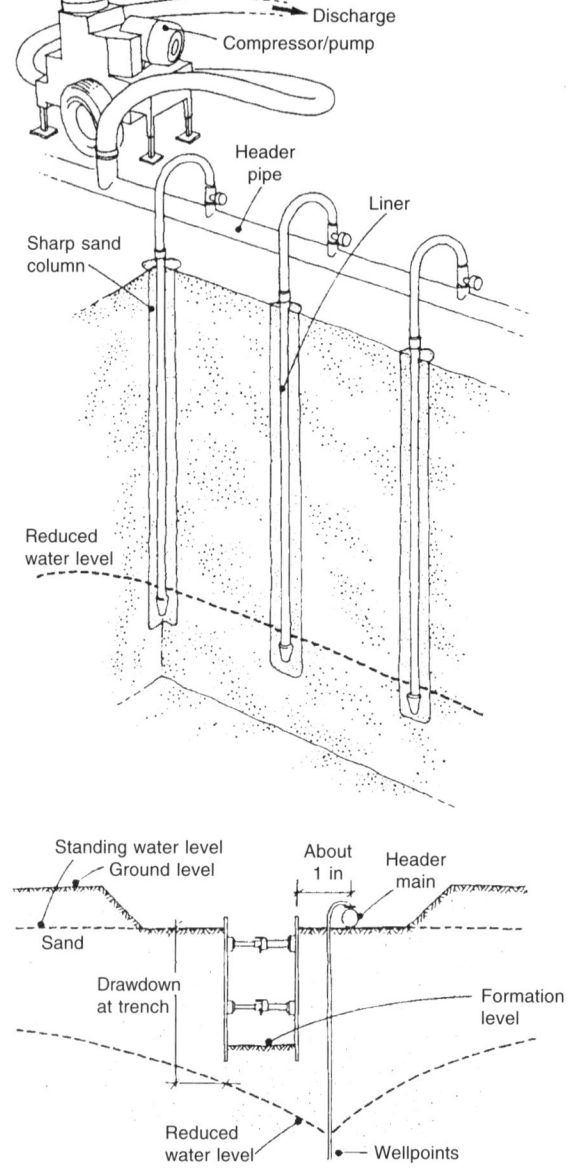

Figure 7.12   The well-point system.

7.2   Pipe laying

Each pipe should be examined carefully on delivery and any damaged in transit must be clearly marked and removed from the site. Where the pipes are to be laid directly on the trench bottom, this should be trimmed to the correct level and gradient immediately before the pipes are laid. Socket holes should be formed at each socket position, leaving the maximum length of support for the barrels. Such holes should be as short as practical and should be scraped or cut in the formation deep enough to give a minimum clearance of 50 mm between the socket and the formation. To ensure that the pipes are laid to the correct gradient each pipe is then set to line and level, using the sight rails or laser. If the formation has been overexcavated and does not provide continuous support, low areas should be brought up to the correct level by placing and compacting suitable material. After laying each pipe should be checked and any adjustments to level should be made by raising or lowering the formation, always ensuring that the pipes finally rest evenly on the adjusted formation throughout the length of the barrels. Adjustments should never be made by local packing.

If the pipes are to be set on a granular bed, this must be laid first to an approximate level and gradient. Any mud should be cleared away and soft spots removed or hardened by tamping in gravel or broken stone – hard spots should also be removed. In soft clay, disturbance of the trench bottom should be minimised by placing a layer of blinding material about 75 mm thick. Socket holes are then scooped out and each pipe is bedded into the granular material using the sight rails or laser to obtain correct line and gradient.

Socket holes need to be deep enough to prevent the weight of the pipe and the load on it bearing on the socket or coupling and should be a minimum of 50 mm deep leaving 50 mm of bedding material below – bedding material should not be compacted into these socket holes.

Measures to prevent migration of fine material from pipe bedding should be undertaken where the pipeline is below ground water level.

7.3   Sewers near existing structures

Where a sewer runs near to the foundations of an existing structure and provided that the water table will be below the trench bottom during constuction, the trench bottom should not be normally below the underside of the foundation of the structure by more than the horizontal distance between the near edge of the foundation and the near side of the trench or 1.5 m whichever is least. If the trench needs to be deeper, if the water table is higher and could rise above the trench bottom during construction or if there is likely to be any risk of causing instability to the existing structure, special precautions may be needed to protect the structure and specialist advice should be obtained. Under any such condition, special support may be required, which may have to be left in place and it may be necessary to backfill the trench above the top of the pipe with weak concrete.

Artificial consolidation of strata may have to be carried out prior to construction of sewers in open-cut, heading or tunnel if the strata revealed during the site investigation or during construction are shown to be excessively difficult and non-cohesive but amenable to one of the accepted forms of soil stabilisation.

There are a number of ways in which this can be done over and above the adoption where suitable of dewatering. These techniques are generally patented and include the injection of chemicals which interact to produce a gel, freezing, or injection of suitable stabilising agents. The cost of such methods needs to be carefully considered as they are generally expensive but may be justified where essential work could otherwise be inhibited by soil conditions or the pressure of existing structures.

7.4   Backfilling and reinstatement

7.4.1   INITIAL BACKFILL

In the case of rigid pipes, as soon as possible after completion of the bedding or surround, selected fill should be placed by hand and carefully compacted between the pipes and the trench sides and brought up in layers so there is at least 150 m of compacted material above the crown of the pipe.

The sidefill for flexible pipes should be the same granular material as that used for bedding. It should be taken to the level of the pipe crown and be carefully compacted, selected backfill should then be placed and carefully compacted in layers to give at least one 150 mm layer above the crown of the pipe.

7.4.2   MAIN BACKFILL

The main backfill may follow after the initial backfill has been placed. Backfill should be in even layers not exceeding 300 mm, each layer being thoroughly compacted before further fill is added.

In the case of trenches in gardens, fields, etc., the top soil and turf (which should have been kept separate) should be replaced and properly reinstated. The final surface should in all cases be levelled off in such a manner so as not be a source of danger to animals or agricultural plant and should be attended to from time to time until consolidation is complete.

Backfilling under highways is a much more complex operation – in fact it is a structural operation and should be treated as such.

Until the introduction of the New Roads and Street Works Act of 1991, the permanent restoration of the highway surface was usually carried out by the Highway Authority when it was assessed that settlement had ceased. In the interim temporary reinstatement for the purpose of carrying traffic was carried out by replacing the original materials forming the road crust in the order in which they were laid. The temporary surface of the trench then being sealed with a layer of tarmacadam to prevent loose material being dispersed by traffic and maintained regularly.

The Public Utilities Street Works Act of 1950 (PUSWA) stipulated the necessary control and management responsibilities. This legislation had been introduced in 1950 when there were some 4 million vehicles on the UK highway network. Although it was enacted in 1950 it was actually drafted in the 1930s but delayed by the war. It was therefore framed around traffic levels and utility usage levels which were different to those obtaining in 1950 and vastly different from those of today. These controls had become increasingly inadequate to deal with the dramatic rise in the levels of both traffic and works on a highway network which had only increased by some 20% since 1950 yet was having to carry some 24 million vehicles (an increase of 500%).

The resultant chaos and interference with traffic flow – this was estimated in 1992 to cost the country some £60 million per year – caused by highway openings of all types could no longer be accepted.

The editor represented the Association of Metropolitan Authorities on the Local Authorities Joint Steering Group for the PUSWA Review 1984 which ultimately led to the introduction of the New Roads and Street Works Act 1991 and the publication of the Statutory Code of Practice 'Specification for the Reinstatement of Openings in Highways' (Department of Transport *et al.*, 1992). This legislation had a significant impact on the way highway reinstatements of all kinds are carried out. All statutory undertakers including sewerage authorities are now fully responsible for all aspects of excavation and reinstatement of the highway.

The aim of excavation and reinstatement should be to complete to a permanent standard in one operation achieving consistently high standards of workmanship and close co-operation between sewerage undertakers and the relevant Highway Authority. High standards will reduce the faults occurring in trench construction such as cumulative settlement, edge depression, cracking and surface in regularity.

High on the list of potential problems are compaction procedures including complying with specified layers thickness, the use of suitable plant and the correct number of compaction passes. Recent research has indicated that in most cases the trenched lane had a worse rate of deterioration than the opposite untrenched lane. This could imply that many local highway authorities will be faced with a substantial repair bill at the end of the contractors guarantee period.

In this connection one noted with interest in May 1999 that in Surrey all reinstatements carried out by utility companies are to be tested, in an unprecedented clampdown on shoddy reinstatements. The County Council decided to extend its already extensive coning test programme having found that eight out of ten reinstatements failed to meet the required standard. Previously Surrey had coned one in four trench reinstatements and planned to increase this to 100% within two years.

Future natural research is likely to concentrate on the reprocessing of excavated materials to make them suitable for reuse rather than disposal with the aim of cost reduction and reduction of impact on the environment.

7.5   Heading construction

The term 'heading' is generally used to describe small tunnels where timber frames are used to provide temporary ground support. In the past when a sewer was being constructed in open-cut and a length had to be built under a main road, railway, canal or similar obstruction, construction in heading was necessary. The present tendency in such instances generally utilises one of the trenchless techniques, e.g. microtunnelling, thrust boring, directional drilling, etc. There are still nevertheless instances where even short lengths of heading have to be constructed, e.g. shield-driven on-line replacement where the old sewer is not in a straight line or where connections have to be amended as work proceeds.

Due to the present-day shortage of skilled timbermen such form of construction is expensive, nevertheless it is often inevitable usually in relatively short lengths.

In part it has been quite widely used where sewers have to be constructed at depths where open-cut was uneconomic and in such instances a timbered heading was constructed between two manholes and this was followed by pipe laying and backpacking working upstream so that the labour involved had one way of escape in case of emergencies.

Nowadays headings or small tunnels are usually cut into the sides of trenches, shafts, tunnels, etc. In this form they constitute the most dangerous sector of temporary works as, in addition to the normal dangers associated with trenches or shafts, they introduce the risk of collapse, trapping men with no alternative escape route available and with the avenues of rescue being difficult and dangerous. The cutting of headings requires a very high degree of careful investigation and consideration in design, quality of material excavation techniques and the actual works of the timbermen notwithstanding the most obvious safety and supervision implications. As distinct from open-cut where the temporary works are removed, in headings the temporary works are usually left in position. It is important therefore in headings that the timbers should be naturally resistant to decay or treated to make them decay resistant.

The supervision and routine checks involved in the safe operation of temporary works in headings are basically different from those employed in open excavations in that the routine inspection must be carried out at the point of maximum risks, namely, the advancing face of the excavation.

Another critical factor is that the speed of the excavation has a direct bearing on safety, the integrity of the cut face is time dependent and the rapid installation of support is of paramount importance. A heading is practical in firm, cohesive soil or rock. CP 2005 recommends that the smallest size of heading for proper safe working is about 760 mm width at the bottom.

All timber used in the construction of a heading shall be to strength class SC4 or better. Where the timbers are to remain they will be protected against decay in accordance with BS 5268: Part 5: 1989, Table 5, Section 1a.

Figure 7.13 shows a typical heading of the type used in bad ground and soft conditions. The sides and roof are lined with 150 mm × 38 mm sheeting placed behind the side and head trees set at approximately 1 metre centres. For headings not exceeding 2 metres either in height or width when measured to the outside of the temporary works the recommended minimum dimensions according to *Timber in Excavations* are as follows:

Side trees	225 mm × 75 mm
Head trees	225 mm × 75 mm
Side boards	150 mm × 38 mm
Crown boards	150 mm × 38 mm
Waling boards	150 mm × 38 mm
Stretchers	150 mm × 38 mm
Sills	225 mm × 75 mm

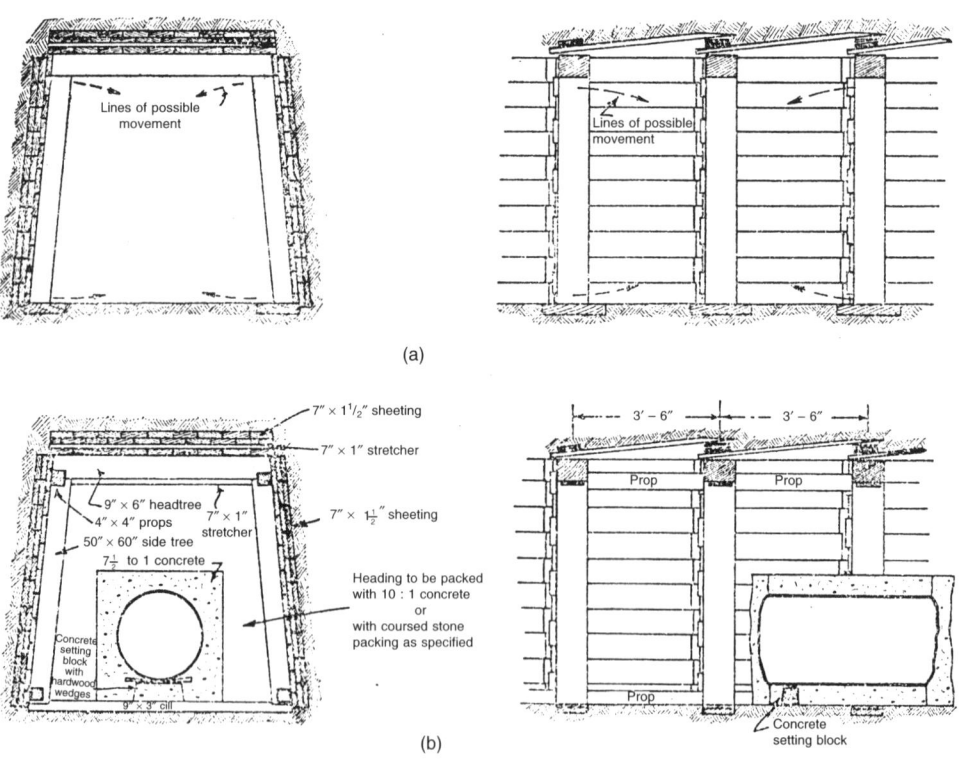

Figure 7.13   (a) Dangerous type of heading. (b) Safe type of heading.

Headings of this type are made large enough only for the men to work in or to construct the sewer.

The sill is always set first and is boxed in from previously set sills which must be checked from time to time to be quite sure as to their accuracy. To enable the timbermen to do their work rapidly and efficiently all the timber should be of standard lengths and sections. The face of the heading should be boarded if the excavation stops for more than two hours.

Adequate temporary electric lighting must be installed in the heading to provide illumination for the proper execution of the work – the voltage employed must not exceed 55 volts to earth. An efficient air blowing system may be required in deep headings and this can be provided with fans and ducting.

Once the sewer has been laid the remaining space in the heading has to be filled in. In bad ground the side sheeting is left in but it must be tightly packed behind and wedged up. In good ground the side sheeting is usually removed as the packing proceeds. The packing is usually done with lean-mix concrete and tightly packed stones.

References

1. *Trenching Practive*. Report 97: Construction Industry Research and Information Association.
2. *Timber in Excavation*: Timber Research and Development Association.

8

Tunnel Construction

Malcolm Chappell BSc (Hons), CEng, FICE, MIMM

Derek Parkin BSc (Hons), MPhil, MBA, CEng, MIMech E, MIMM, MIMgt

8.1 Introduction

All of the water that is supplied to a city or conurbation has to be collected and discharged to a river or sea system through sewers and treatment works. The supply of water not only includes that planned and controlled through feeder pipes, but that which precipitates through rain, snow and hail.

Initially, sewers start as small diameter pipes, a little below ground level, and eventually are collected through interceptors of increasing size to feed the treatment works. These interceptors are sometimes of sufficient size or depth that the sewers have to be constructed using tunnelling techniques. Also surface topography may dictate a tunnelled solution.

There has been a trend over recent years to construct tunnels for use as storage tunnels to hold surface water during periods of storm. It is normally during these periods that treatment works have insufficient capacity to treat everything. It has been normal to bypass the treatment works, but recent EEC directives demand higher levels of treatment before discharge, particularly at coastal locations. Hence the requirement to store stormwater and then treat the effluent during dry periods, e.g. Fylde Coastal Waters Scheme, Clacton-on-Sea.

It is arguable that civil engineers have done more for world health than the medical profession by designing and providing infrastructure to populated areas to remove effluent and to supply fresh water.

The city of Manchester possesses possibly the oldest extensive sewerage system in England. The sewerage network of the city centre was commenced in the late eighteenth century and continued with great rapidity throughout the early nineteenth century. By 1855 this city centre system was fully developed and remains largely unchanged and fully operational today.

It is interesting to note that Manchester Corporation was able to report to the River Pollution Commissioner in 1868 that every street in Manchester was sewered and that a total length of public sewers was 450 km. By comparison Liverpool reported that within its greater, though more dispersed, population, it had 80 km of sewer, Bolton 72 km and Preston 40 km.

Prior to designing London's first sewerage system, which became operational in 1875, Sir Joseph Bazalgette visited Manchester to view the techniques used. His system took 16 years to construct and it contained many brick-lined tunnels running along both sides of the

River Thames. Much of his system is still in use today and parts have recently been strengthened as a consequence of the Jubilee Line Project Extension.

The year 1870 witnessed the use of the first circular shield and segmental cast-iron lining. This method was used to construct the 2.14 m diameter Tower Subway beneath the River Thames. It was built by Barlow and his young assistant James Henry Greathead. Shields and tunnelling machines today are a derivative of this original concept with the basic design commonly known as the Greathead Shield. This invention, together with the segmental lining, was to make tunnelling more economical and thus a viable solution to many civil engineering designs, e.g., London Underground.

The majority of tunnels for sewers are constructed within the diameter range of 1.0 m to >3.0 m, although there are exceptions, e.g. Don Valley Interceptor Sewer (5.2 m diameter) and Brighton (6.0 m diameter). Also the smaller end of the range tends to be constructed by remotely controlled machines rather than by miners working at the face.

Irrespective of the method of construction the primary needs of a sewer are:

- A precise gradient to maintain an adequate velocity to carry solids and prevent settling out of sands and grits
- A smooth internal lining to minimise friction loss through roughness
- Watertightness to stop egress and ingress, either of which could be detrimental to the integrity of the structure
- Resistance to corrosion and erosion.

Taking the above into account a designer will produce drawings and a specification from which a contractor can decide on the best method of construction, and, hence, prepare an estimate for the cost of constructing the works.

On acceptance of the estimate by the client the contractor needs to ensure that the contract is effectively managed to ensure a successful and financially beneficial conclusion.

The execution of tunnelling contracts is becoming more mechanised with more sophisticated *modi operandi*. The reasons for this change are numerous and include:

- An increase in labour costs
- More restrictive legislation in relation to work practices
- Tunnels being driven in ground conditions that were previously thought too difficult
- Greater reliability of electronic components in arduous conditions
- An increase in availability of machines due to more reliable mechanics and materials
- Technological advance of lubricants not only limiting component wear and corrosion but having properties to act as conditioning agents for the excavated material.

A site manager has many aspects to concentrate on, but cannot cope with all matters, especially with existing, wide-ranging and complex legislation. Therefore, the site manager needs to call upon specialists to help share the workload so that he/she can be left to concentrate on the fundamentals of effectively managing a tunnelling contract. This item will be discussed with greater detail in Chapter 11.

The discussion which follows in Chapters 10 and 11 will be limited to current practices within the UK.

8.2   Geological/Topographical Aspects

8.2.1   INTRODUCTION

Tunnelling can trace its roots back to antiquity, where it was often used as a technique to breach fortification. However, while the techniques of making peaceful holes in the ground grew originally from military science they have more latterly evolved due to differing ground conditions and the available technology. There is little doubt that modern tunnelling techniques have grown from the metalliferous mining industry of the fifteenth, sixteenth and seventeenth centuries and it is only during the last 75 years that the tunnelling industry has started to develop its own method and techniques, indeed the development of rock mechanics and soil mechanics is a relatively modern development in tunnelling technology notwithstanding its importance in determining the most appropriate tunnelling methods and techniques.

These methods and techniques adopted within civil engineering vary considerably depending on the geological and topographical aspects (Fig. 8.1) and they can be divided into two broad classifications: rock tunnelling and soft ground tunnelling.

Interpretation of ground information is not a precise science, therefore it needs to be supplemented by information from previous experience of the ground in the area and taking account of special problems such as the presence of flammable gases, deoxygenated air or contaminated ground. When the presence of excessive water is expected, evidence of water tables, artesian head should be noted along with the results of permeability and/or pumping tests and other factors such as abrasivity.

Table 8.1 represents tests which would normally be expected from a typical ground investigation for a soft ground tunnel in the UK. These tests would indicate the geological and topographical aspects as illustrated in Fig. 8.1 which in turn would be used to determine such parameters as:

1. Face stability in cohesive soil
2. Tunnel stability in cohesive soil
3. Ground loading in cohesive soil
4. Calculation of ground closure
5. Conditions for local yielding at the tunnel surface
6. Calculations of long-term settlements
7. Choice of ground treatment and face support
8. Choice of type of excavating process
9. Total jacking forces required (discussed in Chapter 9)
10. Pipe friction and soil parameters (discussed in Chapter 9).

All of these parameters impinge on the fundamental issue of the excavated bore either during or after tunnelling operations.

8.2.2   SITE INVESTIGATION – SOFT GROUND TUNNELLING

Before any practical decisions can be made relating to a tunnelling scheme design, it is essential that an appropriate ground investigation is completed in order to evaluate the basic selection of:

• The type of tunnel excavation method
• Any requirements for ground stability control or conditioning

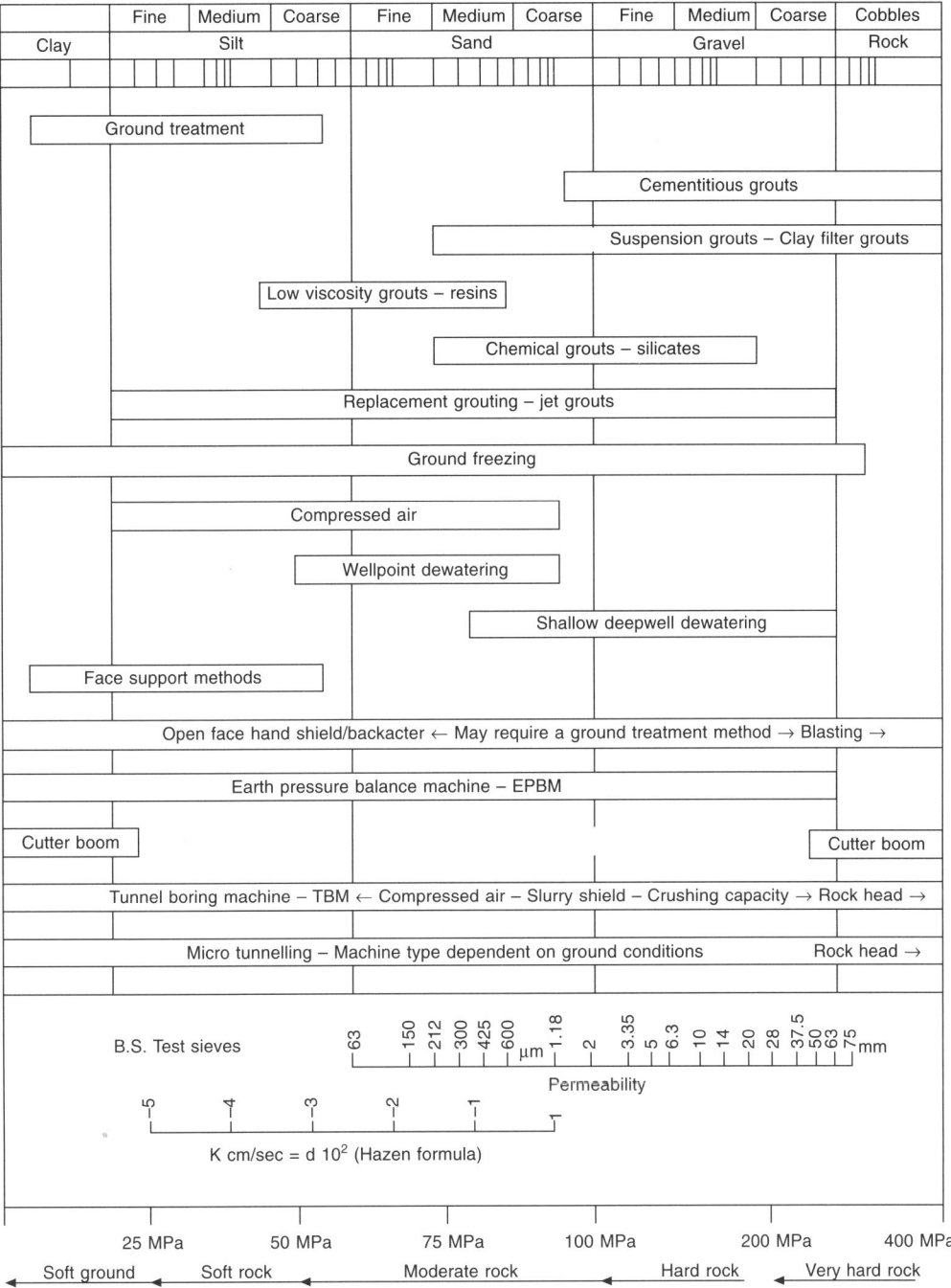

Figure 8.1 Ground treatments and face support methods for varying ground conditions.

Table 8.1   Soil testing for ground investigations for tunnels

	In-situ	Laboratory	
Clays	Standard penetration tests Pressuremeter testing Undrained strength hand vanes Undrained strength pocket penetrometers	Index tests Triaxial tests Undrained strength Particle size distributions Soil chemical tests Swelling Consolidation	Moisture content Liquid limit Plastic limit Undrained, drained Hand vanes, pocket penetrometers Sieving/sedimentation
Soft clays Alluvium	Cone penetration tests Shear vanes *In-situ* permeabilites	Index tests Undrained shear strength Particle size distributions Soil chemical tests	Moisture content Liquid limit Plastic limit Triaxial undrained, drained Sieving/sedimentation
Granular	Standard penetration tests *In-situ* permeabilites Cone penetration tests Friction lines for cohesionless soils	Particle size distributions Soil chemical tests SPT tests	Sieving/sedimentation
Ground water	Drilling observations Piezometers Pumping tests Tracer tests	Water chemical tests	
All types	Good desk study Good design to ground investigation Enough Bhs to build a good 1-D model Ground investigation done in stages Good logging of soils (to relevant British Standards) Good reporting Attention to specific type of site (site peculiarities) Attention to specific type of tunnelling method Attention to shafts and not just tunnels		
Other types	Geophysical methods, e.g. seismic surveys		

- The length, strength and type of tunnel lining
- Tunnel vertical alignment.

The purpose of an effective geotechnical design summary report is to serve as a starting point for measuring the degree of differing site conditions. The British Standard code of practice for site investigations (BS 5930) lists the primary objectives of a site investigation as:

- Suitability – to assess the general suitability of the site and appropriateness for the proposed works
- Design – to enable an adequate and economic design to be prepared
- Construction – to plan the best method of constructions to *foresee* and provide against difficulties and delays that may arise during construction due to ground and other local conditions
- Effect of change – to determine the change that may arise in the ground and environmental conditions, either naturally or as a result of the works and the effects of such changes on the works, on adjacent works and on the environment in general

- Choice of site – where alternatives exist to advise on the relative suitability of different sites or different parts of the same site.

The stability of an excavated bore is of primary importance in any form of tunnelling. Instability can cause damage to surface structures or services and also endanger the miners and excavating machinery apart from the operational difficulties posed. Consequently ground conditions must be carefully assessed to anticipate possible face instabilities, particularly in cohesionless soils below the water table, soft clays, silts and mixed soils. The following examples consider the importance of the interrelationships between the geotechnical parameters listed above and the results derived from the ground investigation shown in Fig. 8.2. They clearly indicate the importance of determining the appropriate properties from a thorough and targeted ground investigation. For example:

1. *Face stability in cohesive soils*: In cohesive soils, the pressure of σ_T required to maintain stability of the tunnel face is given by:

$$\sigma_T > \gamma(H + D_e/2) - T_c S_u \tag{8.1}$$

Where γ = unit weight of soil
S_u = undrained shear strength of soil
T_c = stability ratio (derived from plot after Atkinson and Mair, 1981)
H = depth below surface
P = unsupported lengths
D_e = diameter ratio

To prevent blowout due to excessive face pressure:

$$\sigma_T > \gamma(H + D_e/2) - T_c S_u \tag{8.2}$$

In both cases a factor of safety of 1.5 to 2.0 on S_u is needed to limit heave and settlement in soft clays (Mair, 1987).

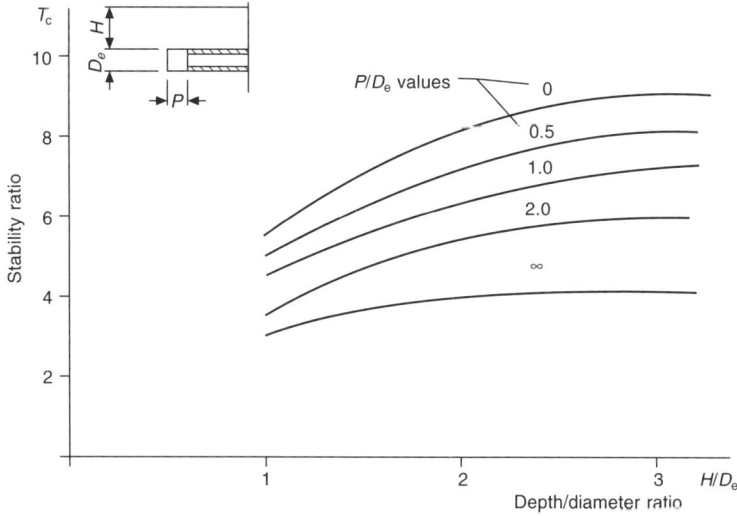

Figure 8.2   Face stability in cohesive soils.

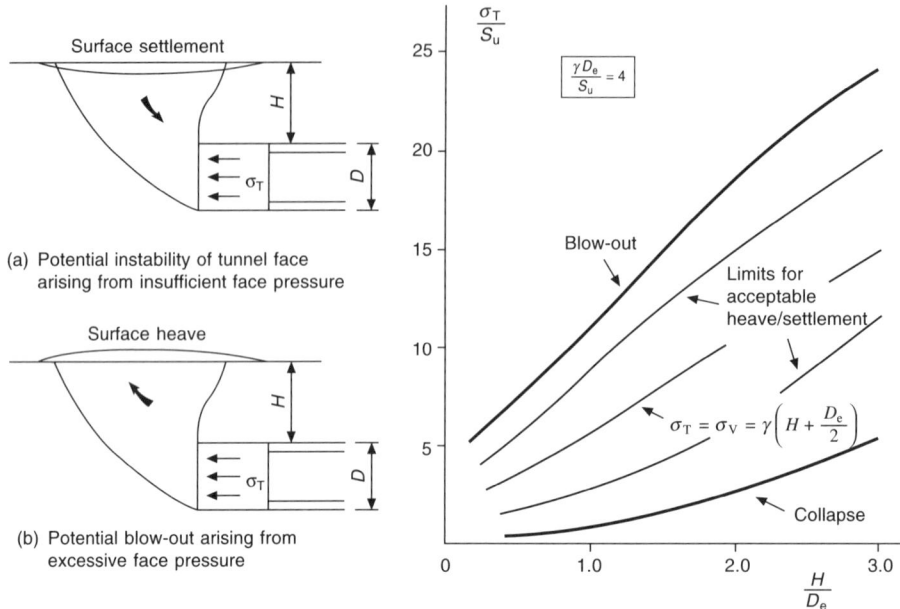

Figure 8.3 Limits of acceptable heave/settlement.

2. *Tunnel stability in cohesive soils*: For the tunnel behind the shield, the conditions correspond to the case in equation (8.1):

$$\sigma_T = \gamma(H + D_e/2) - T_c S_u$$

Which may be rearranged to give:

$$\sigma_T = \gamma D_e\left(H - D_e + \frac{1}{2}\right) - T_c$$

Which in turn gives rise to the plot shown in Fig. 8.4.

3. *Tunnel stability in cohesionless soils*: For tunnels in cohesionless soils without a surcharge on the surface, the required support pressure is independent of the cover depth and is given by:

$$\sigma_T = \gamma D_e T\gamma$$

Where T_γ is the stability number derived from the plot below, it is a function of ϕ, the friction angle of the soil.

Alternatively, if the tunnel is at a shallow depth and a large surcharge σ_s acts on the surface, the weight of soil may be neglected and then:

$$\sigma_T = \sigma_s T_s$$

With the stability number T_s as given by the plot in Fig. 8.6.

However, both solutions apply to dry soils; water pressure, if present, must be added to σ_T and the buoyant weight of the soil used in equation 8.1.

Also the means of generating σ_T must be considered since in the case of a slurry mode of operation, under optimum operational conditions a filter cake forms on the

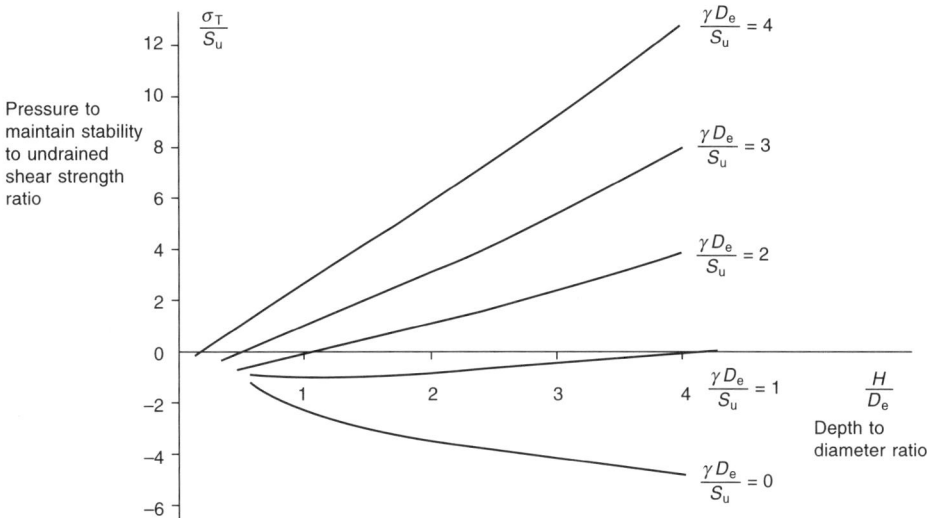

For example, for $H/D_e = 2$, and $\gamma D_e/S_u = 4$. $T_c = 4$ then $\sigma_T/S_u = 4 \times 2.5 - 4 = 6$, or directly from plot above, i.e. for $S_u = 10$ kPa, $\sigma_T = 60$ kPa.

Figure 8.4   Tunnel stability in cohesive soils.

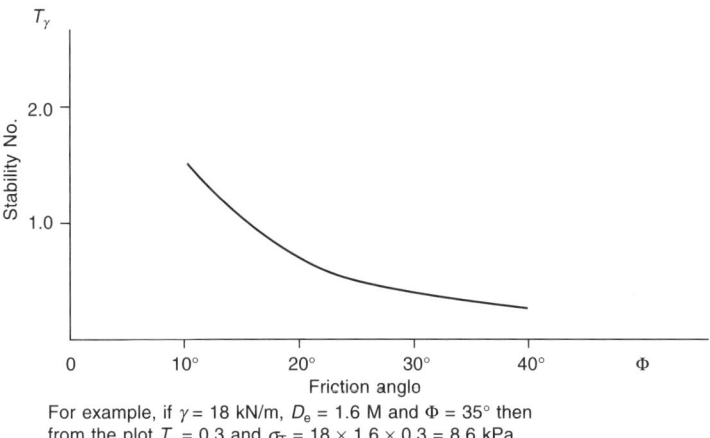

For example, if $\gamma = 18$ kN/m, $D_e = 1.6$ M and $\Phi = 35°$ then from the plot $T_\gamma = 0.3$ and $\sigma_T = 18 \times 1.6 \times 0.3 = 8.6$ kPa

Figure 8.5   Tunnel stability in cohesionless soils.

tunnel face acting like a membrane and inhibiting the infiltration of the suspension into the ground. However, for soils with exceptionally high permeability or when the shear resistance of the slurry is low, the bentonite could penetrate the ground to a certain degree. The final distance of infiltration e_{max} can be estimated by an empirical formula:

$$e_{max} = \frac{\Delta P d_{10}}{2 T_f}$$

Where ΔP = excess slurry pressure
          d_{10} = characteristic grain size
          T_f = yield strength of slurry

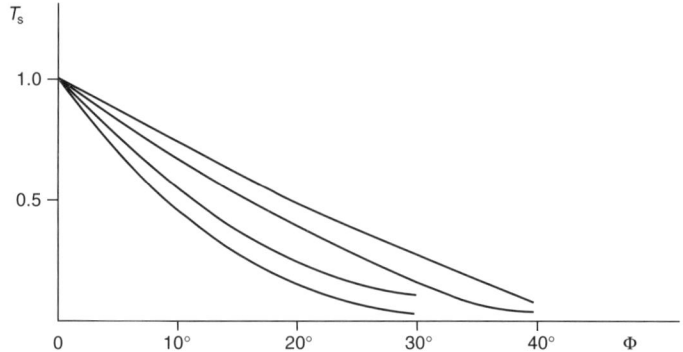

Figure 8.6   Tunnel stability in cohesionless soils – large surface surcharge.

Accordingly, the extent of the slurry infiltration is governed by the soil particle fraction and the yield strength of the suspension depends essentially on the bentonite concentration.

4. *Ground closure*: For initial vertical and horizontal stresses in the ground δ_v and δ_h, the reduction in vertical diameter of the tunnel bore due to elastic stress relief is given by:

$$\delta_v = \frac{(1 - v^2)}{E_s} D_e (3\sigma_v + \sigma_h) \tag{8.3}$$

and similarly the reduction in the horizontal diameter is given by:

$$\delta_h = \frac{(1 - v^2)}{E_s} D_e (3\sigma_h + \sigma_v) \tag{8.4}$$

Where E_s and v are Young's modulus and Poisson's ratio for the soil.

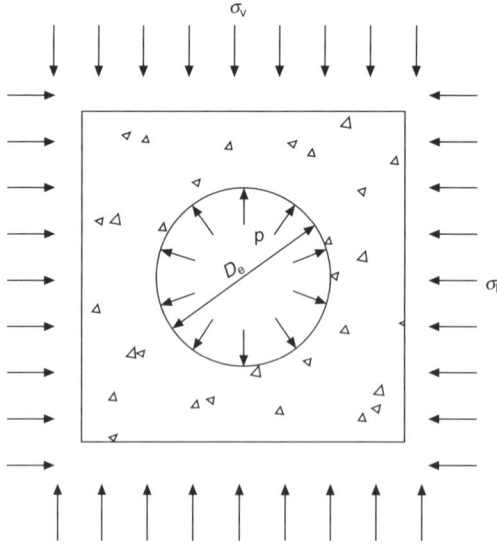

Figure 8.7   Ground closure.

5. *Ground closure in cohesionless soils (after Terzaghi, 1943).*

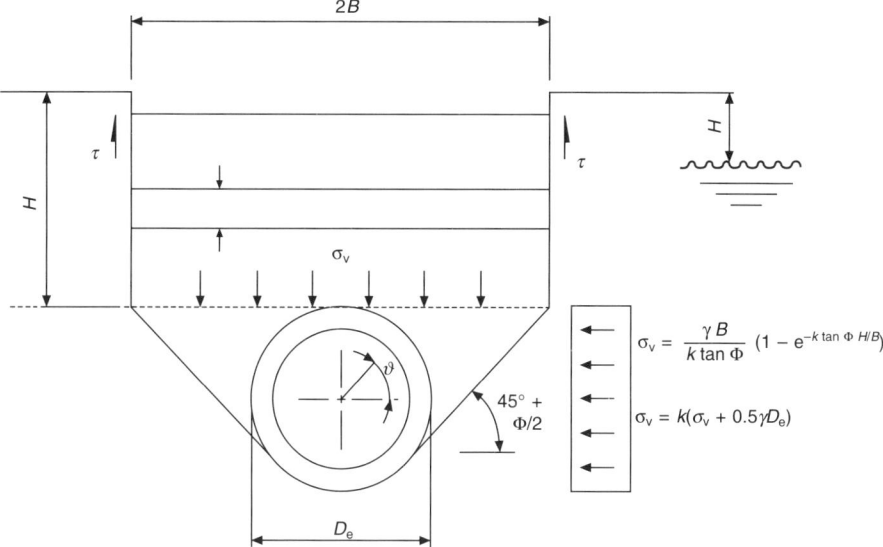

Figure 8.8   Ground closure in cohesionless soils.

The radial stress around the pipe is:

$$\sigma_p = \frac{(\sigma_v + \sigma_h)}{2} + (\sigma_v + \sigma_h)\cos\phi$$

and the total frictional resistance is:

$$F = \frac{\pi D_e}{2}(\sigma_v + \sigma_h)\tan\delta$$

Where π is the angle of internal friction of the soil and δ is the angle of friction between the pipe and the soil.

When a water table is present at depth H, the expression for σ_v becomes:

$$\sigma_v = \sigma_{v1}e^{-k\tan\phi(H-H_1)/B} + \frac{\gamma'B}{k\tan\phi}(1 - e^{-k\tan\phi(H - H_1)/B})$$

Where:

$$\sigma_v = \sigma_{v1} = \frac{\gamma B}{k\tan\phi}(1 - e^{-k\tan\phi(H - H_1)/B})$$

and:

$$\sigma_h = k(\sigma_v + 0.5\gamma'D)$$

Note that γ is bulk unit weight (above water table) and γ' is submerged unit weight (below water) and:

$$B = \frac{D_E}{2}\tan(45° - \phi'/2) + \frac{D_E}{2\sin(45° + \phi'/2)}$$

6. *Surface settlement*: After O'Reilly and New (1982) surface settlement S at a given point is:

$$S = S_{max}e^{y^2/2i^2} = \frac{V_s e^{-y^2/2i^2}}{i\sqrt{2\pi}}$$

derived from Fig. 8.9.

The grading of non-cohesive soils will indicate the compaction of looseness of the soil and the presence of fines which may cause instability.

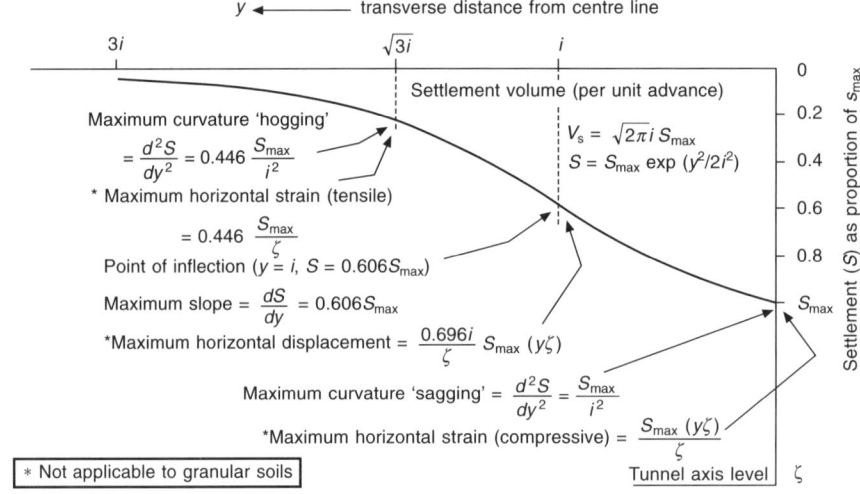

Figure 8.9 Long-term surface settlement.

The view of the *Drifter* article, 'Tunnels and Tunnelling', September 1996, was quite clear:

It is the geotechnical engineer's responsibility to educate owners regarding risk, geotechnical exploration, base line reports and differing site conditions and all the anomalies associated with construction in a geotechnical medium. Solutions can only develop in response to identification of a problem. All too often the signs of impending catastrophe have been ignored and the blame culture predominates. However, blame is equivalent to the avoidance of responsibility and therefore cannot produce an identification of the problem.

This view as to where the responsibility lies was challenged within the Harding Memorial Lecture, 1996, 'Risk: A Tunneller's View' by D.B. Parkes, which recognised that risk in tunnelling begins and often ends with the physical risk from the ground and the way in which engineers deal with it and the plant and equipment they employ in the process. The fact that often quite small variations in ground conditions affect the economy of operations in a disproportionally large way makes it vitally important to identify those factors which have the potential to affect the tunnelling process and then to elicit their presence from the ground investigation. However, consideration of the parameters touched on already in this chapter suggests that it is unlikely that a geotechnical engineer could achieve this in isolation.

Getting the wrong perception of ground conditions can be due to a number of factors including: no proper desk study, inappropriate field work, *in-situ* testing or lab testing inadequacies or poor geological interpretation. Ground conditions receive recognition in the Institute of Civil Engineers (ICE) Conditions of Contract where ground conditions which cause problems are often called 'unforseen', as they were not predicted. This raises the question as to whether the conditions were genuinely unforeseeable, or the ground investigations were inadequate. Of course to the purist the definition of an adequate ground investigation is one from which all the discoverable conditions, which could affect the tunnel driving economy, were identified and determined. However, the law of diminishing returns places a practical limit on this and we are left with a requirement of a prudent and equitable risk management strategy, which includes a properly targeted investigation which is solving the right problems with the right balance between the theoretical ground investigation and the empirical experience.

8.2.3   ROCK CLASSIFICATION AND SITE INVESTIGATION

One of the basic differences between hard and soft ground tunnelling is that in the harder rock conditions the main objective is to excavate the ground. That is, the prime objective and the major use of installed mechanical power is the excavating of the rock. Consequently some parameters used for the assessment of soft ground tunnelling are not as important during a hard ground tunnelling investigation. From a geological perspective an often used parameter, as used by Hoek and Bray in 1977, is unconfined compressive strength (UCS). However, since the development of the philosophy from the simple slabbing theories of cutting, which proposed cutting relied predominantly on the crushing of the rock, to the more current crack propagation philosophy of cutting, has meant that parameters other than UCS affect its ability to be excavated. The overall driveability of a hard ground tunnel is influenced by many factors which will be discussed within this chapter.

Hard rock tunnelling

The means of rock classification are derived from the fact that rock as an engineering material is far more complex in behaviour than that of soils. Although fundamental theoretical rock mechanics techniques have adapted the Coulomb law approach of soil mechanics, in general, rock mechanics has developed far beyond this basic concept. Consequently with respect to tunnelling and stability, three major rock deformation processes are recognised, all of which require a different design approach:

1. *Instability resulting from the rock structure:* This includes the natural features of the rock structure such as bedding planes, joints fault planes, in addition to the movement of discrete blocks under the action of gravity.
2. *Stress-induced deformations:* This arises as a result of the stress field changes due to tunnel formation causing localised rock failure and large movements into the opening.
3. *Effect of semi-failed state of ground prior to excavation:* Situations can also exist where the rock mass is highly fractured and weathered and is, in effect, already in a semi-failed state. In these conditions, even with low values of *in-situ* stress, large values of stress-induced closures can occur although the intact strength of the rock may be significantly higher than the levels of *in-situ* stress. In civil engineering tunnels because of the relatively shallow depths it is more common for the *in-situ* stress field to be of low magnitude and the rock mass degradation high. Consequently tunnel instability is more likely to be related to stress-induced movements in conjunction with rock mass of low strength as given by (1) and (2).

Examination of the rock mass structure is a necessary requirement in the design process of any tunnel project irrespective of whether that structure is a dominant factor influencing tunnel stability. Essentially the rock structure may be considered as a rock property of equal importance to the deformation moduli and rock strengths.

The concept of rock mass structure, which has been effectively discussed by Hoek and Bray will affect all aspects of driveage and support of a tunnel, namely:

1. The rock breakage process and face stability
2. The excavated tunnel profile prior to installation of support (if required)
3. The length of unsupported span and maximum unsupported tunnel size
4. The gravity loading of the supports from loose material
5. The *in-situ* stress loading component of supports resulting from rock mass failure and large-scale movements.

A rock mass can vary from containing no or just a few widely spaced closed joints, e.g. granite, to situations where the discontinuities are so frequent that it becomes a dense pulverised mass, e.g. slates and mudstones. Most tunnelling conditions lie between these two extremes whereby consideration of the rock intact strength, rock mass structure and *in-situ* stress levels will determine the form of instability and therefore the stand-up time and appropriate support measures.

In Table 8.2 Terzaghi has presented an attempt at estimating the degree of support required in various rock conditions according to the general character of the rock mass. It is more suited to the use of steel sets as opposed to the more recent techniques of shotcrete and rockbolting and it assumes that the tunnel is of sufficient depth to allow full ground arching action and thus the rock load is practically independent of depth. The minimum depth at which this occurs is stated to be 1.5 times the combined width and height of the tunnel.

Table 8.2 presents Terzaghi's design guidelines to the range of likely values for the rock load height (H_p) and the lining requirements. The general nature of this approach is recognised and consequently other authors, namely, Deere *et al.* and Rose have published adaptations of the method which allow a more quantitative evaluation of the rock mass quantity. Three such methods are:

- Rock quality designation (RQD) which was later used in Lock Mass Rating.
- Rock mass rating (RMR).
- Tunnelling quality index (*Q* system).

Rock quality designation (RQD)

Deere has put forward a qualitative index of the rock mass quality as illustrated in Table 8.3, based upon the core that is recovered from diamond drilling operations. RQD is defined as the percentage of recovered core in intact pieces greater than 100 mm in length in the total length of a core or the thickness of any strata layer.

$$\text{RQD } (\%) = \frac{100 \times \text{Length of core in pieces} > 100 \text{ mm}}{\text{Total length}}$$

Table 8.3 relates rock quantity to a range of values for RQD and Table 8.4 presents support recommendation according to RQD and driveage method.

This method has been widely used in tunnelling and has been found to be very useful in giving guidance on the selection, dimensioning of tunnel supports and method of excavation.

Table 8.2 Rock load classification according to Terzaghi

Rock condition	Rock loading height (H_p)	Comments
1. Hard and intact	Zero	Light lining only if spalling or popping occurs
2. Hard stratified, or schistose	0–0.5B	Light support, mainly against spalling; loading changes likely
3. Massive, moderately jointed	0–0.25B	
4. Moderately blocky and seamy	0.25B–0.35($B + H1$)	Negligible side pressure
5. Very blocky and seamy	(0.35–1.10)($B + H1$)	Little to negligible side pressure
6. Completely crushed	1.10($B + H1$)	Significant side pressure; softening from seepage in lower part of tunnel requires continuous support
7. Squeezing rock, moderate depth	(1.10–2.10)($B + H1$)	Heavy side pressure, with need for invert struts; circular ribs recommended
8. Squeezing rock, significant depth	(2.10–4.50)($B + H1$)	
9. Swelling rock	Up to 76 m, irrespective of ($B + H1$)	Circular ribs required and yielding support possible

Table 8.3 RQD index as a qualitative description of the rock mass (after Deere)

RQD (%)	Rock quality description
<25	Very poor
25–50	Poor
50–75	Fair
75–90	Good
90–100	Excellent

However, it has limitations: the use of RQD in the design process poses some difficulties as it is only a measure of the degree of rock mass fracturing and there are other factors which influence tunnel stability. Merrit points out the limitations of RQD when joints contain clay filling or weathered material although high joint spacing and high RQD may be evident. However, unstable conditions may ensue due to the presence and nature of joint filling material. It is currently recognised that although RQD does not provide a complete structural analysis of the rock mass, it is a quick and inexpensive index which can provide a reasonable estimate of the likely rock mass fracturing and as a result is a common factor in the more advanced rock mass classification systems.

Rock mass rating (RMR)

The RMR system was developed by Bieniawski in South Africa during the early 1970s and the classification is based on six basic parameters, namely:

1. Intact UCS of rock: range 0 to 15
2. RQD: range of values 3 to 20

3. Joint spacing (referring to all discontinuities) range of values 5 to 20
4. Joint condition: this considers joint separation surface roughness, continuity, wall condition and any fill material: range of values 0 to 30
5. Ground water: assessed according to either in flow (l/s) per 10 m length of tunnel, joint water pressure to major principal stress ratio or a general description: range of values 0 to 15
6. Discontinuity orientation: the range of values is 0 to 60 according to favourability and type of structure, e.g. tunnel, foundation or slope.

Details of the individual ratings are given by Bieniawski in a series of tables, the sum of which gives an overall RMR which has a maximum value of 100 as illustrated in Table 8.4.

Table 8.4   RMR rating related to rock mass class (After Bieniawski)

Class number	Description	Rating
(i)	Very good rock	81–100
(ii)	Good rock	61–80
(iii)	Fair rock	41–60
(iv)	Poor rock	21–40
(v)	Very poor rock	<20

The RMR concept has been used in a variety of applications including a class description of the rock mass, an average stand-up time and approximate values of cohesion and friction angle.

The NGI tunnelling quality index (Q system)

It is felt by many that the RMR system was developed principally for mining applications as opposed to civil engineering tunnels. Conversely, the 'Q' system was developed by the Norwegian Geotechnical Institute (NGI) and was first proposed by Barto *et al.* based on the evaluation of some 200 case histories of civil engineering tunnels in Scandinavia. The concept upon which the 'Q' system is based depends upon three fundamental requirements:

1. Classification of the relevant rock mass quality
2. Choice of the optimum dimensions of the excavation with consideration given to its intended purpose and the required factor of safety (not considered in RMR)
3. Estimation of the appropriate support requirements for that excavation.

The 'Q' system takes numerical account of the following parameters:

1. RQD: The absolute value is employed
2. Joint set numbers (J_n): This is the measure of the number of joint sets within the rock mass and has a range of values of 0–5 (massive deposits/few joints) to 20 (crushed rock)
3. Joint alteration number (J_a): This takes account of infilling and has a range of values from 0.75 (no infilling) to 20 (thick band of crushed rock infilled with clay material)
4. Joint roughness number (J_r): The range or values is 0.5 (smooth, planar, slicken sided joints) to 4 (rough, undulating and discontinuous joint)
5. Joint water reduction factor (J_n): This takes into account the presence of water under pressure affecting the shear strength of joints and has a range of valves 0.5 (high pressure) to 1 (zero pressure)

6. Stress reduction factor (SRF): This takes account of several factors, the principal ones being:
   - Loosening of rock mass as a result of shear zones and clay-bearing rocks
   - Rock stress problems in competent rock
   - Loads induced by squeezing and swelling ground conditions. The range of valves for SRF cover 1 to 15 and depend upon the nature of the problem although actual valves for the parameters 1 to 6 can be assigned using tables provided by Barton *et al.*

The NGI index (Q) is given by the expression:

$$Q = (RQD/J_n)(J_r/J_a)(J_w/SRF) \tag{8.5}$$

The expression is in effect a function of only three parameters which are approximations of:

Block size $= RQD/J_n$
Interblock shear strength $= J_r/J_a$
Active stress $= J_w/SRF$

The value of Q can be related to the support requirements of a tunnel by considering the equivalent dimensions (D_e) of the excavation and this is defined by the expression:

$$D_e = \frac{\text{Excavation span, diameter or height}}{\text{Excavation support ratio (ESR)}} \tag{8.6}$$

The ESR is a function of the operational duty of the tunnel and relates to the stability by such an opening. A full description is given by Barton *et al.* with its range being from 3 to 5 for a temporary mine opening to 0.8 for an underground nuclear power station opening. Using these values for Q and D_e Barton *et al.* defined a series of 38 support types designed to encompass a full range of excavation requirements and rock mass conditions.

Bieniawski has examined the use of rock mass classification systems in tunnelling and he has pointed out a number of shortcomings in relation to inappropriate use of such systems resulting in the following recommendations regarding their application to tunnel design:

- Classification schemes are not rigid guidelines or a substitute for engineering judgement
- Consideration must be given to alternative schemes
- Check applicability – classification schemes are not always appropriate to particular situations
- Classification schemes are generally conservative in nature
- A system with a proven record should be used – the Q system and RMR tend to be superior
- At least two classification schemes should be applied.
- Their application with respect to interaction with the theoretical concept, modeling results or field measurements should be examined
- Their incorporation into appropriate expert systems should be considered
- A complete record or database of previous experience with the classification systems should be kept.

Generally the use of rock mass classification schemes in tunnelling not only provides a quantitative empirical guide to support requirements for a given tunnel but also provides other significant benefits:

1. They allow a tunnel route to be subdivided into areas requiring different support measures.
2. They initiate the systematic collection and recording of geological data.
3. They provide an estimate of the unsupported span of the ground allowable and consequently the type/design of excavation equipment.
4. The unsupported span of the ground in conjunction with stand-up time estimates allow phasing of the various support erection operations to be appropriate for the likely ground behaviour.

Notwithstanding these benefits designers should not be afraid to change their minds on the support requirements if the *in-situ* rock mass behaves differently from those envisaged.

8.3   Linings

One of the fundamental requirements for a tunnelled sewer is an impermeable lining which is as smooth as possible to maintain the hydraulic head. This is normally quite low due to the long distances involved. Thus any effect causing a reduction in the gradient, e.g. roughness, would be detrimental to the scheme. Thus, it is highly desirable for the designer to specify linings constructed from high quality materials.

Most tunnelled sewers in the UK are constructed with a diameter of 3 metres or less. There are exceptions, e.g. Brighton and Don Valley, but these are few and far between. Therefore, the linings discussed in this chapter will be restricted to the most common sizes.

The discussion will also be limited to the permanent lining to the tunnel as the temporary works required during excavation are discussed under the relevant tunnelling method. Temporary works in soft ground could involve the erection of an initial segmental lining to provide temporary support. This is known as the primary lining. Normally this is a panelled segment and therefore 'rough' in hydraulic terms, thus requiring a secondary or permanent *in-situ* lining. Pipejacking is not included as it warrants a separate chapter of its own.

Thus, taking account of the above, the following discussion will concentrate on:

- *In-situ* concrete linings
- Segmental 'one-pass' linings.

8.3.1   DEVELOPMENT

The majority of urban areas are located on river plains or coastal areas. Thus the predominant geology will fall in the soft ground range of clays, sands, silts and gravels and will be invariably linked to a ground water regime. Consequently the stand-up time of these strata is so small that special tunnelling techniques are required to facilitate construction.

Before these techniques had been developed all sewers were, and continued to be for some time, brick lined. These were mainly constructed using open-cut techniques, but where tunnels were necessary elaborate timber supports were required and progress was slow.

The first segmental lining was used in 1869 on the Tower Subway tunnel in combination with a circular shield. The principles behind this technique are still relevant and applied in modern-day tunnelling. Also when this technique was complemented with low pressure compressed air to balance ground water pressure, many geotechnical problems for the tunneller were solved. As the cast-iron linings were panelled a brick, concrete or combined lining was used to provide the hydraulic requirements.

Although concrete segments had been experimented with it was the advent of World

War I which accelerated their development due to scarcity of cast iron for civilian requirements. The original concrete segments were cast to the same dimensions as those adopted through experience for cast iron. Eventually the concrete segments standardised on a width of 2 feet (610 mm). Also, as these segments were also panelled, i.e. reflecting the design of the cast-iron ring, an *in-situ* concrete lining was required for smoothness, i.e. a two-stage operation. In a constant effort to win work in a fiercely competitive market, the contractor prompted the development of a single pass grouted smooth lining which had financial benefits to both himself and his client.

A smooth bore lining had been developed for the Metropolitan Water Board in London which became known as the 'wedgeblock' lining since it used a key in the crown of the tunnel to expand the segments against the tunnel walls, thus eliminating the grouting annulus. This lining was developed from the 'Donseg' lining and standardised on 100 inch (2.54 m) diameter although other diameters have since been produced.

This lining is only applicable in a medium which allows a sufficient stand-up time, e.g. London clay. It is an unreinforced lining as it acts in pure compression once the load has built up on it. Also as the lining is erected quickly, against the ground, and because it restresses the ground, the settlement results from this method are very good.

The development of the grouted smoothbore lining came with the development of fixings to take the heavier loads during erection and were based on the sizes adopted for the standard panelled ring, although these were sometimes increased to suit the adopted method, e.g. use of an erector should the length of tunnel be sufficient to cover the cost of new moulds for the rings.

Despite the developments of the smoothbore ring, problems still remained in water-bearing fine-grained strata, preventing ingress of water and fine particles. Grouting was not always effective.

The development of hydrophillic and elastomeric gaskets have led to the latest type of ring – the trapezoidal segment ring. Gaskets are fitted to each individual segment, thus creating 'star' joints when adjacent plates and rings are connected. This applies particularly when rings, tapered to suit the tunnel curvature are used, cannot be rolled and thus creating a potential weakness in the sealing of the ring. The solution was the development of the trapezoidal ring which eliminates a four corner junction. This lining was used successfully in the recent Fylde Coastal Waters Interceptor Sewer (2.85 m diameter) which had up to 3.5 bar of ground water pressure. These rings were made 1.0 m long to eliminate a substantial number of joints and fixings in the 11.7 km tunnel as a hydraulic erector on the tunnelling machines was necessary in order to provide the necessary pressure to squeeze the cross joint gasket to allow the joint to be fitted. Without this process the ring cannot be built.

8.3.2 DESIGN

In-situ concrete lining

The *in-situ* concrete lining is applicable to both soft and hard ground tunnels although the requirements can vary as follows:

1. To provide a smooth impermeable lining
2. To provide structural support.

It is unusual for (1) to be the sole requirement.

The smooth impermeable lining is required when a tunnel has been built with a primary lining which is capable of meeting the structural and exposure criteria, but not the hydraulic roughness, e.g. a tunnel can be built with a primary standard bolted lining which is capable of meeting structural design criteria, but cannot meet the hydraulic requirements of the system unless lined with an internal lining.

The second case is used when the life expectation of the primary lining will not meet the specification or when the long-term loading is fully applied the primary lining has insufficient load capacity. Accordingly the *in-situ* lining will have to be designed using the long-term load case together with precautions to maintain the integrity of the lining, e.g. exposure condition and early age thermal cracking according to BS 8110 and BS 8007, as applicable.

The requirement is not so severe for the first case as the structural integrity of the lining is not affected. Thus, the secondary lining can be unreinforced as crack control is not important and certainly the lifespan of the lining is increased by eliminating that which is corrosive. Precautions against chemical attack of concrete should be taken in both cases.

Internal linings to tunnels are becoming less frequent, the smoothbore type lining being more economical to install in most cases. However, the *in-situ* lining has one advantage for the designer over the segmentally lined tunnel as the internal lining can be used to regrade a tunnel where construction tolerances have been exceeded. This is extremely difficult to do otherwise.

The designer should accordingly take cognisance of the ground conditions and alignment, and then specify an internal lining if he/she believes the contractor could have problems meeting the construction tolerances and/or that the hydraulic requirements are paramount for the system to work.

Segmental linings

Segmental linings are designed taking the following criteria into account:

- Exposure conditions: The lining should be designed to withstand corrosion attack from both the effluent and the ground water and strata.
- Service life: This is determined by the designer and may reflect the repayment period to justify the capital cost. However, it should be borne in mind that detailed analyses of concrete do not exist and, therefore, designs can only be undertaken to best practice.
- Handling and method of erection: From the time when a segment is lifted from its mould it is repeatedly handled until it reaches the tunnel face before erection into its final position. The stresses generated by this process are often more onerous than the overburden requirements.

  Each one of these processes needs to be addressed so that fixings can be cast-in accordingly. It is quite common to double up the use of rings, e.g. the cast-in fixing for grout injection can be used for segment erection.
- Structural: The lining needs to be designed to support ground loadings and ground water pressures. Other than in exceptional circumstances the latter are time dependent, but this is a function of geology, diameter and depth. Obviously, the latter is instantaneous unless the ground water is being lowered artificially.

  The segments are normally analysed using the Muir–Wood method, but as amended by Curtis. Design life is applied equally to the fixings as well as the segments as corrosion of the former could seriously affect the integrity of the segments, even though they are not providing any structural purpose.
- Shove forces: These are the longitudinal forces which are transmitted into the lining when a tunnel is constructed using a shield or boring machine. The magnitude of the

total shove force can reach 4000 kw or 5000 kw but, due to the alignment of cylinders to the lining, the forces are eccentric and are most pronounced in the invert of the tunnel.

- Hydraulic design: The final circular cross-section produced by a tunnel lining does not always provide the ideal solution, particularly for dry weather flow. Therefore, a channel may be cast into the invert to facilitate such requirements.

Most segment manufacturing will have the moulds for most commonly used segments and will provide designs to suit the actual circumstances. Marketing brochures listing these rings are normally available on request.

Non-standard designs are expensive and time consuming to produce as new master steel and concrete production moulds have to be manufactured. The cost of this process is normally only financially viable when a long drive, say greater than 1 km, is under consideration.

Having taken all the above into consideration, the ring designer must ensure that his design is practical by producing trial segments. When the problems with these have been eliminated, full production of the units can commence.

8.3.3   MANUFACTURE OF SEGMENTS

Segments are manufactured in purpose-designed factories specifically equipped for segment production. The segments are cast inside face down into concrete moulds. The latter are generated from a master mould which is manufactured to high tolerances from steel.

Each mould is individually marked and each segment can be identified by:

- The manufacturer
- The date of casting
- The mould number
- The diameter of the ring
- The type of segment.

Cube strengths are available for each batch to verify that the design criteria have been met. Reinforcement can also be traced to manufacturers.

Segments are normally left to cure overnight in their moulds, although inserts for bolt holes, etc. will have been removed at initial set. They will then be lifted from the moulds and stored outside for 28 days. It is possible to speed up the manufacturing process by using steam curing so that each mould can be used twice per day instead of once. Of course, there is a cost penalty to pay.

8.3.4   CONCLUSION

Although an acceptable method, the use of *in-situ* linings to tunnels is becoming less common due to the popularity of the watertight smoothbore segment and its economic advantages.

From the foregoing, the choice of lining system is intimately linked to the construction method. However, the purchase of an off-the-shelf lining alone will not make the system work. There is much minor detail in the design and manufacture of segments – which there has not been room to discuss – which is essential to the success of a segmentally lined tunnel whether as primary or permanent.

It also needs to be pointed out that while tunnel linings are described as smoothbore, there are still bolt hole and grout hole pockets and joints that need to be filled after the erection of the lining to provide a completely smooth tunnel bore.

8.4   Soft ground tunnelling

8.4.1   INTRODUCTION

Deposits which form soft ground conditions include clays, silts, sands and gravels. These deposits are typical of plains and valleys where the majority of towns and cities are located. By necessity, these have a requirement for large capacity sewerage systems which could include tunnelled sections.

The above deposits can be seriously affected by the presence of ground water and, accordingly, special precautions have to be taken to facilitate tunnel construction. These may range from the application of simple dewatering techniques to the adoption of closed face tunnel boring machines (TBMs).

Whichever method of tunnelling is employed, the fundamental operations remain the same for a successful conclusion, namely:

- Excavation of ground
- Immediate support of ground
- Permanent support of ground
- Management of water.

In the following chapters each of the above stages will be discussed in more detail.

8.4.2   EXCAVATION

Excavation of soft ground tunnels can be undertaken by hand or with mechanical assistance. In making the choice the following commercial and logistical aspects must be taken into consideration:

- Capital expenditure on plant and segment moulds
- Length of tunnel to be constructed
- Procurement period – TBM can take up to 20 weeks for delivery
- Labour costs – mechanisation usually reduces the labour content
- Stand-up time of the face
- External water pressures
- Programme requirements
- Topographical features which may impose settlement limitations.

This list is not exhaustive as each contract is different and has its own characteristics. Handwork can be undertaken with or without the assistance of a tunnel shield. The latter has benefits of providing immediate support and providing a safe area for miners to work. However, traditional timbering techniques are still used when the configuration of the tunnel does not warrant the investment in a shield, e.g. driving TBM/shield launch chambers from the shaft.

Mechanical excavation has become more sophisticated over the last 10–15 years as more difficult ground conditions are negotiated in an attempt to minimise on expensive mining labour. Types of mechanical excavation vary from backhoes and roadheaders through TBM and slurry machines to earth pressure balance machines.

With the exception of slurry machines which use charge and discharge pipes, muck removal from the remaining machines is undertaken by narrow gauge rolling stock within the tunnel. Skips are emptied at the shaft either by tipping into a pit bottom bunker arrangement or by winding to the surface. It is essential in the design of any spoil removal system that it must have sufficient capacity to cope with the anticipated outputs from the excavation method employed.

8.4.3   IMMEDIATE SUPPORT

Fundamental to any soft ground excavation system is the immediate support of the surrounding ground. Traditionally this was undertaken using timbering techniques which became more costly with an increase in diameter. Rates of advance on these drives were slowed by the requirement for masons to install the permanent block or brick lining before the next cycle could commence.

This method was not suitable for water-bearing ground or very soft clays which required immediate support.

This situation was rectified initially by Marc Isambard Brunel, who designed, built and employed the first tunnel shield for the Wapping Tunnel (1825–43). This idea was further developed by Barlow and Greathead in 1875 to facilitate the construction of the Tower Subway. They patented the circular shield and used it with a cast-iron sectional lining. The principles of the Greathead shield (as simple shields are known by) are employed in all modern-day tunnelling machines.

The Greathead shield is fabricated from steel plate and provides:

* Protection for the miners
* Support to the surrounding ground
* Safe place to erect the tunnel lining.

The main components of the shield are:

* Hood and cutting edge
* Main body with diaphragms and containing the propulsion cylinders
* Tailskin.

Classically the hood and tail skin are designed to be compatible with the length of the segmental lining.

The normal operation of the shield in soft ground is as follows:

1. Initially the hood and cutting edge of the shield are forced by the cylinders into the tunnel face to provide overhead protection for the miners during excavation.
2. As the shield is thrust forward, the tailskin moves over the last ring erected until sufficient room is created to build the next ring, hence the length of the cylinder stroke and the rink width are also compatible.
3. The next ring can be erected in the tailskin area created by 'shoving' the shield forward. However, before erecting this ring the annulus between the extrados of the ring and the excavated surface has to be filled.

8.4.4   PERMANENT SUPPORT

The permanent support of the tunnel can be divided into two components: primary and secondary linings. The primary lining is normally erected within the tailskin immediately following excavation.

This void is created by space taken by the tailskin and by the room required to build a ring within the tailskin. The annulus is filled with grout pumped under pressure through injection sockets cast into the segments.

Traditionally, to provide primary support, these rings were standard bolted panelled rings mainly manufactured of pre-cast concrete. This lining was then completed with an *in-situ* cast concrete lining to provide the necessary hydraulic requirements, see section 8.3.

With an improvement in concrete technology and an increased accuracy in constructing tunnels to the required gradients, it is now common practice to construct smoothbore segments as the combined primary and permanent lining and the increased cost of the segments is offset by many other factors (see section 8.3).

8.4.5   MANAGEMENT OF WATER

When working below the water table, control of the ground water is vital to provide a successful outcome. The addition of water can dramatically and adversely affect the properties of soft ground strata. Soils exhibiting this condition can be treated by several differing methods. These will be discussed in a later chapter.

The process of introducing means to the tunnel to control the ground water can be extremely expensive and, as they are not guaranteed, are a potential risk. Consequently, tunnelling machines were developed to contain the water and spoils under balanced conditions. The balancing is achieved by either pumping bentonite slurry into a plenum chamber, or by pressure exerted from excavated earth in a plenum with its pressure controlled by a screw conveyor (see section 8.6.5).

These machines, together with a well-designed lining, allow tunnels to be constructed through extremely difficult ground conditions which could not be tackled using more conventional techniques.

8.5   Ground treatment

8.5.1   INTRODUCTION

Water contained within soft ground strata, such as sands and gravels, can have a serious detrimental effect on the properties of these strata and make tunnelling operations extremely difficult, if not impossible. As it is not always financially effective to introduce modern tunnelling machine technology, more traditional techniques must be adopted.

Before choosing a ground treatment method consideration must be given to the consequences of the application, e.g. a reduction in the water table locally could have a serious effect on the integrity of sensitive adjacent buildings/structures and also might encourage migration of polluted water at a nearby fill/tip. The latter environmental issues are becoming a sensitive subject with the Environmental Agency who closely monitor such matters, particularly with regard to discharge into adjacent water courses or sewer systems. Often the water drawn must be treated prior to discharge.

The consequential and environmental effects must be considered in tandem with the method applicable. Grading curves, water pressure, depth to strata and thickness of strata must be taken into account along with a written risk assessment before the final technical decision is taken. Obviously the final decision is based on technical and commercial respects.

The most commonly used techniques for managing water to permit the construction of soft ground tunnels are as follows:

- Low pressure compressed air
- Ground water lowering
- Chemical/grout injection
- Ground freezing.

Each of the above techniques will be discussed in turn during the remainder of this chapter.

8.5.2 LOW PRESSURE COMPRESSED AIR (LPCA)

This method was, 10–15 years ago, the most widely used technique for ground water control. However, a review of the physiological effects that working in a compressed air environment can have on a tunnel worker has changed this view; it is now a last resort if alternative solutions cannot be demonstrated to be viable. This is particularly applicable at pressure ≥ 1 bar when depressurisation procedures are mandatory. Strict codes of practice and recent legislation make the application of LPCA very onerous.

The method requires the counterbalancing of ground water pressure at the tunnel face with air pressure contained within the tunnel. Bulkheads built into the tunnel lining and containing access doors are used to create an airlock. This can be located either within the access shaft (vertical airlock) or within the tunnel (horizontal airlock).

Although stated earlier that the water pressure has to be balanced, this is very difficult to do in practice as there is a variation in pressure vertically in the face. The differential in pressure is dependent on the excavation diameter of the face.

Thus, if the tunnel pressure was raised to equal the ground water pressure at the tunnel invert, then the crown of the excavation would be overcompensated and 'air losses' will occur. The tunnel engineer has to strike a balance between these criteria to achieve the best results.

The compressed air for the tunnel is generated by high volume low pressure compressors with either a diesel or electric prime mover. Before being fed the working air is passed through a cleansing system, commonly known as 'scrubbers' to eliminate pollutants which may be present from the process, e.g. oil. After cleansing, the air is stored in accumulators/receivers to await automatic instructions from the control panel to discharge into the workings.

A low pressure compressor is time consuming and difficult to install. It is also expensive to run, not only from the energy consumed and plant hire costs, but from the extra labour required to implement the system, e.g. lock attendants, compressor attendants. If the system is raised above 1 bar then extra plant is required (standby compressors, generators, medical locks), extra resources (medical lock attendants, doctors) and time for acclimatisation (build-up of exposure times) of the pressure side tunnel workers and engineers must be added to the programme.

Although onerous in its application, low pressure compressed air has its advantages compared to other systems in that its applications minimise surface disruption as compared to deep wells or freezing.

It also has a specialist application – being required in some modern closed face TBMs to provide a safe access for the tunnel face.

8.5.3   GROUND WATER LOWERING

This system involves the lowering of the ground water table by extraction through wells to provide stable conditions at the tunnel face. The wells can either be deep or vacuum wells. The latter only has a maximum lift of 6 metres, and thus is normally associated with shaft sinking where it can be used in stages.

Deep wells are also associated with shaft sinking, but are more expensive to install as they require drilling in contrast to jetting.

To be effective in tunnel construction the wells have to extend below tunnel invert to give the borehole pumps sufficient room to extract the ground water. The spacing of these wells longitudinally is also important.

As most tunnels are driven within an urban environment the associated problems with using this system limit its application.

Deep and ejector wells can be used, not only for dewatering, but for depressurising strata to reduce base heave problems in shafts.

8.5.4   CHEMICAL/GROUT INJECTION

This process is applicable mainly to granular materials when others are eliminated through environmental constraints (e.g. settlement), cost and access. One advantage of this method is that it is capable of being undertaken from the tunnel face.

The injection takes place through tubes which are drilled or pushed into the required treatment zone at predesigned centres. The most common form of injection pipe is known as tube-à-manchette, which allows injection at predetermined points along the pipe by the use of internal packers.

The injection medium has the advantage of consolidating the ground conditions besides excluding the ground conditions. Thus it is important to understand the extant soil condition prior to determining an injection regime.

Dependent upon the void ratio in the granular strata the injection process is normally in two stages – the first using a grout to fill the majority of the void followed by the more expensive chemical injection to fill the smaller voids.

This is an expensive process and its success is dependent on the vagaries of the ground.

8.5.5   GROUND FREEZING

This process is the most positive of all the treatments and offers a greater degree of success than the other methods. However, it is a costly process to install and run and, therefore, its application must be assessed through a risk assessment.

Artificial freezing of the ground is performed by inducing ice growth around predrilled freeze pipes (either vertically or horizontally) for shafts and tunnels respectively until the ice growth is sufficient to form a continuous structure. The ice wall is normally designed to be a structure having the ability to support external loads. Therefore it only needs to be applied to the periphery of the structure as long as sufficient cut-off is allowed for.

The ground water temperature is lowered to below freezing by pumping brine or liquid nitrogen at temperatures of approximately – 30°C through the freeze tubes. The latter takes less time to take effect but is more expensive. A brine freeze will take up to eight weeks to take effect after completion of drilling.

With subzero temperatures, special precautions must be taken for the safety of operatives.

Care should be taken with the insulation of the freeze wall and design of the lining which can be cast against the ice wall.

8.5.6   CONCLUSION

The guarantee of the success of water management is proportional to the establishment and the application cost which can be a high proportion of the project cost.

Before adopting one of the above methods, alternative solutions should be investigated, e.g.:

- For tunnels – earth pressure balance or slurry machines
- For shafts – sheet piled cofferdams, bored pile, or diaphragm wall solution.

8.6   Mechanism of the tunnelling process

8.6.1   INTRODUCTION

Soft ground tunnelling has a definite beginning certainly in Northern Europe. In the early 1840s a French engineer resident in England called Marc Isambard Brunel invented what he called his tunnelling shield, after failed attempts by others to dig a subaqueous tunnel under the Thames through London clay with ground and water difficulties. It consisted of a series of boxes about 1.8 m × 1.2 m that divided up the tunnel face. A miner was established in each box and excavated forward 6″ by removing one plank at a time while supporting the ground by tightening screwjacks. As the shield moved forward 6″ the masons filled the gap behind with a brick arch.

The completion of this tunnel at Wapping in 1843 was not without its disastrous experiences to such an extent that finding a regular contractor to undertake construction of the Tower Subway could not be found.

James Henry Greathead successfully tendered for the construction which was begun early in 1869. It was during this contract that a new form of cylindrical shield was developed which proved to be the nucleus of the mechanisation of the tunnelling process.

The shield was circular steel tube with a strengthened bulk head which was shoved forward by six $2^{1}/_{2}″$ jacks worked by men inside the shield. The tunnel lining of cast iron was composed of 18″ wide rings, each consisting of three segments and a key piece bolted together which was erected manually. During the 1870s Greathead further developed the shield to solve such problems as tunnelling in sands below the water table with the diaphragm shield (1874). This shield being a prototype of a much more recently developed slurry machine in the way in which it removed the spoil from the face and a flood door mechanism for sealing the face. Further developments of the shield such as hydraulic jacks and the use of grout pans to improve the effectiveness of filling the cavity between the lining and the ground are examples of a period in time where ideas and concepts were translated into working engineering systems which form the basis of the mechanised tunnelling process of today.

8.6.2   GREATHEAD/OPEN SHIELDS

Open shields can be used whenever the ground is stable enough to maintain the face, but where ground conditions are very loose or water bearing then some form of face support and/or ground treatment may be necessary. Face support is normally achieved by the use of timber supports or more recently breasting plates. However, if water is present some form of ground treatment may be necessary. During the late 1800s the technique of using low pressure compressed air as a means of treatment, by pressuring the whole tunnel in order to balance the water pressure, was developed. The same technique is still in use today, but there are various health hazards and a recent revision of the work in Compressed Air Special Regulations 1958 will introduce constraints for its use in certain applications.

Another technique for supporting ground while operating a hand shield consists of face trays and sand tables which use the internal friction angle of granular materials such as dry sands as the basis of offering support to the face.

With a few exceptions, up until 1940, virtually all soft ground tunnels were hand excavated using miners working within variations of the open shield, breaking down the face with pneumatic clay spades and loading by hand into waiting skips. An effort to increase productivity and thus reduce tunnelling costs along with the increasing need to construct tunnels in increasingly challenging conditions has meant an inevitable drive towards mechanising the tunnelling process.

8.6.3   BACKACTER/CUTTERBOOM

Since 1950, various methods have been used for mechanising a basic shield including backhoe excavators and drum differ type rotary cutting machines, used first in the London Underground in the 1930s and further developed during the construction of the Victoria Line in the 1960s. In the early 1970s a development in shield mechanisation occurred which was taken from the mining industry – the road header boom mounted within a tunnelling shield. However, these machines are essentially open-faced shields with a mechanical means of excavation which may affect its ability to deal with certain ground conditions. The backacter type is suitable in semi-stable to stable soils up to strong cohesion values while the cutterboom is more suitable in higher strength soils marls and some weaker rock types.

8.6.4   SLURRY MACHINES

Up to quite recently it was extremely difficult, costly and risky to consider constructing a tunnel in water bearing granular and silt materials.

A leap forward in soft ground tunnelling technology during the last 20 years has been the development of what are known generally as closed face tunnelling machines. The concept is, by many, attributed to John Bartlett who in the 1960s was looking for a method of reducing the dependence on compressed air. The original idea became known as the bentonite slurry shield, as it had a closed compartment, plenum chamber at the front of the machine filled with a slurry made from bentonite and water. Pressure is kept on the fluid in the closed face of the machine to balance the ground water pressure and the cutting wheel excavates the ground at the same time, mining the spoil in the face with the fluid which is then pumped out of the machine. A surface plant removes the spoil from the bentonite slurry which is then recirculated. This system works particularly well in soil and gravel but less well in the benign clays. After much experimentation in the early 1970s development was continued in Japan where the need to construct many miles of main sewers in the water-bearing granular soils to Tokyo Bay and other coastal conurbations provided the impetus.

8.6.5 EARTH PRESSURE BALANCE MACHINE (EPBM)

In certain ground conditions and in areas where there was an environmental probem in disposing of the bentonite contaminated material slurry machines were thought by many to be not suitable. A subsequent Japanese development was a variation on the slurry shield where instead of introducing bentonite, these shields use the excavated material to provide pressure to contain the ground face load. This is achieved by the material from the tunnel face and is broken down by cutters into a chamber where it is kept under pressure via the rate of passage of excavated material through a balanced auger or pressure relieving valves on a screw conveyor. Water and other conditioning agents such as foams and polymers can be added at some considerable cost, if necessary, and the spoil once conditioned can then be extracted through a screw conveyor or belt conveyor into a tunnel transport system within the tunnel. The foams are generally used to lubricate sticky clays to prevent clogging in the chamber and the polymers act as a dewatering agent in wet conditions.

The earth pressure balance shield has proven to be highly successful in poor and variable ground conditions such as high permeability, low cohesion, high ground water pressures and also in benign clays with boulders with sand and gravel interfaces, i.e., Fylde Coastal Waters Interceptor Sewer. One of the main advantages of an EPBM, apart from its ability to deal with variable ground conditions, is also that it uses conventional rail-mounted transport systems.

A variation of a TBM is one which contains a compressed air chamber which can be used to balance the water pressure in the excavation chamber, either during excavation or during maintenance/repair of the TBM.

8.6.6 MICROTUNNELLING

The earth pressure balance and slurry techniques have also been scaled down in size to form microtunnelling machines, which are steerable, remote control pipejacks that are able to install pipes of internal diameter less than that permissible for man entry (900 mm) with the minimum diameter generally in use being about 400 mm. They are normally remote controlled and the system is steered from a control cabin on the surface by the use of active target systems or close circuit television (CCTV). These machines are more associated with pipejacking which is discussed in Chapter 9. However, equipment capable of installing 100 mm–200 mm internal diameter pipes over a length of 30 m from within a 2 m diameter shaft have been developed in Europe and used elsewhere and include directional drilling, augering and impact moling. All of these techniques enhance the minimum disruptive benefits of pipejacking systems generally and offer the following benefits:

1. Health and safety benefits: The highly mechanical nature of the systems ensures that manpower requirements are very low, direct contact with the ground is minimised and the remote operation ensures that the influence of restrictive legislation is all but eliminated.
2. Social cost benefits: As a result of the factors mentioned in (1) social factors such as noise, dust, heat, pollution and surface disturbance are minimised.
3. 'Short-cut' pipeline routes in comparison with open-cut minimise the disruption to services (e.g., gas, water, cables).
4. No problems related to back fill material in comparison with open-cut.
5. Exceptionally small launch pit which minimises intrusion when compared to open-cut or other means of tunnelling.

6. Potential for utilising microtunnelling techniques and equipment for larger tunnels.
   Benefits
   - Reduced manpower
   - Cost reduction
   - Less constrained by legislation
   - Less affected by ground conditions
   - Less environmentally intrusive.

   Deficits
   - Perceived limitation of drive lengths
   - High power consumption – particularly slurry pumps
   - Environmental impact of slurry disposal
   - Cost of spoil treatment.

   Notwithstanding the deficits, the best aspects of microtunnelling in terms of its remote operation and semi-automation can only offer benefits to the tunnelling process.

7. Development of the microtunnelling system (by Iseki pipe replacer) to replace existing sewer pipes while still live. With our ageing sewer system, non-disruptive systems of pipe replacement must be the key to maintaining existing sewer systems.

8.6.7   TUNNEL BORING MACHINES (TBM)

Tunnel boring machines are probably the most common mechanical face excavation method in use in varying face conditions. The support to the face is achieved either by the full cutting head, full face or by the cutter arms only.

Tunnel boring began in the latter part of the nineteenth century with limited success. The imagination of scientists and engineers was too advanced for the technology required to turn concepts into practical reality.

A second generation of tunnel boring started in North America during the 1950s continuing into the 1960s with considerable success in soft rocks. The use of tunnel boring has increased during the 1980s and 1990s to the extent that more than 200 km of tunnels are constructed each year by new and used TBMs worldwide (Fig. 8.10).

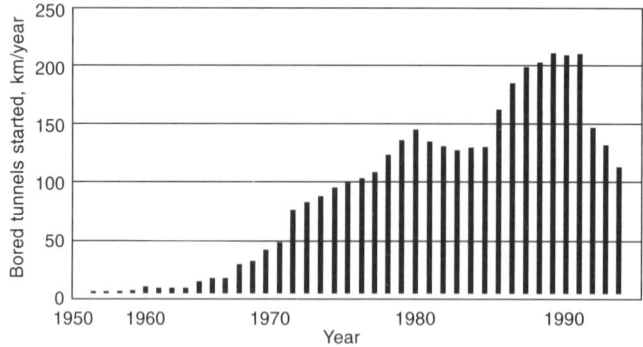

Figure 8.10   TBM tunnels built annually.

Technological advances of the 1980s encouraged an increased worldwide use of TBMs even in labour-rich countries. International TBM contractors started to compete with local non-mechanical and drill and blast excavation to such an extent that tunnel boring currently accounts for over 90% of all civil tunnel excavation in North America and an ever-increasing proportion of tunnels worldwide.

Rock hardness greater than 400 Mpa has been successfully bored in various parts of the world in the past 10–15 years which has significantly expanded the scope of available works. Hard rock boring has become more economical as a result of these increases in technology which have in turn improved such factors as:

- Increased cutter diameter (by a factor of 1.5)
- Cutter head gauge velocities (by a factor of 2)
- Cutter load capacity (by a factor of up to 9)
- Improvement in cutter geometry and metallurgy.

A recent comparison between TBMs with drill and blast excavation (after Tarkoy, 1995) suggests that TBMs have an advantage in the following ways:

1. Performance: Excavation rates 4–6 times higher than drill and blast; overbreak – reduced cost of delivered material and labour to fill (typically drill and blast minimum of 10% overbreak up to 25%)
2. Costs: Reduced support requirement; reduced temporary construction structures; reduced crews and skill levels
3. Safety: According to a US Bureau of Reclamation study during the 1960s and 1970s the average accident rate for TBM excavation is nearly half that of drill and blast; however, this has more recently been contradicted by Bevan and Parkes (1991) who suggested that mechanical excavation is more prone to accidents because of the confined spaces around TBMs.

8.7 Surveying/alignment

A tunnel recently excavated in the area where archaeologists suspect Babylon was located found many unexplained shafts. Various theories were proposed to try to determine the reason for these shafts. These included the possibility of sewer systems and water supplies or maybe that the shafts (which appeared to be raised from underneath) were because the Old Testament miners kept coming up to see where they were, as did the eighteenth century canal tunnellers.

Engineering surveying is generally required to ensure that the civil engineering works are in the corect place and of the correct size. It has been said with some justification that correcting differences between as constructed positions and those acceptable to the client can be the most costly activity on a project. This is particularly so with tunnels especially when the surrounding ground is unstable and water bearing.

8.7.1 ALIGNMENT AND LEVELLING

Control survey

Access to construct tunnels is usually via vertical shafts at or near the ends with possibly some intermediary. Survey stations are required convenient to each shaft which can be accurately related to each other and the control survey.

The relative positions of shaft centres are usually established by traversing as illustrated in Fig. 8.11 – a closed traverse is preferable, but as tunnel surveys are frequently long and narrow this is not always practical. It is often convenient to relate the survey to the national grid. In any case, a co-ordinate grid is necessary if the tunnel is other than a straight line between two intervisible shafts. If an orbiting grid is used, it is essential to ensure that the origin and orientation are chosen so that all the co-ordinates are positive. It is sometimes possible to obtain national grid co-ordinates of ordnance survey points convenient to the ends of the traverse and use these as the closure. With the advent of the Global Positioning System (GPS), it is now possible to use an accurate alternative to traversing if the scale of the tunnel works is such as to justify the cost and the required accuracy is achievable.

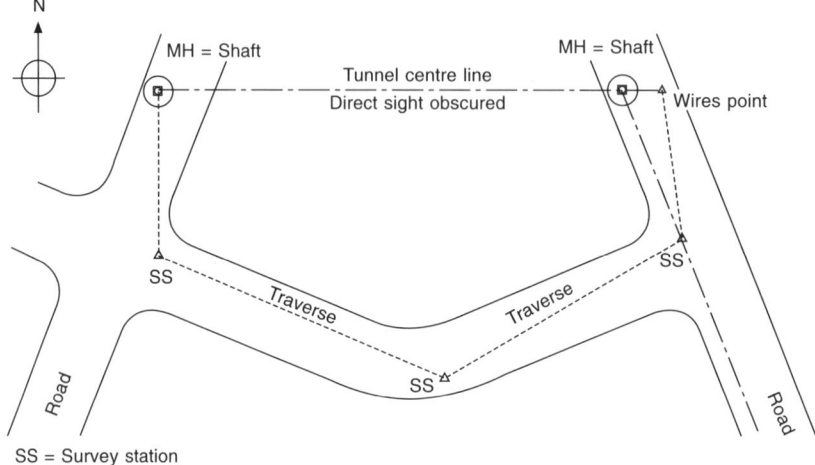

Figure 8.11   Illustration of control survey.

The position of a tunnel relative to surface features is usually given on layout drawings which should show the position and direction of straights, intersection points between straights and the location of access shafts. It is necessary to transfer these onto the ground so that the survey can be made. Intersection points have a habit of being situated in inaccessible positions. In this case sufficient points must be set out for the co-ordinates of the intersection points to be calculated. The centres of shafts should be set out and included in the survey, although they will disappear the minute the shaft excavation starts.

As soon as possible after compiling the survey a 'wires point' must be set on the line of each drive out of each shaft preferably not more than a shaft diameter from the side of the shaft. The connection from the wires point to the main survey must be as long a line as possible since this line will control the accuracy of the drive. Clearly this wires point must be as stable as possible and connected to the main survey by direct measurements.

Shaft sinking

Access shafts can be related to the line of drive in one of three ways. The shaft centre can be:

1. On the line of drive
2. Offset from the line of drive
3. Offset from the line of drive by so much that an addit is required to reach the line of drive.

For all these situations it is necessary to control the position, orientation, level and plumb of rings accurately. In many cases the 'eye' or breakout shaft suppport structure design calls for a particular ring to be at a specific level and orientation.

Whether a shaft is to be sunk by underpinning or as a caisson the initial requirements are similar. That is, a hole is required some 300 mm larger in diameter than the outside of the rings to be built and 1500 mm or 2000 mm deep, depending on ground conditions. A centre peg is required in the bottom of this hole so that the first ring can be built as accurately as possible.

Before excavation starts, four pegs should be driven clear of the proposed excavation such that strong lines between points intersect at the shaft centre, roughly at right angles (Fig. 8.12 – stage 1). It is advisable to extend these lines to secure points so that they can be re-established if disturbed. When the hole has been excavated the centre peg can be set out by plumbing down from the intersection of these two lines (stage 2). A separate level peg, a convenient distance above/below the level of the bottom of the first ring, should be set out at the same time. From this peg can be set support blocks for building the first ring. At this time the position of the centre of the key segment is also required (stage 3).

The first ring is very important as it sets the standard for the rest of the shaft. Once it has been set to the allowed tolerance it should be securely locked in position while sufficient rings are added on top to project above ground. Setting out for different types of shaft sinking methods differs from this point on.

Breaking out

Five things are needed for controlling a successful breakout from shafts:

1. The centre of the tunnel marked on the shaft wall
2. The direction of the tunnel opening
3. 'Square marks' to ensure the correct orientation of the first tunnel rings set
4. Some means of ensuring that the gradient of the tunnel centre line follows the required vertical profile
5. 'Look-up' or 'overhang' for the setting of the first rings to ensure correct gradient.

Items (1) and (2) require the establishment of a vertical plane on the line of drive. Initially this can be done by fastening a string line between points set to the top of the shaft and plumbing down the shaft at each end of the line (Fig. 8.14). By extending the line of these plumbs, short vertical lines 'a' can be established near the axis level at the back of the shaft and on the drive side. In addition, a line 'b' above the scope of the opening will be needed to re-establish the line of drive after the breakout. The appropriate axis level 'e' can then be marked on the two short vertical lines. At this stage provisions should be made for a tight line to be stretched between these two axis points, from which the opening, if circular, can be marked on the shaft lining by trammelling.

Item 3: It is important to set the first ring(s) so that the plane of the ring is at right angles to the line of drive. A point is fixed each side of the shaft such that a line stretched between them is at right angles to the line of drive (L & R). It is advisable for these square marks to be at axis level and most convenient if it passes near the shaft centre (Fig. 8.14).

Item 4 is needed in order to re-establish the gradient once the breakout is started, C_2 having been destroyed (Fig. 8.14). Points are established each side of and clear of the opening axis level GL and GR. A tight line between these two points will guide the level of a line from

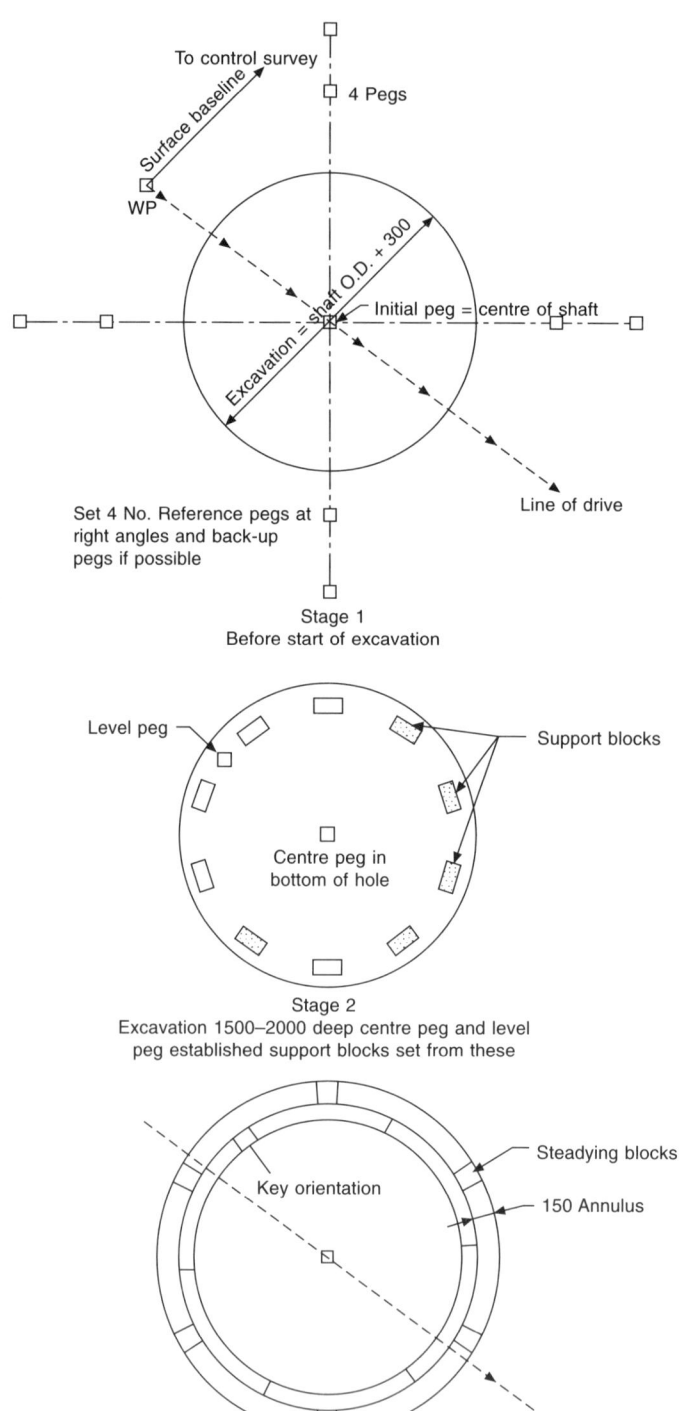

To control survey

4 Pegs

Surface baseline

WP

Excavation = shaft O.D. + 300

Initial peg = centre of shaft

Line of drive

Set 4 No. Reference pegs at
right angles and back-up
pegs if possible

Stage 1
Before start of excavation

Level peg

Support blocks

Centre peg in
bottom of hole

Stage 2
Excavation 1500–2000 deep centre peg and level
peg established support blocks set from these

Steadying blocks

Key orientation

150 Annulus

Line of drive

Stage 3
First ring built

Figure 8.12

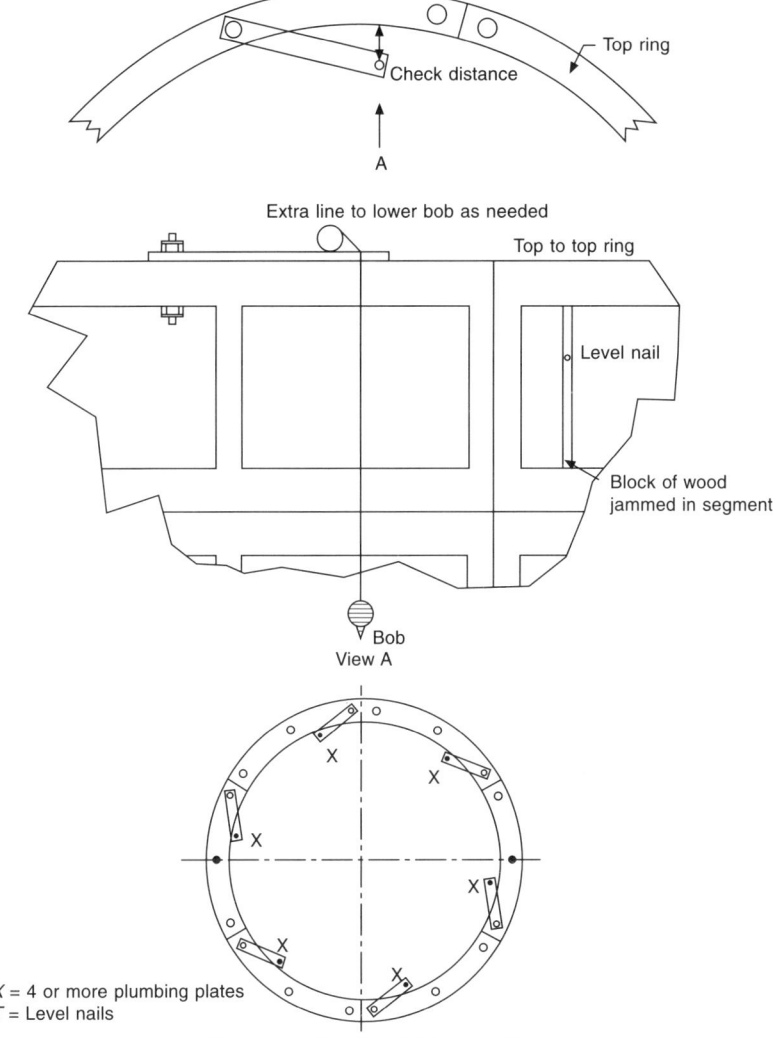

Extra line to lower bob as needed

Top ring

Check distance

A

Top to top ring

Level nail

Block of wood
jammed in segment

Bob

View A

X = 4 or more plumbing plates
T = Level nails

Plan of top of shaft as it is being built

Figure 8.13 Shaft sinking and level nails for underpinning.

the back of the pit (C_1). For this line to be of maximum use the grade line needs to be as far as practicable from the back of the shaft. It is advisable for the grade line to be set square to the line of drive. This can be done from the square marks already positioned.

Item 5 (Fig. 8.15): If the tunnel is to be on a rising gradient the first rings must be set with 'look-up'. The amount of look-up can be determined readily from the tunnel gradient given and the diameter of the tunnel. These first rings are built resting on a support in the invert set firmly to level and gradient. Once they are completed and bolted up, look-up can be checked and, if necessary, adjusted. Falling gradient and 'overhang' are treated similarly.

Much of the foregoing setting out can be done by experienced engineers using a theodolite with a diagonal eyepiece instead of plumb lines and string; however, it is of doubtful accuracy where the depth of the shaft exceeds four times the diameter.

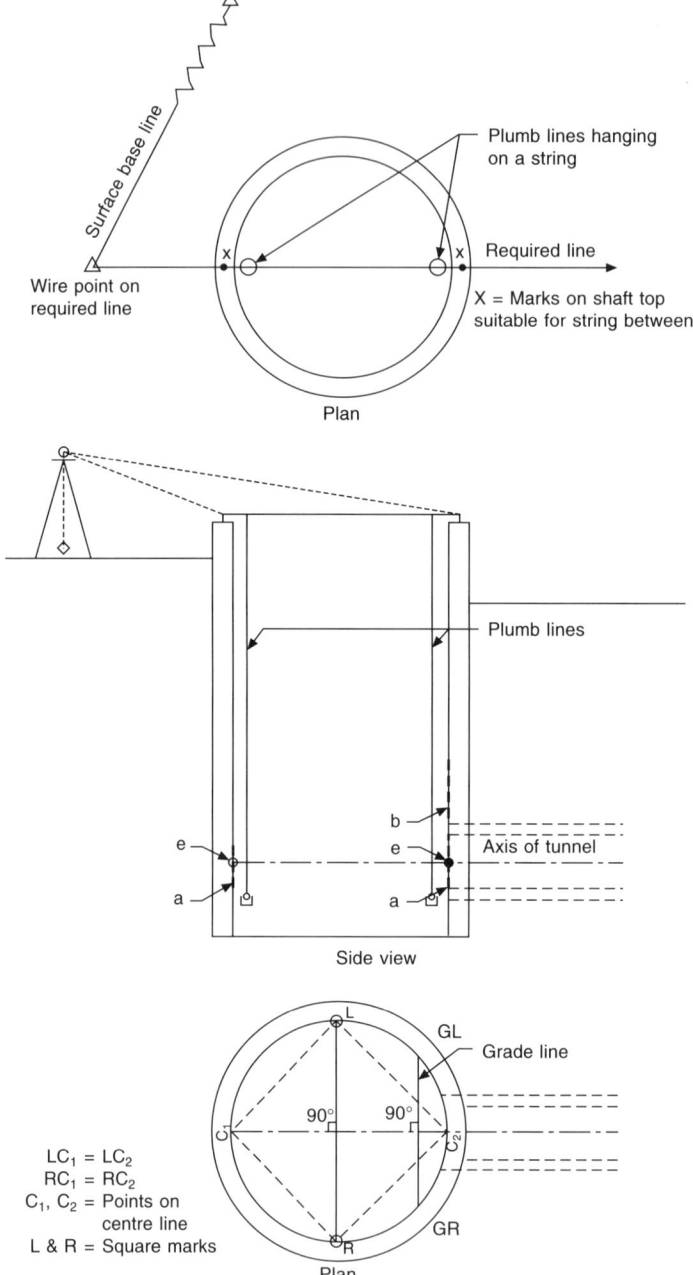

Figure 8.14   Breakout.

Breakout – shield driving: In this case a cradle is built in the shaft bottom to the outside diameter of the shield after the side of the shaft has been broken out. The shield is then lowered onto the cradle which is fixed to the required breakout markings. The position is fixed quite simply by axis marks or points at the back of the shaft and at the breakout and a tight string between them represents the axis of the shield.

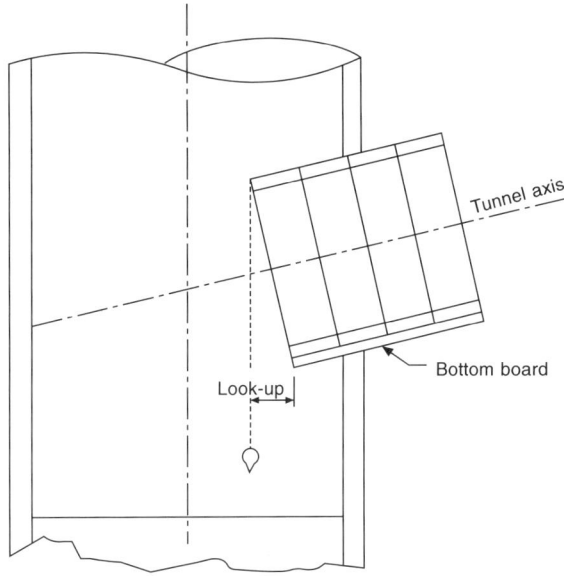

Figure 8.15 Establishing the grade

Co-planing

Once the breakout and a few metres of tunnel are completed by 'string' methods, a more accurate direction control is required using coplaning.

The theoretical sequence for coplaning to establish direction underground is as follows:

1. At the surface two plumb bobs are set in the shaft as far apart as is practical, on the line of the drive. They are lowered so that a theodolite set in the tunnel can observe them.
2. Underground a theodolite is set in line with the two plumb bobs.
3. The line between bobs and the theodolite is transferred forward either by transit or turning off 180°. The bearing obtained by this process is established by roof stations placed on this line for further tunnel driving.

Again much of this work can be done by experienced engineers using a theodolite and diagonal eyepiece; however, if a tunnel is of substantial length or complicated by circular or transition curves the aforementioned process is more reliable (Fig. 8.16).

8.7.2   UNDERGROUND CONTROL SURVEY – WEISBACH

The establishment of direction underground is normally in two stages if the length of the drive is substantial. The first stage is a 'setting-out' process by coplaning. If a drive is short it is usually sufficient to use coplaning processes to achieve a satisfactory junction. If the drive is longer than can be satisfactorily achieved by coplaning, a second stage based on surveying processes as against setting-out processes is employed. The basis of this surveying method can best be described as the *Weisback process*.

The mathematics of the process depends on triangles having two very small angles which are proportional to the lengths of the opposite sides within practical limits. One of the small angles is measured, the other can be deduced from this value and the lengths of the sides (Fig. 8.17).

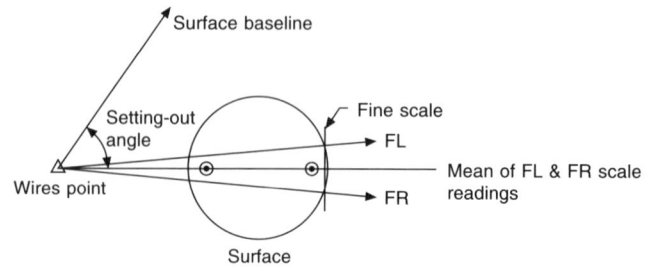

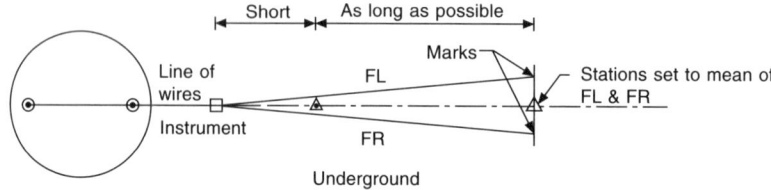

Figure 8.16   Coplaning diagrams.

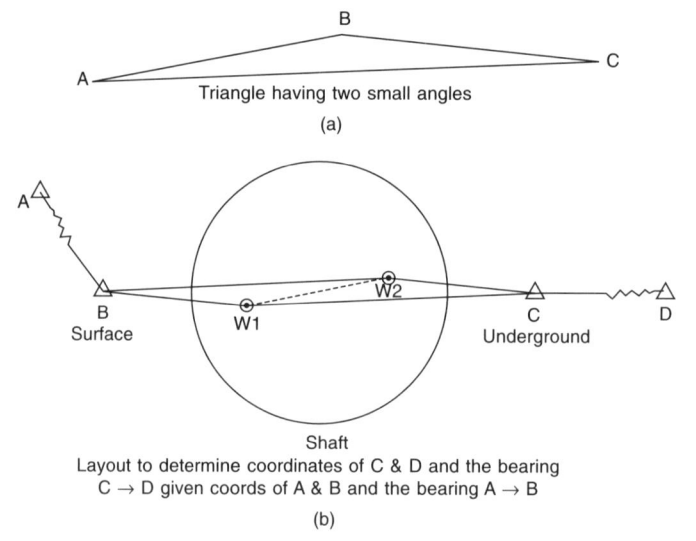

Figure 8.17   Weisbach diagrams.

From Fig. 8.18, in the triangle shown, if angle BAC and lengths AB and BC are measured:

   Angle ACB = AB × angle BA

      Let AB = surface baseline

   W1W2 = plumb lines hanging in shaft

      CD = underground baseline

arranged such that AB and CD are as long as practicable and B-W1, W1-W2 and W2-C. If the bearing A → B and the coords of B are known then the bearing W1 → W2 can be formed by measuring angles ABW and W1 B W2.

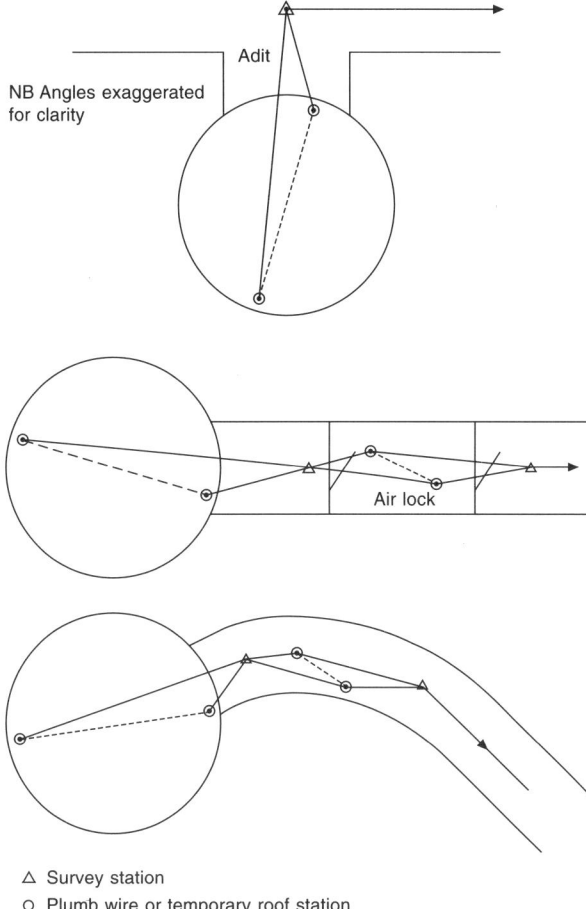

△ Survey station
○ Plumb wire or temporary roof station

Figure 8.18 Application of Weisbach principle.

The following comments are relevant to this operation:

1. If BW is greater than unity, any inaccuracy of the angle W1 B W2 is multipled by W1 W2.
2. Provided the angles between the wires are less than 15 minutes of arc the angles are proportional to the lengths of the two sides to an accuracy of nine significant figures.
3. The error in bearing introduced by setting up inaccuracies is a function of the length of baseline, not the short distances between the instruments and wires.

The initial underground baseline is, if the circumstances are favourable, suitable for extension in a straight line to an adequate length whose bearing can be established to an accuracy relevant to the length of drive. The setting out of changes of direction from this baseline needs follow-up surveying to establish a continuing relationship between as set out and theoretical.

When coordinates of a forward station have been established, the bearing between the station and the next theoretical point should be calculated and the necessary setting-out angle determined. This process needs repeating regularly so that actual positions do not differ significantly from theoretical.

When line of drive is not directly from the access shaft the Weisback principle, as indicated in Fig. 8.19, can be employed to obtain directional control.

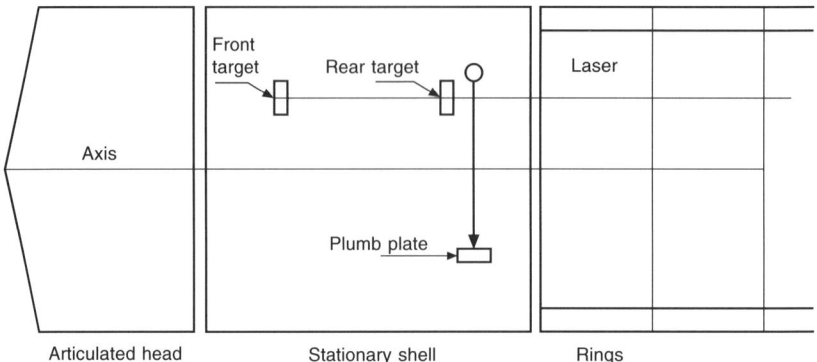

Figure 8.19   TBM line and level control.

8.7.3   ALIGNMENT

Straight drive

The simplest type of tunnel to control is one that goes directly from a shaft and continues in a straight line. Whatever means are used to set the first linings at breakout, it is essential to follow with more alignments as the tunnel is extended. From the heading of these stations the bearings to the next theoretical point are calculated to give the setting-out angle. It is prudent and good practice to reference all survey stations so that they can be positively identified when used. A further advantage of referencing is as a check against a station being damaged or moved. Helium–neon lasers are commonly used to provide a narrow straight beam of light set to a required direction. Plainly, the projector must be fastened to a substantial mounting to reduce the chance of accidental displacement. A further precaution is usually taken by fixing at a distance a target having a hole slightly smaller than the laser beam so that the beam passes through when correctly aligned which ensures continuing accuracy. When lasers are used they are normally seen to give both horizontal and vertical alignment; however, because it is rarely practicable to set the laser on the tunnel axis due to obstructions of sight, a vertical and horizontal offset is needed. When a shield or TBM is being used a target is fixed to it apparently offset so that the line and level position can be checked at all times and that allowance is made for any 'roll' the machine may have at the time of checking.

A tunnel boring machine has three parts:

1. Cutting head: Which is the steering part of the TBM and is articulated by rams located between the head and the main can. Reading dials are situated at the back end of the head so that the articulation can be controlled and monitored.
2. Main stationary can: This is the main body of the TBM within which are two laser targets (front and back) from readings taken on these targets the following information can be obtained.
   - Horizontal position of stationary can
   - Vertical position of stationary can
   - Square and plumb of stationary can.

Table 8.5 Support recommendations for large diameter (6–12 m) tunnels based on RQD after Deere *et al.*

RQD rating	Tunnelling method	Alternative selection of support systems sets	Rockbolts	Shortcrete
Excellent	TBM	None/Occ. LT.S RL: (0–0.2)B	None/Occ.	None/Occ.
	Con	None/Occ. Lt. S RL: (0–0–3) B	None/Occ.	None/Occ. (50–75 mm)
Good	TBM	Occ.Lt.S/Pat 1.5–2 m ctr: RL: (0–0.4) B	Occ/Pat 1.5–2 m ctr	Occ. (50–75 mm)
	Con	Lt.S. 1.5–2 m ctr	Pat 1.5–2 m ctr	
Fair	TBM	Lt/Med. S 1.5–2 m ctr RL: (0–4–1) B	Pat 1.5–2 m ctr	
	Con	Lt/med. S 1–2–1.5 m ctr RL: (0.6.1.3) B	Pat 0.9.15 m ctr	100 mm or more crown and sides
Poor	TBM	Med. CS, 0.9–1.2 m ctr	Pat 0.9–51.5 m ctr	100–150 mm
	Con	Med/Hv sets 0.6–1.2 m ctr RL: (1.3–2) B	Pat 0.6–1.2 m ctr	crown and sides and bolts
Very poor (excluding squeezing ground)	TBM	Med/Hv CS 0.6 m ctr RL: (1.6–2–2) B	Pat 0.6–1.2 m ctr	150 mm or more, whole section; and med set 150 mm or more whole section
	Con	Hv. Cs 0.6 m ctr RL: (1.6–2–2) B	Pat 0.9 m ctr	
Very poor (squeezing or swelling ground)	TBM	V.Hv. CS, 0.6 m ctr RL: up to 75 m	Pat 0.6–0.9 m ctr	150 mm or more, whole section; and heavy sets 150 mm or more, whole section and heavy sets
	Con	V.HvCS 0.6 m ctr RL: up to 75 m	Pat 0.6–0.9 m ctr	

Symbols:
TBM – tunnel boring machine; Con – conventional tunnelling method; None/Occ – none to occasional; Occ/Pat – Occasional to pattern bolting; Pat – specified pattern support system; RL – rock load; B – tunnel width; ctr – centre; Hv.CS – heavy circular sets; Med. CS – medium circular sets: Lts – light set; V – very.

Table 8.6 Rock load classification according to Terzaghi

Class	UCS	Typical rock type
Very soft rock	<25 MPa	Chalk, rock salt
Soft rock	25–50 MPa	Siltstone, schist
Moderate hard rock	50–100 MPa	Sandstone, schist
Hard rock	100–200 MPa	Marble, granite, gneiss
Very hard rock	>200 MPa	Quartzite, gabbro, basalt

Using these readings together with the articulation readings the position of the cutting head can be deduced and remedial action taken if necessary. In addition to the laser targets a plumb plate is located somewhere near the operator's seat so that a constant plumb check can be maintained when the TBM is moving forward. The plumb plate is also used to monitor the 'roll' of the machine which affects the position of the targets relative to the central axis of the TBM.

3. Tailskin: The tailskin is where the segmental rings are built square to the machine to prevent an 'iron bound' situation. Additional square checks on the TBM and the rings can be made using a rotating squaring prism which is held in the laser beam and rotated giving a square plane across the tunnel.

Curves

Tunnels of any lengths usually have changes in direction made necessary by the constraints brought about by the need to avoid underground obstructions and/or sensitive surface features. Such changes can sometimes be made at manhole locations or access shafts, but sudden changes are not always acceptable. The simplest way to change direction smoothly is by a circular curve or, alternatively, with added complexity, transition curves, both of which are beyond the scope of this chapter; however, apart from the theoretical means of achieving the curves, operational consideration must be given to 'creep' – that is, the progress greater than the set length of prefabricated lining units is greater on curves than on straights unless tapered rings are used. Also there is a requirement for a greater number of intersection points and entry and exit transitions into shafts will be affected.

8.8 Temporary works

8.8.1 Introduction

Temporary works are used, almost certainly, on every tunnelling contract. They are an essential, but often neglected, aspect of shaft and tunnel construction.

In modern tunnelling, which can utilise highly sophisticated machinery, temporary works have become an interaction between civil, mechanical and electrical engineers to find the optimum solution. It is important that all these facets come together and, therefore, it is necessary to hold a meeting as early as possible after the award of contract so that the estimator can inform the respective heads of department about the contract and the assumptions he/she has made. This is commonly known as the 'handover' meeting because from thereon the contract mangement team assume responsibility.

It is a common situation in contracting to find the award of a contract delayed by a client for a variety of reasons. Yet on award, the client demands an immediate presence on site and an early start to construction.

This is a very poor state of affairs as it leads to mistakes and, hence, arguments with the client because insufficient time is allocated to the thought process. Paper is cheap and the early delays can be pulled back by good planning and detailed control of the temporary works. It is often a failure of clients to recognise the importance of temporary works and the time required to implement them successfully.

Therefore, the start of the execution of the works should be delayed a sufficient period to suit the contract conditions under which the contract is awarded.

8.8.2 SCHEDULES

On award, the contractor must first plan and programme his proposed sequence of working. This programme is known as the contract programme and is very important contractually as, besides, sequencing, it indicates rates of progress.

As the contract programme only highlights work activities it does not, unless broken down into minor detail, define the associated activities which must come together for a successful outcome. The associated activities include:

- Permanent works design
- Temporary works design
- Procurement of permanent and temporary materials
- Procurement of plant and equipment.

In order to plan and co-ordinate these activities, it is good practice to produce schedules and distribute them to the relevant managers, even if these managers are external to the company.

8.8.3   DESIGN BRIEFS

Temporary works are the responsibility of the contractor who employs specialist engineers as designers. It is normal, other than on major projects, for these to be employed at the head or regional office. This system has the advantage of acting as a centre of knowledge for all solutions which have been implemented on other contracts which probably would not be known to the contract staff.

It is important also to remember in modern mechanised tunnelling that temporary works are not limited to civil design but must, by necessity, to include electrical and mechanical requirements.

To ensure that all these activities come together at the requisite time, it is necessary for the company chief engineer to appoint a temporary works co-ordinator. This person must be, by definition, competent and experienced to undertake the role. On mechanised tunnelling contracts there exists a strong argument to appoint two temporay works co-ordinators – one for civil and structural aspects, the other for mechanical and electrical matters.

No design should be undertaken without a written design brief accompanied by the hazard awareness/risk assessment analysis. It is only through the brief that the designer (civil, mechanical and/or electrical) can undertake a purposeful design. An informal meeting between the temporary works co-ordinator and the designer is recommended as the brief is written. This meeting can be extremely valuable in saving abortive work.

8.8.4   CHECKS/PERMITS

On receipt of a design and risk assessement, the temporary works co-ordinator must ensure that the assumptions made by the designer are implemented through the method statement. The latter should include hold points that should not be passed unless the necessary actions have been taken.

A hold point should be used for the issue of a permit to load by an appointed person, e.g. scaffolding, compressed air-lock doors, electrical circuits.

8.8.5   MONITORING

Having used a system to ensure the adequacy of the design, it is a legal requirement to monitor temporary works and record the results in a register.

All mechanical equipment must be subject to a planned preventative maintenance (PPM) scheme which requires daily, weekly and monthly checks.

To ensure that these checks are being implemented correctly and conscientiously, it is essential that audits are undertaken at regular intervals. These audits should trace an event through the system and raise corrective actions for non-compliance. Auditors should also offer advice.

8.8.6   CONCLUSION

Temporary works have been, and are, an important part of tunnelling. While civil and structural works would dominate a hand-driven tunnel, electrical and mechanical requirements for mechanised tunnels must be recognised and treated with equal importance.

The negative side to undertaking the above is that it requres time and resources. While the conscientious contractor undertakes the above at a cost, clients are not convinced and are still willing to take the lowest tender bid from unscrupulous tenderers at their peril, rather than accept a more expensive but better balanced tender.

It is also important in terms of time and cost to try to combine the temporary works within the permanent works design.

8.9   Future

With ever-expanding populations, urban communities are bound to grow and to maintain reasonable living standards demand infrastructure developments such as sewage disposal. Also as society becomes more environmentally conscious there will be an increasing demand to treat all sewage to a high standard before disposal to water courses.

Accordingly, tunnels are going to be required due to topographical restrictions, particularly in urban areas. However, tunnels have always been dangerous environments in which to work and with machinery becoming more powerful and complicated automation gradually eliminates the need for expensive labour at the face.

The contractors for the Tokyo Harbour project in Japan have successfully implemented a fully automated segment handling system where a segment is not touched from off-loading to erection in its final position.

9

On-line Sewer Replacement in Tunnel

Geoffrey F. Read MSc, CEng, FICE, FIStructE, FCIWEM, FIHT, MILE, FconsE, Fcmi, MAE

9.1 Introduction

In certain instances, after detailed engineering consideration of all the known options for repair and renovation, a decision has to be made to replace the defective sewer.

If the old sewer is relatively shallow, say up to 6 m below ground level, it will probably be economic to replace it by excavating from the surface, although in a busy city centre street the indirect or social costs of such an operation may well be the deciding factor leading to an alternative solution being preferable.

9.2 Procedure – open-cut

Excavation would no doubt proceed utilising either one of the specialised trench support systems, or in certain instances traditional labour-intensive timbering may be necessary. A buried concrete pipeline, complete with its bedding, is considered as a structure and it is necessary to calculate the type of bedding required, with due regard to the crushing strength of the pipes, the depth, the subsoil, the superimposed traffic loading and the width of the trench. Complete information on the subject, including recommendations for laying and testing concrete pipelines, is contained in the booklets issued by the Concrete Pipe Association, the Clay Pipe Development Association, etc. Fig. 9.1 and Table 9.1.

9.3 Heading

In the case of deeper sewers, or where the indirect costs justify the decision, it may be necessary to consider carrying out the work in heading – the method that was often adopted for the original construction of the older sewers.

Headings are practical in firm, cohesive soil or in rock and it is generally recommended that the smallest size of heading for proper and safe working is about 1140 mm clear height by 760 mm width. In all cases of work in heading, great care is necessary with backfilling and consolidation around the pipes.

A pneumatic stowage system is acceptable, but where manual backpacking with crushed stone is employed, each pipe must be surrounded before commencing to lay and joint the next pipe. Lean concrete is often specified for backfilling.

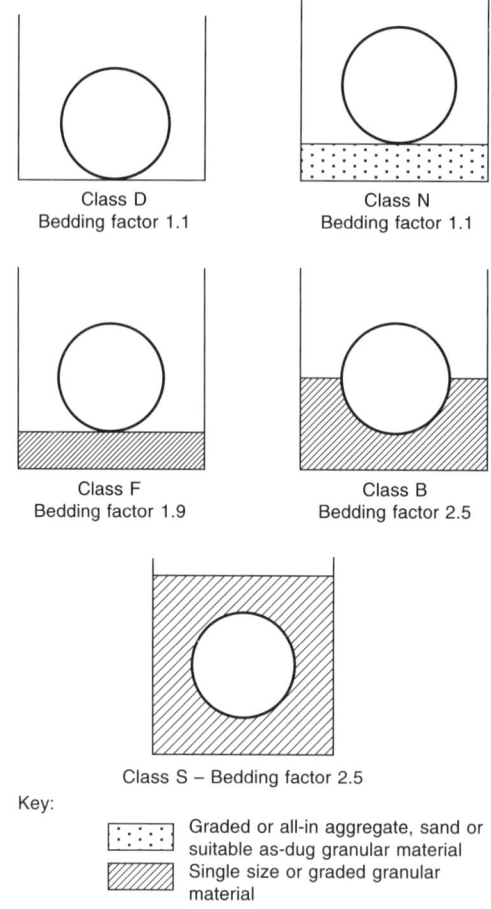

Figure 9.1 shows bedding classes with the following labels:

Class D — Bedding factor 1.1
Class N — Bedding factor 1.1
Class F — Bedding factor 1.9
Class B — Bedding factor 2.5
Class S — Bedding factor 2.5

Key:
Graded or all-in aggregate, sand or suitable as-dug granular material
Single size or graded granular material

Figure 9.1 Bedding factors for pipe bedding classes (Extracts from the specification, design and construction of drainage and sewerage systems using vitrified clay pipes (Clay Pipe Development Association Ltd).

Table 9.1 Strengths, beddings and depths of cover.

Nominal diameter (mm)	Bedding construction class	Crushing strength (kN/m)	Main roads (m)	Fields and gardens (m)
100	D or N	28	0.4–5.7	0.4–6.0
		40	0.4–8.5	0.4–8.6
	F	28	0.4–10.0+	0.4–10.0+
		40	0.4–10.0+	0.4–10.0+
	B or S	28	0.4–10.0+	0.4–10.0+
		40	0.4–10.0+	4.0–10.0+
150	D or N	28	0.7–3.4	0.6–4.0
		40	0.6–5.5	0.6–5.8
	F	28	0.6–6.9	0.6–7.1
		40	0.6–10.0+	0.6–10.0+
	B or S	28	0.6–9.3	0.6–7.1
		40	0.6–10.0+	0.6–10.0+

Table 9.1 Strengths, beddings and depths of cover.

Nominal diameter (mm)	Bedding construction class	Crushing strength (kN/m)	Main roads (m)	Fields and gardens (m)
225	D or N	36	0.9–2.6	0.6–3.5
	F	36	0.6–5.9	0.6–6.2
	B or S	36	0.6–8.0	0.6–8.2
300	D or N	36	–	0.6–2.5
		48	0.8–2.7	0.6–3.5
	F	36	0.6–4.2	0.6–4.6
		48	0.6–6.0	0.6–6.2
	B or S	36	0.6–5.9	0.6–6.2
		48	0.6–8.1	0.6–8.3
375	D or N	36	–	0.9–1.9
		45	–	0.6–2.7
	F	36	0.7–3.2	0.6–3.9
		45	0.6–4.5	0.6–5.0
	B or S	36	0.6–4.8	0.6–5.2
		45	0.6–6.3	0.6–6.6
400	D or N	38	–	0.9–1.8
		48	–	0.6–2.6
		64	0.8–2.9	0.6–3.7
	F	38	0.8–3.0	0.6–3.8
		48	0.6–4.4	0.6–4.9
		64	0.6–6.3	0.6–6.5
	B or S	38	0.6–4.6	0.6–5.1
		48	0.6–6.2	0.6–6.5
		64	0.6–8.5	0.6–8.6
450	D or N	43	–	0.8–1.9
		54	–	0.6–2.7
	F	43	0.7–3.2	0.6–3.9
		54	0.6–4.5	0.6–5.0
	B or S	43	0.6–4.8	0.6–5.2
		54	0.6–6.3	0.6–6.6
500	D or N	48	–	0.8–1.9
		60	–	0.6–2.7
	F	48	0.7–3.2	0.6–3.9
		60	0.6 4.5	0.6 1.9
	B or S	48	0.6–4.8	0.6–5.2
		60	0.6–6.3	0.6–66
600	D or N	48	–	–
		57	–	0.8–1.9
	F	48	1.0–2.1	0.6–3.2
		57	0.7–3.2	0.6–3.9
	B or S	48	0.6–3.8	0.6–4.4
		57	0.6–4.8	0.6–5.2

As evidence of the strength of clay pipes, Table 9.1 shows that, under main traffic roads, DN100 clay pipes of 40 kN/m strength can be laid at up to 8.5 m cover on a trimmed trench bottom and at over 10.0 m cover on a flat bed (Class F) of granular material. DN 600 clay pipes of 57 kN/m strength can be laid at up to 3.2 m depth on a flat bed (Class F) of granular material. It is evident that the majority of clay pipe drains and sewers do not structurally require large amounts of granular bedding material, unlike flexible pipes, which need appropriate surrounding material in order to limit deformation and prevent buckling (BS EN 1295-1 B.2.12.1.4)

Generally, due to the shortage of skilled timbermen, this form of construction is no longer a possibility except in isolated cases.

9.4   Segmental Tunnelling

Traditional tunnelling methods, in which the tunnel is built approximately to the required size of the finished sewer or to a minimum safe construction size, is generally now the standard technique for other than the smaller sewers – the built-up conduit having the dual role of temporary ground support and then permanent pipeline, the method relying on the intrinsic passive resistance of the ground for its stability. Although it is possible to use smaller size tunnels (such as mini-tunnels) the most common size of tunnel for this type of work is the traditional 1520 mm diameter pre-cast reinforced concrete (usually shield driven) bolted segmental tunnel unit with smooth invert, in either free or compressed air.

9.5   Pre-cast concrete tunnel linings

Circular pre-cast concrete segmental linings are nowadays a standard tool of civil engineering tunnellers. The use of pre-formed elements to line tunnels is nothing new – the first linings recorded being brick or stone cut to shape, jointed with cement mortar.

There is little experience of their use in the weak laminated coal measure strata in the United Kingdom but elsewhere there has been considerable use made over many years of pre-cast concrete segments to form the main roadway support systems in mining work. For instance, in Belgium the first concrete linings were formed of rings made up of between 46 and 90 tapered unreinforced concrete blocks up to 500 mm thick and up to 450 mm wide with crushable chipboard packings between 20 and 40 mm in thickness. The internal diameters of the roadways were up to 5.4 m. Although rather slow and expensive, this system has proved very successful and has been used to line 500 km of roadway in the Belgian coalfields. In the United Kingdom cast-iron segmental lining came into use during the Industrial Revolution and proved a strong and reliable system. Its success carried it all over the world together with the related expertise of British tunnelling engineers and miners.

Today it is difficult to envisage in soft ground tunelling, as is usually the case for sewerage schemes, another material that will rival concrete in terms of cost, strength, engineering flexibility or the possibility – where appropriate – of manufacturing on any chosen site.

In the early nineteenth century cast-iron segmental linings were regarded as the ultimate prefabricated lining structures for tunnels excavated in soft ground. Sir Marc Isambard Brunel had anticipated the use of cast-iron segmental linings in an 1818 patent for a circular iron tunnelling shield, though half a century elapsed before both ideas were put into triumphant practice when Greathead built the Tower Subway beneath the Thames in 1869.

It took exceptional circumstances of rapidly increasing costs and shortages of raw materials, at the approach of the Second World War, to provide the impetus for using materials that were cheaper and then more readily available. The idea of substituting reinforced concrete was embodied in a design and patented. A test programme was developed to compare 8′6″ ID cast iron and reinforced concrete linings. Linings of both types were constructed in a tunnel which was loaded to destruction. Both linings failed at the same load but the reinforced concrete segmental linings recovered during unloading. The quantity of reinforcement was apparently at first a matter of guesswork and test rather than calculation and in order to achieve an economic construction of the required strength, the quantity of steel was built up deliberately to the required level.

The first concrete bolted segment tunnel lining was used for the London Central Underground line extension to Ilford in 1937.

It was quickly established that reinforced concrete was a satisfactory substitute for cast-iron segments – a close enough match to allow continuity in systems of excavation, erection and bolting together while providing the necessary longitudinal strength to resist the thrust of shield rams without damage.

After the war, there was a steady development in the use of concrete bolted linings for sewer construction.

The range of diameters grew with demand and a number of odd sizes appeared as a result of sewers still being brick lined. As the use of brickwork lining declined as a result of cost and lack of skilled labour, *in-situ* secondary concrete lining was constructed using ribs and laggings 'in lieu'. Later units with smooth inserts followed and secondary lining was no longer specified.

9.6   Sewer replacement tunnels

As mentioned earlier the most common size for a segmental tunnel for on-line sewer replacement is 1520 mm diameter.

These tunnels are generally oversized from a hydraulic viewpoint, but permit the existing structure to be encompassed within the face excavation and enable side connections to be readily located and satisfactorily re-made.

Where the existing sewer wanders from the tunnel face – which is not uncommon – it is a relatively easy matter to remove side segments and head out to pick up any isolated connections.

On completion full man access to the system then becomes available and because of this improved accessibility the need for manholes at junctions with minor connecting sewers is obviated. Segmental tunnel and shaft linings are grouted by forcing opc/pfa grout (1:7 by mass) through the grouting holes in the segments at least once per shift.

9.7   Access shafts

By modern standards access to the majority of sewers in the UK is deficient in some aspects. Access to the earliest sewers (those constructed before the latter part of the nineteenth century) is often non-existent and access to later sewers is frequently irregular and restricted for personnel use only. Consequently prior to replacing a sewer in tunnel or heading it will usually be necessary to construct appropriate access manholes.

Manholes may be classified by the material used or form of construction adopted. Four main types are used:

- Pre-cast concrete chamber rings
- Brick
- *In-situ* concrete
- Shafts constructed from pre-cast reinforced concrete ribbed segments bolted together.

The availability of pre-cast concrete units and a standardised design allows for economic construction particularly when combined with open trench construction of sewers.

Because of its incremental nature and easily manageable component materials, brick construction is often used for modifications to existing structures or new manholes in existing sewers in urban locations.

In-situ concrete is more likely to be used on open sites for large chambers at relatively shallow depths.

These three types of manhole structures are built from the base upwards. Construction is carrried out with an oversize excavated area which must provide sufficient space on the outside of the structure to enable work to be undertaken externally. Generally constructon is carried out from within a pit, with the earth walls supported by appropriate timbering but nowadays steel sheets supported by hydraulic frames and struts are generally adopted.

Shaft manholes are now stabilised by filling the annulus between them and the surrounding ground with grout – these are constructed from the top downwards and are lined as excavation continues in order to provide continuous ground support.

The minimum shaft size from which heading works can comfortably and safely be undertaken is 2.44 mm and where a shield-driven tunnel is to be constructed a 3.05 m diameter shaft is necessary for entry and removal of a 1.5 m diameter tunnelling shield.

The selection of shaft size should also take into account the degree of certainty ascribed to the sewer records available or the existence of manholes on site. A larger size than that required to accommodate the shield allows a greater 'margin for error' in the incorporation of an 'off-line' sewer at the base, as well as changes to location at surface level to suit other mains. Clearly in busy urban areas there are conflicts of interest. Potential shaft locations, especially those at sewer junctions, frequently coincide with major traffic junctions and plant of other statutory undertakers, so that compromised solutions ultimately seem inevitable.

Shafts formed of pre-cast bolted segments are now the general format for other than shallow sewers (Fig. 9.2). They may be constructed by underpinning or underhanging methods or sunk as caissons (Figs 9.3a–d, 9.4). The prime consideration affecting the method is the type of ground to be excavated. The underpinning method requires ground

Figure 9.2   Typical shaft formed of pre-cast bolted segments.

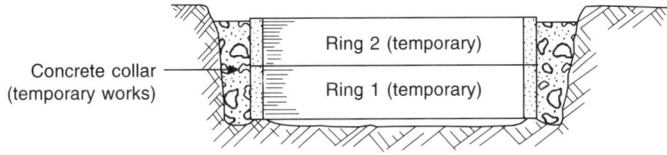

Stage 1. Establishment of support collar

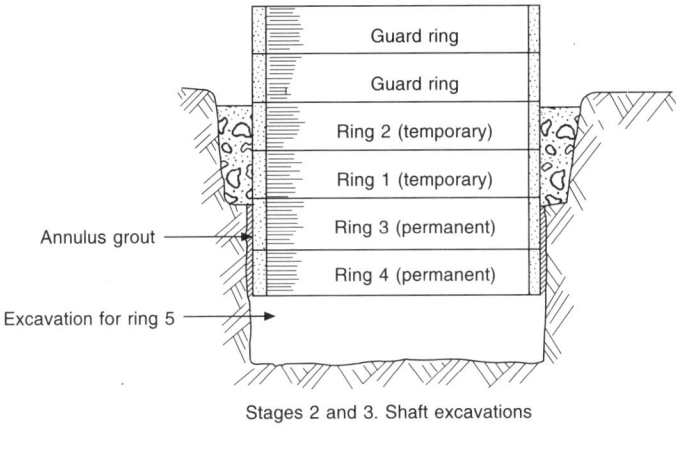

Stages 2 and 3. Shaft excavations

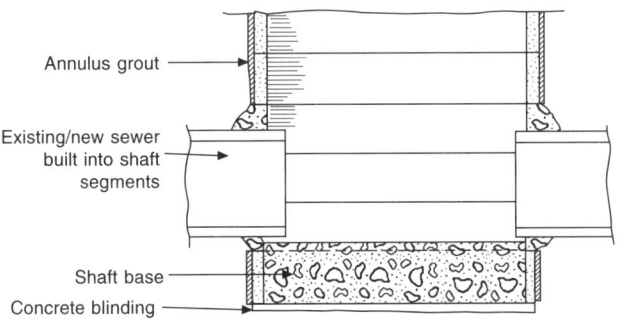

Stages 4 and 5. Shaft base

Figure 9.3 Shaft construction by underpinning.

that is capable of remaining virtually unsupported for a height of at least 700 mm while a 610 mm deep segmental ring is added to the shaft lining. It also requires an absence of ground water or at least conditions where ground water can be controlled either by pumping from within the shaft or by dewatering of the surrounding ground.

Conversion of the shaft is the last stage in the construction. Shafts may be provided with secondary concrete lining (100–150 mm in thickness) to provide additional strength or as a safeguard against water ingress, or for the protection of the bolts when storage or surcharge is anticipated. Alternatively, the segment panel may be infilled with pre-cast concrete 'pellets' to give a smooth finish. Pre-cast concrete landing slabs and roof slabs lessen the time required for construction. Reinstatement of the ground surface requires removal of the top temporary segmental rings which have served as a guard and construction of the small access shaft on the roof slab which is then capped with the manhole cover and frame and finally breaking out part if not all of the temporary concrete collar at ground level.

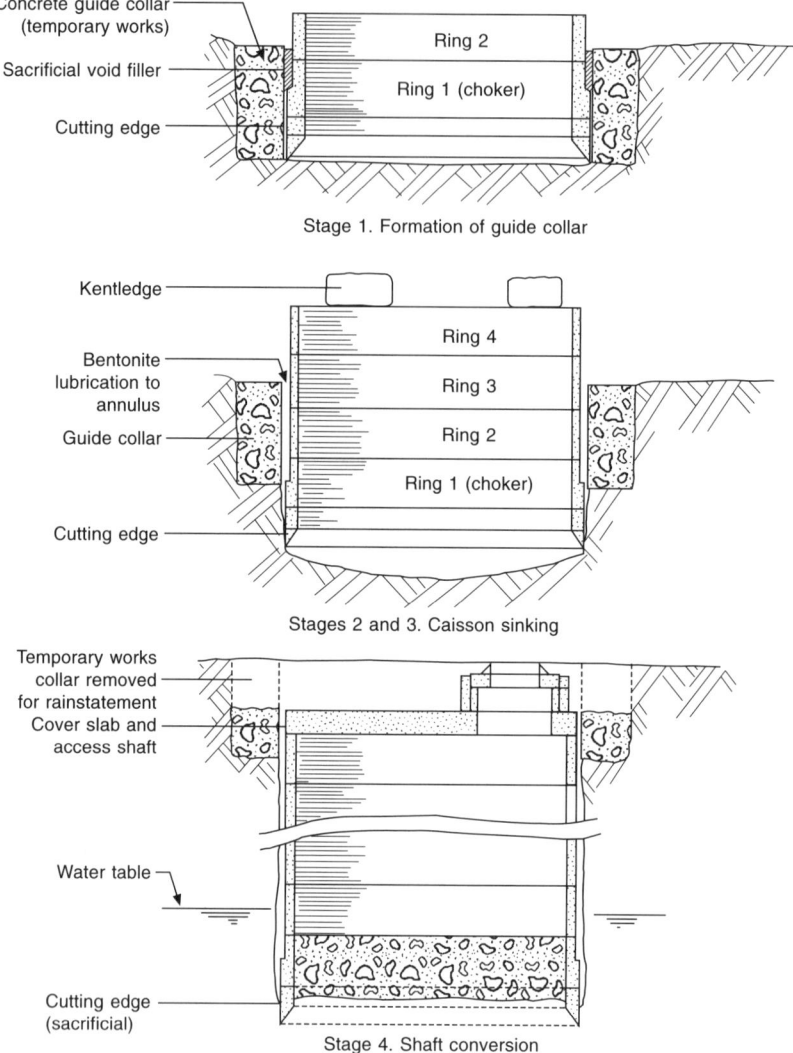

Stage 1. Formation of guide collar

Stages 2 and 3. Caisson sinking

Stage 4. Shaft conversion

Figure 9.4   Shaft construction as a caisson.

Caisson methods are best suited for less stable ground conditions where ground water may also be present. The underpinning method using pre-cast bolted segments is a versatile form of construction. The incremental nature of the method allows variations in soft and hard ground conditions to be accommodated and allows obstructions to be overcome. For example, services and mains running through the shaft (Fig. 9.2) can be supported and maintained, shafts can be constructed around existing inadequate manhole structures – quite a common problem – and existing sewers can be built into the shaft base.

By comparison, caisson construction is almost always restricted to new shafts for new sewers. Existing features cannot easily be accommodated and though there may be scope in some circumstances for using a caisson for shaft construction through difficult ground conditions, where these are encountered as high level strata it is then possible to convert to underpinning methods at a depth to incorporate an existing sewer, although such situations are relatively uncommon.

Construction of a caisson shaft is carried out at ground level. Excavation of the shaft core is made and as spoil is removed the shaft lining sinks under its own weight or, if necessary, further weight (Kentledge) is added. More segmental units are then bolted to the top rungs to extend the caisson as the lining continues to sink below ground level and excavation of the shaft core continues. The lower edge of the first ring is provided with a sacrificial cutting edge and arrangements are made for lubricating the outside face of the rungs with bentonite.

Once sinking is complete and the shaft base constructed, shaft conversion follows – the operation being similar to those described earlier for shaft construction by underpinning. Grouting of the segments is not usually necessary bcause the nature of the ground being excavated precludes the possibility of void formation in the surrounding ground but it may be specified in some circumstances so that bentonite remaining in the annulus is removed by displacement.

Relatively small diameter shafts may also be sunk as caissons constructed from single piece pre-cast concrete rings bolted together by long bolts passing through the full depth of the concrete ring. The first ring consisting of a combined cutting edge/choker unit complete with an integral steel cutting shoe. The sinking of the caisson is generally assisted by using the excavator machine to push down the caisson units once this core has been excavated. At present these shafts are available in four sizes ranging from 2 to 3 m in diameter. Limitation on depth for this form of construction is around 10 m.

Shaft construction by underpinning requires less space and fewer operations to be carried out at ground level and is likely therefore to be better suited to urban locations where working space is limited and general disruption has to be minimised.

9.8 City centre access problems

The earliest sewers with their limited access are generally concentrated in city centres. It is here where the tight network of streets, the repeated cycles of urban development and changing site uses have led to a large number of sewer junctions and connections, duplication of sewerage systems and abandonment of others. But although the need for good access to city centre sewers is clearly apparent, that which presently exists is frequently inadequate.

Physical constraints in city centres combine to make provision of access difficult. Streets are often narrow, buildings large and foundations and basements encroaching below ground into the width of the street require that sewers are deep, while at the same time severely restricting access from the surface.

These constraints of course apply to the other statutory undertakers, but because sewers are generally the deepest and the first laid, access is further constrained by the other services laid across and above.

The final constraint on city centre working is the *raison d'être* of the city itself – people, trade and commerce, manufacturing and service industries, culture, entertainment and shopping, traffic and pedestrians – all the necessary constituents of city life.

An important point to be recognised when considering sewer works in the city centre is that costs are unavoidably high. This applies to all forms of construction activity in such locations; costs may be orders of magnitude higher than in rural areas and several times that of suburban areas. Construction work in cities is restricted by proximity of traffic, buildings and the presence of public utility apparatus, all of which could limit the type of construction plant which may be used. Similarly, the frequency of vehicle movements for material deliveries and spoil removal may be restricted, and, finally, social and environmental constraints may impose limits on working hours. These factors together contribute to lower productivity.

Inevitably compromise has to be sought to satisfy the competing claims of achieving all the objectives of a shaft construction programme in terms of access provisions and longer-term uses for the shafts, against the direct costs which have to be met by the owners of the sewers and the disruption (social costs) which has to be borne by people in the city.

A considerable proportion of the efforts made to reduce disruption has very little effect on direct costs, but can have significant effects on the social costs. These would include:

- A comprehensive study of relevant records to ensure all information for location of manholes has been assessed.
- Careful examination of the records of statutory undertakers to minimise conflict at proposed locations.
- Prior consultation with police, highway authorities, emergency services and safety advisers to ensure all preparations and signing for road closures and traffic orders (including parking restrictions, etc.) have been agreed. Bearing in mind the time required for posting of statutory notices of temporary traffic orders and the consequential delays if the legal requirements are not satisfied exactly, this factor is probably the greatest influence on successful co-ordination of work in city centres and is dealt with more fully in another chapter.
- Detailed programming of works to maximise efficiency and minimise site occupation. The implication here is that time is of the essence, careful planning of the works will reduce disruption, but this may be considered to be an extension of good site management practice, which is desirable in any situation.

An obvious way to reduce disruption further is to carry out work during those periods when fewer people will be inconvenienced. This may entail restricting working to off-peak traffic hours during the day, say after 9.30 am or before 3.30 pm, or both, and ensuring the site is clear to allow uninterrupted traffic flow at other times. A more extreme solution is to permit night-time working only, say after 7.00 pm, together with Sunday working.

Direct costs for this type of working can increase significantly. Fragmented working days suggested for non-peak-hour working are not favoured by contractors since productivity is inhibited. Night-time working, although traditionally associated with tunnelling work and to a lesser extent with shaft sinking, would only achieve daytime productivity rates if continuity of work over periods of whole weeks can be ensured.

However, even night-time working can be disruptive in some city centre locations. Cinemas and theatres are busy until near midnight, restaurants and night-clubs until the early hours. Streets around these places are busy with people and traffic, and sometimes the cessation of daytime parking restrictions means that the proposed night-time working areas are even more congested than during the day. Although temporary traffic orders can be introduced to limit night-time parking, ensuring compliance with orders is even more difficult than for the normal daytime regulations.

Noise, one of the main disruptive features of daytime work, can be equally disruptive at night. Restaurants, theatres and cinemas are noise-sensitive locations and it should not be forgotten that as well as hotel residents, there are other people, such as residents of flats and permanent residents in city centres. Even some hospitals are situated in city centres.

How should the lesser inconvenience of the many be measured against the greater disruption of the few? Guidelines could be expected to cover all situations; however, the importance of the planning stage, which would include advance observations of night-time conditions, will be apparent.

The choice of working method can also influence the level of disruption. The advantages of concrete segmental shaft construction in confining working areas and reducing plant requirements have already been mentioned. The method also has advantages wherever off-peak or night-time working is specified because shafts can be 'plated' over either by using steel beams and sheet steel road plates (placed in sections), or even by adapting the heavy duty, pre-cast concrete shaft cover slab of the permanent works as a temporary cover. In either event, design checks are needed to ensure that the covers are capable of carrying the appropriate traffic loadings and that seatings do not move under the repeated impact of heavy traffic. Attention to the details at the edge of the covers and the matching of road camber are essential for avoiding hazards to cyclists.

The key to the success of using covers is the effectiveness of the initial set-up. In the first session the road surface construction has to be broken out and, ideally, the concrete collar and the two top rings built before the plates are seated. This is more than can comfortably be accomplished in a single night-time session working to a tight schedule, and therefore it is preferable for the night-time session to provide more time. Care and attention at this stage of the operation avoid potential problems the following morning. If all is not well when the road is reopened to traffic, full emergency procedures may be required for a new road closure to apply corrective measures. The consequences for this course of action could be even more disruption than the night-time work was intended to avoid.

A further complication for night-time and Sunday working is obtaining ready-mix concrete supplies for the concrete collar. Other materials can be stockpiled locally during the day and brought to the working area as required. Parking meter bays, temporarily closed off, can provide a suitable daytime storage area for materials and barriers.

While present-day citizens have perhaps come to expect that change and renewal are an integral part of life in a city, there remains an obligation on those responsible for maintaining the working environment to avoid disruption if possible.

What appears to be of most concern to the citizens is the need to reduce the perceived level of disruption in terms of the noise, dust, restriction on access and visual intrusion. Thus subsurface solutions to sewer work that can be confined to easily maintainable compound areas negotiable by pedestrians and traffic are more readily assimilated into city life. The initial confusion when work areas are established in the city centre usually lasts only a day or two and by the end of the first week travel patterns have been reestablished to accommodate them.

9.9 Tunnel construction

Where complete on-line replacement is necessary, the method adopted is the traditional 1520 mm diameter pre-cast reinforced concrete (usually shield driven) bolted segmental tunnel generally with smooth invert constructed on the mean line of the existing sewer, bearing in mind the wandering alignment so common in sewers of this period. The built-up conduit has the dual role of temporary ground support and permanent pipe line, the method relying on the intrinsic passive resistance of the surrounding ground for its stability. Generally the segmental tunnel is constructed within a steel tunnel shield which is moved forward by a number of internal hydraulic jacks and protects the operatives excavating within it (Fig. 9.5). The bolted segmental units are erected within the shield after the jacks have been retracted and the latter then thrust off the newly erected units as the shield is

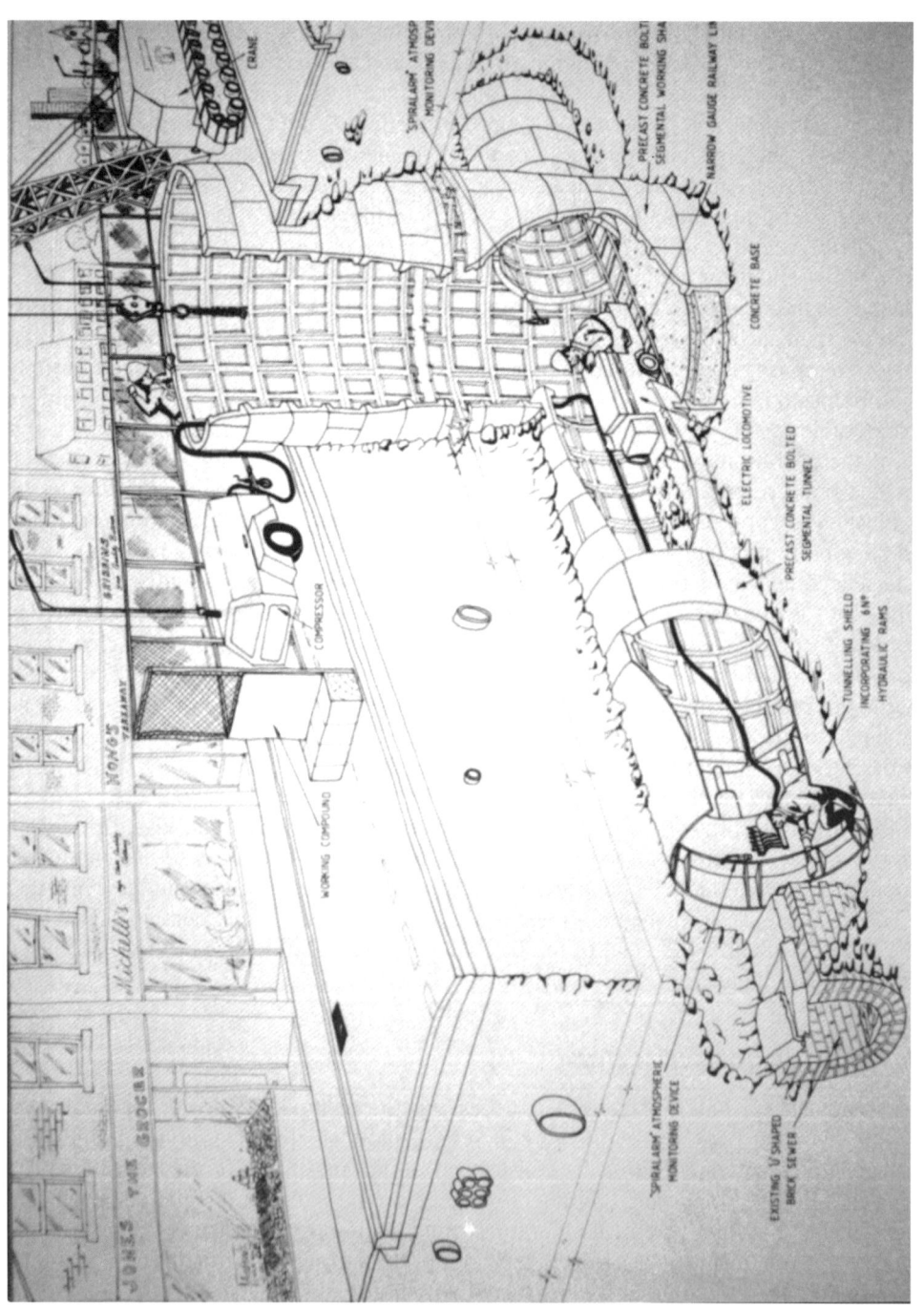

Figure 9.5 Schematic drawing of shield driven tunnelling project.

advanced. The smooth invert units are similar to the standard bolted rings but the lower segments have fewer recessed panels and consequently the invert section ring when built is completely smooth so as not to retard dry weather flow. The annular space outside the new conduit is grouted by forcing cement and pulverised fuel ash (1:7 by mass) grout through the grouting holes in the segments at least once per shift.

Pipejacking (man access) is being used on an increasing scale both in this country and abroad for new construction producing a high quality internal finish with few joints. Using this technique, specially designed thick-walled reinforced concrete pipes can be jacked underground from one jacking station to another behind a steel jacking shield within which manual excavation is carried out – without any necessity for breaking the ground surface between (Fig. 9.6).

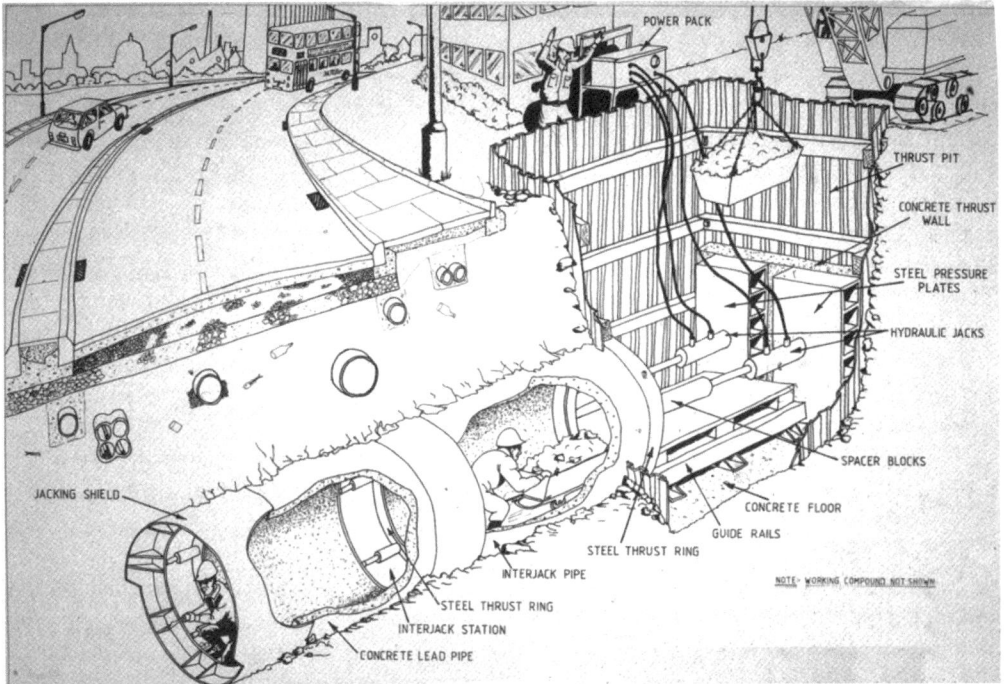

Figure 9.6 Schematic drawing of pipejacking project.

In certain instances this technique has been used for sewer replacement, but where there are a number of lateral connections it is not always economic, in view of the special arrangements which have to be made to deal temporarily with the incoming flow until the new pipeline has been completed and the subsequent reinstatement of connections.

Normal pipejacking is suitable for diameters above 0.9 m (the lower limit for man entry) but a recent development enables smaller diameters to be constructed still involving pipejacking methods but with remote controlled excavation techniques utilising tunnel boring machines (TBMs). This technique is generally referred to as microtunnelling and has successfully involved the use of specially designed concrete pipes. Much of the development of these small bore TBMs has been carried out in Japan, particularly by Iseki.

Iseki equipment is now available to totally replace existing damaged or undersized sewers with new pipelines using 'no-dig' technology; clayware, concrete, asbestos cement, grp and reinforced concrete pipes. The same systems can also carry out trenchless installation of new pipelines.

Iseki's Pipe Replacer is based on their Unclemole TBM and it crushes the existing pipe with an eccentric-motion core crusher. It permits 'on-line' pipe replacement – the sewer flow being pumped through the shield while installation is carried out thus removing the need for overpumping but still leaving the connections to be dealt with separately. Slurry technology permits the replacement or new construction to take place in unstable waterlogged soils. This equipment has a maximum working depth of 30 m and can deal with ground water up to 30 m head pressure.

9.10   Dealing with existing flow

9.10.1   OVERPUMPING

While replacing a sewer on-line the temporary removal of at least some of the sewage normally flowing through it will be required. Consideration of health and safety is of paramount importance in these operations. Some form of diversionary pumping is clearly required and this is generally referred to as overpumping and involves blocking off the flow at a manhole upstream of the length being treated and pumping the sewage 'up and over' to a convenient downstream manhole. An overpumping system must be designed to allow sufficient time for workmen to exit from the system if the flow should suddenly increase for any reason such as there being storms higher up the system, burst water mains or a sudden inflow of industrial effluent.

The importance, both in terms of safety and cost, of the overpumping operations should not be underestimated. Flow rates in sewers can change dramatically in a short space of time, and equipment, materials and lives can be at great risk. The cost of overpumping may well be considerable and should never be minimised when estimating the cost of a project. It is necessary to provide estimates of the dry weather and peak flows in the contract documents. As well as isolating the length of sewers being replaced by overpumping it is necessary to deal with the flow from connections within it. To accommodate this flow as the work proceeds a temporary UPVC pipeline is constructed leading back through the completed section to the downstream manhole where it is discharged into the existing pipeline. Appropriate junction pipes are built in as the various connections are reached.

Local information about the behaviour of any pipe network is invaluable in any overpumping situation. For example, a large-diameter pipe may only run at a small percentage of its capacity if the land use in the area has changed since the pipe was laid. Also the availability of alternative systems into which flow may be diverted may reduce the length of main required and hence the cost. A further consideration, if the opportunity is available, may be to divert the flow into a parallel storm or foul system, returning the flow further downstream.

Expected rates of flow may be derived from flow monitoring exercises over a suitably extensive period of time during which a variety of rainstorms are experienced. Alternatively, flow rates may be calculated using established hydrological methods. The most convenient and reliable methods are now computer based, and there are a variety of sewer flow simulation models which can provide accurate information on anticipated flows at all points in the sewer network for a wide range of possible rainstorms. If such a model is available for the area where the work will be carried out, a very good estimate of flows based on short-term forecasts of meteorological data can be made. Further details of hydraulic modelling in connection with sewer rehabilitation work can be found in Volume I.

Flow information may be conveniently scheduled in the contract as shown in Table 9.2 or covered by more general clauses such as:

> When the drawings or bill items indicate work on, into, or by an existing live sewer the contractor must take all measures necessary to deal with and maintain the flow in such sewers without impediment including the provision of stanks, night and weekend working.
> and
> The peak rate of dry weather flow which the contractor may expect to have to deal with in any length of the proposed pipeline is 'A' l/s.

> In addition the peak rate of storm flow which the proposed pipeline is expected to convey is 'B' l/s. These are estimated rates of flow for a storm with a return period of 'X' years.

Table 9.2

Sewer length reference
Min. overpumping capacity (l/s)
Dry weather flow (l/s)
One year . . . minute storm
Peak flow (l/s)
Time to reach peak flow (min.)

As well as the normal dry weather flow, large industrial discharges should be identified and the nature (content, volume, temperature, etc.) described in the contract documents. Significant flows from laterals should also be indicated as these may cause problems with some renovation techniques.

An alternative contract arrangement would be to state, in the preamble to the documents, that separate items for keeping the works free from flow have not been included. This would then leave the tenderer free to price appropriately in the method-related charges, or to make allowance in the unit rates. Such rates would normally make allowance for any delays caused to the programme as a result of unexpectedly high flows. As these are difficult to predict, this alternative contract arrangement cannot be viewed as totally satisfactory.

As well as dealing with the anticipated flows, it will be necessary to specify the permitted location of the temporary rising mains, particularly in relation to property access and pedestrian and vehicle movement. It may in fact be necessary in some city centre locations to require them to be buried beneath the road surface. In this connection, the method dealing with the suction hoses and small pumps for controlling lateral connections on the length of sewer being treated should also be clearly defined.

It should also be appreciated that when a large sewer is being treated, cases may arise where there is a need for a temporary underground pumping station, complete with all the necessary switchgear and flow control mechanisms.

In most sewer replacement work it is common that all the overpumping pumps, hoses and ancillary equipment will be provided by a specialist pump hire company. Although the main contractor for the work may expect to get specialist advice from the pump supply company, it is clearly important for him to have a sound understanding of the relative advantages and drawbacks of the various different pumping and pipe systems available. It may be that the client or contractor will subcontract the entire overpumping operation to include not only the supply but also the operation of the pumps. However, such an arrangement is comparatively rare, at least in the UK, as the contractor would normally prefer to retain complete control.

9.10.2   PUMP CHOICE

The selection of a suitable pump for a particular job will be dictated not only by the flow rate but also by the suction and delivery heads, the length and diameter of the delivery pipes, the availability of suitable power sources, the access to the sewer, environmental and other legislation (particularly that relating to noise), and the overall cost of hiring and operating the equipment.

9.10.3   SUCTION AND DELIVERY PIPES

While suction hoses are almost universally wire armoured, delivery pipes may be steel, rigid plastic, wire armoured or the flexible lay-flat type. The choice depends on the length of time a main may be in position, the need to bury part or all of it, and the anticipated flow rate and pressure. A further consideration in pipe choice is the type of connection to be used. Smaller diameter hoses often come with quick-release couplings, while larger steel mains often have bolted flanged connectors, which are more troublesome to connect.

For low flows and pressures over short periods, lay-flat hosing is probably the best choice. It is cheap, easy and quick to place and, should vehicular access be required, simple low ramps may be provided over the hose. Lay-flat hose is available in sizes up to approximately 200 mm diameter and, if the pump has to be frequently relocated, the hose is simple to roll up and move. In lay-flat operations it is usual for each pump to have its own hose.

For many operations rigid plastic pipes are the most common choice. The individual lengths are easily handled by two people and, with quick release couplings, they can be quickly connected. The main problem arises in maintaining access to property or in crossing roads. For pipes of up to 75 mm diameter, ramps may be used, but the larger sizes require temporary burial if adequate access is to be maintained without damaging the pipes. If high pressures are to be maintained it may be preferable to use wire-armoured hose, though this is not at all common.

For larger projects with high flows, the provision of a steel pumping main may be the best solution. Several pumps may be connected through a manifold arrangement into one line, thus reducing the total length. As with rigid plastic pipes there is often a problem in maintaining vehicular access, and several excavations may be required to bury the pipe at access points to individual properties or for road crossings. The drop and return bends on the pipeline will result in further head loss and consequential additional pumping costs. In addition the ground disturbance caused by the excavation can give rise to other indirect costs including the disruption of other services, such as water, gas and telecommunications.

Besides the type of hose or pipe to be used, a further consideration is the size of the delivery main, and it is almost invariably false economy to select one that is too small. It should be remembered that, for a given flow rate, reducing the diameter of a pipe by 50% will result in a 32-fold increase in the frictional head loss. The additional cost of a larger diameter pipe will almost always be outweighed by the savings in energy costs resulting from the reduction in the total head.

9.10.4   PRACTICAL CONSIDERATIONS

A major practical consideration is the ease with which the pump unit and associated pipework may be handled. Pumps up to 150 mm nominal size are usually trailer mounted and are easily moved by towing. Pumps above this size are normally skid mounted and require either a free-standing or lorry mounted crane. In addition the larger diameter hoses, especially steel, are heavy and difficult to manoeuvre.

With regard to suction hoses, the amount of space available in the suction manhole may preclude the use of large diameter hoses. To cope with a large flow it may be necessary to excavate onto the sewer to provide a larger opening to allow the larger suction hoses into the sewer. In this case provision of several smaller pumps connected through a manifold to the delivery pipe may be a preferable alternative.

A suitable reservoir for the pumps to extract from and to prevent water from passing into the working area will be required. The most common method is to provide a stank, or small temporary dam, to hold back the flow. For low flows on smaller jobs this may simply be a sandbag wall about half the depth of the sewer. On larger diameter sewers it may be necessary to make special provision to hold back the flow by means of a brick dam or timbers seated into channel sections bolted to the sewer wall. If 24-hour overpumping is not required provision must be made to release the flow.

Bibliography

1. The specification, design and construction of drainage and sewerage systems using vitrified clay pipes – Clay Pipe Development Association Ltd.
2. Design tables for determining the bedding construction of vitrified clay pipe lines – Clay Pipe Developments Association Ltd.
3. Loads on buried pipe lines in trench – simplified tables – The Concrete Pipe Association.

10

Pipejacking

Malcolm Chappell BSc (Hons), CEng, FICE, MIMM

Eur Ing Derek Parkin BSc (Hons), MPhil, MBA, CEng, MIMechE, MIMM, MIMgt

10.1 Introduction

Pipejacking has been used for many years but its increasing worldwide popularity and use over the last decade has meant it has increased its UK market share from 10% in 1985 to 30% in 1990 in preference to trenching or segmental tunnelling methods. The major difference between pipejacking and segmentally lined tunnels is the latter requires erection of the lining in the tunnel, normally behind the shield, whereas the former is installed in the pit bottom.

Pipejacking is basically a construction technique used to install pipes into a near horizontally excavated hole. It is considered to be a 'minimum disruption' method in terms of its interference with surface traffic, services and the environment. The technique is particularly popular in urban areas to provide pipelines, conduits or access for sewers, gas and water mains, electric and telephone cables, sewer relining, subways and land drainage.

Installation of pipes by jacking requires a large force, normally provided by hydraulic rams, which is uniformally distributed through a thrust ring around the circumference of the pipe being jacked. The jacking equipment is accommodated in a thrust pit shaft which incorporates a thrust wall to provide a reaction against which jacking forces can be applied. The excavation is normally carried out by a shield in its various forms as discussed in Chapter 8, in front of the pipeline being jacked.

The spoil is then transported along the pipeline via a conventional muck clearance system or through pipes in a slurry mode to the surface for disposal.

Pipes are advanced as the ground is mined. Continuity is achieved by lowering a pipe into the thrust pit area vacated by the previously jacked pipe once the hydraulic rams have been retracted.

Corrections to the alignment of the pipes are made using steering hydraulic rams in the shield in conjunction with frequent surveying to fixed reference points.

10.1.1 HISTORY OF PIPEJACKING

The earliest recorded use of the pipejacking method was in America in about 1910, (Richardson and Mayo, 1941) and the basic principles of the technique have been presented in detail by the American Pipe Jacking Association (1960) and others. Over the last two decades, the increasing use of modern technology and a competitive tendering market have led to many innovative methods being introduced particularly in Japan and West Germany, where pipes from 200 mm and 400 mm diameter and above are jacked in almost any ground conditions and over ever-increasing distances. Pipejacking was first introduced into the UK in the early 1960s and was used as an alternative tunnelling method to small segmental tunnels and timber headings for providing short crossings under obstructions such as railways, rivers and canals. The technique developed and it became practical to use the method for longer tunnel drives and so pipejacking found a market in the sewer and surface water sectors.

The introduction of mechanised excavation and associated equipment has increased jacking distances. In the UK, Wallis (1982) reports on a 1800 mm diameter pipejack, 460 m long in London clay and Byles (1983) reports an increase in length to 560 m in water-bearing sands and gravels. These two contracts significantly lifted the profile of pipejacking within the UK. Notwithstanding these improvements there is a distance at which it becomes more economic to excavate a thrust shaft and commence a new pipejack. This is primarily due to increased muck haulage distances and costs due to lower production and the required shove pressures. However, improvements in techniques and technology are constantly challenging this distance.

Indeed, new methods of construction and new materials, which will be discussed within this chapter, are continually being developed and implemented.

The most recent British Standard for pipejacking pipes ensures that the pipes are designed to higher standards and strengths for withstanding more onerous loading requirements. As the method of excavation improves (discussed in Chapter 8) then so must the means of dealing with the spoil on the surface to minimise the environmental impact while maintaining the performance capabilities of the excavating process. Consequently processes for the treatment of slurry which are reducing the need to decant will significantly improve performance and costs.

The so-called 'Electronic Revolution' has had an earlier impact on pipejacking than conventional tunnelling techniques. Due predominantly to the tunnel dimensions and the shorter tunnelling distance it is already commonplace to see remotely operated shields using CCTV and laser guidance systems linked back to a control cabin situated on the surface. This has been supplemented by microprocessor-based data loggers which by and large monitor the operating parameters but which have the intelligence to actually operate within these parameters. The days of the fully automated pipejack are not too far in the future.

10.1.2   EFFECTIVE MANAGEMENT

In 1973 six contractors who were using the pipejacking technique formed the Pipe Jacking Association (PJA), to promote and market the method through a code of conduct and good practice.

The PJA in conjunction with other associations such as the Concrete Pipe Association, the Construction Industry Research and Information Association (CIRIA), the Science and Engineering Research Council (SERC) and Oxford University have published a number of design guides, case histories, newsletters and findings of research projects. As a direct outcome of these associations and the support of several major water companies a guide to best practice for the installation of pipejacks has been compiled to provide a code of conduct and good practice in the art and practice of pipejacking.

The guide lays down the following parameters to be considered to produce the best pipejacking scheme design.

Design

- Thorough and focused soil investigation and interpretation
- Suitability and selection of pipejacking excavation method
- Working shaft design and thrust wall
- Surface establishment, muck disposal and treatment.

Operational

- Surveying and alignment (steering control)
- Geological aspects and its effects on pipe installation (ground movement, loading control)
- Pipejacking techniques – selection and suitability taking account of drive lengths, jacking and friction loads, intermediate jacking stations and lubrication.

These parameters are considered in greater detail within this chapter in conjunction with Chapter 8.

10.2   Technical aspects of pipejacking

10.2.1   INTRODUCTION

The purpose of an effective geotechnical design summary report is to serve as a starting point for measuring the degree of differing site conditions. The same parameters discussed in Chapter 8 apply in pipejacking with the added consideration of the *prediction of jacking forces* which is influenced by the geological circumstances.

Pipejacking contractors generally use empirical methods to predict jacking forces which allow for pipe/soil friction forces in different ground conditions. Typical values of friction loads are listed in Table 10.1 (Craig, 1983).

Table 10.1   Typical values of friction loads on pipes (after Craig, 1983).

Ground type	External load (kN/m^2)
Rock	2–3
Boulder clay	5–18
Firm clay	5–20
Wet sand	10–15
Silt	5–20
Dry loose sand	25–45
Fill	45

Production of jacking forces is derived from multiplying the figure from Table 10.1 by the total external surface area of the pipe being installed. This process would indicate a possible maximum jacking force which could be reduced by the use of injected lubricants or the installation of intermediate jacking stations.

A more analytical prediction of jacking forces is detailed by Auld (1982) and more recently outlined by Millian and Norris (1993) and others based on standard soil properties and theories. These predictions of jacking loads are based on ground pressures, the angle of internal friction of the surrounding soil and an 'experienced guess' based on a number of factors which influence jacking loads, namely:

1. Resistance at excavation face
2. Amount of overcut during excavation
3. Variations in ground conditions
4. Recommencement of jacking after a prolonged stoppage
5. Steps at joints
6. Joint deformation
7. Misalignment of jacking pipe
8. Jacking around curves
9. Injection of lubricant into the overbreak void
10. The use of interjack stations.

Haslam (1986) and O'Reilly and Rogers (1987) have presented papers proposing new approaches to predictions of jacking forces based on the analysis of field data from selected pipejacks. However, differences between predicted jacking forces and those encountered on the study were often found, which were unexplainable or at least quantifiably so.

The commonly used empirical approach has a track record of predicting jacking forces with reasonable accuracy with errors encountered using this approach due most likely to variations in soil properties or large misalignments of the pipejack, whereas the presentations of Haslam and O'Reilly only allow for small misalignments of the pipejacks and for consistent ground conditions.

Arguably, the analysis of each author is relevant to the ground conditions and contractor of the specific contract.

Variations between jacking loads may be as much to do with the method of working, labour employed, amount of supervision, jacking arrangements, quantity and pressure of lubricant injection and excavation method than the actual ground conditions themselves.

However, more recent research carried out by Milligan and Norris (1993) suggests that the parameters which are particularly pertinent to the pipejacking process are those soil parameters that have their influence on:

- pipe/soil friction
- total jacking forces

which are considered in the following section.

10.2.2   CONSIDERATION FOR PIPEJACKING

Once an appropriate soil investigation has been completed and interpreted, practical decisions can be made with regard to the pipejacking scheme design. This scheme will take into account the following factors:

1. Requirements for ground stability control
2. Type of tunnel excavation method
3. Drive lengths and interjack requirements/pipe friction and soil parameters and total jacking forces
4. The working shafts and thrust wall construction
5. Surface layout and the treatment/disposal of spoil.

1. Requirement for ground stability control: Ground stability may be assessed from consideration of Chapter 8. Where any possibility of ground collapse exists, consideration should be given to the use of ground water lowering or removal techniques, grouting or

chemical stabilisation of the soil, or the use of specifically designed shaft construction (discussed in (4)) and excavation methods to accommodate these problems.

2. Type of tunnel excavation method: In ground conditions where movement above the pipeline not only causes damage to surface structures or services but also leads to increased resistance to jacking and endangering miners and machinery at the face, then the use of earth pressure balance or slurry support tunnelling machines must be considered. However, in cohesive soils the face pressure developed by these specialist tunnelling machines is required to be controlled to ensure that neither excessive settlement nor heave occur.

   Prior to selection consideration needs to be given to the drive length requirements (3), the working shafts and thrust wall requirements (4) and the surface layout that is allowable.

3. Drive lengths and interjack requirements – pipe friction/soil parameters and total jacking forces: Even when the excavated tunnel is stable, the ground may close onto the pipe due to the 'elastic' unloading of the ground around the tunnel. These reductions in vertical and horizontal diameter of the opening, if they exceed the initial overbreak, will ensure that contact between the soil and pipe will occur, radial stresses will develop and resistance to jacking will start to increase. Work carried out by Milligan and Norris (1993) from results obtained of the radial (normal) and shear stresses at the interface between the pipe and ground, suggests that the relation between shear and total normal stresses appears to be frictional in all the ground materials, i.e., that shear stresses increase more or less linearly with normal stresses. However, in the cohesive soils, at higher stresses the shear stresses seem to tend towards a limiting value which is probably a function of the undrained strength and an adhesion factor. For this frictional behaviour, the apparent interface friction angles are given in Table 10.2.

Table 10.2   Measured local interface friction (after Milligan and Norris, 1993)

Soil type	Friction angle (degrees)
Glacial clay	19
Madstone	17
London clay	12.7
Silty sand	38
Sandy silt	30

However, misalignments in the pipeline must inevitably induce contact stresses between pipe and ground.

   Also the time factor is a well-known phenomenon in that in cohesive soils the force needed to restart a jack after a stoppage is usually higher than that needed to maintain subsequent motion. This mechanism is probably because the pore pressures generated during pushing dissipate during a stoppage and hence the effective stress increases. Relations between increase in jacking load and duration of stoppage amounted to around 25% in the first hour. Effective lubrication of a pipeline also affects the pipe and soil interface. Lubrication of any kind can only work if a discrete layer of the lubricant is maintained between the two sliding surfaces. If this is achieved the interface shear stresses can be reduced considerably.

   Research carried out (Craig, 1983 and others on the extent of total jacking forces, measured by load cells on the main jacks, indicates that the increase in jacking forces, as you would expect, is an indication of resistance at the face. This is relatively small in cohesive soils but large in mudstones. The resistance was also closely related to the excavation and trimming process at the face. Generally when the excavation was slightly larger than the outside diameter of the shield the face resistance was very small. However, if the shield is being used to trim the face or it is being used either in earth pressure balance or slurry mode, then the face resistance increases significantly.

In stable ground conditions, it is considered that only the bottom of the pipeline is really in contact with the soil, consequently it is reasonable to assume in such cases that the average resistance should be related simply to the weight of the pipe. Use of the measured local interface friction coefficients given in Table 10.3 has indicated reasonable agreement between theory and measurement, with the measured values being somewhat higher,

Table 10.3 Pipe self-weight friction (after Milligan and Norris, 1993)

Scheme	Field skin friction (δ)	W ton (δ)	A_u friction (kN/m)
1	19	6.1	7.2 (dry)
2	17	7.0	8.0
3	38	18.7	23.1

probably as a result of increased contact stresses due to misalignment (normally 25%). However, in softer clays a more appropriate model may be that of Haslam (1986) in which the undrained adhesion between pipe and soil is multiplied by a contact width determined from elasticity theory.

In cases where the ground closes onto the pipeline, the resistance will increase considerably which is discussed in 7.2.2(4).

The factors discussed with regard to soil conditions and pipejacking can be summarised by:

- The contact stresses between pipe and ground depend on the stability of the tunnel bore, the initial stresses in the ground and the stiffness of the soil.
- The pipe to soil interface sliding behaviour is frictional in nature even in cohesive soils albeit the undrained strength provides an upper limit.
- In stable tunnels the resistance to sliding is related simply to the self-weight of the pipes.
- The resistance to sliding is related to pipe alignment and the steering process adopted to correct misalignment.
- Effective lubrication requires complete filling of the overbreak for all of the annulus and furthermore in cohesionless soils at sufficient pressure to maintain the stability of the tunnel bore.
- Face loads are likely to be relatively high with slurry or earth-pressure-balance tunnelling machines and in strong soils especially where the shield performs the trimming.

When considering the factors discussed within this section and Chapter 8 it re-emphasises the need for achieving the points laid down by the British Standard Code of Practice for site investigations. That is, that the extent of the ground investigation is determined by the character and variability of the ground. The extent of ground water and its effect on the surrounding environment needs to be identified to determine the economic operation of the pipejack. To achieve this it is important that the general character and variability of the ground be established and due account taken of the fact that the greater the natural variability of the ground, the greater will be the extent of the ground investigation required.

In conclusion, theoretically, there is no limit to the drive length that can be achieved by a pipejack, providing that all the elements that make up a pipejack have no limitations. The working shaft and thrust wall will have to be designed and constructed to withstand the thrust requirements but this can be offset by the use of interjack stations. Obviously the number of interjacks in use has to be balanced against their cost and the time required to operate the interjacks relative to the speed of the excavation shield and the overall friction load.

In terms of overall friction load, the practical lubricant injection limitations together with the time dependent settlement of the ground around the pipe will ultimately determine the drive length. A major practical consideration is the cost due to production being constrained by spoil disposal limitations. Both mechanical and pumping disposal will limit economic production. Consequently, from a practical perspective, the PJA guide to pipejacking recommends the practical and economic lengths with current technology to be 500 m for pipes in the 1200 mm to 2500 mm range.

4. Working shafts and thrust wall construction: The working shaft requirements for pipejacking are dependent on the following factors:

- Adequately sized to accommodate the shield/TBM and jacking equipment selected and provision for an adequately *designed* and installed thrust wall and the necessary temporary works to accommodate entry and exit seals (discussed in 8.6).
- A facility for removing spoil and space for lowering and jointing the jacking pipes and interjack stations.
- In the case of a reception shaft, space to remove the shield via a pipejack eye.
- Its suitability to support the ground pressure and water intrusion particularly during entry and exit of shields.
- The practicability of positioning the working shaft together with the installation plant required in the available site area.
- The presence of adjacent structures, existing services, traffic considerations, vibration problems, environmental and noise limitations.
- Working shaft construction method adopted which could include:
  - Segmental linings
  - Pre-cast or cast *in-situ* caissons
  - Sheet piling or secant piling
  - Shallow trench sheeted or timber supported excavation
  - Battered excavation
  - Ground anchorages.

The method chosen must take cognisance of the best shaft size and its ability to support the ground applicable to that scheme and also their suitability for converting into manholes or other permanent works. It must be properly designed to withstand not only the stresses imposed by the subsoils and water pressures but also the thrust wall loads and local concentrations caused by construction of entry and exit eyes for the excavation shield.

Thrust walls have for many years been designed empirically by taking a conservative ground passive resistance stress and determining the area of the thrust wall against a calculated maximum jacking load. It is advisable that standard structural design practice is used or each specific scheme as the design parameters of depth, soil strata and thrust load will be unique to each project (see 8.6).

It should be recognised that a thrust wall is a temporary structure, which is likely to be aborted after use. Therefore notwithstanding the design requirements and the far reaching consequences of the failure of a thrust wall, due account should be taken in terms of factors of safety and the economics of the structure.

5. Surface layout and the treatment/disposal of spoil: The surface establishment is a support service of equipment which is required to feed the main pipejacking operation. It must take into account surface structures (including overhead obstruction), location of shafts, traffic requirements, environmental noise requirements and the available space.

The main support services generally include: cranage or gantry capable of handling the shield, pipes and ancillary equipment; a storage facility for pipes and other materials and a lubrication mixing and injection plant based on the anticipated daily performance and haulage limitations; a control station for the main and interjack power units and services such as electricity, communications and sufficient water supply. The spoil handling facilities need to take account of traffic constraints and the available space for stockpiling and the nature of the material to be disposed of. If a slurry system is envisaged then consideration must be given to slurry disposal and either settlement areas or spoil treatment equipment required to deal with the type and characteristics of the spoil. There is a growing environmental and economic advantage in the use of more sophisticated process engineering in separating the spoil from the slurry with the use of centrifuges and flocculents as opposed to slurry disposal and replenishment.

10.3 Pipe design and manufacture

10.3.1 INTRODUCTION

Pipes for use in jacking operations can be manufactured from various materials including clay, steel, glass reinforced plastic (GRP) and glass reinforced concrete (GRC). However, the most commonly used material is steel reinforced concrete. The pipes manufactued using concrete are cast in factories by specialist companies under controlled conditions.

This section will restrict the discussion to steel reinforced concrete pipes but will explain application of other materials.

Jacking pipes are normally manufactured to give an effective length of 2500 mm. However, shorter pipes are used occasionally, with a consequential increase in cost, when a tight curve needs to be negotiated, or when the surface 'footprint' available is so confined that shorter pipes have to be used because of insufficient plan area in the access shaft.

10.3.2 DESIGN

Jacking pipes are designed in a different way to segmental lining. The latter has joints which limit the build-up of bending moments as deflection can occur at the cross joints. However, there is a similarity is that the segmental lining has to withstand the push forces of a tunnel boring machine; a jacking pipe must be capable of withstanding the jacking loads which are required to overcome pipe weight and friction along the extrados of the pipeline.

Pipes are designed to meet the requirements of BS 5911; Part 120 for installation by jacking. Manufacturers will normally quote a maximum jacking load which the pipes can contain. Designs are normally based on a minimum concrete cube strength of 50 N/mm^2.

For long-term loading pipes must be designed for expected vertical loads. This will vary with each geotechnical condition, but as most pipejacks are reasonably shallow, i.e. less than 20 m bgl, full overburden should be designed for. The exception to this would be a pipejack through rock. However, the dominant factors are the jacking forces and handling of pipes.

Transference of the jacking load is not always with full contact at the ends of the adjacent pipes. The method of excavation for practical reasons requires an overcut to allow the pipes to follow through. Athough this overcut is small, it allows the pipe to wander within the bore and thus create angular deflection at the joints.

This deflection must be taken into account in the design of the pipe to allow for the concentration of load and to provide rigidity so that the pipe will not deform when subject to non-uniform axial loading.

The maximum angular displacement taken into account within the design is normally 0.25° in combination with a suitable packing material. An increase in this angle will result in a corresponding reduction in the maximum allowable jacking load.

The packing between the pipes is an important component in the performance of the pipes as it must transmit jacking forces from pipe to pipe and be sized to contain these forces within the reinforcing cage at the pipe end. Further aspects of the pipe packing are discussed under section 10.5.

It can be seen from the foregoing that the joint is very important to the success of the pipejack. Basically there are two types of joint that are commonly used.

- Rebated
- Butt joint.

The latter is the more recent introduction and was adopted because it maximises the area available for the transference of the jacking load. Consequently, it has become the most common joint in the UK and Europe. With the rebate joint there is always the risk that if the deflection of pipe became too great then the outside leg of the joint could shear. This problem is difficult to detect but can seriously affect the progress of the pipejack. Similarly, it could happen to the butt joint but as the sections are so much smaller the consequences are not so significant. Having decided on the type of joint to adopt it is important to keep the joint clean, i.e. free form debris from within the tunnel and from material ingressing from the surrounding ground conditions. This latter point is particularly important in non-cohesive soils, particularly when these are negotiated below the water table. Any water seepage could carry with it fine particles. The joints must be kept clean to avoid point loads on the pipe ends.

Sealing of the joint is normally achieved by a single seal along the pipe joint. However, in poor ground conditions it is recommended that a secondary seal is used in addition.

The design of pipes can be varied to suit the prevalent conditions, e.g. the reinforcing cage can be amended to give an increased cover to increase the exposure condition.

10.3.3   MANUFACTURE OF PIPES

Manufacture of reinforced concrete jacking pipes is undertaken at purpose-built factories under controlled conditions to ensure a high quality product is consistently produced. The pipes are also cast to very high tolerances to ensure square ends for uniform load transference.

There are two distinct methods of manufacturing pre-cast concrete pipes. Both methods utilise a low water cement ratio which produces a concrete of high strength with good durability properties to comply with BS 5911: Part 120. Briefly, the methods are:

1. Centrifugal spinning: Semi-dry concrete is fed into a steel mould rotating horizontally on a spinning bed at high speed. Compaction is a combination of vibration and centrifugal action.
2. Vertical casting: Semi-dry concrete is fed into a vertical steel mould and compacted by high frequency vibration through vibrators attached to the form walls.

The reinforcement cage for both types is manufactured by welding an internally and externally wound spiral around the longitudinal bars. The welding is undertaken by an automatic welding machine.

10.3.4 SPECIAL CONSIDERATIONS

There are occasions when the prevalent conditions, either from the ground conditions and/or from the effluent containing chemicals in sufficient quantities, cause corrosion of the concrete.

In these circumstances there are two alternatives (at least):

1. Provide alternative materials resistant to chemical attack, e.g. GRP/GRC/clay. The latter is not manufactured to man entry sizes, i.e. greater than 1000 mm diameter.
2. Alternatively, the concrete pipes and collars could be protected by various measures including:
   - Epoxy coated reinforcement
   - Stainless steel bands for the butt joints
   - Coating the concrete with an epoxy paint or similar
   - Casting an internal or external integral lining to act as a pipe wall.

In addition to standard pipes specials have to be cast for the system to work. These include:

- A lead pipe which has a long rebate on the external wall to fit in the tailskin of the tunnelling machine or shield.
- Lead pipes are also required for interjack stations. The body skins for these units can be independent or can be cast as part of a pipe, i.e. the overall length of the unit is the same as that for a standard pie, but on closing up after removal of the interjack rams, etc. the effective length will only be approximately 1.25 metres.

10.3.5 CONCLUSION

Pipes are a very effective method of producing a tunnel as they produce a high quality, durable product which provides a smooth, almost blemish-free finish. Best results are produced on straight lines, but in the right conditions curves are possible, though the radius is governed by the deflection angle between adjacent pipes.

10.4 Surveying and alignment

10.4.1 SURVEYING

Much of that discussed in section 8.8 applies to pipejacking. The main difference is that in a pipejacking operation the tunnels are constructed by jacking from an access pit and the whole tunnel moves forward as excavation proceeds. Each time the tunnel has been jacked forward one pipe length, the jacks are retracted ready for a further length of pipe to be placed in the pit bottom. At this point the pit bottom can be made relatively clear for surveying work. As the tunnel is continually being moved forward, survey stations attached to it are not normally stable enough to be used for directing the drive. The simplest way to control the direction of the tunnel is to set a laser in the pit bottom as soon as a short length has been constructed by 'strings' methods similar to those indicated in 'Breaking out' under section 8.7.1.

This is satisfactory for the completion of the many jacked tunnels that are short and straight. It should be borne in mind that the forces exerted by the jacks onto the shove wall are considerable and consequently much of the pit bottom structure could move when they are operated. The most stable place for mounting the laser is likely to be the concrete plug of the pit bottom but care must be taken to site the mounting to avoid disturbance of the laser during tunnelling operations.

Self-adjusting lasers can be used, set to the tunnel gradient. This can reduce the number of level checks on the laser but plainly the alignment must be checked regularly by instruments. Having in mind that work in the tunnel must stop while these routine checks are made, it is essential to use methods that take as little time as possible.

Developments in the use of gyroscopes for maintaining direction have meant that where the cost can be justified their employment can increase the interval between precise checks and assist in the negotiation of curves. There are also devices available for the control of curved steering which is comparable in accuracy to the laser for straight steering. A development of the guidance system is the active target system which measures X and Y co-ordinates and the shield's roll, incline and yaw and it calculates the anticipated position of the shield.

10.4.2 ALIGNMENT

The degree of misalignment between pipes within the pipe string is of considerable significance apart from the requirement of the completed pipeline. For instance, the generating of additional interface friction between pipes and soil increases the total jacking load which has to be transmitted safely through the pipe joints. *Misalignment in this context means the angular deviation between the central axis of successive pipes.* In an ideal pipejack, no such deviations would exist but in practice irregularities in ground conditions, excavation methods, etc., will inevitably cause the pipe to stray from the ideal course. In the case of a mechanised method of excavation, corrections are continually made with the steering jacks. The normal practice is to specify limits to the allowable errors in line and level at any point along the tunnel, typically 50 or 75 mm. While these may be necessary to maintain adequate clearance from obstructions or other services, or to provide correct hydraulic flow conditions, they may be quite inadequate as a means of controlling the angular deviation between successive pipes within acceptable limits for transmission of large axial forces during pipejacking operations.

Likewise, the allowable angular deviations specified in BS 5911 relate only to the satisfactory performance of the joint sealing arrangements, not the transmission of longitudinal load.

Research carried out by the University of Oxford (Milligan and Norris, 1993) showed that the alignment of the pipeline (in its unloaded state) did not change significantly as the pipeline was extended. Thus, local curvatures once established remain throughout the drive. This is particularly important during the early stages of a drive where if the alignment control is poor and gives rise to serious joint angles, this is the location at which the highest pipe loads will be generated later in the drive.

Conventional line and level measurements at the shield are sufficient for practical purposes to determine joint angles, assuming that no significant change in alignment occurs due to the application of jacking loads or the passage of successive pipes.

Normal measurements of line and level may be used in a simple method of making decisions on steering adjustments to keep angular deviations within acceptable limits.

10.5 Temporary works

As with tunnelling contracts, temporary works are an integral part of the success of the contract. Again, temporary works include civil, electrical and mechanical engineering designs.

While there is a lot of commonality between tunnel construction and pipejacking temporary works, *one* major difference is in the thrust requirement. Thrust rams and power packs must be supplied to overcome frictional requirements, but most importantly the thrust wall has to be designed to transfer the loads. *Pipejacking – A Guide to Good Practice* almost ignores the fact that the thrust wall must be designed. To the authors this is one of the most important aspects of pipejacking. If the thrust wall is incorrectly designed, then the pipejack will fail.

There are two ways in which the thrust forces can be distributed through the shaft. The first is to mobilise the passive resistance of the ground surrounding the shaft wall. This is the normal method, but when pipelines are in shallow and weak ground they cannot always be accommodated using this method. The second is to increase the mass of the shaft by increasing the depth of the base slab, then by using drains and ties connected into the base slab the forces can be distributed into this point. In soft water-bearing ground it is important that the shaft temporary works are thought of and designed prior to construction of that shaft commencing. The temporary works have to be designed such as to allow the breakout without losing any ground, i.e. through seals, and that all the pipejacking equipment and machinery can be accommodated within the shaft bottom.

The seals are very important in wet flowing ground conditions and in particular when slurry type machines are used as the excavation process. In these circumstances the seals do three things:

1. Minimise the loss of water and material surrounding the shaft
2. Allow the use of a pressurised slurry system (without seals bentonite would pour into the shaft)
3. Allow the injection of pipeline lubrication.

Seals are normally fabricated from neoprene, which is a flexible material, backed up with a steel keeper plate which is bolted into the shaft wall. With the neoprene seal being flexible it will bend towards the face as the machine moves forward, but the keeper plate will counteract any reverse pressure from bentonite injection.

In these sort of ground conditions, the shaft will have to be sunk as a caisson, either using kentledge or jacks.

Pipejacking machines cannot tunnel through reinforced or high strength concrete. Consequently, a substitute low strength concrete segment or form has to be introduced. The soft eye or segment must be positioned at the right level. Therefore, it is important to know what depth of base plug is required in the shaft before the caisson commences. The depth of the base plug could be governed by the antiflotation requirement or by the jacking wall regrout. There are normally two or more pipes which enter and exit into a shaft.

To accommodate the temporary works for the first drive is normally quite straight-forward. The second or subsequent drives can be more difficult. This is dependent on the intersection angle of the two drives. If the drive is straight through, i.e. an intersection angle of 180°, then the first pipe can be used to test the reaction to the second pipeline. The major difficulty comes when the thrust wall is part way across the first pipeline.

Care should be taken when planning pit bottom arrangements because, besides fitting in the major pipejacking components, including ground seals around the shaft opening, cognisance must be taken of the relaxation of the pipes when the push rams are released. This relaxation is caused by the packings between pipes being crushed and then recovering to their normal thickness.

Having taken care to design the thrust wall and pipe bottom arrangement, it is essential that the thrust is transferred down the length of the pipe to the shield and machine.

To some extent this has been dealt with in section 10.3.2, but did not discuss the material. The material must be elastic in order to recover from deformation, but have sufficient modules to withstand the transfer of loads through the pipes.

The material normally used is a chipboard with a specific density of 700 kg/m^3.

With a system designed correctly for load transfer it is now necessary to reduce frictional forces along the tunnel walls by the injection of a material which gives good lubrication qualities. In the majority of pipejacks this material is bentonite and is injected into the excavated annulus as close to the face as possible.

However, in cohesionless soils where frictional forces can be very high, it will be necessary to inject further quantities of bentonite. Therefore, injection ports need to be allowed for during the casting of the pipes. Normally every third pipe has one bentonite injection post at crown and each axis.

A further problem posed in water-bearing, cohesionless, fine soils is the ingress of fine particles into joints, not only pipe to pipe, but pipe to shield and pipe to interjack skins.

The pipe to pipe joint has been dealt with under the section on design. The latter two only matter when the skin is moving relative to the pipe. To date an effective seal has not been developed.

10.6   The finished product

Although both are manufactured under factory controlled conditions, the pre-cast concrete pipe has one distinguishing feature over a tunnel lined with a segmental lining. The pipejack has far fewer joints and, therefore, has less potential for leakage.

Ground water ingress into a tunnel can be controlled by use of elastomeric gaskets which are crushed together under the force from the erector or from the shield rams. This cannot be done in a pipejack due to the constant forward and backward motion of the pipe joints. Accordingly an elastomeric seal is positioned underneath the steel bands to help control water ingress into the pipeline.

As with the tunnel the inner face of the pipe at the joints is sealed with a mortar. When a pipeline has been installed by jacking the first thing to do is to grout the annulus between the external face of the pipe and the ground. Ports in the pipe which have been used for bentonite injection can now be used for grout injection, the grout actually replacing the bentonite which has acted as a lubricant up to this point.

Having the minimum number of joints in the actual finished product, the pipeline is extremely smooth giving good hydraulic qualities and a very low smoothness factor. One of the interesting points of the pipejacking code is that the concrete in the pipe only has to be impermeable to external pressure, or internal pressure, up to 2 bar. The alternative impermeable qualities of tunnel segments are far higher than this and that is probably why the gaskets are required to maintain this impermeability. This does not mean that pipes should only be put into 2 bar of external water pressure. Indeed there are several instances where this has been done. The authors believe that the code needs to be revised accordingly.

In conclusion pre-cast concrete pipes offer this high quality finish to a tunnel and their structural integrity is not compromised by poor workmanship when erecting segments as part of a tunnel. Besides the joints, all bentonite/grout supports must be mortared as well.

11

Management of Construction

Malcolm Chappell BSc (Hons), CEng, FICE, MIMM

Eur Ing Derek Parkin BSc (Hons), MPhil, MBA, CEng,
MIMechE, MIMM, MIMgt

11.1 Introduction

It is the authors' intention through this chapter to make readers aware of some of the essential requirements to run a tunnel contract successfully. It will only briefly touch on financial control and commercial awareness as there is a plethora of information available on this topic.

Although reference will be made throughout to a medium sized (£3m–£5m) tunnelling contract using sophisticated machinery such as an earth pressure balance machine, much contained herein is applicable to all construction contracts, including the less complicated tunnelling and pipejacking contracts. Each contract has its variations in specification and conditions of contract, etc. but it should not be forgotten that Her Majesty's Government legislation still has to be adhered to.

The modern construction site manager cannot be expected to be conversant with all aspects of construction as it is too diverse for a single individual to fully comprehend. Therefore, while the site manager must be made aware of his/her responsibilities he/she must be able to call upon advisers and delegate responsibilities to specialist engineers to offer and undertake specific tasks outside his/her normal contract team.

It has been a recent feature of site managers that they have become overloaded with legislation to which they must adhere, such that it dictates their time rather than management of the contract. This is probably as a result of a lack of staff to meet needs. Although legislation has become more time consuming, this has not been recognised by some organisations who still have their 'heads in the sand' dreaming of the 'good' old days and pertinent staff levels. However, this situation is changing for the better.

11.2 Types of contract

There are numerous conditions of contract available which are applicable to tunnelling contracts. These vary through the traditional 5th and 6th editions of the Institution of Civil Engineering, through the New Engineering Contract to Institution of Chemical Engineers (Green Book) and target cost. Dependent on the conditions, a contractor's role can vary from constructor to designer.

There are distinct advantages to the latter approach in that a contractor can, through his normal business, accumulate plant of a specialist nature, particularly tunnelling machines. These are expensive and time-consuming commodities to procure. Knowing the availability of such equipment, the contractor can design the permanent works to suit and, in theory, offer the client a saving in both time and cost.

The obvious extension to this is for the contractor to be taken on board at the initiation of the scheme such that the scheme can be designed around the plant available and the construction techniques favoured, i.e. getting it right first time, thus minimising the requirement for expensive redesigns.

11.3   CDM Regulations

The Construction (Design and Management) Regulations 1994 (CDM) came into force on 31 March 1995. They are intended to provide minimum health and safety standards at temporary or mobile construction sites. In addition, the Regulations put into place provisions which are the result of an extensive review of existing construction-related legislation.

The CDM Regulations are founded on the general duties of the Health and Safety at Work Act 1974 as extended and made more explicit by the Management of Health and Safety at Work Regulations 1992.

The CDM Regulations place specific duties upon clients, designers and contractors to rethink their approach to health and safety so that it is taken into account and then co-ordinated and managed effectively throughout all stages of a construction project from conception, design and planning through to the execution of works on site and subsequent maintenance and repair, even to final demolition and removal.

The CDM Regulations are based on the following principles:

* Systematic consideration of safety through every stage of a project
* Involvement of all who can make a valid contribution to health and safety on the project
* Throughout the project, proper planning and co-ordination are to be included
* Provision of health and safety is to be undertaken by adequate competent persons
* Safety and health must be planned and managed
* Communication between all parties is essential
* Formal record of safety and health information for future use.

To provide a framework for the allocation of responsibilities to implement the above fundamental principles the following key personnel are necessary during the construction phase and are referred to in the Regulations:

* Clients and their agents
* Developers
* Planning supervisor
* Designers
* Principal contractors
* Contractors.

Each of these is given a range of dutes: they may be asked to demonstrate undertaking of their legal responsibilities to avoid criminal sanctions and possible civil actions.

Of the above probably the most onerous role is the planning supervisor. This is a new role introduced within the Regulations, whereas the others were parties in the development and construction of a project. This important role through all phases of a project is now explained further.

The appointment of a planning supervisor (PS) is made by the client. The PS has to co-ordinate the project design and planning from a health and safety viewpoint with the aim of ensuring that risks relating to the construction of the project are identified. In addition those risks which are associated with maintenance or eventual demolition of a structure are eliminated or at least minimised.

To undertake this role an appropriate level of design and construction experience will be required and be particular to the project.

After appointment a health and safety plan will be prepared during the pre-tender period. This plan will be developed during all stages of the project with designs, principal contractors supplying relevant information for inclusion.

The planning supervisor function includes co-ordination with all designers to ensure all relevant information is exchanged and that health and safety issues are properly considered and addressed accordingly at this important stage of project development. It should be noted that the design of a project is not limited to detailed design but includes many decisions taken prior to this stage.

It can be appreciated that for an individual to undertake this role an appropriate level of design and construction experience will be required particular to the project.

Further information regarding CDM Regulations is available from guides published by the Construction Industry Advisory Committee (CONIAC) and the Health and Safety Commission. In addition each company should take cognisance of these Regulations through company procedures.

11.4   Contract management plan

At the start of a contract the site management term will have (or need to) develop a plan to describe, not only to the client, but internally, how the contract is to be organised, managed and administered. Legislation now imposes this requirement as mandatory, but any contractor that is quality registered through one of the recognised institutions will need to produce this document.

This plan is known as the contract management plan. It incorporates safety, environmental, quality and programme as these are intimately linked and thus can be easily cross-referenced. The plan should be readily available for access by all personnel involved with the management of the contract and should be reviewed and audited at regular intervals to ensure that the contents reflect the prevalent contract circumstances which can vary significantly from those originally perceived.

In turn, this plan should be referenced to the company safety, environmental and quality procedures and other relevant documentation.

11.5   Site organisation

The normal route of command is through the contract director and the contract manager to the site manager/agent. This person is normally the resident figurehead on the site. It is worth noting that under the ICE 5th and 6th editions of the Conditions of Contract only the agent is recognised as the contractor's representative.

On a modern construction site the manager has to manage many diverse facts to effectively control the contract. Also due to the recent plethora of safety legislation and the demands of quality assurance systems effective management can easily disappear under red tape.

It is without doubt a physical and mental impossibility for one person to fully comprehend in detail everything required. Therefore, beside delegating job responsibilities to staff who are resident on site, e.g. quantity surveyors, office manager, etc. the manager has to be supported by advisers and engineers who have specialist knowledge.

On a medium size tunnelling contract this support would extend to (though not be limited to) the following fields:

- Electrical engineering
- Mechanical engineering
- Civil engineering
- Safety
- Quality assurance, e.g. auditors
- Permanent and major temporary works design
- Purchasing
- Planning programming.

The site manager expects the on-site engineer and the above specialists to give advice to the best of their ability. Of course, this is best achieved if the advisers and engineers required in the support functions are directly employed by the contractor, i.e. company employees.

It is only through this support that site managers can be left to do what they do best – manage. Also without this support there is an ever-present danger that the contract will not comply with prevalent safety legislation in its manifest forms.

11.6 Training

Having set, through the previous sections, the standards required to undertake a tunnel/pipejack tunnel project, it is necessary to implement a training programme to ensure that all sections of the site team are fully capable and aware of their roles and responsibilities. Thus every member of the site team, including the workforce, must be included in the training programme.

It is also mandatory, through safety legislation, for all members of the site team to receive instruction on any method statement which is pertinent to their work.

There are two types of training – general and specific. General to improve the overall performance of the individual, and specific to aim at particular aspects of individual assignments.

General training for the workforce could include recently developed NVQ qualifications, e.g. mining, general operatives, which are based on work experience, whereas training for engineers could also lead to a suitable qualification.

It is becoming more and more a prerequisite for contractors to pre-qualify for a tender list to demonstrate not only the training undertaken but also the recognised qualifications achieved.

Specific training should include but is not limited to the following.

11.6.1 SITE SAFATY INDUCTION

It is a mandatory requirement of legislation that all new starters must attend a general site induction before commencing work. This induction is to explain about the contract, the management organisation, welfare facilities, specific safety requirements, emergency procedures and arrangements for personnel matters.

Dependent upon the size and complexity of the contract this induction can range between one and three hours. If longer than three hours then people's interest starts to wane and the important message fails to make the desired impact.

On completion of the safety site induction each new employee will be directed to the section/department leader who is required to give a specific induction on the new employee's workplace. This induction will introduce specific locations, e.g. fire and emergency doors, welfare facilities, departmental procedures, current method statements, etc. After completion the new employee is allowed to start work.

11.6.2   METHOD STATEMENT INDUCTION

Each work activity must be accompanied by a method statement and an appropriate risk assessment as required by the Construction (Design and Management) Regulations. This must be prepared by a competent person, which is normally the section engineer, and then approved by the site manager.

Following approval, and prior to the work activity commencing, each person involved in the method statement must attend a briefing. The briefing is given by the section engineer and will outline the method with its associated risks and highlight plant, materials, permits and each person's responsibilities to him/herself and to each colleague.

11.6.3   OPERATIVE TRAINING

There are particular jobs which require specific employee training before being considered for this post. These particular posts include:

- Banksman
- Slinger
- Circular saw operator
- Crane driver
- Signaller
- Employees working according to the Highways and Street Works Act
- First aider
- Compressed air lock attendants
- Construction vehicle drivers
- Locomotive drive.

It is imperative that employees in these posts receive specific training and that this training is recognised as being appropriate. Accordingly the training must be recorded on the employee's personnel record so that it can be transferred from contract to contract as the employee is transferred.

11.6.4   COMPANY LICENCES

There are some areas of activity which are covered by specific legal requirements – some of these have been listed in the previous section of operative training. However, there are other areas which are not subject to legislative training but to the tunneller are equally, if not more, important.

In particular, these pertain to the tunnel boring machine drivers. These employees are placed in charge of very expensive and intricate pieces of machinery.

It is the authors' belief that each tunnel boring machine operative should be licensed for each particular operation. In other words, the licence applies only for the duration of that contract. When transferred to a new location a new licence will be issued by senior management once they are satisfied that the operator fully comprehends the operation of the machine through the geological strata expected on the designed alignment and, accordingly, the limits within which he/she can operate. If these limits are exceeded then work must be suspended until a senior, experienced member of staff is consulted to make a decision on the next sequence of events.

11.6.5   TOOLBOX TALKS

Safety is culture which has to be preached on site until all members of staff and workforce work together to provide a safe environment in which to work. However, it is important that complacency does not creep in.

This can be achieved by bringing together groups of employees and giving toolbox talks. These are reiterations of particular subjects which are packaged into 30 minute talks. With this length, these talks can be given at the end of a shift without much interruption to the normal running of a contract.

It may be the impression of the reader that the authors are advocating the training of the operatives only. This is certainly not the case. All members of staff and workforce must be trained in the roles they are undertaking.

Nowadays it is incumbent on each chartered engineer, irrespective of the Institution, to undertake continued professional development. This is set at a minimum of five days per annum. Also continued professional development is recognised by the Engineering Council and consequently all engineers should strive to improve their knowledge throughout their career.

11.7   Site support

11.7.1   PLANNING

The contract programme normally has to be submitted to the client for approval 21 days after the award of contract. This is an extremely important document as it explains in a graphical form how a contractor intends to sequence the works and the duration which he anticipates to undertake his activities.

Time and care must be taken on the planning which goes into its production as it can be used in later stages of a contract to substantiate or disprove claims for additional payments or extensions of time. Therefore it must be a fair reflection of the estimate, otherwise monetary substantiation can go seriously awry.

Having produced a programme it is then necessary to produce the planning documents which will assist in ensuring that the programme is adhered to or bettered if possible. These include:

1. Materials schedule: A schedule which lists the major permanent and temporary items and lists the dates when each item is required. Having drawn the document it is then issued to the site manager and the buyer.

   Boxes are available within the document to record delivery periods, requisition and order placing.

Having raised the document, to be effective it needs to be monitored at regular intervals and outstanding items brought to the attention of the site manager for action.

2. Plant schedule: This is a similar document to the materials schedule and is intimately linked to the programme. It is used to decide where plant is to be secured from and whether the plant should be hired or purchased.

3. Temporary works schedule: This document is prepared to highlight the temporary works which are required to undertake construction of the works. Each scheme must then be categorised or describe the extent and importance of it. Within tunnelling these can range from major through general to traditional. The latter would normally involve the use of timbering for excavation support in a tunnel or heading. General designs can normally be handled by the site engineer or head office designs. Major designs may be handled by the in-house design team, but if outside their scope/experience/ability then an outside consulting engineer should be appointed.

   The document should also record how and who is to undertake the design check and, again, a consulting engineer may be required to issue a design check certificate.

   Whoever undertakes the design will require a brief which fully explains the problem and includes enough information for the designer to undertake the brief in the correct manner. To comply with CDM Regulations a hazard awareness sheet should accompany the brief to highlight potential or prevailing problems which could affect the design or the execution thereof. It must be remembered that almost all temporary works designs are linked to a method of operation.

4. Permanent works schedule: A document very similar to the temporary works schedule but highlights the permanent structures. This document is extremely important on design and construct contracts.

The significance of these documents fades if they are not monitored at regular intervals and the site manager is not made aware of potential problems. It is suggested that this function is undertaken by a specialist planning engineer.

11.7.2   DESIGN – CIVIL, ELECTRICAL AND MECHANICAL

It is the opinion of the authors that the above disciplines should be treated with equal merit. For example: *the temporary electrical distribution around the compound and into the tunnel is as important as any major temporary works scheme.*

   Accordingly a design brief should be formally issued to the electrical engineer and the design should be returned with a schematic layout drawing. This is recorded on the temporary works schedules, and distributed accordingly.

11.7.3   MECHANICAL AND ELECTRICAL SUPPORT

With the ever-increasing mechanisation of tunnelling sites and added complexity of the equipment used there is a growing need for the support of mechanical and electrically competent staff. The civil engineering process of driving the tunnel must not overshadow the means by which the majority of the work is achieved, by mechanical and electrical equipment. Work equipment, its provision and use are the subject of a set of regulations (1992) in addition to the Health and Safety at Work Act of 1974 and the Electricity at Work Regulations of 1989. The importance of these resources and assets in achieving the aims of the contract should not be underestimated. Their effective management is often the difference between success and failure. Effective management refers to appropriately qualified people involved in the planning stage of contracts to ensure that equipment is suitable for the purpose for which it is to be provided both in terms of the safety implications of its selection and its ability to perform the required task. Often the latter prerequisite takes precedence based on decisions made by individuals with limited knowledge of the former prerequisite.

Effective management also refers to the safe and effective operation of work equipment. From its provision through to its installation, commissioning, operation and recovery, to achieve this requires the appointment of suitably trained experienced and professionally qualified electrical and mechanical engineers who are given the responsibility and the means to contribute towards this end. We will take these issues in turn.

Provision

Work equipment and the supply of electrical power should be an integral part of the temporary works schedule and the contract management plan. The programming of the contract needs to identify a plant schedule which makes decisions on how close a fit the plant requirements are to the tender and where plant is to be secured from and whether the plant should be hired or purchased. All too often equipment is paid for several times over through a hire agreement than if the rational decision to purchase had been taken at the outset. The fundamental requirement is that every employer must ensure that work equipment provided is suitable for its intended use, having regard to working conditions and any additional risks.

Installation/recovery

Once the decision to hire or purchase has been taken it must be ensured that the equipment arrives with certain legally required information:

1. If the equipment is bought/hired from within the European Community it shall be CE marked which indicates that it satisfies the essential health and safety requirements which have been agreed by member states. In this context the most important product directive is the machinery directive (89/392/EEC as amended by 91/368/EEC). This does not remove the *users'* responsibility in the use of such equipment under the provision and use of Work Equipment Regulations 1992 (PUWER) or employees' responsibility under sections 7 and 8 of the Health and Safety at Work Act (HSWA).
2. Information and instructions: The employer has to provide to its employees adequate health and safety information which will include (Reg. 8, PUWER):
   - The conditions under which and the methods by which the work equipment may be used and installed
   - Any foreseeable abnormal situation and action to be taken if such a situation were to occur.
   During the installation and recovery operations, risk assessments need to be completed and method statements discussed with and provided to all those concrened with the procedure.

Operation

Once the equipment is operating every employer shall ensure that all persons who use work equipment have received adequate training (Reg. 9, PUWER) when using and installing the equipment.

There is also the requirement that employers shall ensure that work equipment is maintained in an efficient state, in efficient working order and in good repair. Furthermore, where any machinery has a maintenance log, the log should be kept up to date, and in circumstances where inadequate maintenance could cause machinery to fail in a dangerous way, a formal system of planned preventative maintenance (PPM) may be necessary (Reg. 6, PUWER).

Although all maintenance is preventative in some respect, the primary aim of PPM is to prevent failures occurring while the equipment is in use. This not only reduces the possibility of accidents associated with equipment, but also reduces unplanned and expensive stoppage times and increases the longevity of equipment life and operational reliability.

This is achieved by establishing the following:

1. The type of examinations required: Taking cognisance of manufacturers' recommendations and the experience of operators and craftsmen who have worked on the equipment.
2. The frequency of examination: Taking account of:
   - The conditions in which the equipment is being used
   - The requirements of the customer
   - The duty of the equipment and the expected duration
   - Any statutory laid down frequencies, i.e. insurance examinations.
3. Recording of the examination: An often quoted adage by the Health and Safety Executive (HSE) is an examination not recorded is an examination not done. Therefore it is important that there is a reporting mechanism that:
   - Prompts the user as to when an examination is due
   - Records the completion of the examination or otherwise
   - Recognises a defect and closes the reporting loop with a corrective action
   - Allows a statistical analysis of the condition of the equipment over time to enable an assessment of the effectiveness of the PPM scheme and the suitability of the equipment.

The PPM scheme along with the overall standard of mechanical and electrical installation should then be audited in the same way as quality and safety to ensure maintenance of the highest standards.

11.7.4   SAFETY

Safety is a very important part of construction management. It is one of the few areas where failure to comply with current legislation may lead to personal legal proceedings, particularly for those in senior management positions. Her Majesty's Health and Safety Executive has the power to prosecute both companies and individuals and do so when there is evidence of negligence.

Safety legislation is a minefield for the uninitiated. It is extremely wide ranging covering a vast range of topics, particularly on construction sites involved with mechanised tunnelling or pipejacking.

While each company must make all individuals working on the site aware of their own and others' responsibilities through specifically targeted training, the support of a specialist adviser is essential to ensure that the work being undertaken and the systems being implemented are compliant with current legislation and company policy.

On major sites, this person will be resident. In fact, the contract may be so large that more than one person is necessary. On smaller contracts the safety adviser will visit the contract on a regular basis to inspect the workplaces and advise on method statements and risk assessments under preparation.

Although deemed advisers, they must have, through the safety manager, a direct route to the company managing director to recommend immediate implementation of measures which have been ignored by the contract management. It is very important that this link exists as safety must begin at the top of the management tree and be seen to do so. It also provides senior management with an independent report into the management of safety on a particular contract.

The safety adviser has an important role at the contract initiation by being involved in the production of the contract safety plan which sets the scene for the contract.

11.7.5 QUALITY ASSURANCE

A formal recognition of a company operating to a system to provide its customer with a recognised quality product is registration to BS 5950, Part 1, and/or Part 2. To ensure that a company continues to comply with the requirements of BS 5950, the registration company visits each company at regular intervals each year.

The system which has been developed by the company, and approved, has to be available for each employee to refer to as and when required. This normally takes the form of written documentation split into sections which should cover all aspects of company and contract administration.

As stated earlier in this chapter, at the start of each contract a contract management plan should be prepared by the site manager and his team to explain to the client how and when the requirements of the contract will be undertaken. A quality adviser will assist with the preparation of this document to ensure it complies with the company system.

11.7.6 LAND SURVEYING

One of the most costly aspects of tunnelling and pipejacking is rebuilding due to misalignment in either line and/or level. The sections in the tunnelling and pipejacking chapters have explained the control of survey for construction purposes.

However, prior to construction starting it is essential to connect the starting and finishing points of the tunnel over the surface with a closed survey to eliminate angular and elevation errors.

To undertake a survey to an accuracy required needs a qualified and experienced surveyor. The surveyor's ability is extremely important to the company and each individual contract as it is an extremely rare occurrence for the land surveyor to be checked by the client or his representatives.

11.7.7 PROCUREMENT

A specialist buying department can make substantial savings to the contract, but only if the correct process is followed to allow the buyer sufficient time to achieve the best deals. To achieve this end he/she needs to be familiar with the contract through the bill of quantities, the materials schedule and specification (see section 11.7.1).

It is an important part of the process for both suppliers and subcontractors to answer relevant questionnaires regarding their attitude towards safety and quality matters. It is essential that the contract is not saddled with inferior and/or substandard products as replacement and/or an alternative could prove to be expensive in terms of duplicated cost, wasted time and unnecessary labour costs.

At the end of each contract, the management team should compile a report on suppliers and subcontractors and their performance. These comments should be recorded on a central register so that poor performers are excluded from further enquiries until proof of an improvement is established.

The latter is normally undertaken by a member of the quality department and is known as a vendor assessment. It is used for new suppliers as well as existing ones that have a poor rating from past performance.

11.7.8   PERSONNEL AND TRAINING

Recently, there has been a plethora of new regulations pertaining to both an employee's and employers's rights and responsibilities. Added to existing legislation, a site manager will constantly need advice and support from a personnel department to deal appropriately with the workforce under his/her control. It should be remembered that many employees on a worksite will be, due to the nature of the construction industry, employed for relatively short periods and possibly itinerant.

Personnel and training departments are interlinked through the necessity to record on an individual's records training received. With the advent of Investors in People those companies who apply the principles of this scheme are required to record training undertaken. Training records should include membership of recognised national schemes.

Clients are asking more frequently for details not only of staff but also of workforce training as it demonstrates a commitment to personnel development.

Thus a training department is required to develop and organise courses to ensure employees undertaking roles and responsibilities are competent in these, which includes the necessary understanding of current health, safety and environmental requirements.

11.8   Audits

Together with current legislation it is necessary to ensure that company departments and contracts are complying with the approved quality assurance system. This compliance is undertaken through audits which follow a trail through a particular process, e.g. design development, materials procurement, plant supply, etc.

Areas that are audited are listed below:

1. Safety: In addition to regular safety inspections by the appointed safety adviser, and less frequently by the Health and Safety Executive inspector, it is strongly advised to take an in-depth audit of the site safety plan and the way it is implemented.

   The audit should be undertaken by the safety manager in conjunction with the site manager and the managing director (or a representative director).
2. Quality: As with safety, audits are necessary to ensure that contracts and departments are 'doing what they have said they will do'. These audits are undertaken in two ways:
   * External audits by representatives of the registered agency
   * Internal audits by trained nominated persons from within the company.

Both are to establish compliance with the department procedures and/or contract management plans. The former normally happen once per year on departments and representative contracts, while the latter are undertaken at more frequent intervals e.g. every six months for departments and every four months for contracts.

Each type of audit can raise corrective action requests (CARs) which must be responded to immediately and then cleared at the next audit.

11.9 Conclusion

It must be appreciated that the foregoing is an ideal and that contracts are not always of a size and complexity to warrant the individual roles mentioned. However, for the smooth management of the contract the work must be undertaken and thus certain individuals have to take on more than one role and its accompanying responsibilities.

It is incumbent on both contracting and client organisations, particularly the later, to recognise the prevalent situation and appreciate that there is a minimum level of resources required to comply satisfactorily rather than being dominated by cost and hoping to get away with it. Safety legislation and quality assurance should be essential elements of today's contracting world.

It is a pity all parties to the contract do not aspire to the same goals. Perhaps invitations to tender should contain a statement of intent from the employer/client to explain their approach to managing the contract.

12

Impact Moling

Norman Howell

12.1 Introduction

The impact mole is probably the most commonly used of all the items of trenchless installation equipment with many thousands of machines in operation worldwide. It has been extensively adopted by utility companies and their contractors as it offers a very cost-effective method of installing small to medium diameter pipes, ducts and cables for a broad range of utilities which include gas, water, electricity, telecommunications and sewerage. A standard set of impact moling equipment is shown in Fig. 12.1. The technique is generally unsteered and is therefore most suitable for short drive lengths such as simple road crossings and the installation of service connections from the distribution network to the end user.

Figure 12.1 Standard set of impact moling equipment.

Due to the limitations on its directional accuracy it is primarily utilised for installing services that do not depend on precise line and level requirements. However, in favourable ground conditions using modern equipment impact moling can be successfully utilised for the installation of small diameter sewerage pumping mains and gravity sewer pipes. It is particularly useful for making lateral connections from properties to main sewer pipelines.

The origins of the present-day impact moles can be traced to 1955 when Wiktor Zinkiewicz, who was employed by the Polish Ship Rescue Company, devised the idea of pulling a rope under a sunken ship on the seabed using a pneumatic hammer device. A small number of machines were built to Zinkiewicz's design by Rzeszow's Transport Equipment Factory. Tests conducted on some of these prototypes at the Technical University of Gdansk showed several shortcomings; however, publicity of these machines known as 'Krets' (the Polish word for moles) had generated widespread interest. So a research and development project was established at the Technical University of Gdansk under the leadership of Professors Gerlach and Zygmunt to improve the device to make it suitable for general commercial use. After careful redesign the first model was put into large-scale production with more than 4000 examples of this model being produced in Poland at Gniezno and Skarzysko. This model was patented by Professor Gerlach in 1958 and sold in many different parts of the world including the USA, Russia, Japan and Europe. A similar machine known as the 88KZ was developed and patented by Professor Zygmunt to produce bores greater than 88 mm in diameter. From 1966 onwards the 88KZ model was manufactured under licence by the Schramm West Chester Company of Pennsylvania under the name of 'Pneuma GopherKS'.

In 1964 the Soviet Union began to take a keen interest in the 'Kret' technology and machines were developed and tested on projects in Siberia and Moscow during that year. This resulted in the Ministry of Medium Machine Building authorising large-scale production and a major factory was established in Odessa. Over the ten-year period between 1970 and 1980, 26 000 machines of this type were produced. As the requirement for larger machines developed in the 1970s the Soviet Union built up a substantial industry manufacturing these machines. Two further factories were established at Minsk and Omsk. Large-scale production continues at all of these factories and machines are also made under licence in several other countries, notably Germany and the USA.

The early impact moles produced in Poland and Russia were heavy for the size of the intended bore and frequently gave problems such as large deviation from the proposed route or the unit being lost in the ground. They were, however, simple in design and rugged in manufacture and met with acceptance in Eastern European countries. The introduction of the equipment to Western Europe and the USA in the mid-1960s initially met with an enthusiastic reception but the problems with directional stability and reliability resulted in the discrediting of moling techniques.

During the late 1960s the Russian equipment was successfully modified independently by German companies Essig and Tracto-Technik to produce a reliable mole which provided consistent results. The successful introduction of these machines overcame the reluctance to use the moling technique and helped to establish it as a viable alternative to conventional open trench methods. Since this time numerous manufacturers have entered the market and the equipment has been continually developed. Impact moles now have reversing features, are capable of being tracked at the surface using electronic transmitters and can also be steered. A full account of the development of impact moles is given in papers by Etherton (1985) and Flaxman (1999).

Since its inception the impact mole has proved to be a very versatile piece of equipment and has been adapted to form the basic element for other more sophisticated trenchless systems. In particular the following two techniques originate from the impact mole concept:

1. The pipe ramming technique is a non-steerable system of forming a bore by driving a steel casing, usually open ended, using a modified impact mole.
2. The pipebursting technique involves breaking an existing pipe by brittle fracture, using mechanical force from within, the pipe fragments being pushed into the surrounding ground. At the same time a new pipe, of equivalent or larger diameter, is drawn in behind the bursting machine. The pneumatic machine variant of this technique is based on a modified impact mole, which converts forward thrust into a radial bursting force.

This often leads to some confusion for engineers who are inexperienced in trenchless pipeline technology as they have problems in differentiating between these various systems. Table 12.1 provides a description of each of these systems and a guide to the capabilities and applications for each technique.

12.2   The technique

Impact moling is the formation of a bore by the use of a tool which comprises a percussive hammer within a torpedo-shaped cylindrical steel body; Fig. 12.2 shows a range of typical pneumatically powered impact moles. The hammer may be hydraulically or pneumatically powered. The technique is usually associated with non-steered or limited steering devices, which have no rigid attachment to the launch pit and rely upon the internal hammer action for forward movement to overcome the frictional resistance of the ground. During operation the soil is displaced, not removed and normally no lubrication or fluids are used to assist in the creation of the bore.

Dependent on the ground conditions an unsupported bore may be formed or a pipe or cable may be drawn or pushed in immediately behind the impact mole.

Impact moles are known by several other names including virgin ground moles, earth piercing tools, soil replacement hammers, percussive moles and Grundomats depending on the term used by the manufacturer and the region of the world where the equipment is used.

12.2.1   THE EQUIPMENT

The basic mechanism of the impact mole is the reciprocating action of the pneumatically or hydraulically powered hammer within the cylindrical steel body. The piston is driven forward and on striking the forward end of the unit, imparts its kinetic energy to the body which is driven forward. The energy of the piston for the return stroke is regulated so as to reposition it for the next forward stroke, rather than reversing the unit out of the bore. Fig. 12.3 shows a cross-section of an impact mole, which illustrates the working parts. Repeated impacts of the hammer piston advance the whole unit through the ground. As forward movement takes place, the soil in front of the mole is forced aside and compacted by the conical or stepped nose to form the walls of the bore. The power of the unit is often used to pull the product pipe, cable or duct through the bore at the same time as the impact mole advances.

A distinction needs to be drawn between impact moles presented in this chapter and hydraulic devices that operate by expansion rather than hammer action. Expanding hydraulic machines are generally used for pipebursting applications rather than new installations.

Table 12.1 Comparison of trenchless techniques where impact hammers are used to provide the motive force

	Technique		
	Impact moling	Pipe ramming	Pipebursting
Description	Used for the installation of new small diameter pipelines over short distances	Used for the installation of new pipelines over short distances. Normally a sleeve is installed into which the carrier pipe is inserted	Used for the on-line replacement of existing pipelines which are constructed from brittle materials such as clayware, concrete, asbestos cement, cast iron, etc.
Ground conditions	Most compressible and displaceable soils such as most clays, sands and gravels. Isolated cobbles and boulders may deflect the mole. Difficulties may arise in soft soils such as wet peaty or silty soils	Most soil conditions ranging from top soil to semi-soluble rock. It is not suitable for use in peaty or marshy soils. Difficulties may be encountered if the ground conditions change over the length of the bore	Most cohesive and granular soil conditions. The performance of the device will be affected by the presence of rock, boulders or artificial obstructions such as concrete adjacent to the pipeline being replaced
Length of drive	Normally up to 15 metres. Capable of achieving up to 70 metres in favourable conditions	Varies up to 30 metres for 200 mm diameter to 80 metres for 1200 mm diameter	Varies normally between 20 and 120 metres
Replacement pipe size and type	20 to 180 mm diameter PE, UPVC, steel	200 to 2000 mm diameter steel	75 to 1000 mm diameter PE, UPVC, steel, clayware
Access requirements	Variable typically 1.5 metres long by 0.5 metres wide	Variable 3 metres long by 1 metre wide minimum	Variable 3 metres long by 1 metre wide minimum
Steering and accuracy	Not steerable within 1% of bore	Not steerable within 1% of bore	Line and level similar to existing pipeline. Controlled by winch cable or rods within the existing pipe
Applications	Property service connections, short road, river and railway crossings	Short strategic crossings under roads, railways and waterways	Size for size or upsize replacement of existing gas, water and sewer pipelines

Hydraulically powered impact moles are available but are not widely used, as pneumatically powered moles are favoured for the following reasons:

1. Pneumatic moles are easier to use, as they are lighter and require only one feed hose. Hydraulic moles need both a feed and return hose.
2. Hydraulic moles tend to be constructed with a greater mechanical complexity which makes them more expensive and difficult to service and maintain. They are, however, in general more powerful than pneumatic moles.

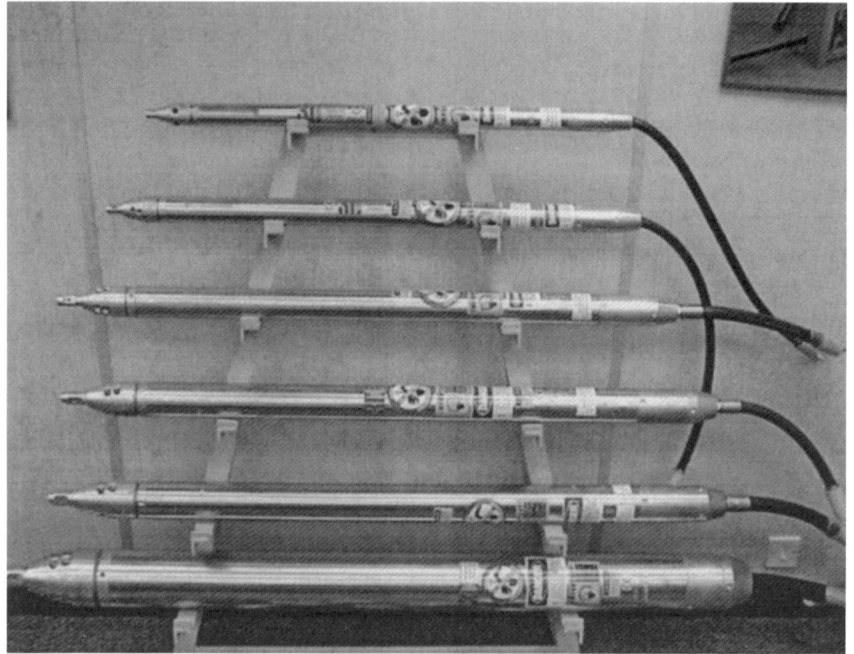

Figure 12.2   Range of typical impact moles.

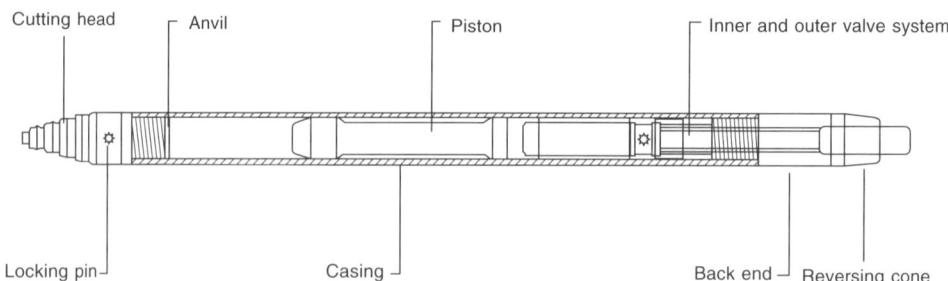

Figure 12.3   Cross-sectional diagram of an impact mole.

3. Due to the greater weight and complex mechanical construction hydraulic moles tend to be less responsive to flexing during operation which can lead to directional stability problems.
4. Pneumatic moles have a tendency to freeze up when operated at low ambient temperatures; hydraulic moles do not suffer from this problem.

Hydraulic moles have only found acceptance in countries such as Sweden, Norway and certain parts of the USA where there is a need to operate them in cold climatic conditions for long periods.

The impact moling technique is capable of installing pipes, ducts and cables in the diameter range 25 mm to 200 mm and Table 12.2 provides details of the various pneumatically powered moles which are capable of undertaking this installation work. Pneumatically powered impact moles are usually driven by compressed air at a pressure of between 6 and

Table 12.2   Technical details for impact moles

Mole dia. (mm)	45	50	65	75	95	110	130	145	160	180	200
Boring dia. (mm)	45	50	65	75	95	110	130	145	160	180	200
(inch)	1.8	2.0	2.6	3.0	3.75	4.3	5.1	5.75	6.3	7.0	8.0
Length (mm)	920	1122	1290	1300	1690	1890	1730	1809	1950	2150	2563
(inch)	36.0	44.0	51.0	51.2	67.0	73.0	68.0	71.2	77.0	85.0	101
Weight (kg)	8	11	25	29	67	96	127	138	216	290	408
(lb)	18	24	55	65	148	212	279	305	476	639	900
Air consumption (m^3/min)	0.45	0.6	0.7	0.9	1.2	1.6	2.4	3.7	4.2	4.5	8.7
(ft^3/min)	16	22	25	32	42	56	85	132	148	160	308
Operating pressure (kPa)	590–690	590–690	590–690	590–690	590–690	590–690	590–690	590–690	590–690	590–690	590–690
(psi)	85–100	85–100	85–100	85–100	85–100	85–100	85–100	85–100	85–100	85–100	85–100
Oil consumption (1/hour)	0.1	0.1	0.15	0.15	0.25	0.35	0.55	0.65	0.65	0.65	0.75
(pt/hour)	0.18	0.18	0.26	0.26	0.44	0.62	1.0	1.14	1.14	1.14	1.32
No. of strokes/min	570	470	470	480	315	280	350	300	320	275	223

7 bar. The efficiency of the performance of the mole will be adversely affected if the pressure is lower than this and the mole will be damaged if excessive pressure is used.

A pneumatically powered impact mole requires constant lubrication to ensure that it performs to its potential and therefore an in-line lubricator should always be used with this equipment. In cold weather during operation the compressed air expands in the mole and as a result the mole may cool down to –30°C. As a consequence water in the soil around the mole freezes and forms an ice coat. This ice coat brings the mole to a standstill even at full impact power. To overcome this problem it is necessary to use a lubricating oil with antifreeze properties and in extreme cold a compressed air heater will be needed.

12.2.2   IMPACT MOLE CONSTRUCTION

The construction of impact moles is relatively simple, as they comprise a main body casing which contains a piston hammer and a compressed air control valve. The body casing is attached to a conical mole head that can be either a simple cone or a chisel head which is effectively a stepped cone, examples of these heads can be seen in Fig. 12.4. During operation the simple cone head pierces the ground and pushes the soil aside and the chisel head also operates in a similar way when it is used in normal conditions as the spaces between the steps fill with soil to form a simple cone. However, when the chisel head strikes an obstacle, the stepped edges concentrate the impact energy against the obstruction. Whereas a smooth cone would tend to be deflected by an obstacle, the stepped shape may

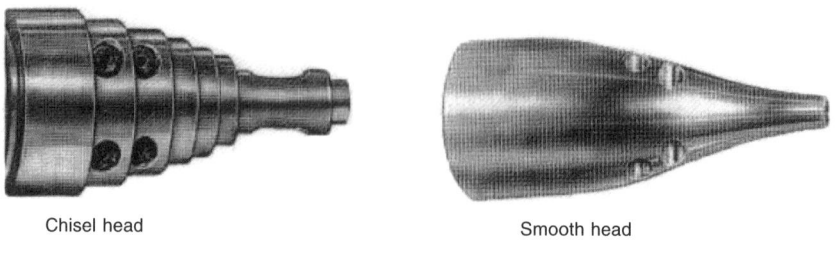

Chisel head   Smooth head

Impact mole fitted with expanders for installing drainage pipes

Figure 12.4   Examples of impact mole heads.

apply sufficient longitudinal force to move the obstruction or shatter it reducing the risk of going off-line.

Impact moles are available with either a fixed or moving head configuration, both types of design exhibit different forms of characteristics through varying ground conditions. It is generally considered that the moving head equipment is more stable and therefore provides greater accuracy for the bore, but there are many thousands of fixed head machines in operation worldwide which are forming bores with a degree of accuracy sufficient for the purpose. With a fixed head mole the head becomes an integral part of the mole body once the unit is assembled and when the piston operates, it acts on the whole of the mole body propelling it forward. The moving head is a spring loaded reciprocating chisel head which is not directly attached to the main mole body, but floats on a shaft passing through the front of the mole, Fig. 12.5 provides an illustration of the moving head configuration. The rear of this shaft is an anvil against which the piston strikes. When the piston accelerates forward and strikes the anvil the impact moves the head forward independently of the main casing to form a small pilot bore. The inertia of the piston returning to its start position together with the expansion of the pretensioned spring, causes the casing to be pulled forward in the pilot hole. The body of the mole acts as an initial directional anchor to the head as it drives forward, giving better directional control. This system offers higher impact energy to penetrate harder ground and allows obstacles to be broken up.

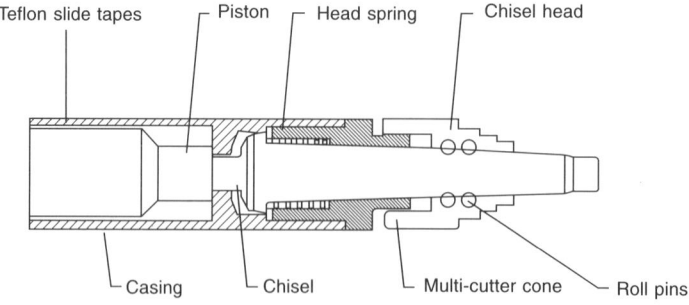

Figure 12.5   Moving head configuration.

If an obstruction is encountered that cannot be penetrated then most moles can be switched to reverse mode and brought back to the launch point.

12.2.3   ACCESSORIES

Accessories for alignment and starting

Once an impact mole is in the ground it will run in a straight line and its direction cannot be changed or corrected. It is therefore extremely important to align the impact mole as accurately as possible prior to commencing a bore and monitor its alignment as it enters the ground. To assist in the precise alignment of the impact mole to its target, a starting cradle is used as shown in Fig. 12.6. This starting cradle is secured to the base of the launch pit

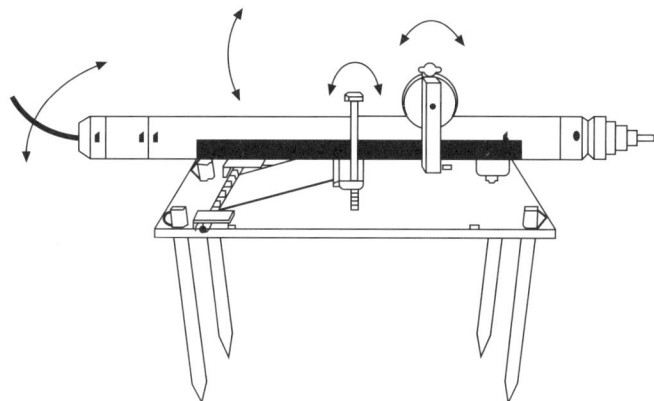

Figure 12.6   Starting cradle.

using earth anchors and can be adjusted vertically, horizontally and laterally by means of adjusting screws. If the soil in the starting pit is very soft or loose, the cradle should be supported on boards or fixed timbers. The starting cradle should be fixed as close as possible to the face of the pit and this face should also be made vertical. After the impact mole has been placed on the starting cradle, to complete the alignment process, an aiming frame fitted with a telescopic sight is used to take aim on a staff in the reception pit, as shown in Fig. 12.7. When the impact mole first enters the ground the roller mechanism mounted on the launch cradle absorbs the impact recoil, until such time as the casing friction builds up and takes over.

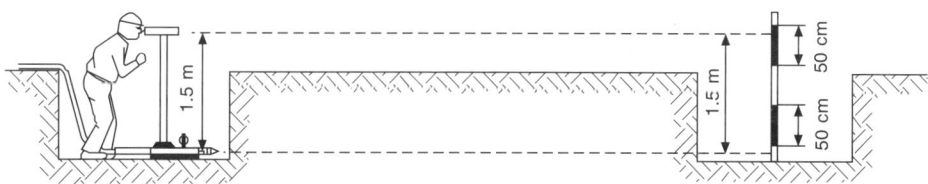

Figure 12.7   Schematic diagram of a site set-up showing the telescope and aiming staff in use.

Add-on cones

Impact moles can be easily adapted to operate in varying ground conditions by changing the design of the conical head of the machine. This is usually achieved by attaching a differently shaped displacement cone over the front end of the mole and some examples of add-on cones are shown in Fig. 12.8 and a description of their uses is given below:

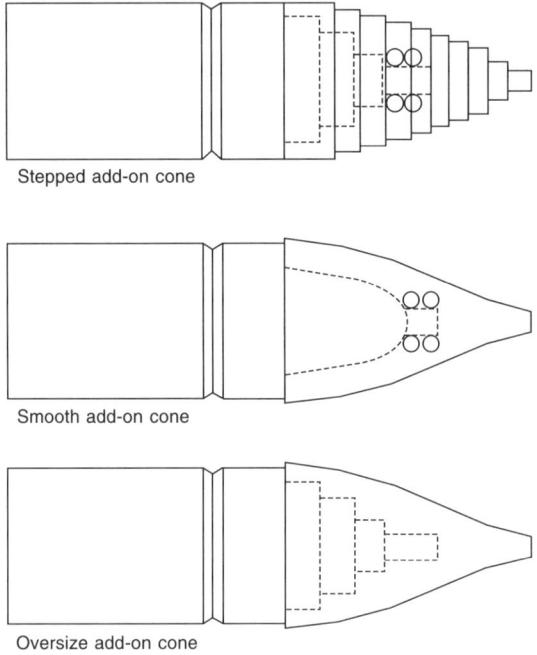

Stepped add-on cone

Smooth add-on cone

Oversize add-on cone

Figure 12.8 Examples of add-on cones for impact mole heads.

1. A stepped cone is used in soils that contain numerous stones and hard inclusions, as these heads are capable of either displacing or breaking up this material. They are also effective in sand and gravel soils. In homogeneous soils with only small amounts of stone inclusions a cone with a finer series of steps can be used to give optimum performance.
2. A smooth cone helps to reduce the friction between the surrounding ground and the casing of the machine and is used to improve installation speeds in sand and hard clay soils.
3. Oversize cones are used in soils that tend to return to their original position after initial compaction by the impact mole. Depending on the soil type the degree of shrinkage of the bore can be up to 10% and this causes the soil to grip the body of the impact mole preventing it from moving forward. These cones are slightly bigger than the mole casing diameter and allow the casing to move more freely in the ground so that propulsion speeds can be increased. Generally they are used in medium stiff clay and dense sand soils and are available in different diameters to suit most conditions. The use of these cones also prevents damage to the impact mole such as casing fractures due to idling strokes when the mole is stationary.

Pipe installation equipment

Accessories are available to allow impact moles to install pipes, cables or ducts in the following ways:

1. A pipe or duct can be installed by directly towing it in behind the impact mole during the bore creation process. This involves replacing the rear reversing cone on the impact mole with a towing adapter and attaching a pulling cable to this adapter. This cable and the air hose from the rear of the impact mole are then threaded through the pipe to be installed and the pipe is then attached to the impact mole using a clamping plate as shown in Fig. 12.9. Where larger diameter pipes such as sewers are to be installed over longer distances Fig. 12.10 shows how assistance to the impact mole's progress can be given using a specially adapted clamping plate which incorporates a pulley wheel and hand operated winches. This installation method is favoured for sewer pipeline installation and is especially favoured where the soil in the borehole has a tendency to collapse. It is important that a pneumatically powered impact mole is always capable of exhausting either through a stable borehole or a towed pipe as a blocked exhaust passage will cause the impact mole to stall underground.

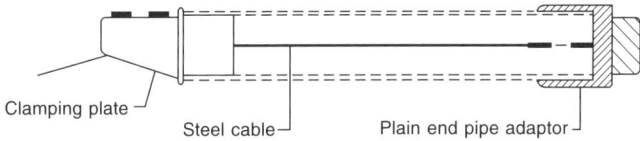

Figure 12.9 Clamping plate arrangement to hold a pipe in the rear of an impact mole.

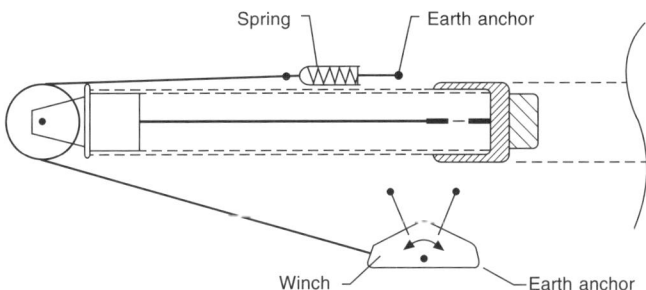

Figure 12.10 Clamping plate with pulley wheel tensioning device.

2. After a bore has been completed the new pipe or cable can be attached to the nose of the mole and towed in by the mole operating in reverse mode. This system is favoured for the installation of small diameter service pipes and cables.
3. On completion of a bore the mole can be disconnected from the air hose and using the adapters shown in Fig. 12.11 the new pipe can be connected to the air hose. The pipe is then installed by pulling back the air hose, through the bore, to the launch excavation. This method is again generally used for installing small diameter pipes and cables.

Monitoring systems

Most impact moles can now be fitted with radio sondes that allow the progress of the mole to be monitored closely both in direction relative to planned course and in depth. This is

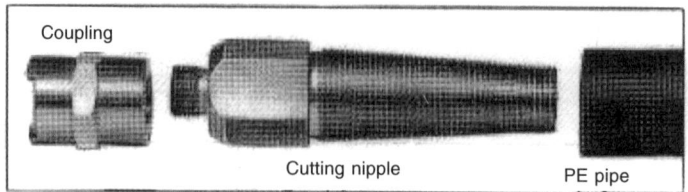

Figure 12.11 Adapter to allow the compressed air supply hose to be used to pull the new pipe into the bore created by the impact mole.

achieved by using a receiver at the surface which picks up the radio signal emitted by the sonde and translates this into data relating to depth, position and angle of inclination of the mole in the ground. The sonde can be fitted either to the rear or within the front end of the mole.

Although rear-mounted sondes give an indication of progress, they provide less useful information than front-mounted units. Depending on the mole size and length, the sonde can be some distance from the penetrating end of the tool, and therefore responds much later than a front-mounted sonde to changes in bore path. Front- or nose-mounted sondes react immediately to changes in direction and pitch, and so give the operator more time to halt the bore and assess what action to take. However, front-mounted sondes have to be more robust and well protected, as they must withstand the shock of the impact forces applied to the front of the unit by the hammer action.

While most impact moles are non-steerable, there are a few steerable machines which normally use steering vanes outside the body to apply corrective action. The latest development is the introduction by the Tracto-Technik Group of the steerable impact mole called the Grundosteer. This machine has a head that is independent of the main body of the mole and can be rotated by turning the air hose. This provides rotation of the machine body by using a hydraulic tensioning unit to turn the machine clockwise into the required position. A friction sleeve located behind the slanted head of the mole reduces the friction and eases the rotation of the main casing. Changes in direction are achieved due to the head having a slanted profile, which is similar to the drill heads used for directional drilling. A sonde is incorporated into the head and this allows monitoring to be undertaken at the surface, using a radio receiver system like those already described.

12.3 New pipes

The impact moling technique can be used to install a wide range of pipes manufactured from materials such as polyethylene, PVC and steel. To ensure the most favourable conditions for pipe installation the pipes should have an in-wall jointing system with no external protrusions. Standard spigot and socket joints will cause drag friction to build up during installation especially in soils, which are susceptible to bore collapse as this may prevent the bore from being completed. It is recommended that the outside diameter of the pipe should be 10% smaller than the bore diameter. Where there is no alternative to a spigot and socket jointed pipe this clearance should be increased to 15%.

Polyethylene pipes can be supplied to site either in continuous coils, which can be cut to length to suit the particular project requirements, or in 6 or 12 metre straight lengths, which can be jointed using butt fusion techniques to form a pipe with smooth inner and outer surfaces of the required length.

PVC pipes are usually supplied to site plain ended in various lengths to suit the particular application. They are jointed using an internal sleeve connector as detailed in Fig. 12.12. These connectors are suitable for use when pressure is exerted from behind, as is normal for sewer pipeline installation, since they ensure that the walls of the pipes are butted up to each other which allows the forces to be transferred across the pipe joints.

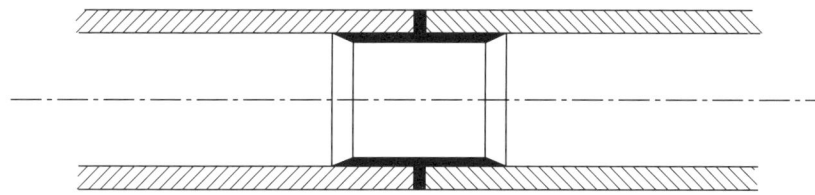

Figure 12.12   Internal sleeve connector.

Steel pipes can be supplied to site in various lengths and may have plain ends for jointing using butt welding methods or pre-formed screw threaded joints. To reduce friction on the outer surface of the steel pipe it is advisable to use a lubricant where bores are undertaken over long distances or in heavily compacted soils where high surface friction is anticipated. A shock absorber is usually positioned between the impact mole and the steel pipes when a threaded jointing system is used as this absorbs the percussive blows and the recoil action of the mole and thus prevents damage to the pipe threads. This pipe system is often chosen for the installation of sewer pipes below the water table as these joints can prevent the ingress of water.

12.4 Ground conditions

Impact moles can only be used in soils that are capable of being compressed or displaced such as clay, sand and some gravels. This is because the compacting action of the impact mole must be able to displace the soil particles into the voids within the soil structure to allow a bore to be created. Table 12.3 provides a summary of suitable ground conditions for impact moling. Obstacles along the bore path can deflect or stop a mole, so a thorough site investigation is essential prior to work commencing, in order to establish a clear route. This should include not only a knowledge of existing utilities, but also soil sampling to ensure that cobbles and boulders are unlikely to impede progress.

The type of ground and soil conditions in which an impact mole will operate has a major influence on the overall performance and accuracy of the machine. The following factors should be taken into consideration when assessing potential mole performance at any particular site before the bore is started:

1. The type of soil and its degree of homogeneity: These are major factors in determining the course the mole will tend to follow. As a general rule homogeneous soils of wet and/ or hard clays will cause the mole to rise upward towards the surface; homogeneous sandy soils will cause the mole to run a fairly level course; and homogeneous sand and gravel soils will cause the mole to run downwards. Difficulties may arise when impact moles are used in soft soils such as running or moist sands, made-up ground, peaty or silty soils and mud or in soils which contain cavities and water pockets. In these soils there may not be sufficient cohesion between the ground and the mole to provide forward motion and even if there is forward motion the borehole may collapse behind

Table 12.3   Suitable ground conditions for impact moling

Soil type	Suitability for impact moling	Comment
Peat	No	Mole may deviate downwards due to its sheer weight. Progress will be slow due to lack of friction between the ground and the mole. The bore produced is likely to collapse directly behind the mole
Silt	No	Mole may deviate downwards due to its sheer weight. Progress will be slow due to lack of friction between the ground and the mole. The bore produced is likely to collapse directly behind the mole
Soft clay	Yes	Care should be taken to ensure that the mole does not deviate downwards
Firm/hard clay	Yes	Mole may rise upwards towards the surface as the bore progresses
Dry sand	Yes	Mole should run to a fairly level course. Bore may collapse behind the mole, it is advisable to install a pipe directly behind the mole
Wet sand	No	These soils do not provide sufficient cohesion between the ground and the mole to provide forward motion. Even if there is forward motion the bore may collapse behind the mole. Mole will also deviate downwards due to its weight
Dry/wet gravel	Yes	Large particles may deflect the mole off course. The presence of cavities and water pockets may cause the mole to deviate

the mole as it penetrates. To overcome these problems it is advisable to install rigid pipes behind the mole as these will maintain the bore opening and allow a force to be applied to the rear of the machine to combat the low cohesion forces. In addition in soft soils the mole may deviate downward due to its sheer weight or vibration; however, to limit this effect the air supply should be reduced by 50%. If the soil conditions change within the course of the bore, the mole will tend to run towards the more easily displaceable soil thus causing directional problems.

2. The degree of compactness of the soil: The degree of compactness and moisture content of the soil are factors that govern the rate of penetration of an impact mole. As the mole penetrates into the ground it displaces the soil both forward and laterally. It can be obviously concluded that in well-compacted soils, this displacement will be slower than in loosely compacted soils. Therefore the rate of penetration is slower in hard compacted soils and increases in soils of lesser compaction.

3. The moisture content of the soil: This influences the soil's ability to be deformed and displaced and consequently affects the rate of penetration of the mole. Wet soils will flow and deform more readily than dry soils and therefore the rate of penetration is greater in wet soil conditions. The highest speeds are obtained in moist sandy soils with a moisture content of 12 to 17% while the lowest speeds are obtained in clay and loam type soils with 4 to 6% moisture content. The moisture content of the soil also affects the frictional force that is generated between the mole and the soil. This force is the principal requirement for the operation of an impact mole to provide forward motion and as the soil moisture content increases, the rate of penetration decreases due to greater slippage occurring.

4. The depth of the bore: If the bore depth is insufficient then the mole will either tend to deviate upwards or cause the soil to be forced upward giving heave at the surface. To avoid these problems the depth of the bore should be at least ten times the diameter of the mole. A survey of the contours of the ground above the line of the bore should be undertaken prior to commencing the bore to ensure that the minimum depth of cover is maintained over its full length.
5. Obstructions: The impact moling technique cannot be used where the ground conditions are predominantly rock or in soils which contain large rocks and boulders or other large obstructions.

12.5 Pipe installation

12.5.1 SITE PREPARATION

Prior to any work commencing at a site it is important to undertake a thorough site survey with the object of determining the following information:

• The prevailing ground conditions
• The position of other utility plant in the vicinity of the proposed line of the moling operation
• The position of any physical or artificial obstructions along the proposed route of the bore.

Once it has been established that it is possible to create a bore using the impact moling technique it is necessary to determine the best positions for the launch and reception pits. To reduce the risk of damage to other utility plant it is always best to bore from the congested plant areas to the non-congested side and an illustration of this can be seen in Fig. 12.13 which shows a schematic diagram of a typical site set-up. The access pits should be kept to the minimum required size for the mole that is to be used but allow sufficient safe working space for the operators. The base of the launch pit should be flat and parallel to the line of the bore. In soft ground conditions it may be necessary to use timbers to form a stable base to operate from. Provision should be made at the reception pit for the removal of the mole and any requirements, such as lead in trenches, for the new pipe.

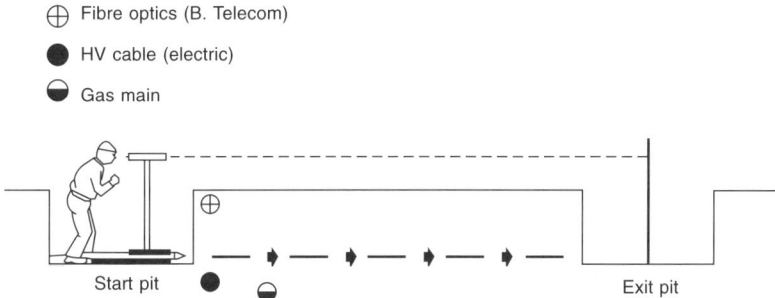

Figure 12.13 Schematic diagram showing correct launch arrangement in relation to existing utility services.

12.5.2   *INSTALLATION METHODS*

There are two main procedures generally adopted for the installation of new pipes using the impact moling technique, the choice of which is dependent on the diameter of the bore to be created, although ground conditions can also influence the decision.

Standard procedure

This method is the most extensively used and involves the installation of pipes and cables up to 75 mm in diameter over relatively short distances. An impact mole is used to create the bore and then the new service is normally installed after the bore has been completed. While this system is not normally used to install a sewer pipe directly it can be used to complete a pilot bore that can be subsequently enlarged using a bigger second mole to allow pipes of greater diameter to be installed.

Initially the impact mole is connected to a compressed air supply via an in-line lubricator which is fitted with a control switch. A distance marker should be fixed to the machine supply hose so that the position of any obstruction can be located and the progress of the mole can be monitored. Additional markers may also be fixed to indicate the expected positions of other underground plant.

A launch cradle is installed in the launch excavation and secured to the ground using earth anchors. The impact mole is then placed on the cradle and clamped to it using the roller mechanism. The level and alignment of the mole is checked either by eye or by using a sighting telescope and ranging rods. Precise horizontal and vertical alignment is achieved by using the adjustment screws on the launch cradle. The impact mole is now started and run into the ground at reduced power as can be seen in Fig. 12.14. During this procedure the mole's line and level should be checked regularly as the accuracy of the bore depends on correct starting alignment. When the mole first enters the ground the roller mechanism

Figure 12.14   Impact mole being launched into the ground.

on the launch cradle absorbs the impact recoil, until such time as the casing friction builds up and takes over. Special care should be taken during the launching operation to ensure that the operator is protected from direct metallic contact with the equipment to avoid accidents, which may result from unintentional cable strikes. It is important that there is permanent supervision of the mole during its boring progress. If the machine runs too fast (more than 15 metres/hour) or the air hose starts twisting back and forth in the launch pit (this is normally due to the impact mole operating without soil friction), there is a risk of deviation from the given course. In both cases the air supply must be reduced using the control switch on the lubricator.

If during the course of the bore an obstruction is met which cannot be penetrated by the mole then by engaging reverse mode the mole can be brought back to the launch pit.

When the mole arrives at the reception pit, the air supply must be reduced in order to avoid idle strokes. Idle strokes may cause damage to the mole if they are allowed to occur frequently. The operator should not stand in the reception pit during the boring operation as there is a risk of being struck by the mole as it arrives and the ground may also collapse as the mole enters the pit.

On completion of the bore the new service can then be installed either by:

- Attaching the service to the nose of the mole and pulling it with the mole in reverse mode
- Disconnecting the mole from the air supply hose and then using the hose with special adapters to pull the service back through the bore.

When the bore is to be enlarged to accommodate a larger diameter pipe, such as a sewer pipe, a winch is established at the reception point and its cable is installed within the pilot bore. This cable is then attached to the mole, which is to be used for the upsizing. The winch acts as a guide for the second mole and also provides additional forward pulling force to assist in the installation of the new sewer pipe. The new pipe is normally installed directly behind the mole during the bore enlargement process.

Procedure for the direct pulling in of pipes

The second system, which is used to install pipes in soft ground conditions or larger diameter pipes over longer bore distances, involves the simultaneous towing in of the new pipe directly behind the mole during the bore formation.

To operate this system the impact mole is fitted with a special adapter that allows the new pipe to be fixed to its rear end. The pipes used normally have an in-wall spigot and socket joint so that they have a smooth outer surface with no external protrusions. The individual pipes are held together in tension by the use of a cable which is attached to the rear of the mole. This cable is tensioned by means of a clamping plate, which is located behind the string of pipes. Once the air hose and the tensioning cable have been installed within the new pipes and the cable has been tensioned the installation procedure followed is the same as that previously described.

In soft or wet soil conditions such as sands and gravels, where there is a danger of the soil collapsing, additional static pressure may be applied to the rear of the pipe being installed. To achieve this a clamping plate with a guide pulley is used instead of the ordinary clamping plate in conjunction with a manually operated hand winch with a steel cable and spring assembly. The winch is anchored at one side of the launch cradle and the steel cable is passed around the pulley and secured to the spring, which is anchored at the other side of the launch cradle. During installation pressure must be exerted continuously via the pulley. The spring overcomes overload of the winch tension and prevents subsequent damage to the pipes and joints.

Additional rear pushing force can also be applied by the use of a pipe pushing unit and a diagrammatic representation of this method being used for the installation of a sewer service connection pipe can be seen in Fig. 12.15

Figure 12.15   Schematic diagram showing the construction of a sewer connection using impact moling.

12.6   Applications

Since the impact moling technique is generally unsteered, it is normally used for short length bores as the accuracy of the bore decreases significantly as the length of the bore increases. Normally the maximum length of bore attempted is 30 metres; however, it is more usual for the system to be used for bores of up to 15 metres in length. At this length of bore an accuracy in line and level of ±150 mm can generally be achieved. Moles of 90 mm to 200 mm diameter can be used over longer distances as their greater length to diameter ratio provides better directional stability. In ideal conditions lengths of 70 to 100 metres have been achieved with these larger moles.

The technique is used to install pipes in the diameter range 45 to 200 mm for short strategic crossings under railways, roads and water courses. When longer bores are required, to achieve greater accuracy, these can be split into two or more shorter bores or a pilot bore can be driven which can be subsequently enlarged as described earlier.

Impact moling is used extensively in Germany and North America to complete new connections from properties to the main sewer in the road. This technique avoids the need to disturb established gardens, footways, drives and landscaped areas and allows the connection to be completed quickly in a cost-effective manner. For the successful installation of sewer pipelines using this technique it is essential that the pipeline to be laid has a gradient greater than 1 in 75. This will provide an allowance to accommodate for any deviations in the line and level that may occur during the installation process due to the inability to steer the mole. If impact moles are used to install pipelines with shallow gradients then it is possible that an unacceptable result will be produced as the pipeline may have backfall and low points. The development of moles with steerable capabilities will improve the results that can be achieved and should result in wider use of the technique for sewer pipeline construction.

Impact moles are relatively easy to use, monitor and maintain in the field and despite their limitations they are successfully used to install a broad range of utility services.

References

Etherton, P. T. (1985) Developments of impact moling techniques in the UK – design, use and cost considerations. Paper given at the No Dig International Conference, London.

Flaxman, E. W. (1999) The development of the percussive mole. *No Dig International*, March 1999 issue.

13

The Pipe Ramming Technique

Norman Howell

13.1   Introduction

Pipe ramming is a non-steerable system of forming a bore by driving a steel casing, usually open ended, using a percussive pipe pushing machine from a drive pit. These machines are similar to the pneumatically powered soil displacement hammers which are used for impact moling; however, in this process they remain located within the launch excavation and drive the steel pipe by dynamic action. When they are used for impact moling the hammer moves through the ground creating a bore by soil displacement. A new cable or pipe may then be drawn or pushed into the bore immediately behind the impact mole. A full description of the impact moling technique is provided in Chapter 12. Since these percussive hammers are used as the power source for a number of trenchless techniques such as pipebursting, pipe ramming and impact moling confusion sometimes arises as to the differences between these methods. To clarify these differences Table 12.1 provides a summary of the essential features of each of these techniques.

In the pipe ramming process steel pipes can be pushed or driven horizontally or at a desired angle. As the bore proceeds the soil is pressed inside the pipe to form a plug and this can be removed either continuously or intermittently during the boring process or as a whole once the bore has been completed. In appropriate ground conditions a closed casing may be used. Figure 13.1 illustrates a typical site set-up for the technique.

Once a drive has commenced it is not possible to steer the pipes, so accurate alignment of the steel pipes at the launch point is needed to achieve the required directional stability.

Normally this technique is used to install a new sacrificial steel casing into which new utilities can be placed. However, it is possible to install product pipes such as insulated gas pipes directly without damaging the insulation or welding joint surfaces.

Pipe ramming is often favoured for short strategic crossings and is capable of installing pipes in the diameter range of 200 to 2000 mm over distances of between 20 and 80 metres.

13.2   The type of pipes to be installed

13.2.1   PIPE CHARACTERISTICS

Due to the method of installation only steel pipes are suitable for pushing as no other material is strong enough to withstand the forces generated by the ramming hammer. The pipes to be driven do not need to possess high mechanical or stress properties. Generally they can be welded longitudinally or spirally or could be seamless with or without an outer

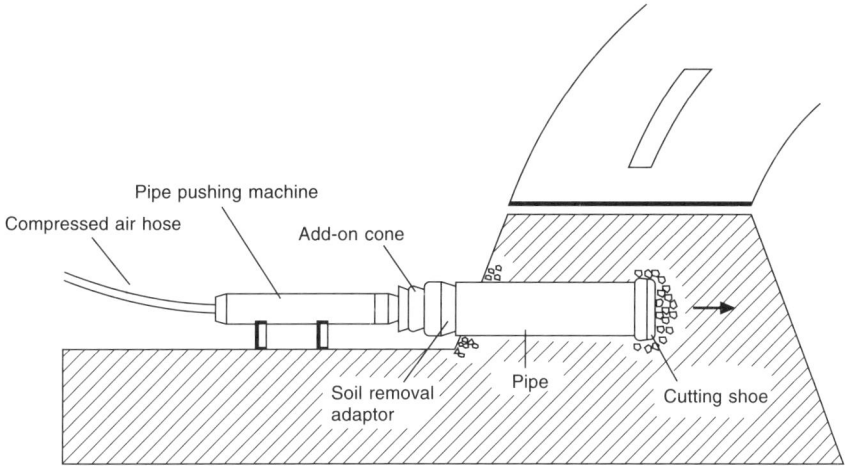

Figure 13.1   General site arrangement for pipe installation using the pipe ramming technique.

protective coating. The wall thickness of the pipe depends on the diameter and length of pipe to be installed. It is important that the correct choice is made, since the impact energy has to be transferred to the front of the pipe length to overcome peak resistance and casing friction. It is not possible, for example, to push a 400 mm diameter pipe with a wall thickness of 5 to 6 mm over a length of 60 metres as instability would result which would have an effect on the propulsion and also the accuracy of the bore. Table 13.1 provides recommendations for pipe wall thickness in relation to pipe diameter and length of bore.

The diameter of the pipe to be installed is initially determined by the function it is required to perform and where it is to be used as a sleeve an allowance must be made for the product pipe installation. There are, however, other reasons to select a pipe of larger diameter than may be needed for the chosen purpose and these include:

- For long drives where a larger diameter pipe may offer greater stability than a smaller diameter pipe which might be subject to deviation
- When the steel pipe is being used for a product pipe that requires a precise gradient such as a sewer pipeline. A larger pipe can be installed to permit adjustment of the product pipe within the sleeve. Table 13.2 provides recommended lengths of bores in relation to the pipe diameter. As a rule of thumb for accuracy the maximum length of bore for smaller pipes up to 800 mm diameter can be derived by dividing the pipe diameter in mm by 10, e.g. for a pipe diameter of 250 mm the maximum bore length is 25 metres.

13.2.2   THE CUTTING SHOE

A specially prepared cutting shoe is attached to the leading end of the first length of steel pipe and this has the following basic functions:

- Strengthens the pipe cross-section at the leading edge and enables obstacles to be penetrated and broken up
- Helps to avoid deformation of the leading edge which can cause steering bias
- It both cuts and displaces the soil and serves to reduce the surface friction and scuffing on both the internal and external pipe walls by producing a slight overbreak
- Protecting coatings or insulation on the pipe to be driven.

Table 13.1 Recommended pipewall thickness in relation to pipe diameter and length of bore

Nominal pipe diameter (mm)		100	150	200	250	300	350	400	500	600	700	800	900	1000	1100	1200	1400	1600
Minimum pipe wall thickness (mm)	For bore lengths up to 25 metres	6–7	6–7	6–7	6–7	6–7	6–7	7–8	8–10	10–12	12–15	12–16	12–16	15–18	15–18	15–18	18–20	20–24
	For bore lengths up to 50 metres	–	–	–	–	12–14	12–14	12–14	12–14	12–14	12–14	14–16	16–18	16–18	18–20	18–20	20–24	24–26

Table 13.2 Recommended lengths of bore in relation to pipe diameter

Nominal pipe diameter (mm)	100	150	200	250	300	400	500	600	700	800
Recommended maximum bore length (metres)	10	15	20	25	30	40	50	60	70	80

Cutting shoes can be constructed in various forms but they are all designed to produce an oversize cut. They can consist of an outer annular cutter with or without an inner cutter and the shape of the cutting shoe has an important influence on the accuracy of the bore. The cutting shoe should have a conical internal surface to reduce soil displacement to a minimum. The design of a typical cutting shoe and how it cuts the soil during pipe installation is shown in Figure 13.2. For light and water retentive soils, the outer cutting ring should encompass only the top two-thirds of the pipe's circumference to avoid a cutting action at the bottom. This reduces the risk of pipes sinking due to the weight of pipe and earth.

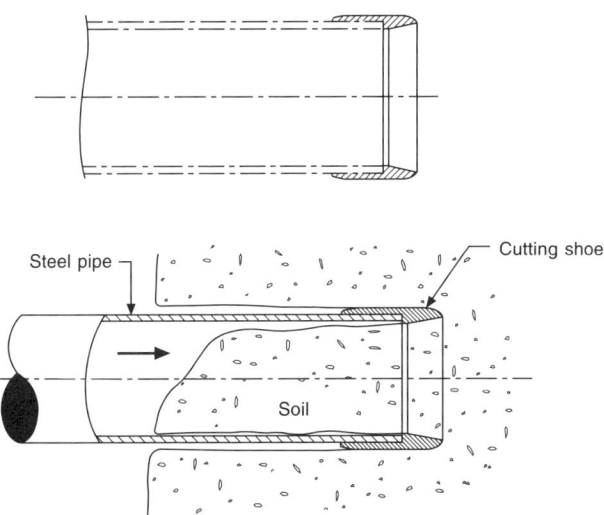

Figure 13.2   Design of a standard cutting shoe and how it produces an oversize cut during the pipe installation operation.

Additional outer annular cutters can be welded on at spacings of approximately 5 to 6 metres to assist the installation process when frictional resistance is high.

Cutting shoes can be either purpose made factory produced units or manufactured on site. When they are made on site they should reach around the whole of the pipe circumference and have a cutting edge to overcome peak resistance. Irrespective of the type of shoes used it must always be secured in place by welding before the pipe driving commences.

13.2.3   PIPE PREPARATION AND WELDING

The steel pipes are jointed by using a standard full penetration butt weld and the welding preparation for the pipe ends can be carried out either on site or the pipes can be supplied with a chamfer for welding. To ensure good accuracy for the bore the individual pipes must be welded competently and in exact alignment with one another. It is also essential to ensure that the entire string of pipes is absolutely airtight when the compressed air/water method of soil removal is to be used.

If taper locking ram cones or adaptors are used to position the impact hammer into the end of the pipe there is a possibility, during the ramming operation, that these cones may cause the pipe end to flare or in extreme cases to split. This belling is due to the impact force on the conical connection between these centralising accessories and the pipe. When this occurs it is necessary to cut off approximately 10 cm from the pipe end to allow a welded butt joint to be formed with the next pipe to be installed. This flaring can be eliminated by simply welding a steel reinforcing ring to the pipe end prior to the ramming operation commencing, although a more satisfactory method of overcoming this problem is to use segmental inserts with a cylindrical outer surface, in preference to tapered accessories, to engage the impact hammer into the pipe.

An alternative joining method that can be adopted is to use pipes that have been prepared with a socket attached. This system requires very little weld preparation since the socketed pipe end is pushed over the end of the previously installed pipe and a fillet weld to the outer surface of the pipe is used to complete the joint. While this jointing method is quicker and thus reduces installation times, greater care is needed to ensure that no deviation of the bore occurs.

13.3   Equipment

13.3.1   PIPE PUSHING MACHINE

The pipe ramming technique utilises a pneumatic driving hammer which is rigidly taper locked into the pipe end. It is a cylindrically shaped machine, which is powered by air from a standard mobile compressor operating at 6 to 7 bar pressure. The machine consists of only three major components: a main casing or body incorporating a conical front striking head or anvil, a high impact piston and an air control device. Figure 13.3 shows a standard driving hammer in use on site.

Compressed air causes the heavy piston to reciprocate several hundred times a minute in the main casing, impacting against the front end anvil on each forward stroke. The impact force that is generated produces a forward movement of the steel pipe through the

Figure 13.3   A standard pipe ramming machine.

ground. The energy of the piston on the return stroke is regulated so as to reposition it for the next stroke rather than producing a reversing action. Forward ramming speed is directly influenced by the number of impact strokes per minute but where high ground resistance and friction exist a low stroke frequency with a higher single impact is required.

Ramming hammers are designed to have sufficient impact power to overcome the ground resistance at the leading edge of the pipe together with the dead weight of the soil inside the pipe while maintaining high ramming speeds.

There is a wide variety of machines available, ranging in diameter from 95 to 600 mm with impact thrusts of between 90 to 2000 tons, to suit any combination of pipe diameter and length of pipe to be installed. In addition to the standard machines a range of mini ramming hammers are produced to allow the technique to be used from within small access pits. These machines are between 1 and 1.2 metres long and are designed to incorporate a rear cone attachment so that they can be almost entirely inserted into the pipe, as illustrated in Figure 13.4, and this significantly reduces the pit length required. It is also possible to use them as conventional ramming hammers with the machine sitting behind the pipe and connected to it using ram cones. Table 13.3 provides technical data on some of the pipe pushing machines that are available together with their air consumption so that the most appropriate compressor can be selected. It is essential for the efficient operation of the pipe pushing machine that the compressor chosen delivers at least the minimum recommended volume of air. In practice it is normal to use a compressor that has a substantially greater capacity than is required to power the impact hammer as large volumes of air are usually required at a later stage in the operation to evacuate soil from the pipe bore. Figure 13.5 provides a guide to the choice of the correct pipe pushing machine in relation to pipe diameter and bore length for average mixed granular soil types.

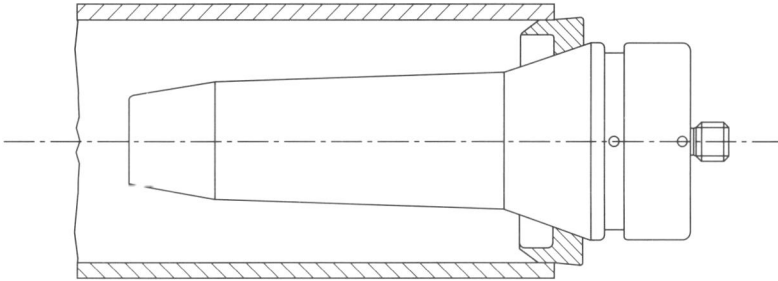

Figure 13.4   A mini pipe pushing machine inserted into a pipe using a rear cone attachement.

During the operation it is important to provide a supply of lubricant to the pipe pushing machine as failure to do so will have an adverse effect on its performance. Lubricant is fed into the machine via an in-line lubricator, this prevents freezing of the compressed air moisture and ensures permanent lubrication. For extreme cold conditions a special compressed air heater may be required to prevent freezing.

13.3.2   RAMMING ACCESSORIES

Ramming cones

To obtain accurate alignment and maximum efficiency the pipe pushing machine must be in solid contact with the pipe and sit axially and centrally behind the pipe. This can be

Table 13.3 Pipe pushing machine technical data

Machine diameter (mm)	95	130	Mini 130	145	180	Mini 180	220	270	Mini 270	350	Mini 350	450	600
Overall length (mm)	1460	1455	905	1545	1690	1080	1915	2010	1230	2345	1850	2855	3465
Weight (kg)	60	95	60	140	230	175	370	615	460	1180	940	2465	4800
Operating pressure (bar)	6–7	6–7	6–7	6–7	6–7	6–7	6–7	6–7	6–7	6–7	6–7	6–7	6–7
Number of strokes/minute	345	320	580	310	280	500	340	310	430	220	300	180	180
Air consumption m^3/minute	1.2	2.7	1.7	4.0	4.5	3.5	8.0	12.0	10.0	20.0	16.0	35.0	50.0
Minimum diameter of pipe that can be installed (mm)	50	50	50	100	100	100	120	200	200	280	280	380	380

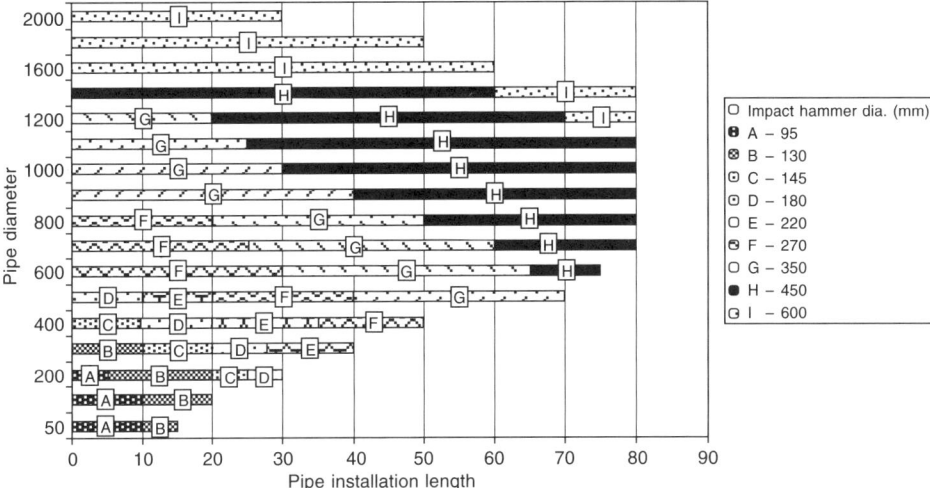

Figure 13.5 Guide to the selection of pipe pushing machines in relation to pipe diameter and bore length.

achieved by positioning the conical head of the machine directly into the pipe but more commonly a conical driving cone is used which corresponds to the diameter of the pipe. The cone is placed in the rear of the pipe and the head of the pushing machine is centrally locked into it. Using interlocking add-on cones as shown in Fig. 13.6 the ramming hammer can be adapted to match the diameter of the steel pipe to be installed. These cones ensure effective impact transmission onto the pipe.

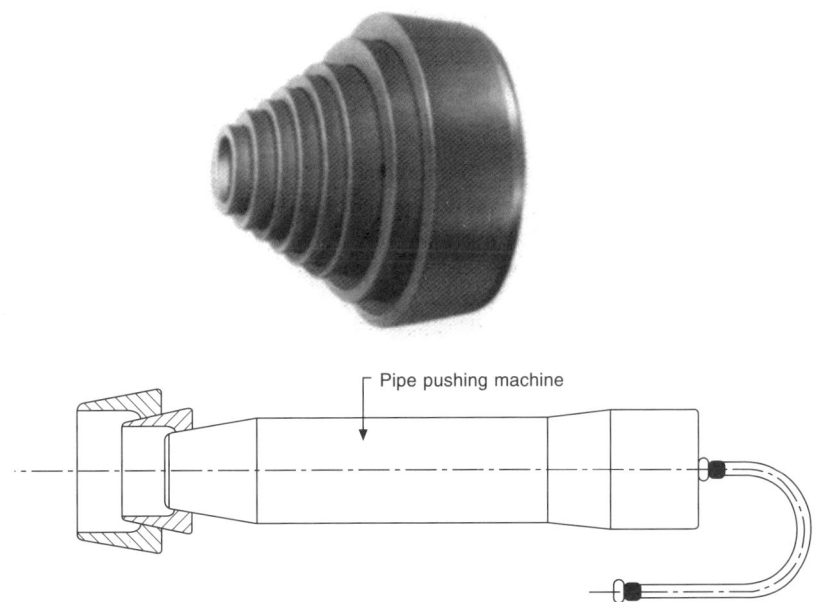

Figure 13.6 Interlocking ram cones.

Segmental ram cones

Segmental ram cones, which are illustrated in Fig. 13.7, are cylindrical at their outer surface and not conical like ramming cones. They fit inside the end of the pipe and due to their shape they cannot fall out. Their purpose is to prevent pipe end flaring which avoids the need to cut off deformed pipe ends prior to joint welding.

These cones can be combined with tapered ram cones to allow different sizes of ramming hammer to be used.

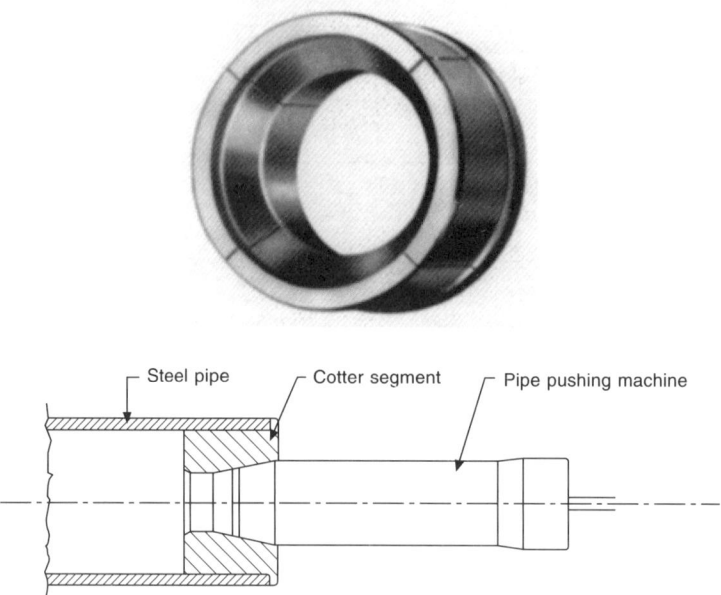

Figure 13.7   Segmental ram cones.

Soil removal adaptor

A soil removal adaptor can be incorporated between the driving cone and the add-on cones to allow the accumulated soil to be removed from within the pipe during the ramming operation. Figure 13.8 provides details of a typical soil removal adaptor. The adaptor is a cone-shaped pipe section with openings on either side through which the soil is continuously discharged; this also relieves the internal pressure within the pipe. Therefore with this arrangement there is no need to interrupt the installation operation in order to remove the soil. The success of this system of soil removal is dependent on the soil characteristics and the degree of soil compaction within the pipe.

13.4   Site preparation works

13.4.1   GENERAL

The pipe ramming technique does not require any fixed abutments or thrust walls for its operation as the dynamic forces produced by the impact hammer are transmitted directly to the steel pipe. The system can therefore be used above ground to install pipes through

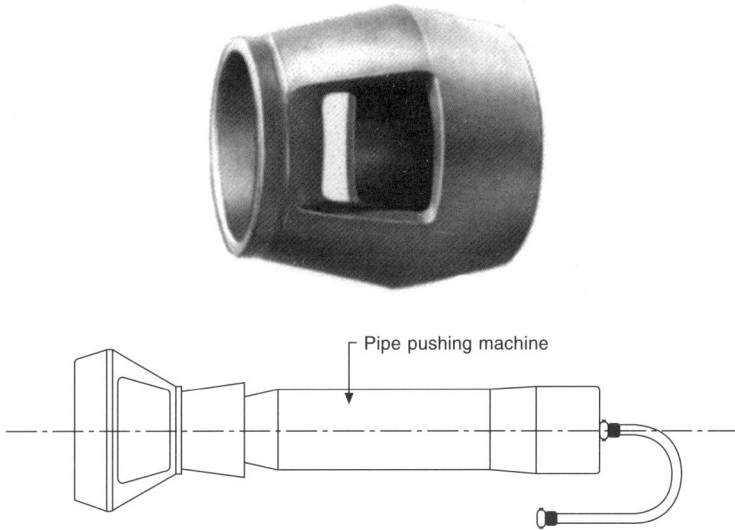

Figure 13.8 Soil removal adaptors.

embankments or slopes as well as below ground from within excavations. At the launch site a large space is normally required as this allows the maximum length of steel pipe possible to be used. This helps to minimise problems with bore alignment as fewer joints are needed so the potential for angular deflection is reduced and also the installation time is substantially lowered since fewer welded joints are required. The length of pipe which is generally used varies from 3 metres to 8 metres with 6 metres being the most favoured length. The technique is not restricted to the use of long lengths of pipe where the site areas are restricted, for example in urban locations, pipes with a length of only 1 metre can be used in conjunction with special short pipe pushing machines. This allows the system to be operated from within small excavations; however, great care must be taken with the joint welding during installation to ensure accurate alignment is achieved.

A typical ramming operation does require the establishment of a solid base on the launch side of the installation and this working surface can be constructed with imported granular backfill or more usually with concrete. The base will be located either against the side of a slope or within a launch pit.

13.4.2 ACCESS EXCAVATIONS

Launch pit

The system can be operated from either an excavation that is fully supported or from within one that has the side walls battered back. To provide the necessary working space at the base of the excavation the width of the pit varies dependent on the size of pipe to be installed and Fig. 13.9 provides recommendations for the minimum widths needed for different pipe diameters. The length of excavation required varies dependent on the length of the individual pipes to be driven and which pipe pushing machine is to be used. Therefore to calculate the length of pit required it is necessary to take these factors into consideration together with the length of the ram cones to be used and include an allowance for working space at both ends of the pit.

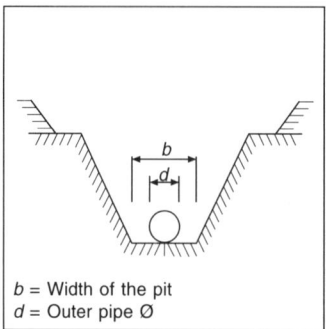

b = Width of the pit
d = Outer pipe Ø

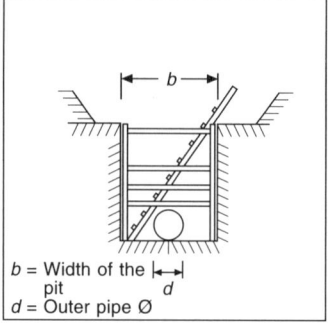

b = Width of the |←→|
pit   d
d = Outer pipe Ø

Outer pipe diameter (d) in mm	Minimum width (b) in mm			
	Pit with lining		Pit without lining	
	Normal	Special	B ≦ 60°	B > 60°
Up to 400	b = d + 400	b = d + 700	b = d + 400	
400–800	b = d + 700		b = d + 400	b = d + 700
800–1400	b = d + 850			
More than 1400	b = d + 1000			

Figure 13.9   Recommended minimum launch pit widths for different pipe diameters.

Reception pit

There are no fixed requirements for reception sites as only a small area is required. The minimum size is usually dictated by the need to provide a safe working environment. To minimise excavation and disruption it is possible to utilise existing manholes or chambers as the reception point.

Launching cradle

Since the system is non-steerable the accuracy of the bore is determined by precise alignment of the pipes at the launch site. In particular the set-up of the first length of pipe to be installed is critical to achieving the desired result. It is therefore important to provide support at the launch pit for the pipes to prevent vertical and horizontal drift. To provide this a launch cradle is installed at the launch point and this can take the form of purpose-built tracking with bogies or be constructed from a combination of steel beams, sheet piles and timber baulks as shown in Fig. 13.10. The cradle should be set back approximately 1 metre from the front face of the pit to leave enough space for a welding hole and should be secured with concrete. A firm alignment base is especially important for long bores.

   A machine support is normally incorporated into the construction of the launch cradle and this allows level adjustment between the pipe pushing machine and pipe to be undertaken. This can be a simple static support or can include a compressed air cushion which provides easy level adjustment. A typical launch cradle set-up is illustrated in Fig. 13.11.

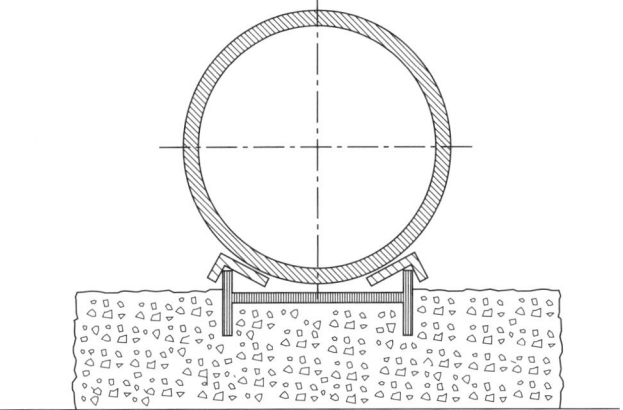

Figure 13.10   Launch cradle alignment arrangement.

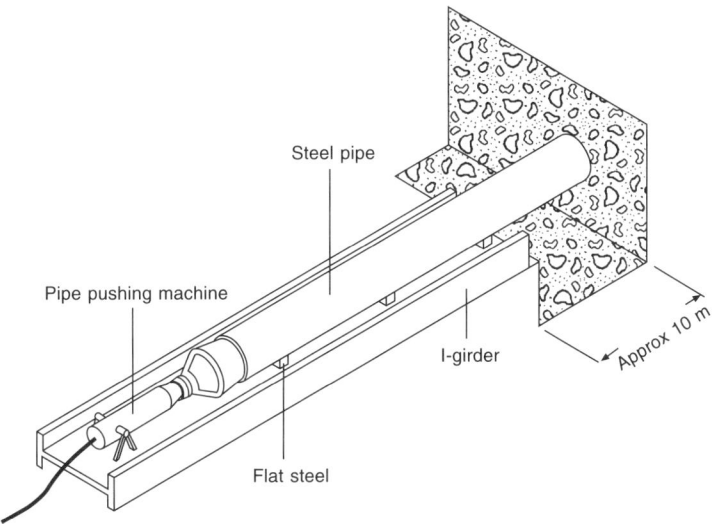

Figure 13.11   Launch cradle general arrangement.

13.5 Pipe installation

13.5.1 GENERAL

The pipe ramming technique can be undertaken using either an open-ended or close-ended pipe and the choice of which method to use is dependent on the prevailing ground conditions. Open-ended pipe ramming is generally preferred as it has several advantages which include:

- Lower reaction against the ramming force, since only the cutting edge is pushed against the ground
- Harder ground can be penetrated as the soil does not have to be compressible
- As the surface area of pipe presented to an obstruction is far less, there is also less likelihood of the pipe deflecting.

For open-ended ramming the ground has to be relatively self-supporting, otherwise there may be loss of ground ahead of the cutting edge as soil moves into the open pipe and flows along it to the launch pit. In severe cases, this could cause surface subsidence or loss of support to adjacent pipelines. Close-end ramming may be effective under such conditions, as soil is displaced around the wall of the bore. However, this method is only usually suitable for smaller pipes up to 300 mm in diameter due to the substantial amount of soil compaction that is required to complete the installation. There is also a risk of surface heave during a compacting bore and to avoid this the depth of cover should be at least ten times the diameter of the pipe.

When using an open-ended system, the cylinder of ground within the circumference of the cutting edge stays inside the pipe during the bore. Over the short distances normally undertaken with pipe ramming, this accumulation of spoil is not usually a problem. However, for longer bores it should be remembered that the spoil adds to the weight of the pipe string being rammed and will therefore affect advance rates. In some instances it may be beneficial to clean out spoil from the pipe during pipe string extension works, to limit the extra burden on the ramming hammer.

If the soil offers considerable resistance, it is advisable to adopt a variation to the standard ramming technique which is shown in Fig. 13.12 and involves initially driving a small diameter pilot bore along the line of the proposed pipe installation. This allows a pulling force to be exerted in conjunction with the jacking force provided by the ramming hammer which improves both the overall efficiency of the operation and guidance of the main bore. The pilot bore is also a cost-effective means of determining accurate ground condition information including the presence of obstructions within the course of the bore.

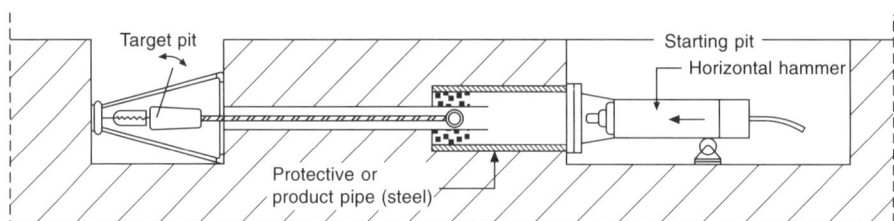

Figure 13.12   Adaption of the standard pipe ramming technique using a pilot bore and winch.

Pilot bores are generally driven using either soil displacement hammers or guided boring rigs with the pulling force being provided by a winch located at the reception point. The winch cable is established within the pilot bore and attached to the front of the leading pipe. This winching equipment supports the pipe ramming operation by exerting a pulling force of up to 20 tonnes.

13.5.2   INSTALLATION PROCEDURE

Prior to installation commencing the pipes are prepared and a cutting shoe is attached to the first length of pipe to be installed. This section of pipe is placed on the launch cradle and is adjusted for direction and gradient. Ramming cones corresponding to the diameter of the pipe and to the pipe pushing machine to be used are placed in the rear of the pipe. The pipe pushing machine is then aligned on its support and pressed into the cones. To form a solid connection with the pipe the machine is attached to the pipe by means of textile strops which have tensioning ratchets. The mobile compressor is connected and the pipe pushing machine is started. The pushing machine forces the steel pipe into the ground along the line dictated by the guide rails on the launch cradle.

Since a controlled start is important to the accuracy of the bore, the ramming should begin at a reduced rate of impact with constant observation being kept on the level and alignment as the direction cannot be altered at a later stage. This degree of control, to give a slow, smooth start, is achieved by regulating the supply of compressed air to the pipe pushing machine. When the pipe first enters the soil, the required surface friction is not on the pipe and this may prevent it from moving forward. Therefore at this phase of the bore it will be necessary to replace this friction by introducing additional thrust by using a winch tensioning device or an excavator bucket.

Once the first pipe has been installed, the pipe pushing machine and cones are removed and the end of the pipe is prepared for jointing. The next pipe is then aligned on the launch cradle and welded to the previously installed pipe. The ram cones and pipe pushing machine are reconnected to the pipe and the installation process is recommenced. This procedure is repeated until the leading edge of the first pipe arrives at the reception point.

The dynamic action transferred from the pipe pushing machine to the front of the pipe ensures that the soil formation and any stone inclusions or obstructions are effectively broken up. This prevents obstacles from being displaced laterally or pushed ahead of the pipe which could influence the accuracy of the bore. Provided the correct machine is chosen for the prevailing conditions its impacting force will overcome casing friction during the boring process and when starting from a standstill after a new pipe has been added. If problems are experienced with the installation progress it is a simple task to change the ramming machine for a more powerful model since the equipment is readily accessible at the launch site. Should an obstruction be encountered along the length of the bore which cannot be penetrated or removed then normally this leads to the installation being aborted. However this only results in the loss of the steel casing that has already been installed and usually it is possible to attempt a new bore in close proximity to the original one.

13.5.3 SOIL REMOVAL

In the majority of cases the accumulation of soil within the installed pipe is removed upon completion of the ramming operation. For longer bores or in soft soil conditions, the earth core enters the pipe at a rate faster then the propulsion speed and compacts to the front of the pipe pushing machine. This can cause the pipe to come to a standstill so in these circumstances intermittent soil removal is undertaken. When a certain jacking resistance is reached the operation is stopped and the pipe pushing machine and pipe are separated. The earth plug inside the pipe is removed either by injecting compressed air or water under pressure or by using mechanical scrapers. This causes interruptions to the installation process and to avoid this a soil removal adaptor can be incorporated between the ramming cones and the pipe pushing machine.

Soil removal after the completion of the pipe pushing operation can be carried out in one of the following ways:

- Removal of the soil core as a whole using compressed air or water
- Breaking up and removal of the soil with an auger, mechanical scraper or high pressure jetting device
- For pipes over 900 mm in diameter the soil can be manually removed.

The first technique is successful in the majority of cases and is also the most economic since the main items of equipment needed, such as the compressor, are already on site. This process does have an element of danger as air compressed at 7 to 8 bar exerts substantial pressure on the earth core. The expulsion of the soil results in a sudden release of pressure which can be explosive. Therefore adequate precautions must be put in place around the reception site to ensure the safety of people and avoid damage to adjacent properties.

The procedure involves either welding a pressure plate on the end of the pipe or placing a pressure sealing plug within the end of the pipe which is secured by two steel dowel bars. Compressed air is blown into the pipe through a valve installed in the pressure plate or plug to fill all the voids and then compress the soil further. This action results in the expulsion of the earth core. Figure 13.13 illustrates the set-up for this system. If the earth does not move this is likely to be due to the compressed air escaping past the earth core preventing sufficient build-up of air pressure. To overcome this a hard foam plug as inserted in front of the sealing plate and this enables pressure to build up as it prevents the forward escape of air. Should this still prove unsuccessful then the soil must be held inside the pipe either by strong internal friction or by the presence of wedged stones. If the problem is due to strong internal friction then it is likely that the power of the compressed air is insufficient to move the soil. To resolve this the pipe should be pressurised by using water or a combination of water and compressed air. Where jamming in the pipe occurs due to larger rocks locking against each other static pressure alone will not overcome the wedging effect. In fact increasing the pressure only serves to increase the wedging effect of the rocks. Jamming can be released only by axial and radial vibration being applied to the pipe. Figure 13.14 illustrates how radial vibration can be provided by an excavator bucket hammering on the outside of the pipe and how axial vibration can be achieved by offsetting the pipe pushing machine from the centre line of the pipe.

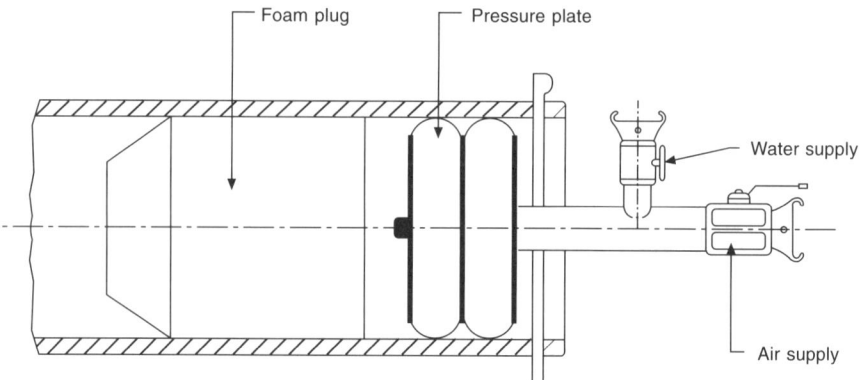

Figure 13.13   Pressure sealing plug arrangement for soil removal.

If having tried all of these techniques the soil has still not come out then the complete process should be attempted from the reception site. Only if all of these attempts have been unsuccessful should soil removal be undertaken using more expensive conventional systems such as auger boring in cohesive soils and pressure jetting in loose sandy soils or gravels.

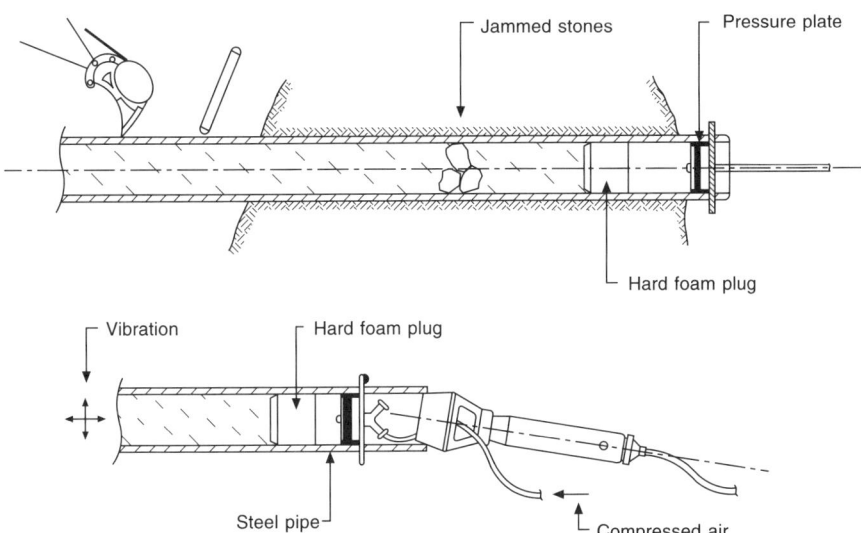

Figure 13.14 Radial vibration provided by a soil displacement hammer or an excavator bucket and axial vibration provided by the pipe pushing machine.

13.5.4 CARRIER PIPE INSTALLATION

Once the bore of the steel casing has been cleaned out the carrier pipe can be easily installed using slip-lining techniques. Spacers can be attached to the outside of the carrier pipe to provide level adjustment to ensure that specific gradients can be met. The treatment of the annular gap between the casing and the carrier pipe is dependent on the design requirements of the carrier pipe. If it has been designed as a standalone pipe the annulus can be left empty or filled with an inert material such as bentonite. This has the added advantage in that the carrier pipe can be removed in the future. The annulus can also be filled with a cementitious or chemical grout which can act simply as a filler or can provide structural enhancement to the carrier pipe if required.

13.6 Applications

It is possible to use the pipe ramming technique in a wide variety of soil types ranging from top soil to semi-soluble rock which includes most granular and cohesive soils. It is not suitable for use in marshy or peaty soils as the steel casing will tend to deflect downwards due to its sheer weight. The system is also not capable of penetrating solid rock. The dynamic impact propulsion is particularly advantageous for dry loose soils such as sand and gravel mixtures and non-homogeneous soils with a high stone content. If the soil conditions change within the course of a bore the casing will tend to move towards the more easily displaceable soil thus causing directional problems. Therefore a thorough ground investigation is an essential requirement for pipe ramming projects. This survey should not be restricted to establishing soil conditions, the presence and position of any obstructions should also be determined. These can include naturally occurring obstacles such as large boulders and manmade items such as old foundations or other utility services. Since there is no means of monitoring the direction of the pipe during a bore it is vital to establish a clear bore path prior to works commencing.

When the open-ended pipe system is used only minimum ground movement occurs as only the wall thickness of the pipe displaces the soil. This allows pipes to be installed with minimum ground cover and avoids disruption at the surface resulting from ground heave. Care should be taken when siting a bore adjacent to buildings or other services as the vibration from the impacting action may affect these structures.

Whilst the pipe ramming technique is generally used for pipeline installation it can also be used for a variety of other construction purposes. These include ramming vertical steel pile casings, providing raft foundations for structures by ramming a series of casings horizontally in parallel and providing arch or circular support structures for tunnelling beneath embankments by ramming a series of interlocking casings in a predetermined pattern.

Figure 13.15   Installation of a small diameter pipe using the pipe ramming technique.

The principles of pipe ramming are relatively simple which makes the transfer of knowledge of the system easy and has resulted in its widespread uptake throughout the world. For sewer projects it offers a highly cost-effective solution for short strategic crossings beneath railways, roads and waterways and is used extensively for this purpose in Europe and North America. It has also been adopted for the installation of new small diameter house sewer connections and this application is particularly favoured in Germany. Figure 13.15 shows the installation of a small diameter pipeline for this application and Fig. 13.16 shows the bore achieved and highlights the excellent line and level that can be gained by using this method to install short length services. In other parts of the world the system is being used to avoid disruption to major arterial routes when new sewerage systems are being constructed. The versatility of the technique, which allows a carrier pipe to be installed within the sacrificial steel sleeve pipe, helps to provide adjustment of the gradient of the service pipe and accommodate shallow gradients. This has been a major factor in the increased use of this technique for the installation of both gravity sewer pipelines and sewerage pumping mains.

Figure 13.16   Completed installation showing the excellent line and level that can be achieved.

14

The Pipebursting Technique

Norman Howell

14.1 Introduction

The pipebursting technique is an on-line trenchless pipeline replacement method, which allows existing underground pipes to be completely replaced on a size for size or upsize basis. It was developed jointly by D. J. Ryan & Sons Ltd and British Gas plc in the late 1970s and patents for both the method and equipment were granted in many countries worldwide between 1980 and 1985. The system has the trade name PIM which is an abbreviation of Pipeline Insertion Method but it is often referred to by several other names and some of the more commonly used terms include on-line replacement, size for size replacement, upsizing, mains bursting and pipe cracking.

The technique was initially used in the UK for the replacement of ageing small diameter cast-iron gas mains and proved to be an extremely efficient replacement option. It was soon identified by other sectors of the utility industry as a system that could offer considerable financial and environmental advantages when compared with conventional methods of pipeline replacement as well as providing the social benefits of minimum disruption during the installation phase of projects. This resulted by 1985 in its widespread adoption in many countries with well-established pipeline networks for the renewal of small diameter gas and potable water mains.

Its potential for the replacement of larger diameter pipes such as sewers was also recognised at this time; however, further research was necessary to develop the technique for this application. In particular, it was necessary to adapt the system to cope with the different problems that are encountered when replacing large diameter pipes. Primarily there was a need to make the system more powerful to increase its ability to fracture the larger pipes, to overcome the greater ground forces which are present when pipes are located at greater depth and to achieve longer replacement lengths in one operation. By the late 1980s successful formats, based on the use of both pneumatically and hydraulically powered machines, had been introduced and since then the system's usage for the replacement of large diameter pipes has grown significantly. This increase in utilisation has provided opportunities to continuously improve the system and extend its capabilities to allow pipes in excess of 1 metre diameter to be replaced and substantial upsizing of existing pipes to be achieved. The technique has now evolved into a proven mainstream pipeline replacement system for old and undersized sewers, which is used extensively throughout the world.

14.2   Principle of the system

Pipebursting is used to replace defective or inadequate underground pipes with a completely new pipe of equivalent or larger diameter with only minimum excavation required. The main purpose of the technique is to replace pipes by utilising the void in the ground occupied by the existing pipe and avoid the need to excavate a new trench.

Essentially the system is based on the relatively simple concept of introducing a device which is usually conical in shape into a defective pipeline and moving it through the pipeline by either pulling or pushing by independent means, its own motive power or a combination of these. The device which can be of various designs such as a static cone or a pneumatically/hydraulically powered machine, fractures and fragments the existing pipe and displaces the fragments radially outwards into the surrounding ground. The soil adjacent to the pipeline is compressed to provide a void of sufficient size to accept the new pipe which is introduced directly behind the pipebursting device.

14.3   Scope of the system

To achieve successful pipe replacement using this method the existing pipe must have a brittle structure and be surrounded by a soil which is capable of being compacted and displaced. Typically the technique can be used to replace pipe which is constructed from fracturable materials such as cast iron, spun iron, clayware, unreinforced concrete, pitchfibre, asbestos cement, etc.

The method is not suitable for use on existing ductile pipes, for example those constructed of ductile iron, steel or polyethylene. Pipes that have a tendency to fail by splitting in one horizontal plane such as those constructed from PVC can cause problems to the pipebursting technique. These pipes will fold back around the replacement pipe as the bursting machine moves forward and this will eventually create sufficient drag to stop the replacement operation. To overcome this, a specially designed bursting head is required which induces more than one fracture in the existing pipe during the bursting phase.

The wall thickness of the pipe to be replaced is not generally an inhibiting factor. All fracturable pipes are susceptible to the method, but exceptional wall thickness may require the use of more powerful pipebursting devices. The addition of steel reinforcement to concrete pipes will normally render most pipebursting devices ineffective in such pipes. Similarly, steel banding which may have been used as a repair clamp may prevent pipebursting by most devices other than ones specifically adapted to deal with these circumstances. The presence of substantial quantities of concrete around the existing pipe in the form of pipe surrounds, at the positions of lateral connections or at the site of previous repairs is likely to prevent the progress of the replacement operation.

14.4   Replacement capabilities

Table 14.1 provides details of the capabilities of the system for the replacement of sewer pipelines and highlights both the wide range of pipe diameters to which pipebursting can be applied for size for size replacement and the substantial upsizing that can be achieved, for example 225 mm to 450 mm provides a 300% increase in pipe capacity. This table should only be used as a guide as there are many factors that can influence the degree of upsizing attainable. The two major influences are the ground conditions and the depth of the pipeline; however, other contributing factors include the degree of compactness and compressibility of the soil adjacent to the pipeline, the moisture content of the soil, ground water movement, construction of the existing pipe and width of the original trench. Since these factors alter from site to site it is important to evaluate each replacement on its own merits.

Table 14.1 Capabilities of the pipebursting technique for the replacement of sewer pipelines

Existing pipe size		New pipe (OD) in mm													
(Inch)	(mm)	90	125	180	200	250	315	355	400	450	500	560	630	710	800
3	75	♦	♦	♦											
4	100	♦	♦	♦											
6	150			♦	♦	♦	♦								
9	225					♦	♦	♦	♦	♦					
12	300						♦	♦	♦	♦	♦	♦			
15	375								♦	♦	♦	♦	♦		
18	450									♦	♦	♦	♦		
21	525											♦	♦	♦	
24	600												♦	♦	♦

There is no theoretical limit on the size of pipe that can be replaced. The method has been used in pipe diameters up to 1200 mm but normally is used for the replacement of pipes within the diameter range of 75 mm up to 600 mm.

The lengths of pipe that can be installed in one operation between entry and exit points will vary according to site conditions, the size of pipe being installed, the degree of upsizing, the power and type of equipment being used and the number of lateral connections. For sewer pipelines the system has been designed to be a manhole to manhole system with the normal length of replacement being approximately 100 metres. In optimum conditions it is possible to install up to 300 metres in one operation.

14.5 Ground conditions

There is always an element of ground movement when using the pipebursting method due to the requirement, at the very least, to slightly enlarge the void taken up by the existing pipe to permit the replacement pipe to be installed. The technique can be operated successfully in a wide variety of soil conditions but there must be scope for the existing void to expand outwards. An existing pipe that was originally installed in a narrow rock cutting will therefore only be capable of being renewed on a size for size basis at best; also an existing pipe encased in a substantial surround of concrete is unlikely to be a suitable candidate for pipebursting.

Table 14.2 provides details of the performance of different pipebursting methods in various ground conditions. Cohesive soils provide the best conditions for all pipebursting devices with firm to soft clays offering ideal ground to allow optimum replacement progress and major upsizing. Very soft clays and silts relax quickly after they have been compressed and tend to ease back onto the replacement pipe as it is being installed. This action produces additional frictional forces between the pipe wall and the soil increasing the drag factor that the pipebursting system has to overcome. Pneumatically powered or static cone systems have the necessary balance of bursting power and external motive force to operate successfully in this type of soil. Where hard clays are encountered a large amount of energy is required to move the soil radially to achieve the required soil compression. A very powerful pipebursting device such as a hydraulically powered or large pneumatically powered machine is normally used to provide the force needed to maintain adequate installation progress.

Table 14.2 The performance of different pipebursting systems in various ground conditions

Ground conditions surrounding the pipeline	Type of pipebursting equipment		
	Pneumatically powered machine	Hydraulically powered machine	Hydraulically powered rod system
Rock	Suitable Size for size only	Suitable Size for size only	Suitable Size for size only
Hard clay	Suitable High powered machine required for upsizing	Suitable For size for size and upsizing	Suitable For size for size and upsizing
Firm clay	Suitable For size for size and upsizing	Suitable For size for size and upsizing	Suitable For size for size and upsizing
Soft clay	Suitable For size for size and upsizing	Suitable For size for size and upsizing	Suitable For size for size and upsizing
Wet sand	Limited Flooding of the machine must be prevented	Not suitable Machines operate poorly in these soils	Suitable For size for size and upsizing
Dry sand	Suitable Upsizing capability may be limited in well-graded deposits	Limited Machine operation may be restricted due to ingress of soil	Suitable Upsizing capability may be limited in well-graded deposits
Wet gravel	Limited Flooding of the machine must be prevented. Not suitable where large cobbles and boulders exist	Not suitable Machines operate poorly in these soils	Suitable Presence of large cobbles and boulders will restrict the operation of the technique
Dry gravel	Suitable Operation will be limited in well-graded deposits and where large cobbles and boulders exist	Suitable Operation will be limited in well-graded deposits and where large cobbles and boulders exist	Suitable Presence of large cobbles and boulders will restrict the operation of the technique

Sandy soils provide the most difficult conditions as their lack of cohesion generally prevents an enlarged void being maintained during pipe installation and as a result the replacement pipe is subjected to additional frictional forces. The degree of compaction that can be achieved in sands is very much dependent on the particle size distribution within the soil structure and the moisture content of the soil. Wet, poorly graded sands have sufficient void spacing to allow the necessary compaction to be achieved and the water within the soil assists in the new pipe installation by acting as a lubricant. Clearly well-graded, dry, compacted sands will offer very little movement and are the most difficult ground conditions in which to undertake pipe replacement using the pipebursting technique.

Sand soils and in particular running sand conditions can be a problem for hydraulically powered machines as their design allows the ingress of particles into the machine, which prevents their efficient operation. Static cone or pneumatically powered machine systems cope best in these conditions. Pipebursting at levels below the water table can cause pneumatic devices to become flooded and prevent their operation. In these conditions static cone systems provide the most efficient replacement option, as the operation of the bursting device is not affected by the presence of water because it has no moving parts.

Care must be exercised in making the proper selection of the most appropriate device for the prevailing conditions.

14.6 Ground movement

The ground movement that takes place during the construction process when the pipebursting technique is used can give rise to concerns for the integrity of the overlying ground surface pavement structures and crossing or adjacent utility services. It is therefore important to determine what ground movement is likely to occur when the method is being proposed for a pipeline replacement project. This can be a particular issue when either large diameter or upsize pipe replacements are being considered where the pipe is located at a shallow depth. As the depth of the pipe increases the potential for surface disruption decreases; however, consideration must still be given to the effect on adjacent utilities. A comparison of the ground movement that occurs when open trench and pipebursting techniques are used for pipe renewal projects is shown in Fig. 14.1. This clearly shows the greater degree of disturbance to the surrounding ground that results from open trench works and as a consequence adjacent services will be at considerably more risk due to ground settlement than on projects where pipebursting is used.

Ground displacements caused by pipebursting result from an outward expansion of the existing pipe followed by an inward relaxation of the ground. This occurs because the pipebursting machine needs to be larger than the existing pipe to facilitate installation of the new pipe. The magnitude and extent of the displacements are dependent on the characteristics of the surrounding soil, degree of soil saturation, the pipeline depth, degree of surface containment and the volume of soil to be displaced. If the pipe being replaced is deep and well defined, as is normally the case with sewer pipelines, then the movements will tend to be radial, whereas in shallow unconfined cases they will be predominantly upward.

Most soils have the ability to undergo considerable compression and for this reason the outward movements tend to diminish rapidly away from the pipebursting machine. In addition the relaxation stage will often subsequently reduce or eliminate the movements that occur away from the pipebursting machine. This process can be accelerated by traffic movements and ground water fluctuation leaving the zone of significant residual displacement within the immediate vicinity of the pipebursting machine. Therefore for normal pipebursting operations consideration of temporary displacement to an adjacent service or structure need only be made if it lies within approximately two to three diameters of the replaced pipe, with significant permanent displacements occurring only closer to the pipe.

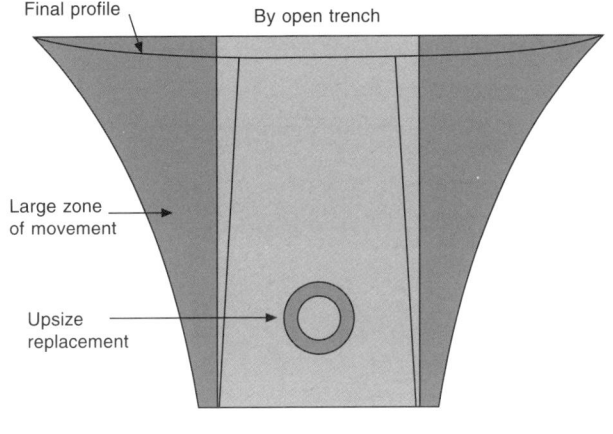

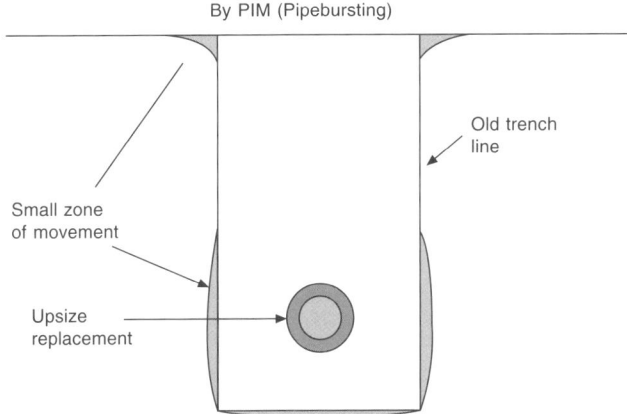

Figure 14.1 Diagrams showing the comparison of ground movement on pipeline renewal projects.

Where a large degree of ground movement is needed to form the void for the new pipe, as is the case with upsizing or large diameter pipe replacement, then more careful consideration will be required. However, significant movements are still likely to be restricted to an acceptable distance from the pipe and can be judged simply using soil mechanics principles. For shallow pipeline replacements, where the ground movements will tend to be mainly upwards, the influence of these at the surface in terms of uplift will be dependent on the soil type and the construction of any overlying structures. Soft to firm clays and poorly graded sands absorb some of the movement through compression and where there is a well-confined surface structure the amount of surface displacement is minimal. Hard clays and well-graded sands have a tendency to transmit all of the ground movement to the surface and uplift can then be a problem if the surface structure is weak. Generally where the pipeline being replaced is at a depth greater than 2 metres there is unlikely to be any significant problem with surface displacement. Figure 14.2 provides examples of the surface movement that was measured on two projects where the pipebursting technique was used for pipeline replaced. These projects involved substantial upsizing of 80% and 320% and the results are displayed diagrammatically, the maximum figures were obtained as the pipebursting machine passed under the measurement station. The final results showed only a few millimetres of movement and were completely acceptable as no damage was caused to the road surface.

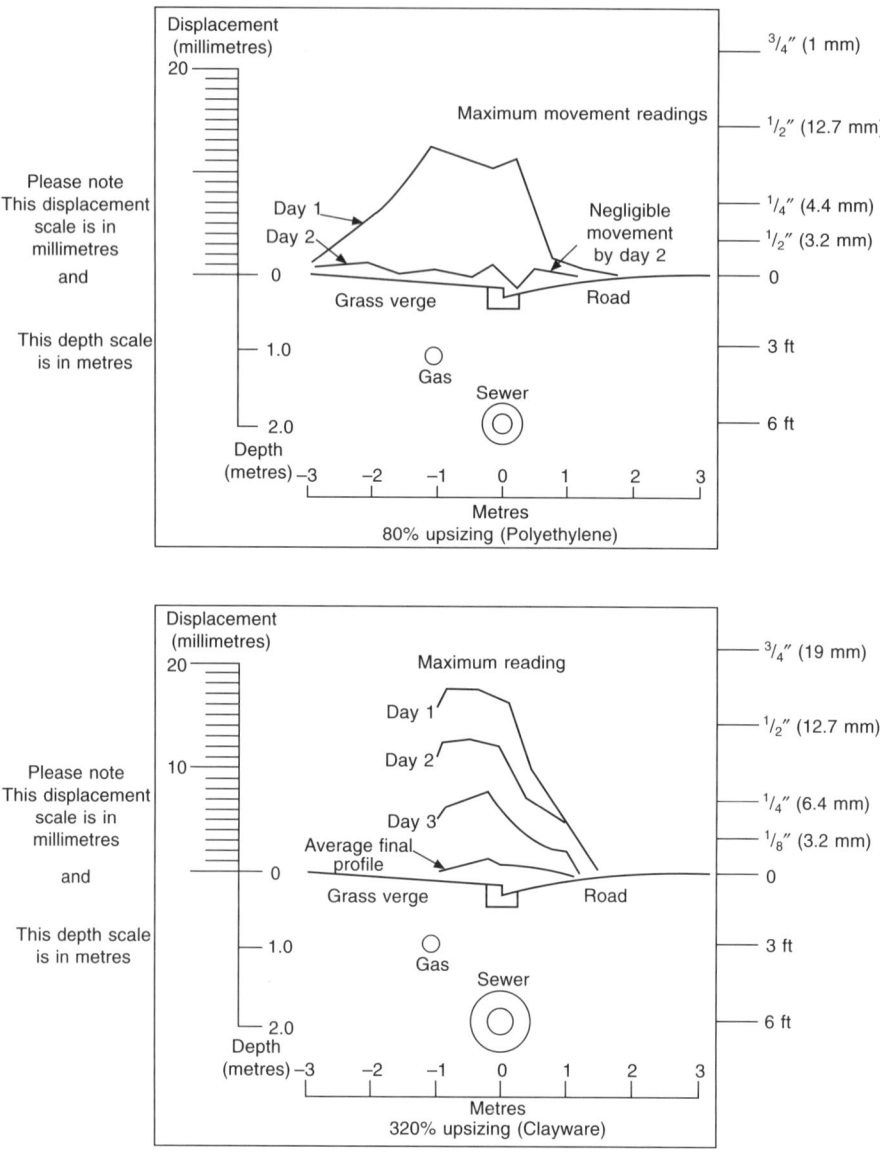

Figure 14.2 Diagrams showing recorded maximum surface movement on pipebursting projects where 80% and 320% upsizing was achieved.

To avoid any potential damage to services which cross over in close proximity to the pipeline being replaced it is normally good practice to expose the service within a small excavation and isolate it from the effects of the pipebursting operation by removing the soil from between the two pipes.

14.7   Replacement pipe options

The technique can be utilised to install pipes constructed of several different materials and employing various jointing systems and details of these are provided in Table 14.3.

Table 14.3   Sewer replacement pipe options for pipebursting

Type of replacement pipe	Format	Jointing system
Polyethylene	Continuous pipe string	Welded joint using butt fusion techniques
Polyethylene	Short length	Mechanical snaplock type jonts with an 'O' ring seal
UPVC	Continuous pipe string	Glued spigot and socket joint
UPVC	Short length	Mechanical locking joint with 'O' ring seal
Polypropylene	Short length	Mechanical screw threaded joint with 'O' ring seal
Clayware	Short length	Mechanical in wall joint with outer stainless steel sleeve and 'O' ring seal
Steel	Continuous pipe string	Butt welded
Steel	Short length	Butt welded

The pipebursting method was initially developed to install long continuous pipe strings of either UPVC or polyethylene pipes as this suited the technique's original application for the replacement of small diameter pipelines located at shallow depth. However, when the system was adapted for sewer pipeline replacement the use of long pipe strings restricted its use on some projects due to the prevailing site conditions. In particular the need for a lead-in trench to provide access to an existing pipeline located at great depth involved substantial excavation, which could in some cases negate the advantages of using a trenchless replacement method. Other problems involved accommodating the pipe string on restricted sites and maintaining property accesses during the replacement works. To overcome these problems short length pipes with mechanical joints were introduced and these are normally manufactured from polyethylene with either snaplock or screw threaded 'O' ring seal joints. Clayware pipes are also available and these pipes have stainless steel collars to provide enhanced shear strength at the joints. They are capable of withstanding higher jacking forces than most polymeric materials, although they are heavier, which makes them more difficult to handle on site.

When short length pipes are used in conjunction with hydraulically powered machine pipebursting equipment it is possible to operate the system from within existing chambers without the need to excavate access pits as can be seen in Fig. 14.3. Careful consideration of the ground conditions should be made before selecting the short length pipe option as the following problems can occur:

- In very wet soft conditions, such as running sand, joint sag may occur due to the pipes sinking under their own weight.
- In conditions where high skin friction is expected, such as very soft clay, there is the potential for damaging the joints due to the exertion of greater jacking forces during pipe installation.

Figure 14.3   A pipebursting operation using short length replacement pipes.

Generally the continuous pipe string option is favoured, as the pipeline is easier to install because there is no need to interrupt the replacement operation for pipe jointing. Since there are no joints along the pipeline length, infiltration, exfiltration and root ingress are eliminated. A typical site arrangement using string welded polyethylene pipe can be seen in Fig. 14.4.

Figure 14.4   Pipebursting operation using string welded polyethylene pipes.

While it is normal practice to install polyethylene pipes when using the pipebursting technique, it has been employed, when required, to install pipes manufactured from steel, UPVC and polypropylene.

14.8 Lateral and service connections

In sewers, due to the varying and sometimes novel configuration in which lateral connections are laid, no successful remote method for reconnection of laterals or services has as yet been devised. Reconnection is completed from within suitable excavations, which are normally completed prior to the replacement operation commencing to avoid damage to the lateral pipework. The connection of small diameter pipes into the new carrier pipe can be made with either a specially designed electrofusion saddle unit or a friction fit saddle unit. Large diameter pipes can be connected by cutting out a section of the carrier pipe and installing a junction pipe using either electrofusion or mechanical couplings to complete the jointing. The connection between the saddle and the existing lateral is completed with conventional clayware or UPVC pipes and mechanical couplings.

Sewer pipelines that have a substantial number of connections may prove uneconomic to replace using the pipebursting technique due to the large number of deep excavations that will be required. An alternative approach is to redirect the laterals at a shallow level to the nearest manhole and the remaining sections of pipework connecting into the original sewer can be abandoned.

Intermediate manholes can easily be accommodated during pipebursting operations, even when upsizing is being carried out. Only minor works are required to remove existing benching at the base of the manhole, thereby allowing the pipebursting machine to pass through the manhole, with the benching being made good following the installation of the new pipeline.

For pressure pipelines, a full range of polyethylene electrofusion and ductile iron mechanical fittings are readily available and as these pipelines are normally located at shallow depth the reconnection can be made quickly and easily from within a small excavation.

14.9 Pipebursting system and equipment options

Since the introduction of the original pneumatically powered machine system the pipebursting equipment has been continuously developed and by using alternative designs and power sources three different system options have evolved. While they all basically achieve the same end result, they all have specific advantages which provide optimum system performance when used in certain site and ground conditions. Details of these operating and equipment options are given below.

14.9.1 OPERATING SYSTEMS

Pneumatically powered machine system

This is the most widely used system throughout the world for the replacement of both pressure pipelines and sewers. For small diameter pipeline replacement where string welded pipes are installed the installation system comprises two main equipment elements: the pipebursting machine and a winch. However, for large diameter or large upsizing projects and where short length replacement pipes are to be installed a pipe pushing machine is used to assist in the pipe replacement operation.

Figure 14.5 shows a typical set-up for the replacement of either a pressure main or small diameter sewer. The pipebursting machine is powered by a compressor located adjacent to the launch excavation and is connected to the machine by umbilical hoses. Figure 14.6 shows a pneumatically powered sewer pipebursting machine. A constant tension winch is located at the reception point with its rope established inside the existing pipeline, which is connected to the nose of the machine. While some pulling force is given by the winch its main function is to give directional stability to ensure that the replacement pipe is installed to the required line and level. The pipe replacement operation is commenced by the introduction of the pipebursting machine into the existing pipeline. The primary pipebursting function is undertaken by the pipebursting machine, as it travels forward it breaks the existing pipe into small fragments which are dispersed into the surrounding ground as a void of sufficient size is created to accommodate the new pipe. The new pipe,

Pipebursting traditional pneumatic PIM

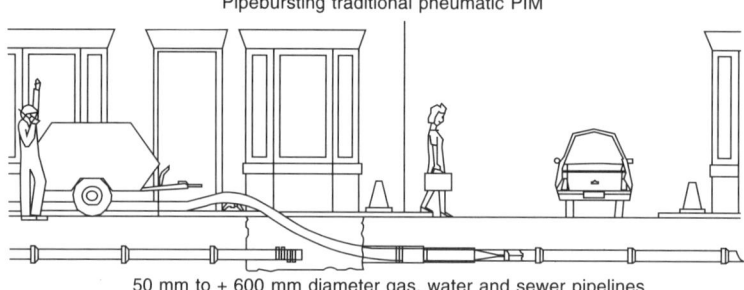

50 mm to + 600 mm diameter gas, water and sewer pipelines

Figure 14.5　Schematic diagram of the pneumatically powered pipebursting system.

Figure 14.6　A pneumatically powered sewer pipebursting machine.

which is attached to the rear of the pipebursting machine, is installed simultaneously as the machine progresses along the old pipeline. To assist the passage of the new pipe a hydraulically powered pusher can be used and this is located at the launch point. These pushers are designed to either grip the outside wall of the new pipe or supply an end thrust load and provide additional power to overcome the frictional forces between the surrounding ground and the new pipe making the installation operation quicker and easier. They are also required for completing the mechanical joints, when short length pipes are employed.

Hydraulically powered machine system

The development of the hydraulically powered pipebursting machine was instigated in the mid-1980s as a result of the failure of the pneumatically powered machines to burst through the heavy Dresser repair fittings which are used extensively in the USA to repair small diameter gas and water mains. They also provided the added benefit of reduced impact on adjacent services, foundations and paved surfaces as the percussion effect of the pneumatically powered machines was removed. The use of hydraulic power at the bursting head provided a substantial lateral force, which was successfully utilised to overcome the problem presented by these fittings. Despite this success the system was not popular for small diameter mains replacement as it proved to be much more time consuming to complete the pipeline replacement operation than its pneumatic counterpart. This was due in part to the fact that hydraulic machines have no forward motive power and rely on the winch to move them along the pipeline during the pipe renewal process. As more powerful pneumatically powered machines evolved the use of hydraulically powered machines declined and are now only seldom used for this application.

Hydraulically powered machines were, however, quickly identified as having considerable potential for the replacement of sewer pipelines. Since they are usually considerably shorter than pneumatically powered machines they are capable of being used for size for size and upsize replacement from existing chambers without the need for launch and reception pits. This offers considerable advantages for the replacement of deep sewers located in congested urban areas and has led to the system being adopted for this purpose throughout the world.

Figure 14.7 shows a schematic representation of a typical sewer project using a hydraulically powered machine. The machine is normally operated in conjunction with a winch and pipe pusher and although pipe strings can be used with this system it is more common for short

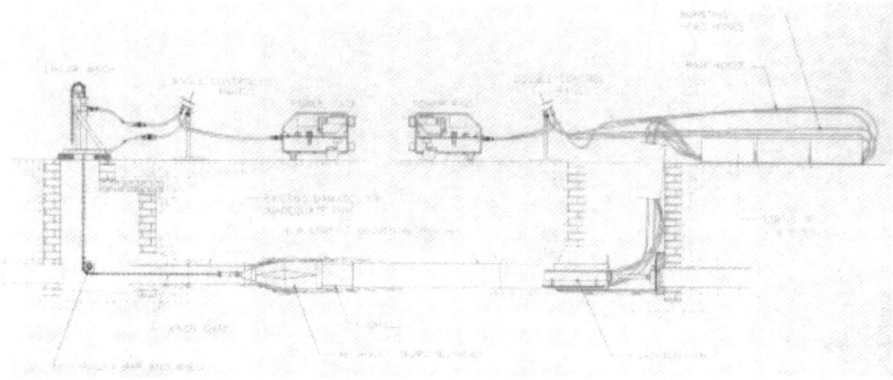

Figure 14.7   Schematic diagram showing a site arrangement for pipeline replacement using a hydraulically powered pipebursting machine.

length pipes to be installed as these are more suitable for installation from existing chambers. The Clearline Expandit machine which can be seen in Fig. 14.8 is an example of a hydraulically powered pipebursting machine. The power source for the pipebursting machine is a hydraulic power pack, which is located adjacent to the launch point and connected to the machine using umbilical hoses. As with the pneumatic system, the winch is located at the reception point and can be either a conventional steel rope capstan winch or a hydraulic chain pull system. In operation, the bursting head is first expanded to crack the old pipe, and is then retracted. The hydraulically powered pusher acting on the new pipe string is used to push the string forward, while tension is applied to the nose of the burster by the winch to maintain directional stability. The process is then repeated, adding further pipes to the end of the string as work progresses. During the installation process the mechanical pipe joints are made by the pipe pusher. The replacement pipes are attached to the rear of the pipebursting machine using a locking ring and this provides a rigid connection that does not allow the differential movement, which is available with the pneumatically powered system.

Figure 14.8   A Clearline expandit hydraulically powered pipebursting machine.

Hydraulically powered rod pipebursting system

This third variant of pipebursting was developed in the early 1990s being derived from soil displacement horizontal directional drilling equipment. It offers considerable advantages over the other systems particularly when used to replace small diameter water mains and these include:

- There are no umbilical lines involved so no precautionary measures are required to prevent contamination of the inside of the pipe during the replacement operation.
- The system's bursting head has no moving parts so this removes the problem of delays due to equipment failure.
- The overall operation process is far quicker improving site efficiency.

Since its introduction there has been a widespread uptake in its use in Europe for small diameter mains replacement and it has now eclipsed pneumatically powered machines for this application. In North America further development of the equipment has taken place and rigs are now available which are capable of replacing pipelines over 1 metre in diameter. The rigs have also been scaled down to allow the system to be used for the replacement of small diameter sewer connections.

In operation, this system differs from the others in that it uses a hydraulically powered push/pull rig, as the one shown in Fig. 14.9, which is located at the launch point. The replacement process can be seen in Fig. 14.10 with the rig used in push mode to introduce a string of steel rods inside the pipeline to be replaced. Once the reception point has been reached the bursting head, which is a static steel expander cone as shown in Fig. 14.11, is fixed to the lead rod. The replacement pipe is then attached to the bursting head and the rods are drawn back to the launch point with the rig in pull mode. This action fragments the existing pipe, compresses the surrounding ground to create the required void and simultaneously installs the new pipe. Generally, due to the installation forces involved, this

Figure 14.9   A hydraulically powered rod pipebursting rig.

Pipe bursting – hydraulic rod system – pull mode

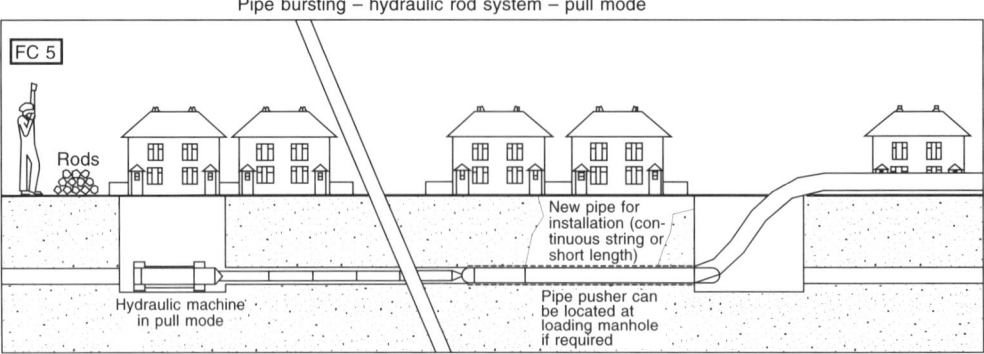

Figure 14.10 Schematic diagram for the hydraulically powered rod pipebursting system.

Figure 14.11 Static bursting cone used with a hydraulically powered rod pipebursting machine.

system is used to install continuously welded polyethylene pipe strings. However, it is possible to install short length pipes but these normally have to be manufactured from rigid materials such as clayware or concrete to withstand the pulling forces exerted on the joints.

The versatility of this system is demonstrated by its adoption by developers of specialist bursting heads for the replacement of pipes and repair section manufactured from non-brittle materials such as steel, ductile iron and plastic. Whilst this new application is principally targeted at gas and water pressure pipelines it will have some relevance for the replacement of sewerage pumping mains manufactured from these materials.

14.9.2 EQUIPMENT

Pneumatically powered pipebursting machine

Pneumatically powered pipebursting machines are generally bullet-shaped devices that are driven by compressed air supplied through air hoses from a compressor, and which fracture the pipeline by forward pressure and impact. A typical pneumatically powered sewer pipebursting machine for replacing 600 mm diameter pipes can be seen in Fig. 14.12.

The construction of these machines is relatively simple as they comprise a pneumatically powered impact hammer, which is contained within a steel expander shield of the appropriate size for the pipe replacement that is to be attempted. The pneumatic impact hammer provides the motive force of the pipebursting machine and is similar to those described in the previous chapters on impact moling and pipe ramming. For comparison of this equipment reference should be made to Table 12.1 in Chapter 12.

It is important for the success of the replacement operation that the correct impact hammer is selected and that it is powered by a compressor that supplies the required volume of compressed air. Table 14.4 provides details of the size and power of the impact hammers available together with recommended size of operating compressor. A guide to the selection of impact hammer size for different pipe replacement projects is given in Table 14.5.

The expander shields are manufactured either to slip over the outside of the body of the impact hammer or to accommodate the hammer within the shield's construction. Each shield is usually built to install one particular pipe size so a range of shields are required to allow a full replacement service to be offered. At the rear of the shield there is a locking attachment for the replacement pipe, which prevents the replacement pipe from becoming detached during the pipebursting operation. On some shields there is an area of float behind the conical head, usually up to 1 metre in length, that allows the replacement pipe to be pushed independently of the pipebursting machine. This provides for a more efficient pipebursting operation since the machine is only carrying out the pipe breaking and soil displacement without having to use additional power to tow in the replacement pipe.

To operate the system effectively it is necessary to use in-line compressed air water separators and lubricators to ensure optimum performance and avoid stoppages due to the impact hammer freezing up. In very cold conditions it may also be necessary to use a compressed air heater to achieve normal working conditions. An in-line control valve is normally incorporated into the compressed air supply line and is located as close as possible to the rear of the impact hammer. This will give a shot of air at the required pressure to restart the impact hammer in the event that it has to be stopped during the replacement operation

Hydraulically powered pipebursting machine

All hydraulically powered pipebursting machines are purpose designed and are based on the principle of an expanding conical head with plates that open and close under hydraulic pressure. This movement of parts of the device is sufficient to fracture the existing pipeline

Figure 14.12 600 mm diameter pneumatically powered sewer pipebursting machine.

and to expand the fragmented main into the surrounding ground. The expansion of the machine is achieved in several different ways such as the use of cams in the Clearline Expandit, a wedge configuration in the Ryan ERS Berster and an umbrella type mechanism in the Express system. An express pipebursting machine for the replacement of 450 mm pipe is shown in Fig. 14.13. Since the expansion of the machine is mechanically induced at right angles to the pipe wall fabric it is extremely powerful and effective for pipebursting. Each machine is generally designed for a particular replacement pipe size so a number of different machines are required to cover the full diameter range. None of these machines

Table 14.4   Technical details of pneumatically powered impact hammers

Impact hammer dimensions and operating parameters	Impact hammer diameter (mm)						
	145	180	220	270	350	460	600
Length (mm)	1550	1750	2050	2160	2400	2900	3470
Weight (kg)	150	240	375	610	1175	2500	4800
Working pressure (bar)	6 to 7	6 to 7	6 to 7	6 to 7	6 to 7	6 to 7	6 to 7
Air consumption (m^3/min)	4 to 5	5 to 6	8	12	20	35	50
Force per unit length (n/m)	420	490	750	1550	2200	6500	9000
Oil consumption (l/h)	0.6 to 0.7	0.7 to 0.8	1.0	1.2	1.5	2.0	3.0
Air hose diameter (mm)	32	32	37.5	50	50	50	50
Delivery of compressed air required (m^3/min)	5.25	7.5	10.5	13.5	22.0	34.75	55.0

Table 14.5   Recommended pneumatically powered impact hammers for different pipeline replacement projects

Existing pipe size inch	mm	New pipe (outside diameter) in mm														
		125	180	200	250	315	355	400	450	500	560	630	710	800	900	1000
4	100	A	B													
6	150		B	B	C	C										
9	225				C	C	D	E	E							
12	300					C	D	D	E	E	E					
15	375							D	E	E	E					
18	450								D	D	E	E				
21	525										E	E				
24	600											E	E	F		
30	750													F	G	G
36	900														G	G

Key							
Reference letter	A	B	C	D	E	F	G
Impact hammer diameter (mm)	145	180	220	270	350	460	600

has any forward motive power and this is normally provided by a winch and pipe pusher operating in tandem with the machine.

These machines are particularly useful in hard ground conditions and for very thick-walled pipe where their mechanical expansion power is essential. They are not as versatile as the pneumatic machines since they are normally only capable of replacing pipes on a size for size or one upsize basis. It is possible to achieve greater upsizing by using machines of different sizes in tandem.

The operation of hydraulically powered machines is heavily influenced by the prevailing ground conditions and some examples of the problems encountered are detailed below:

- Running sand type conditions can cause these machines problems since the mechanical movement design results in open joints allowing ingress of debris and particles into the machine, which eventually prevents its operation.

Figure 14.13   The Express hydraulically powered pipebursting machine.

- The lack of forward motive power can reduce the effective upsizing that can be achieved in hard ground.
- The use of polyethylene pipes with mechanical joints can limit the length of replacement possible due to the maximum pushing force that can be applied to the joint. This is a particular problem in wet clays when additional power is needed to overcome the skin friction forces that develop between the surrounding ground and the new pipe.

Hydraulically powered rod pipebursting equipment

The equipment that is used for this system of pipebursting consists of three main components and is a very powerful hydraulically powered pushing and pulling machine which acts on high tensile steel rods that are connected to a static bursting head.

The main component is the push/pull machine, which consists of a rigid steel frame that contains a series of hydraulic cylinders, which provide the thrust required for the bursting operation. These cylinders are connected to a rod clamping device and this device can be of various designs, which include parallel rollers, jaw gripping blocks or a rack and pinion system. Its purpose is to achieve sufficient gripping force on the rods to ensure the equipment operates efficiently during the replacement process. The rig is operated from a control console and this can either be incorporated into the rig or can be a separate device that is located at the surface adjacent to the launch point. There are a wide variety of rigs available with pulling capacities ranging from 20 to 250 tonnes capacity; Fig. 14.14 shows a TRS 250 tonne push/pull rig. Generally rigs with a capacity up to 50 tonnes are used for the replacement of pipes up to 300 mm in diameter with the more powerful rigs being used for larger diameter pipes such as sewer pipelines.

The rods that are used with these rigs are usually 1 metre long and vary in diameter from 35 to 200 mm dependent on the force that is to be applied to them. They can have either parallel or tapered screw threaded joints, which are similar to those used on directional drilling rods. A recent development is the introduction of quicklock bursting rods, which

Figure 14.14 TRS Hydra Haul 250 tonne capacity hydraulically powered rod pipebursting rig.

are forged with a rectangular ladder shape and hook type joint. These rods improve operating efficiency as they are used with a rack and pinion gripping device, which provides a more positive hold during pipe replacement and the time-consuming screw jointing process is also eliminated.

The standard bursting head used with these rigs is a static cone-shaped steel expander which is drawn through the existing pipe by the rods normally with the replacement pipe attached at the rear. Each head is designed to replace a pipe of a particular diameter. A significant advantage of this system over the pneumatically and hydraulically powered methods is that a series of different sized heads can be used with a single push/pull rig making them very cost effective especially on projects where there are several different pipe diameters to be replaced.

For small diameter pressure pipes a cutting blade which is in the shape of an arrowhead is normally attached in front of the bursting head. This helps to assist the pipebursting process by fracturing the pipe prior to the expander coming into contact with the pipe.

Specialist heads have also been developed for use with this system for the replacement of pipes manufactured from ductile materials and these include:

- The McElroy Bullet which has cutting wheels incorporated into the bursting head
- The Con-Ed Consplit and Ryan Clampburster where the bursting head has an eccentric cone design incorporating a single cutting blade. The Consplit device also includes a pneumatically powered impact hammer to assist the pipe splitting operation. The Clampburster head is shown in Fig. 14.15

Winches

Winches of various designs can be used in conjunction with pipebursting equipment and these include trailer mounted steel rope capstan types and hydraulically powered steel rope and chain pullers. The winches used vary in capacity dependent on the type of replacement being attempted. For pipes in the diameter range of 75 to 200 mm a winch with a capacity of 5 tonnes is normally adequate; however, for larger diameter pipe replacement winches with a capacity of 20 tonnes may be required. A larger capacity winch is the normal choice

Figure 14.15   A clampburster pipebursting head.

for sewer replacement work, as greater winching loads are usually needed due to the depth at which the pipes are located. Chain and rope pullers tend to be more compact and portable units than the trailer mounted winches and are therefore favoured on projects where access to the pipeline is limited.

Pipe pushers

The pipe pushers used with pipebursting equipment to assist in the installation of the replacement pipe are usually hydraulically powered and have two distinct designs. The first is used when long pipe strings are to be installed and is designed to grip the outside of the replacement pipe and thrust it forward behind the pipebursting machine; an example of this type of machine is shown in Fig. 14.16. It comprises a rigid steel frame, into which is incorporated a travelling clamp and a pair of hydraulic thrusting cylinders. The second type is employed with a short length pipe system and is used to make the pipe joints as well as assisting with the installation process. Normally these pushers apply the thrusting force to the rear of the pipe string and comprise a series of hydraulic cylinders acting on an end plate. A stationary clamping mechanism is placed in front of the end plate to allow the pipe joints to be completed prior to installation.

   Pipe pushers are available in various sizes and with a range of capacities from 5 to 40 tonnes to suit most applications. Each pusher can normally be adjusted to allow it to be used for different sizes of pipe.

14.10   Site operational requirements

A great deal of planning and preliminary work is required prior to the commencement of the pipebursting operation. Programming of the works is essential given that, unlike replacement by open trench methods, connecting services or laterals will be out of operation while the replacement pipe is being installed. It should be understood that pipebursting is an accelerated method of pipe replacement and as such this accelerated activity requires additional planning. In particular it is important for the following items to be considered:

1. The position of other utility services in relation to the pipeline being replaced and how they may be affected by the pipebursting operation. Liaison meetings with representatives of all interested parties should be arranged.

Figure 14.16 A pipe pushing rig installed within a launch excavation.

2. Establish the position of any old repairs to the pipeline that may prevent the pipebursting operation from being successful. If substantial quantities of concrete have been used when making these repairs then this will need to be removed prior to the pipebursting commencing.
3. Determine the position of all service or lateral connections and establish the location of all excavations especially if launch and reception pits are required.
4. Determine what traffic and pedestrian management systems are needed to maintain traffic flows and public safety.
5. Establish a system of handling existing flows within the pipeline while the pipeline replacement is taking place.

The preparation required to operate the pipebursting technique at any particular site will vary and is dependent on the prevailing site conditions, the type of pipebursting system to be used and the replacement pipe chosen. While it is advantageous to utilise where possible existing chambers or manholes to gain access to the pipeline, on many occasions this is not possible due to the size of the equipment that has to be accommodated within these entry and exit points. In these cases, to allow the system to be operated, it is necessary to excavate launch and reception pits. These can normally be sited at convenient locations on the site to avoid unnecessary excavation such as where manholes are to be replaced or at the points where lateral connections join the pipeline. The objective should always be to minimise the amount of excavation without prejudicing the operation. Where possible a launch pit should be used twice, to launch the pipebursting machine in opposite directions in two separate operations. When this is not possible it should be the objective to utilise the first reception pit as the next launch pit.

The size of the excavations vary significantly according to the diameter of the replacement pipe, depth of the existing pipe, the type of replacement pipe system and the pipebursting system to be used. For small diameter pipe replacement with the pipe located at a depth up to 2 metres the launch and reception pits would be approximately 3 metres long by 1 metre wide. A typical launch pit for pipelines up to 600 mm in diameter located at greater depth is 4 metres long by 2 metres wide on plan area and this can increase to 5 metres by 4 metres for replacements up to 1000 mm diameter. Reception pits are usually 3.5 metres long by 1.5 metres wide; however, these dimensions will increase for large diameter replacement. These pits are normally excavated to a depth of between 250 and 400 mm below pipe invert and the bases are blinded with lean mix concrete to provide a solid working surface. When a continuous welded pipe string replacement system is used then a lead-in trench will be required at the rear of the launch pit to provide access to the pipeline for the new pipe. These are excavated to the minimum required width for new pipe and normally have a slope of 3 to 1 down to pipe invert.

Generally it is necessary to provide launch and reception pits when operating the pneumatically powered machine and hydraulically powered rod systems as the equipment is either too long or too bulky to be operated from within existing chambers. A number of innovative solutions have been devised to remove the need for access pits and examples of these are as follows:

1. Some pneumatically powered machines have been designed to allow recovery of the bursting head from within a chamber with the impact hammer power unit being drawn back to the launch pit through the newly installed pipe. This eliminates the need for a reception pit.
2. A technique that allows a pneumatically powered pipebursting machine to be launched from the surface without the need for a launch excavation is shown in Fig. 14.17. This involves using an impact mole to provide a slanted pilot bore from the surface down to the invert of the pipeline. The pilot bore is normally located adjacent to an existing manhole to provide access to the equipment during the replacement operation. The pipebursting machine can then be launched from the surface using the pilot bore to guide it down to the existing pipeline.

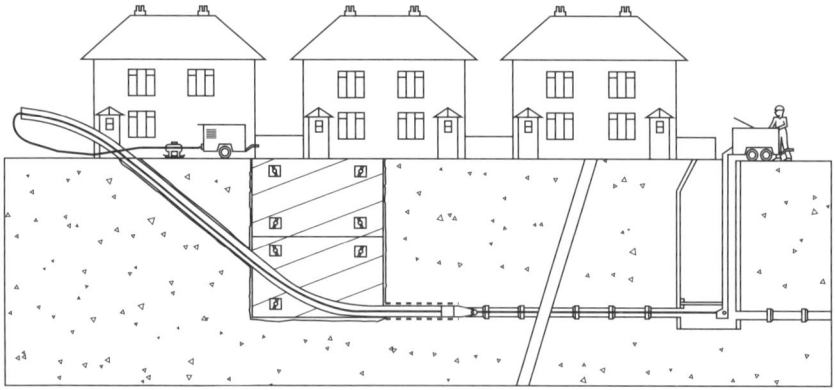

Figure 14.17 Schematic diagram of the windowing method of pipebursting.

3. Compact hydraulically powered rod pipebursting machines have also been developed which can be operated from within chambers with a minimum diameter of 1.2 metres; however, a reception pit or lead-in trench is still required to accommodate the replacement pipe.

Hydraulically powered pipebursting machines are shorter and more compacted than either of the other two pipebursting equipment options and can therefore be operated from within existing chambers with a minimum internal dimension of 1 metre. This has led to the widespread use of this equipment for the replacement of sewer pipelines since it is sufficiently compact for use in locations with difficult and limited access such as city centres, gardens, under buildings, etc. To operate the machine it is necessary to undertake preparatory works within the chamber and this involves breaking out locally the manhole benching, wall and surround to provide access to the pipeline. This is easily reinstated once the pipe replacement has been completed.

It is possible for all types of pipebursting machines to pass through intermediate manholes without the need for excavation; however, similar preparatory work is necessary to that described above for working within chambers.

14.11 Application of the system

Like most trenchless pipeline replacement techniques pipebursting offers to the pipeline owners and the general public many considerable advantages when compared with conventional open trench methods and these include:

- Substantial direct cost savings due to the reduction in excavation, backfill and reinstatement
- Social and traffic disruption is minimised
- Road construction integrity is preserved
- Natural resources and the environment are conserved with the reduced requirement for aggregates and quarried material
- Reduced risk of damage to other utility plant when compared to open trench methods
- Improved safety for both operators and the general public due to reduced open excavation
- Substantial time savings can be achieved on project completions.

However, it is the ability of the pipebursting technique to be used to replace existing pipelines with a pipe of equivalent or larger capacity while requiring only minimal excavation for its operation that has been the major attraction in its continued adoption throughout the world. The popularity of the system has grown as confidence in its use has been gained together with the continuous improvement of the equipment and operating methods. This has led to a greater range of applications for the system with more challenging projects being attempted.

The technique is now used throughout Europe for the replacement of small diameter pressure pipelines with both the pneumatically powered machine and hydraulically powered rod systems being extensively utilised for this purpose. This application was pioneered in the UK where a widespread replacement programme of ageing cast-iron distribution systems is taking place and it has become established as a mainstream replacement technique with over 1200 kilometres of mains being renewed per annum. This level of usage has helped to promote the system for sewer replacement particularly in locations where it is difficult to use other pipeline renewal methods such as the centres of cities, gardens and environmentally sensitive areas. Substantial success has been achieved using the pneumatically powered machine system with both large diameter and upsizing projects being completed. The development of the hydraulically powered machine system created further opportunities for pipebursting, as it is more compacted and can be operated from within existing chambers. As a consequence it is ideally suited for sewer replacement as it avoids the need for deep excavations and allow disruption to be kept to an absolute minimum. The introduction of hydraulically powered rod systems has provided a more efficient method of replacing small diameter pressure pipelines and this equipment has also been adapted to allow small diameter sewer connections to be replaced. This equipment can be installed within existing chambers and provides a simple way of replacing pipes that are often located on private property without the need to disturb driveways and gardens. The use of this process is becoming extremely popular in both the UK and Germany. Large capacity hydraulically powered rod systems are also becoming established as an effective means of replacing large diameter sewer pipelines; however, their use is often restricted in busy urban environments due to the size of the access pits required for their operation. This has limited the adoption of the system in Europe, as many of the sewer replacement projects are located in the older cities where it is not possible to physically accommodate large equipment. The latest generation of the push/pull rigs that are used with this system have been designed to be more compacted for operation from within chambers. However, it will still be necessary to have a lead-in trench when a continuous pipestring is to be installed.

Both hydraulically powered and pneumatically powered machine systems are now well established for sewer pipeline replacement throughout Europe. While the hydraulically powered rod system is used extensively for small diameter pipe replacement in Europe it has not yet achieved the same level of acceptance for sewer renewal as the other systems.

Having become a proven technique in Europe all of the pipebursting systems have now been successfully introduced into many other parts of the world such as the Middle East, Australasia, South Africa, South America, Mexico, India, Hong Kong and China. The technique has proved to be extremely versatile in that it has been easily adapted to suit local conditions. It has been particularly effective for pipe replacement in congested urban areas where traffic chaos would result if open trench methods were employed.

In North America the poor performance of the early pneumatically powered machines, which were generally not capable of breaking the heavy repair clamps placed on gas distribution mains, made engineers reluctant to utilise the system for trenchless pipeline replacement. The implementation of the technique was further limited by the preference for steel and ductile iron pipes for new pressure distribution mains and so it has not been used as extensively as in Europe for this application. Development of more powerful equipment coupled with the realisation of the potential of the system, following success in other regions of the world, has seen a rapid rise in the use of the technique primarily for sewer replacement. The need to replace large diameter pipes has resulted in the capabilities of the system being continually extended as its popularity has grown with many landmark projects being completed. These have included the replacement of pipes in excess of 1000 mm in diameter and replacement lengths of over 200 metres being achieved. The preference within the North American region is to use either pneumatically powered machine or hydraulically powered rod systems as they offer greater bursting power and the road and existing utility layouts can normally allow sufficient space for access excavations.

Innovation, research and development are continuing to consistently improve the pipebursting technique in terms of its operating capabilities and equipment efficiency and reliability. The technique has now evolved into a proven, environmentally friendly, cost-effective alternative to traditional methods for the replacement of structurally inadequate and undercapacity pipelines which is being utilised throughout the world.

15

Horizontal Directional Drilling

Eur Ing Brian Syms BSc, CEng, MICE

15.1 Introduction

Historically, horizontal directional drilling technology was developed for the installation of pressurised pipelines for the oil and gas industry, usually over large distances, much of the early development being in the USA. The capabilities of the technology led to its development for the smaller gas and water distribution mains and later for the cable industry.

The advantages of directional drilling as a trenchless technique for pipeline and cable installation are becoming more widely understood and accepted. This has led to projects which have become more challenging and diverse, including sewerage schemes. However, whenever there are doubts on the geology of the installation area, the use of directional drilling should be discussed with experts particularly if the provision of a gravity pipeline is being considered.

The type of equipment, its categorisation and capability range are subject to debate. Consequently, the description of plant within this chapter, particularly sections 15.1.2 and 15.3 below, is a representation of the author's findings and may not be universally agreed, particularly by some contractors and manufacturers.

The information and equipment detail is given in good faith and is generally taken from supplied information and published articles, consequently any omissions or errors will not, therefore, be the responsibilty of the author. Much of the technical information was derived from US publications, consequently some of the terminology may not be as familiar to readers from Europe.

15.1.1 USE

Cables and pipelines

In the UK, horizontal directional drilling techniques are usually associated with cable and pressure main installation. The development of the technique by the UK electricity industry in tandem with 3-D mapping, particularly Midlands Electricity plc and Northern Electric plc, has benefited their industry to such an extent that the installation of cables using directional drilling units is as great, if not greater, than by open-cut excavation, saving millions of pounds per year in the process. The gas and oil industry has, arguably, led to the development of pressure pipe installation such that its advantages have been recognised by the water industry for water main and sewage pumping main installation.

In the current environment of recognising social or indirect costs, this technique is particulaly useful for operating from a small *footprint*, thus minimising the effect on the public and traffic.

Generally, sewerage projects using this method of construction have been for pressure pipelines where line and level, for instance, are not so critical. However, the accuracy that can now be achieved has encouraged the construction of gravity sewers, although caution cannot be emphasised too highly since there are many factors that will influence the satisfactory completion of a bored pipeline.

Caution

It should be noted that the likelihood of finding suitable ground conditions and a scheme for this technology will be rare since the line and level of the bore has a greater tendency to be deflected by stone or geological variations, for instance. Consequently, since there are a number of potential problems when considering an even gradient, there are many, including equipment suppliers, who do not consider directional drilling techniques appropriate to install gravity sewers.

15.1.2   TERMINOLOGY

Horizontal directional drilling is usually associated with the practice of installing a cable or a pipeline by drilling a hole in a generally horizontal direction. The terms 'guided boring' and 'directional drilling' are, for the general purposes of this chapter, interchangeable although the latter term is frequently used to describe the heavier end of the market, as described further below. There are, however, possibly four types or groups of equipment for carrying out such work which reflect the size and scale of the application. There is some debate regarding the power grouping for the differing size descriptions of machine however, for the purposes of this chapter, the four size groups considered are:

1. Guided moling: Micro system, short, small diameter pipe and cable installation, thrust capability up to 3 tonne (t)
2. Guided boring: Mini system, technique used by most cable and pressure main (<300 mm dia.) installation, thrust capability 5 t to 17 t
3. Horizontal directional drilling: Midi system, larger diameter bores used for pressure main and medium sized pipeline (<600 mm dia.) installation, thrust capability 20 t to 40 t
4. Horizontal directional drilling: Maxi system, large diameter bores historically used for the offshore oil industry, thrust capability 55 t to 400 t.

A more detailed description (but not exhaustive) of the types of machine available and their relevance to sewerage projects is given in Table 15.1

These systems should not be confused with other pipe and cable installation sytems such as auger boring, microtunnelling, etc. which form a bore by removal of the spoil. Auger boring uses a rotating cutting head, the spoil being moved back to the drive pit by helical auger flights and is usually regarded as having limited steering capability. Microtunnelling is a steeerable, remote controlled pipejacking system for the installation of non-man entry pipelines and is capable of operating within tight line and level tolerances.

Table 15.1 Indication of types of directional drilling machine available for sewer and pumping main installation

Manufacturer	Model	Thrust (t)	Pilot bore OD (mm)	Max. backream (mm)	Max. bore distance (m)	Drill pipe length (mm)
Micro upto 3t						
Errut	'Roadent' 900	2	40	150	10	225
Steve Vick International	'Pitmole'	3	90–125	200	40	500
Ditch Witch	PT620	2	87	100	65	1350
Mini 5 t–17 t						
American Augers	DD-15	7	100–125	350–450	230–330	3000
	DD-25	11	100–125	500–600	300–400	3000
Ditch Witch	JT 2320	8	125	350	165	3000
	8/60 JT	10	125	350	200	3500
Steve Vick International	'Rotamole'	5	87	225	170	2000
Straightline	1010	4	75–80	350	200	3300
	2062	9	80	400	270	3300
Terra	1010C	10	75–100	150	170	1500
	2510C	10	75–100	350	330	1500
T.T. Technologies	Grundodrill	6–65	75	250	200	3300
	Grundo drill 12G with percussion	6–12	80	450	400	630
Midi 20 t–40 t						
American Augers	DD-50	22	100–150	600–750	500–670	5000
	DD-90	40	150–200	900–1200	1000–1350	9000
Advanced Directional Drilling Systems	PB70	18–31	75–150	750	1000	3300–5000
	PB90	27–40	75–200	900	1200	5000–10000
Hutte	HBR 206D-40	26–38	94	600	425	3300
Straightline	5152	20	106–112	750	330	3300
Vermeer	D50X100	22	120			4575
Maxi 50 t +						
Power Bore		31–110	28–125	900	1000	
Titan			96–270	1000	600	
				760	1200	
v-Eijk		60	96–270	1120	1700*	
				1620	850	
Wirth		80	28–125	900	1000	

15.1.3 APPLICATION TO DRAINAGE

Since the technology is new to the installation of sewerage pipes, there is relatively little historic knowledge and expertise to describe. The chapter will, therefore, concentrate on the equipment available and how it can be best applied to sewerage projects. The ground condition is usually the most influential parameter when considering any trenchless technique and this topic is no different, hence there is detailed discussion of geological conditions and an indication of how differing ground types should be approached. Finally, the chapter will consider the likely uses for directional drilling for the installation of sewerage systems and describe a number of case studies.

15.2 Directional drilling means what?

15.2.1 *PROVISION OF A HORIZONTAL BORE HOLE*

In simple terms, the technique involves the steering of a small diameter drill pipe between two points along a planned route. If a larger pipe or duct is required, the pilot bore is enlarged with a reamer, the required pipeline simultaneously being pulled into the reamed hole. Very large diameter pipelines may require a number of reaming stages to reach the required bore. The basic principle is shown in Fig. 15.1.

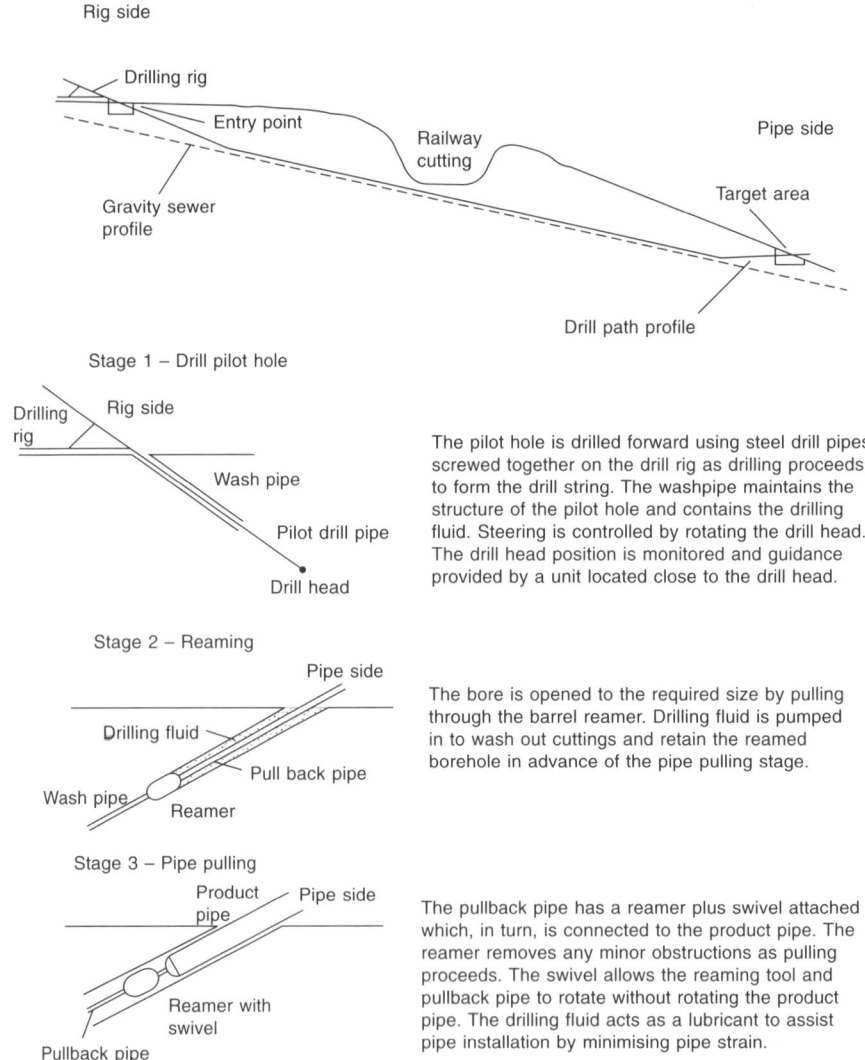

Figure 15.1 Typical horizontal directional drilling location and procedure.

15.2.2   LOCATION

The technique has become more popular for pipeline and cable crossings of rivers, canals, highways, railways, sea defences and even rough terrain or areas with difficult access. The technology suits both welded steel and polyethylene (PE) pipelines. Welded steel pipelines are less flexible than PE and tend to be used for the longer, large diameter projects where a stronger pipeline is required. PE pipelines, usually HPPE, are used for the smaller diameter (less than 600 mm dia., say) projects where the micro and mini systems are utilised, the wall thickness being increased by 10% to meet the demands of its installation. For both materials, pipe lengths are usually welded together into strings on the opposite side of the project site to the drilling rig in advance of the pullback stage – the two sides being referred to as the *rig* side and the *pipe* side respectively.

Some consider that directional drilling can be used in most ground conditions although there is some debate on its success where gravels exist. Some systems are able to cope with a gravel content of more than 30%; however, due to added complexities of such a bored pipeline, it may be more appropriate to try an alternative method or route.

The technology should be treated no differently than any other *tunnelling* type of pipe installation and its success is dependent on the ground conditions. *Stony soils* may deflect the drill head and instant decisions have to be made whether to continue the bore or withdraw to try a slightly different route. Such instances will, therefore, have a greater impact on a gravity sewer scheme than cable installation. Rock outcrops will also cause drilling problems. As described above, drilling through gravel will prove difficult and in such cases a different soil type layer should be sought or a different technique used.

In conclusion, directional drilling should not be attempted unless a good knowledge of the ground conditions is available.

15.3   Equipment and requirements

15.3.1   GENERAL DESCRIPTION

Pilot Bore

A directional or guided drilling system basically comprises a drilling rig, a supply of pipes and a mud or slurry balancing system, usually a bentonite mix, if the latter is required to complete the pipeline installation.

The drilling rig sits on an inclined carriage, which can be adjusted to between 5° and 30°, and bores a small diameter hole using a pilot drill. For large systems, the steering of the drill head is made by rotating a small bend, the bent housing, positioned behind the drill head. For smaller systems, steering as achieved by the movement of a fixed spade or chisel head or by high pressure water jets. Swivel joints allow the drilling head to cut through the soil without rotating the pilot string.

A transmitter device is located behind the drill head to allow the continuous monitoring of the head position and its orientation. The tracking is monitored at ground level at frequent intervals, usually when each drill rod is added and the profile of the drill path is logged. Additional problems arise, therefore, when the bore is under a body of water, for instance, and other tracking methods may be preferred. Tracking and guidance systems are described in greater detail in section 15.5 below. If the line or level exceeds the permitted tolerances, the drill string can be withdrawn sufficiently to redrill the pilot hole on a more acceptable route.

Pre-ream

When a bore in excess of the drill pipe diameter is required to allow the installation of a larger diameter pipe, a (barrel) reamer is rotated and pulled along the drilled path, enlarging the hole by up to one and a half times the diameter of the required pipeline. When the slurry system is used, drilling mud is continuously pumped down the drill pipe to the reamer to power the jet (when appropriate), flush cuttings away and provide support to the reamed hole.

Pipeline installation

The assembly for the pipeline installation, or pullback, comprises a reamer, followed by a universal joint and a swivel to prevent rotation of the pipeline being installed. The reamer and pullback assembly are pulled and rotated from the drill rig using the drill pipe, thus pulling the pipeline into the reamed hole. Again, drilling mud is continuously pumped to keep the hole open, flush out the ground material and provide lubrication to the walls of the pipeline being installed.

In most ground conditions, the bored hole will relax to provide support to the pipeline. When drilling in rock, the oversized borehole should be sufficient to allow the pipeline to be pulled through and does not need to be increased by 50%, say, as for clay.

Naturally, there are variations to the basic principle, particularly when the larger capacity units are involved and these are described in greater detail below.

15.3.2   GUIDED MOLING – MICRO SYSTEM

The machines used for the smaller end of the technology can be regarded as small versions of the standard equipment, such as the Ditchwitch PT620 rod pusher, or pit launched equipment such as the Errut 'Roadent' 900 or Steve Vick International 'Pitmole'. In terms of 'tonnes thrust', the machines are capable of up to 3 t thrust with an increased pullback capability.

The pit or chamber launched machines are hydraulic powered, using short 225 mm to 500 mm long rods in a compact, self-contained unit that can also include trench support. The launch pit can be as small as 900 mm × 600 mm for working in pavements, narrow locations and within buildings and can install pipelines up to 150 mm diameter and from 10 m to 40 m in length, depending on the equipment used and ground conditions. A small power pack is located adjacent to the launch pit.

The rack mounted rod pusher uses rods of 1350 mm length and generally operates from the surface to provide pipelines up to 100 mm diameter over a distance of up to 65 m, again depending on ground conditions.

Both systems are usually operated in dry conditions, without the use of a mud or slurry system.

15.3.3   GUIDED BORING – MINI SYSTEM

Set-up

Mini systems have a thrust and pullback range of 5 t–17 t and are capable of installing pipelines of up to 250 mm diameter over distances of up to 200 m. These systems tend to use high pressure water, or drilling fluid, to cut through the ground and in some models to assist steering. The drilling is carried out from the surface, although shallow starting and reception pits are often required to obtain cover depth, access for making connections and to reduce the entry and exit bend radius.

Drilling process

The drilling rig consists of a drilled rod unit as described in general terms in 15.3.1 and is usually aligned along the centre line of the proposed route and set up at the required entry angle. For safety reasons, the rig should be anchored with pins and an earth bonding system installed, forming a *Faraday Cage,* the latter to protect the operator in case the drill strikes a live cable. In some soil conditions, e.g. gravels, a pneumatic percussion mole with or without a drilling fluid system may be required to successfully install a pipeline.

The pre-ream operation utilises the same source of water or drilling fluid to clean and lubricate the hole opened assembly and provide support to the borehole.

Percussive action

There are at least two steerable boring systems which incorporate a percussive action to assist difficult bores. The 'Grundodrill' utilises a hammer assisted bentonite slurry system which is capable of drilling bores in soils up to 30% stone or gravel content. The hammer action acts at the rig and some of the effect is, therefore, lost through the drill rods. The 'Rotamole', developed by Steve Vick International and British Gas plc uses a hammer version of a rock drill which is fitted to the *sharp* end of the drill assembly. Consequently, it is capable of drilling through more severe ground conditions.

Typical machines in this category found in UK are: American Augers; Advanced Directional Drilling Systems Ltd; Ditch Witch (including Jet Trac systems); Steve Vick International 'Rotamole'; Straightline; Terra; T.T. UK Ltd (Grundodrill systems).

15.3.4   HORIZONTAL DIRECTIONAL DRILLING – MIDI SYSTEM

Set-up

These drilling rig systems are generally rated as having a thrust capability of 20 t to 40 t with a pullback capability somewhat higher, depending on the type and manufacture of the machine itself. Midi systems are usually capable of installing pipelines up to 600 mm diameter for distances up to 800 m. Most midi horizontal drilling systems consist of two parts: a drilling rig and a bentonite mixing system. It is usual to manufacture a self-contained motorised drilling rig to eliminate the need for an additional hydraulic power unit, thus keeping the required plant to two units.

Drilling rig

Again the basic principles are followed except on a larger scale and the provision of mud or slurry tanks and pumps make the site a little more congested. The pilot hole drill is larger, from 75 mm to 100 mm diameter, and a bent housing is used for steering. A change in direction is obtained by rotating the drill string so that the bent housing is in the required direction before advancing the drilling procedure.

The location and level of the drill bit is usually taken at intervals of 3 m, say, i.e. or when another drill rod is added to the rig carriage. It is common practice to store the drill pipes on a loader which, for operation efficiency, is located adjacent to the drilling rig. If deviations, in excess of the allowed tolerances occur, the drill string is withdrawn until a suitable, alternative route can be attempted.

Reaming and pipe installation follow similar procedures as previously described with the drilling fluid being used to assist cleaning away the cuttings and lubricating the drill string.

Typical systems in this range found in the UK are: American Augers; Advanced Directional Drilling Systems Ltd; Ditch Witch; Hutte; Straightline; Terra; T.T. UK Ltd (Grundodrill systems); Vermeer.

Bentonite mixing system

The technical capabilities of a steerable boring or drilling system are self-explanatory but the key to a successful steerable bore is actually the correct use of the bentonite slurry if and when used. Well drillers have maximised the potential of bentonite in vertical boring applications: however, its use for horizontal drilling has yet to be fully realised in the pipe and cable laying industry.

A well-designed bentonite mixing system will usually include the following attributes:

- Venturi system that injects the bentonite powder into the mixing system
- A filtering system to keep out wind blown debris, e.g. sand
- A mixing and circulation pump which mixes the slurry to the correct viscosity required for each project requirement
- A high viscosity pump to keep the bentonite slurry in permanent circulation
- Freshwater tanks to supply the automatic drill stem cleaning system, alternative water supplies can sometimes be used depending on quality, quantity etc.
- Excess bentonite slurry is returned into the mixing system where it is usual practice to provide a recycling process.

It should be noted that the disposal of drilling slurry/mud must comply with the requirements of the Environmental Protection Act 1990 and *The Duty of Care* procedures. Since the disposal is likely to be to a licensed tip, the cost may be significant.

15.3.5   HORIZONTAL DIRECTIONAL DRILLING – MAXI SYSTEM

The drilling process

The drilling rig and bentonite mixing system are much the same as described above but on a much larger scale, say using a 127 mm drill string to provide a 300 mm dia pilot hole. The machines are rated as having 50 t to 400 t thrust. The larger units being able to use drill strings up to 250 mm dia.

Drilled depths are usually 10 to 30 metres, with lengths up to 1800 metres and reamed diameters up to 1200 mm being achievable, as is frequently required by the oil and gas offshore industry. Directional control is by rotating the drill string to move the bent housing into the required direction.

The standard drilling approach of installing a pilot string followed by the washpipe until the pilot string exits in the target area is followed. The pilot string is then removed by pulling back to the drill rig, leaving the washpipe in place as a drawstring for the reaming operation.

In addition to the usual reaming assembly, a 10 m length of heavy drill collar plus a hole opener is located behind the reamer. The heavy drill collar is used to absorb the cutting assembly forces as the hole opener enlarges the bore to the required diameter. As the reamer is pulled through the bored hole and additional lengths of drill pipe are added on behind, so that when the reaming is complete, there is a full length drill string left in the hole to act as a pullstring for the next operation, pulling in the pipeline.

Typical systems in this range found in the UK are: American Augers; Advanced Directional Drilling Systems Ltd; Hutte; Power Bore; Sharewell; Titan; Vermeer; Wirth.

Site arrangements

The area required for the drilling rig and its associated equipment can be as large as 50 m × 50 m, although for restricted sites it is possible to fit the plant into a 30 m × 35 m footprint.

A typical drilling rig footprint, as shown in Figure 15.2, comprises:

- Drill rig weighing up to 42 tonnes
- Two skid mounted power packs, which provide all power and fluid flow/pressure requirements
- Drilling fluid (bentonite) tank with up to 40 000 litres capacity, say
- Control unit, which contains the driller's control console and survey instrumentation
- Skid mounted pipe rack, positioned alongside the rig for storing pilot string and washover pipe.

The welded pipeline is usually strung out in a single length or a series of long lengths placed on piperollers as usually found with a sliplining contract.

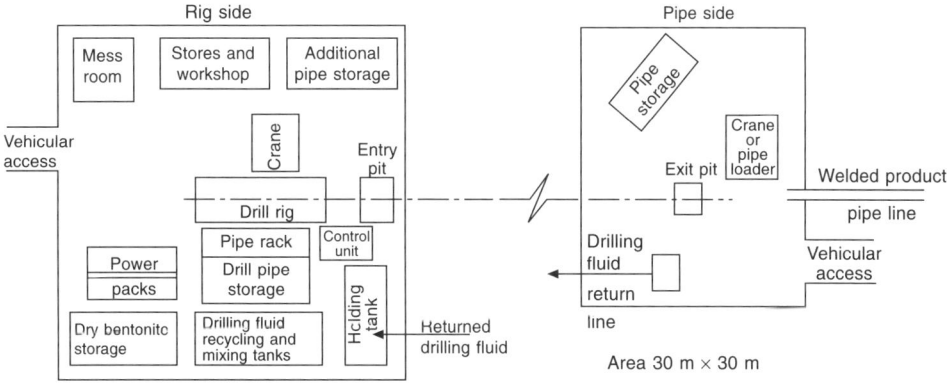

Figure 15.2 Typical maxi rig drilling operation footprint.

15.4 Planning a directional drilling project

15.4.1 *INFORMATION*

As for any project, each underground installation must begin with a plan, which should be comprehensive but concise and then followed.

The first step in planning is to review all available project details and job site information. Utility asset plans need to be obtained and studied. Information about existing or proposed above and below ground structures should also be considered. As shown by some parts of the electricity industry, the transfer of such information onto a computerised 3-D mapping system will bring additional benefits to the planning and execution of a directional drilling project.

The geology of the area is the most important aspect of information gathering and no directional drilling project should be considered unless the data is comprehensive.

15.4.2   SAFETY

In addition to general site safety, which obviously cannot be overlooked, it is important to employ a competent drilling operator, who knows emergency procedures since there are a number of situations which can cause accidents, such as striking an existing live electrical cable. The location of existing utilities should be identified and marked at ground level and, to aid recognition, it may be appropriate to colour code the different services. Trial holes may be required to identify the more accurate location of services, particularly in the vicinity of the entry and exit pits. Despite the smaller working footprint, when compared to an open-cut contract, for instance, there is still a need to comply with the requirements of the New Roads and Street Works Act 1991 and the CDM Regulations 1994.

To help identify possible safety issues in advance it is always advisable to complete a site inspection before the contract starts. The following detail should be checked during the site visit:

• Overall land gradient or slope
• Changes in elevation such as hills, cuttings or open trenches
• Obstacles such as buildings, rail crossings or watercourses
• Signs of existing buried utilities
• Traffic conditions and restrictions
• Consider access to all areas of the site, particularly the working areas
• Soil type and condition if observed
• Possible sources of interference to the guidance system such as overhead electricity cables electrical railway tracks, etc.

15.4.3   THE ROUTE

Identify the best end of the route to use as a starting point, taking into consideration the amount of space available to properly set up the machine, including the unit's electrical safety strike system. If drilling mud is proposed, the fluid system should be located on level ground. In addition, vehicular and pedestrian traffic should be kept at least 3 m away from the equipment.

The full route must be planned before drilling begins and it is good practice to plot the proposed horizontal route on the ground with marking paint.

The drilled bore, hence the route, is usually limited by four measurements: recommended bend limits; entry pitch; minimum set-back; minimum depth.

Recommended bend limits

The recommended bend limits of the pipeline must be considered throughout any bore, not just during the bore entry. Although the drill pipe is designed to bend, bending beyond recommended limits will cause unseen damage which can lead to pipe failure.

Entry pitch

This is measured as the comparison of the slope of the drilling rig to the slope of the ground and can be considered from two extremes. A shallow entry pitch allows the drill pipe to achieve the horizontal plane sooner with less bending of the pipe while increasing the entry pitch makes the bore path longer and deeper.

Minimum set-back

This is the distance from the entry point to the horizontal plane. If the set-back is too small, bend limits are exceeded and the pipe can be damaged.

Minimum depth

Design standards determine the minimum depth for installation of pipelines and these must be considered when the route and position of the drilling rig is being considered.

It is clear that the recommended bend limits and the entry pitch are the two key factors when considering the route of the bore. Pipes must bend gradually for best results, the entry pitch and recommended bend limits of the drillstring and the pipe material determine how deep the pipe can be when it reaches the proposed pipeline depth. To reduce depth the entry pitch should be reduced; conversely, to increase depth, the entry pitch and set-back must be increased. An indication of the set-back and entry pitch for a desired depth is shown in Fig. 15.3

The first pipe section of the installed pipeline must be straight and not at an immediate bend to prevent bending or straining the pipe. To achieve this requirement, the drilling or

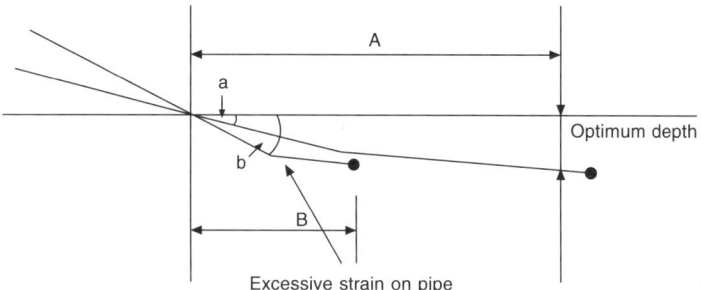

a = entry pitch for proper set-back distance, A
b = entry pitch for incorrect set-back distance, B, which is too short

Typical configurations

Entry pitch (degrees)	Minimum set-back (m)	Minimum depth (m)
16	6.80	0.70
18	7.40	0.84
20	7.90	0.98
22	8.50	1.14
24	9.10	1.31
26	10.10	1.66
30	10.70	1.86

Figure 15.3   Set-back and entry pitch configurations.

boring unit should be set up for a straight entry and this is helped by digging a small entry hole so that the first rod length is easily bored into the surface, thus avoiding bending.

15.4.4 ANCHORING SYSTEMS

Anchoring a machine determines the amount of power the unit can utilise during thrust and pullback operations. A machine can have thousand of pounds of force but if it is not anchored properly and subsequently moves, a number of problems can occur, such as:

- Loss of downhole power
- Full power remains at the unit and major machine damage and/or personal injury could result
- Reduced steering capabilities
- Bent or broken drill pipe
- Failure to complete the bore.

Anchor pins are straight spikes driven into the ground to secure the drilling rig. Naturally, tight dry soil conditions provide the better security and for softer soil conditions, auger anchors may be required. Larger machines can be supplied with front 'dozer' type blades which are also used to assist the anchoring.

15.4.5 CHOICE OF DRILL BIT

Site and ground conditions will determine the choice of drill bit and a general indication of bit choice against ground conditions can be given as

- Silty clay: A wide drill bit is required to make directional changes, steering may be relatively slow
- Dry soft clay: A medium-size bit is preferred, steering may be fast due to the dry soil
- Hard clay: A smaller bit works well but the casing must be at least 12.5 mm larger than the guidance sonde housing.

In addition, the performance of the bit is dependent on the choice and operation of drilling fluid, when used. The pressure of the drilling mud or bentonite should be minimal and for most soil conditions the continuous flow pattern is more influential than its pressure.

A more detailed guide to ground types and drill bits is given in Table 15.2

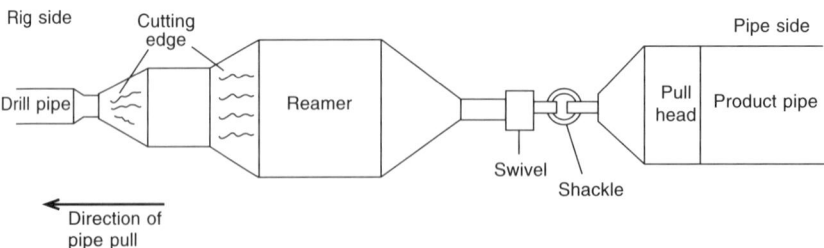

Figure 15.4 Pipe pullback assembly.

Table 15.2 Ground type and drilling bit type combinations

Ground type	Bit type	Comments
Silty clay	Wide	Wider drill bit required to make directional changes, may also be necessary to use a larger bit or a 10° (dogleg).
Dry soft clay	Medium	Directional change may be fast due to dry soil.
Hard clay	Small	Bit should be 12.5 mm larger than the sonde housing. Thicker, blunt-faced bits difficult to push through ground, making steering more difficult.
Caliche (gravel/sand/nitrates)	Smallest/Thinnest	Difficult to thrust forward, chipping technique may be necessary to change direction. Good anchoring essential.
Sugar sand	Mid-size dogleg	Welded carbide narrow bit preferred. Anchoring and drilling fluids important for successful bores.
Sandy silt	Medium to wide	In softer ground, a square point bit or 10° (dogleg) radius bit required to give steering capability. If ground gets hard during progress of the bore, increased torque required to rotate drill head.
Compacted sand	Small/Tapered	Bit must be wider than sonde housing. Forward thrust difficult but steering rapid. Machine anchoring a key issue.
Gravel	Small	Welded carbide bit favoured.
Cobblestone	Steep/Tapered	Difficult conditions but achievable if enough cohesive soil intermixed with boulders. Drilling fluid and welded carbide bit essential. Back reaming will prove difficult.
Solid rock	Rock tool	Sharp changes in direction not possible with favoured standard downhole rock tool. Slight changes can be achieved by experienced operator.

15.5 Guidance systems

15.5.1 INTRODUCTION

Successful guided boring or drilling requires the ability to control the pilot string and monitor its position in space.

At present, many directional drilling projects are for shallow, cable installation, consequently most guidance systems are based on the overland radio transmission location technology. Where large water crossings are encountered, tracking and guidance is often provided by a wireline steering device which is located within the drill string. This method has a number of disadvantages and the development of a cableless system has produced wireless measurement-while-drilling (MWD) tools. Since these systems are expensive to set up and operate, the development in USA of a low-cost cableless steering tool has provided the benefits of the MWD tool but at a much lower cost.

15.5.2 RADIO TRANSMITTER

This method relies on the detection, at ground level, of signals which are generated from a transmitter located near to the drill head. These systems are usually cheap but can be high tech and expensive and are used successfully for short, shallow, easy access projects. As depth increases, the signal resolution diminishes and accuracy reduces. Depths greater than 4 m or 5 m, depending on ground type, should be avoided, although product manufacturers may claim otherwise. In addition, tracking and guiding through buildings, waterways, motorways, etc. are difficult.

There are numerous devices available, too many to describe in detail but an example of the types available follows. Seba Dynatronic provides a simple, solid, inexpensive system which provides the position and depth of the sonde. The newer Radiodetection Drill Track and Subsite Total Tracking System are more sophisticated systems which also provide data on the roll and pitch of the drill head, battery condition and temperature; however, the sondes are consequently more expensive to purchase. DigiTrak provides a tracking system capable of monitoring bores up to 10–15 m depth over a distance of 40 m. Since it is not necessary to monitor over the full route, the system is ideal for motorway, rail track and narrow river crossings where walking over the location is dangerous or impossible. It should be noted that all basic magnetic sonde systems are affected by overhead or underground high voltage cables.

15.5.3   WIRELINE

The wireline system of location and guidance was developed for the oil and gas industry where the monitoring of horizontal bores over long distances and at large depths had to be overcome. The position and attitude of the drill head are transmitted through a wire or trace line which is positioned through the centre of a drill pipe. The main disadvantage is, therefore, the threading of the wire through the pipeline and its protection during drilling activities. The signals are transmitted through the wire to a monitoring system which is usually housed close to the drill rig operator so that changes to direction and pitch can be made without delay or the use of an additional resource.

Sharewell's magnetic guidance system (MGS) is such a system which, when combined with the TruTraker system, can operate in depths over 30 m to an accuracy of 2%.

15.5.4   MWD

The impracticalities of the wireline led to the development of MWD systems for larger drill pipes, i.e. greater than 119 mm dia. The MWD technique uses the principle of transmitting electromagnetic signals from within the drill string. Although a 3 m long device, it is approximately 40 mm dia and fits into the drill string without hindering the drilling process and works either in air or with slurry drilling systems. The technique has been used for more than 15 years in oilfield locations but is thought to be too expensive for many trenchless technology methods for the installation of the smaller diameter pipes and ducts. MWD transmits electromagnetic signals which, when analysed by the receiver, give the position of the drill head in three co-ordinates and tool face attitude. MWD systems tend to be expensive to install and operate, often requiring a specialist operator, although developments, such as AccuNav (USA), are in hand to provide simpler, cheaper units.

15.5.5   CABLELESS STEERING

The cableless steering tool does everything a wireline steering tool does but without the need of the wire fixed inside the drill string. The technology has been designed so that it is relatively cheap to produce and simple to operate, thus avoiding the expense of a specialist operator. In addition, the use of appropriate software, such as TruTracker, operating on a laptop PC, provides user friendly operation and easy to understand graphic representation of the drilling operation.

One system is Boregyde's Arc-gyde, a downhole unit which contains sensors, electronics and batteries and is, consequently, on the large side (two models: one being 2.1 m long with dia. range 50–119 mm; the other 3.6 m long with dia. range 87–190 mm). The new sensor technology overcomes the earth's magnetic and gravity fields, providing information on the attitude of the drill string with respect to magnetic north and local variations. All signals are transmitted to the surface, via the drill string, to a single receiver and data processor. Manufacturing and maintenance costs are low which should help provide higher productivity at reduced cost which will benefit the use of guided boring systems for utility apparatus installation.

15.5.6   LATERAL CONNECTIONS

For making connections to main sewers, a combination of systems to include a wireless system operating from within a sewer manhole, allows a drill rig operator to know when contact has been made with the sewer pipe. The two instruments feed data to a single receiver unit making overall control simple, efficient and effective.

15.5.7   DESIGN SOFTWARE

In the USA, guided horizontal drilling has become one of the most popular trenchless techniques for the installation of utility cables and mains, and a number of design software packages have been developed to aid the design of a guided drilling project. One such package was the result of a joint development involving the Gas Research Institute (GRI), Jason Consultants and Maurer Engineering to produce DrillPath (marketed by Infrasoft). The software is aimed to take out the 'unknowns' of a pipeline design by identifying the drill path coordinates (and in 3-D graphics) and calculating the construction loads for plastic and steel pipes to be installed by mini and midi drilling systems.

15.6   Drainage applications

15.6.1   PUMPING MAINS

The installation of pumping mains has been the most common use of directional drilling techniques for drainage projects. Since the line and level of a pumping main is not as critical as for a gravity sewer, the technique lends itself well to gas, oil and water main installation. Any obstacles found en route during the drilling cycle can be overcome by withdrawing the drill head a short distance, then making a detour around the obstruction in either a vertical or horizontal plane, or both. Naturally, sharp bends should be avoided for the reasons discussed in section 15.4.3 above.

Using a 90 mm diameter drill, the provision of small diameter mains can usually be installed without the need to ream out to a larger bore. Any finished outside diameter (OD) of 90 mm or greater is likely to require reaming out to a bore of one and a half times the required finished size. Ground conditions will determine the need for the use of a bentonite fluid balancing system.

15.6.2   GRAVITY SEWERS

Understandably, the provision of gravity pipelines using this technique is the least common, to date. A gravity sewer relies on a self-cleansing velocity to keep solids moving in order to avoid deposition and the operational difficulties that subsequently arise. It is acknowledged that the gradient within existing sewers varies due to settlement of the pipeline over the years but it is normal practice to construct a new sewer with an even gradient between manholes. Since the level of a drilled bore cannot be guaranteed, there is a natural tendency for design staff to avoid the method.

However, as shown in the case studies there are criteria when the use of directional drilling is acceptable and even preferred where access, for instance, is difficult. Sewers with a good gradient, i.e. steep enough such that a localised change in level of, say 25–50 mm will not adversely affect the overall flow characteristics, can be installed – particularly a surface water sewer where the solids content should be minimal. Guidance and steering systems are becoming accurate enough to install a pipeline to within acceptable tolerances as a number of the case studies demonstrate. Obviously, the best results will be achieved if the depth of the bore is shallow and the pipe run is short.

15.6.3   LATERAL CONNECTIONS

The use of the smaller drilling systems can be used, as pneumatic moling systems are used for water and gas service pipes, for connecting domestic and industrial properties to the public sewerage system or another outfall.

15.6.4   VACUUM SEWERAGE SYSTEMS

Vacuum sewerage systems are usually installed in flat areas where gravity systems are difficult to establish and cause many operational problems and where a pumping regime can prove expensive. The vacuum system makes efficient use of a *pull and push* action which offers a number of advantages in terms of construction criteria and to the nature of the sewage flow.

Vacuum sewerage systems, therefore, arguably offer an ideal scenario for the use of directional drilling techniques since most of the advantages are present – shallow depth, small diameter pipes, suitable sewer lengths, small excavations (for chamber construction) are available for drive and reception pits, etc. Shallow depth cannot be regarded as any depth, since boring a hole too close to the ground surface will cause heave and disturbance which is not good practice in the highway for instance! For the smaller pipe bores e.g. for pipelines up to 225 mm dia. say, a minimum cover depth of the standard 1.20 m for working in the highway should be satisfactory. Assuming that ground conditions are favourable, directional drilling will prove to be even cheaper than narrow trenching due largely to the reduced reinstatement requirements. This is particularly advantageous when working in the highway since road surface reinstatement is subject to high standards and reliability at the asset owner's risk, as required by the New Roads and Street Works Act 1991.

15.6.5   ENVIRONMENTAL PROJECTS

Contaminated land

The cleaning up of contaminated land has received much publicity as governments worldwide, attempt to put right some of the industrial ills of the past. Wherever contaminated land or waste dumps are found, there is every likelihood that ground water has been polluted, which then discharges to a nearby water course. The type of pollutant is quite varied and often hazardous, requiring treatment to bring the quality to an acceptable standard for disposal.

Nineteenth century landfill

Contaminated land is not restricted to identified urban or rural disposal sites. In the nineteenth century, many old cities and towns had water courses culverted (many of which subsequently became public sewers) and the ravines and valleys filled in to provide land for development. The landfill often included industrial waste from chemical, glass, iron and coke works and from mining, consequently there can be a lack of real knowledge about the possible risks associated with such projects until boreholes, for example, are completed and analysed. It is common to find gases such as methane, carbon monoxide and pollutants including arsenics and cyanides in such locations.

A possibly extreme example of this can be found on Tyneside where the banks of the River Tyne were littered with industries which created chemical wastes which in turn were rarely processed further. Typical industries included lead, glass, clay, lime, alkali (e.g. caustic soda for the textile industry), explosives, soap and detergent, pulp and paper and metal works. As a result wastes such as calcium sulphate, calcium sulphide, lead acetate, lead chlorate, arsenic sulphide, cyanide and gypsum were dumped without much thought of the likely environmental impact in years to come. There will be similar and worse situations in other parts of UK, some of which have been addressed by the Development Corporations.

Draining of pollutants

Where pollutants, polluted water and gases are identified, the use of open-cut excavation to drain such areas will create additional risks for which specific safety procedures will need to be identified, addressed and implemented. The use of directional drilling to install a land drainage system for such an area, for instance, offers many advantages since any excavations to start off the drilling process are likely to be shallow thus avoiding the polluted layers. In addition, the risks associated with the physical excavation of pipe laying in contaminated ground are also avoided. The method therefore offers a low risk solution to the installation of a pipeline and reduces the need and associated risks of removing contaminated soil from the site.

In order to flush out an area of contaminated land, a high level perforated pipeline can be similarly installed through which clean water flows into the contaminated soil to the lower land drain, thus assisting the removal of polluted water and gases.

15.7 Advantages and disadvantages

15.7.1 ADVANTAGES OF DIRECTIONAL DRILLING

Horizontal directional drilling offers a major alternative to conventional methods of installing pipelines in urban and rural locations and for the crossing of obstacles such as rivers and roads. There are four basic advantages.

Figure 15.5   Typical Roadent 900 (top) and Pitmole installations.

Cost

It is generally accepted that for most pipelaying schemes, construction by open-cut techniques is more economically viable. However due to the increasing number of other factors now affecting such construction, including the New Roads and Street Works Act 1991 and environmental considerations (sustainable development, use of resources, etc.), the construction of simple pipelines is often cheaper using directional drilling, depending on

Figure 15.6 Typical mini drilling system site set-up for Ditch Witch 8/60 Jet Trac.

ground conditions, access, size of project, etc. Savings can be made due to the shorter working period, e.g. days instead of a few weeks to construct a pipeline of a few hundred metres and the reduced volume of imported materials, e.g. trench fill and reinstatement products.

For harbour crossings, installation costs are significantly lower than methods such as open-cut dredging or the construction of a cofferdam – possibly by as much as one-third. Again installation time is a major factor with schemes taking several months when more conventional methods are used, being completed within a few weeks or even days.

Environmental

When considering pipeline construction, the use of a trenchless technique has the advantage of reducing the imported material content, particularly in terms of trench fill and reinstatement. Consequently, there is a reduced environmental impact on resources and also a reduction in quarrying or mining activities, in turn reducing air pollution.

When crossing a water course or waterway, the operation generally creates no environmental disturbance to the bed or banks and no interruption to river traffic. Similarly for rail track crossings, there is little or no interruption to rail traffic depending on the requirements of the rail track owner. Activities within the highway will be subject to the usual highway restrictions but these will be applied to much smaller areas around the working sites and reception pits and will, therefore, have a much reduced impact on road traffic.

Protection

Due to the relative simplicity of operation and the lesser importance of depth on the operation of a directional drilling system, significant depths of burial can easily be achieved with little additional cost. Consequently, for harbours and busy waterways, protection from anchors, future dredging or excavation works can be ensured.

As demonstrated in Case Study No. 7, an outfall was installed at a depth such that the sealing clay layer of an old chemical dump was not disturbed.

Pipelines can also be installed in ground with high ground water levels without the risk of pipeline flotation.

Weather

Since the method and equipment used are simple in principle, there are usually no seasonal restrictions to directional drilling, although green field sites in wet weather need to be considered in detail and possibly reprogrammed. However, as for any construction method using fluids and hydraulic controls, severe temperature conditions may affect normal practice and require slight adjustments to achieve a satisfactory conclusion.

15.7.2   DISADVANTAGES

Records

Accurate utility records are essential so that other cables (especially electricity and telecommunications) and pipes are avoided. Some survey companies, such as Aegis, have developed 3-D mapping which has benefited the cable industry but comes at a cost.

Records of old, possibly demolished, buildings need to be investigated to ensure that there is no danger of striking foundations or cellars.

Location and gradient

It is extremely unlikely that the line and level of a pipeline can be kept even over the full length of an installed pipeline, since the drill pipe can be deflected off-line by any stone or geological irregularity along its route thus creating local or extensive low or high spots.

Figure 15.7 Typical midi drilling system site set-up for 40 t Powerbore 40 and 10 000 litre mud mixing and recycling system.

As previously identified, when the route crosses open water, the *overland* location procedure must be carried out from a non-metallic boat, which will add its own problems particularly safety precautions and the effects of the tide when in estuarial or coastal waters.

Figure 15.8   Typical maxi drilling system site set-up for Stockton Pipelines 100 t unit.

Ground conditions

A thorough knowledge of the local geology and topography is essential, the information regarding ground conditions needing to be extensive and possibly more detailed than for other tunnelling techniques. Typical borehole requirements are:

Figure 15.9 Stockton Pipelines 100 t maxi drilling system plus 500 m of welded steel pipeline.

- Borehole location:
  - within 10 m of proposed centre line of route
  - to a depth 5 m to 10 m below expected drilling depth
  - at vicinity of both entry and exit points and additional locations if bore length is greater than 150 m
  - additional locations if additional strata are expected

    – every 300 m for bores of 1000 m or more.
- Borehole data:
  - accurate soil and rock classification
  - sieve analysis for cohesiveless soils
  - compressive strength of rock samples
  - water pressure.

Water supply

A suitable and reliable source of water is required for slurry mixing where used.

Drilling fluids

When drilling fluids are used it is essential that the correct fluid is used for the project and ground conditions in question. Projects can fail due to fluid problems such as: incorrect water pH; incorrect viscosity; use of wrong drilling mud on pilot bore and pipeline pullback; incorrect mixing process; use of fluid before it has yielded; insufficient lubrication to external surface of pipeline; high fluid loss to ground.

15.8 Case studies

15.8.1 APPLICATION TO SEWERAGE PROJECTS

In comparison to the cable and water supply industries, for instance, the use of directional drilling techniques for sewerage applications are relatively few and the majority of these are for small diameter pumping mains. As previously stated, its use for gravity systems has yet to be really developed and it is clear from the case studies reported that the industry has yet to be convinced of its suitability.

15.8.2 UK EXPERIENCE

The description of a number of case studies from UK and other European locations has been added to the end of this chapter together with photographs of relevant equipment.

15.9 Future development

As with most technology, advances are being developed for the good of both contractor and client. Some recent and other areas of proposed development are suggested below:

- Design software:
  - expansion of the computerised software approach by companies experienced in mapping and guidance systems to make the concept financially viable for clients and drilling contractors.
- Control techniques:
  - fully automated, computerised control system
  - virtual reality guidance system, extension of existing 3-D mapping and simulator training facilities.
- Drill pipe:
  - new friction welded drill pipe, with low internal hydraulic resistance (LIHR) offers

a drill pipe that combines high mechanical performance and lower hydraulic resistance which are aimed at reducing drilling times and costs.

* Drilling systems
  – greater availability of drill bits and systems with the capability of drilling through gravel-based soils for the smaller mini and midi units.
* Other uses:
  – sewer renovation system such as the Nowak Pipe Reaming Innerream pipe replacement and enlarging system.
* Acceptance:
  – accepted as conventional technique for gravity sewer installation.

Acknowledgements

The author would like to thank the following people for their advice and assistance in the writing of this report:

Chris Barlow, D & C Pipelines
Dave Cook, North West Water Ltd
Dave Gill, Mike Johnson, Mark Horsley, Northern Utility Services Ltd
Dave Harris, Advanced Directional Drilling Systems
Paul Hayward, Paul Hayward Associates
Alwyn Morgan, Midlands Electricity plc
Tony Newton, R.E. Docwra
Nick Taylor, Aegis Survey Consultants Ltd
David Toms, T.T. UK Ltd.

In addition, the following contractors and manufacturers have provided information, technical data and photographs to help form the substance of this chapter:

Bendigo Directional Drilling Ltd
Errut Products Ltd
Euro Equipment Ltd.
Pipe Equipment Specialists Ltd
Stockton Pipelines Ltd
Steve Vick International.

Bibliography

Directional Drilling Magazine – Various.
Branson, Audrey. Planning a directional boring job. *Directional Drilling Magazine*, 1996.
Harrison, W. H. and Rubin, L. A. *New Low Cost Cableless Steering Tools for Directional Drilling.*
New Civil Engineer – Various.
No-Dig International – Various.
Syms, Brian. An overview of trenchless technology. ICE/UKSTT Joint meeting, Newcastle upon Tyne, 1997.
Szczupak, J. R. Horizontal directional drilling. Asian Trenchless Technology '93, Dubai, 1993.
Trenchless Technology Guidelines. International Society for Trenchless Technology. London, 1998.

Appendix

CASE STUDY NO. 1

Project	Off-site foul and surface water for housing development, Barnet, UK.
Client	London Borough of Barnet for Thames Water.
Contractor	Bendigo Directional Drilling.
Equipment	Straightline 20 t DL4010 guided boring rig with 17 t thrust and pullback capability and spindle torque of 5154 Nm at 100 rpm.
Project Detail	Installation of 2 no. pipe runs, each of 70 m to provide foul and surface water gravity sewers for development of 38 dwellings. The 250 mm OD MDPE pipe was butt fusion welded following the pilot bore. For each pipe installation, the pilot bore took an hour and the back reaming 2.5 hours. The drill head entered the reception pit exactly on line and only 20 mm higher than the designed drill path.
Comment	Following a detailed consideration of trenchless techniques, directional drilling was chosen to avoid major disruption to well-established allotment gardens. The drilling was completed through stiff London clay following a small design change at the upstream end to avoid a lens of water-bearing gravel. This part of the scheme took two days rather than the two week period likely for open-cut excavation and avoided major compensation and reinstatement costs.

CASE STUDY NO. 2

Project	New Development, Knaresborough, UK.
Client	Howarth Associates, Harrogate.
Contractor	Northern Utility Services Ltd.
Equipment	Jet Trac 4/40 with 4.25 t thrust capability.
Project Detail	Installation of 100 m of 150 mm diameter MDPE SW gravity sewer for housing development. The 3.5 m deep 75 mm pilot bore was reamed out in three stages, 150 mm, 200 mm and 250 mm in order to install the 180 mm OD pipeline using a cone and cutter tool.
Comment	Probably the first gravity sewer installation in UK. The pilot bore through stony clay proved difficult due to deflections by stones. The second bore proved successful. The choice of this technique was influenced by the narrow road, number of utility services and high reinstatement costs. The project was completed in one and a half days.

CASE STUDY NO. 3

Project	New Development, Kirkby Malzearth, Ripon, UK.
Client	Howarth Associates, Harrogate.
Contractor	Northern Utility Services Ltd.

Equipment	Jet Trac 8/60 with 10 t thrust capability.
Project detail	Installation of 30 m of 315 mm OD MDPE SW gravity sewer for housing development. The 2.5 m deep 125 mm diameter pilot bore was reamed out to 360 mm in one pass using a cone and cutter tool.
Comment	Gravity sewer installation in narrow lane between old buildings through stony, sandy, silty clay. Open-cut excavation to a depth of 2.5 m would undermine the building foundations.

CASE STUDY NO. 4

Project	Sewage Pumping Main, Heathrow Airport, UK.
Client	Walter Lawrence.
Contractor	Steve Vick International.
Equipment	Steve Vick International 'Pitmole'.
Project Detail	Construction of three 12 m bores through concrete and very hard ground for the installation of a 90 mm diameter PE foul sewage pumping main.
Comment	Directional drilling was chosen to construct the main under two airside roads and one security fence to avoid the obvious disruption that excavation would have brought. A dry guided 'Pitmole' boring rig was used, launched from a 1.65 m × 0.75 m pit.

CASE STUDY NO. 5

Project	Sewage Pumping Main, Norwich, UK.
Client	Norwich City Council as Highways Agent.
Contractor	Claret Streetworks Ltd.
Equipment	Steve Vick International 'Pitmole' and 'Pulvic 4' hammer + 'Rotasteer' guidance system.
Project Detail	Installation of 35 m of 90 mm diameter PE pipe under A47 dual carriageway.
Comment	Directional drilling was used to install this sewage pumping main under a busy trunk road in order to avoid traffic disruption. The pilot bore was completed in 5 hours and reamed out with a 132 mm dia. expander to pull back the 90 mm pipe.

CASE STUDY NO. 6

Project	Sewage Pumping Main, Isle of Wight, UK.
Client	Southern Water
Contractor	
Equipment	
Project Detail	Installation of a 570 m long, 400 mm diameter sewage pumping main across the mouth of the River Medina and Cowes Harbour.
Comment	The use of directional drilling allowed the pipeline to be installed across a heavily used waterway at a depth 16 m below the harbour bed. Conventional construction methods would have caused extreme

disruption to harbour traffic, marine anchors and future dredging and significantly reduced the construction period. The ground conditions varied from stiff silty clay to very soft clay, with the occasional band of fine sand.

CASE STUDY NO. 7

Project	Storm Sewer Outfall, Widnes, UK.
Client	Halton Borough Council for North West Water.
Contractor	
Equipment	
Project Detail	Installation of a 200 m long, 800 mm diameter PE gravity storm sewer outfall into the River Mersey at a depth 12 m below ground level through ground conditions varying from firm clay to soft silt to fine silty sand.
Comment	The use of directional drilling permitted the installation of this outfall pipe below an old chemical dump at a depth which kept the pipeline below the clay sealing layer. It was also necessary to avoid disturbance to recently landscaped recreational area.

CASE STUDY NO. 8

Project	Gravity Sewer, Gothenburg, Sweden.
Client	
Contractor	Styrud AB.
Equipment	Vermeer D50X100 Navigator plus DigiTrak Location Equipment.
Project Detail	Installation of 2 no. 50 m lengths of 460 mm diameter gravity sewer at a depth of 1.5 m under a civil/military runway. Project completed in one and a half days.
Comment	The lack of cover meant that soil compaction could not be tolerated due to the risk of distortion to the runway surface. A fly cutter reamer was used with low speed pullback and specialised drilling fluids.

CASE STUDY NO. 9

Project	Gravity Sewer, Mori, Northern Italy.
Client	Trento Province.
Contractor	Sharewell Horizontal Systems for Pato Perforazioni.
Equipment	Vermeer D50X100 Navigator.
Project detail	Installation of 178 m of 225 mm HDPE gravity sewer through fractured limestone using a $4^3/_4$ Tungsten carbide insert bit and a wireline guidance system.
Comment	The rig was set up on a platform to drill through the hillside to allow a straight line bore without entry or exit curves. The pilot bore was completed in 30 hours within 80 mm of the target elevation. The bore was enlarged in another 50 hours using a Sharewell Lo

torque Hole Opener. The product pipe was satisfactorily installed to the delight of the local authority which intends to use the technology on other projects within its region.

CASE STUDY NO. 10

Project	Sewage Pumping Main, Kustelberg, Germany.
Client	Stadt Medebach.
Contractor	Fa Feldhaus.
Equipment	Grundodrill.
Project Detail	Installation of the 90 mm diameter HDPE main in two drives of 90 m and 130 m to connect a small housing estate to the main sewerage system.
Comment	The pipelines were installed through parkland and a tree nursery at a depth of 1.6 m through loamy soil heavily mixed with slate. The whole project was completed in a few days to the full satisfaction of the local inhabitants.

CASE STUDY NO. 11

Project	Sewage Pumping Main, Dresden, Germany.
Client	Local authority.
Contractor	Fa Lavinger.
Equipment	Grundodrill 6.5.
Project Detail	Installation of 500 m of 280 mm diameter HDPE foul sewage pumping main adjacent to A13 motorway to connect the village of Ortrand to a treatment works.
Comment	The installation at the shallow depth of 1.5–1.6 m through running sand, the ground water level being between 0.6 and 1.0 m below ground level. In addition, there was a water main at 1.8 m depth to avoid. The bores were completed from six working pits. Open-cut excavation was overruled due to the proximity of the motorway and to avoid damage to some valuable 'Saxon age' oak trees.

CASE STUDY NO. 12

Project	Sewage Pumping Main, Weilburg, Germany.
Client	Wastewaterboard Weilburg.
Contractor	Fa Haab GmbH.
Equipment	TT.
Project Detail	Installation of 135 m of 180 mm diameter HDPE foul sewage pumping main under the River Lahn to connect the village of Selters to the local treatment works.
Comment	An 8 m deep shaft was required to complete the main at a depth of 3 m below the river bed. The ground conditions were poor with significant ground water flow into the working pit. The pilot bore took one day to complete and the pipeline was installed behind a 250 mm diameter reamer in another one and a half days.

CASE STUDY NO. 13

Project	SW Sewer, Graz, Austria.
Client	Styrian Hospitals Ltd.
Contractor	
Equipment	Di-Drill Power Bore 100.
Project Detail	Installation of 215 m of 400 mm diameter DI SW sewer up to 10 m in depth.
Comment	Congestion of inner city road crossing and main hospital entrance avoided.

CASE STUDY NO. 14

Project	Gravity Drain, Bradford, UK.
Client	Princes Soft Drinks.
Contractor	Northern Utility Services Ltd.
Equipment	Errut Roadent 900.
Project Detail	Installation of 6 m of 100 mm diameter HDPE gravity main within factory unit under strict health and hygiene controls. The bore was at a depth of 1.3 m through clay and infill materials.
Comment	The pilot bore was driven from a 900 mm × 600 mm chamber and arrived at just above the benching of the receiving, live manhole. The whole project was completed in 9 hours.

OTHER CASE STUDIES OF INTEREST

Project	Waste Water Sewer, Schongau, Germany.
Client	Highland Dairy, Schongau.
Contractor	
Equipment	Di-Drill Power Bore 100.
Project detail	Installation of 150 m of 140 mm diameter HDPE pipeline up to 15 m deep under lake.
Comment	Difficult ground conditions – coarse gravel/marl.

Project	Waste Water Sewer, Kempten, Germany.
Client	Waste Water Union, Kempten.
Contractor	
Equipment	Di-Drill Power Bore 100.
Project Detail	Installation of 680 m of 400 mm diameter HDPE sewer up to 36 m. deep under a river in rock.

Project	Waste Water Sewer, V.
Client	Municipal City of V.
Contractor	
Equipment	Di-Drill Power Bore 100.
Project detail	Installation of 900 m of 300 mm diameter HDPE sewer in four bores up to 3.55 m deep.
Comment	The pipeline was installed under a steep, cobbled road.

16

Vacuum Sewerage

Geoffrey F. Read MSc, CEng, FICE, FIStructE, FCIWEM,
FIHT, MILE, FconsE, Fcmi, MAE

16.1 Introduction

In view of the significant flows involved a surface water sewerage system is usually designed as a gravity system, pumping being kept to a minimum. For a foul sewerage scheme it is often necessary to compare the alternative engineering considerations and economics of gravity sewer against a pumping scheme. However, the latter alternative should also include consideration of vacuum sewerage particularly in very flat areas.

Vacuum drainage systems are not an innovation. It is recorded that a Dutch engineer, Captain Liernur, invented a pneumatic system as early as 1868 where sewers were laid at uniform depth regardless of gradient and sewage was drawn through cast-iron pipes under half an atmosphere of vacuum. This system was used in Amsterdam, other continental towns and Stansted in Essex. No record can be found now of the Stansted system.

Following on from this a Swedish engineer. Mr Liljendahl, invented a vacuum sewerage system which was first used in 1969, and installations were completed in Scandinavia, Mexico, Israel and the Bahamas. This system was eventually acquired by AB Electrolux of Sweden who installed it in several countries worldwide. The Electrolux schemes mainly utilised the two-pipe system of separating the 'black' and 'grey' water, i.e. WC wastes were carried in separate pipes to those taking sullage wastes. Dependent on circumstances only black or both black and grey water would be transported under vacuum. The black water system incorporated a vacuum toilet as opposed to a conventional WC while baths and basins were fitted with vacuum traps.

In the mid-1960s National Homes in the USA marketed the Electrolux system through their Airvac division who made considerable developments in the one-pipe system to convey all foul wastes in the same pipe. Later National Homes sold out their interface valve in 1970. From that date onwards they have developed and marketed their own Airvac sewerage system based on the one-pipe system evacuating from collection sumps which accept gravity flows from normal sanitary fittings. The Airvac system in the USA has been used extensively in industry and also to sewer small townships, large estates consisting predominantly of weekend leisure homes, and lately yacht marinas. The Airvac system at marinas is utilised mainly to remove sewage from holding tanks contained in the yachts.

In vacuum sewerage a system of vacuum pipes is used to collect sewage from properties in place of a conventional system of pipes operated by gravity. The idea was first taken up in this country for a public sewerage system by Anglian Water who installed two systems in the Spalding area in the early 1980s and this was followed in 1982 by Severn Trent Water at Four Crosses, Powys.

The village of Four Crosses lies in the Welsh borders some 25 km west of Shrewsbury, in a flat low lying area near to the confluence of the River Severn and the River Vyrnwy. Ground investigation confirmed a high water table and poor subsoils, consisting of a thin substratum of impervious clays underlain by bands of weak silty material with some sands and gravel. As a result of these conditions the septic tanks in the area were functioning poorly, resulting in stream pollution and a smell problem. These conditions together with the lack of main drainage facilities were preventing further development in the village.

Two solutions were considered: a vacuum system and a conventional system of gravity sewers and pumping stations. Because of the flatness of the area and the scattered nature of the community, drainage of the whole village would have called for up to five pumping stations with excavations as deep as 5 m. Because of the poor ground conditions, the construction of this option was thought likely to be difficult and costly.

A vacuum sewerage system, however, does not rely on gravity falls, with the vacuum mains being laid parallel to the ground surface at nominal cover. Deep excavations are avoided and there is no need for intermediate pumping stations. The system therefore promised savings in maintenance times and cost which led to the vacuum scheme receiving approval.

The design consisted of a single vacuum pumping station (the collection station) receiving flows from two MDPE vacuum mains serving different parts of the village. Flows from individual properties were to be led to collection sumps at intervals along the two mains. Periodically the contents of these sumps would be sucked into the vacuum main and along to the collection station. The collection station was to be sited immediately alongside a package rotary contactor sewage treatment plant. Householders would connect their private foul drainage to collection sumps in the same way as they would connect to an inspection chamber in a conventional system, so householders' connection costs remain the same for either system.

16.2   Basic considerations

Vacuum technology has a flexibility which is generated by the ever-present availability of atmospheric pressure. Unlike pressure systems where power has to be provided at every entry point on the system, a vacuum can be generated at a single point in a system thus requiring only one motor point. This not only simplifies power sourcing but in many cases, also reduces significantly the ongoing running costs.

Much of the present-day viability of vacuum sewerage has come with the development of modern piping materials and pipe installation techniques.

In most instances if it is possible to install a gravity-based sewer system then this should be done as the lower lifetime cost covering system installation plus simple ongoing maintenance would far outweigh the cost of pipe installations and ongoing running costs of a vacuum system. However, where it is not possible to install a gravity system, due to difficult or wet ground or accessibility difficulties, then the shallow, small diameter, lower ongoing running cost installations of a vacuum system may well offer a viable alternative.

16.3   System operation

The vacuum pipes feeding a collection station are generally small bore in the range of 90 to 200 mm diameter thick polyethylene pipes which would be installed using minimum disruption techniques such as moling, directional drilling or narrow high speed trenching. Thick plastic is used for vacuum systems because of the negative pressure inside the pipes. The vacuum inside the pipes makes it virtually impossible for rats to live in these sewers and the valves prevent rodents from entering via the collection chamber or sump. Thick plastic also means that the pipes will last longer than the conventional thin-walled ones.

Polyethylene pipes are favoured because their flexibility enables the use of the minimum disruption techniques mentioned earlier, avoidance of existing obstacles and services and the use of sealed watertight joints through the use of plastic welding techniques. There is also a total omission of manholes associated with a gravity system. These vacuum sewers are linked to collection sumps or chambers, which are situated close to the buildings served by the sumps being gravity fed from the buildings served in the conventional way (Fig. 16.1). In most cases the sump can serve several buildings.

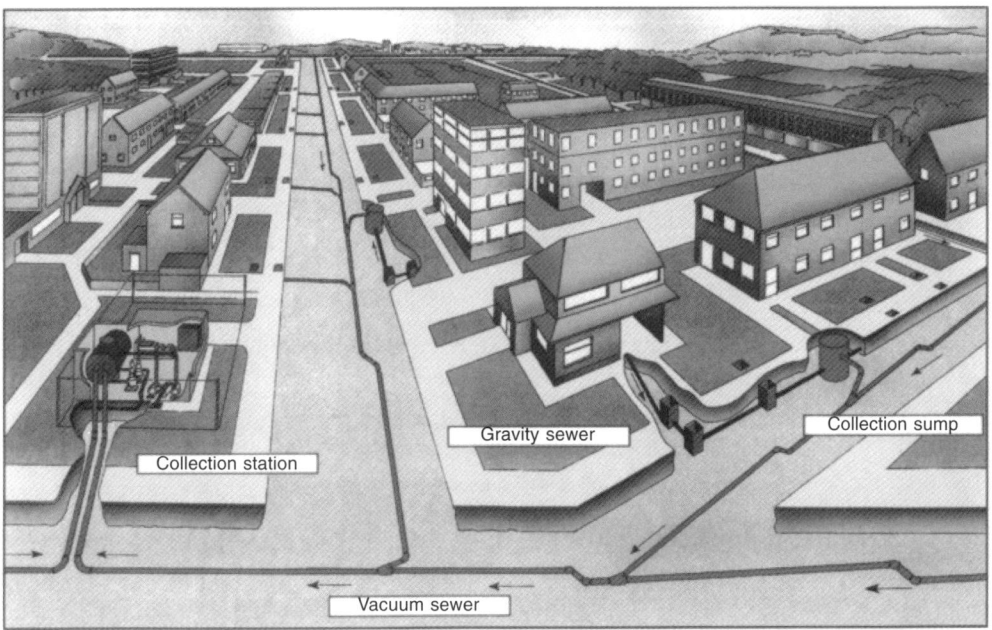

Figure 16.1 Typical vacuum sewerage system layout.

These sump chambers serve two purposes:

1. To collect the effluent discharge from the connecting properties
2. To allow the collected sewage to enter the sewer networks via the interface valve.

A typical sump is 1 m in diameter, 2 m deep and divided into two sections vertically. The top section is the vacuum valve housing with the bottom section being the effluent collector and gravity/vacuum transfer. The 90 mm vacuum valve is designed to handle and cope with any size of effluent material normally carried in a sewer.

As the level of the effluent within the wet sump rises, air is trapped in a pipe called a 'sensor pipe', the pressure of which increases as the effluent level continues to rise (Fig. 16.2). This increase in air pressure is subsequently transferred via flexible tubing to the top section of the interface valve which is known as the controller. Eventually this pressure

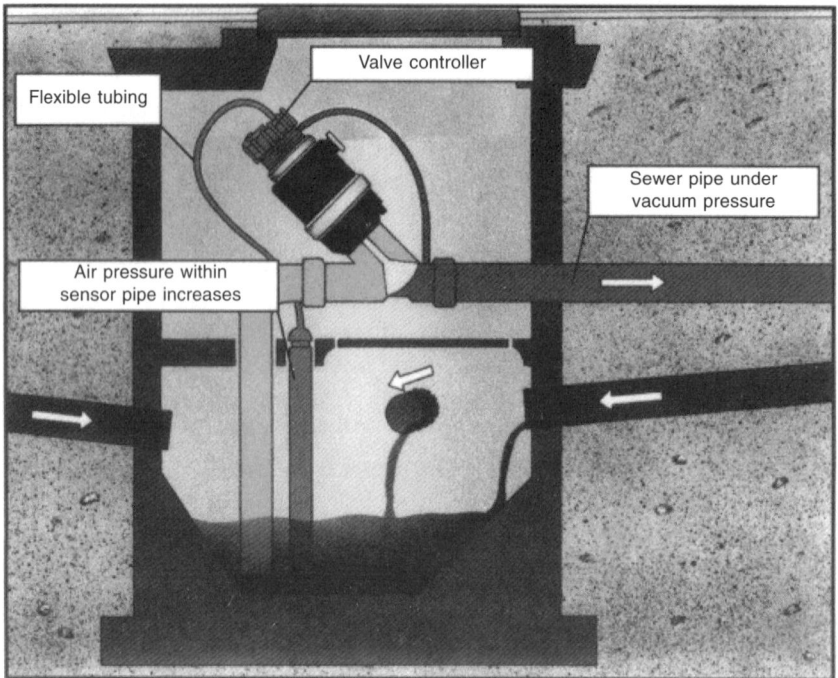

Figure 16.2 Typical sump chamber

becomes great enough to operate a switch within the controller which then allows vacuum pressure to be transferred to the main body of the valve causing it to open (Fig. 16.3).

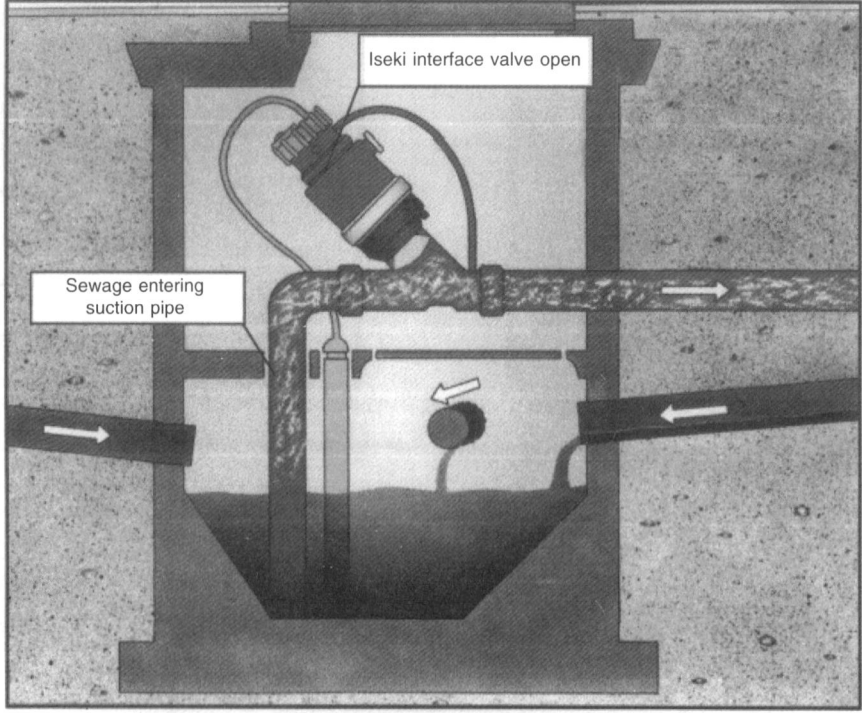

Figure 16.3 Typical sump chamber with interface value open.

When the valve is in the open position, air at atmospheric pressure acting on the surface of the effluent within the wet sump then forces the sewage into the suction pipe, past the interface valve and onward into the sewer pipe network.

Once all the sewage has been removed from the wet sump, the valve remains open for a short period to allow air at atmospheric pressure to enter the sewer pipe network (Fig. 16.4).

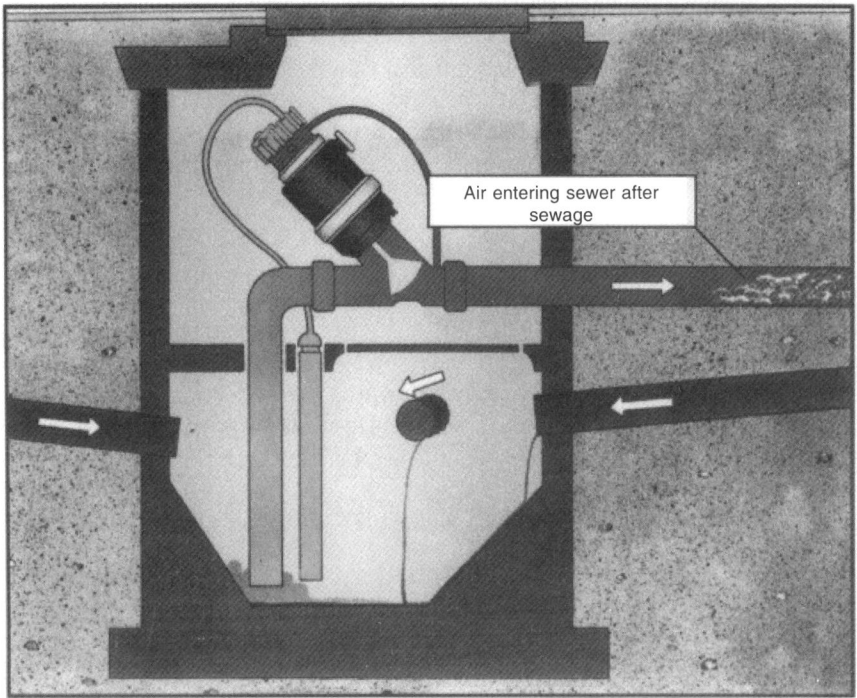

Air entering sewer after sewage

Figure 16.4   Typical sump chamber with valve open, allowing air at atmospheric pressure to enter sewer pipe.

Each vacuum valve is monitored at the vacuum station which allows any faults in this system to be picked up early and easily rectified.

The collection or vacuum station is the heart of the system and is where the vacuum pressure is generated for the whole of the sewerage network being developed. It allows the effluent to be collected and forwarded to the sewage treatment plant.

Each vacuum sewer system tends to be custom designed for the situation where it is to be used. However, according to Iseki Utility Services Ltd, who market this technology worldwide, a system based around a single vacuum generation station can normally handle pipe lengths of up to a 3 km radius on flat ground, and with in-pipe flow velocities of up to 6 m/s the system is generally self-cleansing and therefore low maintenance levels are an inherent feature.

Under this system sewage should never escape into the environment. Even if a pipe splits, which is unlikely as there are fewer joints than in conventional sewers, the negative pressure will only lead to ground water infiltration.

16.4 Interface valve monitoring system

As mentioned earlier each vacuum valve is monitored at the vacuum station. This is an essential operational tool as it pinpoints valves which have failed to close properly and are therefore causing a loss of pressure within the vacuum system. A loss of pressure generally results from a foreign object preventing the plunger inside an interface valve from properly sealing the sewer pipe and it is at this stage that the monitoring system becomes invaluable in locating the troublesome valve.

A switch attached to the body of each interface valve detects whether the valve is open or closed and relays the information to the vacuum station via a signal cable installed alongside the vacuum sewer pipes. Within the vacuum station a light emitting display indicates the open and closed status of each valve.

Examinations of the display panel will indicate which interface valve has failed to close – a visit to the appropriate collection chamber should then allow the problem to be rectified.

16.5 Design details

Although vacuum sewers can be laid 'uphill' in practice they should always be constructed with a minimum slope of 0.2% between lifts.

When a vacuum valve cycles and admits sewage into the system, the flow initially moves in two directions. About 80% moves towards the collection station and the other 20% in the opposite direction. When the 'reverse flow' slows down, as a result of friction, the sewage now moves under gravity towards the collection station, hence the reason for the minimum 0.2% inclination.

Should the natural slope of the ground exceed 0.2% the pipe is laid to the actual gradient (Fig. 16.5).

16.6 Piping design

The design of the vacuum proposals is carried out in a similar manner to that of a heating circuit where it is only necessary to check out the index circuit for the summation of static and friction bases. Static lifts in the system are limited to 300 mm at any one time and it is recommended that with such lifts it is only necessary to take into account 50%, i.e. 150 mm, when summating the losses.

The friction head loss created by the two-phase flow of air and water in a pipe is greater than that created by the water component alone flowing at the same volumetric flow rate in the same pipe. The Hazen–Williams formula for full bore flow is used to arrive at the friction loss and this is set out below:

$$h = \frac{112.8 \times 10^9}{d^{4.87}} \times \frac{(Q)}{(C)} \, 1.85$$

Where h is frictional head loss in metres/100 m
Q is the flow in cubic metres per hour
d is the internal diameter of the main in millimetres
C is a coefficient of friction (150 for UPVC pipes)

Without going into theoretical detail it is general to multiply the above head loss figure by 2.75 to arrive at the actual loss due to the flow in the pipe.

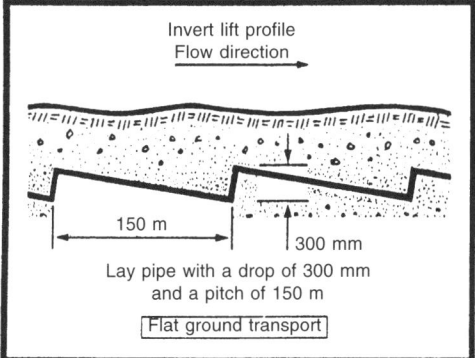

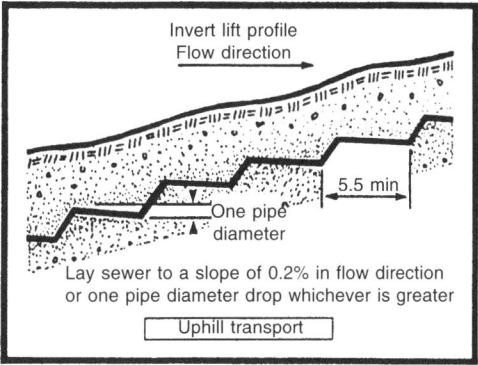

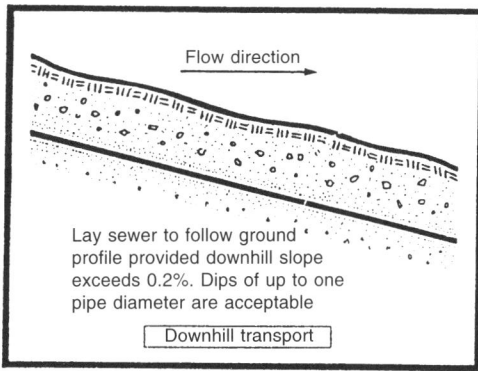

Figure 16.5   Pipe configurations for varying level conditions.

16.7 Station design

Vacuum sewers are designed to cope with peak flows and these are normally assumed to be 4DWF, bearing in mind there will be no infiltration.

The following is a design which was used for an actual scheme some years ago at Chelmsford for the Essex Area Health Authority:

1. DWF 21 gpm (1.6 litres/second)
2. Peak flow 4 × DWF = 84 gpm (6.4 litres/second)
3. Number of valves 38
4. Vacuum consumed by each valve controller 0.15 cfm
5. Therefore total vacuum consumed by valve 38 × 0.15 = 5.7 cfm
6. Vacuum pump capacity required:

$$Q_{vpc} = \frac{5 \times \text{peak flow cfm}}{6.25} = 67.2 \text{ cfm (effective)}$$

Add vacuum consumed by controller   5.7 cfm

Total   72.9 cfm

Use 100 cfm pumps (2 no. duty and standby)

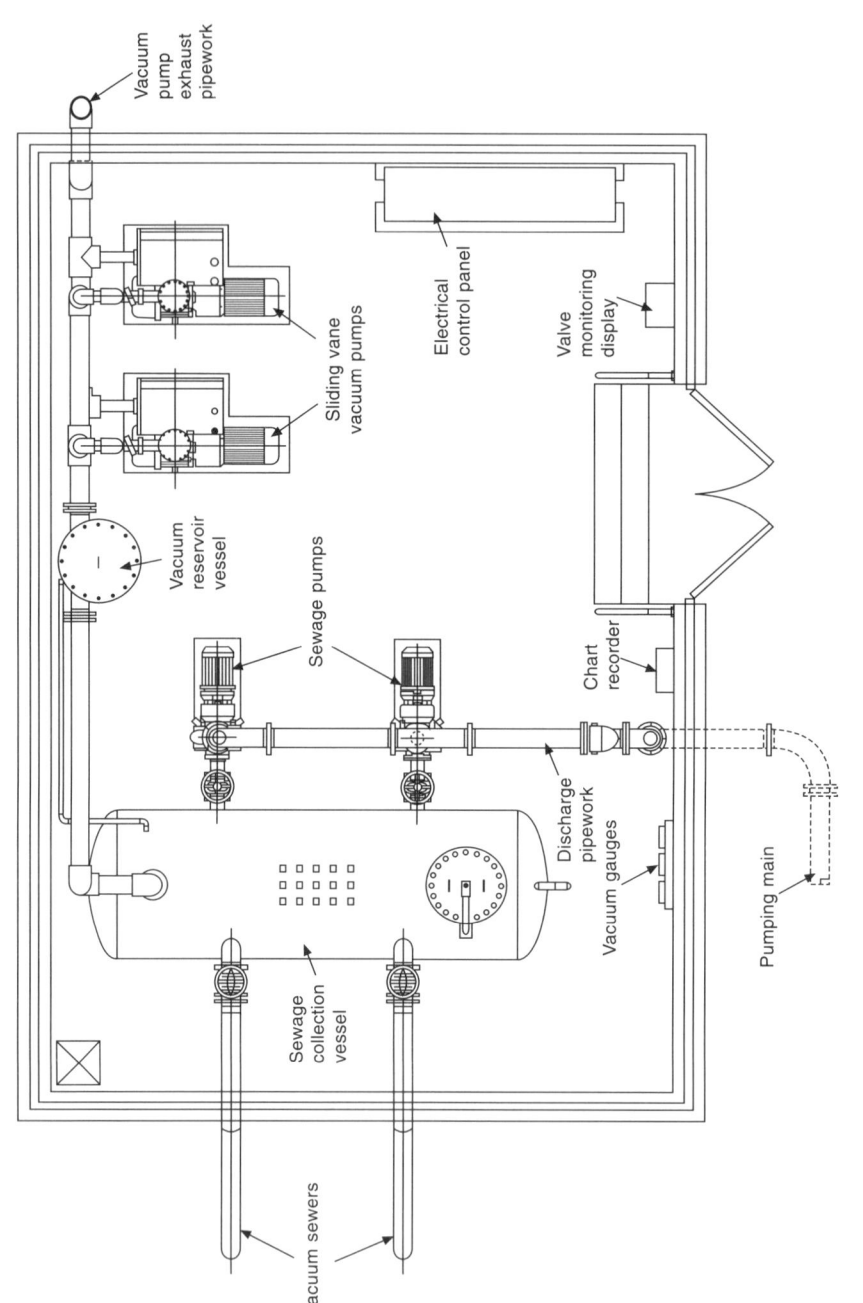

Figure 16.6 Typical vacuum station.

7. Discharge pump capacity $= 1.2 \times$ peak flow

$$= 1.2 \times 84$$

$$= 100 \text{ gpm}$$

8. Sewage collection tank operating volume. This is based on 30 minute cycle at DWF

$V_o =$ operating volume of tank

Now $30 = \dfrac{V_o}{\text{DWF}} + \dfrac{V_o}{\text{Discharging pump DWF}}$

i.e. $30 = \dfrac{V_o}{21} + \dfrac{V_o}{100 - 21}$

$\therefore$ $V_o = \dfrac{30 \times 21 \times 79}{100} = 498$ gallons say 500 gallons

Volume of collection tank $= V_{\text{CT}}$

$V_{\text{CT}} = 1.5 \times V_o = 1.5 \times 500$

$\therefore$ $V_{\text{CT}} = 750$ gallons

Tank used was 775 gallons

9. Volume of vacuum required in systems $= V_t$, i.e. vacuum in storage reservoir in system + vacuum stored in pipework + vacuum storage in collection tank.
It was recommended that the volume in the system should be brought up from 16–20 inches of mercury vacuum in 1.5 minutes using the effective pump capacity.
Hence,

System storage $V_t = 3 \times Q_{\text{vpc}} \times 1.5$

$$= 4.5 \times 67.2 \text{ cu.ft}$$

$\therefore$ $V_t = 302.4 \times 6.25 = 1890$ gallons

Volume of storage reservoir tank V_{rt}

$V_{\text{rt}} = V_t = V_p$ piping vol $- (V_{\text{ct}} - V_o)$

In this instance the piping volume was 1260 gallons

Now $V_{\text{rt}} = 1890 - V_t \times 1260 - (775 - 500)$

$\therefore$ $V_{\text{rt}} = 1190$ gallons

Tank used had capacity of 1200 gallons

10. Diesel standby generator
This was rated at 50 kVA continuous running with a 10% overload running for 1 hour. The generator had a capability to cope with both sewage pump and vacuum pumps running together plus the electrical loading within the station.

16.8   Recent vacuum sewerage projects in the UK

16.8.1   *WRANGLE VILLAGE, ENGLAND*

Client:	Boston Borough Council
Country:	UK
Length of pipe:	3500 m
No. of Valves:	76
Volume of flow:	10.5 litres/second
Specialist feature:	Very flat, high water table, ribbon development

The village of Wrangle is situated in the Fenlands of Lincolnshire, not far from the traditional seaside holiday resort of Skegness. The village suffered from poor drainage and a very limited sewerage system – only the council estates had mains drainage, all other properties were served by septic tanks which in the high ground water of this area often failed to function properly.

Water courses within the village often became polluted with foul sewage and the local council were concerned about a potential health hazard. Therefore, the decision was taken to install a mains drainage system that would serve the needs of the existing villagers and also make allowance for the future expansion of this popular village.

The very difficult ground conditions, the long ribbon development and the flat topography meant the council drainage engineers were faced with a difficult sewerage project. The most efficient and cost-effective gravity/pumped sewerage system required several pumping stations and some deep sewers. The potential of a vacuum system was assessed and found most suitable as the village could be served by only a single vacuum station, shallow sewers and there would be an overall cost saving of some 40%.

Installation of the vacuum system was carried out efficiently and quickly, with narrow trenches excavated using a trenching machine.

The existing council estates were served by the installation of dual interface valve chambers which intercepted the gravity pipes running from the estates toward the redundant sewage treatment works, which were to be demolished.

The vacuum sewerage system at Wrangle illustrates the traditional use of vacuum sewerage systems in areas with flat topography, high water table and widespread housing.

16.8.2   *ROSEDALE ABBEY VILLAGE, ENGLAND*

Client:	Ryedale District Council
Country:	UK
Length of pipe:	200 m
No. of valves	28
Volume of flow:	10 litres/second
Specialist feature:	Steep gradient, rocky ground, stream crossings

The picturesque village of Rosedale Abbey lies deep within the North York Moors National Park – an area of great natural beauty and environmental sensitivity. Very popular with holidaymakers, the village contains peaceful caravan parks and several restaurants and pubs. The population varies greatly between winter and summer.

The village is situated alongside the River Severn and is in a steep sided valley. In the past, inadequate sewerage within the village meant risks of pollution were always present, more so in the summer time and in particular in the clear water of the river.

Having made the decision to install a mains sewerage system to serve the village, Ryedale District Council designers were faced with some difficult problems to overcome; the river and a tributary stream had to be crossed by the sewers and the environmental impact of the constructed works had to be kept to an absolute minimum. This meant that fixing pipes to the existing bridges was not acceptable – the pipes had to cross under the deeply cut stream and river.

To use conventional sewerage techniques would have meant pumping stations at each crossing together with some deep trenching works in difficult ground. Consideration was accordingly given to using a vacuum system and the designers were able to offer a cost-effective and well-engineered sewerage system that in addition kept the environmental impact to a minimum with shallow, narrow trenches and a single pumping station located away from the attractive village centre. The system is sufficiently flexible to cope well with the wide seasonal variety in sewerage flows.

This project demonstrates the applicability of vacuum sewerage in situations where its use may not be an immediately attractive option. Its ability to cross under streams without the need for separate pumping stations, the shallow trenches and the steep hills to be coped with offered specific advantages in this instance.

16.8.3   *ALL ENGLAND LAWN TENNIS CLUB, ENGLAND*

Client:	All England Lawn Tennis Club
Country:	UK
Length of pipe:	600 m
No. of valves:	54
Volume of flow:	120 litres/second
Specialist feature:	Roof rainwater drainage, steep lifts from gutter

As part of an ongoing development of the excellent facilities at Wimbledon, the All England Lawn Tennis Club commissioned the design and construction of a new Number 1 Court. The design chosen included a symetrical-cantilevered roof with no supporting columns within the seating area, thus ensuring that all spectators will be able to enjoy unobstructed views of the tennis. However, this aesthetically pleasing design included a horizontal gutter around the inner and outer circumferences of the roof. Since it was not desirable to have downpipes from the high level gutter to ground level, the designers looked for an alternative system. A vacuum system proved an ideal choice, with the interface valves located high in the roof space, draining both the inner and outer gutters via dual rain mains back to a single collection station, together with the system's ability to cope with a full range of rainfall conditions.

This particular project demonstrates the versatility of a vacuum system for collecting effluents other than sewage. The roof had the difficult geometry of the gutter being some metres below the crown of the roof, which was where the main drainage pipes had to run for aesthetic reasons. This meant that a powered system of some kind had to be used and a vacuum system offered the most cost-effective and efficient solution.

16.8.4   *SPECIAL PRE-CAST UNITS FOR VACUUM SEWER SYSTEMS*

Milton Pipes Ltd is supplying McNicholas Construction with special pre-cast collection chambers for an extensive sewerage system being installed at Wraysbury, on behalf of Thames Water and the Royal Borough of Windsor and Maidenhead.

The system will provide first drainage to some 700 houses in Wraysbury, comprising 19 km of rising mains and sewers. Flows along the 200 mm sewers, buried to 3 m, will be driven by two vacuum pumping stations and 130 intermediate pre-cast concrete vacuum chambers. Milton was commissioned to construct the specially designed vacuum chambers. and provide 150 standard 1050 mm gravity manholes.

Vacuum systems can provide an efficient method of sewage transfer on very flat sites and sites with a high water table or with a complex network of existing utilities. They can be laid around existing services, and sewage can be transported over bridges and other constructions.

Each chamber consists of two sections including watertight joints and a specially developed pre-cast concrete conical base. The entire unit is designed with holes for access and pipe connections for the vacuum network.

The lower chamber, where sewage is collected, incorporates the conical section. The upper chamber consists of a standard 750 mm deep × 1050 mm diameter manhole chamber ring.

A specially cast landing slab forms a separate 'clean' area in the top chamber for the vacuum valve system and associated monitoring equipment.

16.9   General observations

While used worldwide only three UK water companies have so far installed these systems in preference to conventional systems – Anglian, Thames and Southern. Water company engineers may have sometimes been accused of being conservative in their approach to new technologies. This might explain why there has not been greater use of the system. This may also be true of house builders. Yet paradoxically they have perhaps been more willing to use the new thin-walled plastic pipes when renovating or building conventional gravity systems.

17

Social or Indirect Costs of Public Utility Works

Geoffrey F. Read MSc, CEng, FICE, FIStructE, FCIWEM, FIHT,
MILE, FconsE, Fcmi, MAE

Ian Vickridge BSc(Eng), MSc, CEng, MICE, MCIWEM

17.1 Introduction

While we all benefit to varying degrees from the provision of public utilities, their construction and maintenance in urban areas inevitably causes some social and environmental costs to arise. These might include disruption to traffic and the local economy, and loss of amenity as a result of noise, dirt and smell, for example. These costs are those which are not borne by the promoting authority but fall directly or indirectly on the public at large. Such costs can, dependent on the techniques employed, amount to several times the direct cost of the works. In urban areas, open-cut is no longer generally acceptable, even where the depth is relatively shallow, because of the consequential social and environmental impacts.

There is a vast network of underground infrastructure in the UK.

The total length of underground utility mains exceeds that of the road network nearly fivefold; 1.69 million kilometres and 0.37 million kilometres respectively (Table 17.1). The bulk of this network of pipes and cables lies beneath highways and is concentrated in urban areas. Most urban roads conceal two wired mains – electricity and telephone – and at least three piped mains – gas, water, and sewerage. Table 17.2 shows the utilities' typical annual workload of excavations involving nearly three million openings each year. Sewerage is an important element within the overall utility network.

Table 17.1 Number of utility customers and lengths of mains

	Number of customers served by underground mains	Length of underground mains (kms)
Electricity	20 million	4 00 000
Gas	15.9 million	2 30 000
Telephone	20 million	4 30 000
Water	21 million	3 25 000
Public sewers	21 million	3 01 000
Total		16 86 000

Table 17.2   Utilities' annual workload of excavations

	New and replacement mains (km)	New and replacement services (no.)	Small openings (no.)
Electricity	4000	2 00 000	2 17 000
Gas	5457	7 67 000	5 44 000
Telecom	3150	4 67 000	74 000
Water	3200	2 30 000	5 36 000
Sewers	2500		
Total	18 307	16 64 000	13 71 000

The development of trenchless technology has only started to have an impact over the last few years. Previously, traditional open-cut trenching has been the almost universal choice for urban and semi-urban locations, on the grounds that it was the cheapest and most practical – a view no longer accepted by the general public!

Civil engineering project appraisal is generally concerned with determining both the direct costs and the benefits of a proposed scheme in order to ascertain which of a number of alternatives is most likely to produce the greatest public benefit for the lowest cost.

The concept of overall costing, where environmental costs are also included in the project appraisal for consideration alongside economic costs, was highlighted in Professor David Pearce's report to the government. In this report it is suggested that an 'economic value' could be assigned to environmental benefits or disbenefits, based on what the public is prepared to pay in order to have them preserved or removed. However, in many cases, there is either insufficient data presently available or the techniques are too cumbersome and time consuming for them to be used routinely by either highway authorities or utility companies to assess the magnitude of social and environmental costs. Relatively simple techniques for estimating social costs are required.

Utilities' works generally can create a wide variety of social and financial costs borne by a number of bodies and individuals who generally do not have any input into the decision-making process.

The authors have been involved in a research programme, based in the Department of Civil and Structural Engineering at the University of Manchester Institute of Science and Technology, which initially examined the social or indirect costs resulting from various sewerage rehabilitation techniques and has been reported at various conferences (See Bibliography nos 1, 2, 3, and 8). The scope of the research has since widened, in view of the interaction between the various elements, to examine the social costs of public utility works in general.

The social costs arising out of any public utility works are not borne by the promoting authority, but fall directly or indirectly onto the public at large. They may be actual cash losses or less tangible losses, such as environmental damage or loss of personal amenity. The main elements of these costs typically include:

- Delays and diversions to traffic, increased accidents and highway maintenance, including any consequential environment damage
- Damage to other underground mains and adjacent buildings
- Disruption of the local economy
- Direct environmental damage including loss of amenity as a result of noise, vibration, visual intrusion, dirt and smell.

There is no direct financial incentive for the promoter to minimise these indirect costs. However, they can be internalised through the adoption of the appropriate legislative mechanisms, as discussed later in this chapter.

In the public utility field of operation, there is as yet no statutory requirement to give consideration to the total project cost to the community although, in the Horne Report on Roads and Utilities 1985 it was estimated that the financial loss to road users, nationally caused by utility works, might amount to £35 million at 1983 prices. Recommendation 55 in this report suggested that the utilities should study the 1985 standards for the design and location of their underground apparatus with a view to minimising their whole-life costs including social costs.

It is the public who pay for the works on the country's infrastructure – either directly through charges, or indirectly. Should the general public therefore not be entitled to demand that preventive maintenance, replacement, renovation, and renewal be carried out with the overall economy in mind – including social and environmental considerations – and not that of one particular undertaker alone? General recognition of this view is growing, albeit somewhat slowly!

The owners of underground plant beneath the highways have the use of the space rent free and exist in a protected, maintained and sheltered environment – courtesy of the highway authority. It seems not unreasonable therefore to suggest that, for the very small part of their life cycle, when plant is installed or maintained, the work should impinge on the users of the highway and the general public as little as possible.

17.2 Social and environmental effects of public utility works

17.2.1   TRAFFIC

Although car ownership levels in the UK are among the lowest in Europe, the UK roads are among the most densely trafficked in Europe. In 1996, for example, the UK had 65 vehicles per km of road, only slightly less than the Netherlands which had the highest density in Europe of 67 vehicles per km.

Bearing in mind the clearly overloaded nature of the UK highway network, traffic disruption is normally the most significant factor in the social cost equation. Disruption to traffic arises primarily because the road space available during the works, and the methods of traffic control adopted for the management of its use, are generally unable to provide adequately for the traffic wishing to use the road. The social and economic consequences of traffic disruption as a result of public utility works therefore typically includes the following effects:

- Delays caused by a reduction in the capacity of the road
- Diversions leading to additional travel time and distance
- Difficulties of vehicle access to premises near the works
- Additional road accidents where works obstruct visibility or the smooth flow of traffic and on diversion routes which may carry changes in volume or types of traffic
- Damage to the road structure or to underground mains and services on diversion routes which are unsuited to heavy vehicles
- Environmental effects caused by unusual traffic on diversion routes and due to extra traffic congestion, wherever it may arise

- Problems for bus operation resulting from traffic delays, diversions and unreliability of journey times.

All these problems can be quantified to some extent; in most cases overall results are a product of the severity of the effect for each vehicle and the relevant number of vehicles involved – the nature of the effects tending to be more complex in the vicinity of major road junctions than elsewhere.

Traffic diversions can also have a serious effect on the commercial and business life in and around the particular area.

Values for the costs of delays and extra distances travelled can be determined using appropriate traffic assignment models (e.g. QUADRO 2) and the Department of Transport's published figures for vehicle operating costs and delay costs.

17.2.2 DIVERSION ROUTE EFFECTS

Where road works occupy a large amount of road space, or a small but strategic sector of space it may be necessary to divert most or all of the existing traffic onto other routes. Even where there is still room for some vehicles to pass, high traffic flows may require diversionary routes in order to avoid excessive delays. Suitable designated and signed diversion routes are normally prescribed but motorists will quickly explore 'rat runs' if it is likely that they will minimise delay. Any diversion is likely to involve a greater cost to the vehicle operator in terms of mileage, time and fuel consumption in comparison with the original route – assuming the latter to have been chosen rationally.

So we have two diversion route variations, namely, the official one which is usually via a similar type of highway to the one affected or the unofficial network utilising the highways of a lower category with resultant effects on the local environment as well as the highway structure itself. The latter unofficial diversion routes are commonly completely unsuitable for the class and density of diverted traffic which endeavours to use them. The layout is often inadequate for large vehicles and footway damage particularly at junctions with small radii kerbing is a common feature which in turn usually leads to damage to statutory undertakers' plant. In addition there is generally a local increase in road accidents as well as rapid deterioration of the carriageway surface itself. Environmental factors such as noise, smell and dust quickly become apparent as a result of greater traffic usage particularly when this will be slow moving controlled by frustrated and irritable drivers. Vehicles will spend more time in queues due to undercapacity of the selected route and fuel consumption will rise as vehicles operate less efficiently leading to increased operating costs.

As part of the research programme previously referred to a study was carried out to examine some of the effects resulting from unofficial traffic diversions in an area of south Manchester consequent upon the execution of sewerage works. As part of this sewerage project it was necessary to close Dickenson Road between Wilmslow Road and Birchfields Road except for access. The official diversion being via Wilmslow Road, Hathersage Road and Anson Road and this was signed to a high standard. The labyrinth of streets between Hathersage Road and Dickenson Road were studied to examine the effects of 'rat-running' traffic.

Carriageway deterioration

This particular study was undertaken using the MARCH system (Maintenance Assessment Rating and Costing for Highways) which is the technique commonly used in urban areas.

Table 17.3 Highway damage from diverting traffic on unofficial diversion routes
(June 1987 to August 1988)

Damage identified by MARCH survey	Diversion routes (unofficial)	Control sites
Major deterioration	1063 sq.m (+19.7%)	1200 sq.m (9.2%)
Minor deterioration	4655 sq.m (+35%)	163 sq.m(16.4%)

∴ Diverted traffic associated with an additional 10.5% increase in major deterioration and 18.5%
increase in minor deterioration.

Estimated additional maintenance costs:		£
Major deterioration	565 Sq.m patching @ £12.90/sq.m	7294
Minor deterioration	2185 Sq.m S/dressing @ £0.90/sq.m	1966
	∴ Total extra maintenance cost (1988 prices)	£ 9260

It was found that the increased traffic usage resulted in an additional repair cost of
£9260 (1988 prices) as a result of 10.5% more major deterioration and 18.5% more minor
deterioration than the control site, the closure operating for some 9 months in total (Table
17.3).

Casualty accidents

The road accident statistics for the area under review in comparison with a similar control
area are tabulated in Table 17.4 in respect of the comparable period prior to the closure of
Dickenson Road and during the closure.

Table 17.4

	Before	During	% change	Actual change
Dickenson Road	29	23	−20.7	−6
'Rat-run' roads	66	73	+10.6	+7
Official diversion	42	59	+40.5	+17
Area total	137	155	+13.1	+18
Control area	143	124	−13.3	−19

Table 17.4 shows an increase in accidents over 'rat-runs' = 7, or an increase over
expected level as a result of accident reduction in control area = +26, or a total cost to
community of £1 41 700.

Damage to public utility equipment

As a result of the additional traffic density on the minor roads used as 'rat-runs' it was
decided to determine whether there was a consequential increase in gas escapes during the
relevant period.

The results showed, however, that there had been no obvious increase in gas escape
activity on the diversion routes in this instance.

17.2.3　NOISE

Noise will be generated by site operations which may include continuous operation of pumps or generators producing noise even when normal working has ceased. Where the traffic flows can still pass freely despite some loss of road width, diversions are unlikely and traffic and works noise will be concentrated on the site. At the other extreme, where a heavily trafficked road is completely blocked by the works, its usual traffic will be forced to divert taking its associated noise problem with it.

17.2.4　OVERPUMPING

Overpumping, in the case of sewerage works, causes continuous noise, and service hoses and pipes can often seriously restrict access to premises and side roads. The degree of overpumping, and hence the social costs incurred, will be dependent on the technique adopted.

17.2.5　VIBRATION

Vibration will be produced by the works particularly during digging or moling operations. In addition to these ground-borne vibrations, there may be very low frequency vibration (infra sound) in the range 0.1 to 1 Hz which can have a psychological impact on people exposed to it.

17.2.6　AIR POLLUTION

Particulate air pollution around the site will include dust and dirt disturbed by the workings as well as carbon smut and lead compounds discharged from plant on site and slow moving traffic in and around the site. Gases will be produced by both diesel engine plant and slow moving vehicular traffic, the main constituents being carbon monoxide, hydrocarbons and nitrogen oxides. In addition sewerage works may often be responsible for a smell problem.

17.2.7　DUST, DIRT AND MESS

Any construction site will produce some mess and disturbance. With utilities work involving excavation, the main problem is likely to be related to the excavated material and imported backfill.

17.2.8　VISUAL INTRUSION

The level of visual intrusion will be directly affected by the size and nature of the machinery and the hoardings around the site.

17.2.9　PLANT AND MATERIALS

The particular construction technique used will dictate the amount and nature of the surface plant required and the materials to be stored on the carriageway and delivered to the access points. Even those techniques which do not need large access shafts may occupy extensive areas of the carriageway with all the plant necessary to service underground operations. A careful examination of the expected area, location and duration of occupation of the carriageway is required in order to assess the likely social costs of any technique.

17.2.10   SAFETY

Accidents and injuries to workers and the general public are a social cost and thus the on-site risk or hazard associated with specific techniques should be considered in any social cost analysis. Any technique which removes people from hazardous situations and replaces them with machines must – all things being equal – lower the risk of serious injuries or fatalities. Hence many of the emerging no-dig technologies should help in alleviating such risks.

17.2.11   GENERALLY

There are a number of other disadvantages which may result from public utility openings of the highway, such as temporary changes in local pedestrian movement, bus stopping places, local shopping patterns, etc.

Reproduced in Appendix A is a paper written by the editor some years ago describing how the traffic element in the social cost equation was used to justify a heading scheme where the direct cost was likely to exeed an open-cut solution until account was taken of some of the social cost implications.

Although many of these effects can be quantified by physical measurements, people's perceptions must also be taken into account when considering overall costs. Indeed some effects can only be identified through people's responses elicited by in-depth social surveys and complaints. Usually the traffic disadvantages result in the most significant social costs.

17.3   The environmental impact of public utility works

Public utility works will inevitably have an impact on the environment in one form or another and each technique will incur its own unique pattern and magnitude of impacts. Some techniques will inevitably cause severe traffic disruption while others may cause more air pollution, noise or vibration. However, the way in which a project is managed can, in many cases, have more influence on the disruption caused to the surrounding area than the choice of technique itself. Nevertheless, it is important to recognise the relative magnitudes of the most significant impacts if a rigorous appraisal is to be carried out when selecting the most appropriate technique for a particular job.

Open trench replacement or repair will almost certainly cause some traffic disruption, which may, in some cases, result in social costs amounting to several times the direct cost of the work. On the other hand, those techniques that require little excavation of the road surface and which require manholes to be enlarged or access shafts to be excavated, will obviously cause some disturbance but they should generally result in less disruption than open trench work. However, the siting of access shafts and the number open at any one time will have a critical influence on traffic flow through and around the site. Access shafts located at or near busy junctions may disrupt traffic more severely than an open trench, particularly if those shafts are open for long periods of time.

Any construction which involves excavation or the removal and transportation of loose and fine fill material will generate wind blown dust in the vicinity of the work, resulting in additional cleaning of cars, furniture and clothing. In addition, wind blown dust may have more serious implications in particularly sensitive locations, such as hospitals and laboratories, where airborne particulates may damage sensitive equipment and machinery, notwithstanding the obvious health hazard. Further, any method which results in longer traffic journey lengths or slower traffic movement will contribute to the overall pollution load on the atmosphere due to the additional pollutants emitted by the vehicles.

The problem of damage caused to other underground services, the highway, and buildings by the construction of public utility works using traditional open trench methods is well recognised. In many cases this is accounted for through compensation payments and the costs incurred by any necessary repairs or diversionary work. However, the question of long-term damage to buried assets has not yet been fully addressed, particularly for those beneath minor roads carrying heavy diverted traffic. Similarly, the long-term damage, caused by moling or boring techniques, to other services, building foundations and the highway surface is still not fully understood. Although these impacts cannot therefore be accurately evaluated, an engineer should be able to make subjective comparisons between methods if it was felt that there was reasonable cause for concern.

All civil engineering works have a visual impact on the environment during the construction phases of a project. Although many completed structures may be regarded as aesthetically pleasing and thus enhance the environment, there can be few people who would argue that the construction of public utility works has anything but a negative visual impact on the landscape, albeit generally temporary.

Some impacts such as noise, health and safety are controlled to some extent through legislation. However, although restrictions related to these factors may be imposed on contractors, it is still true to say that some methods of working will be less noisy or more safe than others, and there will therefore be a difference in the impact on the community. There are two separate target groups to consider: the contractor's employees on site, and the public at large. It is clear that any method which seeks to remove operatives from a dangerous underground environment by using remotely controlled machines instead must be intrinsically safer. Those techniques which are inherently less safe will not only directly affect those implementing them, but it must be remembered that the cost of treating injuries ensuing from hazardous working conditions are borne by society as a whole. It should further be noted that any technique which leads to large volumes of traffic being diverted onto narrow residential streets could result in an increase in road accidents, particularly on normally quiet routes taken by children or the elderly.

Some, but not all, of the environmental impacts and the way in which they might vary in nature and magnitude with the different techniques currently available have been outlined above.

17.4 Environmental impact assessment

To many people the concept of an impact on the environment is restricted to physical effects, such as an increase in noise levels, more air pollution, or the degradation of the quality of the water in our rivers and seas. However, all development activity will also have an impact on the social and economic life of a community.

Concern about the adverse effects of development projects on the natural world and on human health and happiness has a long history. Modern times have seen this concern rise to the forefront of public consciousness. The UK was early in the field of environmental investigation and legislation, notably in the fields of public health (clean air, sewage treatment, etc.), and in development planning. However, other countries have subsequently led the way in the introduction of formal, obligatory, environmental assessments.

In accordance with the European Directive on Environmental Assessment the UK Government introduced regulations to ensure that, for projects of major potential impact, comprehensive environmental data would have to be provided by the developer when seeking planning approval. To date the regulations do not require a mandatory assessment to be submitted for infrastructure projects, these being included in Schedule 2, indicating, however, that they may be required in certain instances.

When carrying out an environmental assessment in respect of the likely effects of development projects a number of basic questions should be addressed:

1. What is the existing environment in the area affected by the proposed development?
2. What effects will the proposed scheme have on the existing environment?
3. What measures can be taken to mitigate any adverse effects of this proposed development?
4. What would happen if the proposed development did not proceed?

It is accordingly suggested that these four basic categories should in fact be the primary ones considered when reviewing the environmental impact of proposed public utility works.

17.5 Legislative mechanisms to control social costs

17.5.1 DIRECT PROHIBITION OR CONTROL

One approach to controlling the social costs arising from underground work is to introduce a complete ban on open trench work, as has been done in Singapore, or in Tokyo where open-cut construction is not normally permitted unless it is the only option, in which case, it is required that the work be carried out at night in a manner that will enable the area to be plated over or backfilled to allow unrestricted movement next morning. Although this will result in lower social costs in many situations, it will not necessarily always be the case.

There are as yet no truly 'no-dig' techniques, as all require some occupation of road space for a certain amount of time in order to gain access to the underground asset and to bring equipment and materials to the access point. If the chosen point is at a heavily trafficked junction in the highway network, for example, albeit occupying a relatively small area, severe traffic disruption will be caused, thus incurring high social costs. Noise, vibration, dust, loss of access to properties, and loss of trade may also be important and significant social costs, which the banning of open trench work does not necessarily minimise.

A refinement on the total prohibition approach is to allow open trench work only at specific times of the day and/or in certain locations. Such times and locations are chosen with a view to avoiding heavily used roads particularly at peak flow hours.

Even if a total ban of open trench work were to minimise social costs, the arguably more desirable objective of minimising social costs would not necessarily be met. The prohibition approach is thus a blunt instrument which does not always result in the lowest overall cost to the economy.

17.5.2 PUBLIC INVESTMENT

The use of public funds to promote the welfare of the general public and reduce adverse environmental and social impacts has long been recognised as a legitimate use of such funds, although this has become less well regarded by recent UK governments. In the past this has meant, among other things, the promotion of the health of the general public through the construction and operation of sewerage systems, sewage treatment works and water supply systems. Even among the commercially oriented utilities, a residual social role can be seen to remain.

There are precedents for the use of public money to minimise social costs, examples being the award of grants for the thermal and noise insulation of domestic properties. Although unlikely, it would be possible for a government to legislate for the award of grants to those agencies and authorities who used 'no-dig' techniques in preference to open trench methods. However, the setting of the level of grant, the processing of applications, and the monitoring of the system would be administratively difficult.

17.5.3   COMPENSATION PAYMENTS

Expenditure that under some circumstances may be regarded as public investment can, with slightly different administrative or legislative arrangements, take the form of compensation payments. For example, a statutory grant system currently operates in the UK providing noise compensation payments for increased traffic noise suffered by properties adjacent to major new highways.

In the case of the public utilities, under present legislation in the UK, compensation claims for loss of trade may be made against water companies undertaking sewerage work if it can be shown that the work has indeed affected trade and resulted in a loss of business. This does not apply to work carried out by the other public utilities, nor does it apply to bus companies who may have increased costs as a result of traffic delays and diversions caused by sewerage works. An extension of the right to compensation to a wider section of the general public and to cover losses and costs incurred not only as a result of sewer works, but also other utilities work, would be a further incentive for the adoption of no-dig techniques.

17.5.4   ENVIRONMENTAL TAXES

The proposition that environmental impacts might best be controlled through a system of environmental taxation has gained credence over the last few years. These ideas can be applied to the reduction of traffic delays through the application of the road lane and road space rental concepts.

In the UK, the 'lane rental' mechanism has been successfully used to reduce traffic delays during motorway maintenance contracts. The lane rental scheme has been employed extensively by the Department of Transport since it was first used in the UK in 1987. It was introduced in an attempt to reduce the time taken by contractors to complete maintenance contracts. The contract can specify, or the contractor can tender, a maximum time that the contractor will be allowed to occupy the particular section of motorway undergoing maintenance. Should the contractor fail to complete within this time, penalties are imposed, and in some cases, bonuses are payable for early completion.

The important point to note, in the context of this chapter, is that the penalties or bonuses are related to the total delay costs incurred by the motoring public as a result of the works being carried out. In terms of reducing delays, the results have been encouraging. However, it has been reported that other less quantifiable social costs have arisen, notably increased stress on construction personnel, who are often obliged to work long hours in close proximity to a fast moving and noisy stream of traffic.

It is generally agreed that the lane rental concept is most appropriate to motorway and rural dual carriageway maintenance work, and is unlikely to be directly applicable to public utilities work, especially in urban areas. However, a modification of this concept, which we term 'road space rental' (RSR) and others, since referred to as highways rental fee, may well be the necessary and structured incentive required to encourage project promoters to adopt the less disruptive trenchless technologies now available for the installation and rehabilitation of underground assets.

17.6   Developing the concept of road space rental to minimise social costs

It is clear that many of the delays caused by utility street works could be avoided. In a paper to the 1988 No-Dig conference in Washington, USA (see Bibliograph no. 13), we proposed a system of road space rental which would involve charging different rents for road space, depending on the sensitivity of the location with respect to the inconvenience and disruption caused.

Under such an arrangement, anyone occupying road space for construction or maintenance purposes would be required to pay a rent to the highway authority. This rent could be based on the amount and traffic significance of the highway occupied at any given time. The charges should reflect the degree of disruption caused, and would thus vary with the precise location of the occupied space in the road network. Work near a significant road junction, for example (Fig. 17.1). or work during periods of peak traffic flow would attract a premium rate. This idea should provide a real incentive to a contractor to use quicker, less disruptive techniques by reducing the financial civil engineering cost advantage enjoyed by open trench techniques compared with many of the no-dig techniques.

Figure 17.1   Work near a significant road junctions.

It must be appreciated, however, that road space rental could aggrevate the impact of other social costs in certain cases. Speeding up the works may require longer working hours which could exacerbate the environmental problems caused, especially for residents (Fig. 17.2). So there is a need for an evaluation technique that can readily assess all the costs and benefits as described in a paper we presented to the 1991 No-Dig Conference in Hamburg (see Bibliography No. 9).

Figure 17.2 RSR could involve more night-time working.

17.7 The New Road and Street Works Act 1991

The Government's reform of street works legislation via the New Road and Street Works Act of 1991 has been widely welcomed by all sides of the industry. The control and management of highway openings of all types had previously relied on the Public Utilities Street Works Act of 1950 introduced when there were some 4 million vehicles on the UK highway networks. Although it was enacted in 1950 it was actually drafted in the 1930s but delayed by the war; it was therefore framed around traffic levels and utility usage levels which were different to those obtaining in 1950 and vastly different from those of today. These controls had become increasingly inadequate to deal with the dramatic rise in the levels of both traffic and works on a highway network which had only increased by some 20% since 1950 yet was having to carry some 24 million vehicles (an increase of 500%). The resultant chaos and interference with traffic flow (currently costing the country an estimated £60 million per year) caused by highway openings of all types could no longer be accepted.

The New Road and Street Works Act provides highway authorities with powers to charge utilities where works exceed a prescribed period.

The 1991 Act was intended to improve co-ordination between utilities and others opening the highway with the aim of reducing the consequent traffic disruption (Fig. 17.3). In

Figure 17.3 Typical traffic disruption as a result of public utility works.

addition, the Government's intention, spelt out in the Citizen's Charter, to reduce the annoyance and inconvenience caused by streetworks, promises well for the future. Local authorities are given powers to designate-traffic-sensitive streets and limit the times at which work can be undertaken in such locations. This alone could lead to the wider adoption of trenchless techniques. The Act includes a number of other major changes designed to minimise the impact of public utility works with particular benefit to the overloaded highway network. Limited powers to introduce the concept of 'highway rental' have been included in the new legislation, but only as an option. The consensus being at the time of enactment that utilities would not prolong their works needlessly, but there is now a growing feeling that this does not obtain in practice. It had been hoped that a more definite and comprehensive approach would have been included, which would no doubt have resulted in greater use being made of trenchless techniques. Nevertheless it is clear that greater emphasis is being given to social and environmental considerations and this is likely to continue.

17.8 Preliminary conclusions

Generally trenchless construction will mean:

- Less disruption to the environment, traffic flows and other public utility equipment
- Less risk to the public operatives and adjacent structures
- Less damage to roads, footpaths and the substrata.

Nevertheless it has been confirmed that 'trenchless' techniques are not in themselves a complete panacea to all social or environmental cost problems. Access shafts or working compounds to accommodate them at or near major road junctions and kept open for unnecessarily long periods may cause as much or more disruption as an open trench, for example.

One of the reasons for ignoring social costs and benefits in project appraisal has been the perceived difficulty in evaluating them. However, we believe that through a rational and structured examination of a hierarchy of social costs, these hitherto ignored components of project costs could be relatively easily incorporated into the project appraisal process.

First, the direct costs of each of the selected potential construction methods would be evaluated in the normal way, and a 'guesstimate' of the worst case disruption costs arising from a complete road closure would be made. At this stage it should be adequate to carry out an assessment of the diversion costs, assuming a complete road closure for the duration of the contract, and using the formula shown below:

$$TC = VPD \times (VOC \times AL + VOT \times AT) \times T$$

Where TC = total cost to road users (£)
  VPD = number of vehicles per day
  VOC = vehicle operating cost (£/km)
  AL = additional distance
    = diversion distance – original distance (km)
  VOT = value of time (£/vehicle per hour)
    = Average value of time per occupant (£/hour) × average number of occupants per vehicle
  AT = additional time (hours)
    = (diversion distance (km)/diversion speed (km/hr))
    – (original distance (km)/original speed (km/h))
  T = construction time (days)

Vehicle operating and value of time costs are, of course, country dependent, but in the UK these are evaluated and published on a regular basis by the Department of Transport, and based on research by the Transport and Road Research Laboratory (TRRL) and others. The number of vehicles using the road in question should be available from the local road or transport planning authority, and the length of diversion and the construction time can be estimated from the project proposals. 'Guesstimates' of journey speeds could be made by local highway engineers familiar with the road network.

If this preliminary assessment indicated that traffic costs were less than a certain proportion (say 20%) of the direct costs, as may be the case in rural areas or small residential urban streets, there would be no need to proceed with any more complex traffic analysis, although other normally less significant social costs might then be considered in cases where they are exceptionally high.

Such other significant social costs might include, for example, disruption of access to critical facilities (fire stations, hospitals, etc.) or noise and vibration in sensitive locations. A worst case assessment of these costs could be made by evaluating the costs of mitigating these impacts.

The social cost of noise nuisance, for example, might be determined by estimating the cost of installing noise control devices or insultation to construction machinery or to buildings in the vicinity of the work. These preliminary estimates could be carried out fairly easily and would indicate their level of significance relative to the direct costs.

Social costs can thus be considered in the selection of technique to be employed on any particular job, even though in many cases final selection might still be based on the least direct cost. If preliminary estimates indicate that the social costs are likely to be significant, it would be necessary to assess these costs for each method under consideration. By far the most significant social cost in the majority of cases is that arising from traffic disruption, but even at this stage of the appraisal a full traffic assignment analysis may not be necessary. A detailed study of the route of the proposed underground utility and the proposed method of working for each option should reveal whether or not high traffic costs would be likely.

Works on a traffic significant route or crossing such a route would almost inevitably result in some traffic costs, but these will depend on the width and length of road space to be occupied at any one time. Any method which leaves less than a 3 metre width of carriageway available for traffic will result in the same disruption as a complete closure, and disruption costs could therefore be evaluated using the simple method previously described. It should be noted here that the exact location of the trench or access shafts within the carriageway is required – in some circumstances access shafts required for trenchless techniques can effectively cause a complete road closure!

If more than 3 metres but less than 5.5 metres of carriageway are left available for traffic, traffic in one direction must be diverted, or shuttle working, controlled by temporary traffic lights, must be introduced. In such a case an estimate of the traffic costs could therefore be made by assuming no delays to traffic in one direction, and evaluating the costs of diverting traffic travelling in the other. If no suitable diversion route is available and shuttle working has to be introduced, any method which requires a shuttle working length in excess of 150 metres or where the works are carried out on a road normally carrying a two-way traffic volume greater than 1300 vehicles per hour, should be deemed to incur significant social costs and might therefore be discounted at this stage, unless restricting the works to off-peak traffic times is a feasible alternative.

Where more than 5.5 metres of carriageway are left available, two-way traffic can be maintained and the social costs incurred will be due to delay rather than diversion. In such cases a more complex traffic analysis would be required if simpler procedures had not already eliminated the other options.

The above procedures would normally apply to works on traffic significant routes, but even works on other roads can incur high social costs if they cross a traffic significant route. Roadworks within 50 metres of a junction with a traffic significant route can causes severe congestion due to tailbacks blocking the major road. It is therefore important to identify the position of trenches and shafts in relation to such junctions. Any methods requiring road openings in such locations should be discarded at this stage, or a more comprehensive analysis of the likely traffic delays should be carried out.

The preceding paragraphs indicate that, for the majority of cases, social costs arising from traffic disruption can be incorporated into the project appraisal process without resorting to complex and expensive traffic assignment studies. We would maintain that for most underground utility projects, a simple consideration of traffic costs as outlined above would be sufficient. In only a few cases, probably located in particularly environmentally sensitive areas, would a consideration of other social costs, such as those arising from noise, dust, vibration, and visual intrusion, be necessary. These costs are more difficult to evaluate in financial terms as we discuss later in this chapter.

By presenting social cost analysis in the overall framework of project appraisal, we hope to have shown that, in many cases, only a rough and preliminary assessment of social costs is required to demonstrate that they have been taken into consideration, and that the most cost-effective option for the community as a whole has been chosen. However, in many cases, there is neither incentive nor legislative instrument to encourage or force promoters of projects to consider social costs at all. The new legislation in the UK has been discussed in previous sections of this chapter, where reference has been made to the concept of charging for the occupation of road space as a legislative instrument to promote the consideration of social costs in project appraisal. One of the difficulties of implementing a charging scheme for road space occupation is that of setting appropriate charges which reflect the social costs incurred by that occupation. The following section is an attempt to show how this might be done.

17.9   The determination of road space rental charges

A number of legislative instruments which could be used to minimise social costs were reviewed in our paper presented at 1991 No Dig conference in Hamburg (see Bibliography no. 9). In that paper we stated that any road space charging scheme would necessarily be more complex than a simple road lane rental or nominal charge per unit area. However, it is also recognised that any scheme must be sufficiently straightforward to keep administrative costs to a minimum.

For such a scheme to be implemented, it would first be necessary for the authority responsible for the road network to designate certain roads as 'traffic significant' and certain junctions as 'critical'. This information would then be made available to the utilities and their contractors. The traffic significance of a road would depend on a number of factors including the capacity of the road, the normal traffic volume on that road and the availability of alternative routes. It is envisaged that within a normal urban network, about 10% to 20% of all roads would be classified as traffic significant. 'Critical' junctions would be those junctions where the traffic significant route did not have automatic and continuous right of way, or where there was a substantial volume of right turning traffic on the traffic significant route.

For any work on a traffic significant road, the level of charge for road space occupancy would be based on a number of factors relating not only to the road itself but also to the proposed method of working. From an understanding of the method of working for any particular construction or maintenance technique, an assessment should first be made of whether or not two-way traffic could be maintained along the route under consideration.

If it can, a charge related to the delay costs could be estimated from graphs similar to that shown in Fig. 17.4a and labelled as graph type 1. It can be seen from this graph that traffic costs are related to both traffic flow and the extent of the obstruction, and that these costs will rise dramatically after a certain 'threshold', value of obstruction extent is exceeded. Although 'extent' in this context refers to both length and width of the obstruction, it is the width that is generally the dominating factor. It should be noted that the graph presented is for illustration purposes only, and that the actual shape of the curves will be dependent on the number and width of the carriageways. Any road openings in the vicinity of 'critical' junctions would attract a supplementary charge.

If two-way traffic cannot be maintained, it would be necessary to either close the road in one or both directions, or introduce shuttle working, all of which would normally incur greater traffic costs, and should therefore attract a higher rental charge. As a rule of thumb, shuttle working on a road normally carrying two-way traffic in excess of 1300 vehicles per hour and requiring a shuttle working length greater than 150 metres will generally cause severe disruption. In such cases closure of the road to traffic in one direction is likely to be the preferred option. Where shuttle working can be employed, the delay costs can be assessed from graphs similar to that shown in Fig. 17.4b and labelled graph type 2. Once again, a supplementary charge would be made for any works near critical junctions.

In situations where two-way traffic cannot be mantained and suitable diversions are

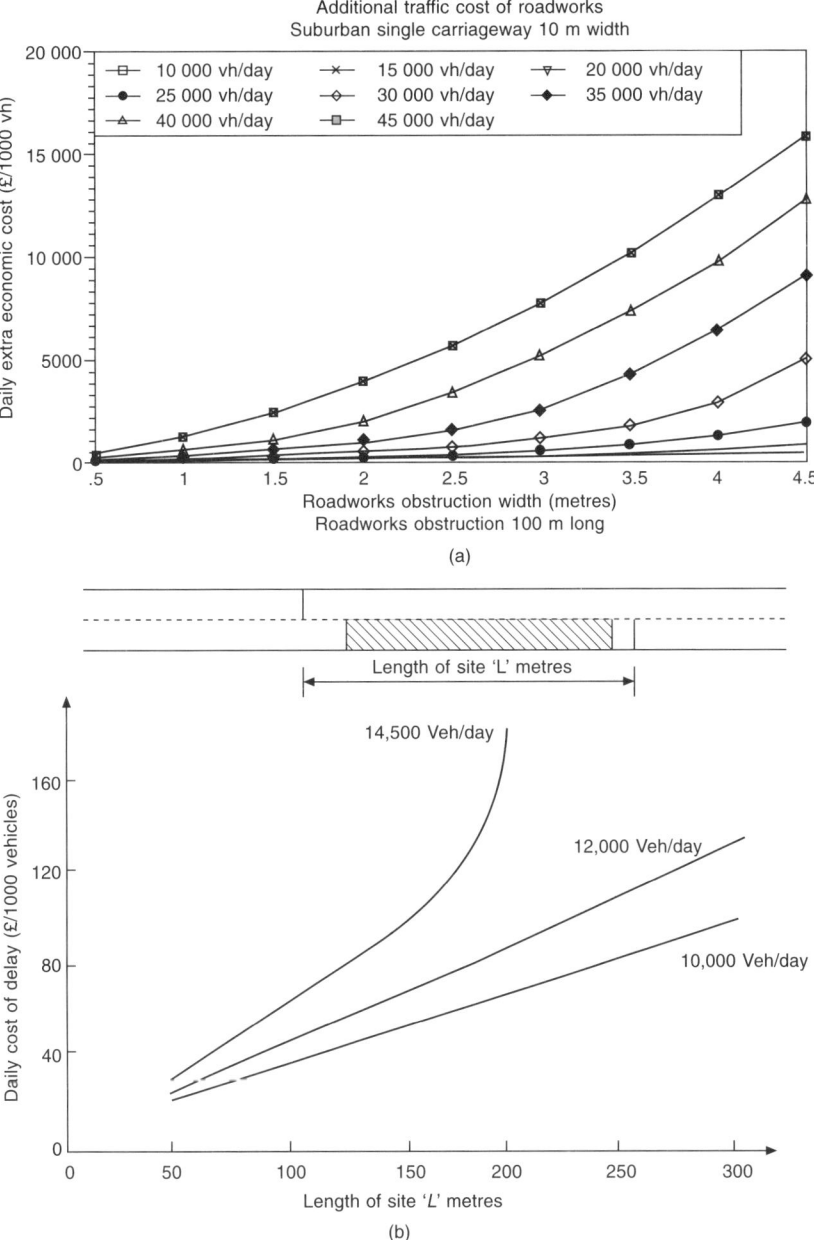

Figure 17.4 (a) Graph type 1; (b) Graph type 2.

available, an alternative to shuttle working is to allow for passage of traffic in one direction only while diverting traffic travelling in the opposite direction. The local pattern of the road network and traffic flow will dictate whether or not this is preferable to shuttle working. If this option is chosen, the one-way diversion costs would be calculated. In most such cases it would then be unnecessary to calculate the costs of traffic delay on the route through the works, as these are likely to be small in comparison to the one-way diversion costs. As with the other situations previously described, a surcharge for any road occupation near critical junctions should also be made.

When a full road closure is required, traffic travelling in both directions must be diverted and diversion costs for both directions must then be assessed, resulting in a consequentially high road space rental charge.

17.10   Problems of assigning monetary values to noise, vibration and air pollution

The purpose of presenting the above sections of this chapter has been to partially allay the anticipated concerns of engineers faced with the complex task of evaluating costs and charges which they often do not regard as 'real'.

Although in most projects involving trenchless technology, social cost issues will be adequately dealth with through a consideration of the traffic costs alone, there will be a small number of situations where other social costs should be considered. These costs, arising from the noise, dust, and visual intrusion caused by the construction process and by any changes in the pattern of traffic flow around the site, are much more difficult to evaluate than the traffic disruption costs previously discussed.

Although difficult to evaluate, they are nevertheless real costs and some attempts have been made to assign monetary value to them (4,5). For example, Pearce (4) reports that studies indicate a 0.6% reduction in house prices for every unit increase in noise exposure frequency (NEF). This figure is taken as the average of several studies carried out in the USA, UK, Canada, and Australia. Similarly, several studies carried out in the USA suggest approximately a 0.1% fall in property value for every 1% increase in concentration of pollutants in the air. The overall annual costs of pollution damage from noise and air pollution in Germany were estimated to be in excess of US$30 billion in 1985 – equivalent to 6% of Germany's GNP at that time!

It is clear that in some cases these less easily evaluated social costs may be the major consideration. At the present time, these factors can generally only be considered in a qualitative way, although the costs of mitigation might indicate their relative monetary value. However, we envisage that in the future, as the results of research become available, more accurate assessments of these costs could be included in the overall appraisal of projects and the setting of road space rental charges.

17.11   Overall effect of road space charging on the economy

Road space rental charges may increase the overall cost of a project and these costs must inevitably be passed to the consumers in the form of increased unit charges. However, these charges would normally be increased across the board, and hence borne by all consumers according to their consumption. This is in contrast to the present situation where most consumers pay the standard charge, but a minority in effect pay an additional 'environmental' levy – that is the cost of enduring the disruption caused by the project. Road space charging would not increase the total economic cost of a project – it would merely redistribute those costs more equitably such that all those receiving the benefits of the project (the consumers) paid their fair share for them.

The main purpose of road space rental is thus to reduce the total economic costs to the community as a whole through the adoption of less disruptive construction techniques where this is appropriate.

17.12 Working shaft location

It will be appreciated that working compounds, if they are situated within the highway, could well cause serious traffic problems, particularly if located near a busy traffic junction.

Away from junctions, traffic delays depend primarily on the extent to which normal traffic flow can be maintained past the works. This will depend mainly on the width of cariageway remaining to traffic, relative to traffic volume. Where less than 3 metres of usable carriageway remain available, the road must be closed completely and traffic diverted to an alternative route.

Where more than 3.0 metres, but less than 5.5 metres of usable carriageway remain, traffic cannot pass in both directions simultaneously and therefore either traffic in one direction must be diverted, or shuttle working must be introduced controlled by temporary traffic signals. The cost of traffic congestion is a major factor in the consideration of social costs and hence the obvious social benefits obtained by the use of trenchless technology can easily be cancelled out by the effects of badly sited working compounds. It is accordingly necessary in the overall equation at the design stage to consider alternative pit locations, even if the civil engineering costs of such may not be the cheapest but where the effect on traffic is less significant (Fig. 17.5).

17.13 The future

It is clear that the ever-increasing public concern for the environment will ensure that the 'green' vote will continue to exert greater influence on development/planning policies and related legislation generally.

Within the European Community the environment ranks as second most important political problem perceived by the electorate – second only to unemployment – ahead of inflation and arms control. Already our industry is moving towards a greater recognition of the social and environmental consequences of its work and hence a more thorough appraisal of all the associated costs and benefits.

Neverthless, there is still a pressing need to put substantially more resource into the detailed planning of street openings, and the co-ordination between utilities and highway authorities, to ensure that traffic disruption is minimised. Only if the necessary skilled human resources are made available for these vital preparatory stages will we finally bid farewell to those long-lived and apparently deserted street openings.

Well-managed and designed public utility works will help to reduce the temporary environmental impact of works designed to give long-term environmental benefit to the population in general.

It is interesting to note that the British Automobile Association (AA), which represents some 7 million interests, has stated that 'Trenchless technology should be encouraged as a means of reducing congestion and delay in traffic sensitive streets' which it identifies as 'virtually all urban and inter-urban routes'. Clearly there is now a high level of conflict between the needs of the utilities on the one hand and the needs of the road users on the other.

The message to the industry is clear – maximise the use of trenchless techniques but still ensure that social cost considerations are given priority at both design and construction stages.

Bibliography

1. UMIST Sewer Rehabilitation Group, John Wood and Colin Green, Current *Research into the Social Costs of Sewerage Systems.* Proceedings of NO-DIG 87, ISTT, 1987.
2. Read, Geoffrey F. and Vickridge Ian. *The Environmental Impact of Sewerage Replacement and Renovation.* Proceedings of NODIG 90, ISTT, Rotterdam, 1990.
3. Vickridge, I. G., Ling, D. J., Letherman, K. M., Read G. F. and Bristow A. L. Social costs of sewerage rehabilitation – where can no-dig techniques help? *Tunnelling and Underground Space Technology,* Vol. No. 4, pp. 495–501, 1989.
4. Pearce, David, Markandya Anil and Barbier, Edward B. *Blueprint for a Green Economy.* Earthscan Publications Limited, 1989.
5. Bristow, A. L. and Ling, D. J. *Reducing congestion from utilities roadworks: whose costs count.* PTRC Summer Annual Meeting, Seminar K – Highway Construction and Maintenance, University of Sussex, 1989.
6. *Trenchless construction for new pipelines: a review of current methods and development.* WRc Engineering Report ER 167E, 1985.
7. Read, Geoffrey F. Social cost implications in sewerage rehabilitation. *Civil Engineering,* Aug. 87, pp. 8–13.
8. Horne Professor Michael. *Roads and the Utilities.* Department of Transport, 1985.
9. Ling, David, Read, Geoffrey F. and Vickridge, Ian. *Road space rental – a structured incentive for the adoption of no-dig technologies.* NO-DIG 91, Hamburg, International Society for Trenchless Technology, October 1991.
10. The New Road and Street works Act 1991.
11. Munasinghe, Mohan and Lutz, Ernst. Environmental–economic evaluation of projects and policies for sustainable development. Environment Group Working Paper No. 42, The World Bank, January 1991.
12. Vickridge, I. Counting the Social Costs of Civil Disruption. *New Civil Engineer,* 19 October 1989.
13. Vickridge, Ian, Read, Geoffrey and Ling, David. Evaluating the social costs and setting the charges for road space occupation. NO-DIG 1992, Washington, USA, April 1992.

Appendix A

Some Social Cost Implications in a Sewerage Rehabilitation Scheme at Cheetham Hill Road, Manchester

Geoffrey F. Read MSc, CEng, FICE, FIStructE, FCIWEM, FIHT, MILE, FconsE, Fcmi, MAE

Concern about the environment has of late become an increasingly common topic of conversation within the Water Industry and has led urban civil engineers to 'go underground' in their thinking rather than proceed in open-cut, even if the depth of the sewer is relatively shallow. It was perhaps not surprising therefore at the interest generated amongst the delegates, at the recent 'No dig 87' conference in relation to the Social Costs of Sewerage Rehabilitation and Flooding Studies currently being carried out at U.M.I.S.T. and Middlesex Polytechnic which were dealt with in a Paper introduced by Geoffrey F. Read and Ian Vickridge of U.M.I.S.T. and John Wood of W.R.C.

In his article Geoffrey F. Read who was formerly City Engineer, Manchester, where he was responsible amongst other things for an extensive Sewerage Rehabilitation programme, describes a project which was mentioned in the Conference Paper in which some Social Costs were taken into account at the design stage – albeit in broad concept – in order to justify the works being carried out in heading rather than, open–cut – the latter being cheaper had civil engineering costs, alone been the only criterion.

Production

No longer is open-cut sewerage rehabilitation acceptable in our towns and cities because of consequential disruption and delay to traffic, damage to existing mains and services, increased highway maintenance and traffic accidents in addition to the environmental impacts which include noise and vibration, air pollution and effects on the amenity and aesthetic value of the areas affected.

Engineering works do not last for ever and remedial works or collapses if they are not found out – eventually materialise. At this stage the public reluctantly accepts the need for keeping the system operating but is now less than enthusiastic about the consequential impact on the normal quality of life in the local environment.

Since the late seventies Manchester has headed the 'Sewer Collapse League' having had over 75 collapses in the City Centre since 1975. Although social costs, often at a high level, are incurred through these emergencies, sewerage rehabilitation work itself causes disruption but in this case the level of social costs can be estimated and will vary dependent upon the particular rehabilitation techniques adopted.

The problem

The 500 × 375 mm non-man entry, butt jointed, egg shaped earthenware sewer in Cheetham Hill Road at depths varying from 4 to 8.9 metres had been constructed in 1858 and was in imminent danger of collapse. It was in poor condition with displaced and open joints and four recent collapses had taken place as a result of ground loss etc., consequent upon regular surcharging. Owing to the grossly displaced joints a full CCTV survey had not been possible.

The sewer ran along the centre line of Cheetham Hill Road (A665). This area contains much light industry and wholesaling activity, largely housed in Victorian property but with some new buildings. It has generally a high concentration of commercial activity. The highway above the sewer is a main arterial traffic route heavily used by both commuters and commercial vehicles into and out of the City centre.

Cheetham Hill Road also has beneath it a large amount of other statutory undertakers' plant including 12″, 9″ and 8″ diameter gas mains together with a 10″ diameter old gas main having rigid joints.

It was considered that the high traffic loading would exacerbate the poor condition of the sewer, increasing the risk of collapse and resultant disruption. Renovation of the existing sewer was not a feasible proposition and in any case it was under capacity often surcharging up to 2 m head. On-line replacement was necessary involving 740 metres of 600 mm diameter pipe and 10 new manholes.

A number of alternative solutions were considered but these were ultimately condensed, in general terms, to two options:

an open-cut solution (A), where the upstream length of 490 m would be in open-cut and the remaining 250 m in heading, consequent upon the greater depth of this section, with significant disturbance to traffic and business;

or a complete heading solution (B), with minimum disturbance.

The estimated costs of the civil engineering works alone were:

| Option A | (Open-Cut/Heading) | £4 71 040 |
| Option B | (Heading) | £5 37 600 |

However, I was deeply concerned at the likely impact of the 'open-cut' solution on the continued economic well-being of this part of the City and recommended Option B. In order to substantiate this recommendation, arrangements were made for a detailed examination of the likely social costs of both options.

This preliminary examination, which was able to take account of some but not all of the social costs that were likely to arise, confirmed that there was in fact even then a significant saving, in terms of the total cost to society, by adopting the heading solution which I had recommended. The results of this examination are summarised in Table 1.

It was accepted that the assessment was of necessity an approximation. In any case traffic figures vary daily as indeed will a driver's choice of diversion route and hence extent of traffic delays.

There are many factors which it will be seen were not taken into account in view of the need to reach early confirmation of the recommendation to proceed in heading before a further collapse situation occurred. It is believed that these would have reinforced the decision to adopt the complete heading solution.

Traffic considerations

As will be seen from Table 1, the main factor which was evaluated, in addition to loss of profits to businesses was the cost of traffic delays and diversions. The general nature of such traffic disruption is similar to that encountered with any roadworks in urban areas. Considerable advance has been made in the measurement or prediction of such traffic disruptions, including the economic assessment of delays and other costs to road users and reference is made under Table 1 to the programme used in this instance.

Cheetham Hill Road contains four traffic lanes and carries an estimated flow of 26 000 vehicles per day.

Table 1 Summary of estimated costs

Nature of costs	Option A Combined open cut/ heading solution	Option B Heading solution
Direct Costs:		
(a) Civil Engineering Works	£4 71 040	£5 37 600
(b) Compensation for loss of business profit*	£1 30 000	£70 000
(c) Compensation for damage as an unavoidable consequence of the works	Unknown However claims resulting from open cut works are likely to be far greater than those resulting from heading works	Unknown
Social Costs:		
(d) Delays and diversions to traffic (1979 prices); (RPI)** Inflation	£7 91 340	£3 58 212
$\dfrac{351.1 - 223.5}{223.5} = 57.1\%$	£4 51 790	£2 04 509
(e) Long term damage to: Buildings Carriageways Services	All intangible but again likely to be far greater for open cut than heading.	
(f) Disruption to local economy and loss of amenity	All intangible but again likely to be far greater for open cut than heading	
Grand Totals	£18 44 170	£11 70 321
	Estimated overall saving of Option B = £18 44 170 − £11 70 321 = £6 73 849	

*Supplied by the Director of Finance NWW and was based on current levels of settlement of claims for loss of profit.
**Based on Department of Transport 'COBA 9' Manual, May 1981 Annex II Highway Economics Note No. 2 (July 1980) 'Values of Time and Vehicle Operating Costs for 1979' (HEN 2).

Three alternative diversion routes exist although the shortest one (Route B) is unsuitable for large vehicles due to tight radii and low headroom under a railway bridge. Accordingly the calculations were based on the average net additional length of the two more attractive routes viz. 1.9 km referred to later as route *x*.

Working arrangements

(a) Option A (Fig. A1)

This would involve a pipe laying 'train' approximately 100 m long occupying 3 of the 4 traffic lanes for some 30 weeks. Following (or preceding) this there would be two phases of heading work occupying the two off-side traffic lanes at the Southern end lasting approximately 8 weeks each.

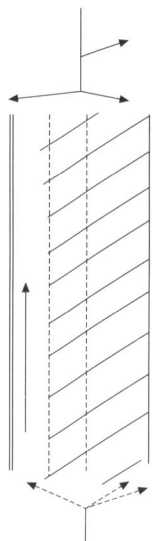

Figure A1   Option 'A' open-cut/heading. Elements in traffic distruption costs: (1) Cost of *time* and extra *distance* for all inbound vehicles to be diverted via route 'x' or in the case of PSVs route C. (2) Cost of *time* for traffic in excess of capacity. (3) Cost of *time* and extra *distance* for outbound traffic in excess of capacity to travel route 'x'. *Note:* this diagram relates to the open-cut part of the option, the heading part being similar to Option B.

During the open-cut work traffic on Cheetham Hill Road would be restricted to one lane outbound (Northward). During the peak period vehicle flows would exceed the Capacity of the one lane and advisory diversion routes for both directions would be signed. All inbound (Southward) traffic would be diverted during this 30 week period.

During the two phases of heading work, two traffic lanes (one in each direction) would be maintained but peak period flows would be advised to use the signed diversion routes.

(b) Option B (Fig. A2)

To accommodate working/reception shafts traffic would be restricted to one lane (nearside) in each direction – peak flows being advised to use diversion route.

Elements contributing to traffic disruption costs

The comparison of the traffic disruption costs associated with the two alternative options has been summarised on Table 2.

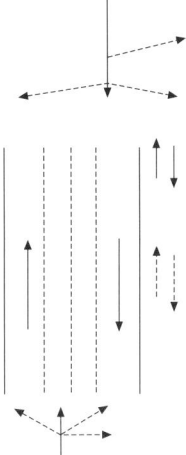

Figure A2 Option 'B' heading only. Elements in traffic distruption costs: (1) Cost of *time* in excess of capacity. (2) Cost of *time* and extra *distance* for traffic in excess of capacity to travel route 'x'.

Table 2 Elements contributing to traffic disruption costs

	Option A Combined Open-Cut/Heading Solution	Option B Heading only solution
	1. The cost of time and distance for all *inbound* (southward) vehicles only to travel Route × (excluding PSVs) during the open-cut element of the works.	H 1. The cost of time and distance for all *inbound* (Southward) and *outbound* (Northward) vehicles, in excess of the 2 single lane capacities to travel Route × at peak periods (excluding PSVs which will remain on Cheetham Hill Road).
	2. The cost of time and distance for all *inbound* PSVs to travel Route C during the open-cut element of the works.	H 2. The cost of time only for all other *inbound* and *outbound* vehicles (including PSVs) which will remain on Cheetham Hill Road.
Open Cut	3. The cost of time and distance for *outbound* (northward) vehicles, in excess of single lane capacity, to travel Route × at peak periods (excluding PSVs which will remain on Cheetham Hill Road), throughout the contract period.	
	4. The cost of time only for all other vehicles (including PSVs) which remain on Cheetham Hill Road, throughout the contract period.	
	5. The cost of time and distance for *inbound* vehicles, in excess of single lane capacity, to travel Route × at peak periods (excluding PSVs which will remain on Cheetham Hill Road) during the heading element of the works.	
Heading	6. The cost of time only for all *inbound* vehicles, including PSVs, which will remain on Cheetham Hill Road during the heading element of the works.	

Note
x = imaginary diversion route distance i.e. average net length of the two most attractive alternative routes.

Calculating traffic disruption costs

HEN2 goes into considerable detail to enable the various contributors to costs to be quantified. From the expression:

$$\text{value p/km} = a + b/v + cv^2$$

given in HEN2, Table 6, where v = vehicle speed in km/hour and a, b and c are parameters also given in HEN2, Table 6, the following values of distance were calculated. These values include, inter alia:

> Wear and tear on vehicle
> Fuel consumption
> Value of time of occupants
> Indirect taxation

Vehicle category	Value of time and distance in p/km (at 30 km/hour)	Value of time in p/hour
NWC (non-working car)	6.5	122.1
WC (working car)	21.5	480.2
LGV (light goods vehicle)	20.0	335.9
OGV (other goods vehicle)	32.0	360.0
PSV (public service vehicle)	73.0	1310.7

The above also includes the value of time only (HEN2, Table 3) in order to quantify the cost of delays to the traffic, on Cheetham Hill Road, that does not divert.

One of the traffic elements (1) if Option A were adopted would be to determine the cost of time and distance for all inbound vehicles to travel Route x during open cut work (excluding PSVs which would use Route C).

This was calculated as follows:

Vehicle category	Estimated number	Value p/km	Cost in £/km/day
NWC	8990	6.5	584
WC	1802	21.5	387
LCV	1197	20.0	239
OGV	718	32.0	230
			1440 £/km/day

Sub total = 1440 × 1.9 km × 6 days* × 30 weeks £ 492 480

* 6 days per week assumes that total traffic on Saturday is equivalent to 1 days traffic Monday to Friday

Similar calculations were made for the other elements in Table 2 which in turn led to the estimated traffic delay and diversion costs given in Table 1.

Generally

Although this assessment was of necessity an approximation it was nevertheless abundantly clear that the apparently more expensive 'Heading Option' was in reality the cheapest to

'Mr Great Britain', being much less disruptive to the community. North West Water accepted the reasoning and the works have proceeded on the basis of Option B. It was interesting to note at the recent No-Dig '87 conference when presenting their paper that Messrs. Rosbrook and Reynolds of Southern Water also believed that social costs were a direct cost to its customers.

It would seem that the day of the cost/benefit analysis for sewerage and other, similar highway openings is approaching and perhaps sewer rental contracts, similar to the procedures operated by Department of Transport on motorways, is not too far away in this environmentally sensitive area, bearing in mind the 'No-Dig' techniques now available to minimise the consequential impact.

This is probably one of the first sewerage rehabilitation schemes where social costs were taken into account at the design stage and it is hoped that the Research Programme at U.M.I.S.T. will, in due course, enable more accurate assessments of this significant element to be made.

Appendix B   Cheetham Hill Road Sewer Replacement Scheme, Manchester

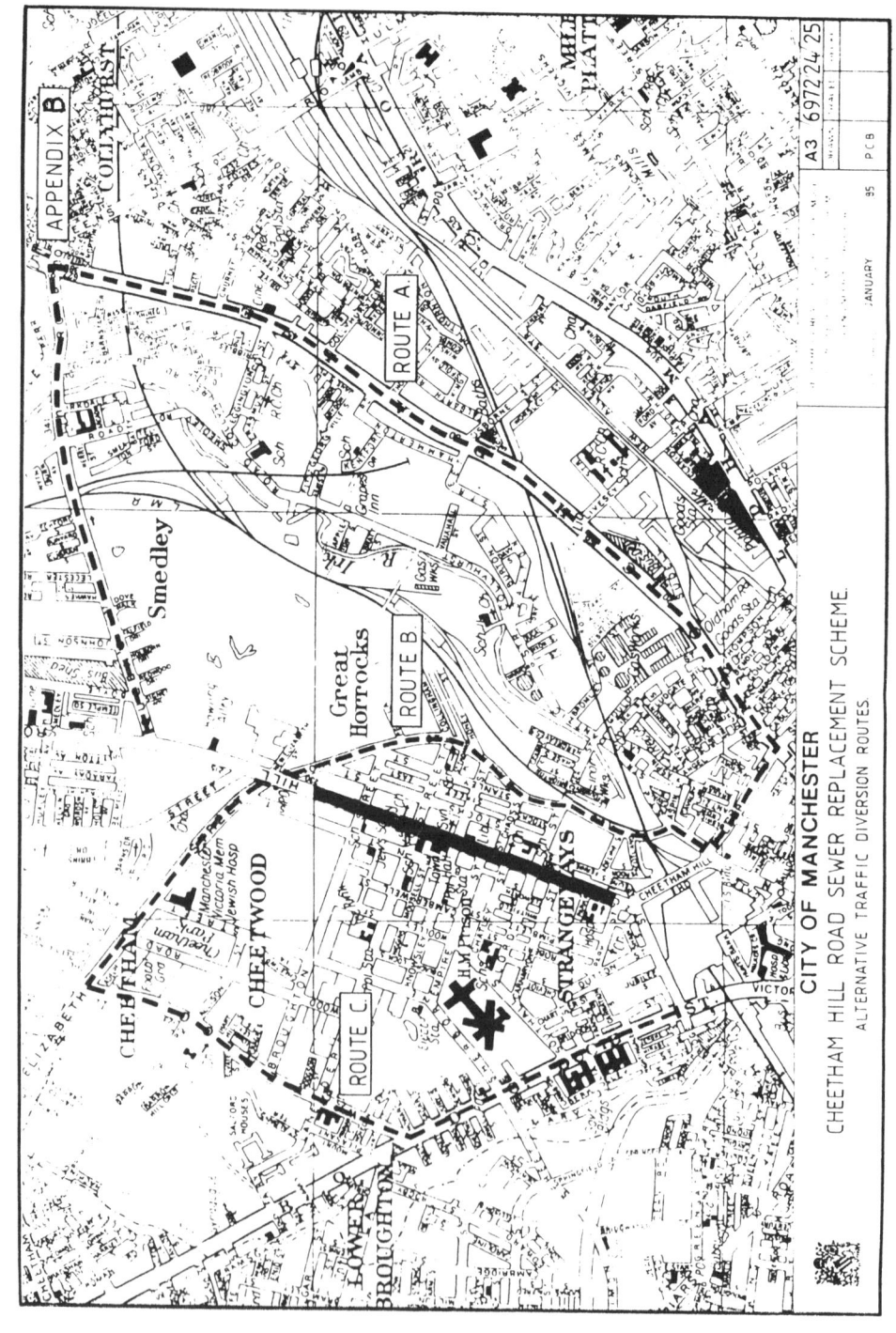

CITY OF MANCHESTER

CHEETHAM HILL ROAD SEWER REPLACEMENT SCHEME.
ALTERNATIVE TRAFFIC DIVERSION ROUTES

18

Factors Affecting Choice of Technique

Geoffrey F. Read MSc, CEng, FICE, FIStructE, FCIWEM, FIHT, MILE, FconsE, Fcmi, MAE

Ian Vickridge BSc(Eng), MSc, CEng, MICE, MCIWEM

18.1 Introduction

A number of different methods for installing, constructing, and replacing sewer pipelines have been reviewed and discussed in other chapters of this book. Each method has its relative advantages and disadvantages, which will be dependent on the specific environment and conditions. The engineer must choose the most appropriate method for any given situation. Although the various factors and parameters affecting the choice of method are much the same as for any engineering work, they must be fully understood in the context of installing and constructing sewers. The selected method must be *technically feasible* and produce an end result that is 'fit for the purpose'. It must also be one that allows for completion of the work at *reasonable cost*, in a *reasonable time*, with due regard to the *safety* of the workforce and the general public, and with minimum *environmental impact* and *social disruption*.

18.2 Technical feasibility

In selecting an appropriate technique, the first criterion must be technical feasibility – will the technique be technically capable of doing the job in the specific prevailing conditions and environment? In the past, the only technically feasible methods for installing sewers were open trench construction or tunnel. Today, the range of options has widened to include pipejacking and microtunnelling, pipe ramming, directional drilling and, in the case of on-line replacement, pipebursting and pipe splitting.

A primary technical requirement for a sewer is that it must be installed to a uniform grade or slope, which is governed by the design flow rate and minimum self-cleansing velocity. In addition, the pipe material should be resistant to any corrosive and abrasive action. The pipe, together with the enhancement from the surrounding soil support, should be capable of withstanding both internal and external imposed loads. All these technical requirements must be met and have to be assessed before considering any other evaluation criteria such as cost, safety, and environmental and social impact.

18.2.1   *LINE AND LEVEL*

Maintaining a constant grade is normally perfectly feasible with conventional open trench construction, although obstacles such as hard ground, buried objects, and other utility installations may result in construction difficulties if not anticipated beforehand. Line and slope of sewers can be accurately controlled using pipejacking and microtunnelling, but unforeseen buried objects or major variations in ground conditions can cause significant difficulties. If a machine gets stuck part way through a drive between two shafts, the only solution may be a very expensive rescue operation involving excavation from the surface, which may not even be possible in some cases.

Pipe ramming, on the other hand, is a non-steerable technique and inconsistent ground conditions and buried obstacles can cause the pipe to deviate from the planned path. Therefore where there is such a risk, pipe ramming should not be considered, except for pressure mains, where grade is not critical.

Although directional drills are of course steerable, until recently it was generally regarded that the line and level could not be controlled sufficiently accurately for gravity sewer installation. However, the accuracy of directional drilling has improved greatly and there have been several reports of gravity sewers being installed successfully with this technique.

Maintaining line and level may also be a problem when attempting to use pipebursting or pipe splitting if the existing sewer has settled and is no longer straight. It is therefore important to ensure that the existing sewer has a constant and uniform gradient before considering bursting type operations. Although there are techniques available for dealing with this problem, these will inevitably increase the cost and time required to complete the job.

18.2.2   *LENGTH OF DRIVE*

Another factor to consider is the maximum length of drive that is possible with each technique. In most sewerage schemes, it is common to have access chambers every 100 metres or so, and as this distance is well within the range of pipejacking and microtunnelling operations, it is feasible to locate jacking shafts at those positions where access chambers will be required. Although there have been reports of successful pipe ramming operations of over 100 metres, it is generally accepted that around 60 metres is a more practical upper limit, particularly as the direction cannot be controlled. For long distance drives of pumping mains under obstacles such as rivers, directional drilling must be the preferred option, as pipes of up to 1 metre diameter have been installed over distances of around 1.5 kilometres without any intermediate shafts. In soft ground, large diameter pipes can also be pipejacked over a similar distance using interjacks at suitable intervals.

18.2.3   *DEPTH*

The costs and technical difficulties of installing sewers using open trench methods increase significantly with depth. Excavation itself becomes more difficult, requiring machines with greater power and reach, and more complex trench support structures and dewatering facilities are needed. In narrow streets, it becomes practically impossible to install deep sewers using open trench methods and this was one of the factors that instigated the development of microtunnelling in Japan. On the other hand, when using trenchless methods, it is only the shafts that become more complex with increased depth, and there is the added benefit of reduced risk of interference with other utilities at greater depth.

18.2.4 GROUND CONDITIONS

Ground conditions and the level of the water table will have an impact on the technical feasibility and cost of whatever installation technique is to be considered. It is therefore essential that good geotechnical information is available when choosing the most appropriate installation technique. The location and amount of rock to be excavated will have an important bearing on open trenching as well as on tunnelling, pipejacking, and microtunnelling. What is often more critical in the case of pipejacking, microtunnelling, and directional drilling is the variability of the ground conditions. This is because the type of machine and the cutting face or drill type must be chosen very carefully to suit the particular ground and ground water conditions. Any changes encountered in the nature of the ground will inevitably affect the rate of work and may even prevent further progress altogether. Although it is claimed that some machines can work in a variety of soil and rock conditions, in general a machine designed for one type of soil or rock will not work well in another.

When considering directional drilling, the ground conditions will not only affect the type of drill and backreamer to be used, but will also have a considerable influence on the nature of drilling fluids to be used. Ground conditions will also dictate whether a mud motor arrangement, rather than a simple fluid assisted drilling technique, is required.

Open trenching below the water table can entail significant extra work and cost, in terms of sinking the well points and running the pumps necessary for a dewatering operation. Unless a sealed head earth pressure balancing machine is used, there may also be a need for dewatering, stabilisation, or ground freezing when pipejacking or microtunnelling is used at depths below the water table.

A geotechnical concern related to pipeline installation is that of settlement or heave at the surface, particularly if the surface carries a road or a railway, or where there is potential for affecting other underground structures, utility installations, or building foundations. In all tunnelling and drilling operations there will be a certain amount of 'overbreak', where the bore is excavated slightly larger than that required to install the pipe. This may cause a problem of subsidence if there is little cover and the pipe is to be installed close to the surface. It is generally less of a problem at greater depths. With trenchless methods, limiting the diameter of the cutting tool to, say, 25 mm more than the outside diameter of the installed pipe is one way to specify the maximum permissible amount of overbreak. However, it should be noted that this approach is not possible with directional drilling because the technique relies on having a bore diameter that is at least 1.25 times the outside diameter of the installed pipe. Having said this, heave is likely to be more of a problem than subsidence with directional drilling. This is because the annular space between the pipe and the bore is filled with drilling fluid (generally bentonite slurry). During installation this fluid is used to seal and maintain the bore and carry the cuttings out of the bore. If for any reason the fluid is prevented from escaping from the bore, pressure can build up and cause heave at the surface. Once the pipe is installed the fluid in the annular space congeals as a consequence of its thixotropic nature, thus minimising the risk of subsidence.

Heave can also occur during pipejacking and microtunnelling operations, particularly when using earth pressure balance machines. Operators of these machines must maintain a delicate balance between the speed of progress, the pressure at the face, and the flow rate of slurry to and from the face. The consequences of getting it wrong can be either subsidence due to a collapse of soil at the face, or heave due to too much pressure.

Subsidence occurs during tunnelling and drilling due to the fact that soil is removed from one location in the soil mass, thus presenting a potential for soil migration into the void that has been formed. This can of course be prevented by adequate and reliable support around the bore. When an open-ended pipe is used, pipe ramming displaces very little soil during the ramming operation, as most of the excavation takes place from within the pipe once it has been rammed into place. Therefore, it would appear that this method offers the lowest risk of subsidence or heave, particularly when the amount of cover is limited.

Any method that does not actually remove the soil, but compacts it around the installed pipe instead, must present the greatest risk of heave at the surface. Pipe ramming operates on this principle, but only a relatively small amount of soil has to be compacted into the surrounding soil mass, when an open-ended pipe is used. Pipebursting, on the other hand, may displace a significant amount of soil, especially when 'upsizing', and at shallow depths, this can present a real risk of heave and consequent damage of the surface.

18.2.5   *INTERFERENCE WITH OTHER UTILITIES*

As well as good geotechnical information, an accurate and comprehensive knowledge of the position, size and nature of utilities is also essential. Unfortunately this is not always readily available and each year there are around 75 000 interference incidents in the UK. Some of these may be minor and have little impact, but fractured gas mains and damaged high voltage cables can be highly dangerous, for operatives as well as the general public. Any broken utility pipe or cable will lead to some disruption to service (and hence the general public, commerce, and industry), and repair or replacement can be very expensive, especially for fibre optic cables. It is therefore worth considering the interference risks when assessing the technical feasibility of any installation technique.

Critics of trenchless methods suggest that there is a greater risk of interference incidents when using trenchless methods than when conventional open trench methods are employed. However, proponents of trenchless methods estimate that greater use of trenchless methods could reduce interference damage by as much as 90%. The justification for this lies to some extent in the fact that very much less ground is disturbed when using trenchless techniques. For example, an open trench to lay a 100 mm diameter pipe across a road will involve the excavation of approximately 6 m^3 of soil, whereas installation of the same pipe using directional drilling only disturbs about 0.2 to 0.3 m^3 of soil – 3% to 5% of that when using open trench. It is therefore argued that there must be much less risk of encountering other utilities as well. In the case of sewers, which are almost always at a greater depth than other utilities, there is bound to be a greater risk of encountering other pipes and cables in a trench than when using pipejacking, microtunnelling, or directional drilling.

Pipebursting and pipe splitting techniques enlarge an existing bore and compress the soil around it, and therefore constitute a risk of damage to other nearby utilities. If other utilities are too close, bursting may not be the most appropriate technique to use. Another factor to consider when assessing whether pipebursting is a feasible option is whether there is any concrete surround at any point on the pipe and, in the case of cast-iron pipes, whether there are any ductile iron or steel clamps or joints. Although tools are now available for dealing with such clamps, any concrete surround is likely to make pipebursting

18.2.6 LATERALS

In conventional open trench work, lateral connections can be made as the work progresses. The open trenches necessary for the lateral connections do not cause any significantly greater disruption than that caused by construction of the main line. However, making lateral connections using trenchless methods is generally not feasible and the open trenches required to connect a large number of laterals may almost completely negate the benefits that may be gained from using a trenchless method to install the main line. In assessing the feasibility of a trenchless technique, the number of laterals to connect may well be a significant factor. However, this could also influence the initial design of the scheme as alternative pipe layouts may be considered. Where microtunnelling is used to install the main line, the possibility of connecting all laterals directly to the access chambers, rather than to the main line sewer, could be considered. In this way, laterals could also be installed using trenchless methods by drilling or tunnelling from the access chambers to the house gulleys. Although this would result in slightly longer runs for the laterals, it would have the added advantage of preserving the structural integrity of the sewer.

A large number of closely spaced laterals will also militate against the use of pipebursting or pipe splitting, as all laterals must be cut and disconnected prior to the bursting operation and reconnected after the new sewer has been installed. This can result in a large number of small excavations being open throughout the bursting operation, and can lead to as much disruption as with a complete open trench job.

18.3 Cost

Although cost is often the overriding criterion for selection of technique, it should be recognised that the cost of any sewer installation project can be split into three distinct elements: direct costs, indirect costs, and social and environmental costs. The direct costs are those borne by the client or project promoter and include planning, design, land acquisition, and construction costs together with any other payment to contractors and suppliers. The indirect costs are also borne by the client and include those not directly attributable to construction but related to it, such as traffic management and compensation payments to owners of businesses, buildings and land affected by the work. The social and environmental costs, on the other hand, are not borne by the client, but by the general public, individuals, groups and organisations affected in one way or another by the work. They include the costs of traffic delays and diversions, increased traffic accidents, loss of amenity, and general mess and pollution.

In most cases the social costs arising from trenchless methods are considerably less than those incurred by open trench construction, and this was regarded by many as the driving force in the development of trenchless methods. However, evidence now suggests that trenchless methods are now generally lower in direct costs as well.

In selecting a sewer installation technique, most client organisations will only consider the direct and indirect costs that they will actually have to bear. They will tend to ignore the social costs because, apart from a consideration of good public relations, it is not in their interest to minimise them. In the UK, this situation is changing as a consequence of legislation, such as the New Road and Street Works Act, and more recently the implementation of section 74, which allows local authorities to impose financial penalties for occupying road space for longer than a specified period. In other parts of the world legislative restrictions may be more severe and can include a complete ban on open trench work in heavily trafficked areas.

18.3.1   DIRECT COSTS

The cost of the installation or repair of underground pipes is perhaps even more difficult to estimate than other civil engineering work, because there are so many variables and unknowns. These include the method of excavation, disposal of excavated material and importation of new fill, the ground and water table conditions, the accessibility to the site and any traffic management implications, the number of access chambers, the number and location of lateral connections, any dewatering requirements, environmental controls, any compensation payments, and the cost of diverting or repairing other services. Each installation will present its own set of challenges and constraints, and it is therefore very difficult to give an overall cost per metre without evaluating these specific requirements.

Although it is not possible to give definitive guidance on the cost of various forms of sewer installation, some general points can be made to aid in preliminary decision making. First of all, although costs will increase with depth, trenchless methods will not be as affected by depth as open trench methods. There is a 'break-even' depth at which trenchless methods will be the same as open trench, and less costly than open trench at greater depths. As a rough guide, although trenchless construction will generally be the most expensive method, it will be cheaper than open trench at depths greater than around 3 metres (Fig. 18.1). Directional drilling and pipe bursting are both even less affected by depth considerations and there is now sufficient evidence to suggest that these techniques are cheaper than open trench in all situations except shallow installations below fields in rural areas, where reinstatement costs are minimal.

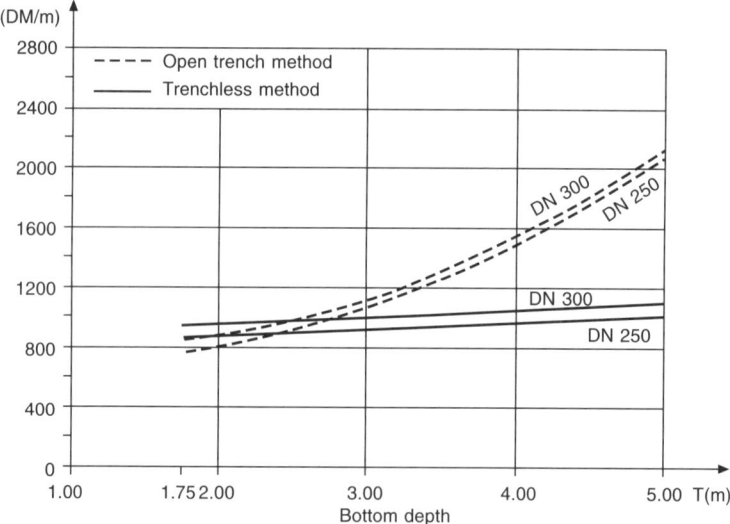

Figure 18.1   Comparison of cost involved with installing size DN 250 and DN 300 sewers by the open and trenchless method in Berlin

18.3.2   INDIRECT COSTS

The amount of indirect cost that the client or utility will be required to pay depends to a large extent on the legislation pertaining in the country or region in which the project is set, and the respective contractual obligations of the client and the contractor. For example, traffic management costs can be regarded as direct costs if the contractor is required to cover this within the terms of the contract. If the client is required to organise and pay for traffic management, it will incur indirect costs. On the other hand, if police and traffic authorities are provided to manage traffic at no cost to the client, the costs can be regarded as social costs.

The right to compensation will also vary. In many countries, householders may claim for damage to their land or buildings caused by underground work and legitimate claims may be brought against the client or his contractor. In the UK, businesses may claim for lost profits arising from disruption caused by sewer works if they can prove that profits have been lost.

18.3.3   *SOCIAL AND ENVIRONMENTAL COSTS*

Social and environmental costs are those borne by the general public or parties other than the contractor or promoter. They arise from traffic disruption, uncompensated loss of trade, disruption to other services, flooding, and pollution. Many of these are difficult to quantify, but a number of studies have been carried out to estimate the costs of traffic delays and diversions. These studies have shown that traffic disruption costs are of the same order as the direct costs of works carried out in the street. In other words, for every pound spent by the client organisation in carrying out the works, the public pay another pound in additional costs of running their vehicles.

In some cases, these traffic disruption costs can be as much as ten times the direct costs and should therefore have a substantial effect on the choice of construction method. Although trenchless methods, such as microtunnelling and directional drilling, do not eliminate traffic disruption, they can significantly reduce it in two ways. First of all, there is less overall occupation of the road space and second, the work is often completed much more quickly than when using open trench methods.

Damage to other underground services has already been mentioned and those services damaged during excavation must be repaired, thus incurring an extra cost. However, less obvious damage and increased stress on exposed pipes may accelerate the incidence of future failures and reduce the life expectancy of services exposed during open trench work. Additionally, heavy traffic diverted onto minor roads during sewer construction will inevitably cause damage to the road structure of the diversionary routes and any services buried below them.

It should be appreciated where opon-cut construction is being considered that breaking up a highway carriageway is a serious interference with the homogeneity of the road pavement and however good the reinstatement some settlement must be expected after back filling of the trench. The reason for this is that the installed base courses are not homogeneously bonded with the existing road pavement either on account of the use of a material mixture whose quality differs from that of the original pavement or because of the customary manual placing of the material and the compacting problems that usually occur in the process. This often results in damage at the seams of the restored road part and could impair the road's useful value and service life. Beyond this the damage may affect the traffic safety (surface water runoff), driving comfort, and appearance of the road pavement. The consequences of such damage are extra servicing and premature repair jobs whose cost must be viewed as past of the indirect cost involved with the open-cut method.

18.3.4   *WHOLE LIFE COSTS*

Even if considering only the direct costs, the initial capital cost of a sewerage scheme is only one element of the full cost. For a comprehensive evaluation and assessment of the relative merits of different techniques their whole life costs should be considered. In addition to the initial capital costs of the scheme, the whole life costs will include the operational and maintenance costs over the whole life of the facility, and the decommissioning and replacement costs that will be incurred when the useful life comes to an end.

Estimating the life expectancy of any structure or facility is difficult. Obviously the strength and durability of the materials used will have a major impact, as will the quality of the workmanship during initial construction, and it is a credit to Victorian engineers and constructors that there are many sewers over 150 years old still in active service.

The Water Research Centre in the UK provided guidance on the estimated life of newly installed sewers as 40 years for plastic materials up to 125 years for large diameter concrete sewers. However, these figures were based on the pipe material only and did not take the method of installation into account. Sewers, like any other underground pipeline, are composite structures consisting of not only the pipe, but also the soil surrounding it. A pipe surrounded by a well-compacted soil will be more durable than one surrounded by poorly compacted soil, and could therefore be expected to have a greater life expectancy. In open trench construction, it is extremely difficult to obtain the same degree of compaction as the original undisturbed soil. Trenchless methods do not disturb the soil in the invasive way that open trenches do and, although there is no definitive research to support this, it is reasonable to suggest that sewers installed in this way will have a greater asset life.

18.4   Time

Although the time to complete a sewer installation project will depend on a number of factors, not least of which are the depth of installation and the diameter of the sewer, it is now generally recognised that most trenchless methods will be quicker than open trench, particularly for smaller diameter installations. However, it should also be noted that significant delays could occur if something goes wrong with a trenchless method. Stuck machines, collapsed pipes, heave, subsidence, and damage to other utilities will not only cause delays but also the inevitable claims and disputes.

A typical drive rate for microtunnelling is about 1 to 2 metres per hour, but could be up to 4 metres per hour in favourable conditions. In some conditions pipe ramming may progress at a similar rate, but in good conditions much higher rates of up to 10 metres an hour can be achieved. Shaft construction and set-up for these methods will add significantly to the total time needed to complete a project. In good ground, small diameter pipes, of up to around 400 mm, can be installed by directional drilling at a rate of around 6 to 12 metres per hour.

18.5   Safety

It is well recognised that open trenches present a danger to both those working in or near them and to the general public, and trench support, protective fencing, and other measures to protect those at risk, must be implemented assiduously. Any technique that eliminates these risks altogether by removing people from open excavations must improve overall safety.

Apart from the excavations required for shafts or access pits, trenchless methods eliminate the need for people to enter or be near a trench and should therefore lead to fewer accidents from soil collapses and pipe falls. However, trenchless methods present their own set of hazards. There is always a risk of striking a high voltage cable and suitable earthing precautions need to be taken to ensure that operatives are protected against such a risk. Another issue related to directional drilling is the fact that operators can be exposed to rotating machinery in the form of the drill string and attachments such as reamers. This is an issue that is currently being pursued by the Health and Safety Executive in the UK, who insist that rotating drill strings present a hazard and need to be fully guarded.

A further factor to consider is the risk associated with narrow traffic lanes that may be necessary when open trench work is under way. There is evidence to indicate that traffic accidents increase under these narrow lane situations and while some of these may result in minor bumps and scratches to vehicles, others will result in injury, or worse, to drivers and pedestrians.

18.6 Environmental impact

The issues of environmental impact and environmental management have grown in importance over recent years and are now important elements of feasibility studies and the selection of appropriate construction methods and processes. International agreements, such as the Rio and Kyoto conventions, and legislation are forcing the construction industry worldwide to recognise and embrace the concepts of sustainability and environmental protection. Many clients are requiring their contractors to prepare and implement environmental management plans for their projects, and contractors are developing environmental management systems in accordance with the international ISO 14000 series of standards.

Environmental effects of alternative construction methods need to be considered at an early stage of any project and excavation of any kind has an environmental impact. It has the potential to cause noise and vibration, dust and other air pollution, and water pollution. The excavation operation will also require energy, produce waste excavated material, and require imported fill material, often in the form of freshly quarried aggregates.

All sewer installation methods create noise, vibration, air and water pollution to varying degrees, which may influence the choice of technique to be used. Pipe ramming generates a lot of noise and vibration, which may be difficult to dampen and control, so this may preclude this method in noise-sensitive areas. Although the static engines used to power the hydraulics for microtunnelling machines and directional drills can also generate a good deal of noise, it is possible to contain these in a noise insulated housing and thus mitigate the problem. Of course noise will also be generated from conventional construction equipment such as excavators, trucks, road saws, and pneumatic breakers used in open trench work.

Although all construction equipment will create some air pollution, perhaps the greatest air pollution nuisance is that from particulates arising from loose wind blown excavated soil particles and dust. A great deal more material is excavated, moved, stored, and transported when open trench methods are used than when trenchless methods are. There is therefore a greater likelihood of airborne dust arising from open trench operations, and this should be given consideration in sensitive areas close to housing, schools, and hospitals.

Sewer installation operations can pollute both ground and surface water sources in a number of ways. Temporary overpumping of sewage will be necessary if pipebursting is used for on-line replacement of a sewer, and this will constitute a water pollution hazard. It is therefore important to ensure that temporary pipes or hoses and joints are watertight and in good condition. Trenchless methods, such as microtunnelling and directional drilling, often use a bentonite- or polymer-based drilling fluid to transport cuttings to the surface. Slurries from such operations should never be discharged to the natural water courses as they can cause serious environmental damage. In many cases these slurry systems are closed and incorporate settling tanks and pumps for separating the cuttings and recycling the fluid, but even in such cases disposal of settled slurries must be carried out in a controlled manner. Water pumped from dewatering operations, which may be required for deep open trench installations and shaft construction, must be discharged and disposed of in a manner that will cause no environmental damage. This will be a particularly important consideration when excavating in contaminated land.

In the UK, the New Road and Street Works Act placed much greater responsibility for long-term integrity of reinstatement on the utility companies and their contractors. This has resulted in a change in trench excavation and reinstatement practice as, in order to avoid any penalties associated with defective reinstatements, it is now common practice to remove excavated material to land disposal and reinstate the trench with quarried aggregates or other material such as foamed concrete. This has two effects related to sustainability. First, it exerts pressure on landfill disposal sites, an already diminishing resource, and second, it increases the demand for quarried aggregates, which are also a diminishing and non-renewable resource. The cost of disposal to landfill is increasing as haulage distances to landfill sites get longer and charges and taxes for dumping also rise. There is therefore a growing financial incentive to find and use sewer installation methods that minimise the amount of excavation required.

Table 18.1   Summary of factors affecting choice of technique

	Technical feasibility	Time, cost, quality	Environmental, safety and social impacts
Open trench	• Deep sewers difficult, particularly in narrow streets • Conventional readily available equipment • Not particularly high skill levels needed • Potential for service interference • Laterals can be connected • Dewatering necessary below water table • Accurate control of line and level • Careful excavation can minimise service interference	• Cheap in fields • Cost rises significantly with depth • Slow progress • Compaction may be problematic • High social costs • Backfilling and reinstatement costs increasing	• Safety hazards to workers and public • Noise and vibration • Dust, mess • Traffic disruption • Loss of trade • Problems with disposal of excavated material • Damages tree roots
Microtunnelling Pipejacking	• Depth not a major problem • Laterals need to be connected at shafts	• Medium rate of progress • High set-up costs • High machine costs	• Careful siting of access shafts needed to minimise traffic disruption

Table 18.1 (*Contd*)

	Technical feasibility	Time, cost, quality	Environmental, safety and social impacts
	• Can be below water table • Accurate control of line and level • Potential for subsidence or heave • Good operator skills required	• Expensive unless high rates of machine use can be guaranteed	• Removes operatives from dangerous deep excavations • Disposal of slurry is an issue
Pipe ramming	• Non-steerable • Can deviate from line and level • Low risk of heave or subsidence	• Fast progress	• High noise and vibration • Dangers of expelling spoil by compressed air
Directional drilling	• Control of line and level not as good as microtunnelling but improving • Accurate information on ground and utility services essential • Good operator skills required	• Fast progress • Relatively low cost	• Electrical strike • Rotating parts • Disposal of drilling fluid and slurry is an issue • Noise may be an issue, but can be contained
Pipebursting and pipesplitting	• Can only use if there is an existing pipe • Bursting not possible with ductile materials • Control of line and level depends on existing pipe • Problems with concrete surround and clamps • Laterals must be cut prior to bursting	• Fast progress • Relatively low cost	• Noise

19

Civil Engineering Contract Management

Geoffrey F. Read MSc, CEng, FICE, FIStructE FCIWEM,

FIHT, MILE, FconsE, Fcmi, MAE

Geoffrey S. Williams CEng, FICE, FCIArb, MAE

19.1 Introduction

19.1.1 TRADITIONAL ATTITUDES

The Institution of Civil Engineers is dedicated to the purpose of 'promoting . . . the art of directing the great resources of power in Nature for the use and convenience of Man'. Engineers have been striving towards this romantic and inspiring purpose since the beginnings of human history. Ours is the inheritance which runs from the monumental achievements of distant time, the walls of China and of Zimbabwe, the irrigation and aqueducts of Egypt and Rome, to the railways, canals, highways, sewers, tunnels and bridges which opened up our modern world. The deliberate designation of civil engineering as an art gives emphasis to an obscured truth; most of us entered this profession for the joy of accomplishment in walking onto a green field site and using science, practicality and trades expertise in the manipulation of huge resources to produce an economic, functional, and geometrically pleasing result; a tangible, enduring testament to our skill and endeavour.

As a profession focused on the demonstration of this art, the end object is seen as a justification in itself. We incline to be dismissive of paperwork and records. We like to 'get things done' while others such as quantity surveyors and lawyers trail in our wake, waving papers. Our designs have responded to the economic parameters, and contractual considerations are seen to be less important than imperatives in the onward progress of the project; commerce is seen as a matter for reliance on equity and good sense, and every honest contractor conscientiously striving to deliver a good product may think himself entitled to a proper reward.

There was indeed a time, remotely within the memory of older professionals, when the continuous availability of profitable work enabled the more relaxed view to prevail. In those days a profit rising to 10% on turnover enabled contractors to turn away from old contentions in order to accept new work which in administrative terms was more rewarding than the prolongation of existing disputes. Final account negotiation was relatively restricted. It was not unknown for smaller contracts to be settled for the tender sum without remeasurement. Good relations and the progressive escalation of the contractor to higher tender listings were then the primary object. Contractors in the main operated with their own plant fleet and directly employed operatives, and sought to foster their own domestic subcontractors by providing them with a continuity of employment.

In those mostly forgotten days, dispute and particularly arbitration were regarded as an ultimate breakdown of relationships. A 'claims-conscious' contractor was likely to have his opportunities quickly curtailed. It is evident that the exact provisions of the ICE contract, formulated to give a fair distribution of risk, an absolute right of claim and reasonable recovery of extra cost, were largely ignored. There could be few industries with less basic awareness of their fundamental 'rules of engagement'. The thought of turning to a third party for an enforceable solution was offensive, particularly to the engineer and employer, even though arbitration was specifically provided by the contract as a means of gaining early relief from disagreement or abuse of the engineer's powers.

19.1.2   INCREASED COMPETITION

Now, the world has turned to a more cynical view. Standards of professional objectivity are thought to have declined. There are numbers of employers (including statutory authorities) who regard professionalism as a liability and prefer their employees and professional advisers to be 'managers' wholly accountable for the employer's interests. The boardrooms of contracting organisations are filled with accountants and legal executives who (rightly, on their own terms) believe that every contractual entitlement must be applied, and that every claim potential must be exaggerated in order to maximise the company balance sheet and give confidence to banks and creditors.

The watershed between old and new attitudes probably occurred around 1973, when a sudden increase in oil prices coincided with a peak of construction activity. Thereafter an increasing proportion of the industry's resources seems always to have been surplus to UK national requirements, and inflation and high interest rates compelled increasing attention to the disciplines of rationalised company resources and regulation of cash flow (a term almost unknown before 1970).

In the ensuing period lower profit margins combined with excess construction resources led companies to search for a higher turnover in a diminishing market. Encouraged by a tender system which (with questionable logic) invariably rewards the lowest bidder, work has for the past generation been bought in at prices which would previously have been regarded as unrealistic. Fee bidding has reduced the effective performance of consultants and has exposed them to negligence claims by the employer. Rationalisation has produced widespread casualties, both in main contractors and in the ghostly band of failed subcontractors least able to defend their rights against contractual legality and outright exploitation. It is estimated that subcontractors underwrote £1 billion per year of the UK construction industry's losses in the decade to 1996.

As long ago as 1979, Max Abrahamson in his definitive work Engineering Law and the ICE Contracts 4th edition) wrote:

> It is very sad that the situation has been reached where it can even be suggested of a great industry, as it has been, that efficiency in the pursuit of payment pays better than efficiency in site work. And this problem should not be represented as a conflict between the interests of contractors and the interests of engineers and employers. Neither side of the industry benefits. Contractors do not make more profit overall, because competition ensures that in seeking contracts they allow in their starting prices for the money to be earned at the end in claims.

How true and how self-destructive that has proved to be!

19.1.3   THE GROWTH OF ADVERSARIAL ATTITUDES

In the effort to avoid the ultimately destructive impact of this downward spiral the relationships between the contractor, the engineer and employer have redefined themselves. With profit set in the range of 2.5% and little or no allowance for contingency within the best programme expectation, commercial considerations dictate that contractors must place an entire reliance on the full range of rights conveyed by the standard forms. These claims representations have to be documented and negotiated through site staffs reduced on both sides to minimal levels. The tension which frequently results from this situation calls for interpersonal skills rarely found in the basic levels of an underrewarded profession, and the reports going back up the line to the senior management of each side may be so subjectively directed as to place the parties on opposite poles.

Employers too may regard the developing contentions between the contractor and the engineer with some suspicion. Contractors have always been ready to consider, when a decision is made against them, that the engineer is refusing to admit the consequence of his own oversight or error, but now employers, too, perceive that the engineer may concede claims without contractual need in order to deflect pressure from matters where he may be at fault. It has to be said that while such instances do occur, they are very much in the minority, and the evidence of probity by the great majority of engineers lies in the infrequency with which their initial determinations of claim are tested by a formal request for an engineer's decision precedent to arbitration. However, it must be true that any lowering of regard from either side for the engineer's professional independence of decision places him in a defensive position which encourages procrastination and the prolongation of dispute.

19.1.4   THE RIGHT TO CLAIM

There is no doubt of the rights of claim expressed in all of the ICE standard forms. Arbitrators and the courts will give full effect to them, and it is purposeless for an employer to protest at the volume of submission and record involved in claims applications, when they have encouraged the contractor to bid the lowest conceivable price (by any objective standard, the wrong price on the consideration of the tenders received) in order to 'win' the contract. The terms of the ICE standard forms are regarded as very fair and attempts in the courts to challenge and push back their boundaries have mostly ended in failure. It is no longer relevant that parties were formerly willing to concede or compromise their express rights in favour of goodwill and more constructive pursuits. There are no more claims-unconscious contractors.

19.1.5   RELUCTANCE TO REQUEST THIRD PARTY HELP

A particularly sad relic of former times is the reluctance of the parties to invite a completely independent third party resolution by an arbitrator who will know the contract forms and the legal precedents inside out. Viewed correctly, the arbitrator is the natural and integral court of appeal provided by the contract. The engineer's sometimes embarrassed position between the contractor and employer would be greatly eased, and contentions would be much abbreviated if the parties would use the service thus provided. There is no reason why an arbitrator should descend from a cloud surrounded by a mighty panoply of legal jurisdiction to hear representations by experts and leading counsel for each side, he is to be regarded as one step up from the engineer, with the advantage of patent impartiality and final decision. If the parties have agreed to resolve their differences by these means, why should they feel offended or aggrieved by an early reference which sets the dispute aside? And if the engineer can make a decision unattended by legal counsel and huge expense, why should the representations for the same issue before an arbitrator assume an extent and expense beyond the pockets of the parties and out of all proportion to the issue?

19.2   Claims under the main contract

19.2.1   WHAT IS A CLAIM?

The word 'claim' has an uncertain ring to it; as if proof was necessarily lacking and a questionable pretension was involved. A 'claims consultant' is thought of in some quarters as a species of ambulance chasing promoter of disputes. However, the apparent distaste for claims is a misconception. In order to view claims in their correct perspective, it is desirable to consider the effect of the particular contract terms and the way in which they supplement or vary the basic common law position between the parties.

The purpose of express terms

In the simplest form of contract, the buyer undertakes to purchase a specified article from the supplier for a fixed price. The supplier must provide the article in exact accordance with the specification, and the buyer must pay the agreed price. If there are no terms to the contrary, the law will imply that the goods when provided will be of reasonable quality and fit for their purpose, and that the contract will be discharged by both parties within a reasonable time. The buyer does not warrant the feasibility of manufacture or supply of the goods he has ordered; the supplier takes the whole risk of providing the article when he binds himself to do so and may set about that in any lawful way he pleases. The buyer must not interfere, since the law further requires that neither party shall obstruct nor hinder the other in the performance of their contracted duty. Circumstances of this kind might apply where a householder contracts for some minor work on a 'handshake' agreement with a local builder.

In any contract agreement the parties are free (subject to illegality) to amend or to augment the basic common law position in any way which they agree is desirable. It is necessary to make sense of the often vexing interaction of changing requirements or unanticipated ground conditions, and to provide for control and inspection of works in which most of the endeavour will be covered and hidden by subsequent activity. The various forms of agreement give expression to the respective rights and duties of the parties, and expressly provide for the right of the employer (through the agency of the engineer) to interfere in and modify the works as they proceed. The imposed rights of interference and variation are controlled by these terms, and are accompanied by prescribed remedies. These generally provide for the contractor to recover for cost including overheads, plus profit on extra works plus additional time warranted by delay. The engineer has an absolute duty to remeasure the works, value variations and certify extension of time for performance; in other respects, the contractor must still claim his additional entitlements as he would in common law. Such claims are thereby made in accordance with the contract agreement rather than in breach of it.

Apportioning risk

In further variation of the common law position, the agreements apportion the risks of construction between the contractor and the employer. Best known of all contractual rights in the civil engineering conditions, the contractor's responsibility to deal at his own cost with physical conditions and artificial obstructions, is limited to the extent of that foresight which, taking account of pre-tender site examination and the information provided by the employer, is reasonably attributable to an experienced contractor. Successive editions of the ICE conditions have moved this provision progressively in favour of the contractor, easing the severity of notice requirements, imposing a duty on the employer to disclose all information in his possession, and adding the entitlement to profit on the full amount of any costs which may reasonably have been incurred.

Although it is questionable whether the employer overall is advantaged by a provision which enables the contractor to diverge entirely from the competitive level of his accepted tender to a recovery based on 'cost plus', one has to take into account the potentially penal effect of a single serious occurrence of unforeseen adverse conditions on the tenuous financial security of a contractor who may find himself with all his eggs in one basket and no profit-making alternatives. Therein lies a commercial justification for the risk-sharing provision. However, the range of remedial options available to the engineer when such conditions are encountered and the breadth of the payment provisions give particular importance to the application of the notice and record requirements placed upon the contractor. The engineer, and if necessary an arbitrator, is entitled to look sceptically at notice and costs produced long after the problem was identifiable. The engineer's duty is to withhold from the certification of costs which by the passage of time cannot be established as being properly and necessarily identified with the causal event.

Claims – a necessary function of express entitlements

It can be seen that where the right to a flexible control of the works is traded against the previously agreed scale of remedy which is triggered by the contractor's duty of notice, the employer, on the one hand, should not complain or pillory the contractor for making the claims or submerging him in records. On the other hand, contractors should not object to a machinery which requires notice if that gives access to work measured generally on a cost-plus basis.

19.2.2   MEASUREMENT CLAIMS

Re-measurement of the works

The ICE contracts are measure and value contracts in which the contract price is defined as the sum to be ascertained and paid in accordance with the contract provisions. The price is determined by the sum of the remeasured quantities extended at the unit tender rates and prices, but adjustments to the price also arise as a part of the machinery of remeasurement on each occasion when work occurs which either is not envisaged by the measurement provisions of the bills of quantity, or where, by significant change of quantity or circumstance of construction, an existing unit rate or price becomes unreasonable or inapplicable.

Variations

In the ICE 6 contract, Clause 52 for the valuation of variations has two subclauses making separate provisions. Subclause 52(1) requires the engineer to value the variations which he has ordered, either

(a) at existing Bill rates (where the circumstances and character of the varied works are substantially the same as those covered by existing rates), or
(b) at new rates based upon the existing rates (where the existing rates do not completely reflect the circumstances or character of the varied work), or
(c) at a 'fair' rate in those cases where the work is entirely different from that covered by existing rates and prices.

Consultation

In making his determination of the rate under this subclause, the engineer has an express duty to consult with the contractor.

Principles involved in rate fixing

Rate fixing is sometimes represented as a kind of black art, but this distorts the intention of the valuation clauses. All of the engineer's determinations call for a high level of probity and experience and must be exercised with a complete absence of prejudice. Beyond that we cannot shrink from the fact that they finally represent an exercise of opinion on the rights of the parties; a principled exercise of professional duty. Provided that the opinion has taken proper account of the representations made by the contractor and is the product of sensible reasoning and professional good faith, it should be supported, and unless irrationality or prejudice is evident, the courts wisely refrain from interference in the process. There is provision for review on notice of the contractor's objection. There is provision for further review on a request for an engineer's decision. The final review is that of an arbitrator, but the determination is still a matter of informed opinion.

When determining a rate adjustment, the engineer may not vary for better or worse the commitments given by the parties to each other, and rate comparability will apply to variations 'so far as may be reasonable'. In general it is not unreasonable to apply rates which the contractor has warranted as sufficient, simply because they prove to be mistaken or uneconomic. But comparability is to be achieved from existing rates 'as the basis for valuation.' 'Basis' is variously defined in the dictionary as a beginning, a main ingredient, and common ground for negotiation, and this leaves the engineer some sensible scope for the exercise of his discretion.

Rate fixing for a change in working method

Complicated considerations arise when the variation involves a change in working method, and existing rates have to be broken into their basic components for purposes of comparison. The adjustment which the engineer should seek to make is the addition or deduction of that sum which represents the difference between what the work was going to cost, and what it now is going to cost (with appropriate adjustment of overhead and profit), thus leaving the original contract commitment unaltered.

Fair rates

In deciding what is fair the engineer is entitled to and should look at the general relationship of rates and costs in all or any group of rates which are most nearly related to the varied work. If no such rates exist he may look for proof of a general market rate for comparable work.

Valuing site instructions

Variations also result from site instructions given by the engineer under Clauses 7 and 13, and from confirmations of instruction issued by the contractor under Clause 2(6)(b). The instructions referred to in these clauses are commonly issued as immediate measures and are sometimes followed by confirming formal variation orders at a later date. The linkage with Clause 51 is the only prescribed means of recovering the costs arising from site instructions issued under Clause 7, and it follows that the provisions of Clause 52 for the valuation of variations will apply. Rate-fixing calculations may be thought cumbersome for the generality of minor instructions; these frequently carry a cost in mobilisation execution and supervision out of all proportion to the considerations underlying the pricing of items in the bill of quantities. In such instances it may be preferable to consider the costs as a 'price' subject to a fair valuation, and to include an allowance for any disruptive effect which falls under the description of *change to specified sequence method or timing of construction* in Clause 51.

Secondary effects or variations – Clause 52(2)

Clause 52(2) provides for the revaluation of rates as the result of the contingent effect of an ordered variation on other activities. Many variations will have a wider impact than that which can be valued by rate comparability in the immediately affected items.

Notice requirements

Unlike the requirement of subclause (1), the engineer has no duty of his own volition to ascertain and value the secondary effects of a variation. The operation of subclause (2) depends entirely upon the service of notice, one to the other, by the engineer or contractor. Such notice is required to be given 'before the varied work is commenced or as soon thereafter as is reasonable in all the circumstances.'

The courts have confirmed that the provision of notice is a condition precedent to consideration of the claim, but there is some doubt as to the extent of time which may be considered 'reasonable in all the circumstances'. Since the clause refers to 'any rate or price contained in the contract [is] by reason of such variation rendered unreasonable or inapplicable', and no further provision is made for consultation or submission by the contractor of proposed adjustments to the rates, the notice must be specific in identifying the rates or prices in question. The engineer is left to determine the issue, but it is difficult to understand how any engineer could have sufficient knowledge of the contractor's pricing and arrangements as to make a decision without inquiry and discussion.

As with subclause (1) no time limit is fixed within which the engineer must communicate his decision, but the contractor may apply pressure by the inclusion of his claims within the monthly statement. Clause 2(1)(a) states that 'the Engineer shall carry out the duties specified in or necessarily to be implied from the contract'. It is suggested that, as a matter of necessary implication, the engineer must deal with the matters put to him by the contractor within a reasonable time and if he fails to do so he commits a breach of duty. The contractor is entitled to know at an early date the value to be placed upon the work he is called on to perform, if only to have the right of objection contained in Clause 52(4)(a).

Sequence of events in valuation of variations

The sequence of events which is derived by examination of Clauses 52(1), (2) and (4) is:

- Engineer instructs a variation (Cl. 51)
- Engineer consults with contractor to ascertain value of variation (Cl. 52(1))
- Engineer or contractor to give earliest notice of contingent effects making other rates unreasonable or inapplicable (Cl. 52(2))
- Engineer determines new rates and prices by agreement or otherwise (Cl. 52(1))
- If disagreed, contractor gives notice of claim within 28 days (Cl. 52(4)(a))
- Upon receipt of notice, Engineer may instruct the contractor to keep contemporary records material to the claim (Cl. 52(4)(c))
- Contractor sends first interim account giving full and detailed particulars of the amount claimed, and grounds of claim (Cl. 52(4)(d))
- Engineer makes interim certification of such amount as he considers due (Cl. 52(4)(f)).

Concluding the determination process

It seems that the only way of ending the pressure applied to the engineer is for the employer to serve a notice of dispute on the engineer requiring his final and binding decision under Clause 66(1). The contractor's only means of setting such a decision aside is to give notice of arbitration, and he will be obliged to contemplate the possibility that the arbitrator may reduce the rates determined by the engineer.

Documentation

Instructions and variations issued by the engineer will normally carry a direction as to the proposed payment provision. If the provision is for the application of existing bill rates and the contractor is dissatisfied, he should immediately reply objecting under Clause 52(4)(a) to the determination of rate, point to the absence (if this is the case) of consultation, and notify (if necessary) the necessity to adjust rates for related items under Clause 52(2). He should keep records and notify their availability.

If the provision of the ordered variation is for 'rates to be agreed' (a frequent occurrence), this signals the engineer's availability for the consultation required in Clause 52(1) and opens the door to the reappraisal of rates in both subclauses (1) and (2). Although the contractor is not called upon to respond (he may simply leave the initiative with the engineer), it is in his interests to do so forthwith by proposing the revised rate under subclause (1) and defining any related rates under subclause (2). This is valid as a response in consultation, defines the scope of the contractor's claim to revision of affected rates, and (in lack of a prompt reply) confines the engineer to an agenda established by the contractor.

If in accordance with Clause 52(3) the engineer has ordered the varied works to be carried out on a daywork basis, no problem is likely to ensue, but the contractor may still have regard to the possibility of contingent effects and if appropriate give notice under Clause 52(2).

From the engineer's viewpoint, a failure to determine the rates may impede any subsequent objection which he might wish to raise in connection with the contractor's submission and records, for the engineer cannot logically complain of an inability to verify records which he did not require the contractor to maintain in accordance with Clause 52(4)(c). The contractor is left with an optional presentation: If the new rates were capable of determination from rate comparability, why did the engineer not respond to the submission? If the new rates were *not* capable of determination from rate comparability, the various issues may have to be resolved from uncontested records or by adoption of uncompetitive rates taken from one of the standard price books.

Variations to quantity – Clause 56(2)

Closely connected with the provisions for variation is the operation of Clause 56(2) for rate adjustment resulting from a fluctuation in quantity where the change does not result from a variation ordered by the engineer. The subclause provides that:

> If the actual quantities executed in respect of any item are greater or less than those stated in the Bill of Quantities and if in the opinion of the Engineer such increase or decrease of itself shall so warrant the Engineer shall after consultation with the contractor determine an appropriate increase or decrease in any rates or prices rendered unreasonable or inapplicable in consequence thereof and shall notify the contractor accordingly.

The operation of the clause is similar in effect to the provisions of Clause 52, but clearly it will generally operate retrospectively to the execution of the work. No order is required to operate the clause (see Clause 51(4)). No notice provision is made, and the responsibility is placed upon the engineer as a duty to consult with the contractor and to determine a new rate (or new rates), consequent only upon his having formed an opinion that the 'increase or decrease of itself shall so warrant'. A reduction in quantity could make plant use uneconomic, and in the extreme case could result in expensive hand work. Conversely an increase in excavation disposed from the site could exhaust the tip provisions.

The increase or decrease in rate is required to be 'appropriate'. There is no express linkage to the rate-fixing provisions of Clause 52, and there are no directions establishing the parameters of what is 'appropriate'. However, the dictionary definition of 'appropriate' is 'belonging, peculiar, suitable, proper', and this indicates that the adjustment should be carried out by reference to the provision of the original rate, so that the precepts of Clause 52 will apply. Clause 56(2) will *only* operate if the change in quantity necessarily involves a change in working method and/or unit expense. As well as opportunities for recovery of increased costs, there are clear circumstances in which it is appropriate to reduce a rate for economies of scale, provided that a beneficial change in the operation is made possible in due time. Most projects could be more economically constructed in hindsight, and that would not be an appropriate basis for rate reduction.

Distinction between treatment of changed work and changed quantity

The distinction between changed work and changed quantity which underlies the expression 'of itself shall so warrant' was underlined in the case of Dudley Corporation *v.* Parsons and Morrin Ltd (1959) where a contractor who had priced an item for excavating 575 cubic yards of rock at 2 shillings a cubic yard (when the economic unit price should have been £2) experienced the necessity of excavating an additional 1125 cubic yards of rock. This was an RIBA contract, and the architect exercised sympathy by valuing the original quantity at 2 shillings per cubic yard and the additional quantity at £2 per cubic yard. It seems that there had been no change in the operation of excavation. The Court of Appeal held that the architect was wrong and that the contractor was only entitled to 2 shillings per cubic yard for the whole quantity excavated, thus upholding the principle that the contractor was responsible for the sufficiency of his rates to discharge the increased quantity of any work in which the character of the operation remained unchanged. It is suggested that the Court would have made the same ruling had the error in rates operated in the reverse direction.

Notice provisions in Clause 56(2)

The contractor's first opportunity to give formal notice for a reconsideration of rate under Clause 56(2) occurs in Clause 52(4)(a), after the execution of the work described by the relevant item.

In practice it is in the contractor's interest to take this matter into his own hands and give notice to the engineer as soon as he believes that a change of quantity has warranted a revision of rate. He should follow with details in the monthly statement under Clause 60(1)(a) for the estimated contract value of the permanent works executed up to the end of that month, and since in Clause 60(2)(a) the engineer must exercise his opinion, and by Clause 60(9) must now (under the 6th edition) supply 'such detailed explanation as may be necessary', the contractor will have the basis for a notice under Clause 52(4)(a) claiming a higher rate or price.

'Lost' overheads and other charges not proportional to quantity

Clause 55(1) states that:

> The quantities set out in the Bill of Quantities are the estimated quantities of the work but they are not to be taken as the actual and correct quantities of the Works to be executed by the Contractor in fulfilment of his obligations under the Contract.

Very few bills of quantity for civil engineering works are wholly accurate, and contractors who have the time and energy during the tender period spend a lot of it searching for those items whose inaccuracy can be exploited by rate loading, either to sharpen the competitive edge of the tender, or to deliver a concealed profit. In stating that the quantities are 'estimated' the clause both implies that they will not be completely accurate, and warrants that they will not be wildly inaccurate. A professional estimate offered for pricing purposes should be as reasonably near to the mark as is permitted by the character of the work described in the itemisation.

The contractor is entitled to distribute his site and overhead allowances and profit across the rates for the works in the limited anticipation that, though the quantities may vary in response to site conditions, the commercial basis of the contract will substantially be the same. It sometimes occurs that nervous employers (or engineers) build an extra contingency allowance into the project by artificially inflating the unit quantities. This is basically dishonest, and a misrepresentation to the contractor.

The contractor is required in Clause 11(3) to satisfy himself in respect of the correctness and sufficiency of his rates and prices which shall (unless otherwise provided in the contract) cover all his obligations under the contract. If the final valuation of the works produces a figure substantially less than the tender figure, the contractor may claim for the loss which results from the underrecovery of overhead allowances. Where the loss arises from a significant reduction in contract quantities, authority for a retrospective adjustment of this kind is found in Clause 56(2), where the operation of the clause requires no notice and has no time limit.

THE EFFECT OF THE STANDARD METHOD OF MEASUREMENT

If the Civil Engineering Standard Method of Measurement (CESSM) is used in accordance with Clause 57, different considerations may affect the case. The CESSM arrangements offer the contractor the opportunity of using the system of method-related charges to identify and separate, on a fixed charge or time-related basis, all those costs within his tender which do not relate to fluctuation of quantity. It is universally recognised that civil engineering works are subject to variations and quantity adjustment; indeed the ICE 6 contract specifically states that the tender quantities are not to be taken as the actual and correct quantities of the works to be executed. The contractor provides unit rates in order that the works can be remeasured by the engineer.

The contractor who fails to grasp the opportunity of establishing a title to the separated measurement of the fixed and time-related charges which are most likely to be at risk from quantity fluctuation may be said to have signalled that he does not wish to recover costs which are not proportional to quantity.

Summary of provisions for valuing variations

The arrangements prescribed by Clauses 51, 52, 55, and 56 make a basis on which variations to civil engineering works can be accommodated and valued under the contract as the work proceeds, without the necessity of delay and the eventual submission of contentions claims. As commonly applied, they are allowed to collect and lie unresolved until they become distinctly unwieldy provisions. There really is no reason why this should be so, if the parties will move away from a mistaken preoccupation with the *cost* of the works, to the intended progressive updating of the contract rates on a basis simulating the contractor's position at the submission of the tender.

19.2.3   THE ESTABLISHMENT OF 'COST'

Before proceeding to remaining claims, the quantum of which is largely based upon 'cost', it is desirable to consider the effect of the definition which arises at Clause 1(5) of the ICE edition:

> The word 'cost' when used in the Conditions of Contract means all expenditure properly incurred or to be incurred whether on or off the Site including overhead finance and other charges properly allocatable thereto but does not include any allowance for profit.

Basic 'cost'

This definition has the benefit of simplicity and directness, and makes clear that the contractor's cost is not to be limited to consideration of the expense incurred directly at the workface.

The cost of plant and equipment

The basis of 'cost' as a damage was emphasised in the case of Bernard Sunley *v*. Cunard White Star (1940). There a machine lying in Doncaster had to be transported for contract use in Guernsey, but it was delayed in transit by a period of one week. The plaintiff sought to charge damages at the customary hire rate, but the Court of Appeal held that only costs in depreciation, interest on investment, and costs in maintenance and wages thrown away were admissible. Loss of profit migth have been accepted, but proof of opportunity for the use of the machine in the delay period was lacking, and Sunley were awarded the sum of £30 in respect of a claim for £577. It is thought that Sunley were penalised by their failure to provide any reasonable proof of their case, upon which the court refused to make guesses. Therein lies a general lesson for all those making claims!

Cost arising from subcontract liability

How is this principle to be applied to the circumstance where the contractor's principal cost is that arising from a subcontractor? In subcontracting the work, the contractor is securing a vicarious performance of his own obligations. Those obligations include (in Clause 52(4)) the creation and communication of records and updated costs in connection with any claim. The subcontract costs are thus as susceptible to proof as those of the contractor's direct employment. The engineer is entitled to proof of the contractor's liability in the scale of costs under the terms of the subcontract and will apply the records (to which he has full entitlement) to the evidence of subcontract costs. Where he is deprived of notice or receives a late claim he has his remedy in Clause 52(4)(e), for he may reject or mitigate the claim on grounds that he has been prevented or prejudiced from proper investigation by absence of notice. It is suggested that simple presentation of subcontract invoices is insufficient proof of cost to satisfy the terms of the main contract.

Overheads on subcontract liabilities

A critical factor in cost may be the 'doubling up' of overhead losses, where those of the subcontractor are compounded by the application of the contractor's own head office overheads. Inexpert assessment of these might lead to a gross escalation of the employer's liabilities, and it is necessary to approach the question in two stages.

The engineer on behalf of the employer has no power to interfere in the contractor's commercial arrangements. The contractor's liability towards the subcontractor is clearly the basis of his cost, and the subcontractor's overheads will be a legitimate element in any claim against the contractor, whether it is brought under the subcontract or in common law.

Thus overhead in the subcontract must be accepted by the employer on the same basis of liability as applies to the contractor. Profit must also be included if there is such a liability in the subcontract, for then it would be a 'cost' which the contractor could not avoid. If the contractor has paid the cost, there should be evidence of the basis of payment, for the cost must be 'properly incurred'. If the claim has not been paid, there is room for proof of liability, and that demands evidence.

The further application of the contractor's overheads to the cost liability incurred towards the subcontractor is not chargeable as a percentage addition. Evidence must be given of the particular overhead costs which are truly attributable to the causal event. Profit by the contractor is excluded – it is not a cost.

The combination of these factors is the 'cost' referred to in Clause 1(5).

19.2.4   UNFORESEEN PHYSICAL CONDITIONS AND ARTIFICIAL OBSTRUCTIONS

Clause 12 of the ICE 6th edition

The contractor must give notice under this clause whenever he encounters physical conditions (other than weather conditions) or artifical obstructions which could not in his opinion reasonably have been foreseen by an experienced contractor. The service of notice is a duty placed on the contractor separate from that required for recovery of extra cost or an award of extension of time, and this would allow the engineer to deal with the situation by taking those steps provided in subclause (4) which require no further notice from the contractor (investigation or variation) or by rejection.

The clause has the effect of transferring a risk, which in common law would belong to the contractor, onto the employer. Of course, often it is not readily possible for either party to establish the degree of risk in advance of civil engineering works. The clause does not remove from the contractor those risks which might reasonably have been thought probable or likely and for which he is held to have made provision by contingency factors in his rates. The operation of the clause is reserved to those circumstances which fall short of common or even infrequent occurrence within the foresight attributed to an experienced contractor, taking into account the information provided by the employer and the possibilities to be considered from the contractor's pre-tender examination of the site.

The ICE 6th edition introduced a requirement for the employer to disclose 'all information on the nature of the ground and sub-soil including hydrological conditions obtained by or on behalf of the Employer from investigations undertaken relevant to the Works'. The contractor is entitled to rely upon a reasonable interpretation of this information, and to have based his tender upon it.

The test of reasonable foresight

The test is not what the contractor *claims* he foresaw, but what a notional experienced contractor would have foreseen in the circumstances of the tender. The expectation of Clause 11 that the contractor should satisfy himself about the ground and subsoil (so far as is practicable and reasonable), and 'in general to have obtained for himself all necessary information as to risks, contingencise and all other circumstances', is likely to be construed generously by arbitrators, having regard for the realities of the tender situation.

As the notional contractor would be aware that all manner of unexpected physical conditions and artificial obstructions might affect the works; a line has to be drawn between those which occur with reasonable frequency (such as pockets of weak ground), and the freak discovery of an unexploded bomb.

Defence to 'lack of foresight' by the engineer

Contractors who are pressing a claim under Clause 12 frequently point to an absence of prior investigation or information from the engineer, and contend that the foresight of those preparing the contract set a standard which the contractor should not be required to exceed. Why, they ask, should the contractor be credited with greater foresight than those who conducted the site investigations and prepared the designs? The contention is compelling and is rebuttable only by the assertion that the investigations were made for the purpose of designing the works, not for the construction of them, that the contractor is entirely responsible for his own assessment of those resources necessary to carry out and complete the works, and that the information made available for interpretation is to be considered on its own merits exclusive of any assumption as to completeness or as to the foresight exercisable by the engineer.

The scope of the provision

The clause is most frequently applied to the occurrence of ground conditions which affect or obstruct the contractor's method of works. The unanticipated occurrence of rock, running sand, soft areas, springs and ground water at higher levels than expected, all send the site agent hurrying to consult the site information to check if a claim can be made. In sewer trenches, a changed ground characteristic may critically affect trench width with penal consequences related to imported filling and disposal. Claims have been founded on rock which was harder than anticipated, and on the unsuspected presence of chemically contaminated materials which required special means of disposal.

The physical conditions do not have to be static in nature, it is immaterial whether the applications of stress in ground behaviour were active or passive provided that the condition meets the criteria for lack of reasonable foresight. Neither are the conditions limited by those existing prior to contract.

Existing service apparatus

By custom, details of existing services apparatus are given without warranty. Technical means of tracing the presence of mains and services have to a great extent diminished the remedy available in this clause to the contractor who claims for such unsuspected obstructions. It should be noted that the wording in subclause 12(1) refers simply to 'conditions or obstructions [which] could not in his opinion reasonably have been foreseen'. It is in the different provisions of Clauses 13 and 14 that the wording is extended to continue 'at the time of tender'. Many circumstances which vary from the tender information could reasonably be detected in advance of the relevant activity by any sensibly diligent contractor, allowing sufficient time to avoid or limit the contingent expense and delay, this will apply particularly to the detection of service apparatus.

Weather conditions

The immediate effects of weather conditions on the works are clearly excluded as a basis of claim. But of course weather conditions may bring secondary effects. Water tables may rise to a level inconsistent with the data given in the site information. All that can be said is that every case will turn upon its own facts, including the extremity of the weather condition, identification of causation, the extent of the tender information, and the reasonable conclusions to have been drawn from that information.

Notices of claim for unforeseen conditions, etc

It is clear that the engineer is to be put on notice as early as possible in order that he can act to limit the consequences for the employer, by instructing those measures which (in the terminology of Clause 51) are necessary or desirable for the satisfactory completion and functioning of the works. Notice is required in two separable circumstances, first that the contractor considers that a Clause 12 event has been encountered, and second that he intends to make a claim.

Decision and payment

If the engineer decides that the condition could 'in whole or in part' have been reasonably foressen, he has a duty to inform the contractor in writing. No time limit is given, but the engineer's duty is to respond within a reasonable time, if only to allow the contractor his right of dispute.

The provisions for payment under this clause are particularly attractive to contractors, comprising 'the amount of any costs which may reasonably have been incurred by the Contractor by reason of such conditions or obstructions together with a reasonable percentage addition thereto in respect of profit'. This seems to invite highly constructive accounting by the contractor, and the engineer should be alert to monitor costs with great throughness, to act promptly in issuing instructions and variations, and to look for those factors in the rates of concurrent works which offset elements attributed to the claim.

19.2.5 DELAYS AND EXTENSIONS OF TIME

The contractor is entitled to the whole of the contract period in which to complete the works, and may programme as be wishes within that period, consistent with sectional completions and any express contract requirements for the order and timing of the works.

The contractor's right to early completion

In practice, every contractor will make his own assessment of the balance of commercial advantage which lies in applying large resources over a shortened programme, or more modest resources over the full period. The shortened programme is a legitimate gamble for commercial gain. Unless such a programme is made a term of the contract at acceptance, the employer cannot hold the contractor to it, neither can the contractor apply it as a term of the contract (*Glenlion Construction Ltd v. The Guinness Trust (1987)*).

Employer not liable for contractor's 'gamble' on early completion

Exclusive of intervening circumstances, if the contractor overruns a shortened programme, but finishes within the contract period, there is no real 'loss' – only a failure to make a profit for which he made a voluntary arrangement. A contractor claiming under the contract for the cost of delay must clearly demonstrate that the cost which he seeks to recover has a causal link to, and is confined to the direct consequence of, some event for which remedy is provided by the contract and for which the contractor gave notice in accordance with the contract (in ICE 6, such causes are listed in Clause 44).

Although the contractor may voluntarily operate an accelerated programme, the employer's obligations for the engineer's response time in giving instructions and variations continue to be construed in the context of the contract period (though they must be issued in accordance with the contractor's reasonable notice and the accepted programme). The duty of co-operation does not imply that the employer has to join in the contractor's gamble by overresourcing the engineer's staff or requiring them to work generalised overtime.

Extension of time and liquidated damages

Delays and extensions of time are separate though related issues. The purpose of extending time is to protect the employer's right to liquidated damages in accordance with the contract. In brief, liquidated damages are recoverable by the employer for each day or week (as described in the contract) by which the contractor fails to deliver the works in time. If by some act of prevention, such as his own delay, interference, failure in giving access, variation or suspension, the employer impedes the contractor in the discharge of his duty, then if commensurate time is not added to the contract period the liquidated damages provision ceases to operate, and time is said to be 'at large'. This simply means that the contractor must deliver the works within a 'reasonable' time, and failure beyond that would entitle the employer to common law damages.

By an odd quirk of the ICE 6 contract, the express provision for the notification of delay compels the contractor to apply for an extension of time. Must the engineer extend time if the contractor is delayed while ahead of programme? The accepted programme is not necessarily the basis for the award of an extension of time; and indeed it may be that an accepted programme is not yet in being, since it is not unknown for programme submission and rejection to continue in 21 day notice cycles until much of the work is in fact completed. The obverse of the employer's inability to enforce a shortened programme which he has accepted from the contractor (and his absence of liability for it), means that if the engineer does not extend time where warranted by a cause of delay, he denies the contractor's right to the full contract period. 'If in the Contract one finds the time limited within which the builder is to do the work, that means, not only that he is to do it in that time, but it means also that he is to have that time in which to do it'. (*Wells v. Army and Navy Co-operative Society* (1902)).

The costs of delay

Extension of time of itself carries no provision for payment. Those provisions are contained in, and should be identified to those, various clauses which provide for recovery of cost, as set out below. Cost is a matter of record in the event, but where time-related costs such as maintenance of offices and facilities and head office overheads are concerned, the preserved rights of the parties to the full time period mean that if contract rates and prices are to be extended, then the incremental weekly costs attributable to delay should continue to be assessed from the base of the weeks in the contract period rather than the programme period. Otherwise 'cost' must be proved by evidence of loss. In the relevant clauses, ICE 6 speaks generally of 'cost' rather than 'necessary cost', but the contractor has a common law duty to mitigate the employer's loss, and if he does not want to be judged by hindsight to have overspent the loss resulting from delay or suspension, the wise contractor will seek the engineer's instructions in respect of standing resources, or at least notify his intentions for the engineer's approval.

Concurrent delays – the 'dominant' or effective cause

The common experience of construction works is that delays do not occur singly, in a nicely separated and sequential manner. The engineer can be faced with extremely difficult problems in the analysis of the proper cause and effect of overlapping questions of prevention (employer's liability), the contractor's culpable delay (contractor's liability) and weather (neutral; contractor's financial risk). Various approaches have their own logical appeal, and in dealing with concurrent causes of delay it might appear correct to adopt the first cause which arises chronologically, or alternatively the 'dominant' cause (though that may not have been 'first in line'), or to make an apportionment between the periods in which the causes competed and those periods, preceding and following the overlap, in which the causes stood alone.

Apportionment between delays

The law seeks where possible to avoid apportionment between competing causes and it seems to be the general case that extensions of time should be treated by the dominant cause principle. However, there is judicial support for the view that in construction cases the engineer or arbitrator has the duty of separating and allocating the extension of time to the variously competing heads of claim in a manner which does *commercial* justice to the facts. In *H. Fairweather and Co.* v. *London Borough of Wandsworth* (1988), it was stated: 'an arbitrator has the task of allocating, when the facts require it, the extension of time to the various heads. I do not consider that the dominant test is correct'. Thus although the effects of competing delay events may and do become confused and inextricable, it remains important to give separate notice of each event as it occurs, and to establish so far as possible the causal link to any related damage, such that the eventual position can be considered on the basis of contemporaneous records. From the engineer's viewpoint it is equally important to record the existence and effect of any culpable delay by the contractor.

Apportionment of cost is given express contractual sanction in ICE 6 under the provision for the care of the works, in the event of liabilities arising jointly from matters of employer's risk and contractor's risk (Clause 20(3)(c)). However, it is thought that in general delay claims, and determination which rested upon some kind of arbitrary overall apportionment between the competing causes would be unsupported in law.

Contractual provisions for claims involving delay

The causes of delay expressly listed by ICE 6, Clause 44 are:

(a) any variation ordered under Clause 51(1)
(b) increased quantities referred to in Clause 51(4)
(c) any cause of delay referred to in these conditions
(d) exceptional adverse weather conditions
(e) other special circumstances of any kind whatsoever which may occur.

The contractual provisions at (c) giving title to reimbursement for extra cost expressly or impliedly including delay are:

Cl. 7(4)(a) Delayed issue of instructions or drawings.
• such cost as may be reasonable.
Cl. 12(6) Adverse physical conditions and artificial obstructions.
• the amount of any costs which may reasonably have been incurred . . . together with a reasonable percentage addition thereto in respect of profit.
Cl. 13(3) Unforeseen instructions concerning the mode, manner and speed of construction
• the amount of such cost as may be reasonable . . . Profit shall be added . . . in respect of any additional permanent or temporary work.
Cl. 14(8) Unavoidable delay or cost arising from unreasonable delay in approval by the engineer of working method, or imposition of design criteria legitimately unforeseen at tender.
• such sum in respect of the cost incurred as the Engineer considers fair in all the circumstances plus profit on additional permanent or temporary work.
Cl. 17(2) Setting out – incorrect data supplied in writing by the engineer.
• the cost of rectification.
Cl. 27(6) Delay due to variation affecting work in controlled land or prospectively maintainable highway.
• such additional cost as the Engineer shall consider to have been reasonably attributable to such delay.
Cl. 31(2) Facilities for other contractors – delay or cost beyond reasonable foresight at the time of tender.
• the amount of such cost as may be reasonable plus profit on additional permanent or temporary work.
Cl. 40 Suspension of works (for reasons other than express provisions, weather conditions, contractor's default, or safe working).
• extra cost incurred plus profit on additional permanent or temporary work.
Cl. 42(3) Failure to give possession of or access to the site.
• the amount of any additional cost to which the contractor may be entitled plus profit in respect of additional permanent or temporary work.
Cl. 50 Searches, tests and trials to detect defects in the works (if contractor found not liable).
• cost of the work carried out. (see also Cl. 38)
Cl. 59(e) Loss following forfeiture exercised with the engineer's consent against a nominated subcontractor (but only following failure by the contractor to recover from the subcontractor).
• unrecovered reasonable expense.

Contractual limitations

The term at Clause 44(e) above providing for the contractor to claim extension of time for 'other special circumstances of any kind whatsoever' is a catch-all which the contractor may use to cover any form of delay which he can claim was unforeseen and is beyond his control. It might be used in connection with Clause 20(3)(b) (Rectification of damage arising from excepted risks), or other neutral events. The catch-all cannot be used in connection with preventions by the employer for which there is no prescribed remedy in the contract. The engineer is restricted to certification in accordance with the contract, and any claim by the contractor which cannot be brought within the contract terms will lie outside the certification and payment provisions and has to be brought separately against the employer.

19.2.6   CLAIMS ON THE PRINCIPLE OF 'PREVENTION' BY THE EMPLOYER

The basis in general terms which would be implied by law

In all of the following cases the employer (or the engineer) is given power to exercise a control on and interference in the performance of the works as they originally stood at the acceptance of the tender. The law holds that there is a necessarily implied term in any contract that neither party may unilaterally amend or prevent the other from performing it; indeed the parties are attributed a qualified duty to do whatever may be necessary to enable the other party to perform in accordance with the contract.

The express terms of the contract

In the ICE contracts the implied terms are generally supplanted by express terms such as those which follow below.

In all of these clauses of the ICE 6th edition, the contractor must comply with the notice provisions of Clause 52(4) to claim his remedy for amendment or interference. The remedy is based upon the common law principle: 'He who alleges must prove', and the burden of proof is 'more likely than not', but increases as the passage of time distances and clouds the event and its consequences. Prompt compliance with the contract provisions greatly eases negotiation and almost certainly maximises the value of the claims.

In all of the following considerations, it is taken that the contractor will be able quite readily to identify and agree the immediate direct costs which arise from the contract terms. This assumes that proper notices and records are made and supplied to the engineer. The stricture is repeated; all rights defer to the necessity of reasonable proof, the best time to do that is when the incident occurs, but if there is delay in realising that an entitlement has arisen, there is no particular difficulty or prejudice of the engineer's ability to ascertain and determine the cost if there has existed a routine and sensibly detailed procedure for recording site attendance and performance.

Clause 7 – issue of further drawings and instructions

The above implied terms are not held to bind the employer and engineer beyond the scope of their reasonable duty.

A contractor claiming under this clause must show first that the drawings or instructions which he required were 'necessary for the purpose of the proper and adequate construction and completion of the Works', second that he gave adequate notice in writing that the relevant drawings or specifications were required, third that the engineer failed to respond with the information at a time 'reasonable in all the circumstances', and fourth that he suffered directly consequential delay and/or extra cost. The timing and adequacy of the contractor's notice requiring information is extremely important, and constitutes a condition precedent to the right to claim. In the building case of *Merton* v. *Leach* it was accepted that a list of information requirements set out upon the contractor's programme was sufficient to establish notice; however, this practice is not to be recommended as the most obvious way of putting the engineer on notice, and the contractor would be well advised to draw attention to such requirements in correspondence, and to update the requirements in writing at every review or amendment of the programme.

Clause 13 – satisfaction of the engineer

The contractor has an absolute duty to construct and complete the works in accordance with the contract except in the instance of a legal or physical impossibility.

Clause 64 entitles the engineer to issue instructions controlling the mode manner and speed of execution of the works, and the contractor must comply with such instructions on any matter in connection with the works 'whether mentioned in the contract or not'.

The contractor is entitled to 'the amount of such cost as may be reasonable' plus profit on additional permanent or temporary work. In many instances the contractor may find it more financially advantageous to press for measurement of the consequences of the instructions as variations.

Clause 14 – programme and method statements

The contractor is obliged to submit a programme of works within 21 days following the award of the contract, together with a general description of the methods of construction which he proposes to adopt. The engineer must respond within 21 days, to accept or reject the proposals, failing which the programme is deemed to have been accepted. Thereafter the employer must give the contractor possession of and access to the site in accordance with the accepted programme, (Cl. 42(2)(b)). If the engineer is not satisfied with the adequacy of the information initially provided to him, he is entitled within 21 days to demand further details, and the contractor must provide the required information within a further 21 days, or the programme is deemed to be rejected.

The programme accepted under Clause 14 binds only in respect of the access and possession to be given by the employer. In other respects it is not a part of the contract agreement between the parties. Different considerations apply to programmes or method statements directed to be supplied with the tender and incorporated into the contract agreement. Departure from such a commitment is either a technical breach by the contractor, or a variation by the employer, to be measured under Clause 52.

Clause 17 – setting out errors

The contractor is responsible for the accuracy of his setting out to the data supplied to him by the engineer. This clause is clear in apportioning liability to the employer only where the data supplied in writing to the contractor was wrong. Even here, though the clause does not provide it, there may be a legitimate escape route for the engineer. It is common experience that errors in supplied information arise because of conflict between drawings, or because amendments to a particular detail are made on one drawing without being carried to a related drawing. Reinforcing details are particularly susceptible, and highway general arrangement modifications in conflict with separate statutory undertaker's information have produced dramatically penal results.

A contractor who holds and continues to operate from conflicting information some part of which he must have known to be in error is arguably in breach of a duty reasonably to be implied to draw the inconsistency to the engineer's attention for correction under Clause 5 of the contract. Although the provision of Clause 17 is very direct, it is thought that particular circumstances and the general duties of the parties might moderate the effect of the clause.

Remedy under the clause is restricted to cost. There is no provision for profit on additional works. There is no express mention of delay, but express provision arguably lies within the terms of Clause 13.

Clause 27 – delay due to variation involving work in controlled land or prospectively maintainable highway

The Public Utilities Street Works Act 1950 referred to in the clause has been replaced by the New Road and Street Works Act 1991, and it is anticipated that this clause will be redrafted in the near future. The effect of the changed legislation upon the clause is not thought to affect the contractual obligation.

The employer is responsible for obtaining all necessary consents and for notifying the contractor before commencement of the works as to which parts are affected by the statutory legislation. If the employer has failed to obtain necessary consents, or has failed to notify the contractor in writing before the commencement of the works, then the engineer must give instructions 'including the ordering of a variation under Clause 51 as may be necessary'.

Clause 31 – facilities for other contractors

The clause is clear that the contractor does not hold exclusive possession of the site. Where his own endeavours may have created the primary means of access in access roads and scaffolding, he is obliged to make these available to others upon the direction of the engineer or the engineer's representative. The expression 'reasonable facilities' is imprecisely defined. It could be extended from access and working space to include site amenities, but is likely to stop short of the supply of water and power unless there is express contract provision.

Clause 40 – suspension of the works

REASONS FOR SUSPENSION
There are a variety of reasons why the engineer may feel compelled to suspend the progress of the works in whole or in part. Disregarding those suspensions which are provided in specifications for the order of the works, the reasons, divided between the responsibilities of the parties, may include:

1. The contractor:
   - fails to comply in the provision of programmes and method statements, such that the engineer is unable properly to control and ensure the safe construction of the project
   - persistently fails to observe specification requirements
   - is working without proper regard for safety
   - is endangering or causing undue damage to surrounding property or paving surfaces

- is wasting contract materials (e.g. rock or suitable materials in earthworks) or is removing them from site without authority
- persists in working in unfavourable weather (rain or low temperature) to the prejudice of the works
- refuses a proper instruction, e.g. to exclude from the site subcontractors or individuals to whom the engineer has properly objected.

2. The employer (or the engineer as his agent):
   - is delayed in conveying access or possession
   - needs time to re-evaluate conditions which have emerged on the site (e.g. Cl. 12)
   - wishes to consider major changes to the works
   - is unable to provide materials identified in the contract for his supply
   - is delayed in the preparation of drawings, or uncovers a fault requiring redesign (also an excepted risk under Cl. 20)
   - is delayed in denomination following valid termination of employment of the previous nominated subcontractor.

This is unlikely to be a comprehensive list. The engineer may be pressed by the employer to suspend the works simply because the employer is not at that time able to pay on further certificates. The authority of the clause in such circumstances would surely be very doubtful as being an improper use of the engineer's powers. However, the costs of construction under the ICE contracts can run greatly beyond the anticipation of the employer in reliance on the tender figure, and it is hard to see how the contractor would take advantage from objection, since he is entitled to the costs of the delay.

THE CONTRACTOR'S RIGHTS
The power of suspension belongs exclusively to the engineer. The contractor has no right of suspension under the conditions, and his common law right is effectively limited to the acceptance of repudiation in those extreme circumstances where the employer can be shown no longer to be willing to uphold the contract. No contractor should even contemplate this without taking legal advice.

CONCURRENT CAUSES
There may of course be more than one cause contributing to a period of suspension, the parties could be jointly responsible for differing reasons. The clause as written indicates a primary assumption that the contractor will be paid 'except to the extent that' there is alternative contract provision, weather conditions, or contractor's default. The words 'to the extent that' indicate that, in circumstances of concurrent joint liability, there should be an apportionment of the resulting cost. Since there might be consecutive followed by overlapping causes, the engineer might best clarify the situation by issuing separate notices of suspension for the competing causes. This would also reduce the risk that the compounded period of the competing causes would exceed four months.

The contractor is not entitled to benefit in a suspension if he could not otherwise have claimed a delay; that is to say, for example, if he experiences inclement weather or frost conditions for which he should have allowed in the tender, he cannot obtain a benefit by continuing the attempt to work and forcing the engineer to suspend. The difference between a delay and a suspension lies in the contractor's title in a delay situation to limp along with whatever work is possible at his own discretion. If the delay is critical, it could fairly be said that he has a duty to do this, as in a Clause 12 situation where he awaits the engineer's response to his initial notice. It does not always follow that work will come to a complete halt. Under an instruction to suspend, the contractor must cease work entirely, except for protection and security operations.

Excepting the listed exclusions, the remedy in either case amounts to cost, plus profit on additional permanent or temporary work, plus extension of time if appropriate. The engineer is never obliged to issue a notice of suspension. It seems that the only real advantage which the contractor might have in persuading the engineer to exercise this power is to formalise the delay, to bring pressure on the engineer and employer to limit it, and to have the doubtful benefit of an option to accept an omission or abandonment if the delay exceeds three months plus 28 days.

CLAUSE 42 – POSSESSION OF THE SITE AND ACCESS TO THE WORKS
In common law failure by an employer to give possession of the site on the date named gives the contractor a right of election to a determination or 'rescission' of the contract (*Freeman v. Hensler* (1900)). The ICE contracts recognise the reality which frequently exists in large amounts of contract space partially or continuously dedicated to manufacture or traffic.

If the contract agreement is unqualified by provisions for partial or sequential possession, the contractor is entitled to be given immediate occupation of the whole site, but in practice, there will be no breach or damage if the employer is able to offer sufficient possession for the works to commence without unreasonable restriction.

DELAYS IN GIVING INSTRUCTIONS TO COMMENCE
Invitations to tender frequently state that the contractor's offer is to be held open for a period of three or more months; irrespective of this the contractor is entitled to withdraw his offer at any time prior to acceptance. Few contrtactors would wish to do this, but employers may overlook the fact that the contractor is compelled to gamble on the date by which he may be given instructions to commence the works. A tender for a six month contract offered in May and immediately accepted could enable a commencement in June with over four months of activity clear of winter conditions. The same contract accepted in August and commenced in September would contain four months of winter activity. In similar circumstances an 18 month contract could effectively fall into two summers and a winter, or less agreeably, two winters and a summer.

FAILURE TO GIVE POSSESSION
The position with regard to a failure to give possession is entirely different. The provisions for partial possession, the order of availability, and the nature of access, where prescribed in the contract, present constraints for which the contractor has tendered. Clause 42 seems to operate upon the assumption that all of these constraints will apply sequentially starting from the date of commencement, but practical considerations may dictate that fixed dates have to be pre-allocated to possession of certain areas, and in such instances the contractor would need to be notified of the intended commencement date at the time of tender.

PROGRAMME RESPONSIBILITIES
The contractor must accommodate the prescribed constraints in his Clause 14 programme, and upon the engineer's acceptance of the programme, the employer is bound to deliver access and the possession in accordance with that programme. Is the contractor similarly bound by that pattern of possession? The programme is an order of works, and circumstances may dictate that the arrangements are altered, but although the contractor is not bound by

his programme, he cannot enforce a right to have the employer change fixed arrangements for possession. It should be noted the Clause 14 provides for revised programmes to be produced when required by the engineer if it appears to him that the actual work is out of conformity with the existing accepted programme. In requiring 'such modifications to the original programme as may be necessary to ensure the completion of the Works or any Section within the time for completion' it might be a reasonable requirement of such a revision that the arrangements of the access and possession provided by the existing programme should be met. The engineer has no power to require the employer to vary the arrangement to suit the contractor, but must in accordance with Clause 51 order any variation which is necessary for the completion of the works. The contractor has no right to payment in the event of his own default, and may find himself put to great expense unless and until the engineer is able to secure the employer's compliance in a revised programme.

THE CLAIMS ENTITLEMENT

The good sense of notifying all those constraints which the contractor is to price at the outset carries the penalty that on breach of the employer's duty to provide possession or access, the contractor is entitled to claim.

RESTRICTIONS ON ACCESS AND POSSESSION

The clause provides that the contract may prescribe 'the availability and nature of the access which is to be provided by the employer' and makes clear 'the employer shall give . . . possession of so much of the Site and access thereto as may be required to enable the contractor to commence and proceed with the construction of the Works'.

Clauses 38 and 50 – searches, tests and trials

These clauses authorise the engineer to require the contractor to uncover or open up any work, either during the progress of the works, or during the defect liability period, where he suspects (or where events have shown) that there is a defect. Perhaps it is questionable that the exercise of this power is an act of prevention. However, a contract devoid of express terms would not provide for a contractor to be required to dig up his completed work for inspecton, especially where it may already have been witnessed and approved by the employer's agents.

COMPLETENESS OF THE CONTRACTOR'S OBLIGATION

The contractor's obligation to construct and deliver the works in entire conformity with the contract is a complete one. It makes no difference that the work may have been given approval by the engineer, his representative or assistants in the course of construction, or that it may previously have been subjected to tests or trials. By Clauses 8 and 13 the contractor undertakes to provide all things necessary for the proper execution and completion of the works, and (except in the event of impossibility) undertakes to construct the works in strict accordance with the contract. By Clause 2(1)(c) the engineer has no authority to relieve the contractor of any of his obligations under the contract, and by Clause 39(3), failure by the engineer or his staff to disapprove work at the time of execution does not prejudice his right at a later date to order the investigation, and if necessary the rectification of any work which is found not to comply with the contract.

BALANCED RESPONSIBILITIES FOR COST

If a defect is found for which the contractor is liable, the contractor carries out the investigation, repair and reinstatement at his own cost. If the works are found to have been constructed in accordance with the contract (Cl. 38) or if the defect is one for which the contractor is not liable (Cl. 50), the employer is to pay the 'cost' of the opening up and reinstatement.

BASIS FOR CLAIM

Neither clause envisages payment to the contractor on a basis other than 'cost', which excludes profit, and would not permit daywork rates to be applied. Neither clause expressly requires a notice of claim under Clause 52(4). In the event that the investigations prove the employer to be liable, the contractor can extend the basis of claim to cover delay and disruption by giving notice of claim under Clause 13(3) on the basis that the engineer has issued instructions which 'involve the contractor in delay or disrupt his arrangements or methods of construction so as to cause him to incur cost beyond that reasonably to have been foreseen by an experienced contractor at the time of tender'.

Clause 59(4)(e) – costs resulting from forfeiture in a nominated subcontract

THE BASIS OF NOMINATD SUBCONTRACTS

In a nominated subcontract the employer not only tells the contractor what work is to be done and specifies to what standard it will be delivered, he instructs the contractor as to who will carry it out, and requires the contractor to accept a responsibility for that subcontractor's performance in accordance with the contract.

THE SEPARABILITY OF THE CONTRACTS, AND ACCEPTABLE TERMS OF SUBCONTRACT

The employer is not a party to the contract into which he requires the contractor to enter with the subcontractor and does not warrant that subcontractor's performance. However, under the main contract, while the contractor 'shall he as responsible for the work executed . . . by a nominated sub-contractor employed by him as if he had himself executed such work', the contractor is unable to determine the subcontractor's employment without the engineer's consent.

It seems sensible that if the employer is to thrust upon the contractor the subcontractor whom he chooses to do his work, the least he can do in justice to the main contractor is to offer some protection against the obvious consequence of failure in performance by the subcontractor. The result is first to offer to the contractor the acceptability of all those clauses in the subcontract which, with the possible exception of security for performance, most main contractors would in any case regard as standard for their domestic subcontracts.

The contractor is therefore given rights of reasonable objection to the appointment of the subcontractor, and is only required to enter into arrangements with a subcontractor who will indemnify the contractor against 'all claims, demands and proceedings, damages costs charges and expenses whatsoever arising out of' the failure by the subcontractor 'in undertaking towards the contractor such obligations and liabilities as will enable the contractor to discharge his own obligations and liabilities towards the employer under the terms of the Contract'. The contractor is entitled to demand unspecified security for proper performance, and to include determination provisions equivalent to those of the main contract.

If for any reason the contractor wishes to determine the subcontractor's employment he must first obtain the engineer's consent in writing. If the engineer refuses, the contractor is entitled to 'appropriate' instructions under Clause 13. If the engineer consents, the subcontract is determined and alternative arrangements are made to discharge (or to omit) the subcontract works. The contractor is then primarily responsible for the recovery of all resulting costs and damages from the subcontractor, including those of any additional expense which has been incurred by the employer.

INDEMNITY GIVEN BY THE EMPLOYER

The price of surrendering to the engineer the power to authorise the determination of the nominated subcontractor's performance is an undertaking by the employer in subclause 59(4)(e) that, if and to the extent that the contractor fails to recover from the subcontractor all his reasonable expenses of completing the subcontract works and all his proper additional expenses arising from the termination, the employer will reimburse the contractor his unrecovered expenses. The clause contains no requirement for notice of claim.

CONTRACTOR HAS NO RIGHT OF CLAIM FOR SUBCONTRACTOR'S DEFAULTS

The absence of a claim provision in this clause demonstrates the fact that the contractor's proper right of recovery exists within his contract with the nominated subcontractor. There will be no extension of time, extra cost or delay and disruption claim under the main contract in respect of subcontract default. This is because the employer is not in a contractual relationship with the subcontractor and has no right of recovery from him. If the contractor were to be compensated directly by the employer there would be no residual cause left for action between the contractor and the subcontractor.

Thus assuming that, in the fullness of time, the contract overruns the time for completion, the employer will be at liberty to deduct liquidated damages, the properly attributable element of which will be added by the contractor to his action against the subcontractor, together with delay and disruption and the extra cost of completing the subcontract works.

The operation of the indemnity given by the employer to the contractor is conditioned by the duty placed on the contractor if necessary to pursue the subcontractor by arbitration or litigation, or if he was insolvent to await a receiver's report or winding-up in liquidation. It is only when the contractor fails in enforcing his legal remedies that he can call on the employer to pay. It is suggested that the clause must contemplate the contractor's legal costs as 'proper additional expenses arising from the termination', as also are interest and financing charges. The penalty for the employer can be very heavy.

19.2.7   THE ALTERNATIVE BASIS – CLAIMS IN BREACH OF THE CONTRACT

Continuing common law rights

The conditions governing delay and prevention matters which give the contractor a right to extra payment relate to circumstances which may otherwise be breaches of the contract by the employer, but no set of conditions can reasonably provide for all the circumstances in which the parties seek redress from each other, and claims which cannot be identified with the express terms of the contract must be made as the assertion of common law rights.

It is necessary to distinguish very clearly between the claims brought under the contract from those brought in common law, since the express terms define and limit (or sometimes extend) the means of recovery, whereas the common law gives entitlement to damages sufficient to restore the claiming party to the position he would have been in if the breach had not occurred. The position of the engineer is also affected.

Where the express conditions of contract deal with common law breach, it has been held that the express terms do not displace the common law rights and duties.

While it is frequently asserted that the common law rights of the parties are not (without clear words) extinguished by contractual provisions, it seems that the law will regard the provisions as sufficient to limit the liability. It is suggested that it is unacceptable in law that a contractor should claim sums in common law which greatly exceed those to be quantified by the limitation of the relevant contract provisions, unless those provisions are found to be unfair or are otherwise ineffective to limit the claimant's rights. The following matters merit consideration:

1. Every contractor tenders in full knowledge that the limiting provisions of the contract will take effect when, entirely foreseeably, the complexities of the work lead the employer into technical breach. His rates and prices include for conformity with the provisions and for elements of risk. Arbitrary departure from the provisions in favour of common law settlement could involve a measure of double recovery.
2. The measure of damage between the contractual provision and the common law remedy is not generally significant, except whereas in *Milburn Services Ltd* v. *United Trading Group (UK) Ltd* (1995), the contractor is asserting the particular rights applicable to common law determination, or as in *Clarksteel Limited* v. *Birse Construction Ltd* (1996), is seeking an escape route from a particularly onerous commitment.
3. The contractual provisions exist, not only to provide a measure of recovery, but to give an administrative structure to a measure and value contract where the end cost is uncertain and the parties need to know where they are. The engineer in particular requires early notice of claim such that he can, if necessary and possible, act to limit the damage for the employer.
4. A contractor's enthusiasm for common law solutions is most likely to be governed by escape from non-compliance with express provisions for notice and record. Where the contractor has entered and executed works and has received payment under particular provisions upon which the employer has relied as defining and limiting his liabilities, the contractor may be subject to the legal doctrine of promissory estoppel.

THE ENGINEER AND COMMON LAW CLAIMS
The authority of the engineer is derived exclusively from the conditions of contract which simultaneously define and limit his duties and powers. Where the conditions say 'the Engineer shall' this denotes a duty which must be exercised in the given circumstance. Where they say 'the Engineer may' this denotes a power which the engineer can exercise at his own discretion. Beyond this, the engineer has neither duties nor powers, and wherever claims cannot be brought within the express terms of the contract, the contractor must deal with the employer on the basis of the rights and duties implied by common law or enforceable by statute.

Thus the engineer must ascertain the basis of any claim put to him. If it lies under the contract, he must determine the liability and value accordingly. If it lies outside the contract, or upon a compounded basis of contractual terms and common law, he must refer it to the employer. He will deal with it only if he receives the employer's authority to do so. If that authority is received, his subsequent negotiations with the contractor are those of opposing parties seeking to settle a difference or dispute; the matter is outside the terms of the contract, and the duties of impartiality are inapplicable.

THE 'ENGINEER'S DECISION' PRIOR TO ARBITRATION
It may be noted that in Clause 60(4) (payment) the engineer is to certify the amount which in his opinion is finally due 'under the contract' (thus excluding claims brought in breach of the contract). In Clause 66 (the arbitration clause), the engineer is required to 'settle' the matter referred for his decision in writing. In every other applicable clause the engineer is required to 'ascertain' or 'determine'. The word 'settle' has a wider meaning (dictionary:

deal effectually with, dispose or get rid of) which enables the engineer to express his decison additionally on ex-contractual representations of claim which he is not otherwise empowered to determine.

Terms implied by common law

THE EMPLOYER
Most claims raised in common law by the contractor are based upon the implied terms of co-operation and non-interference in the performance of the works. The cases of *Merton* v. *Leach* (1985) and *Perini* v. *Commonwealth of Australia* (1969) serve to emphasise that:

- The employer is bound to take all steps reasonably necessary to enable the contractor to discharge his contractual obligations and to execute the works in a regular and orderly manner
- The employer must not hinder or prevent the contractor from carrying out his obligations in accordance with the terms of the contract.

Under these implied terms the contractor may require the employer to compel the engineer *inter alia* to issue drawings and instructions in good time, attend for inspection and measurement, value variations and assess delays, and certify payment. The employer himself must give possession and access to the site and supply all materials components and services at the time and in the manner necessary to the contract, including those of a replacement engineer if the original appointment is unable or fails to exercise his duties.

Thus the employer is bound by a term necessarily implied in law that the engineer will carry out the duties undertaken in the contract. Most, but not all of the potential breaches by the engineer of the contract conditions are addressed by contractual remedies.

THE CONTRACTOR
The duties of co-operation imply conversely and less obviously that the contractor must perform his side of the contract bargain by providing proper and adequate contract supervision, and by bringing all appropriate matters to the engineer's attention and supplying temporary and permanent designs which are the product of skill and care, and the work records which are properly required for the management and valuation of the work. Contractors are less willing to accept that the implied terms cut both ways, and that in the event of failure to supply these services, they are not only in breach of contract terms, but of implied terms as well, which equally entitle the employer to damages in the appropriate case.

Implication of rights and duties distinguished

Unlike the application of the continuing rights to common law damages described above, the courts are generally reluctant to consider the implication of general duties where the express terms are directly relevant.

COURT RELUCTANT TO 'IMPROVE THE CONTRACT AGREEMENT'
The courts will not use implied terms to rectify or improve the balance of responsibility between parties, even where by probable oversight or bad drafting the contractual terms are clear, but have proved to be unexpectedly onerous.

19.2.8   COMBINED CAUSES – DAMAGES ARISING FROM MIXED CONTRACTUAL AND COMMON LAW BASES

The ICE contracts are set out in a manner which supposedly would enable every cause of loss envisaged by the conditions to be separately identified and addressed, and this is administratively and contractually desirable wherever it is possible of achievement. In reality the starting point for serious claims negotiation is frequently delayed until in the preparation of the final certificate application the contractor seeks a peg on which to hang his claimed losses, and a full retrospective analysis of the impact and interaction of the various causes is possible. The interim assessment of claims is readdressed to take account of duplications of measured effects and reassessment of the site and head office overheads appropriate to the claims, taking into account the overall delays and the increase or decrease of measured works. At this point it is practical and sensible to ensure that there is neither over- nor underrecovery in the contingent effects of the liabilities. Claims determinations on the contractual basis must be considered alongside any made in common law in order to finalise the appropriate measure of liability.

Rolled-up claims

There are now a number of cases authenticating the right of a contractor to 'roll-up' issues which are, or have progressively become, genuinely inseparable in their effect. This does not mean that a contractor can aggregate all his grumbles, tot up his costs, add overhead and profit, and value his claim by a simple deduction of the measured value. Neither does it mean that he can rely on the *Chaplin* v. *Hicks* (1911) judgement that where damage is uncertain in scope the tribunal must 'do its best' to arrive at an appropriate figure. Every engineer is justified in throwing out claims on any basis which fails to establish proper liability in each of the connected issues, and to show a proper link to the occurrence of damage. It is only where there is a genuine uncertainty of scale in damage that a best objective assessment can be applied, wherein the balance of any doubt must be ruled against the claimant. If any part of the claim is not shown to be the employer's liability, there is the danger that the whole of the claim could be lost.

The rationale for the consideration of a rolled-up claim is:

- That all matters which reasonably can be separated should so be separated off and settled on their merit
- Liability and causal link having been established in the remainder, regard must be had to concurrent and supervening causes which may have taken priority in effect; the contractor's own culpable actions, other settled claims, and neutral events such as weather which may have been the dominant or contributory causes in all or part of the alleged damage
- In the remainder, the valuation of the damage must (in view of inseparability) be made on the 'lowest common denominator' of the contract entitlements
- There is no allowance of profit.

Thus, while the law recognises that the complexity of construction work gives rise to incidents of damage which cannot readily be isolated, and that the contractor in that situation should not be deprived of redress, proof is still the primary requirement. The conservative nature of the eventual result in most rolled-up claims should be an incentive to contractors to maximise their rights in the use of prompt notice, full records, and claims separation wherever possible.

Primary and secondary effects

It is convenient and practical to separate those losses of expense which are directly attributable and ascertainable in the identified causes (*the splash*) from those which have interacted in secondary effect in the remainder of the contract (*the ripple*), and to ascertain and determine the costs of the secondary effects by a collective analysis. Disruption, loss of productivity, and overhead loss claims often arise from multiple causes, the particular allocation of which is not readily distinguishable and not all of which may be the employer's liability. It is accepted in law that there is little point in attempting to make a totally hypothetical apportionment of collective damage between competing causes.

Before proceeding further, three points must be made in respect of interactive claims for secondary effects.

1. THE NECESSITY OF A CAUSAL LINK

There is no sustainable claim unless the claimant can establish a breach (either of the contract or common law), a damage (which falls to be ascertained) and what is known as a *causal link* between the two. The facts of accepted liability and incidental loss on their own are insufficient. A connection between liability and damage must be established. Very often, where a damage is all too evident, any available breach is seized on in order to claim liability, and the connection between the two is overlooked.

2. SECONDARY EFFECT OF VARIATIONS NOT SEPARATELY CLAIMABLE

No consolidation can be made of the effects of multiple variations. The two subclauses of Clause 52 form a complete framework for the valuation of variations and their contingent financial effects. The pricing is based on existing rates and prices which contain overhead and risk elements, or on 'fair' rates which, while still based within the framework of the tender, are final in taking account of all relevant circumstances. Variations are not retrospectively assessed and are not breaches by the employer to be treated as damages.

3. INCLUSION OF COMMON LAW CLAIM ELEMENTS INVALIDATES THE ENGINEER'S POWER OF DETERMINATION

There are no separate provisions in the ICE contracts for the consolidated assessment of financial losses. There is no separate provision for disruption, lost productivity, or unrecovered overheads. The normal coverage of such claims arises under the individual identifications of liability in the contract clauses. To the extent that all such liabilities have been subject to the contractor's notices issued in accordance with the contract (and accepted), interacting effects which are otherwise inseparable can be ascertained and certified by the engineer in any way which is consistent with his contractual duty. But if the consolidation is broadened to include common law breaches which lie outside the express terms of the contract the engineer loses the power of determination, and may proceed in negotiation only with the permission of the employer.

Claims for disruption and loss of productivity

A claim for disruption is possibly the most difficult of any to substantiate. Everyone knows that the effects of delay, stop-start, failure to meet programme targets, displacement of subcontract entry and performance dates, and inability to deploy major units of company owned plant at the internally planned dates, can be demoralising and contributive to reductions of unit productivity which go well beyond the factual estabilishment of standing time in periods of delay.

Every precept of good site management dictates that the contractor puts together the best team available, offers them bonus incentive and continuity of work, and looks for rising productivity as the team settles beyond the initial period of the 'learning curve'. Work which is disrupted so that the team is jumping from job to job around the site, or from one site to another with repeat mobilisation costs and continually changing personnel, is unlikely to achieve the production targets set by the estimator or the bonus clerk. Although the situation is relatively familiar, to the extent that it is attributable to matters for which the employer is responsible, it is excluded from the risks which the contractor is obliged to cover within his rates and prices.

DIFFICULTY IN SUBSTANTIATION

Knowledge and proof are different animals. The contractor's claim is likely to look like an implausible cost founded on an improbable target. The alleged costs of the alleged disruptions may alternatively derive from any of the following hypotheses:

- The contractor overestimated the obtainable production rates
- The contractor misdirected the work, with an imbalance or inadequacy in the provision of labour, plant or materials
- Labour or plant operatives were less competent than envisaged
- Working conditions or access were less satisfactory than expected, but not beyond reasonable foresight
- Weather conditions reduced productivity
- There was culpable delay on the contractor's part
- There was concurrent disruption from authorised variations which were separately accounted.

Obviously the contractor faces an uphill battle to prove his case His best policy is to disarm the objections so far as possible by admitting defects in performance and pointing out that site operation is seldom ultimately efficient; a common level of culpable error was priced into his rates through costings derived from similar works. He will represent that all work could be perfectly executed with the benefit of hindsight, and that the engineer is not entitled to form his opinion from that basis. The basis he seeks is the degree by which the notional reasonable output consistent with the working condition was diminished by the influence of the factors for which the employer was responsible. This preserves the causal link, but in all but the most direct circumstances carries the problem of comparing two hypothetical conditions.

DETERMINATION OF COST IN THE ABSENCE OF DETAILED PROOF

The engineer may recognise the liability but remain sceptical as to the quantum of damages. All that can be said is that he must maintain his impartiality and leave any prejudices outside the door in forming an objective decision as to the extent of any entitlement. Provided that requirements as to liability and causal link are satisfied, it is held that uncertainty of extent is not a barrier to the award of damage; where the claimant is unable to prove that he would necessarily have achieved any benefit but for the breach, he may still recover damages for the loss of a chance and the certifier must 'do the best he can' with the available evidence.

Claims for the recovery of overhead costs

THE ENTITLEMENT TO 'OVERHEADS'
The definition of 'cost' in ICE 6 includes the cost of overhead finance and other charges properly allocatable thereto whether on or off the site. Overheads are thus recoverable within any provision which gives the contractor entitlement to 'cost'. Potentially the overheads will include the relevant costs of site operation, administration and supervision, together with admissible levels of head office charges. The claim is limited to those costs which are 'properly incurred' and is subject to reasonable proof. Unlike variations where the contractor is entitled to rates which include overhead and profit at the levels anticipated in his tender pricing, it is suggested that evidence is required of the costs actually experienced, and that the basis of tender is irrelevant.

SUBSTANTIATION OF OVERHEADS
In the majority of cases there will be no difficulty in the substantiation of on-site overheads, but the practices of contractors in tendering varies widely, so that head office overheads may be added as a reduced percentage to the gross of on-site costs and administration, while the contractor's balance sheets may present more substantial overheads which combine expense truly attributable to head office with staff and other charges which arise on site. There is obvious potential for double accounting.

APPORTIONMENT BETWEEN SITE AND HEAD OFFICE OVERHEADS
The contract conditions give no help in directing when this apportionment is applicable, or how it is to be made. There are many occasions when claims events on the site give rise to little or no expenditure at head office, and others in which head office expenditure is substantially involved. Whether engineering administrators, planners, buyers, designers, and quantity surveying staff are employed on the site or in a remote office is entirely a matter for the contractor.

NECESSITY FOR PROOF OF LOSS
The engineer may be required to exercise knowledge and ability beyond the expectation of his profession, and might be compelled to call on the services of accountants. However unsatisfactory that may appear in terms of promoting quick claims settlements, there can be little doubt that the engineer who certifies overhead additions without checking for proof and double accounting is in principle liable to the employer for a breach of duty.

EXTENDED OVERHEADS DUE TO CONTRACT PROLONGATION
Individual delays ascertainable from the claims made under the various separate provisions of the contract sustain claims for overhead charges as a legitimate part of 'cost' as set out above, and are allowable at any time during the works.

EXTENDED OVERHEADS – THE FALSE PREMISE
The general case for recovery of head office overheads due to prolongation is commonly taken upon the propositions:

- That the rates and prices collectively included provision of an *allowance* for a planned level of support to head office services in the duration of the contract
- That the overhead *cost* incurred by the site works was expended by the contractor substantially on a linear basis unrelated to site performance
- That the duration of the works was extended beyond the original contract period by

instructions or delays for which the employer was responsible, such that the *allowance* was exceeded
- That the excess of *cost* over *allowance* is the measure of damage.

It is therefore sometimes assumed simplistically that, to establish the value of a claim, the number of weeks awarded in extension of time can be multiplied by the original proportionate weekly overhead charge. This is entirely false. Extension of time is not in itself a ground for claim; time-based claims are based on the proper association of culpable delay with proof of damage.

Questionable assumptions in the 'linear extension' approach to overhead loss

Even having reduced the equation to culpable delay, there remains the requirement to prove damage, and this is not the easy matter it appears. It is a false assumption that the necessary loss equates to an extension of the head office charges priced into the contract. In all normal circumstances, most head office costs will continue unaffected whether the contract winds up on time or not; it cannot therefore be said that they are a 'loss' attributable to the employer. Neither can it be said that percentages derived from company balance sheets are an appropriate basis. So far as the contract is concerned, these include non-recurring costs such as estimating, promotion, professional fees, subcontract and hire arrangements, which will not be repeated as a result of delay. Irrespective of time, there will be only one expense in the negotiation of final account. Most insurance cost is related to turnover rather than time. Depreciation of capital assets clearly will continue unaffected.

The alternative causes of overhead 'loss'

There are in fact two differing bases for this claim. The first is that the contractor was compelled to allocate more overhead expenditure to the contract than was to have been contemplated at the time of the contract, and is subject to proof that the alleged sum was in fact expended as a consequence of the delay. The second rests on the premise that, as a result of uncertainty in the period of delay, the contractor could take no steps to reduce head office expenditure, and could not accept other additional work to support that expenditure because the available resources were tied to the existing contract.

In the construction market, at the time of writing, successful claims of the second character will be few and far between, since contractors are competing on falling turnover for a diminished share of the national market and are releasing underemployed staff of long establishment. There can surely be few instances, for example, where a major company can sustain the argument that, as a result of delay on the Glasgow contract, it refused major work offered in Cardiff. However, the argument might have greater application for a specialist subcontractor whose pattern of work was not adapted to carrying out more than a few contracts at a time.

Overhead recovery based on formula calculations

The difficulties of proper assessment of loss in a claim for overhead recovery led contractors and engineers towards the easy and widespread solution of applying a formula to approximate the figure of supposed 'loss'.

The '*Hudson*' *formula*, taken from *Hudson's Building and Engineering Contracts* (10th edition), page 599, is:

$$\frac{\text{HO/Profit percentage}}{100} \times \frac{\text{Contract sum} \times \text{Period of delay (weeks)}}{\text{Contract period (in weeks)}}$$

The formula is subject to the objections expressed above, but additionally includes profit (to which the contractor is not entitled in this claim), uses the percentage included in the contract (which is not necessarily the figure appropriate to operating loss) and contains mathematical inaccuracy, since the proper isolation of the overhead figure demands that the contract sum is divided by 100 plus HO% (and not 100 as shown). So far as is known, the 'Hudson' formula, though widely used in claims submission, has never been endorsed by the courts. In the key case of *J.F. Finnegan Ltd* v. *Sheffield City Council* (1988), the difficulty of ascertaining the overhead loss was met by the use of the 'Emden' formula.

The *'Emden' formula* is taken from *Emden's Building Contracts and Practice*, 8th edition, vol 2:

$$\frac{h}{100} \times \frac{c}{cp} \times pd$$

Where h = head office percentage obtained by dividing the total overhead cost and profit of the contractor's organisation as a whole by total turnover

c = the contract sum

cp = contract period in weeks

pd = period of delay in weeks

This formula at least operates upon the balance sheet figures for the organisation as a whole, but is similarly exposed to foregoing objections, and includes profit.

The formulae are subject to the qualifications made by the authors (and confirmed by the courts) which make it clear that they are solely applicable to the second case recovery above, in which work at the same level of overhead recovery was simultaneously available during the overrun period. This should (but frequently does not) restrict their use, and their frequent adoption in inappropriate circumstances without the requisite level of corroborating evidence can only be ascribed to the unwillingness of the parties to 'get down to brass tacks'.

THE EFFECT OF CONTINGENCY SUMS
Both the above formulae, as applied to civil engineering forms, involve the presumption that the contract sum, as established in lump sum building contracts, is equivalent to the tender price of a measure and value contract in civil engineering. In the latter, the contractor has no title for instruction in the use of provisional sums inserted for contingency, and no claim in the event of non-use.

THE EFFECT OF INCREASED MEASUREMENT IN THE CONTRACT
The theory that the contractor has lost the recovery of overheads from the relationship between time and earning capacity rests in either case on the surmise that the increase in time has not in any case been offset by extra earning capacity on the contract. All of the measurements and claims settled under the contract carry overhead allowance, so if it happens that the increase of the contract value over the tender sum is proportional to or exceeds the gross of delay for which the employer is liable, there is no 'loss'.

DANGERS FOR THE ENGINEER
There are consequently dangers for the unwary or uninformed engineer. As previously noted, these claims arise in common law and fall beyond the boundaries of the engineer's

authority under the contract. Even where the employer gives authority to negotiate, there will exist a reliance situation in which the employer is entitled to be appropriately and professionally guided by the engineer, and a breach of his duty to exercise the requisite level of skill and knowledge in advising the employer can give rise to his liability for any resulting economic loss.

Accelerated programmes – the effect of the 'Glenlion' case

For many years, contractors believed that by producing a programme showing a saving in time to the contract period, they gave themselves two possible advantages: first they could base claims for extended overheads upon an overrun to the shorter period, and second, they could claim that the weekly unit overhead commitment had that higher level derived by dividing the reduced number of weeks into the tender sum. These arguments were effectively shut out by the judgement in the case of *Glenlion Construction Ltd* v. *The Guinness Trust* (1987).

In that contract, the tender invitation specified a contract period of 104 weeks. This was held to be irrelevant, since Glenlion offered a reduction of £23 570 for a period of 114 weeks and the contract was accepted on that basis. Following acceptance, Glenlion submitted a programme showing completion in 101 weeks, and disputes arising from the contract having been referred to an arbitrator, an interim award was obtained from which Glenlion appealed and asked the court *inter alia* to decide whether there was an implied term that, if the programme showed a completion date before the date for completion, the employer should so perform the agreement as to enable the contractor to carry out the works in accordance with the programme and to complete the works on the said (earlier) date. The court held:

1. That the contractor was entitled to complete the works before the contractual completion date
2. That the contractor was entitled to early completion whether or not he followed his programme, and whether or not the production of a programme was a contractual requirement
3. Despite having made such a programme, there was no contractual obligation on the contractor to perform so as to enable early completion. It was not apparent why a unilateral absolute obligation should be placed on an employer. There was no implied term requiring the employer to perform the agreement so as to enable the contractor to carry out the works in accordance with the programme or to achieve early completion.

Glenlion was a contract under the JCT form of contract. In the ICE 6 contract, the employer must in fact give the contractor possession and access in conformity with the accepted programme, but there his programme commitment ends, so that the *Glenlion* judgement otherwise applies equally in civil engineering works. The contractor programmes for early completion entirely at his own risk and commercial discretion. The employer is not bound by the contractor's choice, and his only liability for the general case of extended overheads arises only in relation to delay beyond the contract period named in the agreement.

However, this insistence upon the sanctity of the contract period potentially carries an additional basis for a contractor's claim. What of the condition in which the contract measurement is reduced, and the contractor (correspondingly) completes on an earlier date (whether he programmed for it or not)? Logic suggests that he may claim for a loss of contribution to head office overheads, even though the period of on-site expenditure was curtailed.

19.2.9   ACCELERATION CLAIMS

There are no means within a contract whereby either party can unilaterally amend the contracted time for completion. The engineer's powers to award an extension of time are in reality a means of maintaining the time for completion, taking account of specific intervening events. Clause 46(3) of ICE 6 allows for a request to be made by the employer or engineer that the contract period be reduced, but the contractor is entitled to refuse, and reduction takes place only if the parties are able to agree 'any special terms and conditions of payment'. The contractor is thus in a free bargaining position.

19.2.10   DELAYED PAYMENT, NON-PAYMENT AND INTEREST CHARGES

Nothing boils the blood of a contractor or subcontractor faster than the absence of the payment to which he considers himself entitled. When attempting to drive the labour force and subcontractors to increased overtime effort in uncongenial conditions to meet a contract target, and simultaneously wondering how the next week's wage bill can be met, the mental image of the responsible official unconcernedly sipping a gin and tonic in the bosom of his family is prone to incline annoyance towards paranoia. When, the following day, third and fourth telephone calls from anxious creditors have again to be discretely avoided, the intention forms: 'something must be done'.

The claim for interest charges on overdue sums whcih lies in Clause 60(7) is at such a time of little service to the contractor. It is, however, a very fair clause, which gives title, not only to interest on delay in payment of certified sums, but also to interest on sums which 'should have been certified and paid'.

Thus if the engineer is proved to have withheld any element of due payment from his certificate, it attracts interest from the date when payment would properly have fallen due in accordance with the contract, and this appears to be a complete answer to the common law claim of financing charges, established in the cases of *F.G. Minter* v. *Welsh Health Technical Services Organisation* (1980) and *Rees and Kirby* v. *Swansea City Council* (1985) – the interest is to be compounded monthy at a rate of 2% above the base rate of a designated bank. There is no provision for a notice of claim, but the contractor's applications for interim and final account payment are to include 'the extimated amounts to which the contractor considers himself entitled in connection with all other matters for which provision is made under the contract', and the engineer is required to certify from the basis of this monthly or final statement.

It might be argued that, depending upon the economic climate, many employers would find this rate of interest no penalty to delayed payment. Similarly, a contractor must look for a return on his capital much above the given rate, and it is bad business to borrow at 3% above base and lend at 2% above base. Interest could not be provided as a penalty, but perhaps it might have been better for the interest rate to have been left for negotiation and insertion in the appendix. Few employers at the date of agreement would have the boldness to contend that the matter was of importance to them.

What other avenues are open to the contractor who finds himself suffocated by an adverse cash flow? The problem of course is not unique to the construction industry, and in similar circumstances most traders exercise a right to withdraw supply. The contractor may be tempted to do the same, for desperate situations call for desperate remedies. Can he claim rescission or repudiation of the contract, or suspend the work in order to apply pressure to the employer?

The answer very positively is – whether or not it is lawfully justified, do not do it. Instead, proceed directly to a specialist construction solicitor, and follow his advice. The problems of repudiation and rescission are forested swamplands where only the foolhardy tread without a guide.

It seems evident that the contractors who successfully navigate this area kept their heads, gave notice of their complaint and the consequences for their performance, and maintained service at minimum levels until they were able to gain the assistance of the law. The 'rapid response' which will shortly be available from contract adjudication should greatly help in these situations, as will other provisions of the Housing Grants, Construction and Regeneration Act 1996, when this comes into force. This legislation gives the contractor a right, subject to seven days' notice, to suspend the works if payment in full is not made on the due date. The difficulty will arise in determining within any complication of interim payment exactly what was the appropriate sum in the circumstance of an interim certification.

19.3 Subcontract claims

19.3.1 INTRODUCTION

Basic problems

The complexities of contract provisions bear very hard upon subcontractors, whose terms of engagement usually reflect the superior bargaining position of the main contractor. The main contractor's concern is to establish a control and flexibility in the management of the works as closely similar to direct employment as possible, in order to facilitate compliance with main contract responsibilities and necessary programme adjustments. This conflicts with the equal concern to minimise management input and pass on the problems of labour control and commercial risk. The creation of subcontract agreements with responsibilities 'back to back' with those of the main contract, implying a contractual awareness which rarely exists, compounds with interruption of those communications, down the chain from the engineer or upwards from the subcontractor, which are essential to the proper operation of the contract provisions.

The engineer and the subcontractor are each excluded from the other by their respective agreements with the contractor, and acting as middleman it would be unnatural if the contractor did not put upon every representation, up or down the line, the interpretation most commercially advantageous to himself. As will be seen in the examination of the subcontract clauses, the ability of the contractor to shelter behind the decisions and determinations of the engineer is more restricted than is commonly realised.

Except in the limited circumstances of nominated subcontract, there have existed no practical means short of the discovery process in arbitration whereby either engineer or subcontractor can test the accuracy or integrity of the representations made to the other party by the contractor.

Within this unhappy situation the subcontractor must perform all the contractual obligations of prompt compliance, notice, record, and claim which enable the contractor to meet the terms of the main contract. In other words, the subcontractor must not only be equally as aware of his rights and duties as the contractor, but he must be more diligent to perform them in a reduced timescale which allows for the contractor's administration. If he does not so perform he is (under the FCEC form of subcontract and most domestic forms) liable for any consequent loss to the contractor.

The effect of commercial pressures

The assumptions of contractual knowledge by the subcontractor and commerical altruism by the contractor have always been unrealistic and have proved unequal to the commercial pressures of overcompetition in the construction market. Contractors will select the lowest credible subcontract offer and renegotiate it to reduced levels. The lowest bid frequently carries minimal administrative back-up and expertise. Often the main contract, though offered for inspection, was never seen, and frequently the subcontract terms are incorporated by reference only, and were never read.

Irrespective of a general sympathy which is felt by experienced people throughout the industry, arbitrators and the courts will continue to apply the terms which the parties have made for themselves, and some help is at hand in the new provisions to outlaw 'pay-when-paid' contracts, and to introduce ready access to adjudication. Thus it should follow that even where the subcontract agreement reflects an imbalance of bargaining power between the parties, misuse of that power will be curtailed, and the subcontractor will be better able to press home his rights, provided that he complies with the contract.

Forthcoming amendments

Standard and domestic forms of contract existing at the time of writing will if necessary have been amended to comply with the provisions of the Housing Grants, Construction and Regeneration Act of 1996 in respect of payment and adjudication. The Federation of Civil Engineering Contractors' Form of Subcontract is certainly one of those affected, but although the FCEC became defunct in November 1996, the form in most respects is likely to remain as the primary basis for subcontract agreements for some time, and accordingly is taken as the model for the following comments. Regard must be had for the terms which may exist in contractor's own domestic forms which may vary subcontractors' duties and rights of claim.

Alternative bases of claim, availability of the main contract terms

The subcontractor may claim either on the express terms of the subcontract, or where main contract conditions are incorporated by reference, he may claim under these through the incorporating term, all as previously described for main contract claims, or in common law.

19.3.2   COMMENCEMENT AND COMPLETION – THE PROGRAMME RESPONSIBILITIES

It is the common observation of subcontractors that their interaction with each other and with the changing programme requirements of the main contractor is a vexatious and unpredictable matter in which the contractor may be compelled to force them into a pattern of work quite unforeseen in the genial accord of the pre-acceptance meeting. Methods vary; subcontractors may be subject to the big stick of contractual threats or the carrot of bonuses for accelerated completion. The carrots are prone to wither before consumption, as further interactive events make the set targets unattainable. Wherever the subcontract work lies on the critical path, the cry is invariably for the supply of more resources; when it is not, or other works create delay, the subcontractor may find himself posted to a pigeon-hole.

Difficulties in establishing grounds of claim

Wherever the subcontract agreement specifies in general terms that the subcontract works shall be carried out in an order or at a time to be instructed by the contractor, it is extremely difficult for a subcontractor to establish claims for interruption of progress due to programme variation, fragmentation of the work opportunity, or unexpected sharing of access and workspace, or for waiting time for the use of essential facilities and supplies pledged by the contractor without warranty.

The effect of express terms

Generally the courts seem to have taken the natural and pragmatic view that he who buys a pig in a poke must not grumble that it is an ugly pig when it is uncovered.

Civil engineering works are not usually so vulnerable to the interactive effect of differing trades subcontract performances as is general in building work, but the problem still exists. It is extremely difficult for a contractor, at the date of agreement with the subcontractor, to tie himself by commitment to dates of attendance and unobstructed access or working space in particular areas. It is therefore to be expected that any standard subcontract form of agreement will convey control to the contractor in very general terms which protect him from subsequent claim, leaving the parties open to agree any more particular commitments. It is up to the subcontractor to ensure that those commitments in which he has a vital interest are addressed in the process of contract negotiation, and are incorporated into the agreement.

FCEC form – commencement and programme provisions

Clause 5(2) of the FCEC form states:

> The Contractor shall from time to time make available to the Sub-Contractor such part or parts of the Site and such means of access thereto within the Site as shall be necessary to enable the Sub-Contractor to execute the Sub-Contract Works in accordance with the Sub-Contract, but the Contractor shall not be bound to give the Sub-Contractor exclusive possession or exclusive control of any part of the Site.

And at Clause 6(1):

> Within 10 days, or such other period as may be agreed in writing, of receipt of the Contractor's written instructions so to do, the Sub-Contractor shall enter upon the Site and commence the execution of the Sub-Contract Works and thereafter shall proceed with the same with due diligence and without any delay, except such as may be expressly sanctioned or ordered by the Contractor or be wholly beyond the control of the Sub-Contractor. Subject to the provisions of this clause, the Sub-Contractor shall complete the Sub-Contract Works within the Period for Completion specified in the Third Schedule hereto.

The importance of supplementary binding agreement

Both parties to the subcontract therefore have a strong interest in seeing that the schedule or other document appended to the subcontract properly defines the terms of the agreement which has been concluded between them. If work is to proceed at a particular rate, or if particular minimum resources or outputs are required from the subcontractor, they should be specified. Provision should be made for any separate visits or phased completions. The subcontractor should take the opportunity to qualify the limitations as to shared possession and control, and to the shared use of access and/or constructional plant provided by the contractor, to the minimum standards essential to the progress demanded for his works.

19.3.3   *SUBCONTRACT INSTRUCTIONS AND DECISIONS*

The FCEC provisions for instruction and decision

Clause 7(1):

> The subcontractor is to comply with all instructions and decisions of the Engineer (as passed forward by the Contractor) and has like rights of payment as the Contractor has under the main contract. If the instructions or decisions are invalidly or incorrectly given by *the Engineer* [emphasis supplied], the sub-contractor is entitled to such costs as may be reasonable (if any).

Clause 7(2):

> The Contractor shall have the like powers in relation to the Sub-Contract Works to give instructions and decisions as the Engineer has in relation to the Main Works under the Main Contract and the Sub-Contractor shall have the like obligations to abide by and comply therewith and the like rights in relation thereto as the Contractor has under the Main Contract. The said powers of the Contractor shall be exercisable in any case irrespective of whether the Engineer has exercised like powers in relation thereto under the Main Contract.

The powers of the main contractor

This means that in putting the duties and rights of the subcontractor back to back with his own, the contractor has assumed all the powers which, under the main contract, the engineer is expressly required to exercise impartially. No such commitment to impartiality is express in the subcontract agreement. The engineer's powers under the main contract are balanced by express duties, but this clause is silent as to the duties of the contractor in the comparable circumstances. Moreover, the engineer is required to be a named chartered engineer, while 'the contractor' may include any servant or agent of the contracting company who is given authority, or who acts with ostensible authority.

It seems that the careful balance between right and duty established in the main contract documents is only available to the subcontractor at the contractor's discretion; he may exercise powers of decision for or against the subcontractor irrespective of those received impartially from the engineer.

The respective rights of the sub- and main contractors may well be quite different, and the subcontractor has little knowledge and no immediate right of access to the engineer's decisions regarding liability or extent of damage.

Duties of a party vested with power of determination

It is the general rule that, where the agreement provides for valuation to be determined by an impartial third party (e.g. the engineer), the courts will refuse to reopen and 'go behind' the third party decision, provided that it was arrived at in good faith, excepting the agreement of both parties.

Access to claims under the main contract terms

Clause 7 is the only avenue in the FCEC form through which the subcontractor can claim under the incorporated terms of the main contract for those matters of instruction by the engineer which arise under the main contract Clauses 7(4), 13(3), 14(8), 27(6), 31(2), 40(1), and 42(3), which constitute interference or prevention and involve delay and extra cost.

Subclause (1) states that the subcontractor has like rights to payment from the contractor as the contractor has against the employer in the given circumstance. Like rights do not imply that the payment will be pro rata to that handed down to the contractor by the engineer. There is no provision in the clause for the contractor to take any reasonable steps to obtain benefits from the employer, and the matter is restricted to direct settlement between the contractor and the subcontractor.

Claims based upon the contractor's own instructions

Subclause (2) refers to a right of recovery in connection with those instructions given by the contractor which did not originate under the main contract from the engineer as 'like rights in relation thereto as the Contractor has under the Main Contract'. This makes little sense if the contractor has no such right under the main contract, as frequently occurs. Perhaps the provision can be taken to intend that the subcontractor's claim will be limited to the extent of the provisions which in similar circumstances apply between the contractor and employer, but that is not what it says. The subcontractor in any event is not to be shut out of his rights without very clear words, and may claim on a common law basis for any act of prevention or interruption of his works by the main contractor which was not envisaged by the agreement.

19.3.4   SUBCONTRACT VARIATIONS

Notices by the subcontractor

In the main contract situation the duty to value variations falls as a duty upon the engineer, and there is no necessity for the contractor to give notice of claim, except in the instance where the effect of the variation extends to related operations which in themselves are not varied, or in event of dispute to the engineer's determination of rate.

In the subcontract, the obligation for notice is similarly limited; the contractor has a liability in good faith to pay for any extra work which he instructs (*Molloy* v. *Liebe* (1910)), and has assumed the position of certifier under the terms of the subcontract (Cl. 7(2)). Subcontract circumstances make it good commercial sense for the subcontractor to assert his contractual title, irrespective of the express duties, and at the earliest opportunity to put the contractor on notice of the full likely effect of the instructions he has issued. In real terms, the contractor and subcontractor are likely to approach the value of measured entitlement as equals on common law principles of ordinary fairness. In that case, as when the intricacies of the subcontract terms are introduced to the arguments, early appreciation and opportunity to record events will serve the parties best. For that reason, and because the contractor is not an independent certifier, variations are treated here on 'claim' principles.

FCEC Provisions for ordered variations

Clause 9(2):

> The value of all authorised variations shall be ascertained by reference to the rates and prices (if any), specified in this Sub-Contract for the like and analogous work, but if there are no such rates and prices, or if they are not applicable, then the value shall be such as is fair and reasonable in all the circumstances. In determining what is a fair and reasonable variation, regard shall be had to any valuation made under the Main Contract in respect of the same variation.

The clause embraces both the variations ordered by the engineer under the main contract, and those separately ordered by the contractor, and gives a common framework for valuation which varies substantially from the terms of Clause 52 of the main contract. The absence of reference to the provision of the main contract constrains the subcontractor within the strict terms of Clause 9.

Thus if the varied work is not similar to that covered by existing rates and prices, there is no mechanism for consultation or the fixing of new rates based upon those existing. The clause moves directly from applicable rates to those which are 'fair and reasonable in all the circumstances', and the absence of any duty upon the contractor to negotiate on the subcontractor's behalf also suggests that alternative considerations to those of the main contract are envisaged.

Basis of variation differs from that of main contract

The bases upon which the engineer will determine rates, and those which should govern the contractor in relation to the subcontractor, are distinctly different. The rate revision determined by the engineer may contain significant elements which are absent from the comparable subcontract rate. It will exclude any contingent effect which would alternatively be claimed in the main contract under Clause 52(2). In the instance of omission, it will deal with a 'measure and value' situation, whereas the subcontract clauses envisage a 'lump sum' subcontract. If will involve examination of the contractor's pricing note submissions, and provides for notice of objection. It is therefore difficult to see how the 'fair and reasonable' rate to be determined for the subcontractor should be much influenced by a third party decision by a certifier to whom he had no right of representation. Any regard which is to be had to the determination in the main contract must be seen to be in respect of the 'same' variation – it is suggested that this is to be read more literally than simply the occurrence of the same set of circumstances.

Secondary effects of ordered variations

Whereas the JCT subcontract (Dom 1 (1980) Cl. 16.3.1.2) and the ICE main contract form (Cl. 52(2)) permit a refixing of rates additional to those which are the immediate subject of the variation, this subcontract form says nothing about the resulting effect when dependent work has to be carried out in more difficult conditions, or where the working method is affected by a change of quantity. It would be iniquitous if the contractor should be given rights to recover costs in the main contract which he had no obligation to pass to the subcontractor, despite it being the subcontractor's loss. In this regard, the principles accepted in the engineer's determination of secondary effect under the main contract may be relevant to subcontract variation. Deprived of a common access to recovery, the subcontractor could claim, through Clause 7(1) of the subcontract in reliance on the referred rights under Clause 13 of the main contract, to be compensated for 'instructions which involve the (sub) contractor in delay or disrupt his arrangements or methods of construction so as to cause him to incur cost beyond that reasonably to have been foreseen . . . etc.'. Alternatively, there are common law rights of claim.

'Measuring' the variation

The FCEC form states at Clause 9(3):

> Where an authorised variation of the Sub-Contract Works, which also constitutes an authorised variation under the Main Contract is measured by the Engineer thereunder, then provided that the rates and prices in this sub-contract permit such variation to be valued by reference to measurement, the Contractor shall permit the Sub-Contractor to attend any measurement made on behalf of the Engineer and such measurement made under the Main Contract shall also constitute the measurement of the variation for the purposes of this Sub-Contract and it shall be valued accordingly.

The key word repeated in this subclause is 'measurement'. On no account must this be confused with 'valuation' or rate-fixing exercises. On any practical analysis (and by comparison with Clauses 52 and 56 of the main contract) 'measurement' simply means physical measurement on the site or from the drawings. The thought of the contractor trailing his subcontractors through the engineer's office to negotiate rates is charming but unlikely.

Variation of subcontract quantity

FCEC Clause 9(4):

> Save where the contrary is expressly stated in any bill of quantities forming part of this Sub-Contract, no quantity stated therein shall be taken to define or limit the extent of any work to be done by the Sub-Contractor in the execution and completion of the Sub-Contract Works, but any difference between the quantity so billed and the actual quantity executed shall be ascertained by measurement, valued under the clause as if it were an authorised variation and paid in accordance with the provisions of the Sub-Contract.
>
> The effect of 'measure and value' and 'lump sum' provisions.

A great confusion in the FCEC document results from the assumption in the clauses that the work is to be carried out on a lump sum contract price. The Third Schedule (at 'A') gives an option of price or measure and value, but measure and value principles are not evident in the drafting of the clauses. It is thus directed here that any *difference* between the quantity billed and the actual quantity executed shall be ascertained by measurement, valued as if it were an authorised variation and paid in accordance with the provisions of the subcontract. Clause 9(1) requires the value of variations to be 'added to or deducted from the price'. The payment provision at Clause 15(5) directs that the final payment shall be 'the Price and or any other sums that may have become due under the sub-contract'.

The effect of having a measure and value main contract with a lump sum subcontract is that the measurement and valuation provisions are not truly 'back to back'. The effect upon rates of a changed quantity are treated in the main contract (where appropriate) by revaluation of the whole measured item under Clause 56(2). In the subcontract, only the excess or shortfall from the billed quantity is revalued as a variation and added to or subtracted from the subcontract price. Fluctuations of quantity are not of themselves considered to be variations in the main contract, and their valuation by the engineer cannot, on either logical or arithmetical bases, be applied directly to the subcontract within the terms of Clause 9. The subcontractor is left with a title to a revised rate which is 'fair and reasonable in all the circumstances', and that would surely include all the costs of any changed method of work which was necessitated by the variation.

A second characteristic of a lump sum priced contract is that the contractor is both obliged and entitled to carry out all of the work scheduled in the agreement. In the instance of a measure and value contract, there is no allowable claim for loss of profit resulting from an ordered omission. In ordering the omission of work from a lump sum subcontract, either directly or by treating quantity fluctuation as a variation, the contractor is depriving the subcontractor of the entitlement on which his tender relied, and is liable to compensate him by placing him in the position in which he would have been had the contract been performed. He is therefore liable to pay for a loss of profit.

The 'fair and reasonable valuation'

Where variations cannot be valued by reference to contract rates and prices, does 'such value as shall be fair and reasonable in all the circumstances' specified for variations to the subcontract conform to the same requirements as those of the main contract, where 'the rates and prices in the Bill of Quantities shall be used as the basis for valuation so far as may be reasonable failing which a fair valuation shall be made'?

Certainly, a connection is made by the reference in the subcontract, 'In determining what is a fair and reasonable valuation, regard shall be had to any valuation made under the main contract in respect of the same variation.' But there is room for doubt that, in view of the possible divergence of terms between main and subcontracts, 'have regard to' is effective to put the valuations on a like basis.

A 'fair' valuation in a measure and value contract, preceded by a qualification requiring that, where work differs in character or condition from the basis of existing rates, still those rates shall create the framework for valuation so far as may be reasonable, puts the engineer in the shoes of the contractor's estimator at time of tender, and by reasonable extension this influences the establishment of the 'fair valuation'. It will react to the level of pricing, high or low, represented in the contract. No such qualification exists in this lump sum contract situation. The sub-contractor here has established the price of a fixed quantity of work, and on the occurrence of further work for which a price was not agreed the courts would if necessary determine a 'fair and reasonable' price.

19.3.5   *ADVERSE PHYSICAL CONDITIONS AND ARTIFICIAL OBSTRUCTIONS, AND THE EFFECT OF CLAIMS DETERMINATIONS IN THE MAIN CONTRACT*

FCEC Clause 19(2)

> Subject to the Sub-Contractor's complying with this sub-clause (e.g. notices and records), the Contractor shall take all reasonable steps to secure from the Employer such contractual benefits, if any, as may be claimable in accordance with the Main Contract on account of any adverse conditions or artificial obstructions or any other circumstances that may affect the execution of the Sub-Contract Works and the Sub-Contractor shall in sufficient time afford the Contractor all information and assistance that may be requisite to enable the Contractor to claim such benefits. On receiving any such contractual benefits from the Employer (including any extension of time) the Contractor shall in turn pass on to the Sub-Contractor such proportion thereof as may in all the circumstances be fair and reasonable.

Limits of main contractor's duty to represent subcontract issues to the employer

Until recently, it was widely thought that the reference 'adverse conditions or artificial obstructions or any other circumstances that may affect the execution of the Sub-Contract Works' required the contractor (subject to notice from the subcontractor) to 'take all reasonable steps to secure from the Employer such contractual benefits, if any, as may be claimable in accordance with the Main Contract' and 'in turn pass on to the Sub-Contractor such proportion thereof as may in all the circumstances be fair and reasonable'. There thus appeared to be a contractual chain which linked damages incurred by the subcontractor through the representations of the contractor to the liabilities of the employer. The subcontractor would be assured that his claims (where appropriate under the main contract) would be passed upwards, and there would be returned to him a proper proportion of the sum determined by the engineer.

The construction placed upon the clause by the Court of Appeal in the cases of *Mooney* v. *Henry Boot Construction Ltd* and *Balfour Beatty* v. *Kelston Sparkes Contractors Limited* (heard on the same day in July 1996) differs from this perception. Each case was directed to the proper application of Clause 10(2). In the first case, it had been found in the lower court that Mooney was not entitled to the sum of £28 601 which Henry Boot had recovered from the employer, for two reasons:

1. A subcontractor had to show an independent entitlement under some other provision of the subcontract, because Clause 10(2) was merely procedural. Mr Mooney had no such independent entitlement so his claim failed.
2. Clause 10(2) applied only to contractual benefits that the main contractors were *entitled* to receive from the employer, and it was unlikely that the main contractors were in fact entitled to recover the £28 601. So the windfall, as it was called, remained with the main contractors.

The subcontractor appealed. On the question, 'Can a claim be made by the sub-contractor on the main contractors by virtue of Cl. 10(2) alone, and without relying on any other term of the contract as entitling him as to the money?' it was held:

> There does not have to be a ground for recovery beyond that contained in Cl. 10(2) itself. The clause says that the main contractors shall pass on such proportion of the benefit as is fair and reasonable. In plain English that in our view imposes a legal obligation on the main contractors.

It was said that if the second sentence of the clause only dealt with procedure for recovering a share of benefits already owed by some other clause, it had little or no content. There was no provision elsewhere for the subcontractor to be compensated for adverse physical conditions and artificial obstructions. Mr Mooney had the award of £28 601 restored, and the question of 'any other circumstances' fell to be considered in the *Balfour* appeal.

In this second case, the lower court had held that, where the engineer under the main contract gives instructions by way of variation consequent on the occurrence of unforeseen ground conditions, then the subcontractor was entitled to a fair proportion of sums recovered by the main contractor in respect of those instructions. Both parties appealed, Balfour because the judgement did not go far enough, and Kelston (*inter alia*) because they argued that *all* contractual benefits fell within Clause 10(2) of the subcontract – the belief perhaps shared by most in the industry.

The Court of Appeal identified 'claimable on account of any adverse physical conditions or artificial obstructions or any other circumstances which may affect the execution of the sub-contract works' as 'the descriptive words'. The contractual benefits referred to were to be those which answered to the descriptive words.

Which main contract clauses gave rise to such benefits? Clause 12 obviously did, since the 'unforeseen conditions' were expressed in Clause 10(2). But claims under Clause 13 or 51 only came within the descriptive words if they derived from 'any other circumstances which may affect the execution of the sub-contract works'. This phrase was to be identified generically with the preceding 'adverse physical conditions or artificial obstructions', to the exclusion of other claims. Further, matters dealt with in Clauses 13 and 51 of the main contract (instruction and directions of the engineer and main contract variations) did not affect the execution of the subcontract works within subcontract Clause 10(2), 'They alter or add to what the sub-contract requires, but they do not affect the performance of it.'

Variations ordered in response to unforeseen conditions, etc.

Of the situation which arises when the engineer authorises variations in respect of 'unforeseen conditions', the Court held that these too did not give rise to benefits in Clause 10(2) of the subcontract:

Clauses 7, 8 and 9 make specific provision for the payment of extra money to the sub-contractor as the result of an instruction, direction or variation…When there has been an instruction, direction, or variation it is the regime of clauses 7, 8 and 9 that applies, and not the somewhat vague assessement of a fair and reasonable proportion under clause 10(2).

A very limited duty

Thus it must be recognised by subcontractors that the only duty which is placed on the contractor to pass forward and secure contractual benefits claimed under the subcontract arises in the narrow context of claims for adverse conditions and artificial obstructions, and that even there, if the engineer responds by issuing a variation order, any benefit to the subcontract arising from the variation must be settled between subcontractor and contractor alone.

Separation of claims considerations

In every other respect of the claims which arise under the subcontract the principle of the separability of contracts applies, and the contractor's duties and liabilities to the subcontractor under Clauses 7, 8 and 9 are to be exercised unprejudiced by any necessity of representation to the employer under the main contract. Clause 10(2) in respect of 'unforeseen conditions' is the only express term within which benefit for the subcontractor depends upon the supply of 'all information and assistance that may be requisite to enable the Contractor to claim such benefits' (from the employer). Even there, in the final sentence of the subclause, 'nothing in this Clause shall prevent the Sub-Contractor claiming for delays in the execution of the Sub-Contract Works solely by the act or default of the Main Contractor on the ground only that the Main Contractor has no remedy against the Employer for such delay.'

The subcontractor must of course supply the necessary substantiation of *any* claim he makes, and the Contractor may use it in evidence on his own behalf to the employer, but the respective rights and duties of the main and sub-contract remain separate except as above. An Engineer's decision on a main contract claim may be good evidence in respect of obligations owed in the sub-contract, but it is not conclusive and cannot be used to excuse liability if liability exists.

19.3.6   SUBCONTRACT PAYMENT

The existing provisions of the FCEC form of subcontract became unlawful when the provisions of the Housing Grants, Construction and Regeneration Bill were enacted, and 'pay when paid' clauses are prohibited. From that time, the subcontractor has unrestricted right to the sums properly certifiable on an interim basis, and in the event of disputes it is intended that he will be able to demand adjudication to correct and enforce any balances properly due to him under the terms of the subcontract.

The effects of non-payment

Non-payment or significant underpayment of subcontractors is the most frequently terminal source of site disputes. Subcontractors tend to think that non-payment constitutes a fundamental breach; an immediate and good reason to suspend the works or to terminate the subcontract.

This is most dangerous ground, and any subcontractor who contemplates suspension or withdrawal should not do so, but should immediately take specialist legal advice. In general, and contrary to popular supposition, breach of interim payment terms has not necessarily been regarded in law as sufficient ground for the abandonment of a civil engineering contract, and suspension or withdrawal may present the employing contractor with the opportunity and necessity to take over the works and to charge all additional expense or loss as damages against any sums which lie unpaid.

The Housing Grants, Construction and Regeneration Bill (1996), made provision at Clause 109 for a statutory right to suspend work where a sum is due, is not paid in full by the final date for payment, and where no effective notice to withhold payment is given. Where the contractor or subcontractor seeks to exercise this right he must first ensure that all of the preconditions have been met, and must give seven days' notice specifying the grounds upon which suspension is based. If it is later shown that the non-payment was justified, the subcontractor may be in repudiatory breach by suspending the work. The suspension will end when the defaulting party makes payment in full, and the period of suspension is added as an extension of the contract period.

Interest on overdue sums

The subcontractor will be able to claim interest on late certification or payment in accordance with the terms of his agreement; or as financing charges as a special damage under common law.

The FCEC provision at Clause 15(3)(f) and (g) creates the following conditions:

- The subcontractor receives interest from the due date of payment only if he makes a written claim within seven days of the date upon which payment should have been made
- The subcontractor alternatively receives interest from the date of the written claim
- Notwithstanding absence of notice, the subcontractor is to be paid any interest paid by the employer attributable to money due to the subcontractor
- Interest is at the rate provided by the terms of the main contract.

Subcontractors will note that in the absence of any express provision for the payment of interest on overdue sums their entitlement is restricted to recovery where they obtain judgement and apply for interest under section 35A of the Supreme Court Act 1981, or under section 49 of the Arbitration Act 1996. There is no room for reliance on custom or on implied obligations drawn from the contractor's rights under the main contract.

The terms of the Arbitration Act 1996 give complete discretion as to the award of interest, including consideration of whether it should be simple or compounded. As such, although the common law right to financing charges as a head of damage remains in common law, it is probably covered by section 49(3) of the Act, which states 'The tribunal may award simple or compound interest from such dates, at such rates and with such rests as it considers meet the justice of the case'

Cash discounts

A frequently occurring source of dispute arises in the deduction by the main contractor of discount from payments delayed beyond the period prescribed in the subcontract. Where the agreement provides for a cash discount, and subject to any express term to the contrary, it is held to be implied that the purpose of such discount is secure payment on the due date.

19.3.7   *CUT-OFF POINT FOR NOTICE OF SUBCONTRACT CLAIM*

The FCEC contract and many other domestic forms contain a cut-off provision purporting to exclude any claim by the subcontractor in which notice is delayed beyond the date of the maintenance certificate in respect of the main contract. This is obviously founded on the necessity to the contractor of being able to include all relevant claims within the final certificate application to the engineer. In fact, the statutory limitation period on claims in contract is six years, or 12 years for contracts under seal.

As a limitation of liability clause, the provision would be open to challenge under the Unfair Contract Terms Act 1977, which restricts the extent to which liability for breach of contract, or for negligence or other breach of duty, can be avoided by limiting terms of this kind. The courts look more kindly upon limitations than outright exclusions, and the success of an application to set aside the term would depend very much upon the circumstances of the case.

19.4   Claims by the contractor against the subcontractor

The contractor is entitled to all of those claims which arise in common law for breaches by the subcontractor of the obligations undertaken by him in accordance with the subcontract agreement. Few of these are left uncovered by express terms in the FCEC form, and the contractor has the normal rights of set-off against any sums properly payable to the subcontractor in respect of the subcontract works.

The main contractor's concerns

The contractor's position arises from four major concerns:

1. Ineffective or delayed performance by a subcontractor can have effect upon the main contract out of all proportion to the limited nature of the original default
2. No contractor will wish to be seen by the employer as being complacent in the acceptance of substandard, slow, or inefficient work for which he holds entire responsibility under the contract
3. Interaction between subcontractors can give rise to cross-claims which being unrecoverable either in the respective subcontracts, or under the main contract, may also compromise parallel legitimate claims which the contractor wishes to press
4. He wants notice and records for all claims which can be replicated in the main contract without encouraging claims consciousness in respect of his own defaults.

These concerns are met in the FCEC subcontract agreement as follows.

Damages sustained in the main contract

Clause 3(4):

> The Sub-Contractor hereby acknowledges that any breach by him of the Sub-Contract may result in the Contractor's committing breaches of and becoming liable in damages under the Main Contract and other contracts made by him in connection with the Main Works and may occasion further loss and expense to the Contractor in connection with the Main Works and all such damages loss and expense are hereby agreed to be within the contemplation of the parties as being the probable results of any such breach by the Sub-Contractor.

The extent of liability in damages

Damages for a breach of contract in common law work on the principle that the offending party is to restore the injured party, so far as money can do it, to the position he would have been in if his rights had been observed. There must be limits to this principle, since otherwise the injured party could claim damages so distant and conjectural as to make the other liable to an unforeseen extent entirely out of proportion to lack of care which gave rise to the breach.

The immediate effect of a breach of the subcontract may be to cost the contractor, *inter alia*, a week of progress, with resulting costs in attendance, site overheads, standing equipment, temporary works, etc. These are damages which arise 'naturally' under the first 'limb' of *Hadley* v. *Baxendale* (1854). Beyond this are the contractor's liabilities for the commitments and charges of other subcontractors and/or suppliers held up by the delay, and the deduction of liquidated damages under the main contract. In order to secure such liabilities against a defence of 'remoteness', all charges of this nature are stated to be held 'in contemplation', thus satisfying the second limb of the rule by putting the subcontractor on written notice. A claim by the contractor that, by some offence given to the employer arising from the breach, he lost the opportunity to tender for further profitable work would be beyond the rule of 'remoteness'.

Proof of damage

In order to apply the damages, the contractor has not only to prove liability, but show that the damage arose solely and necessarily from the breach. Direct damage should be relatively straightforward, provided that it is properly based on 'cost', and that mitigation is allowed in respect of other beneficial use of the 'lost' time. Indirect damage may not be easy to prove.

Apportionment between subcontractors

In *Banque Keyser Ullman* v. *Skandia* (1987) Judge Steyn said:

> There is no more difficult area of our law than causation, Scientific precepts and philosophical notions are frequently invoked. Ultimately, it seems to me, a judge is on safer ground if he puts his trust in precedent, or in its absence, in common sense.

This was later quoted in the judgement in the case of *Fairfield-Mabey Ltd.* v. *Shell UK Trading Ltd, Metallurgical Services (Scotland) Ltd, third party* (1989):

> Fairfield-Mabey alleged that the actions of two subcontractors resulted in delay, each of which was claimed to be sufficient to cause the termination which Shell had imposed on their contract for the construction of an oil rig. They also acknowledged that one finding of the court might be that the loss was in part caused by their own culpable delay, but submitted that if both subcontractors had responsibility, it was not a defence for either to say that the delay was caused by the other, and they were entitled to proceed against either of them for the whole of the damages.

> It was held by Judge Bowsher that there should be a finding of contributory negligence on the part of the plaintiffs and an apportionment of the damages.

It seems to follow that unless the default of one subcontractor is sufficient to identify liability sufficient for the whole of a particular alleged loss (in which case the concurrent liability of others would be immaterial) and there is no contributory negligence by the contractor, it would be extremely difficult to provide definitive proof of the extent of individual liability. This places the contractor in a difficulty which contributes to the frequent practice of charging almost everything against every possibly culpable target, and waiting to see what happens. The contractor might try to apportion the damage without prejudice to his right to hold each of the defaulting subcontractors individually responsible for the whole of concurrent defaults, but this is obviously a tortuous and dangerous territory in which the contractor would only justify himself by explaining everything to everybody, and nothing could be finalised until the whole of his liability was known.

That situation is so unlikely as to be commercially unrealisable. Consequently, unless contractors turn to the application of liquidated damages, as discussed below, they will continue to treat their subcontractors on the demerits of their individual performance, and, since it is difficult to envisage how adjudication in the individual case would totally resolve the problem, the possibility of gross over recovery by the contractor must be lived with as a fact of commercial life unless and until a particular excess is curbed by reference to arbitration. Subcontractors must demand the proper measure of proof and use the adjudication provisions to test and at least limit the case.

Subcontract liquidated damages

It might be more practical for contractors to create realistic sectional liquidated damage provisions in each of their subcontracts to deal with the combined effects of delay. The occurrence of a gross undeserved overcompensation will not then be open to challenge, provided that the weekly increment of liquidated damage is sustainable as a genuine pre-estimate of potential loss, and is not so high as to be construed as a penalty. This will only take out the contentious time-related expense; other direct damage from breach will be recoverable on the ordinary basis of proof, but must be carefully defined to exclude those elements which fall to be considered within the liquidated damage provision.

The right of set-off

There is no express provision in the FCEC form of subcontract for set-off by the contractor of damages arising from breach of the subcontract by the subcontractor. However, the contractor has a right of set-off in common law and may deduct the damages from interim payments due to the subcontractor. The common law right of set-off is avoidable only by express contract provision (*Modern Engineering* v. *Gilbert-Ash* (1974)), but to be effective, must be quantified in detail and notified to the subcontractor.

Normally a payment which is made due under a certification is capable of enforcement by the subcontractor under a summary judgement application to the court, but provided that the contractor can show a triable issue, the subcontractor must submit to the deduction until the matter is resolved by the appropriate tribunal. This can lead to a comfortable margin of time during which the subcontractor must continue with its contracted obligations.

Misuse of set-off power

The right of set-off has been widely misused by employers as well as contractors to delay or frustrate contracted obligations of payment, sometimes on the flimsiest of pretexts. To some extent the circumstances can be understood. A contractor's agent, project manager or quantity surveyor is often left by the estimator with an improbable task of making something out of nothing, and the temptation to balance the contract books by the invention or exaggeration of claims, both upwards to the employer and downwards to the subcontractors, is a powerful lever under the balance of integrity when careers are on the line.

The control of substandard slow or defective work

Default by the subcontractor is dealt with in Clause 17 of the FCEC Agreement, which expressly provides for the determination of the subcontract on the occurrence of

(a) failure to proceed with due diligence.
(b) failure to execute the works in accordance with the Sub-Contract, or to perform other due obligations.
(c) refusal or neglect to remove defective materials or make good defective work, and
(d) commit an act of bankruptcy or go into liquidation or receivership.

The question of due diligence has been discussed above, under the programme rights of subcontractors. Contractors would be well advised to tighten the terms of their forms of subcontract to ensure that they have established the principles upon which to base any assertion which they may need to make regarding failures in diligence. Otherwise their rights may be unclear up to the point where completion in due time is impossible and they can confidently apply the sanctions of determination and take over the subcontract works in whole or part. However, this does not imply that an incompetent subcontractor must be supported until the site work is faced with disaster. The contractor will be supported in removing the subcontractor where it has become clear that the subcontractor, by incompetence or lack of resource, is unable to discharge the subcontract according to its terms.

Low rates or prices not an excuse

The acceptance by contractors of subcontract tenders at exceptionally low values does not mean that any degree of tolerance can be exercised toward dilatory attendance or substandard works. No generalised sympathy for the commerical difficulties of subcontractors will be extended by engineers or arbitrators in connection with bad work, particularly where the contractor has been careful in honouring his own obligations to the subcontract.

'Temporary disconformity'

Every contractor from time to time experiences some failure of intention, and in the nature of site work errors occur which have to be corrected. The contractor or engineer would be unjustified in applying sanctions towards those breaches of performance which were described by Lord Diplock in *Kaye Ltd* v. *Hosier and Dickinson Ltd* (1972): 'Provided that the contractor puts it right timeously I do not think that any temporary disconformity should of itself amount to a breach of contract by the contractor.' Any competent contractor will very quickly recognise the relative degree of expertise evident in the subcontractor's performance. If this falls short of the required reasonable level a reserve position has to be established, since the determination provisions of this or any other agreement will only be upheld if the express requirements for written notice have been observed.

Notice requirements of determination provisions

Any failure by the subcontractor to deal immediately with disconformities with the specification should be made subject to written notice requiring compliance within a reasonable time. The reasonable time will depend upon the circumstances, but usually seven days will be a minimum unless the work in question is to be covered over or put into use. The clause then provides that 'in any such event and without prejudice to any other rights or remedies, the Contractor may by written notice to the Sub-Contractor forthwith determine the Sub-Contract' with all the rights and consequences described. A written notice is effective when it is received, and it is desirable to serve notices in respect of determination by recorded delivery to the subcontractor's principal address.

The consequences of contractual determination are so major that the justification is likely to be disputed. If an adjudicator is already appointed under the contract, it might be well to have him view the works before taking the ultimate step, and this will help to ensure that the contractor is not drawn into administrative and legal expense which he may be unlikely to recover. In any event, before proceeding with the works, the contractor should make careful records of the defects by photograph and measurement, and record the existing measured work, together with every nut and bolt of the equipment he proposes to take over, and seek agreement of those records with the subcontractor.

Clause 17(2) – the contractor's rights on 'deemed repudiation' of the subcontract by the subcontractor

> Upon such determination, the rights and liabilities of the Contractor and the Sub-Contractor shall, subject to the preceding sub-clause, be the same as if the Sub-Contractor had repudiated this Sub-Contract and the Contractor had by his notice of determination under the preceding sub-clause elected to accept such repudiation.

The classic definition of repudiation of a contract is taken from *Chitty on Contracts*, p. 1601: 'The test [of repudiation] is to ascertain whether the actions of the party in default are such as to lead a reasonable person to conclude that he no longer intends to be bound by its provisions.' The common law right exists in parallel with the express terms of determination set out in the subcontract, and arises upon the occurrence of what is called a 'fundamental breach' of the subcontract. Abandonment of the works, direct refusal to construct in accordance with the specification, or defiance of legitimate instruction would be fundamental breaches by the subcontractor. Unlawful exclusion from the works or total failure to provide materials or services necessary to the works would be fundamental breaches by the contractor.

Failure in payment is a fundamental breach only where the contractor demonstrates that he will not certify or pay the sums properly due in accordance with the agreement; delayed payment is not usually of itself a fundamental breach, and the subcontractor's claim to specifically ascertained sums is usually sufficiently uncertain to cloud any representation that, upon proof of the debt, the contractor did not intend to pay. There is authority to support the view that, where the contractor deliberately omits to make interim payment altogether, this may be accepted as a repudiatory breach. (*D.R. Bradley (Cable Jointing) Ltd* v. *Jefco Mechanical Services Ltd* (1988)).

Danger of reliance on the common law right to accept repudiation

Much as the original contract was established by offer and acceptance, it is now extinguished by repudiation and acceptance. The party who 'accepts' the other's repudiation brings about 'rescission', and a discharge from further performance in accordance with the contract. The liabilities accrued between the parties prior to the rescission remain enforceable in accordance with the contract, but outstanding liabilities are not governed by the extinct terms, and the contractor is entitled to the excess of his costs and damages in completion of the subcontract works over the equivalent cost to him of outstanding works, had the subcontract been performed and completed.

Contractor's liability on wrongful operation of the determination provisions

Where the contractor is held to have repudiated the subcontract, the subcontractor may claim for the full value of outstanding measure and claims up to the date of rescission in accordnce with the contract, plus damages thereafter to include loss of profit on outstanding contracted work, and overhead loss on the basis that no immediate opportunity was available for the reapplication of his resources. If the contractor has wrongfully taken possession of materials and construction equipment belonging to the subcontractor, the damages within reasonable contemplation may run very widely, and most contractors are sensibly reluctant to apply those aspects of the contractual determination provisions without absolute necessity.

Bibliography

1. Abrahamson, Max W. *Engineering Law and the ICE Contracts.*

20

Project Management

Eur Ing Brian Syms BSc, CEng, MICE

20.1 Introduction

20.1.1 PROJECT MANAGEMENT

In order to ensure the successful construction of a scheme, the whole project requires close but efficient project management. While most civil engineering projects are prepared with a standard approach, this chapter will emphasise the needs of a sewerage project. Consequently, project management has been considered with the intention of providing a guide to avoid some of the problems and pitfalls that have occurred on sewerage schemes.

Many of the topics covered are very extensive and warrant a chapter or, in some cases, a complete publication to cover the subject in detail. This chapter will give guidelines only on an approach to project management for sewerage projects and further reading will be required to enhance this approach.

The details provided are typical of the requirements for the UK but are also applicable to other international projects, particularly those covered by European legislation.

20.1.2 PROJECT LIFECYCLE

The general concept of *project lifecycle*, as shown in Fig. 20.1, is followed although the chapter emphasises some of the more crucial topics such as environmental impact, risk and health and safety.

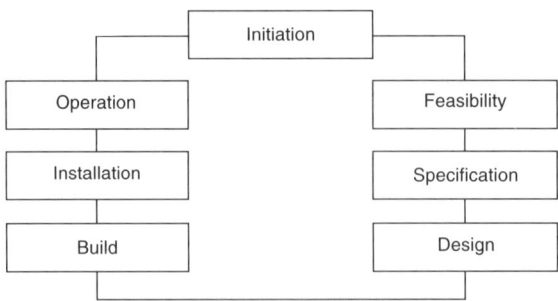

Figure 20.1 The project lifecycle.

20.1.3   CONTENT

The chapter will identify, define and detail the following topics.

Project evaluation

The evaluation of a project must identify why an investigation is required, what is required and how it is to be progressed. Once the need is established, the steps to achieve successful completion should be stated in a formal procedure which must be maintained and updated on a regular basis.

In addition, the current legislation will be examined to identify the implications on the project as a whole and on the construction phase in particular.

Feasibility

This section will look at the options available to satisfy the removal of the immediate problem. However, it is also necessary to identify other aspects which may influence the final solution such as future development. The feasibility study is the formal, written report that defines the situation and identifies the likely solution required to overcome the specific problem before describing the design criteria. At the end of the feasibility stage, the preferred option should be apparent and the conceptual design identified.

The concepts of *value management* and *value engineering* are also introduced at this stage.

The design process

The design process will identify the progression of the feasibility study from conceptual design to detailed design objectives. At an early stage, it should also be established exactly who will carry out the various stages. This decision may also influence the type of contract produced.

At this stage it is also important to identify the other services which may influence the route of a sewer, for instance, or even the type of solution, such as on-line and off-line storage. Consequently, contacts and communication links with utility services have to be established.

Risk

Risk and safety issues have become major influences during the life of a project. The risk of failure can be applied to both the financial and problem-solving aspects of a project.

CDM Regulations

Health and safety requirements have far reaching consequences both in terms of the design of a scheme and the construction of the final works. Health and safety issues must be recognised at the design stage with a forward consideration of the construction stage and to the future operation and maintenance of the completed scheme. Indeed, the proposed construction method can greatly influence the design of a sewerage pipeline or installation.

Procurement

The procurement of services usually follows an established format within a company and the larger sewerage schemes will be subject to European legislation and advertising. It is important to establish quality assurance procedures and whether formal accreditation is required for the service or product being supplied.

The contract documents should be such as to ensure good contract procedures and a good final product. A standard approach is usually made, consequently the documents do not need to explain every detail but must refer to standard or accepted procedures and products so that a potential tenderer can understand all the conditions that apply.

Construction and site management

One of the most important aspects of site management is the communication between the supervision personnel and the contractor's staff. However, in today's more inquisitive and educated society, contact with the general public must not be ignored or undervalued.

Resources cannot be underestimated and roles should be clearly defined.

All aspects of contract procedure must be identified and established early into the construction phase so that poor workmanship and trends are avoided. The commissioning of the finished product and its effect on operational procedures must be discussed with the operations group in good time to avoid delays or unacceptable conditions.

Other contract types

There are a number of ways of procuring the services to complete the requirements of a project. The admeasurement contract remains the most popular but there is gaining interest and recognition of other contract types such as design and build, turnkey, partnerships and term contracts.

Summary

A summary of the main points has been included by concentrating on the role of the project manager and the interaction of this role with all of the stages identified throughout the project.

20.1.4   HEALTH AND SAFETY LEGISLATION

The health and safety legislation detailed, particularly that for the CDM Regulations, reflects the position in 1998. Readers must be aware of continual changes to operational and construction practice that may subsequently occur as a result of EU and UK legislation.

20.1.5   WHO WILL BENEFIT?

The chapter is aimed at all those involved in the successful completion of a sewerage project. Consequently, engineers and technicians involved in the client's project management, feasibility and design stages and on-site supervision will all benefit from the content. In addition, the detail included will assist the contractor's understanding of the needs of a scheme and should help the tender bid preparation and on-site establishment.

20.2   Project evaluation

20.2.1   WHAT IS REQUIRED?

The need for a project is often fairly clear as service failures occur in a sewerage system; however, it is not always so easy to facilitate or decide what is required to overcome the problem. Perhaps 'what is required' cannot be decided until the questions 'why?' and 'how?' are considered. In addition, the end user's requirements should not be overlooked, i.e. the needs of the operator and of the operatives.

An example of looking further than the immediate problem can be emphasised by the real example of the flooding of a single garden, which upon the completion of simple manhole inspections revealed a surcharged system within a housing estate of some 200 properties. Following flow measurement and system modelling, a major scheme to the value of £1.2m resulted.

Whenever a project is being considered, there is a chance to look outside the immediate problem and identify what else can be achieved in order to improve the system or its operation within the bounds of *value for money*. Sewer operatives as well as management should be approached so that the project team can understand the existing problems as well as the future needs of operations. There is little point wasting time on the design and construction of a particular type of plant if operational experience suggests otherwise. The involvement of the user/operator must not be avoided as the practical aspects of most solutions will need to be considered. Where new technology is considered, its *value* must be explained to the user in order to win over the concept. A number of project management initiatives include the project manager/user relationship within the official project cycle.

It is important to appreciate and understand the long-term finances of a project. Taking this a stage further, *whole life costing* is a look at the cost of the project over its life and therefore includes operational costs, maintenance costs, periodic plant renewal, energy costs, etc.

It is frequently found that a new sewerage scheme is required for a number of reasons. It is important, therefore, that as with any other civil engineering project, the scheme receives the usual attention to detail in the three areas of evaluation:

- Planning and feasibility studies
- Project assessment during implementation
- Project performance review.

20.2.2   PROJECT PLANNING AND ASSESSMENT

At this stage, an evaluation of the problem should establish the way forward and identify the procedures to be addressed. In particular, whether to carry out a feasibility study, identify objectives and basic rules to follow, etc. It is often helpful to develop and follow a flow diagram. If a standard diagram is used, it is possible to omit some stages when they are not needed for a particular scheme. A typical approach as shown in Fig. 20.2 can be used for appraisal and feasibility studies.

20.2.3   FEASIBILITY

A feasibility study is required to define the problem which has been identified; investigate the real issues; evaluate alternatives and promote a recommended solution.

The steps required during such a study, as shown in Fig. 20.2, are fully defined in a number of publications and papers and will be subject to company project management procedures. However, each step should ensure that, as a minimum, the following topics are covered:

- Objectives: Define the project objectives, identify the client

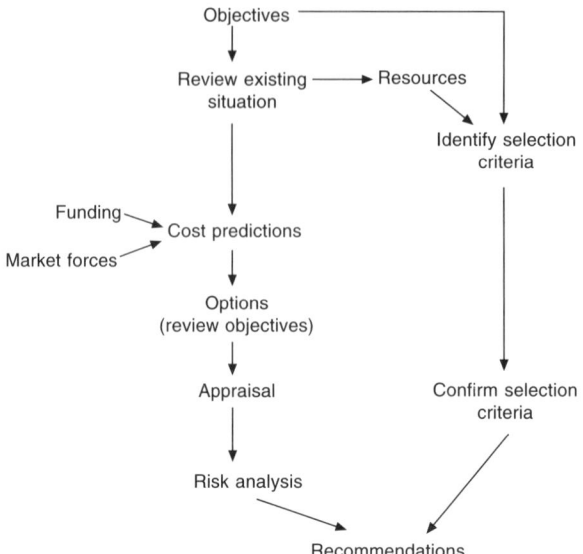

Figure 20.2 Flow diagram for a typical, standard approach to a feasibility study.

- Examine existing conditions: System failure, operator needs, customer perception
- Resources: Financial (capital investment and revenue expenditure), personnel, materials
- Identify selection criteria: Value for money, operational economics, financial return, environmental impact, social costs, political pressure
- Cost predictions: Project funding, service demands, running costs, current system performance
- Options: Identify options to meet current and future demands/legislation, effect of resources (financial, human, materials, power, etc.), review objectives
- Appraisal: Assessment of technical, financial and economic benefits, environmental impact of each option, practical issues (social, operational, planning)
- Risk analysis: Identify possible contractual risks, identify risk of failure, identify contingencies
- Confirm selection criteria: Have they been met by identified options
- Recommendations: Summarise advantages and disadvantages of each option, identify preferred option or options if further analysis or investigation is required.

It should be noted that value management is a procedure that can be applied at various stages of a project: value analysis being undertaken at the feasibility stage; value engineering studies being held during the design stage. The process will, therefore, assist the determination of the preferred option.

20.2.4 PROJECT VIABILITY

Definition

The viability of a project is generally accepted as the ability of the project to meet its predetermined objectives. It is usually financial viability that is of most importance but technical, economic and social targets also need to be addressed.

Financial viability

The financial viability of a project can be measured in a number of ways but the most common approach is a discounting technique of which there are arguably two main types: net present value (NPV) and internal rate of return (IRR); both of which use the principle of discounted cash flow. NPV and IRR are both acceptable methods for considering the financial viability of a scheme, the choice being a matter of company preference.

Discounted cash flow

The discounted cash flow (DCF) approach considers the present-day value of a scheme over the life of the asset and whether, or not, it is a long-term economical solution. DCF uses the principle that £1 today is worth more than £1 in a year's time, taking into account earned interest but ignoring inflation. While manufacturing projects use a 15–20 year project life and retailing a 5–10 year period, it is more common for sewerage projects to use a figure of 20–40 years to reflect the minimum asset life. However, when considering long-term operating schemes (see section 20.9) a figure of 15–25 years may be more appropriate, i.e. the life of the operating contract and the exercise will compare income against capital and revenue expenditure.

DCF techniques use compound interest principles to convert cash flows in differing time periods to a present-day base or value. The general formula $X = P(1 + r)^n$ calculates the value of today's 'cash' in a number of years' time where:

X = sum at the end of the investment period
P = principal sum invested
r = rate of interest
n = number of years
$(1 + r)^n$ = compound factor

Example: for an interest rate of 8%:

n	$(1 + r)^n$	Compound factor
1	1.08	1.08
2	1.08^2	1.17
3	1.08^3	1.26
4	1.08^4	1.36
5	1.08^5	1.50

The term discount factor is defined as giving the present value of a sum of money paid in the future and is the reciprocal of the compound factor defined above, consequently the formula is $\dfrac{1}{(1 + r)^n}$.

Example: for a fixed income of £10 per annum for the next five years at a fixed interest rate of 8%:

n	Income	Discount factor	Discounted income
1	£10.00	0.926	£9.26
2	£10.00	0.857	£8.57
3	£10.00	0.794	£7.94
4	£10.00	0.735	£7.35
5	£10.00	0.681	£6.81

The present-day value of the total is the sum of the discounted income and is calculated by the formula:

$$\text{Present day value} = \sum_{i=1}^{n} \frac{Ci}{(1 + r)^i} \text{ which in this example is £39.93}$$

Where r = annual interest rate
n = no. of years (project life)
Ci = cash flow or income in period i
i = cash flow period

Net present value

NPV, or Net present cost (NPC), is the method most frequently used by the sewerage industry; however, unlike the examples given above, and apart from the turnkey type of contract, there is generally no income. Again, NPV is a method of considering the value of £1 today at a date in the future or today's value of £1 from a date in the future. That is, if £1 today is the equivalent of £1 $(1 + R)^n$ in n years' time where R is the required rate of return, then £1/$(1 + R)^n$ today is equivalent to £1 in n years' time.

For appraisal purposes, the factors $(1 + R)^n$, i.e. the future value of £1, and $1/(1 + R)^n$, i.e. the present value of £1 in n year's time, are tabulated for general use, thus simplifying financial calculations.

An NPV calculation converts the annual cash flows (expenditure and income, where appropriate) and sums them to give the value of the project's earnings or savings, less the present value and should be, therefore, a positive figure. In theory, a project resulting in a negative value should be reassessed and probably abandoned. However, since most sewerage schemes will have negative NPV, unless a specific income stream can be identified, the option with the lowest NPV must be chosen. If environmental issues were considered and social/indirect costs included, the NPV would result in a positive figure.

$$NPV = -Io + \sum_{i=1}^{n} \frac{Ci}{(1 + R)^i}$$

Where Io = the initial cash outlay
n = no. of years (project life)
Ci = cash flow or income in period i
i = cash flow period
R = rate of return

Figure 20.3(a)(i) gives an example of an NPV calculation.

Internal rate of return

When an investment is made, the question 'what rate of return does this investment offer?' is usually asked. The cost of implementing a new sewerage project or scheme should be treated no differently, with the same question asked. IRR is only appropriate for income generating projects and its value is that which satisfies the requirement that the present value of the cash inflows equals the value of the cash outflows. It can be shown that the IRR is the required rate of return when NPV is zero and is given by the equation:

$$Io = \sum_{i=0}^{n} \frac{Ci}{(1+k)^i} \quad \text{or} \quad 0 = -Io + \sum_{i=0}^{n} \frac{Ci}{(1+k)^i}$$

Where Io = the initial cash outlay
$\quad n$ = no. of years (project life)
$\quad Ci$ = cash flow or income in period i
$\quad i$ = cash flow period
$\quad k$ = unknown return

Figure 20.3(a)(ii) gives the IRR for the same example as used for NPV above.

It is generally agreed that if a real discount rate on the cost of raising capital is 8%, an acceptable IRR should be greater than 10%; schemes giving an IRR of less than 10% are unlikely, therefore, to be approved.

Conclusion

The use of NPV for comparing two or more solutions to a project will provide an insight to the financial liability of a project. The method can be used to compare a high capital cost

---

Consider the net present value and internal rate of return approaches to the project where:
Initial cash outlay, I_0 = £500 000
Cash income in year 1, C_1 = £250 000
Cash income in year 2, C_2 = £200 000
Cash income in year 3, C_3 = £150 000
Rate of return, R = 8%

1. $NPV = -I_o + \dfrac{C_1}{(1+R)^1} + \dfrac{C_2}{(1+R)^2} + \dfrac{C_3}{(1+R)^3}$

$\quad = -500{,}000 + \dfrac{250{,}000}{(1.08)^1} + \dfrac{200{,}000}{(1.08)^2} + \dfrac{150{,}000}{(1.08)^3}$, from tables

$\quad = -500{,}000 + 250{,}000 \times 0.926 + 200{,}00 \times 0.857 + 150{,}000 \times 0.794$
$\quad = +22{,}000$

Therefore project has a positive NPV over three years.

2. For IRR = 0.

$\quad 0 = -I_o + \dfrac{C_1}{(1+k)^1} + \dfrac{C_2}{(1+k)^2} + \dfrac{C_3}{(1+k)^3}$

where k = unknown return
Try k = 10%

$NPV = -500{,}000 + \dfrac{250{,}000}{(1.10)^1} + \dfrac{200{,}000}{(1.10)^2} + \dfrac{150{,}000}{(1.10)^3}$, from tables

$\quad = -500{,}000 + 250{,}000 \times 0.909 + 250{,}000 \times 0.826 + 150{,}000 \times 0.751$
$\quad = +5100$

Try k = 11%
$NPV = -500{,}000 + 250{,}000 \times 0.901 + 250{,}000 \times 0.812 + 150{,}000 \times 0.731$
$\quad = -2700$

Linear interpolation is generally satisfactory for these applications, giving IRR = 10.65%

Figure 20.3 (a) NPV and IRR calculations.

scheme against a low capital but high revenue cost, e.g. a tunnelled pipeline through a hill versus a pumped solution around or over the hill. However, it must be remembered that a pumped solution adds a number of additional costs, apart from the day-to-day maintenance, such as:

- Electrical and mechanical overhaul at 10 and 30 years, which for budget purposes can be assumed to be valued at 25% of the original value of the equipment
- Complete electrical and mechanical refurbishment/replacement at 20 years.

Although the simplified example shown in Fig. 20.3(b) shows that Option B, the pumped solution, is less than half of the capital cost of Option A, the tunnelled solution, the NPVs are so close that Option A cannot be ruled out on cost grounds alone. The pumped option will require consideration of other operational issues, such as the consequences of pump

A very basic look at using NPV to consider the viability of two options is illustrated in the following example.

Option A Construct a 1 km tunnel to divert flow from one catchment to another. Assumptions made for budgeting purposes:

- design costs   = £40,000
- overall tunnelling rate, inclusive of shafts, etc. = £1000 per lin. m.
- construction period   = 2 years
- discount factor   = 8%
- future annual operations cost   = £1000 (nominal)

	Year	costs × £1000	Income/savings × £1000	Discount factor	NPV
Capital investment	0	(40)		1.000	(40)
	1	(750)		0.926	(695)
	2	(250)		0.857	(214)
			NIL		
Labour	2–40	(1) pa		10.999	(11)
Total					(960)

Option B Construct a pumping station and 1.2 km pumping main to divert flow from one catchment to another. Assumptions made for budgeting purposes:

- design costs   = £60,000
- estimated cost of pumping station   = £250,000
- overall pipe laying rate   = £125 per lin. m.
- construction period   = 1 year
- discount factor – general; power   = 8%; 6%
- future annual operations cost   = £2000 (nominal)
- power   = £30,000 pa

	Year	costs × £1000	Income/savings × £1000	Discount factor	NPV
Capital investment	0	(60)		1,000	(60)
	1	(400)		0.926	(370)
E&M refurb	10	(50)		0.463	(23)
E&M replmt	20	(200)	NIL	0.215	(43)
E&M refurb	30	(50)		0.099	(5)
Power	1–40	(30) pa		14.103	(424)
Labour	1–40	(2) pa		10.999	(22)
Total					(947)

Figure 20.3   (b) Using NPV principles to compare two project options.

failure, for instance, which may subsequently promote the tunnel as the preferred option.

A more accurate comparison will identify the most financially viable solution but operational views and construction techniques (practicalities, environmental impact, etc.) must be considered before making the final choice.

20.2.5 LEGISLATION

Health and safety

Health and safety legislation has far reaching consequences. Client/owner responsibilities cannot be ignored, site safety is no longer 'the responsibility of the contractor' alone and such topics are described in greater detail in section 20.6 below. The Health and Safety at Work Act 1974 places obligations upon an employer to establish safe working conditions for employees, e.g.:

- A safe place to work
- Safe systems of work
- Safe equipment
- Sufficient information and training to permit safe working.

As a result of EC legislation, regulations have placed a greater obligation and greater level of responsibility on the client/owner over a contractor. Six codes of practice, sometimes referred to as the '6 pack', were introduced, effective from 1 January 1993:

> The Management of Health and Safety at Work Regulations 1992
> The Workplace (Health, Safety and Welfare) Regulations 1992
> The Provision of Use of Work Equipment Regulations 1992
> The Personal Protective Equipment at Work Regulations 1992
> The Manual Handling Operations Regulations 1992
> The Health and Safety (Display Screen Equipment) Regulations 1992.

An example of the changes in responsibility can be given by the safe entry into a confined space. This is not just an operations issue since the construction of new sewerage installations usually requires confined space entry at some stage, particularly during commissioning works. A confined space is defined in 1974 Act as a space in which there is *liable* to be a dangerous atmosphere. The new regulations require the client to: identify potential confined spaces and whether dangerous substances could be discharged to the sewer; approve the contractor's health and safety policy document; confirm contractor personnel have received confined space training; ensure that the correct entry procedures are pursued.

Matters were taken further for installation and construction works with the Construction (Design and Management) Regulations 1994, generally referred to as the CDM Regulations. These Regulations placed obligations on the client, the designer and the planning supervisor as well as on the principal contractor and are discussed in greater detail in section 20.6 below.

Traffic management

Traffic management is met by the requirements of the 'New Roads and Street Works Act 1991' (NRASWA). The legislation covers signing, guarding and lighting, road surface reinstatement, traffic sensitivity (which can affect *normal* working hours) and whereabouts notices, all of which can cause delay and/or financial implications.

The training and accreditation of the workforce for the signing, guarding and lighting of a site is now a legal requirement for all supervisors and operatives working in the highway. It should be recognised that once established, the signing, guarding and lighting of a site has to be maintained 24 hours a day.

Utilities are required to give a guarantee for the integrity of the road surface to the highway authority according to the depth of the required excavation. Road surface reinstatement must be guaranteed for a minimum period of two years for trenches up to 1.5 metres and three years for depths greater than 1.5 metres. Sewers are generally laid at depths well in excess of 1.5 metres! Should the reinstatement fail an inspection, the renewed surface then follows a further guarantee of the same duration.

Minimum working widths and provision of pedestrian walkways are required when working in the highway, which for sewer construction invariably means traffic control facilities and even road closures with traffic diversions. The latter now being charged for by most highway authorities.

Some roads, or sections of streets, are defined by the highway authority as being *traffic sensitive* and as such may impose restricted working hours, usually off-peak.

Notices regarding the construction of most capital investment schemes are co-ordinated and monitored by quarterly meetings of the local HAUC, consequently there is a need to ensure early identification of a scheme to avoid a negative response from this committee. The highway authority is also responsible for monitoring all road openings, which will ultimately be recorded on a national, computerised register.

Entry onto private land

The need for land acquisition, entry and easements should be identified early and acted upon to avoid delays and disruption to progress. Section 159 of the Water Industry Act 1991 provides the formal notice to lay sewers in private land but it should be remembered that three months' notice for most construction (i.e. other than emergencies and sewer requisitions) is required.

Environmental considerations

Environmental legislation could include anything that has an effect on the environment, the latter usually being regarded as physical surroundings that are common to all of us: air, space, waters, land, plants and wildlife. Indeed, Einstein suggested: 'the environment is everything that isn't me'.

The methods of translating general legislative statements into specific provisions can be summarised as regulations, directives, decisions, recommendations and opinions. Regulations, being directly applicable laws, are the most forceful while directives specify policy objectives but leave the means for achieving them to each member state. Decisions are binding only upon those specified, whereas recommendations and opinions carry no mandatory obligations. In the field of environmental policy, directives have been the dominant legislative device.

European environmental law has centred on the EC Directive 85/337, from which environmental impact assessment (EIA) has evolved and developed to integrate into member state domestic legislation. Consequently, it is often forgotten because domestic law is usually cited as the applicable law. 'EC Directive 85/337 on the assessment of the effects of certain public and private projects on the environment' identified the need for a statement to outline the effects of certain projects on the environment – in the UK this is implemented in town and country planning legislation.

European environment policy has over 200 items of environment legislation, most aspects being reactive provisions aimed at curing existing 'ills'. On the other hand, EIA is a preventative policy to anticipate and resolve potential problems in advance and is detailed further in section 20.3.5 below.

Historically, the 'cost' of compliance was cheap, for example sewage disposal was often direct to water courses or the sea. The introduction of integrated pollution control, for instance, has resulted in: increased controls applied to discharges of prescribed substances to sewers, higher effluent standards required from sewage treatment works and the increasing control for the disposal of sewage sludge which is therefore more expensive, e.g. dumping at sea ceased in 1998; disposal to land is restricted; higher emission standards are required for incineration.

Today the cost of non-compliance can result in

- Criminal liabilities
- Administration sanctions (regulator)
- Clean-up costs
- Civil liability
- Adverse publicity.

The Environmental Protection Act 1990 has led to *The Duty of Care* regulations for the collection and removal of waste materials. Sewer 'arisings' are regarded as a controlled waste and the removal and transfer of such waste requires the completion of transfer notices and disposal at licensed sites. Similarly, the removal and transfer of excavated material and demolition waste is covered by this legislation. In addition, noise regulation can restrict working hours and is monitored by local authority environmental health officers.

The Control of Pollution Act 1974 provides, among many other objectives, the requirements to prevent the pollution of watercourses. This is not an obvious section of legislation to cover but when commissioning works that involve a CSO, for instance, the discharge of flows outside of the discharge consent parameters is not permitted. Failure to meet these requirements usually leads to prosecution and fining of all responsible parties, which includes the client as well as the contractor and is guaranteed when fish are killed as a result of such pollution.

Sustainable development

Sustainable development is not legislation as such but reflects the *recommendations* or *opinions* of others, particularly environmentalists; however, the principles have been incorporated by some governments into domestic legislation.

This term was coined in response to the environmental and development challenges that confronted the international community during the last decade and continues to feature in human activities worldwide. Ozone depletion and global warming received most attention but it was also identified that depletion of natural resources was of equal importance. One definition of sustainable development has been given as:

> Development that meets the needs of the present without compromising the ability of future generations to meet their own needs. (Mrs Gro Bruntland, Norwegian Prime Minister)

In other words, we need to continue developing but we also need to keep an eye on the resources that we use so that future generations are not deprived of them. More appropriate to sewer construction, the areas where the principles of sustainable development can be applied include:

- Traffic delays and associated atmospheric pollution
- Surplus excavated material requiring tipping space
- Use of quarried products for backfill, reinstatement, etc.
- Fuel
- Noise, dust, etc.

20.2.6   STANDARD APPROACH

There is a general leaning towards a standard approach to most services and supplies and contract management is no exception. For instance, most sewerage undertakers and large industrial companies have established methods for the procurement of services. However, those new to the procedures or wishing to improve their methods can consider the guidelines given in the British Standard BS 6079 'Guide to Project Management (1996)' and assistance from the Association for Project Managers (for address see Bibliography).

20.3   Feasibility stage

20.3.1   NEED

Sewerage projects are generally identified to resolve a basic problem or combination of problems, such as the following:

- Hydraulic: Surcharge, flooding
- Development: New pipelines, effect of flow on existing system
- Operational: Service failure, failure prevention, improved efficiency
- Legislation: Removal of flows from inland and coastal waters, pollution.

Flooding

Flooding due to insufficient capacity is probably the most obvious sign (and least acceptable by the public) that something is not right underground. The capacity of a sewer is finite and in time will reach its potential by increased water usage, additional new flows, system failure or the recent changes to atmospheric conditions. The change in hydraulic stability is not always obvious, due to surcharged conditions and/or leakage. Systems have been discovered to have survived in a surcharged condition for years with the hydraulic head in chambers pressurising the pipeline, thus forcing more flow through for prolonged periods. Flooding occurs only once the lowest point is breached. Today, we can use sophisticated methods to use the surcharged condition to its full potential and is far removed from the 'spare' capacity originally available in older combined systems.

Development

Many areas of the country are being developed with the emphasis on inner city redevelopment, out of town retail and sporting facilities and suburban housing. Inner city redevelopment often utilises the existing sewerage system which, depending on the change in runoff area, can usually cope with the change. Out of town retail parks and residential schemes provide a different challenge. Off-site drainage is required by law – the question being 'where to take the flow?' In the past, the construction of a new treatment plant may well have been the favoured solution but today, water companies are reluctant to adopt and operate new sewage treatment works, consequently pipeline construction and the subsequent effects of the additional flows on the established system are more likely.

Operational failure

Operational failure can lead to the need for new pipeline installation particularly where ground movement or subsidence has occurred. Such failure usually manifests itself in siltation, blockages and pollution but can lead to structural failure if the pipeline support is removed. The use of CCTV methods to inspect strategic sections of the sewerage system, for instance, will allow preventative action, thus reducing the likelihood of service failures. The analysis of inefficient systems often leads to sewer diversions, abandonment of treatment works, etc.

Pollution

Both UK and EU legislation are addressing the impact of sewage pollution on inland and coastal waters. There are many schemes currently being considered to remove river and coastal discharges, all of which will involve new pipelines and often pumping regimes. However, there is also a need to consider the position of surface water discharges due to wrong or illegal connections and the illicit disposal of waste as the effect can be equally devastating on the environment.

20.3.2   FEASIBILITY ISSUES

A return to the main observation 'what is required by the user' is necessary when considering the feasibility or options stage. In general, the client will require the most cost-effective solution to the immediate problem. The user/operator will usually require a simple to operate, low energy consumption solution, the operation of which can be added to the normal day-to-day operations schedule.

Flooding problems are usually overcome by the provision of a new larger diameter sewer or storage facilities. Flow measurement and the construction of a hydraulic model become an essential part of the feasibility and design stages. Flow measurement, correctly established, is not cheap and relies on a number of rainfall events to verify the model and predict storm flows accurately and hence assist the design of the alternative solutions. In UK, the climate complements the needs of flow measurement (until flow monitors are installed!) and it is normal practice to leave the monitors in place for a minimum period of five or six weeks in order to gain an appreciation of the flow characteristics of a system. However, during dry periods, the survey period may be extended significantly and the flow survey budget can easily be exceeded.

20.3.3   CONSTRUCTION OPTIONS

The construction of a new, long, large diameter sewer may be appropriate in a relatively shallow, green field site but would not be appreciated in a city centre. Construction techniques other than open-cut excavation must be considered, not only to congested areas. The development of tunnelling techniques (pipejack, microtunnel, directional drilling, etc.) make them viable in many locations.

In addition, the use of on-line and off-line storage facilities are more popular in today's *customer aware* environment. Both types of storage can include tank or oversized pipeline solutions although the former is usually used for off-site storage and the latter for on-line solutions. Self-cleansing velocities and the deposition of silt become real issues in storage systems as the basic concept is to hold up the flow and restrict its entry into the downstream system. Desilting operations can be built in as an automatic facility or it can be left as an operational function in which case the programming, monitoring and resource impact must be established and maintained. A badly maintained storage facility can quickly lead to repetition of flooding which, from a public relations viewpoint, is an embarrassment and could lead to prosecution and a fine.

Another construction option which requires further consideration is vacuum sewerage system technology. Developed largely in the USA, this technology is relatively new to the UK and little use has been made apart from in flatter areas such as East Anglia. The basic principle of *push and pull* allows a shallower, smaller diameter pipeline and therefore offers immediate savings on capital expenditure. The provision and maintenance of the equipment required at both ends and often at intermediate stages will need to be assessed against the more common approach of a gravity and/or pumped system. It is claimed that a vacuum system can work uphill, although a small gradient is preferred. Another useful 'by-product' is created by the collision of sewage waste as it moves through the pipeline, the resulting smaller solid sizes then benefiting the sewage treatment process.

20.3.4   RATIONALISATION

There is a tendency to reduce the number of sewage treatment works to lessen the risk of consent failure and the pollution of water courses. Due to the very nature that a treatment works serves a *gravity catchment*, sewage flows have to be diverted *over* or *around* a hill. The diversion of flows via gravity sewers would be preferable but a pumped solution is often cheaper than a tunnelled solution even when the operation and maintenance costs are examined.

20.3.5   ENVIRONMENTAL IMPACT ASSESSMENT

Definition

Environmental impact assessment (EIA) is the term given to the whole process, prepared by a 'developer', in the collection and systematic examination of the environmental effects of proposed development leading to approval or refusal by a local planning authority. The study is called the 'environmental statement' and contains the information gathered and assessed by the developer and submitted with a planning application, whether a mandatory requirement or not.

In the UK, information can be gathered from a number of sources in addition to the developer, such as the Environment Agency, Ministry of Agriculture, Fisheries and Food, Health and Safety Executive, conservation groups, the public and planning authorities.

Which projects?

Projects where EIA is compulsory are given in Annex I of the Directive and Annex II provides a list of projects where it is discretionary, these lists being summarised in Fig. 20.4. Compulsory projects include those processes that clearly have a major impact on the environment, e.g. refineries, combustion installations, large transport schemes, disposal of 'nasty' wastes (radioactive, asbestos, toxic), etc.

```
┌─────────────────────────────────────────────────────────────────────┐
│              Annex I – Projects which require an EIA                   │
│                                                                       │
│ crude oil refineries;                                                 │
│ gasification & liquefaction of coals and shales;                      │
│ thermal power stations & other large combustion installations;       │
│ radioactive waste storage & disposal facilities;                      │
│ integrated steel & cast-iron melting works;                           │
│ asbestos extraction, processing, products etc.;                       │
│ integrated chemical installations;                                    │
│ motorways and express roads;                                          │
│ long distance railways;                                               │
│ airports with runways greater than 2100 m.;                           │
│ trading ports, inland waterways & ports;                              │
│ waste disposal for the incineration, chemical treatment or land fill of│
│ toxic and dangerous wastes.                                           │
│            Annex II – Projects where an EIA is recommended             │
│ agriculture;                                                          │
│ extractive industry;                                                  │
│ energy industry;                                                      │
│ processing of metals;                                                 │
│ manufacture of glass;                                                 │
│ chemical industry,                                                    │
│ food industry;                                                        │
│ textile, leather, wood, & paper industries;                           │
│ rubber industries;                                                    │
│ infrastructure projects;                                              │
│ other projects e.g. holiday, sport & leisure complexes, waste disposal,│
│ storage of scrap iron, manufacturing etc.; modifications etc. to Annex I│
│ projects.                                                             │
└─────────────────────────────────────────────────────────────────────┘
```

Figure 20.4   Extract from EC Directive 85/337.

EIA is required on Annex II projects which 'are likely to have significant effects on the environment by virtue of factors such as their nature, size or location'. The three general criteria are:

- Local significance in terms of size and physical scale
- Sensitivity of location, e.g. adjacent to national park, site of special scientific interest (SSSI), etc.
- Polluting effects.

As there have been wide discrepancies between the quality and quantity of assessments carried out by member states, it is recommended that the formal screening procedure (see below) be applied to all Annex II projects.

Detail

Annex III of the Directive describes the information required in an environmental statement and should include:

- Description of proposed development (site, design, size and scale)
- Data to identify and assess likely main effects on the environment
- Likely direct and indirect effects on human beings, fauna, flora, the environment and material assets and the cultural heritage
- Description of measures taken to overcome or reduce significant adverse effects.

Additional information should include:

- Physical charactcristics of proposed development and land use required

- Characteristics of production process
- Estimated type and quality of emissions
- Alternatives considered
- Direct/indirect effects on the environment resulting from: use of natural resources; emission of pollutants; creation of nuisances; the elimination of waste.

In addition, a non-technical summary should be included for non-experts to understand the proposal.

To assist the production of an environmental statement, the Department of the Environment publication 'Guide to Environment Assessment' has a simple checklist:

- Information describing the process
- Information describing the site and its environment
- Assessment of the effects
- Mitigating measures and their likely effects
- Risk of accident.

There are three main stages of an assessment to complete before the formal assessment is submitted:

1. Screening:
   - which impacts are likely to be significant
   - magnitude and significance of these impacts, identifying which issues need to be addressed for each impact.
2. Scoping:
   - preliminary meetings with planning authorities, consultees, etc.
   - enables EIA to acquire objectivity
3. Consultation:
   - developers' responsibility to approach statutory consultees which can charge for information
   - non-statutory consultees also useful.

Commitment

Gaining commitment is arguably the most important aspect towards setting up an environmental assessment. It enables a company to take the lead and understand the basic issues, such as training. The advantages of good environmental management are:

- Awareness of legislation
- Market forces 'one step ahead'
- Avoiding problems, e.g. project delays
- Staff recruitment and morale
- Financial, e.g. waste reduction, appropriate design
- Self-interest and selflessness benefit all.

EIA is a committed phase within the feasibility stage and as such demands the correct approach. The detail required is dependent on the type and size of project involved and hence the cost of the study. However, if EIA is ignored or undervalued the consequences can result in delays, additional cost and possible failure to meet timescales. Commitment is essential at the highest level, only then can all staff be persuaded of the importance and the benefits.

20.3.6   *PROBLEM RESOLVED?*

Usually, there are many options and possible solutions and cost is frequently the most important factor. For a particular scheme, each proposed solution will have to be assessed in terms of:

- Original problem addressed
- Estimated cost, value for money
- Original problem resolved
- Operational impact
- Drainage area plan strategy
- Environmental impact
- Discharge consent criteria
- Risk assessment
- Economic benefits
- Legal issues.

The final choice is usually governed by all of the above and tends to be the most favourable economic solution (capital and revenue costs) which resolves the problem to the satisfaction of the user.

Since a great deal of the 'design' stage of a scheme has been achieved by this stage, the favoured option is referred to as the *conceptual design*. This is then used to progress to the detailed design stage.

20.3.7   *VALUE MANAGEMENT*

Concept

Value management is a multidisciplinary team-based or workshop process used for group problem solving designed to maximise the value of a project. The technique was developed in the USA over 50 years ago from what was at first a novel procurement method where items were specified solely by performance.

Value management is the generic term used to describe the whole process and although the same term is often used for the earlier interventions, from inception through to outline design, it is also referred to as value analysis in some quarters to avoid confusion. The term *value engineering* is applied to later interventions – from commencement of the detail design phase through to construction/manufacture.

Early in the life of a project, the technique can be used to gain a shared understanding of a problem or an opportunity and to give direction to a feasibility study. Later in the project lifecycle, the process can be used to examine the detailed design with the aim of reducing costs or to ensure the client/user understands exactly what the project can and cannot deliver.

Value management can be introduced at any stage in the life of a project as follows:

1. Value analysis
   - understand and define the problem
   - develop objectives
   - define the feasibility brief and consider all options
   - select preferred option and develop conceptual design

2. Value engineering:
  - consider design criteria
  - detailed design
  - pre-award of contract.

The number of workshops employed in any project will depend on its scale and complexity. It is usual to apply the technique to all projects over a specific value or to those of a complex nature.

It is generally accepted that the greatest potential for adding value is early in the life of a project. Often it is difficult to determine the scale of the benefits of early intervention whereas in the value engineering phase, it is usual to generate savings in the order of 5% to 10% of the construction cost. One must not forget, however, the cost of the process itself. Both the direct costs of engaging a trained facilitator from a specialist consultant and the cost of an off-site venue together with the indirect costs of taking senior people out of the business for a day or more (since it is recommended that a maximum of three to five meetings are held) must be balanced against the likely gains from a workshop. Usually, the number and complexity of the workshops will be a direct consequence of the scale and *nature* of the project.

The technique brings together all the interested parties, the stakeholders, under the guidance of a trained facilitator. The workshop is run in a structured manner, designed to encourage participation, to generate ideas and to achieve the agreement or buy-in of every stakeholder. In order to be successful, the workshop must include everyone who has a significant influence on the project, from the person who most understands the problem to those who will be responsible for operating the solution. Stakeholders will therefore include representatives from operations, engineering, finance and commissioning as well as technical experts and contractors as appropriate.

Structure

The process typically follows seven distinct stages:

1. Information stage: Gathering and presentation of all data, facts, etc.
2. Function analysis: Clarify project, remove technical jargon
3. Idea generation: Fast-track or brainstorming session
4. Idea evaluation: Evaluate relevant ideas inside and outside workshop
5. Idea development: Develop technical detail
6. Decision building: Relevance of each option considered, preferred solution agreed
7. Implementation: Final solution is disassembled and individual aspects reviewed to ensure objectives have been met.

The workshop is the core of the value management process but it must be preceded by a brief meeting attended by the facilitator and two or three key stakeholders to set the workshop objectives and to agree on the presentations needed for the information stage. The workshop participants will be identified at this meeting and all domestic arrangements agreed. The workshop is followed by a debriefing session to check that the actions and recommendations resulting from the workshop have been implemented and the added value realised.

Value can be expressed as a function of performance over cost as either increased performance or reduced cost can be equated to increased value. More often than not the enhanced value of a scheme through the value management process is achieved by a combination of improved performance and reduced cost. The value management process breaks down a project to allow systematic consideration of each element. It results in consensus which allows the project to go forward to its next phase while ensuring that value is maximised at every stage.

The costs of employing the value management technique can be significant but so can the savings. For instance, some regard £6000 to be an acceptable average cost. However, significant savings have been identified, for instance a £600k saving in a £4m sewerage scheme has been achieved by increasing tunnel drives to reduce the number of deep access shafts (at £50k each say) and altering access road construction to allow a cheaper method of surface water drainage and discharge to soakaways rather than to the public surface water system.

Since the decision making is widespread but collective, everyone 'buys into' the solution.

20.4   The design process

20.4.1   WHO?

In the water industry, the design process has evolved since the major changes that the reorganisation of 1974 brought. Water authorities in England and Wales were given the responsibility for the water, sewerage and sewage treatment services but due largely to the historic involvement of the local authority environmental health departments, the operation, maintenance and construction of the public sewerage system remained with the local authority under an Agency agreement. Later, many water authorities created in-house design teams as Sewerage Agency agreements were gradually terminated. Today, most Agencies have been terminated as the privatised UK water companies have taken stock of their responsibilities and schedules of work. One consequence of the vastly increased investment programme is that the water companies and the Scottish Water Authority perform project management roles while most design work is completed by consultants or consortia.

Consultants are used, increasingly, for most major sewerage and sewage treatment schemes and in many cases for packages of smaller schemes including sewer renovation. A major benefit to the client is manpower – the client can establish and maintain a manning level that will manage the peaks and troughs of a long-term capital investment programme without the disruption of detailed design and meeting time schedules. It is the consultant who has to meet programme demands although the client will still be legally bound to meet EU Directives, for instance. It is the project manager's role to ensure that all targets, i.e. time, financial and objectives, are met.

The client/owner can also dictate its involvement in the construction phase by considering a variety of forms:

- Turnkey
- Lump sum
- Measured works
- Fixed fee and costs
- Cost plus.

For example, for turnkey contracts, the owner has little involvement, risk, etc., while the cost plus contract requires the most involvement, risk and control. Similarly, project management arrangements can be covered in a number of ways from fixed fee, a variety of percentage options to full reimbursement. The degree of client/owner involvement and risk will dictate the final method.

20.4.2 DESIGN OBJECTIVES

The key objective is to meet the client's requirements and also solve the problem efficiently. The selection criteria must, therefore, be carefully identified and adhered to when appointing the *designer*. The client must, therefore, select a design team that is expected to meet the demands of the project, indeed in some cases there may be a restricted number of organisations that can supply the appropriate expertise, e.g. long, off-shore sea outfalls. Consequently, in some cases, there may be a need to subcontract the detailed design to specialists; however, overall responsibility should remain with the principal design team rather than being co-ordinated by the client's project management team.

20.4.3 UNDERSTANDING THE PROBLEM

An understanding of the existing problem is essential for the design team and should be clearly defined in the feasibility brief. Loose descriptions of apparent service failure are not acceptable therefore. It is important for the client to provide all relevant, available information and also information that may not be so obvious. Operations reports on system failure are useful for the designer to understand and *get a feel* for the problem. Sewer maps (presumed to be reasonably accurate!) are necessary, together with any known knowledge of other utility apparatus in the vicinity and other third party interference.

Information regarded as essential for most projects includes: system records, existing sewer pipe data, normal flow conditions, predicted flow conditions, local geography, location of other underground and overground utility services, other underground structures and local knowledge. Other information that will prove useful includes: manhole surveys, traffic flows, highway status, ground conditions, ground water levels, tidal reach (if appropriate) and local issues/policies/politics.

20.4.4 PROJECT MANAGEMENT

Project management should be established at the very start of a scheme and continued until the finished works have been commissioned. Progress must be maintained and monitored throughout so that actual performance against projection can be recorded. When establishing the project management team, responsibilities and delegated duties must be identified particularly for those involving decision making and budgeting.

One approach to project management is the user/project sponsor/project manager interface. Definitions for these roles are given by Northumbrian Water Limited *Project Management Manual* as:

User: 'an individual who is responsible for providing operational and maintenance advice and assistance throughout the life of the project and for ensuring that the completed facility is operated as specified following the handover'.

Project sponsor: 'the individual who is responsible for justifying the Company's investment in a defined project and is accountable for the achievement of the project's business aims'.

Project manager: 'the individual responsible to the Project Sponsor for the planning, organising, resourcing, directing and controlling of the project to meet the objectives set'

It is imperative that an idea and understanding of the construction techniques being considered is appreciated by the project management and design teams. Knowledge is unlimited and the site investigation is required at an early stage to evaluate the advantages and disadvantages of different technologies. In order to construct a pipeline using microtunnelling, for instance, recognition of the location required for working shafts and reception pits must be established. Similarly, directional drilling may require a lead-in trench to allow the pipe material to reach the required depth without inducing excessive stress/strain within the pipe walls. It is also essential to know the capabilities of each technique being considered, e.g. how far can a microtunnel be efficiently driven in differing ground conditions.

20.5   Risk

20.5.1   WHAT IS RISK?

Risk is a small word that covers a multitude of activities and responsibilities. A capital investment project can be completed successfully only if the risk has been identified and its management has been established. Risk assessment commences during the project feasibility stage and continues through its commissioning and future operation.

What is risk? Risk can be the consequences of safety, finance and problem solving and is often subjective. Risk analysis is required to identify the potential risks that can affect the project's completion and reduce the possibility of failure. This is important from all aspects since any failure could lead to financial and possible legal penalties.

20.5.2   HEALTH AND SAFETY

Operational risk

Health and safety issues are important risk factors both from a construction and operation aspect. The risk of constructing a gravity sewer in open-cut excavation at a depth of 5 m and in running sand must be compared to pipejack and pumped solutions – both from a safe working environment situation and the possibility of delays and equipment failure which incur financial penalties to the client and/or the contractor. The maintenance of a deep sewerage system adds additional risk to operational practices and hence revenue expenditure. A long pumping main adds risk at the discharge point to a gravity sewer or treatment works due to the likely presence of hydrogen sulphide – additional operational practices having to be introduced.

Machinery now has to meet the requirements of an EU directive which is generally *recognised* as the CE mark. Naturally there is a cost implication and it is accepted that machine prices have increased as a direct consequence of the legislation. However, in today's health and safety conscious Europe, it is an expense that neither the supplier/contractor nor the client/purchaser can afford to overlook. Failure to comply with the directive can lead to imprisonment and/or heavy fines.

There are health and safety implications if a project is on or passes through other private land. On major industrial sites, there may be even stricter safe working procedures and permit to work systems which will override legislation. For example, contamination of land on petrochemical sites can lead to a relatively shallow excavation being classed as a confined space, requiring a breathing apparatus entry procedure.

Construction

The CDM Regulations identified in section 20.2.5 above have far reaching consequences and as such requirements are described separately in section 20.6 below. An example of the new requirements is the 'health and safety' file. This *file* does not have a particular format but it must exist in a *permanent form*, consequently photocopied papers, faxes and even computerised (digital) details are not permitted. The planning supervisor is responsible for its preparation, upkeep and amendment and once the project is completed, the file is handed on to the user/operator.

Confined spaces

The construction of new installations will, in general, avoid *confined spaces.* However, most sewerage schemes will connect to a 'live' part of the sewerage system or be an upgrade of a system, when confined spaces will occur.

Risk assessment, if carried out correctly, will determine whether an area of work is a confined space and must ensure safe working conditions. As indicated in section 20.2.5 above, a confined space is *any location* where there is a potentially dangerous (hazardous) atmosphere. The hazard can be brought into the location by the operator, e.g. welding equipment, use of resinous materials.

Confined spaces are usually categorised into three areas of risk:

1. High risk installations are locations where isolation of any potentially dangerous materials or equipment is essential to make the space safe to enter and work in. For sewerage schemes, such installations include: sewers or wet wells where large amounts of sewage can enter rapidly; in pumping mains; culverts or tunnels which may fill rapidly. These systems require a formal 'permit to work' system which ensures, by the use of written documentation (see Fig. 20.5(a)), the risks are minimised or, preferably, removed.
2. Medium risk locations are those where flow level and atmospheric conditions normally remain safe but where precautions are required to ensure that all conditions are safe before and during entry. For sewerage schemes, such installations include: *all* sewers; submersible pumping station wet wells; inside any wet well or tank; *any* chamber 2.0 m deep or greater. A typical entry procedure is shown in Fig. 20.5(b).
3. Low risk locations are those that are generally accepted as safe under normal operating conditions and unlikely to be a danger, such as: dry wells; chambers less than 2.0 m deep; open wet wells and tanks; pipe tunnels.

One area of concern is within a sewerage system located downstream of an industrial estate where a number of chemical processes are safely used in isolation but could cause a *cocktail of effluents*, from which hazardous conditions may arise. The degree of potential danger will determine which category of risk applies.

Entry into confined spaces procedures will be required as part of a contract pre-tender appraisal and a contractor will be required to submit a procedure equal or better than that of the client or owner of the apparatus in question.

Note that legislative changes were due to be enforced from early 1998 which will change the approach to risk assessment for entry into confined spaces.

20.5.3   *FINANCIAL*

It is imperative that from the feasibility stage all present and future costs are identified and evaluated. The risk of getting the 'sums' wrong can be a problem, particularly to the

---

PERMIT TO WORK FOR ENTRY INTO CONFINED SPACES

1 Precise site details: ....
....

2 Work to be done: ....
....

3 Plant/Equipment withdrawn from service: This ....
.... has been withdrawn from service. Relevant personnel have been informed.
Time .... Date .... Signed .... Title ....

4 Isolation: The following equipment has been isolated (describe equipment and form of isolation)
....
....
....
Time .... Date .... Signed .... Title ....

5 Cleaning and Purging: The following plant has been cleaned of all dangerous materials.
Dangerous materials: ....
Method: ....
....
....
Time .... Date .... Signed .... Title ....

6 Testing for dangerous materials: Results
....
....
....
Time .... Date .... Signed .... Title ....

7 Issue: I certify that I have personally examined the plant detailed above and I am satisfied that the above particulars are correct.
*1) The confined space is safe for entry without breathing apparatus.
*2) Breathing apparatus must be worn.
*3) Safety harness must be worn.
*4) Other safety precautions necessary are ....
....
Time of issue .... Date of issue ....
Time of expiry .... Date of expiry ....
Signed .... Authorised person ....
* Delete as necessary

8 Acceptance: (1) I have read and understood this permit and all persons I supervise will work in accordance with it.
            (2) I have read and understand the 'Safe Working Procedures – Entry into Confined Spaces (Current Edition)'
Time .... Date .... Signed .... Supervisor of Work

9 Completion: The work has been completed, all persons, materials and equipment removed from the area.
Time .... Date .... Signed .... Supervisor of Work

10 Return to service: I have examined all the plant detailed and I am satisfied that it is **safe** to return it to service and cancel this Permit to Work
Time .... Date .... Signed .... Authorised person

---

Figure 20.5 (a) Typical permit to work for confined spaces entry documentation.

project management team and consistent errors can lead to personal and company dilemmas. Project estimating can be straightforward but one eye has to be kept on the market for the services required. Consequently, when a specific type of construction is required and there is a lot of that type of expertise required nationwide, market forces are likely to increase the cost of that work.

It is as important to correctly predict the future running costs of any new sewerage project both for general operations and maintenance and other costs such as energy (power) consumption or the use of other materials, such as chemicals. The water industry is actively

---

CERTIFICATION FOR MEDIUM RISK CONFINED SPACES

Location .... Date .... Time ....

Gas tests	Pass	Fail

Oxygen deficiency
Flammable gas
Hydrogen
Other ....

If atmosphere and physical conditions are safe, proceed with entry using:

Safety Harness
Safety Ropes
Carry Escape Breathing Apparatus
Continuous Atmosphere Monitoring
Complete Works

If there is an ALARM – EVACUATE confined space immediately using ESCAPE BREATHING APPARATUS and report to Supervisor.

If GAS is present and it is essential to complete the work in the confined space, longer duration working breathing apparatus MUST be used by TRAINED PERSONNEL and entry APPROVED by the authorised person.

Name of Responsible Person .... Signed ....

Figure 20.5   (b) Typical medium risk confined space documentation.

making efforts to reduce operations and maintenance costs, consequently new apparatus requiring high revenue expenditure is not welcomed unless it is unavoidable.

The cost of providing legal, planning and land acquisition services should not be underestimated particularly if a public inquiry is envisaged.

20.5.4   *LEGISLATION*

A risk factor of growing importance is that of keeping to legal time schedules. Much of the work to coastal and estuarial waters is due to EU legislation (Urban Waste Water Directive – UWWD) and as such is controlled by legal deadlines rather than finance. Much of this work was due to be completed by the year 2000 although agreement with the regulators extended some project timetables in certain cases. However, the failure to meet these basic schedules can lead not only to domestic prosecution but also action from the European courts.

20.5.5   *PROJECT SUCCESS*

To be successful, the completed scheme has to meet the original requirements. Typical questions once the scheme has been commissioned are:

- Has the risk of flooding been removed or reduced to an acceptable level?
- Has the pollution of a Class 1 river been removed?
- Is the system more effective and efficient?

It should be concluded, therefore, that risk assessment is a complicated exercise and that no single element of risk is really more important than another. The success of risk assessment is in the recognition of all elements of risk associated with a project and the subsequent management and, hopefully, reduced chance of failure.

20.6   The Construction (Design and Management) Regulations 1994

20.6.1   CONCEPT

The Regulations, referred to as the CDM Regulations, place duties on clients, designers and contractors and also on a new role and duty holder, the planning supervisor. The Regulations also introduced new documents – the health and safety plan and the health and safety file. As a consequence, the duty holders all have roles and responsibilities to ensure that health and safety requirements are identified and managed correctly throughout all stages of the project.

The health and safety plan is considered in two stages – pre-tender and construction. The pre-tender plan collates the health and safety information available to the client and the designer and helps the selection of the main or principal contractor. The plan produced for the construction phase will identify how the construction work will be managed to ensure health and safety throughout the duration.

The health and safety file is a record of the appropriate information for the client and must address risks during construction and future maintenance, including repair and cleansing. The Construction Industry Advisory Committee (CONIAC), for the Health and Safety Commission, produced guidelines (1995) which approached the topic with the following five stages:

- Concept and feasibility
- Design and planning
- Tender/selection stage
- Construction phase
- Commissioning and handover.

20.6.2   DEFINITION

The importance and impact of these Regulations on construction work cannot be emphasised enough. The definition of *construction work* within the legislation is far reaching and includes many day-to-day activities not normally associated with new installation.

The definitions refer to work carried out on a structure and from a sewerage aspect include:

- Shafts and tunnels
- Bridges
- Sewers
- Sewage works
- Drainage works
- Formwork and falsework
- Scaffolding
- Work on any plant where there is a risk of falling 2 metres.

Within the confines of sewerage the activities included are:

- Construction and commissioning
- Renovation and repair
- Demolition of a structure

- Installation, commissioning, maintenance, repair or removal of services
- High pressure water cleaning (jetting).

20.6.3 APPLICATION OF THE REGULATIONS

Clearly, CDM affects a wide ranging field of works but not all activities, the main trigger really being associated with size or scale of a project. CONIAC (1995) provided flow diagrams to help determine whether CDM applies (partially or wholly) and whether the Health and Safety Executive (HSE) has to be notified of the works in question. Figures 20.6(a) and 20.6(b) are reproductions of these diagrams and Fig. 20.6(c) shows the form of notification.

The CDM Regulations apply to most construction work, including:

- New build construction
- Alteration, maintenance and renovation of a structure
- Site clearance
- Demolition and dismantling of a structure
- Temporary works
- All associated design work.

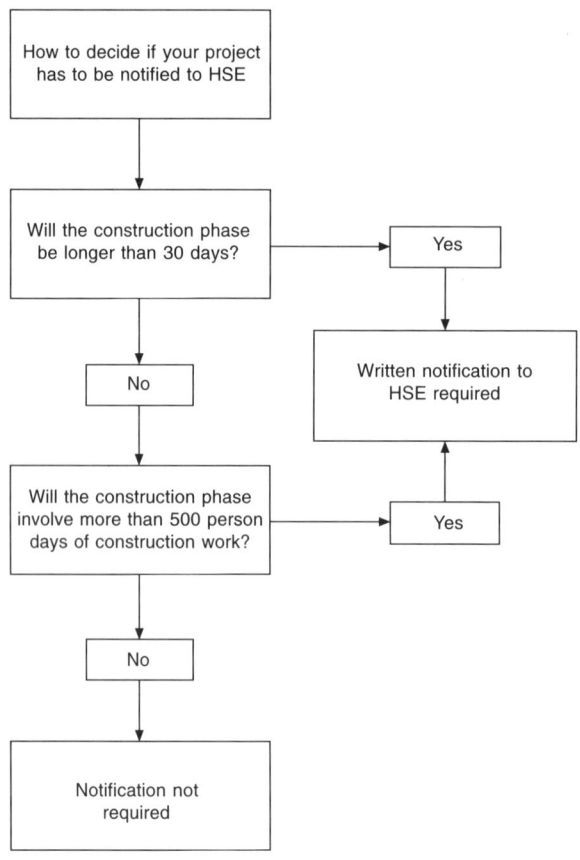

Figure 20.6 (a) HSE written notification flow diagram.

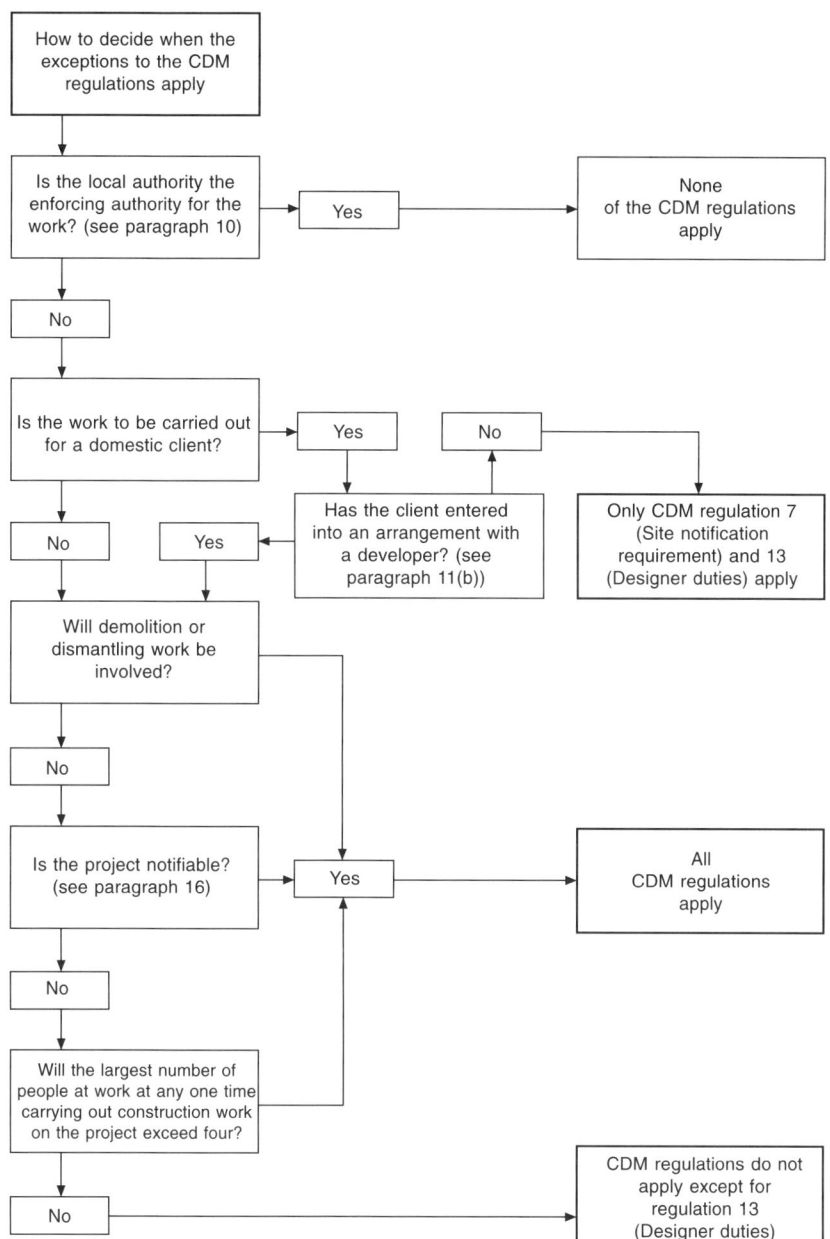

Figure 20.6 (b) HSE CDM application flow diagram.

Executive
Notification of project

Note

1. This form can be used to notify any project covered by the Construction (Design and Management) Regulations 1994 which will last longer than 30 days or 500 person days. It can also be used to provide additional details that were not available at the time of initial notification of such projects. (Any day on which construction work is carried out (including holidays and weekends) should be counted, even if the work on that day is of short duration. A person day is one individual, including supervisors and specialists, carrying out construction work for one normal working shift).

2. The form should be completed and sent to the HSE area office covering the site where construction work is to take place. You should send it as soon as possible after the planning supervisor is appointed to the project.

3. The form can be used by contractors working for domestic clients. In this case only parts 4–8 and 11 need to be filled in.

HSE – For official use only

Client	V	PV	NV	Planning Supervisor	V	PV	NV
Focus serial number				Principal contractor	V	PV	NV

1. Is this the initial notification of this project or are you providing additional information that was not previously available.

   Initial notification   Additional Notification

2. **Client:** name, full address, postcode and telephone number (if more than one client, please attach details on separate sheet).

   Name: Telephone Number:

   Address:

   Postcode:

3. **Planning Supervisor:** name, full address, postcode and telephone number

   Name: Telephone Number:

   Address:

   Postcode:

4. **Principal Contractor:** (or contractor when project for a domestic client) name, full address, postcode and telephone number.

   Name: Telephone Number:

   Address:

   Postcode:

5. **Address of site:** where construction work is to be carried out.

   Address:

   Postcode:

Figure 20.6   (c) Notification of project.

6. **Local Authority:** name of the local government district council or island council within whose district the operations are to be carried out.

---

7. **Please give your estimates on the following:** Please indicate if these estimates are original   revised

(a) The planned date for the commencement of the construction work:

(b) How long the construction work is expected to take (in weeks):

(c) The maximum number of people carrying out construction work on site at any one time:

(d) The number of contractors expected to work on site:

8. **Construction work:** give brief details of the type of construction work that will be carried out.

---

9. **Contractors:** name, full address and postcode of those who have been chosen to work on the project (if required continue on a separate sheet). (Note this information is only required when it is known at the time notification is first made to HSE. An update is not required).

Declaration of planning supervisor
10. I hereby declare that (name of organisation) has been appointed as planning supervisor for the project.

    Signed by or on behalf of the organisation _____
    (print name)

    Date _____

Declaration of principal contractor
11. I hereby declare that (name of principal contractor) has been appointed as principal contractor for the project (or contractor undertaking project for domestic client).

    Signed by or on behalf of the organisation _____
    (print name)

    Date _____

Figure 20.6   (c) (*Contd*)

However, the CDM Regulations do not apply if the work is due to last 30 days or less and involves four or less people on site at any one time or the work is to a private residence. In addition, they do not apply when the local authority is the enforcing health and safety authority and work is carried out within shops and offices or involves work to heating and water systems.

20.6.4 FEASIBILITY

During this phase the client's main tasks are to:

- Determine if the project falls within the scope of the Regulations (Fig. 20.6(b))
- Appoint a planning supervisor
- Ensure competency of planning supervisor and designer
- Provide planning supervisor and designer with information relevant to the health and safety of the project.

Planning supervisor

The client must appoint the Planning Supervisor at an early stage, the role being carried out by the client or a consultant or contractor specialising in the role.

The selection of the planning supervisor is critical to the successful co-ordination of the health and safety aspects of project planning and design and the duties can be summarised as follows:

- Assess the competence and allocation of resources of the designer
- Ensure that the designer complies with the duties, i.e. the identification, avoidance and reduction of risk
- Ensure co-operation between designers (where appropriate)
- Prepare the health and safety plan prior to the appointment of the contractor(s)
- Assess the competence and allocation of resources of the contractor
- The project is notified to the HSE (if applicable)
- The health and safety file is prepared, maintained and given to the client at the end of the project.

Designer

The designer plays an important role during the project to ensure that the health and safety of all those involved with the scheme during construction and future maintenance is addressed during the design phase, i.e. to avoid risk to contract labour and user operatives. An example of a situation to be avoided being the design of a sewer or pumping main which, due to inadequate velocity or the length of a main, increases time retention and, as a consequence, either methane and/or hydrogen sulphide (the latter at the discharge point of a pumping main) is encouraged. At all stages and where different teams of designers are employed, all involved have a contribution to make.

The main duties include:

- Make client aware of its duties
- During the design phase, identify hazards and risks that may arise during construction and future maintenance
- Design to avoid risk to health and safety
- Reduce risk at source if avoidance is not possible
- Consider risk control measures to protect operatives if avoidance or reduction of risk to a safe level is not possible
- Ensure the design includes adequate information on health and safety issues for those who require it for inclusion in the health and safety plan
- Co-operate with the planning supervisor and any other designers involved with the project.

20.6.5   DESIGN

During this stage the client must provide the planning supervisor and the designer with appropriate details to the health and safety of the project and be satisfied that the designer is competent to continue with the design. In addition, the planning supervisor must advise the client when appropriate, continue the supervision and co-ordination of the designer(s) and ensure the health and safety plan and files are prepared. The designer, meanwhile, must identify the significant health and safety hazards and risks, consider risk control and co-operate with the planning supervisor and other designers.

20.6.6 TENDER PHASE

The appointment of the main contractor, or principal contractor, is critical to the overall success of the construction phase. As previously discussed, pre-tender qualification procedures can eliminate those contractors which do not meet the health and safety requirements by considering a company's health and safety policy.

Due to the demands of the CDM Regulations, it is important to allow tenderers sufficient time to consider the health and safety implications of the tender documentation. The documentation should ensure that all known risks, etc. are identified and indicate how the contractor may be expected to deal with potential hazards.

The pre-tender stage health and safety plan should address at least the following detail:

- A general description of the work including the programme of works
- The significant health and safety risks
- The standards to be applied to control significant health and safety risks
- Information to demonstrate competence to the client
- Other specific details required by the client, e.g. monitoring systems
- Information for preparing the construction safety plan.

The prospective contractor should provide the following documentation:

- Company health and safety policy (if not already submitted at the pre-qualification stage)
- An outline of the health and safety provision required by the tender documents
- Identify the risk control and management methodology
- Supply evidence of competence with regard to the Regulations.

It is important that the post-tender health and safety review includes careful consideration of the rates and programme of works as low rates and/or a very short contract period could indicate a relaxed approach to the CDM Regulations.

20.6.7 CONSTRUCTION PHASE

The first stage, and arguably the most important, is the appointment of the principal contractor who is responsible for the safety of the site. The client must ensure that the contractor does not commence on-site activities until its health and safety plan has been approved and has demonstrated competency and compliance with the legislation. The planning supervisor and the designer will continue the duties already identified but directed at this phase.

The contractor is now faced with complying with the Regulations and its key duties are:

- Develop, implement and maintain the health and safety plan
- Ensure resources are adequate and competent, including any subcontractor's personnel
- Ensure the co-ordination and co-operation of other contractors including risk assessments and method statements for any high risk operations
- Ensure site workers have received the appropriate training, are subsequently properly informed and are regularly consulted
- Ensure that all site workers comply with the rules set out in the safety plan
- Monitor health and safety performance
- Ensure that only authorised people are allowed on site
- Display the notification details submitted to the HSE

- Pass on information to the planning supervisor for the health and safety file.

The health and safety plan for this phase must address the following issues:

- Arrangements for ensuring the health and safety of all who may be affected by the works
- Managing health and safety of the construction and monitoring of compliance with legislation
- Information about welfare arrangements
- Collation of relevant information for future use including *as built* drawings, construction methods and materials used, details of equipment and maintenance facilities and procedures, operation manuals and the identification and location of all other utilities and services found during the contract period.

20.6.8   *COMMISSIONING AND PROJECT REVIEW*

Commissioning

Many of the tasks carried out during the construction phase are continued through commissioning, which for some projects may be lengthy and require an additional contract. The risks associated with the commissioning of a new sewer installation will invariably involve *live* flows and require particular safe working procedures to address the *snagging lists* and the commissioning works. The commissioning of a pumping station must be carefully planned with the operations group as it will immediately affect the operation of the receiving sewerage system, this is particularly important if large flows are involved and the discharge is to man-entry sewers where sewer gangs could be working.

Once again, sufficient time must be allowed for the commissioning works to be completed safely particularly where electrical and mechanical systems are involved.

All information gathered at this stage must be passed on to the planning supervisor for inclusion in the safety file and will include the operation manuals referred to above. The completed safety file is then handed on to the client or system user.

Project review

The feedback from all the parties identified in this section is important to evaluate the adequacy of existing health and safety management systems and procedures and to identify amendments which can influence subsequent projects and management systems. Learning from experience will allow the client to manage risks more reliably and effectively.

It has been found in some cases that if standards are addressed at an appropriate level, risk control can identify financial savings when health and safety management systems are introduced.

20.7   Procurement

20.7.1   *PROCEDURES*

Once a project has been identified and designed, how do we 'buy' it? Today, procurement procedures are applied to all goods and services required for project completion and should define the aims, criteria and restrictions of a company's strategy. It is usual for sewerage contracts that the successful contractor is responsible for the purchase of all equipment and materials unless there are items that are required from a specific supplier. This usually simplifies contractual responsibilities although it is not always the case, particularly when complicated pumping regimes and/or specialised equipment are required.

Procurement of services can involve a large number of contracts: materials and supplies; equipment; contractor; subcontractor; contract packages (including design, contract management, etc.). The procurement procedure must also: identify the type of contract preferred; provide all appropriate information; define the requirements of the contract; identify any pre-qualification stage; identify any special conditions. The estimated value of the contract will also determine how the tender will be advertised to comply with the EU regulations for the purchase of services.

20.7.2   TENDER LISTS

It is common practice to have identified a number of suppliers, for both materials and services. These select lists should not remain static – material supplies and service contractors should face a continuing appraisal system to ensure that value for money and quality are maintained. The assessment should include not only price but also quality, delivery punctuality and reliability.

Quality assurance is a process through which a supplier can demonstrate its reliability to consistently provide a service or a product to a required standard. Many clients, therefore, now have tender lists which are dependent on the upkeep of contractor assessment with information covering both financial status and capability. These lists are usually kept by the client or it may prefer to use the services of a separate consultant or a central organisation say, which consequently reduces the involvement/workload of the client. Quality assurance plays a major part of the selection process and those suppliers, contractors and consultants with BS 5750 (ISO 9001) accreditation will have a head start against those which do not. However, the award of the accreditation should also be assessed as it may not be for the product or service required. In ISO 9001 the quality system is required to provide a means for ensuring that products conform to specified requirements. It does not link the quality policy to the quality objectives and system except by definition identified in other standards.

Due to the recognition of environmental issues in modern-day construction, it is as appropriate to consider the supplier's environmental policy and whether BS 7750 (ISO 14001) is being pursued. This standard requires the system to provide a means of ensuring that the effects of the organisation's activities conform to its environmental policy and associated objectives and targets. It avoids, therefore, any misunderstanding by requiring the management system to ensure conformity to its policies and objectives.

Quality assurance is a major topic in itself and cannot be considered half heartedly. The main requirement is commitment and this must start at the top and be supported throughout all levels of a business.

20.7.3   PROCUREMENT PLAN

The procurement plan must identify the conditions of purchase; ordering procedures; inspection and quality control procedures. It should also warn prospective suppliers of any political restraints, e.g. avoidance of products from a country faced with a trade ban. Currency issues, particularly exchange rates, are most important when any tender/contract advertisement is international.

Care must be taken when considering the advertising of a tender. Within the EU, privately owned utilities must advertise construction works in the *European Journal* if the project value is Euro 5m (approximately £4m). There are two issues to be aware of:

- The value of the £ against the Euro
- If the project value is £4m, i.e. inclusive of feasibility and design costs, the actual contract value may well be between £3.0m and £3.5m.

In order to comply with the legislation, it may be appropriate, therefore, to advertise all contracts valued above £3.5m in the *European Journal*. However, company procurement rules will identify a de-minimus value, above which all projects will be advertised.

20.7.4   TENDER DOCUMENTS

Tender documents are usually produced to a standard particularly within the water industry and major industrialists. However, there has been a change from dictating to the contractor how a scheme should be completed and identifying every single item in the bill of quantities. For example, the construction of a manhole is usually a single item which is easier for all when compared to shuttering items that used to include the *bullnose* finish to the incoming and outgoing pipes!

Tender documents should ensure, in clear English, that the tenderer is fully aware of what is required and every imposition that may inhibit progress which should reduce the likelihood of claim situations. It is also common practice to ask the tenderer to include a *method statement* which is particularly useful when considering safe working procedures and new technology.

A visit by each tenderer to the site with the project manager is highly recommended, if not compulsory, as practical issues for both parties can be observed, discussed and usually resolved before the appointed contractor arrives on site.

20.7.5   TENDER ANALYSIS AND AWARD OF CONTRACT

Tender appraisal

Tender analysis can be an extensive exercise but it is now common practice to include some criteria at the pre-tender selection stage. Items such as insurance cover, available equipment and company safety policy should be included at the select list stage to ensure that the companies being asked to tender are not going to waste time for both client and contractor.

In larger companies the procurement methodology will include a standard approach to tender analysis or appraisal. This will ensure that the arithmetic is correct, that all items have been addressed (not all items may be priced as contractors frequently add *inclusive rates* for a product) and that the contractor has complied with the tender requirements.

Cash flow

It is also important to consider the cash flow. An even cash flow throughout the contract period is unlikely as progress and the type of work tends to be slow at the beginning of a contract, often rising to high spend for the middle half of a contract. Again it is usual to provide the client's finance department with a spend profile and it is obviously preferable to avoid large peaks and surprises.

Method statement

The method statement must be examined closely – the cheapest tender may not always be the most appropriate submission. Very low rates should trigger alarm bells. A contractor must be allowed to make a reasonable profit, consequently the client should not necessarily be looking solely for the cheapest option. Certainly, the client will want value for money and the lowest rates will tend to identify which contractor has this to offer. However, rates that are regarded as too low should be avoided as the contractor will be under financial pressure from the start of the contract and may become claim conscious. It may be necessary to interview a number of contractors to discuss their proposals particularly where health and safety issues or the public are concerned. Two or more contractors may have made similar financial bids and the client can look for an opening, even a rate discount, in order to identify the preferred submission.

Award of contract

The award of the tender should not be delayed even if the start of construction is some time away. It may be preferable to delay the start of a scheme, such as after the cutting of crops in a field or following the popular period for a tourist attraction but that should not affect the award. A contractor needs to know as soon as possible if it has been successful and when it is expected on site as it has a major exercise to organise the labour, plant and materials and sufficient time should be allowed by the client for this mobilisation.

Value engineering

Value management techniques can be applied at any stage of a project, including the post-tender period where the contractor who submitted the most acceptable tender is invited to contribute to a workshop where the aim is to reduce costs but not the contractor's profit!

The team would include the earlier workshop team with the tenderer, the quantity surveyor and any other technical experts as required.

The concentration of ideas will be towards specific parts of the project where savings can be made and/or the functions can be improved. Examples of these include:

- The positioning of new construction with reference to existing apparatus, particularly when considering commissioning works
- Siting of working compounds and storage of materials
- Maximise tunnel drives to reduce the number of access shafts and also to minimise the environmental impact (e.g. sporting facilities, allotments, etc.)
- Use of 'one-pass' tunnel linings
- Use of caissons rather than bolted segmental access shafts
- Evaluation of specialist services provided.

20.7.6   CONTRACT DOCUMENTS

Good quality contract documents will undoubtedly help to ensure a satisfactory conclusion. There have been many advances in standardisation and an approach to quality. The majority of companies and authorities will follow the principles of the ICE Conditions of Contract (5th or 6th edition), Institution of Chemical Engineers contract documents or others as considered appropriate to the scheme, together with the use of the Civil Engineering Standard Method of Measurement (3rd edition) (CESMM3) for the schedule of rates. All standards will require specific clauses to accommodate the needs of the sewerage business. For sewerage contracts, specifications are identified in the Civil Engineering Specification for the Water Industry (latest edition) (CESWI).

The I.Chem.E. documents have been introduced to sewerage projects although they are often misused. The *Green Book* is the 'model form of Conditions of Contract for process plant for cost reimbursable contracts' and is a flexible approach to a variety of the different types of contract detailed in section 20.9, particularly the partnering type of contract. It avoids confrontation and is regarded as *user friendly* although there should be a significant rewriting to be job specific. The *Red Book* is the 'model form of Conditions of Contract for process plant for lump sum contracts'. It was really developed for production or process plant contracts but has been used for mechanical and electrical works on sewerage or sewage treatment schemes. It is not regarded as user friendly and should, arguably, be avoided for sewerage schemes since the clauses are not really appropriate for civil engineering works.

Contract documents should cover all aspects of the project, not necessarily in fine detail but refer to the health and safety issues, legal requirements, traffic requirements, available land, available records, etc. Reimbursement of compensation claims and contractual variation claims can become costly and the early attention to detail will prove beneficial.

20.8 Construction and site management

20.8.1 SCOPE OF SITE MANAGEMENT

Site management offers its own list of activities and priorities and a pre-contract meeting should address a number of issues such as:

- Site set-up
- Resources and roles, communications
- Value engineering (see 20.7.5 above)
- Programme of works, recording and monitoring
- Quality control and inspection
- Safe working procedures, health and safety
- Payment
- Commissioning, review.

20.8.2 CONSTRUCTION SITE

The construction site must be large enough to accommodate the works without endangering those on site but not so large as to make it difficult to manage. Site offices should be as close to the working site as possible, preferably within the same site boundaries. However, this will not be possible when sewers are constructed over long distances, although it is recommended that additional offices for contract and site supervision staff are provided.

Projects which have a high profile may require the provision of permanent public relations presentation areas, hence safe access by the general public must also be provided.

The storage of materials requires some thought. If on-site concrete batching is involved, there must be sufficient room to store and prepare materials so that the process is safe from contamination and avoids environmental *stress* to the surrounding *space*. Materials have value, e.g. manhole covers and frames, stainless steel products, electrical cable and components, fuel oil, etc., and must be stored safely and securely. Material storage areas must not inhibit the progress of work on site. The transport of materials to distant areas of the project should be considered particularly if it involves the use of public highways.

20.8.3 STAFFING RESOURCES

The site establishment should reflect the magnitude, nature and importance of the project. This is of equal importance for the supervisory staff as the contractor's on-site resources. Employment efficiency policies are commonplace but the underestimation of on-site resources can have dire consequences. There is a need to provide not only the right numbers of staff but they should also be appropriately trained and experienced. Yes, there is the opportunity to give invaluable training to younger engineers and technicians but only if experienced people are on site to provide it. Arguably the most important site supervisor is the clerk of works role while the contractor's general foreman and leading hands are most crucial.

Roles are important as are the points of contact and communication. At the pre-contract stage, it is necessary for both parties to establish who is responsible for what and how communications are to be exchanged. For instance, all contractual changes or amendments to the programme must be in a written form and evaluated as soon as possible. These relationships are also important to ensure cost control systems are maintained and monitored correctly.

20.8.4 PROGRAMME OF WORKS

A programme of works will need to be established but once agreed, continual and regular monitoring and updating are required. Monitoring and updating rely on the assessment of daily reports and records, consequently the establishment of systems needs to be quick and simple to operate and maintain. Daily records will include the usual items for progress and use of materials but more specialised contracts will require more specific records. On tunnelling contracts, attention to wriggle diagrams should give an indication of workmanship and give advanced warning of problem drives. Similarly, the monitoring of grout used for filling voids or abandoned sewers on a measured basis is imperative to keep abreast of trends and the avoidance of shocks – if larger than predicted volumes are being used other apparatus or facilities may be interfered with and there may be a need to readdress that part of the contract.

20.8.5 INSPECTION OF WORKS

Quality control systems help to ensure that the completed works will achieve the required standard(s). Such systems can take a variety of forms to cover the works and techniques involved to complete a project from simple material quality checks to the inspection of *high tech* control panels. As previously identified in section 20.7.2 above, the use of companies accredited to BS 5750 and/or BS 7750 will give added security and faith in a product or service.

Responsibilities not to be ignored include those required by *The Duty of Care* and NRASWA as previously identified, as the consequences are unwanted and are, by and large, avoidable. It should not be forgotten that the client has responsibilities for these issues and usually 'blame' cannot be attributed totally to the contractor.

20.8.6   SAFE WORKING PROCEDURES

On-site health and safety issues, as described in section 20.5.2 above, need to be identified and discussed at the pre-contract meeting so that on-site procedures can be established from day one and maintained throughout the contract period. There is little excuse in the eyes of the Health and Safety Executive if identified safe working procedures are ignored! Again, it should be remembered that it is the legal duty of the client and the contractor to adhere to and monitor the agreed procedures.

20.8.7   PAYMENT

Regular invoicing and payment are important matters, particularly to the contractor but also to the client for cash flow purposes. A contract is not established to make a contractor struggle – a well-run contract requires the right approach from both contractor and client for the scheme to be completed cost effectively for a sensible price and enable the contractor to gain reasonable profit.

   The formal presentation of invoices on an agreed, regular basis and the prompt payment as identified in the contract documents will avoid confrontation and help towards maintaining a good working relationship. Agreement of measured quantities prior to the submission of an invoice is a good practice for the supervisory and contractor staff to achieve as it assists the smooth progress of the payment procedure.

20.8.8   COMMISSIONING AND REVIEW

The commissioning of the new works should involve the operator, particularly if there is an effect on the existing drainage system. Attention to confined space entry is likely as most sewerage contracts will involve the diversion of flows to and/or from the new works.

   Project review is included in this part of the chapter as a reminder that it should be part of the general management of a project. For the majority of sewerage schemes, a project review team should include representatives from the project manager, design team, site staff and operations group. In this way, the concept can be addressed by all participants with the obvious question 'has the problem been resolved for the approved budget' being the main topic on the agenda.

   Sensible, constructive comment will allow all parties to learn from successes and failures. For large contracts or where major problems or failures have occurred, it may be appropriate to use an independent group, possibly from an external source, to avoid direct hostile criticism and embarrassment within an organisation. However, the group must include an understanding of the problem, the systems used, the appropriate technology and the aims of the client or owner. Lessons learned can then be incorporated into the company policy and objectives, where appropriate, to make future contract procedures more efficient.

20.9   Other contract types

20.9.1   Introduction

The move away from the more conventional contract procedures which rely on a tight specification, schedules of work and a tendering procedure has gathered momentum in recent years. Due to the large volume of work required in a short period of time, there is a willingness for sewerage undertakers to enter into contractual arrangements which will increase 'productivity' and in doing so, share the risk.

The term *turnkey* is used to identify a provider of a service, consequently the contractual arrangements found within this category are varied in detail and complexity. The types of arrangement vary from a simple *design and build* format to the *private finance* initiatives that have been promoted by international projects such as some European nations' municipality-owned sewerage systems which have been operated and maintained by private finance and, as more recently highlighted, by major projects in Central and South America and Asia. The contract types identified in the following section are not exclusive nor extensively described. Most have complications and must be approached with some thought and probably require assistance from experts.

Most of the arrangements are based on the design and build principles which are then 'extended' to introduce the operation of a system or process. Risk is therefore increased from the usual technical and quality standards to include income, which is the key attraction to all operational type contracts. If the income element to a contract is low, then the profit to a prospective contractor will be limited and the likelihood of a good product and service diminishes. The skill is to identify a potential project by identifying a problem or need and then promoting a viable solution that will solve the problem or satisfy the need while rewarding the product or service provider.

20.9.2   *DESIGN AND BUILD*

Consortium

Some of the larger projects involve design and build contractual arrangements. In such cases, it is usual for a number of specialists to form a consortium for the duration of one or more projects. A consortium could consist of a design capability, project management expertise and a specialist contractor for instance. When electrical and mechanical expertise is required, additional members of the consortium should be included.

Design and build can be associated with conventional style or turnkey contracts, often to overlap engineering with construction. Hence it is more likely to be applied to a sewage treatment facility rather than straightforward sewer construction, although the sewerage transfer scheme may be part of the treatment contract.

Client's role

The client can issue a design and build contract on conventional lines although the statement of requirements can vary. For example, the documents could describe a similar project or be in the form of a specification, together with the conceptual design, either of which can then be developed within the contractor's scope of work and expertise. Alternatively, the client can invite competitive proposals, the contractor being required to submit a proposal based on conception statements plus an indication of the quality being offered within an agreed contract sum.

Both of these options have relatively narrow definitions and are the type more commonly used in building and road construction. However, there is a trend towards treatment facilities and transfer works being serviced in this manner.

Costs

Construction contracts usually have a cost ceiling set by the contract price but with some flexibility to which 'quality' is attributed. However, 'quality' is subjective as the design element is usually subcontracted to another party which will have its own ideas on quality.

A design and build contract will include some or all of the following criteria:

- Predetermined cost
- Encouragement of new technology
- Environmental impact reduction
- Construction risks
- Use of comparable project services or supplies
- Timescales.

The basic theory assumes that the type of project, which may have been completed elsewhere, can be improved upon, completed sooner and at a lower cost, while maintaining or improving on the required quality and standards.

Risk

Most of the risk is held by the contractor, although the client may be subjected to an inquiry if the project does not meet legislative standards and timescales. Assuming that the contract documents are fair and the timescales reasonable, the contractor will shoulder the risk element of not meeting these targets.

Risk is associated with the design of the scheme as referred to above. It is likely that the design will be subcontracted to a specialist design consultant or contractor (e.g. pumping facilities) and risk will be attributed to technical capabilities and quality of the product or service. In addition, there is a risk of exceeding the contract period and/or budget as a result of a design issue.

Confidence

Due to this change in approach, client confidence takes time to achieve an acceptable level. The client, designer and contractor(s) must work closely together to develop a working relationship that relies on trust throughout the lifecycle of a project. As has already been identified for good project management, there needs to be a common understanding and agreement of the management, design, construction, cost and time objectives from the start of the project to its conclusion. Within these objectives, the use of subcontractors or specialist designers should be addressed when the design strategy is defined. Finally, the involvement of the user or operator during final inspections and commissioning is imperative.

20.9.3   *BUILD, OWN AND OPERATE*

The design and build concept can be taken a stage further into *build, own and operate* (BOO) or *build, own, operate and transfer* (BOOT) arrangements. As the names suggest, the objective is for the contractor to own and operate the system, usually for a finite period of time before transferring ownership to the client organisation. Many sewerage applications of this type are those township operation and maintenance contracts located in Central and Southern America and Asia. However, the same principles can be applied to major sewerage systems and/or sewage treatment facilities in Europe. In order to be successful, there has to be an income for the contractor. In the case of operation and maintenance contracts, the income is usually from billing the public who receive the service. A treatment facility can gain income, for providing the service, on an effluent strength and volumetric basis and could, therefore, be particularly profitable where chemical or toxic waste processing is involved.

Risk

With these arrangements, the risk factor lies with the service provider. For the client, infrastructure investment is secured by giving the opportunity to a concession or service company to provide the design, construction, operation and maintenance of a service for a specific period of time, at the end of which the infrastructure and service is transferred to the client, usually free of charge. There may be a case for residual value to be considered but in the case of a long-term concession, say 15 to 25 years, this is likely to be nil.

The contractor is faced with a complex task in which there is a mixture of financial and contractual arrangements to identify, establish and manage. The promoter of the project therefore has a much greater role than for a conventional contract as the project is likely to include:

- Feasibility study
- Planning approval
- Financial backing
- Quality standards
- Construction
- Operation and maintenance
- Income.

20.9.4 PARTNERING

Concept

The basic success of partnering lies with the desire of professional bodies, client and contractor, to complete projects as efficiently and cost effectively as possible within identified standards. It also leads away from the 'claim' and 'extras' issues that are invariably associated with conventional construction arrangements, thus reducing additional cost to both client and contractor.

Partnering arrangements rely on trust, which is shown by the client by sharing knowledge with the contractor(s). In return, the contractor demonstrates trust by completing the job properly without risk to the arrangements. In the water industry, partnering is relatively new but a number of water companies are using this approach with specialist contractors for major contracts, e.g. long sea outfalls. Similarly, partnering for term contract activities such as minor works and operation and maintenance contracts are becoming more popular.

Benefits

The client has the knowledge and satisfaction that for a certain period, say three to five years, programmes of work can be addressed without the need for individual project or annual tendering procedures. In addition, short-, medium- and long-term cash flows can be predicted with the aim of reducing peaks and troughs.

Continuity and a planned programme of work offer the contractor the ability to make efficient use of resources, plant and labour. Specialist equipment can be purchased or hired, knowing that the investment is 'safe'.

Selection of a contractor

The selection of the right contractor is critical to establish successful partnering. Since trust is the key to this success, the client must be sure that he has selected the appropriate contractor(s) for the service or product required. There is a need to ensure that the service can be provided throughout the period of the agreement. For instance, a sewerage infrastructure partnering agreement may include the installation and renovation of pipelines as well as the more general operation and repair activities. Installation will include techniques other than open-cut excavation such as microtunnelling or pipebursting, consequently the tenderer needs to demonstrate that he has the facility to meet these demands or has identified a subcontractor who can. Similarly, sewer renovation may involve a number of techniques over a period of a few years.

Term contracts

Term contracts have been established for many years in the drainage business to provide a number of services, the most common being CCTV survey, manhole survey, minor works, operation and maintenance, etc. although the principle was also used by North West Water Limited for microtunnelling works. However, the contracts usually followed a schedule of works and tendered rates to cover an expected annual turnover. There is now a current trend to move towards partnering as the client offers security of work in return for a share of the risk.

20.9.5   PRIVATE FINANCE INITIATIVE

Consortium

The *private finance initiative* (PFI) is probably the ultimate in allowing the public sector to procure services rather than assets over a long-term period, using the principles of the various arrangements described above, from design and build to partnering.

In the vernacular of the previous descriptions, PFI is a DBFO (design, build, finance and operate) arrangement and therefore requires the input of a number of experts.

PFI will almost certainly involve a consortium of at least three members for the design, financing and construction of the scheme. The 'design' group may also be required to undertake the full feasibility study and, depending on the solutions identified, may need the assistance of specialist designers. Many of these schemes necessitate huge financing (e.g. the £50m project by East of Scotland Water on the Esk Valley) and the consortium will need the expertise and services of a banking organisation. The contractor again will be required to introduce significant resources, both in terms of people and plant, and may also need to sublet some of the work to specialists.

Advantages

As previously stated, entering into an arrangement such as PFI requires careful consideration and is certainly not universally appropriate. The advantages to the public sector, in particular, will include:

- Reduced cost, therefore increasing value for money
- Much of the risk element is transferred to the service provider
- Improved level of service is provided to the customer.

However, the advantages do not appear 'overnight'. In order to achieve these, especially the improved level of service, the contract needs to be such that a profit can be achieved by the contractor while the documents will need to ensure that the customer does receive a better service, hence improved value for money.

A number of guidelines have been produced by HM Treasury, which include details of risk sharing and how 'payment mechanisms' can be optimised to achieve value for money.

Comparison

A conventional contract will require the client or user to:

- Design the scheme or employ a consultant to complete the detailed design
- Produce a specification and complete the tender stage
- Finance, manage and supervise the construction stage
- Operate the new asset.

With PFI arrangements, all of these roles are taken on by the consortium:

- The client specifies its service requirements to the consortium, which is then responsible for the design, specification, etc.
- The consortium manages the construction
- The consortium finances the whole project which it recovers through income arrangements for the service being provided
- The asset is operated and maintained by the consortium.

Risk sharing

It is clear that sharing the risk is beneficial to the client and can be advantageous to the contractor. Risk should be allocated to whoever is best able to manage it, hence some risks may be retained by the client. Naturally, tenderers may offer an acceptance of increased risk and an indication of the areas usually involved are:

- Design and construction: Responsibility lies with the contractor to complete the scheme on time and to budget, although the client may retain statutory risks.
- Operation: A longer commissioning period is the responsibility of the contractor and therefore at its cost, similarly increased operating costs will be at the cost of the contractor.
- Residual value: As far as the client is concerned, a PFI service contract will not have a residual value at the end of the contract since it is a service agreement rather than an asset purchase; if there is an asset value at the end of the contract, its value will be included in the following contract assuming that it has a value and that it is suitable for the following service provision period.
- Technology: Usually the client's risk since technical progress may provide a better but more economical solution in a few years' time; PFI promotes innovation, consequently there are occasions when new technology will reward the contractor that considers some risk by investing in a new product (or service).
- Income: For sewerage and sewage treatment schemes, the volume of effluent 'serviced' will provide the main source of income, although the transfer of all risk associated with volume would make the project non-viable, hence volume criteria will have to be identified; similarly, the quality of the influent will require specification since a higher risk approach will provide a higher income.
- Legislation: Generally the client's responsibility, although the contract must specify the

standard criteria, regulations, consents, etc. that apply, for which failure to meet will be penalised. It should not be forgotten, however, that in the case of an illegal discharge from a combined sewer overflow (CSO) any resulting prosecution will involve the owner of the asset as well as the operator.

20.10   Summary

20.10.1   *PROJECT MANAGER*

The success of a project lies directly with the appointment of the project manager. This key role is responsible for the project to be completed at cost, within the time schedule and to the technical specification identified. The project manager is required to plan, organise, resource, direct, co-ordinate and control all activities from scheme justification to its final construction, commissioning and handover to the user. To accomplish this, the project manager must be able to motivate, delegate, communicate and lead the project team, integrating all aspects to a successful completion. These skills are not all usually second sense and require a varying degree of training in order to enhance and maximise the skills to the benefit of the post holder and the employer and hence the project.

20.10.2   *PROJECT MANAGER RESPONSIBILITIES*

Understand the problem

The client or user must be confident that the appointed project manager understands the basic problem(s) being addressed, otherwise there may be some hesitancy when a preferred solution is offered to the client. The project manager must, therefore, gain the trust of the user or sponsor of the project by allowing sufficient opportunity and depth of discussion to understand the main issues. In addition, the business needs and objectives of the client must be fully appreciated in order to approach the project correctly in terms of finance, time, legislation (e.g. consent criteria) and technical aspects.

Communication

As previously identified, communication skills are essential to this post. Communication at all levels of the project – evaluation, feasibility, design, procurement, construction, commissioning and handover – to include all the appropriate parties, cannot be overemphasised. Regular meetings with the project sponsor, for instance, should keep the user aware of progress. During the feasibility and design stages, communication is equally important to ensure that a satisfactory solution is being established. There is an advantage for the project manager to act as the central, focal point for all communications between the various participants of a project (except on-site communications, say) and must also report progress, on a regular basis, to those responsible for contract monitoring.

Communication is not restricted to speaking and writing. The efficient reporting, administration and filing of conversations, meetings and written correspondence must be established and maintained from an early stage.

Feasibility stage

The project manager is responsible for planning the feasibility stage and the external consulting engineers, if required, in accordance with the plan to meet the project objectives.

Once into the feasibility study, there is a need to appraise all contributors to the project of their involvement and interaction with others and then motivate these participants to achieve the set objectives.

Changes to the brief must be agreed with the project sponsor and can only be authorised, registered (preferably through a formal procedure) and monitored by the project manager in order to ensure that the project objectives are maintained. Such changes will include cost, time period, technical aspects and operational implications.

The project manager will review the options considered and the effect of each on the aims of the project, environmental impact, risk, etc. and then identify the recommended solution for submission to the user or sponsor for acceptance and report. Once agreed, the proposal proceeds through project appraisal and financial viability.

Value management should also be addressed during this stage.

Design

Since each suitable option has had some degree of design applied in order to consider its suitability, the accepted preferred solution will have a 'conceptual' design. This is developed with the designer so that full design can follow and be compared with the project objectives, performance requirements and business standards. This may involve the input from a specialist contractor. The design stage therefore takes the conceptual design forward to where procurement may start. The final design is then recommended and offered for final approval by the project manager.

Risk

Risk assessment is not confined to any particular stage of a project, nor to the health and safety aspect alone, as the issues relating to all areas of risk must be continuously addressed and monitored. However, the two main areas that do required attention are, arguably, related to health and safety – safe working procedures and CDM Regulations. Failure to meet these requirements could lead to the termination or protracted delay of a contract, which would incur significant cost both in time and money and could lead to prosecution and a fine.

Procurement

The project manager will need to be aware of procurement procedures, particularly for large schemes where advertising in keeping with European legislation is required. The latter is a lengthy procedure which necessitates early consideration and action within the lifecycle of a project, particularly as the procedure will also include an 'invitation' to tender stage. The project may require the same procedure to the design stage as well as the construction phase, when an external consultant or contractor is required.

The choice of contract documentation will also require attention as it may be advantageous to use contract types other than 'ICE 6th Edition' such as 'green book' or 'red book' contracts, as indicated in section 20.7.6 above.

Tender appraisal is managed by the project manager, evaluation of the technical and financial submissions being the main tasks. Care must be taken when a cheap, out of line tender is received as there may be a danger of substandard workmanship and/or a negligent safety approach, or a claim-conscious contractor (or, in fact, a combination of any of these). Clear, precise assessment procedures, including meetings with the tenderers, should identify any potential problems and whether the lowest cost submission should be accepted.

Value engineering may be considered at this stage with the preferred tenderer to examine the contractor's ideas and views on cost savings and/or improved efficiencies.

Construction

The *engineer to the contract* has overall control for the running of the contract but usually delegates the powers to the resident engineer. However, the project manager must retain overall responsibility for the project. Consequently, regular meetings to discuss progress are essential. The project manager will also be required to ensure that the project develops as predicted and that the contractor performs as defined by the method statement and agreed procedures.

The project manager will also manage the commissioning and handover procedures. At this stage, user operational employees and the contractor(s) must understand their roles and responsibilities during the commissioning works and the formal handover of the completed project to the client. Any training for operatives must be planned and completed in advance of the handover, unless a specialist operator is to be employed.

Project completion and review

At the completion of the project, i.e. handover to the client or user, there is a tendency to 'relax'. However, all project documentation must be finished (e.g. as-built drawings), gathered together and suitably stored in an accessible location.

Project review is essential as a method to learn from the experience of completing the project. The review team will include representation from operations, project management, the design team and construction site staff. The review team can be led by the project manager or by an independent source if the scheme is particularly large, complex or if relationships are strained.

20.10.3   CONCLUSION

Project management is the key to the successful implementation of a scheme or project. The responsibilities, particularly those of the project manager, are great and need to be harnessed correctly. Not all good engineers or technicians make good project managers.

To help matters, there are standard, structured ways of completing each stage of a project, without which the complications faced would become more difficult to identify, address and manage and a typical activity summary is shown in Table 20.1. Above all, a sensible approach to contract procedure should be encouraged and a simple but close working relationship between client, consultant and contractor will enable the completion of a project, in line with its objectives, to be accomplished without fuss or aggravation.

Acknowledgements

The author wishes to thank former colleagues at Northumbrian Water Limited for their assistance, comments, proof reading, use of company manuals and patience in the production of this chapter.

The Special Projects Group of the former East of Scotland Water provided helpful information on PFI project management.

Thanks also to Ian Vickridge, former Senior Lecturer, UMIST, for general advice.

Table 20.1 Typical project lifecycle activity summary

Stage	Purpose	Main activities	Responsibility
Project identification and evaluation	Define problem.	Identify problem and need for investment solution.	User/Sponsor
		Define active personnel.	Sponsor
		Scope feasibility study.	Sponsor
		Prepare feasibility study brief.	Sponsor
Feasibility	Consider alternative options.	Identify and evaluate all recognised options.	Project manager
	Identify conceptual design.	Select preferred solution and develop conceptual design.	Project manager
		Risk analysis.	Project manager
Design	Take forward preferred option.	Design consultant selection procedure.	Project manager
	Develop design and confirm final preferred option.	Ensure preferred option is acceptable by operations.	Sponsor
		Ensure preferred option can be constructed in area considered.	Project manager
		Monitor design process.	Project manager
		Promote preferred option.	Project manager
		Tender document preparation.	Project manager
		Tender list selection.	Project manager
		Tender appraisal.	Project manager
Construction	Award contract.	Appoint contractor.	Project manager
	Site management.	On-site project management.	Resident engineer
		CDM Regs.	Planning supervisor/
		• safety plan	Principal contractor
		• safety file.	
Commission and project review	Commission new works.	Commission and hand over to user.	Project manager
			Contractor/User
	Project completion.	Completion of project documentation and storage.	Project manager
	Review project achievement.	Post-completion review.	All
		Act on relevant issues.	Project manager

Bibliography

Association for Project Managers, 85 Oxford Road, High Wycombe, Bucks HP11 2DX.

Brown, Mark. *Successful Project Management – in a week*. British Institute of Management. Sevenoaks, Kent, Hodder & Stoughton Ltd, 1992.

Corrie, R. K. *Project Evaluation*. Engineering Management. London, Thomas Telford Ltd, 1991.

Health and Safety at Work Act 1974.

Health and Safety Executive. *A Guide to the New Regulations*. McKenna & Co., 1993.

Health and Safety Executive. *A Guide to Managing Health and Safety in Construction*, 1995.

Hirst, I. R. C. *Business Investment Decisions*. Hemel Hempstead, Herts, Philip Allan, 1988.

HM Treasury – Private Finance Panel. *Practical Guidance on the Sharing of Risk and Structuring of PFI Contracts*. Risk and Reward in PFI Contracts Series, 1996.

Joyce, Raymond. *The CDM Regulations Explained*. London, Thomas Telford Ltd, 1995.

New Roads and Street Works Act 1991.

Northumbrian Water Limited. *Project Management Manual*, 1994.

Northumbrian Water Limited, *Safe Working Procedure No. 1: Entry into Confined Spaces*, 1989.

Norton, Brian. *Practical Guide on Value Management in Construction*. MacMillan Press, 1995.

Syms, Brian. Environment Impact Assessment – Undergraduate Course Notes. UMIST and 1996.

21
Sewer Safety

M.J. Ridings

21.1 Introduction

The subject of sewer safety has been discussed briefly as part of *Sewers Rehabilitation and New Construction*, Volume I. It is a subject however, which requires a more detailed study. Due to the dangers associated with confined spaces, of which sewers and manholes are just an example, there are many safety issues which must be addressed prior to any person entering such an area. As regards the legal framework in place to protect persons entering these areas, the legislation is somewhat outdated, confusing and in certain situations impractical. As you will discover on reading the section on current legislation the use of safety equipment was originally more widespread within the factory environment than it was within the construction sector. During the course of this chapter I will endeavour to outline the principles why persons required to enter such systems must develop and operate safe working practices. This should ensure that the task required can be carried out with minimal risk to those forming the working group.

21.2 Current legislation

Although certain legislation has been mentioned in Volume I, I have indicated below a more comprehensive listing together with a summary of their requirements:

1. Health and Safety at Work, etc. Act 1974
2. Factories Act 1961
3. Breathing Apparatus (Report on Examination) Order 1961
4. Construction (Lifting Operations) Regulations 1961 No. 1581
5. Construction (Head Protection) Regulations 1989
6. Management of Health and Safety at Work Regulations 1992
7. Provision and Use of Work Equipment Regulation 1992
8. Personal Protective Equipment at Work Regulations 1992
9. Control of Substances Hazardous to Health Regulations 1994
11. Construction (Design and Management) Regulations 1994
12. Construction (Health, Safety and Welfare) Regulations 1996

The above listing covers only the sewer environment and further legislation is in place for other types of confined spaces.

21.2.1   HEALTH AND SAFETY AT WORK, ETC. ACT 1974

The requirements under this legislation as regards sewer safety are covered under sections 2, 6, and 7.

Section 2 places duties upon the employer as regards the provision of:

- A safe working environment
- Safe access and egress from the working environment
- Equipment, information, training and supervison as regards safety
- Adequate welfare facilities.

Section 6 places duties upon the manufacturer and designers as regards the production and installation of safety equipment:

- Equipment provided to be fit for the purpose intended
- Provision of information as required to operate and maintain equipment.

Section 7 places a duty upon the employee as regards:

- Co-operating with his/her employer
- Responsibility for themselves and others by their acts or omissions.

21.2.2   FACTORIES ACT 1961, SECTION 30

Although strictly speaking the Factories Act 1961 protects only the workers as defined in the definition of a 'factory' it is used as a guidance document for workers who are in the construction industry. One reason for this being that the definitions of a confined space contained in the above Act and the Construction (General Provisions) Regulations 1961 are similar. The nature of the factory environment also demands from time to time that persons enter these confined spaces wearing breathing apparatus. The construction legislation, however, does not lend itself to entering environments where this may be required. Construction legislation from around this time is not as specific as regards other equipment and entry procedures.

The Factories Act 1961, section 30 concerns 'dangerous fumes and lack of oxygen' and requires that persons entering confined spaces have the following equipment:

- Suitable breathing apparatus
- Belt (although a full body harness has generally replaced the belt)
- Lifeline
- Testing equipment
- Reviving apparatus – oxygen resuscitation.

In addition to the above-mentioned requirements, there must also be provided a person or persons keeping watch outside the confined space and capable of pulling the person entering out. According to the legislation this person would be holding the free end of the rope. A person capable of administering first aid and trained in the use of reviving apparatus and oxygen must also be available. In relation to the present method of working many of the above requirements are in my experience not adhered to.

One part of section 30 of the Factories Act does allow for entry without breathing apparatus subject to certain criteria being met. However, this relates to the isolation of the area from the ingress of dangerous gases together with the removal of substances liable to give off dangerous gases. These points although possible to achieve in the factory environment are almost certainly not feasible within a sewer or tunnel environment. The exeption to this being while working in compressed air which in itself carries its own inherent dangers.

21.2.3   *BREATHING APPARATUS (REPORT ON EXAMINATION) ORDER 1961*

The purpose of this order is to ensure that breathing apparatus, harnesses and lifelines, together with other associated equipment, are thoroughly examined and maintained and the record of the findings kept in a register for inspection. This examination must be carried out at least once a month by a competent person. Details of this inspection include the following:

- Name and address of the owner of the equipment and where it is kept
- Serial number or distinguishing mark of the equipment
- Condition of the equipment
- Cylinder pressure
- Signature of the person carrying out the examination
- Date of the examination.

This examination is covered to some extent by the Control of Substances Hazardous to Health Regulations 1994.

A more detailed interpretation of the above is covered by section 21.7

21.2.4   *CONSTRUCTION (LIFTING OPERATIONS) REGULATIONS 1961 NO. 1581*

Although similar to the requirements detailed in the Factories Act 1961 these regulations support further additions including the fitting of safe working load indicators to most jib cranes with 'safety' hooks being fitted with a safety catch. The age for operators and banksmen is also contained within the legislation.

21.2.5   *CONSTRUCTION (HEAD PROTECTION) REGULATIONS 1989*

The purpose of this legislation is to ensure that where there is a risk of head injury then appropriate head protection is worn. This may take into account working in manholes, sewers, excavations, in the vicinity of lifting machinery and the erection of scaffolding. Industrial safety helmets are now in common use as the main form of head protection on construction sites. The use of chin straps being of significant use in the situation where bending or the donning of breathing apparatus during an emergency evacuation from confined spaces is required.

21.2.6   *MANAGEMENT OF HEALTH AND SAFETY AT WORK REGULATIONS 1992*

As previous legislation was qualified by the term 'reasonably practicable' the ability to interpret the requirements of older legislation lent itself to a wide variance in safety standards set within the construction industry. The Management of Health and Safety at Work Regulations 1992, however, placed an 'absolute' duty on employers to identify any significant foreseeable risk to which their employees were to be exposed. They are then obliged to identify the hazard and prevent or control the exposure by either removing the hazard or adequately controlling it. For example, the locking off of electrical and mechanical devices together with the use of suitable presonal protective equipment are forms of control.

21.2.7 *PROVISION AND USE OF WORK EQUIPMENT REGULATIONS 1992*

The main objective of the above legislation is to ensure the provision of safe work equipment and its safe use. The equipment should not give risk to the health and safety of its users by either its age or origin.

The regulations are set summarised thus:

- Regulations 1–3:   Date of entry of regulations, scope, etc.
- Regulation 4:   Defines who has duties
- Regulations 5–10:   Selection, maintenance, instruction, training, etc.
- Regulations 11–24:   Use of equipment to control hazards
- Regulations 25–27:   Exemption certificates, repeals and revocations.

21.2.8 *PERSONAL PROTECTIVE EQUIPMENT AT WORK REGULATIONS 1992*

The requirements for the provision of personal protective clothing at work depend on its interpretation and this is quite simple. Any equipment that is required to protect persons from one or more risks to their health and safety including hazards such as weather come within its scope. As regards breathing apparatus, gloves, safety helmets, etc. these are obvious articles that are required.

21.2.9 *CONTROL OF SUBSTANCES HAZARDOUS TO HEALTH REGULATIONS 1994*

These regulations replace earlier legislation and ensure that no employer can carry out work that would expose an employee to substances that are hazardous to their health, unless a suitable and sufficient assessment of the risks associated with that work and product is carried out. Not only must this assessment be carried out prior to the work being undertaken but it must also be reviewed from time to time especially when circumstances surrounding the task or product change. This assessment must be carried out by a competent person who must obtain all the relevant product data sheets. Where the hazardous substance cannot be practically removed then the employer must provide suitable control measures. The employer must also ensure that maintenance, examination and tests of the control measures take place. The workplace must also be monitored and this will be discussed later in the chapter. As in other health and safety legislation the employer must provide adequate and suitable information, instruction and training for its employees.

21.2.10   *CONSTRUCTION (DESIGN AND MANAGEMENT) REGULATIONS 1994*

The above legislation makes provision for the production of a health and safety plan together with the keeping of a health and safety file. This file must remain available for inspection by any party requiring information contained within. This information can relate to the experience and qualifications of the employees engaged on the project, risk assessment details and generic COSHH assessments. Projects coming under the scope of the above regulations are detailed within it.

21.2.11   *CONSTRUCTION (HEALTH, SAFETY AND WELFARE) REGULATIONS 1996*

These regulations revoke many earlier legislation and contain details for the safe working on construction sites. The regulations include safe areas of work, falls, use of explosives, training and welfare facilities.

   As can be seen from the vast array of legislation it is not difficult to understand why many companies have difficulty in complying with all the statutory requirements. However, there are draft proposals to bring confined space entry legislation under one regulation which should simplify the situation. The draft document issued by the Health and Safety Commission addresses not only the existing legislation but in my opinion tackles the issue relating to types of breathing apparatus currently being used and the need to implement suitable and sufficient arrangements for the rescue of persons in the event of an emergency.

21.3   Types of confined spaces

As defined in the Factories Act 1961, section 30 a confined space can be taken as a 'chamber, tank, vat, pit, pipe, flue or similar confined space in which dangerous fumes are liable to be present to such an extent as to involve risk of persons being overcome thereby'. This definition therefore includes sewers, tunnels and culverts as the risk from the types of effluent passing through them may result in an irrespirable atmosphere. This is also true with regard to certain ground types. It is also possible for certain areas to be affected by hazardous substances being introduced into them thus creating a dangerous area.

21.4   Hazard identification

Once an area has been classified as a confined space it is essential that a comprehensive assessment of the hazards associated with the task in hand is undertaken.

   This hazard identification process once completed will result in an evaluation of the risk involved in entering the manhole and the necessary equipment, personnel and procedures that need to be in place to carry out the task safety. This evaluation may also result in the abandonment of the task due to a high risk becoming evident.

   As regards the manhole in Fig. 21.1 the hazards may be of the following nature:

* Location: As a result of the location of the manhole, the hazard may be from traffic, livestock or incoming tides.
* Gases: The gases associated with the manhole could be generated from a variety of

sources, e.g. breakdown of organic material within the sewage, trade effluent discharges, ground conditions (water filtering through chalk), exhaust fumes, etc.

- Access: Although the Factories Act 1961 section 30 gives minimum size of access as 18″ Diameter or 18″ × 16″ this does result in sizes above these dimensions being safe to enter. This would depend on the size of person entering together with the type of safety equipment carried in the case of an emergency.
- Falls: If a person is required to descend from the top of the manhole to the bottom they may, if unaided, fall. This could not be classed as safe access and egress and as such is governed by the requirements set out under the Health and Safety at Work, etc. Act 1974.
- Flow conditions: The content of the effluent (e.g. corrosive), depth and speed all affect the safety of the person entering.
- Biological: Persons entering underground systems are at risk from diseases which find their way into that environment by being carried in the flow, present in the ground or from rats (in the form of leptospirosis).

Other hazards exist within such an environment including structural instability, hypodermic needles, weather conditions, mechanical and electrical features. The above sample is not exhaustive but gives an indication of the hazards that need to be controlled. If they are not controlled then a risk exists which has to be evaluated prior to a decision as to whether the task should or should not be undertaken.

In relation to controlling the hazards we can look at the earlier samples:

- Location: Traffic can be diverted, signing and guarding measures implemented. Livestock can be removed or secure fencing erected to prevent ingress into the work area by the animals.
- Gases: The gases concerned need to be identified and monitored. Ventilation systems need to be considered to dilute gases that may build up within the area.
- Access: An initial measurement of the access points is required and then an assessment of the size of personnel and equipment involved needs to be carried out.
- Falls: Falls can be prevented by the use of suitable access equipment including tripod, winch, fall-arrest lines, harnesses, etc.
- Flow conditions: Obtaining flow rates from the water company, local authority or client may assist followed by a visual inspection.
- Biological: Observing correct hygiene precautions prior to entry and on exiting the system including after handling equipment.

You can see that many of the hazards identified can be controlled although some cannot. The hazards that remain uncontrollable result in the 'risk'. An example may be gases being given off by a landfill site or spillage from a tanker containing hazardous liquids. These hazards cannot be identified and therefore the monitoring of the environment in which persons are required to work becomes impractical. The requirement therefore for suitable and sufficient emergency procedures to be in place becomes a serious consideration. As I mentioned earlier this is now under consideration by the Health and Safety Commission in the draft proposals for confined space entry.

21.5 Personal protective equipment

The legislation regarding the requirement to be provided with and wear personal protective equipment has already been discussed earlier in the chapter. The types of protective equipment need now to be identified. A typical list of equipment may be as follows:

- Safety helmet: Safety helmets must comply with the relevant standards and be free from damage and not be defaced by stickers unless the adhesive does not cause deterioration of the fabric of the helmet.
- Protective overalls: These overalls should protect the person from the effects of exposure to chemicals, bacteria, corrosives or other harmful substances. The suits should also not be a static risk.
- Safety boots (including steel midsole and toecaps): The boots should protect the wearer from falling objects and puncturing from objects such as glass and hypodermic needles.
- Waterproof gloves: Gloves should prevent the wearer from coming into contact with substances similar to those listed for protective overalls.
- Full body harness: The harness should be so designed as to enable the wearer to be lifted vertically out of a shaft or in the case of a narrow tunnel, horizontally. This is usually achieved by the locating of 'D' rings at strategic points on the harness.
- Eye protection (if required): This may take the form of a full face mask to prevent spray from hitting the face.
- Barrier cream: A water resistant barrier cream should be used.

The above equipment is essential to provide protection from many of the hazards that exist in the sewer environment.

In addition to the above the following will also be required prior to entry:

- Appropriate gas monitor: These monitors need to be capable of monitoring the gases that may be encountered and incapable of creating a source of ignition of flammable gases, e.g. intrinsically safe. This topic will be covered later in the chapter.
- Escape breathing apparatus: This equipment must be suitable and sufficient for the type of environment and the length of time needed to escape in the event of an emergency. This type of equipment will be covered later in the chapter.
- Oxygen resuscitation equipment: All resuscitation equipment must only be operated by suitably qualified persons, e.g. first aiders. It is essential also that the equipment is suitable for the needs of the potential casualty.
- Tripod winch, safety lines: The choice available in the above type of equipment sometimes leaves the user with many options when hiring or buying.

The user must bear in mind that current legislation requires that employees not only have a safe working environment but also safe access and egress from that environment. The method of entry not only for a single person but multiple occupants needs to be considered. Other factors affecting equipment choice would be:
  - Numbers of intermediate landing platforms, e.g. ability to set up fall-arrest systems
  - Depth of entry, e.g. speed of evacuation
  - Weather conditions, e.g. water may affect efficiency of rope systems
  - Ground conditions, e.g. grit may affect efficiency of rope systems
  - Physiology of winchman, e.g. relates to size of possible casualty
  - Intrinsically safe lighting.
All electrical equipment for use within potentially explosive atmospheres must be certified intrinsically safe. The lighting must also be sufficient to allow the person entering the area to be able to see any potential hazards. This lighting may come in the form of handlamps, torches or larger festoon type air-driven turbo lights.
- Traffic control equipment (if required): This would include signs, barriers, cones, etc.
- Manhole cover lifting equipment: It was quite common in the past to use pick axes to

lift manhole covers and sometimes they are still used. This practice is dangerous and either manhole keys or lever type systems should be used.

- Ventilation equipment (air movers/extractors) (if required): The use of compressor driven ventilation systems is a common method for introducing air into an area lacking a respirable atmosphere. However, care must be taken in introducing air which may be itself contaminated. Ideally a forced ventilation system which also extracts should be used. This is discussed later in the chapter.
- Method of communication: Communications from the outside of the confined space can be achieved simply by voice contact or by more sophisticated means. There are intrinsically safe wire type communication systems such as the 'Diktron' and 'Talking Rope' methods. A series of whistles or blasts on fog horns is also possible.

21.6   Gases and gas detection

The atmosphere that we breathe consists of mainly 21% oxygen and 79% nitrogen. To enable us to carry out work within confined spaces it is essential to maintain this balance within set tolerances. We achieve this by the monitoring of the atmosphere with appropriate gas detection instruments. As our bodies' respiratory system uses oxygen there is a resultant waste product in the form of carbon dioxide. This gas excludes oxygen while in a confined space and therefore it is essential to provide adequate ventilation and continuous monitoring while the space is occupied. The build-up of carbon dioxide is also possible through the action of water filtering through chalk. As a person's sense of judgement is affected by a lack of oxygen, it becomes apparent that the evacuee's inability to react effectively to an emergency situation could result in a serious injury.

Table 21.1 indicates a variety of average breathing rates.

Table 21.1

Exertion	Oxygen consumed per min. (litres)	Air breathed per min. (litres)	Volume of air at each respiration (litres)	No. of respiration's per min.
Rest in bed	0.24	7.7	0.46	16.8
Rest, standing	0.33	10.4	0.61	17.1
Walking – 2 mph	0.78	18.6	1.27	14.7
Walking – 3 mph	1.07	24.8	1.53	16.2
Walking – 4 mph	1.6	37.3	2.06	18.2
Walking – 5 mph	2.54	60.9	3.14	19.5

21.6.1   RESPIRATORY SYSTEM

To the average person, respiration is an effortless process consisting of breathing in and breathing out. You breathe in fresh air and breathe out 'stale' air. It is known as oxygen inhalation (or inspiration) and carbon dioxide exhalation (or expiration).

Not many of us know or care much more than that. It is taken for granted along with other vital functions of the body and keeps us alive without our even being conscious of it.

The job of the lungs is to get oxygen into your blood and carbon dioxide out. The lungs are constructed on the same pattern as a tree, with hollow trunk, branches, twigs and leaves. The trunk is the wind pipe (trachea), the branches and twigs are progressively finer air tubes, of Bronchi and bronchioles. These tubes penetrate all parts of your lungs, deep inside your chest. The leaves are the alveoli. These are minute air sacs which together form a very large area in close contact with the blood that flows all around them. Here the oxygen enters the blood and carbon dioxide leaves.

The air in the alveoli is kept fresh by breathing and oxygen from the air is carried to all parts of your body, dissolved in your blood. (It is actually combined with haemoglobin, the substance that makes your blood red.)

Breathing may be considered the most important of all the functions of the body for, indeed, all the other functions depend upon it. Humans may exist some time without eating, a shorter time without drinking, but without breathing, their existence may be measured by a few minutes.

21.6.2 RESPIRATION

Respiration or breathing is the means by which the tissue or organs of the body are supplied with oxygen which is essential if they are to continue to live and function efficiently.

It is a spontaneous action which unless some physical cause intervenes is performed automatically by the human body some 15 to 18 times every minute from the moment of birth to death, this average rate being increased by exercise or excitement. The body performs this action because it needs oxygen which is normally obtained only from the atmosphere.

To obtain this, air must be drawn into the lungs, held for a sufficient time for the oxygen required to be absorbed and then expelled. This is normally effected by the action of the diaphragm and the ribs and consists of three phases: inspiration, expiration and a pause.

21.6.3 INSPIRATION

On inspiration or inhalation the diaphragm contracts and its dome-shaped centre becomes flattened thereby increasing the capacity of the chest from above downwards. The ribs, which are normally inclined downwards and forwards, rise outwards and upwards thereby increasing the capacity of the chest from front to back and from side to side.

The muscles responsible for this are the outer of the two layers of muscles occupying the spaces between the ribs. This would not in itself cause air to be drawn in, were it not for the elasticity of the lungs and the fact that for physical reasons the two layers of pleura keep in close contact and thus the lungs themselves expand the air that is drawn into them.

21.6.4 EXPIRATION

In expiration or exhalation the reverse process takes place owing to the relaxation of the diaphragm and of the action of the inner of the two layers of muscles occupying the spaces between the ribs as a result of which the lungs contract by their own elasticity. The muscles of the abdominal wall also play a part in respiration; when the diaphragm descends they relax to allow room for the displaced abdominal organs; when the diaphragm ascends they contract.

Table 21.2 indicates the effect of oxygen starvation on the human body.

Table 21.2 How the body reacts when the proportion of oxygen in the air drops below 20.8 per cent

Concentration (per cent)	Effects on health
17	Judgement impairment commences.
16	Signs of oxygen starvation of tissues appear.
12	Breathing rate increases, pulse becomes faster, and muscular co-ordination impairment starts.
10	Abnormal fatigue felt, breathing difficulties encountered, emotions upset, but victim remains conscious.
6	Vomiting takes place and nausea is felt. Consciousness may be lost. Freedom of movement impaired and victim may experience physical collapse, while still mentally aware of surroundings.
<6	Extreme breathing difficulties experienced. Breathing stops. Heart action ceases after a few minutes.

While symptoms are grouped opposite a specific concentration, some may occur at concentrations higher or lower than those against which they are shown.

21.6.5 HAZARDOUS GASES

As regards gases associated with sewer and tunnel work, the list of commonly encountered gases can generally be narrowed down to the following.

Methane

Formerly known as lightcarburetted hydrogen, it is a compound of carbon and hydrogen. Its composition is denoted by terming it a hydrocarbon.

The name 'marsh gas' was applied to the gas when it was obtained by stirring the mud of stagnant pools and marshes.

Methane is the flammable constituent of fire-damp. Some samples of fire-damps consist almost entirely of this gas, while others consist of it in a mixture with carbon dioxide and with nitrogen.

Methane is readily kindled yielding carbon dioxide and steam as the product of combustion. When the gas is kindled in a ventilated, confined space, the flame burns and scorches seriously. If the supply of air to the flame is restricted, highly poisonous carbon monoxide is formed.

The lightness and buoyancy of methane cause it to rise rapidly to the roof and it will take the highest place possible.

Methane if kindled during its escape or immediately after its escape will burn with a flare. If, however, it has time to diffuse and become mingled with air, its behaviour in contact with a flame will vary with the proportion of air with which it is mingled. Mixture with small proportions of air tends to hasten the burning of the gas. When the mixture is diluted until 16% of the methane is present, it begins to be explosive. But as the proportion of methane in the air approaches 10% the mixture begins to burn with greater rapidity, and when fired in large quantity becomes explosive. When the proportion reaches 10% the explosion is the most violent possible. When the proportion of methane in the air falls to below 5% the gas cannot be fired. The lower explosive level of methane is 5% (LEL 5). The main properties of methane are shown in Table 21.3

Hydrogen sulphide

Hydrogen sulphide, or as it is better known, sulphurated hydrogen, is one of the more commonly encountered gases that persons working within the sewerage environment are likely to encounter. It is a naturally occurring gas that results from the decaying of organic matter. It is therefore found naturally in sewers within the sludge deposits and also in mines. The hydrogen sulphide is released upon agitation of the deposits.

There is also a danger from this gas within the oil industry and this is due to the high toxicity levels as a result of the gas being under pressure. The gas is used within industry for a variety of products including pharmaceuticals and sulphide production. As regards its effect on the human body it acts as an irritant to the respiratory system and when levels reach approximately 600 parts per million can result in coma and death.

The gas also has a peculiar effect on the sense of smell which in effect is paralysed temporarily. The result therefore may lead the unsuspecting person to think that the danger has passed. If the person fails to use suitable gas monitoring equipment then a false sense of security may exist.

In addition to its toxic properties hydrogen sulphide has a flammable range. This means that under certain conditions it will burn and may also ignite to form an explosive reaction. The main properties of hydrogen sulphide are shown in Table 21.3.

Carbon dioxide

Carbon dioxide is the natural stimulant to breathing and this is automatically carried out so as to keep the percentage in the lungs at about 5.5%. When more carbon dioxide is produced by oxidative processes in exertion, then we breathe harder to get rid of the excess and keep the normal amount in the depth of the lungs in the blood. The breathing of small percentages of carbon dioxide, e.g. 1–2%, in the inspired air only deepens the breathing so as to keep the percentage in the alveolar air normal. When 5% is present the breathing becomes very laboured and if a person breathes 10% they may become unconscious through the excess of this gas in the blood.

The breathing centre in the brain is controlled by carbon dioxide in the depths of the lungs so that the percentage of this gas is left constant therein (5–6%). The respiratory centre of the brain, which controls the muscles used for breathing, does so under the chemical stimulus of a percentage of carbon dioxide in the blood. When the percentage rises, as it does after exertion, the brain instructs the muscles to make the owner breathe deeply and make the heart pump blood more rapidly. The carbon dioxide stimulant will cause a patient to breathe deeper. However, a patient who is not breathing may not commence breathing because of the introduction of a stimulant to the lungs and heart. Respiration would have to be mechanically produced with the aid of artificial respiration.

The question often asked is, why cannot the body function efficiently when only 16% or less of oxygen is present in view of the fact that only one-fifth of that oxygen is used up by the body. There is a surplus of four-fifths which one would have thought would overcome the deficiency. The explanation is involved and cannot be dealt with in this context but the fact must be accepted that in the alveolar air the percentage of the 5–6% carbon dioxide and 13–14% oxygen is maintained as a constant during inspiration and expiration. The inhaled air can be regarded as a tidal wave which is flushing out the pool of alveolar air.

The flood tide maintains the required 14% of oxygen which is being continually consumed by the matters in the body which require oxidation. Upon being oxidised carbon dioxide is produced which if it were not for the reflushing flood tide of inhaled air, would exceed the

Table 21.3

Gas	Properties	Hazard
Carbon dioxide OES 5000 ppm	Colourless Odourless Heavier than air Non-flammable	Displacement of Oxygen Mildly toxic
Carbon monoxide OES 50 ppm	Colourless Odourless Lighter than air Flammable Toxic	Toxicity Flammability
Chlorine OES 0.5 ppm	Yellow/Green colour Choking odour Much heavier than air Non-flammable Toxic	Toxicity
Hydrogen sulphide OES 10 ppm	Colourless Rotten-egg odour Heavier than air Flammable Toxic	Toxicity Flammability Impairment of the sense of smell by high concentration
Methane (constituent of natural and reformed gas)	Colourless Odourless Lighter than air Flammable	Flammability Explosion
Petroleum Diesel vapour OES 500 ppm	Colourless Paraffinic odour Much heavier than air Toxic Flammable	Toxicity Flammability Explosion

operation values of 5–6% which, in turn, would entail a decrease in the operative 13–14% of oxygen. The ebb tide, i.e. the exhaled air, carries away the surplus carbon dioxide and nitrogen which are present in the pool by inhalation and by oxidation.

One can conclude that to maintain these readily critical values of 13–14% of oxygen and 5–6% of carbon dioxide in the pool of the alveolar air, the lungs have considerable work to perform and that the lungs are insufficient to the extent that within the period of time involved in one inhalation they are hard put to abstract one-fifth or 20% of the available oxygen. It is vital that they should abstract the available oxygen and one has to accept that because of the time/volume factor involved in the breathing process no more than one/fifth of the available oxygen can be abstracted. One-fifth of the volume of oxygen available in air inhaled amounts to a definite quantity of oxygen. If there is a deficiency in the oxygen content of the air being inhaled, i.e. 16% instead of 20%, it is harder passing air and so the lungs become overworked, i.e., breathing is laboured. The labouring will increase as the oxygen content of the air diminishes and if it is diminished to a great extent then the lungs cannot labour sufficiently to secure the necessary volume of oxygen which is required in the pool. It will be apparent then that it is not so much a matter of how much oxygen is available but rather how much work will the lungs have to perform in order to secure the vital content of oxygen in each inhalation. The main properties of carbon dioxide are shown in Table 21.3.

Carbon monoxide

Carbon monoxide is generally produced as a waste gas from internal combustion engines. It is also produced during combustion of liquid and solid fuels when there is not enough oxygen in the atmosphere to allow complete combustion, e.g. carbon monoxide is produced instead of carbon dioxide.

To prevent this an adequate means of ventilation is required. Within manholes, sewers and tunnels certain activities give rise for concern. Two of these are the use of petrol driven generators and STIHL saws within a confined area. The use of such machinery gives rise to the production of carbon monoxide and should be avoided. The problem as a result of the production of this gas is not restricted to the confined spaces mentioned but may include car parks and car ferries.

The effect on the human body is one of a toxic nature. As oxygen passing through the lungs combines with haemoglobin in the blood to form oxyhaemaglobin it is carried through the arterial circulation to the cell structure. The deoxygenated haemoglobin then transports carbon dioxide which is a waste product of respiration back to the lungs where it is exhaled. The cycle is then repeated. The introduction of carbon monoxide causes a severe disruption of this process as the haemoglobin 'prefers' this to oxygen. The haemoglobin therefore combines preferentially with the carbon monoxide to form carboxyhaemaglobin. The continuous inhalation of carbon monoxide results in the uptake of oxygen into the bloodstream being severely reduced and the cell structure of the body being deprived of this essential oxygen.

The symptoms of exposure to various concentrations may be as follows:

Up to 400 parts per million   – severe headache and nausea
400–800 parts per million   – collapse and unconsciousness after two hours
10 000 parts per million   – death.

As regards treatment the following action should be taken:

• The victim should be removed from the contaminated area
• If heart has stopped then apply cardiac massage
• Give oxygen as soon as possible
• Refer to hospital.

Preventative measures are essential due to the dangerous nature of the gas and may include the following:

• Removal of the hazard
• Controlling of the hazard if unable to remove
• Ventilation
• Use of respiratory protective equipment

21.6.6   GASEOUS ATMOSPHERES

Most atmospheres encountered with sewerage and sewage functions will contain some or all of the following gases, dependent upon trade waste conditions – this list is not exhaustive.

The gases fall into three distinct categories – some gases, however, occur in each – due to their properties, i.e. flammable, explosive, toxic and suffocating.

Flammable:
- Methane: Usually from sewage sludge, natural gas
- Ammonia: From sewage, or refrigerating plants, may be liberated in alkaline sewage
- Petroleum vapours: Trade wastes, spillages, etc.
- Carbon monoxide: Sewage – trade waste.

Toxic:
- Hydrogen sulphide: From sewage – acid conditions make liberation easier.
- Cyanides: Usually trade wastes
- Ammonia: See above
- Chlorine: Bleaches, hospital wastes
- Carbon monoxide: Sewage – trade wastes.

Suffocating:
- Ammonia: See above
- Carbon dioxide: From sewage – outside
- Hydrogen sulphide   See above
- Chlorine: See above
- Steam: Laundry waste – factory cooling water
- Methane: See above

Thus any gas monitoring system will require a capability to detect and evaluate the presence of each type of hazardous gas.

21.6.7   ATMOSPHERE MONITORING SYSTEMS

Gas detector tubes

There are various types of gas detector tube systems currently in use. These systems are used to analyse specific gases which are not monitored using conventional electronic gas detectors.

The aim is to determine very small concentrations with the maximum reliability in the shortest time. Laboratory methods are not always suitable, since the time expenditure is very high and it can sometimes take hours to analyse the results. The development of the gas detector tube system enables the air to be sampled to be drawn up through a glass tube containing a reagent by means of a bellows type pump.

A typical instrument uses one stroke of the bellows to provide a 100 cubic cm sample. The gas detector thus not only supplies a sample but measures it as well. The pump bellow may be made of neoprene and will open automatically after compressing and releasing the bellows. The end of the stroke is determined by the limit of the attached chain becoming taut. Many types of detection tube system now utilise automatic pump and stroke counters to assist in the measurements. The shelf life of the tubes tend to be no longer than two years at room temperature.

OPERATION

A typical tube system is operated by drawing a sample of air into the tube by compressing the bellows a number of times as determined by the choice of tube. This figure is indicated in the handbook and is specific to individual gases. There is, however, a cross-sensitivity between some gases and the same tube can be used in certain instances. To draw the sample up the tube the ends of the tube must first be broken before being placed in the main detector body. The tubes can come as short-term exposure or long-term exposure and can be manually operated or battery powered.

Electronic gas detectors

There are many electronic gas detectors currently in operation. The types used for monitoring for oxygen, methane and hydrogen sulphide are usually of the multigas type, e.g. used for the monitoring of all these gases at the same time. A typical monitor, the Neotronics Minigas 4, is shown in Fig. 21.1.

Figure 21.1   Neotronics Minigas 4 electronic gas detector.

The instrument is lowered into the confined space prior to entry to allow readings to be taken and recorded on the entry log. The instruments usually have alarm settings to warn of high and low oxygen levels which are measured as a % vol. in air. The time weighted average (TWA), short-term exposure level (STEL) and instantaneous alarm levels for toxic gases are usually measured in parts per million and follow standards as set out in the EH40 publication. Flammable gases are measured as a % LEL, e.g. methane alarm settings being 20% LEL (1%) vol. in air. It is usually the instantaneous alarm that warns the person entering that a hazard exists. As regards the operation of the instruments, they should be accompanied by a set of instructions.

This is of utmost importance as instruments vary considerably in the way they are calibrated. As well as a service date of usually six months they must be checked and adjusted daily to enable relative readings between the normal atmosphere and conditions below ground level. This is often overlooked when the instrument is in practical use.

Where there exists a known gas that cannot be monitored using the above type instrument then either a further sensing cell must be inserted into the instrument and calibrated or another specific monitor for the gas in question used. The Minigas 4 as described earlier has the ability to monitor for other gases such as carbon monoxide and chlorine. Where chlorine coexists with hydrogen sulphide the cross-sensitivity between the two gases means that they are not suitable for measurement within that environment.

The other problem surrounding the use of electronic gas detectors relates to the contamination of the explosive sensor. Certain gases or substances may affect the sensors of the Minigas 4:

- Silicones
- Halogens(>100 ppm)
- Halogenated hydrocarbons
- Phosphorus-containing compounds
- Sulphur-containing compounds (>100 ppm)
- Volatile organometallics.

21.7 Breathing apparatus

Although we have covered the legislation relating to safety in sewers, tunnels, etc. earlier an important piece of legislation still remains and that is the Breathing Apparatus (Report on Examination) Order 1961. Many of the requirements under this document are repeated within the terms of the Control of Substances Hazardous to Health Regulations 1994.

21.7.1 BREATHING APPARATUS (REPORT ON EXAMINATION) ORDER 1961

All breathing apparatus, safety harnesses, lifelines, reviving apparatus and any other equipment provided for use in, or in connection with, entry into confined spaces, and for use in emergencies, must be properly maintained and thoroughly examined at least once a month, and as soon as possible after every occasion on which it has been used. The manufacturer's advice should be followed regarding regular maintenance and servicing requirements. When, and where appropriate, spare cylinders of air and/or oxygen should be kept.

The monthly thorough examination of the equipment must be made by a competent person, who should sign a report containing at least the following particulars:

- The name of the occupier of premises where the equipment is stored
- The address of the premises
- In the case of breathing apparatus or reviving apparatus, the particulars of the type of apparatus and of the distinguishing number or mark, together with a description sufficient to identify the apparatus, and the name of the marker
- In the case of a safety harness, belt or rope, the distinguishing number or mark and a description sufficient to identify the safety harness, belt or rope
- The date of the examination and by whom it was carried out

- The condition of the apparatus, safety harness, belt or rope, and particulars of any defect found at the time of the examination
- In the case of a compressed-oxygen apparatus, a compressed-air apparatus or a reviving apparatus, the pressure of oxygen or air, as the case may be, in the supply cylinder.

Such reports must be kept available for reference and inspection purposes and may be kept in the form of a register.

21.7.2   BREATHING APPARATUS TYPES

Resuscitation equipment has to be kept available for dealing with emergency situations in all instances where the use of breathing apparatus is required.

Types of breathing apparatus in most use are as follows:

- Fresh air breathing apparatus (open circuit)
- Compressed air breathing apparatus (open circuit) including air-line breathing apparatus and self-contained breathing apparatus
- Self-rescuers – oxygen rebreathers (closed circuit).

Fresh air type

This equipment can usually only be used at distances in the order of 10 metres from a supply of naturally available fresh air. The idea behind the equipment is to enable the wearer to breathe fresh air by the action of his breathing drawing the air from a non-contaminated site. This may mean siting the inlet hose at a point where exhaust gases or other contaminants cannot enter.

Compressed air breathing apparatus (open circuit)

GENERAL

The common feature of the various types of compressed air breathing apparatus is that the wearers receive a supply of pure air under pressure at a rate sufficient to meet their highest respiratory demands. This is achieved by a valve fitted to the face mask of the breathing apparatus. The two main types of valve in use in the water industry are:

1. Demand valve:
   - Limitations: The wearer must obtain an air-tight seal between the contours of his face and the face seal of the breathing mask or he will be at risk from inward leakage from the atmosphere when inhaling.
   - Description: A demand valve is a spring loaded non-return valve in the compressed air line and fitted onto the breathing mask. The act of inhalation by the wearer causes the valve to open and admit compressed air into the breathing mask. As soon as the wearer stops inhaling the valve shuts cutting off the air supply until the wearer's next inhalation cycle
2. Positive pressure demand valve:
   - Limitations: If the wearer is using a self-contained breathing apparatus and has not obtained an airtight seal between the contours of his face and the face seal of the breathing mask the duration of his set will be reduced due to the outward leakage of air from the mask. It must also be borne in mind that at very high work rates the level of positive pressure in the face mask will be diminished or fail.
   - Description: The positive pressure demand valve is a development of the demand

valve described above but in this case a positive pressure in the mask is achieved by the continuous action of a spring in the diaphragm such that the demand valve always remains open until pressure in the mask increases. This increase in pressure arises when the mask is worn and its magnitude is controlled by the cracking pressure of the exhaust valve.

Air-line system

This system utilises either a supply of air from compressed air cylinders mounted on a frame with the air supply being reduced in pressure through a series of valves before being fed to the wearer, the supply may come from an on-site fixed supply, e.g. as on a chemical site. The supply may also be from a compressor which passes the air supply through a series of filters.

This system, however, requires that another person is in attendance at all times while the system is in operation.

Self-contained breathing apparatus (open circuit)

LIMITATIONS

Air consumption rates: under normal working conditions the average consumption rate of air per person is taken as 40 litres per minute.

However, in heavier work such as working in confined spaces or ascending ladders consumption of air may be as high as 80 litres per minute or even more for short periods. In each case the duration of the set would be correspondingly less.

It is therefore essential that when assessing a task consideration is given to the time required to complete the task in relation to the capacity of the cylinder, allowing for the escape factor.

Thus the limitations of this type of equipment are:

- Time: This is dependent on the capacity of the air cylinder carried, the distance of the workplace from a safe atmosphere and the required work rate of the wearer.
- Weight: A working set weighs between 9 kg and 13.5 kg (30 lb and 40 lb), and is carried on the wearer's back, which increases fatigue in the wearer.
- Size: The size of the wearer plus the bulk of the breathing apparatus on his back limits him to the size of openings he can get through.

APPLICATIONS

There are two main types of self-contained compressed air breathing apparatus both of which can be used in toxic or oxygen deficient atmospheres as follows:

1. Escape breathing apparatus:
   - Limitations: This type of apparatus can be put on by personnel to escape to safe atmosphere should an incident occur involving the contamination of the atmosphere.
   - Applications: For escape purposes only.
   - Description: this type of apparatus may supply a limited quantity of respirable air depending on the demand rate of the wearer but may not have an audible alarm to warn the wearer of the depletion of his air supply.
2. Short duration working or rescue breathing apparatus:
   - Limitations: Duration according to the capacity of the air cylinder carried and the demand rate of the wearer. Cylinders for working or rescue sets range in capacity from a nominal half hour to a nominal one hour.

- Applications: For use in routine inspection or maintenance work in oxygen deficient or toxic atmospheres or entry into untested atmospheres for rescue work.
- Description: This type of apparatus is worn as a backpack and must be fitted with a contents gauge and an audible low air alarm. They can also be fitted with a coupling to connect into an air line if required.

AIR PURITY

Air being supplied to the wearer should not contain impurities in excess of the following limits:

Carbon monoxide 5 ppm	(5.5 mg/m^3)
Carbon dioxide 500 ppm	(900 mg/m^3)
Oil mist	(0.5 mg/m^3)
Water vapour	(300 mg/m^3)

STORAGE OF CYLINDERS

It is essential that full and empty cylinders are unmistakably identifiable as such and should be stored separately.

USE OF EQUIPMENT

There are common limitations in the use of respiratory protective equipment. The safe use of all types of respiratory equipment depends on the wearer obtaining an airtight seal between his face and the mask to prevent inward leakage from the atmosphere. This can be affected by facial contours and features (e.g. beards, sideboards and spectacles) which may prevent an airtight seal being obtained.

ROUTINE BREATHING APPARATUS OPERATIONS

When it is considered necessary to have airline or short duration working and rescue breathing apparatus available for use in routine operations, these operations should be carried out under permit-to-work conditions.

Particular attention should be paid to the following points:

- Breathing masks, hoods, and blouses: These should not be cracked, punctured or otherwise damaged or distorted. Attachments should be secure.
- Harnesses: Stitching should be sound and elastic bands should have good elasticity.
- Inlet and outlet valve mountings: These should be undamaged and make a perfect joint with the mask, hood or blouse.
- Inlet and outlet valves: Valves should not be damaged or distorted, should move freely on the mountings and seat properly.

HOSES AND AIR LINES

Hoses and air lines used in conjunction with respiratory protective equipment should be undamaged and free from deterioration that might cause leakage. Means of attachment should be secure.

BREATHING APPARATUS CYLINDERS

These must be examined and tested according to statutory and manufacturer's requirements. The date of test must be stamped on the valve end of the cylinder.

CLEANING AND DECONTAMINATION

Equipment that is used regularly should be collected and cleaned as frequently as necessary to ensure that proper protection is provided for the wearer.

The breathing masks and tubes should be dismantled and cleaned by washing with soap and warm water and then thoroughly rinsed. Each worker should be briefed in the cleaning procedures and always be issued with clean equipment.

SERVICING

Servicing should be carried out only by authorised persons using parts designed for the particular make of respiratory protective device. No attempt should be made to replace components or to make adjustments or repairs beyond the manufacturers' recommendation.

Self-contained compressed air breathing apparatus

Immediately after use, used and partly used cylinders or other containers should be replaced with correctly charged ones so that the breathing apparatus is ready for immediate use with a full nominal duration availability.

It should be remembered that a breathing apparatus team working duration is limited to the duration of the team member who has the lowest air reserve. Therefore, all team members should start with correctly charged cylinders.

DURATION OF CYLINDER

The duration of a specific cylinder and its contents can easily be identified and worked out by the markings and stampings on the cylinder casing.

Example
Medical air
Colour: French grey body
       Black and white quarters around top
Capacity and details
WC: Water content in litres (stamped on neck or top of cylinder)
TP: Test pressure
WP: Working pressure (BAR/ATS/PSI)
Locate WC on cylinder
Locate WP on cylinder

$$\text{Duration of cylinder} = \frac{\text{WC} \times \text{WP (ATS)}}{40}$$

Average consumption = 40 litres/min
e.g. WC = 2 litres
    WP = 200 ATS

$$\text{Duration} = \frac{2 \times 200}{40} = 10 \text{ minutes}$$

21.8 Ventilation

The requirement to provide adequate ventilation is covered under sections 2 and 3 of the Health and Safety at Work Act 1974. These are general duties and require employers to ensure that the premises are ventilated sufficiently so they do not place their employees at risk from dangerous gases or lack of respirable air. Further legislation covers the subject of ventilation in more detail such as the Factories Act 1961, Workplace (Health, Safety and Welfare) Regulations and the COSHH Regulations. As regards the COSHH Regulations, they require the initial control of harmful substances by means of other than personal protective equipment. The use of local exhaust ventilation or LEV is the most common method.

We can look at the subject of ventilation in two ways. First, there is a requirement for the employees to work in comfort. This means that a source of respirable or fresh air is provided to the workplace. Second, dust, vapours and fumes are removed from the workplace by means of extraction ventilation. There may also be a requirement for the removal of heat from the workplace. This can also be performed by extraction ventilation. Extraction ventilation may also be local or general.

If we were to look at comfort ventilation certain areas need to be addressed. As we are looking for the provision of a supply of respirable or fresh air the obvious criteria are that of rate of air movement and also the measurement of the number of air changes per hour. The air supply should be relatively dry and cool. As the supply of air coming into the work area is required to enable employees to breathe comfortably it should be from a clean source and in certain cases the air may need to be filtered.

In relation to extraction ventilation systems we can divide these into three main categories of receptor, captor and low volume/high velocity systems:

- Receptor system: A receptor system uses a fan to provide an air flow to carry contaminants from a type of hood to a collection point. These systems are used for highly hazardous contaminants.
- Captor system: This system again uses a hood but it is designed to collect the contaminants at a point outside the hood and encourage the flow of the contaminants into the hood and through the system. The requirements to capture different contaminants vary widely with some particles requiring only a capture velocity of 0.25 m/s while others require a velocity of 10 m/s.
- Low volume/high velocity: This system requires the collection hood or cowl to be placed very near to the source of contaminant. If a source of contaminant is through a small aperture and at a high speed then if the hood is placed close to the aperture very low air flow rates would be needed.

Another possibility would be to use dilution ventilation to lower the concentration of harmful contaminants. This can be done by inducing large volumes of air into the area required.

21.9   Safe working procedures

Employees whose duties involve working in confined spaces should give consideration to their state of health. Defects such as hearing, eyesight disorders together with a lack of sense of smell are all problems that may give rise to accidents while working. Other problems include physical deformities, claustrophobia, mental disorders and weaknesses which may increae the risks involved in this type of physical activity. Systems should be established which regularly monitor the health of such employees.

21.9.1   TRAINING

Training will vary depending on the requirements of the operational need. In view of the potentially high risk to employees when entering and working in sewers and confined spaces, personnel should be competent in the following aspects:

- Be capable of recognising a confined space
- Have a knowledge of current legislation relating to confined spaces
- Be able to identify potential hazards
- Understand permit to work systems
- Be able to interpret safe working procedures
- Recognise a need for hygiene precautions
- Understand and select appropriate protective equipment
- Have a thorough knowledge of gas detection equipment
- Be capable of exiting a confined space in an emergency
- Exit a confined space wearing suitable breathing apparatus
- Be able to use access equipment
- One team member would also require first aid at work training.

In certain situations the person may be required to enter as part of a rescue team in which case further training would be required.

21.9.2   MEDICAL INSTRUCTION CARD

A medical instruction card should be given to every employee likely to come into contact with sewage. This card should list the basic precautions to be taken by an employee against the risk of contracting leptospirosis jaundice.

In line with the Management of Health and Safety at Work Regulations employers should provide employees with information regarding known operational hazards, how they can be avoided and the most effective methods of dealing with them, e.g. a method statement.

21.9.3   PHYSICAL INJURIES

Many injuries can be caused to persons descending through manholes and working inside sewers from falls, slipping, cuts, scratches, implements dropped from above or faulty tools. Failure to report defects by the first person entering the system may result in other persons being injured.

Exiting from the workplace immediately to apply first aid is a priority in this situation. Injuries underground may result in a major rescue operation.

21.9.4   DANGEROUS ATMOSPHERES

Permission to enter a confined space should be by an authorised competent person who has relevant local knowledge of the system. Entry should only be attempted once the area has been tested and appropriate equipment is available. Continuous monitoring while persons are in the confined space must be carried out at all times as the conditions within the confined space may change. Alternatively if the confined space requires entering while the atmosphere is harmful this must only be done using suitable breathing apparatus, e.g. in the case of rescue.

Other gases as a result of accidental spillage or discharge of trade effluents either accidental or consented may cause hazards as a result of toxic or explosive fumes. It should be noted that gases or liquids that may be safe on their own may become volatile when mixed with other substances. Industrial areas where sewer work is to be carried out should be studied so as to assess the likelihood of this type of event happening. However, accidental spillages can occur almost anywhere and therefore persons working in underground environments should not only make correct use of the monitoring systems available, but should use the benefit of their own senses to monitor for the unexpected.

21.9.5   TYPICAL PROCEDURES FOR MANHOLE, SEWER OR TUNNEL ENTRY

Written procedures should take account of all local conditions. Some examples of principal procedures that may be developed are given in Table 21.4

At the depot

Receive the task, location, plans, records and known hazards – this enables the employer to assess the risks involved in carrying out the task. It should also give an indication of the manpower, equipment and control measures that need to be in place prior to entry.

Brief other members of working party – every member of the working party should be involved with the safety of the task and therefore needs to be fully aware of the procedures.

Checking of equipment – the equipment required needs to be inspected for suitability, serviceability and also that the right quantity is available. It is often a problem that insufficient equipment is provided for the number of persons entering.

Transporting of equipment – safety equipment can be not only expensive but also fragile and therefore the handling and storage are very important. The equipment is also very dangerous with regard to the compressed air cylinders, therefore during transportation the cylinders should be secure.

Communications – a form of communication between site and base should be established and this would be continued between the control point outside the confined space and the persons entering.

Arrival on site

Inform control of arrival – this allows the control to log the whereabouts of the entry team should there be more than one location to be visited.

No-smoking area – essential when there is the possibility of explosive gases.

Erect barriers and road signs – these should be erected to the relevant standards to prevent unauthorised access and protect third parties.

Opening of manhole covers – this may be for access purposes and also ventilation.

Ventilate sewer – although ventilation may only dilute toxic and flammable gases in the confined space, it may also make the atmosphere more acceptable as regards unpleasant odours. The use of gas monitoring systems is the key to assessing whether the confined space is safe for entry.

Table 21.4   Confined space work record/entry check list

If Unsatisfactory Contact Supervisor			Signature								
Equipment			Do not Smoke within the Vicinity of a Confined Space								
Depot	Waterproofs 1 set/person		Division								
Personal	PVC Gloves 1 pair/man		Location								
	Safety Boots 1 pair/man		Team Names								
	Waders/Thigh Boots 1 pair/man		1. 2.								
	Safety Helmet 1 per man		3.								
	Gas Detector 1 No Test		4. 5.								
Depot	Full Harness 1 No										
Safety	Lifeline 1 No		Sites/Locations visited								
	Winch/Tripod 1 No										
	B.A. 1 Per Man										
	Cap/Hand Lamps 1 Per Man										
	Resuscitator 1 No										
Depot	First Aid Kit										
Vehicle	Soap/Cleaner										
	Radio/Phone Working										
	Check Weather Conditions										
	Check Personal Equipment										
Site Checks	Check Gas Detector OK Yes/No										
	Check B.A. Equipment										
	Check Winch/Tripod										
	Check Harness/Lifelines										
	Check Hand/Cap Lamps										
Ventilate Space											
Assemble Safety Equipment											
Inspect Entry											
Isolate Machinery											
Entry	Lower Equipment										
	Gas Test 5 minutes	Hydrogen Sulphide Satis Flammable Gases Satis Oxygen Levels Satis									
	If applicable	Visib/Struct									
	Enter if checks OK Yes/No										
Complete Work/Recover Equipment											
Exit from Space											
Secure Site											
Clean Equipment											
Report alarms/defects/etc. to supervisor											

Check for unusual smells – there may be certain gases that cannot have been foreseen during the risk assessment that will not be detected by the available monitoring system but can be detected by the sensory organs.

Visual check – the condition of ladders, step irons, platforms and landings may be assessed from the surface.

Depth and velocity of flow – this may affect the ability of the persons entering to stand within the flow of a pipe, without being washed away.

Gas monitoring – a test of the atmosphere at all entry and exit points should be undertaken prior to entry. If unsafe to enter further tests should be obtained. Continuous ventilation should assist in this matter.

Rescue equipment – all appropriate rescue equipment should be set out prior to entry.

Personal safety equipment – the person in charge together with the person entering should check that their equipment is in good order before entering.

Entry

Briefing – as many top, intermediate and bottom persons as may be necessary must be briefed as to their duties.

Entry – all persons should descend on a safety line

First person – the first person down on discovering any defect must inform the person in charge.

Arrival – on arrival at landing or benching the person should stand clear of ladder and call the next person down.

Access – only one person on a ladder or step irons at a time.

Dangerous flows or falls – fix safety chain or barriers and where considered necessary running lines.

Traversing of sewer systems

Communications – confirm communication arrangements.

Traversing of sewer – inform person in charge that working party is ready to move off. Members of the working party should be in visual or aural contact.

Equipment checks – atmosphere monitoring equipment should be checked and tested at predetermined intervals.

Movement – walk slowly and carefully taking care not to disturb any sludge which may be present.

Responses – before answering pre-arranged calls from top man, check each gang member and atmosphere before answering.

Arrival at exit point – on arrival at exit manhole inform the person in charge.

Leaving the sewer

Persons to leave – members of sewer team not required for recovery of working equipment should leave the sewer. Fix safety barriers or chains.

Ascent – ascent by ladder or step irons should be carried out one person at a time.

Falling objects – do not stand under ladder or step irons while any person is ascending.

Head counts – when all working party, bottom personnel and any intermediate personnel are withdrawn conduct a head count.

Equipment check – all safety and working equipment must be checked for serviceability and quantity. Missing equipment left in a sewer system may cause operational problems in the future.

Control – control must be informed that all men are out of sewer.

Site safety – all manhole covers must be replaced, barriers and road signs removed.

Load vehicle – care must be taken when loading equipment onto vehicle. This equipment may well be contaminated and strict hygiene precautions should be observed.

Hygiene – carry out personal hygiene procedures.

Return to the office

Reports – report to control any abnormal occurrences, engineering defects, personal injury and defects in safety equipment.

Hygiene – carry out further hygiene procedures as you may have handled equipment from the vehicle on your return.

In the event of an accident in the sewer

Collapse of personnel – if a person collapses in a sewer, don breathing apparatus and prop the person up if possible. Leave the sewer immediately.

Recovery – do not attempt to recover collapsed person unless rescue breathing apparatus is available.

Emergency services – the person in charge should inform the emergency services and control immediately with details of accident.

Fire services – if fire services are used to recover injured person then the top person should give details of sewer layout, position of injured person and offer use of specialised equipment if required. The fire services, however, should have received a copy of the location prior to commencement of the works.

Emergency procedures

Gas alarm – if atmosphere monitors indicate a dangerous atmosphere don escape breathing apparatus and leave the sewer immediately.

Increase in flow – if depth or velocity of flow increases then leave the sewer immediately.

Calls from person in charge – if person in charge calls for sewer team to leave the sewer then leave the sewer immediately.

Recovery of equipment – do not try to recover working equipment during an emergency evacuation.

21.10   Rescue

When considering the rescue facilities for sewer entry work several questions need to be identified and answered. These may be as follows:

- Have all the emergency situations been identified?
- What is the risk of an emergency arising?
- Is it necessary to have rescue equipment and trained operatives on site during the course of the works?
- Are the emergency services, e.g. the fire service, able to respond quickly should a given situation arise?
- Are there adequate communications on site?
- Is there easy access to the site should the emergency services be called?

The above are just some of the topics that need addressing when setting up a safe system of work involving rescue.

Historically, reliance on the fire service has been high on the agenda of the majority of local authorities and water companies. Having carried out a variety of exercises with the fire service, the author has found that the equipment available to them and the lack of familiarity with the sewer environment leave them at a disadvantage, if compared to trained personnel who have worked in this type of environment. With reference to the equipment if would appear that the equipment used to combat fires is the type available to enter underground systems and may be suitable for certain tunnel systems where not only the size of the tunnel is large enough to allow entry with large self-contained breathing apparatus but the cover slab may also be absent. Should the rescue be attempted through a standard 610×610 mm manhole opening the restriction may prevent rescue personnel getting through with their breathing apparatus being worn.

Before considering the use of an outside organisation for rescue the response time should be assessed. Factors for the response times should be distance and access. If you consider that once the human brain is starved of oxygen for a certain period of time, it will start to deteriorate. If the response and setting-up time on site is of such a period that this is exceeded then the need for an on-site rescue team must be considered.

Once an on-site rescue team has been decided on then all personnel should be adequately trained and be provided with suitable and sufficient equipment. This equipment would include:

- Breathing apparatus
- Oxygen resuscitation
- Stretcher
- First aid kit
- Lifelines
- Drag sheets
- Communication equipment
- Etc.

The above would be in addition to the usual confined space entry equipment.

21.11 Hygiene

People are very familiar with everyday hygiene precautions in the home in relation to bodily waste. This waste is passed into the sewer system for the local water companies to deal with. Sewer workers therefore are at risk of contracting various ailments as a result of this waste product. Sewers are also a habitat for an increasing rat population. These rodents and in particular *Rattus norvegicus* are a source of a disease commonly known as weil's disease or leptospirosis (Fig. 21.2)

Zoonose – disease passed from animal to humans

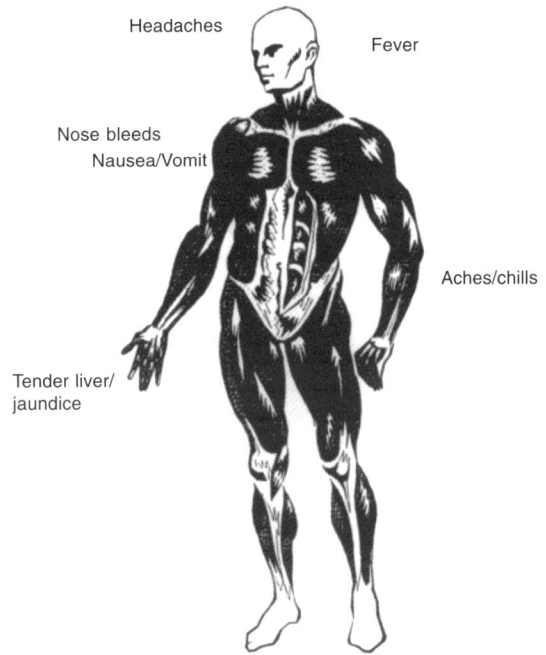

Figure 21.2 Weil's disease.

This disease, although recognised as a distinct clinical entity in 1886, did not have its organism identified until 1916.

Inadequate hygiene precautions can result in death to the unwary. It is a potentially fatal disease which is a notifiable condition caused by bacteria in rats' urine which can contaminate fresh water, including rivers, ponds reservoirs, etc.

The condition can affect not only sewer workers but also farmers and water sports enthusiasts. It is not only a disease associated with the rat but also with cows and to a lesser degree dogs. Although worldwide there are over 200 strains of the disease only three serogroups are associated with human infection:

- Icterohaemorrhagiae: From rats and can cause liver or kidney failure, conjunctivitis and meningitis. It has a fatality rate of 5–10%.
- Canicola: From dogs and can cause jaundice, conjunctivitis and meningitis with a fatality rate of less than 1%.

- Hardjo: From dairy herds and can cause flu-like symptoms with severe headache and meningitis but is rarely fatal.

As far as detection in humans is concerned antibodies are not always detectable in the blood for 7–10 days after contraction.

It is essential that persons who come into contact with the rat population take adequate measures. This may involve the covering of cuts and abrasions. Persons at risk should refrain from rubbing their eyes or touching their mouths and noses after coming into contact with potentially contaminated substances.

21.12 Summary

This chapter is meant as an in-depth guide to safe working in confined spaces. In the final analysis it is essential to plan activities that involve the safety of personnel working underground in potentially hazardous areas. This planning must be thoroughly documented, and cater for the provision of adequate information, instruction, training and equipment. As regards persons being unsure of the requirements of a safe system of work they should remember 'If in Doubt Don't Enter'.

Bibliography

Croners, *Health and Safety at Work,* 1996.

Mepham, P. Infectious Diseases at Work – *The Safety Practitioner*, July 1988.

National Joint Health and Safety Committee for the Water Services. Guidelines for the selection maintenance and training in the use of respiratory protective equipment, 1979.

National Joint Health and Safety Committee for the Water Services. Safe Working in Sewers and at Sewage Works, 1979.

Neotronics. *Minigas User Manual,* 1995.

Sampson, Bill. *Safe to Enter – Working in Confined Spaces,* 1993.

22

Contract Supervision

Barry H. Lewis* CEng, MICE

22.1 Introduction

For too long contract supervision has meant the separate supervision of the various aspects of the contract so that planning and designing the project, drawing up the contract, the tendering procedure, phasing the project, site supervision and the final account were often dealt with in groups independent of each other. 'Safety' was an item that was given consideration from time to time, as finances and time permitted.

As a consequence many major projects gain publicity for the unforeseen time delays and budget overspends due to lack of preparation in design, risk assessment and drawing up the contract. At the same time many of these projects suffered fatal or near fatal accidents and the safety record of the industry became a serious concern see Table 22.1. It was not before time, therefore, that the government and the profession brought about new regulations and codes to instigate a more holistic conception of contract supervision. The Construction (Design and Management) Regulations 1994 (CDM) and the New Engineering Contract

Table 22.1 Likely severity of on-site accidents

	High	Medium	Low	
Falls from height	•			More than 2 metres
		•		2 metres or less
Being struck by mobile plant	•			
Tripping			•	
Collapse	•			
Manual handling		•	•	Depending on the object handled
Moving objects	•			
Electricity	•			Normal voltage and above
			•	110 V and below
Contact with moving machinery	•			
Fire	•			
Harmful substances	•	•	•	Depending on the substance concerned
Noise and vibration	•	•	•	Depending on exposure levels

'High'   – Fatality; major injury or illness causing long-term disability.
'Medium' – Injury or illness causing short-term disability.
'Low'   – Other injury or illness.

---

*Consultant Engineer – Support Services. Formerly Divisional Engineer with the City Engineer's Department, Manchester.

(NEC) published in 1993, now published in its second edition as the NEC Engineering and Construction Contract (ECC), give emphasis to the integration of all aspects of a contract at the development and design stage and not just considered as they arise.

While the ICE Conditions of Contract 5th edition, and subsequently the 6th edition, have been employed on sewer construction contracts for many years the water industry has found that as an alternative the I. Chem. E. Model Form of Conditions of Contract for Process Plant, Reimbursable Contracts, 2nd edition (i.e. the I. Chem. E. *Green Book*) has certain advantages. In the confines of a chapter it will not be possible to give more than an outline of each of the above listed aspects of contract supervision and greater import will be given to the more recent developments in this field, as mentioned in the previous paragraph.

A successfully concluded contract is more likely to come about if *all* parties achieve their preset targets. The contractor aims to complete the work within the programmed time while making the assessed profit. The engineer aims to hand over to the client a structure suitable for its purpose and to the required standards at a cost within the allowed budget. Those interested in the environs to the project require that minimal disruption is caused and that the locality is reinstated to, at least, the same standard as prior to the commencement of the work. All parties are concerned that the operation of the works will be carried out at minimal risk to site personnel, the public and adjoining structures and properties.

To achieve all these aims, detailed planning is required. Safety planning is now a requirement of the law but planning construction at as early a stage as possible will pre-empt the unforeseen. Variations and delays cause friction between the parties to the contract and tend towards the short cutting of procedures, with the resultant reduction in prime issues such as safety and quality. Sewer replacement and the construction of new sewers are no different in principle to other major civils works; however, they do incorporate some of the most interesting and difficult aspects of civil engineering expertise.

22.2   Contract responsibilities

Clarity is the key quality of any contract and much of the 6th revision of the ICE Conditions of Contract was motivated by the need to produce a more 'user friendly' document. The responsibilities of each party to the contract must be thoroughly understood as anything to the contrary leads to confusion and disputes with the resultant loss of co-operation between the parties at the expense of efficiency and achievement of the programme timing. Clause 1.1 of the Conditions of Contract gives the parties to the contract the titles of employer engineer and contractor and defines the overall responsibility of each role.

22.2.1   *THE EMPLOYER*

The employer is defined as the person, firm or organisation named in the Appendix to the Form of Tender and includes the employer's personal representatives, successors and permitted assigns. He is the client or promoter and is the main party to the contract as it is he who commissions the project and provides the finance for the design, supervision and construction of the project. However, when the contract has been entered into then employer and contractor assume equal if different responsibilities. The employer's principal responsibility is to define the functions that the project is to perform, to issue to the engineer the relevent information and data held by him, to obtain all necessary legal authority and approvals for construction and commissioning of the project, to finance the project and pay for the other parties to the contract and to acquire the land necessary for the project and its construction. His other responsibilities are usually delegated to the engineer.

22.2.2 THE ENGINEER

The employer must nominate the engineer in the contract documents and any change of personnel must be notified in writing to the contractor. The engineer's function is to plan and design the project, draw up the contract, implement the tender procedure let the work and then to supervise the work, authorise payments and issue completion certificates. The engineer's responsibilities and powers are clearly defined in the ICE Conditions of Contract and must be thoroughly understood prior to agreeing the contract.

The engineer acts on behalf of the employer and must ensure that he exercises scrupulous integrity and employs due professional skill as any instructions issued by him are legally binding on the employer. However, the engineer must be impartial as he is not a party to the contract and gains no benefit from its execution. Although the engineer is bound to protect the employer he is not to promote the employer's interests and he must not lose sight of the contractor's rights and entitlements.

The decisions of the engineer form a 'contract' with the contractor which is only enforceable between the contractor and the employer so that any breach of the 'contract' may result in legal action against either party. The engineer is not empowered to alter any of the contract conditions or to relieve the contractor of his responsibilities unless so specifically stated in the contract, nor is he authorised, without the employer's approval. Here it is worth quoting from Clause 2, subclause (2) of the ICE Conditions of Contract 6th edition:

> Where the Engineer . . . is not a single named Chartered Engineer the Engineer shall . . . notify to the Contractor in writing the name of the Chartered Engineer who will act on his behalf and assume the full responsibilities of the Engineer under the Contract.

Employers with a continuous programme of civil engineering projects, such as large local and regional authorities, usually delegate the duties of the engineer to their chief technical officers and under such circumstances it is difficult for the engineer to act fairly and without bias. However, he must always bear in mind that any action or instruction may ultimately be judged by an arbiter.

22.2.3 THE PLANNING SUPERVISOR

The appointment of the planning supervisor is another critical duty of the employer and/ or the engineer, which appointment must be carried out in compliance with the Construction (Design and Management) Regulations 1994, commonly referred to as the CDM Regulations.

This is a statutory appointment and should be made as soon as the client knows enough about the project to decide who would be best suited to the role. Should the post fall vacant during the project an immediate replacement must be found in order to comply with the law.

The Regulations do not stipulate whether the planning supervisor need be independent of the other parties to the contract only that he be competent and adequately resourced for health and safety.

The duties of the planning supervisor include the following:

- Ensure that the Health and Safety Executive is notifed

- Supervise a pre-tender *health and safety plan* and introduce a *health and safety file*.
- Oversee the design of the project to ensure safety considerations are foremost at the design stage
- Ensure that measures are in place so that individual designers liaise on health and safety matters
- Advise the client on the competence and resources of designers and contractors and on the suitability of the safety plan.

The *safety plan* should be developed as the design proceeds and must be ready at the tender stage so that it can be issued with the tender documents. It should include all information on particular risks and critical operations that require particular consideration of working methods (Table 22.2(a)). An example of the design and construction information which may be included in the safety plan is shown in tabular form (Table 22.2(b)).

The *health and safety file* should include relevant information as follows:

- 'Record' or 'as-built' drawings and plans used and produced throughout the construction process.
- The design criteria
- General details of the construction methods and materials used
- Details of the equipment and maintenance facilities in the structure
- Maintenance procedures and requirements for the structure
- Manuals produced by specialist contractors and suppliers which outline operating and maintenance procedures and schedules for plant and equipment installed as part of the structure
- Details of the location and nature of utilities and services, including emergency and fire-fighting systems.

22.2.4 ENGINEER'S REPRESENTATIVE AND STAFF

It is usual for the engineer to appoint a representative, usually termed the resident engineer, to watch and supervise the works. Subclause (3) states:

(a) The Engineer's Representative shall be responsible to the Engineer who shall notify his appointment to the Contractor in writing.
(b) The Engineer's Representative shall watch and supervise the construction and completion of the Works. He shall have no authority (i) to relieve the Contractor of any of his duties or obligations under the Contract nor except as expressly provided hereunder (ii) to order any work involving delay or extra payment by the Employer or (iii) to make any variation of or in the works.

The engineer may delegate any of his duties (except those listed below) to the engineer's representative provided that such be in writing and that prior written notification is also given to the contractor. Any of these delegated duties may be revoked at any time, again subject to written notification. The duties which may *not* be delegated by the engineer are those which relate to the following clauses of the ICE Conditions of Contract 6th edition:

- Clause 12(6) 'Delay and extra cost': Giving power to determine payment for unforeseen adverse conditions or obstructions.
- Clause 44 'Assessment of delay': Entitling the engineer to assess any claim for delay and grant any extension of time to the contract period that he considers reasonable.
- Clause 46(3) 'Provision for accelerated completion': Enabling the engineer to agree modified payments for any required acceleration of the works.

Table 22.2 (a) The pre-tender stage health and safety plan

1. Nature of the project	Name of the client Location Nature of construction work Timescale for completion
2. The existing environment	Surrounding land uses and restrictions, e.g. shops, schools, businesses, etc. Relevant planning restrictions Existing underground services and overhead lines Existing traffic systems and restrictions, e.g. access for fire appliances, delivery times, parking requirements, etc. Existing structures, e.g. special health problems from existing materials on site, any fragile materials or instability problems Ground conditions, e.g. contamination, gross instability, possible subsidence, old mines, underground obstructions
3. Existing drawings	Available drawings of structures to be demolished or incorporated in proposed structures (may include a health and safety file held by the client)
4. The design	Significant hazards or work sequences identified by designers which cannot be avoided or designed out plus precautions assumed for dealing with them The principles of the structure's design and any precautions that might be needed or sequences of assembly that need to be followed during construction Detailed reference to specify problems where contractors will be required to explain their proposals for managing them
5. Construction materials	Health hazards arising from construction materials where particular precautions are required, due to their nature or their use, which the designer cannot design out. They should be specifically specified to enable a competent contractor who may be assumed to know the precautions that suppliers are legally required to provide
6. Site-wide elements	Position of the site access and egress points for deliveries and emergencies Location of temporary site accommodation Location of unloading, layout and storage areas. Traffic/pedestrain routes
7. Overlap with client's undertaking	Consideration of the health and safety issues which arise when the project is to be located in premises occupied or partly occupied by the client
8. Site rules	Specific site duties which the client or planning supervisor may lay down as result of points 2 to 7 or for other reasons, e.g. specific permit-to-work rules, emergency procedures
9. Continuing liaison	Procedures for considering the health and safety implications of design elements of the principal, and other, contractor's packages Procedures for dealing with unforeseen eventualities during the project execution resulting in substantial design change which might affect resources

Table 22.2 (b) Safety plan – design/construction information

Hazard	Location	References	Site-specific hazard control
Interface with pedestrians	Around compounds	Code of Practice for Safety at Street Works and Road Works	Signs advising of pedestrian routes required. Particular care required in vicinity of hospital.
Interface with traffic	Access to compounds	*Traffic Signs Manual*, Chapter 8	Contractor's vehicles to comply with temporary traffic orders. Reflective jackets to be worn.
Services	All excavations	Recommendations on the Avoidance of Danger from Underground Electricity Cables Model Consultative Procedure for pipelines in deep excavations	Actual positions of services to be determined using all known detection methods. Procedures to be developed for the protection of services exposed in excavations. Special requirements of statutory undertakers to be followed
Mechanical plant operations	Within compounds		Contractor to develop safe working procedures for all plant operations. Plant to be operated only by trained personnel.
Manual handling operations	During construction	Manual handling operations	Minimise the requirement for handling materials. Train personnel in safety techniques.
Lifting operations	During construction		Protection to be provided where risk to damage to underground services exists. Slewing of plant outside compounds to be prohibited.
Excavations	During construction		Contractor to design suitable temporary support measures. Services to be located before excavation commences.
Confined spaces	All manholes and excavations	NWW SSW No. 13 – Entry into Confined Spaces	Permit to work/enter system required.
Disease	Live sewers		Train personnel regarding risks and necessary precautions.
Needles/ syringes	Sewers/manholes		Contractor to develop safe working procedure for dealing with discarded hypodermic needles/syringes. Advice may be obtained from environmental health dept.
Hazardous substances	During construction	COSHH regulations	COSHH assessments to be carried out. Personnel to be trained in risks and precautions to be taken.
Flooding	Live sewers		Contractor to provide adequate overpumping. No working in

Table 22.2(b)   Safety plan – design/construction information

Hazard	Location	References	Site-specific hazard control
			live sewers during periods of heavy rainfall. Connections to lined sewers to be reopened as soon as possible.
Pollution	During overpumping		Contractor to ensure that discharge points for overpumping are to foul or combined sewers.
Noise	During construction	Prior consent	Contractor to use and maintain suitable plant to reduce noise levels to minimum. Contractor to carry out noise level assessments and provide hearing protection as necessary.
Waste disposal	During construction		All waste to be disposed of by registered carriers to licensed tips.
Security	All compounds during construction	HSE Guidance Note GS7 – Accidents to Children on Construction Sites	Compounds to be fenced in accordance with the guidelines in HSE Note GS7. Compounds to be secured when lift unattended.

- Clause 48 'Certificate of substantial completion': Enabling the engineer to issue a certificate of substantial completion for any section of the work.
- Clause 60(4) 'Final account': Placing responsibility on the engineer to issue a certificate of final payment within three months of receipt of the contractor's final account.
- Clause 61 'Defects correction certificate': Empowering the engineer to issue a defects correction certificate to the employer stating the date on which the contractor shall have completed his obligations.
- Clause 63 'Determination of contractor's employment': Empowering the engineer to issue a certificate to the employer in respect of the contractor's non-compliance.
- Clause 66 'Settlement of disputes'.

Commonly the duties which the engineer elects to delegate may include most or all of the responsibilities defined by the following clauses of the ICE Conditions of Contract 6th edition (Table 22.3).

Subclause (5) allows for the engineer, or his representative, to 'appoint any number of persons to assist the Engineer's Representative'. 'Such assistants shall have no authority to issue any instructions to the contractor 'other than those necessary for them to carry out their duties and any such instructions may be referred by the contractor to the Engineer's representative for confirmation'. The representative's staff may comprise assistant resident engineers, each responsible for a separate section of the work with clerks of work and inspectors responsible to each assistant resident engineer. The number of staff employed will clearly depend on the size of the project and the extent of the site and where different engineering skills are being carried out simultaneously it may be appropriate to designate particular staff to each section of the work. This is particularly advantageous where the contractor also has different staff supervising separate sections of the project. An example

Table 22.3 Engineer's delegated duties

Clause no.	Clause title	Clause no.	Clause title
13(1)	Work to be to satisfaction of Engineer	38	Examination of work before covering up
13(2)	Mode and manner of construction	39	Removal of unsatisfactory work and materials
14(1)	Programme to be furnished	45	Night and Sunday work
14(2)	Action by Engineer	49(1)	Work outstanding
14(3)	Provision of further information	49(2)	Execution of work of repair etc.
14(4)	Revision of programme	49(3)	Methods of construction
14(5)	Design criteria	49(4)	Remedy on failure to carry out work required
14(6)	Method of construction	50	Contractor to search
14(7)	Engineer's consent	51(1)	Ordered variations
14(8)	Delay and cost	51(2)	Ordered variations to be in writing
15(1)	Contractor's superintendence	52(3)	Daywork (ordering work)
18	Boreholes and exploratory excavation	52(4)	Notice of claims
19(1)	Safety and security	53(1)	Vesting of Contractor's equipment
20(3)	Rectification of loss or damage	56(1)	Measurement and valuation
31(1)	Facilities for other contractors	56(3)	Attending for measure
32	Fossils etc.	56(4)	Daywork (payment)
33	Clearance of site on completion	58(1)	Use of Provisional Sums
35	Returns of labour and contractor's plant	58(2)	Use of Prime Cost Items
36	Quality of materials and workmanship	60(1)	Monthly statements
37	Access to site	62	Urgent repairs

of such division of duties might be where a new sewer is being laid partly in trench work and partly by tunnelling and both sections of the work are programmed to run simultaneously. Alternatively the pipe laying functions may be supervised by one dedicated group of staff with numerous accommodation works being dealt with by another (Fig. 22.1).

All contracts should have appropriately experienced staff supervising each stage of the project production. The engineer must appoint an engineer's representative (resident engineer)

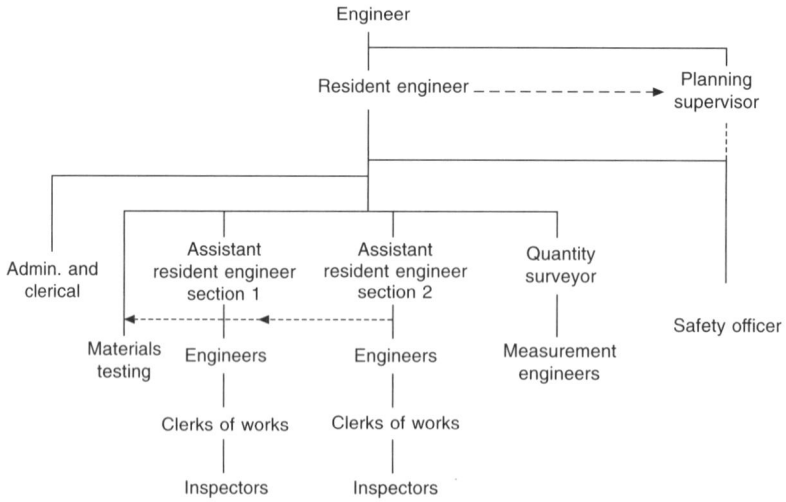

Figure 22.1 Engineer's site staff.

and other site staff with the right amount of construction experience to supervise the construction works. Ideally the resident engineer's function should start well before the on-site start date and his involvement at the design and feasibility stage is essential if the proper risk assessment process is to be carried out. Too often senior site staff are recruited from the design team with the view that, as they understand the design concepts, they are best suited to supervise the works, irrespective of their on-site contract experience. This can lead to dogmatic insistence on the construction of design features which, with a more detached and flexible approach, might lead to time and cost savings, without detraction from the quality or suitability of the construction.

Involvement of an experienced resident engineer at the design stage can result in the detailing of the design being more tailored to the site conditions and producing a more pragmatic approach at the site supervision stage. This form of liaison between site and design staff will avoid differences of opinion occurring at the start of the site work. On contracts where there has been little time for the resident engineer to ponder the design the relationship between designer and the resident engineer can become strained as minor adjustments to design are requested in order to facilitate construction. Consequently, there then becomes a conflict of interests where the designer resists what he comes to consider as attempts to compromise his design criteria. Such differences are best aired prior to the completion of the design in a spirit of co-operation.

To achieve this 'ideal partnership' situation the engineer must be certain that he has selected a highly experienced representative and that he is supported by staff with the necessary abilities.

22.2.5 THE CONTRACTOR

The contractor, that is the person, firm or company to whom the employer awards the contract, undertakes under Clause 8 to construct and complete the works and to provide all labour, materials and everything whether of a temporary or permanent nature required for completion of the works. He shall not be responsible for the design or specification unless it is expressly provided in the contract and then only in as far as to the use of reasonable skill, care and diligence. This latter limitation on Clause 8 was introduced in the 6th edition of the ICE Conditions of Contract. The contractor must also take full responsibility for the adequacy, stability and safety of all site operations and methods of construction. Clause 3 precludes both contractor and employer from transferring all or part of the contract or assign any benefit or interest without the written consent of the other party.

The ICE Conditions of Contract have been revised in its 6th edition to recognise that subcontractors and the self-employed now constitute the major proportion of the labour force in the construction industry. Subletting the whole of the works requires the employer's consent; subletting part of the works requires that prior notification is given to the engineer. The employment of labour-only subcontractors does not require prior notification. The contractor is still fully responsible for all subcontracted work (see Clause 4(4)) and the engineer has the specific right after due warning is given in writing to require the contractor to remove any subcontractor from the site who misconducts himself (Clause 4(5)). This is in addition to the powers given to the engineer under Clause 16 to require the removal of any employee of a subcontractor who misconducts himself.

Within 21 days of the award of the contract the contractor shall submit to the engineer a programme of works, taking into account any condition regarding possession of the site as prescribed by Clause 42(1), together with a general description, in writing, of the arrangements and method of construction which the contractor intends to utilise. Within 21 days of receipt of such proposals the engineer shall accept or reject them, failure to respond within this timescale shall be deemed as acceptance. In the event of the engineer's rejection, the contractor must submit a revision of the programme within a further 21 days.

It is the contractor's responsibility to provide the appropriate superintendence of the works either personally or by appointing a representative to constantly attend the work. This person is usually called the 'agent' or 'project manager' and his main duties are to ensure the proper, efficient management of the construction works in accordance with the contract and within the contractor's programme and financial targets Fig. 22.2)

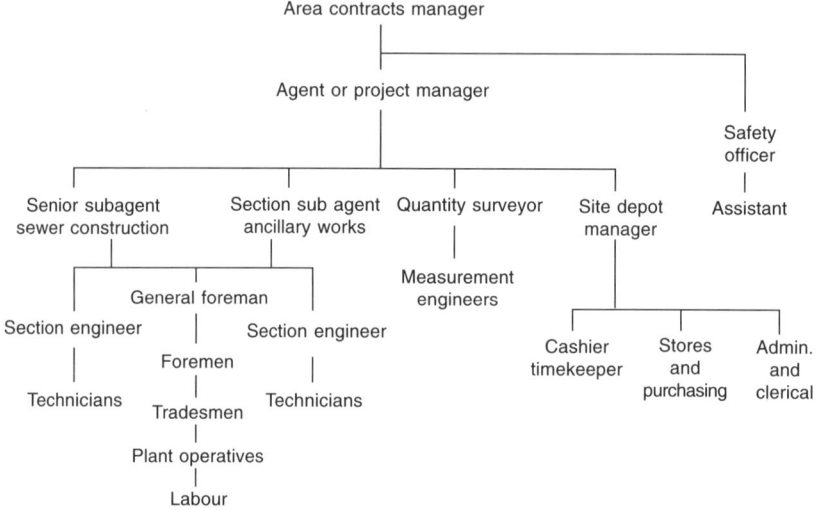

Figure 22.2   Contractor's contract and site staff.

It is essential that the Agent is experienced and technically competent so that he and the Resident Engineer form a working relationship by which they may, together, deliver the project on time and to the required standard of workmanship.

22.3 Programming the work

The employer's requirements will more or less determine the start and completion dates of the contract. The date of legal possession of the site needs to be assured before the date for commencement of the works can be firmly fixed. Many a contract has fallen behind the initial programme before site work has commenced due to delays with land acquirement. Consequently this aspect of the planning and programming of a project is as important as any aspects of the construction process and I shall elaborate further on site possession later in this chapter.

Thorough and accurate programming of the work at the earliest possible opportunity is a major factor in determining whether or not the contract will be achieved satisfactorily, i.e. to specification and design, on time, within the target costs and at a profit to the contractor. Therefore the sooner the parties of the contract can be drawn together to discuss the programming issues the more likely are the above to be achieved. The ability to bring the contractor into the planning of the project at an early stage will depend on the form of contract entered into and these have been expanded on earlier in this chapter. It is also essential that the engineer's representative or resident engineer is consulted throughout the design stage as this will facilitate the smooth supervision of the contract and will ensure that many minor anomalies are resolved prior to commencement on site.

Clause 14 of the ICE Conditions of Contract states that the contractor has 21 days after the award of the contract to submit his programme, together with details of the mode of construction, in writing to the engineer (Table 22.4). The engineer may then accept or reject these proposals, within a further 21 days, or request further clarification or substantiation of the details or question the reasonableness of the proposals in relation to the contractor's obligations. If the engineer rejects the programme the contractor must submit a revision within 21 days of the receipt of the notification of rejection or within a period agreed jointly. At any stage if the engineer does not respond within the stipulated 21 days the last submitted proposals will be deemed to have been accepted. If during the progress of the work the engineer considers that the contractor is not conforming to the agreed programme he may require that a revised programme be submitted showing modifications that will ensure the completion of the work or any section of the work within the time specified by the conditions of contract.

The engineer may instruct the contractor to supply him with the details of proposed methods of construction, including temporary works and use of equipment and for calculations of any loadings that may affect the permanent works. This is clearly essential if the engineer is to ensure that the proposed methods of construction and use of equipment comply with the contract and are not to have a detrimental effect on the completed structures. Again the engineer has 21 days to approve or reject the contractor's proposals or require changes to the proposals. Once consent has been received the contractor may not amend the methods agreed by the engineer without the further approval of the engineer. Needless to say all of the above communications and approvals must be in writing and shall not relieve the contractor of any of his obligations or responsibilities under the contract. It is worth noting that Clause 14 has been amended under the 6th edition of the ICE Conditions of Contract to enable the contractor to claim costs with a profit margin added.

In complying with Clause 14 the engineer's consent or acceptance of any proposals, submitted by the contractor in relation to the above, may not be 'unreasonably' withheld. Should the engineer's consent be 'unreasonably' delayed or should the engineer or the design criteria impose limitations to the extent that the contractor incurs unavoidable delay or cost, then the engineer shall take these into account when the contractor submits a claim for extension of time under Clause 44 and additional payment under Clause 52(4). However, the above entitlement to claim is subject to the provision that the 'limitations' could not 'reasonably' have been foreseen by an experienced contractor at the time of tender.

This of course is very clear, apart from the small matter of what shall be considered 'reasonable' and 'unreasonable' and here we have what frequently becomes a bone of contention on many contracts and polarises opinion between the client's view and the contractor's view on the legitimacy or otherwise of a claim for delay and cost. No text book can give definitive guidance on such a matter as the details and conditions arising in each case are different, hence the need for highly experienced staff to be employed by the engineer and the contractor. I have raised the issue at this point rather than leave its discussion to a chapter on contractural claims and arbitration in order to emphasise the importance of the employer ensuring that all information is available to all parties, at as early a stage as possible, and that all parties combine to agree a fully developed and highly detailed works programme.

Too often a contractor is expected to submit an outline programme with his tender and

Table 22.4

Year 1991 Month	Jan		Feb				March					April				May				June					July				Aug					Sept				Oct	
Week ending	19	26	2	9	16	23	2	9	16	23	30	6	13	20	27	4	11	18	25	1	8	15	22	29	6	13	20	27	3	10	17	24	31	7	14	21	28	5	12
Project week no.	1	2	3	4	5	6	7	8	9	10	11	12	13	14	15	16	17	18	19	20	21	22	23	24	25	26	27	28	29	30	31	32	33	34	35	36	37	38	
Establish compound S1																																							
Establish compound S2																																							
Foul water P.																																							
Foul water P.S. building											E																												
Surface water pump stn											A																												
Shaft F5 finishes											S																												
Shaft F4 finishes											T																												
Shaft F3 finishes											E																												
Pipejack to F5 drive			strip								R																												
ditto grout		set up								set up										strip																			
ditto foul											H																												
Pipejack F4 to F3 drive											O																												
ditto grout										strip																													
ditto foul											L																												
Pipejack F5 to F4 drive											S																												
ditto grout																																							
ditto foul																																							
Shafts P1 and P2 finish																																							
300 mm dia. rising main																																							
Heading F5 to exist P.S.																																							
Heading FWPS-incSWPS																																							
Hotel connections																																							
Amendments to exist P.S.																																							
Abandon exist outfall																																							

Complete section one   Complete section Two   Substantial completion this is accepted as complying with Clause 14. This programme may need continual amendment as further design details and site conditions become evident so that the engineer has to trust the interpretation of the conditions of contract to ensure that he maintains some control of the programme. If the contractor fails to comply with the time requirement, when providing the requested information, there is no penalty stipulated in the conditions but, as the contract period will have been fixed, he is likely to incur an additional time constraint. While the above clauses do empower the engineer to obtain a detailed programme this may well lead to a confrontational relationship between him and the contractor, a situation from which nobody benefits.

22.4   Service diversions and accommodation works

Apart from the construction works the programme has to include accommodation work and service diversions. Unless the sewer is to be constructed on a green field site then it is certain that any or all of gas, water, electric, telecom, and cable services will be encountered as well as vehicle and pedestrain traffic. These may require temporary or even permanent diversion and complications to the programming of works will be encountered when the service companies carry out their statutory right to remove, divert or protect their own duct, cable or pipeline. Here the project is immediately in the hands of other statutory authorities over which the employer has no control.

The managing of works of this nature requires full knowledge of all services at the preliminary stages of the project investigation process so that the service companies can be given the optimum notice of the requirement to deal with their respective services. Lack of such notice invariably leads to delays in the contract programme as the service company is given too tight a schedule to carry out its work, bearing in mind the emergency functions to which it has a statutory duty to attend.

Unfortunately, due to the age of services in this country, service information is often unreliable. It is necessary therefore that site investigation work is carried out thoroughly and that information from the service authorities is checked physically on site, as far as is feasible, by means of trial holes and trenches. Again the criteria is to leave as little as possible to chance as time expended prior to the commencement of the work on site can save time and money as the contract progresses.

Service mains will often be located running across excavations and need to be supported, with the approval and under the supervision of the respective statutory undertakers. Although such services mains inevitably hamper progress, diversion of the mains may be considered too costly or too disruptive from the service customer's point of view and also as far as the contract programme is concerned. Too often bits of rope, wire and timber are used to support pipes and ducts running across trenches; however, purpose-made slings and chains are far more effective and take up less room and can also be easily repositioned when necessary to facilitate the construction works (Fig. 22.3). Needless to say such service pipes and ducts have to be treated with supreme care as one careless act by a machine driver can cost a small fortune in cash terms and, while the contractor must be fully insured for such eventualities, he will not have anticipated the disruption to his works schedule and the knock-on effect that this may have on his performance. The experienced resident engineer will be only too aware of the adverse effect damage to a major service main can cause, not only in terms of the contract programme but also to the service customers and therefore the reputation of the employer. Subcontractors are prone to such accidents due to coming to the site some time after the start date so that they may not have been fully

Figure 22.3   Service pipes supported across excavation.
Photographs by courtesy of Manchester City Council – Environment and Development Design Consultancy.

briefed by the main contractor who may already be under pressure on his own works and therefore affords only perfunctory superintendence on his subcontractor.

22.5   Possession of the site

Having clarified all legalities with regard to the land upon which the project is to be constructed, the timing of the possession of the site needs to be determined. On a large sewer contract the overall length of the structure may be considerable and it is inevitable that the possession of 'the site' needs to be phased. Clause 42 allows for the extent of portions of the site of which the contractor may take phased possession to be stipulated, giving the order of possession of such portions and the availability and nature of the access to be provided. Also the order in which the sections of the work is to be undertaken may be prescribed.

The work's commencement date is that date specified in the appendix to the form of tender or, in accordance with Clause 41, if no date is specified, a date within 28 days of the award of the contract as notified in writing by the engineer. Alternatively, the commencement date may be by mutual agreement of the parties. The clause requires the contractor to start work on site as soon as is reasonably practicable after the commencement date. However, once again no clear guidelines can be given as to what constitutes 'reasonably practical' thereby leaving a further issue for possible dispute. Once commenced the contractor is obliged, under the terms of Clause 41(2) to proceed with the works with due expedition and without delay as required by the terms of the contract.

For his part the employer is obliged under Clause 42(2) to give the contractor, on the work's commencement date, possession of as much of the site, together with access to it, as may be required to enable the contractor to start and continue with the work. Also, during the course of the contract the contractor shall be given possession and the necessary access to it, of further portions of the site as is required so that the contractor may undertake his programme of works, as was accepted by the engineer under Clause 14.

Failure on the part of the employer to give possession of the site in accordance with the clause will render him liable to meet any resultant costs incurred by the contract (subject to Clause 52(4)) or a possible claim for an extension of time (in accordance with Clause 44). The engineer is given the responsibility by this clause to determine the quantum of such a claim, which shall include a profit element, and shall notify the contractor with a copy to the employer (Table 22.5).

Table 22.5   Checklist of actions required by the engineer

	Refer to:
Programme	
1 Check that the land for the works and rights of access have been acquired.	* Action by Promoter.
2 Check that the promoter has obtained planning, financial and other approvals to proceed.	* Action by promoter.
3 Notify the contractor of date for commencement.	ICE Cl. 41
4 Get the contractor to submit a programme and description of arrangements and methods of construction. Inform the contractor of consent or of requirements for changes and resubmission.	ICE Cl. 14 and Cl. 13
5 Agree with the contractor a programme for issue of further drawings.	(ICE Cl. 7
6 Agree with the promoter a programme for decisions required from the promoter.	
Organisation	
7 Appoint the planning supervisor	CDM Regs
8 Select, train and arrange induction of the engineer's representatives and staff.	
9 Delegate to representatives by letter – copied to the contractor.	ICE Cl. 2
10 Approve the contractor's agent.	ICE Cl. 15
Communications	
11 Agree with the promoter a system and the scope of reports for promoter and others.	
12 Agree with the promoter responsibilities for relationship with the press.	
13 Arrange start-up meeting with the contractor's agent and senior management, on-site relationships, contract administration, safety, security and welfare. Invite the chief officers of the local authorities, health and safety inspectorate, and others as appropriate.	
14 Agree a system and scope of meetings, reports and instructions with the contractor.	ICE Cl. 13 and Cl. 68
Bonds and insurance	
15 Call for bonds from the contractor, if contract requires them (for the promoter).	ICE Cl. 10
16 Call for insurance certificates from the contractor (for the promoter).	ICE Cl. 23
Subcontracts	
17 Call for list from contractor of proposed subcontractors.	ICE Cl. 4
18 Nominate or state dates when subcontractors will be nominated for provisional sums and prime cost items.	

22.6 Site supervision

22.6.1 *SUPERVISION AND SUPERINTENDENCE OF THE CONSTRUCTION*

Supervision and superintendence are two separate and distinctive roles which should not be confused. The latter is the function of the contractor's agent and relates to the direct organisation of the construction activity. The main components of this function involve the programming of the works and cash flow targets and the direct organisation of the labour resource, the use of plant/transport and purchase of materials.

The resident engineer as the engineer's representative is responsible for the on-site supervision of the works. His function is to examine and approve the contractor's methods, materials and workmanship and to assess and measure the work for payment, including any claims for additional payment or extension of contract period. As detailed elsewhere in this chapter, the ICE Conditions of Contract specifies which decisions, arising from these assessments, are delegated to him and which he must refer to the engineer.

The resident engineer has powers under the contract to decide which of the contractor's methods, materials and work he will accept and which he will reject. However, he should look to use these powers sparingly and seek to discuss the contractor's proposals before they are made formal so that he obtains the agent's compliance through consultation and persuasion. If the resident engineer demonstrates his knowledge and experience he will not need to evoke his contractual powers formally on many occasions and the atmosphere of co-operation will be conducive to a well-run contract. An inexperienced resident engineer will tend to oppose the agent when he proposes any alternative methods of construction than those stipulated in the contract documents, even though they are quite sound, purely through lack of confidence in his own judgement. This can be the cause of considerable frustration to the contractor. The contractual situation is that it is basically for the contractor to determine how the works should be executed, provided he complies with the contract and design specification, and he is entitled a degree of freedom to prove his point.

During the early days of the contract the contractor will need to obtain the resident engineer's approval of a number of issues. Apart from the programming of the work and the mode of operation there will be samples of materials that will need approval. It may be a requirement that the source of materials will have to be inspected, for example pre-cast concrete pipe suppliers or stone quarries and that sample panels of particular finishes have to be prepared for inspection. The materials testing engineer, in conjunction with the resident engineer, will take on many of these duties and both will prepare reports giving their findings for submission to the engineer.

The resident engineer's staff need to be organised so that their manning resources are used to the best effect. Their duties are to watch the work under construction to see that correct engineering practices are adopted. They use a combination of experience and the requirements of the specification to oversee that the work is carried out to the necessary standards. Many of the operations they inspect will be governed by strict specification and design criteria, for example the size and spacing of reinforcement steel, the minimum concrete cover to reinforcement bars or the maximum depth of each layer of compacted fill material in a trench.

On the other hand, experienced staff will be expected to use a great deal of discretion in determining what is good engineering construction practice. Work not carried out to specification may not be automatically unacceptable and more often than not a clerk of works or inspector will accept that the result is more important than the means of achieving it. It may not be physically possible for resident engineer's staff to supervise all operations on site and here it is necessary for the resident engineer to develop a clear strategy so that his staff know what the priorities are and what is expected of them. To do so close co-operation from the contractor is required so that due notice is given to the resident engineer of the significant activities on site so that the resident engineer can deploy the staff resources effectively. The contract can be drawn up to include regular programmes detailing such operations to be issued by the contractor particularly in relation to out of normal working hours proposals, such as weekend concrete pours. The resident engineer will also be able to assess which areas of the contractor's expertise needs close monitoring and to which he can confidently give less rigorous attention.

The resident engineer is responsible to the engineer for the quality and accuracy of the construction and his staff will be assigned to the various quality inspection duties. Materials brought to site will need to be sampled and tested. The inspectors may deliver materials to the test laboratory or alert the testing technicians that material has been delivered or when *in-situ* testing is required. In co-ordination with the inspectors, the test laboratory will ensure that concrete is sampled and cubes crushed, stone is sampled and graded and asphalt and bituminous materials are within specification and correctly laid. Many of the checks for compliance with standards are of a day-to-day nature and the resident engineer must set up recording systems whereby he can be satisfied that inspectors are undertaking the necessary inspections and tests at each stage of the operation to which they have been delegated. Formwork, shutters and reinforcement need inspection prior to a concrete pour. Inspectors need to be present at the pour and when the formwork is struck and shutters stripped. Pipelines are to be checked for line and level and for correctly formed and sealed joints prior to backfilling of trenches. Backfill material must be sampled and tested and close supervision is required of the compaction process.

It is in no one's interest if errors are discovered too late to prevent the completion of work which the resident engineer will then have to reject. Consequently, a speedy response is needed by inspectors and laboratory technicians when they are made aware that certain operations are imminent. A clear understanding must be reached with the agent's staff so that they are under no illusions as to what work must be carried out in the presence of an inspector and how far that work can progress without the need for satisfactory test results to be produced. The resident engineer must organise inspections and tests so that they are carried out quickly and do not delay further stages of the programme. If testing and the wait for results are likely to affect the progress of the work then this should be built into the programme or alternative faster methods of checking and testing employed. This is where highly experienced site staff can be invaluable in assessing exactly when testing is essential and when more empirical methods are all that are required. Both contractor's and engineer's staff must be aware as to what tolerances are to be permitted on tests so that the contractor can undertake a proper assessment of what time and effort should be invested in each operation.

No matter what level of supervision or checks and tests the engineer's staff undertakes the contractor cannot absolve himself from liability for any errors in his work even when expressly approved by the engineer. Neither the engineer nor his staff carries any legal responsibility for any such errors and owe no duty to the contractor to prevent or even warn him from making mistakes. There is, of course, an ethical and professional obligation upon the engineer and his staff to act openly and fairly and to use his skill and judgement for the good of his profession.

22.6.2   *SPECIALIST SUPERVISION AND SAFETY*

The construction or replacement of pipelines underground presents a number of engineering techniques which require specialist knowledge and experience, together with rigorous safety procedures. Consequently the respective resident engineer's and agent's staff need to be aware of all the possible problems and hazards that such operations can create. Some of these operations are briefly summarised here to give a sample of the complexity of sewer replacement or construction contracts.

Excavation in any form of ground, apart from rock, will require expert trench support techniques. When deep excavations are dug in unstable gound close boarded timbering is extremely effective and the flexibility that timbering gives enables very poor and uncertain ground conditions to be combated. This form of support also ensures that any unknown services, or known ones for that matter, are not damaged as the shaft is sunk (Fig. 22.4). However, skilled timber men are a dying breed and many contractors prefer the speed which sheet piling can provide. For shallow trenches, trench sheeting may be used to line the trench, with the toe of the sheet being driven below the foot of the trench (Fig. 22.5). This photograph shows how pre-formed metal expanding frames can be utilised to speed up the process of supporting excavated trenches. This photograph also shows the use of a well point dewatering system which may be the only effective way of combating serious ground water problems which threaten the stability of the trench.

Figure 22.4   Close boarded timbering of deep shaft.
Photographs by courtesy of Manchester City Council-Environment and Development Design Consultancy.

For deeper trenches or where the ground is too unstable for the trench to be dug to any significant depth without support or where ground water is a serious problem, interlocking steel sheet piles are required. The only safe and accurate way of driving these piles is by the use of a piling frame and a piling hammer and many an attempt to speed up the process

Figure 22.5 Steel sheet piling to trench with dewatering drainage.

has resulted in severe injury to an operative, fingers being particularly vulnerable in these circumstances (Fig. 22.6). Sheet piles can only be used when the precise position of all services is known. For relatively shallow trenches metal pre-formed adjustable trench formers are ideally suited to long trenches where repeated use of each former makes their hire an extremely viable expense. The cost of hire may well be offset by the speed of construction and, if used correctly, they provide a level of safety second to none (Fig. 22.7).

Tunnelling techniques are an essential aspect of sewer construction and it may be necessary, or desirable, for a number of reasons to construct the sewer in heading. Tunnelling

Figure 22.6   Piling rig and frame for steel sheet piling.

is often the only feasible way of negotiating hard rock but tunnels are also economically viable where the sewer line crosses roads, railways, waterways or passes under or close to buildings. This form of construction can only be placed in the hands of an experienced contractor and supervision by the resident engineer and the agent must be of the highest order. Safety is of course the main consideration, as it must be on all construction sites, but due to the great potential for major disaster and loss of life the margins for error are very small. Where the sewer is to pass through solid rock no timbering is required and the tunnel will be formed by blasting. Again a great deal of experience and expertise is necessary to ensure such an operation is conducted safely and effectively. Safety procedures for all site personnel must be rigorously controlled and no risks must be taken which may jeopardise public safety, either in the blasting process or in the storage of the explosives. Seamed rock will require bracing support of the tunnel and softer or loose ground will need timber support to differing degrees from frames and soffit lagging to close boarded, fully braced timbering of walls and soffit (Fig. 22.8).

Soft rock may be effectively drilled by the use of specialist boring equipment and in soft ground the more traditional method of digging out at the forward edge of a steel shield may be employed (Fig. 22.9). Where the material is wet and ground water is likely to be a problem the shield will have an enclosed end to form a caisson and compressed air will be used to maintain a pressurised atmosphere to keep out the water. To maintain the air pressure specialist compressed air plant has to be maintained in best mechanical condition and the workings may only be entered or exited via an air lock (2.10). This again presents particular health and safety problems as operatives are working in a pressurised atmosphere and have to be held in the air lock while the pressure is adjusted slowly to prevent what is commonly called 'the bends', a particularly debilitating and life threatening disease.

Figure 22.7 Adjustable pre-formed trench shuttering.

The use of concrete segments has revolutionised large diameter sewer and manhole construction. These effectively provide ground support to the sewer or manhole structure in one operation and when used appropriately can expedite the construction time dramatically. Concrete segments are expensive and their use can eliminate any potential problems that are frequently encountered using more traditional methods and, consequently, may ultimately prove the more economical process (Figs 22.11 and 22.12).

Figure 22.8   Tunnel heading in solid ground with timber support.

22.6.3   *SETTING OUT*

Setting out is the responsibility of the contractor and Clause 17 establishes these responsibilities as:

1. The true and proper setting out of the works
2. The correctness of the position, levels, dimensions and alignment of all parts of the works
3. The provision of all instruments, appliances and labour in connection therewith
4. The rectification of any error in (2) above to the satisfaction of the engineer at his own cost; unless such error is based on information from the engineer or his representative when the cost falls on the employer
5. The protection and preservation of all benchmarks, sight rails, pegs, etc. used in the setting out.

It would of course be naive and unprofessional of the engineer's representative not to undertake thorough checks of the contractor's calculations and setting out; however, any such checks by the Engineer's staff cannot relieve the contractor of his responsibilities, under Clause 17, for the accuracy of his seeting out or of any of his lines or levels.

It is usual for the contract to specify that the contractor establishes survey station points and benchmarks from which the setting out of the works on site will be carried out. The engineer's representative's staff will check these plus any level pegs, traverse stations and lines to establish that the contractor has been utilising the correct data. The engineer's representative needs to set up a formal system whereby his staff are notified by the corresponding member of the contractor's staff of any setting out that has been carried out so that checks can be carried out prior to any structural work being undertaken.

Figure 22.9 Tunnel boring tool breaking into manhole pit from tunnel.

Although any checks carried out by the engineer's staff do not absolve the contractor from his Clause 17 responsibilities, the importance of these checks cannot be stressed too highly as any errors in works, due to inaccurate setting out, can have serious cost and safety implications and can be a severe embarrassment to the engineer's site staff concerned.

Frequently, in order to expedite the process the setting out and checking are carried out simultaneously by a member of the contractor's staff and a member of the engineer's staff. In other words theodolites, levels and tapes etc. are read simultaneously by representatives of both sets of staff. Although tempting to the resident engineer's staff, this is poor professional practice as it will fail to pick up any errors caused by inaccurately calibrated or defective equipment or by a negligent chainman holding a staff, pin or tape in the wrong position. A

Figure 22.10 Air-lock chamber access to tunnel in compressed air amosphere.

Figure 22.11 Shaft being constructed using pre-cast concrete segments.

Figure 22.12 Sewer pipeline tunnelling using pre-cast concrete segments.

completely independent check with different equipment and chainmen is essential. This can be particularly important when setting out line, and the level of pipelines needs to be particularly accurate as any slight deviation off line or level may be perpetuated and exacerbated as the pipe laying progresses.

Particular challenges to accurate setting may be caused by pipe laying in deep trenches or in headings and tunnels. The transference and maintenance of line and level from the surface to the bottom of the trench will require scrupulous accuracy and thorough checking. In such instances the use of laser beams can ensure a high degree of accuracy but the operatives of the equipment must be highly trained and experienced in its use.

22.6.4 THE SITE AND THE ENVIRONMENT

We are not concerned here with how the project will affect the environment when it has been constructed and is operational but with the inevitable problems caused by the construction works to the locality in which they are sited. The former will have been careful considered and addressed by the project manager and designer.

One of the primary functions of the resident engineer is to ensure that the works and associated operations are phased and controlled to cause as little disruption to the locality and its residents as possible and he must work in close liaison with the relevant departments of the local authority. The spirit of co-operation which is necessary between the site staff and the local authority can soon evaporate if the latter first hear of a contract activity from the mouth of an inconvenienced and irate resident or council member.

Whether the site is in an urban or rural area then aspects of the control of noise, vibration, dirt, waste and traffic are inevitable to some degree or other. The resident engineer will want to ensure that the contractor has obtained all necessary approvals and has served the appropriate notices. These may include:

• Notifying the Health and Safety Executive where the works are to last in excess of six weeks. Usually this will have been undertaken by the planning supervisor
• Contacting the environmental health officer in relation to hours of work and type of equipment to be used in order to ensure noise and vibration levels do not exceed those permitted
• Obtaining planning permission for site compounds which are to remain in place for more than 12 months
• Serving notice under the New Roads and Street Works Act and ensuring that the information received from the statutory undertakers is acted upon
• Undertaking a close liaison with the highway authority in relation to the roads and streets adjoining the site. This may relate to plant and transport accessing the site or the need for temporary traffic regulation orders to be implemented so that temporary diversion of road traffic can be set up to facilitate the works.
• Advising emergency service and transport services when the work is likely to significantly affect the local highway network

The resident engineer will also ensure that the contractor carries out consultation exercises with local residents groups or individual residents well in advance of any section of the work causing disruption to them. He will be very aware that the reputation of the employer as a caring developer may rest on how the concerns and complaints of the local residents and businesses in the area are addressed.

The choice of the location for a site compound is likely to be limited and determined by the environment rather than practical construction factors. By their very nature sewer construction sites are inevitably going to extend over a considerable distance and more than one site will be necessary for a compound. Either two or more compounds will be needed simultaneously or a single compound will be resited a number of times during the duration of the works.

An urban environment, particularly a city or town centre, will pose far greater problems when attempting to find a suitable location than will be experienced in a rural setting. In the latter location the environmental issues will revolve around preservation of top soil and trees together with protection of any livestock on adjacent land (Fig. 22.13). Storage and stacking of plant and equipment is not likely to present many problems so long as a degree of sensitivity is used and consultation with the local authorities and residents is undertaken meaningfully.

The town centre site will test the ingenuity of all concerned in the contract and the engineer and resident engineer will play a major role in how a compound may be slotted into a particular location.

Whether in the rural or urban setting, the contract will need at least one sizeable compound to store plant and materials, particularly large deliveries of pipes, concrete segments and stone bedding material and fill. For this cranage or lifting gantries will be required and adequate space to ensure off-loading and storage are carried out safely (Fig. 22.14).

Traffic regulation orders will be necessary on street compounds and in order to divert traffic from streets blocked by the works and special signs may have to be manufactured and erected to advise the public that businesses and shops are still open and how access may be gained (Figs 22.15 and 22.16). Particular attention must be paid to providing safe

Figure 22.13   Sewer construction in green field site.

Figure 22.14   Loading gantry at pipe storage compound.

Figure 22.15   Site compound, Manchester city centre.

Figure 22.16   Barriers and specialist sign for commerce.

walkways for pedestrians and secure fencing needs to be erected around working areas to ensure the public does not inadvertently enter the site. Fencing of the works in an urban area will have to be particularly secure as vandalism and theft are a daily occurrence (Fig. 22.17).

Figure 22.17 Provision of safe pedestrian walkways.

22.6.5 SUBCONTRACTS

Subcontracts are entered into in two ways. Often a main contractor sublets part of his work, either to provide specialist work or to 'hire in' additional labour resources. However, such a facility can only be employed with the consent of the engineer. Alternatively the client or his engineer determines that a particular element of the project needs to be undertaken by a specialist contractor and 'nominates' the subcontractor. Such a nominated subcontractor will be selected by means of a tendering process and the main contractor is instructed to enter into a subcontract. The contractor, however, shall no be under an obligation to enter into a subcontract if he has a reasonable objection against the subcontractor. If a valid objection is made or the main contractor terminates the employment of the subcontractor Clause 59(2) specifies that the engineer may:

- Nominate an alternative subcontractor
- Vary the work under Clause 51
- Omit all or part of that work so that the employer can undertake the work
- Instruct the contractor to find a subcontractor of his own choosing
- Invite the contractor to undertake the work directly.

The subcontractor must agree to be bound by the conditions of the general contract, that is to the same conditions to which the main contractor is bound and the contractor is responsible for the nominated subcontractor in all matters apart from his design and specification. This means that the resident engineer must have no direct contact with the subcontractor unless in the presence of the main contractor. The exceptions to this will be when issues of emergency or safety arise and the resident engineer assesses that immediate action is necessary. As in the case of all significant instructions emergency instructions must be confirmed in writing to the agent immediately and a full report of the incident kept on record. Should the subcontractor be in default the contractor may terminate the nominated subcontract but only with the engineer's written consent. If the Engineer does not consent then he must instruct the contractor in accordance with Clause 13. Where the engineer agrees to the termination of the nominated subcontract, he has to undertake one of the above listed options. The contractor must recover from the subcontractor all additional expenses incurred, including the employer's costs resulting from the termination.

22.6.6   *MEETINGS, REPORTS AND RECORDS*

There are many forms of meeting which will be held during the running of a long contract but only formal meetings held and chaired by the resident engineer will be dealt with here. A precontract meeting should be called as soon as the programme has been agreed and a start date known and from then on regular, probably monthly, progress meetings will be held.

At these meetings the engineer will be present to act on behalf of the employer and the agent will usually invite his contracts manager or equivalent. However, the resident engineer will preside over the meeting as he is independent of the parties to the contract. Apart from resolving details of construction and matters of progress and payment, such meetings enable the resident engineer to keep the engineer informed of progress and how he is supervising the works.

It is essential that an agenda is issued prior to the meeting and this may be based on the subheadings in the bill of quantities. Minutes must be taken and it is necessary for the minute taker to confirm to the meeting his understanding of any decisions that are reached. It should be borne in mind that a set of full and accurate minutes can curtail many a dispute at a later stage when individual recollections differ.

Between the progress meetings there will be plenty of opportunity for either party to request a meeting to discuss a particular issue or series of issues. All such meetings should be chaired by the resident engineer or his representative and the conclusion committed to writing, even if it is purely a note in the diary, and confirmed in writing to the contractor. Subcontractors may be called to or requested to attend meetings relating to their work and this is clearly to be encouraged as they will have first hand knowledge which can be invaluable or conversely may inadvertently show both agent and resident engineer where problems may arise. The resident engineer should not hold meetings with the subcontractor save in the presence of the main contractor's representative and so the resident engineer must not permit the main contractor to absolve himself of the direct control of his subcontractor.

All the resident engineer's site staff should keep a daily diary and record his actitivies for the day and particularly any minor instructions or warnings of poor or unsafe practices to the contractor's staff or workforce. Such information can prove invaluable to the resident engineer when considering claims made by the contractor or extension of time due allegedly to unforeseen conditions but possibly due to inefficient or abortive work (Table 22.6).

Table 22.6 Relief sewer contract – progress meeting no. 5 to be held at Resident Engineer's Office at 9.30 am Wed. 12th June 1997

Agenda
Item
1. Minutes of last meeting.
2. Matters arising.
3. Safety:
    (i) Review of safety plan.
    (ii) Excavation support and access.
    (iii) Subcontractor training.
4. General progress:
    (i) Delayed possession section 3
    (ii) Progress review.
5. General items:
    (i) Engineer's offices – outstanding items.
    (ii) Testing pipework.
    (iii) Traffic management.
6. Ground investigation:
    (i) Trial pits No. 7–11.
7. Site clearance:
    (i) Tree removal parkland verge.
8. Statutory undertakers:
    No items.
9. Earthworks:
    (i) Topsoil for reuse.
    (ii) Excavation.
10. Pipework:
    (i) Pipework defects M.H. 5 to 7
    (ii) Report from supplier.
    (iii) Testing proposals.
    (iv) Bedding concrete test results.
11. Ancillaries:
    (i) M.H. 3, 4 and 9 remedial work.
    (ii) Subcontract performance.
    (iii) M.H. ironwork sample.
12. Accommodation works:
    (i) Wall to bridge house.
13. Interim payments:
    (i) Payment 4.
14. Any other business:
    (i) Complaint from local residents.
15. Date of next meeting.

Resident Engineer

Apart from receiving copies of minutes of progress meetings the employer will need to be appraised on a regular basis of contract expenditure. This will be in the form of a report detailing costs on the following items:

- Prime cost
- Provisional sums
- Bill of quantity items
- Variation orders
- Dayworks
- Claims.

The employer will require further information as to the reason for the claim and the resident engineer's estimate as to the probable size of the settlement or the reason for rejection of the claim. The engineer, in conjunction with the resident engineer, will issue a periodic report outlining a technical appraisal of any items within the contract that are thought appropriate. These may include:

- Details of any statutory undertaker's work, its cost and progress
- Required amendments to the design
- Site conditions affecting progress and/or cost
- Suppliers' performance
- Quality of materials
- Contractor's superintendence
- Advanced information on likely final cost
- Information on probable completion date.

The resident engineer will be under pressure to favour the client/employer in his dealings but must remember that his contractual duty is to supervise the contract impartially and to uphold the rights of both employer and contractor.

22.7 Measurement and payment

22.7.1   THE BILL OF QUANTITIES CONTRACT

This form of contract is considered the best suited to civil engineering works where it is usual for a detailed design and specification to have been carried out prior to letting the contract. For many years this form of contract has been the most widely used for sewer construction works which are frequently subject to variation as unforeseen site and ground conditions are met. The design drawings are broken down into a detailed list of items each of which are quantified by abstraction of measurements from the design drawings. At tender stage the contractor prices each item and the total tender price is stated. During the performance of the works the actual quantity of work carried out under each item is measured and costed at the tendered rate for that item.

Generally the final account is subject to a remeasurement at completion. Measurements are carried out in accordance with the Civil Engineering Standard Method of Measurement (CESMM), now in its 3rd edition, published by the Institution of Civil Engineers and provision is made under the ICE Conditions of Contract for rates to be adjusted, or for new rates to be drawn up by the engineer, for additional or varied items of work.

This form of contract enables tenders from individual contractors to be compared on equal terms and provides for a common ground competitive assessment to be made. It also furnishes the employer with an estimated overall contract cost which allows further consideration of the budgetary implications. Other advantages are that it provides a facility to pay for works on an interim basis and for a final measurement to be made prior to settlement of the final account. The financial implications of any variations to the works can be readily assessed. In addition the bill of quantities contract enables contractors to obtain prices of materials and quotations for works to be sublet to other contractors as well as calculating labour payment bonuses. The contractor is also able to assess whether the contract covers all items of work required.

Payment for the work is the issue which accentuates the differing pre-occupations of the resident engineer and the contractor's agent. The former is primarily concerned with the quality of the work and that the manner of its completion accords with the specification and safety requirements. Finance becomes a secondary issue, provided the overall budget is not significantly exceeded and provided the engineer is kept appraised of the financial situation.

On the other hand, the contractor, while keen to maintain a good reputation, cannot afford to do so for very long if his costs are not being met so that his predominant interest in any particular item of work is, how much he is to be paid for its completion. The experienced resident engineer recognises this fact and works with the contractor to make sure he is paid promptly and accurately for work completed. Such an arrangement results in a better performance than will be produced by a claim conscious agent under financial pressure.

Clauses 55 and 56 of the ICE Conditions of Contract state that the quantities set out in the bill of quantities are only estimated and not to be taken as the actual quantities to be undertaken by the contractor and the engineer shall determine the value of the actual work completed by means of measurement. Should the quantities differ significantly from those set out in the bill then the engineer, after discussion with the contractor, should determine an increase or decrease in any rates rendered unreasonable by the change.

If the bill is found to contain any error in description or omission the contractor shall not be relieved from his obligations and the error or omission shall be corrected by the engineer and paid for, in accordance with Clause 52.

22.7.2 VARIATIONS AND DAYWORKS

Clause 51 allows for variations to the contract as deemed necessary by the engineer for the completion or the improvement of the works. The clause includes the following list of possible variations additions, omissions, substitutions, alterations, changes in quality, form, character, kind, position, dimension, level or line and changes in any specified sequence, method or timing of construction required by the contract and may be ordered during the defects correction period.

The contract should stipulate who will be responsible for the issue of variation orders and that they must be ordered in writing and referenced with a sequential number so that each variation order can be costed under the terms of the contract and must state under which clause of the contract the order is issued. It should also state the basis for pricing the variation order. Clause 52 specifies that where work is of similar nature and carried out under similar conditions to work already contained in the bill of quantities then the variation order shall be valued at the same rate or price quoted for that work.

Even where the variation work does not correspond directly to any existing rates in the bill or where the work is ordered during the defects correction period, the rates in the bill of quantities are to be used as a basis for arriving at a reasonable value. Where this not applicable a 'fair valuation shall be made'. If the engineer and the contractor fail to agree on a price then the engineer will determine the value of the variation work in accordance with the above principles. It is clear therefore that bill of quantity rates may not be applicable after the substantial completion date.

Large contracts may contain a schedule of 'daywork' rates which may be used by the engineer to pay for additional work ordered by the engineer which bears little relation to items in the bill of quantities. Daywork rates are based on the cost per hour for the use of items of plant and labour, plus a percentage for overheads and profit margin. Where a schedule has not been included in the bill, then Clause 56(4) determines that the 'Schedule of Dayworks carried out incidental to Contract Work', issued by the Federation of Civil Engineering Contractors, shall be used to value the appropriate work. The materials to be used on work ordered as 'daywork' shall be paid for at purchase price with a predetermined percentage on cost. The engineer may require the contractor to submit quotations for materials for his approval.

Clearly it is important that accurate records of time spent by labour and plant on such works are diligently kept and submitted by the contractor to the engineer's representative. In any event the resident engineer should already be receiving labour and plant returns on a weekly basis, in accordance with Clause 35, which may be utilised to check daywork claims. The contractor must give due notice to the resident engineer of his intention to commence any work which the resident engineer has ordered to be carried out under the dayworks schedule so that the resident engineer may ensure that a clerk of works closely supervises the work and records, with times, all plant and labour usage. These records need to include the class and trade of each operative employed and type and size of each item of plant used and whether it has been hired with or without an operator.

All quantities of materials should be noted and copies of delivery tickets retained. It is evident that while all unit costs and rates may be known prior to the commencement of the daywork, the major unknown element in the ultimate cost of the work is the efficiency of the manner in which the contractor undertakes the task.

The cynical reader may suggest that dayworks are heaven sent for a contractor to 'offload' his poorer workers who, by their slow progress, may prove highly profitable. Time lost due to ineptitude will be booked against the daywork and the method of payment may prove a disincentive to efficient working. The contractor, however, will be aware that excessive delays may adversely affect his programme and he will need to be able to substantiate any claim for delay due to a significant volume of additional work. Also it is incumbent upon the resident engineer to have the work closely supervised by site staff experienced in the type of work ordered so that they have the ability to pass on constructive criticism of any inappropriate modes of working and to place on record the contractor's refusal, or inability, to comply. In practice many contractors are willing to allow the resident engineer to determine many details of the method of operation, without compromising the contractor's responsibilities under the contract. When works is ordered at daywork rates it is well for the resident engineer to consider that he is effectively hiring the contractor's labour and plant which is to be charged by the hour. Consequently he must take a more proactive interest in the daywork, one which would not be appropriate for bill of quantity items.

22.7.3 *PROVISIONAL SUMS AND PRIME COST ITEMS*

Clause 58 provides for the employer to include provisional sums and prime cost (PC) items. A provisional sum is basically an estimated cost of a contingency item which the resident engineer may order in whole or in part, if he considers it necessary to complete the contract effectively.

The provisional item may be for work to be carried out or materials or services to be supplied by the contractor and the value determined in accordance with clause 52. The contract may specify that the work, material or service be supplied by a nominated subcontractor or supplier.

Prime cost items may be contained within the contract whereby the engineer orders specialist work, materials or services and may nominate a subcontractor or may request the contractor, with the contractor's agreement, to undertake the item. Payment is normally in accordance with the terms of a lump sum quotation submitted by the contractor, normally the subcontractor's price with a percentage on-cost to the main contractor for his attendance.

Clause 59, however, relieves the contractor from any obligation to enter into a subcontract with the nominated party if he has reasonable grounds not to do so. In such an event the engineer may instruct the contractor to submit his own nomination for approval and, failing agreement may instruct the contractor to undertake the work directly, in accordance with Clause 58(1)(a) or 58(2)(b).

22.7.4 *INTERIM PAYMENTS*

Most contracts allow for interim payments to be made to the contractor and Clause 60 gives the terms which enable such an arrangement. The contractor is to submit a monthly statement which usually, unless specified otherwise, includes the estimated value of completed work up to that date, the value of materials stored on the site awaiting use plus an estimated cost of agreed extras or claims. On large contracts both agent and resident engineer will employ quantity surveyors, respectively, to submit and to check the valuation which should contain the following items:

- Preliminaries and temporary works included in the bill of quantities
- Measured work carried out by the contractor and his subcontractors (including materials from nominated suppliers)
- Payment for work by nominated subcontractor
- Insurances
- Dayworks
- Adjustment for fluctuation (if included in the contract)
- Materials on site not yet used.

The employer retains a percentage of the interim valuation, predetermined in the contract but not exceeding 10%. The usual way of agreeing the measure is for the resident engineer's and the agent's staff to undertake the task together. Clause 56 requires the contractor to attend at the resident engineer's request to assist in the measurement and on a large contract the respective site engineers or quantity surveyors will undertake the measurement as a continual process.

22.7.5   *FINAL ACCOUNT AND CLAIMS*

At the end of the maintenance period and upon completion of the works and remedial works, to the satisfaction of the engineer, the engineer shall issue the maintenance certificate and enter into 'discussion' with the contractor to settle the final account. At this stage the value of the resident engineer's staff having carried out a continual measure throughout the term of the contract will become apparent. Within three months of the issue of the maintenance certificate the contractor shall present a statement, with supporting information, of the value of measured work and any other items he considers he is entitled to payment for. Within three months of receiving this information the engineer must issue a final certificate stating the sum due to the contractor.

Few contracts reach completion without the contractor making a claim for additional payment due to unforeseen circumstances or due to breaches in the Contract. He will need to establish that such circumstances were not reasonably foreseeable by an experienced contractor. This will depend on the extent and quality of the information given to the contractor at the tender stages.

Claims from a contractor may arise as a result of a liability specified under the following clauses of the 6th edition of the ICE Conditions of Contract:

- 5 Documents not mutually explanatory – Ambiguities or errors result in delay/cost.
- 7(4) Delay in issue – Failure of Engineer to issue documents in time causing delay/cost.
- 12 Adverse conditions and obstructions – Unforeseen conditions resulting in delay/cost.
- 13(3) Delay and extra cost – Unforeseen delay/cost due to Engineer issuing instructions.
- 14(8) Methods of construction – Engineer delays consent or imposes unforeseeable conditions.
- 17(2) Setting out – Costs of rectifying errors due to incorrect data supplied by Engineer.
- 20(2) Excepted risks – Events/damage for which contractor is not liable.
- 22(2) Damage to persons and property – Due to Employer's negligence or contractual liability.
- 26(1) Payment of fees – Statutory fees paid by contractor to be reimbursed by the Employer.
- 27(6) Delays due to variations – Public utility street works resulting in delay.
- 31(2) Facilities for other contractors – Other contractors and utilities causing delay/cost.
- 32 Fossils etc. – Delay due to precautions needed to protect articles of value or antiquity.
- 36(3) Costs of tests – Costs of tests not specified in the contract (but not tests which fail).
- 38(2) Uncovering and making openings – to examine work (not if work out of specification).
- 40(1) Suspension of work – Engineer instructs suspension, not due to contractor's failure.
- 42(1) Failure to give possession – Employer fails to give possession causing delay.
- 49(3) Cost of repair work – Payment for repairs where contract is not at fault.
- 50 Contractor to search – Searches or tests for defects (not if defect is contractor's liability).
- 52(4) Notice of claims – The contractor considers payments for variations insufficient.
- 56(2) Increase of rate – Rates rendered unreasonable due to increase or decrease in quantities.
- 59(4) Nominated sub-contractors default – Contractor to recover expenses incurred.
- 60(7) Interest on overdue payments – Engineer fails to certify or Employer fails to pay on time.

- 64 Payment in event of frustration – Employer to pay value of work completed at the time.
- 65 War clause – Employer to pay for work completed prior to contract determination.
- 69 Labour tax fluctuations – Adjustment of contract price due to change of labour tax rates.
- 70 Value added tax – Payment of V.A.T. by Employer to contractor.
- 71 Special conditions – Contract price fluctuation payments where included in contract.

All claims must relate to a specific clause in the contract as the engineer is not authorised to accept a claim made otherwise. Clause 52(4) specifies the terms under which a claim is to be made and the timescales within which claims must be made.

A number of the above clauses relate to delay as well as additional costs and should significant delays occur the contractor may request an extension of time. The engineer may grant such an extension without a corresponding additional payment and this is purely to determine the date of the contractor's liability for deduction of liquidated damages from payments. However, the contractor is likely to link the granting of an extension in time with justification for a claim for additional costs. The most frequent occurrences which lead to delays and subsequent claims from contractors for an extension are:

- Exceptionally inclement weather
- Delay in issue of documents (Clause 7)
- Adverse physical conditions and artificial obstructions (Clause 12)
- Industrial dispute
- Problems in obtaining labour or materials
- Fire or flood.

This is by no means a complete list and the contractor, to be successful, will have to show that any event given as a reason for a claim for extension could not have been foreseen by an experienced contractor.

22.8 Completion

The engineer, after considering the contractor's claims for an extension of time should determine the revised completion date and an interim award should be granted immediately. There may be a circumstance where the engineer may think fit to grant an extension of time even though no application has been made by the contractor. In either case the contractor is to be notified of the revised date.

During the course of the contract the engineer may become concerned at the slow rate of progress, for reasons other than to entitle the granting of an extension of time, to such a degree that he considers that substantial completion will not be achieved by the time for completion. In this event he should write to the contractor instructing him to submit for approval measures which will expedite the works. The contractor shall not be entitled to claim additional costs for such provision.

The contractor may consider night work or Sunday work as a means of catching up with the programme but, unless this provision is already in the contract, he will require the permission of the engineer. Exceptions to the need for approval may arise when the work is 'unavoidable or absolutely necessary' for the saving of life or property or for the safety of the works, in which case the engineer or the resident engineer must be advised immediately. Apart from the above exceptions Clause 45 states that none of the works may be carried out on Sundays or at night unless it is work customarily carried out outside normal working hours or as part of shift working.

When the contractor has substantially completed the whole of the works, or a section of work where the contract is so divided and it has been tested satisfactorily, he may notify the engineer in accordance with Clause 48 and confirm in writing that he will complete any remnants of the work within the agreed time. Within 21 days of the notification of substantial completion the engineer must issue a certificate of substantial completion stating the completion date. He shall inform the contractor of all work required to be completed prior to him issuing the certificate. Within 21 days of the contractor having completed all outstanding work to the satisfaction of the engineer, the engineer shall issue a certificate of substantial completion. Unless the certificate expressly states as much, it is not deemed to certify any ground or surfaces to be reinstated.

22.9   Contract strategy

22.9.1   CONTRACT

When drawing up a contract there are key decisions which need to be made to ensure the appropriate form of contract is chosen for the project in hand. The factors affecting the selection relate to how much information concerning the design is available at the time. Clearly the promoter/client needs to be aware of the likely total expenditure on the project before entering a contract; however, the time available for pre-contract deign will be a major issue in what form of contract is most suitable. Conversely, there may be circumstances when the promoter needs a degree of flexibility to amend the design as the work progresses and certain forms of contract give this facility.

There are a number of different types of contract each with particular features the effect of which must be assessed carefully, prior to selecting the form of contract to be adopted. The most widely utilised are *bill of quantities contract, schedule of rates contract, lump sum contract* and *cost reimbursement contract*.

The *Bill of Quantities contract* is considered the best suited to civil engineering works where it is usual for a detailed design and specification to have been carried out prior to letting the contract and this has formed the basis for the information contained in the greater part of this chapter. However, where the work to be executed is limited and the quantity and specification of the work are not, or unlikely to be, subject to variation then a *lump sum contract* may be considered appropriate. Under such a contract drawings and specification will be issued to the contractor who agrees to fulfil the specification and then prices the work without the facility of a bill of quantities.

The contractor undertakes to complete all the works for the tender sum and payment may be made via a series of interim payments based on completed work or at predetermined stages of the contract.

To cover the contingency that additional or varied work may be required, the contract may include a schedule for items of work which is to be priced by the contractor and which may be used by the engineer for the ordering of variations to the contract.

Cost reimbursement contracts are most usefully entered into where the work is of an emergency nature and there is little time available to carry out a detailed design or draw up a detailed schedule of measured items. They are also employed when the extent or the progress of the work cannot be forecast with any reliable accuracy. Such contracts are sometimes called 'cost plus contracts', the contractor being reimbursed his actual incurred cost plus overheads and profit margin. The contract takes the form of a detailed schedule of expenditure headings, from wages and materials to insurance premiums. The contractor is required to obtain competitive quotes for all plant and materials prior to obtaining agreement that he may place the orders.

Cost reimbursement contracts have the disadvantage of discouraging efficiency or economy, particularly where an agreed percentage of the agreed cost is paid to the contractor. This is called a '*cost plus percentage contract*' and sometimes referred to as a 'daywork contract'.

Alternatively a lump sum fee may be paid (*cost plus fixed fee contract*) which has the advantage of ensuring the contractor is not rewarded for delaying the progress of the works.

A further and more advantageous variation, from the employer's viewpoint, is where the contractor is rewarded for efficiency and reduction in costs (*cost plus fluctuating fee contract*). This is effected by paying the contractor his actual costs plus a fee which is inversely proportional, in accordance with a predetermined scale, to the cost submitted for payment by the contractor.

The fourth variation of cost reimbursement contract is the '*target contract*' where time to develop a detail design is limited and the tendering process has to be commenced before the design is complete. A bill of quantities on the known elements of the design is drawn up to enable contractors to estimate the total costs and a more accurate figure agreed as the work progresses and the design is finalised. On completion the contractor is paid his costs plus a fee based on a percentage of the estimated target. If the contractor completes the work within the 'target' figure he receives an additional sum related to the 'savings' made, this is over and above his actual cost plus the agreed fee. Should the 'target' be exceeded the effect is a reduction in the fee payment. Facility needs to be made under such a contract for adjustment of the target cost to allow for variations to the design and measurement and changes in costs of resources. The advantage of this contract is that it provides an incentive to the contractor to expedite the programme and to effect efficiencies in use of labour plant and material resources.

The *New Engineering Contract*, or NEC Engineering Construction Contract as it is now named, will be discussed in some detail later in this chapter. However, engineers are more likely to come in contact with the traditional forms of contract as prescribed by the ICE.

22.9.2 CONDITIONS OF CONTRACT

While *the ICE Conditions of Contract* is now in its 6th edition some water authorities have favoured the *IChemE Model Form of Conditions of Contract for Process Plant, Reimbursable Contracts*, 2nd edition (i.e. the I.Chem.E.*Green Book*) for use on sewer and water process projects, e.g. the Thames Water Ring Main Project Phase 2 by Thames Water plc.

The ICE Conditions of Contract 6th edition produces a contract where design and construction are quite separate entities and have distinct and separate management processes. There is no mechanism for the contractor to be involved at the design stage of a project. This has brought about an ethos were price differentiation is the overriding factor during the selection process. This can often be at the expense of quality, performance and innovation.

A third option, which is being favoured on extensively serviced projects is the *IEE/ IMechE Model Form of General Conditions of Contract 1988 (MF/1)*, where the contract places the responsibility directly on the contractor to produce a detailed design, to the engineer's outline plan, and then to undertake the supervision of the construction. The drawback to this form of contract is that the selection of a contractor is decided on contract price and performance criteria. This means that the effect of any possible variations to the contract cannot be assessed and costs of individual items cannot be taken into account.

However, the I.Chem.E. form of contract, known as the *Green Book*, is increasingly receiving favour among water authorities for use on sewer and water main contracts. Under this form of contract the client/purchaser and contractor enter into a partnership where the contractor agrees to undertake the design, manufacture and purchase and install while the client agrees to pay for these services. The contractor is empowered to submit proposals for variation and alteration to the client and between them they share responsibilities for problem solving by team working. Without careful planning and complete openness between contractor and client this might lead to what amounts to the client having to accept underwriting costs he had not been made aware of at the outset.

When comparison is made between the three above forms of contract the most notable differences are:

1. The ICE 6th edition provides for contract payment related to measurement of the works
2. The MF/1 type contract provides for a lump sum payment
3. The I.Chem.E. (*Green Book*) is a cost reimbursement contract

However, there are other significant provisions which are worth comparison as follows:

4. There are differences in the completion process in that MF/1 and I.Chem.E. provide for testing after handover and when the structure is operational
5. Each contract enables the engineer or the project manager to issue variation orders, instruct methods of construction, award extensions to the contract period, certify payment and agree extra payments
6. Clause 61(2) of the ICE Conditions 6th edition stipulates that the contractor and the employer are not relieved from any liability under the contract, even when the defect works have been completed and a certificate issued by the engineer. The contractor retains responsibility for the quality of the work when the contract has been completed.
7. Performance is the main criterion under MF/1 and I.Chem.E., which is to be measured through testing procedures, and liability in the long term is of lesser consequence
8. MF/1 puts clear responsibility upon the engineer in relation to the issuing of the final certificate which is deemed to provide conclusive evidence that all work complies with and has been completed in accordance with the contract and substantiates the value of the work
9. Similarly, under I.Chem.E. the only contractual liability residual upon the contractor, at the end of the defects liability period, is where performance guarantees were entered into as a part of the contract. Otherwise the purchaser becomes liable for the structure and its performance and maintenance
10. Arbitration and liquidated damages are facilitated by each of the three forms of contract (Table 22.7).

22.9.3   NEW ENGINEERING CONTRACT

In 1996 the Institution of Civil Engineers published the second edition of its New Engineering Contract (NEC), having first introduced it in 1993. By 1997 over 5 billion pounds worth of projects had been procured worldwide incorporating this form of contract.

In order that an informed decision may be made as to how best to utilise the NEC it is worthwhile considering the thought process that went into the development of this form of

Table 22.7 Equivalent terms used in standard forms of contract

	ICE Conditions of Contract 6th edition	IEE/IMechE model form of general conditions (i.e. MF/1)	IChemE model for reimbursable contracts (i.e. the *Green Book*)
Type of contract	Measurement contract	Lump sum contract	Cost reimbursement contract
Purchaser	Employer	Purchaser	Purchaser
Manager of the project	Engineer	Engineer	Project manager
Manager's obligation	To act 'impartially'	Must 'exercise such discretion fairly'	Must 'exercise (his) discretion . . . to the best of his skill and judgement as a professional engineer'
Supervision	Engineer's representative (resident engineer)	Engineer's representative (resident engineer)	Project manager's representative
Contractor's obligations	To construct and complete to the satisfaction of the engineer	To design, manufacture, deliver to site, erect and test plant, execute the work and carry out the tests on completion to the reasonable satisfaction of the engineer	To execute the work to the reasonable satisfaction of the project manager
Payments	Monthly, within 28 days of the certificate less retention	As per on schedule, usually at intervals of 25%, 50%, 75%, etc.	Effectively in advance, i.e. two weeks after receipt of a certificate predicting the following month's expenditure
Retention	2% retention deducted from above	15% retention deducted from above	No retention deducted from above
Variations	No financial limit	Total must not exceed 15% of the contract price unless by joint agreement	Contractor can object if the total exceeds 25% of the first agreed estimate
Substantial completion	Substantial completion certificate	Tests on completion	Construction completion report signed by the project manager and the contractor
Maintenance period	Defects correction period typically 6 months	Defects liability period typically 12 months	Defects liability period typically 365 days
Final certificate	Defects correction certificate	Taking over certificate with performance tests after taking over	Taking over procedures and test followed by taking over certificate
Completion	Final certificate	Final certificate	

contract. It had become apparent that the large number of available forms of contract, some of which are outlined previously within this chapter, was creating complications in the industry and that a more flexible and simpler form of contract was highly desirable. Promoters of projects expressed the need for a much broader series of choices when planning a contract than was available among the various existing forms of contract, particularly in relation to the disparate apportionment of risk between parties to the contract. Also there was felt to be a need to review the manner in which contracts were managed and how payments were made. In general improvements were requested to ensure that the project was delivered in a manner that would more likely guarantee that the purchaser obtained the required return for his investment. This could best be achieved by promoting greater co-operation between purchaser, engineer and contractor and reducing the confrontational elements within a contract that inevitably lead to disputes and claims. It was also considered that a new form of contract was now required to eliminate the unnecessary variations in contract procedure between the differing professional disciplines employed on many projects and that this would provide a more flexible arrangement appropriate for national and international use.

As a result the Institution undertook 'a fundamental review of alternative strategies for civil engineering design and construction with the objective of identifying the needs of good practice'. In 1988 Dr Martin Barnes and Dr John Perry drew up parameters for an innovative form of contract presented to a working group, chaired by Mr Robin Wilson and made up of members from consultant engineers, contracting engineers, engineers from organisations in the private and public sectors, together with representation from the legal profession. A consultation document was published in 1991 and the initial response was extremely favourable with much constructive feedback from many quarters of the industry. Consequently in 1993 the Institution of Civil Engineers published the first edition of its New Engineering Contract (NEC) with a second edition following in 1997 under the title of the NEC Engineering and Construction Contract (NEC/ECC).

The basic aims of the NEC/ECC are to provide a contract which can easily be adapted to design and construct any building or engineering project, to create a clear and straightforward contract administration and to motivate management of the contract, such that the objective of the project is attained while the contractor also achieves a reasonable return with the minimum of disputation.

The NEC/ECC outlines several options for consideration when the employer is drawing up the contract in order that a degree of flexibility is incorporated:

- The choice of a priced contract, reimbursable contract, target contract and management contract enables the employer to adopt the appropriate strategy for the overall project and best contract payment methods for each contract
- The chosen strategy may be supplemented by any of a series of secondary options as required, these covering such issues as inflation adjustment, retention sums, penalties for delay or underperformance of the delivered asset and early completion incentive bonuses
- The limit of the contractor's responsibilities for design of the works, if any at all, may be specified by use of the 'works information'.
- Adopting appropriate terminology and test procedures common to the respective engineering and building components of a contract and using the 'works information' for other specialist specification
- Deciding what degree of subcontractor involvement in the management contract is appropriate, i.e. from nil to total involvement
- Drawing up a schedule of contract data and specifying which law pertains to which contract, facilitating use on international contracts.

Clarity and simplicity are essential ingredients of the NEC/ECC and these are achieved by adopting:

- Economy of language rather than convoluted technical jargon and clear concise sentences.
- A user friendly format so that the available options are easily identified and simple to employ
- Specification of procedures which have been fully assessed and examined by means of flow charts.

The third objective of the NEC/ECC is to motivate good management by the following means:

- The contract to be a 'working document' for day-to-day guidance to ensure the work is to the programme and specification and not merely for reference when conflict arises
- The contract is to establish the grounds for all decisions and actions taken so that misconceived decisions are avoided
- The clear definition of the roles of project manager, engineer, designer and supervisor as appointed by the employer. Similarly the role of the adjudicator, as appointed by the employer and contractor, is to be defined and his powers specified
- A stimulus being created for the contractor to provide an accurate programme of works and to issue revisions as appropriate
- A clause which lays down all compensation events, i.e. those events which entitle the contractor to compensation if they affect costs and programming. Also the identification of those risks for which the contractor must take liability
- The NEC/ECC incorporates a schedule of cost components which defines the components of actual cost and those not included are to be covered by the contractor's fee with a profit element
- In order to reduce claims and disputes to a minimum, a major objective of the NEC/ECC, the employer assesses the compensation events based on actual cost and time and the contractor is required to submit alternatives for coping with them
- The introduction of a process by which the project manager and the contractor have a joint responsibility to warn of any possible increase in cost, delay in the programme or impairment in performance of the asset and to enjoin in remedial action to mitigate the effect.

A second option which has been successfully employed as a contract strategy is the use of the priced contract with a bill of quantities. Alternatively some employers have decided to adopt a target cost strategy and then selected the NEC/ECC as the only standard contract form available of its type. Target costs contracts are more often employed where there is a high level of risk or where there is little time before commencement of work on site to produce a detailed bill of quantities. In such cases it is likely that the contractor has to carry out much of the design function and so the decision may be made to use an activity schedule.

Since the conception of the NEC/ECC practice has found that it is at its most effective when a combination of priced contract with activity schedule is used. Under such an arrangement design is a possible item of the activity schedule and consequently any design required to be carried out by the contractor is readily incorporated. Unless the design is fully complete prior to tendering, a complete bill of quantities is not a possibility.

Because the contractor has to draw up an activity schedule he has to plan the job in detail and consequently a far more detailed tender is produced. This enables the contractor to have a complete 'feel' for the project so that more accurate and effective pricing is produced.

Payment for the work is made in stages as various activities are completed and therefore cash flow is tied closely to the programme and is more easily predicted. The more accurately the contractor can preprogramme the work the better are his use of resources and the more accurate his financial estimates of expenditure and income.

Note: Photogrraphs by Courtesy Manchester City Council – Environment and Development Design Consultancy.

Bibliography

Clarke, R. H. *Site Supervision*. Thomas Telford, London, 1988.

ICE Conditions of Contract 5th and 6th editions Compared. Thomas Telford London, 1991.

Civil Engineering Procedure 4th edition. The ICE, Thomas Telford London, 1986.

Stephen, Wearne. *Civil Engineering Contracts*. Thomas Telford, London, 1989.

Atkinson, A.V. *Civil Engineering Contract Administration* 2nd edition. Stanley Thornes (Publishers) Ltd 1992.

Markes, R. I., Marks, R. J. E., Rosemary, E. Jackson. *Aspects of Engineering Contract Procdure* 3rd edition, Pergamon Press. 1985.

Best Practice with the New Engineering Contract. Broome, C. paper 11154 ICE Proceedings.

Payne, A. C. *An Introduction to Management for Engineers*. Wiley, Chichester, 1995.

Designing for Health and Safety in Construction. Health and Safety Commission 1995.

Nicholson, T. H. N. *The New Engineering Contract*. paper in ICE Proceedings November 1992.

The New Engineering Contract: A Promising Start. Martin Barnes, paper in ICE Proceedings Augest 1994.

Designing for Health and Safety in Construction. Health and Safety Commission 1995.

Ballantyne, J. K. *The Resident Engineer* 2nd edition Works Construction Guide, (Thomas Telford Ltd, London, 1986.

Madge, P. *Civil Engineering Insurance and Bonding*, Thomas Telford Ltd, London 1987.

INDEX

Entries are in letter-by-letter alphabetical order. Entries in *italics* refer to publications and legal cases. Reference locators in *italics* indicate figures or tables when they appear outside a reference range.

AA (Automobile Association), 105, 113, 357
Abrahamson, Max, 379
Access
   ambulances, 103, 105
   chambers, 116–18
   city centres, 193, 201–3, 210, 532, 534
   contractors, 397, 399–400
   pipe ramming, 263–5
   safety, 482, 483, 484
   shafts, 33, *35*, 179–85, 197–201, 351
   to manholes, 35–6, 482, 483, 500–4
   traffic disruption, 103, 104–5
Accidents, 132–3, 343, 507
Adversarial attitudes, 379–80
Air
   purity, 496
   respiration, 485–7, 488–90, 498
   see also Breathing apparatus; Ventilation
Airline breathing apparatus, 495, 496
Air pollution
   monetary values, 356
   public utility works, 344
   technique choice, 375
Airvac sewerage system, 327
Air valves, 45
Alignment
   impact moling, 243
   pipejacking, 210, 214, 220–1, 233
   tunnel construction, 179–85, 188–90, 203, 233
      curves, 190
      straight drive, 188–90
Ambulance access, 103, 105, 108
American Society of Civil Engineers (ASCE), 51, 91
Ammonia, 491
Anchoring systems, 308
Arbitration Act 1996, 423

Asbestos–cement pipes, 38
ASCE (American Society of Civil Engineers), 51, 91
Asia, 469, 470
Association for Project Managers, 442, 477
ASTM standards
   cone penetration test, 79
   laboratory tests, 82
Audits, 192, 234–5
Augers, 64–5, *66*, 298
Austria, 326
Automobile Association (AA), 105, 113, 357

Backacters, 140–1, 176
Backdrop manholes, 33, 35, *36*
Backfilling
   boreholes, 67
   carriageway reinstatement, 114–15
   initial, 145–6
   main, 146–7
   on-line sewer replacement, 193
   pipe laying, 129–30, 145–7
Balancing reservoirs, 29
Balfour Beatty v. Kelston Sparkes Contractors Limited (1996), 420–1
Banque Keyser Ullman v. Skandia (1987), 425
Barlow, Peter William, 151, 171
Bazalgette, Sir Joseph, 150
Bedding factors, 127–8, 193, *194*, *195*
Belgium, 196
Bending
   flanged pipes, 41, *42*
   horizontal directional drilling, 306, 307
   semi-rigid and flexible pipes, 129
Bentonite, 176, 301, 303, 304, 369, 376
Bernard Sunley v. Cunard White Star (1940), 388

Bill of Quantities contract, 387, 419–20, 536, 538–9, 544, 549
Blind persons, 110, 112
Blue Book, 101, 109, 110, *512*
Bolting
  flanges, 36
  segments, 9, 13–14, *18*, 197, 198, 203–5
BOO (build, own and operate), 470–1
BOOT (build, own, operate and transfer), 470
Boreholes
  backfilling, 67
  ground water, 80
  horizontal directional drilling, 300
  in-situ testing, 76–9
  location and depth, 60–1, 310
  trial pits, 47–8, 61, 67
  water balance, 63, *64*
Boring
  horizontal directional drilling, 302–3
  methods, 61–7
Breathing, 485–7, 488–90
Breathing apparatus, 479, 480, 484, 493–7, 503, 504
  cylinders, 496, 497
  routine operations, 496–7
  seals, 494, 496
  types, 494–7
  valves, 494–5, 496
Breathing Apparatus (Report on Examination) Order 1961, 478, 480, 493–4
Brick sewers
  London, 150–1
  manholes, 198
  rehabilitation, 12, *14*, 15
  'U'-shaped sewer type, 8–9, *18*
British Gas plc, 16, 272, 303
British Standards
  borehole diameter, 65
  buried pipeline design, 125
  cone penetration test, 79
  construction investigation, 89
  environmental policy, 463
  excavation support, 133
  ground hazards, 49
  pipejacking, 211, 215, 217, 220
  pipes, 37–9, 217
  project management, 442
  quality assurance, 233, 463, 467
  sample quality, 72
  semi-rigid and flexible pipes, 129
  site investigation, 57, 59, 154–5, 215
  soil chemical tests, 89
  standard penetration test, 77
  timber supports, 148

tunnel linings, 168
Brunel, Marc Isambard, 171, 175, 196
BS *see* British Standards
Build, own and operate (BOO), 470–1
Build, own, operate and transfer (BOOT), 470
Buried assets
  mapping, 99
  pipes, 6, 124–7
  site investigation, 95–9
  tracing, 96–8
Businesses, traffic disruption, 103–8, 109–10, 361, 532, *534*
Butt-jointed pipes, 8, 218, 257–8, 360

Cables
  buried, 96–8
  horizontal directional drilling, 297–8
  impact moling, 250–1
Caissons, 198, 200–1, 221–2
Campaign for the Renewal of Older Sewerage Systems (CROSS), *10*, 19
Cantilever supports, 135
Carbon dioxide, 485, 488–90, 491, 496
Carbon monoxide, *489*, 490, 491, 496
Carriageways
  deterioration, 342–3
  reinstatement, 114–15, 146–7, 373, 440
  restricted width, 109, 110, 112, 353, 357, 375
Carrier pipe installation, 269
CARs (corrective action requests), 235
Cascades, 35
Cash flow, 435–6, 464, 468, 550
Casings
  gas cylinders, 497
  impact moles, 241–2
  light cable percussion drilling, 62–4
  pipe ramming, 254, 269, 270
Cast iron
  linings, 166, 196, 197
  pipes, 38
Caulking, 36
CCTV (close circuit television), 177, 211, 360, 443
CDM *see* Construction (Design and Management) Regulations 1994
CE marks, 231, 451
Cement
  asbestos–cement pipes, 38
  mortar, 36
Central and South America, 469, 470
Centrifugal spinning, 219
CESMM (Civil Engineering Standard Method of Measurement), 387–8, 465, 538
Cesspools, 3, 5

CESWI (Civil Engineering Specification for the Water Industry), 466
Chaplin v. Hicks (1911), 405
Cheetham Hill Road, Manchester, 345, 359–66
Chemical injection, 145, 174
Chemical process effluents, 452
Chemical tests, 89
Chitty on Contracts, 428
Chlorine atmospheres, *489*, 491
CI (consistency index), 83–4
Circular letters, 104, 108
CIRIA (Construction Industry Research and Information Association), 211
Citizen's Charter, 351
City centre access, 193, 201–3, 210, 532, 534
Civil engineering, 17, 190, 191, 192, 230, 340
    contract management, 227, 378–429
    see also Engineers
Civil Engineering Specification for the Water Industry (CESWI), 466
Civil Engineering Standard Method of Measurement (CESMM), 387–8, 465, 538
Civilisations, 2–5, 7, 8, 9
Claims
    adversarial attitudes, 379–80
    causal links, 406
    contracts, 379, 382, 391, 400, 401, 542–3
        apportionment, 425–6
        breach, 402–5, 422–3, 424, 425, 542
        combined causes, 405–11, 422
        contractor against subcontractor, 379, 388–9, 402, 424–9
        cut-off point, 424
        main contract, 381–413, 414, 416–17, 420–2, 424–9, 542–3
        proof, 425
        subcontracts, 388–9, 413–24
    delay, 394, 408–9, 543
    disruption, 406–7
    liquidated damages, 392, 402, 426
    measurement claims, 382–8, 410, 418–19
    overheads, 408–11
        Emden formula, 410
        Hudson formula, 409–10
    primary and secondary effects, 406, 418
    rolled-up, 405
    site investigation, 51, 91
    time extension, 395, 543
    traffic disruption, 103–4
Clarksteel Limited v. Birse Construction Ltd (1996), 403
Classification and index tests, 82, 83–4
Clay Pipe Development Association (CPDA), 127–8, 132, 193, 194

Clayware pipes, 8–9, 15, 37, 127, *194*, *195*, 219, 279, 281
Cleaning, breathing apparatus, 497
Clearline Expandit machine, 284, 288
Clients, 469, 470, 471, 473, 474, 508–9
Cloacina (goddess of sewers), 3
Close circuit television (CCTV), 177, 211, 360, 443
Coal mining, 196
Codes of practice
    '6 pack', 439
    Blue Book, 101, 109, 110, *512*
    New Road and Street Works Act 1991, 101, 115
    pipejacking, 211–12, 215
    Specification for the Reinstatement of Openings in Highways, 115, 146
Cofferdams, 139
Cohesionless soils, 156–61, 222, 275
Cohesive soils, 87, 155–6, *157*, 214–15
Colebrook–White formula, 119–20
Collapse of sewers, 17–19, 359–60
Combined sewerage systems, 1, 7, 23, 27, 32, 473
Combustion, 490
Commissioning, 462, 468, 476, *477*
Common law rights, 381–2, 389, 399, 402–11, 425, 428
Communication
    confined spaces, 485, 500, 502
    see also Information provision
Compensation
    NEC/ECC contracts, 549
    public utility works, 346, 348, 373
Completion, 392, 400, 402, 411–12, 416–17, 454, 476, *477*, 543–4, 546
Compressed air
    breathing apparatus, 494–5, 496, 497
    impact moling, 250, 251
    pipebursting, 282, 287
    pipe ramming, 258–9, *260*, 267–9
    soft rock tunnelling, 526, *530*
Compressed Air Special Regulations 1858, 176
Concrete
    glass reinforced, 217, 219
    in-situ
        manholes, 198
        tunnel linings, 167–8
    pre-cast
        manholes, 33, *34*, 197–9
        pipes, 37, 218–19, 222–3
        shafts, 33, *35*
        soakaways, 29
        tunnel linings, 9, 13–14, *18*, 196–7, 527, *530*, *531*
    reinforced, 37, 196–7, 217, 218–19
Concrete Pipe Association, 193, 211

Cone penetration test (CPT), 78–9, *89*
Cones
  impact mole heads, 244
  pipebursting, 273, 276, 285, *286*, 287, 291, *292*
  pipe ramming, 257, 259, 261–2, 266–7
  traffic management, 112, 113, 114, 147
Confined aquifers, 61
Confined spaces, 439, 451, 452, 478–506, *512*
CONIAC (Construction Industry Advisory Committee), 226, 455, 456
Consistency index (CI), 83–4
Consortia, 469, 472–4
Constant head tests, 78
Construction
  ground hazards, 49–54
  heading, 147–9, 193–6, 345, 359–66, 363, 364–5, 525
  open-cut, 132–44, 193
  technique choice, 367–77
  tunnels, 150–92
  see also New construction
Construction (Design and Management) Regulations 1994 (CDM)
  application, 456–9
  construction management, 225–6, *477*
  contract supervision, 507–8
  definitions, 455–6
  health and safety files, 452, 455, 462, 482, 510
  health and safety plans, 226, 455, 461–2, 482, 510, *511–13*
  horizontal directional drilling, 306
  method statement, 228
  planning supervisors, 225–6, 452, 455, 459–60, 509–10
  principles, 225, 461–2
  project management, 431, 432, 439, 452, 455–62, *477*
    commissioning and project review, 462, 468, 476, *477*
    construction phase, 461–2
    design phase, 460
    feasibility phase, 459–60
    tender phase, 461
  risk assessment, 452, 475
  sewer safety, 478, 482
Construction (Head Protection) Regulations 1989, 479, 480
Construction (Health, Safety and Welfare) Regulations 1996, 478, 482
Construction Industry Advisory Committee (CONIAC), 226, 455, 456
Construction Industry Research and Information Association (CIRIA), 211
Construction investigations, 89–90

Construction (Lifting Operations) Regulations 1961 No. 1581, 478, 480
Construction management, 224–35
  audits, 192, 234–5
  CDM Regulations, 225–6, *477*
  procurement, 233–4
  project management, 432, 466–8
  safety, 232–3
  site management, 226–7
  site support staff, 227, 229–34
  training, 227–9
  see also Contracts
Construction sites *see* Site
Consultation
  contracts, 383
  New Road and Street Works Act 1991, 101–5
  public relations, 105–8, 532
  sewage treatment schemes, 449
Contaminated land, 312–13, 451
Contour mapping, 68, 116
Contractors
  appointment of subcontractor, 401–2, 535–6
  claims against subcontractors, 379, 388–9, 402, 424–9
  contract responsibilities, 225–6, 404, 515–16, 523, 528–9, 546, 548
  principal, 461
  programme of works, 396, 399–400, 414–15, 467, 515–19, 549–50
    accelerated, 411–12
  rights, 398, 404
  selection, 472, *477*
  setting out errors, 396–7, 400–1, 528–9, 531
  site access, 397, 399–400
  staffing resources, 467
  superintendence, 522–3
  suspension of work, 397–9, 412–13, 422–3
Contracts
  adjudication, 414, 549
  apportioning risk, 382, 389, 473–4, 548
  arbitration, 380–1, 403–4, 423
  award of, 465, *477*
  Bill of Quantities, 387, 419–20, 536, 538–9, 544, 549
    changes, 386–7
  breach of, 402–5, 422–3, 424, 425, 542
  build, own and operate, 470–1
  commencement, 399, 414–15, 520–1
  completion, 392, 400, 402, 416–17, 476, *477*, 543–4, 546
  conditions, 545–6, *547*
  contingency sums, 410
  cost plus fixed fee, 545
  cost plus fluctuating fee, 545

cost reimbursement, 544–5, 546

daywork rates, 540, 545

defective work, 400–1, 427

delays and extensions of time, 391–5, 397, 399, 406–7, 408–9, 540, 543, 549

design and build, 469–70

documents, 229–30, 385, 465–6

express terms, 381, 395, 400, 415

flow information, 207

foresight, 390

ICE, 224, 382–412, 418, 465, 475, 508, 545, 546, *547*

   claims, 542–3

   contractors' responsibilities, 515–16

   dayworks, 540

   engineers' duties, 509, 510, 513–14, 522

   payments, 541–3

   possession, 520–1

   programme of works, 517, *518*

   remeasurement, 538–9

   setting out, 528–9

   site organisation, 226

   subcontracts, 536

   unforeseen conditions, 161, 542

   variations, 539

IChemE, 224, 465, 466, 508, 545, 546, *547*

IEE/IMechE, 545–6, *547*

JCT, 411, 418

lump sum, 449, 466, 544, 546

management, 378–429

   plans, 226

   standard approach, 442

measurement, 382–8, 410, 418–19, 538–40

NEC, 224, 546–8

   NEC/ECC, 507–8, 548–50

notice requirements, 384–5, 386–7, 389–90, 392, 396, 397

overheads, 387, 389, 408–11

payment, 422–4, 468, 541–3, 548

pay-when-paid, 414, 422

rescission, 399, 412–13, 428

responsibilities, 508–16

safety plans, 233

strategy, 544–50

supervision, 507–50

target, 545, 549

temporary works, 190–1, 227

traditional attitudes, 378–9

tunnelling, 151

turnkey, 449, 469

types, 224–5, 432, 449, 468–74, 475

unforeseen conditions, 161, 389–91, 420–2, 542

valuation, 382–8, 419–20, 471, 476, 540

variations, 382–6, 397, 406, 417–20, 421–2, 539–40

   see also Claims; Procurement; Subcontracts

Contracts term contracts, 472

Control of Pollution Act 1974, 441

Control of Substances Hazardous to Health Regulations 1994 (COSHH), 478, 480, 481, 482, 493, 498, *512*

Co-planing, 185, *186*

Core barrels, 66

Corrective action requests (CARs), 235

Corrugated metal

   pipes, 38

   unflanged liner plates, 38

COSHH (Control of Substances Hazardous to Health Regulations 1994), 478, 480, 481, 482, 493, 498, *512*

Cost plus fixed fee contracts, 545

Cost plus fluctuating fee contracts, 545

Cost plus percentage contract (daywork), 540, 545

Cost reimbursement contracts, 544–5, 546

Costs

   contract management, 388–9

   defective work, 400–1

   delay, 393, 407, 408–9, 543, 549

   design and build, 469–70

   desk study, 56–7

   direct, 372

   horizontal directional drilling, 298, 314–16

   indirect, 298, 339–66, 372–3

   NEC/ECC contracts, 549

   New Road and Street Works Act 1991, 109, 314

   procurement, 463–4

   project management, 447, 448–9, 469–70

   public utility works, 340

   rate fixing, 383–7

   risk management, 452–6

   road space rental, 349–50, 354–6, 356–7

   sampling, 76

   sewers

      collapses, 17–18, 360

      construction, 9, *11*

      rehabilitation, 9, 11–12, 13, 15, 17, 360–1

   site investigation, 51, *53*, 54

   social costs, 202–3, 298, 339–66, 371, 373

   suspension of work, 398–9

   technique choice, 371–4

   tunnel boring machines, 179

   vchicle operation, 352, 364

   whole life costs, 374, 433

CPDA (Clay Pipe Development Association), 127–8, 132

CPT (cone penetration test), 78–9, *89*

Crack propagation, 161
Critical sewers, *10*, *11*, 12, 19
CROSS (Campaign for the Renewal of Older
　　Sewerage Systems), *10*, 19
Crushing strength, 127–8, *194*, *195*
Culverts, 6, 8
Cut and cover construction, 52, 54
Cutterbooms, 176
Cutting
　　pipe ramming shoes, 255, 257
　　rock, 161
Cyanide atmospheres, 491

Damages *see* Claims
Daywork rates, 540, 545
DCF (discounted cash flow), 435–6
DDB (double decker buses) factor, 18
Delivery pipes, 208
Department of Transport
　　geotechnical reports, 90–1
　　'lane rental' mechanism, 348–9
　　Traffic Signs Manual, 109
　　vehicle operating costs, 352
Dereliction, 16–18
Design
　　civil, electrical and mechanical, 230
　　engineer consultation, 517
　　horizontal directional drilling, 311
　　inverted syphons, 30–2
　　objectives, 450
　　pipejacking, 212
　　　　pipes, 217–18
　　project management, 431, 449–51, 460, 475, *477*
　　sewerage systems, 6, 20, 24–7
　　sewers, 116–31
　　　　external forces, 124–9
　　　　flexible pipes, 125–6, 129
　　　　hydraulic design, 118–19
　　　　preliminary design, 116–18
　　　　rates of flow, 118–23
　　　　rigid pipes, 125–6, 127–9
　　　　semi-rigid pipes, 125, 129
　　　　structural design, 123–30
　　　　survey and scoping, 116
　　software, 311, 320
　　temporary works, 191
　　tunnel linings, 167–9
　　vacuum sewerage, 328, 332–5
　　　　stations, 333–5
　　see also Construction (Design and Management)
　　　　Regulations 1994
Design and build contracts, 469–70

Desk study
　　existing plans, 95–6
　　reports, 59–60, 89, 90
　　site investigation, 48, 49, 56–60
Dewatering, 141–4, 172, 369, 376
　　well-point system, 142–4
Diameter
　　manholes, 33, *34*
　　pipes, 37, 118–23, 128, 141, 250, 251, 252, 254–
　　　　5, *256*, 273–4
　　　　large, *299*, 300, 301–2
　　　　small, *270*, 271, 272, 283, 295, 296, 444
　　shafts, 198, 201
　　tunnels, 151, 166, 197, 203, 205
Diesel vapour, *489*
Direct costs
　　Cheetham Hill Road, Manchester, *361*
　　public utility works, 352
　　technique choice, 372
Dirt, dust and mess, 344
Disabled people, 109, 110
Discharges
　　chemicals, 452
　　foul sewage, 1–5, 21, 441
Discounted cash flow (DCF), 435–6
Diseases, 3, 7, 483, 505, *514*
Display screen equipment, 439
Diversion routes, 101, 103, 105, 107–8, 109–10,
　　342–3, 362–3, 532
　　unofficial, 342
Documentation
　　contracts, 229–30, 385, 465–6, 475
　　health and safety files/plans, 226, 452, 455, 461–
　　　　2, 482, 510, *511–13*
　　tenders, 461, 464
DoE/NWC Working Party, 25
Domestic waste water, 1–5, 21–3
Donseg lining, 167
Double decker buses (DDB) factor, 18
DPT (dynamic probing), 79
Drag shovels, 140–1
Drainage
　　horizontal directional drilling, 299, 311–13
　　pollutants, 313
　　trenches, 142–4
　　see also Surface water sewers
Drained tests, 87
Drawings, 116–18, 395–7
D.R. Bradley (Cable Jointing) Ltd v. Jefco
　　　　Mechanical Services Ltd (1988), 428
Drilling
　　fluids, 320
　　methods, 61–7

sample disturbance, 72
see also Horizontal directional drilling
Drive length, 368
Dry weather flow (DWF), 21–3, 27, 122, 123, 207
Dry wells, 43
Ductile iron pipes, 38
Dudley Corporation v. Parsons and Morrin Ltd (1959), 386
Dust, 344
Duty of Care procedures, 304
DWF (dry weather flow), 21–3, 27, 122, 123, 207
Dynamic probing (DPT), 79

Earth pressure balance machine (EPBM), 177, 214, 215
ECC (Engineering and Construction Contract), 508, 548–50
Effective stress tests, 87
EIA (environmental impact assessment), 346–7, 440–1, 444–6
Electrical engineering, 190, 191, 192, 227, 230–2
Electrical resistivity surveys, 70
Electricity supply
    contract management, 231
    horizontal directional drilling, 297, 303
    pumps, 43
    safety, 375, 484, *512*
    underground mains, 339–40
Electricity at Work Regulations 1989, 230
Electrolux schemes, 327
Electromagnetic location (EML), 97, 99
Electromagnetic surveys, 68, 69–70
Electronic gas detectors, 492–3
Embedment
    classes, 129
    stiffness, 126, 127
Emden formula, 410
Emergencies, 103, 105, 202, 503–4, 532, 544–5
EML (electromagnetic location), 97, 99
Employees
    contract responsibilities, 234
    health, 498, 505–6
Employers
    contract responsibilities, 234, 401, 404, 508–9, 520–1
    reports, 536–8
    indemnity, 402
    principle of 'prevention', 395–402
    suspension of work, 397–9
Engineering Council, 229
Engineers
    common law claims, 403–4

contract responsibilities, 396, 398, 401–2, 410–11, 413, 416, 476, 546
    delegated duties, 513–15, 522
    environmental impact, 531–2
    programme of works, 517, 519
    supervision, 509, *521*, 541–4
representatives and staff, 227, 510, 513–15, 517, 522–3, 528–9, 536–8
see also Civil engineering
Entry procedures, 500–4
ENV7: Geotechnical Design (BSI 1995), 49
Environment Agency, 444
Environmental costs
    public utility works, 340, 348–9, 356
    rehabilitation, 13
    technique choice, 373
Environmental impact
    contract supervision, 531–5
    green field sites, 532, *533*
    horizontal directional drilling, 316
    international agreements, 375
    legislation, 346, 375, 440–1
    protection, 304, 375
    public utility works, 341–7
    sewer collapses, 17–18
    sustainable development, 375, 376, 441–2
    taxes, 348–9
    technique choice, 375–9
Environmental impact assessment (EIA), 346–7, 440–1, 444–6
Environmental policy, 463
Environmental projects, horizontal directional drilling, 312–13
Environmental Protection Act 1990, 304, 441
EPBM (earth pressure balance machine), 177, 214, 215
Escape breathing apparatus, 484, 495
ESR (excavation support ratio), 165
European Community
    coastal and estuarial waters, 454
    environmental impact assessment, 347, 357, 440–1, 443, 444–6
    health and safety legislation, 439
    machinery directive, 231, 451
    sewage treatment schemes, 449
European Journal, 464
European Standards
    buried pipeline design, 125
    semi-rigid and flexible pipes, 129
Excavations
    depth, 368–9
        up to 6 m depth, 136–7, 193
    dewatering, 141–4, 172

environmental impact, 345–6
fatal accidents, 132–3
manual, 170, 176, 197
mechanised, 170–1, 176–9, 191, 211, 230
minimal, 140–1
pipe ramming access, 263–5
safety, *512*
soft ground tunnelling, 152–61, 170–5, 196, 526,
    530
standard solutions, 135–6
support, 132–40, *141, 142*, 171–2, 524–5
trench, 138–9, 140–1
trial pits, 47–8, 61, 67
see also Open-cut construction; Tunnels
Excavation support ratio (ESR), 165
Existing flow, 206–9, 450
Existing plans, 95–6, 450, 519
Existing structures
    contract management, 391
    rehabilitation, 13–14
    sewer construction, 145
    site investigation, 48, 292–3
Expiration (exhalation), 485, 486–7, 488–90
Extraction ventilation, 498

Factories Act 1961, 478, 479–80, 482, 483, 498
Factors of safety, 127–8
Fairfield–Mabey Ltd v. Shell UK Trading Ltd,
    Metallurgical Services (Scotland) Ltd, third
    party (1989), 425–6
Falling head tests, 77–8
Faraday Cage, 303
Fatal accidents, 132–3, 507
FCEC (Federation of Civil Engineering Contractors'
    Forum of Subcontract), 413, 414, 415–29
Feasibility
    project management, 431, 433–4, 442–9, 459–
        60, 474–5, *477*
    technique choice, 367–71
Federation of Civil Engineering Contractors' Forum
    of Subcontract (FCEC), 413, 414, 415–29
F.G. Minter v. Welsh Health Technical Services
    Organisation (1980), 412
Finance
    risk factors, 452–4
    sewer rehabilitation, 19
    viability, 435–9
Fire services, 103, 105, 108, 503, 504
First aid, 479, 484, 499, 503, 504
Flanged pipes, 40–1
Flexible pipes, 36, 38
    backfilling, 146

buckling capacity, 129
design, 125–6, 129
Flooding of sewers, 442, 443, 512
Flood plains, 29
Flow rates *see* Rates of flow
Flushing, 32
Footway reinstatement, 114–15
Foul sewage
    discharge, 1–5, 21, 441
    pumping stations, 41–3
    rates of flow, 6, 20, 21–3, 118
    sewer design, 6
    vacuum sewerage, 327
Francis, John, 8
Freeman v. Hensler (1990), 399
Freezing, ground treatment, 174–5
Fresh air breathing apparatus, 494
Friction
    angle, 214, 215
    load, 212, 214, 215, 275
    losses, 45–6, 332
Funding *see* Finance

Gantries, 532, *533*
Gas detector tubes, 491–2
Gases, 485–93
    atmospheres, 482, 491, 499–500
    hazardous, 479, 482, 484, 487–91, 499–500
    monitoring, 484, 485, 491–3, 499–500, 502, 503
        exposure levels, 492
    respiration, 485–7
Gaskets, 167, 222, 223
Gas pipelines, 16, 297, 304, 339–40
Geographical Information Systems (GIS), 96
Geology
    horizontal directional drilling, 306
    pipejacking, 212
    tunnel construction, 152–66
Geophysical investigation, 60, 67–71
    induced methods, 68, 70–1
    planning, 71
    potential field methods, 68–70
Geotechnical Baseline Report (ASCE), 51, 91
Geotechnical parameter tests, 82–3, 85–9
Geotechnical reports, 51, 90–2, 154, 212
    commentary, 91–2
    post construction, 91
Germany
    horizontal directional drilling projects, 325, 326
    impact moling, 237, 253
    pipebursting, 296
    pipejacking, 211

pipe ramming, 271
technique choice, *372*
GIS (Geographical Information Systems), 96
Give and take system, 111
Glass reinforced concrete (GRC), 217, 219
Glass reinforced plastic (GRP), 38, *39*, 217, 219
Glenlion Construction Ltd v. The Guinness Trust
(1987), 392, 411
Global Positioning System (GPS), 180
GPR (ground probing radar), 68, 69, 71, 98
Gradient
horizontal directional drilling, 316
sewers, 6, 33, 35, 116, 118–23, 144–5, 367–8
tunnels, 181, 182–5
vacuum sewerage, 332, *333*, 444
Gravity sewers, 6, 327, 444
above ground, 39–41
horizontal directional drilling, 298, 312, 368
inverted syphons, 29–32
Gravity surveys, 69
GRC (glass reinforced concrete), 217, 219
Greathead, James Henry, 151, 175, 196
Greathead shields, 151, 171, 175–6, 196
Green Book (IChemE), 224, 466, 475, 508, 545,
546, *547*
Ground anchors, 134
Ground conditions
closure, 158–9, 214, 215
desk study, 48, 49, 56
heave, 155, *156*, 249, 312, 369–70
horizontal directional drilling, 298, 301, 308–9,
318–20
impact moling, 247–9
pipebursting, 274–8
pipe ramming, 267–9
site investigation, 49
technique choice, 369–70
unforeseen, 161, 390, 420–3
Ground exploration
borehole location and depth, 60–1
geotechnical reports, 90, 212
ground water, 48, 79–82
hazards, 48, 49–54
longitudinal section, 118
main ground exploration, 48–9, 60–89
methods, 61–7
planning, 60–1
records, 89
sampling and sample quality, 72–6
site investigation, 47, 48–9, 60–89
soft ground tunnelling, 152–61, 170–5
testing
in-situ, 76–9, *154*
laboratory, 82–9, *154*

Ground Investigation Price Index, 54
Ground movement, 274, 276–8, 443
Ground probing radar (GPR), 68, 69, 71, 98–9
Ground treatment, 172–5, 213–14
chemical/grout injection, 145, 174, 205, 222
freezing, 174–5
see also Soils
Ground water
lowering water table, 141–4, 174
reports, 90
site investigation, 48, 79–82
tunnel construction, 51, 141, 170, 172–5
Grout injection, 167, 174, 205, 222
GRP (glass reinforced plastic), 38, *39*, 217, 219
Guarding public utility works, 109
Guidance systems, 301, 302–3, 309–11, 320
Gyroscopes, 220

Hadley v. Baxendale (1854), 425
Hard rock
drilling, 61, 65–7, 369
tunnelling, 161–6, 179, 526
Hazard identification
confined spaces, 482–3
control, 483
health and safety plan, *514*
site investigation, 48, 49–54, 91
Hazardous substances, 493, 498, *512*, *513*
gases, 479, 482, 484, 487–90, 499–500
legislation, 478, 480, 481, 482
Hazen–Williams formula, 45, *46*, 332
HDPE (high density polyethylene), 38
Heading construction, 147–9
on-line sewer replacement, 193–6, 345, 359–
66, 363, 364–5, 525
Head losses
inverted syphons, 30, 32
pumps, 45–6
vacuum sewerage, 332
Head protection, 479, 480
Health and safety
employees, 498–500
files, 452, 455, 462, 482, 510
hygiene, 2, 3, 505–6
legislation, 432, 439, 478–82
plans, 226, 455, 461–2, 482, *511–13*
procedures, 468, 475, 498–504
risk, 232, 451–2
tender phase, 461
work equipment, 231
see also Safety
Health and Safety at Work, etc. Act 1974 (HSWA),
225, 230, 231, 439, 478, 479, 483, 497

Health and Safety Commission, 226, 455, 482
Health and Safety (Display Screen Equipment) Regulations 1992, 439
Health and Safety Executive (HSE), 232, 234, 375, 444, 468, 513
  notification of projects, 456–9, 509, 532
Heave, 155, *156*, 249, 312, 369–70
Helium–neon lasers, 188
H. Fairweather and Co. v. London Borough of Wandsworth (1988), 393
High density polyethylene (HDPE) pipes, 38
Highway authorities, 101–2, 105, 109, 114, 115, 146, 202, 440, 532
Highways Act 1980, 110
History
  buried asset maps, 96
  roads, 100
  sanitation, 2–5, 7, 8–9
  tunnelling, 179
HM Treasury guidelines, 473
Horizontal directional drilling, 297–326
  advantages, 313–16
  anchoring systems, 308
  applications, 297–8, 299, 311–13, 320, 321
  bend limits, 306, 307
  bit choice, 308, *309*, 321
  cableless steering, 310–11
  case studies, 320, 322–6
  control techniques, 320
  design software, 311
  disadvantages, 316–20
  drainage applications, 299, 311–13
  drilling process, 303, 304–5
  entry pitch, 307
  environmental projects, 312–13
  equipment, *299*, 301–5
  future development, 320–1
  ground conditions, 298, 301, 308–9, 318–20, 369
  guidance systems, 301, 302–3, 309–11, 320
  information provision, 305–6
  lateral connections, 118, 311, 312
  maxi system, 298, *299*, 304–5, *318*, *319*
  micro system, 298, *299*, 302
  midi system, 298, *299*, 303–4, *317*
  minimum depth, 307–8
  mini system, 298, *299*, 302–3, *315*
  pilot bore, 301
  planning, 305–9
  reaming, *299*, 300, 302, 304
  route, 306–8
  safety, 306
  sewerage projects, 320
  technique choice, *377*
  terminology, 298–9

Horne Report on Roads and Utilities (1985), 341
Hoses
  breathing apparatus, 496
  impact moling, 245, *246*
  suction, 207, 208–9
House of Commons' Environment Subcommittee, 17
Housing Grants, Construction and Regeneration Act 1996, 413, 414, 422, 423
HSE *see* Health and Safety Executive
HSWA (Health and Safety at Work, etc. Act 1974), 225, 230, 231, 439, 478, 479, 483, 497
Hudson formula, 409–10
Hydraulic design, 118–19
Hydraulic frames, 138, *139*
Hydraulic machines
  horizontal directional drilling, 302
  impact moling, 238–41
  pipebursting, 275–6, 279, 283–91, 295–6
  pipejacking, 210
Hydraulics Research Station, 25
Hydrogen sulphide, 488, *489*, 491, 492
Hydrological cycle, 2
Hygiene, 2, 3, 505–6

ICE *see* Institute of Civil Engineers
IChemE (Institution of Chemical Engineers), 224, 465, 466, 508, 545, 546, *547*
IEE/IMechE contracts, 545–6, *547*
Impact moling, 236–53
  accessories, 243–6
  applications, 252–3
  equipment, 238–41
  ground conditions, 247–9
  launch arrangement, 249, 250–1
  mole construction, 241–3
  new pipes, 246–7, 250–2
  pipe installation, 249–52
    direct pulling in, 251–2
    standard procedure, 250–1
  radio sondes, 245–6
  site preparation, 249
  starting, 243, 250
  technique, 238–46
Indirect costs
  horizontal directional drilling, 198
  public utility works, 339–66
  technique choice, 372
Induction training, 227–8
Industrial Revolution, 7
Industrial waste water, 1, 2, 7–8, 21, 23, 207
Information provision
  circular letters, 104, 108

contract management, 395–6
  subcontracts, 416–17
  supervision, 536–8
desk study, 57, *58*
feasibility brief, 450–1
ground hazards, 50
health and safety files/plans, 226, 452, 455, 461–2, 482, 510, *511–13*
horizontal directional drilling, 305–6
medical instruction cards, 499
project management, 474
site investigation, 89–92
Infrastructure
  build, own and operate, 471
  existing hole in the ground, 9, 11, 12
  partnering agreements, 472
Injuries, 17–18, 499
In-situ concrete
  manholes, 198
  tunnel linings, 167–8
In-situ testing, 76–9, *154*
Inspection
  breathing apparatus, 478, 480, 493–4
  confined spaces, 500–3
  contract management, 400–1, 467, 523
  walk-over, 57, 59
Inspiration (inhalation), 485, 486, 488–90, 494
Installation
  horizontal directional drilling, 302
  impact moling, 249–52
  pipebursting, 279–80
  pipe loading, 130
  pipe ramming, 254–8, 265–9
  remote, 15
  technique choice, 367–77
  work equipment, 231
Institute of Civil Engineers (ICE), 378
  Conditions of Contract, 224, 382–412, 418, 465, 475, 508, 545, 546, *547*
  claims, 542–3
  contractors' responsibilities, 515–16
  dayworks, 540
  engineers' duties, 509, 510, 513–14, 522
  payments, 541–3
  possession, 520–1
  programme of works, 517, *518*
  remeasurement, 538–9
  setting out, 528–9
  site organisation, 226
  subcontracts, 536
  unforeseen conditions, 161, 542
  variations, 539
  standard forms, 380, *547*

Institute of Hydrology, 25
Institution of Chemical Engineers (IChemE), 224, 465, 466, 508, 545, 546, *547*
Institution of Public Health Engineers, 9
Interest charges, 412–13, 423
Internal pressure, 123–4
Internal rate of return (IRR), 435, 437–9
International Standards
  environmental management, 375, 463
  quality assurance, 463
Intrusive investigation methods, 99
Inverted syphons, 29–32
Investment
  rehabilitation, 16–17, 19
  sewage treatment schemes, 449
  site investigation, 51, *53*
  social cost minimisation, 347–8
Investors in People, 234
Iron, 38, 166, 196
IRR (internal rate of return), 435, 437–9
Iseki Pipe Replacer, 178, 205–6
Italy, horizontal directional drilling projects, 324–5

Jacking *see* Pipejacking
Japan, 176–7, 192, 205–6, 211, 347, 369
JCT contracts, 411, 418
J.F. Finnegan Ltd v. Sheffield City Council (1988), 410
Joints
  butt, 8, 218, 257–8, 360
  flexible, 37, 123
  mechanical, 279, 281, 290
  rebated, 218
  rigid, 36, 37
  rock mass, 164

Kay Ltd v. Hosier and Dickinson Ltd (1972), 427

Laboratory testing, 82–9, *154*
Ladder access, 35–6
Land
  contaminated, 312–13
  entry onto private land, 440, 451
Land drains, 8–9, 15, 313
Landfill
  horizontal directional drilling, 313
  material disposal, 376
  nineteenth century, 313
Land surveying *see* Surveying
Lane marking, 112–13

Laser guidance systems, 188, 211, 220
Lateral connections
    horizontal directional drilling, 118, 311, 312
    pipebursting, 281, 293
    rates of flow, 207
    sewer networks, 117–18
    technique choice, 371
Lateral pressures, 136
Launching cradles, 250–1, 264, 264–5, *265*, 266–7
Launch pits, 249, 250–1, 263–4, 294, 302
Lead pipes, 219
Legislation
    environmental, 346, 375, 440–1
    health and safety, 225–6, 232, 432, 439, 478–82
    private finance initiatives, 473–4
    project management, 432, 439–42
    risk factors, 454
    sewer safety, 478–82
    social costs, 202, 347–9
    traffic management, 101–5, 350–1, 439–40
Leptospirosis, 505
Levelling of tunnels, 179–85
LEV (local exhaust ventilation), 498
Licences, 228–9
Liernur, Captain Charles T, 327
Life cycle management, 9, *10*, *11*
Lifting operations, 479, 480, *512*, 532, *532*
Light cable percussion drilling, 62–4, *66*, 85, *86*,
    88
Lighting, 149, 484
Liljendahl, Joel, 327
Linings
    brickwork, 9, 166, 197
    concrete
        in-situ, 167–8
        pre-cast, 9, 13–14, *18*, 196–7, 527, *530*, *531*
    design, 167–9
    development, 166–7
    segmental, 9, 13–14, *18*, 151, 166, 168–9, 196,
        197
    shafts, 199
    smoothbore, 167, 168, 169–70
    tunnel construction, 166–70, 196–7
Liquidated damages, 392, 402, 426
Liquid limit, 83–4
Lloyd–Davies storm formula, 23, 24–6
Loads
    external forces, 124–7
    pipes
        impact loading, 130
        installation, 130
        operational, 123–30
        transportation, 130
Local authorities, 449, 504

Local exhaust ventilation (LEV), 498
Local radio traffic news, 108
London
    sewerage systems, 150–1
    tunnel construction, 167, 171, 175, 176, 196–7
London County Council, 5
Low pressure compressed air (LPCA), 166, 173,
    176
Lubrication
    impact moling, 250
    pipejacking, 212, 214, 215, 222
Lugeon (packer) tests, 78
Lump sum contracts, 449, 466, 544, 546

Magnetic surveys, 68, 69
Main ground exploration, 48–9, 60–89
    see also Ground exploration
Maintenance
    breathing apparatus, 497
    certificate, 542
    planned preventative maintenance, 191, 232
    traffic signs, 113–14
    work equipment, 232
Maintenance Assessment Rating and Costing for
    Highways (MARCH), 342–3
Management
    construction, 224–35
    contracts, 226, 378–429
    environmental, 446
    pipejacking, 211–12
    projects, 430–77
    risk, 54–5
    see also Construction (Design and Management)
        Regulations 1994; Traffic management
Management of Health and Safety at Work
    Regulations 1992, 225, 439, 478, 480–1, 499
Manchester
    historical development, 7–8, 150
    Paving and Soughing Committee, 8
    sewerage rehabilitation scheme, 345, 359–66
    unofficial traffic deversions, 342–3
Manchester Literary and Philosophical Society, 17
Manchester through the Ages (David Rhodes), 7
Manholes
    access, 35–6, 482, 483, 500–4
    backdrop, 33, 35, *36*
    cover lifting equipment, 484–5
    diameter, 33, *34*
    hazard identification, 482–3
    new construction, 32–6
    public utility works, 345
    rehabilitation, 14–15
    sewer layout, *117*

small sewers, 32–3
spacing, 32–3
tunnels, 197
types, 197–201
Manning equation, 120–2
Manometric head, 44–6
Manual Handling Operations Regulations 1992, 439
Manual labour, 15, 170, 176, 197, *514*
Manufacture
    pipejacking pipes, 218–19
    tunnel lining segments, 169
Maps
    buried assets, 96, 99
    sewer layout, 116–17, 450
    three-dimensional, 297, 306, 320
MARCH (Maintenance Assessment Rating and Costing for Highways), 342–3
Materials
    public utility works, 344–5
    records, 540
    schedules, 229–30
    storage, 466–7, 532, *533*
    testing, 523
Measurement
    claims, 382–8, 410
    contract supervision, 538–40
    remeasurement, 382, 538–9
    standard method, 387–8, 465, 538
    variations, 418–19
Mechanical engineering, 190, 191, 192, 227, 230–2
Medical instruction cards, 499
Meetings, 536–8
Merton v. Leach, 396
Meteorological Office, 25
Methane, 487, *489*, 491, 492
Method statement, 396
    induction, 228
    safety advisers, 232
    temporary works, 191
    tender analysis, 464, 465
Michael Angelo Taylor's Act 1817, 5
Microtunnelling, 177–8, 298
    layout design, 118, 177–8, 298
    pipes, 37
    technique choice, 368, 369, 371, 374, *376*
Middle East, 12
Millburn Services Ltd v. United Trading Group (UK) Ltd (1995), 403
Ministry of Agriculture, Fisheries and Food, 444
Ministry of Health storm formula, 23–4
Modern Engineering v. Gilbert–Ash (1974), 426

Moles
    horizontal directional drilling, 298, 302
    impact moling, 236–53
        clamping plates, 245, 251
        head configurations, 241–2, 244
Molloy v Liebe (1910), 417
Monitoring
    gaseous atmospheres, 484, 485, 491–3, 499–500, 502, 503
    horizontal directional drilling, 301
    impact moling, 245–6
    temporary works, 191–2
    vacuum sewerage valves, 332
Mooney v. Henry Boot Construction Ltd (1996), 420–1
MWD (wireless measurement-while-drilling), 309, 310

Narrow trenching, 140–1
NEC *see* New Engineering Contract
Neoprene seals, 221
Neotronics Minigas 4, 492–3
Net present value/cost (NPV), 435, 436, 437–9
New construction, 20–46, 443
    ground hazards, 51–4
    impact moling, 246–7
New Engineering Contract (NEC), 224, 507–8, 545, 546–8
    Engineering and Construction Contract (NEC/ECC), 508, 548–50
New Road and Street Works Act 1991 (NRASWA)
    environmental impact, 532
    inspection, 467
    reinstatement practice, 312, 314, 376, 397
    safety, 306
    traffic management, 101–5, 109, 114, 115, 146, 350–1, 371, 439
NGI tunnelling quality index (Q system), 164–6
Night–time working, 202–3, 543–4
Noise
    control, *513*
    house prices, 356
    monetary values, 352, 356
    public utility works, 344, 352
    shaft construction, 202
    technique choice, 375
Non-intrusive investigation methods, 96–9
North America
    impact moling, 253
    pipebursting, 284–5, 296
    pipe ramming, 271
    sewerage rehabilitation, 12

tunnel boring machines, 178–9
see also USA
Norwegian Geotechnical Institute (NGI), 164–6
NPV (net present value), 435, 436, 437–9
NRASWA *see* New Road and Street Works Act 1991
NVQ qualifications, 227
Oedometer tests, 88, 89
Off-line sewer replacement, 13, 15
OFWAT
    implied critical sewer life, *10*, *11*, 19
    sewerage network funding, 19
Oil mist, 496
Oil pipelines, 297, 304
One-way working, 111, 353, 355, 357
On-line sewer replacement, 12, 13, 15, 193–209
    see also Pipebursting technique
Open-cut construction, 132–44, *376*
    ground conditions, 369
    on-line sewer replacement, 13, 193, 359, 360, 362
    prohibition and control, 347–8
    sewer layout, 118
    sloping sides, 132
    support systems, 132–41, *142*
        standard solutions, 135–6
    up to 6 m depth, 136–7, 193
    see also Excavations; Tunnels
Open shields, 175–6
Operational loads on pipes, 123–30
Operative training, 228
Overbreak, 369
Overflows, 23–4, 27–9, 474
Overpumping, 206–9, 344, *513*
Oxford University, 211, 220
Oxygen
    monitoring, 492
    respiration, 485–7, 488–90

Packer (Lugeon) tests, 78
Partially separate sewerage systems, 1, 2, 32
Particle size distribution, 84
Partnering, 466, 471–2
Passenger transport authorities, 105, 108
Payment
    cash discounts, 423
    contractors, 468
    contract supervision, 541–3, 548
    delayed, 412–13
    final account and claims, 542–3
    interest charges, 412–13, 423
    interim, 423, 539, 541, 550
    non-payment, 412–13, 422–3

private finance initiatives, 473
provisional sums and prime cost items, 541
unforeseen conditions, 391, 542
Peak flow, 118, 119–22, 207
Pearce, Prof. David, 340, 356
Pedestrian safety, 100, 109, 110, 112, 113, 440, 534, *535*
Penetrometers, 67
PE (polyethylene), 38, 246, 279–81, 286, 290, 301, 329
Percussion drilling, 62–4, *66*, 85, *86*, *88*, 303
Permanent works, 227, 230
Permeability
    soil, 23, *24*, 25
    testing, 77–8, 88–9
Permits to work, 191, 452, *453*, *454*, 496, *512*
Personal protective equipment, 481, 483–5, 502
    see also Breathing apparatus
Personal Protective Equipment at Work Regulations 1992, 439, 478, 481
Personnel records, 234
Petroleum vapour, *489*, 491
PFI (private finance initiative), 469, 472–4
Piezometers, 80–2, *83*, *89*
Piling, 139–40, *142*, 524–5
PIM *see* Pipebursting technique
Pipebursting technique, 238, *239*, 272–96
    applications, 295–6
    cones, 273, 276, 285, *286*, 287, 291, *293*
    equipment, 281, 287–92
    Express system, 288–9, *290*
    ground conditions, 274–6
    ground movement, 274, 276–8
    lateral and service connections, 281, 292–3
    launch pits, 294
    operating systems, 281–7
    pipe strings, 279–80, 283, 286, 292
    pushers, 292, *293*
    replacement capabilities, 273–4, 279–81
    rod pipebursting, 284–7, 290–1, *292*, 296
    site operational requirements, 292–5
    technique choice, 368, *377*
    utility interference, 370
Pipejacking, 205–6, 210–23
    history, 210–11
    interjack requirements, 214–16
    lead pipes, 219
    loads, 212–13, 214–15, 217–18, 222
    lubrication, 212, 214, 215, 222
    management, 211–12
    pipe design and manufacture, 37, 217–19
    seals, 218, 221–2
    shafts, 216

spoil disposal, 214, 215, 217
surveying and alignment, 210, 212, 214, 219–21, 233
technical aspects, 212–17, 368
technique choice, *376*
temporary works, 221–2
thrust walls, 216–17, 221
Pipejacking – A Guide to Good Practice, 221
Pipe Jacking Association (PJA), 211
Pipe laying, 144–5
backfilling, 129–30, 145–7
Pipelines *see* Pipes
Pipe pushing machines, 258–9, 292
Pipe ramming technique, 210, 238, *239*, 254–71
accessories, 259–62
applications, 269–71
carrier pipe installation, 269
close-ended, 265–6
cones, 257, 259, 261–2, 266–7
cutting shoes, 255, 257
environmental impact, 375
equipment, 258–62
installation procedure, 265–7
launch arrangements, 263–5, 266–7
open-ended, 265–6
pipe preparation, 257–8
site preparation, 262–5
soil removal, 262, *263*, 267–9
technique choice, 368, 374, *377*
types of pipes, 254–8
welding, 254, 257–8, 267
Pipes, 36–41
asbestos–cement, 38
buried, 6, 96–8
carrier pipes, 269
cast and ductile iron, 38
characteristics, 254–5, *256*
clayware, 8–9, 15, 37, 127, *194*, *195*, *219*, *279*, 281
concrete, 9, 37, 193
corrugated metal, 38
delivery, 208
design, 217–18
diameter, 37, 118–23, 128, 141, 250, 251, 252, 254–5, *256*, 273–4
large, *299*, 300, 301–2
small, *270*, 271, 272, 283, 295, 296, 444
external forces, 124–7
flanged, 40–1
flexible, 36, 38, 125–6, 129, 146
friction loads, 212, 214, 215
gas supply, 16
internal pressure, 123–4
joints, 8, 36, 37, 123, 164, 218, 257–8, 279

laying, 144–5
lead, 219
length referencing system, 119
manufacture, 218–19
operational loads, 123–30
ovoid, 37
pipebursting, 279–81
pipejacking, 217–19
pipe ramming technique, 210, 254–8
polyethylene, 246, 279–81, 286, 290, 301, 329
high density, 38
pressurised, 297–8
PVC, 39, 206, 246, 247, 279, 281
rigid, 36–9, 125–6, 127–9, 145–6, 248
roughness, 120
semi-rigid, 125, 129
sizing, 25–7
spigot and socketed, 8, 38, 40, 144–5, 246, 251, 258
steel, 38, 40–1, 246, 247, 254–71, 301
stiffness, 126, 129
strength, 126, 127–9
strings, 279–80, 282, 283, 286, 301
vacuum sewerage, 328–9, 332, *333*
Pipe splitting, 368, 370, *377*
PI (plasticity index), 84
Piston samplers, 74, 75
PJA (Pipe Jacking Association), 211
Planned preventative maintenance (PPM), 191, 232
Planning
contracts
management plan, 226
safety plans, 233
geophysical investigation, 71
health and safety plans, 226, 455, 461–2, 482, 510, *511–13*
horizontal directional drilling, 305–9
new construction, 20–1
procurement, 463–4
project management, 433, *434*
site support staff, 227, 229–30
Planning supervisors (PS), 225–6, 452, 455, 459–60, 509–10
Plant, 231–2, 481
cost, 388
public utility works, 344–5
schedules, 230, 231
storage, 532
Plasticity index (PI), 84
Plastic limit, 83–4
Pneumatic machines
impact moling, 238–41
pipebursting, 275–6, 281–3, 287, *288*, *289*, 294, 295

pipe ramming, 258–9, *260, 261*
stowage systems, 193
Poland, 237
Pole mounted signs, 112–13
Police, 105, 108, 114, 202
Pollution
    air, 344, 356
    drainage, 313
    water, 7, 17, 27, 313, 441, 443, *513*
Polyethylene (PE) pipes, 38, 246, 279–81, 286, 290, 301, 329
Polyvinyl chloride (PVC) pipes, 39, 206, 246, 247, 279, 281
Population density, 22–3
Portable traffic signals, 112
Potential field methods, 68–70
Power augering, 64–5, *66*, 298
PPM (planned preventative maintenance), 191, 232
Pre-cast concrete
    collection chambers, 337–8
    manholes, 33, *34*, 197–9
    pipes, 37, 218–19, 222–3
    shafts, 33, *35*
    soakaways, 29
    tunnel linings, 9, 13–14, *18*, 196–7
Preliminary sources survey *see* Desk study
Press releases, 104, 108
Pressure
    internal, 123–4
    lateral, 136
Pressure mains, 297–8
Prevention by employer, 395–402
Prime cost items, 541
Priority signs, 111
Private culverts, 8
Private finance initiative (PFI), 469, 472–4
Private land entry, 440, 451
Probes, 67, 78–9
Procurement
    construction management, 233–4
    plan, 463–4
    procedures, 462–3
    project management, 431–2, 462–6, 475–6
    tender lists, 463
Productivity, 406–7
Programme of works, 396, 399–400, 414–15, 467, 515–19, 549–50
    accelerated, 411–12
Project lifecycle, 430, *477*
Project management, 430–77
    CDM Regulations, 225–6, 431, 432, 452, 455–62, *477*
    completion, 454, 476, *477*

construction and site management, 432, 466–8, *477*
contracts
    supervision, 507–50
    types, 432
definitions, 450
design process, 431, 449–51, 460, 475, *477*
evaluation, 431, 432–42, *477*
feasibility phase, 431, 433–4, 442–9, 459–60, 474–5, *477*
handover procedures, 476
health and safety, 432, 451–2
managers' responsibilities, 474–6, 476
notification to HSE, 456–9, 509, 532
procurement, 431–2, 462–6, 475–6
progress, 450–1
project review, 462, 468, 476, *477*
risk, 431, 451–4, 475
standard approach, 442
value management, 433, 447–9, 465
viability, 434–9
Project Management Manual (Northumbrian Water Limited), 450
Proprietary support systems, 138–9
Protective equipment *see* Personal protective equipment ...
Provision and Use of Work Equipment Regulations 1992 (PUWER), 231, 439, 478, 481
PS (planning supervisors), 225–6, 452, 455, 459–60, 509–10
Public Health Act 1936, 103
Public health engineering, 2, 3, 7–9, 17
Public relations
    construction sites, 466
    consultation, 105–8, 532
    traffic management, 100–5
Public safety
    excavations, 374–5
    sewer collapses, 17–18
    traffic management, 108–9
Public Utilities Street Works Act 1950 (PUSWA), 101, 146, 350, 397
Public utility works
    case study, 345, 359–66
    environmental impact, 340, 341–9, 356
    future development, 357–8
    road space rental, 349–50, 354–6, 356–7
    safety, 345
    social or indirect costs, 339–66
        investment, 347–8
    working shaft location, 357
    see also Traffic disruption; Traffic management; Utilities

Pumping
  dewatering, 141–4
  foul sewerage systems, 6
Pumping mains
  horizontal directional drilling, 311
  internal pressure, 123
Pumping stations, 41–6
  standby pumping capacity, 43
Pumps
  manometric head, 44–6
  overpumping, 206–9
  power, 43–5
  selection, 208
  types, 42
PUSWA (Public Utilities Street Works Act 1950), 101, 146, 350, 397
PUWER (Provision and Use of Work Equipment Regulations 1992), 231, 439, 478, 481
PVC pipes, 39, 246, 247
  unplasticised, 39, 206, 279, 281

Q system (NGI tunnelling quality index), 164–6
Quality
  rock quality designation, 65, 162–3, 164
  samples, 72, *73*, 75, 523
Quality assurance, 227, 233, 234, 463, 467, 470, 523

RAC (Royal Automobile Club), 105, 113
Radar *see* Ground probing radar
Radio transmitters
  horizontal directional drilling, 301, 309–10
  impact moling, 245–6
Rainfall intensity, 1, 6, 23–6, 118, 120
Raking supports, 134
Rate fixing, 383–7
Rates of flow
  depth of flow, 120–2
  feasibility issues, 443
  flooding, 442, 443
  foul sewage, 6, 20, 21–3, 118
  new construction, 20–1
  on-line sewer replacement, 206–9
  safety, 483, 502
  sewer design, 118–23
  surface water sewers, 21, 23–7
Rat-runs, 342–3
Rats, 505
Reaming, *299*, 300, 302, 304
Rebate joints, 218
Reception pits, 249, 250–1, 264, 294
Records
  construction investigations, 89–90

contract supervision, 536–8
  dayworks, 540
  horizontal directional drilling, 316
  streetworks register, 101–2
  training, 234
  see also Reports
Red Book (IChemE), 466, 475
Rees and Kirby v. Swansea City Council (1985), 412
Rehabilitation, 9–19
  cost effectiveness, 9, 11–12, 13, 15
  pipebursting, 272–96
  risk, 15, 50–1
  sewer layout, 118
  social cost implications, 13, 345, 359–66
  strategy, 16–17
Reinforced concrete, 37, 196–7, 217, 218–19
Reinstatement
  backfilling, 145–7
  carriageway and footway surfaces, 114–15, 146–7, 373, 440
  manholes, 199
Remeasurement, 382, 538–9
Remote control, 177, 205
Renewal *see* Rehabilitation
Renovation *see* Rehabilitation
Repair *see* Rehabilitation
Replacement
  borehole location and depth, 61
  on-line in tunnel, 193–209
  pipebursting, 273–4, 279–81
  risk factors, 50–1
  technique choice, 367–77
  tunnels, 197
Reports
  contract supervision, 536–8
  desk study, 59–60
  feedback, 89–90
  geotechnical reports, 51, 90–2, 154, 212
  see also Records
Repudiation of subcontracts, 428–9
Rescission of contracts, 399, 412–13, 428
Rescue, 495–6, 502, 504
Respiration, 485–7, 488–90, 498
Respiratory protective equipment *see* Breathing apparatus
Retail Price Index, 54
Retention tanks, 27, 29
Rigid pipes
  backfilling, 145–6
  design, 125–6, 127–9
  impact moling, 248
Rising head tests, 77–8
Rising mains, 41–6, 207

Risk
    apportioning in contracts, 382, 389, 473–4, 548
    assessment, 451–4, 475, *477*
    build, own and operate, 471
    design and build, 470
    excavations, 374–5
    financial, 452–4
    hazard identification, 48, 49–54, 91, 482–3
    health and safety, 232, 451–2
    legal deadlines, 454
    management, 54–5
    Management of Health and Safety at Work
        Regulations 1992, 480–1
    operational, 451
    private finance initiatives, 473–4
    project management, 228, 431, 451–4, 470, 471,
        473–4, 475
    rehabilitation, 15, 50–1
    sewer collapses, 17–18
    site investigation, 47–55, 90–2
    tunnel construction, 160
RMR (rock mass rating), 162, 163–4
Road space rental, 349–50, 354–6, 356–7
Road Traffic Regulation Act 1984, 104, 109
Road Traffic (Temporary Restrictions) Act 1991,
    104
Roadworks *see* Traffic ...
Rock
    classification, 65, 161–6
    cores, 65–6
    drilling, 61, 65–7, 369
    hardness, 179
    tunnel construction, 161–6, 179, 526
Rock mass rating (RMR), 162, 163–4
Rock mechanics, 161–4
Rock quality designation (RQD), 65, 162–3, 164
Roman civilisation, 2–3, *4*, *5*, 7, 8, 9
Rotary drilling, 65–7
Royal Automobile Club (RAC), 105, 113
RQD (rock quality designation), 65, 162–3, 164
Runoff, 1, 2, 5, 23–5, 27–9, 118, 131
Russia, 237
(D.J.) Ryan & Sons Ltd, 272, 288, 291

Safety
    advisers, 232–3
    audits, 234
    Faraday Cage, 303
    heading construction, 147–9
    horizontal directional drilling, 306
    induction training, 227–8
    pedestrians, 100, 109, 110, 112, 113, 440, 534,
        535

    public utility works, 345
    sewers, 478–506
        factors of safety, 127–8
        legislation, 478–82
        working procedures, 468, 475, 498–504
    site organisation, 226–7
        support staff, 227, 232–3
    supervision, 524–8
    technique choice, 374–5
    toolbox talks, 229
    tunnelling, 179, 526
    see also Health and safety
Safety at Street Works and Road Works (*Blue Book*),
    101, 109, 110, *512*
Sampling
    cores, 65–6
    cost, 76
    disturbed samples, 72–5
    drive methods, 75
    inside clearance ratio, 75
    laboratory testing, 82–9
    quality levels, 72, *73*, 75
    samplers, 72, 74
    site investigation, 72–6
    size of samples, 75, *76*
    undisturbed samples, 72
    window core sampling, *87*
Sanitation, history, 2–5
Scandinavia, 164
Science and Engineering Research Council (SERC),
    211
Scoping, 116
Seals, 218, 221–2, 494, 496
Seepage, 80, 82
Segmental tunnel linings, 9, 13–14, *18*, 151, 166,
    168–9, 196, 197, 527
    gaskets, 167, 222, 223
    handling, 192
    segment manufacture, 169
Seismic surveys, 68, 70
Self-cleansing
    inverted syphons, 32
    pumps, 43
    sewers, 119, 121, 444
Self-contained breathing apparatus, 494, 495–7
Self-contained packaged pumping station, 42
Semi-rigid pipe design, 125, 129
Separate sewerage systems, 1, 2, 23
SERC (Science and Engineering Research Council),
    211
Services *see* Public utility works; Utilitites
Setting out, 528–9, 531
Sewerage Rehabilitation Manual (WRC), 11, 12,
    13, 15, 120

Sewerage systems
  definition, 1
  development effects, 442–3
  failure, 443
  flooding, 442, 443, 512
  horizontal directional drilling, 320
  outline design, 6, 20, 24–7, 116–18
  planning, 20–1
  pumping, 41–6
  rehabilitation, 9–19
  types, 1–6
  underground mains, 339–40
  see also Vacuum sewerage
Sewers
  above ground, 39–41
  access shafts, 33, *35*, 197–201
  collapses, 17–19, 359–60
  critical, *10*, *11*, 12, 19
  dereliction, 16–18
  design, 116–31
    external forces, 124–9
    flexible pipes, 125–6, 129
    hydraulic design, 118–19
    layout, 116–17
    preliminary design, 116–18
    rates of flow, 118–23
    rigid pipes, 125–6, 127–9
    semi–rigid pipes, 125, 129
    structural design, 123–30
    survey and scoping, 116
  entry procedures, 500–4
  gradients, 6, 33, 35, 116, 118–23, 144–5, 367–8
  near existing structures, 145
  new construction, 20–46, 443
  on-line replacement in tunnel, 193–209
  safety, 478–506
  size determination, 119–23
  small, 32–3
  stepped, 33, 35
  survey and scoping, 116
  see also Gravity sewers
Sewers (Bevan and Rees 1937), 100
Shafts
  borehole location and depth, 60–1
  breaking out, 181, 183–5, 203
  caissons, 198, 200–1, 221–2
  diameter, 198, 201
  pipejacking, 216
  public utility works, 345, 351, 353, 357
  sewer access, 33, *35*, 197–201
  sinking, 180–1, *182*, 201
  timber support, 524
  tunnel construction, 179–85
Sheeting excavations, 133–5

Sheet piling, 139–40, *142*, 524–5
Shields
  excavation support, 138–9, *142*
  pipebursting, 287
  tunnel construction, 151, 166, 170, 171, 175–6, 196, 203–5, 526, *529*
Short duration breathing apparatus, 495–6
Shuttle lanes, 111, 112, 353, 354–5, 357
Side weirs, *28*
Signing
  covered signs, 113
  fixing methods, 113
  lane marking, 112–13
  maintenance, 113–14
  pole mounting, 112–13
  portable traffic signals, 112
  priority signs, 111
  stop/go boards, 111–12
  traffic management, 108–10, 532, *534*
Silting
  inverted syphons, 32
  manhole spacing, 33
  pumps, 43
  sewers, 116, 119, 444
Singapore, 347
Single-pass reinstatement, 114–15
Site inspection survey (walk–over), 57, 59
Site investigation, 47–94, 249
  aims and structures, 48–9
  buried assets, 95–9
  construction investigations, 89–90
  existing services, 519–20
  flowchart, 54, *55*
  geophysical investigations, 60, 67–71
  geotechnical reports, 90–2, 212
  ground hazards, 49–54
  hard rock tunnelling, 161–6
  intrusive methods, 99
  non-intrusive methods, 96–9
  preliminary sources survey, 48, 49, 56–60
  soft ground tunnelling, 152–61
  trial holes, 47–8
  see also Ground exploration
Site management, 226–7
  facilities, 397
  health and safety legislation, 439
  horizontal directional drilling, 305
  offices, 466
  pipebursting, 292–5
  possession, 399–400, 411, 516, 520–1
  project management, 432, 466–8, *477*
  scope, 466
  security, *513*
  superintendence, 522–3

supervision, 522–38
support staff, 227, 229–34
Sizing, 25–7, 119–23
Slurry machines, 176, 214, 215, 217, 301, 303, 304, 320
Soakaways, 29
Social costs
    Cheetham Hill Road, Manchester, *361*
    horizontal directional drilling, 198
    legislative control, 202, 347–9
    public utility works, 339–66
    rehabilitation, 13, 345, 359–66
    road space rental, 349–50, 354–6, 356–7
    sewer collapses, 360
    shaft construction, 202–3
    technique choice, 371, 373
Soft ground
    boring, 61–5, 170
    tunnelling, 152–61, 170–2, 196, 526
        ground treatment, 172–5
Software
    horizontal directional drilling, 311, 320
    hydraulic design, 119, 131
Soil
    cohesionless, 156–61, 222, 275
    cohesive, 87, 155–6, *157*, 214–15
    compaction, 88, 126, 130, 147, 247, 248, 275–6
    consolidation, 88, 145
    friction forces, 214, 215, 275
    ground closure, 158–9, 214, 215
    horizontal directional drilling, 308–9
    impact moling, 247–9
    load on pipes, 124–30
    permeability, 23, *24*, 25
    pipejacking, 212–17
    pipe ramming, 267–70
    settlement, 155, *156*, 160–1, 369
    stability, 155–61, 213–14
    strength, 85–6
    subsoil survey, 47, 48
    testing
        in-situ, 76–9, *154*
        laboratory, 82–9, *154*
Soil water, 48, 248, 276
Soviet Union (formerly), 237
Specification for the Reinstatement of Openings in Highways, 115, 146
Spigot and socketed pipes, 8, 38, 40, 144–5, 246, 251, 258
SRF (stress reduction factor), 165
Standard penetration tests (SPT), 76, 77
Standards *see* ASTM standards; British Standards; European Standards; International Standards

Standpipes, 80–2, *83*
Static probing, 78–9
Steam, 491
Steel
    pipes, 38, 40–1, 246, 247, 254–71, 301
    sheet piling, 139–40, *142*, 524–5
Steering
    horizontal directional drilling, 300, 303, 310–11
    impact moling, 246
Stop/go boards, 109, 111–12
Storage
    gas cylinders, 496
    materials, 466–7, 532, *533*
    tanks, 27, 29, 443–4
    tunnels, 150
Storm water, 1, 6, 23–5, 150, 207
    sewage overflows, 23–4, 27–9
Street authorities, 102–3, 105, 108
Streetworks *see* Traffic management
Strength tests, 85
Stress reduction factor (SRF), 165
Structural design, 123–30
Strutting excavations, 133–5, 136
Subcontracts
    adverse physical conditions, 420–2
    claims, 388–9, 413–24
        by contractors, 379, 388–9, 402, 424–9
        deemed repudiation, 428–9
        due diligence, 415, 427
        set-off, 428
        temporary disconformity, 427
    FCEC, 413, 414, 415–29
    ICE, 515
    instructions and decisions, 416–17
    NEC/ECC, 548
    nominated, 401–2, 413, 535–6
    notice requirements, 417, 424, 427–8
    payment, 422–4
    supervision, 535–6
    supplementary binding agreements, 415
    tendering, 535–6
    variations, 417–20, 421–2
    see also Contracts
Subdrains, 142, *143*
Submersible pumps, 42
Subsidence, 370, 443
Suction hoses, 208–9
Sunday working, 202–3, 543–4
Superintendence, 522–3
Supervision
    contracts, 507–50
        meetings, 536–8
        programming the work, 515–19, 549

service diversions and accommodation works, 519–20
  setting out, 528–9, 531
planning supervisors, 225–6, 452, 455, 459–60, 509–10
safety, 524–8
site management, 522–38
specialists, 524–7
staffing resources, 467, 514–15
traffic management, 440
Support systems
  boxes, 138–9, *140*, *142*
  cantilever, 135
  double–sided, 133–4
  ground anchors, 134
  hydraulic frames, 138, *139*
  immediate, 171
  open–cut construction, 132–40, *141*, *142*
  permanent, 171–2
  proprietary, 138–9
  raking, 134
  rock, 162, *163*, 526
  service pipes, 519–20
  sliding panels, 138, 139, *141*
  standard solutions, 135–6
  steel sheet piling, 139–40, *142*, 524–5
  see also Shields
Surface water sewers, 1, 5
  design, 6
  rates of flow, 21, 23–7, 118
Surrey County Council, 147
Surveying
  pipejacking, 212, 219–20, 233
  setting out, 528
  sewer design, 116
  tunnels, 179–90, 233
    Weisback process, 185–8
Sustainable development, 375, 376, 441–2
Sweden, 324
Syphons, 29–32

Target contracts, 545, 549
Taxation, 348–9
Taxi Drivers Associations, 105, 108
TBM (tunnel boring machines), 170, 177, 178–9, 188–9, 205–6, 229
Technical feasibility of techniques, 367–71
Telephone mains, 339–40
Temporary traffic orders, 102, 104, 105, 108, 202, 532
Temporary works
  pipejacking, 221–2
  tunnel construction, 147–9, 190–2

design, 191, 227
  linings, 166
  monitoring, 191–2
  schedules, 190–1, 230
Tendering
  analysis, 464–5, 475
  cash flow, 464
  CDM Regulations, 461
  competition, 379, 414, 475
  documents, 461, 464
  instructions to commence, 399, 520–1
  procurement, 463, 464–5
  site investigation, 47
  subcontracts, 535–6
  see also Contracts
Term contracts, 472
Testing
  contract management, 400–1, 523
  in-situ, 76–9, *154*
  laboratory, 82–9, *154*
Thermographic surveys, 70
Thrust walls, 216–17, 221
Timber in Excavations, 133, 138, 148
Timbering supports, 52, 132–9, *141*, *142*, 148–9, 166, 170, 196, *527*, *528*
Time
  extensions, 391–5, 543
  slow work, 427, 543–4
  technique choice, 374
Toolbox talks, 229
Topography
  sewer network design, 117
  surface water sewers, 6
  tunnel construction, 150, 152–66, 192
Town and country planning, 440
Toxicity of gases, *489*, 490, 495–6, 500
Traffic authorities, 104, 109, 110
Traffic disruption
  Cheetham Hill Road, Manchester, 361–5
  claims, 103–4
  public utility works, 339, 341–3, *351*, 353–4
  road traffic significance, 354
  sewer collapses, 18, 360
  shaft construction, 202–3, 345, 351, 353, 357
  social and environmental costs, 373
Traffic loading of pipes, 124, 127–8
Traffic management, 100–15
  barriers, 113, 114
  calming measures, 105
  carriageway restriction, 109, 110, 112
  cones, 112, 113, 114, 147
  diversion routes, 101, 103, 105, 107–8, 109–10, 342–3, 362–3, 532
  give and take system, 111

lane marking, 112–13
New Road and Street Works Act 1991, 101–5, 109, 114, 115, 146, 350–1, 371, 439
one-way working, 111, 353, 355, 357
pipebursting, 293
portable traffic signals, 109, 112
priority signs, 111
public relations, 105–8
Public Utilites Street Works Act 1950, 101, 146
rat-runs, 342–3
shaft construction, 202
signing and statutory requirements, 108–10, 532, *534*
statutory requirements, 108–10
stop/go boards, 109, 111–12
surface reinstatement, 114–15, 373, 440
temporary traffic orders, 102, 104, 105, 108, 202, 532
two-way traffic, 109, 110, 353, 354
traffic Signs Manual (Department of Transport), 109, *512*
Training, 227–9, 467
courses, 234
method statement induction, 228
New Road and Street Works Act 1991, 101
NVQ qualifications, 227
operatives, 228, 440
records, 234
sewer safety, 499
site safety induction, 227–8
toolbox talks, 229
work equipment, 231
workshops, 448
Transmitter devices, 245–6, 301, 309–10
Transportation of pipes, 130
Transport and Road Research Laboratory (TRRL)
hydrograph method, 24, 25, 119
vehicle operating costs, 352
Trench excavation, *239*, 345–6, 524–5
proprietary support systems, 138–9
Timber in Excavations, 133, 138
width of trench, 125, 127–8, 132, 143
narrow trenches, 140–1
Trenching Practice, 136, 138
Trenchless methods
pipe operation loads, 129–31
public utility works, 340, 351, 357
rehabilitation, 15
sewer layout, 118, 131
utility interference, 370
see also specific techniques
Trial pits, 47–8, 61, 67, 77
TRRL (Transport and Road Research Laboratory), 24, 25, 119, 352

Tube samplers, 74
Tunnel boring machines (TBM), 170, 177, 178–9, 188–9, 205–6
operative licences, 228–9
Tunnels
alignment, 179–85, 188–90, 203, 233
bored tunnels, *178*
borehole location and depth, 60
construction, 150–92, 203–6
contracts, 151, 224–35
supervision, 525–8
co-planing, 185, *186*
diameter, 151, 166, 197, 203, 205
entry procedures, 500–4
gradient, 181, 182–5
ground hazards, 51, *52*, *53*
hard rock tunnelling, 161–6, 179, 526
heading construction, 147–9, 193–6, 525, *528*
levelling, 179–85
linings, 166–70, 196–7
manholes, 197
microtunnelling, 118, 177–8, 298
on-line sewer replacement, 193–209
shafts, 179–85
shields, 151, 166, 170, 171, 175–6, 196, 203–5, 526, *529*
soft ground tunnelling, 152–61, 170–5, 196, 526
stability, 155–62
surveying, 179–90
temporary works, 190–2, 227
tunnelling process, 175–9
see also Excavations
Turnkey contracts, 449, 469
Two–way traffic, 109, 110, 353, 354

UCS (unconfined compressive strength), 161
UHF/VHF signals *see* Ground probing radar
UK
Citizen's Charter, 351
environmental impact assessment, 346–7, 440–1, 443
ground water, 79–80
horizontal directional drilling, 297, 303, 304, 305
projects, 320, 322–4, 326
'lane rental' mechanism, 348–9
pipebursting, 272, 295, 296
pipejacking, 210, 211
sewers
access, 197
rehabilitation, 12, 196
social cost compensation, 348, 373
traffic disruption, 341–2

vacuum sewerage, 327–8, 333–5, 444
   projects, 336–8
   water industry, 449
Uncertainty *see* Ground exploration; Risk
Unconfined compressive strength (UCS), 161
Underground control survey, 185–8
Underpinning, 198–200
Undrained tests, 85–6
Unfair Contract Terms Act 1977, 424
Unflanged corrugated liner plates, 38
Unplasticised PVC pipes, 39, 206, 279, 281
USA
   buried pipeline development, 6
   Geotechnical Baseline Report, 51, 91
   horizontal directional drilling, 297, 309, 310, 311
   impact moling, 237, 253
   pipebursting, 283
   pipejacking, 210
   sewer rehabilitation, 12
   tunnel ground hazards, 51, *53*
   vacuum sewerage, 327, 444
   see also North America
Utilities
   buried, 95–9, 339–40
   diversions and accommodation works, 519–20
   existing plans, 95–6, 292–3
   interference, 370
   sewerage systems, 6
   surface reinstatement, 440
   tracing, 96–8
   see also Public utility works

Vacuum sewerage, 6, 327–38, 444
   case studies, 336–8
   design, 328, 332–5
   horizontal directional drilling, 312
   piping, 328–9, 332, *333*
   pre-cast units, 337–8
   stations, 333–5
   sumps, 329–31, 337–8
   system operation, 328–31
   valves, 329–31
      monitoring, 332
Value engineering, 431, 434, 447–9, 465, 476
Value management, 382–8, 419–20, 433, 447–9, 465
Valves
   breathing apparatus, 494–5, 496
   rising mains, 45
   vacuum sewerage, 329–31
      monitoring, 332
Vehicle operating costs, 352, 364

Ventilation, 485, 490, 497–8, 500
   manhole spacing, 33
Vertical casting, 219
Vibration
   monetary values, 356
   public utility works, 344
   technique choice, 375
Victorian era, 3, 7–8, 15, 17, 37, 374
Visual intrusion, 344

Walings, 136, *137*, 138, *139*
Walk-over, 57, 59
Wallingford Procedure, 25
Washboring, 65
Wash-out valves, 45
WASSP–SIM simulation program, 25, 29
Waste water, 1–5
Water
   consumption, 20, 118
   demand, 21–2, 131
   levels, 48
   mains, 141
   storage, 27, 29, 150, 443–4
   supply, 2–3, 5, 7, 16–17, 20, 320, 339–40
   table levels, 141–4, 145, 172, 174, 369, 391
   vapour, 496
Water closets, 5
Water cycle, *21*
Water Industry Act 1991, 440
Water meadows, 29
Water pollution, 7, 17, 27, 313, 441, 443, *513*
   technique choice, 375–6
Water Research Council (WRC), 11, 12, 13, 15, 374
Weather
   contract management, 391
   dry weather flow, 21–3, 27, 122, 123, 207
   horizontal directional drilling, 316
Wedgeblock lining, 167
Weil's disease, 505
Weisback process, 185–8
Welding
   pipe ramming technique, 254, 257–8, 267
   pipes, 36, 254, 301, 305
Welfare, 439, 478, 482
Well–point dewatering, 142–4
Wells v. Army and Navy Co-operative Society (1902), 392
Wet wells, 43
Wheelchairs, 110
Whole life costs
   project management, 433
   public utility works, 341
   technique choice, 374

Winches, 484
    impact moling, 251–2
    pipebursting, 281–2, 283, 291–2
    pipe ramming, 266
Wireless measurement–while–drilling (MWD), 309, 310
Wireline system, 309–10
Work equipment, 231, 439, 478, 481

Working procedure safety, 468, 475, 498–504
Workplace (Health, Safety and Welfare) Regulations 1992, 439, 498
Workshops, 448
WRC (Water Research Council), 11, 12, 13, 15, 374

Zinkiewicz, Wiktor, 237

Methods in Cell Biology

VOLUME 100

The Zebrafish: Cellular and Developmental Biology, Part A

Third Edition

Series Editors

Leslie Wilson
Department of Molecular, Cellular and Developmental Biology
University of California
Santa Barbara, California

Paul Matsudaira
Department of Biological Sciences
National University of Singapore
Singapore

Methods in Cell Biology

VOLUME 100

The Zebrafish: Cellular and Developmental Biology, Part A
Third Edition

Edited by

H. William Detrich III

Department of Biology, Northeastern University, Boston, MA, USA

Monte Westerfield

Institute of Neuroscience, University of Oregon, Eugene, OR, USA

Leonard I. Zon

Division of Hematology/Oncology, Children's Hospital of Boson,
Department of Pediatrics and Howard Hughes Medical Institute, Harvard Medical School, Boston, MA, USA

ELSEVIER

AMSTERDAM • BOSTON • HEIDELBERG • LONDON
NEW YORK • OXFORD • PARIS • SAN DIEGO
SAN FRANCISCO • SINGAPORE • SYDNEY • TOKYO
Academic Press is an imprint of Elsevier

Academic Press is an imprint of Elsevier
30 Corporate Drive, Suite 400, Burlington, MA 01803, USA
525 B Street, Suite 1900, San Diego, CA 92101-4495, USA
32, Jamestown Road, London NW1 7BY, UK
Linacre House, Jordan Hill, Oxford OX2 8DP, UK

Third edition 2010

ISBN–13: 978-0-12-384892-5
ISSN: 0091-679X

For information on all Academic Press publications
visit our website at elsevierdirect.com

Printed and bound in USA
10 11 12 13 10 9 8 7 6 5 4 3 2 1

DEDICATION

We dedicate the Third Edition of *Methods in Cell Biology: The Zebrafish* to Wolfgang Driever, Mark C. Fishman, Charles Kimmel, and Christiane Nüsslein-Volhard. Through their foresighted embrace of the zebrafish as a model vertebrate and their pursuit of genetic screens to illuminate vertebrate development, they fostered the emergence of the vibrant zebrafish research community.

8. Physiological Recordings from Zebrafish Lateral-Line Hair Cells and
Afferent Neurons

Josef G. Trapani and Teresa Nicolson

    I. Introduction   220
    II. Zebrafish Mounting and Immobilizing   221
    III. Microphonics   223
    IV. Action Currents   225
    V. Summary   228
    VI. Discussion   228
    References   229

9. Zebrafish Kidney Development

Iain A. Drummond and Alan J. Davidson

    I. Introduction   234
    II. Structure of the Zebrafish Pronephros   236
    III. Formation of the Pronephros   238
    IV. Methods to Study Pronephros Function   245
    V. Conclusions   256
    References   256

10. Molecular Regulation of Pancreas Development in Zebrafish

Robin A. Kimmel and Dirk Meyer

    I. Introduction   262
    II. Pancreas Development   262
    III. Analysis of Beta-Cell Migration and Proliferation   268
    IV. Future Directions   276
    References   276

11. Monitoring Sleep and Arousal in Zebrafish

Jason Rihel, David A. Prober, and Alexander F. Schier

    I. Introduction   282
    II. Behavior, Genetics, and Pharmacology of Zebrafish Sleep   282
    III. Methods for Monitoring Sleep/Wake Behavior in Zebrafish   288
    IV. Conclusion   291
    References   291

12. Use of Flatbed Transparency Scanners in Zebrafish Research: Versatile and
Economical Adjuncts to Traditional Imaging Tools for the *Danio rerio* Laboratory

Charles A. Lessman, Michael R. Taylor, Wilda Orisme, and Ethan A. Carver

    I. Introduction   296
    II. Scanner Basics   297
    III. Motility Analysis   302

IV. Oocyte and Egg Assays   309
V. Using Scanners to Count and Measure   316
VI. Other Potential Applications of Scanners   319
VII. Summary: Inexpensive Adjunct to Microscopy   321
References   321

Subject Index   323

Volumes in Series   339

CONTRIBUTORS

Numbers in parentheses indicate the pages on which the authors' contributions begin.

Andrei Avanesov, (153) Division of Craniofacial and Molecular Genetics, Tufts University, Boston, Massachusetts

Ethan A. Carver, (295) Department of Biological and Environmental Sciences, The University of Tennessee-Chattanooga, Chattanooga, Tennessee

Prisca Chapouton, (73) Institute of Developmental Genetics, Helmholtz Zentrum München, German Research Center for Environmental Health, Neuherberg, Germany

Chi-Bin Chien, (3) Department of Neurobiology and Anatomy, University of Utah, Salt Lake City, Utah

Alan J. Davidson, (233) Center for Regenerative Medicine, Massachusetts General Hospital, Boston, Massachusetts

Iain A. Drummond, (233) Departments of Medicine and Genetics, Harvard Medical School and Nephrology Division, Massachusetts General Hospital, Charlestown, Massachusetts

James M. Fadool, (205) Department of Biological Science, Florida State University, Tallahassee, Florida

John A. Gaynes, (3) Department of Neurobiology and Anatomy, University of Utah, Salt Lake City, Utah

Leanne Godinho, (73) Lehrstuhl für Biomolekulare Sensoren, Institute for Neuroscience, Technische Universität München, Munich, Germany

Paul D. Henion, (127) Center for Molecular Neurobiology and Department of Neuroscience, Ohio State University, Columbus, Ohio

Yunhan Hong, (55) Department of Biological Sciences, National University of Singapore, Singapore

Sumio Isogai, (27) Department of Anatomy, School of Medicine, Iwate Medical University, Morioka, Japan

Makoto Kamei, (27) Program in Genomics of Differentiation, National Institute of Child Health and Human Development, Bethesda, Maryland; Cell Biology of Diseases Group, Sansom Institute for Health Research, School of Pharmacy and Medical Science, University of South Australia, Adelaide, South Australia, Australia

John P. Kanki, (127) Department of Pediatric Oncology, Dana-Farber Cancer Institute, Harvard Medical School, Boston, Massachusetts

Robin A. Kimmel, (261) Institute of Molecular Biology, University of Innsbruck, Innsbruck, Austria

Martina Lachnit, (127) Department of Oncological Sciences, Huntsman Cancer Institute, University of Utah, Salt Lake City, Utah

Mei-Yee Law, (3) Department of Neurobiology and Anatomy, University of Utah, Salt Lake City, Utah

Jeong-Soo Lee, (127) Department of Pediatric Oncology, Dana-Farber Cancer Institute, Harvard Medical School, Boston, Massachusetts

Charles A. Lessman, (295) Department of Biological Sciences, The University of Memphis, Memphis, Tennessee

A. Thomas Look, (127) Department of Pediatric Oncology, Dana-Farber Cancer Institute, Harvard Medical School, Boston, Massachusetts

Jarema Malicki, (153) Division of Craniofacial and Molecular Genetics, Tufts University, Boston, Massachusetts

Dirk Meyer, (261) Institute of Molecular Biology, University of Innsbruck, Innsbruck, Austria

Teresa Nicolson, (219) Howard Hughes Medical Institute, Oregon Hearing Research Center and Vollum Institute, Oregon Health and Science University, Portland, Oregon

Wilda Orisme, (295) Department of Chemical Biology and Therapeutics, St. Jude Children's Research Hospital, Memphis, Tennessee

Weijun Pan, (27) Program in Genomics of Differentiation, National Institute of Child Health and Human Development, Bethesda, Maryland

Brian D. Perkins, (205) Department of Biology, Texas A&M University, College Station, Texas

Fabienne E. Poulain, (3) Department of Neurobiology and Anatomy, University of Utah, Salt Lake City, Utah

David A. Prober, (281) Division of Biology, California Institute of Technology, Pasadena, California

Jason Rihel, (281) Department of Molecular and Cellular Biology, Harvard University, Cambridge, Massachusetts

Alexander F. Schier, (281) Department of Molecular and Cellular Biology, Harvard University, Cambridge, Massachusetts; Division of Sleep Medicine, Harvard University, Cambridge, Massachusetts; Center for Brain Science, Harvard University, Cambridge, Massachusetts; Harvard Stem Cell Institute, Harvard University, Cambridge, Massachusetts

Cornelia Stacher Hörndli, (3) Department of Neurobiology and Anatomy, University of Utah, Salt Lake City, Utah

Rodney A. Stewart, (127) Department of Oncological Sciences, Huntsman Cancer Institute, University of Utah, Salt Lake City, Utah

Michael R. Taylor, (295) Department of Chemical Biology and Therapeutics, St. Jude Children's Research Hospital, Memphis, Tennessee

Josef G. Trapani, (219) Howard Hughes Medical Institute, Oregon Hearing Research Center and Vollum Institute, Oregon Health and Science University, Portland, Oregon

Brant M. Weinstein, (27) Program in Genomics of Differentiation, National Institute of Child Health and Human Development, Bethesda, Maryland

PREFACE

The publication of the hundredth volume of *Methods in Cell Biology* is a significant milestone for the Series, indicative both of the success of the volumes in providing state-of-the-art technical protocols and of the commitment of the many contributors, editors, and publisher, Elsevier/Academic Press, to ensuring their widespread dissemination. Two individuals in particular, the Series Editors Leslie Wilson and Paul Matsudaira, deserve our gratitude for fostering the technical and scientific excellence of *Methods in Cell Biology*. Les became the Series Editor beginning with Volume 27 in 1986 and Paul joined to create the "Dynamic Duo" with Volume 37 in 1993. Monte, Len, and I salute Les and Paul for their signal accomplishments in developing the Series into the foremost "go-to" resource for cell and developmental biologists.

Building on the foundation of our first (1999) and second (2004) editions of *Methods in Cell Biology: The Zebrafish*, Monte, Len, and I are pleased to introduce this Third Edition, beginning with *Methods in Cell Biology* Volume 100, *Cellular and Developmental Biology, Part A*. In this volume (and its soon-to-be-released companion, *Part B*), our contributors present the latest technical advances in the Cell, Developmental, and Neural Biology of the zebrafish that have appeared since the second edition. One theme that clearly emerges from these chapters is that the zebrafish is the preeminent vertebrate model for mechanistic *cellular* studies of developmental processes *in vivo*. Subsequent volumes on *Genetics, Genomics, and Informatics* will cover new technologies in Forward and Reverse Genetics, Transgenesis, The Zebrafish Genome and Mapping Technologies, Informatics and Comparative Genomics, and Infrastructure. The Third Edition will also introduce new volumes on *Disease Models and Chemical Screens*, two rapidly emerging and compelling applications of the zebrafish. We trust that these volumes will prove valuable both to seasoned zebrafish investigators and to those who are newly adopting the zebrafish model as part of their research armamentarium.

We thank Les, Paul, and the staff of Elsevier/Academic Press, especially Zoe Kruze, for their enthusiastic support of our Third Edition of *Methods in Cell Biology: The Zebrafish*. Their help, patience, and encouragement are profoundly appreciated.

H. William Detrich, III
Monte Westerfield
Leonard I. Zon

PART I

Cellular Biology

CHAPTER 1

Analyzing Retinal Axon Guidance in Zebrafish

Fabienne E. Poulain, John A. Gaynes, Cornelia Stacher Hörndli, Mei-Yee Law, *and* **Chi-Bin Chien**

Department of Neurobiology and Anatomy, University of Utah, Salt Lake City, Utah

Abstract

I. Introduction
II. Visualizing Retinal Axons
   A. Transgenic Lines
   B. Labeling with Antibodies
   C. Labeling with Lipophilic Dyes
   D. Transiently Expressing DNA Constructs
   E. *In Vivo* Single Cell Electroporation
   F. Time-Lapse Imaging
   G. Protocols for Labeling Methods
III. Perturbing the Retinotectal System
   A. Retinotectal Mutants
   B. Injecting DNA or Morpholinos
   C. Using Heat Shock to Induce Misexpression
   D. Transplanting to Test Cell Autonomy of Gene Function
   E. Protocols for Transplants
IV. Future Directions
   References

Abstract

How neuronal connections are established during development is one of the most fascinating questions in the field of neurobiology. The zebrafish retinotectal system offers distinct advantages for studying axon guidance in an *in vivo* context. Its accessibility and the larva's transparency not only allow its direct visualization, but

978-0-12-384892-5
DOI: 10.1016/B978-0-12-384892-5.00001-3

also facilitate experimental manipulations to address the mechanisms of its development. Here we describe methods for labeling and visualizing retinal axons *in vivo*, including transient expression of DNA constructs, injection of lipophilic dyes, and time-lapse imaging. We describe in detail the available transgenic lines for marking retinal ganglion cells (RGCs); a protocol for very precise lipophilic dye labeling; and a protocol for single cell electroporation of RGCs. We then describe several approaches for perturbing the retinotectal system, including morpholino or DNA injection; localized heat shock to induce misexpression of genes; a comprehensive list of known retinotectal mutants; and a detailed protocol for RGC transplants to test cell autonomy. These methods not only provide new ways for examining how retinal axons are guided by their environment, but also can be used to study other axonal tracts in the living embryo.

I. Introduction

Axon guidance is an essential process for proper formation of neuronal connections during development. This is certainly true in the visual system, where retinal axons must interpret a large variety of signals to navigate to their brain target and establish precise and ordered connections reflecting our perception of the environment. The accessibility of the visual system not only allows its easy visualization, but also facilitates experimental manipulations to test the mechanisms of its development. Many studies have taken advantage of this accessibility to give a precise description of the visual system's anatomy and identify important factors required for its formation. In the past decade, the zebrafish retinotectal system has drawn attention for its distinct advantages. The optical transparency of zebrafish embryos allows direct visualization of retinal axons and is particularly suited for high-resolution imaging, including time-lapse analysis. Chimeric embryos with retinal neurons of different genetic backgrounds can be easily generated by cell transplants. Finally, the short generation time of zebrafish as well as the recent characterization of its genome are especially suited for genetic analysis and have allowed the generation and identification of many mutants with retinotectal defects. These properties establish zebrafish as an excellent model for studying retinal axon guidance and, more generally, for studying cell biology in an *in vivo* context, as many *in vivo* experiments not possible in other systems can be performed.

Retinal ganglion cells (RGCs) are the primary cell type in the innermost cellular layer of the retina, responsible for carrying visual information from the eye to the brain. In zebrafish, the first RGCs are born at 28 h post-fertilization (hpf) (Hu and Easter, 1999; Masai *et al.*, 2005) and immediately extend axons that then must pass several landmarks (Fig. 1A). Retinal axons first grow within the retina to the optic disc, where they exit (30–32 hpf). They then join the optic nerve and elongate toward the ventral midline of the diencephalon, where nerves coming from both eyes meet to form the optic chiasm (34–36 hpf). In zebrafish and other species lacking binocular vision, all axons cross the midline. Retinal axons then

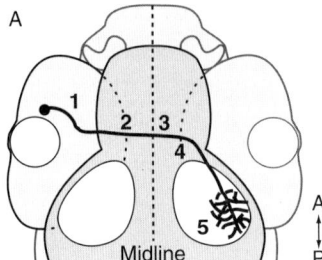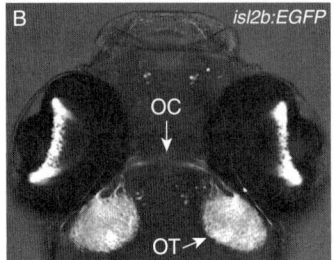

Fig. 1 The zebrafish retinotectal projection. (A) Diagram of the retinal axon pathway. Retinal axons navigate to the optic nerve head (1), pass through the optic nerve and exit the eye (2), cross the midline at the chiasm (3), and grow dorsally along the optic tract (4) to reach the tectum (5). (B) Dorsal view of a Tg(*isl2b: EGFP*)^{zc7} transgenic embryo, which specifically expresses EGFP in all RGCs, allowing a direct visualization of retinal projections. Courtesy of A. Pittman. A: anterior; P: posterior; OC, optic chiasm; OT, optic tectum. *Maximum intensity projection, confocal microscopy.* (A, B): dorsal views, anterior up.

navigate dorsally through the optic tract to reach their main target, the optic tectum (48 hpf), where they establish a topographic map, making connections according to their position in the retina (Fig. 2C–D). Axons originating from the more rostral retina project to the more posterior tectum, and axons from the dorsal retina project to the ventrolateral tectum. Interestingly, this ordering in the tectum can already be observed along the dorso-ventral axis in the optic tract: dorsal axons grow through the ventral branch of the tract, and ventral axons through its dorsal branch. Once in the tectum, retinal axons mature, arborize, and form synapses with their tectal targets.

Retinal axons encounter many guidance decision points along their pathway and respond to various attractive or repulsive cues to choose the right track. Many factors acting as road signs have been identified, but how retinal axons respond to them *in vivo* still remains poorly understood. Many laboratories, including ours, have developed tools for visualizing retinal axons during their navigation and modifying their nature or their environment to test specific functions. We describe here the different methods used for labeling and visualizing retinal axons, as well as several approaches for perturbing the retinotectal system. Many of these methods are also applicable to nonretinal axons. We finish with an overview of methods likely to be important in the future.

II. Visualizing Retinal Axons

Understanding how retinal projections develop requires specific labeling and precise visualization of retinal axons *in vivo*. Several methods can be used, depending on which part of the retinotectal pathway is studied, how many axons are observed, and whether the axons are observed live. Thanks to the optical transparency of zebrafish

embryos, transgenic lines expressing fluorescent proteins in RGCs can be used to visualize retinal axons as they develop. Lipophilic dyes are particularly useful to label specific groups of axons. Finally, transient expression of DNA constructs and *in vivo* electroporation are specially suited for labeling single axons and imaging them as they elongate. For all these approaches, precise imaging is best achieved using confocal microscopy.

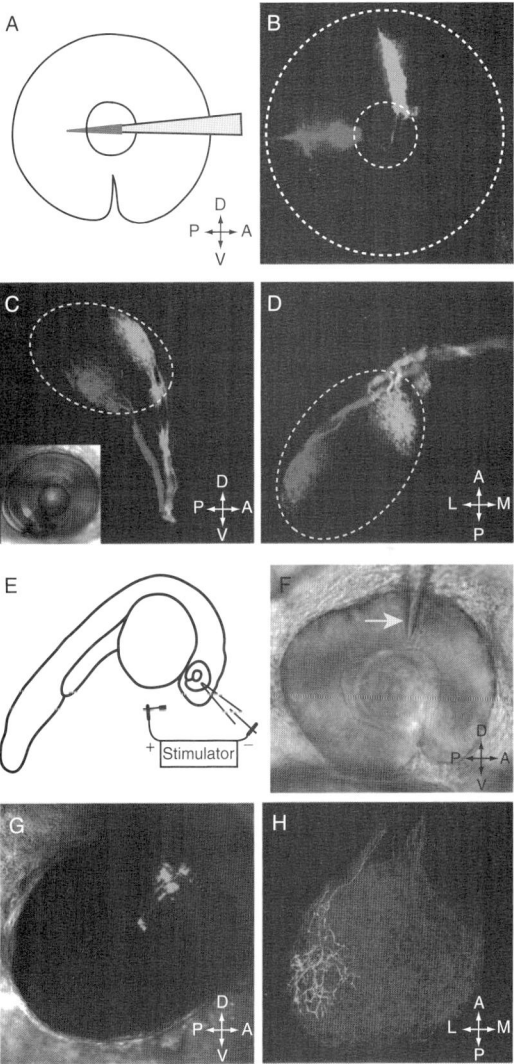

Fig. 2 *(Continued)*

A. Transgenic Lines

Several transgenic lines that express fluorescent proteins (FPs) under the control of RGC-specific promoters have been developed (Table I). Their main advantage is to allow clear and direct visualization of retinal projections in live embryos. Labeled embryos are simply obtained by crossing transgenic carriers. Depending on the promoter used, all RGCs or a subset of them are labeled. Promoters from the *isl2b* and *atoh7* (previously named *isl3* and *ath5*) genes drive transgene expression in all RGCs, allowing the visualization of all retinal axons (Fig. 1B, Masai *et al.*, 2003; Pittman *et al.*, 2008). In contrast, promoters from the *pou4f3* (previously named *brn3c*) gene can be used to label a subset of RGCs (Neumann and Nüsslein-Volhard, 2000; Xiao *et al.*, 2005). For instance, the *pou4f3* promoter drives expression in RGCs that project mainly into one of the four retinorecipient layers of the tectum, allowing characterization of laminar targeting of retinal axons (Xiao *et al.*, 2005).

Different FPs can be expressed to label RGCs. Enhanced green fluorescent protein (EGFP) is the most frequently used, as it is stable and particularly bright. RGCs can also be labeled in red using TagRFP or mCherry. Adding specific tags to the FP coding sequence allows labeling of specific cellular compartments such as the nucleus or the plasma membrane. For instance, the N-terminal palmitoylation sequence from GAP-43 (Moriyoshi *et al.*, 1996) or the CAAX consensus motif from Ras (Choy *et al.*, 1999) can target FPs to the plasma membrane, giving better labeling of axonal arbors.

Finally, other transgenic lines express the strong transcriptional activator Gal4-VP16, which drives the expression of DNA constructs containing a UAS (upstream activation sequence) control element (Köster and Fraser, 2001). These lines can be

---

Fig. 2 Methods for visualizing retinal axons. (A–D) Focal injection of dyes in the retina allows visualization of retinal axons exiting from the retina and making topographic connections in the tectum. (A) After removing lens, a dye-coated glass micropipette is briefly inserted in a peripheral direction into the RGC layer (method described in detail in Section II.G.1). (B) Lateral view of a 48 hpf eye focally injected with DiI (red) and DiO (green). Labeled retinal axons can be observed exiting from the retina. *Maximum intensity projection, confocal microscopy.* (C) Lateral view of a 4 dpf embryo topographically injected with DiI and DiO into the dorsonasal (DN) and ventrotemporal (VT) retina, respectively, using a vibrating-needle injection apparatus (Baier *et al.*, 1996). Inset shows the sites of injection in the retina. DN (red) and VT (green) retinal axons navigate through the ventral and dorsal branches of the optic tract, respectively, and terminate topographically in the tectum. Yellow dashed line: tectal border. *Maximum intensity projection, confocal microscopy.* (D) Dorsal view of the projections showed in C. DN axons terminate in the posterolateral tectum, whereas VT axons innervate the antero-medial tectum. Yellow dashed line: tectal border. *Maximum intensity projection, confocal microscopy.* (E–H) *In vivo* single cell electroporation allows visualization of retinal arbors in the tectum. (E) Schematic representation of the electroporation setup: a 22–28 hpf embryo is mounted laterally under a compound microscope. A negatively charged glass microelectrode is filled with DNA solution and placed in the retina, with a positively charged ground electrode placed near the head. (F) DIC picture of the microelectrode (arrow) placed into the DN retina just prior to electroporation. *40× water immersion objective, compound microscope.* (G) Electroporated RGCs expressing GAP43-EGFP (green) in a live 5 dpf embryo mounted laterally with the lens removed. The EGFP image has been merged with a DIC image of the head. *Maximum intensity projection, confocal microscopy.* (H) Dorsal view of the contralateral tectum of the same embryo, with tectal neuropil visualized by *isl2b:mCherry-CAAX* transgene [red; Tg(*isl2b:mCherry-CAAX*)^{zc23}] and electroporated RGC axons and arbors visualized with GAP43-EGFP (green). *3D projection from Fluorender software, 40× water immersion objective, confocal microscopy.* A: anterior; P: posterior; D: dorsal; V: ventral; L: lateral; M: medial. (See Plate no. 1 in the Color Plate Section.)

Table I
Transgenic Lines that Label RGCs

Transgenic line	Previous names, ZFIN allele number	Retinal expression	Other expression	References
atoh7:GFP	ath5:GFP, rw021	New born RGCs	Forebrain, tectum	Masai et al. (2003), Poggi et al. (2005)
atoh7:mGFP	ath5:mGFP, cu1	"	"	Vitorino et al. (2009), Zolessi et al. (2006)
atoh7:mRFP	ath5:mRFP, cu2	"	"	Vitorino et al. (2009), Zolessi et al. (2006)
atoh7:Gal4-VP16	zf138	"	"	Maddison et al. (2009)
pou4f1-hsp70:GFP	brn3a-hsp70:GFP, rw0110	RGCs (likely a subset)	Tectum, habenula, cranial sensory ganglia	Aizawa et al. (2005), Sato et al. (2007)
pou4f3:mGFP	brn3c:mGFP, s273, s356t	Subset of RGCs	Inner ear, lateral line neuromasts	Del Bene et al. (2008), Xiao et al. (2005)
pou4f3:Gal4VP16	s311t	"	"	Xiao and Baier (2007)
isl2b:GFP	isl3:GFP, zc7	All RGCs	Cranial ganglia Rohon-Beard neurons, a few cells in forebrain dorsal midbrain	Pittman et al. (2008)
isl2b:mGFP	zc20	"	"	Law and Chien (unpublished)
isl2b:mCherryCAAX	zc23, zc25	"	"	Pittman et al. (2008)
isl2b:Gal4VP16	zc60	"	"	Ben Fredj et al. (2010)
chrnb3b:GFP	jt0021	RGCs	Trigeminal ganglion, Rohon-Beard neurons, some tectal cells	Matsuda and Mishina (2004), Tokuoka et al. (2002), Yoshida and Mishina (2003)
-2.7shh:GFP	t10	RGCs	Amacrine cells, notochord, floor plate, pharyngeal arch endoderm, ventral forebrain	Neumann and Nüsslein-Volhard (2000), Nevin et al. (2008), Roeser and Baier (2003)

mGFP, Membrane-targeted GFP; RGCs, retinal ganglion cells.

particularly useful for expressing DNA constructs at high levels in a few RGCs (described in Section II.D).

B. Labeling with Antibodies

Alternately, antibodies can be used to label retinal axons. Although they cannot be employed for live visualization, they provide strong staining that can be useful to examine details or specific aspects of retinal axon navigation. Several antibodies have been widely used to label retinal axons using standard whole-mount antibody staining techniques. Anti-acetylated tubulin (Sigma, St. Louis, Missouri) recognizes a form of tubulin found in stable microtubules, and thus labels all axons. This staining has been used to visualize the earliest axons crossing the chiasm (Karlstrom et al., 1996) and to label axon bundles within the retina (Li et al., 2005). Zn-5 and zn-8 (Zebrafish International Resource Center, Developmental Studies Hybridoma Bank, Iowa City, Iowa) are two monoclonal antibodies, likely derived from the same hybridoma, that recognize the cell surface adhesion molecule Alcam-a (previously named neurolin/DM-GRASP, Laessing et al., 1994). Alcam-a is expressed by newly born RGCs that are added in successive peripheral rings around the retina, but turns off in central RGCs by 48 hpf (Laessing and Stuermer, 1996). Consequently, zn-5/8 staining is particularly appropriate to label retinal axons navigating within the retina to the optic nerve head. Finally, anti-GFP (Invitrogen, Carlsbad, California), anti-DsRed (which also recognizes mCherry, Clontech, Mountain View, California) and anti-TagRFP (Evrogen, Moscow, Russia) antibodies can be used to amplify the signal from FPs.

C. Labeling with Lipophilic Dyes

While transgenic lines and antibodies are appropriate for labeling a large population of axons, they cannot be used to visualize spatially specific sub populations of RGCs. Lipophilic carbocyanine dyes such as DiI, DiO, DiA, or DiD (Invitrogen) offer the great advantage of being easily injected in specific locations within the retina. Structurally, they consist of a fluorophore attached to two long aliphatic alkyl tails responsible for their insertion within membranes. Carbocyanine dyes are highly fluorescent in lipid bilayers, but weakly fluorescent in water. Once applied, they become incorporated into the plasma membrane and diffuse laterally, labeling the entire cell. These properties have made lipophilic dyes the tool of choice for anterograde and retrograde tracing of neurons in both live and fixed tissues (Honig and Hume, 1989).

DiI (red) and DiO (green) are the most commonly used. They can be applied using several methods. The first is to inject DiI or DiO dissolved in chloroform into the eye, which labels the entire projection ("whole eye fills"). This technique is particularly useful for studying guidance at the chiasm, as each eye can be labeled with a different color. It has been described previously (Hutson et al., 2004) and is not repeated here.

DiI and DiO can also be delivered into specific regions of the retina, so that only a subset of RGCs is labeled (Fig. 2A–D). In the second method, dyes dissolved in dimethylformamide are focally injected using a vibrating-needle injection apparatus (Baier et al., 1996; Trowe, 2000). DiI or DiO is loaded in a reservoir through which

passes a tungsten needle. Fixed larvae are mounted in an agarose form. A small loudspeaker vibrates the needle, transporting dye to its tip, where the dye precipitates in the embedded tissue. This method has the advantage of labeling many embryos reproducibly and has been used to analyze projection topography in the tectum and axon ordering in the tract (Karlstrom *et al.*, 1996; Lee *et al.*, 2004; Trowe *et al.*, 1996). However, the custom-built apparatus is not widely available. The third technique uses a dye-coated microneedle to focally deposit dye into the retina. The needle is coated with dye and can be reused several times. It does not require any specialized apparatus and can be used to label very few cells. We describe this method in Section II.G.1. A final method is to focally inject DiI along the retinal pathway to retrogradely label RGCs. Although it is difficult to inject dye precisely enough, this technique can be used to visualize RGC morphology and organization within the retina (Mangrum *et al.*, 2002).

D. Transiently Expressing DNA Constructs

Whereas lipophilic dyes can easily label a subset of RGCs, they are more difficult to use for single axons. These can be better visualized by transiently expressing DNA constructs encoding FPs. Plasmids injected at the one cell stage are expressed mosaically, labeling a few cells randomly. Expression can be targeted to RGCs using the *atoh7* or *isl2b* promoters (Masai *et al.*, 2003; Pittman *et al.*, 2008). This method has been used to visualize single retinal arbors (Campbell *et al.*, 2007) and RGC dendritic outgrowth (Mumm *et al.*, 2006). Alternatively, constructs containing a UAS element upstream of an FP coding sequence can be injected into transgenic embryos expressing Gal4-VP16 in RGCs (Table I). The Gal4/UAS system amplifies FP expression and gives better labeling. While DNA methods are very useful for labeling single axons, they cannot yet be used to target specific RGC subtypes or RGCs in particular locations, since the required enhancers have not yet been identified.

E. *In Vivo* Single Cell Electroporation

Another way to label individual axons is *in vivo* single cell electroporation (Fig. 2E–H). Although technically demanding, this powerful approach offers the possibility of delivering DNA constructs or dextran-coupled indicators to individual RGCs or RGCs in specific locations in the retina. We have used it to visualize projections and arborizations of individual dorsonasal RGCs (Pittman *et al.*, 2010). In this approach, an applied voltage generates an electric field across cells in the retina, breaking down the plasma membrane and creating transient pores through which negatively charged DNA molecules move into the cell. Briefly, embryos are mounted laterally on a glass slide in agarose that is windowed to expose the eyes, covered with medium, and viewed under a 40× water immersion objective. A glass microelectrode filled with DNA or tracer solution is poked into the retina with a micromanipulator, and a voltage train applied. Embryos are then unmounted and raised. This approach allows coelectroporation of several indicators or constructs into the same cell, allowing both visualization and perturbation experiments. We describe this technique in detail in Section II.G.2.

F. Time-Lapse Imaging

Time-lapse imaging of RGC axons is crucial to understand their response to the environment. It has been used by many investigators to monitor axons' behavior (Campbell *et al.*, 2007; Hutson and Chien, 2002; Kaethner and Stuermer, 1992; O'Brien *et al.*, 2009; Schmidt *et al.*, 2000) and can be used with all the labeling techniques described above, except for antibody labeling. Confocal or two-photon microscopy is most appropriate for time-lapse imaging and can be performed with an upright or inverted microscope. Several protocols have been previously described, so we do not discuss them here (Campbell *et al.*, 2007; Hutson and Chien, 2002; Hutson *et al.*, 2004; Meyer and Smith, 2006).

G. Protocols for Labeling Methods

Here we describe detailed protocols for focal injections of lipophilic dyes in the retina and for *in vivo* single cell electroporation.

1. Method 1: Precise Labeling with Intraretinal Injection of Lipophilic Dyes

This method uses glass microneedles coated with lipophilic carbocyanine dyes to focally deposit dye into the retina (Fig. 2A–B). It can be used to target specific locations in the retina and label very few cells. It was originally developed by Torsten Trowe (2000).

a. Solutions Needed

- DiI or DiO crystals (Molecular Probes)
- 4% PFA (4% paraformaldehyde in 0.1 M phosphate buffer, pH 7.4)
- 1% low-melt agarose in PBS (phosphate-buffered saline)
- 50% and 80% glycerol in water

b. Protocol

1. Fix zebrafish embryos at required stage in 4% PFA at 4°C for at least 12 h. For growth cone labeling, fix at room temperature for the first 2 h.
2. Use glass capillary with an outer diameter of 1.0 mm and an inner diameter of 0.58 mm to prepare the micropipette for injections. Pull the capillary to a final taper length of 9.0 mm and a tip size of 2 μm. To coat the micropipette with dye, place a few dye crystals on a cover glass and melt them at 100°C on a hot plate. Dip the tip of the micropipette horizontally into the dye paste and roll it to cover the tip equally on all sides. Wipe off as much dye from the tip as possible onto the cover glass.
3. Prepare 30 ml of 1% low-melt agarose and keep on heating block at 45°C to prevent from solidifying. Use a Petri dish lid to embed embryos for dye injection. Coat bottom with a thin layer of 1% low-melt agarose and let solidify. Transfer embryos with as little PFA as possible onto the agarose. Cover embryos with a drop of 1% low-melt agarose and orient them in a lateral position.
4. When the agarose covering is solid, use a sharpened tungsten needle to remove the top-facing lens by carefully cutting the skin covering the eye in a circle along the

border between lens and retina. The lens will become loose and can now be easily removed. The resulting hole should be refilled with 1% low-melt agarose. The embryos are now ready to be injected with the dye.

5. Use a standard pipette holder and three-axis micromanipulator to hold the dye-coated micropipette. Insert it into the RGC layer by placing it in the empty lens cup and advancing in a peripheral direction at a roughly 45° angle (Fig. 2A). Leave the needle in the eye for not more than 2 s to ensure a small injection site and labeling of only a few axons. The coated micropipette can be reused for several injections before it has to be coated again with fresh dye.

6. After finishing the injections, cover embedded embryos with 1× PBS or water to avoid drying. This step also washes off excessive dye. Store the embryos for a few hours at room temperature for fast diffusion of the dye, or keep them at 4°C overnight if slower diffusion is desired. Long incubation times can result in nonspecific diffusion of the dye within the eye, which can prevent clear imaging results later on.

7. Recover embryos from the agarose bed using forceps. Place them in a microfuge tube and wash them in 1× PBS. Transfer embryos to 50% glycerol/H_2O and incubate them for 3 h at 4°C with agitation. Change the medium to 80% glycerol/H_2O, and store embryos at 4°C overnight. Now that they are cleared, embryos can be mounted for confocal imaging in 80% glycerol between two coverslips (Fig. 2B).

2. Method 2: Single Cell *In Vivo* Electroporation

In vivo focal electroporation is used to deliver tracers or transgenes into single RGCs. It can target several or individual cells in precise topographic positions within the retina (Fig. 2E–H). An electric field applied across an RGC progenitor creates transient pores in the plasma membrane through which negatively charged DNA molecules move into the cell. We have used the protocol detailed here to image single RGC arbors in the tectum (Pittman *et al.*, 2010); it was slightly modified from a previous method for imaging habenular neurons (Bianco *et al.*, 2008).

a. Solutions Needed

- E2 medium (15 mM NaCl, 0.5 mM KCl, 1 mM $CaCl_2$, 1 mM $MgSO_4$, 0.15 mM KH_2PO_4, 1.7 mM $NaHCO_3$)
- 0.1 mM phenylthiourea (PTU) in E3 embryo medium (5 mM NaCl, 0.17 mM KCl, 0.33 mM $CaCl_2$, 0.33 mM $MgSO_4$)
- tricaine stock (0.4% tricaine, 10 mM HEPES, pH 7.4)
- 1% low-melt agarose in E2/GN/tricaine (10 μg/ml gentamicin in E2 medium, 0.02% tricaine)

b. Protocol

1. Raise embryos at 28.5°C in E3 medium containing 0.1 mM PTU to inhibit pigment formation, and dechorionate them between 22 and 28 hpf. Anesthetize embryos by adding tricaine to a final concentration of 0.02%. Mount laterally in a drop of 1% low-melt agarose in E2/gentamycin/ tricaine, in wells built with quick-hardening

epoxy on a glass microscope slide. Expose the eye by cutting a small window in the agarose with forceps, and cover the embryo with E3-PTU + 0.02% tricaine.

2. After mounting the embryo, place the glass slide under a 40× water immersion objective on an upright compound microscope. Place an Ag/AgCl cathode in the overlying buffer near the head of the embryo. Backfill a glass microelectrode (1–3 μm diameter tip) with 2 μl of solution containing the tracer or DNA (final concentration of 1–3 μg/μl in water or 10 mM Tris-HCl, pH 8.5), and place it in the retina using a micromanipulator. Use a stimulator to deliver 1 s trains of 2 ms negative-going square pulses at 200 Hz, 30–50 V (reverse polarity and use 3–5V for positively charged tracers). An effective train will cause a visible rippling effect (tissue response) in the tissue surrounding the microelectrode tip when the voltage train is applied. A clogged needle will result in a less pronounced tissue response. A "pop" will occasionally appear in the tissue in response to a voltage train, resulting in an ineffective electroporation. While the exact cause of the pop is not known, it occurs less often with a lower DNA concentration and a lower voltage. Each cell is targeted with 3–5 trains, and several cells can be targeted per eye. After electroporation, the embryo is removed from the agarose and raised in E3 + PTU at 28.5°C

3. FP expression can be seen in electroporated RGCs in the eye under a fluorescent dissecting microscope by 12 h after electroporation. Corresponding axons can be visualized in the contralateral optic tract and tectum under a 40× water objective on a compound microscope, or by confocal microscopy. Labeled axons are best observed from a dorsal view in the contralateral tectum, or from a lateral view in the contralateral optic tract after removal of the contralateral eye. Time-lapse imaging can also be performed.

III. Perturbing the Retinotectal System

Experimental manipulations perturbing axons or their environment are crucial to understand how and by which molecular mechanisms retinotectal projections develop. Many important factors have been discovered through the generation and characterization of mutants with retinotectal defects isolated in large-scale genetic screens. In addition, several approaches including DNA or antisense morpholino oligonucleotide (MO or "morpholino") injections, heat shock experiments, or transplants can be used to assess the function of a particular protein.

A. Retinotectal Mutants

The first mutants with retinotectal defects were obtained from a large genetic screen performed in Tübingen in the 1990s (Karlstrom et al., 1996; Trowe et al., 1996). Topographic injections of DiI and DiO in the retina were used as an assay to identify mutants with defects in retinal axon pathfinding, sorting in the tract, and topography in the tectum. Almost all the genes affected in these mutants have now been identified, allowing the discovery of crucial regulators of axon guidance or brain patterning, including the receptors *robo2* and *patched1*, the transcription factor *lhx2*, and the

Table II
Retinotectal Pathfinding Mutants

Mutant name (abbreviation)	Region in which pathfinding affected	Gene	Brain defect?	References
acerebellar (ace)	Chiasm, anterior projection, optic tract, topography	fgf8	Yes	Picker et al. (1999), Shanmugalingam et al. (2000)
astray (ast)	Chiasm, anterior projection, optic tract, tectum aroborization	robo2	No	Campbell et al. (2007), Fricke et al. (2001), Hutson and Chien (2002), Karlstrom et al. (1996)
bashful (bal)	Retinal exit, anterior projection	laminin α1	Yes	Karlstrom et al. (1996), Paulus and Halloran (2006)
belladonna (bel)	Midline crossing	lhx2	Yes	Karlstrom et al. (1996), Seth et al. (2006)
beyond borders (beyo)	Confinement to tectal neuropil	?	Yes	Xiao et al. (2005)
blind date (blin)	Tectum innervation	?	No	Muto et al. (2005), Xiao et al. (2005)
blowout (blw)	Midline crossing eye shape	patched 1 (ptc)	Yes	Karlstrom et al. (1996), Lee et al. (2008)
blue kite (bluk)	Tectum innervation	?	No	Xiao et al. (2005)
blumenkohl (blu)	Expanded terminations	slc17a6b (glutamate transporter)	No	Smear et al. (2007), Trowe et al. (1996)
bogus journey (boj)	Midline crossing	?	?	Muto et al. (2005)
boxer (box)	Tract sorting, crossing in posterior commissure	extl3	No	Karlstrom et al. (1996), Lee et al. (2004), Trowe et al. (1996)
breaking up (brek)	Confinement to tectal neurop1	?	No	Xiao et al. (2005)
chameleon (con)	Retinal exit, midline crossing	dispatched homolog 1 (dips1)	Yes	Karlstrom et al. (1996), Nakano et al. (2004)
clueless (clew)	Tectum innervation	?	No	Xiao et al. (2005)
coming apart (coma)	Optic tract, tectum innervation	?	No	Xiao et al. (2005)
cyclops (cyc)	Midline crossing	nodal related-2 (ndr2)	Yes	Karlstrom et al. (1996), Rebagliati et al. (1998), Sampath et al. (1998)
dackel (dak)	Tract sorting crossing in posterior commissure	ext2	No	Karlstrom et al. (1996), Lee et al. (2004), Trowe et al. (1996)
dark half (darl)	Ventral branch of the optic tract missing, topography	gdf6a	No	Gosse and Baier (2009), Muto et al. (2005)
detour (dtr)	Midline crossing	gli1	Yes	Karlstrom et al. (1996, 2003)
dragnet (drg)	Laminar specificity in the tectum	collagen IVa5 (col4a5)	No	Xiao and Baier (2007), Xiao et al. (2005)
esrom (esr)	Midline crossing, termination	MYC binding protein 2 (mycbp2) or PAM	No	D'Souza et al. (2005), Karlstrom et al. (1996), Trowe et al. (1996)
excellent adventure (exa)	Targeting defect in the tectum	?	?	Muto et al. (2005)

Mutant (abbreviation)	Defect	Gene	Known	References
fizz wuzzy (fuzz)	Confinement to tectal neuropil	?	No	Xiao et al. (2005)
gnarled (gna)	Tectal entry, tectal misrouting	?	Yes	Trowe et al. (1996), Wagle et al. (2004)
grumpy (gup)	Anterior projection, midline crossing	laminin β1	Yes	Karlstrom et al. (1996), Parsons et al. (2002)
iguana (igu)	Midline crossing	DAZ interacting protein 1 (dzip1)	Yes	Karlstrom et al. (1996), Sekimizu et al. (2004), Wolff et al. (2004)
late bloomer (late)	Delayed innervation of the tectum	?	No	Xiao et al. (2005)
no isthmus (noi)	Chiasm, anterior projection, tectal bypass	pax2a	Yes	Brand et al. (1996), MacDonald et al. (1997), Trowe et al. (1996)
macho (mao)	Expanded terminations	?	No	Gnuegge et al. (2001), Trowe et al. (1996)
michikusa (mich)	Ectopic arbor after crossing the midline	?	?	Muto et al. (2005)
missing link (miss)	Pretectal targets (AF4, AF9) absent or reduced	?	?	Muto et al. (2005)
nevermind (nev)	Tract sorting, D-V topography	cyfip2	No	Pittman et al. (2010), Trowe et al. (1996)
odysseus (ody)	Intraretinal guidance defects	cxcr4b	No	Knaut et al. (2003), Li et al. (2005)
parachute (pac)	Ipsilateral projection; entering chiasm area	N-cadherin	Yes	Lele et al. (2002), Masai et al. (2003)
pinscher (pic)	Tract sorting, crossing in posterior commissure	papst1 (sulfate transporter)	No	Clément et al. (2008), Karlstrom et al. (1996), Trowe et al. (1996)
shirli-myrli (shir)	Delayed innervation of the tectum	?	No	Muto et al. (2005)
sleepy (sly)	Anterior projection; midline crossing	laminin γ1	Yes	Karlstrom et al. (1996), Parsons et al. (2002)
smooth muscle omitted (smu)	Midline crossing	smoothened (smo)	Yes	Chen et al. (2001), Varga et al. (2001)
sonic-you (syu)	Retinal exit, midline crossing	sonic hedgehog (shh)	Yes	Brand et al. (1996), Schauerte et al. (1998)
space cadet (spc)	Retinal exit, midline crossing	?	No	Karlstrom et al. (1996), Lorent et al. (2001)
tarde demais (tard)	Delayed innervation of the tectum	?	No	Xiao et al. (2005)
umleitung (uml)	Midline crossing	?	Yes	Karlstrom et al. (1996)
vertigo (vrt)	Delayed innervation of the tectum	?	No	Xiao et al. (2005)
walkabout (walk)	Pretectal target AF4 overinnervated	?	?	Muto et al. (2005)
who cares (woe)	Tract sorting, D-V topography	?	No	Trowe et al. (1996)
you-too (yot)	Midline crossing	gli2	Yes	Karlstrom et al. (1996, 1999)

? = not known

adhesion molecule *N-cadherin* (see Table II for a complete listing of these mutants). While some genes such as *astray (robo2)* primarily affect axon navigation, others such as *ace (fgf8)* disrupt brain patterning, resulting in mispresented axon guidance cues. More recently, a new screen has been performed using the *pou4f3:mGFP* transgenic line expressing membrane-targeted GFP (mGFP) in a subset of RGCs (Xiao *et al.*, 2005). This approach allowed the identification of novel mutants with various defects in tectum innervation (Table II). Two mutants from this screen have been cloned, revealing new functions for *gdf6a* and *collagenIVa5* in regulating eye dorso-ventral patterning and tectum laminar targeting, respectively (Gosse and Baier, 2009; Xiao and Baier, 2007). Finally, a recent screen using behavioral assays identified mutants with disrupted response to visual motion and/or impaired background adaptation (Muto *et al.*, 2005). Some of these mutants also have abnormal retinotectal projections or a lack of RGCs that are likely responsible for their phenotype. Identifying the mutations generated in these newer screens will give new clues about the factors involved in retinal axon guidance.

B. Injecting DNA or Morpholinos

A common approach to characterize protein function in zebrafish is to inject stable MOs into one-cell stage embryos. MOs inhibit either protein translation when targeted near the start codon of mRNAs (Nasevicius and Ekker, 2000) or splicing of the pre-mRNAs when they are targeted to exon–intron or intron–exon boundaries (Draper *et al.*, 2001). Under good conditions, MOs can quickly reveal required functions for a targeted gene, though their use is subject to several caveats, including loss of efficacy as they are diluted during development (Eisen and Smith, 2008). We took advantage of this dilution with an MO against the transcription factor *atoh7* to specifically block differentiation of early- but not late-born RGCs, allowing the functional analysis of isotypic interactions between pioneer and follower axons during navigation (Pittman *et al.*, 2008).

Alternatively, DNA constructs encoding dominant negative forms of the protein of interest can be transiently or stably expressed. Temporal or spatial control can be provided by the *hsp70l* heat shock promoter (see following section) or cell-specific promoters, respectively. Similarly, gain-of-function experiments can be performed by misexpressing genes of interest at specific times or locations. For greater precision, DNA constructs or MOs can be delivered to individual RGCs by *in vivo* cell electroporation (described in Section II.E), allowing functional studies at single-cell resolution (Pittman *et al.*, 2010).

C. Using Heat Shock to Induce Misexpression

A powerful technique to misexpress genes in a temporally or spatially controlled manner is to use heat shock. This approach is particularly useful for studying genes with both early and late roles during development. Heat shocks can be performed after transient injection of DNA constructs or on stable transgenic lines. The *hsp70l* promoter

is an inducible element that drives strong gene expression in response to a temperature shift from 28.5°C (normal rearing temperature) to 37–40°C (Halloran *et al.*, 2000). Global heat shocks have been widely used to induce ubiquitous gene expression in embryos at specific times. The exact heat shock duration and temperature depend on the age of the embryo, the transgene to be expressed, and the level of expression desired. For instance, raising the temperature to 42°C for 5 min can induce detectable transgene expression in 20 hpf embryos (Thummel *et al.*, 2005).

Recently, we developed a technique using a sharpened soldering iron to induce focal heat shocks in restricted regions of the embryo (Hardy *et al.*, 2007). For this approach, a copper soldering iron tip with a diameter of 15 μm is heated to 60°C and put directly in contact with the embryo for 3 min. A perfusion chamber keeps fluid flowing over the embryo during heat shock, thereby preventing heating of the medium and restricting the area of activation. This method is rapid and easy, allows the targeting of ~100 μm patches of tissue, and can be used in a variety of tissues and stages. A detailed protocol has been described (Hardy *et al.*, 2007). Even more recently, Rolf Karlstrom's group developed another focal heat shock method using an optical fiber to deliver energy to a localized region (Placinta *et al.*, 2009).

D. Transplanting to Test Cell Autonomy of Gene Function

Transplanting cells or tissues is a powerful approach to test cell autonomy of gene function. Different types of transplant can be performed depending on the question (e.g., transplanting all RGCs, or RGCs in specific parts of the retina; labeling donors, hosts, or both labeled). A tricky but elegant approach is to transplant entire eye primordia, yielding mosaic embryos in which the whole eye comes from the donor while the rest of the embryo is derived from the host. A main advantage of this approach is that all retinal axons coming from the transplanted eye share the same genotype and are not influenced by interactions with host retinal axons, as these have been removed. We used eye transplants to demonstrate that *robo2* acts eye-autonomously to regulate retinal axon guidance (Fricke *et al.*, 2001). A detailed protocol has been previously described (Hutson *et al.*, 2004).

Alternatively, early transplants at blastula stage can be used to test cell autonomy (Ho and Kane, 1990). These are easy to perform and allow quite effective targeting of the retina (Moens and Fritz, 1999). Cells are removed from donor embryos between 4 and 6 hpf and replaced into the animal pole of host embryos. The resulting mosaic embryos display clones of RGCs in the retina, as well as some clones of cells in the brain. An abbreviated protocol is given below. While the presence of donor cells in the brain may make results harder to interpret, this approach is the easiest way to generate mosaic embryos with RGCs from different genetic backgrounds. However, it cannot be employed to target RGCs from or to specific regions within the retina.

Instead, transplants at a later stage are required. We have recently begun to use a technique for transplanting RGCs in a topographic manner (Fig. 3; inspired by Masai *et al.*, 2003). Donor and host embryos labeled with different transgenes are grown to 30–33 hpf, when the first RGCs are specified. After mounting embryos laterally in

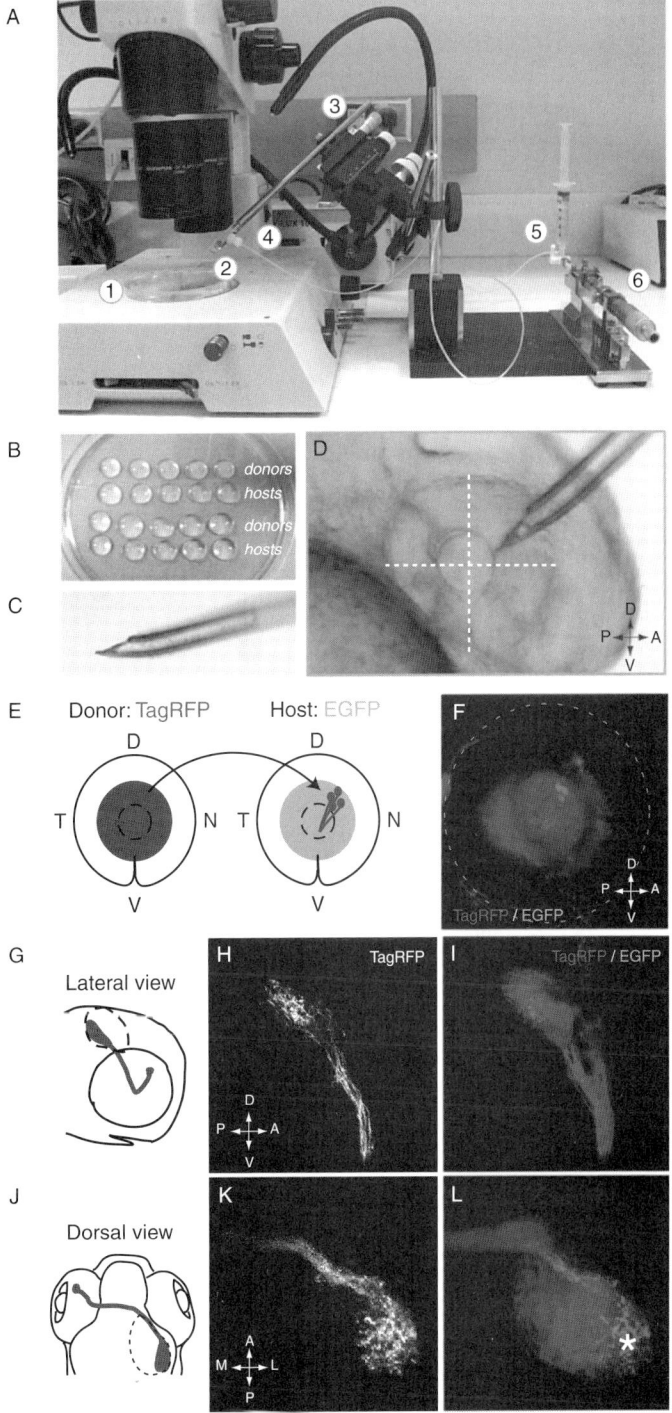

Fig. 3 *(Continued)*

agarose, donor RGCs are precisely removed from a specific location in the retina with a 40 μm glass micropipette and replaced at the same position in the host retina. The transplant is considered successful if, after raising the host, the transplanted RGCs are observed in the correct area of the retina (from a lateral view), and if in control conditions their arbors terminate in the appropriate part of the tectum (from a dorsal view). We obtain ~25% successful transplants with this approach and provide a detailed protocol below.

E. Protocols for Transplants

1. Method 3: Blastula Transplants

Since blastula stage transplants have been explained in detail elsewhere (Ho and Kane, 1990; Kemp et al., 2009), only a succinct description of the method is provided here.

1. Donor embryos are injected at the one-cell stage with 5% Alexa-488 dextran or rhodamine dextran (10,000 MW) as a lineage marker. The light color from the dextran helps to distinguish donors from hosts during later steps. We use agarose-groove dishes for the injections (mold TU-1, Adaptive Science Tools, Worcester, Massachusetts; 1% agarose w/v in E2 or E3 embryo medium). Donor and host embryos are raised at 28.5°C until the sphere (4 hpf) or shield stage (6 hpf).

2. While waiting for the embryos to develop, pull and bevel standard wall, non-filament capillaries for use as transplant needles and prepare an agarose transplant dish (single-well mold; mold PT-1, Adaptive Science Tools; Kane and Kishimoto, 2002).

3. Dechorionate donor and host embryos. Use a clean fire-polished large-bore Pasteur pipette to transfer one donor and four hosts into each row of the transplant dish using an air-filled syringe and fire-polished transplantation pipette. Remove cells

---

Fig. 3 Perturbing the retinotectal system with late topographic transplants. (A) Embryos are mounted laterally in drops of low-melt agarose deposited on a dish lid that is then placed under a dissecting microscope (1). The transplant needle is mounted in a micropipette holder (2), itself mounted onto a three-axis micromanipulator (3) placed next to the microscope. The micropipette holder is connected via a tube filled with mineral oil (4) to an oil-filled Hamilton syringe with a micrometer drive (6). The syringe is attached by a three-way stopcock to a reservoir filled with mineral oil (5). (B) Donor and host embryos mounted laterally in low-melt agarose drops. Embryos are arranged so that each donor is close to its respective host. (C) The transplant needle has a 40 μm diameter opening with a sharp tip that is slightly bent (around 20°). (D) The transplant needle is inserted into the dorsonasal retina, close to the lens, at a 45° angle. The bend of the needle tip is facing upward, so that ventral RGCs cannot be drawn up. (E) Dorsonasal (DN) RGCs from an isl2b:TagRFP donor are isotopically transplanted into the DN retina of an isl2b:EGFP host between 30 and 33 hpf. Their axonal projections are then visualized at 4 dpf by live confocal microscopy. (F) Lateral view of a WT isl2b:EGFP host eye in which WT TagRFP-positive RGCs have been transplanted. GFP is shown as blue for the best visualization. (G, J) Projections of DN donor axons observed in transplants in lateral (G) and dorsal (J) views. (H, I) Lateral view of TagRFP-positive projections at 4 dpf. DN donor axons navigate along the ventral branch of the tract to reach the tectum. (K, L) Dorsal view of the same projections. DN donor axons project to the posterolateral part of the host tectum (asterisk). F, H, I, K, L: confocal maximum intensity projections. (See Plate no. 2 in the Color Plate Section.)

from each donor embryo and transplant 20–50 cells into the animal pole of each corresponding host. While the origin of the transplanted cells is not important, the location where they are placed into the host is crucial. A fate map of the 6 hpf embryo can be used as a reference (Woo *et al.*, 1995).

4. After transplantation, transfer the agarose dish carefully to the 28.5°C incubator. During gastrulation, donor cells will spread out and form a mosaic patch of fluorescently labeled cells; choose those in which this patch includes cells in the eye. Once embryos have developed to bud stage, it is safe to remove them from the transplant dish and put them in 4-well or 24-well dishes. For experiments in which mutant cells are transplanted, donors should be kept together with their respective hosts until genotyped, either by PCR or by mutant phenotype. If necessary, hosts can be genotyped as well.

2. Method 4: Late Topographic Transplants

While blastula transplants are useful for testing functional cell autonomy and can be easily performed, they cannot target RGCs within specific regions of the retina. Testing the roles of genes specifically expressed in the dorsal or ventral retina, for instance, requires transplanting at later stages in a topographic manner. Here, we describe a detailed protocol for transplanting dorsonasal RGCs into the host dorsonasal retina. These transplants are performed between 30 and 33 hpf, when the first RGCs are specified and have acquired their positional identity within the retina. Donor and host embryos are labeled with *isl2b:TagRFP* and *isl2b:EGFP* transgenes, respectively, so that axons of transplanted RGCs and their projections can be easily visualized by live confocal miscroscopy at 4 days post-fertilization (dpf).

a. Solutions Needed

- E2 medium (15 mM NaCl, 0.5 mM KCl, 1 mM $CaCl_2$, 1 mM $MgSO_4$, 0.15 mM KH_2PO_4, 1.7 mM $NaHCO_3$)
- 0.1 mM phenylthiourea (PTU) in E3 embryo medium (5 mM NaCl, 0.17 mM KCl, 0.33 mM $CaCl_2$, 0.33 mM $MgSO_4$)
- tricaine stock (0.4% tricaine, 10 mM HEPES, pH 7.4)
- 1% low-melt agarose in E2/GN/tricaine (10 µg/ml gentamicin in E2 medium, 0.02% tricaine)

b. Protocol

1. The transplant needle is prepared in advance and can be reused several times. The quality of its preparation is the most important parameter for successful transplants. Pull standard wall, non-filament capillaries and polish them using a microforge, so that the tip displays a 20° angle with a 40 µm diameter opening (Fig. 3C).

2. Raise embryos at 28.5°C in E3 medium containing 0.1 mM PTU to inhibit pigment formation, and dechorionate them between 22 and 28 hpf. At 30 hpf, anesthetize embryos by adding tricaine to a final concentration of 0.02%. Mount laterally in a

drop of 1% low-melt agarose in E2/gentamycin/tricaine deposited on the lid of a Petri dish (Fig. 3B). Donors and hosts should be arranged in lines, so that each donor is close to its respective host. Once the drops have solidified, fill the Petri dish with PTU-E3/tricaine and position it under a dissecting scope.

3. Prepare the transplant setup (Fig. 3A): an oil-filled Hamilton syringe with a micrometer drive is connected by a three-way stopcock to a reservoir filled with mineral oil and to a micropipette holder through flexible plastic tubing. It is important to fill the system completely with mineral oil and ensure that air bubbles have been eliminated (air bubbles impair the ability to control suction and pressure). The transplant pipette is mounted in the micropipette holder, itself mounted onto a three-axis micromanipulator positioned next to the dissecting scope.

4. Using the micromanipulator, bring the transplant pipette near the dorsonasal retina, with a 45° angle (Fig. 3C). Make sure that the needle opening is facing upward, so that ventral RGCs cannot be drawn up into the needle. Insert the needle into the dorsonasal retina close to the lens, and slowly and carefully suck up 40–100 cells into the needle. At this stage, the fluorescence of the transgene expressed in RGCs is not yet visible, so the fraction of RGCs among the removed cells can vary. After cells have been taken up, reverse the pressure in the needle to stop the suction, and remove the needle from the donor eye. Insert the needle into the host retina in a similar way, and slowly expel the cells with as little medium as possible. After transplantation, let embryos recover for few minutes, remove them from the agarose, and raise them in E3 + PTU at 28.5°C in 24-well plates. Axons of transplanted RGCs can then be observed after 48 hpf by live imaging.

IV. Future Directions

The approaches developed over the past decade have greatly improved our ability to label and visualize the retinotectal projection *in vivo,* as well as to perform functional assays for understanding the molecular mechanisms that control its development. Nevertheless, novel techniques will be required for observing retinal axons in greater detail and to ask new biological questions.

Three methods already used in other systems are currently being adapted to study new aspects of zebrafish retinotectal system development. The Brainbow approach, initially developed in mice, allows labeling and mapping of neurons with a wide range of colors by randomly varying the levels of red, green, and blue FPs expressed in individual neurons (Livet *et al.*, 2007). It has been used to reconstruct the architecture of neuronal circuits in different systems and will be a powerful tool for analyzing sorting of retinal axons in the tract as well as topographic mapping in the tectum. A second approach is to use enhancer trap (ET) screens to isolate lines expressing transgenes in specific subsets of neurons. For instance, new lines with interesting expression patterns in the tectum have recently been produced with a Gal4 ET screen (Scott and Baier, 2009). Such an approach will potentially allow the identification of new lines driving expression in specific regions of the retina (Picker *et al.*, 2009) or RGC subtypes. Finally, calcium

imaging has been used *in vitro* to measure growth cone responses to guidance cues (Guan *et al.*, 2007; Tojima *et al.*, 2010). Adapted to zebrafish, it will give the ability to monitor, *in vivo*, the activity of retinal axons as they elongate. Combined together, these emerging techniques will improve our ability to examine retinal axons as they navigate, shedding new light on axon guidance *in vivo*.

References

Aizawa, H., Bianco, I. H., Hamaoka, T., Miyashita, T., Uemura, O., Concha, M. L., Russell, C., Wilson, S. W., and Okamoto, H. (2005). Laterotopic representation of left-right information onto the dorso-ventral axis of a zebrafish midbrain target nucleus. Curr. Biol. **15**, 238–243.

Baier, H., Klostermann, S., Trowe, T., Karlstrom, R. O., Nusslein-Volhard, C., and Bonhoeffer, F. (1996). Genetic dissection of the retinotectal projection. Development **123**, 415–425.

Ben Fredj, N., Hammond, S., Otsuna, H., Chien, C. B., Burrone, J., and Meyer, M. P. (2010). Synaptic activity and activity-dependent competition regulates axon arbor maturation, growth arrest, and territory in the retinotectal projection. J. Neurosci. **30**, 10939–10951.

Bianco, I. H., Carl, M., Russell, C., Clarke, J. D., and Wilson, S. W. (2008). Brain asymmetry is encoded at the level of axon terminal morphology. Neural Dev. **3**, 9.

Brand, M., Heisenberg, C. P., Jiang, Y. J., Beuchle, D., Lun, K., Furutani-Seiki, M., Granato, M., Haffter, P., Hammerschmidt, M., Kane, D. A., Kelsh, R. N., Mullins, M. C., *et al.*, (1996). Mutations in zebrafish genes affecting the formation of the boundary between midbrain and hindbrain. Development **123**, 179–190.

Campbell, D. S., Stringham, S. A., Timm, A., Xiao, T., Law, M. Y., Baier, H., Nonet, M. L., and Chien, C. B. (2007). Slit1a inhibits retinal ganglion cell arborization and synaptogenesis via Robo2-dependent and - independent pathways. Neuron **55**, 231–245.

Chen, W., Burgess, S., and Hopkins, N. (2001). Analysis of the zebrafish smoothened mutant reveals conserved and divergent functions of hedgehog activity. Development **128**, 2385–2396.

Choy, E., Chiu, V. K., Silletti, J., Feoktistov, M., Morimoto, T., Michaelson, D., Ivanov, I. E., and Philips, M. R. (1999). Endomembrane trafficking of ras: The CAAX motif targets proteins to the ER and Golgi. Cell **98**, 69–80.

Clement, A., Wiweger, M., von der Hardt, S., Rusch, M. A., Selleck, S. B., Chien, C. B., and Roehl, H. H. (2008). Regulation of zebrafish skeletogenesis by ext2/dackel and papst1/pinscher. PLoS Genet. **4**, e1000136.

Del Bene, F., Wehman, A. M., Link, B. A., and Baier, H. (2008). Regulation of neurogenesis by interkinetic nuclear migration through an apical-basal notch gradient. Cell **134**, 1055–1065.

Draper, B. W., Morcos, P. A., and Kimmel, C. B. (2001). Inhibition of zebrafish fgf8 pre-mRNA splicing with morpholino oligos: A quantifiable method for gene knockdown. Genesis **30**, 154–156.

D'Souza, J., Hendricks, M., Le Guyader, S., Subburaju, S., Grunewald, B., Scholich, K., and Jesuthasan, S. (2005). Formation of the retinotectal projection requires Esrom, an ortholog of PAM (protein associated with Myc). Development **132**, 247–256.

Eisen, J. S., and Smith, J. C. (2008). Controlling morpholino experiments: Don't stop making antisense. Development **135**, 1735–1743.

Fricke, C., Lee, J. S., Geiger-Rudolph, S., Bonhoeffer, F., and Chien, C. B. (2001). Astray, a zebrafish roundabout homolog required for retinal axon guidance. Science **292**, 507–510.

Gnuegge, L., Schmid, S., and Neuhauss, S. C. (2001). Analysis of the activity-deprived zebrafish mutant macho reveals an essential requirement of neuronal activity for the development of a fine-grained visuotopic map. J. Neurosci. **21**, 3542–3548.

Gosse, N. J., and Baier, H. (2009). An essential role for Radar (Gdf6a) in inducing dorsal fate in the zebrafish retina. Proc. Natl. Acad. Sci. U.S.A. **106**, 2236–2241.

Guan, C. B., Xu, H. T., Jin, M., Yuan, X. B., and Poo, M. M. (2007). Long-range Ca2+ signaling from growth cone to soma mediates reversal of neuronal migration induced by slit-2. Cell **129**, 385–395.

Halloran, M. C., Sato-Maeda, M., Warren, J. T., Su, F., Lele, Z., Krone, P. H., Kuwada, J. Y., and Shoji, W. (2000). Laser-induced gene expression in specific cells of transgenic zebrafish. Development **127**, 1953–1960.

Hardy, M. E., Ross, L. V., and Chien, C. B. (2007). Focal gene misexpression in zebrafish embryos induced by local heat shock using a modified soldering iron. Dev. Dyn. **236**, 3071–3076.

Ho, R. K., and Kane, D. A. (1990). Cell-autonomous action of zebrafish spt-1 mutation in specific mesodermal precursors. Nature **348**, 728–730.

Honig, M. G., and Hume, R. I. (1989). DiI and diO: Versatile fluorescent dyes for neuronal labelling and pathway tracing. Trends Neurosci. **12**, 333–335, 340–341.

Hu, M., and Easter, S. S. (1999). Retinal neurogenesis: The formation of the initial central patch of postmitotic cells. Dev. Biol. **207**, 309–321.

Hutson, L. D., Campbell, D. S., and Chien, C. B. (2004). Analyzing axon guidance in the zebrafish retinotectal system. Methods Cell Biol. **76**, 13–35.

Hutson, L. D., and Chien, C. B. (2002). Pathfinding and error correction by retinal axons: The role of astray/robo2. Neuron **33**, 205–217.

Kaethner, R. J., and Stuermer, C. A. (1992). Dynamics of terminal arbor formation and target approach of retinotectal axons in living zebrafish embryos: A time-lapse study of single axons. J. Neurosci. **12**, 3257–3271.

Kane, D. A., and Kishimoto, Y. (2002). Cell labelling and transplantation techniques. In "Zebrafish, Practical Approach" (C. Nüsslein-Volhard, R. Dahm, eds.) No. 261, pp. 95–120, Oxford University Press, Tubingen, Germany.

Karlstrom, R. O., Talbot, W. S., and Schier, A. F. (1999). Comparative synteny cloning of zebrafish you-too: Mutations in the Hedgehog target gli2 affect ventral forebrain patterning. Genes Dev. **13**, 388–393.

Karlstrom, R. O., Trowe, T., Klostermann, S., Baier, H., Brand, M., Crawford, A. D., Grunewald, B., Haffter, P., Hoffmann, H., Meyer, S. U., Muller, B. K., Richter, S., et al. (1996). Zebrafish mutations affecting retinotectal axon pathfinding. Development **123**, 427–438.

Karlstrom, R. O., Tyurina, O. V., Kawakami, A., Nishioka, N., Talbot, W. S., Sasaki, H., and Schier, A. F. (2003). Genetic analysis of zebrafish gli1 and gli2 reveals divergent requirements for gli genes in vertebrate development. Development **130**, 1549–1564.

Kemp, H. A., Carmany-Rampey, A., and Moens, C. (2009). Generating chimeric zebrafish embryos by transplantation. J. Vis. Exp. Jul 17;(29). pii: 1394. doi: 10.3791/1394.

Knaut, H., Werz, C., Geisler, R., and Nusslein-Volhard, C. (2003). A zebrafish homologue of the chemokine receptor Cxcr4 is a germ-cell guidance receptor. Nature **421**, 279–282.

Koster, R. W., and Fraser, S. E. (2001). Tracing transgene expression in living zebrafish embryos. Dev. Biol. **233**, 329–346.

Laessing, U., Giordano, S., Stecher, B., Lottspeich, F., and Stuermer, C. A. (1994). Molecular characterization of fish neurolin: A growth-associated cell surface protein and member of the immunoglobulin superfamily in the fish retinotectal system with similarities to chick protein DM-GRASP/SC-1/BEN. Differentiation **56**, 21–29.

Laessing, U., and Stuermer, C. A. (1996). Spatiotemporal pattern of retinal ganglion cell differentiation revealed by the expression of neurolin in embryonic zebrafish. J. Neurobiol. **29**, 65–74.

Lee, J. S., von der Hardt, S., Rusch, M. A., Stringer, S. E., Stickney, H. L., Talbot, W. S., Geisler, R., Nusslein-Volhard, C., Selleck, S. B., Chien, C. B., and Roehl, H. (2004). Axon sorting in the optic tract requires HSPG synthesis by ext2 (dackel) and extl3 (boxer). Neuron **44**, 947–960.

Lee, J. S., Willer, J. R., Willer, G. B., Smith, K., Gregg, R. G., and Gross, J. M. (2008). Zebrafish blowout provides genetic evidence for Patched1-mediated negative regulation of Hedgehog signaling within the proximal optic vesicle of the vertebrate eye. Dev. Biol. **319**, 10–22.

Lele, Z., Folchert, A., Concha, M., Rauch, G. J., Geisler, R., Rosa, F., Wilson, S. W., Hammerschmidt, M., and Bally-Cuif, L. (2002). Parachute/n-cadherin is required for morphogenesis and maintained integrity of the zebrafish neural tube. Development **129**, 3281–3294.

Li, Q., Shirabe, K., Thisse, C., Thisse, B., Okamoto, H., Masai, I., and Kuwada, J. Y. (2005). Chemokine signaling guides axons within the retina in zebrafish. J. Neurosci. **25**, 1711–1717.

Livet, J., Weissman, T. A., Kang, H., Draft, R. W., Lu, J., Bennis, R. A., Sanes, J. R., and Lichtman, J. W. (2007). Transgenic strategies for combinatorial expression of fluorescent proteins in the nervous system. Nature **450**, 56–62.

Lorent, K., Liu, K. S., Fetcho, J. R., and Granato, M. (2001). The zebrafish space cadet gene controls axonal pathfinding of neurons that modulate fast turning movements. Development **128**, 2131–2142.

Macdonald, R., Scholes, J., Strahle, U., Brennan, C., Holder, N., Brand, M., and Wilson, S. W. (1997). The Pax protein Noi is required for commissural axon pathway formation in the rostral forebrain. Development **124**, 2397–2408.

Maddison, L. A., Lu, J., Victoroff, T., Scott, E., Baier, H., and Chen, W. (2009). A gain-of-function screen in zebrafish identifies a guanylate cyclase with a role in neuronal degeneration. Mol. Genet. Genomics **281**, 551–563.

Mangrum, W. I., Dowling, J. E., and Cohen, E. D. (2002). A morphological classification of ganglion cells in the zebrafish retina. Vis. Neurosci. **19**, 767–779.

Masai, I., Lele, Z., Yamaguchi, M., Komori, A., Nakata, A., Nishiwaki, Y., Wada, H., Tanaka, H., Nojima, Y., Hammerschmidt, M., Wilson, S. W., and Okamoto, H. (2003). N-cadherin mediates retinal lamination, maintenance of forebrain compartments and patterning of retinal neurites. Development **130**, 2479–2494.

Masai, I., Yamaguchi, M., Tonou-Fujimori, N., Komori, A., and Okamoto, H. (2005). The hedgehog-PKA pathway regulates two distinct steps of the differentiation of retinal ganglion cells: The cell-cycle exit of retinoblasts and their neuronal maturation. Development **132**, 1539–1553.

Matsuda, N., and Mishina, M. (2004). Identification of chaperonin CCT gamma subunit as a determinant of retinotectal development by whole-genome subtraction cloning from zebrafish no tectal neuron mutant. Development **131**, 1913–1925.

Meyer, M. P., and Smith, S. J. (2006). Evidence from in vivo imaging that synaptogenesis guides the growth and branching of axonal arbors by two distinct mechanisms. J. Neurosci. **26**, 3604–3614.

Moens, C. B., and Fritz, A. (1999). Techniques in neural development. Methods Cell Biol. **59**, 253–272.

Moriyoshi, K., Richards, L. J., Akazawa, C., O'Leary, D. D., and Nakanishi, S. (1996). Labeling neural cells using adenoviral gene transfer of membrane-targeted GFP. Neuron **16**, 255–260.

Mumm, J. S., Williams, P. R., Godinho, L., Koerber, A., Pittman, A. J., Roeser, T., Chien, C. B., Baier, H., and Wong, R. O. (2006). In vivo imaging reveals dendritic targeting of laminated afferents by zebrafish retinal ganglion cells. Neuron **52**, 609–621.

Muto, A., Orger, M. B., Wehman, A. M., Smear, M. C., Kay, J. N., Page-McCaw, P. S., Gahtan, E., Xiao, T., Nevin, L. M., Gosse, N. J., Staub, W., Finger-Baier, K., *et al.*, (2005). Forward genetic analysis of visual behavior in zebrafish. PLoS Genet. **1**, e66.

Nakano, Y., Kim, H. R., Kawakami, A., Roy, S., Schier, A. F., and Ingham, P. W. (2004). Inactivation of dispatched 1 by the chameleon mutation disrupts Hedgehog signalling in the zebrafish embryo. Dev. Biol. **269**, 381–392.

Nasevicius, A., and Ekker, S. C. (2000). Effective targeted gene "knockdown" in zebrafish. Nat. Genet. **26**, 216–220.

Neumann, C. J., and Nuesslein-Volhard, C. (2000). Patterning of the zebrafish retina by a wave of sonic hedgehog activity. Science **289**, 2137–2139.

Nevin, L. M., Taylor, M. R., and Baier, H. (2008). Hardwiring of fine synaptic layers in the zebrafish visual pathway. Neural Dev. **3**, 36.

O'Brien, G. S., Rieger, S., Martin, S. M., Cavanaugh, A. M., Portera-Cailliau, C., and Sagasti, A. (2009). Two-photon axotomy and time-lapse confocal imaging in live zebrafish embryos. J. Vis. Exp. Feb 16;(24). pii: 1129. doi: 10.3791/1129.

Parsons, M. J., Pollard, S. M., Saude, L., Feldman, B., Coutinho, P., Hirst, E. M., and Stemple, D. L. (2002). Zebrafish mutants identify an essential role for laminins in notochord formation. Development **129**, 3137–3146.

Paulus, J. D., and Halloran, M. C. (2006). Zebrafish bashful/laminin-alpha 1 mutants exhibit multiple axon guidance defects. Dev. Dyn. **235**, 213–224.

Picker, A., Brennan, C., Reifers, F., Clarke, J. D., Holder, N., and Brand, M. (1999). Requirement for the zebrafish mid-hindbrain boundary in midbrain polarisation, mapping and confinement of the retinotectal projection. Development **126**, 2967–2978.

Picker, A., Cavodeassi, F., Machate, A., Bernauer, S., Hans, S., Abe, G., Kawakami, K., Wilson, S. W., and Brand, M. (2009). Dynamic coupling of pattern formation and morphogenesis in the developing vertebrate retina. PLoS Biol. **7**, e1000214.

Pittman, A. J., Gaynes, J. A., and Chien, C. B. (2010). nev (cyfip2) is required for retinal lamination and axon guidance in the zebrafish retinotectal system. Dev. Biol. **344**, 784–794.

Pittman, A. J., Law, M. Y., and Chien, C. B. (2008). Pathfinding in a large vertebrate axon tract: Isotypic interactions guide retinotectal axons at multiple choice points. Development **135**, 2865–2871.

Placinta, M., Shen, M. C., Achermann, M., and Karlstrom, R. O. (2009). A laser pointer driven microheater for precise local heating and conditional gene regulation in vivo. Microheater driven gene regulation in zebrafish. BMC Dev. Biol. **9**, 73.

Poggi, L., Vitorino, M., Masai, I., and Harris, W. A. (2005). Influences on neural lineage and mode of division in the zebrafish retina in vivo. J. Cell Biol. **171**, 991–999.

Rebagliati, M. R., Toyama, R., Haffter, P., and Dawid, I. B. (1998). cyclops encodes a nodal-related factor involved in midline signaling. Proc. Natl. Acad. Sci. U.S.A. **95**, 9932–9937.

Roeser, T., and Baier, H. (2003). Visuomotor behaviors in larval zebrafish after GFP-guided laser ablation of the optic tectum. J. Neurosci. **23**, 3726–3734.

Sampath, K., Rubinstein, A. L., Cheng, A. M., Liang, J. O., Fekany, K., Solnica-Krezel, L., Korzh, V., Halpern, M. E., and Wright, C. V. (1998). Induction of the zebrafish ventral brain and floorplate requires cyclops/nodal signalling. Nature **395**, 185–189.

Sato, T., Hamaoka, T., Aizawa, H., Hosoya, T., and Okamoto, H. (2007). Genetic single-cell mosaic analysis implicates ephrinB2 reverse signaling in projections from the posterior tectum to the hindbrain in zebrafish. J. Neurosci. **27**, 5271–5279.

Schauerte, H. E., van Eeden, F. J., Fricke, C., Odenthal, J., Strahle, U., and Haffter, P. (1998). Sonic hedgehog is not required for the induction of medial floor plate cells in the zebrafish. Development **125**, 2983–2993.

Schmidt, J. T., Buzzard, M., Borress, R., and Dhillon, S. (2000). MK801 increases retinotectal arbor size in developing zebrafish without affecting kinetics of branch elimination and addition. J. Neurobiol. **42**, 303–314.

Scott, E. K., and Baier, H. (2009). The cellular architecture of the larval zebrafish tectum, as revealed by gal4 enhancer trap lines. Front Neural Circuits **3**, 13.

Sekimizu, K., Nishioka, N., Sasaki, H., Takeda, H., Karlstrom, R. O., and Kawakami, A. (2004). The zebrafish iguana locus encodes Dzip1, a novel zinc-finger protein required for proper regulation of Hedgehog signaling. Development **131**, 2521–2532.

Seth, A., Culverwell, J., Walkowicz, M., Toro, S., Rick, J. M., Neuhauss, S. C., Varga, Z. M., and Karlstrom, R. O. (2006). belladonna/(lhx2) is required for neural patterning and midline axon guidance in the zebrafish forebrain. Development **133**, 725–735.

Shanmugalingam, S., Houart, C., Picker, A., Reifers, F., Macdonald, R., Barth, A., Griffin, K., Brand, M., and Wilson, S. W. (2000). Ace/Fgf8 is required for forebrain commissure formation and patterning of the telencephalon. Development **127**, 2549–2561.

Smear, M. C., Tao, H. W., Staub, W., Orger, M. B., Gosse, N. J., Liu, Y., Takahashi, K., Poo, M. M., and Baier, H. (2007). Vesicular glutamate transport at a central synapse limits the acuity of visual perception in zebrafish. Neuron **53**, 65–77.

Thummel, R., Burket, C. T., Brewer, J. L., Sarras, M. P., Jr., Li, L., Perry, M., McDermott, J. P., Sauer, B., Hyde, D. R., and Godwin, A. R. (2005). Cre-mediated site-specific recombination in zebrafish embryos. Dev. Dyn. **233**, 1366–1377.

Tojima, T., Itofusa, R., and Kamiguchi, H. (2010) Asymmetric clathrin-mediated endocytosis drives repulsive growth cone guidance. Neuron **66**, 370–377.

Tokuoka, H., Yoshida, T., Matsuda, N., and Mishina, M. (2002). Regulation by glycogen synthase kinase-3beta of the arborization field and maturation of retinotectal projection in zebrafish. J. Neurosci. **22**, 10324–10332.

"Analyse von Mutationen mit Einfluss aud die topographische Ordnung von Axonen im retinotektalen System des Zebrabärblings, Danio rerio. PhD Thesis, Eberhard-Karls-Universität Tübingen, Tübingen, Germany.

Trowe, T., Klostermann, S., Baier, H., Granato, M., Crawford, A. D., Grunewald, B., Hoffmann, H., Karlstrom, R. O., Meyer, S. U., Muller, B., Richter, S., Nusslein-Volhard, C., *et al.*, (1996). Mutations disrupting the ordering and topographic mapping of axons in the retinotectal projection of the zebrafish, Danio rerio. Development **123**, 439–450.

Varga, Z. M., Amores, A., Lewis, K. E., Yan, Y. L., Postlethwait, J. H., Eisen, J. S., and Westerfield, M. (2001). Zebrafish smoothened functions in ventral neural tube specification and axon tract formation. Development **128**, 3497–3509.

Vitorino, M., Jusuf, P. R., Maurus, D., Kimura, Y., Higashijima, S., and Harris, W. A. (2009). Vsx2 in the zebrafish retina: Restricted lineages through derepression. Neural Dev. **4**, 14.

Wagle, M., Grunewald, B., Subburaju, S., Barzaghi, C., Le Guyader, S., Chan, J., and Jesuthasan, S. (2004). EphrinB2a in the zebrafish retinotectal system. J. Neurobiol. **59**, 57–65.

Wolff, C., Roy, S., Lewis, K. E., Schauerte, H., Joerg-Rauch, G., Kirn, A., Weiler, C., Geisler, R., Haffter, P., and Ingham, P. W. (2004). iguana encodes a novel zinc-finger protein with coiled-coil domains essential for Hedgehog signal transduction in the zebrafish embryo. Genes Dev. **18**, 1565–1576.

Woo, K., Shih, J., and Fraser, S. E. (1995). Fate maps of the zebrafish embryo. Curr. Opin. Genet. Dev. **5**, 439–443.

Xiao, T., and Baier, H. (2007). Lamina-specific axonal projections in the zebrafish tectum require the type IV collagen Dragnet. Nat. Neurosci. **10**, 1529–1537.

Xiao, T., Roeser, T., Staub, W., and Baier, H. (2005). A GFP-based genetic screen reveals mutations that disrupt the architecture of the zebrafish retinotectal projection. Development **132**, 2955–2967.

Yoshida, T., and Mishina, M. (2003). Neuron-specific gene manipulations to transparent zebrafish embryos. Methods Cell Sci. **25**, 15–23.

Zolessi, F. R., Poggi, L., Wilkinson, C. J., Chien, C. B., and Harris, W. A. (2006). Polarization and orientation of retinal ganglion cells in vivo. Neural Dev. **1**, 2.

CHAPTER 2

Imaging Blood Vessels in the Zebrafish

Makoto Kamei[*,‡], **Sumio Isogai**[†], **Weijun Pan**[*],
and **Brant M. Weinstein**[*]

[*]Program in Genomics of Differentiation, National Institute of Child Health and Human Development, Bethesda, Maryland

[†]Department of Anatomy, School of Medicine, Iwate Medical University, Morioka, Japan

[‡]Cell Biology of Diseases Group, Sansom Institute for Health Research, School of Pharmacy and Medical Science, University of South Australia, Adelaide, South Australia, Australia

Abstract
I. Introduction
II. Imaging Vascular Gene Expression
III. Non-vital Blood Vessel Imaging
  A. Micro-Dye and Micro-Resin Injection
  B. Alkaline Phosphatase Staining for 3 dpf Embryos
IV. Vital Imaging of Blood Vessels
  A. Microangiography
  B. Imaging Blood Vessels in Transgenic Zebrafish
V. Conclusion
References

Abstract

Understanding on the mechanisms of vascular branching morphogenesis has become a subject of enormous scientific and clinical interest. Zebrafish, which have small, accessible, transparent embryos and larvae, provides a unique living animal model to facilitating high-resolution imaging on ubiquitous and deep localization of vessels within embryo development and also in adult tissues. In this chapter, we have summarized various methods for vessel imaging in zebrafish, including in situ hybridization for vascular-specific genes, resin injection- or dye injection-based vessel visualization, and alkaline phosphatase staining. We also described detail protocols for live imaging of vessels by microangiography or using various transgenic zebrafish lines.

I. Introduction

The circulatory system is one of the first organ systems to begin functioning during vertebrate development, and its proper assembly is critical for embryonic survival. Blood vessels innervate all other tissues, supplying them with oxygen, nutrients, hormones, and cellular and humoral immune factors. The heart pumps blood through a complex network of blood vessels comprised of an inner single-cell thick endothelial epithelium surrounded by outer supporting pericyte or smooth muscle cells embedded in a fibrillar matrix. The mechanisms of blood vessel growth and morphogenesis are a subject of intensive investigation, and a large number of genes important for blood vessel formation have been identified in recent years. This has been achieved through developmental studies in mice and other animal models. However, our understanding of how these genes work together to orchestrate the proper assembly of the intricate system of blood vessels in the living animal remains limited, in part because of the challenging nature of these studies. The architecture and context of blood vessels are difficult to reproduce *in vitro*, and most developing blood vessels *in vivo* are relatively inaccessible to observation and experimental manipulation. Furthermore, since a properly functioning vasculature is required for embryonic survival and major defects lead to early death and embryonic resorption in amniotes, genetic analysis of blood vessel formation has been largely limited to reverse-genetic approaches.

The zebrafish provides a number of advantages for *in vivo* analysis of vascular development. As noted elsewhere in this book, zebrafish embryos are readily accessible to observation and experimental manipulation. Genetic and experimental tools and methods are available for functional manipulation of the entire organism, vascular tissues, or even single vascular- or non-vascular cells. Two features in particular make zebrafish especially useful for studying vascular development. First, developing zebrafish are very small—a 2 dpf embryo is just 2 mm long. Their embryos are so small, in fact, that the cells and tissues of the zebrafish receive enough oxygen by passive diffusion to survive and develop in a reasonably normal fashion for the first 3–4 days of development, even in the complete absence of blood circulation. This makes it fairly straightforward to assess the cardiovascular specificity of genetic or experimental defects that affect the circulation. Second, zebrafish embryos and early larvae are virtually transparent. The embryos of zebrafish (and many other teleosts) are telolecithic—yolk is sequestered in a single large cell separate from the embryo proper. The absence of obscuring yolk proteins gives embryos and larvae a high degree of optically clarity. Genetic variants deficient in pigment cells or pigment formation are even more transparent. This remarkable transparency is probably the most valuable feature of the fish for studying blood vessels, facilitating high-resolution imaging *in vivo*.

In this chapter we review some of the methods used to image and assess the pattern and function of the zebrafish vasculature, both in developing animals and in adults. First, we briefly touch on visualizing vascular gene expression (*in situ* hybridization, immunohistochemistry). In the next section we detail methods for imaging blood vessels in fixed developing and adult zebrafish specimens (resin and dye injection, alkaline phosphatase (AP) staining). In the final section we describe several methods

for imaging blood vessels in living animals (microangiography, time-lapse imaging of transgenic zebrafish with fluorescently tagged blood vessels). Collectively, these methods provide an unprecedented capability to image blood vessels in developing and adult animals.

II. Imaging Vascular Gene Expression

Experimental analysis of blood vessel formation during development requires the use of methods for visualizing the expression of particular genes within blood vessels and their progenitors. There are two general methods available to visualize endogenous gene expression within zebrafish embryos and larvae, *in situ* hybridization and immunohistochemistry. Neither of the methods is specific to the vasculature, and detailed protocols for these methods are available elsewhere (Hauptmann and Gerster, 1994; Westerfield, 2000). *In situ* hybridization is used routinely to assay the spatial and temporal patterns of vascular genes. A variety of different probes are available, some of which are listed in Table I. The *fli1a* and *scl* genes are early markers of vascular and hematopoietic lateral mesoderm. The expression of the *fli1a* becomes restricted to endothelial cells, a subset of circulating myeloid cells, and cranial neural crest derivatives (Brown *et al.*, 2000; Thompson *et al.*, 1998), while *scl* expression becomes restricted to the hematopoietic lineage at later stages (Gering *et al.*, 1998). The *tie2* and *cdh5* genes are zebrafish orthologs of angiopoietin-1 receptor and VE-Cadherin, respectively, and are expressed in a vascular-specific manner (Lyons *et al.*, 1998; Sumanas *et al.*, 2005). The *kdrl* and *flt4* genes (Fouquet *et al.*, 1997; Sumoy *et al.*, 1997; Thompson *et al.*, 1998) are zebrafish orthologs of mammalian endothelial-specific tyrosine kinase receptors for the important vascular signaling molecule

Table I
Common Marker Genes Used in Zebrafish Vasculature Research

Marker genes	Expression pattern	Reference
Fli1a	Pan-endothelial	Brown *et al.* (2000); Thompson *et al.* (1998)
tie2	Pan-endothelial	Lyons *et al.* (1998)
Kdrl (flk1)	Initially pan-endothelial, enriched in arteries at later stages	(Bussmann *et al.*, 2008); Sumoy *et al.* (1997)
cdh5	Pan-endothelial	Sumanas *et al.* (2005)
efnb2	Artery only	Lawson *et al.* (2001); Zhong *et al.* (2000)
Grl	Artery only	Zhong *et al.* (2000)
notch5	Artery only	Lawson *et al.* (2001)
dll4	Artery only	Herpers *et al.* (2008)
tbx20	Artery only	Szeto *et al.* (2002)
dab2	Vein only	Herpers *et al.* (2008)
ephb4	Vein only	Lawson *et al.* (2001)
flt4	Initially pan-endothelial. Later restricted to vein only.	Thompson *et al.* (1998)
prox1	Lymphatic vessel	Yaniv *et al.* (2006)
lyve1	Lymphatic vessel	Hogan *et al.* (2009)
Scl	Hematopoietic	Gering *et al.* (1998)

vascular endothelial growth factor (vegf). They are initially expressed in hemangio-genic lateral mesoderm then become restricted to angioblasts and endothelium. In the axial vessels of the trunk (dorsal aorta and posterior cardinal vein) *kdrl* becomes preferentially expressed in the aorta while *flt4* becomes preferentially expressed in the posterior cardinal vein (similar expression patterns of the corresponding orthologs are observed in mouse) (Kaipainen *et al.*, 1995; Thompson *et al.*, 1998). Other genes such as *efnb2, grl, Dll4, Tbx20,* and *notch5* are useful as markers of specification of arterial rather than venous endothelium, although all of these markers also exhibit substantial expression in non-vascular tissues, particularly the nervous system (Herpers *et al.*, 2008; Lawson *et al.*, 2001; Szeto *et al.*, 2002; Zhong *et al.*, 2000). The most commonly used venous marker is *ephb4,* although *flt4* has also been used to identify venous endothelium as noted above. *Disabled-2 (Dab2),* a cytosolic adaptor regulating endocytosis, also localizes very specifically to venous but not arterial endothelium in both *Xenopus* and *Zebrafish* embryos (Cheong *et al.*, 2006; Herpers *et al.*, 2008). The lymphatic endothelial markers Prox-1 and Lyve-1 are both conserved between mammals and fish, and both are useful zebrafish markers of lymphatic specification (Hogan *et al.*, 2009; Yaniv *et al.*, 2006). *Prox-1* functions as a key transcriptional factor for the differentiation of lymphatic endothelial cells. *Lyve-1* has been identified as a major receptor for HA (extracellular matrix glycosaminoglycan hyaluronan) on the lymph vessel wall, although its function in lymphangiogenesis is unclear.

III. Non-vital Blood Vessel Imaging

A number of methods are available for visualizing the pattern of blood vessels in fixed specimens. Micro-dye injection and micro-resin injection can be used to delineate the patent vasculature (lumenized or open blood vessels connected to the systemic circulation). Both of these methods rely on injection to fill blood vessels with dye or plastic resin that can be visualized in detail following the procedure. Dye injection methods are most useful in embryos and larvae up to a few weeks old. At later juvenile stages, and in adults, tissue opacity and thickness interfere with dye visualization in deeper vessels and resin injections can be more useful. Resin injections are difficult to perform on small specimens (such as embryos) but could be used to visualize vessels at almost any stage of development. While technically challenging, resin injection provides excellent visualization of the adult vasculature, since tissues surrounding the plastic resin are digested away and do not interfere with vessel observation. In addition to these two injection methods for lumenized vessels, staining for the endogenous AP activity of vascular endothelium can also be used to visualize vessels in fixed specimens. This method is useful for easy, rapid observation of vessel patterns, and does not require the vessel be patent, but it cannot be used effectively prior to approximately 3 days post-fertilization due to low signal and high background staining. Even at 3 dpf the method gives a relatively high background and is not particularly useful for visualizing cranial vessels. This method is also less useful at later stages due to increasing background. We describe the procedures for all of these methods in detail below.

A. Micro-Dye and Micro-Resin Injection

Since the 19th century, the dye injection method has been the most widely used tool for visualizing the developing circulatory system. Pioneering vascular embryologists such as Florence Sabin carried out their groundbreaking descriptive studies by injecting India ink into blood vessels of vertebrate embryos to reveal their patterns (for example, Evans, 1910; Sabin, 1917). In the 1970s the corrosive resin casting method, previously employed to visualize larger adult blood vessels, was combined with scanning electron microscopy to permit its use for visualizing vessels on a microscopic scale, such as in the developing renal vasculature (Murakami, 1972). Although micro-angiography and vascular-specific transgenic fish have now become the tools of choice in most cases for visualizing vessels in living zebrafish embryos (see below), these newer methods have limited usefulness in later stage larvae, juveniles, and adult fish. At these later stages the "classical" dye or resin injection methods still provide the best visualization of the majority of blood vessels (via direct injection into the dorsal aorta or caudal artery) or lymphatic vessels (via direct injection into the thoracic duct) (Yaniv *et al.*, 2006). The resin casting method involves injection of a plastic resin that is allowed to harden *in situ*, followed by etching away of tissues to leave behind only the plastic cast. The cast is rotary shadowed and visualized by scanning electron microscopy. The dye injection method described below involves injection of Berlin Blue dye followed by fixation and clearing of the embryos or larvae and whole-mount microscopic visualization.

1. Resin Injection Method

a. Materials
- Paraffin bed (see Fig. 2)
- Injection apparatus for circulating saline buffer (x2; see Fig. 4)
- Injection apparatus for fixative
- Injection apparatus for resin injection (one apparatus per sample to be injected)
- Physiological saline buffer suitable for bony fish
- 2% glutaraldehyde solution in saline buffer (Sigma cat# G6403, 50% solution in water)
- Methacrylate resin components

  Methyl methacrylate monomer (Aldrich cat# M55909 or Fluka cat# 03989)
  Ethyl methacrylate monomer (Aldrich cat# 234893 or Fluka cat# 65852)
  2-Hydroxypropyl methacrylate monomer (Aldrich cat# 268542 or Fluka cat# 17351)

b. Protocol

Preparation of the apparatus
1. The *paraffin bed* is made in a 9 cm glass petri dish by pouring molten paraffin wax (Fig. 1). While the wax is solidifying, tilt the dish approximately 15° to create a gentle slope. A depression is made in the middle of the bed for settling a fish.

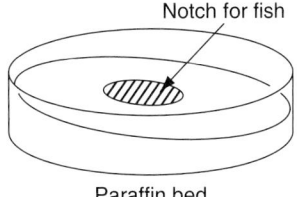

Fig. 1   Paraffin bed used for holding adult zebrafish.

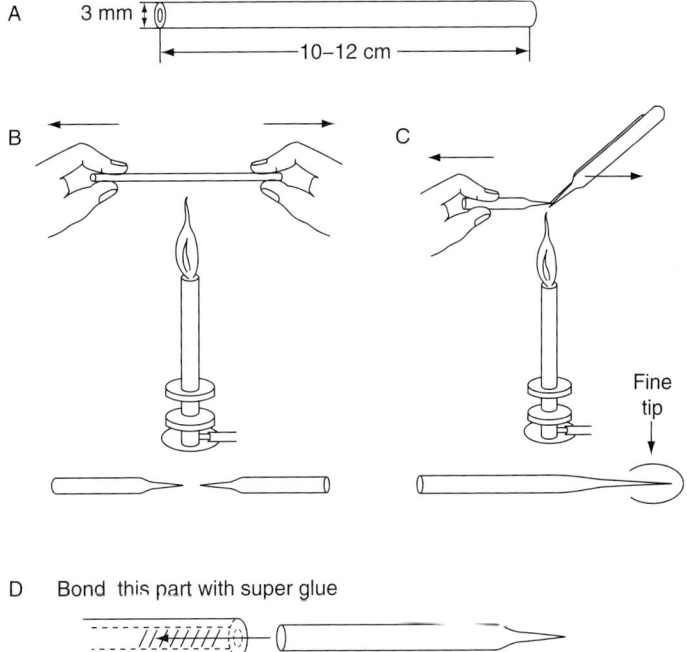

Fig. 2   Preparation of glass needles for injection of adult zebrafish. Glass stock (A) is pulled on a Bunsen burner (B), the tips are re-pulled (C), and then the needles are bonded to vinyl tubing with super glue (D).

2. The *glass needles* are made from stock glass tubing (3 mm outside diameter). The tubes are cut into 10–12 cm lengths (Fig. 2). The needles are pulled from the tube by heating the middle of the tube with a Bunsen burner. When the color of the glass tube is changed to red and the tubes feel soft, remove the tube from the heat, and pull on both ends. This should produce two injection needles with length of 5–6 cm. Let the needles cool down. Then holding the thick end of the needle by hand, and the sharp end of the needle by a pair of forceps, re-heat the sharp end on a Bunsen burner, and pull as before. By pulling the needles twice, it is possible to create

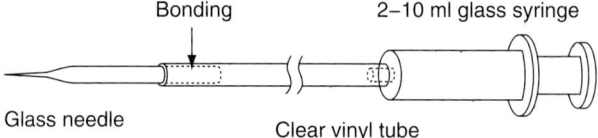

Fig. 3 Apparatus for injecting adult zebrafish with saline and fixative. For injecting resin, the glass syringe is replaced with a disposable plastic syringe.

needles with very fine points. The tip of the needles made in this manner is closed, and the tip needs to be broken open just before use.

3. The *apparatus for injecting physiological saline buffer* is made by attaching a glass needle (see step 2 above) to a clear vinyl tubing (3 mm inside diameter, 20 cm in length). When plugging the needle in, a small amount of superglue is applied to reinforce the attachment. For injection of the buffer a 2–10 ml glass syringe is attached as shown (Fig. 3).

4. The *apparatus for injecting fixative* is prepared as described in step 3 above.

5. The *apparatus for injecting resin* is prepared as described in step 3, except that a 10 ml disposable syringe is attached instead of a glass syringe and heat-resistant silicone tubing is used.

**** IMPORTANT CAUTION ON RESIN USE ****

As resin polymerizes, heat is generated and the viscosity of the resin increases. The heated vinyl tube can detach from the syringe suddenly, causing pressurized, viscous hot resin to splatter. The resin is harmful to skin and mucous membranes, and personal safety measures should always be taken when performing this procedure (e.g., use of goggles, face masks, gloves, and other protective clothing). In addition, make sure that the end of the vinyl tubing is securely attached to the syringe and use heat-resistant silicone tubing for the resin injection apparatus.

6. *Preparing resin (methyl methacrylate and ethyl methacrylate monomers).* Commercially available methacrylate monomers contain monomethyl ether hydroxyquinone to prevent polymerization, and it is necessary to remove this before use. Prepare 500 ml of 5% NaOH. Pour 100 ml methacrylate monomer and 50 ml 5% NaOH in a separating funnel and shake (Fig. 4A). Wait until the two solutions separate, and then remove the lower 5% NaOH layer (should be brown in color). Repeat until the NaOH solution remains clear, then remove the NaOH by extracting the methacrylate monomer (upper layer) with distilled water. Pour 100 ml of distilled water in the separating funnel containing the methacrylate monomer, shake, and wait until the two layers separate. Remove the lower distilled water layer. Repeat three to four times. Filter methacrylate monomer using double filter paper (Whatman) and incubate the monomer in a 150 ml air-tight container with sodium sulfite overnight (Fig. 4B). Place this container within a desiccator containing silica-gel, and store in a refrigerator at 4°C.

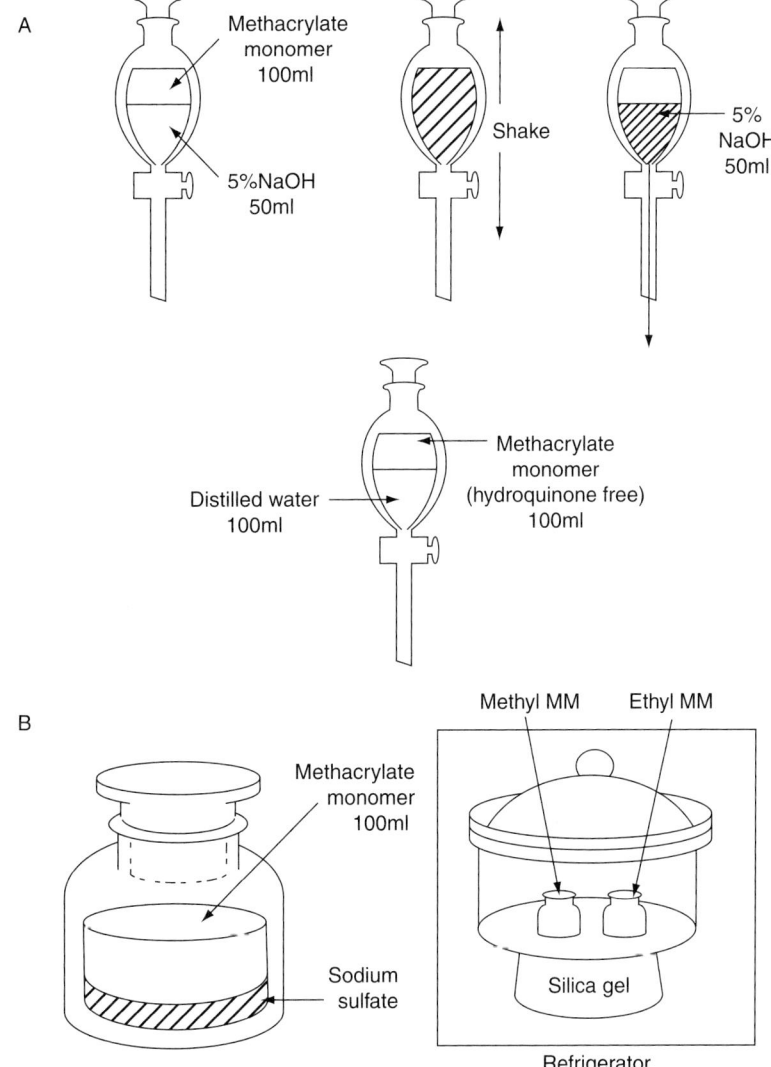

Fig. 4 Preparation of resin for injection. Commercial resin is supplied with monomethyl ether hydroxyquinone to prevent polymerization. This must be extracted before use (A) as described in the text. After extraction resin is stored refrigerated over sodium sulfate in a desiccator (B).

Experimental procedure

Steps 1, 2, and 4 are to be carried out under a dissecting microscope.

1. *Washing the circulatory system with saline buffer.* Place anesthetized adult zebrafish on the depressed part of the paraffin bed ventral side up. Use a pair of watchmaker's forceps and a pair of fine surgical or iridectomy scissors to remove the outer skin

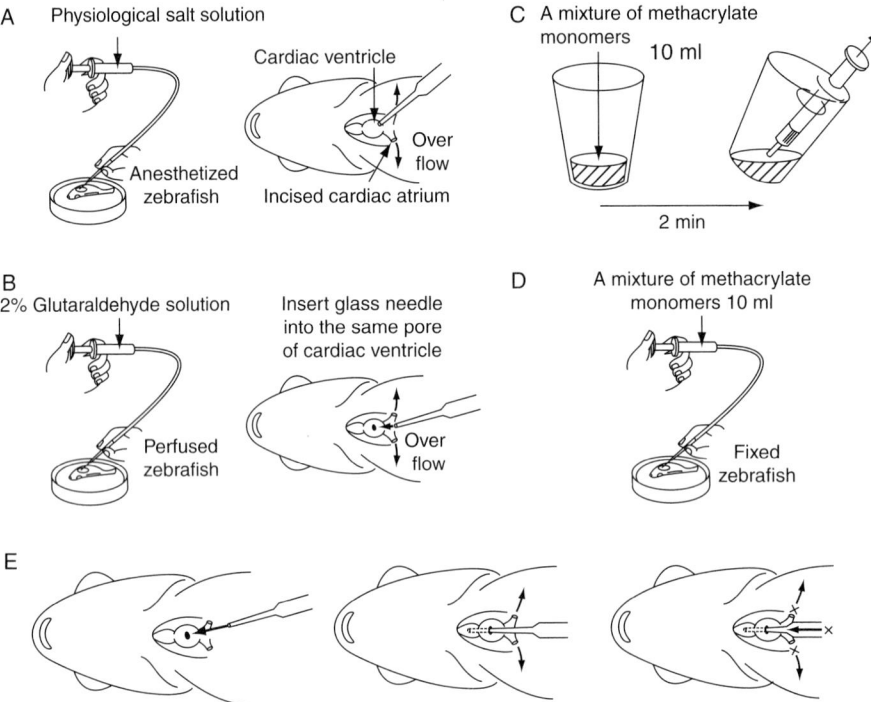

Fig. 5 Resin injection of adult zebrafish. Anesthetized animal is thoroughly flushed with physiological salt solution (A), then with glutaraldehyde fixative (B). Mixed resin is taken up into plastic syringe (C) which is attached to the rest of the resin injection apparatus and injection is performed (D). Injection should be stopped when resin hardens and flow ceases (E). See text for details.

and pericardial sac surrounding the heart. Use the forceps to sever the sinus venosus to allow blood to drain. Break the tip of the needle to the size of the ventricle and attached to a glass syringe containing saline buffer. Stab the glass needle into the ventricle in the direction of the head, and apply pressure on the syringe to flush the circulatory system with buffer (Fig. 5A). Do not stop until the system is very well flushed out; flushing should be continued well after the flow from the sinus venosus has become clear saline.

2. *Fixation with 2% glutaraldehyde.* Break a glass needle as above, attach to a syringe containing the 2% glutaraldehyde solution, and start circulating the fixative in the same manner as for the saline buffer (Fig. 5B). Fix well (overnight preferable).

3. *Mixing the resin.* Mix together 3 ml methyl methacrylate monomer, 1.75 ml ethyl methacrylate monomer, and 5.25 ml 2-hydroxypropyl methacrylate monomer (to make 10 ml final volume) in a disposable plastic 100 ml cup (Fig. 5C). To this mixture, add 0.15 g benzoyl peroxide (catalyst), 0.15 ml *N,N*-dimethylaniline (polymerization agent), and Sudan III (dye), then mix well and sonicate for 2 min.

4. *Resin injection*. Immediately after sonication, aspirate the resin mixture into a 10 ml disposable plastic syringe, and attach a glass needle. This time break the needle so that it will have a slightly larger bore. Make sure the vinyl tubing is attached to the end of the syringe firmly. Push the needle in the direction of the head of the fish through the same hole used for washing and fixing (Fig. 5D). If possible, push the tip of the needle all the way into the arterial cone (back of the ventricle) before beginning the injection. At first the injection will be easy since the viscosity is low, but after 3–5 min, the viscosity will increase, and the resin will start to heat up. Keep pushing down on the syringe. After 10–12 min, the resin hardens sufficiently for resin flow to stop, at which point the injection should be stopped (Fig. 5E). Wait about 10 min more for resin to fully harden.

5. *Digestion of tissue*. The injected adult zebrafish is digested with 10–20% KOH for a few days, and gently washed with distilled water to remove tissue. The resin cast is dissected out using watchmaker's forceps and small scissors. Sometimes sonication is used to remove bones and hard-to-remove tissues, but care is required not to destroy the cast itself. For SEM (scanning electron microscopy) observation, each local vascular system is divided and trimmed. These procedures must be performed under water using dissecting microscope. Each block is frozen in distilled water then freeze-dried.

6. *Scanning electron microscopy*. The dried block is mounted on a metal stub and coated with osmium or platinum. Observations are performed with a scanning electron microscope using an acceleration voltage of 5–10 kV.

2. Dye Injection Method

a. Materials
- Berlin blue dye solution
- Injection apparatus for embryos and early larvae (as described for microangiography, below) *OR*
- Injection apparatus for juvenile and adult fish as described for resin injection, above)
- Paraffin bed (for juvenile and adult injections—see resin injection above)

b. Protocol

Dye injection of embryos and early larvae

1. Prepare Berlin blue dye solution by adding 0.5–0.75 g Berlin blue powder (Aldrich cat# 234125, Prussian blue) to 100 ml distilled water and dissolving thoroughly. Filter solution through double layers of Whatman 3MM filter paper and store in an air-tight bottle.

2. Glass microneedles are prepared as for microangiography (see below).

3. Agarose embed embryos/larvae as follows (Fig. 6A): Immobilize embryo in 1X tricaine methanesulfonate (MS-222) in embryo media (Westerfield, 2000). Place embryo on slide in a drop of tricaine embryo media and them remove as much of liquid as possible with a pipette. Place a single drop of 1% molten low-melting

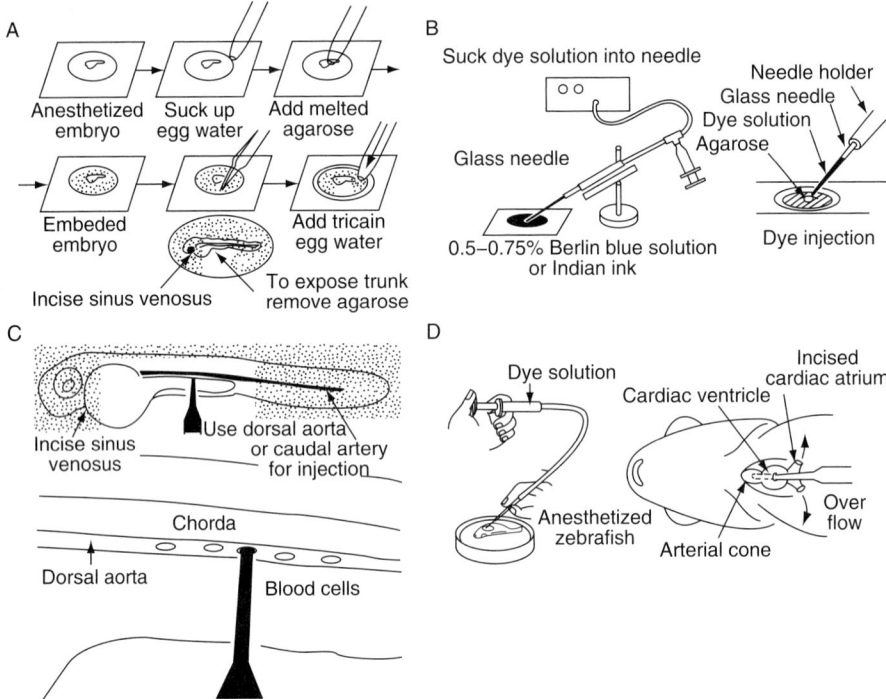

Fig. 6 Mounting and dye injection of developing and adult zebrafish. Embryos and larvae are embedded in agarose on a glass slide for dye injection (A). Injection is performed into the dorsal aorta or caudal artery, with the sinus venosus incised to permit dye to flow through the vasculature (B). The apparatus used for dye injection is shown in (C). For dye injection of adult zebrafish, an apparatus similar to that used for saline and fixative injection prior to resin injection is employed (D). See text for further details.

temperature agarose on the embryo, allowing it to harden and embed the embryo. Attempt to orient the embryo before the agarose hardens such that either its left or its right side is facing up.

4. In order to allow blood drain, sever the sinus venosus using the watchmaker's forceps.

5. Remove the agarose covering the caudal half of the trunk or cranial half of the tail with the forceps, and then add a drop of 1X tricaine.

6. Break the fine tip of needle to the size of the dorsal aorta or the caudal artery, and, from the needle tip, aspirate enough Berlin blue solution to cannulate all the vessels thoroughly (Fig. 6B).

7. To pierce precisely the dorsal aorta or the caudal artery with the fine tip, the point just beneath the notochord must be targeted. To make sure the tip is in the correct position with respect to these vessels, inject the dye for 0.1 s using a picopump. The blood cells move according to the pumping if the tip is in the right position. Inject the dye for 1–2 s, and continue the procedure until all the vessels are cannulated

thoroughly (Fig. 6C). For lymphangiography, attempt to pierce the thoracic duct, which lies ventral to the dorsal aorta above the posterior cardinal vein, instead of the dorsal aorta. This vessel is more difficult to visualize and more challenging to inject than the dorsal aorta, so repeated attempts and patience will likely be needed. One can distinguish when the lymphatic rather than the blood vascular system has been injected because the dye will fill the thoracic duct but will not be visible or highly diluted in blood vessels such as the cardinal vein and dorsal aorta.

8. Fix the dye injected embryos.

Dye injection of juvenile and adult zebrafish

Note: This method is similar to the resin injection method described above.

1. The injection of dye is carried out under a dissecting microscope. Place an anesthetized adult zebrafish on the depressed part of the paraffin bed ventral side up.

2. Use a pair of watchmaker's forceps and a pair of fine surgical or iridectomy scissors to remove the outer skin and pericardial sac surrounding the heart. Use the forceps to sever the sinus venosus to allow blood to drain during injection (Fig. 6D).

3. Break the tip of the needle to the size of the ventricle and attach to a glass syringe containing 0.5~0.7% Berlin blue solution buffer (apparatus is very similar to that used for saline injection in the resin injection method described above). Stab the glass needle into the ventricle in the direction of the head, and apply pressure on the syringe to flush the circulatory system with buffer. Continue injection well after dye begins to flow from sinus venosus; make sure that the dye is thoroughly injected.

4. Fix the dye injected sample immediately in either 4% paraformaldehyde or 10% neutral formaldehyde; store in fixative.

5. For observation lightly wash the samples and then clear them by passing through 50%, 70%, 80%, 90%, and then 100% glycerol solution (1 solution change per day for a total of 5 days). Image samples under a dissecting microscope, dissecting away tissues as needed for observation of deeper vessels.

B. Alkaline Phosphatase Staining for 3 dpf Embryos

Zebrafish blood vessels possess endogenous AP activity. Endogenous AP activity is not detectable in 24 hpf embryos, but is weakly detectable by 48 hpf and strongly at 72 hpf. Staining vessels by endogenous AP activity is useful for easy and rapid visualization of the vasculature in many specimens, but provides less resolution than many of the other methods. We use a protocol modified from Childs *et al.* (2002)

a. Materials

• Fixation Buffer: 10 ml 4% paraformaldehyde + 1 ml 10% Triton-X100; makes 11 ml, scale up or down as needed.

• Rinse Buffer: 10 ml 10X PBS + 5 ml 10% Triton-X100 + 1 ml normal horse serum + 84 ml distilled water; makes 100 ml, scale up or down as needed.

- Staining Buffer: 1 ml 5M NaCl + 2.5 ml 1M MgCl2 + 5 ml 1M Tris pH 9.0 – 9.5 + 500 μl 10% Tween-20 + 41 ml distilled water; makes 50 ml, scale up or down as needed.
- Staining Solution: 10 ml staining buffer + 45 μl NBT + 35 μl BCIP; scale up or down as needed.
- NBT 4-Nitro Blue Tetrazolium (Boehringer-Mannheim Cat# 1-383-213), 100 mg/ml in 70% dimethylformamide.
- BCIP X-Phosphate or 5-Bromo-4-Chloro-3-indolyl-phosphate (Boehringer-Mannheim Cat# 1-383-221), 50 mg/ml in dimethylformamide.

b. Protocol

1. Fix embryos at room temperature for 1 h in fixation buffer.
2. Rinse 1× in rinse buffer.
3. Wash 5× 10 min at room temperature in rinse buffer **or** leave washing in rinse buffer at 4°C for up to several days. If doing the latter, wash again at RT for 10 min before going on to the next step.
4. Wash 2× 5 min in staining buffer.
5. Stain in 1 ml of staining solution. Color development takes about 5–30 min.
6. To stop reaction, wash 3X in rinse buffer without horse serum, then fix in 4% paraformaldehyde for 30 min, and store in fixative at 4°C.

c. Important Notes

1. Avoid putting the embryos in methanol (this destroys endogenous AP activity). If embryos been placed in methanol, some AP activity can be reconstituted by washing embryos in PBT overnight or even over a weekend before starting the staining procedure, although staining will be weaker than in non-methanol-exposed embryos.
2. If combined AP staining and antibody staining (e.g., anti-EGFP) staining is desired, antibody staining should be done first. After the DAB (3,3′-diaminobenzidine) staining, do a quick post-fix and go straight into the washes for the AP protocol. AP staining works very well after an antibody stain. Alternatively, stain for exogenous AP phosphatase at a time point when the endogenous vascular form is not active (24 hpf).

IV. Vital Imaging of Blood Vessels

While the methods described above are useful for visualizing vascular patterns in fixed zebrafish specimens, particularly at later developmental stages and in adults, the zebrafish is perhaps best known for its accessibility to vital imaging methods. A number of vascular imaging methods are available that take advantage of the optical clarity and experimental accessibility of zebrafish embryos and larvae. Confocal microangiography can be used for imaging blood vessels with active circulation and to detect defects in their patterning and/or function. Confocal microangiography is

Table II
Zebrafish Transgenic Lines for Time–Lapse Vascular Imaging

Transgenic zebrafish line	Reference
$Tg(fli1a:EGFP)^{y1}$	Lawson and Weinstein (2002)
$Tg(flii1a:nEGFP)^{y7}$	Yaniv *et al.* (2006)
$Tg(flii1a:EGFP–cdc42wt)^{y48}$	Kamei *et al.* (2006)
$Tg(mTie2:GFP)$	Motoike *et al.* (2000)
$Tg(kdrl:G-RCFP)$	Cross *et al.* (2003)
$Tg(kdrl:memCherry)^{s896}$	Chi *et al.* (2008)
$Tg(kdrl:EGFP)^{s843}$	Jin *et al.* (2005)
$Tg(fli1a: DsRed)$	Geudens *et al.* (2010)
$Tg(fli1a:EGFP; kdrl:ras-cherry)$	Hogan *et al.* (2009)
$Tg(flt1:YFP, kdrl:mCherryRed)$	Hogan *et al.* (2009)
$Tg(stabilin:YFP)^{hu4453}$	Hogan *et al.* (2009)
$Tg(gata1:DsRed)^{sd2}$	Traver *et al.* (2003)

performed by injecting fluorescent microspheres into the circulation of living embryos, and then collecting 3-dimensional image "stacks" of the fluorescently labeled vasculature in the living animal using a confocal or (preferably) multiphoton microscope. This method can be used from the initiation of circulation at approximately 1 dpf out to 10 dpf or even older larvae, although injections become progressively more technically challenging to perform after about 2 dpf. Increasing tissue depth makes high-resolution imaging of deep vessels increasingly difficult at later stages. Repeated microangiographic imaging of the same animal and microangiography on animals with impaired circulation may be difficult or impossible to perform.

Numerous transgenic zebrafish lines have been generated with endogenous fluorescent labeling of blood or blood and lymphatic vessels (Table II). These animals facilitate high-resolution imaging of the vasculature *in vivo* and make possible long-term time-lapse imaging of the dynamics of vessel growth and remodeling. Unlike dye or resin injection methods, transgenic animals can be repeatedly re-imaged over an extended period of time with continued normal development of the imaged vessels, particularly when multiphoton imaging is employed.

Zebrafish *fli1a:EGFP* (Lawson and Weinstein, 2002), *kdrl:EGFP* (Cross *et al.*, 2003) and other transgenic zebrafish lines have already been widely used in studies of vasculogenesis and angiogenesis in the zebrafish. Since these lines permit imaging of the endothelial cells themselves, rather than vessel lumens, they can be used to image vessels that are not carrying circulation, cords of endothelial cells lacking a vascular lumens, or even isolated migrating angioblasts. The *fli1a:EGFP* line permits visualization of both blood and lymphatic vessels, while only blood endothelial cells are marked in the *kdrl:EGFP* line. The *fli1a:EGFP* line has already been used in a very large variety of studies, including examination of the mechanisms of cranial (Lawson and Weinstein, 2002) and trunk (Isogai *et al.*, 2003) blood vessel formation, lymphatic vessel formation (Yaniv *et al.*, 2006), and tumorigenesis (Stoletov *et al.*, 2007). The *fli1a* and *kdrl* promoters have also both been used to generate additional lines expressing other fluorescent proteins. Transgenic lines expressing nuclear-targeted EGFP

(nEGFP) or membrane-targeted EGFP (EGFP-cdc42wt) have been derived. Multiphoton time-lapse imaging using $Tg(fli1a:nEGFP)^{y7}$ transgenic embryos has been used to trace the migration and lineage of individual endothelial cells and to quantify their proliferation (Yaniv et al., 2006). The $Tg(fli1a:EGFP–cdc42wt)^{y48}$ line has been used to visualize the dynamics of endothelial vacuoles and their contribution to vascular lumens formation (Kamei et al., 2006). By combining microangiographic imaging of functional vascular lumens using red quantum dots, and imaging of forming lumenal compartments using the $Tg(fli1a:EGFP–cdc42wt)^{y48}$, the details of progressive formation of vascular lumens (via formation and intracellular/intercellular fusion of endothelial vacuoles) and their connection to the rest of the vascular circulation could be followed dynamically in vivo for the first time (Kamei et al., 2006).

More recently, transgenic lines have been established with a variety of fluorescent proteins expressed under the control of a number of different endothelial promoters (Table II). Taking advantage of differences in the preferential labeling of different vascular promoters, double transgenic animals were generated for the specific observation on blood/lymph vessels, like $Tg(fli1a:EGFP; kdrl:ras-cherry)$, or artery/venous sprouts $Tg(flt1:YFP; kdrl:mCherryRed)$ (Hogan et al., 2009).

In addition to endothelial promoter-driven transgenic lines, hematopoietic-specific transgenic lines such as $Tg(gata1:DsRed)^{sd2}$ have proven useful for applications in vascular analysis. Motion-based imaging of blood flow has been used to characterize lumens formation, distinguish lymphatic vessels from blood vessels, and examine which vessels are patent and carrying blood flow in ($Tg(fli1a:EGFP; gata1:DsRed$) double transgenic animals (Tong et al., 2009; Yaniv et al., 2006).

Methods for confocal microangiography and time-lapse multiphoton imaging of transgenic animals are described in more detail below. Finally, we describe some of the novel insights into in vivo vessel formation processes that have already been obtained through use of time-lapse imaging methods.

A. Microangiography

Confocal microangiography is useful for visualizing and assessing the patent circulatory system; blood vessels that are actually carrying blood flow or that at least have open lumens connected to the functioning vasculature. The method facilitates detailed study of the pattern, function, and integrity (leakiness) of vessels. The method is relatively easy to perform, particularly on younger animals, and does not require that the animal be of any particular genotype (although animals with impaired circulation may be difficult or impossible to infuse with fluorescent microspheres).

a. Materials

- 0.02–0.04 μm fluoresceinated carboxylated latex beads, available from Invitrogen. The yellow-green (cat # F8787), red-orange (cat # F8794), or dark red (cat # F8783) beads are suitable for confocal imaging using the laser lines on standard Krypton–Argon laser confocal microscopes. Other colors may be used for when

multiphoton imaging is employed. Quantum dots (QD) are fluorescent semiconductors that are especially suited for multiphoton imaging with their high quantum yield, as well as the fact that the same excitation wavelength could be used for obtaining multiple different colors depending on the QD used. For microangiography, PEG-coated non-targeted QDs are available from Invitrogen. Qtracker 565, 655, 705, and 800 (cat# Q21031MP, Q21021MP, Q21061MP, and Q21071MP, respectively) are suitable for multiphoton confocal imaging. The fluorescent bead suspension as supplied is diluted 1:1 with 2% BSA (Sigma) in deionized distilled water, sonicated approximately 25 cycles of 1" each at maximum power on a Branson sonifier equipped with a microprobe, and subjected to centrifugation for 2 min at top speed in an Eppendorf microcentrifuge. The quantum dots in supernatant are used as supplied.

- 1 mm OD glass capillaries (World Precision Instruments, Cat # TW100-4 without filament or TW100F-4 with an internal filament) for preparing holding and microinjection pipettes.
- 2 Coarse micromanipulators with magnetic holders and base plates.
- 30% Danieu's solution (1X Danieu's: 58 mM NaCl, 0.7 mM KCl, 0.4 mM MgSO4, 0.6 mM Ca(NO3)2, 5 mM Hepes, pH 7.6).
- Holding and microinjection pipettes.
- 6 cm culture dish (Falcon).
- Micromanipulator and microinjection apparatus.
- Dissecting microscope equipped with epifluorescence optics.

b. Protocol

Preparation of the apparatus

1. The *glass microinjection needles* are prepared from 1 mm capillaries with internal filaments using a Kopf vertical pipette puller (approximate settings: heat $= 12$, solenoid $= 4.5$; see Fig. 7A for desired shape of microneedle). Needles are broken open with a razor blade just behind their tip to give an opening of approximately 5–10 μm in width.
2. The *holding pipettes* are prepared from 1 mm capillaries WITHOUT filaments by partially melting one end of the capillary with a Bunsen burner, such that the opening is narrowed to approximately 0.2 mm (slightly smaller for younger embryos, slightly larger for older larvae). A photographic image of the end of the tip of a holding pipette is shown in Fig. 7B.
3. The *apparatus for microinjection* is made by attaching a glass microinjection needle (step 1) to a pipette holder (World Precision Inst Cat # MPH6912; adapter for holder and tubing to attach to picopump, WPI Cat # 5430). The pipette holder is attached to a controlled air pressure station such as World Precision Instruments Pneumatic Picopump (catalog # PV820).
4. The *apparatus for holding embryos* is made by attaching a glass holding pipette (step 2) to a pipette holder (World Precision Inst Cat # MPH6912). The holding pipettes and their holders are attached via mineral-oil filled tubing (Stoelting

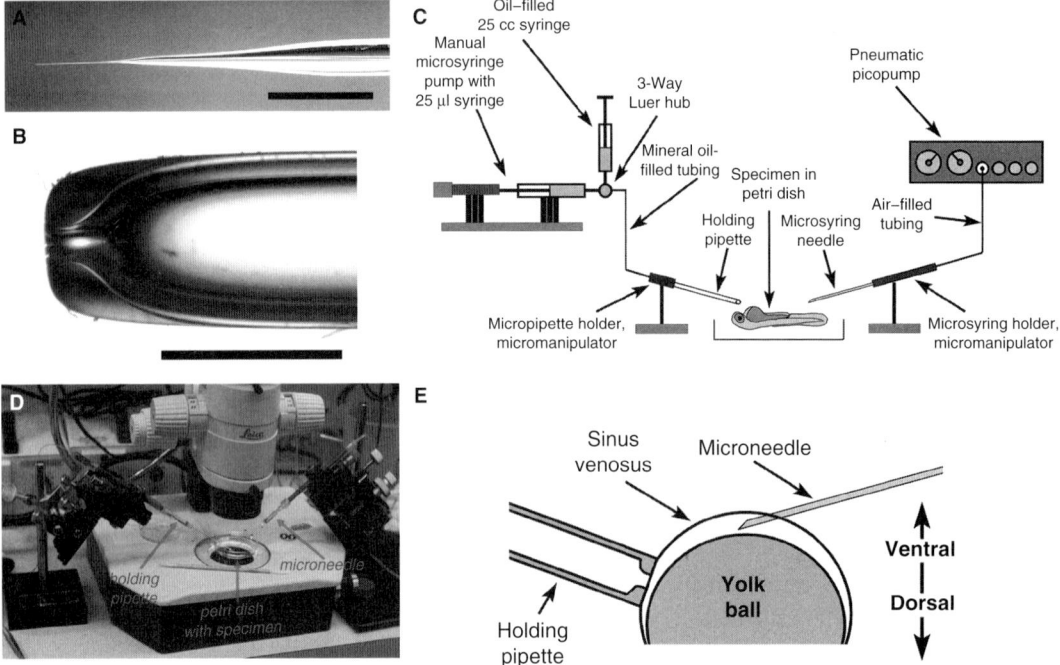

Fig. 7 Microangiography of developing zebrafish embryos and larvae. The desired configurations for injection needles (A) and holding pipettes (B) are shown. A schematic diagram of the apparatus used is shown in (C) and a photographic image of an actual setup is shown in (D). For injection, an embryo is held ventral side up with suction applied through the holding pipette and injected obliquely through the sinus venosus (E). Older larvae are injected by direct intracardiac injection. See text for details. Scale bars are 3 mm (A) and 1 mm (B).

Instruments, Clay-Adams cat # 427415) to a manual microsyringe pump (Stoelting Instruments cat # 51222, with 25 µl syringe).

5. Holding pipettes and microneedles and their associated holders and other equipment are arranged on either side of a stereo dissecting microscope as diagrammed in Fig. 7C. A photographic image of a typical arrangement is shown in Fig. 7D.

Experimental procedure

1. Embryos are collected and incubated to the desired stage of development. Use of albino mutant lines or PTU (1-phenyl-2-thiourea) treatment improves visualization of many vascular beds at later stages (see Westerfield 2000 for PTU treatment protocol).

2. A few microliters of fluorescent microsphere suspension are used to backfill a glass microneedle for injection. The tip should be broken off to the desired diameter just before use.

3. Embryos are dechorionated and anesthetized with tricaine in embryo media.

4. 1- to 3-day-old embryos and larvae are held ventral side up for injection using a holding pipette applied to the side of the yolk ball (Fig. 7E), with suction applied via a microsyringe driver. Care should be taken to not allow the holding pipette to rupture the yolk ball. 4- to 7–day-old larvae are held ventral side up for injection by embedding in 0.5% low-melting temperature agarose.

5. For 1- to 3-day-old embryos a broken glass microneedle is inserted obliquely into the sinus venosus (as diagrammed in Fig. 7E). For 4- to 7-day-old larvae a broken glass microneedle is inserted through the pericardium directly into the ventricle.

   For lymphangiographic labeling of the lymphatic vasculature injections are performed directly into the thoracic duct, located between the dorsal aorta and the cardinal vein. Lymphangiographic injections are generally significantly more difficult than angiographic injections of comparable stage embryos.

   Labeling of the lymphatic vasculature in living animals can also be accomplished by subcutaneous or intramuscular injections after insertion of the broken glass microneedle through the tough outer periderm of the developing animal. The dyes or microspheres are preferentially taken up by and drained through the lymphatic vessels. Injections should be performed at a site relatively distant to (preferably caudal to) the area to be imaged and the animal monitored after injection to choose the optimal time for imaging. This is when the dye has filled the lymphatic vessels, but has not yet diffused far from the site of injection through other tissues (which creates a high background labeling outside the lymphatics). Subcutaneous dye injection labeling of the lymphatics is most effective in older larvae (2 weeks +) in which the lymphatic vasculature is well developed.

6. Following microneedle insertion, many (20+) small boluses of bead suspension are delivered over the course of up to a minute. Smaller numbers of overly large boluses can cause temporary or permanent cardiac arrest. The epifluorescence attachment on the dissecting microscope can be used to monitor the success of the injection.

7. Embryos or larvae are allowed to recover from injection briefly (approximately 1 min) in tricaine-free embryo media, then rapidly mounted in 5% methylcellulose (Sigma) or low-melt agarose (both made up in embryo media with tricaine). For short-term imaging (generally one stack of images) methylcellulose is applied to the bottom of a thick depression well slide. The rest of the well is carefully filled with 30% Danieu's solution containing $1\times$ Tricaine, trying not to disturb the methylcellulose layer below. The injected zebrafish embryo is placed in the well, moved on top of the methylcellulose, and then gently pushed into the methylcellulose in the desired orientation to fully immobilize. Methylcellulose is only useful for short-term mounting because the embryo gradually sinks in the methylcellulose (which also loses viscosity by absorbing additional water over time). For longer term or repeated imaging animals can be mounted in agarose, using methods such as that described below.

8. Injected, mounted animals are imaged on a confocal or multiphoton microscope using the appropriate laser lines/wavelengths. Although the fluorescent beads are initially distributed uniformly throughout the vasculature of the embryo, within minutes they

begin to be phagocytosed by and concentrated in selected cells lining the vessels (cf. "tail reticular cells" in Westerfield 2000). Because of this, specimens must be imaged as rapidly as possible, generally within 15 min after injection. Generally between 20 and 50 frame-averaged (5 frames) optical sections are collected with a spacing of 2–5 μm between sections, depending on the magnification (smaller spacing at higher magnifications). Three-dimensional reconstructions can be generated using a variety of commercial packages (see below).

B. Imaging Blood Vessels in Transgenic Zebrafish

Confocal microangiography is a valuable tool for imaging developing blood vessels, but it has limitations. The method is well suited for delineating the luminal spaces of functional blood vessels, but those that lack circulation, vessels that have not yet formed open lumens, and isolated endothelial progenitor cells are essentially invisible. Much of the "action" of early blood vessel formation occurs prior to the initiation of circulation through the relevant vessels. The first, major axial vessels of the zebrafish trunk, the dorsal aorta and cardinal vein, coalesce as defined cords of cells at the trunk midline with distinct molecular arterial-venous identities many hours before they actually begin to carry circulation. Later-developing vessels generally form by sprouting and migration of strings of endothelial cells or even individual cells that are likewise undetectable by angiography until well after their initial growth has been completed. Furthermore, because of leakage of low molecular weight dyes, or pinocytotic clearance of microspheres, injected animals can only be imaged for a short time (up to 1/2 h) after injection and repeated imaging requires re-injection of microspheres with different excitation and emission spectra. Thus, for most practical purposes dynamic imaging of blood vessel growth using this method is not possible. What is needed is a specific and durable fluorescent "tag" for endothelial cells and their angioblast progenitors.

As already described, autofluorescent proteins such as GFP have been used to mark a variety of tissues in transgenic zebrafish embryos and larvae. Methods for generating germline transgenic zebrafish are now widely used and their application and resulting lines have been thoroughly reviewed elsewhere (Lin, 2000; Udvadia and Linney, 2003). Tissue-specific expression of fluorescent (or other) proteins in germline transgenic animals is achieved through the use of tissue-specific promoters, and a number of different promoters have been used to drive fluorescent protein expression in zebrafish vascular endothelium. Murine *Tie2* (a tyrosine kinase receptor expressed in endothelium activated by angiopoietin ligands) promoter constructs drive GFP expression in endothelial cells in mice and zebrafish, and stable germline transgenic lines have been prepared in both species (Motoike *et al.*, 2000). The usefulness of mTie2-GFP has been limited in fish by the fact that germline transgenic zebrafish show substantial non-vascular expression of GFP in the hindbrain and more posterior neural tube. The overall level of expression from the murine promoter is also relatively low in fish compared to mice. Germline transgenic zebrafish have also been generated expressing EGFP (enhanced green fluorescent protein) in the

vasculature under the control of the zebrafish *fli1a* promoter (Lawson and Weinstein, 2002). The *fli1a* is a transcription factor expressed in the presumptive hemangioblast lineage, and later restricted to vascular endothelium, cranial neural crest derivatives, and a small subset of myeloid derivatives (Brown *et al.*, 2000). These lines express abundant EGFP in the vasculature, faithfully recapitulating the expression pattern of the endogenous *fli1a* gene, and permitting resolution of very fine cellular features of vascular endothelial cells *in vivo*. The *fli1a:EGFP* transgenic lines have become the most widely used resource for transgenic visualization of blood vessels and have already been used in a variety of different published studies to examining developing trunk and cranial vessels (Isogai *et al.*, 2003; Lawson and Weinstein, 2002; Roman *et al.*, 2002) and regenerating vessels in the adult fin (Huang *et al.*, 2003). Most recently, transgenic zebrafish with fluorescently labeled blood vessels have also been generated by using the promoter for the *kdrl* receptor tyrosine kinase to drive EGFP expression in endothelium (Cross *et al.*, 2003).

Here, we review methods for exploiting what is perhaps the most important feature of these transgenics, that they permit repeated and continuous imaging of the fluorescently labeled blood vessels. This has made it possible, for the first time, to image the dynamics of blood vessel growth and development of vascular networks in living animals. We describe methods for mounting embryos and larvae for long-term observation and for time-lapse multiphoton microscopy of blood vessels within these animals. The mounting of animals for time-lapse imaging is much more difficult and in some ways more critical to the success of the experiment than the actual imaging, which is relatively straightforward to set up on most imaging systems.

1. Long-Term Mounting for Time-Lapse Imaging

For time-lapse imaging of blood vessels in transgenic fish over the course of hours or even days, the animals must be carefully mounted in a way that maintains the region of the animal being imaged in a relatively fixed position, yet keeps the animal alive and developing normally throughout the course of the experiment. This task is complicated further by the fact that developing zebrafish are continuously growing and undergoing morphogenetic movements and this must be accommodated in whatever scheme is used to hold them in place. We describe below a relatively simple mounting method that is adaptable to imaging different areas of embryos or larvae and holds them in place over the course of hours. For time-lapse experiments that run up to a day or more imaging chambers with buffer circulation are employed.

a. Materials
- Imaging chambers (see below)
- 2% low-melting temperature agarose made up in 30% Danieu's solution containing 1X PTU (if non-albino animals are used) and 1.25X Tricaine (if non-paralyzed animals are used)

- 30% Danieu's solution with or without 1× PTU and 1.25× Tricaine (see above)
- Fine forceps (Dumont #55)

b. Method

Preparation of imaging chambers

1. Imaging vessels are prepared from 6 cm polystyrene culture dishes (Falcon 3002) and 14 ml polypropylene tubes (Falcon 2059). Model cement is also required for assembly.
2. The polypropylene tube is sliced into 5 mm segments (rings) using a heated razor blade. One ring is glued to the bottom plate of the culture dish using model cement. Care should be taken to glue the slice of the polypropylene tube to the center of the dish and to avoid smearing the glue inside the polypropylene ring (to avoid obscuring the optical clarity). See Fig. 8A.
3. The glue should be allowed to dry overnight before use.
4. Just before use the polypropylene ring in the imaging chamber should be slightly overfilled with the low-melting temperature agarose to make a slightly convex dome (Fig. 8A).
5. Time-lapse imaging for longer duration (more than 1 day) requires a modified version of this chamber. Instead of using a 6 cm polystyrene culture dish, a T-25 tissue culture flask (Nalge Nunc cat# 163371) is used. A 1–1/2" square is cut from the upper side of the flask with a coping saw. The opening should stretch from one side of the flask to the other, and be large enough to accommodate the objective lens to be used in the observation. Portions of the adjoining walls of the flask should also be cut down as shown in (Fig. 8B) in order to allow the objective lens to enter the imaging chamber. A second opening is made in the bottom of the flask opposite the cap by drilling a 3/32" hole using an electric drill. This second opening is to be used for the inflow port, and it should be large enough to accommodate the female Luer bulkhead (Small Parts, Inc., 3/32" barb, cat# LCN-FB-093-25). A silicone O-ring (Small Parts, Inc., 1/4" ID × 3/8" OD × 1/16" wide, cat# ORS-010-25) is inserted over the bulkhead before its attachment to the flask, and then secured onto the flask with a locking nut (Small Parts, Inc., cat# LCN-LN0-25). A 5 mm-wide slice of a 14 ml Falcon tube is glued to the center of the chamber opposite the large opening using Z-Poxy resin (Small Parts, Inc., cat# EPX-PT38). Baffles are made from a lid of a 14 ml Falcon tube, and also glued on inside of the chamber, one over the inlet port and another one before the outlet. Another small hole on the top of the chamber was made by drilling to allow the insertion of a temperature probe into the chamber. The cap of the flask needs to be modified so that the Luer bulkhead can be attached. The center portion of the lid is excised and discarded. The Luer female bulkhead is then screwed tightly onto the lid with an O-ring. To prevent flooding of the stage should accidental overflow of the chamber occur a safety drain is installed into the side of the chamber ("Gutter" in Fig. 8B). The safety drain is constructed from a plastic drinking straw connected to a cutoff disposable

pipette. The bulb and first 1" of the pipette are removed. A quarter of the wall of the main pipette body was also removed to form a catchment opening for chamber overflow. This piece is attached to the side of the chamber at an angle of 25–30° using Z-Poxy resin. After drying, a plastic drinking straw is connected to the drain to lead the overflow to a container. The completed chamber is air-dried for overnight to allow all resin to cure, and then tested for any leakage by filling it with distilled water before use.

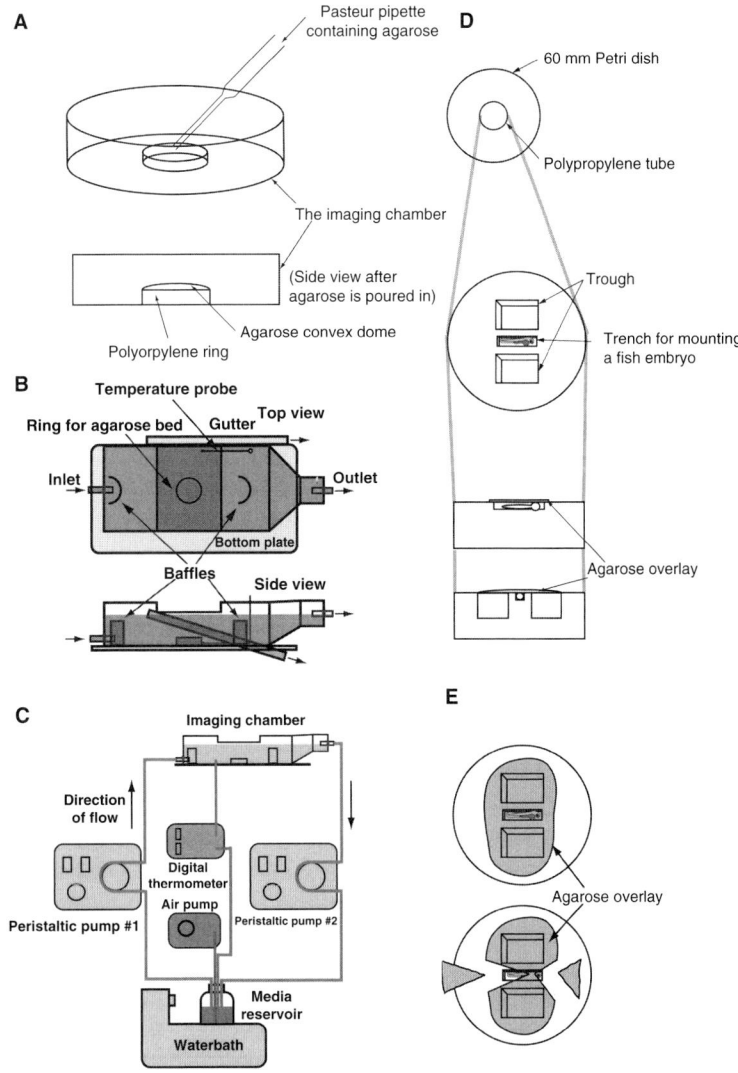

Fig. 8   *(Continued)*

6. The imaging chamber forms part of a loop of fluid circulation connected together with silicone tubing (Fig. 8B and C). Embryo media warmed in a bottle in a waterbath is carried through silicone tubing to the chamber by a peristaltic pump. Excess media is continuously removed from the chamber and returned to the media reservoir by a second peristaltic pump. The embryo media in the reservoir is aerated by an aquarium air pump. The temperature of the waterbath is calibrated to warm the media sufficiently to maintain a constant temperature (28.5°C) in the imaging chamber (an electronic temperature probe measures the temperature in the imaging chamber). The required waterbath temperature should be determined empirically and depends on a number of factors including the ambient temperature of the room, the length and diameter of the tubing between the reservoir and imaging chamber, and the flow rate. In a cool room with long tubing the temperature of the waterbath may need to be 37°C or even higher.

Mounting animals in imaging chambers
These procedures are done on a dissecting microscope.

1. Dechorionate and select embryos for mounting. Only a single embryo is generally mounted per imaging chamber.
2. Fill the imaging vessel with the 30% Danieu's solution to just below the rim of the petri dish. If pigment-free albino mutant embryos are used, the PTU can be left out, and if paralyzed mutant embryos are used the tricaine can be left out of the Danieu's solution.
3. Pick an embryo for mounting, and drop it in the middle of the agarose bed (it may begin to roll of because of the convex surface).
4. Using the fine forceps blades, make a shallow, narrow trench in the center of the agarose dome. This trench should be slightly wider than the dimensions of the embryo in its desired orientation for imaging, with the animal below the surface of the agarose but not too deep (Fig. 8D). It is critical that the trench be carved out carefully to make a space that holds the animal relatively motionless at rest. For imaging most portions of later stage embryos and larvae the animal should lie on its side in the trench, for lateral view. A larger cavity should be carved out posterior to

---

Fig. 8 Mounting zebrafish embryos and larvae for time-lapse imaging. For shorter-term imaging, imaging chambers are prepared from a ring of polypropylene tube glued to a 60 mm petri dish (A). For long-term (greater than 1 day) time-lapse imaging, an imaging chamber is constructed from a modified tissue culture flask (B). The approximate areas filled with water in an operating chamber are noted in blue. For details on construction, see text and Kamei & Weinstein (2005). The imaging chamber is a key part of the apparatus used for long-term time-lapse imaging of developing zebrafish, diagrammed schematically here (C). Tubing carrying water is noted in blue, wires for temperature probe are shown in green, and the air line is shown in red. The inlet and outlet ports of the imaging chamber are each connected via silicone tubing through two separate peristaltic pumps to a heated, aerated reservoir of embryo buffer, forming a continuous circuit of fluid. Temperature of the imaging chamber and fluid reservoir are both monitored using separate temperature probes, and the reservoir is continuously aerated with an aquarium air pump. See text and Kamei & Weinstein (2005) for additional details. In either short- or long-term imaging chambers the polypropylene ring in the center is filled with low-melt agarose and cavities carved out to accommodate the animal and to act as anchor points for top agarose (D). After covering the animal with top agarose much of the agarose over the animal is carved away in wedges (E). See text for further details. (See Plate no. 3 in the Color Plate Section.)

the tail to accommodate additional increases in trunk/tail length. Additional space should also be left around the head to accommodate shifting and growth, particularly in younger animals.

5. Two additional large cavities should also be carved out perpendicular to the trench on either side (Fig. 8D). These cavities will act as anchor points for the agarose layer that is overlaid on top of the embryo.

6. Place the embryo in the trench, and slowly overlay with molten agarose. It should be warm enough to freely flow in the buffer, but not too hot to kill the embryo. We typically use glass Pasteur pipette for this since it offers more precise control. Start from one well next to the trench. Apply the agarose at steady rate, and once it filled the well, move over to the other well by moving the pipette over the embryo. Once positioned over the other well, fill up this well also (Fig. 8E). This should create an agarose bridge over the embryo and should hold the embryo down.

7. Cut excess agarose away by using the blades of a pair of fine forceps. For imaging the trunk, we slice away a triangle of agarose over the trunk and tail and over the rostral region, leaving a "bridge" of agarose over the yolk sac, posterior head, and anterior-most trunk sufficient to hold the embryo firmly in place (Fig. 8E). This ensures the optical clarity of the trunk vessels. These cuts are necessary, since the embryo is growing in anterior-posterior axis, and straightening. Without removing these wedges of agarose, continued growth and straightening of the embryo/larva could not be accommodated.

2. Multiphoton Time-Lapse Imaging

Once animals are properly mounted they can be imaged by relatively straightforward time-lapse imaging methods. It is strongly recommended that a multiphoton microscopy system be used for this rather than a standard confocal microscope. The advantages of multiphoton microscopy and its use in developmental studies have been reviewed elsewhere (Denk and Svoboda, 1997; Weinstein, 2002). Multiphoton imaging reduces photodamage over the course of long imaging experiments, improves resolution of fluorescent structures deep in tissues, and improves the "three dimensionality" of resulting image reconstructions. Most imaging systems designed for or adaptable to multiphoton imaging (Leica, Zeiss, Olympus, etc.) have software interfaces that allow simple implementation of "4-D imaging" (*x, y, z,* time) experiments. Below we provide a cursory experimental description with some of the important parameters and experimental considerations. The melanophores in developing embryos interfere with multiphoton imaging when 976 nm excitation (EGFP two-photon excitation wavelength) is used. This interference can be avoided either by imaging techniques (moving the field of view), by treating embryos with PTU, or by using genetic mutants lacking or with reduced melanophores (such as *albino, nacre,* or *casper*) (Lister *et al.,* 1999; White *et al.,* 2008). PTU should be used carefully in later stage embryos because it may cause developmental defects. Use of tricaine (MS-222) for extended periods of time can also be problematic as the proper dosage for the drug becomes difficult to control, and excess tricaine can easily kill the animal. For imaging

experiment lasting longer than 1 day, the use of paralyzed *nic1* mutant animals (Westerfield *et al.*, 1990) is strongly encouraged.

1. Transfer the imaging chamber with mounted animal very carefully to the stage of a multiphoton microscope, taking care not to dislodge the animal. If the animal is very easily dislodged in transfer then it was likely not well enough mounted and should be more securely held when re-mounted.

2. After locating the field to be imaged, the time-lapse parameters should be set. Below are listed some guidelines for a few of the important parameters for multiphoton transgenic blood vessel imaging:

Maximal imaging depth:	approximately 250 µm for best image quality
Objectives:	10–100X, must pass long wavelength light
Spacing between planes:	1–5 µm, depending on magnification
Number of planes imaged:	10–60, depending on region and magnification
Interval between time points:	1–15 min (5 min is most typical)
Length of timelapse:	Up to 24 h, longer if a chamber is used with circulation of warmed buffer. With such a chamber, we have successfully imaged an animal for 5 days (Kamei and Weinstein, 2005). The chamber without flow described above is mainly used for shorter time-lapse experiments, and some developmental delay may be noted in longer runs.
Laser power setting:	Minimum necessary to obtain good images; if possible increase power with greater depth and use sensitive detectors to permit further decreases in required power.
Frames averaged:	5 frames averaged/plane

3. Once an imaging run has been initiated the images being collected should be checked frequently for shifting of the field being imaged. Often some shifting occurs due to growth or morphogenetic movements of the developing animal. Some shifting is also sometimes seen at the beginning of a time-lapse run as the animal "settles in." The field being imaged will sometimes need to be adjusted several times during the course of an experiment to maintain the vessels being imaged within the field. The stage can be adjusted to reset the X and Y positions. The Z positions (bottom and top of the images stack) may also need to be reset, usually by stopping and restarting the time-lapse program. If excessive shifting occurs due to the embryo being improperly mounted a new animal should be re-mounted.

V. Conclusion

Recent evidence suggests that genetic factors are critical in the formation of major vessels form during early development. Understanding the mechanisms behind the emergence of these early vascular networks will require the use of genetically and

experimentally accessible model vertebrates. The accessibility and optical clarity of the zebrafish embryo and larva make it particularly useful for studies of vascular development. Studies of developing vessels are likely to have far-reaching implications for human health, since understanding mechanisms underlying the growth and morphogenesis of blood vessels has become critical for a number of important emerging clinical applications. Pro- and anti-angiogenic therapies show great promise for treating cancer and ischemia, respectively, and a great deal of effort is currently going into uncovering and characterizing factors that can be used to promote or inhibit vessel growth *in vivo*. Since many of the molecules that play key roles in developing vessels carry out analogous functions during postnatal angiogenesis, it seems likely that the zebrafish will yield many important clinically applicable insights in the future.

References

Brown, L.A., Rodaway, A.R., Schilling, T.F., Jowett, T., Ingham, P.W., Patient, R.K., and Sharrocks, A.D. (2000). Insights into early vasculogenesis revealed by expression of the ETS-domain transcription factor Fli-1 in wild-type and mutant zebrafish embryos. *Mech. Dev.* **90**, 237–252.

Bussmann, J., Lawson, N., Zon, L., and Schulte-Merker, S. (2008). Zebrafish VEGF receptors: A guideline to nomenclature. *PLoS Genet.* **4**, e1000064.

Cheong, S.M., Choi, S.C., and Han, J.K. (2006). Xenopus Dab2 is required for embryonic angiogenesis. *BMC Dev. Biol.* **6**, 63.

Chi, N.C., Shaw, R.M., De Val, S., Kang, G., Jan, L.Y., Black, B.L., and Stainier, D.Y. (2008). Foxn4 directly regulates tbx2b expression and atrioventricular canal formation. *Genes Dev.* **22**, 734–739.

Childs, S., Chen, J.N., Garrity, D.M., and Fishman, M.C. (2002). Patterning of angiogenesis in the zebrafish embryo. *Development* **129**, 973–982.

Cross, L.M., Cook, M.A., Lin, S., Chen, J.N., and Rubinstein, A.L. (2003). Rapid analysis of angiogenesis drugs in a live fluorescent zebrafish assay. *Arterioscler. Thromb. Vasc. Biol.* **23**, 911–912.

Denk, W., and Svoboda, K. (1997). Photon upmanship: Why multiphoton imaging is more than a gimmick. *Neuron* **18**, 351–357.

Evans, H. (1910). "Manual of Human Embryology." J. B. Lippincott & Company, Philadelphia, PA and London.

Fouquet, B , Weinstein, B.M., Serluca, F.C., and Fishman, M.C. (1997). Vessel patterning in the embryo of the zebrafish: Guidance by notochord. *Dev. Biol.* **183**, 37–48.

Gering, M., Rodaway, A.R., Gottgens, B., Patient, R.K., and Green, A.R. (1998). The SCL gene specifies haemangioblast development from early mesoderm. *EMBO J.* **17**, 4029–4045.

Geudens, I., Herpers, R., Hermans, K., Segura, I., Ruiz de Almodovar, C., Bussmann, J., De Smet, F., Vandevelde, W., Hogan, B.M., Siekmann, A., *et al.* (2010). Role of Dll4/Notch in the Formation and Wiring of the Lymphatic Network in Zebra Fish. *Arterioscler. Thromb. Vasc. Biol.* **30**, 1695–1702.

Hauptmann, G., and Gerster, T. (1994). Two-color whole-mount in situ hybridization to vertebrate and Drosophila embryos. *Trends Genet.* **10**, 266.

Herpers, R., van de Kamp, E., Duckers, H.J., and Schulte-Merker, S. (2008). Redundant roles for sox7 and sox18 in arteriovenous specification in zebrafish. *Circ. Res.* **102**, 12–15.

Hogan, B.M., Bos, F.L., Bussmann, J., Witte, M., Chi, N.C., Duckers, H.J., and Schulte-Merker, S. (2009). Ccbe1 is required for embryonic lymphangiogenesis and venous sprouting. *Nat. Genet.* **41**, 396–398.

Huang, C.C., Lawson, N.D., Weinstein, B.M., and Johnson, S.L. (2003). reg6 is required for branching morphogenesis during blood vessel regeneration in zebrafish caudal fins. *Dev. Biol.* **264**, 263–274.

Isogai, S., Lawson, N.D., Torrealday, S., Horiguchi, M., and Weinstein, B.M. (2003). Angiogenic network formation in the developing vertebrate trunk. *Development* **130**, 5281–5290.

Jin, S.W., Beis, D., Mitchell, T., Chen, J.N., and Stainier, D.Y. (2005). Cellular and molecular analyses of vascular tube and lumen formation in zebrafish. *Development* **132**, 5199–5209.

Kaipainen, A., Korhonen, J., Mustonen, T., van Hinsbergh, V.W., Fang, G.H., Dumont, D., Breitman, M., and Alitalo, K. (1995). Expression of the fms-like tyrosine kinase 4 gene becomes restricted to lymphatic endothelium during development. *Proc. Natl. Acad. Sci. U.S.A.* **92**, 3566–3570.

Kamei, M., Saunders, W.B., Bayless, K.J., Dye, L., Davis, G.E., and Weinstein, B.M. (2006). Endothelial tubes assemble from intracellular vacuoles *in vivo*. *Nature* **442**, 453–456.

Kamei, M., and Weinstein, B.M. (2005). Long-term time-lapse fluorescence imaging of developing zebrafish. *Zebrafish* **2**, 113–123.

Lawson, N.D., Scheer, N., Pham, V.N., Kim, C.H., Chitnis, A.B., Campos-Ortega, J.A., and Weinstein, B.M. (2001). Notch signaling is required for arterial-venous differentiation during embryonic vascular development. *Development* **128**, 3675–3683.

Lawson, N.D., and Weinstein, B.M. (2002). *In vivo* imaging of embryonic vascular development using transgenic zebrafish. *Dev. Biol.* **248**, 307–318.

Lin, S. (2000). Transgenic zebrafish. *Methods Mol. Biol.* **136**, 375–383.

Lister, J.A., Robertson, C.P., Lepage, T., Johnson, S.L., and Raible, D.W. (1999). Nacre encodes a zebrafish microphthalmia-related protein that regulates neural-crest-derived pigment cell fate. *Development* **126**, 3757–3767.

Lyons, M.S., Bell, B., Stainier, D., and Peters, K.G. (1998). Isolation of the zebrafish homologues for the tie-1 and tie-2 endothelium-specific receptor tyrosine kinases. *Dev. Dyn.* **212**, 133–140.

Motoike, T., Loughna, S., Perens, E., Roman, B.L., Liao, W., Chau, T.C., Richardson, C.D., Kawate, T., Kuno, J., Weinstein, B.M., *et al.* (2000). Universal GFP reporter for the study of vascular development. *Genesis* **28**, 75–81.

Murakami, T. (1972). Vascular arrangement of the rat renal glomerulus: A scanning electron microscope study of corrosion casts. *Arch. Histol. Jap.* **34**, 87–107.

Roman, B.L., Pham, V.N., Lawson, N.D., Kulik, M., Childs, S., Lekven, A.C., Garrity, D.M., Moon, R.T., Fishman, M.C., Lechleider, R.J., *et al.* (2002). Disruption of acvrl1 increases endothelial cell number in zebrafish cranial vessels. *Development* **129**, 3009–3019.

Sabin, F.R. (1917). Origin and development of the primitive vessels of the chick and of the pig. *Carnegie Contrib. Embryol.* **6**, 61–124.

Stoletov, K., Montel, V., Lester, R.D., Gonias, S.L., and Klemke, R. (2007). High-resolution imaging of the dynamic tumor cell vascular interface in transparent zebrafish. *Proc. Natl. Acad. Sci. U.S.A.* **104**, 17406–17411.

Sumanas, S., Jorniak, T., and Lin, S. (2005). Identification of novel vascular endothelial-specific genes by the microarray analysis of the zebrafish cloche mutants. *Blood* **106**, 534–541.

Sumoy, L., Keasey, J.B., Dittman, T.D., and Kimelman, D. (1997). A role for notochord in axial vascular development revealed by analysis of phenotype and the expression of VEGR-2 in zebrafish flh and ntl mutant embryos. *Mech. Dev.* **63**, 15–27.

Szeto, D.P., Griffin, K.J., and Kimelman, D. (2002). HrT is required for cardiovascular development in zebrafish. *Development* **129**, 5093–5101.

Thompson, M.A., Ransom, D.G., Pratt, S.J., MacLennan, H., Kieran, M.W., Detrich, H. W.3rd, Vail, B., Huber, T.L., Paw, B., Brownlie, A.J., *et al.* (1998). The cloche and spadetail genes differentially affect hematopoiesis and vasculogenesis. *Dev. Biol.* **197**, 248–269.

Tong, E.Y., Collins, G.C., Greene-Colozzi, A.E., Chen, J.L., Manos, P.D., Judkins, K.M., Lee, J.A., Ophir, M.J., Laliberte, F.M., and Levesque, T.J. (2009). Motion-based angiogenesis analysis: A simple method to quantify blood vessel growth. *Zebrafish* **6**, 239–243.

Traver, D., Paw, B.H., Poss, K.D., Penberthy, W.T., Lin, S., and Zon, L.I. (2003). Transplantation and *in vivo* imaging of multilineage engraftment in zebrafish bloodless mutants. *Nat. Immunol.* **4**, 1238–1246.

Udvadia, A.J., and Linney, E. (2003). Windows into development: Historic, current, and future perspectives on transgenic zebrafish. *Dev. Biol.* **256**, 1–17.

Weinstein, B. (2002). Vascular cell biology *in vivo*. A new piscine paradigm? *Trends Cell Biol.* **12**, 439–445.

Westerfield, M.(2000). "The Zebrafish Book. A Guide for the Laboratory Use of Zebrafish (Danio Rerio)," 4th ed. University of Oregon Press, Eugene.

Westerfield, M., Liu, D.W., Kimmel, C.B., and Walker, C. (1990). Pathfinding and synapse formation in a zebrafish mutant lacking functional acetylcholine receptors. *Neuron* **4**, 867–874.

White, R.M., Sessa, A., Burke, C., Bowman, T., LeBlanc, J., Ceol, C., Bourque, C., Dovey, M., Goessling, W., Burns, C.E., *et al.* (2008). Transparent adult zebrafish as a tool for *in vivo* transplantation analysis. *Cell Stem Cell.* **2**, 183–189.

Yaniv, K., Isogai, S., Castranova, D., Dye, L., Hitomi, J., and Weinstein, B.M. (2006). Live imaging of lymphatic development in the zebrafish. *Nat. Med.* **12**, 711–716.

Zhong, T.P., Rosenberg, M., Mohideen, M.A., Weinstein, B., and Fishman, M.C. (2000). gridlock, an HLH gene required for assembly of the aorta in zebrafish. *Science* **287**, 1820–1824.

CHAPTER 3

Medaka Haploid Embryonic Stem Cells

Yunhan Hong

Department of Biological Sciences, National University of Singapore, Singapore 117543

Abstract
I. Introduction
   A. Haploidy in Evolution
   B. History Toward Haploid ES Cell Culture
   C. Rationale
II. Methods
   A. Production of Haploid Embryos
   B. Cell Culture Derivation
   C. Culture Condition
   D. Generation of Stable Haploid ES Cell Lines
   E. Characterization of Haploid ES Cells
   F. Stable Growth and Genetic Stability
III. Applications of Haploid ES Cells
   A. Semicloning
   B. Analysis of Gene Function
   C. Haploid Genetic Screens
IV. Summary
Acknowledgments
References

Abstract

The appearance of diploidy, the presence of two genomes or chromosome sets, is a fundamental hallmark of eukaryotic evolution and bisexual reproduction, because diploidy offers the basis for the bisexual life cycle, allowing for oscillation between diploid and haploid phases. Meiosis produces haploid gametes. At fertilization, male and female gametes fuse to restore diploidy in a zygote, which develops into a new life. At sex maturation, diploid cells enter into meiosis, culminating in the production of haploid gametes. Therefore, diploidy ensures pluripotency, cell proliferation, and

functions, whereas haploidy is restricted only to the post-meiotic gamete phase of germline development and represents the end point of cell growth. Diploidy is advantageous for evolution. Haploidy is ideal for genetic analyses, because any recessive mutations of essential genes will show a clear phenotype in the absence of a second gene copy. Recently, my laboratory succeeded in the generation of medaka haploid embryonic stem (ES) cells capable of whole animal production. Therefore, haploidy in a vertebrate is able to support stable cell culture and pluripotency. This finding anticipates the possibility to generate haploid ES cells in other vertebrate species such as zebrafish. These medaka haploid ES cells elegantly combine haploidy and pluripotency, offering a unique yeast-like system for *in vitro* genetic analyses of molecular, cellular, and developmental events in various cell lineages. This chapter is aimed to describe the strategy of haploid ES cell derivation and their characteristics, and illustrate the perspectives of haploid ES cells for infertility treatment, genetic screens, and analyses.

I. Introduction

Our interest in creating haploid embryonic stem (ES) cells comes from the curiosity and need. It has been well known for centuries that haploidy, the presence of a single genome or a complete set of chromosomes or genes within a cell, is an excellent system for genetic analyses of molecular events, because any recessive mutations of genes essential, for example, tumor suppressors and pluripotency will show a clear phenotype in the absence of a second gene copy. However, haploidy in eukaryotes is usually restricted to single-celled organisms, among them is the yeast (*Saccharomyces cerevisiae*), which has been best employed in genetic analyses (Botstein and Fink, 1988). Multicellular organisms, in particular vertebrates, are usually diploid and contain two genomes, one from the father and the other from the mother. The absence of haploid vertebrates has raised a fundamental question as to whether haploidy can support normal development in these organisms. Curiously, it is completely unknown about the absence of haploid vertebrates.

ES cells are an excellent system for experimental analyses of cellular and developmental events *in vitro*. ES cells are derived from early fertilization embryos (Evans and Kaufman, 1981), and induced pluripotent stem (iPS) cells can even be derived by reprogramming from various differentiated types of cells (Takahashi and Yamanaka, 2006). These are diploid fertilization ES or iPS cells, because they originate directly from fertilization embryos or indirectly from their differentiated derivatives and contain two chromosome sets. Although these diploid ES cells have been widely used for studying vertebrate development *in vitro*, a direct analysis of recessive phenotypes has been hampered by their diploidy. Therefore, it is highly desirable to create haploid ES cells to combine haploidy and pluripotency for developmental genetic analyses in biomedicine, because any recessive and disease phenotypes of essential genes can easily be dissected *in vitro* either in an undifferentiated state or under various differentiation schemes to follow molecular, cellular, and developmental events during the

restriction of various cell lineages and differentiation of various cell types. Similarly, genes essential for normal and diseased cellular metabolism can also be investigated on a well-defined genetic background.

In this chapter, we will discuss the practice for our recent success in the generation of first haploid ES cells from the fish medaka (*Oryzias latipes*) (Yi *et al.*, 2009). We will focus on the rationale and strategies we have used to help readers develop an intuitive understanding of the possibility and approach to derive haploid ES cells from other organisms. Detailed experimental protocols will be published elsewhere (Yi *et al.*, 2010). We will also discuss problems and future directions.

A. Haploidy in Evolution

In most bisexual animals, haploidy oscillates with diploidy. Diploidy ensures mitotic divisions and pluripotency throughout life, whereas haploidy exists only in the post-meiotic germline and lacks the mitotic ability, and thus represents the dead end of the life unless fertilization to restore diploidy in a zygote. Haploidy is the ancestral status of evolution. In a broad sense, all viruses and prokaryotic organisms are haploid. They often possess a single RNA or DNA molecule as their genome. In single-celled eukaryotic organisms such as yeast, the genome usually comprises several DNA molecules that are compacted with proteins into individual chromosomes. In these single-celled organisms, haploidy prevails, and cell divisions by fission or budding lead to individual propagation. This is asexual reproduction. Under defined conditions such as starvation, these haploid organisms fuse together to form transient diploid cells, which immediately undergo meiosis and become haploid cells. In plants, haploid callus and pollens in culture can form plantlets. It is well known in certain invertebrates such as the honeybee (Heimpel and de Boer, 2008) that parthenogenetic haploid embryos can develop into adult animals. Therefore, haploidy in these unicellular organisms, in plants, and in invertebrates is associated with the intrinsic ability for continuous cell divisions, pluripotency *in vivo*, fertility, and heredity. However, haploid organisms have so far been absent in vertebrates.

B. History Toward Haploid ES Cell Culture

Driven by the enormous potential for genetic analyses in biomedicine, haploid cell cultures have attracted considerable interest. Pioneer work was done in frog, where androgenetic embryos at the late tail-bud stage produced two cell lines, which consisted of diploid and near-haploid populations but could not give rise to pure near-haploid cultures (Freed and Mezger-Freed, 1970). Later on in *Drosophila*, several cultures were derived from haploid embryos, which initially had a high percentage of 80–90% of haploid cells but were quickly overgrown by diploidy and aneuploidy in serial culture (Debec, 1978, 1984). In human, a heterogeneous culture called KBM7 was obtained from near-haploid chronic myeloid leukemia, which was unstable and quickly overgrown by diploid cells (Andersson *et al.*, 1987). By making use of their

expertise in the derivation of first mouse ES cells, Kaufman *et al.* (1983) repeatedly attempted to derive mouse haploid ES cells. In their experiments, they used 7% ethanol to activate eggs to develop into parthenogenetic embryos, which predominantly (>80%) contained only haploid mitoses when examined at the morula stage. When reaching the blastocyst stage, these embryos were used for cell initiation, resulting in several stable ES cell lines. "However, chromosome analysis at early passage cell lines revealed that all were diploid with a modal number of 40 chromosomes" (Kaufman *et al.*, 1983). Therefore, this work demonstrated the ability to derive homozygous diploid cell lines of parthenogenetic origin but the inability to generate haploid ES cells in mouse. These studies provided valuable data on haploid cell cultures but also led to a generally accepted hypothesis that haploidy is associated with inferior growth and genetic instability due to diploidization and aneuploidy.

C. Rationale

As outlined above, there are two pros for the feasibility to derive haploid ES cells in a vertebrate: (1) haploidy has the start point of evolution into multicellular organism and must be able to support cell division and (2) haploidy *per se* can even sufficiently support embryonic and adult development in plants and invertebrate animals. There are also two cons, which are (1) the lack of report on any haploid organism in vertebrates and (2) the unsuccessful attempts in *Drosophila*, frog, and mouse. The curiosity to understanding these controversial aspects, the great potential of haploid ES cells and our expertise in generating stable cell lines of diploid ES cells and adult germ stem cells have been the driving force for us to create haploid ES cells.

Technically, we envisioned that the successful generation of haploid ES cells might depend on three critical factors (Yi *et al.*, 2009): (1) a pure or highly inbred strain in which many—if not all—deleterious (recessive) alleles have been deleted; (2) conditions most conducive to stem cell proliferation rather than differentiation; and (3) ability to maintain the genetic stability/integrity. These considerations have led us to choose the medaka as a test organism. This laboratory fish is an excellent model for analyzing vertebrate development (Wittbrodt *et al.*, 2002) and has given rise to several diploid lines of ES cells (Hong and Schartl, 2006; Hong *et al.*, 1996, 1998b) and male germ stem cell spermatogonia (Hong *et al.*, 2004b). This fish also has many inbred strains known to be permissive to ES cell derivation (Hong *et al.*, 1998a). Experience of nearly 20 years working with medaka stem cells has been an encouraging force for haploid ES cell derivation.

II. Methods

Detailed procedures are the topics of several papers. The focus here is on those that are particularly important for haploid ES cells.

A. Production of Haploid Embryos

One exception is the need to obtain haploid embryos as the start point. There are two approaches. One is haploid androgenesis, in which the oocyte is genetically inactivated before fertilization with a normal sperm. The other is haploid gynogenesis, in which sperm are inactivated before they are used for insemination with normal eggs. We adopt gynogenesis to obtain haploid embryos (Fig. 1). Sperm are small, and their nuclei can easily be inactivated by UV-light irradiation without affecting their ability to trigger egg development. To ensure haploid gynogenesis, we made use of two inbred albino medaka strains, i^1 and i^3, which produce genetically pigmented hybrids. In the absence of one nucleus, namely the sperm nucleus, the embryo will be haploid and albino. The haploidy can also be judged by the so-called haploid syndrome. Direct evidence comes from chromosome analysis demonstrating the presence of 24 chromosomes for the haploid set in medaka. Once 100% efficiency of haploid gynogenesis is achieved, embryos thus obtained can be used for cell culture initiation.

B. Cell Culture Derivation

The strategy to obtain haploid ES cells is similar to that for diploid ES cells. Blastomere cells were dissociated from gynogenetic embryos at the midblastula stage and seeded in gelatin-coated 48-well plates for culture under feeder-free conditions (Hong and Schartl, 2006). Blastomeres moved around by forming pseudopodia and underwent rapid divisions during the first hours of culture, and reached the final

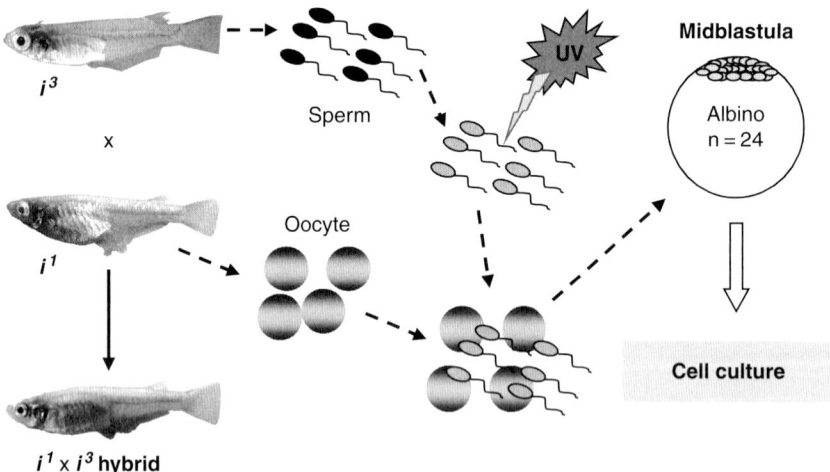

Fig. 1 Flow chart illustrating how to produce haploid embryos. Two different albino strains are used to phenotype successful induction of haploid gynogenesis. Solid lines depict normal fertilization, resulting in hybrids with black pigmentation as clearly seen in the eye. Broken lines depict steps in producing gynogenetic haploid embryos for cell initiation. The haploid embryos are genetically albino and possess a haploid set of 24 chromosomes.

small size and attached by day 1 of culture. Serial passages led to stable cultures. We initiated five primary cultures and obtained three ES cell lines designated as HX1, HX2, and HX3, because they predominantly comprise haploid X-chromosome-carrying cells of gynogenetic origin. Albinism, haploid syndrome, and haploid metaphases demonstrated uniparental haploidy for the embryos from which the cell lines were derived. Direct evidence for the parental origin came from genotyping. In all the three cell lines we failed to detect sperm-derived autosomal tyrosinase allele (Hong et al., 1998b), convincingly demonstrating a gynogenetic origin. In addition, the Y-chromosomal dmrt1y (Matsuda et al., 2007) is absent, confirming a female genotype in accordance with the maternal origin.

All the three cell lines had similar stable growth and phenotype following more than 120 passages during more than 400 days of culture. Chromosome examination revealed that 77% cells were haploid. These haploid ES cells have a round or polygonal shape, little cytoplasm, and large nuclei with prominent nucleoli. Importantly, the number of nucleoli serves as a marker to monitor haploidy in living cell cultures. The medaka karyotype has a single nucleolar organizer region (Uwa and Ojima, 1981), indicating that the haploid genome can form only one nucleolus. The nucleolus is easily visible in living cells and its number delineates ploidy levels in culture. During the nucleolar cycle, a haploid medaka cell may have one (interphase), no (metaphase and anaphase), or two nucleoli (telophase with two daughter nuclei). A diploid medaka cell displays a similar picture except for up to a doubled number of nucleoli. In culture, the small cells indeed have one nucleolus per nucleus, whereas the large cells have one (fused from two) to two nucleoli per nucleus. Nucleolar silver staining of fixed cells confirmed these observations. We found that haploid blastula-derived cultures invariantly have ~20% diploid cells (Fig. 2).

C. Culture Condition

Cell dissociation and feeder-free culture conditions for the derivation of gynogenetically haploid ES cells are similar to those established for fertilization diploid ES cells (Hong and Schartl, 1996). One big difference is a more demanding culture medium for haploid ES cell initiation. We found that a mixture of MES1-conditioned medium and MO1-conditioned medium enhanced the efficiency of haploid ES cell initiation and early passage culture up to 100 days. Afterward, the parental haploid ES cell lines and their pure haploid clones can easily be maintained in normal medaka ES cell media. MO1 is an uncharacterized cell culture derived from the adult medaka ovary. MES1 is one of the three medaka diploid ES cell lines derived from fertilization midblastula embryos of strain HB32C (Hong et al., 1996). HX cells originated from medaka strain i[1]. It remains to be determined whether the requirement for conditioned medium for initial culture is due to differences in ploidy and/or medaka strain, as we have found that under conditions for MES1 cell derivation, only certain strains are permissive for ES cell derivation, whereas others are not (Hong et al., 1998a).

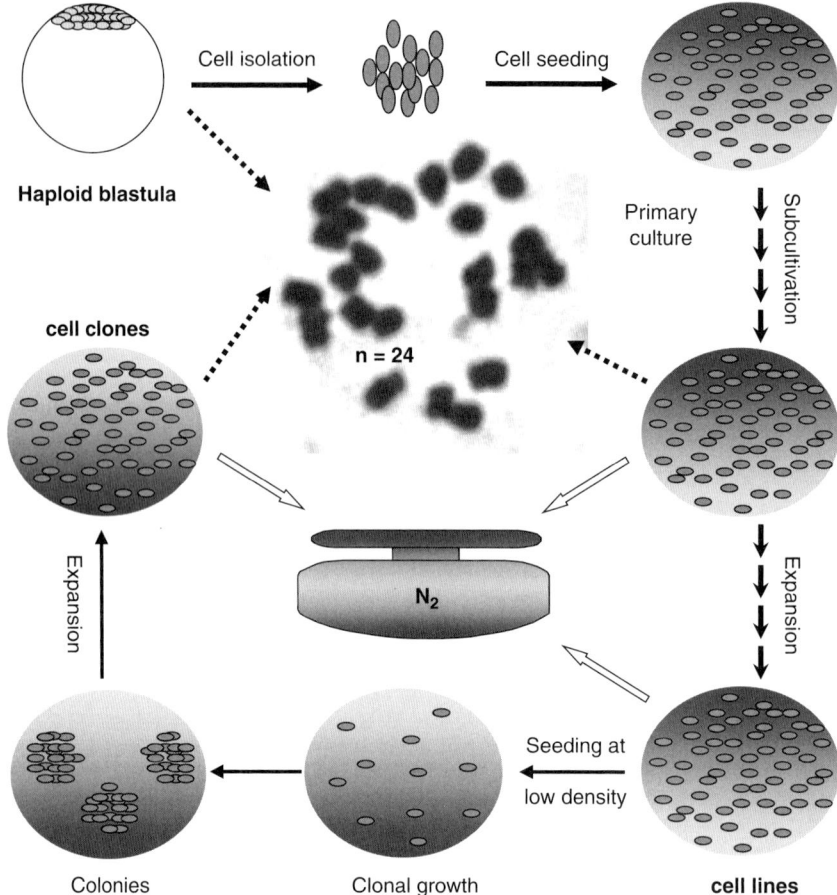

Fig. 2 Flow chart illustrating how to create pure haploid ES cells. Solid arrows depict major steps to derive haploid ES cells and pure ES cell clones. Broken arrows depict major steps for characterization by, e.g., chromosome analysis. Open arrows depict when cells are frozen for cryostorage. Cell initiation and subculture are in multiwell plates. Clonal growth and expansion are often in 10-cm dishes.

D. Generation of Stable Haploid ES Cell Lines

Fortunately, medaka haploid ES cells can easily be dissociated into single cells. When seeded at a low cell density, they can form distinct colonies during clonal growth. Importantly, colonies comprising undifferentiated cells were capable of expansion to stable clones of homogeneous cells exhibiting high alkaline phosphatase activity, a general marker for undifferentiated stem cells (Hong and Schartl, 2006; Hong et al., 1996, 1998b, 2004b). Medaka gynogenetic ES cells have the intrinsic ability for stable growth. In our experiment, 100 colonies were randomly picked up and seeded in 96-well plates. We obtained 60 clones and expanded them for cytogenetic analyses. Chromosome examination revealed that 45 clones were haploid with the

remainder being diploid. This ratio is again in accordance with the haploid/diploid ratio observed in parental cell cultures. Flow cytometry analyses revealed that the parental cell lines indeed had haploid and diploid peaks, whereas haploid clones essentially had only the haploid peak (Yi *et al.*, 2009). Taken together, both haploid and diploid clones of HX1 comprise essentially pure haploid and diploid cells capable of stable growth.

E. Characterization of Haploid ES Cells

Besides stable growth and phenotype, there are several other experiments to characterize the pluripotency of diploid ES cells. These procedures apply also to haploid ES cells. These include pluripotency gene expression profile, induced differentiation in suspension culture to form embryoid bodies followed by phenotypic monitoring and immunostaining, and chimera formation by cell transplantation into early developing embryos. Interestingly, upon transplantation into fertilization embryo hosts, haploid ES cells are not different from diploid counterparts in the ability to take part in chimeric embryogenesis, generating functional cells that contribute to several organ systems, including the heart, blood cells, skeletal muscles, and fin epithelia (Yi *et al.*, 2009).

F. Stable Growth and Genetic Stability

It is widely accepted that three factors are against haploid cell culture (Andersson *et al.*, 1987; Debec, 1978, 1984; Freed and Mezger-Freed, 1970). Diploidization has been thought to be a major cause to prevent haploid cell derivation. Conceivably, somatic cells in culture may undergo spontaneous fusion and/or endomitosis. The reverse situation, somatic haploidization to convert diploidy to haploidy, has not been reported, because haploidization has so far been found only in the meiotic division of germ cells. Consequently, diploid and even polyploid cells will gradually accumulate during serial culture, ultimately overgrowing haploid cells. Haploid cells in culture may intrinsically have genetic instability and become aneuploid cells, abolishing in part haploid cell cultures. Inferior growth has also been ascribed to haploid cells in culture and accepted as an important factor for outgrowth by diploid cells. Stable growth and the retention of a constant ratio between haploidy and diploidy in all the three HX cell lines indicate that these factors are essentially absent in medaka haploid ES cells. The availability of pure haploid ES cell clones, diploid gynogenetic ES cell clones, and fertilization diploid ES cell lines enabled us to address the stability issue. We found that haploid and diploid ES cells are not different in doubling time and proliferation during attachment and suspension cocultures. Interestingly, we found that di- and polyploidization events rarely occur in self-renewing haploid ES cells, but do occur during differentiation. These di- and polyploidized differentiation products usually disappear by cell death, without overgrowing haploid ES cells (Yi *et al.*, 2009). Conceivably, the three parental cell lines maintained a dynamically stable proportion of haploid cells during serial culture because under standard culture conditions, spontaneous differentiation is a rare event (≤1%), and occasional diploidization

does not accumulate. These results demonstrate that the haploidy retains pluripotency-dependent genetic stability *in vitro*.

The ploidy levels in the parental HX cell lines and their haploid clones remain fairly constant during continuous culture. However, we noticed that diploid cells will increase when cell cultures are re-initiated from frozen vials. But we do not know when diploidization occurs (unpublished data). Therefore, it is a routine to regularly monitor the ploidy level by the nucleolus number and cytogenetic analyses. In the case that pure haploid cells are required, clonal growth and expansion can lead to new haploid ES cell clones.

III. Applications of Haploid ES Cells

A. Semicloning

In normal reproduction, fertilization combines two haploid meiotic nuclei of the sperm and egg, and ensures a high efficiency of normal development and genetic diversity in progeny. By mimicking fertilization, intracytoplasmic sperm injection has been developed and increasingly used for treating infertile men who have germ cells but defects in post-meiotic progression. Embryos by diploid somatic cell nuclear transfer to enucleated oocytes produced viable offspring in frogs, fish, and mammals (Campbell *et al.*, 1996; Gurdon and Melton, 2008; Lee *et al.*, 2002; Wakamatsu *et al.*, 2001), providing a powerful tool for animal cloning, human ES cell derivation, and analyzing nuclear reprogramming. However, the application of this cloning strategy to human-assisted reproduction has widely been debated because of low efficacy and ethical concern about identical progeny to the nuclear donor. Recently, a new assisted reproductive technology on the basis of creating mosaic oocytes has been proposed to treat infertile patients lacking any germ cells. This approach, called semicloning, uses nuclear transfer to combine a haploid somatic nucleus and a haploid gamete nucleus in the oocyte. Semicloning has remained hypothetical because viable offspring has not yet been obtained (Tsai *et al.*, 2000).

Haploid ES cells and sperm are similar to each other in terms of a complete single genome. However, they are different in terms of mitotic or meiotic products. In addition, haploid ES cells lack the ability of sperm to penetrate eggs. It is curious to know whether haploid ES cells can replace sperm for reproduction. The availability of haploid ES cells enabled us to directly test semicloning by producing mosaic oocytes. To this end, i[1]-derived HX1 cells were labeled with nuclear GFP and their nuclei were transplanted into non-enucleated matured oocytes of i[3] albino. The nuclear transplants exhibited black pigmentation (Fig. 3), similar to the fertilization hybrid between i[1] and i[3] albinos (Fig. 1), demonstrating the functional contribution from both the oocyte and the HX1 nuclei. Furthermore, nuclear transplant embryos also displayed nuclear GFP in all tissues (Fig. 3B and C). One of the nuclear transplant fry grew into a pigmented fertile female (Fig. 3D), namely NT1, which also exhibited continuous GFP expression from the hES genome. Most importantly, NT1 showed normal fertility and germline

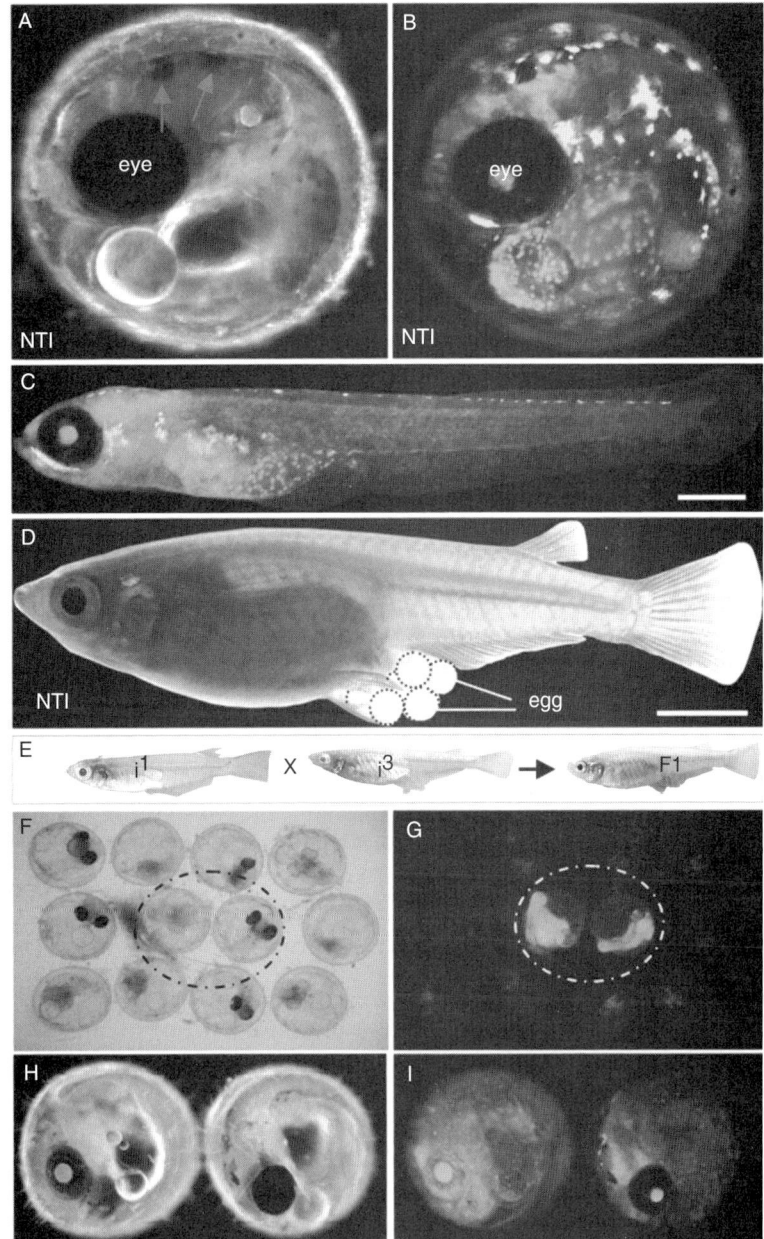

Fig. 3 *(Continued)*

transmission upon test-crosses to both i^1 and i^3 males (Fig. 3E), producing four types of F1 progeny, namely albino or pigmented and GFP-positive or GFP-negative (Fig. 3F–I). Pigmented and albino progeny were segregated at the Mendelian 1:1 ratio (Fig. 3F–I), demonstrating that the HX1-derived and oocyte genomes were not different in germline transmission. We termed this first semicloned medaka Holly. Holly lived for 18 months, and her albino and transgene expression has been passed over four generations. Since embryos with genomic abnormalities cannot reach advanced stages of development in mouse and human (Yanagimachi, 2005), germline transmission demonstrates the retention of genetic stability and integrity in hES cultures. Most importantly, mosaic oocytes comprising a haploid mitotic nucleus from haploid ES cells and a haploid meiotic nucleus from the oocyte can generate viable and fertile offspring, demonstrating the feasibility of semicloning for reproductive medicine.

Compared to cloning through diploid somatic nuclear transfer, semicloning would not only ensure the biparental contribution to the progeny but also create a new and unpredictable combination of genetic traits from both parents similar to normal fertilization (Fig. 4). However, semicloning at its present form can treat only male infertility and produce female individuals. Future work is needed to determine whether male haploid ES cells can also be obtained.

B. Analysis of Gene Function

Gene targeting by homologous recombination in ES cells is a powerful routine in mouse developmental genetics. This complicated approach starts with targeted gene disruption in ES cells, production of whole animals by germline chimera formation from gene-targeted ES cell clones, and breeding the founder to produce F1 heterozygous animals, which in turn are intercrossed to obtain heterozygous and, more importantly, homozygous null embryos/animals for phenotypic analyses. This series of steps are aimed at generating a null locus at the homozygous state. Although this approach is tedious and experimentally demanding, it remains the standard in creating homozygous null vertebrates, because precise genetic alterations mediated by homologous recombination in mouse ES cells occur at an exceedingly low efficiency of $\sim 10^{-6}$. The frequency for independent and simultaneous disruption within a particular cell of both copies of a gene in question would be $\sim 10^{-12}$, which is well beyond the

---

Fig. 3 Semicloning and germline transmission. (A, B) Nuclear transplant founder NT1 embryo at day 7. The i^1-derived HX1a was stably labeled with nuclear GFP, and its nuclei were transplanted into non-enucleated oocytes of strain i^3. Wild-type pigmentation is seen in the eye and body surface (arrows; A), and nuclear GFP is throughout the embryo (B). (C) Fry NT1 showing GFP expression and eye pigmentation. (D) NT1 fertile female laying eggs. (E) Crossing between albino strains i^1 and i^3. F1 hybrids have wild-type black pigmentation in the eye. (F–I) Germline transmission. (F, G) Cluster of embryos between NT1 and i^1 male showing germline transmission into four classes of progeny: GFP-positive albino or pigmented and GFP-negative albino or pigmented. (H, I) Higher magnification of GFP-positive albino or pigmented progeny as circled in (F, G). Embryos are 1 mm in diameter. Bars, 0.2 mm (C) and 0.5 mm (D). (reproduced with permission from the Science Publisher).

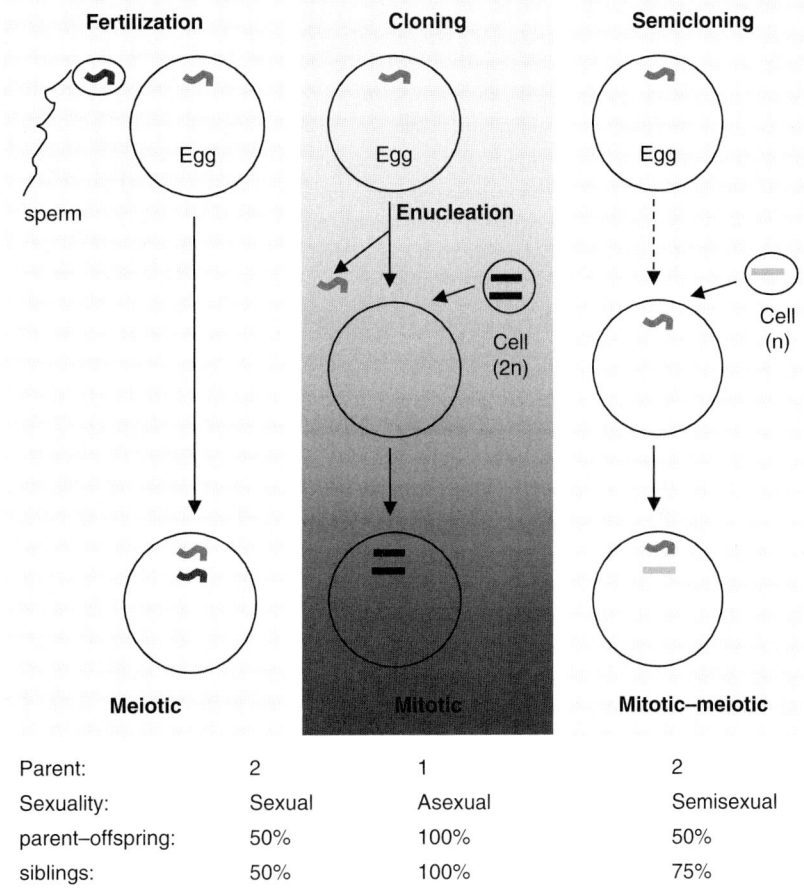

Fig. 4 Genetic consequences of semicloning. (Top) Comparisons between fertilization, cloning by diploid somatic nuclear transfer, and semicloning by haploid nuclear transfer. (Bottom) Genetic consequences of the three reproduction modes. Semicloning omits the enucleation step and mimics fertilization in the production of diversity in an unpredictable way. Curved line, meiotic genome; straight line, mitotic genome.

practical limit of screening for putative homozygous null mutants. In the case of haploid ES cells, the approach becomes straightforward, because the disruption of a single gene copy will directly show the loss-of-function phenotype at molecular and cellular levels. If the gene concerned is involved in pluripotency maintenance or lineage restriction and differentiation, the role of the gene can also be easily investigated by induced ES differentiation under defined culture conditions.

In theory, genes encoding cell cycle repressors and tumor suppressors such as p53 are perfect candidates for gene targeting in haploid ES cells. Conceivably, disruption of such genes alone or in combination will not compromise but promote haploid ES cell growth. In fact, the work with fish gene targeting began ~20 years ago, when I and

Manfred Schartl (Würzburg, Germany) initiated ES cell derivation and attempted gene targeting by using p53 as a model. Although the medaka ES cell work has been a success, germline chimera production from ES cell cultures has remained open. Recent work by Schartl and colleagues led to the notion that the medaka germline is cell-autonomously specified, pointing to the possible inaccessibility of this organism to germline contribution by ES cell cultures. Recently, we showed that p53 is expressed in both diploid and haploid medaka ES cells. By using a homologous recombination vector we previously designed for gene targeting in diploid medaka ES cells (Chen et al., 2002) and drug selection conditions established for diploid medaka ES cells (Hong et al., 2004a), we showed that p53 disruption in haploid ES cells can easily be detected by genomic PCR for the targeted locus and RT-PCR for the loss of the endogenous p53 expression, thus its absence. Importantly, the p53-disrupted haploid ES cells indeed exhibit a clear phenotype, including the loss of sensitivity for induced cell cycle arrest and senescence (unpublished data), a well-conserved role of this gene in diverse organisms examined so far. In medaka haploid ES cells, a typical gene-targeting experiment starting with available vectors can often be achieved within 3 months to offer targeted cell clones in large quantity for phenotypic analyses.

C. Haploid Genetic Screens

Mutagenesis screening is a powerful tool for generating, identifying, and analyzing genes involved in a particular process or pathway. Such screens have so far been performed in several model organisms including *Drosophila*, zebrafish, and medaka. These are diploid genetic screens because they are also based on the production of homozygous animals containing a particular mutation. In this regard, haploid ES cells offer a paradigm for high-throughput haploid genetic screens for random mutations (e.g., chemical mutagenesis), designed genetic modifications (gene targeting), and endogenous genes involved in particular pathways and processes. Recently, it has been reported that haploid genetic screens in human cells identify host factors used by pathogens (Carette et al., 2009). Similar approaches toward identifying host genes responsible for sensitivity to infection by viruses will shed light on new strategies to disrupt the virus susceptibility/sensitivity in aquaculture species by gene transfer technology.

IV. Summary

We have developed a strategy for generating haploid ES cell lines from gynogenetic blastula embryos of a highly inbred medaka strain. We can reproducibly obtain and stably cultivate haploid ES cells capable of forming pure haploid clones. We have demonstrated that haploidy in medaka has the intrinsic ability to support stable growth and pluripotency in culture. Most importantly, we have shown that haploidy after long-term culture and even genetic engineering is capable of producing fertile animals, demonstrating that haploidy retains the genetic integrity, and providing a powerful tool

for germline transmission in this organism. Most strikingly, we have generated oocyte creatures comprising a haploid mitotic nucleus and a haploid meiotic nucleus and demonstrated their ability to develop into fertile offspring. Most recently, we have also shown that the medaka haploid ES cells are excellent for gene-targeting experiments. Future work will determine whether these haploid cells are also a paradigm for genetic screens.

Acknowledgments

The author thanks Professors Manfred Schartl and Jochen Wittbrodt (Germany) for encouragement, and the members of his Developmental Genetics Lab for discussions and unpublished data. This work was supported by the Biomedical Research Council of Singapore (R-08-1-21-19-585 and SBIC-SSCC C-002-2007) and the National University of Singapore (R-154-000-153-720).

References

Andersson, B. S., Beran, M., Pathak, S., Goodacre, A., Barlogie, B., and McCredie, K. B. (1987). Ph-positive chronic myeloid leukemia with near-haploid conversion *in vivo* and establishment of a continuously growing cell line with similar cytogenetic pattern. *Cancer Genet. Cytogenet.* **24**, 335–343.

Botstein, D., and Fink, G. R. (1988). Yeast: An experimental organism for modern biology. *Science* **240**, 1439–1443.

Campbell, K. H., McWhir, J., Ritchie, W. A., and Wilmut, I. (1996). Sheep cloned by nuclear transfer from a cultured cell line. *Nature* **380**, 64–66.

Carette, J. E., Guimaraes, C. P., Varadarajan, M., Park, A. S., Wuethrich, I., Godarova, A., Kotecki, M., Cochran, B. H., Spooner, E., Ploegh, H. L., and Brummelkamp, T. R. (2009). Haploid genetic screens in human cells identify host factors used by pathogens. *Science* **326**, 1231–1235.

Chen, S., Hong, Y., and Schartl, M. (2002). Development of a positive–negative selection procedure for gene targeting in fish cells. *Aquaculture* **214**, 67–79.

Debec, A. (1978). Haploid cell cultures of *Drosophila melanogaster*. *Nature* **274**, 255–256.

Debec, A. (1984). Evolution of karyotype in haploid cell lines of *Drosophila melanogaster*. *Exp. Cell Res.* **151**, 236–246.

Evans, M. J., and Kaufman, M. H. (1981). Establishment in culture of pluripotential cells from mouse embryos. *Nature* **292**, 154–156.

Freed, J. J., and Mezger-Freed, L. (1970). Stable haploid cultured cell lines from frog embryos. *Proc. Natl. Acad. Sci. U. S. A.* **65**, 337–344.

Gurdon, J. B., and Melton, D. A. (2008). Nuclear reprogramming in cells. *Science* **322**, 1811–1815.

Heimpel, G. E., and de Boer, J. G. (2008). Sex determination in the Hymenoptera. *Annu. Rev. Entomol.* **53**, 209–230.

Hong, Y., and Schartl, M. (1996). Establishment and growth responses of early medakafish *(Oryzias latipes)* embryonic cells in feeder layer-free cultures. *Mol. Mar. Biol. Biotechnol.* **5**, 93–104.

Hong, Y., Chen, S., Gui, J., and Schartl, M. (2004a). Retention of the developmental pluripotency in medaka embryonic stem cells after gene transfer and long-term drug selection for gene targeting in fish. *Transgenic Res.* **13**, 41–50.

Hong, Y., Liu, T., Zhao, H., Xu, H., Wang, W., Liu, R., Chen, T., Deng, J., and Gui, J. (2004b). Establishment of a normal medakafish spermatogonial cell line capable of sperm production *in vitro*. *Proc. Natl. Acad. Sci. U. S. A.* **101**, 8011–8016.

Hong, Y., and Schartl, M. (2006). Isolation and differentiation of medaka embryonic stem cells. *Methods Mol. Biol.* **329**, 3–16.

Hong, Y., Winkler, C., and Schartl, M. (1996). Pluripotency and differentiation of embryonic stem cell lines from the medakafish (*Oryzias latipes*). *Mech. Dev.* **60**, 33–44.

Hong, Y., Winkler, C., and Schartl, M. (1998a). Efficiency of cell culture derivation from blastula embryos and of chimera formation in the medaka (*Oryzias latipes*) depends on donor genotype and passage number. *Dev. Genes Evol.* **208**, 595–602.

Hong, Y., Winkler, C., and Schartl, M. (1998b). Production of medakafish chimeras from a stable embryonic stem cell line. *Proc. Natl. Acad. Sci. U. S. A.* **95**, 3679–3684.

Kaufman, M. H., Robertson, E. J., Handyside, A. H., and Evans, M. J. (1983). Establishment of pluripotential cell lines from haploid mouse embryos. *J. Embryol. Exp. Morphol.* **73**, 249–261.

Lee, K. Y., Huang, H., Ju, B., Yang, Z., and Lin, S. (2002). Cloned zebrafish by nuclear transfer from long-term-cultured cells. *Nat. Biotechnol.* **20**, 795–799.

Matsuda, M., Shinomiya, A., Kinoshita, M., Suzuki, A., Kobayashi, T., Paul-Prasanth, B., Lau, E. L., Hamaguchi, S., Sakaizumi, M., and Nagahama, Y. (2007). DMY gene induces male development in genetically female (XX) medaka fish. *Proc. Natl. Acad. Sci. U. S. A.* **104**, 3865–3870.

Takahashi, K., and Yamanaka, S. (2006). Induction of pluripotent stem cells from mouse embryonic and adult fibroblast cultures by defined factors. *Cell* **126**, 663–676.

Tsai, M. C., Takeuchi, T., Bedford, J. M., Reis, M. M., Rosenwaks, Z., and Palermo, G. D. (2000). Alternative sources of gametes: Reality or science fiction? *Hum. Reprod.* **15**, 988–998.

Uwa, H., and Ojima, Y. (1981). Detailed and banding karyotype analyses of the medaka, *Oryzias latipes* in cultured cells. *Proc. Jpn. Acad.* **57**, 39–43.

Wakamatsu, Y., Ju, B., Pristyaznhyuk, I., Niwa, K., Ladygina, T., Kinoshita, M., Araki, K., and Ozato, K. (2001). Fertile and diploid nuclear transplants derived from embryonic cells of a small laboratory fish, medaka (*Oryzias latipes*). *Proc. Natl. Acad. Sci. U. S. A.* **98**, 1071–1076.

Wittbrodt, J., Shima, A., and Schartl, M. (2002). Medaka—a model organism from the far East. *Nat. Rev. Genet.* **3**, 53–64.

Yanagimachi, R. (2005). Intracytoplasmic injection of spermatozoa and spermatogenic cells: Its biology and applications in humans and animals. *Reprod. Biomed. Online* **10**, 247–288.

Yi, M., Hong, N., and Hong, Y. (2009). Generation of medaka fish haploid embryonic stem cells. *Science* **326**, 430–433.

Yi, M., Hong, N., and Hong, Y. (2010). Derivation and characterization of haploid embryonic stem cell cultures in medaka fish. *Nat. Protoc.* **5**, 1418–1430.

PART II

Developmental and Neural Biology

CHAPTER 4

Neurogenesis

Prisca Chapouton* *and* Leanne Godinho[‡]

*Institute of Developmental Genetics, Helmholtz Zentrum München, German Research Center for Environmental Health, Neuherberg, Germany

[‡]Lehrstuhl für Biomolekulare Sensoren, Institute for Neuroscience, Technische Universität München, Munich, Germany

Abstract
I.   Introduction
II. Neurogenesis During Development and in the Adult Brain
     A. Morphogenesis: Formation of the Neural Plate and Neural Tube
     B. Molecular Determinants and Patterning of the Neural Plate
     C. Proneural Domains, Lateral Inhibition, and Neurogenic Cascade
     D. Zones of Delayed Differentiation
     E. Signaling Centers and Downstream Neuronal Specification
     F. Generation of Glial Cells
     G. Neurogenesis at Juvenile and Adult Stages
III. Methods for Studying Neurogenesis in the Developing and Adult Brain
     A. Molecular Markers
     B. Methods to Label Live Cells
     C. *In Vivo* Imaging
     D. Methods to Follow Cell Cycle Events
     E. Manipulating the Expression of Genes Involved in Neurogenesis
     F. Methods for Targeted Cell Ablations
IV. Specific Protocols to Study Adult Neurogenesis
     A. Fixation of the Adult Brain for Immunohistochemistry and *In Situ* Hybridization
     B. Immunohistochemistry on Vibratome Sections
     C. *In Situ* Hybridization on Whole Mount Adult Brains
     D. *In Situ* Hybridization on Gelatine–Albumin Sections
     E. Intraperitoneal Injections of BrdU
     F. Lipofections/Electroporations of the Adult Brain *In Vivo*
     G. Slice Culture
V.   Conclusion
     Acknowledgements
     References

METHODS IN CELL BIOLOGY, VOL. 100
Copyright © 2010 Elsevier Inc. All rights reserved.

73

978-0-12-384892-5
DOI: 10.1016/B978-0-12-384892-5.00004-9

Abstract

For more than a decade, the zebrafish has proven to be an excellent model organism to investigate the mechanisms of neurogenesis during development. The often cited advantages, namely external development, genetic, and optical accessibility, have permitted direct examination and experimental manipulations of neurogenesis during development. Recent studies have begun to investigate adult neurogenesis, taking advantage of its widespread occurrence in the mature zebrafish brain to investigate the mechanisms underlying neural stem cell maintenance and recruitment. Here we provide a comprehensive overview of the tools and techniques available to study neurogenesis in zebrafish both during development and in adulthood. As useful resources, we provide tables of available molecular markers, transgenic, and mutant lines. We further provide optimized protocols for studying neurogenesis in the adult zebrafish brain, including *in situ* hybridization, immunohistochemistry, *in vivo* lipofection and electroporation methods to deliver expression constructs, administration of bromodeoxyuridine (BrdU), and finally slice cultures. These currently available tools have put zebrafish on par with other model organisms used to investigate neurogenesis.

I. Introduction

The mechanisms by which undifferentiated neural progenitor cells generate mature, functioning neurons are referred to as neurogenesis. Because neural progenitors also generate glial cells (oligodendrocytes and astrocytes), we also include gliogenesis in this chapter. The earliest step in neurogenesis (and gliogenesis) begins with the specification of neural progenitors, followed by proliferative cell divisions that amplify the progenitor pool, cell fate specification and determination, exit from the cell cycle, and finally terminal differentiation. Each of these steps is precisely orchestrated to generate multiple cell-types that ultimately will populate the mature central nervous system (CNS).

For more than a decade, the vertebrate organism zebrafish has been highly productive for studies of neurogenesis, because it harbors the most advantageous features of several different model organisms. Like mice and *Drosophila*, zebrafish are amenable to genetic analysis and like *Xenopus*, fertilization and subsequent development occurs externally, making them particularly accessible for examination and experimental manipulation.

Although neurogenesis during development has been the subject of the vast majority of studies conducted thus far, neurogenesis in the adult zebrafish brain has also recently received attention. Our goal in this chapter is to provide an overview of the tools and techniques currently available to study neurogenesis in zebrafish both during development and in adulthood. We begin, however, with a brief description of how

neurogenesis is initiated during development and how it occurs in the adult brain of zebrafish.

II. Neurogenesis During Development and in the Adult Brain

A. Morphogenesis: Formation of the Neural Plate and Neural Tube

As early as the 16-cell stage of development, it is possible to identify the cells (4 central blastomeres) that will give rise to the future CNS (Helde *et al.*, 1994). The descendents of these four cells are located in the dorsal blastoderm and become assembled into a two-dimensional neural plate by the end of gastrulation. A detailed fate mapping study (Woo and Fraser, 1995) that traced dorsal blastula cells at the shield stage to their corresponding locations in the neural plate revealed that cells located at the animal pole are fated to anterior locations, whereas more laterally located cells are fated to more posterior brain regions. Territories along the anterior–posterior (A–P) axis of the neural plate are fated to give rise to distinct brain regions (retina and telencephalon, diencephalon, mesencephalon, hindbrain, and spinal cord) and can be labeled by specific patterning markers (see II-B and Fig. 1A). At the 6–10 somite stages, the neural plate becomes a three-dimensional keel and subsequently a rod (Kimmel *et al.*, 1995). During this morphogenetic process, medial cells of the neural plate reach ventral locations, while lateral cells become located in dorsal midline domains (Papan and Campos-Ortega, 1997; see Fig. 1B–E). Cells in the rod gradually become polarized, with the apical localization of ZO1 and N cadherin, followed by Lin7c and Nok (Yang *et al.*, 2009). This gradual polarization, as well as mirror symmetric type divisions that occur medially in the neural rod, with the localization of Par3 at the cleavage furrow, induces the formation of a lumen (Tawk *et al.*, 2007; Yang *et al.*, 2009), thereby forming a neural tube. Thus, the establishment of apico-basal polarity by neuroepithelial cells is critical to neural tube formation. This polarity, which involves the apical localization of structural proteins [ZO-1, Par3 (von Trotha *et al.*, 2006), Nok (Yamaguchi *et al.*, 2010), lin7 (Wei *et al.*, 2006)] and of enzymatic proteins [aPKC (Alexandre *et al.*, 2010; Rohr *et al.*, 2006; Tawk *et al.*, 2007)], and the basolateral localization of Numb (Reugels *et al.*, 2006) and lgl2 (Reischauer *et al.*, 2009) is also essential at later stages of development for progenitor maintenance and neuronal fate specification throughout the CNS (Roberts and Appel, 2009). Mutants exhibiting defects in apico-basal polarity are listed in Tables V and VI. As another aspect of tissue polarization, the nuclei of dividing progenitors in the neural tube move between the ventricular (apical) and basal (pial) surfaces in a process called interkinetic nuclear migration (INM), with nuclear position correlating with specific phases of the cell cycle. INM continues to occur at later stages of neurogenesis throughout the CNS and has been well characterized in the developing zebrafish retina where the extent of nuclear movement during INM was shown to correlate with the fate of daughter cells: the more basal a progenitor cell nucleus migrates, the more likely it is to generate a postmitotic daughter cell (Baye and Link, 2007b).

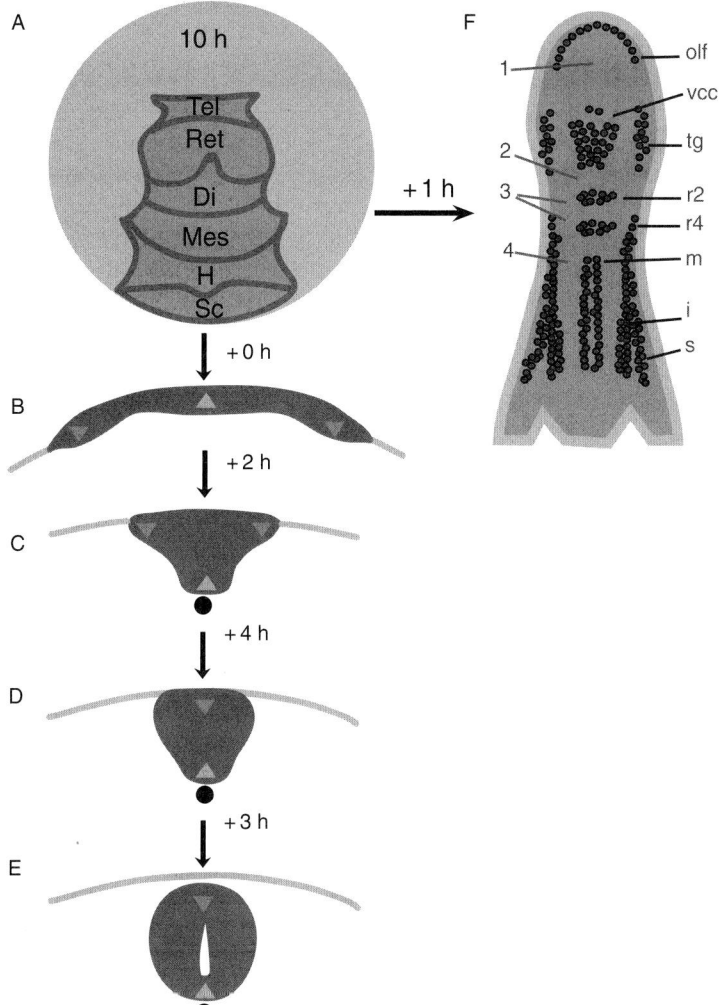

Fig. 1 Fate map, neurogenesis, and early proneural clusters. (A) Fate map of the neural plate at tail-bud stage (10 h post-fertilization, dorsal view). Schematization adapted from Woo and Fraser (1995). The different neural territories are organized in partially overlapping domains that align along the A–P axis during gastrulation. (Tel, telencephalon; Ret, retina; Di, diencephalon; Mes, mesencephalon; H, hindbrain; Sc, spinal cord). (B–E) Morphogenesis of the neural tube at tail-bud (B), 5 somites (C), 14 somites (D), and 20 somites (E) after (Papan and Campos-ortega, 1994; Schmitz *et al.*, 1993). Lateral neural plate bulges converge toward the midline, which folds inward leading to the formation of a neural keel (C) and rod (D). A process of cavitation then shapes a lumen, and thus a tube (E). Prospective dorsal neural tube cells originate from lateral neural plate domains (indicated by red triangles) while ventral neural tube cells originate from medial neural plate domains (indicated by a green triangle). At the keel/rod stage, progenitors undergo mirror symmetric divisions and contribute to cells on both sides of the future neural tube. (F) Early proneural clusters at the 3-somite stage, visualized on a schematically represented flat mounted embryo (anterior to the top) after *neurog1* whole mount in situ hybridization. The following proneural clusters are visible: olf, olfactory neuron progenitors; tg, trigeminal ganglia; vcc, ventro-caudal cluster; r2, motoneuron progenitors of rhombomere 2; r4 motoneuron progenitors of rhombomere 4; m, spinal motoneuron progenitors; i, spinal interneuron progenitors; s, spinal sensory neuron progenitors. These clusters are separated by prepatterned zones of delayed differentiation (1, ANP, anterior neural plate; 2, intervening zone at the mid-hindbrain boundary; 3, longitudinal stripes in r2 and r4; 4, longitudinal stripes in the spinal cord). (See Plate no. 4 in the Color Plate Section.)

B. Molecular Determinants and Patterning of the Neural Plate

The neural territory is marked and specified by members of the SoxB1 family, *sox1a/b, sox2, sox3,* and *sox 19a/b* (Dee *et al.*, 2008; Okuda *et al.*, 2006; Vriz *et al.*, 1996), which are already expressed at blastula stages and in the neural plate. The SoxB transcription factors regulate, both positively and negatively, several transcriptional programs (Okuda *et al.*, 2010) implicated in the patterning of the neural plate (*bmp2b/7*) and in neural differentiation (*her3* and *neurog1*, positively regulated and *ascl1*, negatively regulated). As in other vertebrates, SoxB1 family members are involved generally in imparting neural fate and preventing further differentiation (Bylund *et al.*, 2003). Upstream and in combination with Sox2, the Pou transcription factor pou5f1/Oct4/pou2/Spg (Belting *et al.*, 2001; Reim and Brand, 2006) also activates repressors of differentiation that specify neural progenitors, such as *hesx1* and *her3* (Onichtchouk *et al.*, 2010).

The anterior neural plate (ANP) border is delimited by the placodal field territory and expresses *dlx3b, dlx5a,* and *foxi1* (Akimenko *et al.*, 1994; Nissen *et al.*, 2003; Quint *et al.*, 2000; Solomon and Fritz, 2002; Solomon *et al.*, 2003; Whitlock and Westerfield, 2000), whereas posteriorly the neural plate is delimited by the future neural crest and Rohon-Beard neurons and is marked by *snail1b, foxd3, sox9b,* and *sox10* (Dutton *et al.*, 2001; Li *et al.*, 2002; Odenthal and Nusslein-Volhard, 1998; Thisse *et al.*, 1995). Along the A–P axis of the neural plate, patterning genes subdivide presumptive brain regions and are involved in their growth and specification. An early ANP marker is *otx2*, whereas *hoxb1b* marks the hindbrain territory, from rhombomere 4 to more caudal regions (Koshida *et al.*, 1998; Li *et al.*, 1994; Prince *et al.*, 1998). The anterior neural border, expressing the Wnt inhibitor Sfrp (Houart *et al.*, 2002), is a crucial organizing center that inhibits posterior fates and induces the formation of the telencephalon in the ANP. Within the ANP, *rx3* expression specifies the eye field territory (Stigloher *et al.*, 2006). Along the dorso-ventral (D-V) axis, BMP signaling is an important regulator defining the territories of distinct neurons (Barth *et al.*, 1999). Markers subdividing domains of the developing anterior brain from 17 somites to 1 day have been described in (Hauptmann and Gerster, 2000a, c; Hauptmann *et al.*, 2002).

C. Proneural Domains, Lateral Inhibition, and Neurogenic Cascade

At the 3 somite stage, the first proneural clusters can be identified in the neural plate by their expression of several markers (see below and Fig. 1F). Within these clusters, a few progenitors are specified toward the neurogenic cascade to exit the cell cycle and differentiate as neurons. This complete neurogenic program takes place for a few cells very early in development so that the first nuclei and axonal tracts can be traced as early as 18 h post-fertilization (hpf) using HRP, diI, or acetylated tubulin antibody staining (Wilson *et al.*, 1990). This early wave of neuron production results in the establishment of a basic functional neuronal network that allows newly hatched fish to exhibit escape responses as early as 2 days post-fertilization (dpf) (O'Malley *et al.*, 1996).

Within the proneural clusters, specified progenitors express proneural genes, such as *neurog1, ascl1a*, and *coe2* (Allende and Weinberg, 1994; Bally-Cuif *et al.*, 1998; Blader *et al.*, 1997) (other proneural genes are listed in Tables I and II), neurogenic genes (*deltaA, deltaB, deltaD*, and *deltaC, jagged 1a, 1b, and 2*) (Haddon *et al.*, 1998; Julich *et al.*, 2005; Yeo and Chitnis, 2007), and, as they eventually exit the cell cycle, the early postmitotic neuronal marker *elavl3* (Park *et al.*, 2000b). A dynamic process of lateral inhibition allows some cells within the proneural clusters to follow the neurogenic cascade while keeping their neighbors in a progenitor state. The expression of the genes named above in these clusters is therefore visible as a "salt and pepper" pattern.

Lateral inhibition involves the interaction between cells expressing Notch ligands (Delta A, B, C, D, Jagged1a, 1b, 2) and their neighbors expressing Notch receptors (Notch1a, Notch1b, Notch2, Notch3) (see Tables I and II). Upon ligand binding, Notch is cleaved intracellularly and transported to the nucleus, where it acts together with the transcription factor Su(H) and its coactivator Mastermind to induce the transcription of the family of Hairy/Enhancer of Split genes, such as *her4* (Bray and Furriols, 2001; Takke *et al.*, 1999). Her proteins, in turn, inhibit the expression of proneural genes. Notch activation requires not only binding of the ligand to the receptor but also an internalization of the ligand (Matsuda and Chitnis, 2009), which is intracellularly bound by the Mindbomb ring ubiquitin ligase (Itoh *et al.*, 2003). The dAsb11 ubiquitin ligase is also required in the ligand-expressing cells for correct Notch activity (Diks *et al.*, 2008). Thus, lateral inhibition ensures the maintenance of progenitors within the proneural clusters while permitting the generation of different neuronal phenotypes, owing, in most cases, to a temporal difference of cell cycle exit (Aizawa *et al.*, 2007; Appel *et al.*, 2001; Cau *et al.*, 2008; Yeo and Chitnis, 2007). Pathways and markers involved in the coordination of cell cycle exit and differentiation have been shown to involve Cdkn1c (Park *et al.*, 2005) and histone deacetylase (Yamaguchi *et al.*, 2005) but are in general not well characterized in the zebrafish.

D. Zones of Delayed Differentiation

Zones of delayed neurogenesis, or non-neurogenic zones, are found around the proneural clusters (areas marked 1–4 in Fig. 1F). In several of these non-neurogenic zones, members of the hairy/enhancer of split family genes have been shown to prevent neural progenitors from entering the neurogenic cascade (Stigloher *et al.*, 2008). The intervening zone around the mid-hindbrain boundary which expresses *her5* and *her11* (Geling *et al.*, 2003, 2004; Ninkovic *et al.*, 2005, 2008); the longitudinal domains in the spinal cord, which express *her9* and *her3* (Bae *et al.*, 2005; Hans *et al.*, 2004) and separate the columns of motoneurons, interneurons, and Rohon-Beard sensory neurons; and the ANP (containing the prospective telencephalon, diencephalon, and eyes), by the action of Foxg1 (Bourguignon *et al.*, 1998), constitute zones in which neurogenesis is inhibited. Notably, not only in the neural plate but also later in brain development and at adult stages, several regions of the zebrafish brain maintain pools of undifferentiated neural progenitors that express *her* genes (Chapouton et al., in revision).

E. Signaling Centers and Downstream Neuronal Specification

Positive signals that induce the first proneural clusters and ongoing neurogenesis involve sources of signaling activity within or adjacent to the neural plate, and at later stages within the developing brain. These signals not only induce the neurogenic cascade (i.e., Shh induces *neurog1* expression Blader *et al.*, 1997) but also specify neuronal identity and the size of the neuronal clusters, for example, the role of Wnt in the diencephalic dopaminergic population (Lee *et al.*, 2006; Russek-Blum *et al.*, 2008). Such signaling centers include the floor plate which secretes Shh and Twhh (Ribes *et al.*, 2010; Strahle *et al.*, 2004), the roof plate with Wnt signaling activity (Amoyel *et al.*, 2005), the interrhombomeric boundary cells in the hindbrain conveying Wnt1 signaling activity and inducing *deltaA* expression in rhombomeres (Amoyel *et al.*, 2005), rhombomere 4 conveying FGF3 and FGF8 signaling activity (Maves *et al.*, 2002), the zona limitans intrathalamica (ZLI), with Shh and Twhh (Scholpp *et al.*, 2006, 2009), and the isthmic organizer at the mid-hindbrain boundary (Raible and Brand, 2004; Wurst and Bally-Cuif, 2001) with FGF and Wnt signaling activity orchestrating the growth of the midbrain and rhombomere 1. Downstream of these signaling pathways, transcription factor combinations are activated in a temporal and region-specific manner and define specific differentiation programs, thereby producing a variety of neuronal phenotypes. One example illustrating this diversity of programs is the specification of three types of interneurons in the ventral spinal cord, V3 neurons, and the two classes of Kolmer–Agduhr cells (Ka″ and Ka′) (Yang *et al.*, 2010). V3 and KA″ interneurons are generated by lateral floor plate progenitors and are intercalated on the same ventral level in the spinal cord. The slightly more dorsally located KA′ interneurons are generated by the *olig2*-expressing motoneuron progenitor pool. Initially all three neuron types depend on Shh. Subsequently, the activation cascade generating the V3 identity goes through the transcription factors Gli, Nkx2.2a, Nkx 2.2b, Nkx 9, Olig2, and Sim1 (a leucine zipper transcription factor) whereas the generation of the KA″ identity goes through the transcription factors Gli, Nkx2.2a, Nkx 2.2b, Nkx 9, Olig2, Gata2, Tal2, and Gata3. The generation of KA′ neurons involves Olig2, Gata3, and Tal2. Interestingly, although both KA″ and KA′ inhibitory interneurons express the enzyme Gad67 at the end of their differentiation program, the expression of *gad67* is acquired via two distinct transcriptional cascades. Tables I and II list some of the markers specific to diverse neuronal populations generated in the neural tube and developing brain, Tables III and IV list transgenic lines, and Tables V and VI mutants affecting neurogenesis and signaling pathways.

F. Generation of Glial Cells

Neural progenitors give rise not only to neurons but also to oligodendrocytes and to radial glial cells. The generation of oligodendrocytes has mostly been studied in the spinal cord (Park *et al.*, 2002) and in the hindbrain (Zannino and Appel, 2009) and takes place after motoneurons have been generated, sharing the same progenitors and some of the molecular determinants of motoneuron progenitors, such as Shh and Olig 2 (Park *et al.*, 2002, 2004). Molecular markers of radial glial cells such as GFAP and *nestin* are expressed in neural plate progenitors (Kim *et al.*, 2008; Mahler and Driever,

2007; Tawk *et al.*, 2007). The radial glial morphology, however, characterized by the long processes spanning from the ventricle to the pial surface, becomes visible from juvenile stages onward (Kim *et al.*, 2008; Tawk *et al.*, 2007). Radial glial cells function as neural progenitors (Bernardos *et al.*, 2007; Chapouton *et al.*, 2010; Kim *et al.*, 2008; Lam *et al.*, 2009; März *et al.*, 2010; Pellegrini *et al.*, 2007b) and may also subserve an astrocytic function in the mature brain. A parenchymal astrocytic population, in addition to radial glia, may also exist in the zebrafish brainstem. Evidence for their existence comes from a study in which immunostaining using a monoclonal antibody revealed cells with a stellar morphology reminiscent of parenchymal astrocytes in other vertebrates (Kawai *et al.*, 2001). Interestingly, these cells also express GFAP.

G. Neurogenesis at Juvenile and Adult Stages

Zebrafish grow throughout life with growth occurring in all organs, including the brain. In accordance with this, neurons continue to be generated within the adult fish brain. This continued neurogenesis is made possible by the maintenance of undifferentiated progenitors in specific niches throughout the brain that can be recruited into the neurogenic cascade even at adult stages. Indeed, many more zones of proliferation and neurogenesis have been identified in the zebrafish brain than in mammals (Adolf *et al.*, 2006; Chapouton *et al.*, 2006, 2010; Grandel *et al.*, 2006; Ito *et al.*, 2010; Kaslin *et al.*, 2009; März *et al.*, 2010) (reviewed in Chapouton *et al.*, 2007; Kaslin *et al.*, 2008; Fig. 2). The generation of neurons of distinct neurotransmitter phenotypes (GABAergic, glutamatergic, Tyrosine hydroxylase (TH)-expressing, serotonergic) as well as the generation of oligodendrocytes has been reported. Except for the cerebellum (Chaplin *et al.*, 2010; Kaslin *et al.*, 2009), dividing progenitors in the adult CNS are located in ventricular regions. These dividing progenitors express radial glial markers and/or early differentiation markers (Adolf *et al.*, 2006; Berberoglu *et al.*, 2009; Chapouton *et al.*, 2010; Ganz *et al.*, 2010; März *et al.*, 2010; Menuet *et al.*, 2005; Pellegrini *et al.*, 2007b; Raymond *et al.*, 2006). Efforts to characterize progenitors (radial glial cells) in the adult telencephalon have revealed that they can exist in a quiescent or proliferating state and can switch back and forth between these two states (Chapouton *et al.*, 2010; März *et al.*, 2010). Additionally, BrdU tracing and marker analysis have revealed that radial glial cells generally proceed along the following lineage to generate new neurons: quiescent radial glial cells (state I progenitors) enter the cell cycle, thus becoming dividing radial glial cells (state II progenitors), which in turn generate progenitors expressing early proneural genes such as *ascl1a*, and the neural cell adhesion molecule PSA-NCAM and are thus committed to differentiation (state III progenitors) (Chapouton *et al.*, 2010; März *et al.*, 2010).

The pronounced neurogenic activity in the mature zebrafish CNS translates to a capacity for a strong regenerative response following injury. Thus injury models have also begun to be used to uncover the molecular and cellular features of the regenerative response in the adult CNS particularly in the spinal cord (Reimer *et al.*, 2008, 2009) and in the retina (Bernardos *et al.*, 2007; Fausett and Goldman, 2006; Fausett *et al.*, 2008; Fimbel *et al.*, 2007; Hieber *et al.*, 1998a; Kassen *et al.*, 2007; Thummel *et al.*, 2008a,c, 2010).

III. Methods for Studying Neurogenesis in the Developing and Adult Brain

The tools and techniques currently available to study neurogenesis allow for the identification of progenitors and their progeny, permit dynamic visualization of mitotic divisions and the behavior of proliferating and postmitotic cells, as well as manipulations of cell fate. We discuss each technique briefly and provide tables of available resources (cell-specific markers, transgenic lines, mutants etc.) and references to detailed protocols where appropriate. Finally, although the tools and methods to study neurogenesis during development can, in principle, be used to examine neurogenesis in the adult zebrafish brain, optimization is necessary. We therefore provide detailed protocols for performing a handful of methods including immunostaining, *in situ* hybridization, slice cultures, BrdU injections, lipofections, and electroporations that have been optimized for application in the adult zebrafish brain (see Section IV).

A. Molecular Markers

Comprehensive analysis of how neurogenesis proceeds during development and adulthood requires the ability to identify progenitor cells at different stages of the cell cycle, maturation and commitment, as well as the various types of progeny they generate. Antibodies and antisense RNA probes (listed in Tables I and II) have been indispensable in providing tools for such analyses. For example, antibodies directed against proliferating cell nuclear antigen (PCNA) are used to mark all proliferating cells, while expression of phosphohistone H3 marks only those proliferating cells in the late G2- and M-phase of the cell cycle. Furthermore, the expression of specific genes can be used to identify progenitors committed to the neuronal fate (e.g., neurod) or mark newly generated postmitotic neurons (e.g., HuC and HuD; Kim *et al.*, 1996; Mueller and Wullimann, 2002; Park *et al.*, 2000b), and specific neuronal subtypes (e.g., glyt2 for glycinergic neurons, Higashijima *et al.*, 2004). Whether performed on whole mount embryos or on sections, immunostaining and *in situ* hybridization also provide spatial information about where specific progenitors and cell-types reside in the nervous system. Detailed protocols for performing *in situ* hybridization and immunostaining on embryonic tissue can be found in Hauptmann and Gerster (2000b) and Jowett (1999). A protocol that has been optimized for examining RNA and protein expression in the adult zebrafish brain is provided in Section IV.

B. Methods to Label Live Cells

Although the expression of specific transcripts and proteins can be used to uniquely label and identify cells, they permit analysis only in fixed specimens. Other labeling methods are required to visualize cells dynamically *in vivo*, either to trace lineages or to study the behavior of cells. One of the classical methods to label cells involves iontophoresis or pressure injection of fluorescent dyes. Lineage analysis can be performed when single cells are labeled as the dye is passed on to the ensuing progeny

Table I
Markers of Distinct States and Fates: Antibodies and Antisense Probes—Distinct Cellular States

Name	Ab/RNA	Type of cells labeled	Reference
Cell cycle markers			
Phosphohistone 3	Polyclonal Ab	Cells in late G2- and M-phase	Wei and Allis (1998)
Ccnb1	RNA	Cells in late G2- and M-phase	Kassen et al. (2008)
Cdkn1c	RNA		Park et al. (2005)
PCNA	Ab, monoclonal and polyclonal	All proliferating cells (the antigen remains present for about 24 h)	Wullimann and Knipp (2000); Thummel et al. (2008b)
MCM5	RNA, Ab	All proliferating cells	Ryu et al. (2005)
Histone H1	Ab	Cells in G1-phase	Huang and Sato (1998); Tarnowka et al. (1978)
BrdU	mAb	Cells in S-phase permanently labeled after BrdU treatment	Adolf et al. (2006); Park et al. (2004)
Histone H2A		Visualization of mitotic figures	Pauls et al. (2001)
Undifferentiated progenitor markers			
gfap	RNA, Ab	Radial glial cells	Bernardos and Raymond (2006)
S100ß	Ab	Radial glial cells	Grandel et al. (2006); März et al. (2010)
Glutamine synthetase	Ab	Mueller glia cells	Thummel et al. (2008a)
BLBP/FABP	RNA, Ab	Radial glial cells	Adolf et al. (2006)
nestin	RNA	Radial glial cells	Mahler and Driever (2007)
Cyp19b (AroB)	RNA, Ab	Radial glial cells	Pellegrini et al. (2007a)
pou3 (tai-ji)	RNA	Progenitor cells	Huang and Sato (1998)
pou5f1 (pou2)	RNA	Progenitors	Hauptmann and Gerster (1996)
sox2	RNA, Ab	Progenitors	März et al. (2010)
sox19	RNA	Progenitors	Vriz et al. (1996)
sox31	RNA	Progenitors	Girard et al. (2001)
sox11a	RNA	Progenitors	de Martino et al. (2000)
sox11b	RNA	Progenitors	de Martino et al. (2000)
sox21	RNA	Progenitors	Rimini et al. (1999)
her3	RNA	Progenitors	Bae et al. (2005)
her4	RNA	Progenitors, radial glial cells	Yeo et al. (2007)
her5, her11	RNA	Midbrain progenitors	Chapouton et al. (2006); Ninkovic et al. (2005); Tallafuss and Bally-Cuif (2003)
her6	RNA	Progenitors	Scholpp et al. (2009)
her9	RNA	Progenitors	Latimer et al. (2005)
id2	RNA	Progenitors	Chong et al. (2005)
id3	RNA	Progenitors	Dickmeis et al. (2002)
notch1a	RNA	Progenitors	Mueller and Wullimann (2002); Westin and Lardelli (1997)
notch1b	RNA	Progenitors	Thisse et al. (2001), http://zfin.org
notch2	RNA	Progenitors	Thisse et al. (2001), http://zfin.org
notch3	RNA	Progenitors	Thisse et al. (2001), http://zfin.org
Committed progenitor markers			
dlA	RNA, Ab	Progenitors	Chapouton et al. (2010); Haddon (1998); Mueller and Wullimann (2002); Tallafuss et al. (2009)
dlD	RNA	Progenitors	Dornseifer et al. (1997); Haddon (1998)
dlB	RNA	Subpopulation of deltaA and deltaD expressing cells: singled-out primary neurons	Haddon (1998)

Table I
(*Continued*)

Name	Ab/RNA	Type of cells labeled	Reference
dlC	RNA, Ab	Progenitors, early expression mainly in presomitic mesoderm	Julich *et al.* (2005); Matsuda and Chitnis (2009)
jagged1a, jagged1b, jagged2		Progenitors	Gwak *et al.* (2010); Yeo and Chitnis (2007); Zecchin *et al.* (2005)
onecut1	RNA	Progenitors, early differentiating neurons, except in the telencephalon	Hong *et al.* (2002)
coe2		Progenitors, early postmitotic neurons	Bally-Cuif *et al.* (1998)
neurog1 (ngn1)	RNA Ab	Progenitors, early postmitotic neurons	Blader *et al.* (1997); Mueller and Wullimann (2003); Thummel *et al.* (2008a)
neurod	RNA, Ab		Kani *et al.* (2010); Korzh *et al.* (1998); Mueller and Wullimann (2002)
neurod2	RNA	Progenitors	Liao *et al.* (1999)
neurod4 (zath3)	RNA	Progenitors, early postmitotic neurons	Park *et al.* (2003); Wang *et al.* (2003)
ascl1a	RNA	Progenitors	Allende and Weinberg (1994)
ascl1b	RNA	Progenitors	Allende and Weinberg (1994)
atoh1a, atoh1b, atoh1c	RNA	Progenitors of the rhombic lip	Adolf *et al.* (2004); Chaplin *et al.* (2010)
olig1	RNA	Oligodendrocytes progenitors	Schebesta and Serluca (2009)
olig2	RNA	Oligodendrocytes and motoneurons progenitors; cerebellar eurydendroid cells	Bae *et al.* (2009); Park *et al.* (2002)
Pax6.1, pax6.2	RNA, Ab	Retinal and brain progenitors	Adolf *et al.* (2006); Blader *et al.* (2004); Thummel *et al.* (2010)
Early differentiating markers			
dcc	RNA	First neuronal clusters	Hjorth *et al.* (2001)
ß-thymosin	RNA	Early differentiating neurons	Roth *et al.* (1999)
Polysialic acid (PSA)	Ab	Differentiating neurons, expression on cell bodies	Marx *et al.* (2001); März *et al.* (2010)
cntn2 (tag1)	RNA	Outgrowing and migrating neurons	Warren *et al.* (1999)
L2/HNK1	mAb Zn12	Outgrowing neurons	Metcalfe *et al.* (1990)
elavl3 (HuC)	RNA	Early differentiating and mature neurons. Start at 1 somite	Kim *et al.* (1996)
			Park *et al.* (2000a)
Elavl3 + 4 (HuC+D)	Ab		Mueller and Wullimann (2002)
elavl4 (HuD)	RNA	Subset of postmitotic neurons. Start at 10 somites	Park *et al.* (2000a)
tuba1 (α1-Tubulin)	RNA	Early differentiating and regenerating neurons	Hieber *et al.* (1998b)
β1-Tubulin	RNA	Early differentiating, start at 24 hpf	Oehlmann *et al.* (2004)
gap43	RNA	Postmitotic neurons in the phase of axonal growth and regenerating neurons start at 17 hpf	Reinhard *et al.* (1994)
Eno2	RNA	Neurons	Bai *et al.* (2007)
α2-Tubulin	RNA	Mature neurons	Hieber *et al.* (1998b)
Acetylated Tubulin	Ab	Membrane staining of all differentiated neurons	
Neurofilament	Ab	Mature neurons, reticulospinal neurons (bodies+axons)	Gray *et al.* (2001); Lee *et al.* (1987)

(*Continued*)

Table I
(Continued)

Name	Ab/RNA	Type of cells labeled	Reference
	RMO-44		
Gefiltin (intermediate filament)	Ab	Retinal ganglion cells	Leake *et al.* (1999)
Plasticin (intermediate filament)	Ab	Subset of neurons extending axons	Canger *et al.* (1998)
	RNA		
nadl1.1 (L1.1)	RNA	Reticulospinal neurons during axonogenesis	Becker *et al.* (1998)
E587 (L1 related)	mAb E17	Axons of primary tracts and commissures. start at 17 hpf	Weiland *et al.* (1997)
	CON1	Subset of axons	Bernhardt *et al.* (1990)
	zn-1	All neurons	Trewarrow *et al.* (1990)

Table II
Markers of Distinct States and Fates: Antibodies and Antisense Probes—Neuronal Identity Markers

Neuronal identity	Marker	Ab/RNA	Reference
GABAergic neurons			
		Antibodies	Extensive list in Marc and Cameron (2001); Yazulla and Studholme (2001)
	Gad1		MacDonald *et al.* (2010)
	gad2 (gad65)	RNA	Martin *et al.* (1998)
	gad1 (gad67)	RNA	Martin *et al.* (1998)
Monoaminergic neurons			
	Vmat2	RNA	Wen *et al.* (2008)
Catecholaminergic neurons (dopamine, noradrenalin, adrenalin)			
Dopaminergic	*TH* (tyrosine hydroxylase)	RNA Ab(Chemicon)	Guo *et al.* (1999b)
	Uch-L1	RNA	Son *et al.* (2003)
	nr4a2a (Nurr1)	RNA	Blin *et al.* (2008); Filippi *et al.* (2007); Luo *et al.* (2008)
	Nr4a2b		
	slc6a3 (dat)	RNA	Bai and Burton (2009); Holzschuh *et al.* (2001)
	Otpa,otpb	RNA, Ab	Blechman *et al.* (2007); Ryu *et al.* (2007)
Noradrenergic and adrenergic neurons	*dbh (dopamine β hydroxylase)*	RNA	Guo *et al.* (1999b)
	TH (tyrosine hydroxylase)	RNA Ab(Chemicon)	Guo *et al.* (1999b)
Serotonergic neurons			
	5HT (serotonin)	Ab (Chemicon)	
	tph (tphD1)	RNA	Bellipanni *et al.* (2002)

Table II
(*Continued*)

Neuronal identity	Marker	Ab/RNA	Reference
	tph2 (tphD2)	RNA	Bellipanni *et al.* (2002)
	tphR	RNA	Teraoka *et al.* (2004)
	Pet1	RNA	Lillesaar *et al.* (2007, 2009)
	slc6a4a; slc6a4b	RNA	Norton *et al.* (2008)
Glutamatergic neurons			
		Antibodies	Extensive list in Marc and Cameron (2001); Yazulla and Studholme (2001);
	vglut1 and *vglut2*	RNA	Higashijima *et al.* (2004)
Cholinergic neurons			
		Antibodies	Extensive list in Marc and Cameron (2001); Yazulla and Studholme (2001)
Glycinergic neurons			
	glyt2	RNA	Higashijima *et al.* (2004)
	glra1, glra4a, glrb (glyR subunits)	RNA	Imboden *et al.* (2001)
Histaminergic neurons			
	L-histidine decarboxylase, HA	RNA, Ab	Eriksson *et al.* (1998); Kaslin and Panula (2001b)
Orexin/Hypocretin neurons			
	preproORX	RNA, antibodies	Kaslin *et al.* (2004)
Isotocynergic neurons			
	Otp	Antibody	Blechman *et al.* (2007)
Motoneurons			
Primary and secondary motoneurons	HB9		Flanagan-Steet *et al.* (2005)
Primary motoneurons, early processes	znp-1	mAb	Melancon *et al.* (1997); Trewarrow *et al.* (1990)
Secondary motoneurons during axonal growth	Alcam (DM-GRASP/ neurolin)	Ab zn-5	Fashena and Westerfield (1999); Ott *et al.* (2001)
Primary (RoP, MiP, VaP, CaP) and secondary motoneurons	*lhx3 (lim3)*	RNA	Appel *et al.* (1995)
		pAb	Glasgow *et al.* (1997)
MiP+RoP, secondary motoneurons and cranial motoneurons	*islet1*	RNA	Inoue *et al.* (1994)
		Ab(DSHB, 40.2D6 and 39.5D5)	Korzh *et al.* (1993); Segawa *et al.* (2001)
CaP and VaP	*islet2*	RNA	Appel *et al.* (1995)
Primary motoneurons	*pnx*	RNA	Bae *et al.* (2003)
Primary motoneurons	*olig2*	RNA	Park *et al.* (2002)
Cranial motoneurons	*tbx20*	RNA	Ahn *et al.* (2000)
Occulomotor and trochlear motoneurons	*phox2a*	RNA	Guo *et al.* (1999a)

(*Continued*)

Table II
(*Continued*)

Neuronal identity	Marker	Ab/RNA	Reference
Spinal motoneurons (transiently)+ vagal motor nucleus(nX)	*sst1 (ppss1)* (somatostatin)	RNA	Devos *et al.* (2002)
Reticulospinal interneurons			
	tlx3a (tlxA)	RNA	Andermann and Weinberg (2001)
Mauthner neuron	3A10	Ab	Hatta (1992)
Mauthner neuron	*RMO-44*	mAb	Gray *et al.* (2001); Lee *et al.* (1987)
Mauthner neuron	*glra1, glra4a, glrb (glyR subunits)*	RNA	Imboden *et al.* (2001)
Spinal cord commissural interneurons			
	pax2a (pax2.1)	RNA Ab	Mikkola *et al.* (1992)
Interneurons of the spinal cord			
	pnx	RNA	Bae *et al.* (2003)
	dbx1a (hlx1)	RNA	Fjose *et al.* (1994)
	cxcr4a	RNA	Chong *et al.* (2001)
Small number of interneurons	*islet1*	RNA, Ab	Korzh *et al.* (1993)
Rohon-Beard primary sensory neurons			
	pnx	RNA	Bae *et al.* (2003)
	cxcr4b	RNA	Chong *et al.* (2001)
	islet1	RNA Ab	Korzh *et al.* (1993)
	islet2	RNA	Segawa *et al.* (2001)
	tlx3a (tlxA)	RNA	Andermann and Weinberg (2001); Langenau *et al.* (2002)
	tlx3b	RNA	Langenau *et al.* (2002)
	cbfb	RNA	Blake *et al.* (2000)
	runx3	RNA	Kalev-Zylinska *et al.* (2003)
Subpopulation of RB	*ntrk3a (trkC)*	RNA	Williams *et al.* (2000)
Cerebellar populations			
Upper rhombic lip, granule cells precursors	*atoh1a, atoh1b*	RNA	Adolf *et al.* (2004); Koster and Fraser (2001)
Rhombic lip and granule cells	*zic1* *Ptf1a*	RNA	Chaplin *et al.* (2010)
Eurydendroid cells	*olig2*	RNA	McFarland *et al.* (2008)
Purkinje cells	zebrin2	mAb	Jaszai *et al.* (2003)
	M1	mAb	Miyamura and Nakayasu (2001)
Retinal cells			
	See extensive reviews	Antibodies	Marc and Cameron (2001); Yazulla and Studholme (2001)
	Brn3a		Sato *et al.* (2007a)
Amacrine cells	*pax6a (pax6.1)*	RNA	Kay *et al.* (2001); Raymond *et al.* (2006); Thummel *et al.* (2008b)
Ganglion cells (RGC)	*atoh7 (ath5)*	RNA	Kay *et al.* (2001); Masai *et al.* (2000)
RGC and INL	*islet 1* *islet3*	RNA, Ab	Korzh *et al.* (1993); Masai *et al.* (2000) Kikuchi *et al.* (1997)
RGC and INL	*lhx3 (lim3)*	RNA, Ab	Glasgow *et al.* (1997); Masai *et al.* (2000)

Table II
(*Continued*)

Neuronal identity	Marker	Ab/RNA	Reference
RGC	*cxcr4b*	RNA	Chong *et al.* (2001)
RGC	*dacha*	RNA	Hammond *et al.* (2002)
RGC	zn-5	mAb	Kawahara *et al.* (2002)
Pineal/habenula neurons			
	brn3a		Aizawa *et al.* (2007)
	Serotonin-N-acetyltransferase-2 (aanat2)	RNA	Gothilf *et al.* (2002a)
	islet1	RNA, Ab	Korzh *et al.* (1993)
	lhx3 (lim3)	RNA, pAb	Glasgow *et al.* (1997)
	tph (tphD1)	RNA	Bellipanni *et al.* (2002)
	cxcrb	RNA	Chong *et al.* (2001)
Pituitary gland neurons			
	tbx20	RNA	Ahn *et al.* (2000)
	lhx3 (lim3)	RNA, pAb	Glasgow *et al.* (1997)
Hypothalamic clusters			
	histamine	Ab	Kaslin and Panula (2001a)
	neurog3 (ngn2, ngn3), expression start at 24 hpf	RNA	Wang *et al.* (2003)
	cxcrb	RNA	Chong *et al.* (2001)
	sst1 (ppss1), start at 5dpf	RNA	Devos *et al.* (2002)
	tph (tphD1)	RNA	Bellipanni *et al.* (2002)
	dacha	RNA	Hammond *et al.* (2002)

(Li *et al.*, 2000; Woo and Fraser, 1995). Caged fluorescent dyes can also be injected at the one-cell stage and subsequently uncaged in brain regions of interest at specific developmental times to allow fate-mapping studies. Photolysis of the caging group to reveal fluorescein in small numbers of cells can be achieved by UV irradiation using an epifluorescence microscope fitted with a 4′,6-diamidino-2-phenylindole (DAPI) filter set and a pinhole in the light-path to restrict the illumination to a small area (Kozlowski and Weinberg, 2000; Tallafuss and Bally-Cuif, 2003). Uncaging in individual cells can be accomplished by two-photon excitation (Russek-Blum *et al.*, 2009). Although fluorescent dyes have been successfully used in several fate mapping studies, photobleaching and loss of label can pose problems; successive divisions dilute the dye making it difficult to verify whether all of a cell's descendants have been labeled.

The advent of genetically encoded fluorescent proteins has provided the most effective method thus far to indelibly label cells. Genetically encoded fluorescent proteins are generally non-toxic, do not dilute with cell divisions, and are less prone to photobleaching because new fluorescent proteins are constantly generated by the cell. Fluorescent proteins can be delivered to cells via DNA or RNA expression constructs. The purpose of the experiment and the degree of labeling specificity desired will determine which method should be employed. Ubiquitous labeling may be useful

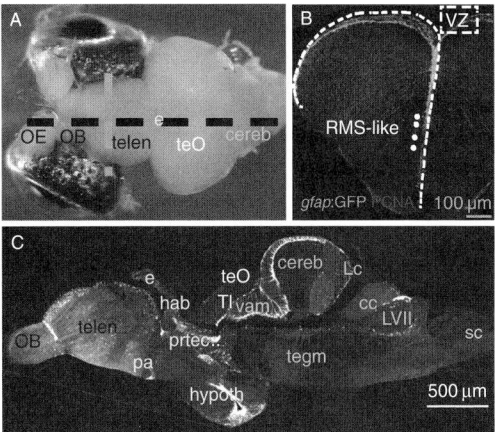

Fig. 2 Neural progenitors in the adult brain. (A) Whole mount preparation of an adult brain of a 3-month-old zebrafish, anterior to the left, depicting the different brain regions and the levels of sections shown in B (green dashed line) and in C (gray dashed line). (B) Cross section of 100 μm thickness through the telencephalon of a *gfap*:GFP transgenic zebrafish depicting in green the cell bodies of GFP-positive radial glial cells located along a single layer at the medial and dorsal ventricular surface (VZ, white dashed line), their processes extending toward the pial surface, and in red the nuclei of dividing cells stained by an anti-PCNA antibody. Dividing cells are scattered along all levels of the VZ and are concentrated in a rostral migratory stream-like stripe (depicted by a white dotted line) situated along the A–P length of the telencephalic midline (see März *et al.*, 2010). (C) Sagittal medial section depicting in white the nuclei of dividing cells stained by a PCNA antibody. Dividing progenitors are located along all ventricular regions of the brain, as well as in the cerebellum (see Kaslin *et al.*, 2009). OE, olfactory epithelium; OB, olfactory bulb; telen, telencephalon; pa, preoptic area; e, epiphysis; hab, habenula; prtec, pretectal nucleus; hypo, hypothalamus; teO, optic tecum; Tl, torus longitudinalis; vam, medial valvula cerebelli; cereb, cerebellum; Lc, lobus caudalis; tegm, tegmentum; cc, crista cerebellaris; LVII facial lobe, sc spinal cord. For more details see the adult brain atlas (Wullimann *et al.*, 1996). (See Plate no. 5 in the Color Plate Section.)

to obtain information about general morphogenesis either during normal development or when things go awry in mutants. However, mosaic labeling provides a powerful way to highlight individual or small numbers of cells allowing their fates and behavior to be studied with greater precision.

1. Transient Expression of RNA and DNA

Ubiquitous labeling of the entire embryo will result when capped RNA is injected at the one-cell stage. To achieve mosaic expression using RNA constructs, injections should be performed at later stages (16–128 cell stages) and into individual blastomeres (e.g., Alexandre *et al.*, 2010). It should be noted that as RNA degrades over time, persistence of the fluorescent label is usually limited to the first 2 days of development. Injecting DNA constructs will result in mosaic expression even when injections are performed at the one-cell stage and when ubiquitous promoters are used. The resulting expression pattern, however, will be stochastic rather than cell specific. When specific cell populations need to be targeted, cell-type-specific promoters and regulatory

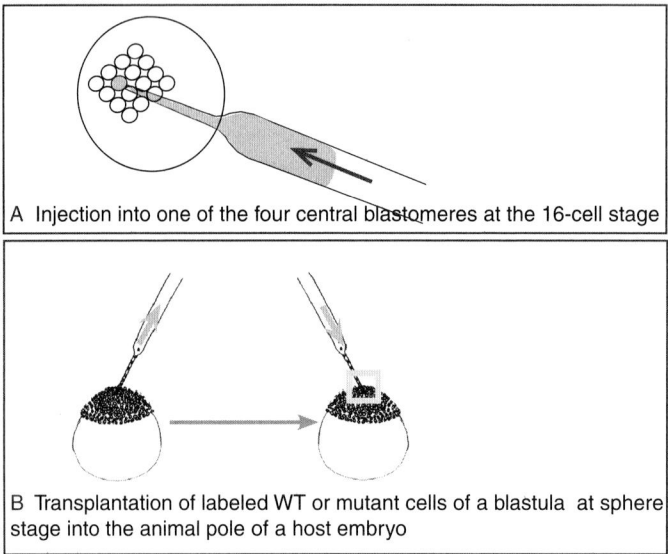

A Injection into one of the four central blastomeres at the 16-cell stage

B Transplantation of labeled WT or mutant cells of a blastula at sphere stage into the animal pole of a host embryo

Fig. 3   Targeting the neural plate by injection into one central blastomere at the 16-cell stage or by blastula transplantation. (A) Injection of DNA or RNA in one of the 4 central blastomeres at the 16-cell stage; (B) Transplantation of blastomeres at sphere stage. Both the injection in one of the 4 central blastomeres and the transplantation into the dorsal side of the blastula lead to a mosaic organization of injected/transplanted cells and their descendants in the neural tube. Prior to transplantation, donor embryos should be injected at the one-cell stage with a fluorescent or biotinylated dextran. Alternatively, donor embryos are transgenically labeled. The dorsal part of the host embryo, depicted by a square, is the targeted area of transplantation.

elements should be used. Although such DNA constructs are used to generate stable transgenic lines, they can also simply be injected at the one-cell stage to generate transient transgenic zebrafish. The resulting mosaic but nonetheless largely cell-specific labeling provides the ability to visualize cells in isolation. It should be noted that transient injections of DNA constructs with cell-specific promoters can sometimes also result in non-specific expression patterns. If it is necessary to restrict expression specifically to the CNS, plasmids can be introduced into one of the 4 central cells at the 16-cell stage, as these cells give rise to the CNS (Helde *et al.*, 1994) (Fig. 3A).

2. Stable Transgenic Lines

To obtain reproducible cell-specific expression patterns for every experiment, stable transgenic lines should be used. A vast array of stable transgenic lines has been generated, where cell-type-specific regulatory elements drive the expression of fluorescent proteins [most commonly green fluorescent protein (GFP)] in progenitors and postmitotic cells of diverse fates (see Table III). Additionally, stable transgenic lines as well as libraries of enhancer trap lines driving the expression of the transcriptional activator Gal4 have been generated (Asakawa and Kawakami, 2009;

Asakawa *et al.*, 2008; Choo *et al.*, 2006; Distel *et al.*, 2009; Ogura *et al.*, 2009; Scott and Baier, 2009; Scott *et al.*, 2007). These Gal4 "driver" lines can be crossed to "effector" transgenic lines expressing fluorescent proteins under the control of the Gal4-specific Upstream Activating Sequence (UAS) promoter (UAS:GFP; UAS: RFP, UAS:Kaede; see Table IV). Alternatively, UAS reporter constructs can be injected into eggs from specific Gal4 driver lines, or vice versa, enhancer-specific Gal4 constructs can be injected into eggs from UAS effector lines to achieve mosaic expression patterns. Expression of the reporter is maintained as long as the cell-specific enhancer and promoter elements drive Gal4 expression. A new UAS effector transgenic line, however, allows continuous expression of a reporter gene once activated. To generate this line, a bicistronic construct was used to express both a reporter gene (GFP) and the Gal4 transcriptional activator, KalTA4, under UAS promoter elements (Distel *et al.*, 2009). Activation of UAS thus results in the expression of both GFP and KalTA4. The expression of KalTA4 allows reiterative activation of UAS and therefore maintenance of expression.

Table III
Transgenic Lines to Visualize Distinct States and Fates

Cell type/brain region	Transgenic line	Reference
Ubiquitous	H2A.F/Z:GFP	Pauls *et al.* (2001)
	H2Afx:H2A-mCherry	McMahon et al. (2009)
	αactin:GFP	Higashijima et al. (1997)
	αactin:M.GFP	Cooper et al. (2005)
	Pax6-DF4:gap43-CFPQ[01]	Godinho *et al.* (2005)
Cell cycle phases	Cecyil double transgenic line	Sugiyama *et al.* (2009)
Progenitors, radial glial cells		
Neural progenitors; radial glial cells	her4:rfp	Yeo *et al.* (2007)
	her4:gfp	
Radial glial cells	Fezf2:GFP	Berberoglu *et al.* (2009)
Radial glial cells	Cyp19a1b:gfp	Tong *et al.* (2009)
Radial glial cells	gfap:gfp	Bernardos and Raymond (2006); Lam *et al.* (2009)
Radial glial cells	nestin:gfp	Kaslin *et al.* (2009); Lam *et al.* (2009)
Mid-hindbrain boundary progenitors	her5:gfp	Chapouton *et al.* (2006); Tallafuss and Bally-Cuif (2003)
Committed progenitors		
Rhombic lip progenitors	Atoh1a:egfp	Kani *et al.* (2010)
	Atoh1a:dTomato	
Neuronal progenitors DRG-sensory neurons	neurog1:gfp	Blader *et al.* (2004); McGraw *et al.* (2008)
Neuronal progenitors	deltaA:gfp	Chapouton *et al.* (2010)
Neuronal progenitors and early differentiating neurons	α1-tubulin:gfp	Fausett *et al.* (2008); Goldman *et al.* (2001)
Differentiating and early postmitotic neurons		
Early differentiating neurons	HuC:gfp	Park *et al.* (2000c); Sato *et al.* (2006)
	HuC:Kaede	

Table III
(*Continued*)

Cell type/brain region	Transgenic line	Reference
Neurons	Eno2:gfp	Bai *et al.* (2009)
	Eno2:rfp	
Differentiating and regenerating neurons	gap43:gfp	Kusik *et al.* (2010)
Oligodendrocyte progenitors	olig1:mgfp	Schebesta and Serluca (2009)
Oligodendrocyte progenitors	sox10:mrfp	Kucenas *et al.* (2008)
Motoneurons and oligodendrocyte progenitors	olig2:GFP	Shin *et al.* (2003)
	olig2:dsred2	Kucenas *et al.* (2008)
	olig2:Kaede	Zannino and Appel (2009)
Motoneurons and oligodendrocyte progenitors	nkx2.2a: EGFP	Kirby *et al.* (2006)
Differentiated neurons and glia		
Oligodendrocytes	plp:gfp	Yoshida and Macklin (2005)
Neuronal subtypes		
Motoneurons	hb9:GFP	Flanagan-Steet *et al.* (2005)
	hb9:mGFP	
Cranial motoneurons	isl1:GFP	Higashijima *et al.* (2000)
Sensory neurons	sensory:GFP	Sagasti *et al.* (2005)
Excitatory interneurons	alx:GFP	Kimura *et al.* (2006)
V2 interneurons	vsx1:GFP	Kimura *et al.* (2008)
Caudal primary neurons	nrp1a:GFP	Sato-Maeda *et al.* (2008)
Spinal cord commissural interneurons	pax2.1:GFP	Picker *et al.* (2002)
Olfactory sensory neurons	pOMP^{2k}:gap-CFP	Sato *et al.* (2005)
	pOMP^{2k}:lyn-RFP	
	pTRPC2^{9k}:gap-Venus	
	pTRPC2$^{4.5k}$:gap-Venus	Sato *et al.* (2007b)
	OR111-7:YFP	
	OR1031:CFP	Feng *et al.* (2005)
	rag1:gfp	
Glycinergic neurons	glyt2:gfp	McLean *et al.* (2007)
Monoaminergic neurons	vmat2:gfp	Wen *et al.* (2008)
TH-positive neurons	TH:Mmgfp	Meng *et al.* (2008a)
Serotonergic neurons	pet1:gfp	Lillesaar *et al.* (2009)
Glutamatergic neurons	vglut2a:egfp	Bae *et al.* (2009)
	vglut2a:dsred2	Kani *et al.* (2010)
	Xeom:gfp	Mione *et al.* (2008)
GABAergic neurons	ptf1a:gfp	Kani *et al.* (2010); Pisharath *et al.* (2007)
	ptf1a:gal4VP16	Zerucha *et al.* (2000)
	1.4dlx5a-6a:GFP	MacDonald *et al.* (2010)
	Dlx1a/2aIG:GFP	
Gonadotropin releasing hormone neurons	gnrh:gfp	Palevitch *et al.* (2007)
Pineal neurons	aanat2:gfp	Gothilf *et al.* (2002b)
Dorsal hindbrain neurons	gad2:gal4VP16	Sassa *et al.* (2007)
Dorsal hindbrain neurons	zic1:gal4VP16	Sassa *et al.* (2007)
Retinal and habenular neurons	Brn3ahsp70:gfp	Aizawa *et al.* (2007)
Telencephalic neurons	Tbr1:yfp	Mione *et al.* (2008)

(*Continued*)

Table III
(*Continued*)

Cell type/brain region	Transgenic line	Reference
Reporter transgenes for specific signaling activity		
Cells containing Notch activity	TP1bglob:gfp	Parsons *et al.* (2009)
	TP1bglob:rfp	
Cells containing wnt activity	Top:dgfp	Dorsky *et al.* (2002)
Cells containing fgf activity	Dusp6:gfp	Molina *et al.* (2007)
Hedgehog-expressing cells	shh:gfp	Ertzer *et al.* (2007)
	twhh:gfp	Du and Dienhart (2001)
Retinal cells		
Rod photoreceptors (pan)	Rhodopsin:GFP	Hamaoka *et al.* (2002)
	Xenopus rhodopsin:GFP	Fadool (2003)
Cone photoreceptors (pan)	TransducinαC:GFP	Kennedy *et al.* (2007)
UV cone photoreceptors (pan)	SWS1:GFP	Takechi *et al.* (2003)
Blue cone photoreceptors (pan)	SWS2:GFP	Takechi *et al.* (2003)
Green cone photoreceptors	RH2-1:GFP, RH2-2:GFP, RH2-3:GFP, RH2-4:GFP	Tsujimura *et al.* (2007)
Horizontal cells (pan)	ptf1a:GFP	Godinho *et al.* (2005)
Bipolar cells	vsx1:GFP	Vitorino *et al.* (2009)
Bipolar cells (subset)	vsx2:GFP	Vitorino *et al.* (2009)
Bipolar cells (subset)	nyctalopin:Gal4VP16 UAS:memYFP	Schroeter *et al.* (2006)
Amacrine cells (pan)	ptf1a:GFP	Godinho *et al.* (2005)
Amacrine cells (subset)	Pax6DF4:memYFP	Kay *et al.* (2004)
	Pax6DF4:memCFP	Godinho *et al.* (2005)
Amacrine cells (subset)	12th:MmGFP	Meng et al. (2008)
Ganglion cells (subset)	brn3c: memGFP	Xiao *et al.* (2005)
Ganglion cells (all or vast majority)	isl2b:mCherry	Pittman *et al.* (2008)
Müller glial cells (pan)	gfap:GFP	Bernardos and Raymond (2006); Kassen *et al.* (2007); Lam *et al.* (2009)
Miscellaneous		
Enhancer trap-viral insertion	yfp	Ellingsen *et al.* (2005)
Enhancer trap	gfp	Parinov *et al.* (2004)
Enhancer trap screen	gal4	Asakawa and Kawakami (2009); Asakawa *et al.* (2008); Choo *et al.* (2006); Distel *et al.* (2009); Ogura *et al.* (2009); Scott and Baier (2009); Scott *et al.* (2007)

Table IV
Uas Effector Lines

Purpose/manipulation	Transgenic line	Reference
Visualization	uas:gfp	Asakawa *et al.* (2008)
Visualization	uas:rfp	Asakawa *et al.* (2008)
Photoconvertible fluorescent protein	uas: kaede	Asakawa *et al.* (2008)
Blocking synaptic transmission	uas:tetxLC	Davison *et al.* (2007)
Overexpression of Notch intracellular domain	uas:nicd	Scheer and Campos-Ortega (1999)

3. Electroporation and Lipofections

When promoter elements that target specific cell populations are unknown, but restricted spatial and/or temporal labeling is still required, DNA or RNA can be transfected into cells either by electroporation or by lipofection. Electroporation methods have been developed for embryonic, juvenile, and adult zebrafish. Following the focal injection of DNA or RNA into the desired brain region to target groups of cells (Bianco *et al.*, 2008; Kera *et al.*, 2010) or using microelectrodes to target individual cells (Cerda *et al.*, 2006; Tawk *et al.*, 2009), a small current is applied. In the adult zebrafish brain, radial glial cells, by virtue of their location near the ventricular surface, can be specifically targeted for lipofection or electroporation by injecting DNA expression constructs into the ventricles (Chapouton *et al.*, 2010) (Fig. 5, see Section IV).

4. Blastomere Transplantation

Mosaic labeling can also be achieved by taking advantage of classical embryological techniques of transplantation (Fig. 3B). Briefly, cells from a transgenically labeled embryo, or from an embryo injected at one-cell stage with rhodamine dextran, are transplanted at sphere stage into unlabeled embryos at the same developmental age, into the region that will develop into the future brain. This region is, however, determined at shield stage (dorsal part of the blastula at shield stage Woo and Fraser, 1995), so that not all transplantations will contribute cells to the brain. Correctly transplanted embryos need to be sorted at later stages. This technique is described in detail in Kemp *et al.* (2009).

C. *In Vivo* Imaging

By permitting direct observations of processes as they occur *in vivo*, time-lapse imaging is proving to be a powerful tool to study neurogenesis. Several characteristics of embryonic and larval zebrafish make them particularly accessible to *in vivo* time-lapse imaging. First, their relative transparency and external development permit imaging without manipulative surgery. Second, genetic tools can be used to label progenitors and their progeny fluorescently (See Section III.B). Third, the speed with which development occurs makes it feasible to follow the fate of individual progenitors over multiple rounds of divisions (Lyons *et al.*, 2003) and until the terminal differentiation of their progeny. Indeed, it has been possible to visualize directly the behavior of neural progenitor cells (including INM), mitotic divisions, where these mitoses occur, and the outcomes of the divisions, whether these are symmetric and proliferative, symmetric and terminal, or asymmetric, with a progenitor and a postmitotic cell being generated (Baye and Link, 2007b; Das *et al.*, 2003; Godinho *et al.*, 2007; Poggi *et al.*, 2005; Norden *et al.*, 2009).

Although largely transparent, zebrafish harbor pigment cells (melanophores and iridophores) which can be problematic for imaging beyond 24 hpf. Melanophores (black pigment cells) are first detectable around 24 hpf, whereas iridophores (silver

pigment cells) can be detected beginning at 42 hpf. Chemical inhibition of melanin formation, by rearing the embryos in *N*-phenylthiourea (PTU) and mutants that lack melanophores (*nacre*; Lister *et al.*, 1999) or iridophores (*roy orbison*; Ren *et al.*, 2002) can be used to circumvent pigment-associated problems. By treating *roy orbison* embryos with PTU, transparency can be maintained until larval stages, whereas breeding *nacre* with *roy orbison* permits transparency to be maintained into adulthood (*mitfa;roy*; White *et al.*, 2008).

In stark contrast to the abundant reports of *in vivo* time-lapse imaging studies of neurogenesis during development, no such reports exist thus far for adult fish. Lack of transparency in adults and the difficulty of immobilizing adult fish are likely contributors. The pigmentation mutants *nacre, roy orbison*, and *mitfa;roy* should, in principle, permit imaging of the intact adult zebrafish CNS. However, using pigmentation mutants to conduct imaging studies, whether during development or at maturity, requires careful verification that physiological processes occur in a normal and timely manner. Although keeping embryonic and larval zebrafish immobilized and alive for imaging is relatively easy, the situation is more challenging for adult fish. Embryonic and larval fish can be embedded in low-melt agarose and continually anesthetized in tricaine for periods of up to 24 h continuously while largely maintaining normal development. For adult fish, it will be necessary to construct special devices to restrain the animal while super-fusing the gills with anesthetic-containing water, methods used in electrophysio-logical recordings of goldfish (Szabo *et al.*, 2008).

The most effective way to indelibly label cells for imaging has been to use genetically encoded fluorescent proteins. Several methods now exist to restrict fluor-escent labeling to individual or a few cells, easing fate tracking during time-lapse imaging (see Section III.B). Despite the vast array of available fluorescent proteins (Rizzo *et al.*, 2009; Shaner *et al.*, 2005, 2008), green fluorescent protein (GFP) remains the most commonly used in reporter constructs and in stable transgenic lines (see Table III). When multicolor imaging is required, either to label different compartments of a cell (e.g., nucleus and cytoplasm) or to label different cell populations concurrently, it is worth noting the fluorescent proteins with non-overlapping spectra (Shaner *et al.*, 2005). Although combinations of GFP and RFP have been the most commonly used (Alexandre *et al.*, 2010), CFP and YFP have also been demonstrated to work well in combination (Mumm *et al.*, 2006). A notable addition to the palette of fluorescent proteins is the growing number of photoconvertible and photoswitchable fluorescent proteins which can serve to highlight one or a few cells among a larger labeled population, thus allowing cells to be tracked more confidently (Lippincott-Schwartz and Patterson, 2009). Kaede and Dronpa are the only examples reported to be used in zebrafish thus far (Aramaki and Hatta, 2006; Hatta *et al.*, 2006; Sato *et al.*, 2006). Kaede converts from green to red fluorescence in response to irradiation with UV light (Ando *et al.*, 2002), while Dronpa can be switched on or off by excitation with 405 and 488 nm respectively (Habuchi *et al.*, 2005). Additionally, two-photon excitation can be used to turn on Dronpa, thus making it possible to target individual cells in a volume (Aramaki and Hatta, 2006).

With its ability to capture images at depths 30–50 µm from the surface and optical sectioning capability, confocal microscopy has been the workhorse of most of the imaging studies conducted in zebrafish so far. Confocal microscopes equipped with the appropriate laser lines to excite the most commonly used fluorescent proteins (440 nm for CFP, 488 nm for GFP, 514 nm for YFP, and 559 or 568 nm for RFP) and several detectors (photomultiplier tubes) allow multichannel imaging. A 405 nm laser line or other source of UV illumination is necessary when working with many photoactivatable fluorescent proteins (e.g., Kaede). Imaging deeper structures, several hundred microns below the surface, can be accomplished only by multiphoton microscopy, which requires a tunable pulsed infrared laser. Additionally, because multiphoton excitation is confined only to the focal plane, repetitive imaging over long time frames produces less phototoxic effects than confocal microscopy (Helmchen and Denk, 2005). When fast acquisition and detection of cellular or intracellular events and optical sectioning are required, spinning disk confocals are more suitable than slow scanning confocal and multiphoton microscopes.

Finally, recent developments in light microscopy that allow imaging at high speeds and at greater depths, while keeping phototoxicity at low levels, have provided unprecedented three dimensional (3D) reconstructions of early zebrafish embryogenesis in its entirety. These new microscopy approaches promise to contribute to studies of neurogenesis by providing not only qualitative data but also quantitative data. As the name implies, digital scanned laser light sheet fluorescence microscopy (DSLM) uses a sheet of laser light to illuminate a transparent, fluorescently labeled specimen in a single plane and a camera to detect images placed at a 90° angle from the axis of illumination. DSLM was used to record the first 24 h of zebrafish embryogenesis with high temporal and spatial resolution allowing the mitotic divisions and movements of every cell to be tracked (Keller et al., 2008). Modifications of DSLM, using structured illumination to overcome the difficulties of imaging older zebrafish embryos which scatter light, allowed imaging up to the third day of development (Keller et al., 2010). Another approach to acquire complete views of zebrafish embryonic development exploited the intrinsic non-linear optical properties of cells to image mitotic spindles and membranes without the need for fluorescent labeling (Olivier et al., 2010). Furthermore, optimizing the scanning mode to capture images of the deepest structures at higher frame rates allowed entire 3D reconstructions of embryogenesis for the first 10 cell division cycles. The vast amounts of data acquired through such imaging approaches have also necessitated the development of software to allow for automated image reconstruction and cell tracking (Keller et al., 2008, 2010; Olivier et al., 2010).

D. Methods to Follow Cell Cycle Events

One of the simplest and yet powerful tools to examine the dynamics of the cell cycle takes advantage of the incorporation of thymidine analogues into newly synthesized DNA. The most commonly used among these analogues is bromodeoxyuridine

(BrdU), followed by chlorodeoxyuridine (CldU), iododeoxyuridine (IdU), and ethynyldeoxyuridine (EdU).

Because DNA is newly synthesized during S-phase, BrdU incorporation, which can be detected by antibody labeling, can be used to mark cells undergoing the S-phase of the cell cycle. BrdU incorporation can also be used to measure other cell cycle parameters. The relative cell cycle speed of a progenitor population can be inferred by calculating the proportion of cells residing in S-phase (BrdU-positive) within the total number of cycling cells (i.e., PCNA-positive), given that the S-phase remains relatively constant, while the G1-phase is more variable. The fates of cells "born" (i.e., that permanently leave the cell cycle) at specific times can be determined by administering BrdU over a short time window and sacrificing the animal after several days or weeks. Because the incorporated BrdU will not be diluted by cells that have left the cell cycle, double labeling with antibodies against BrdU and specific cell-type markers will reveal the fate of the cells. After long chase periods, some BrdU label-retaining cells remain in the stem cell niche without having divided many times, therefore without having diluted BrdU. The cycling status of these label-retaining cells can be determined by a co-staining for PCNA (Adolf *et al.*, 2006; Alunni *et al.*, 2010; Chapouton *et al.*, 2006; Kaslin *et al.*, 2009) or by the incorporation of a second thymidine analogue, such as IdU (Aluni et al., 2010). BrdU and the other base analogues can be administered to the living animal in different ways: embryos can be soaked in a BrdU solution (although the skin can be difficult to permeabilize) (Kim *et al.*, 2008; Park *et al.*, 2004) or BrdU can be directly injected into the brain region of interest (Baye and Link, 2007b). To examine cell proliferation in the adult CNS, BrdU can be injected intraperitoneally (detailed protocol provided in Section IV, Fig. 4) (Adolf *et al.*, 2006) or added to the swimming water (Grandel *et al.*, 2006).

An alternative to using BrdU incorporation to determine the birthdates of cells takes advantage of the photoconversion properties of Kaede. By using transgenic lines in which neurons express Kaede shortly following their birth (Huc:Kaede), selective photoconversion can highlight neurons generated at specific developmental times. The fates that these Kaede-marked new-born cells adopt can be examined by their subsequent expression of cell-type-specific transgenic markers. This method termed BAPTISM (birthdating analysis by photoconverted fluorescent protein tracing *in vivo* combined with subpopulation markers), permits continuous observation of the birthdated cells throughout their development (Caron *et al.*, 2008).

Taking advantage of the upregulation of specific proteins at distinct phases of the cell cycle, a new technique, Fucci (fluorescent ubiquitination-based cell cycle indicator), allows an *in vivo* glimpse into cell cycle progression (Sugiyama *et al.*, 2009). By crossing two transgenic lines in which the ubiquitination domain of Cdt1 is fused to Kusabira Orange 2 and the ubiquitination domain of geminin is fused to Azami Green, respectively, it is possible to distinctly mark the G1-phase (orange) and the S/G2/M-phase (green) of the cell cycle. This new tool permits cycling progenitors to be imaged *in vivo*, while displaying the phase of the cell cycle they are in. It will thus be possible to investigate whether progenitors in specific brain regions or at different developmental times display heterogeneous cell cycle behaviors.

E. Manipulating the Expression of Genes Involved in Neurogenesis

To investigate the genetic mechanisms underlying neurogenesis, loss-of-function and gain-of-function approaches can be used. Gain-of-function approaches typically involve the overexpression or misexpression of genes by RNA injection or transient or stable transgene expression. Restricting gene expression temporally and/or to specific cell-types can be achieved by using cell-type-specific promoters or inducible systems (discussed below). Loss-of-function studies examine the effects of disruptions in gene function. Antisense morpholino oligonucleotides to knock-down gene expression (Knowlton *et al.*, 2008), the use of dominant negative constructs, and analysis of mutant fish can provide insights into the genetic underpinnings of neurogenesis. Mutants have been isolated in several mutagenesis screens (Tables V and VI) generated using TILLING (Targeting induced local lesions in genomes) libraries (Wienholds et al., 2002) or by the more recent technology of zinc finger nucleases (Doyon *et al.*, 2008; Meng *et al.*, 2008b). The use of small interfering RNAs (siRNAs) to knock-down-specific genes in zebrafish has not proved successful so far due to strong developmental side effects. These side effects have been attributed to an interference with the microRNA processing machinery (Fjose and Zhao, 2010), leading to an impairment of microRNA function.

Many loss- and gain-of-function approaches disrupt gene expression early in development and potentially throughout the embryo. Thus it may be difficult to decipher a specific role for a gene in neurogenesis if it also plays a role in early embryonic development. To circumvent this, genetic perturbations that can be spatially and/or temporally restricted are required.

Table V
Mutants and Transgenics Allowing for Manipulation of the Whole Embryo

	Transgenic, mutant, or chemical inhibitor	Affected protein	Reference
Mutants affecting cell cycle			
	mcm5	MCM5	Ryu *et al.* (2005)
	disarrayed	–	Baye and Link (2007a)
	perplexed	carbamoyl-phosphate synthetase2-aspartate transcarbamylase-dihydroorotase (cad)	Link *et al.* (2001b); Willer *et al.* (2005)
	pescadillo	Pescadillo	Allende *et al.* (1996)
	APC	APC	Wehman *et al.* (2007)
	curly fry(cfy)	-	Song *et al.* (2004)
	flotte lotte (flo)	Elys/Ahctf1	Cerveny et al. (2010)
	Transgenic: hsp70:cdkn1c-myc		Park *et al.* (2005)
Mutant of the neurogenic cascade			
	Notch pathway mutants (see below)		
	neurog1	Neurog1	Golling *et al.* (2002)
	ascl1a	Ascl1a	http://zfin.org

(*Continued*)

Table V
(Continued)

	Transgenic, mutant, or chemical inhibitor	Affected protein	Reference
Cell polarity mutants			
	ncad	N-Cadherin	Lele *et al.* (2002); Malicki *et al.* (2003); Yamaguchi *et al.* (2010)
	nagy oko (nok)	Pals1/Stardust	Bit-Avragim *et al.* (2008); Wei and Malicki (2002)
	mosaic eyes (moe)	FERM domain protein	Jensen and Westerfield (2004)
	heart and sould (has)	aPKCλ	Horne-Badovinac *et al.* (2001)
	oko meduzy (ome)	Crumbs2	Malicki and Driever (1999); Omori and Malicki (2006)
	pen/lgl2	Lgl2	Reischauer *et al.* (2009); Sonawane *et al.* (2005)
	ale oko	P50 component of dynactin complex	Jing and Malicki (2009)
Signalling pathway mutants or inhibitors			
FGF-pathway	Hsp:dn-fgfR1		Gonzalez-Quevedo *et al.* (2010); Lee *et al.* (2005)
	Hsp: ca-fgfR		Marques *et al.* (2008)
	Hagoromo FGF upregulation mutant		Amsterdam *et al.* (2009)
	fgf20 mutant		Gonzalez-Quevedo *et al.* (2010)
	fgf8 mutant		Reifers *et al.* (1998)
	Inhibitor	SU5402	Furthauer *et al.* (2001)
Shh-pathway	smoothened (smo)	Smoothened receptor	Chen *et al.* (2001)
	Sonic you (Syu)	Shh	Schauerte *et al.* (1998)
	detour	Gli1	Odenthal *et al.* (2000); Vanderlaan *et al.* (2005)
	you too (yot)	Gli2	Vanderlaan *et al.* (2005)
	broad minded (bromi)		Ko *et al.* (2010)
	Inhibitor: cyclopamine		Chen *et al.* (2001)
Notch-pathway	deadly seven (des)	Notch1a	Gray *et al.* (2001); Liu *et al.* (2003)
	dla^dx2	DeltaA	Appel *et al.* (1999)
	after eight (aei)	DeltaD	Holley *et al.* (2000)
	Beamter (bea)	DeltaC	Julich *et al.* (2005)
	mind bomb (mib)	E3 ubiquitin ligase that is required for Delta endocytosis	Itoh *et al.* (2003)
	Transgenic: hsp70:dnSu(H)	Dominant negative effect on Su(H) and block of Notch activity	Latimer *et al.* (2005)
	Transgenic: hsp:gal4 x uas:nicd-myc	Activation of Notch signalling	Scheer *et al.* (2001)
	Inhibitor: DAPT		Chapouton *et al.* (2010); Geling *et al.* (2002)
BMP pathway	swirl (swr)	Bmp2	Kishimoto *et al.* (1997); Nguyen *et al.* (1998)
	Snailhouse (snh)	Bmp7	Nguyen *et al.* (1998)
	chordino (chd)	Chordin	Schulte-Merker *et al.* (1997)
	Somitabun (sbn)	Smad5	Hild *et al.* (1999)
Nodal pathway	one eyed pinhead (oep)	EGF-CFC class protein	Gritsman *et al.* (1999)
	cyclops	TGFβ of nodal class	Feldman *et al.* (2000)
	squint	TGFβ of nodal class	Feldman *et al.* (2000)

Table V
(Continued)

	Transgenic, mutant, or chemical inhibitor	Affected protein	Reference
Wnt pathway	Inhibitor: SB431542		Ribes *et al.* (2010)
	masterblind	Axin1	Heisenberg *et al.* (1996); Masai *et al.* (1997)
	headless	TCF3	Kim *et al.* (2000)
	GSK3ß Inhibitor: OTDZT		Ninkovic *et al.* (2008)
	Activator: LiCl		Ninkovic *et al.* (2008)
Mutant collections	Viral mutagenesis screen		Wang *et al.* (2007)
	Insertional mutagenesis screen		Amsterdam *et al.* (1999); Golling *et al.* (2002)
	Gene trap screen		Asakawa *et al.* (2008)
	TILLING mutants		Moens *et al.* (2008)

Table VI
Mutants Affecting Neuronal Specification

Mutant name	Abbreviation	Mutated gene	Phenotype	Reference
narrowminded	nrd	*prdm1* (SET/ zinc finger transcription factor)	Loss of Rohon-Beard and decrease in neural crests neurons	Artinger *et al.* (1999); Hernandez-Lagunas *et al.* (2005)
motionless	mot	*Med* (component of the Mediator complex)	Less dopaminergic hypothalamic neurons, lacking brain ventricles, cell death in telencephalon and lens by 50 hpf.	Guo *et al.* (1999b); Wang *et al.* (2006)
foggy	fog	*supt5h (spt5)*	Lack of dopaminergic neurons in hypothalamus, telencephalon and retina, and lack of noradrenergic neurons in the locus Coeruleus (+cardiovascular defects)	Guo *et al.* (1999b); Guo *et al.* (2000)
too few	tof	*fezl*	Reduction of hypothalamic dopaminergic neurons	Guo *et al.* (1999b); Levkowitz *et al.* (2003)
opta	m866	*orthopedia*	Fewer dopaminergic neurons in the hypothalamus and caudal posterior tuberculum	Ryu *et al.* (2007)
soulless	sll	*phox2a*	Loss of locus coeruleus and arch associated catecholaminergic neurons	Guo *et al.* (1999a)
lakritz	lak	*atoh7 (ath5)*	Loss of retinal ganglion cells	Kay *et al.* (2001)
young	yng	*smarca4*	Blocked final differentiation of retinal cells	Gregg *et al.* (2003); Link *et al.* (2000)
perplexed	plx	carbamoyl-phosphate synthetase2-aspartate transcarbamylase-dihydroorotase (cad)	Cell death of retinal cells before exiting the cell cycle	Link *et al.* (2001b); Willer *et al.* (2005)
confused	cfs		Cell death in a subset of retinal postmitotic cells	Link *et al.* (2001a)
ascending and descending	add	Hdac1	Retinal cells do not exit the cell cycle	Yamaguchi *et al.* (2005)

1. Spatial Control of Genetic Manipulations

Blastomere transplantation is one approach to spatially restrict genetic perturbations. Mosaic mutants can be generated by transplanting a few cells at the sphere stage from tracer-labeled (rhodamine dextran or other fluorophore) mutant embryos into the dorsal blastula of unlabeled wild-type embryos. Because the dorsal blastula is the site of the future presumptive brain (Woo and Fraser, 1995), integration of the transplanted mutant cells will be restricted to this region. Genetic mosaic animals have the added benefit of allowing studies of the effect of a mutation in isolated cells and whether a phenotype resulting from a mutated gene is cell autonomous or non-autonomous (Ho and Kane, 1990).

Genetic perturbations can also be targeted to specific brain regions or even single cells at selected times (from early embryos until adulthood) by the electro-poration (Bianco et al., 2008; Cerda et al., 2006; Chapouton et al., 2010; Kera et al., 2010; Tawk et al., 2009) or lipofection (Chapouton et al., 2010) of constructs (e.g., overexpression, dominant-negative). Additionally, electroporation can be used to deliver antisense morpholinos directly into CNS regions of interest even at adult stages (Fausett et al., 2008; Thummel et al., 2008c). The ability to knock-down genes at later ages using this approach circumvents the declining efficacy of morpholinos injected at the one-cell stage. (Morpholinos are usually not effective at knocking down genes beyond the first 2 days of development, presumably due to their dilution with cell divisions and/or due to the synthesis of new mRNA.) Recombinant sindbis and rabies viruses can also be used to deliver transgenic constructs by infecting neurons in larval or adult zebrafish brains (Zhu et al., 2009). CMV-based retro- and lentiviruses can be used to target the adult brain ventricular zones (Rothenaigner and Bally-Cuif, personal communication).

Cell-type-specific promoters can be used to target genetic manipulations to specific cell populations using either transient transgenic approaches or stable transgenic lines. Additionally, one can take advantage of the vast array of available Gal4 driver lines generated through enhancer trap screens, some of whose expression patterns have been characterized. Specific Gal4 driver lines can be crossed with UAS effector transgenic lines that allow for genetic manipulation (e.g., *UAS:NICD* that produces overexpression of the Notch intracellular domain; Table IV). Alternatively, in the absence of UAS effector transgenic lines, UAS constructs can be injected into eggs obtained from Gal4 driver lines.

In a similar manner to the Gal4-UAS bipartite system, but leading to permanent genetic recombination, the Cre-loxP system can be used to achieve spatial control of gene expression. Cre recombinase expressed under a specific enhancer in one transgenic line can be used to excise loxP sites in a second transgenic line carrying a ubiquitously expressed reporter. Recombination will occur in cells expressing Cre and in the progeny they generate, providing a way to mark clones of cells genetically. Conversely, the reporter can be expressed under a specific enhancer and Cre expressed ubiquitously (Seok et al., 2009) (see Table VII).

Table VII
Cell-Type-Specific Manipulations: Simple and Combined Driver + Effector Transgenic Lines

Description	Transgenic line	Reference
Tracing of cells progeny	Tissue-specific:KalTA4 (several enhancer trap lines)uas: gfpT2ATalTA4	Distel *et al.* (2009)
	http://plover.imcb.a-star.edu.sg/~zetrap/ZETRAP.htm	Asakawa *et al.* (2008); Choo *et al.* (2006); Davison *et al.* (2007); Scott *et al.* (2007)
Permanent ubiquitous deletion of gfp and expression of dsred.	EF1α-lox-gfp-lox-dsred2	Yoshikawa *et al.* (2008); Sinha *et al.* (2010)
Permanent ubiquitous deletion of dsred and expression of gfp.	EF1α-lox-dsred2-lox-egfp	Hans *et al.* (2009)
Permanent expression of KRAS in progenitors after Cre- induced recombination.	Nestin-lox-Cherry-lox-EGFP-KRAS	Seok *et al.* (2009)
Permanent ubiquitous expression of KRAS after Cre- induced recombination.	ß-actin-lox-egfp-lox- KRasG12D	Le *et al.* (2007)

2. Temporally Controlled Genetic Manipulations

Approaches that allow inducible gene expression permit temporal control of genetic manipulation. One of the most widely used inducible systems in zebrafish involves the use of the heat shock promoter *hsp70* (Collins *et al.*, 2010; Le *et al.*, 2007; Thummel *et al.*, 2005). Expression of genes cloned downstream of the *hsp70* promoter involves exposure of the embryos to 37°C which can be spatially restricted to single cells by using a laser beam (Halloran *et al.*, 2000). Other methods developed to achieve temporal control of gene expression (Table VIII) use substrate-dependant transcriptional activation. Although some of these methods have not been used specifically in studies of the nervous system, they should in principle work. For example, the expression of a glucocorticoid receptor fused to Gal4 (de Graaf *et al.*, 1998) or of a modified Ecdysone receptor from insect fused to Gal4 (Esengil *et al.*, 2007) induces expression of a UAS-driven target gene only upon administration of a chemical compound (i.e., dexamethasone or tebufenozide, respectively). Similarly, the lexPR operator line is activated by the administration of the steroid Mifepristone, thereby inducing expression of the LexOP-driven effector line (Emelyanov and Parinov, 2008). Furthermore, a recent report suggests that the Tet system, used in mammals (Schonig and Bujard, 2003), has been successfully transferred to zebrafish (Zhu *et al.*, 2009). Here the Tet-Off system was used to conditionally repress transgenes upon doxycycline application (Zhu *et al.*, 2009). Temporal control of recombination using the Cre-LoxP system can also be achieved by using the ligand-inducible form of the Cre recombinase, CreER^{T2}, whose expression occurs only upon administration of the drug tamoxifen or its active metabolite 4-hydroxy-tamoxifen (Hans *et al.*, 2009). CreER^{T2} can also be activated in single cells when two-photon excitation is used to uncage a caged ER ligand, 4-hydroxy-cyclofen (Sinha *et al.*, 2010).

Table VIII
Time Controlled Cell-Type-Specific Manipulation: Driver + Effector Transgenic Lines + Substrate Administration

Description	Transient induction by	Transgenic line or constructs	Reference
	Targeted heat shock on single cells via laser beam	hsp70:egfp	Halloran et al. (2000)
	Heat shock	hsp:cre hsp:egfp-cre	Le et al. (2007)Thummel et al. (2005)
	Heat shock	hsp:gal4	Scheer and Campos-Ortega (1999)
HOTcre	Heat shock	hsp:lox-mcherry-lox-H2Bgfp	Hesselson et al. (2009)
	4OH-Tamoxifen	Pax2.1:cre^{ERT2}	Hans et al. (2009)
	Light-induced uncaging of 4OH-cyclofen	cre^{ERT2} transient ubiquitous expression	Sinha et al. (2010)
Self-excising Cre;Mosaic Clonal analysis	Heat shock	EF1α-L-hsp:cre-L-gal4VP16-uas-nlsRFP	Collins et al. (2010)
Glucocorticoid receptor	Dexamethasone	No stable line-but construct: CMV: GalGR	de Graaf et al. (1998)
Insect-derived Ectysone receptor	Tebufenozide	No stable line, but construct: CMV:EcR-gal4-VP16	Esengil et al. (2007)
Fusion between bacterial repressor/ activator domain/ human progesterone receptor	Mifepristone (Ru486)	Driver+operator line: Krt8:LexPR-LexOP:egfp Krt8:LexPR-LexOP:Kras Operator line: Cry:cfp-LexOP:mCherry	Emelyanov and Parinov (2008)
	Transient inactivation by administration of doxycyclin	Activator lines: HuC:itTA dlx4/6:itTA Effector lines: Ptet:venus Ptet:ChR2YFP	Zhu et al. (2009)
Cell ablation	NTR substrate: Metronidazole	uas:nfsBmCherry uas:nfs-cfp	Davison et al. (2007)Curado et al. (2008)Review Pisharath and Parsons (2009)

F. Methods for Targeted Cell Ablations

Cell ablation studies can be used to reveal the importance of specific cell-types during development, examine lineage relationships, or probe the mechanisms that trigger proliferation and neurogenesis in the regenerative response to injury. Laser-mediated cell ablation (Gahtan and Baier, 2004; Yang et al., 2004) and the use of diphtheria toxin under the control of specific promoters to mediate cell death have both been reported in zebrafish (Kurita et al., 2003; Li et al., 2009; Wan et al., 2006). It should be noted, however, that reports of diphtheria toxin-mediated cell ablation in the zebrafish nervous system are currently lacking.

A relatively recent addition to the array of cell ablation techniques takes advantage of the ability of the bacterial nitroreductase enzyme to convert the prodrug, metronidazole into a cytotoxic agent (Bridgewater *et al.*, 1995). Using cell-specific promoters to express nitroreductase in specific cell populations and exposure to metronidazole allows targeted cell ablation that can be spatially and temporally controlled (Curado *et al.*, 2008; Davison *et al.*, 2007; Montgomery *et al.*, 2009; Pisharath *et al.*, 2007; Zhao *et al.*, 2009). Fusing fluorescent proteins to the nitroreductase coding sequence allows cells to be monitored before being ablated.

Another enzyme prodrug combination, thymidine kinase–ganciclovir, which is used as a therapy for cancer, can also be employed to ablate cells (Springer and Niculescu-Duvaz, 2000). The herpes simplex virus thymidine kinase phosphorylates the drug, ganciclovir, which then acts by inhibiting DNA polymerase and thus arresting the cell cycle and causing subsequent cell death. No report of its use exists in zebrafish to date. Additionally, it should also be noted that, as a result of its mechanism of action, cell ablation would be restricted to proliferating cells. Finally, expression of the fluorescent protein KillerRed has the potential to mediate cell death. When illuminated with green light, KillerRed generates reactive oxygen species thereby killing the cells that express it (Bulina *et al.*, 2006). No reports of cell ablation using KillerRed in zebrafish currently exist.

IV. Specific Protocols to Study Adult Neurogenesis

A. Fixation of the Adult Brain for Immunohistochemistry and *In Situ* Hybridization

Anesthetize adult fish on an ice/water mix for 3–5 min and decapitate with a scalpel. Either dissect the brain immediately in phosphate-buffered saline (PBS) and fix for 4–6 h at 4°C in 4% paraformaldehyde (PFA) or fix the whole head at 4°C in 4% PFA overnight on a shaker and dissect the brain the next day. After washing out the PFA in PBS, proceed to a methanol series (sequentially: 25% MeOH (in PBS+0.1%Tween), 50%, 75%, and 100% MeOH). Freeze the brain in 100% Methanol at –20°C for at least 1 h. The brains can be preserved for several months in MeOH at –20°C.

B. Immunohistochemistry on Vibratome Sections

Brains stored in MeOH at –20°C must be processed through a descending methanol series, back to PBS (sequentially 75% MeOH (in PBS+0.1% Tween=PBT); 50% MeOH, 25% MeOH, PBS). It should be noted that some antigens (e.g., lipid soluble) do not tolerate methanol treatment (recommended above for preparing brains). Some antigens do not require any methanol treatment. Some antibodies require a special pretreatment, such as a second methanol series, treatment with hydrochloric acid (HCl), or a citrate buffer treatment (see antigen retrieval, below).

Embed the brain in 3% agarose in PBS. Cool the agarose and cut a block around the brain.

Cut 70–100 μm-thick sections on a vibratome.

Incubate the sections in a blocking solution (PBS containing 0.5% Triton X-100 and 5–10% normal immune goat serum).

Primary antibody (diluted in blocking solution) incubation should proceed for 2 h at room temperature or overnight at 4°C on a shaking incubator.

Wash 2–3× for 5 min in PBS.

Incubate the sections in secondary antibody (1:1000 dilution, e.g., Alexa Fluor-coupled antibodies from Molecular Probes-Invitrogen) in PBS containing 0.5% Triton and 10% normal goat serum for 30–45 min at room temperature in the dark.

Wash 2–3× in PBS

The sections are mounted on slides in Aqua polymount (Polyscience) or any other water-based mounting medium.

Pretreatments, Antigen Retrieval

Hydrochloric Acid Pretreatment   Detection of BrdU by immunostaining requires HCl pretreatment which denatures DNA and thus exposes the BrdU epitope. HCl pretreatment might destroy other antigens and should therefore be performed after the completion of immunohistochemistry for the other markers. Incubate the sections with freshly made 2M HCl (1:4.4 of 32% HCl solution) for 30 min at room temperature. Wash once quickly, and twice for 5 min each, with PBS. The BrdU antibody should then be diluted in PBS containing 0.5% Triton X-100 but no serum.

Citrate Retrieval   Some antigens require a retrieval step in citrate buffer. Slices are incubated at 85°C for 30 min in 10 mM sodium citrate in PBS, pH 6, and washed three times in PBS.

C. *In Situ* Hybridization on Whole Mount Adult Brains

Fix brains as described above. Incubate the brains in a descending methanol series until PBS, as described above in (B).

Place fixed, methanol-treated brains into a 48-well-plate. Treat with proteinase K (10 μg/ml) for 30 min at room temperature. Proceed through to *in situ* hybridization using the standard protocol for zebrafish embryos (Hauptmann and Gerster, 2000b), until all post-hybridization washes have been performed. Embed the brain in 3% agarose and cut 70–100 μm sections at the vibratome. Block the sections in blocking buffer (PBS; 0.1% Tween; 2% normal goat serum (NGS); 2 mg/ml bovine serum albumin) and incubate them in anti-DIG antibodies and continue following the standard embryo *in situ* protocol.

D. *In Situ* Hybridization on Gelatine–Albumin Sections

For weak RNA probes which do not reveal a strong signal following whole mount *in situ* hybridization, gelatine–albumin sections are better suited. However, it is not

possible to perform fluorescence immunohistochemistry following *in situ* hybridization, due to the high level of autofluorescence resulting from the embedding mixture.

Prepare the gelatine–albumin mixture: To 4.5 g gelatine in a beaker add PBS up to the 250 ml mark and dissolve by heating at 50°C. In a separate beaker, to 270 g albumin and 80 g sucrose, add PBS up to the 750 ml mark and dissolve overnight at room temperature. Filter the solution.

Mix both solutions after the gelatine solution has cooled down. The mixture can be aliquoted and stored at –20°C.

Immediately before embedding the brain, add 200–500 µl glutaraldehyde to 5 ml of the gelatine–albumin mixture. Embed the brain very quickly, as the gelatine–albumin polymerizes within 30 s to 1 min.

Trim a block around the brain and cut 70–100 µm sections on a vibratome. Process the sections through an ascending methanol series and freeze in 100% methanol at –20°C for a minimum of 1 h. Reverse the methanol series to 100% PBS and process for *in situ* hybridization, following the standard zebrafish embryo (Hauptmann and Gerster, 2000b) protocol, without proteinase K pretreatment.

E. Intraperitoneal Injections of BrdU

As described in (Adolf *et al.*, 2006), dissolve BrdU in a saline solution (110 mM NaCl pH 7.2) at a concentration of 2.5 mg/ml. Add a few drops of methylene blue to color the solution. Vortex the solution for about 5 min to ensure that the BrdU is completely dissolved. The BrdU solution can be stored for up to one week at 4°C.

Inject the fish intraperitoneally with BrdU solution (5 µl per 0.1 g body weight). As illustrated in Fig. 4, the required injection volume is pipetted as a drop onto a piece of Parafilm (Fig. 4A), aspirated into a small syringe (insulin syringe). The fish is held on its side in a net on a wet Petri dish (Fig. 4B) and injected right below the skin at the level of the belly (Fig. 4C). The procedure is rapid and does not require anesthetizing the fish.

F. Lipofections/Electroporations of the Adult Brain *In Vivo*

Prepare a piece of plastic foam to fit into in a 6 cm Petri dish, with a slit to hold a fish, as illustrated in Fig. 5A (Chapouton *et al.*, 2010). Fill the Petri dish and foam

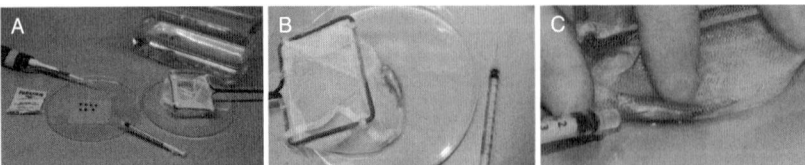

Fig. 4   Intraperitoneal injection of BrdU into the adult fish. (A) The desired volume of BrdU solution (5 µl/ 0.1 g body mass) is dropped onto a piece of Parafilm. (B) The fish is immobilized for a short time in a net in a wet Petri dish. (C) Stabilize the fish on its side and inject the BrdU solution using a small syringe, inserted at an oblique angle, into the peritoneum.

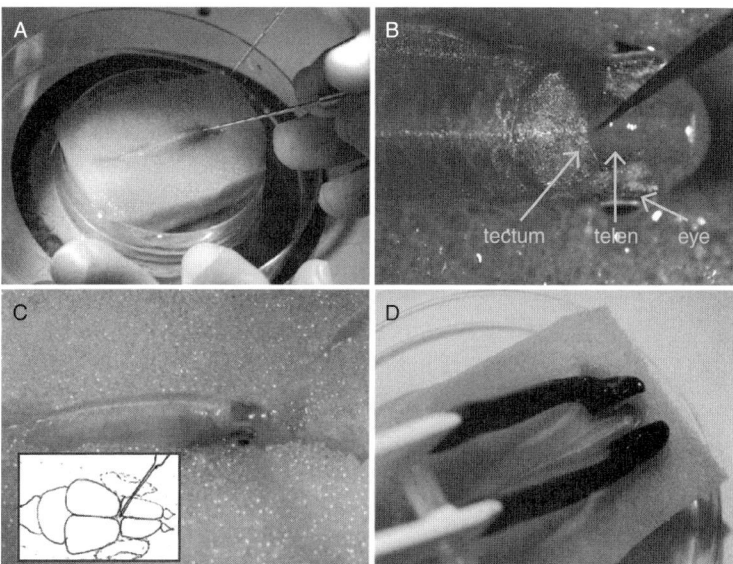

Fig. 5 Lipofection and electroporation *in vivo* in the adult zebrafish brain. (A) The anesthetized fish is placed into a slit in a block of wet foam and a hole is made in its skull at the level of the epiphysis, as shown in (B). (C) The capillary containing DNA (with lipofectamine if lipofecting) is inserted through the hole and the solution is pressure injected into the brain ventricles. The propagation of the solution through the ventricles can be visualized easily due to the presence of a dye (Fast Green). (D) For electroporation, tweezertrodes are placed on both sides of the head and electroshocks are given.

plastic with fish water containing 0.02% tricaine (MS-222). Anesthetize an adult fish in 0.02% tricaine and place it into the slit of the plastic foam so that its dorsal-side faces up (Fig. 5A). With the help of a microsurgery tool (i.e., Fine Science Tools 10055-12) under a stereomicroscope, make a small hole in the skull at the level of the epiphysis (as in Fig. 5A and B). Introduce an injection capillary (as used for injections of embryos at the one-cell stage), containing DNA solution (see below) through the hole at the posterior end of the telencephalon into the diencephalic ventricle and pressure inject the solution (Fig. 5C). The presence of Fast Green should make it relatively easy to visualize the propagation of the DNA solution into all the ventricles. Propagation of the solution is easier to visualize when working with brass fish, and slightly more difficult to detect in the AB, Tü, or EK lines, due to their stronger pigmentation. It should be possible to distinguish between a successful injection that fills the ventricles from an injection which results in the DNA solution only being distributed on the surface of the brain, outside the meninges.

To deliver DNA into the living brain using lipofection, prepare a mixture of DNA and lipofectant (Lipofectamine 2000, Invitrogen). Dissolve Lipofectamine in a ratio of 1:2 in water or Hank's buffered salt solution (HBSS) (solution A). Prepare 2–5 µg of DNA (a single plasmid or two plasmids) in a 5 µl volume (solution B). Mix 5 µl each of solution A and solution B at room temperature and incubate 20 min before use. This

mixture should be used within a few hours of preparation. Prior to injection, add 2 μl of a 2 mg/ml Fast Green solution to the lipofection mix.

For electroporations, inject DNA (2–5 μg/μl) in a solution containing 0.2 mg/ml fast green. Remove the fish from the plastic foam and place it into fish water containing tricaine. Connect electrodes (e.g., Tweezertrodes 520, Harvard Apparatus) to an electroporator (Ovodyne) and current amplifier (TSS20 and EP21; Intracell), and position the electrodes laterally on both sides of the head, slightly anterior to the eyes. Deliver 5 pulses (40 V, with 50 ms pulse width and 1000 ms time interval) (Fig. 5D).

Both lipofection and electroporation work equally well and allow isolated cells located on the ventricular surface to be transduced (about 20–50 cells per telencephalon). Lipofection presents the advantage of avoiding electroshocks to the fish.

G. Slice Culture

Culture medium (modified artificial cerebrospinal fluid, ACSF) to be prepared fresh: for 1 l:

100 mM	NaCl	5.84 g
2.46 mM	KCl	0.183 g
1 mM	MgCl$_2$.6H2O	0.203 g
0.44 mM	NaH$_2$PO4.H2O	0.060 g
1.13 mM	CaCl$_2$	0.166 g
5 mM	NaHCO$_3$	0.420 g
10 mM	Glucose	1.802 g
pH 7.2		

Sterile filter the solution

Brain Embedding, Cutting, and Culture Dissect brains in cold culture medium and embed in 2% low gelling agarose cooled down to 28°C. Cut 270 μm slices on a vibratome. Collect the slices gently using a spatula or a plastic Pasteur pipette with a large bore. Lay the slices on a Millicell-CM (Millipore) culture plate insert placed in a 6-well plate filled with 1.5 ml culture medium (containing penicillin/streptomycin) per well. The slices should not float, they should only be covered by a thin layer of medium. Culture the slices at -28°C, in a normal egg incubator. Change the medium every second day. Slices can be cultured for about 4–5 days.

Fixation Remove the medium and fill the wells with about 3 ml of 4% PFA, drop some more PFA gently onto the slices. Fix for 1 h at 4°C. Wash the slices in PBS and proceed to a methanol series, freeze in 100% methanol at –20°C for at least 1 h, before proceeding for immunohistochemistry.

V. Conclusion

Both as an embryo and as an adult, the zebrafish continues to contribute to our understanding of neurogenesis. Many methods of manipulation, mutants, and transgenic lines are available, as well as genetic methods to manipulate the zebrafish in a conditional (spatially and temporally) manner. Optical methods to image processes *in vivo* and to manipulate neuronal activity have been implemented. In the future, more genetic and optical tools will be developed, making neurogenesis in the zebrafish a rich field to explore.

Acknowledgments

We are very grateful to Laure Bally-Cuif, Philip Williams, and Thomas Misgeld for their helpful comments and suggestions while writing this chapter.

References

Adolf, B., Bellipanni, G., Huber, V., and Bally-Cuif, L. (2004). atoh1.2 and beta3.1 are two new bHLH-encoding genes expressed in selective precursor cells of the zebrafish anterior hindbrain. *Gene Expr. Patterns* **5**, 35–41.

Adolf, B., Chapouton, P., Lam, C. S., Topp, S., Tannhauser, B., Strahle, U., Gotz, M., and Bally-Cuif, L. (2006). Conserved and acquired features of adult neurogenesis in the zebrafish telencephalon. *Dev. Biol.* **295**, 278–293.

Ahn, D., Ruvinsky, I., Oates, A., Silver, L., and Ho, R. (2000). tbx20, a new vertebrate T-box gene expressed in the cranial motor neurons and developing cardiovascular structures in zebrafish. *Mech. Dev.* **95**, 253–258.

Aizawa, H., Goto, M., Sato, T., and Okamoto, H. (2007). Temporally regulated asymmetric neurogenesis causes left-right difference in the zebrafish habenular structures. *Dev. Cell* **12**, 87–98.

Akimenko, M. A., Ekker, M., Wegner, J., Lin, W., and Westerfield, M. (1994). Combinatorial expression of three zebrafish genes related to distal-less: Part of a homeobox gene code for the head. *J. Neurosci.* **14**, 3475–3486.

Alexandre, P., Reugels, A. M., Barker, D., Blanc, E., and Clarke, J. D. (2010). Neurons derive from the more apical daughter in asymmetric divisions in the zebrafish neural tube. *Nat. Neurosci.* **13**, 673–679.

Allende, M. L., Amsterdam, A., Becker, T., Kawakami, K., Gaiano, N., and Hopkins, N. (1996). Insertional mutagenesis in zebrafish identifies two novel genes, pescadillo and dead eye, essential for embryonic development. *Genes Dev.* **10**, 3141–3155.

Allende, M. L., and Weinberg, E. S. (1994). The expression pattern of two zebrafish achaete-scute homolog (ash) genes is altered in the embryonic brain of the cyclops mutant. *Dev. Biol.* **166**, 509–530.

Alunni, A., Hermel, J. M., Heuze, A., Bourrat, F., Jamen, F., and Joly, J. S. (2010). Evidence for neural stem cells in the medaka optic tectum proliferation zones. *Dev. Neurobiol.* **70**, 693–713.

Amoyel, M., Cheng, Y. C., Jiang, Y. J., and Wilkinson, D. G. (2005). Wnt1 regulates neurogenesis and mediates lateral inhibition of boundary cell specification in the zebrafish hindbrain. *Development* **132**, 775–785.

Amsterdam, A., Burgess, S., Golling, G., Chen, W., Sun, Z., Townsend, K., Farrington, S., Haldi, M., and Hopkins, N. (1999). A large-scale insertional mutagenesis screen in zebrafish. *Genes Dev.* **13**, 2713–2724.

Amsterdam, A., Lai, K., Komisarczuk, A. Z., Becker, T. S., Bronson, R. T., Hopkins, N., and Lees, J. A. (2009). Zebrafish Hagoromo mutants up-regulate fgf8 postembryonically and develop neuroblastoma. *Mol. Cancer Res.* **7**, 841–850.

Andermann, P., and Weinberg, E. (2001). Expression of zTlxA, a Hox11-like gene, in early differentiating embryonic neurons and cranial sensory ganglia of the zebrafish embryo. *Dev. Dyn.* **222**, 595–610.

Ando, R., Hama, H., Yamamoto-Hino, M., Mizuno, H., and Miyawaki, A. (2002). An optical marker based on the UV-induced green-to-red photoconversion of a fluorescent protein. *Proc. Natl. Acad. Sci. U.S.A.* **99**, 12651–12656.

Appel, B., Fritz, A., Westerfield, M., Grunwald, D. J., Eisen, J. S., and Riley, B. B. (1999). Delta-mediated specification of midline cell fates in zebrafish embryos. *Curr. Biol.* **9**, 247–256.

Appel, B., Givan, L. A., and Eisen, J. S. (2001). Delta-Notch signaling and lateral inhibition in zebrafish spinal cord development. *BMC Dev. Biol.* **1**, 13.

Appel, B., Korzh, V., Glasgow, E., Thor, S., Edlund, T., Dawid, I. B., and Eisen, J. S. (1995). Motoneuron fate specification revealed by patterned LIM homeobox gene expression in embryonic zebrafish. *Development* **121**, 4117–4125.

Aramaki, S., and Hatta, K. (2006). Visualizing neurons one-by-one *in vivo*: Optical dissection and reconstruction of neural networks with reversible fluorescent proteins. *Dev. Dyn.* **235**, 2192–2199.

Artinger, K. B., Chitnis, A. B., Mercola, M., and Driever, W. (1999). Zebrafish narrowminded suggests a genetic link between formation of neural crest and primary sensory neurons. *Development* **126**, 3969–3979.

Asakawa, K., and Kawakami, K. (2009). The Tol2-mediated Gal4-UAS method for gene and enhancer trapping in zebrafish. *Methods* **49**, 275–281.

Asakawa, K., Suster, M. L., Mizusawa, K., Nagayoshi, S., Kotani, T., Urasaki, A., Kishimoto, Y., Hibi, M., and Kawakami, K. (2008). Genetic dissection of neural circuits by Tol2 transposon-mediated Gal4 gene and enhancer trapping in zebrafish. *Proc. Natl. Acad. Sci. U.S.A.* **105**, 1255–1260.

Bae, Y. K., Kani, S., Shimizu, T., Tanabe, K., Nojima, H., Kimura, Y., Higashijima, S., and Hibi, M. (2009). Anatomy of zebrafish cerebellum and screen for mutations affecting its development. *Dev. Biol.* **330**, 406–426.

Bae, Y. K., Shimizu, T., and Hibi, M. (2005). Patterning of proneuronal and inter-proneuronal domains by hairy- and enhancer of split-related genes in zebrafish neuroectoderm. *Development* **132**, 1375–1385.

Bae, Y. K., Shimizu, T., Yabe, T., Kim, C. H., Hirata, T., Nojima, H., Muraoka, O., Hirano, T., and Hibi, M. (2003). A homeobox gene, pnx, is involved in the formation of posterior neurons in zebrafish. *Development* **130**, 1853–1865.

Bai, Q., and Burton, E. A. (2009). Cis-acting elements responsible for dopaminergic neuron-specific expression of zebrafish slc6a3 (dopamine transporter) *in vivo* are located remote from the transcriptional start site. *Neuroscience* **164**, 1138–1151.

Bai, Q., Garver, J. A., Hukriede, N. A., and Burton, E. A. (2007). Generation of a transgenic zebrafish model of Tauopathy using a novel promoter element derived from the zebrafish eno2 gene. *Nucleic Acids Res.* **35**, 6501–6516.

Bai, Q., Wei, X., and Burton, E. A. (2009). Expression of a 12-kb promoter element derived from the zebrafish enolase-2 gene in the zebrafish visual system. *Neurosci. Lett.* **449**, 252–257.

Bally-Cuif, L., Dubois, L., and Vincent, A. (1998). Molecular cloning of Zcoe2, the zebrafish homolog of Xenopus Xcoe2 and mouse EBF-2, and its expression during primary neurogenesis. *Mech. Dev.* **77**, 85–90.

Barth, K. A., Kishimoto, Y., Rohr, K. B., Seydler, C., Schulte-Merker, S., and Wilson, S. W. (1999). Bmp activity establishes a gradient of positional information throughout the entire neural plate. *Development* **126**, 4977–4987.

Baye, L. M., and Link, B. A. (2007a). The disarrayed mutation results in cell cycle and neurogenesis defects during retinal development in zebrafish. *BMC Dev. Biol.* **7**, 28.

Baye, L. M., and Link, B. A. (2007b). Interkinetic nuclear migration and the selection of neurogenic cell divisions during vertebrate retinogenesis. *J. Neurosci.* **27**, 10143–10152.

Becker, T., Bernhardt, R. R., Reinhard, E., Wullimann, M. F., Tongiorgi, E., and Schachner, M. (1998). Readiness of zebrafish brain neurons to regenerate a spinal axon correlates with differential expression of specific cell recognition molecules. *J. Neurosci.* **18**, 5789–5803.

Bellipanni, G., Rink, E., and Bally-Cuif, L. (2002). Cloning of two tryptophane hydroxylase genes expressed in the diencephalon of the developing zebrafish brain. *Mech. Dev.* **119S**, S215–S220.

Belting, H. G., Hauptmann, G., Meyer, D., Abdelilah-Seyfried, S., Chitnis, A., Eschbach, C., Soll, I., Thisse, C., Thisse, B., Artinger, K. B., *et al.* (2001). spiel ohne grenzen/pou2 is required during establishment of the zebrafish midbrain-hindbrain boundary organizer. *Development* **128**, 4165–4176.

Berberoglu, M. A., Dong, Z., Mueller, T., and Guo, S. (2009). fezf2 expression delineates cells with proliferative potential and expressing markers of neural stem cells in the adult zebrafish brain. *Gene Expr. Patterns* **9**, 411–422.

Bernardos, R. L., Barthel, L. K., Meyers, J. R., and Raymond, P. A. (2007). Late-stage neuronal progenitors in the retina are radial Muller glia that function as retinal stem cells. *J. Neurosci.* **27**, 7028–7040.

Bernardos, R. L., and Raymond, P. A. (2006). GFAP transgenic zebrafish. *Gene Expr. Patterns* **6**, 1007–1013.

Bernhardt, R. R., Chitnis, A. B., Lindamer, L., and Kuwada, J. Y. (1990). Identification of spinal neurons in the embryonic and larval zebrafish. *J. Comp. Neurol.* **302**, 603–616.

Bianco, I. H., Carl, M., Russell, C., Clarke, J. D., and Wilson, S. W. (2008). Brain asymmetry is encoded at the level of axon terminal morphology. *Neural Dev.* **3**, 9.

Bit-Avragim, N., Hellwig, N., Rudolph, F., Munson, C., Stainier, D. Y., and Abdelilah-Seyfried, S. (2008). Divergent polarization mechanisms during vertebrate epithelial development mediated by the Crumbs complex protein Nagie oko. *J. Cell. Sci.* **121**, 2503–2510.

Blader, P., Fischer, N., Gradwohl, G., Guillemot, F., and Strahle, U. (1997). The activity of neurogenin1 is controlled by local cues in the zebrafish embryo. *Development* **124**, 4557–4569.

Blader, P., Lam, C. S., Rastegar, S., Scardigli, R., Nicod, J. C., Simplicio, N., Plessy, C., Fischer, N., Schuurmans, C., Guillemot, F., *et al.* (2004). Conserved and acquired features of neurogenin1 regulation. *Development* **131**, 5627–5637.

Blake, T., Adya, N., Kim, C. H., Oates, A. C., Zon, L., Chitnis, A., Weinstein, B. M., and Liu, P. P. (2000). Zebrafish homolog of the leukemia gene CBFB: Its expression during embryogenesis and its relationship to scl and gata-1 in hematopoiesis. *Blood* **96**, 4178–4184.

Blechman, J., Borodovsky, N., Eisenberg, M., Nabel-Rosen, H., Grimm, J., and Levkowitz, G. (2007). Specification of hypothalamic neurons by dual regulation of the homeodomain protein Orthopedia. *Development* **134**, 4417–4426.

Blin, M., Norton, W., Bally-Cuif, L., and Vernier, P. (2008). NR4A2 controls the differentiation of selective dopaminergic nuclei in the zebrafish brain. *Mol. Cell. Neurosci.* **39**, 592–604.

Bourguignon, C., Li, J., and Papalopulu, N. (1998). XBF-1, a winged helix transcription factor with dual activity, has a role in positioning neurogenesis in Xenopus competent ectoderm. *Development* **125**, 4889–4900.

Bray, S., and Furriols, M. (2001). Notch pathway: Making sense of suppressor of hairless. *Curr. Biol.* **11**, R217–R221.

Bridgewater, J. A., Springer, C. J., Knox, R. J., Minton, N. P., Michael, N. P., and Collins, M. K. (1995). Expression of the bacterial nitroreductase enzyme in mammalian cells renders them selectively sensitive to killing by the prodrug CB1954. *Eur. J. Cancer* **31A**, 2362–2370.

Bulina, M. E., Lukyanov, K. A., Britanova, O. V., Onichtchouk, D., Lukyanov, S., and Chudakov, D. M. (2006). Chromophore-assisted light inactivation (CALI) using the phototoxic fluorescent protein KillerRed. *Nat. Protoc.* **1**, 947 953.

Bylund, M., Andersson, E., Novitch, B. G., and Muhr, J. (2003). Vertebrate neurogenesis is counteracted by Sox1-3 activity. *Nat. Neurosci.* **6**, 1162–1168.

Canger, A., Passini, M., Asch, W., Leake, D., Zafonte, B., Glasgow, E., and Schechter, N. (1998). Restricted expression of the neuronal intermediate filament protein plasticin during zebrafish development. *J. Comp. Neurol.* **399**, 561–572.

Caron, S. J., Prober, D., Choy, M., and Schier, A. F. (2008). *In vivo* birthdating by BAPTISM reveals that trigeminal sensory neuron diversity depends on early neurogenesis. *Development* **135**, 3259–3269.

Cau, E., Quillien, A., and Blader, P. (2008). Notch resolves mixed neural identities in the zebrafish epiphysis. *Development* **135**, 2391–2401.

Cerda, G. A., Thomas, J. E., Allende, M. L., Karlstrom, R. O., and Palma, V. (2006). Electroporation of DNA, RNA, and morpholinos into zebrafish embryos. *Methods* **39**, 207–211.

Cerveny, K. L., Cavodeassi, F., Turner, K. J., de Jong-Curtain, T. A., Heath, J. K., and Wilson, S. W. (2010). The zebrafish flotte lotte mutant reveals that the local retinal environment promotes the differentiation of proliferating precursors emerging from their stem cell niche. *Development* 137, 2107–2115.

Chaplin, N., Tendeng, C., and Wingate, R. J. (2010). Absence of an external germinal layer in zebrafish and shark reveals a distinct, anamniote ground plan of cerebellum development. *J. Neurosci.* **30**, 3048–3057.

Chapouton, P., Adolf, B., Leucht, C., Tannhauser, B., Ryu, S., Driever, W., and Bally-Cuif, L. (2006). her5 expression reveals a pool of neural stem cells in the adult zebrafish midbrain. *Development* **133**, 4293–4303.

Chapouton, P., Jagasia, R., and Bally-Cuif, L. (2007). Adult neurogenesis in non-mammalian vertebrates. *Bioessays* **29**, 745–757.

Chapouton, P., Skupien, P., Hesl, B., Coolen, M., Moore, J. C., Madelaine, R., Kremmer, E., Faus-Kessler, T., Blader, P., Lawson, N. D., *et al.* (2010). Notch activity levels control the balance between quiescence and recruitment of adult neural stem cells. *J. Neurosci.* **30**, 7961–7974.

Chen, W., Burgess, S., and Hopkins, N. (2001). Analysis of the zebrafish smoothened mutant reveals conserved and divergent functions of hedgehog activity. *Development* **128**, 2385–2396.

Chong, S., Emelyanov, A., Gong, Z., and Korzh, V. (2001). Expression pattern of two zebrafish genes, cxcr4a and cxcr4b. *Mech. Dev.* **109**, 347–354.

Chong, S. W., Nguyen, T. T., Chu, L. T., Jiang, Y. J., and Korzh, V. (2005). Zebrafish id2 developmental expression pattern contains evolutionary conserved and species-specific characteristics. *Dev. Dyn.* **234**, 1055–1063.

Choo, B. G., Kondrichin, I., Parinov, S., Emelyanov, A., Go, W., Toh, W. C., and Korzh, V. (2006). Zebrafish transgenic Enhancer TRAP line database (ZETRAP). *BMC Dev. Biol.* **6**, 5.

Collins, R. T., Linker, C., and Lewis, J. (2010). MAZe: A tool for mosaic analysis of gene function in zebrafish. *Nat. Methods* **7**, 219–223.

Cooper, M. S., Szeto, D. P., Sommers-Herivel, G., Topczewski, J., Solnica-Krezel, L., Kang, H. C., Johnson, I., Kimelman, D. (2005). Visualizing morphogenesis in transgenic zebrafish embryos using BODIPY TR methyl ester dye as a vital counterstain for GFP. *Dev. Dyn.* **232**, 359–368.

Curado, S., Stainier, D. Y., and Anderson, R. M. (2008). Nitroreductase-mediated cell/tissue ablation in zebrafish: A spatially and temporally controlled ablation method with applications in developmental and regeneration studies. *Nat. Protoc.* **3**, 948–954.

Das, T., Payer, B., Cayouette, M., and Harris, W. A. (2003). *In vivo* time-lapse imaging of cell divisions during neurogenesis in the developing zebrafish retina. *Neuron* **37**, 597–609.

Davison, J. M., Akitake, C. M., Goll, M. G., Rhee, J. M., Gosse, N., Baier, H., Halpern, M. E., Leach, S. D., and Parsons, M. J. (2007). Transactivation from Gal4-VP16 transgenic insertions for tissue-specific cell labeling and ablation in zebrafish. *Dev. Biol.* **304**, 811–824.

de Graaf, M., Zivkovic, D., and Joore, J. (1998). Hormone-inducible expression of secreted factors in zebrafish embryos. *Dev. Growth Differ.* **40**, 577–582.

de Martino, S., Yan, Y. L., Jowett, T., Postlethwait, J. H., Varga, Z. M., Ashworth, A., and Austin, C. A. (2000). Expression of sox11 gene duplicates in zebrafish suggests the reciprocal loss of ancestral gene expression patterns in development. *Dev. Dyn.* **217**, 279–292.

Dee, C. T., Hirst, C. S., Shih, Y. H., Tripathi, V. B., Patient, R. K., and Scotting, P. J. (2008). Sox3 regulates both neural fate and differentiation in the zebrafish ectoderm. *Dev. Biol.* **320**, 289–301.

Devos, N., Deflorian, G., Biemar, F., Bortolussi, M., Martial, J., Peers, B., and Argenton, F. (2002). Differential expression of two somatostatin genes during zebrafish embryonic development. *Mech. Dev.* **115**, 133–137.

Dickmeis, T., Rastegar, S., Lam, C. S., Aanstad, P., Clark, M., Fischer, N., Rosa, F., Korzh, V., and Strahle, U. (2002). Expression of the helix-loop-helix gene id3 in the zebrafish embryo. *Mech. Dev.* **113**, 99–102.

Diks, S. H., Sartori da Silva, M. A., Hillebrands, J. L., Bink, R. J., Versteeg, H. H., Van Rooijen, C., Brouwers, A., Chitnis, A. B., Peppelenbosch, M. P., and Zivkovic, D. (2008). d-Asb11 is an essential mediator of canonical Delta-Notch signalling. *Nat. Cell Biol.* **10**, 1190–1198.

Distel, M., Wullimann, M. F., and Koster, R. W. (2009). Optimized Gal4 genetics for permanent gene expression mapping in zebrafish. *Proc. Natl. Acad. Sci. U.S.A.* **106**, 13365–13370.

Dornseifer, P., Takke, C., and Campos-Ortega, J. A. (1997). Overexpression of a zebrafish homologue of the Drosophila neurogenic gene Delta perturbs differentiation of primary neurons and somite development. *Mech. Dev.* **63**, 159–171.

Dorsky, R. I., Sheldahl, L. C., and Moon, R. T. (2002). A transgenic Lef1/beta-catenin-dependent reporter is expressed in spatially restricted domains throughout zebrafish development. *Dev. Biol.* **241**, 229–237.

Doyon, Y., McCammon, J. M., Miller, J. C., Faraji, F., Ngo, C., Katibah, G. E., Amora, R., Hocking, T. D., Zhang, L., Rebar, E. J., *et al.* (2008). Heritable targeted gene disruption in zebrafish using designed zinc-finger nucleases. *Nat. Biotechnol.* **26**, 702–708.

Du, S. J., and Dienhart, M. (2001). Zebrafish tiggy-winkle hedgehog promoter directs notochord and floor plate green fluorescence protein expression in transgenic zebrafish embryos. *Dev. Dyn.* **222**, 655–666.

Dutton, K. A., Pauliny, A., Lopes, S. S., Elworthy, S., Carney, T. J., Rauch, J., Geisler, R., Haffter, P., and Kelsh, R. N. (2001). Zebrafish colourless encodes sox10 and specifies non-ectomesenchymal neural crest fates. *Development* **128**, 4113–4125.

Ellingsen, S., Laplante, M. A., Konig, M., Kikuta, H., Furmanek, T., Hoivik, E. A., and Becker, T. S. (2005). Large-scale enhancer detection in the zebrafish genome. *Development* **132**, 3799–3811.

Emelyanov, A., and Parinov, S. (2008). Mifepristone-inducible LexPR system to drive and control gene expression in transgenic zebrafish. *Dev. Biol.* **320**, 113–121.

Eriksson, K. S., Peitsaro, N., Karlstedt, K., Kaslin, J., and Panula, P. (1998). Development of the histaminergic neurons and expression of histidine decarboxylase mRNA in the zebrafish brain in the absence of all peripheral histaminergic systems. *Eur. J. Neurosci.* **10**, 3799–3812.

Ertzer, R., Muller, F., Hadzhiev, Y., Rathnam, S., Fischer, N., Rastegar, S., and Strahle, U. (2007). Cooperation of sonic hedgehog enhancers in midline expression. *Dev. Biol.* **301**, 578–589.

Esengil, H., Chang, V., Mich, J. K., and Chen, J. K. (2007). Small-molecule regulation of zebrafish gene expression. *Nat. Chem. Biol.* **3**, 154–155.

Fadool, J. M. (2003). Development of a rod photoreceptor mosaic revealed in transgenic zebrafish. *Dev. Biol.* **258**, 277–290.

Fashena, D., and Westerfield, M. (1999). Secondary motoneuron axons localize DM-GRASP on their fasciculated segments. *J. Comp. Neurol.* **406**, 415–424.

Fausett, B. V., and Goldman, D. (2006). A role for alpha1 tubulin-expressing Muller glia in regeneration of the injured zebrafish retina. *J. Neurosci.* **26**, 6303–6313.

Fausett, B. V., Gumerson, J. D., and Goldman, D. (2008). The proneural basic helix-loop-helix gene ascl1a is required for retina regeneration. *J. Neurosci.* **28**, 1109–1117.

Feldman, B., Dougan, S. T., Schier, A. F., and Talbot, W. S. (2000). Nodal-related signals establish mesendodermal fate and trunk neural identity in zebrafish. *Curr. Biol.* **10**, 531–534.

Feng, B., Bulchand, S., Yaksi, E., Friedrich, R. W., and Jesuthasan, S. (2005). The recombination activation gene 1 (Rag1) is expressed in a subset of zebrafish olfactory neurons but is not essential for axon targeting or amino acid detection. *BMC Neurosci.* **6**, 46.

Filippi, A., Durr, K., Ryu, S., Willardt, M., Holzschuh, J., and Driever, W. (2007). Expression and function of nr4a2, lmx1b, and pitx3 in zebrafish dopaminergic and noradrenergic neuronal development. *BMC Dev. Biol.* **7**, 135.

Fimbel, S. M., Montgomery, J. E., Burket, C. T., and Hyde, D. R. (2007). Regeneration of inner retinal neurons after intravitreal injection of ouabain in zebrafish. *J. Neurosci.* **27**, 1712–1724.

Fjose, A., Izpisua-Belmonte, J. C., Fromental-Ramain, C., and Duboule, D. (1994). Expression of the zebrafish gene hlx-1 in the prechordal plate and during CNS development. *Development* **120**, 71–81.

Fjose, A., and Zhao, X. F. (2010). Inhibition of the microRNA pathway in zebrafish by siRNA. *Methods Mol. Biol.* **629**, 239–255.

Flanagan-Steet, H., Fox, M. A., Meyer, D., and Sanes, J. R. (2005). Neuromuscular synapses can form *in vivo* by incorporation of initially aneural postsynaptic specializations. *Development* **132**, 4471–4481.

Furthauer, M., Reifers, F., Brand, M., Thisse, B., and Thisse, C. (2001). sprouty4 acts *in vivo* as a feedback-induced antagonist of FGF signaling in zebrafish. *Development* **128**, 2175–2186.

Gahtan, E., and Baier, H. (2004). Of lasers, mutants, and see-through brains: Functional neuroanatomy in zebrafish. *J. Neurobiol.* **59**, 147–161.

Ganz, J., Kaslin, J., Hochmann, S., Freudenreich, D., and Brand, M. (2010). Heterogeneity and Fgf dependence of adult neural progenitors in the zebrafish telencephalon. *Glia* **58**, 1345–1363.

Geling, A., Itoh, M., Tallafuss, A., Chapouton, P., Tannhauser, B., Kuwada, J. Y., Chitnis, A. B., and Bally-Cuif, L. (2003). bHLH transcription factor Her5 links patterning to regional inhibition of neurogenesis at the midbrain-hindbrain boundary. *Development* **130**, 1591–1604.

Geling, A., Plessy, C., Rastegar, S., Strahle, U., and Bally-Cuif, L. (2004). Her5 acts as a prepattern factor that blocks neurogenin1 and coe2 expression upstream of Notch to inhibit neurogenesis at the midbrain-hindbrain boundary. *Development* **131**, 1993–2006.

Geling, A., Steiner, H., Willem, M., Bally-Cuif, L., and Haass, C. (2002). A gamma-secretase inhibitor blocks Notch signaling *in vivo* and causes a severe neurogenic phenotype in zebrafish. *EMBO Rep.* **3**, 688–694.

Girard, F., Cremazy, F., Berta, P., and Renucci, A. (2001). Expression pattern of the Sox31 gene during Zebrafish embryonic development. *Mech. Dev.* **100**, 71–73.

Glasgow, E., Karavanov, A., and Dawid, I. (1997). Neuronal and neuroendocrine expression of lim3, a lim class homeobox gene, is altered in mutant zebrafish with axial signaling defects. *Dev. Biol.* **192**, 405–419.

Godinho, L., Mumm, J. S., Williams, P. R., Schroeter, E. H., Koerber, A., Park, S. W., Leach, S. D., and Wong, R. O. (2005). Targeting of amacrine cell neurites to appropriate synaptic laminae in the developing zebrafish retina. *Development* **132**, 5069–5079.

Godinho, L., Williams, P. R., Claassen, Y., Provost, E., Leach, S. D., Kamermans, M., and Wong, R. O. (2007). Nonapical symmetric divisions underlie horizontal cell layer formation in the developing retina *in vivo*. *Neuron* **56**, 597–603.

Goldman, D., Hankin, M., Li, Z., Dai, X., and Ding, J. (2001). Transgenic zebrafish for studying nervous system development and regeneration. *Transgenic Res.* **10**, 21–33.

Golling, G., Amsterdam, A., Sun, Z., Antonelli, M., Maldonado, E., Chen, W., Burgess, S., Haldi, M., Artzt, K., Farrington, S., *et al.* (2002). Insertional mutagenesis in zebrafish rapidly identifies genes essential for early vertebrate development. *Nat. Genet.* **31**, 135–140.

Gonzalez-Quevedo, R., Lee, Y., Poss, K. D., and Wilkinson, D. G. (2010). Neuronal regulation of the spatial patterning of neurogenesis. *Dev. Cell* **18**, 136–147.

Gothilf, Y., Toyama, R., Coon, S., Du, S., Dawid, I., and Klein, D. (2002). Pineal-specific expression of green fluorescent protein under the control of the serotonin-N-acetyltransferase gene regulatory regions in transgenic zebrafish. *Dev. Dyn.* **225**, 241–249.

Grandel, H., Kaslin, J., Ganz, J., Wenzel, I., and Brand, M. (2006). Neural stem cells and neurogenesis in the adult zebrafish brain: Origin, proliferation dynamics, migration and cell fate. *Dev. Biol.* **295**, 263–277.

Gray, M., Moens, C. B., Amacher, S. L., Eisen, J. S., and Beattie, C. E. (2001). Zebrafish deadly seven functions in neurogenesis. *Dev. Biol.* **237**, 306–323.

Gregg, R. G., Willer, G. B., Fadool, J. M., Dowling, J. E., and Link, B. A. (2003). Positional cloning of the young mutation identifies an essential role for the Brahma chromatin remodeling complex in mediating retinal cell differentiation. *Proc. Natl. Acad. Sci. U.S.A.* **100**, 6535–6540.

Gritsman, K., Zhang, J., Cheng, S., Heckscher, E., Talbot, W. S., and Schier, A. F. (1999). The EGF-CFC protein one-eyed pinhead is essential for nodal signaling. *Cell* **97**, 121–132.

Guo, S., Brush, J., Teraoka, H., Goddard, A., Wilson, S. W., Mullins, M. C., and Rosenthal, A. (1999a). Development of noradrenergic neurons in the zebrafish hindbrain requires BMP, FGF8, and the homeodomain protein soulless/Phox2a. *Neuron* **24**, 555–566.

Guo, S., Wilson, S. W., Cooke, S., Chitnis, A. B., Driever, W., and Rosenthal, A. (1999b). Mutations in the zebrafish unmask shared regulatory pathways controlling the development of catecholaminergic neurons. *Dev. Biol.* **208**, 473–487.

Guo, S., Yamaguchi, Y., Schilbach, S., Wada, T., Lee, J., Goddard, A., French, D., Handa, H., and Rosenthal, A. (2000). A regulator of transcriptional elongation controls vertebrate neuronal development. *Nature* **408**, 366–369.

Gwak, J. W., Kong, H. J., Bae, Y. K., Kim, M. J., Lee, J., Park, J. H., and Yeo, S. Y. (2010). Proliferating neural progenitors in the developing CNS of zebrafish require Jagged2 and Jagged1b. *Mol. Cells* **30**, 155–159.

Habuchi, S., Ando, R., Dedecker, P., Verheijen, W., Mizuno, H., Miyawaki, A., and Hofkens, J. (2005). Reversible single-molecule photoswitching in the GFP-like fluorescent protein Dronpa. *Proc. Natl. Acad. Sci. U.S.A.* **102**, 9511–9516.

Haddon, C., Smithers, L., Schneider-Maunoury, S., Coche, T., Henrique, D., and Lewis, J. (1998). Multiple delta genes and lateral inhibition in zebrafish primary neurogenesis. *Development* **125**, 359–370.

Halloran, M. C., Sato-Maeda, M., Warren, J. T., Su, F., Lele, Z., Krone, P. H., Kuwada, J. Y., and Shoji, W. (2000). Laser-induced gene expression in specific cells of transgenic zebrafish. *Development* **127**, 1953–1960.

Hamaoka, T., Takechi, M., Chinen, A., Nishiwaki, Y., and Kawamura, S. (2002). Visualization of rod photoreceptor development using GFP-transgenic zebrafish. *Genesis* **34**, 215–220.

Hammond, K. L., Hill, R. E., Whitfield, T. T., and Currie, P. D. (2002). Isolation of three zebrafish dachshund homologues and their expression in sensory organs, the central nervous system and pectoral fin buds. *Mech. Dev.* **112**, 183–189.

Hans, S., Kaslin, J., Freudenreich, D., and Brand, M. (2009). Temporally-controlled site-specific recombination in zebrafish. *PLoS One* **4**, e4640.

Hans, S., Scheer, N., Riedl, I., v Weizsacker, E., Blader, P., and Campos-Ortega, J. A. (2004). her3, a zebrafish member of the hairy-E(spl) family, is repressed by Notch signalling. *Development* **131**, 2957–2969.

Hatta, K. (1992). Role of the floor plate in axonal patterning in the zebrafish CNS. *Neuron* **9**, 629–642.

Hatta, K., Tsujii, H., and Omura, T. (2006). Cell tracking using a photoconvertible fluorescent protein. *Nat. Protoc.* **1**, 960–967.

Hauptmann, G., and Gerster, T. (1996). Complex expression of the zp-50 pou gene in the embryonic zebrafish brain is altered by overexpression of sonic hedgehog. *Development* **122**, 1769–1780.

Hauptmann, G., and Gerster, T. (2000a). Combinatorial expression of zebrafish Brn-1- and Brn-2-related POU genes in the embryonic brain, pronephric primordium, and pharyngeal arches. *Dev. Dyn.* **218**, 345–358.

Hauptmann, G., and Gerster, T. (2000b). Multicolor whole-mount in situ hybridization. *Methods Mol. Biol.* **137**, 139–148.

Hauptmann, G., and Gerster, T. (2000c). Regulatory gene expression patterns reveal transverse and longitudinal subdivisions of the embryonic zebrafish forebrain. *Mech. Dev.* **91**, 105–118.

Hauptmann, G., Soll, I., and Gerster, T. (2002). The early embryonic zebrafish forebrain is subdivided into molecularly distinct transverse and longitudinal domains. *Brain Res. Bull.* **57**, 371–375.

Heisenberg, C. P., Brand, M., Jiang, Y. J., Warga, R. M., Beuchle, D., van Eeden, F. J., Furutani-Seiki, M., Granato, M., Haffter, P., Hammerschmidt, M., *et al.* (1996). Genes involved in forebrain development in the zebrafish, Danio rerio. *Development* **123**, 191–203.

Helde, K. A., Wilson, E. T., Cretekos, C. J., and Grunwald, D. J. (1994). Contribution of early cells to the fate map of the zebrafish gastrula. *Science* **265**, 517–520.

Helmchen, F., and Denk, W. (2005). Deep tissue two-photon microscopy. *Nat. Methods* **2**, 932–940.

Hernandez-Lagunas, L., Choi, I. F., Kaji, T., Simpson, P., Hershey, C., Zhou, Y., Zon, L., Mercola, M., and Artinger, K. B. (2005). Zebrafish narrowminded disrupts the transcription factor prdm1 and is required for neural crest and sensory neuron specification. *Dev. Biol.* **278**, 347–357.

Hesselson, D., Anderson, R. M., Beinat, M., and Stainier, D. Y. (2009). Distinct populations of quiescent and proliferative pancreatic beta-cells identified by HOTcre mediated labeling. *Proc. Natl. Acad. Sci. U.S.A.* **106**, 14896–14901.

Hieber, V., Dai, X., Foreman, M., and Goldman, D. (1998). Induction of alpha1-tubulin gene expression during development and regeneration of the fish central nervous system. *J. Neurobiol.* **37**, 429–440.

Higashijima, S., Okamoto, H., Ueno, N., Hotta, Y., Eguchi, G. (1997). High-frequency generation of transgenic zebrafish which reliably express GFP in whole muscles or the whole body by using promoters of zebrafish origin. *Dev. Biol.* **192**, 289–299.

Higashijima, S., Hotta, Y., and Okamoto, H. (2000). Visualization of cranial motor neurons in live transgenic zebrafish expressing green fluorescent protein under the control of the islet-1 promoter/enhancer. *J. Neurosci.* **20**, 206–218.

Higashijima, S., Mandel, G., and Fetcho, J. R. (2004). Distribution of prospective glutamatergic, glycinergic, and GABAergic neurons in embryonic and larval zebrafish. *J. Comp. Neurol.* **480**, 1–18.

Hild, M., Dick, A., Rauch, G. J., Meier, A., Bouwmeester, T., Haffter, P., and Hammerschmidt, M. (1999). The smad5 mutation somitabun blocks Bmp2b signaling during early dorsoventral patterning of the zebrafish embryo. *Development* **126**, 2149–2159.

Hjorth, J. T., Gad, J., Cooper, H., and Key, B. (2001). A zebrafish homologue of deleted in colorectal cancer (zdcc) is expressed in the first neuronal clusters of the developing brain. *Mech. Dev.* **109**, 105–109.

Ho, R. K., and Kane, D. A. (1990). Cell-autonomous action of zebrafish spt-1 mutation in specific mesodermal precursors. *Nature* **348**, 728–730.

Holley, S. A., Geisler, R., and Nusslein-Volhard, C. (2000). Control of her1 expression during zebrafish somitogenesis by a delta-dependent oscillator and an independent wave-front activity. *Genes Dev.* **14**, 1678–1690.

Holzschuh, J., Ryu, S., Aberger, F., and Driever, W. (2001). Dopamine transporter expression distinguishes dopaminergic neurons from other catecholaminergic neurons in the developing zebrafish embryo. *Mech. Dev.* **101**, 237–243.

Hong, S., Kim, C., Yoo, K., Kim, H., Kudoh, T., Dawid, I., and Huh, T. (2002). Isolation and expression of a novel neuron-specific onecut homeobox gene in zebrafish. *Mech. Dev.* **112**, 199–202.

Horne-Badovinac, S., Lin, D., Waldron, S., Schwarz, M., Mbamalu, G., Pawson, T., Jan, Y., Stainier, D. Y., and Abdelilah-Seyfried, S. (2001). Positional cloning of heart and soul reveals multiple roles for PKC lambda in zebrafish organogenesis. *Curr. Biol.* **11**, 1492–1502.

Houart, C., Caneparo, L., Heisenberg, C., Barth, K., Take-Uchi, M., and Wilson, S. (2002). Establishment of the telencephalon during gastrulation by local antagonism of Wnt signaling. *Neuron* **35**, 255–265.

Huang, S., and Sato, S. (1998). Progenitor cells in the adult zebrafish nervous system express a Brn-1-related Pou gene, Tai-ji. *Mech. Dev.* **71**, 23–35.

Imboden, M., Devignot, V., Korn, H., and Goblet, C. (2001). Regional distribution of glycine receptor messenger RNA in the central nervous system of zebrafish. *Neuroscience* **103**, 811–830.

Inoue, A., Takahashi, M., Hatta, K., Hotta, Y., and Okamoto, H. (1994). Developmental regulation of islet-1 mRNA expression during neuronal differentiation in embryonic zebrafish. *Dev. Dyn.* **199**, 1–11.

Ito, Y., Tanaka, H., Okamoto, H., and Ohshima, T. (2010). Characterization of neural stem cells and their progeny in the adult zebrafish optic tectum. *Dev. Biol.* **342**, 26–38.

Itoh, M., Kim, C. H., Palardy, G., Oda, T., Jiang, Y. J., Maust, D., Yeo, S. Y., Lorick, K., Wright, G. J., Ariza-McNaughton, L., et al. (2003). Mind bomb is a ubiquitin ligase that is essential for efficient activation of Notch signaling by Delta. *Dev. Cell* **4**, 67–82.

Jaszai, J., Reifers, F., Picker, A., Langenberg, T., and Brand, M. (2003). Isthmus-to-midbrain transformation in the absence of midbrain-hindbrain organizer activity. *Development* **130**, 6611–6623.

Jensen, A. M., and Westerfield, M. (2004). Zebrafish mosaic eyes is a novel FERM protein required for retinal lamination and retinal pigmented epithelial tight junction formation. *Curr. Biol.* **14**, 711–717.

Jing, X., and Malicki, J. (2009). Zebrafish ale oko, an essential determinant of sensory neuron survival and the polarity of retinal radial glia, encodes the p50 subunit of dynactin. *Development* **136**, 2955–2964.

Jowett, T. (1999). Analysis of protein and gene expression. *Methods Cell Biol.* **59**, 63–85.

Julich, D., Hwee Lim, C., Round, J., Nicolaije, C., Schroeder, J., Davies, A., Geisler, R., Lewis, J., Jiang, Y. J., and Holley, S. (2005). beamter/deltaC and the role of Notch ligands in the zebrafish somite segmentation, hindbrain neurogenesis and hypochord differentiation. *Dev. Biol.* **286**, 391–404.

Kalev-Zylinska, M. L., Horsfield, J. A., Flores, M. V., Postlethwait, J. H., Chau, J. Y., Cattin, P. M., Vitas, M. R., Crosier, P. S., and Crosier, K. E. (2003). Runx3 is required for hematopoietic development in zebrafish. *Dev. Dyn.* **228**, 323–336.

Kani, S., Bae, Y. K., Shimizu, T., Tanabe, K., Satou, C., Parsons, M. J., Scott, E., Higashijima, S., and Hibi, M. (2010). Proneural gene-linked neurogenesis in zebrafish cerebellum. *Dev. Biol.* **343**, 1–17.

Kaslin, J., Ganz, J., and Brand, M. (2008). Proliferation, neurogenesis and regeneration in the non-mammalian vertebrate brain. *Philos. Trans. R. Soc. Lond., B, Biol. Sci.* **363**, 101–122.

Kaslin, J., Ganz, J., Geffarth, M., Grandel, H., Hans, S., and Brand, M. (2009). Stem cells in the adult zebrafish cerebellum: Initiation and maintenance of a novel stem cell niche. *J. Neurosci.* **29**, 6142–6153.

Kaslin, J., Nystedt, J. M., Ostergard, M., Peitsaro, N., and Panula, P. (2004). The orexin/hypocretin system in zebrafish is connected to the aminergic and cholinergic systems. *J. Neurosci.* **24**, 2678–2689.

Kaslin, J., and Panula, P. (2001). Comparative anatomy of the histaminergic and other aminergic systems in zebrafish (Danio rerio). *J. Comp. Neurol.* **440**, 342–377.

Kassen, S. C., Ramanan, V., Montgomery, J.E., C., T.B., Liu, C. G., Vihtelic, T. S., and Hyde, D. R. (2007). Time course analysis of gene expression during light-induced photoreceptor cell death and regeneration in albino zebrafish. *Dev. Neurobiol.* **67**, 1009–1031.

Kassen, S. C., Thummel, R., Burket, C. T., Campochiaro, L. A., Harding, M. J., and Hyde, D. R. (2008). The Tg(ccnb1:EGFP) transgenic zebrafish line labels proliferating cells during retinal development and regeneration. *Mol. Vis.* **14**, 951–963.

Kawahara, A., Chien, C. B., and Dawid, I. B. (2002). The homeobox gene mbx is involved in eye and tectum development. *Dev. Biol.* **248**, 107–117.

Kawai, H., Arata, N., and Nakayasu, H. (2001). Three-dimensional distribution of astrocytes in zebrafish spinal cord. *Glia* **36**, 406–413.

Kay, J., Finger-Baier, K., Roeser, T., Staub, W., and Baier, H. (2001). Retinal ganglion cell genesis requires lakritz, a zebrafish atonal homolog. *Neuron* **30**, 725–736.

Kay, J. N., Roeser, T., Mumm, J. S., Godinho, L., Mrejeru, A., Wong, R. O., and Baier, H. (2004). Transient requirement for ganglion cells during assembly of retinal synaptic layers. *Development* **131**, 1331–1342.

Keller, P. J., Schmidt, A. D., Santella, A., Khairy, K., Bao, Z., Wittbrodt, J., and Stelzer, E. H. (2010). Fast, high-contrast imaging of animal development with scanned light sheet-based structured-illumination microscopy. *Nat. Methods* **7**, 637–642.

Keller, P. J., Schmidt, A. D., Wittbrodt, J., and Stelzer, E. H. (2008). Reconstruction of zebrafish early embryonic development by scanned light sheet microscopy. *Science* **322**, 1065–1069.

Kemp, H. A., Carmany-Rampey, A., and Moens, C. (2009). Generating chimeric zebrafish embryos by transplantation. *J. Vis. Exp.*17; (29). pii: 1394. doi: 10.379/-1394.

Kennedy, B. N., Alvarez, Y., Brockerhoff, S. E., Stearns, G. W., Sapetto-Rebow, B., Taylor, M. R., and Hurley, J. B. (2007). Identification of a zebrafish cone photoreceptor-specific promoter and genetic rescue of achromatopsia in the nof mutant. *Invest. Ophthalmol. Vis. Sci.* **48**, 522–529.

Kera, S. A., Agerwala, S. M., and Horne, J. H. (2010). The temporal resolution of *in vivo* electroporation in zebrafish: A method for time-resolved loss of function. *Zebrafish* **7**, 97–108.

Kikuchi, Y., Segawa, H., Tokumoto, M., Tsubokawa, T., Hotta, Y., Uyemura, K., and Okamoto, H. (1997). Ocular and cerebellar defects in zebrafish induced by overexpression of the LIM domains of the islet-3 LIM/homeodomain protein. *Neuron* **18**, 369–382.

Kim, C. H., Oda, T., Itoh, M., Jiang, D., Artinger, K. B., Chandrasekharappa, S. C., Driever, W., and Chitnis, A. B. (2000). Repressor activity of Headless/Tcf3 is essential for vertebrate head formation. *Nature* **407**, 913–916.

Kim, C. H., Ueshima, E., Muraoka, O., Tanaka, H., Yeo, S. Y., Huh, T. L., and Miki, N. (1996). Zebrafish elav/HuC homologue as a very early neuronal marker. *Neurosci. Lett.* **216**, 109–112.

Kim, H., Shin, J., Kim, S., Poling, J., Park, H. C., and Appel, B. (2008). Notch-regulated oligodendrocyte specification from radial glia in the spinal cord of zebrafish embryos. *Dev. Dyn.* **237**, 2081–2089.

Kimmel, C. B., Ballard, W. W., Kimmel, S. R., Ullmann, B., and Schilling, T. F. (1995). Stages of embryonic development of the zebrafish. *Dev. Dyn.* **203**, 253–310.

Kimura, Y., Okamura, Y., and Higashijima, S. (2006). alx, a zebrafish homolog of Chx10, marks ipsilateral descending excitatory interneurons that participate in the regulation of spinal locomotor circuits. *J. Neurosci.* **26**, 5684–5697.

Kimura, Y., Satou, C., and Higashijima, S. (2008). V2a and V2b neurons are generated by the final divisions of pair-producing progenitors in the zebrafish spinal cord. *Development* **135**, 3001–3005.

Kirby, B. B., Takada, N., Latimer, A. J., Shin, J., Carney, T. J., Kelsh, R. N., and Appel, B. (2006). *In vivo* time-lapse imaging shows dynamic oligodendrocyte progenitor behavior during zebrafish development. *Nat. Neurosci.* **9**, 1506–1511.

Kishimoto, Y., Lee, K. H., Zon, L., Hammerschmidt, M., and Schulte-Merker, S. (1997). The molecular nature of zebrafish swirl: BMP2 function is essential during early dorsoventral patterning. *Development* **124**, 4457–4466.

Knowlton, M. N., Li, T., Ren, Y., Bill, B. R., Ellis, L. B., and Ekker, S. C. (2008). A PATO-compliant zebrafish screening database (MODB): Management of morpholino knockdown screen information. *BMC Bioinformatics* **9**, 7.

Ko, H. W., Norman, R. X., Tran, J., Fuller, K. P., Fukuda, M., and Eggenschwiler, J. T. (2010). Broad-minded links cell cycle-related kinase to cilia assembly and hedgehog signal transduction. *Dev. Cell* **18**, 237–247.

Korzh, V., Edlund, T., and Thor, S. (1993). Zebrafish primary neurons initiate expression of the LIM homeodomain protein Isl-1 at the end of gastrulation. *Development* **118**, 417–425.

Korzh, V., Sleptsova, I., Liao, J., He, J., and Gong, Z. (1998). Expression of zebrafish bHLH genes ngn1 and nrd defines distinct stages of neural differentiation. *Dev. Dyn.* **213**, 92–104.

Koshida, S., Shinya, M., Mizuno, T., Kuroiwa, A., and Takeda, H. (1998). Initial anteroposterior pattern of the zebrafish central nervous system is determined by differential competence of the epiblast. *Development* **125**, 1957–1966.

Koster, R. W., and Fraser, S. E. (2001). Direct imaging of *in vivo* neuronal migration in the developing cerebellum. *Curr. Biol.* **11**, 1858–1863.

Kozlowski, D. J., and Weinberg, E. S. (2000). Photoactivatable (caged) fluorescein as a cell tracer for fate mapping in the zebrafish embryo. *Methods Mol. Biol.* **135**, 349–355.

Kucenas, S., Takada, N., Park, H. C., Woodruff, E., Broadie, K., and Appel, B. (2008). CNS-derived glia ensheath peripheral nerves and mediate motor root development. *Nat. Neurosci.* **11**, 143–151.

Kurita, R., Sagara, H., Aoki, Y., Link, B. A., Arai, K., and Watanabe, S. (2003). Suppression of lens growth by alphaA-crystallin promoter-driven expression of diphtheria toxin results in disruption of retinal cell organization in zebrafish. *Dev. Biol.* **255**, 113–127.

Kusik, B. W., Hammond, D. R., and Udvadia, A. J. (2010). Transcriptional regulatory regions of gap43 needed in developing and regenerating retinal ganglion cells. *Dev. Dyn.* **239**, 482–495.

Lam, C. S., Marz, M., and Strahle, U. (2009). gfap and nestin reporter lines reveal characteristics of neural progenitors in the adult zebrafish brain. *Dev. Dyn.* **238**, 475–486.

Langenau, D., Palomero, T., Kanki, J., Ferrando, A., Zhou, Y., Zon, L., and Look, A. (2002). Molecular cloning and developmental expression of Tlx (Hox11) genes in zebrafish (Danio rerio). *Mech. Dev.* **117**, 243–248.

Latimer, A. J., Shin, J., and Appel, B. (2005). her9 promotes floor plate development in zebrafish. *Dev. Dyn.* **232**, 1098–1104.

Le, X., Langenau, D. M., Keefe, M. D., Kutok, J. L., Neuberg, D. S., and Zon, L. I. (2007). Heat shock-inducible Cre/Lox approaches to induce diverse types of tumors and hyperplasia in transgenic zebrafish. *Proc. Natl. Acad. Sci. U.S.A.* **104**, 9410–9415.

Leake, D., Asch, W., Canger, A., and Schechter, N. (1999). Gefiltin in zebrafish embryos: Sequential gene expression of two neurofilament proteins in retinal ganglion cells. *Differentiation* **65**, 181–189.

Lee, J. E., Wu, S. F., Goering, L. M., and Dorsky, R. I. (2006). Canonical Wnt signaling through Lef1 is required for hypothalamic neurogenesis. *Development* **133**, 4451–4461.

Lee, V. M., Carden, M. J., Schlaepfer, W. W., and Trojanowski, J. Q. (1987). Monoclonal antibodies distinguish several differently phosphorylated states of the two largest rat neurofilament subunits (NF-H and NF-M) and demonstrate their existence in the normal nervous system of adult rats. *J. Neurosci.* **7**, 3474–3488.

Lee, Y., Grill, S., Sanchez, A., Murphy-Ryan, M., and Poss, K. D. (2005). Fgf signaling instructs position-dependent growth rate during zebrafish fin regeneration. *Development* **132**, 5173–5183.

Lele, Z., Folchert, A., Concha, M., Rauch, G. J., Geisler, R., Rosa, F., Wilson, S. W., Hammerschmidt, M., and Bally-Cuif, L. (2002). parachute/n-cadherin is required for morphogenesis and maintained integrity of the zebrafish neural tube. *Development* **129**, 3281–3294.

Levkowitz, G., Zeller, J., Sirotkin, H. I., French, D., Schilbach, S., Hashimoto, H., Hibi, M., Talbot, W. S., and Rosenthal, A. (2003). Zinc finger protein too few controls the development of monoaminergic neurons. *Nat. Neurosci.* **6**, 28–33.

Li, M., Zhao, C., Wang, Y., Zhao, Z., and Meng, A. (2002). Zebrafish sox9b is an early neural crest marker. *Dev. Genes Evol.* **212**, 203–206.

Li, Y., Allende, M. L., Finkelstein, R., and Weinberg, E. S. (1994). Expression of two zebrafish orthodenticle-related genes in the embryonic brain. *Mech. Dev.* **48**, 229–244.

Li, Z., Joseph, N. M., and Easter, S. S. Jr. (2000). The morphogenesis of the zebrafish eye, including a fate map of the optic vesicle. *Dev. Dyn.* **218**, 175–188.

Li, Z., Korzh, V., and Gong, Z. (2009). DTA-mediated targeted ablation revealed differential interdependence of endocrine cell lineages in early development of zebrafish pancreas. *Differentiation* **78**, 241–252.

Liao, J., He, J., Yan, T., Korzh, V., and Gong, Z. (1999). A class of neuroD-related basic helix-loop-helix transcription factors expressed in developing central nervous system in zebrafish. *DNA Cell Biol.* **18**, 333–344.

Lillesaar, C., Stigloher, C., Tannhauser, B., Wullimann, M. F., and Bally-Cuif, L. (2009). Axonal projections originating from raphe serotonergic neurons in the developing and adult zebrafish, Danio rerio, using transgenics to visualize raphe-specific pet1 expression. *J. Comp. Neurol.* **512**, 158–182.

Lillesaar, C., Tannhauser, B., Stigloher, C., Kremmer, E., and Bally-Cuif, L. (2007). The serotonergic phenotype is acquired by converging genetic mechanisms within the zebrafish central nervous system. *Dev. Dyn.* **236**, 1072–1084.

Link, B., Fadool, J., Malicki, J., and Dowling, J. (2000). The zebrafish young mutation acts non-cell-autonomously to uncouple differentiation from specification for all retinal cells. *Development* **127**, 2177–2188.

Link, B. A., Kainz, P. M., Ryou, T., and Dowling, J. E. (2001). The perplexed and confused mutations affect distinct stages during the transition from proliferating to post-mitotic cells within the zebrafish retina. *Dev. Biol.* **236**, 436–453.

Lippincott-Schwartz, J., and Patterson, G. H. (2009). Photoactivatable fluorescent proteins for diffraction-limited and super-resolution imaging. *Trends Cell Biol.* **19**, 555–565.

Lister, J. A., Robertson, C. P., Lepage, T., Johnson, S. L., and Raible, D. W. (1999). nacre encodes a zebrafish microphthalmia-related protein that regulates neural-crest-derived pigment cell fate. *Development* **126**, 3757–3767.

Liu, K. S., Gray, M., Otto, S. J., Fetcho, J. R., and Beattie, C. E. (2003). Mutations in deadly seven/notch1a reveal developmental plasticity in the escape response circuit. *J. Neurosci.* **23**, 8159–8166.

Luo, G. R., Chen, Y., Li, X. P., Liu, T. X., and Le, W. D. (2008). Nr4a2 is essential for the differentiation of dopaminergic neurons during zebrafish embryogenesis. *Mol. Cell. Neurosci.* **39**, 202–210.

Lyons, D. A., Guy, A. T., and Clarke, J. D. (2003). Monitoring neural progenitor fate through multiple rounds of division in an intact vertebrate brain. *Development* **130**, 3427–3436.

MacDonald, R. B., Debiais-Thibaud, M., Talbot, J. C., and Ekker, M. (2010). The relationship between dlx and gad1 expression indicates highly conserved genetic pathways in the zebrafish forebrain. *Dev. Dyn.* **239**, 2298–2306.

Mahler, J., and Driever, W. (2007). Expression of the zebrafish intermediate neurofilament Nestin in the developing nervous system and in neural proliferation zones at postembryonic stages. *BMC Dev. Biol.* **7**, 89.

Malicki, J., and Driever, W. (1999). oko meduzy mutations affect neuronal patterning in the zebrafish retina and reveal cell-cell interactions of the retinal neuroepithelial sheet. *Development* **126**, 1235–1246.

Malicki, J., Jo, H., and Pujic, Z. (2003). Zebrafish N-cadherin, encoded by the glass onion locus, plays an essential role in retinal patterning. *Dev. Biol.* **259**, 95–108.

Marc, R., and Cameron, D. (2001). A molecular phenotype atlas of the zebrafish retina. *J. Neurocytol.* **30**, 593–654.

Marques, S. R., Lee, Y., Poss, K. D., and Yelon, D. (2008). Reiterative roles for FGF signaling in the establishment of size and proportion of the zebrafish heart. *Dev. Biol.* **321**, 397–406.

Martin, S., Heinrich, G., and Sandell, J. H. (1998). Sequence and expression of glutamic acid decarboxylase isoforms in the developing zebrafish. *J. Comp. Neurol.* **396**, 253–266.

Marx, M., Rutishauser, U., and Bastmeyer, M. (2001). Dual function of polysialic acid during zebrafish central nervous system development. *Development* **128**, 4949–4958.

März, M., Chapouton, P., Diotel, N., Vaillant, C., Hesl, B., Takamiya, M., Lam, C. S., Kah, O., Bally-Cuif, L., and Strähle, U. (2010). Heterogeneity in progenitor cell subtypes in the ventricular zone of the zebrafish adult telencephalon. *Glia* **58**, 870–888.

Masai, I., Heisenberg, C. P., Barth, K. A., Macdonald, R., Adamek, S., and Wilson, S. W. (1997). floating head and masterblind regulate neuronal patterning in the roof of the forebrain. *Neuron* **18**, 43–57.

Masai, I., Stemple, D., Okamoto, H., and Wilson, S. (2000). Midline signals regulate retinal neurogenesis in zebrafish. *Neuron* **27**, 251–263.

Matsuda, M., and Chitnis, A. B. (2009). Interaction with Notch determines endocytosis of specific Delta ligands in zebrafish neural tissue. *Development* **136**, 197–206.

Maves, L., Jackman, W., and Kimmel, C. B. (2002). FGF3 and FGF8 mediate a rhombomere 4 signaling activity in the zebrafish hindbrain. *Development* **129**, 3825–3837.

McFarland, K. A., Topczewska, J. M., Weidinger, G., Dorsky, R. I., and Appel, B. (2008). Hh and Wnt signaling regulate formation of olig2+ neurons in the zebrafish cerebellum. *Dev. Biol.* **318**, 162–171.

McGraw, H. F., Nechiporuk, A., and Raible, D. W. (2008). Zebrafish dorsal root ganglia neural precursor cells adopt a glial fate in the absence of neurogenin1. *J. Neurosci.* **28**, 12558–12569.

McLean, D. L., Fan, J., Higashijima, S., Hale, M. E., and Fetcho, J. R. (2007). A topographic map of recruitment in spinal cord. *Nature* **446**, 71–75.

McMahon, C., Gestri, G., Wilson, S. W., Link, B. A. (2009). Lmx1b is essential for survival of periocular mesenchymal cells and influences Fgf-mediated retinal patterning in zebrafish. *Dev. Biol.* **332**, 287–298.

Melancon, E., Liu, D. W., Westerfield, M., and Eisen, J. S. (1997). Pathfinding by identified zebrafish motoneurons in the absence of muscle pioneers. *J. Neurosci.* **17**, 7796–7804.

Meng, S., Ryu, S., Zhao, B., Zhang, D. Q., Driever, W., and McMahon, D. G. (2008a). Targeting retinal dopaminergic neurons in tyrosine hydroxylase-driven green fluorescent protein transgenic zebrafish. *Mol. Vis.* **14**, 2475–2483.

Meng, X., Noyes, M. B., Zhu, L. J., Lawson, N. D., and Wolfe, S. A. (2008b). Targeted gene inactivation in zebrafish using engineered zinc-finger nucleases. *Nat. Biotechnol.* **26**, 695–701.

Menuet, A., Pellegrini, E., Brion, F., Gueguen, M. M., Anglade, I., Pakdel, F., and Kah, O. (2005). Expression and estrogen-dependent regulation of the zebrafish brain aromatase gene. *J. Comp. Neurol.* **485**, 304–320.

Metcalfe, W., Myers, P., Trevarrow, B., Bass, M., and Kimmel, C. (1990). Primary neurons that express the L2/HNK-1 carbohydrate during early development in the zebrafish. *Development* **110**, 491–504.

Mikkola, I., Fjose, A., Kuwada, J. Y., Wilson, S., Guddal, P. H., and Krauss, S. (1992). The paired domain-containing nuclear factor pax[b] is expressed in specific commissural interneurons in zebrafish embryos. *J. Neurobiol.* **23**, 933–946.

Mione, M., Baldessari, D., Deflorian, G., Nappo, G., and Santoriello, C. (2008). How neuronal migration contributes to the morphogenesis of the CNS: Insights from the zebrafish. *Dev. Neurosci.* **30**, 65–81.

Miyamura, Y., and Nakayasu, H. (2001). Zonal distribution of Purkinje cells in the zebrafish cerebellum: Analysis by means of a specific monoclonal antibody. *Cell Tissue Res.* **305**, 299–305.

Moens, C. B., Donn, T. M., Wolf-Saxon, E. R., and Ma, T. P. (2008). Reverse genetics in zebrafish by TILLING. *Brief Funct. Genomic Proteomic* **7**, 454–459.

Molina, G. A., Watkins, S. C., and Tsang, M. (2007). Generation of FGF reporter transgenic zebrafish and their utility in chemical screens. *BMC Dev. Biol.* **7**, 62.

Montgomery, J. E., Parsons, M. J., and Hyde, D. R. (2009). A novel model of retinal ablation demonstrates that the extent of rod cell death regulates the origin of the regenerated zebrafish rod photoreceptors. *J. Comp. Neurol.* **518**, 800–814.

Mueller, T., and Wullimann, M. F. (2002). BrdU-, neuroD (nrd)- and Hu-studies reveal unusual non-ventricular neurogenesis in the postembryonic zebrafish forebrain. *Mech. Dev.* **117**, 123–135.

Mueller, T., and Wullimann, M. F. (2003). Anatomy of neurogenesis in the early zebrafish brain. *Brain Res. Dev. Brain Res.* **140**, 137–155.

Mumm, J. S., Williams, P. R., Godinho, L., Koerber, A., Pittman, A. J., Roeser, T., Chien, C. B., Baier, H., and Wong, R. O. (2006). *In vivo* imaging reveals dendritic targeting of laminated afferents by zebrafish retinal ganglion cells. *Neuron* **52**, 609–621.

Nguyen, V. H., Schmid, B., Trout, J., Connors, S. A., Ekker, M., and Mullins, M. C. (1998). Ventral and lateral regions of the zebrafish gastrula, including the neural crest progenitors, are established by a bmp2b/swirl pathway of genes. *Dev. Biol.* **199**, 93–110.

Ninkovic, J., Stigloher, C., Lillesaar, C., and Bally-Cuif, L. (2008). Gsk3beta/PKA and Gli1 regulate the maintenance of neural progenitors at the midbrain-hindbrain boundary in concert with E(Spl) factor activity. *Development* **135**, 3137–3148.

Ninkovic, J., Tallafuss, A., Leucht, C., Topczewski, J., Tannhauser, B., Solnica-Krezel, L., and Bally-Cuif, L. (2005). Inhibition of neurogenesis at the zebrafish midbrain-hindbrain boundary by the combined and dose-dependent activity of a new hairy/E(spl) gene pair. *Development* **132**, 75–88.

Nissen, R. M., Yan, J., Amsterdam, A., Hopkins, N., and Burgess, S. M. (2003). Zebrafish foxi one modulates cellular responses to Fgf signaling required for the integrity of ear and jaw patterning. *Development* **130**, 2543–2554.

Norden, C., Young, S., Link, B. A., Harris, W. A. (2009). Actomyosin is the main driver of interkinetic nuclear migration in the retina. *Cell* **138**, 1195–1208.

Norton, W. H., Folchert, A., and Bally-Cuif, L. (2008). Comparative analysis of serotonin receptor (HTR1A/HTR1B families) and transporter (slc6a4a/b) gene expression in the zebrafish brain. *J. Comp. Neurol.* **511**, 521–542.

O'Malley, D. M., Kao, Y. H., and Fetcho, J. R. (1996). Imaging the functional organization of zebrafish hindbrain segments during escape behaviors. *Neuron* **17**, 1145–1155.

Odenthal, J., and Nusslein-Volhard, C. (1998). Fork head domain genes in zebrafish. *Dev. Genes Evol.* **208**, 245–258.

Odenthal, J., van Eeden, F. J., Haffter, P., Ingham, P. W., and Nusslein-Volhard, C. (2000). Two distinct cell populations in the floor plate of the zebrafish are induced by different pathways. *Dev. Biol.* **219**, 350–363.

Oehlmann, V., Berger, S., Sterner, C., and Korsching, S. (2004). Zebrafish beta tubulin expression is limited to the nervous system throughout development, and in the adult brain is restricted to a subset of proliferative regions. *Gene Expr. Patterns* **4**, 191–198.

Ogura, E., Okuda, Y., Kondoh, H., and Kamachi, Y. (2009). Adaptation of GAL4 activators for GAL4 enhancer trapping in zebrafish. *Dev. Dyn.* **238**, 641–655.

Okuda, Y., Ogura, E., Kondoh, H., and Kamachi, Y. (2010). B1 SOX coordinate cell specification with patterning and morphogenesis in the early zebrafish embryo. *PLoS Genet.* **6**, e1000936.

Okuda, Y., Yoda, H., Uchikawa, M., Furutani-Seiki, M., Takeda, H., Kondoh, H., and Kamachi, Y. (2006). Comparative genomic and expression analysis of group B1 sox genes in zebrafish indicates their diversification during vertebrate evolution. *Dev. Dyn.* **235**, 811–825.

Olivier, N., Luengo-Oroz, M. A., Duloquin, L., Faure, E., Savy, T., Veilleux, I., Solinas, X., Debarre, D., Bourgine, P., Santos, A., *et al.* (2010). Cell lineage reconstruction of early zebrafish embryos using label-free nonlinear microscopy. *Science* **329**, 967–971.

Omori, Y., and Malicki, J. (2006). oko meduzy and related crumbs genes are determinants of apical cell features in the vertebrate embryo. *Curr. Biol.* **16**, 945–957.

Onichtchouk, D., Geier, F., Polok, B., Messerschmidt, D. M., Mossner, R., Wendik, B., Song, S., Taylor, V., Timmer, J., and Driever, W. (2010). Zebrafish Pou5f1-dependent transcriptional networks in temporal control of early development. *Mol. Syst. Biol.* **6**, 354.

Ott, H., Diekmann, H., Stuermer, C., and Bastmeyer, M. (2001). Function of neurolin (DM-GRASP/SC-1) in guidance of motor axons during zebrafish development. *Dev. Biol.* **235**, 86–97.

Palevitch, O., Kight, K., Abraham, E., Wray, S., Zohar, Y., and Gothilf, Y. (2007). Ontogeny of the GnRH systems in zebrafish brain: In situ hybridization and promoter-reporter expression analyses in intact animals. *Cell Tissue Res.* **327**, 313–322.

Papan, C., and Campos-Ortega, J. (1997). A clonal analysis of spinal cord development in the zebrafish. *Dev. Genes Evol.* **207**, 71–81.

Papan, C., and Campos-ortega, J. A. (1994). On the formation of the neural keel and neural tube in the zebrafish Danio (Brachydanio) rerio. *Rouxs Arch. Dev. Biol.* **203**, 178–186.

Parinov, S., Kondrichin, I., Korzh, V., and Emelyanov, A. (2004). Tol2 transposon-mediated enhancer trap to identify developmentally regulated zebrafish genes *in vivo. Dev. Dyn.* **231**, 449–459.

Park, H. C., Hong, S. K., Kim, H. S., Kim, S. H., Yoon, E. J., Kim, C. H., Miki, N., and Huh, T. L. (2000a). Structural comparison of zebrafish Elav/Hu and their differential expressions during neurogenesis. *Neurosci. Lett.* **279**, 81–84.

Park, H. C., Boyce, J., Shin, J., and Appel, B. (2005). Oligodendrocyte specification in zebrafish requires notch-regulated cyclin-dependent kinase inhibitor function. *J. Neurosci.* **25**, 6836–6844.

Park, H. C., Kim, C. H., Bae, Y. K., Yeo, S. Y., Kim, S. H., Hong, S. K., Shin, J., Yoo, K. W., Hibi, M., Hirano, T., *et al.* (2000b). Analysis of upstream elements in the HuC promoter leads to the establishment of transgenic zebrafish with fluorescent neurons. *Dev. Biol.* **227**, 279–293.

Park, H. C., Mehta, A., Richardson, J. S., and Appel, B. (2002). olig2 is required for zebrafish primary motor neuron and oligodendrocyte development. *Dev. Biol.* **248**, 356–368.

Park, H. C., Shin, J., and Appel, B. (2004). Spatial and temporal regulation of ventral spinal cord precursor specification by Hedgehog signaling. *Development* **131**, 5959–5969.

Park, S. H., Yeo, S. Y., Yoo, K. W., Hong, S. K., Lee, S., Rhee, M., Chitnis, A. B., and Kim, C. H. (2003). Zath3, a neural basic helix-loop-helix gene, regulates early neurogenesis in the zebrafish. *Biochem. Biophys. Res. Commun.* **308**, 184–190.

Parsons, M. J., Pisharath, H., Yusuff, S., Moore, J. C., Siekmann, A. F., Lawson, N., and Leach, S. D. (2009). Notch-responsive cells initiate the secondary transition in larval zebrafish pancreas. *Mech. Dev.* **126**, 898–912.

Pauls, S., Geldmacher-Voss, B., and Campos-Ortega, J. A. (2001). A zebrafish histone variant H2A.F/Z and a transgenic H2A.F/Z:GFP fusion protein for *in vivo* studies of embryonic development. *Dev. Genes Evol.* **211**, 603–610.

Pellegrini, E., Mouriec, K., Anglade, I., Menuet, A., Le Page, Y., Gueguen, M. M., Marmignon, M. H., Brion, F., Pakdel, F., and Kah, O. (2007). Identification of aromatase-positive radial glial cells as progenitor cells in the ventricular layer of the forebrain in zebrafish. *J. Comp. Neurol.* **501**, 150–167.

Picker, A., Scholpp, S., Bohli, H., Takeda, H., and Brand, M. (2002). A novel positive transcriptional feedback loop in midbrain-hindbrain boundary development is revealed through analysis of the zebrafish pax2.1 promoter in transgenic lines. *Development* **129**, 3227–3239.

Pisharath, H., and Parsons, M. J. (2009). Nitroreductase-mediated cell ablation in transgenic zebrafish embryos. *Methods Mol. Biol.* **546**, 133–143.

Pisharath, H., Rhee, J. M., Swanson, M. A., Leach, S. D., and Parsons, M. J. (2007). Targeted ablation of beta cells in the embryonic zebrafish pancreas using E. coli nitroreductase. *Mech. Dev.* **124**, 218–229.

Pittman, A. J., Law, M. Y., and Chien, C. B. (2008). Pathfinding in a large vertebrate axon tract: Isotypic interactions guide retinotectal axons at multiple choice points. *Development* **135**, 2865–2871.

Poggi, L., Vitorino, M., Masai, I., and Harris, W. A. (2005). Influences on neural lineage and mode of division in the zebrafish retina *in vivo. J. Cell Biol.* **171**, 991–999.

Prince, V. E., Moens, C. B., Kimmel, C. B., and Ho, R. K. (1998). Zebrafish hox genes: Expression in the hindbrain region of wild-type and mutants of the segmentation gene, valentino. *Development* **125**, 393–406.

Quint, E., Zerucha, T., and Ekker, M. (2000). Differential expression of orthologous Dlx genes in zebrafish and mice: Implications for the evolution of the Dlx homeobox gene family. *J. Exp. Zool.* **288**, 235–241.

Raible, F., and Brand, M. (2004). Divide et Impera–the midbrain-hindbrain boundary and its organizer. *Trends Neurosci.* **27**, 727–734.

Raymond, P. A., Barthel, L. K., Bernardos, R. L., and Perkowski, J. J. (2006). Molecular characterization of retinal stem cells and their niches in adult zebrafish. *BMC Dev. Biol.* **6**, 36.

Reifers, F., Bohli, H., Walsh, E. C., Crossley, P. H., Stainier, D. Y., and Brand, M. (1998). Fgf8 is mutated in zebrafish acerebellar (ace) mutants and is required for maintenance of midbrain-hindbrain boundary development and somitogenesis. *Development* **125**, 2381–2395.

Reim, G., and Brand, M. (2006). Maternal control of vertebrate dorsoventral axis formation and epiboly by the POU domain protein Spg/Pou2/Oct4. *Development* **133**, 2757–2770.

Reimer, M. M., Kuscha, V., Wyatt, C., Sorensen, I., Frank, R. E., Knuwer, M., Becker, T., and Becker, C. G. (2009). Sonic hedgehog is a polarized signal for motor neuron regeneration in adult zebrafish. *J. Neurosci.* **29**, 15073–15082.

Reimer, M. M., Sorensen, I., Kuscha, V., Frank, R. E., Liu, C., Becker, C. G., and Becker, T. (2008). Motor neuron regeneration in adult zebrafish. *J. Neurosci.* **28**, 8510–8516.

Reinhard, E., Nedivi, E., Wegner, J., Skene, J. H., and Westerfield, M. (1994). Neural selective activation and temporal regulation of a mammalian GAP-43 promoter in zebrafish. *Development* **120**, 1767–1775.

Reischauer, S., Levesque, M. P., Nusslein-Volhard, C., and Sonawane, M. (2009). Lgl2 executes its function as a tumor suppressor by regulating ErbB signaling in the zebrafish epidermis. *PLoS Genet.* **5**, e1000720.

Ren, J. Q., McCarthy, W. R., Zhang, H., Adolph, A. R., and Li, L. (2002). Behavioral visual responses of wild-type and hypopigmented zebrafish. *Vision Res.* **42**, 293–299.

Reugels, A. M., Boggetti, B., Scheer, N., and Campos-Ortega, J. A. (2006). Asymmetric localization of Numb:EGFP in dividing neuroepithelial cells during neurulation in Danio rerio. *Dev. Dyn.* **235**, 934–948.

Ribes, V., Balaskas, N., Sasai, N., Cruz, C., Dessaud, E., Cayuso, J., Tozer, S., Yang, L. L., Novitch, B., Marti, E., *et al.* (2010). Distinct Sonic Hedgehog signaling dynamics specify floor plate and ventral neuronal progenitors in the vertebrate neural tube. *Genes Dev.* **24**, 1186–1200.

Rimini, R., Beltrame, M., Argenton, F., Szymczak, D., Cotelli, F., and Bianchi, M. E. (1999). Expression patterns of zebrafish sox11A, sox11B and sox21. *Mech. Dev.* **89**, 167–171.

Rizzo, M. A., Davidson, M. W., and Piston, D. W. (2009). Fluorescent protein tracking and detection: Applications using fluorescent proteins in living cells. *Cold Spring Harb. Protoc. 2009, pdb top64.*

Roberts, R. K., and Appel, B. (2009). Apical polarity protein PrkCi is necessary for maintenance of spinal cord precursors in zebrafish. *Dev. Dyn.* **238**, 1638–1648.

Rohr, S., Bit-Avragim, N., and Abdelilah-Seyfried, S. (2006). Heart and soul/PRKCi and nagie oko/Mpp5 regulate myocardial coherence and remodeling during cardiac morphogenesis. *Development* **133**, 107–115.

Roth, L., Bormann, P., Bonnet, A., and Reinhard, E. (1999). beta-thymosin is required for axonal tract formation in developing zebrafish brain. *Development* **126**, 1365–1374.

Russek-Blum, N., Gutnick, A., Nabel-Rosen, H., Blechman, J., Staudt, N., Dorsky, R. I., Houart, C., and Levkowitz, G. (2008). Dopaminergic neuronal cluster size is determined during early forebrain patterning. *Development* **135**, 3401–3413.

Russek-Blum, N., Nabel-Rosen, H., and Levkowitz, G. (2009). High resolution fate map of the zebrafish diencephalon. *Dev. Dyn.* **238**, 1827–1835.

Ryu, S., Holzschuh, J., Erhardt, S., Ettl, A. K., and Driever, W. (2005). Depletion of minichromosome maintenance protein 5 in the zebrafish retina causes cell cycle defect and apoptosis. *Proc. Natl. Acad. Sci. U.S.A.* **102**, 18467–18472.

Ryu, S., Mahler, J., Acampora, D., Holzschuh, J., Erhardt, S., Omodei, D., Simeone, A., and Driever, W. (2007). Orthopedia homeodomain protein is essential for diencephalic dopaminergic neuron development. *Curr. Biol.* **17**, 873–880.

Sagasti, A., Guido, M. R., Raible, D. W., and Schier, A. F. (2005). Repulsive interactions shape the morphologies and functional arrangement of zebrafish peripheral sensory arbors. *Curr. Biol.* **15**, 804–814.

Sassa, T., Aizawa, H., and Okamoto, H. (2007). Visualization of two distinct classes of neurons by gad2 and zic1 promoter/enhancer elements in the dorsal hindbrain of developing zebrafish reveals neuronal connectivity related to the auditory and lateral line systems. *Dev. Dyn.* **236**, 706–718.

Sato, T., Hamaoka, T., Aizawa, H., Hosoya, T., and Okamoto, H. (2007a). Genetic single-cell mosaic analysis implicates ephrinB2 reverse signaling in projections from the posterior tectum to the hindbrain in zebrafish. *J. Neurosci.* **27**, 5271–5279.

Sato, T., Takahoko, M., and Okamoto, H. (2006). HuC:Kaede, a useful tool to label neural morphologies in networks *in vivo. Genesis* **44**, 136–142.

Sato, Y., Miyasaka, N., and Yoshihara, Y. (2005). Mutually exclusive glomerular innervation by two distinct types of olfactory sensory neurons revealed in transgenic zebrafish. *J. Neurosci.* **25**, 4889–4897.

Sato, Y., Miyasaka, N., and Yoshihara, Y. (2007b). Hierarchical regulation of odorant receptor gene choice and subsequent axonal projection of olfactory sensory neurons in zebrafish. *J. Neurosci.* **27**, 1606–1615.

Sato-Maeda, M., Obinata, M., and Shoji, W. (2008). Position fine-tuning of caudal primary motoneurons in the zebrafish spinal cord. *Development* **135**, 323–332.

Schauerte, H. E., van Eeden, F. J., Fricke, C., Odenthal, J., Strahle, U., and Haffter, P. (1998). Sonic hedgehog is not required for the induction of medial floor plate cells in the zebrafish. *Development* **125**, 2983–2993.

Schebesta, M., and Serluca, F. C. (2009). olig1 Expression identifies developing oligodendrocytes in zebrafish and requires hedgehog and notch signaling. *Dev. Dyn.* **238**, 887–898.

Scheer, N., and Campos-Ortega, J. A. (1999). Use of the Gal4-UAS technique for targeted gene expression in the zebrafish. *Mech. Dev.* **80**, 153–158.

Scheer, N., Groth, A., Hans, S., and Campos-Ortega, J. A. (2001). An instructive function for Notch in promoting gliogenesis in the zebrafish retina. *Development* **128**, 1099–1107.

Schmitz, B., Papan, C., and Campos-Ortega, J. A. (1993). Neurulation in the anterior trunk region of the zebrafish Brachydanio rerio. *Rouxs Arch. Dev. Biol.* **202**, 250–259.

Scholpp, S., Delogu, A., Gilthorpe, J., Peukert, D., Schindler, S., and Lumsden, A. (2009). Her6 regulates the neurogenetic gradient and neuronal identity in the thalamus. *Proc. Natl. Acad. Sci. U.S.A.* **106**, 19895–19900.

Scholpp, S., Wolf, O., Brand, M., and Lumsden, A. (2006). Hedgehog signalling from the zona limitans intrathalamica orchestrates patterning of the zebrafish diencephalon. *Development* **133**, 855–864.

Schonig, K., and Bujard, H. (2003). Generating conditional mouse mutants via tetracycline-controlled gene expression. *Methods Mol. Biol.* **209**, 69–104.

Schroeter, E. H., Wong, R. O., and Gregg, R. G. (2006). *In vivo* development of retinal ON-bipolar cell axonal terminals visualized in nyx::MYFP transgenic zebrafish. *Vis. Neurosci.* **23**, 833–843.

Schulte-Merker, S., Lee, K. J., McMahon, A. P., and Hammerschmidt, M. (1997). The zebrafish organizer requires chordino. *Nature* **387**, 862–3.

Scott, E. K., and Baier, H. (2009). The cellular architecture of the larval zebrafish tectum, as revealed by gal4 enhancer trap lines. *Front Neural Circuits* **3**, 13.

Scott, E. K., Mason, L., Arrenberg, A. B., Ziv, L., Gosse, N. J., Xiao, T., Chi, N. C., Asakawa, K., Kawakami, K., and Baier, H. (2007). Targeting neural circuitry in zebrafish using GAL4 enhancer trapping. *Nat. Methods* **4**, 323–326.

Segawa, H., Miyashita, T., Hirate, Y., Higashijima, S., Chino, N., Uyemura, K., Kikuchi, Y., and Okamoto, H. (2001). Functional repression of Islet-2 by disruption of complex with Ldb impairs peripheral axonal outgrowth in embryonic zebrafish. *Neuron* **30**, 423–436.

Seok, S. H., Na, Y. R., Han, J. H., Kim, T. H., Jung, H., Lee, B. H., Emelyanov, A., Parinov, S., and Park, J. H. (2009). Cre/loxP-regulated transgenic zebrafish model for neural progenitor-specific oncogenic Kras expression. *Cancer Sci.* **101**, 149–154.

Shaner, N. C., Lin, M. Z., McKeown, M. R., Steinbach, P. A., Hazelwood, K. L., Davidson, M. W., and Tsien, R. Y. (2008). Improving the photostability of bright monomeric orange and red fluorescent proteins. *Nat. Methods* **5**, 545–551.

Shaner, N. C., Steinbach, P. A., and Tsien, R. Y. (2005). A guide to choosing fluorescent proteins. *Nat. Methods* **2**, 905–909.

Shin, J., Park, H. C., Topczewska, J. M., Mawdsley, D. J., and Appel, B. (2003). Neural cell fate analysis in zebrafish using olig2 BAC transgenics. *Methods Cell Sci.* **25**, 7–14.

Sinha, D. K., Neveu, P., Gagey, N., Aujard, I., Le Saux, T., Rampon, C., Gauron, C., Kawakami, K., Leucht, C., Bally-Cuif, L., *et al.* (2010). Photoactivation of the CreER T2 recombinase for conditional site-specific recombination with high spatiotemporal resolution. *Zebrafish* **7**, 199–204.

Solomon, K. S., and Fritz, A. (2002). Concerted action of two dlx paralogs in sensory placode formation. *Development* **129**, 3127–3136.

Solomon, K. S., Kudoh, T., Dawid, I. B., and Fritz, A. (2003). Zebrafish foxi1 mediates otic placode formation and jaw development. *Development* **130**, 929–940.

Son, O., Kim, H., Ji, M., Yoo, K., Rhee, M., and Kim, C. (2003). Cloning and expression analysis of a Parkinson's disease gene, uch-L1, and its promoter in zebrafish. *Biochem. Biophys. Res. Commun.* **312**, 601–607.

Sonawane, M., Carpio, Y., Geisler, R., Schwarz, H., Maischein, H. M., and Nuesslein-Volhard, C. (2005). Zebrafish penner/lethal giant larvae 2 functions in hemidesmosome formation, maintenance of cellular morphology and growth regulation in the developing basal epidermis. *Development* **132**, 3255–3265.

Song, M. H., Brown, N. L., and Kuwada, J. Y. (2004). The cfy mutation disrupts cell divisions in a stage-dependent manner in zebrafish embryos. *Dev. Biol.* **276**, 194–206.

Springer, C. J., and Niculescu-Duvaz, I. (2000). Prodrug-activating systems in suicide gene therapy. *J. Clin. Invest.* **105**, 1161–1167.

Stigloher, C., Chapouton, P., Adolf, B., and Bally-Cuif, L. (2008). Identification of neural progenitor pools by E(Spl) factors in the embryonic and adult brain. *Brain Res. Bull.* **75**, 266–273.

Stigloher, C., Ninkovic, J., Laplante, M., Geling, A., Tannhauser, B., Topp, S., Kikuta, H., Becker, T. S., Houart, C., and Bally-Cuif, L. (2006). Segregation of telencephalic and eye-field identities inside the zebrafish forebrain territory is controlled by Rx3. *Development* **133**, 2925–2935.

Strahle, U., Lam, C. S., Ertzer, R., and Rastegar, S. (2004). Vertebrate floor-plate specification: Variations on common themes. *Trends Genet.* **20**, 155–162.

Sugiyama, M., Sakaue-Sawano, A., Iimura, T., Fukami, K., Kitaguchi, T., Kawakami, K., Okamoto, H., Higashijima, S. I., and Miyawaki, A. (2009). Illuminating cell cycle progression in the developing zebrafish embryo. *Proc. Natl. Acad. Sci. U.S.A.* **106**, 20812–20817.

Szabo, T. M., Brookings, T., Preuss, T., and Faber, D. S. (2008). Effects of temperature acclimation on a central neural circuit and its behavioral output. *J. Neurophysiol.* **100**, 2997–3008.

Takechi, M., Hamaoka, T., and Kawamura, S. (2003). Fluorescence visualization of ultraviolet-sensitive cone photoreceptor development in living zebrafish. *FEBS Lett.* **553**, 90–94.

Takke, C., Dornseifer, P., v Weizsacker, E., and Campos-Ortega, J. A. (1999). her4, a zebrafish homologue of the Drosophila neurogenic gene E(spl), is a target of NOTCH signalling. *Development* **126**, 1811–1821.

Tallafuss, A., and Bally-Cuif, L. (2003). Tracing of her5 progeny in zebrafish transgenics reveals the dynamics of midbrain-hindbrain neurogenesis and maintenance. *Development* **130**, 4307–4323.

Tallafuss, A., Trepman, A., and Eisen, J. S. (2009). DeltaA mRNA and protein distribution in the zebrafish nervous system. *Dev. Dyn.* **238**, 3226–3236.

Tarnowka, M. A., Baglioni, C., and Basilico, C. (1978). Synthesis of H1 histones by BHK cells in G1. *Cell* **15**, 163–171.

Tawk, M., Araya, C., Lyons, D. A., Reugels, A. M., Girdler, G. C., Bayley, P. R., Hyde, D. R., Tada, M., and Clarke, J. D. (2007). A mirror-symmetric cell division that orchestrates neuroepithelial morphogenesis. *Nature* **446**, 797–800.

Tawk, M., Bianco, I. H., and Clarke, J. D. (2009). Focal electroporation in zebrafish embryos and larvae. *Methods Mol. Biol.* **546**, 145–151.

Teraoka, H., Russell, C., Regan, J., Chandrasekhar, A., Concha, M., Yokoyama, R., Higashi, K., Take-Uchi, M., Dong, W., Hiraga, T., *et al.* (2004). Hedgehog and Fgf signaling pathways regulate the development of tphR-expressing serotonergic raphe neurons in zebrafish embryos. *J. Neurobiol.* **60**, 275–288.

Thisse, B., Pflumio, S., Fürthauer, M., Loppin, B., Heyer, V., Degrave, A., Woehl, R., Lux, A., Steffan, T., Charbonnier, X. Q., and Thisse, C.. (2001) Expression of the zebrafish genome during embryogenesis. ZFIN Direct Data Submission (http://zfin.org).

Thisse, C., Thisse, B., and Postlethwait, J. H. (1995). Expression of snail2, a second member of the zebrafish snail family, in cephalic mesendoderm and presumptive neural crest of wild-type and spadetail mutant embryos. *Dev. Biol.* **172**, 86–99.

Thummel, R., Burket, C. T., Brewer, J. L., Sarras, M. P. Jr., Li, L., Perry, M., McDermott, J. P., Sauer, B., Hyde, D. R., and Godwin, A. R. (2005). Cre-mediated site-specific recombination in zebrafish embryos. *Dev. Dyn.* **233**, 1366–1377.

Thummel, R., Enright, J. M., Kassen, S. C., Montgomery, J. E., Bailey, T. J., and Hyde, D. R. (2010). Pax6a and Pax6b are required at different points in neuronal progenitor cell proliferation during zebrafish photoreceptor regeneration. *Exp. Eye Res.* **90**, 572–582.

Thummel, R., Kassen, S. C., Enright, J. M., Nelson, C. M., Montgomery, J. E., and Hyde, D. R. (2008a). Characterization of Muller glia and neuronal progenitors during adult zebrafish retinal regeneration. *Exp. Eye Res.* **87**, 433–444.

Thummel, R., Kassen, S. C., Enright, J. M., Nelson, C. M., Montgomery, J. E., and Hyde, D. R. (2008b). Characterization of Muller glia and neuronal progenitors during adult zebrafish retinal regeneration. *Exp. Eye Res.* **87**, 433–444. Epub 2008 Aug 5.

Thummel, R., Kassen, S. C., Montgomery, J. E., Enright, J. M., and Hyde, D. R. (2008c). Inhibition of Muller glial cell division blocks regeneration of the light-damaged zebrafish retina. *Dev. Neurobiol.* **68**, 392–408.

Tong, S. K., Mouriec, K., Kuo, M. W., Pellegrini, E., Gueguen, M. M., Brion, F., Kah, O., and Chung, B. C. (2009). A cyp19a1b-gfp (aromatase B) transgenic zebrafish line that expresses GFP in radial glial cells. *Genesis* **47**, 67–73.

Trewarrow, B., Marks, D., and Kimmel, C. B. (1990). Organization of hindbrain segments in the zebrafish embryos. *Neuron* **4**, 669–679.

Tsujimura, T., Chinen, A., and Kawamura, S. (2007). Identification of a locus control region for quadrupli-cated green-sensitive opsin genes in zebrafish. *Proc. Natl. Acad. Sci. U.S.A.* **104**, 12813–12818.

Vanderlaan, G., Tyurina, O. V., Karlstrom, R. O., and Chandrasekhar, A. (2005). Gli function is essential for motor neuron induction in zebrafish. *Dev. Biol.* **282**, 550–570.

Vitorino, M., Jusuf, P. R., Maurus, D., Kimura, Y., Higashijima, S., and Harris, W. A. (2009). Vsx2 in the zebrafish retina: Restricted lineages through derepression. *Neural Dev.* **4**, 14.

von Trotha, J. W., Campos-Ortega, J. A., and Reugels, A. M. (2006). Apical localization of ASIP/PAR-3: EGFP in zebrafish neuroepithelial cells involves the oligomerization domain CR1, the PDZ domains, and the C-terminal portion of the protein. *Dev. Dyn.* **235**, 967–977.

Vriz, S., Joly, C., Boulekbache, H., and Condamine, H. (1996). Zygotic expression of the zebrafish Sox-19, an HMG box-containing gene, suggests an involvement in central nervous system development. *Brain Res. Mol. Brain Res.* **40**, 221–228.

Wan, H., Korzh, S., Li, Z., Mudumana, S. P., Korzh, V., Jiang, Y. J., Lin, S., and Gong, Z. (2006). Analyses of pancreas development by generation of gfp transgenic zebrafish using an exocrine pancreas-specific elastaseA gene promoter. *Exp. Cell Res.* **312**, 1526–1539.

Wang, D., Jao, L. E., Zheng, N., Dolan, K., Ivey, J., Zonies, S., Wu, X., Wu, K., Yang, H., Meng, Q., *et al.* (2007). Efficient genome-wide mutagenesis of zebrafish genes by retroviral insertions. *Proc. Natl. Acad. Sci. U.S.A.* **104**, 12428–12433.

Wang, X., Emelyanov, A., Korzh, V., and Gong, Z. (2003). Zebrafish atonal homologue zath3 is expressed during neurogenesis in embryonic development. *Dev. Dyn.* **227**, 587–592.

Wang, X., Yang, N., Uno, E., Roeder, R. G., and Guo, S. (2006). A subunit of the mediator complex regulates vertebrate neuronal development. *Proc. Natl. Acad. Sci. U.S.A.* **103**, 17284–17289.

Warren, J. T. Jr., Chandrasekhar, A., Kanki, J. P., Rangarajan, R., Furley, A. J., and Kuwada, J. Y. (1999). Molecular cloning and developmental expression of a zebrafish axonal glycoprotein similar to TAG-1. *Mech. Dev.* **80**, 197–201.

Wehman, A. M., Staub, W., and Baier, H. (2007). The anaphase-promoting complex is required in both dividing and quiescent cells during zebrafish development. *Dev. Biol.* **303**, 144–156.

Wei, X., Luo, Y., and Hyde, D. R. (2006). Molecular cloning of three zebrafish lin7 genes and their expression patterns in the retina. *Exp. Eye Res.* **82**, 122–131.

Wei, X., and Malicki, J. (2002). nagie oko, encoding a MAGUK-family protein, is essential for cellular patterning of the retina. *Nat. Genet.* **31**, 150–157.

Wei, Y., and Allis, C. (1998). Pictures in cell biology. *Trends Cell Biol.* **8**, 266.

Weiland, U., Ott, H., Bastmeyer, M., Schaden, H., Giordano, S., and Stuermer, C. A. (1997). Expression of an l1-related cell adhesion molecule on developing cns fiber tracts in zebrafish and its functional contribution to axon fasciculation. *Mol. Cell. Neurosci.* **9**, 77–89.

Wen, L., Wei, W., Gu, W., Huang, P., Ren, X., Zhang, Z., Zhu, Z., Lin, S., and Zhang, B. (2008). Visualization of monoaminergic neurons and neurotoxicity of MPTP in live transgenic zebrafish. *Dev. Biol.* **314**, 84–92.

Westin, J., and Lardelli, M. (1997). Three novel Notch genes in zebrafish: Implications for vertebrate Notch gene evolution and function. *Dev. Genes Evol.* **207**, 51–63.

White, R. M., Sessa, A., Burke, C., Bowman, T., LeBlanc, J., Ceol, C., Bourque, C., Dovey, M., Goessling, W., Burns, C. E., *et al.* (2008). Transparent adult zebrafish as a tool for *in vivo* transplantation analysis. *Cell Stem Cell* **2**, 183–189.

Whitlock, K. E., and Westerfield, M. (2000). The olfactory placodes of the zebrafish form by convergence of cellular fields at the edge of the neural plate. *Development* **127**, 3645–3653.

Wienholds, E., Schulte-Merker, S., Walderich, B., Plasterk, R. H. (2002). Target-selected inactivation of the zebrafish rag1 gene. *Science* **297**, 99–102.

Willer, G. B., Lee, V. M., Gregg, R. G., and Link, B. A. (2005). Analysis of the Zebrafish perplexed mutation reveals tissue-specific roles for de novo pyrimidine synthesis during development. *Genetics* **170**, 1827–1837.

Williams, J. A., Barrios, A., Gatchalian, C., Rubin, L., Wilson, S. W., and Holder, N. (2000). Programmed cell death in zebrafish rohon beard neurons is influenced by TrkC1/NT-3 signaling. *Dev. Biol.* **226**, 220–230.

Wilson, S. W., Ross, L. S., Parrett, T., and Easter, S. S. Jr. (1990). The development of a simple scaffold of axon tracts in the brain of the embryonic zebrafish, Brachydanio rerio. *Development* **108**, 121–145.

Woo, K., and Fraser, S. E. (1995). Order and coherence in the fate map of the zebrafish nervous system. *Development* **121**, 2595–2609.

Wullimann, M., and Knipp, S. (2000). Proliferation pattern changes in the zebrafish brain from embryonic through early postembryonic stages. *Anat. Embryol.* **202**, 385–400.

Wullimann, M. F., Rupp, B., and Reichert, H. (1996). Neuroanatomy of the zebrafish brain. *Birkhäuser verlag.*

Wurst, W., and Bally-Cuif, L. (2001). Neural plate patterning: Upstream and downstream of the isthmic organizer. *Nat. Rev. Neurosci.* **2**, 99–108.

Xiao, T., Roeser, T., Staub, W., and Baier, H. (2005). A GFP-based genetic screen reveals mutations that disrupt the architecture of the zebrafish retinotectal projection. *Development* **132**, 2955–2967.

Yamaguchi, M., Imai, F., Tonou-Fujimori, N., and Masai, I. (2010). Mutations in N-cadherin and a Stardust homolog, Nagie oko, affect cell cycle exit in zebrafish retina. *Mech. Dev.* **127**, 247–264.

Yamaguchi, M., Tonou-Fujimori, N., Komori, A., Maeda, R., Nojima, Y., Li, H., Okamoto, H., and Masai, I. (2005). Histone deacetylase 1 regulates retinal neurogenesis in zebrafish by suppressing Wnt and Notch signaling pathways. *Development* **132**, 3027–3043.

Yang, C. T., Sengelmann, R. D., and Johnson, S. L. (2004). Larval melanocyte regeneration following laser ablation in zebrafish. *J. Invest. Dermatol.* **123**, 924–929.

Yang, L., Rastegar, S., and Strahle, U. (2010). Regulatory interactions specifying Kolmer-Agduhr interneurons. *Development* **137**, 2713–2722.

Yang, X., Zou, J., Hyde, D. R., Davidson, L. A., and Wei, X. (2009). Stepwise maturation of apicobasal polarity of the neuroepithelium is essential for vertebrate neurulation. *J. Neurosci.* **29**, 11426–11440.

Yazulla, S., and Studholme, K. (2001). Neurochemical anatomy of the zebrafish retina as determined by immunocytochemistry. *J. Neurocytol.* **30**, 551–592.

Yeo, S. Y., and Chitnis, A. B. (2007). Jagged-mediated Notch signaling maintains proliferating neural progenitors and regulates cell diversity in the ventral spinal cord. *Proc. Natl. Acad. Sci. U.S.A.* **104**, 5913–5918.

Yeo, S. Y., Kim, M., Kim, H. S., Huh, T. L., and Chitnis, A. B. (2007). Fluorescent protein expression driven by her4 regulatory elements reveals the spatiotemporal pattern of Notch signaling in the nervous system of zebrafish embryos. *Dev. Biol.* **301**, 555–567.

Yoshida, M., and Macklin, W. B. (2005). Oligodendrocyte development and myelination in GFP-transgenic zebrafish. *J. Neurosci. Res.* **81**, 1–8.

Yoshikawa, S., Kawakami, K., and Zhao, X. C. (2008). G2R Cre reporter transgenic zebrafish. *Dev. Dyn.* **237**, 2460–2465.

Zannino, D. A., and Appel, B. (2009). Olig2+ precursors produce abducens motor neurons and oligodendrocytes in the zebrafish hindbrain. *J. Neurosci.* **29**, 2322–2333.

Zecchin, E., Conigliaro, A., Tiso, N., Argenton, F., and Bortolussi, M. (2005). Expression analysis of jagged genes in zebrafish embryos. *Dev. Dyn.* **233**, 638–645.

Zerucha, T., Stuhmer, T., Hatch, G., Park, B. K., Long, Q., Yu, G., Gambarotta, A., Schultz, J. R., Rubenstein, J. L., and Ekker, M. (2000). A highly conserved enhancer in the Dlx5/Dlx6 intergenic region is the site of cross-regulatory interactions between Dlx genes in the embryonic forebrain. *J. Neurosci.* **20**, 709–721.

Zhao, X. F., Ellingsen, S., and Fjose, A. (2009). Labeling and targeted ablation of specific bipolar cell types in the zebrafish retina. *BMC Neurosci.* **10**, 107.

Zhu, P., Narita, Y., Bundschuh, S. T., Fajardo, O., Scharer, Y. P., Chattopadhyaya, B., Bouldoires, E. A., Stepien, A. E., Deisseroth, K., Arber, S., *et al.* (2009). Optogenetic Dissection of Neuronal Circuits in Zebrafish using Viral Gene Transfer and the Tet System. *Front Neural Circuits* **3**, 21.

CHAPTER 5

Studying Peripheral Sympathetic Nervous System Development and Neuroblastoma in Zebrafish

Rodney A. Stewart[*], Jeong-Soo Lee[†], Martina Lachnit[*], A. Thomas Look[†], John P. Kanki[†], _and_ Paul D. Henion[‡]

[*]Department of Oncological Sciences, Huntsman Cancer Institute, University of Utah, Salt Lake City, Utah

[†]Department of Pediatric Oncology, Dana-Farber Cancer Institute, Harvard Medical School, Boston, Massachusetts

[‡]Center for Molecular Neurobiology and Department of Neuroscience, Ohio State University, Columbus, Ohio

Abstract
I. Introduction
II. The Peripheral Autonomic Nervous System
    A. Overview
    B. Molecular Pathways Underlying PSNS Development
III. The Zebrafish as a Model System for Studying PSNS Development
    A. Overview
    B. Development of the PSNS in Zebrafish
    C. Mutations Affecting PSNS Development
IV. Zebrafish as a Novel Model for Studying Neuroblastoma
    A. Overview of Neuroblastoma
    B. Using Zebrafish to Study Neuroblastoma
V. Conclusion and Future Directions
VI. Acknowledgments
References

Abstract

The combined experimental attributes of the zebrafish model system, which accommodates cellular, molecular, and genetic approaches, make it particularly

well-suited for determining the mechanisms underlying normal vertebrate development as well as disease states, such as cancer. In this chapter, we describe the advantages of the zebrafish system for identifying genes and their functions that participate in the regulation of the development of the peripheral sympathetic nervous system (PSNS). The zebrafish model is a powerful system for identifying new genes and pathways that regulate PSNS development, which can then be used to genetically dissect PSNS developmental processes, such as tissue size and cell numbers, which in the past haves proved difficult to study by mutational analysis in vivo. We provide a brief review of our current understanding of genetic pathways important in PSNS development, the rationale for developing a zebrafish model, and the current knowledge of zebrafish PSNS development. Finally, we postulate that knowledge of the genes responsible for normal PSNS development in the zebrafish will help in the identification of molecular pathways that are dysfunctional in neuroblastoma, a highly malignant cancer of the PSNS.

I. Introduction

The cellular, molecular, and genetic attributes of the zebrafish system make it particularly well suited for studying mechanisms underlying normal vertebrate developmental processes and disease states, such as cancer. The ability to analyze developing tissues in the optically clear embryo, combined with the unbiased nature of forward genetic screens, allows the identification of genes that function during vertebrate organogenesis. These genes may also contribute to diseases of a specific tissue/organ, providing new drug targets for development of therapies to treat the analogous human disease.

In this chapter, we focus on the advantages of the zebrafish system for analyzing the developing peripheral sympathetic nervous system (PSNS). We provide a brief overview of the genetic pathways involved in the PSNS development, our current understanding of zebrafish PSNS development and discuss the rationale for developing a zebrafish model. We also include examples that illustrate the potential of mutant analysis in zebrafish PSNS research. Finally, we explore the potential of the zebrafish system for discovering genes that are disrupted in neuroblastoma (NB), a highly malignant cancer of the PSNS.

II. The Peripheral Autonomic Nervous System

A. Overview

The internal organs, smooth muscles, skin, and exocrine glands of the vertebrate body are innervated by the peripheral autonomic nervous system (ANS), which comprises the PSNS and the parasympathetic (PAS) and enteric nervous systems (ENS). The three components of the ANS differ structurally and functionally in their characteristic location of the cell bodies, the targets they innervate, the neurotransmitters they utilize, and the molecular pathways controlling their development (Ernsberger and Rohrer, 2009; Young *et al.*, 2010). These structural and functional differences

allow the sympathetic and parasympathetic systems to function largely in complementary opposition to each other in order to maintain organ homeostasis by adjusting vascular tone, heart rate, and endocrine secretion to specific environmental challenges, which can in turn generate the fight-or-flight response.

The ANS in general consists of central preganglionic and peripheral postganglionic neurons that regulate the function of a target organ. In the sympathetic nervous system, the preganglionic neurons are short, mainly cholinergic and their cell bodies are generally found in the thoracic and lumbar areas of the spinal cord. Their axons exit ventrally and innervate long adrenergic postganglia lying near the spinal cord that will innervate the target organs. A preganglionic PSNS neuron can innervate up to 20 postganglionic neurons, resulting in a massive signal amplification and eventual ubiquitous activation of all organ systems. Sympathetic neurons are predominantly adrenergic, producing the neurotransmitter noradrenalin along with one or more other neuropeptides, including adenosine triphosphate. For example, sympathetic neurons that stimulate smooth muscle cells (e.g., cardiac muscle cells) depend on a postsynaptic receptor in the target cell, activation of which can be mimicked by administering appropriate receptor ligand analogous (β-blockers), causing the heart to beat faster and blood vessels to constrict. In contrast, the preganglionic neurons of the PAS are long and their cell bodies lie in the cranial (brain stem) and sacral regions of the spinal cord. They synapse with maximally five, but normally only one short postganglionic neuron, which happens to be located within a specific target organ or its immediate vicinity. Signal transmission therefore results in localized low-dose organ innervations. Both pre- and postganglionic PAS neurons are cholinergic, releasing the neurotransmitter acetylcholine along with other neuromodulators (Thexton, 2001).

Chromaffin cells located in the adrenal medulla represent a unique subset of the PSNS that develop into endocrine cells (instead of postganglion neurons) that release hormones directly into the blood stream, instead of sending out processes. The adrenal medulla may therefore be considered as a modified sympathetic ganglion. Chromaffin cells are the bodies' main source of circulating catecholamines (adrenaline, noradrenaline) and endorphins, which are stored in intracellular granules and released in response to stress. As such they play an important role in the generation of the fight-or-flight response. Chromaffin cells can also be found in fewer numbers in structures such as the carotid aorta, vagus nerve, bladder, and prostate. While it is generally accepted that chromaffin cells and sympathetic neurons are derived from a common precursor population (the sympathoadrenal (SA) cell), there is evidence that these cell populations can also develop independently from the neural crest (NC) in the absence of a common progenitor (Huber, 2006). Importantly, chromaffin cells can develop into malignant tumors, called pheochromocytomas, which produce excessive amounts of catecholamines, usually adrenaline and noradrenaline, which cause severely high blood pressure. Pheochromocytomas are tumors of the multiple endocrine neoplasia syndromes type IIA and type IIB (also known as MEN IIA/B), with up to 25% being hereditary. Mutations in *VHL, RET, NF1, SDHB*, and *SDHD* are associated with familial pheochromocytomas, and recent studies suggest that loss of developmentally regulated apoptosis (resistance to neurotrophic withdrawal) is the mechanism driving tumor formation in these cases (Nakamura and Kaelin, 2006; Schlisio *et al.*, 2008).

The function of the ENS is relatively independent of the CNS and other components of the ANS. In most vertebrates, the enteric neurons form two layers of ganglionic plexuses located along the entire length of the gastrointestinal tract consisting of a microscopic meshwork of ganglia connected to each other by short nerve trunks. The inner myenteric plexus (termed Auerbach plexus), situated between the longitudinal and the circular muscle layers, is mainly responsible for muscle contraction. Neurons in the Auerbach plexus regulate peristaltic waves, which move digestive products. The submucosal plexus (termed Meissners plexus) controls motility, secretion, and microcirculation processes as well as blood flow in the gut. The functions of the ENS are complex, and 17 different neuronal types have been identified, which can be classified into sensory neurons, interneurons, and motor neurons. Each type produces a variety of different neurotransmitters including nitric oxide, ATP, and 5-hydroxytryptamine. Enteric ganglia integrate sensory and reflex signals from the parasympathetic and sympathetic neurons, which innervate the gastrointestinal tract to coordinate peristalsis (Hansen, 2003).

B. Molecular Pathways Underlying PSNS Development

The early development of the PSNS can be divided into four overlapping stages, based on both morphologic and molecular criteria: (i) formation and fate specification of NC cells that will develop into SA progenitors, (ii) bilateral migration of SA cells and their coalescence in regions adjacent to the dorsal aorta, (iii) neuronal and noradrenergic differentiation of SA progenitors, and (iv) maintenance of PSNS neurons in fully developed ganglia and the establishment of their efferent synaptic connections.

Considerable progress has been made elucidating the cellular and molecular mechanisms underlying PSNS development, which represents one of the best described genetic pathways establishing vertebrate neuronal and neurotransmitter identity (Anderson, 1993; Apostolova and Dechant, 2009; Ernsberger and Rohrer, 2009; Francis and Landis, 1999; Goridis and Rohrer, 2002; Young *et al.*, 2010). Briefly, NC progenitors form at the border between the neural and non-neural ectoderm through a process regulated by bone morphogenetic proteins (BMPs), fibroblast growth factors (FGFs), Retinoic acid (RA), and Wnt signaling (Knecht and Bronner-Fraser, 2002). These NC progenitor cells express genes such as *Snai11/2, Tfap2α*, and *Foxd3* that appear to play roles in their induction and/or early specification (Sauka-Spengler and Bronner-Fraser, 2008). Subsequent to the morphogenetic movements that result in neural tube closure, NC progenitors are localized to the most dorsal aspect of the neural tube. The premigratory NC then undergoes an epithelial–mesenchymal transition and migrates away from the neural tube (Acloque *et al.*, 2009). NC cells first migrate ventromedially and later others follow a dorsolateral pathway (Fig. 1A; Goridis and Rohrer, 2002; LeDouarin and Kalcheim, 1999). Ventral-migrating NC cells aggregate in the vicinity of the dorsal aorta, where they form primary sympathetic ganglia, and then migrate to the secondary sympathetic ganglia and adrenal gland, where they undergo neuronal and catecholaminergic differentiation (Anderson and Axel, 1986; Anderson *et al.*, 1991; Huber, 2006).

During the premigratory stage, NC cells may become fate-restricted to specific lineages. Indeed, lineage-tracing studies in mice show that the expression of

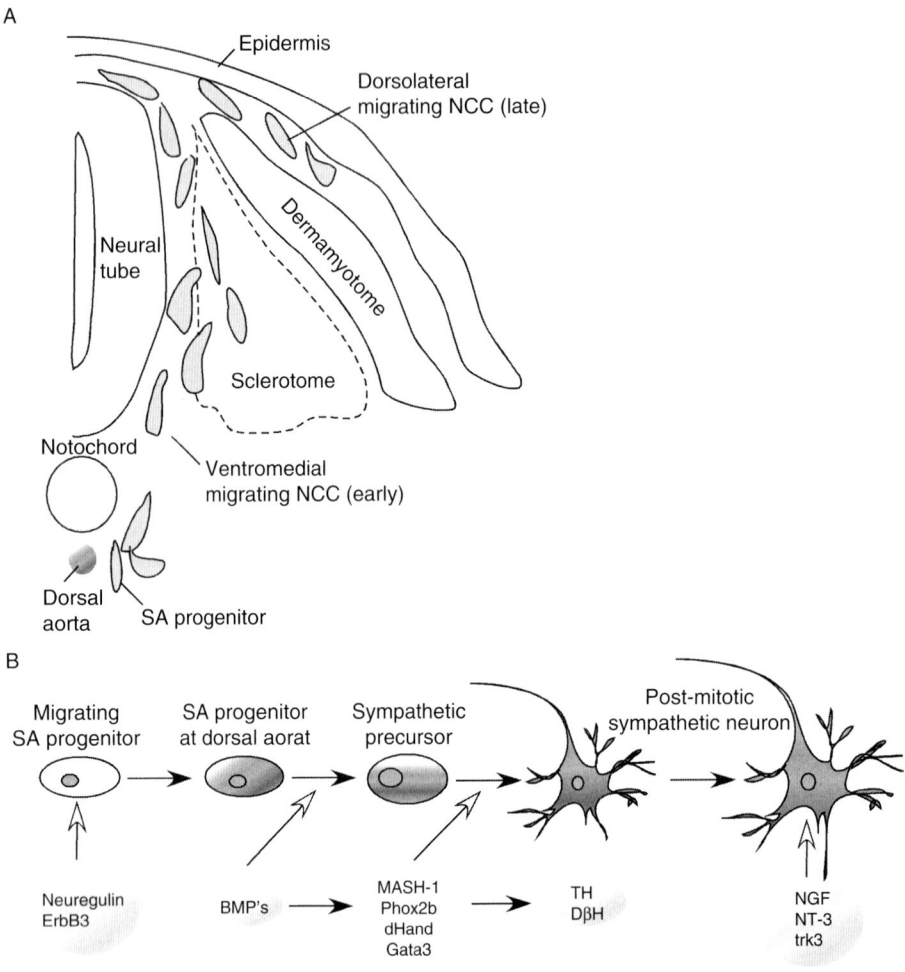

Fig. 1 Neural crest-derived SA progenitors migrate along a ventromedial pathway to bilateral regions adjacent to the dorsal aorta. (A) Schematic diagram of a transverse section through the trunk of a vertebrate embryo (embryonic day 10.5 in the mouse, 2.5 in the chick, and approximately 28 hpf in the zebrafish embryo). In avian and rodent embryos, presumptive SA progenitor cells derived from the neural crest migrate ventromedially within the sclerotome region of the somites and ultimately cease migration in the region of the dorsal aorta. In zebrafish, neural crest-derived SA precursors migrate ventromedially between the neural tube and the somites to the dorsal aorta region. (B) Molecular pathways governing sympathetic neuron development. During migration, signaling via the neuregulin-1 growth factor is required for the development of at least some SA progenitors. Once the SA progenitor cells arrive at the dorsal aorta, BMP signaling activates the transcriptional regulators MASH-1 and Phox2b that ultimately lead to the expression of the transcription factors Phox2a, GATA-3, and dHand. Together, these factors are responsible for differentiation of SA progenitors into noradrenergic neurons. Fully differentiated neurons express biosynthetic enzymes responsible for the synthesis of noradrenalin, such as tyrosine hydroxylase and dopamine-β-hydroxylase. Survival of the differentiated sympathetic neurons is governed by a number of neurotrophic factors, such as NGF and NT-3.

Neurogenin-2 in pre- and early migrating NC cells promotes the differentiation of sensory neurons at the expense of sympathetic neurons (Zirlinger *et al.*, 2002). These results are consistent with single cell labeled studies in zebrafish, which show that the SA lineage is specified at early migratory stages (Raible and Eisen, 1994). Unfortunately, the molecular mechanisms governing these decisions (including what determines *Ngn2* expression in a sub-population of NC) are not known. Instead, most of our knowledge of PSNS development comes from studies during or after NC migration.

During ventrolateral migration, precursors of the SA lineage are exposed to signaling factors from the somites, ventral neural tube, and notochord, such as sonic hedgehog (Shh) and Neuregulin-1, an EGF-like growth factor (Crone and Lee, 2002; Krauss *et al.*, 1993; Patten and Placzek, 2000; Williams *et al.*, 2000b). Neuregulin-1 expression is associated with the origin, migration, and target site of SA progenitors. Mice and zebrafish lacking components of the Neuregulin-1 pathway, such as the ErbB3 receptor, exhibit severe hypoplasia of the primary sympathetic ganglion chain while the migration of cranial NC-derived enteric neurons appears normal (Britsch *et al.*, 1998). Migrating trunk NC cells of *ErbB3* mutants cannot recognize their target location and instead accumulate dorsal to the sites of normal sympathetic ganglion formation (Britsch *et al.*, 1998; Crone and Lee, 2002; Murphy *et al.*, 2002). A number of other signaling molecules also function to restrict the migration path of SA precursors toward the dorsal aorta. For example, trunk NC cell migration is restricted to the anterior somite by EphrinB ligands, which repel early migrating NC cells from the dorsolateral pathway and redirects them toward the ventrolateral migration path (Santiago and Erickson, 2002; Schwarz *et al.*, 2009). Semaphorin-3A and 3F, Neuropilin-1 and -2, and F-Spondin are also involved in restricting migration to the anterior somite (Debby-Brafman *et al.*, 1999; Gammill *et al.*, 2006; Kawasaki *et al.*, 2002). Once the migrating NC cells reach the dorsal aorta, N-Cadherin, Eph/ephrin, CXCL12, and Artemin signaling are required for the subsequent formation and segmental organization of sympathetic ganglia (Honma *et al.*, 2002; Kasemeier-Kulesa *et al.*, 2005, 2006).

As the SA precursors aggregate in the vicinity of the dorsal aorta, a molecular signaling cascade is initiated in response to BMPs secreted by these cells (Fig. 1B). Dose-dependent BMP signaling appears to be essential and sufficient to initiate the development of both noradrenergic sympathetic neurons (*Phox2b/Th*; high BMP concentration) and cholinergic parasympathetic neurons (*Phox2b /Chat*; low BMP concentration) of the autonomic lineage (Huber and Ernsberger, 2006; Morikawa *et al.*, 2009; Muller and Rohrer, 2002). Response to the BMP gradient occurs through the ALK3 receptor (BMP receptor IA) in mice, as conditional deletion of *Alk3* in NC cells caused death of these cells immediately after reaching the dorsal aorta (Morikawa *et al.*, 2009). BMP signaling induces the expression of the proneural gene *Mash-1*, a mammalian *achaete-scute* homolog, and the homeodomain transcription factor PHOX2B in sympathetic neuroblasts (Ernsberger *et al.*, 1995; Groves *et al.*, 1995; Guillemot *et al.*, 1993; Hirsch *et al.*, 1998). PHOX2B is essential for the maintained expression of *Mash-1* and proliferation of sympathoadrenergic precursor cells, as SA precursors fail to proliferate and degenerate before differentiation is induced in *Phox2b* mouse mutants (Huber, 2006; Pattyn *et al.*, 1999, 2000). MASH-1 appears to support sympathoblast differentiation, and *Mash-1* deficient mice

show delay of sympathoadrenergic differentiation that eventually leads to their loss (Pattyn *et al.*, 2006). It is still not clear how BMP signaling results in *Mash-1* or *Phox2b* expression, although it is likely that multiple BMP pathways, including *Smad4*-dependent and-independent pathways are required for proliferation, survival, and differentiation of PSNS precursors (Morikawa *et al.*, 2009).

Several other critical transcription factors are activated downstream or in parallel to MASH-1 and PHOX2B, including the homeobox proteins PHOX2A, bHLH transcription factor HAND2, and the zinc-finger proteins GATA-2/3 (Howard *et al.*, 2000; Huber, 2006; Tsarovina *et al.*, 2004). PHOX2A and PHOX2B are not functionally equivalent, though both appear to bind and activate the noradrenergic marker genes, tyrosine hydroxylase (*Th*) and dopamine-β-hydroxylase (*Dβh*), that are essential enzymes in the production of noradrenaline (Coppola *et al.*, 2005). HAND2 (also known as dHAND) is essential for the differentiation of noradrenergic neurons and is expressed in the early noradrenergic sympathetic chain but not in the cholinergic parasympathetic system (Lucas *et al.*, 2006; Muller and Rohrer, 2002; Schmidt *et al.*, 2009). Expression of HAND2 in terminally differentiated sympathetic neurons maintains their neurotransmitter properties by promoting the expression of noradrenergic genes, while suppressing cholinergic genes (Apostolova and Dechant, 2009; Schmidt *et al.*, 2009). In contrast, HAND1 (also known as eHAND) appears to have no significant impact on *Dbh* expression (Doxakis *et al.*, 2008). GATA 2/3 expression is initiated after MASH-1, PHOX2A, PHOX2B, and HAND2 expression, but before the onset of the noradrenergic marker genes, and is maintained throughout development (Huber, 2006; Tsarovina *et al.*, 2004). However, GATA2/3 overexpression results in the formation of ectopically non-autonomic TH-negative neurons, which may reflect a dependence on additional cofactors for induction of sympathetic neurons (Tsarovina *et al.*, 2004). The transcriptional activity of GATA 2/3 appears to be directly regulated via protein–protein interactions with the transcription factor CREB, which also seems to interact with Phox2a and to influence noradrenergic differentiation (Benjanirut *et al.*, 2006; Hong *et al.*, 2006; Rudiger *et al.*, 2009). Together, these genes drive SA differentiation, which is further modulated by cAMP and MAPK signaling to activate *Dbh* and *Th* expressions (Ernsberger *et al.*, 2000; Kim *et al.*, 2001; Seo *et al.*, 2002).

As mentioned above, a later stage of PSNS development consists of modeling of the sympathetic ganglia. The neurotrophic factors, NGF and NT-3, have been shown to control sympathetic neuron survival and the maintenance of their synaptic connections (Birren *et al.*, 1993; DiCicco-Bloom *et al.*, 1993; Francis and Landis, 1999). In the embryo, NGF is secreted from sympathetic target tissues upon arrival of sympathetic neurons (Chun and Patterson, 1977; Heumann *et al.*, 1984; Korsching and Thoenen, 1983; Shelton and Reichardt, 1984). Analysis of NGF and its high affinity receptor, TrkA, in mouse mutants confirmed their requirement for the *in vivo* survival of sympathetic neurons (Fagan *et al.*, 1996; Smeyne *et al.*, 1994). In their absence, sympathetic neuron development proceeds normally, but is then followed by neuronal cell death. A similar phenotype is observed in NT-3 mutants, although neuronal death occurs at later embryonic stages (Ernfors *et al.*, 1994; Farinas *et al.*, 1994; Francis and Landis, 1999; Wyatt *et al.*, 1997). HAND1 and HAND2 support NGF-dependent

neurotrophic survival by enhancing the expression of *TrkA* in sympathetic neurons (Doxakis *et al.*, 2008; Ma *et al.*, 2000; Schmidt *et al.*, 2009). Interestingly, adrenalin transporter and TrkA co-localize in noradrenergic cells, which suggests that TrkA ligands (NGF or NT3) might indirectly support a positive selection of noradrenergic cells (Brodski *et al.*, 2002). In summary, neurotrophic factors (such as NT-3 and NGF) appear to be largely responsible for establishing and maintaining mature ganglion cell numbers during embryonic or early postnatal development in chick and rodents.

Although many of the inductive signaling pathways affecting different stages of NC development have been identified, the regulatory mechanisms controlling these pathways remain poorly understood. While substantial evidence links BMP signaling with the induction of SA progenitor cell development, the genetic control of SA cell responsiveness to BMP signals and their specification remain unclear. While sympathetic precursors are competent to express MASH1 and PHOX2B in response to BMP signaling near the dorsal aorta, they do not respond to BMPs present in the overlying ectoderm during earlier premigratory stages. Furthermore, the molecular mechanisms regulating the interactions of transcription factors and downstream pathways, specifying neuronal and noradrenergic differentiation, are incompletely understood. Other signaling pathways, such as cAMP (Lo *et al.*, 1999), may also contribute to this process. Finally, proliferation of sympathoblasts and differentiated sympathetic neurons occurs throughout embryogenesis (Birren *et al.*, 1993; Marusich *et al.*, 1994; Rohrer and Thoenen, 1987; Rothman *et al.*, 1978). An inability to control cell proliferation in sympathetic ganglia is of particular medical interest, since it can lead to NB, the most common human cancer in infants less than 1 year of age. While the study of PSNS development in tetrapods will continue to contribute to our knowledge of PSNS development, a goal of this chapter is to demonstrate the power of the zebrafish, *Danio rerio*, exploiting its forward genetic potential and advantages as an embryologic system, for making significant contributions to PSNS research. We also propose that the study of zebrafish PSNS development will contribute to our understanding of both normal and abnormal PSNS development and may ultimately provide a genetic model for human NB.

III. The Zebrafish as a Model System for Studying PSNS Development

A. Overview

One of the most powerful attributes of the zebrafish system is its capacity for large-scale genetic screens due to its rapid embryonic development and high expansion rate. The unbiased nature of phenotype-based genetic screens enables new genes to be identified without prior knowledge of their function or expression in the tissue of interest. This approach is particularly attractive for study of the PSNS, as many signaling components involved in determining sympathetic fate are incompletely understood. Also, most of our current understanding of PSNS development has relied on functional assays on isolated sympathetic cells in culture or gene misexpression analyses (Francis

and Landis, 1999; Goridis and Rohrer, 2002). While these studies can determine whether certain genes are sufficient to direct sympathetic development, they do not address whether those genes are normally required for PSNS development. Murine gene knock-out models have been used in loss-of-function studies to confirm the *in vivo* requirement for particular genes in PSNS development (Guillemot *et al.*, 1993; Lim *et al.*, 2000; Morin *et al.*, 1997; Pattyn *et al.*, 1999). For example, although both *Phox2a* and *Phox2b* can induce a sympathetic phenotype when misexpressed in chick, only the selective knock-out of *Phox2b* eliminates PSNS development *in vivo*, as the PSNS in *Phox2a$^{-/-}$* mutant mice appears relatively normal (Morin *et al.*, 1997; Pattyn *et al.*, 1999).

These studies provide valuable insights into the regulatory pathways directing sympathetic neuron development and emphasize the advantages of using mutants to dissect genetic pathways *in vivo*. The capacity for experimental mutagenesis and manipulation of gene expression in the developing embryo is a major strength of the zebrafish system. Forward genetic zebrafish screens can be especially valuable for identifying genes affecting complex signaling pathways that rely upon interactions between the developing PSNS and the surrounding tissues, which may be impossible to address using *in vitro* assays. Critical roles for extrinsic factors are particularly evident in PSNS development and SA progenitors migrate past a number of tissues expressing different signaling molecules, such as the neural tube, notochord, and floor-plate. In addition, zebrafish mutant embryos often survive for longer periods of time during embryogenesis than knock-out mice lacking orthologous genes due to their *ex-utero* development. This allows the analysis of the PSNS to extend through later stages of sympathetic neuron differentiation and maintenance. Finally, the zebrafish system offers the most amenable vertebrate model for performing large-scale mutagenesis screens to identify novel genes affecting all aspects of PSNS development, as the time, space, and expense associated with mutagenesis techniques in mice can be prohibitive.

Establishing the zebrafish as a useful vertebrate model for identifying new genes important for PSNS development will require (i) an analysis of zebrafish sympathetic neuron development and its comparison with other vertebrates, (ii) an analysis of the genetic programs regulating zebrafish PSNS development and their conservation in other vertebrates, (iii) the generation of efficient mutagenesis protocols and screening assays for the isolation of PSNS mutants. These areas are addressed below.

B. Development of the PSNS in Zebrafish

1. Neural Crest Development and Migration

A number of studies have now analyzed the anatomical and molecular mechanisms underlying the different stages of PSNS development in the zebrafish (An *et al.*, 2002; Raible and Eisen, 1994). The findings show that the morphogenesis and differentiation of sympathetic neurons in zebrafish are qualitatively very similar to other vertebrates. Migration and cell fate specification of trunk sympathetic precursors in zebrafish were initially analyzed by labeling single NC cells with vital dyes and following their subsequent development (Raible and Eisen, 1994). In the trunk, NC migration begins

around 16 h postfertilization hpf, at the level of somite 7 and sympathetic neurons are only derived from NC cells migrating along the ventromedial pathway. Hence, the ventromedial migration of SA precursor cells is conserved in zebrafish. These studies also demonstrated the existence of both multipotent and fate-restricted NC precursors that generate a limited number of derivatives, such as sympathetic neurons, before or during the initial stages of migration (Raible and Eisen, 1994). Although little is known about the molecular mechanisms underlying such fate decisions, the ability to analyze the fate-restriction of sympathetic neurons in zebrafish, using different NC mutants, affords a powerful method to dissect the genetic pathways underlying this process.

2. Gene Expression in Migrating SA Progenitors

Many of the genes capable of inducing the development of SA progenitors in birds and rodents have been identified in zebrafish (Table I). Furthermore, the expression of some of these genes in dorsal aorta cells and in NC-derived cells in its vicinity, where noradrenergic neurons form, is consistent with their role in fish PSNS development (Cheung *et al.*, 2005; Elworthy *et al.*, 2005; Lucas *et al.*, 2006; Sakai *et al.*, 2006; Tsarovina *et al.*, 2004). A number of BMP homologues have been identified in zebrafish and some have been shown to be expressed by the dorsal aorta (Dick *et al.*, 2000; Nguyen *et al.*, 1998). Several zebrafish mutants exhibit midline defects affecting structures that may be responsible for BMP signaling. In *flh* mutant embryos, the notochord and dorsal aorta fail to develop and *bmp4* expression is absent (Talbot *et al.*, 1995). The loss of the local source of BMP signaling corresponds with a failure of sympathetic neurons (Hu$^+$/TH$^+$) to form (P. Henion, unpublished data). Interestingly, NC-derived cells (*crestin*$^+$) continue to populate this region where sympathetic neurons would normally develop. These observations suggest that, like other vertebrates, dorsal aorta-derived BMPs are required for SA development in zebrafish. In

Table I
Expression of Conserved PSNS Genes in Zebrafish

Genes	Expression during PSNSdevelopment	References
ErbB3	N/A	Lo *et al.* (2003)
BMP4	Dorsal aorta	Fig. 2A, Dick *et al.* (2000), Nguyen *et al.* (1998)
Crestin	Migrating SA, nascent SCG	Luo *et al.* (2001)
Zash1a (Ascl1)	SCG	Lucas *et al.* (2006), Stewart *et al.* (2006)
Phox2a	SCG	Guo *et al.* (1999a), Holzschuh *et al.* (2003)
Phox2b	Migrating SA, SCG	Elworthy *et al.* (2005), Lucas *et al.* (2006), Stewart *et al.* (2006)
HuC (Elavl3)	SCG, Trunk sympathetic ganglia	An *et al.* (2002)
Gata2	SCG	Neave *et al.* (1995)
dHand	SCG	Fig. 2C and D, Yelon *et al.* (2000)
Th	SCG, Trunk sympathetic ganglia	An *et al.* (2002), Guo *et al.* (1999b), Holzschuh *et al.* (2001)
Dβh	SCG, Trunk sympathetic ganglia	An *et al.* (2002), Holzschuh *et al.* (2003)
Pnmt	SCG	R. Stewart (unpublished)
Trkreceptors	ND	Martin *et al.*, (1995), Williams *et al.* (2000a)

contrast, floor-plate cells, which are lacking in the *cyclops* mutant, do not appear to be required for dorsal aorta development, BMP expression or SA development, and all of these functions appear to be normal in these mutant embryos (P. Henion, unpublished data). Interestingly, SA development appears normal in *no tail* mutants (Fig. 2A and B) even though dorsal aorta development is impaired (Fouquet *et al.*, 1997) and BMP expression is reduced. It is possible that weak BMP persists in *ntl* due to the continued presence of notochord precursor cells that fail to differentiate properly in this mutant (Melby *et al.*, 1997). However, whether the notochord is directly responsible for BMP expression and dorsal aorta development is unclear.

Most of the described transcription factors known to direct the development of the sympathetic precursors in other species are present in the zebrafish and exhibit appropriate gene expression patterns. The zebrafish *zash1a* gene, a homolog of *Mash-1*, is transiently expressed in cells near the dorsal aorta by 48 hpf (Allende and Weinberg, 1994). Preliminary gene knockdown experiments using antisense morpholinos to specifically target the *zash1a* gene resulted in the loss of *th*-expressing noradrenergic neurons in the developing PSNS (R. Stewart, unpublished data). The *phox2a, phox2b, gata-2/3*, and *hand2* genes have also been cloned in zebrafish, and they are all expressed in cells adjacent to the dorsal aorta by 2 days postfertilization (dpf) (Guo *et al.*, 1999a; Holzschuh *et al.*, 2003; Neave *et al.*, 1995; Yelon *et al.*, 2000). Initial reports of embryos deficient for *phox2b* suggest that, unlike other vertebrates, it is not required for normal PSNS development, but instead is required for development of the ENS (Elworthy *et al.*, 2005). This may be due to the fact that Phox2b is only transiently expressed in nascent

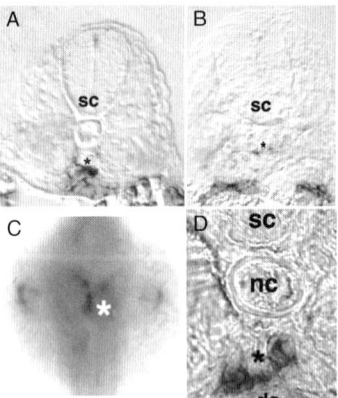

Fig. 2 Expression of *bmp* by the dorsal aorta and dHand by sympathetic neurons. (A) Expression of *bmp4* in transverse sections at the mid-trunk level in wild-type (top) and *ntl* mutant embryos at 48 hpf. *bmp4* expression is evident in cells of the dorsal aorta (asterisk) of wild-type embryos. (B) In *ntl* mutants, the dorsal aorta does not form completely (asterisk) although expression of *bmp4* is present albeit to a reduced extent. (C) Expression of zebrafish dHand at 58 hpf in cervical sympathetic neurons (asterisk). (D) Co-expression of dHand (blue) and TH (red) in sympathetic neurons (asterisk) in a transverse section in the anterior trunk region of a 3 dpf embryo. sc = spinal cord; nc = notochord; da = dorsal aorta. Top is dorsal in (A, B, D) and anterior in (C). (See Plate no. 6 in the Color Plate Section.)

sympathetic ganglia (Lucas *et al.*, 2006) or it could be due to incomplete knockdown of *phox2b* that is commonly observed using antisense morpholinos technologies. Analysis of PSNS development in *phox2b* mutants will be needed to resolve this issue. Noradrenergic differentiation requires *hand2* in zebrafish, as the zebrafish *hand2* deletion mutant, *hands off*, fails to express the noradrenergic genes *th* and *dbh*, even though the pan-neuronal marker *HuC* (*elav13*) and other PSNS precursor markers are normally expressed (Lucas *et al.*, 2006). Further analysis of *Zash1a, Phox2b, Phox2a, Gata-2/3*, and *Hand2* expression, together with the examination of Th and Dβh expression in SA precursors, will provide insight into the functions of these genes with respect to sympathetic neuron differentiation (see below). Importantly, analysis of compound mutants utilizing existing zebrafish mutants such as *soulless* (*phox2a*; Guo *et al.*, 1999a and *hands off*; Lucas *et al.*, 2006) together with other mutants described in this chapter will contribute to our understanding of the functional roles of these genes in sympathetic neuron development. The combination of transient gene knockdown techniques using morpholinos, and the identification of new zebrafish mutants affecting other regulators of sympathetic neuron development, should also contribute to novel insights relevant to PSNS development across vertebrate species.

3. Neuronal Differentiation and Coalescence into Sympathetic Ganglia

The timing of overt neuronal differentiation of sympathetic precursors and their transition to fully differentiated NA-producing neurons has been described in detail in zebrafish (An *et al.*, 2002). The pan-neuronal antibody 16A11 recognizes members of the Hu family of RNA-binding proteins and labels sympathetic precursors located ventrolateral to the notochord and adjacent to the dorsal aorta (An *et al.*, 2002). Sympathetic neurons were found to differentiate at different times in the zebrafish embryo, and two populations of sympathetic ganglion neurons were defined. The most rostral population develops at 2 dpf and comprises the superior cervical ganglion (SCG) complex that consists of two separate ganglia arranged in an hourglass shape. Several days later, more caudal trunk sympathetic neurons develop as irregular, bilateral rows of single neurons adjacent to the dorsal aorta, presumably analogous to the primary sympathetic chain in other vertebrates. These neurons differentiate in an anterior to posterior temporal progression, extending caudally as far as the level of the anus, and eventually form regular arrays of segmentally distributed sympathetic ganglia (An *et al.*, 2002). The reason for the delay in the differentiation of the caudal sympathetic neurons is not known, since the formation of the dorsal aorta (Fouquet *et al.*, 1997) and its expression of BMPs (Martinez-Barbera *et al.*, 1997) occur well before the differentiation of sympathetic neurons is observed. Importantly, ventrally migrating NC-derived cells populate the region adjacent to the dorsal aorta between 24 and 36 hpf. Thus, the delay in caudal PSNS development may be due to a delay in their becoming competent to respond to BMP signaling. Since the expression of some zebrafish BMPs has not been examined in the dorsal aorta, it remains possible that different types of BMPs may be selectively expressed by dorsal aorta cells and/or that SA progenitors exhibit differential responsiveness to different BMPs.

4. Differentiation of Noradrenergic Neurons

One of the key events in PSNS differentiation is the acquisition of the NA-neurotransmitter phenotype indicated by the expression of noradrenalin and genes such as *tyrosine hydroxylase* (*th*) and *dopamine-β-hydroxylase* (*dbh*) that are required for the enzymatic conversion of the amino acid L-tyrosine to noradrenalin (Goridis and Rohrer, 2002; Huber, 2006). In zebrafish, *th* expression has been used as the principal marker for the presence and formation of fully differentiated sympathetic neurons although it is also expressed by other catecholaminergic neurons in the central nervous system (CNS) (Figs. 3A–D; An *et al.*, 2002; Guo *et al.*, 1999b; Holzschuh *et al.*, 2001). Expression of Dβh protein and mRNA is also used as markers of PSNS differentiation because of its requirement for the conversion of dopamine to noradrenaline in sympathetic neurons of the PSNS and a subset of dopaminergic neurons in the CNS (An *et al.*, 2002; Holzschuh *et al.*, 2003). By 10 dpf, all of the sympathetic ganglia contain neurons expressing *th*, although some neurons within the nascent sympathetic ganglia do not express Th protein. However, by 28 dpf, all of the neurons uniformly express Th, suggesting the complete maturation of sympathetic ganglia by this time. Both Th protein and mRNA are detectable in the SCG complex beginning at 48 hpf. Consistent with the expression of Hu proteins, most sympathetic neurons located posterior to the SCG complex do not begin to express *th* mRNA until approximately 5 dpf, in a few of the more rostral trunk segments (Fig. 3D). It should be noted that a second tyrosine hydroxylase gene has been identified in zebrafish, called *th2* (Filippi *et al.*, 2009; Schweitzer and Driever, 2009; Yamamoto *et al.*, 2010). The expression of *th2* overlaps that of *th1* in the CNS; however, it is not known if *th2* also labels PSNS. The expression of *dbh* is generally observed slightly later than *th* in differentiating sympathetic neurons (An *et al.*, 2002). However, *dbh* is expressed along with *th* as early as 2 dpf in the SCG complex (Holzschuh *et al.*, 2003). Once noradrenergic identity has been established, the continued expression of *hand2* appears to be essential for the maintenance and proliferation of sympathetic neurons (Hendershot *et al.*, 2008; Lucas *et al.*, 2006; Morikawa *et al.*, 2007).

Another element of PSNS function is the regulated release of adrenalin and noradrenalin by chromaffin cells of the adrenal gland, which form in and/or around the developing kidney (Liu, 2007; Unsicker *et al.*, 2005). Chromaffin cells represent a specialized component of the PSNS, and most of these cells express an additional enzyme in the catecholaminergic pathway, called phenylethanolamine *N*-methyltransferase (PNMT), which converts noradrenaline into adrenaline (Kalcheim *et al.*, 2002; Schober *et al.*, 2000). As in other species, both noradrenergic and adrenergic chromaffin cells have been described in zebrafish (Hsu *et al.*, 2003; Liu *et al.*, 2003). However, in contrast to mammals, where chromaffin cells are located in the adrenal medulla separated from steroidogenic cells in the adrenal cortex, the chromaffin cells in zebrafish and other teleost fish are interspersed with adrenocortical cells in a specialized region of the kidney called the interrenal gland (Liu, 2007). Initial observations indicated that non-neuronal (16A11-negative), Th and Dβh-positive cells are present in the SCG complex at 2 dpf, which continue to migrate ventrally to the

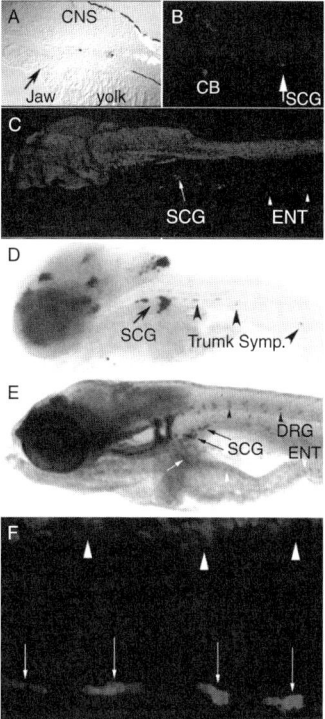

Fig. 3 Development of the peripheral sympathetic nervous system in zebrafish embryos. (A–C) Parasagittal section of 3.5 dpf embryo. High-magnification DIC (A) and fluorescence (B) of the same field showing TH-IR (red) in the SCG (arrow), carotid body (CB), and a group of anterior cells in the midbrain (CNS). (C) Low-magnification view of 3.5 dpf embryo labeled with anti-Hu to reveal all neurons. A subset of cervical sympathetic neurons are indicated by the arrow and enteric neurons (ENT) by arrowheads. (D) Lateral view of whole-mount *TH* RNA *in situ* preparation at 5 dpf. *TH* RNA is strongly expressed in the SCG (arrow) at this stage and is beginning to be expressed in the trunk sympathetic chain (arrowheads). A description of *TH* RNA expression in the head is described in Guo *et al.*, 1999b. (E) Whole-mount antibody preparation of a 7 dpf larvae labeled with anti-Hu to reveal neurons. Black arrows indicate SCG, black arrowheads indicate dorsal root ganglion (DRG) sensory neurons, and white arrow and white arrowheads indicate enteric neurons (ENT). (F) Parasagittal section in the mid-trunk region of a 17 dpf embryo labeled with anti-Hu. Ventral spinal cord neurons are evident at the top (arrowheads). Four segmental sympathetic ganglia (arrows) are located ventral to the notochord adjacent to the dorsal aorta.

interrenal gland (Fig. 4; An *et al.*, 2002). Consistent with these studies, mRNA *in situ* hybridization assays using *pnmt* expression show that double positive *pnmt*[+]/*th*[+] cells are present in the SCG at 3 dpf, suggesting that chromaffin cells can be specified before migration to the kidney (R. Stewart, unpublished). Once chromaffin cells reach the kidney, signaling from the pituitary gland and endothelial cells appears to be required for continued maintenance and development of a functional interrenal gland (Liu, 2007; To *et al.*, 2007).

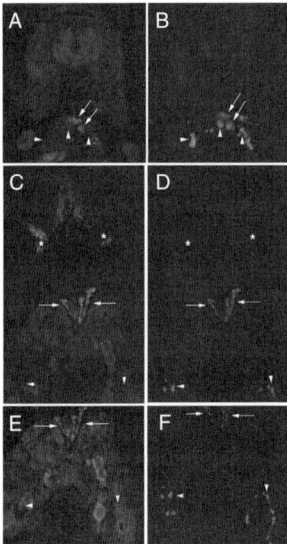

Fig. 4 Sympathoadrenal derivatives in embryonic and juvenile zebrafish. (A, B) Transverse section of a 3.5 dpf embryo double labeled with anti-Hu (green) and DβH (red). Arrows indicate Hu$^+$/DβH$^+$ sympathetic neurons of the cervical ganglion. Arrowheads indicate Hu$^-$/DβH$^+$ presumptive chromaffin cells. (C, D) Transverse section through the mid-trunk region at 28 dpf double labeled with anti-Hu (green) and anti-TH (red). (E, F) Higher magnification of C and D, including a slightly more ventral region. Arrows indicate sympathetic neurons, arrowheads indicate chromaffin cells, and asterisks denote dorsal root ganglia. (See Plate no. 7 in the Color Plate Section.)

5. Modeling of Sympathetic Ganglia

In rodents and birds, neurotrophic factors, such as NGF and NT-3, control sympathetic neuron cell numbers through the regulation of their survival and continued maintenance of their synaptic connections (Francis and Landis, 1999; Schober and Unsicker, 2001). In zebrafish, the ability of these factors to control the survival of sympathetic neurons is unknown. However, NT-3 has been shown to act as a neurotrophic factor regulating cell death in Rohon-Beard sensory neurons (Williams *et al.*, 2000a). In teleost sympathetic ganglia, the proliferation of cells occurs during early development and may possibly continue throughout adult life (An *et al.*, 2002; Weis, 1968). Recent data suggest that Hand2 in combination with Hand1 appears to enhance the NGF response by upregulating *TrkA* expression (Doxakis *et al.*, 2008; Lucas *et al.*, 2006; Ma *et al.*, 2000; Schmidt *et al.*, 2009). Analysis of BrdU incorporation and phospho-histone H3 immunoreactivity indicate that cells proliferate and undergo cell division within the developing sympathetic ganglia (An *et al.*, 2002). Interestingly, some of these cells also expressed the pan-neuronal marker, 16A11, suggesting that pre-existing neuronal cells proliferated within the ganglia, a process that has also been observed in both chick and mouse PSNS (Birren *et al.*, 1993; Cohen, 1974; DiCicco-Bloom *et al.*, 1993; Marusich *et al.*, 1994; Rohrer and Thoenen, 1987; Rothman *et al.*, 1978). However, in chick and rodents sympathetic neurons become

post-mitotic during embryonic development, while they may remain competent to divide throughout life in the zebrafish (Weis, 1968).

C. Mutations Affecting PSNS Development

1. Introduction

PSNS development, from the induction of NC through the overt differentiation of sympathetic ganglia, can be readily observed within the first 5 days of zebrafish development (An *et al.*, 2002). During this time, dynamic changes in both the numbers and the distribution of sympathetic cells within the SCG can be easily visualized by *th* mRNA whole-mount *in situ* hybridization. At 2 dpf, bilateral rows containing approximately 5–10 *th*-positive cells are ventrally located near the dorsal aorta. By 5 dpf, the number of *th*-positive cells have increased fivefold and coalesced into a V-shaped ganglia, including some appearing to migrate ventrally toward the kidney, which may represent putative adrenal chromaffin cells (An *et al.*, 2002). Thus, the evaluation of SCG formation at 3- and 5 dpf represents an excellent assay for early PSNS development that can be used in combination with forward genetic screens to detect novel mutations affecting different stages of PSNS development. Assaying the SCG would also be able to confirm mutations found to affect very early NC development. Such mutagenesis screens have been performed and examples of the different mutant classes that have been isolated thus far are discussed below.

2. Mutations Affecting PSNS Development

Mutations affecting PSNS development can be divided into a number of classes. Mutants can be identified that either fail to (1) form NC precursors, (2) migrate to the dorsal aorta, (3) differentiate into SA progenitors, or (4) proliferate and survive once they differentiate. So far, most of the zebrafish PSNS mutants described disrupt the early phases of PSNS development, notably failure of NC precursors to be specified before cell migration commences. One example is *sympathetic mutant 1* (*sym1*), which was discovered in a diploid gynogenetic screen designed to identify mutations disrupting *th* expression in the SCG complex at 5 dpf (Fig. 5; Stewart *et al.*, 2006). The *sym1* mutation causes a severe reduction or absence of *th*- and *pnmt*-expressing cells in the SCG complex of the PSNS, but *th* expression is not affected in other regions of the CNS. Subsequent cloning of the *sym1* mutation revealed that it was a deletion within the *foxd3* gene (and renamed *foxd3*^{zdf10}), which was previously shown to be essential for early NC development in other vertebrates (Fig. 5A and C; Stewart *et al.*, 2006). Analysis of the *sym1* phenotype showed that early NC survival, migration, and specification is impaired in these embryos, and they exhibit a reduction in the expression of a number of early NC markers, including *snai1b, crestin*, and *sox10* (Luo *et al.*, 2001). Thus, Foxd3 is required at multiple stages in early NC progenitors for the specification and migration of PSNS progenitors.

Another NC mutant that displays defects in PSNS development is the zebrafish *colorless* (*cls*) mutant, which disrupts the *sox10* gene required for development of most

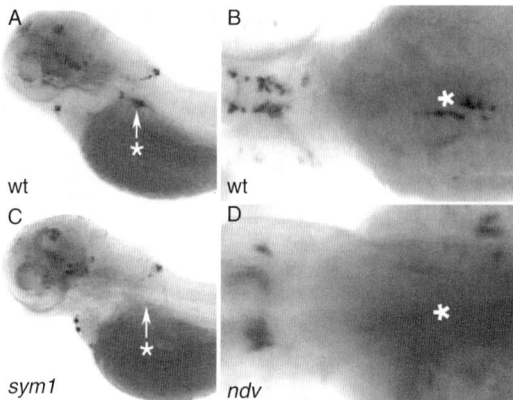

Fig. 5 Isolation of PSNS mutants in zebrafish. Whole-mount *TH in situ* preparation showing expression of *TH* mRNA in wild-type (A, B), *sym1* (C), and *nosedive* (D) mutant embryos. (A, C) Lateral view of *TH* expression at 5 dpf. The *sym1* mutant phenotype (C) was identified in an *in situ* screen at 5 dpf for mutations that specifically lack *TH* expression in the SCG (C, asterisk), but leave others areas of TH expression in the CNS and carotid body unaffected. (B, D) Dorsal view of *TH* expression in the SCG region at 3 dpf in wild-type embryos (B, asterisk). Expression of *TH* is absent in the SCG in *nosedive* mutants (D, asterisk). Analysis of Hu immunoreactivity revealed that the lack of *TH* expression in the cervical region in *nosedive* mutant embryos is due to the absence of sympathetic neurons (data not shown).

non-ectomesenchymal NC lineages, including the PSNS (Dutton *et al.*, 2001; Kelsh and Eisen, 2000). The mutation was originally isolated in a screen for NC mutants affecting pigment cell development (Kelsh and Eisen, 2000). Analysis of the *cls* phenotype shows that Sox10 appears to be required for the early survival and migration of NC-derived cells, as well as for their specification into certain sub-lineages and glial formation at a later developmental stage (Arduini *et al.*, 2009; Dutton *et al.*, 2001; Kelsh and Eisen, 2000; McKeown *et al.*, 2005). *cls* homozygotes have a complete absence of the ENS, glia and pigment cells, and a strong reduction of sensory neurons, although craniofacial derivatives are unaffected (Dutton *et al.*, 2001). Interestingly, the *sym1* and *cls* mutations affect complementary sets of NC derivatives. For example, unlike *cls, sym1* mutants have severe defects in craniofacial cartilage development, while unlike *sym1*, the *cls* mutant lacks pigment cells (Kelsh and Eisen, 2000). These phenotypes suggest that multiple genetic pathways control NC fate specification, including the specification of PSNS progenitors. Indeed, analysis of *foxd3* and *tfap2a* double mutants revealed that these genes act in parallel to regulate cell specification within the premigratory NC, at least in part by controlling the expression of *soxE* group genes, such as *sox10* (Arduini *et al.*, 2009).

Similar to the *foxd3* and *sox10* mutant phenotypes, the *lockjaw* or *mount blanc* mutant (*tfap2a*) also shows a lack of *th-* and *dbh*-positive neurons in the region of the SCG complex (Holzschuh *et al.*, 2003; Knight *et al.*, 2003). The *tfap2a* gene is normally expressed both in the premigratory NC and again in the region of the developing SCG. Recent studies have demonstrated that *tfap2a*, along with *foxd3*, clearly plays an important role in the early specification of all NC lineages (Arduini *et al.*, 2009). Therefore it is likely that *tfap2a* is

involved in both specification of premigratory crest toward the PSNS lineage and differentiation of PSNS progenitors once they reach the dorsal aorta. Indeed, in the absence of *tfap2a* function, a subset of NC cells can still migrate and undergo neural differentiation to SA progenitors at the SCG, as determined by 16A11 immunoreactivity (Holzschuh *et al.*, 2003; O'Brien *et al.*, 2004). Furthermore, the expression of *phox2a* in cells in the region of the SCG in *tfap2a* mutants indicates that the signaling cascade required to induce the initial stages of noradrenergic differentiation at the dorsal aorta is intact in these mutants (Holzschuh *et al.*, 2003). The failure of *tfap2a* mutants to express the noradrenergic differentiation markers *th* and *dbh* is therefore likely due to the requirement for Tfap2a to activate these genes, as Tfap2a has conserved DNA binding regions in both *th* and *dbh* promoters (Holzschuh *et al.*, 2003; Seo *et al.*, 2002). Retinoic acid (RA) signaling pathways function upstream of *tfap2a* in the differentiation of noradrenergic neurons because incubation of wild-type zebrafish embryos in RA induces ectopic *th*-positive cells in the region of the SCG, and this affect is blocked in *tfap2a* mutants (Holzschuh *et al.*, 2003). In addition, mutations in *neckless/rald2*, which disrupt the biosynthesis of RA from vitamin A, have fewer *th*-expressing cells in the SCG (Holzschuh *et al.*, 2003).

The expression of *tfap2a* in PSNS precursors is also controlled by Hand2, as the zebrafish *hands off* mutant (deletion of *hand2)* fails to express normal levels of *tfap2a* and *gata2* in PSNS precursors that have arrived at the dorsal aorta, which in turn causes a strong reduction in the number of *dbh*- and *th*-expressing cells in the SCG (Lucas *et al.*, 2006). The loss of noradrenergic identity in *hands off* mutants may also be due to the direct loss of *dbh* expression in these precursors, as Hand2 has been shown to function with Phox2b to directly activate the *dbh* expression in other vertebrates (Rychlik *et al.*, 2003; Xu *et al.*, 2003).

In a screen for mutations affecting NC derivatives (Henion *et al.*, 1996), three mutants were identified based on the absence of dorsal root ganglion (DRG) sensory neurons, based on the loss of 16A11 immunoreactivity. Subsequently, it has been found that these mutants, one of which is called *nosedive*, also completely lack sympathetic and enteric neurons based on the expression of *th* mRNA and 16A11, respectively (Fig. 5B and D). In contrast, the development of trunk NC-derived chromatophores and glia appears normal in all three mutants. These observations suggest that the genes disrupted by these mutations do not affect all NC derivatives, but selectively function during the development of crest-derived neurons. It is tempting to speculate that the phenotypes of these mutants further indicate an early lineage segregation of neurogenic precursors and/or suggest the function of these genes in the development of multiple NC-derived neuronal subtypes.

IV. Zebrafish as a Novel Model for Studying Neuroblastoma

A. Overview of Neuroblastoma

Neuroblastoma (NB) is an embryonic tumor of the PSNS that affects 650 children in the United States each year, making it the most common extracranial solid tumor and the leading cause of cancer death in children 1–4 years of age (Goodman, 1999). Clinically,

NB most often arises in the adrenal medulla (40% of cases); however, it manifests diverse behavior and can spontaneously regress or differentiate without treatment in infants less than 1-year-old (Maris, 2010; Maris and Matthay, 1999). Unfortunately, older children with advanced disease account for 70% of all NB patients, and their long-term survival rate remains a dismal 20–30% (Maris, 2010; Maris and Matthay, 1999). Cytogenetic analyses of NB have identified a number of aberrant chromosomal regions in NB cells. These regions include allelic losses of 1p36.1, 2q, 3p, 4p, 5q, 9p, 11q23, 14q23-qter, 16p12–13 and 18q –, which have been identified in 15–44% of primary NB (Brodeur, 2003; Brodeur et al., 1997; Maris and Matthay, 1999). Although the majority of critical target genes within these chromosomal regions remain to be identified, the findings suggest that multiple tumor suppressor genes may function during different stages of NB pathogenesis. Extensive interrogation of the 1p36 deletion has identified potential tumor suppressor genes that include *KIF1B* and *miR34a* (Schlisio et al., 2008; Welch et al., 2007; Westermann and Schwab, 2002). Furthermore, increased single nucleotide polymorphisms detected in *BRAD1* (*BRCA1*-associated domain 1 located at 2q35) and the putative gene *FLJ22536* at 6p22, along with copy number variations in a novel NB breakpoint family gene 23 (*NBPF23*), have been recently been implicated in NB (Capasso et al., 2009; Diskin et al., 2009; Maris et al., 2008). However, the functional roles of these genes in the pathogenesis of NB remain unclear.

Amplification of the potent proto-oncogene *MYCN* is found in over 20% of NB patients and is the most reliable marker of poor prognosis. Aberrant *MYCN* expression is believed to mediate NB formation by promoting cell proliferation and survival of NC-derived PSNS cells while blocking cell death (Brodeur, 2003). Other genetic factors in NB include the *PHOX2B* gene, which encodes a homeobox transcription factor that is essential to the development of the sympathoadrenal lineage cells (see above). Heterozygous mutations in *PHOX2B* have been identified in both familial and sporadic NB, albeit at low frequencies (Bourdeaut et al., 2005; van Limpt et al., 2004). Recently, the mutation and amplification of the anaplastic lymphoma kinase (*ALK*) gene were identified in up to 15% of both familial and sporadic NB cases (Chen et al., 2008; George et al., 2008; Janoueix-Lerosey et al., 2008; Mosse et al., 2008). The *ALK* gene encodes a receptor tyrosine kinase that regulates cell proliferation, differentiation, and apoptosis through a number of different signaling pathways including the PI3K/Akt pathway, MAPK pathway, and the STAT3 pathway (Palmer et al., 2009). These recent significant findings are sure to be among many other disrupted genetic pathways yet to be identified, and lack of understanding of their functional roles in this disease poses a major obstacle to understanding the molecular pathology of NB and to the development of effective therapies to treat this devastating disease.

B. Using Zebrafish to Study Neuroblastoma

Previous studies of NB pathogenesis have relied on cell lines derived from NB patient samples that are cultured *in vitro*. Also, an *in vivo* animal model of NB has been established in mice by specifically driving the overexpression of *MYCN* under the control of the tyrosine hydroxylase (Th) promoter, which resembles human NB (Weiss et al., 1997). Another murine transgenic mice expressing H-Ras chimerically under the

dopamine-beta-hydroxylase (DβH) promoter was reported to develop tumors similar to human ganglioneuroma and NB (Sweetser *et al.*, 1997). The development of a new zebrafish model of NB can contribute substantially to the analysis of NB *in vivo* based upon the strengths of the zebrafish as a genetic and tumor model system. These strengths include (1) direct visualization of tumor formation and progression in adult pigment mutants by observing the expression of fluorescently tagged oncogenes in the tumor tissues and (2) the relative ease of maintaining large numbers of animals for tumor sample collection and genetic and chemical modifier screens. In addition, the ability to simultaneously misexpress mRNAs and knockdown genes of interest in zebrafish embryos using morpholinos (Eisen and Smith, 2008) and shRNA techniques (Dong *et al.*, 2009) allow genetic epistasis analyses to be performed that provide insights into the molecular pathways that are dysfunctional in this disease. Furthermore, the zebrafish is especially amenable to the generation of transgenic lines that express human oncogenes, such as *MYCN*, to the analysis of the cellular properties of transformation and to mutagenic techniques, such as TILLING (Moens *et al.*, 2008) and Zinc-Finger Nuclease strategies (Doyon *et al.*, 2008; Foley *et al.*, 2009; Meng *et al.*, 2008), which allow the contribution and cooperation of potential tumor suppressors to be assessed *in vivo*. Importantly, the zebrafish NB model represents a platform that can be used in large high-throughput drug screens to foster the development of therapies to successfully target this deadly cancer.

V. Conclusion and Future Directions

Impressive advances have been made in understanding the genetic mechanisms that regulate PSNS development through studies in birds and rodents. Zebrafish studies that further define the anatomical and morphological aspects of the developing PSNS have accelerated the pace of discovery in this field. Exploiting the forward genetic and imaging capacity of the zebrafish system can contribute significantly to the identification of new genes and pathways that regulate PSNS development, and these mutants also provide the means to genetically dissect PSNS developmental processes *in vivo*. Finally, we postulate that knowledge of the genes responsible for normal PSNS development, such as Phox2b and Alk, will help define the molecular pathways that are affected in NB. Ultimately, second-generation suppressor screens based on established PSNS mutants exhibiting proliferative abnormalities, or on new transgenic NB models themselves, can be used to identify potential genes and pathways that may be relevant to the development of effective therapies for this disease.

VI. Acknowledgments

We would like to thank Hermann Rohrer for comments on the first edition of this manuscript. This work was supported by grant R01 CA104605 (A.T.L.). J.S.L. was supported by the Hope Street Kids foundation. R.A.S is supported by NIH/NINDS award R00 NS058608.

References

Acloque, H., Adams, M. S., *et al.* (2009). Epithelial-mesenchymal transitions: The importance of changing cell state in development and disease. *J. Clin. Invest.* **119**(6), 1438–1449.

Allende, M. L., and Weinberg, E. S. (1994). The expression pattern of two zebrafish achaete-scute homolog (ash) genes is altered in the embryonic brain of the cyclops mutant. *Dev. Biol.* **166**(2), 509–530.

An, M., Luo, R., *et al.* (2002). Differentiation and maturation of zebrafish dorsal root and sympathetic ganglion neurons. *J. Comp. Neurol.* **446**(3), 267–275.

Anderson, D. J. (1993). Cell fate determination in the peripheral nervous system: The sympathoadrenal progenitor. *J. Neurobiol.* **24**(2), 185–198.

Anderson, D. J., and Axel, R. (1986). A bipotential neuroendocrine precursor whose choice of cell fate is determined by NGF and glucocorticoids. *Cell* **47**(6), 1079–1090.

Anderson, D. J., Carnahan, J. F., *et al.* (1991). Antibody markers identify a common progenitor to sympathetic neurons and chromaffin cells in vivo and reveal the timing of commitment to neuronal differentiation in the sympathoadrenal lineage. *J. Neurosci.* **11**(11), 3507–3519.

Apostolova, G., and Dechant, G. (2009). Development of neurotransmitter phenotypes in sympathetic neurons. *Auton. Neurosci.* **151**(1), 30–38.

Arduini, B. L., Bosse, K. M., *et al.* (2009). Genetic ablation of neural crest cell diversification. *Development* **136**(12), 1987–1994.

Benjanirut, C., Paris, M., *et al.* (2006). The cAMP pathway in combination with BMP2 regulates Phox2a transcription via cAMP response element binding sites. *J. Biol. Chem.* **281**(5), 2969–2981.

Birren, S. J., Lo, L., *et al.* (1993). Sympathetic neuroblasts undergo a developmental switch in trophic dependence. *Development* **119**(3), 597–610.

Bourdeaut, F., Trochet, D., *et al.* (2005). Germline mutations of the paired-like homeobox 2B (PHOX2B) gene in neuroblastoma. *Cancer Lett.* **228**(1–2), 51–58.

Britsch, S., Li, L., *et al.* (1998). The ErbB2 and ErbB3 receptors and their ligand, neuregulin-1, are essential for development of the sympathetic nervous system. *Genes Dev.* **12**(12), 1825–1836.

Brodeur, G. M. (2003). Neuroblastoma: Biological insights into a clinical enigma. *Nat. Rev. Cancer* **3**(3), 203–216.

Brodeur, G. M., Maris, J. M., *et al.* (1997). Biology and genetics of human neuroblastomas. *J. Pediatr. Hematol. Oncol.* **19**(2), 93–101.

Brodski, C., Schaubmar, A., *et al.* (2002). Opposing functions of GDNF and NGF in the development of cholinergic and noradrenergic sympathetic neurons. *Mol. Cell. Neurosci.* **19**(4), 528–538.

Capasso, M., Devoto, M., *et al.* (2009). Common variations in BARD1 influence susceptibility to high-risk neuroblastoma. *Nat. Genet.* **41**(6), 718–723.

Chen, Y., Takita, J., *et al.* (2008). Oncogenic mutations of ALK kinase in neuroblastoma. *Nature* **455**(7215), 971–974.

Cheung, M., Chaboissier, M. C., *et al.* (2005). The transcriptional control of trunk neural crest induction, survival, and delamination. *Dev. Cell* **8**(2), 179–192.

Chun, L. L., and Patterson, P. H. (1977). Role of nerve growth factor in the development of rat sympathetic neurons in vitro. I. Survival, growth, and differentiation of catecholamine production. *J. Cell Biol.* **75**(3), 694–704.

Cohen, A. M. (1974).DNA synthesis and cell division in differentiating avian adrenergic neuroblasts. In "Wenner-Gren Center International Symposium Series: Dynamics of Degeneration and Growth in Neurons" (K. Fuxe, L. Olson, and Y. Zotterman, eds) 359–370. Pergamon, Oxford.

Coppola, E., Pattyn, A., *et al.* (2005). Reciprocal gene replacements reveal unique functions for Phox2 genes during neural differentiation. *EMBO J.* **24**(24), 4392–4403.

Crone, S. A., and Lee, K. F. (2002). Gene targeting reveals multiple essential functions of the neuregulin signaling system during development of the neuroendocrine and nervous systems. *Ann. N. Y. Acad. Sci.* **971**, 547–553.

Debby-Brafman, A., Burstyn-Cohen, T., *et al.* (1999). F-Spondin, expressed in somite regions avoided by neural crest cells, mediates inhibition of distinct somite domains to neural crest migration. *Neuron* **22**(3), 475–488.

DiCicco-Bloom, E., Friedman, W. J., *et al.* (1993). NT-3 stimulates sympathetic neuroblast proliferation by promoting precursor survival. *Neuron* **11**(6), 1101–1111.

Dick, A., Hild, M., *et al.* (2000). Essential role of Bmp7 (snailhouse) and its prodomain in dorsoventral patterning of the zebrafish embryo. *Development* **127**(2), 343–354.

Diskin, S. J., Hou, C., *et al.* (2009). Copy number variation at 1q21.1 associated with neuroblastoma. *Nature* **459**(7249), 987–991.

Dong, M., Fu, Y. F., *et al.* (2009). Heritable and lineage-specific gene knockdown in zebrafish embryo. *PLoS One* **4**(7), e6125.

Doxakis, E., Howard, L., *et al.* (2008). HAND transcription factors are required for neonatal sympathetic neuron survival. *EMBO Rep.* **9**(10), 1041–1047.

Doyon, Y., McCammon, J. M., *et al.* (2008). Heritable targeted gene disruption in zebrafish using designed zinc-finger nucleases. *Nat. Biotechnol.* **26**(6), 702–708.

Dutton, K. A., Pauliny, A., *et al.* (2001). Zebrafish colourless encodes sox10 and specifies non-ectomesenchymal neural crest fates. *Development* **128**(21), 4113–4125.

Eisen, J. S., and Smith, J. C. (2008). Controlling morpholino experiments: Don't stop making antisense. *Development* **135**(10), 1735–1743.

Elworthy, S., Pinto, J. P., *et al.* (2005). Phox2b function in the enteric nervous system is conserved in zebrafish and is sox10-dependent. *Mech. Dev.* **122**(5), 659–669.

Ernfors, P., Lee, K. F., *et al.* (1994). Lack of neurotrophin-3 leads to deficiencies in the peripheral nervous system and loss of limb proprioceptive afferents. *Cell* **77**(4), 503–512.

Ernsberger, U., Patzke, H., *et al.* (1995). The expression of tyrosine hydroxylase and the transcription factors cPhox-2 and Cash-1: Evidence for distinct inductive steps in the differentiation of chick sympathetic precursor cells. *Mech. Dev.* **52**(1), 125–136.

Ernsberger, U., Reissmann, E., *et al.* (2000). The expression of dopamine beta-hydroxylase, tyrosine hydroxylase, and Phox2 transcription factors in sympathetic neurons: Evidence for common regulation during noradrenergic induction and diverging regulation later in development. *Mech. Dev.* **92**(2), 169–177.

Ernsberger, U., and Rohrer, H. (2009). Development of the autonomic nervous system: New perspectives and open questions. *Auton. Neurosci.* **151**(1), 1–2.

Fagan, A. M., Zhang, H., *et al.* (1996). TrkA, but not TrkC, receptors are essential for survival of sympathetic neurons in vivo. *J. Neurosci.* **16**(19), 6208–6218.

Farinas, I., Jones, K. R., *et al.* (1994). Severe sensory and sympathetic deficits in mice lacking neurotrophin-3. *Nature* **369**(6482), 658–661.

Filippi, A., Mahler, J., *et al.* (2009). Expression of the paralogous tyrosine hydroxylase encoding genes th1 and th2 reveals the full complement of dopaminergic and noradrenergic neurons in zebrafish larval and juvenile brain. *J. Comp. Neurol.* **518**(4), 423–438.

Foley, J. E., Ych, J. R., *et al.* (2009). Rapid mutation of endogenous zebrafish genes using zinc finger nucleases made by Oligomerized Pool ENgineering (OPEN). *PLoS One* **4**(2), e4348.

Fouquet, B., Weinstein, B. M., *et al.* (1997). Vessel patterning in the embryo of the zebrafish: Guidance by notochord. *Dev. Biol.* **183**(1), 37–48.

Francis, N. J., and Landis, S. C. (1999). Cellular and molecular determinants of sympathetic neuron development. *Annu. Rev. Neurosci.* **22**, 541–566.

Gammill, L. S., Gonzalez, C., *et al.* (2006). Guidance of trunk neural crest migration requires neuropilin 2/semaphorin 3F signaling. *Development* **133**(1), 99–106.

George, R. E., Sanda, T., *et al.* (2008). Activating mutations in ALK provide a therapeutic target in neuroblastoma. *Nature* **455**(7215), 975–978.

Goodman, N. W. (1999). An open letter to the Director General of the Cancer Research Campaign. *J. R. Coll. Physicians Lond.* **33**(1), 93.

Goridis, C., and Rohrer, H. (2002). Specification of catecholaminergic and serotonergic neurons. *Nat. Rev. Neurosci.* **3**(7), 531–541.

Groves, A. K., George, K. M., *et al.* (1995). Differential regulation of transcription factor gene expression and phenotypic markers in developing sympathetic neurons. *Development* **121**(3), 887–901.

Guillemot, F., Lo, L. C., *et al.* (1993). Mammalian achaete-scute homolog 1 is required for the early development of olfactory and autonomic neurons. *Cell* **75**(3), 463–476.

Guo, S., Brush, J., *et al.* (1999a). Development of noradrenergic neurons in the zebrafish hindbrain requires BMP, FGF8, and the homeodomain protein soulless/Phox2a. *Neuron* **24**(3), 555–566.

Guo, S., Wilson, S. W., *et al.* (1999b). Mutations in the zebrafish unmask shared regulatory pathways controlling the development of catecholaminergic neurons. *Dev. Biol.* **208**(2), 473–487.

Hansen, M. B. (2003). The enteric nervous system I: Organisation and classification. *Pharmacol. Toxicol.* **92**(3), 105–113.

Hendershot, T. J., Liu, H., *et al.* (2008). Conditional deletion of Hand2 reveals critical functions in neurogenesis and cell type-specific gene expression for development of neural crest-derived noradrenergic sympathetic ganglion neurons. *Dev. Biol.* **319**(2), 179–191.

Henion, P. D., Raible, D. W., *et al.* (1996). Screen for mutations affecting development of Zebrafish neural crest. *Dev. Genet.* **18**(1), 11–17.

Heumann, R., Korsching, S., *et al.* (1984). Relationship between levels of nerve growth factor (NGF) and its messenger RNA in sympathetic ganglia and peripheral target tissues. *EMBO J.* **3**(13), 3183–3189.

Hirsch, M. R., Tiveron, M. C., *et al.* (1998). Control of noradrenergic differentiation and Phox2a expression by MASH1 in the central and peripheral nervous system. *Development* **125**(4), 599–608.

Holzschuh, J., Barrallo-Gimeno, A., *et al.* (2003). Noradrenergic neurons in the zebrafish hindbrain are induced by retinoic acid and require tfap2a for expression of the neurotransmitter phenotype. *Development* **130**(23), 5741–5754.

Holzschuh, J., Ryu, S., *et al.* (2001). Dopamine transporter expression distinguishes dopaminergic neurons from other catecholaminergic neurons in the developing zebrafish embryo. *Mech. Dev.* **101**(1–2), 237–243.

Hong, S. J., Huh, Y., *et al.* (2006). GATA-3 regulates the transcriptional activity of tyrosine hydroxylase by interacting with CREB. *J. Neurochem.* **98**(3), 773–781.

Honma, Y., Araki, T., *et al.* (2002). Artemin is a vascular-derived neurotropic factor for developing sympathetic neurons. *Neuron* **35**(2), 267–282.

Howard, M. J., Stanke, M., *et al.* (2000). The transcription factor dHAND is a downstream effector of BMPs in sympathetic neuron specification. *Development* **127**(18), 4073–4081.

Hsu, H. J., Lin, G., *et al.* (2003). Parallel early development of zebrafish interrenal glands and pronephros: Differential control by wt1 and ff1b. *Development* **130**(10), 2107–2116.

Huber, K. (2006). The sympathoadrenal cell lineage: Specification, diversification, and new perspectives. *Dev. Biol.* **298**(2), 335–343.

Huber, K., and Ernsberger, U. (2006). Cholinergic differentiation occurs early in mouse sympathetic neurons and requires Phox2b. *Gene Expr.* **13**(2), 133–139.

Janoueix-Lerosey, I., Lequin, D., *et al.* (2008). Somatic and germline activating mutations of the ALK kinase receptor in neuroblastoma. *Nature* **455**(7215), 967–970.

Kalcheim, C., Langley, K., *et al.* (2002). From the neural crest to chromaffin cells: Introduction to a session on chromaffin cell development. *Ann. N. Y. Acad. Sci.* **971**, 544–546.

Kasemeier-Kulesa, J. C., Bradley, R., *et al.* (2006). Eph/ephrins and N-cadherin coordinate to control the pattern of sympathetic ganglia. *Development* **133**(24), 4839–4847.

Kasemeier-Kulesa, J. C., Kulesa, P. M., *et al.* (2005). Imaging neural crest cell dynamics during formation of dorsal root ganglia and sympathetic ganglia. *Development* **132**(2), 235–245.

Kawasaki, T., Bekku, Y., *et al.* (2002). Requirement of neuropilin 1-mediated Sema3A signals in patterning of the sympathetic nervous system. *Development* **129**(3), 671–680.

Kelsh, R. N., and Eisen, J. S. (2000). The zebrafish colourless gene regulates development of non-ectomesenchymal neural crest derivatives. *Development* **127**(3), 515–525.

Kim, H. S., Hong, S. J., *et al.* (2001). Regulation of the tyrosine hydroxylase and dopamine beta-hydroxylase genes by the transcription factor AP-2. *J. Neurochem.* **76**(1), 280–294.

Knecht, A. K., and Bronner-Fraser, M. (2002). Induction of the neural crest: A multigene process. *Nat. Rev. Genet.* **3**(6), 453–461.

Knight, R. D., Nair, S., *et al.* (2003). lockjaw encodes a zebrafish tfap2a required for early neural crest development. *Development* **130**(23), 5755–5768.

Korsching, S., and Thoenen, H. (1983). Nerve growth factor in sympathetic ganglia and corresponding target organs of the rat: Correlation with density of sympathetic innervation. *Proc. Natl. Acad. Sci. U.S.A* **80**(11), 3513–3516.

Krauss, S., Concordet, J. P., *et al.* (1993). A functionally conserved homolog of the Drosophila segment polarity gene hh is expressed in tissues with polarizing activity in zebrafish embryos. *Cell* **75**(7), 1431–1444.

LeDouarin, N., and Kalcheim, C. (1999). "The Neural Crest". Cambridge University Press, Cambridge; New York.

Lim, K. C., Lakshmanan, G., *et al.* (2000). Gata3 loss leads to embryonic lethality due to noradrenaline deficiency of the sympathetic nervous system. *Nat. Genet.* **25**(2), 209–212.

Liu, Y. W. (2007). Interrenal organogenesis in the zebrafish model. *Organogenesis* **3**(1), 44–48.

Liu, Y. W., Gao, W., *et al.* (2003). Prox1 is a novel coregulator of Ff1b and is involved in the embryonic development of the zebra fish interrenal primordium. *Mol. Cell. Biol.* **23**(20), 7243–7255.

Lo, J., Lee, S., *et al.* (2003). 15000 unique zebrafish EST clusters and their future use in microarray for profiling gene expression patterns during embryogenesis. *Genome Res.* **13**(3), 455–466.

Lo, L., Morin, X., *et al.* (1999). Specification of neurotransmitter identity by Phox2 proteins in neural crest stem cells. *Neuron* **22**(4), 693–705.

Lucas, M. E., Muller, F., *et al.* (2006). The bHLH transcription factor hand2 is essential for noradrenergic differentiation of sympathetic neurons. *Development* **133**(20), 4015–4024.

Luo, R., An, M., *et al.* (2001). Specific pan-neural crest expression of zebrafish Crestin throughout embryonic development. *Dev. Dyn.* **220**(2), 169–174.

Ma, L., Merenmies, J., *et al.* (2000). Molecular characterization of the TrkA/NGF receptor minimal enhancer reveals regulation by multiple cis elements to drive embryonic neuron expression. *Development* **127**(17), 3777–3788.

Maris, J. M. (2010). Recent advances in neuroblastoma. *N. Engl. J. Med.* **362**(23), 2202–2211.

Maris, J. M., and Matthay, K. K. (1999). Molecular biology of neuroblastoma. *J. Clin. Oncol.* **17**(7), 2264–2279.

Maris, J. M., Mosse, Y. P., *et al.* (2008). Chromosome 6p22 locus associated with clinically aggressive neuroblastoma. *N. Engl. J. Med.* **358**(24), 2585–2593.

Martin, S. C., Marazzi, G., *et al.* (1995). Five Trk receptors in the zebrafish. *Dev. Biol.* **169**(2), 745–758.

Martinez-Barbera, J. P., Toresson, H., *et al.* (1997). Cloning and expression of three members of the zebrafish Bmp family: Bmp2a, Bmp2b and Bmp4. *Gene* **198**(1–2), 53–59.

Marusich, M. F., Furneaux, H. M., *et al.* (1994). Hu neuronal proteins are expressed in proliferating neurogenic cells. *J. Neurobiol.* **25**(2), 143–155.

McKeown, S. J., Lee, V. M., *et al.* (2005). Sox10 overexpression induces neural crest-like cells from all dorsoventral levels of the neural tube but inhibits differentiation. *Dev. Dyn.* **233**(2), 430–444.

Melby, A. E., Kimelman, D., *et al.* (1997). Spatial regulation of floating head expression in the developing notochord. *Dev. Dyn.* **209**(2), 156–165.

Meng, X., Noyes, M. B., *et al.* (2008). Targeted gene inactivation in zebrafish using engineered zinc-finger nucleases. *Nat. Biotechnol.* **26**(6), 695–701.

Moens, C. B., Donn, T. M., *et al.* (2008). Reverse genetics in zebrafish by TILLING. *Brief. Funct. Genomic. Proteomic.* **7**(6), 454–459.

Morikawa, Y., D'Autreaux, F., *et al.* (2007). Hand2 determines the noradrenergic phenotype in the mouse sympathetic nervous system. *Dev. Biol.* **307**(1), 114–126.

Morikawa, Y., Zehir, A., *et al.* (2009). BMP signaling regulates sympathetic nervous system development through Smad4-dependent and -independent pathways. *Development* **136**(21), 3575–3584.

Morin, X., Cremer, H., *et al.* (1997). Defects in sensory and autonomic ganglia and absence of locus coeruleus in mice deficient for the homeobox gene Phox2a. *Neuron* **18**(3), 411–423.

Mosse, Y. P., Laudenslager, M., *et al.* (2008). Identification of ALK as a major familial neuroblastoma predisposition gene. *Nature* **455**(7215), 930–935.

Muller, F., and Rohrer, H. (2002). Molecular control of ciliary neuron development: BMPs and downstream transcriptional control in the parasympathetic lineage. *Development* **129**(24), 5707–5717.

Murphy, S., Krainock, R., *et al.* (2002). Neuregulin signaling via erbB receptor assemblies in the nervous system. *Mol. Neurobiol.* **25**(1), 67–77.

Nakamura, E., and Kaelin, W. G. Jr. (2006). Recent insights into the molecular pathogenesis of pheochromocytoma and paraganglioma. *Endocr. Pathol.* **17**(2), 97–106.

Neave, B., Rodaway, A., *et al.* (1995). Expression of zebrafish GATA 3 (gta3) during gastrulation and neurulation suggests a role in the specification of cell fate. *Mech. Dev.* **51**(2–3), 169–182.

Nguyen, V. H., Schmid, B., *et al.* (1998). Ventral and lateral regions of the zebrafish gastrula, including the neural crest progenitors, are established by a bmp2b/swirl pathway of genes. *Dev. Biol.* **199**(1), 93–110.

O'Brien, E. K., d'Alencon, C., *et al.* (2004). Transcription factor Ap-2alpha is necessary for development of embryonic melanophores, autonomic neurons and pharyngeal skeleton in zebrafish. *Dev. Biol.* **265**(1), 246–261.

Palmer, R. H., Vernersson, E., *et al.* (2009). Anaplastic lymphoma kinase: Signalling in development and disease. *Biochem. J.* **420**(3), 345–361.

Patten, I., and Placzek, M. (2000). The role of Sonic hedgehog in neural tube patterning. *Cell Mol. Life Sci.* **57**(12), 1695–1708.

Pattyn, A., Goridis, C., *et al.* (2000). Specification of the central noradrenergic phenotype by the homeobox gene Phox2b. *Mol. Cell. Neurosci.* **15**(3), 235–243.

Pattyn, A., Guillemot, F., *et al.* (2006). Delays in neuronal differentiation in Mash1/Ascl1 mutants. *Dev. Biol.* **295**(1), 67–75.

Pattyn, A., Morin, X., *et al.* (1999). The homeobox gene Phox2b is essential for the development of autonomic neural crest derivatives. *Nature* **399**(6734), 366–370.

Raible, D. W., and Eisen, J. S. (1994). Restriction of neural crest cell fate in the trunk of the embryonic zebrafish. *Development* **120**(3), 495–503.

Rohrer, H., and Thoenen, H. (1987). Relationship between differentiation and terminal mitosis: Chick sensory and ciliary neurons differentiate after terminal mitosis of precursor cells, whereas sympathetic neurons continue to divide after differentiation. *J. Neurosci.* **7**(11), 3739–3748.

Rothman, T. P., Gershon, M. D., *et al.* (1978). The relationship of cell division to the acquisition of adrenergic characteristics by developing sympathetic ganglion cell precursors. *Dev. Biol.* **65**(2), 322–341.

Rudiger, R., Binder, E., *et al.* (2009). In vivo role for CREB signaling in the noradrenergic differentiation of sympathetic neurons. *Mol. Cell. Neurosci.* **42**(2), 142–151.

Rychlik, J. L., Gerbasi, V., *et al.* (2003). The interaction between dHAND and Arix at the dopamine beta-hydroxylase promoter region is independent of direct dHAND binding to DNA. *J. Biol. Chem.* **278**(49), 49652–49660.

Sakai, D., Suzuki, T., *et al.* (2006). Cooperative action of Sox9, Snail2 and PKA signaling in early neural crest development. *Development* **133**(7), 1323–1333.

Santiago, A., and Erickson, C. A. (2002). Ephrin-B ligands play a dual role in the control of neural crest cell migration. *Development* **129**(15), 3621–3632.

Sauka-Spengler, T., and Bronner-Fraser, M. (2008). A gene regulatory network orchestrates neural crest formation. *Nat. Rev. Mol. Cell Biol.* **9**(7), 557–568.

Schlisio, S., Kenchappa, R. S., *et al.* (2008). The kinesin KIF1Bbeta acts downstream from EglN3 to induce apoptosis and is a potential 1p36 tumor suppressor. *Genes Dev.* **22**(7), 884–893.

Schmidt, M., Lin, S., *et al.* (2009). The bHLH transcription factor Hand2 is essential for the maintenance of noradrenergic properties in differentiated sympathetic neurons. *Dev. Biol.* **329**(2), 191–200.

Schober, A., Krieglstein, K., *et al.* (2000). Molecular cues for the development of adrenal chromaffin cells and their preganglionic innervation. *Eur. J. Clin. Invest.* **30**(Suppl 3), 87–90.

Schober, A., and Unsicker, K. (2001). Growth and neurotrophic factors regulating development and maintenance of sympathetic preganglionic neurons. *Int. Rev. Cytol.* **205**, 37–76.

Schwarz, Q., Maden, C. H., *et al.* (2009). Neuropilin 1 signaling guides neural crest cells to coordinate pathway choice with cell specification. *Proc. Natl. Acad. Sci. U.S.A* **106**(15), 6164–6169.

Schweitzer, J., and Driever, W. (2009). Development of the dopamine systems in zebrafish. *Adv. Exp. Med. Biol.* **651**, 1–14.

Seo, H., Hong, S. J., *et al.* (2002). A direct role of the homeodomain proteins Phox2a/2b in noradrenaline neurotransmitter identity determination. *J. Neurochem.* **80**(5), 905–916.

Shelton, D. L., and Reichardt, L. F. (1984). Expression of the beta-nerve growth factor gene correlates with the density of sympathetic innervation in effector organs. *Proc. Natl. Acad. Sci. U.S.A* **81**(24), 7951–7955.

Smeyne, R. J., Klein, R., *et al.* (1994). Severe sensory and sympathetic neuropathies in mice carrying a disrupted Trk/NGF receptor gene. *Nature* **368**(6468), 246–249.

Stewart, R. A., Arduini, B. L., *et al.* (2006). Zebrafish foxd3 is selectively required for neural crest specification, migration and survival. *Dev. Biol.* **292**(1), 174–188.

Sweetser, D. A., Kapur, R. P., *et al.* (1997). Oncogenesis and altered differentiation induced by activated Ras in neuroblasts of transgenic mice. *Oncogene* **15**(23), 2783–2794.

Talbot, W. S., Trevarrow, B., *et al.* (1995). A homeobox gene essential for zebrafish notochord development. *Nature* **378**(6553), 150–157.

Thexton, A. (2001). Vertebrate peripheral nervous system. Encyclopedia of Life Sciences. Retrieved 2001.

To, T. T., Hahner, S., *et al.* (2007). Pituitary-interrenal interaction in zebrafish interrenal organ development. *Mol. Endocrinol.* **21**(2), 472–485.

Tsarovina, K., Pattyn, A., *et al.* (2004). Essential role of Gata transcription factors in sympathetic neuron development. *Development* **131**(19), 4775–4786.

Unsicker, K., Huber, K., *et al.* (2005). The chromaffin cell and its development. *Neurochem. Res.* **30**(6–7), 921–925.

van Limpt, V., Schramm, A., *et al.* (2004). The Phox2B homeobox gene is mutated in sporadic neuroblastomas. *Oncogene* **23**(57), 9280–9288.

Weis, J. S. (1968). Analysis of the development of nervous system of the zebrafish, Brachydanio rerio. I. The normal morphology and development of the spinal cord and ganglia of the zebrafish. *J. Embryol. Exp. Morphol.* **19**(2), 109–119.

Weiss, W. A., Aldape, K., *et al.* (1997). Targeted expression of MYCN causes neuroblastoma in transgenic mice. *EMBO J.* **16**(11), 2985–2995.

Welch, C., Chen, Y., *et al.* (2007). MicroRNA-34a functions as a potential tumor suppressor by inducing apoptosis in neuroblastoma cells. *Oncogene* **26**(34), 5017–5022.

Westermann, F., and Schwab, M. (2002). Genetic parameters of neuroblastomas. *Cancer Lett.* **184**(2), 127–147.

Williams, J. A., Barrios, A., *et al.* (2000a). Programmed cell death in zebrafish rohon beard neurons is influenced by TrkC1/NT-3 signaling. *Dev. Biol.* **226**(2), 220–230.

Williams, Z., Tse, V., *et al.* (2000b). Sonic hedgehog promotes proliferation and tyrosine hydroxylase induction of postnatal sympathetic cells in vitro. *NeuroReport* **11**(15), 3315–3319.

Wyatt, S., Pinon, L. G., *et al.* (1997). Sympathetic neuron survival and TrkA expression in NT3-deficient mouse embryos. *EMBO J.* **16**(11), 3115–3123.

Xu, H., Firulli, A. B., *et al.* (2003). HAND2 synergistically enhances transcription of dopamine-beta-hydroxylase in the presence of Phox2a. *Dev. Biol.* **262**(1), 183–193.

Yamamoto, K., Ruuskanen, J. O., *et al.* (2010). Two tyrosine hydroxylase genes in vertebrates New dopaminergic territories revealed in the zebrafish brain. *Mol. Cell. Neurosci.* **43**(4), 394–402.

Yelon, D., Ticho, B., *et al.* (2000). The bHLH transcription factor hand2 plays parallel roles in zebrafish heart and pectoral fin development. *Development* **127**(12), 2573–2582.

Young, H. M., Cane, K. N., *et al.* (2010). Development of the autonomic nervous system: A comparative view. *Auton. Neurosci.* Mar 24. [Epub ahead of print]. PMID: 20346736

Zirlinger, M., Lo, L., *et al.* (2002). Transient expression of the bHLH factor neurogenin-2 marks a subpopulation of neural crest cells biased for a sensory but not a neuronal fate. *Proc. Natl. Acad. Sci. U.S.A* **99**(12), 8084–8089.

CHAPTER 6

Analysis of the Retina in the Zebrafish Model

Andrei Avanesov *and* **Jarema Malicki**

Division of Craniofacial and Molecular Genetics, Tufts University, Boston, Massachusetts

I. Introduction
II. Development of the Zebrafish Retina
    A. Early Morphogenetic Events
    B. Neurogenesis
    C. Development of Retinotectal Projections
    D. Non-Neuronal Tissues
III. Analysis of Wild-Type and Mutant Visual System
    A. Histological Analysis
    B. The Use of Molecular Markers
    C. Analysis of Cell Movements and Lineage Relationships
    D. Analysis of Cell and Tissue Interactions
    E. Analysis of Cell Proliferation
    F. Behavioral Studies
    G. Electrophysiological Analysis of Retinal Function
    H. Biochemical Approaches
    I. Chemical Screens
IV. Analysis of Gene Function in the Zebrafish Retina
    A. Reverse Genetic Approaches
    B. Forward Genetics
V. Summary
    Acknowledgments
    References

Abstract

The zebrafish is one of the leading models for the analysis of the vertebrate visual system. A wide assortment of molecular, genetic, and cell biological approaches is available to study zebrafish visual system development and function. As new techniques

become available, genetic analysis and imaging continue to be the strengths of the zebrafish model. In particular, recent developments in the use of transposons and zinc finger nucleases to produce new generations of mutant strains enhance both forward and reverse genetic analysis. Similarly, the imaging of developmental and physiological processes benefits from a wide assortment of fluorescent proteins and the ways to express them in the embryo. The zebrafish is also highly attractive for high-throughput screening of small molecules, a promising strategy to search for compounds with therapeutic potential. Here we discuss experimental approaches used in the zebrafish model to study morphogenetic transformations, cell fate decisions, and the differentiation of fine morphological features that ultimately lead to the formation of the functional vertebrate visual system.

I. Introduction

The vertebrate central nervous system (CNS) is enormously complex. The human cerebral cortex alone is estimated to contain in excess of 10^9 neurons (Jacobson, 1991), each characterized by the morphology of its soma and processes, synaptic connections with other cells, receptors expressed on its surface, the neurotransmitters it releases, and numerous other molecular and cellular features. Together these characteristics define cell identity. To understand the development of the CNS, multiple steps involved in the formation of numerous cell identities must be determined. One way to approach this enormously complicated task is to study a relatively simple and accessible region of the CNS. The retina is such a region.

Several characteristics make the retina more approachable than most other areas of the CNS. Most importantly, the retina contains a relatively small number of neuronal cell classes, and these are characterized by stereotypical positions and distinctive morphologies. Even in very crude histological preparations, the identity of individual cells can be frequently and correctly determined based on their location. Cajal noted that the separation of different cells into distinct layers, the small size of dendritic fields, and the presence of layers consisting almost exclusively of neuronal projections are fortuitous characteristics of the retina (Cajal, 1893). In addition, the eye becomes isolated from other parts of the CNS early in embryogenesis, and consequently cell migrations into the retina are limited to the optic nerve and the optic chiasm only (Burrill and Easter, 1994; Watanabe and Raff, 1988). Such anatomical isolation simplifies the interpretation of developmental events within the retina. Taken together, all these qualities make the retina an excellent model system for the studies of vertebrate neuronal development and function.

Teleost retinae have been studied for over a century (Cajal, 1893; Dowling, 1987; Malicki, 2000; Muller, 1857; Rodieck, 1973). The eyes of teleosts in general and zebrafish in particular are large and their neuroanatomy is well characterized. An important advantage of the zebrafish retina for genetic and developmental research is that it is formed and becomes functional very early in development. Neurogenesis in the central retina of the zebrafish eye is essentially complete by 60 hours post fertilization (hpf) (Nawrocki, 1985) and, as judged by behavioral responses to visual stimuli, the zebrafish eye detects light surprisingly early, starting between 2.5 and 3.5 days post fertilization (dpf) (Clark, 1981; Easter and Nicola, 1996). Studies of the zebrafish retina benefit from many general qualities of the system: high fecundity, transparency, embryogenesis that

occurs outside the maternal organism, the ease of maintenance in large numbers, the short length of the life cycle, the ability to study haploid development, and most recently the progress in zebrafish genomics, including the genome sequencing project.

The neuronal architecture of the vertebrate retina has been remarkably conserved in evolution. Early investigators noted that even retinae of divergent vertebrate phyla, including teleosts and mammals, display similar organization (Cajal, 1893; Muller, 1857). Gross morphological and histological features of mammalian and teleost retinae display few differences. Accordingly, human and zebrafish retinae contain the same major cell classes organized in the same layered pattern, where light-sensing photoreceptors occupy the outermost layer, while the retinal projection neurons, the ganglion cells, reside in the innermost neuronal layer, proximal to the lens. The retinal interneurons, the amacrine, bipolar, and horizontal cells, localize in between the photoreceptor and ganglion cell layers (Fig. 2). Similarities extend beyond histology and morphology. Pax-2/noi and Chx10/Vsx-2 expression patterns, for example, are very similar in mouse and zebrafish eyes (Liu *et al.*, 1994; Macdonald and Wilson, 1997; Nornes *et al.*, 1990; Passini *et al.*, 1997), and a number of genetic loci display closely related phenotypes in humans and zebrafish alike. These observations stimulated efforts to use the zebrafish as a model of human eye disorders (reviewed in Gross and Perkins, 2008). Consequently, zebrafish eye mutants have been proposed as models of pyruvate dehydrogenase deficiency, choroidemia, achromatopsia, as well as June, Joubert, and Hermansky–Pudlak syndromes (Bahadori *et al.*, 2006; Brockerhoff *et al.*, 2003; Duldulao *et al.*, 2009; Hudak *et al.*, 2010; Krock *et al.*, 2007; Taylor *et al.*, 2004). This is a fortuitous circumstance, considering that throughout the world diseases of the retina affect millions (Cedrone *et al.*, 1997; Dryja and Li, 1995; Hartong *et al.*, 2006; Seddon, 1994). Thus, in addition to being an excellent model for the studies of vertebrate neurogenesis, the zebrafish retina is likely to provide medically relevant insights. In this chapter, following an introduction to zebrafish eye development, we focus on tools currently used to study various aspects of the zebrafish visual system. Since many techniques described in this chapter are also applied to the analysis of other organs, the reader is encouraged to search for more information in other sections of this volume.

II. Development of the Zebrafish Retina

A. Early Morphogenetic Events

Fate-mapping studies indicate that during early gastrulation the retina originates from a single field of cells positioned roughly between the telencephalic and the diencephalic precursor fields (Woo and Fraser, 1995). During late gastrulation, the anterior and lateral migrations of diencephalic precursors are thought to subdivide the retinal field into two separate primordia (Rembold *et al.*, 2006; Varga *et al.*, 1999). Neurulation in teleosts proceeds somewhat differently than in higher vertebrates. First, the primordium of the CNS does not take the form of a tube (the neural tube), and instead is shaped in the form of a solid rod called the neural keel (Fig. 1B and C) (Kimmel *et al.*, 1995; Lowery and Sive, 2004; Schmitz *et al.*, 1993). Consistent with that, optic vesicles are not present, and the equivalent structures are called optic lobes. These first become evident as bilateral thickenings of the anterior neural keel at about 11.5 hpf, and gradually become more and

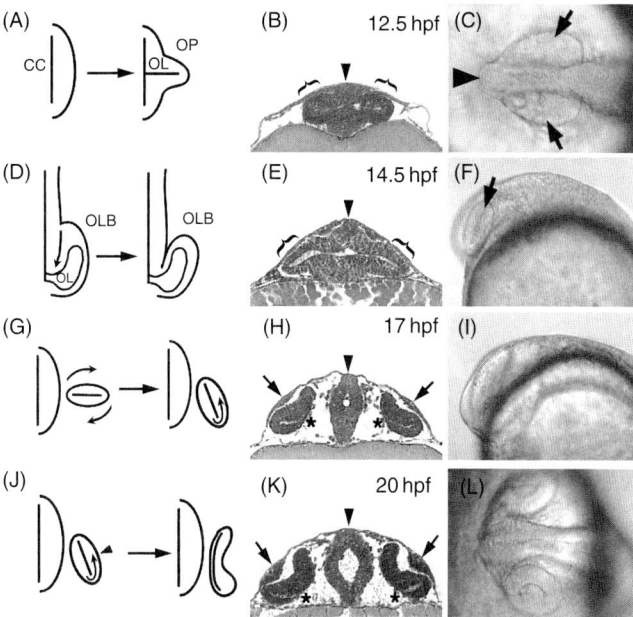

Fig. 1 Early morphogenetic events leading to the formation of the optic cup. (A) A diagram of a transverse section through anterior neural keel illustrating morphogenetic transformation that leads to the formation of optic lobes. Solid horizontal line represents the ventricular lumen (OL) of the optic lobe. (B) A transverse plastic section through the anterior portion of the neural keel and optic lobes (brackets). (C) Dorsal view of anterior neural keel and optic lobes (arrows) at 12.5 hpf. (D) A schematic representation of anterior neural keel (dorsal view, anterior down). Wing-shaped optic primordia gradually detach from the neural keel starting posteriorly (arrow). (E) A transverse plastic section through anterior neural keel and optic lobes (brackets) at 14.5 hpf. (F) Lateral view of anterior neural keel and optic lobe (arrow) at the same stage. (G) A diagram of dorsoventral reorientation of the optic lobe. (H) A transverse plastic section through neural keel and optic lobes during the reorientation at ca. 17 hpf. At about the same time, lens rudiments start to form (arrows) and the medial layer of the optic lobe becomes thinner as it begins to differentiate into the pigmented epithelium (asterisks). The lateral surface of the optic lobe starts to invaginate. (I) A lateral view of anterior neural keel during optic cup formation. (J) A schematic representation of morphogenetic movements that accompany optic cup formation. Cells migrate (arrow) from the medial to the lateral cell layer around the ventral edge of the lobe. Simultaneously, the initially flat lobe invaginates (arrowhead) to become the concave eye cup. (K) A transverse plastic section through the anterior neural tube during optic cup formation at 20 hpf. Lens rudiments are quite prominent by this stage (arrows). Most of the medial cell layer already displays a flattened morphology, except for the ventralmost regions, which still retain columnar appearance (asterisks). (L) A dorsal view of anterior neural tube and optic lobes at 20 hpf. Vertical arrowheads in B, E, H, and K indicate the midline. CC, central canal; OL, optic lumen; OP, optic primordium; OLB, optic lobe; hpf, hours post fertilization. Except D, C and L, in all panels dorsal is up. Panels A, D, G, and J are based on Easter and Malicki (2002). The remaining panels reprinted from Pujic and Malicki (2001) with permission from Elsevier.

more prominent (Fig. 1A–C) (Schmitt and Dowling, 1994). They are initially flattened and protrude laterally on both sides of the brain (brackets and arrows, respectively in Fig. 1B and C). At approximately 13 hpf, the posterior portion of the optic lobe starts to separate from the brain, while its anterior part remains attached (Fig. 1D). This attachment will persist throughout eye development, at later stages forming the optic stalk. As its morphogenesis advances, the optic lobe turns around its anteroposterior axis so that its

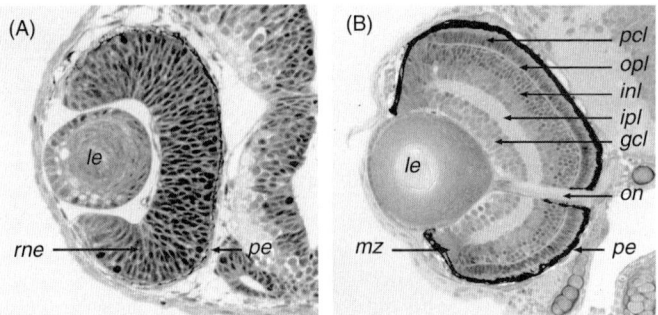

Fig. 2 Histology of the zebrafish retina. (A) A section through the zebrafish eye during early stages of neurogenesis at approximately 36 hpf. At this stage, the retina mostly consists of two epithelial layers: the pigmented epithelium and the retinal neuroepithelium. Although some retinal cells are already postmitotic at this stage, they are not numerous enough to form a distinct layer. (B) A section through the zebrafish eye at 72 hpf. With the exception of the marginal zone, where cell proliferation will continue throughout the lifetime of the animal, retinal neurogenesis is mostly completed. The major nuclear and plexiform layers, as well as the optic nerve and the pigmented epithelium, are well differentiated. gcl: ganglion cell layer; inl: inner nuclear layer; ipl: inner plexiform layer; le: lens; mz: marginal zone; on: optic nerve; opl: outer plexiform layer; pcl: photoreceptor cell layer; pe: pigmented epithelium; rne: retinal neuroepithelium.

ventral surface becomes directed toward the brain while the dorsal surface starts to face the outside environment (Fig. 1G). Cells forming the outside surface will differentiate into the neural retina. Fate-mapping studies suggest that starting at ca. 15 hpf, cells migrate from the medial to lateral epithelial layer of the optic lobe (Li *et al.*, 2000b). The medial layer becomes thinner and subsequently differentiates as the retinal pigmented epithelium (RPE) (asterisks in Fig. 1H and K). At about the same time, an invagination forms on the lateral (upper, before turning) surface of the optic lobe (Schmitt and Dowling, 1994). This is accompanied by the appearance of a thickening in the epithelium overlying the optic lobe: the lens rudiment (arrows in Fig. 1H). Subsequently, over a period of several hours, both the invagination and the lens placode become increasingly more prominent, transforming the optic lobe into the optic cup (Fig. 1J–L). The choroid fissure forms in the rim of the optic cup next to the optic stalk. The lens placode continues to grow and by 24 hpf it is detached from the epidermis. At the beginning of day 2, the optic cup consists of two closely connected sheets of cells: the pseudostratified columnar neuroepithelium (rne) and the cuboidal pigmented epithelium (pe) (Fig. 2A). Starting at about 24 hpf, melanin granules appear in the cells of the pigmented epithelium. In the first half of day 2, concomitant to the expansion of the ventral diencephalon, the eye rotates so that the choroid fissure, which at 24 hpf was pointing above the yolk sac, is now directed toward the heart (Kimmel *et al.*, 1995; Schmitt and Dowling, 1994). Throughout this period, the optic stalk gradually becomes less prominent. In the first half of day 2 as ganglion cells begin to differentiate, the optic stalk provides support for their axons. Later in development, it is no longer present as a distinct structure and its cells may contribute the optic nerve (Macdonald *et al.*, 1997). Lastly, the optic cup rotates around its mediolateral axis (Schmitt and Dowling, 1994). This is the final major morphological transformation in zebrafish eye development.

B. Neurogenesis

At the beginning of the second day of development, the zebrafish neural retina still consists of a single sheet of pseudostratified neuroepithelium. Similar to other epithelia, the retinal neuroepithelium is a highly polarized tissue, characterized by apico-basal nuclear movements, which correlate with cell cycle phase (Baye and Link, 2007; Das *et al.*, 2003; Hinds and Hinds, 1974). Nuclei of cells that are about to divide migrate to the apical surface of the neuroepithelium, where both nuclear division and cytokinesis take place. After the division, the newly formed nuclei move back to more basal locations. Although it has been assumed for a long time that dividing cells lose their contact with the basal surface of the neuroepithelium (Hinds and Hinds, 1974), more recent two-photon imaging studies in zebrafish show that this view is most likely incorrect, as a tenuous cytoplasmic process extends toward the basal surface during nuclear division of the neuroepithelial cell (Das *et al.*, 2003). Interestingly, in the brain neuroepithelium, and possibly in the retina, this process splits into two or more prior to the cytokinesis, and the daughter processes are inherited either symmetrically or asymmetrically by the daughter cells (Kosodo *et al.*, 2008).

In between mitotic divisions, the movement of cell nuclei is stochastic most of the time, so that persistent nuclear movements, directed either basally or apically, occur during less than 10% of the cell cycle (Norden *et al.*, 2009). The maximum depth of basally directed translocation is very heterogeneous, ranging from 10% to 90% of neuroepithelial thickness. Interestingly, deeper nuclear migration correlates with divisions that generate post-mitotic cells (Baye and Link, 2007). Mitotic divisions are observed nearly exclusively at the apical surface of the neuroepithelium until about 1.5 dpf. Following that, between 40 and 50 hpf, ca. 50% of mitoses occur in the inner nuclear layer (INL) (Godinho *et al.*, 2005). Very few mitotic divisions are observed in the central retina at later stages.

Despite its uniform morphology, the retinal neuroepithelium is the site of many transformations, apparent in the changes of cell cycle length and in the dynamic characteristics of gene expression patterns. After a period of very slow cell cycle progression during early stages of optic cup morphogenesis, the cell cycle shortens to ca. 10 h by 24 hpf, and later its duration appears even shorter (Baye and Link, 2007; Hu and Easter, 1999; Li *et al.*, 2000a; Nawrocki, 1985). Imaging of individual neuroepithelial cells between 24 and 40 hpf revealed that their cell cycle varies greatly in length from about 4 to 11 h, averaging ca. 6.5 h (Baye and Link, 2007). The significance of changes in the length of the cell cycle, or the genetic mechanisms that regulate them, are not understood.

In parallel to fluctuations of cell cycle length, the expression patterns of numerous genes display dramatic changes in the retinal neuroepithelium during this time. While the transcription of some early expressed genes, such as *rx3* or *six3*, is downregulated, other genes become active. The zebrafish *atonal 5* homolog, *lakritz*, is one interesting example of an important genetic regulator characterized by a dynamic expression pattern. The *lakritz* gene becomes transcriptionally active in a small group of cells in the ventral retina by 25 hpf, and from there its expression spreads into the nasal, dorsal, and finally temporal eye (Masai *et al.*, 2000). This gradual advance of expression around the retinal surface is noteworthy because it characterizes many other developmental regulators and neuronal differentiation markers (reviewed in Pujic and Malicki, 2004).

Another noteworthy feature of neuroepithelial cells is the orientation of their mitotic spindles. The mitotic spindle position and its role in cell fate determination has been interesting, albeit contentious issue. It has been proposed that in some species the vertical (apico-basal) reorientation of the mitotic spindle characterizes asymmetric cell divisions, which produce cells of different identities: a progenitor cell and a postmitotic neuron for example (Cayouette and Raff, 2003; Cayouette *et al.*, 2001). As such divisions first appear in the neuroepithelium at the onset of neurogenesis, so should vertically oriented mitotic spindles. The analysis of zebrafish neuroepithelial cells found, however, little support for the presence of vertically oriented mitotic spindles: the majority, if not all, of zebrafish neuroepithelial cells divide horizontally (Das *et al.*, 2003).

As the morphogenetic movements that shape and orient the optic cup come to completion, the first retinal cells become postmitotic and differentiate. Gross morphological characteristics of the major retinal cell classes are very well conserved in all vertebrates. Six major classes of neurons arise during neurogenesis: ganglion, amacrine, bipolar, horizontal, interplexiform, and photoreceptor cells. The Müller glia are also generated in the same period. Ganglion cell precursors are the first to become postmitotic in a small patch of ventrally located cells between 27 and 28 hpf (Hu and Easter, 1999; Nawrocki, 1985). The early onset of ganglion cell differentiation is again conserved in many vertebrate phyla (Altshuler *et al.*, 1991). Similar to expression patterns that characterize the genetic regulators of retinal neurogenesis, differentiated ganglion cells first appear in the ventral retina, nasal to the optic nerve (Burrill and Easter, 1995; Schmitt and Dowling, 1996). The rudiments of the ganglion cell layer are recognizable in histological sections by 36 hpf. Approximately 10 h after the first ganglion neuron progenitors exit the cell cycle, cells that contribute to the INL also become postmitotic. Again, this first happens in a small ventral group of cells (Hu and Easter, 1999). By 34–36 hpf, and possibly even earlier, terminal divisions of retinal progenitor cells give rise to pairs of ganglion and photoreceptor cells, indicating that these two cell classes are generated in overlapping windows of time (Poggi *et al.*, 2005).

By 60 hpf, over 90% of neurons in the central retina are postmitotic, and the major neuronal layers are distinguishable by morphological criteria. Cells of different layers become postmitotic in largely non-overlapping windows of time. This is particularly obvious for ganglion cell precursors, most of which, if not all, are postmitotic before the first INL cells exit the cell cycle (Hu and Easter, 1999). This is different from *Xenopus*, where the times of cell cycle exit for different cell classes overlap extensively (Holt *et al.*, 1988). In contrast to mammals, neurogenesis in teleosts and larval amphibians continues at the retinal margin throughout the lifetime of the organism (Marcus *et al.*, 1999). In adult zebrafish, as well as in other teleosts, neurons are also added in the outer nuclear layer. In contrast to the marginal zone, where many cell types are generated, only rods are added in the outer nuclear layer of the adult (Mack and Fernald, 1995; Marcus *et al.*, 1999).

Photoreceptor morphogenesis starts shortly after the exit of photoreceptor precursor cells from the cell cycle (reviewed in Tsujikawa and Malicki, 2004a). The photoreceptor cell layer can be distinguished in histological sections by 48 hpf. The expression of visual pigments, opsins, is necessary for photoreceptor outer segment differentiation. Rods are the first to express opsin around 50 hpf, shortly followed by blue and red cones, and somewhat later by short single cones (Raymond *et al.*, 1995; Robinson *et al.*, 1995; Takechi *et al.*, 2003). Photoreceptor outer segments first appear in the

ventral patch by 60 hpf, and ribbon synapses of photoreceptor synaptic termini are detectable by 62 hpf (Branchek and Bremiller, 1984; Schmitt and Dowling, 1999). The photoreceptor cell layer of the zebrafish retina contains five types of photoreceptor cells: rods, short single cones, long single cones, and short and long members of double cone pairs. The differentiation of morphologically distinct photoreceptor types becomes apparent by 4 dpf, and by 12 dpf all zebrafish photoreceptor classes can be distinguished on the basis of their morphology (Branchek and Bremiller, 1984).

The photoreceptor cells of the zebrafish retina are organized in a regular pattern, referred to as the "photoreceptor mosaic." In the adult, cones form regular rows. The spaces between these rows are occupied by rods, which do not display any obvious pattern. Within a single row of cones, double cones are separated from each other by alternating long and short single cones. Adjacent rows of cones are staggered relative to each other so that short single cones of one row are flanked on either side by long single cones of the two neighboring rows (Fadool, 2003; Larison and Bremiller, 1990). In addition to morphology, individual types of photoreceptors are uniquely characterized by spectral sensitivities and visual pigment expressions. Long single cones express blue light-sensitive opsin; short single cones, ultraviolet (UV)-sensitive opsin; double cones, red-sensitive and green-sensitive opsins; whereas rods express rod opsin (Hisatomi *et al.*, 1996; Raymond *et al.*, 1993). The number of opsin genes exceeds the number of photoreceptor types, as two and four independent loci encode red and green opsins, respectively (Chinen *et al.*, 2003). Each green and red opsin gene is expressed in a different subpopulation of double cones. Of the two red opsin genes, LWS-2 is expressed in the central retina, while LWS-1 in the retinal periphery (Takechi and Kawamura, 2005). Similarly, the expression domains of green opsin genes RH2-1 and RH2-2 occupy largely overlapping areas in the central retina, while RH2-3 and RH3-4 are expressed at the retinal circumference in what appears to be non-overlapping regions (Takechi and Kawamura, 2005).

C. Development of Retinotectal Projections

As this aspect of retinal development is discussed at length in an accompanying chapter (Chapter 1), here we comment on some of the most basic observations only. The neuronal network of the retina is largely self-contained. The only retinal neurons that send their projections outside are the ganglion cells. Their axons navigate through the midline of the ventral diencephalon into the dorsal part of the midbrain, the optic tectum. The ganglion cells extend axons shortly after their final mitosis, already while they are migrating toward the vitreal surface (Bodick and Levinthal, 1980). The projections proceed toward the inner surface of the retina and subsequently along the inner limiting membrane toward the optic nerve head. In zebrafish, the first ganglion cell axons exit the eye between 34 and 36 hpf and navigate along the optic stalk and through the ventral region of the brain toward the midline (Burrill and Easter, 1995; Macdonald and Wilson, 1997). At about 2 dpf, the zebrafish optic nerve contains ca. 1800 axons at the exit point from the retina (Bodick and Levinthal, 1980). Cross sections near the nerve head reveal a crescent-shaped optic nerve. Axons of centrally located ganglion cells occupy the outside (dorsal) surface of the crescent whereas the axons of more peripheral (younger) cells localize to the inside (ventral) surface of the optic nerve. With the exception of

the axonal trajectories of cells separated by the choroid fissure, axons of neighboring ganglion cells travel together in the optic nerve (Bodick and Levinthal, 1980). In addition to ganglion cell axons, the optic nerve contains retinopetal projections, which appear considerably later, after 5 dpf, and originate in the nucleus olfactoretinalis of the rostral telencephalon (Burrill and Easter, 1994). After crossing the midline, the axonal projections of the ganglion cells split into the dorsal and ventral branches of the optic tract. The ventral branch contains mostly axons of the dorsal retinal ganglion cells, and the dorsal branch mostly of the ventral cells (Baier et al., 1996). The growth cones of the retinal ganglion cells first enter the optic tectum between 46 and 48 hpf. In addition to the optic tectum, the retinal axons innervate nine other, much smaller targets in the zebrafish brain (Burrill and Easter, 1994).

Spatial relationships between individual ganglion cells in the retina are precisely reproduced by their projections in the tectum. The exactitude of this pattern has long fascinated biologists and has been a subject of intensive research in many vertebrate species (Drescher et al., 1997; Fraser, 1992; Sanes, 1993). The spatial coordinates of the retina and the tectum are reversed. The ventral-nasal ganglion cells of the zebrafish retina project to the dorsal-posterior optic tectum whereas the dorsal-temporal cells innervate the ventral-anterior tectum (Karlstrom et al., 1996; Stuermer, 1988; Trowe et al., 1996). By 72 hpf, axons from all quadrants of the retina are in contact with their target territories in the optic tectum.

In summary, development of the zebrafish retina proceeds at a rapid pace. By the end of day 3, all major retinal cell classes have been generated and are organized in distinct layers (Fig. 2B), the photoreceptor cells have developed outer segments, and the ganglion cell axons have innervated their target, the optic tectum. It is also about this time that the zebrafish visual system becomes functional (Clark, 1981; Easter and Nicola, 1996). The brevity of eye morphogenesis and retinal neurogenesis is a major advantage of the zebrafish eye as a model system.

D. Non-Neuronal Tissues

In many vertebrates, the retina is intimately associated with some form of the vascular system (Wise et al., 1971). The mature zebrafish retina features two vessel systems: the choroidal and retinal vasculatures. The first of these tightly surrounds the retinal pigment epithelium, while the second differentiates on the inner surface of the retina (Alvarez et al., 2007; Kitambi et al., 2009). The development of the eye vasculature can be efficiently visualized using transgenic lines. Carriers of the fli-GFP and flk-GFP transgenes are suitable for this purpose (Choi et al., 2007; Lawson and Weinstein, 2002). In these strains, GFP-positive cells first appear in the retinal choroid fissure and the retina toward the end of the first 24 h of embryogenesis (Kitambi et al., 2009). By 48 hpf, a vascular bed forms on the medial surface of the lens (Alvarez et al., 2007; Kitambi et al., 2009). Initially, retinal blood vessels appear to adhere tightly to the lens. As the organism matures, however, vasculature appears to progressively lose contact with the lens and starts to adhere to the vitreal surface of the eye (Alvarez et al., 2007). In contrast to many mammals, including primates, blood vessels do not penetrate the neural retina in zebrafish (Alvarez et al., 2007). In addition to the vasculature, several other non-neuronal ocular tissues, such as the cornea, the iris, the ciliary body, and the

lens, have been characterized in the zebrafish in detail (Dahm *et al.*, 2007; Gray *et al.*, 2009; Soules and Link, 2005; Zhang *et al.*, 2009; Zhao *et al.*, 2006).

III. Analysis of Wild-Type and Mutant Visual System

A major goal of eye research in zebrafish is to characterize phenotypes obtained in the course of new generations of forward and reverse genetic studies as well as small-molecule screens. Diverse research approaches are available to study the zebrafish retina. This chapter provides an overview of the available methods. While some techniques are described in detail, the majority are discussed only briefly because of space constraints, and references to sources of more comprehensive protocols are provided. Where applicable, other chapters of this volume are referenced as the source of more complete information. Table I lists some of the most important techniques currently available for the analysis of the zebrafish retina.

After 30 hpf, the observations of retinal development in the zebrafish embryo are hampered by the differentiation of pigment granules in the RPE. In immunohistochemical experiments, for example, the staining pattern is not accessible to visual inspection in whole embryos unless they are sectioned or their pigmentation is inhibited. To inhibit pigmentation, embryos are raised in media containing 1-phenyl-2-thiourea (PTU). Concentrations ranging from 75 to 200 µM are recommended (Karlsson *et al.*, 2001; Westerfield, 2000). Starting between 2 and 3 dpf, embryogenesis is somewhat delayed in PTU-treated embryos, hatching is inhibited, and pectoral fins are abnormal (Karlsson *et al.*, 2001). Appropriate controls have to be included to account for these deviations from normal embryogenesis. An additional disadvantage of using PTU is that it does not inhibit the differentiation of iridophores, which are present on the surface of the eye by 42 hpf, and by 4 dpf are dense enough to impair visualization of retinal cells with fluorescent probes. An alternative to using PTU is to conduct experiments on pigmentation-deficient animals. *albino; roy* double mutant line is the most useful for this purpose as it lacks both RPE pigmentation and iridophores (Ren *et al.*, 2002). As crossing a mutation of interest into a pigmentation-deficient background takes two generations (or about 6 months), this approach is, however, time consuming.

A. Histological Analysis

Following morphological description, the first and the simplest step in the analysis of a phenotype is histological examination. It allows one to evaluate the major cell classes in the retina at the resolution that whole-mount preparations do not offer. Given the exquisitely precise organization of retinal neurons, histological analysis on tissue sections is frequently very informative. Plastic sections in particular offer very good tissue preservation and thus reveal fine detail. Prior to sectioning, tissue samples are usually embedded in either epoxy (epon, araldite) or in methacrylate (JB4) resins (Polysciences Inc.). Epoxy resins can be used for both light and electron microscopy. Several fixation methods suitable for plastic sections are routinely used (Li *et al.*, 2000b; Malicki *et al.*, 1996). For light microscopy, plastic sections are frequently prepared at 1–8 µm thickness

Table I
Techniques Available to Study the Zebrafish Retina and Their Sources/Examples of Use

Protocol	Goal	Sources/Examples of Use
Histological analysis		
Electron microscopy	Evaluation of phenotype on a subcellular level	Allwardt *et al.* (2001), Doerre and Malicki (2002), Kimmel *et al.* (1981)
Light microscopy	Evaluation of phenotype on a cellular level	Malicki *et al.* (1996), Schmitt and Dowling (1994)
Molecular marker analysis		
Antibody staining (whole mount)	Determination of expression pattern on protein level	Schmitt and Dowling (1996)
Antibody staining (sections)	Determination of expression pattern on protein level	Pujic and Malicki (2001), Wei and Malicki (2002)
In situ hybridization—double labeling	Parallel determination of two expression patterns on transcript level	Hauptmann and Gerster (1994), Jowett (2001), Jowett and Lettice (1994), Strahle *et al.* (1994)
In situ hybridization—frozen sections	Determination of expression pattern on transcript level	Barthel and Raymond (1993), Hisatomi *et al.* (1996)
In situ hybridization—whole mount	Determination of expression pattern on transcript level	Oxtoby and Jowett (1993), Thisse *et al.* (2004)
Gene function analysis		
Implantation	Test of function for a factor (most often diffusible) via the implantation of a bead saturated with this substance	Hyatt *et al.* (1996), Martinez-Morales *et al.* (2005)
Morpholino knockdown	Test of gene function based on antisense inhibition of its activity	Eisen and Smith (2008), Nasevicius and Ekker (2000), Tsujikawa and Malicki (2004a)
Overexpression (DNA injections)	Test of gene function based on enhancement of its activity through DNA injections	Koster and Fraser (2001), Mumm *et al.* (2006)
Overexpression (light-mediated RNA/DNA uncaging)	Identification of gene function through enhancement of its activity in selected tissues at specific developmental stages	Ando *et al.* (2001), Ando and Okamoto (2003)
Overexpression (RNA injections)	Test of gene function based on enhancement of its activity through RNA injections	Macdonald *et al.* (1995), reviewed in Malicki *et al.* (2002)
Overexpression (UAS–GAL4 system)	Test of gene function in selected tissues using stable transgenic lines	Del Bene *et al.* (2008), Scheer and Campos-Ortega (1999)
TILLING (targeting induced local lesions in genomes)	Identification of chemically induced mutant alleles in a specific genetic locus	Colbert *et al.* (2001), Wienholds *et al.* (2002)
Zinc finger nucleases	Identification of mutant alleles in a specific locus	Doyon *et al.* (2008), Meng *et al.* (2008)

(Continued)

Table I
(Continued)

Protocol	Goal	Sources/Examples of Use
Embryological techniques		
Cell labeling (caged fluorophore)	Fate determination for a specific group of cells	Take-uchi et al. (2003)
Cell labeling (iontophoretic)	Determination of morphogenetic movements or cell lineage relationships	Li et al. (2000b), Varga et al. (1999), Woo and Fraser (1995)
Cell labeling (lipophilic tracers)	Analysis of ganglion cell development (for example, retino tectal projection)	Baier et al. (1996), Malicki and Driever (1999) Mangrum et al. (2002)
Cell labeling (fluorescent protein transgenes)	Determination of cell fate and fine differentiation features in living animals	Hatta et al. (2006), Mumm et al. (2006), Neumann and Nuesslein-Volhard (2000)
Mitotic activity detection (BrdU)	Identification of mitotically active cell populations; birth dating	Hu and Easter (1999), Larison and Bremiller (1990), Del Bene et al. (2008)
Mitotic activity detection (tritiated thymidine)	Identification of mitotically active cell populations; birth dating	Nawrocki (1985)
Tissue ablation	Functional test for a field of cells via their removal by surgical means	Masai et al. (2000)
Transplantation (whole eye)	Test whether a defect (in axonal navigation, for example) originates within or outside the eye	Fricke et al. (2001)
Transplantation (fragment of tissue)	Functional test for a field of cells via their transplantation to an ectopic position by surgical means	Masai et al. (2000)
Transplantation (blastomere)	Test of cell autonomy of a mutant phenotype by generating a genetically mosaic embryo	Ho and Kane (1990), Jensen et al. (2001), Malicki and Driever (1999), Jing and Malicki (2009)
Behavioral tests		
Optokinetic response	Test of vision based on eye movements; allows for evaluation of visual acuity	Brockerhoff et al. (1995), Clark (1981), Neuhauss et al. (1999)
Optomotor response	Test of vision based on swimming behavior	Clark (1981), Neuhauss et al. (1999)
Startle response	Simple test of vision based on swimming behavior	Easter and Nicola (1996)
Electrophysiological tests		
ERG	Test of retinal function based on the detection of electrical activity of retinal neurons and glia	Avanesov et al. (2005), Brockerhoff et al. (1995)

Table I
(Continued)

Protocol	Goal	Sources/Examples of Use
Biochemical approaches		
Co-immunoprecipitation from embryo extracts	Identification of direct and indirect protein binding partners	Insinna et al. (2008), Krock and Perkins (2008)
Tandem affinity purification from embryo extracts	Identification of direct and indirect protein binding partners	Omori et al. (2008)
Chemical screens		
Screens of small-molecule libraries	Identification of chemicals that affect a developmental process	Kitambi et al. (2008)
Genetic screens		
Behavioral	Detection of mutant phenotypes by behavioral tests	Muto et al. (2005), Neuhauss et al. (1999)
Histological	Detection of mutant phenotypes via histological analysis of sections	Mohideen et al. (2003)
Marker/tracer labeling	Detection of mutant phenotypes via staining with antibodies, RNA probes, or lipophilic tracers	Baier et al. (1996), Guo et al. (1999)
Morphological	Detection of mutant phenotypes by morphological criteria	Malicki et al. (1996)
Transgene guided	Detection of mutant phenotypes in transgenic lines expressing fluorescent proteins in specific cell populations	Xiao et al. (2005)

In this table, we primarily cite experiments performed on the retina. Only where references to work on the eye are not available, we refer to studies of other organs. Most forward genetic approaches such as mutagenesis, mapping, and positional cloning methods do not contain visual system-specific features and thus are not listed in this table. These approaches are discussed in depth in other sections of this volume. Entries are listed alphabetically within each section of the table.

and stained with an aqueous solution of 1% methylene blue and 1% azure II (Humphrey and Pittman, 1974; Malicki *et al.*, 1996; Schmitt and Dowling, 1999).

Following transmitted light microscopy, histological analysis of mutant phenotypes can be performed at a higher resolution using electron microscopy. This allows one to inspect morphological details of subcellular structures, such as the photoreceptor outer segments, cell junctions, cilia, synaptic ribbons, mitochondria, and many other organelles (Allwardt *et al.*, 2001; Doerre and Malicki, 2002; Schmitt and Dowling, 1999; Tsujikawa and Malicki, 2004b). These subcellular features frequently offer insight into the nature of the process being studied (Avanesov *et al.*, 2005; Emran *et al.*, 2010). Electron microscopy can be used in combination with diaminobenzidine (DAB) labeling of specific cell populations. Oxidation of DAB results in the formation of polymers which are chelated with osmium tetroxide and subsequently observed in the electron microscope (Hanker, 1979). Prior to microscopic analysis, cells can be selectively DAB-labeled using several approaches: photoconversion (Burrill and Easter, 1995), antibody staining combined with peroxidase detection (Metcalfe *et al.*, 1990), or retrograde labeling with horseradish peroxidase (HRP) (Metcalfe, 1985).

B. The Use of Molecular Markers

A variety of molecular markers are used to study the zebrafish retina before, during, and after neurogenesis. Endogenous transcripts and proteins are among the most frequently used markers, although smaller molecules, such as neurotransmitters, and neuropeptides can also be used (Avanesov *et al.*, 2005; Cameron and Carney, 2000). During early embryogenesis, the analysis of marker distribution allows one to determine whether the eye field is specified correctly. Several RNA probes are available to visualize the optic lobe during embryogenesis (Table II). Some of them label all cells of the optic lobe uniformly, while others can be used to monitor the optic stalk area (Table II). After the completion of neurogenesis, cell class-specific markers are used to determine whether particular cell populations are specified and occupy correct positions. Some of these markers are listed in Table II. Many transcript and protein detection methods have been described. Detailed protocols for most of these are available and we reference many of them in Table II. Below we discuss in detail the main types of molecular probes used to study the zebrafish visual system.

1. Antibodies

Antibody staining experiments can be performed in several ways. Staining of whole embryos is the least laborious. One has to keep in mind, however, that many antibodies produce high background in whole-mount experiments, and the eye pigmentation needs to be eliminated after 30 hpf as described above. At later stages of development, tissue penetration may become an additional problem. This can be circumvented by permeabilizing larvae via increasing detergent concentration above the standard level of 0.5% (2.5% Triton in both blocking and staining solution works well for anti Pax-2 antibody; see Riley *et al.*, 1999) or by enzymatic digestion of embryos (for example

Table II
Selected Molecular Markers Available to Study the Zebrafish Retina

Name	Type	Expression pattern	References[a]/Sources
Optic lobe, optic stalk markers			
pax2a (pax 2)	RNA probe & Ab (poly)	Nasal retina, optic stalk ($\leq$24 hpf); ON (2 dpf)	Kikuchi et al. (1997), Macdonald et al. (1997), Covance PRB-276P
rx1 (zrx1)	RNA probe	Anterior neural keel, optic primordia ($\leq$11 hpf)	Chuang et al. (1999), Pujic and Malicki (2001)
rx2 (zrx2)	RNA probe	Anterior neural keel, optic primordia ($\leq$11 hpf)	Chuang et al. (1999), Pujic and Malicki (2001)
rx3 (zrx3)	RNA probe	Anterior neural plate ($\leq$9 hpf); optic primordia ($\leq$12 hpf)	Chuang et al. (1999), Pujic and Malicki (2001)
six3a (six3)	RNA probe	Neural keel, optic primordia ($\leq$11 hpf)	Pujic and Malicki (2001), Seo et al. (1998)
six3b (six6)	RNA probe	Anterior neural keel, optic primordia ($\leq$11 hpf)	Pujic and Malicki (2001), Seo et al. (1998)
vax2	RNA probe	Optic stalk ($\leq$15 hpf); optic stalk, ventral retina ($\leq$18 hpf)	Take-uchi et al. (2003)
Ganglion cell markers			
alcam[b] (neurolin)	RNA probe, Ab (mono & poly)	Ganglion cells (28 hpf, RNA; $\leq$32 hpf protein)	Fashena and Westerfield (1999), Laessing et al. (1994), Laessing and Stuermer (1996), Zn-5/Zn-8 DSHB and ZIRC
cxcr4b	RNA probe	Ganglion cells (30 hpf)	Pujic et al. (2006)
gc34	RNA probe	Ganglion cells ($\leq$36 hpf)	Pujic et al. (2006)
L3	RNA probe	Ganglion cells (30 hpf)	Brennan et al. (1997)
Tg (ath5:GFP)	Transgene	Ganglion cells (25 hpf)	Masai et al. (2003, 2005)
Tg (brn3c: gap43-GFP)	Transgene	Ganglion cells (42 hpf)	Xiao et al. (2005)
Amacrine cell markers			
Ap2α	RNA probe	Amacrine cells (1.5–2 dpf)	Pujic et al. (2006)
Ap2β	RNA prove	Amacrine cells ($\leq$36 hpf)	Pujic et al. (2006)
Choline acetyltransferase	Ab (poly)	Subset in INL and GCL, IPL ($\leq$5 dpf)	Avanesov et al. (2005), Millipore, cat# AB144P
GABA	Ab (poly)	Subset in INL and GCL, IPL (2.5 dpf); ON (2 dpf)	Sandell et al. (1994), Millipore, cat# AB131; Sigma, cat# A2052
GAD67	Ab (poly)	Subset in INL and few in GCL, IPL ($\leq$7 dpf)	Connaughton et al. (1999), Kay et al. (2001), Millipore, cat# AB9706
Hu C/D	Ab (mono)	INL and GCL ($\leq$3 dpf)	Kay et al. (2001), Link et al. (2000), Invitrogen, cat# A21271
Neuropeptide Y	Ab (poly)	Subset in INL, IPL ($\leq$4 dpf)	Avanesov et al. (2005), ImmunoStar, cat# 22940
Parvalbumin	Ab (mono)	Subset in INL and GCL, IPL ($\leq$3 dpf)	Malicki et al. (2003), Millipore, cat# MAB1572
pax6a (pax6.1)	RNA probe Ab (poly)	Neuroepithelium (12–34 hpf); INL and GCL (2 dpf); INL (5 dpf)	Hitchcock et al. (1996), Macdonald and Wilson (1997)

(Continued)

Table II
(Continued)

Name	Type	Expression pattern	References[a]/Sources
Serotonin	Ab (poly)	Subset in INL (≤5 dpf)	Avanesov et al. (2005), Sigma, cat# S5545
Somatostatin	Ab (poly)	Subset in INL (≤5 dpf)	Malicki Lab, unpublished data; ImmunoStar; cat# 20067
Substance P	Ab (mono)	Subset in INL (≤5 dpf)	Malicki Lab, unpublished data; AbCam, cat# AB6338
Tyrosine hydroxylase	Ab (mono)	Subset in INL (3–3.5 dpf)	Biehlmaier et al. (2003), Pujic and Malicki (2001), ImmunoStar; cat# 22941; Millipore, cat# MAB318
Bipolar cell markers			
vsx1	RNA probe	Neuroepithelium (31 hpf); outer INL (50 hpf)	Passini et al. (1997)
vsx2	RNA probe	Neuroepithelium (24 hpf); primarily or exclusively in the bipolar cells (50 hpf)	Passini et al. (1997)
Protein kinase C-β1	Ab (poly)	IPL, OPL (2.5 dpf); bipolar cell somata (4 dpf)	Biehlmaier et al. (2003), Kay et al. (2001), Santa Cruz, cat# sc-209
Tg(nyx::Gal4VP16; UAS::MYFP)	Transgene	ON bipolar cells (2.5 dpf)	Schroeter et al. (2006)
Horizontal cell markers			
Cx 52.6	Ab (poly)	Horizontal cells (≤7 dpf?)	Shields et al. (2007)
Cx 55.5	Ab (poly)	Horizontal cells (≤7 dpf?)	Shields et al. (2007)
Horizin	RNA probe	Horizontal cells, weak staining in GCL and inner INL (≤60 hpf)	Pujic et al. (2006)
Photoreceptor markers			
Blue opsin	Ab (poly)	Blue cones (≤3 dpf)	Doerre and Malicki (2001), Vihtelic et al. (1999)
Blue opsin[c]	RNA probe	Blue cones (52 hpf)	Chinen et al. (2003), Raymond et al. (1995), Vihtelic et al. (1999)
Green opsin	Ab (poly)	Green cones (≤3 dpf)	Doerre and Malicki (2001), Vihtelic et al. (1999)
Green opsins (four genes)	RNA probes	Green cones (40–45 hpf)	Chinen et al. (2003), Takechi and Kawamura (2005), Vihtelic et al. (1999)
NDRG1	RNA probe	Photoreceptors (36–48 hpf)	Pujic et al. (2006)
Red opsin	Ab (poly)	Red cones (≤3 dpf)	Doerre and Malicki (2001), Vihtelic et al. (1999)
Red opsins (two genes)	RNA probes	Red cones (40–45 hpf)	Chinen et al. (2003), Raymond et al. (1995), Takechi and Kawamura (2005)
Rod opsin	Ab (poly)	Rods (≤3 dpf)	Doerre and Malicki (2001), Vihtelic et al. (1999)
Rod opsin[c]	RNA probe	Rods (50 hpf)	Chinen et al. (2003), Raymond et al. (1995)
Tg (opn1sw1: EGFP)	Transgene	UV cones (≤56 hpf)	Takechi et al. (2003)
Tg (xops: GFP)	Transgene	Rods	Fadool (2003)
UV opsin	RNA probe	UV cones (56 hpf)	Hisatomi et al. (1996), Takechi et al. (2003)

Table II
(Continued)

Name	Type	Expression pattern	References[a]/Sources
UV opsin	Ab (poly)	UV cones (≤3 dpf)	Doerre and Malicki (2001), Vihtelic et al. (1999)
Zpr1 (FRet 43)	Ab (mono)	Double cones in larvae (48 hpf); double cones &bipolar cell subpopulation in the adult	Larison and Bremiller (1990), ZIRC
Zpr3 (FRet 11)	Ab (mono)	Rods (50 hpf)	Schmitt and Dowling (1996), ZIRC
Zs-4	Ab (mono)	Rod inner segments (adult), onset unknown	Vihtelic and Hyde (2000), ZIRC
Müller glia markers			
cahz (carbonic anhydrase)	RNA probe Ab (poly)	Müller glia (≤4 dpf)	Peterson et al. (1997, 2001)
GFAP	Ab (poly)	Müller glia (5 dpf)	Malicki Lab, unpublished data; *DAKO, cat# Z0334*
Glutamine synthetase	Ab (poly)	Müller glia (60 hpf)	Peterson et al. (2001)
Tg (gfap: GFP)	Transgene	Müller glia (48 hpf)	Bernardos and Raymond (2006)
Plexiform layer markers			
Phalloidin	Fungal toxin	IPL, OPL, ON (≤60 hpf)	Malicki et al. (2003), *Invitrogen, cat# A-12379*
Snap-25	Ab (poly)	OPL, IPL (≤2.5 dpf)	Biehlmaier et al. (2003), *StressGen, cat# VAP-SV002*
SV2	Ab (mono)	IPL, OPL (≤2.5 dpf)	Biehlmaier et al. (2003), *DSHB*
Syntaxin-3	Ab (poly)	OPL (2.5 dpf); faint IPL (5 dpf)	Biehlmaier et al. (2003), *Alamone Labs, cat# ANR-005*

Approximate time of the expression onset is indicated in parenthesis. Sources of commercially available reagents are listed, including catalog numbers where appropriate. Names of markers are listed alphabetically within each section.

DSHB = Developmental Studies Hybridoma Bank (http://dshb.biology.uiowa.edu); ZIRC = Zebrafish International Resource Center (http://zfin.org/zirc/home/guide.php). dpf = days post fertilization; hpf = hours post fertilization; GCL = ganglion cell layer; INL = inner nuclear layer; IPL = inner plexiform layer; OPL = outer plexiform layer; ON = optic nerve.

[a] When references to work performed on zebrafish are not available, experiments on related fish species are cited.

[b] Zn-5 and Zn-8 antibodies both recognize neurolin (Kawahara et al., 2002).

[c] Transcript expression onset was estimated by using goldfish probes (Raymond et al., 1995).

collagenase treatment; see Doerre and Malicki, 2002). When background or tissue penetration is a problem, useful alternatives to using whole embryos is staining of either frozen or paraffin sections. Confocal microscopy of retinal sections reduces the background even further, while also enhancing the details of cell architecture.

For cryosectioning, embryos should be fixed as appropriate for a particular antigen and infiltrated in 30% sucrose/phosphate buffered saline (PBS) solution for cryoprotection. While for many antigens simple overnight fixation in 4% paraformaldehyde (PFA) at 4°C is sufficient, some others require special treatments. For example, anti-gamma aminobutyric acid (GABA) staining of amacrine cells requires fixation in both glutaraldehyde and paraformaldehyde (2% each; see Avanesov *et al.*, 2005; Sandell *et al.*, 1994) (Fig. 3F). Glyoxal-based fixatives (such as Prefer fix supplied by Anatech)

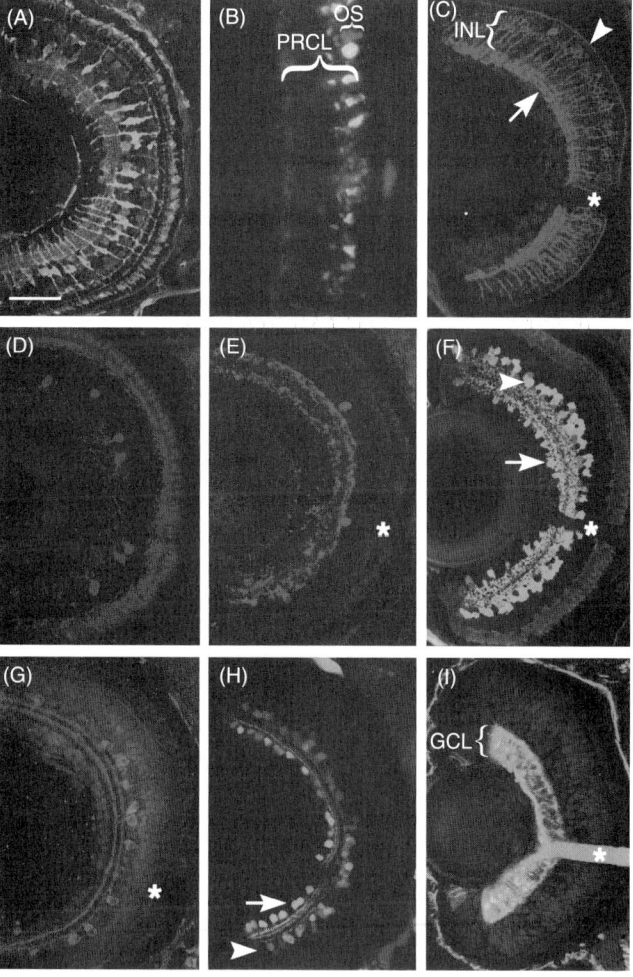

Fig. 3   *(Continued)*

may also be useful when testing new antibodies (Dapson, 2007; Pathak *et al.*, 2007). Fixed specimen can be oriented as desired using molds prepared from Eppendorf tubes that are cut transversely into ca. 3–4 mm wide rings. These are then placed flat on a glass slide and filled with embedding medium (Richard-Allan Scientific Inc.). Embryos are placed in the medium, oriented with a needle, and transferred into a cryostat chamber that is cooled to −20°C. Once the medium solidifies, plastic rings are removed with a razor blade.

Antibody staining can be efficiently performed on 15–30 μm frozen sections, and analyzed by confocal microscopy. For conventional epifluorescence microscopy, thinner sections may be desired. Upon the application of modified infiltration and embedding protocols, 3 μm sections of the zebrafish embryos can be prepared and analyzed using a conventional microscope equipped with UV illumination (Barthel and Raymond, 1990). Some antigens require the application of additional steps during staining protocols, such as antigen retrieval. Sections are immersed in near-boiling solution of 10 mM sodium citrate for 10 min prior to the application of blocking solution. This method significantly improves the labeling of amacrine cell populations by anti-serotonin or anti-choline acetyltransferase antibodies (Fig. 3G and H) (Avanesov *et al.*, 2005). Immersion in cold acetone is another treatment that improves staining with some immunoreagents, such as certain anti-gamma-tubulin antibodies (Pujic and Malicki, 2001).

Alternatively, antibody staining can be performed on plastic sections. Anti-GABA antibodies, for example, work very well with this method. Both epoxy (Epon-812,

---

Fig. 3 Transverse sections through the center of the zebrafish eye reveal several major retinal cell classes and their subpopulations. (A) Anti-rod opsin antibody detects rod photoreceptor outer segments (red), which are fairly uniformly distributed throughout the outer perimeter of the retina by 5dpf. On the same section, an antibody to carbonic anhydrase labels cell bodies of Müller glia in the INL as well as their radially oriented processes. (B) A higher magnification of the photoreceptor cell layer shows the distribution of rod opsin (red signal) and UV opsin (green signal) in the outer segments (OSs) of rods and short single cones, respectively. (C) A subpopulation of bipolar cells is detected using antibody directed to protein kinase C-β (PKC). While cell bodies of PKC-positive bipolar neurons are situated in the central region of the INL, their processes travel radially into the inner (arrow) and outer (arrowhead) plexiform layers, where they make synaptic connections. (D) Tyrosine hydroxylase-positive interplexiform cells are relatively sparse in the larval retina. (E) Similarly, the distribution of neuropeptide Y is limited to only a few cells per section. (F) The distribution of GABA, a major inhibitory neurotransmitter. GABA is largely found in amacrine neurons in the INL (arrowhead), although some GABA-positive cells are also found in the GCL (arrow). (G) Choline acetyltransferase, an enzyme of acetylcholine biosynthetic pathway, is restricted to a relatively small amacrine cell subpopulation. (H) Antibodies directed to a calcium-binding protein, parvalbumin, recognize another fairly large subpopulation of amacrine cells in the INL (green, arrowhead). Some parvalbumin-positive cells localize also to the GCL and most likely represent displaced amacrine neurons (arrow). By contrast, serotonin-positive neurons (red) are exclusively found in the INL. (I) Ganglion cells stain with the Zn-8 antibody directed to neurolin, a cell surface antigen (Fashena and Westerfield, 1999). In addition to neuronal somata, strong Zn-8 staining exists in the optic nerve (asterisk). In all panels lens is left, dorsal is up. A–H show the retina at 5dpf, while I shows a 3dpf retina. Asterisks indicate the optic nerve. Scale bar equals 50 μm in A and C–I and 10 μm in B. dpf, days post fertilization; GCL, ganglion cell layer; INL, inner nuclear layer; OS, outer segments; PRCL, photoreceptor cell layer. Panels D, G, and H are reprinted from Pujic and Malicki (2004) with permission from Elsevier. (See Plate no. 8 in the Color Plate Section.)

Electron Microscopy Sciences Inc.) and methacrylate (JB-4, Polysciences Inc.) resins can be used as the embedding medium. This improves the quality of staining, as plastic sections preserve tissue morphology better, compared to frozen ones. In the GABA staining protocol, primary antibody can be detected using avidin–HRP conjugate (Vector Laboratories Inc.) or a fluorophore-conjugated secondary antibody (Fig. 2F and Malicki and Driever, 1999; Sandell *et al.*, 1994). An extensive collection of antibodies that can be used to visualize features of the retina in the adult zebrafish has been also characterized (Yazulla and Studholme, 2001).

2. mRNA Probes

In situ hybridization with most RNA probes works very well on whole embryos (Oxtoby and Jowett, 1993). Following hybridization, embryos are gradually dehydrated in a series of ethanol solutions of increasing concentration, and embedded in plastic as described above (Pujic and Malicki, 2001). An additional fixation step prior to dehydration reduces the leaching of *in situ* signal (Westerfield, 2000). Expression patterns are then analyzed on 1–5 μm thick sections. Several *in situ* protocols are available to monitor the expression of two genes simultaneously (Jowett, 2001, and references in Table II; Jowett and Lettice, 1994). In the experiment shown in Fig. 4B, expression patterns of two opsins are detected simultaneously using two different chromogenic substrates of alkaline phosphatase (AP) (Hauptmann and Gerster, 1994). *In situ* hybridization can also be combined with antibody staining (Novak and Ribera, 2003; Prince *et al.*, 1998). In embryos older than 5 dpf, *in situ* reagents sometimes do not penetrate to the center of the retina. In such cases, hybridization procedures can be performed more successfully on sections (Hisatomi *et al.*, 1996). Given the small size of zebrafish embryos, *in situ* hybridization experiments can be performed in a high-throughput fashion using hundreds or even thousands of probes to screen for genes expressed in specific organs, tissues, or even specific cell types (Thisse *et al.*, 2004). In recent years, *in situ* hybridization could also be performed using robotic devices that carry out most of the tedious steps, including hybridizations and washes (Intavis Bioanalytical Instruments AG). This approach was also applied to the retina and led to the identification of numerous transcripts expressed in subpopulations of retinal cells (Pujic *et al.*, 2006). Some of these transcripts can be used as markers of specific retinal cell classes (Table II).

3. Lipophilic Tracers

Details of cell morphology can also be studied using lipophilic carbocyanine dyes, DiI, DiO, and others, which label cell membranes (Honig and Hume, 1986; 1989). In the retina, these are especially useful in the analysis of ganglion cells. Carbocyanine dyes can be used as anterograde as well as retrograde tracers. When applied to the retina, DiI and DiO allow one to trace the retinotectal projections (Baier *et al.*, 1996). When applied to the optic tectum or the optic tract, they can be used to determine the position of ganglion cell perikarya, and even to study the stratification and branching

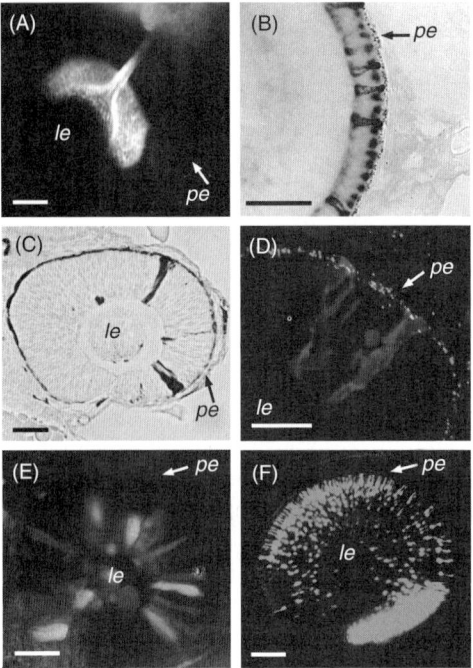

Fig. 4 Selected techniques available to study neurogenesis in the zebrafish retina. (A) DiI incorporation into the optic tectum retrogradely labels the optic nerve and ganglion cell somata. (B) A transverse plastic section through the zebrafish retina at 3 dpf. *In situ* mRNA hybridization using two probes, each targeted to a different opsin transcript and detected using a different enzymatic reaction, visualizes two types of photoreceptor cells. (C) A plastic section through a genetically mosaic retina at ca. 30 hpf. Biotinylated dextran-labeled donor-derived cells incorporate into retinal neuroepithelial sheet of a host embryo and can be detected using HRP staining (brown precipitate). (D) A transverse cryosection through a genetically mosaic zebrafish eye at 36 hpf. In this case, donor-derived clones of neuroepithelial cells are detected with fluorophore-conjugated avidin (red). The apical surface of the neuroepithelial sheet is visualized with anti-γ-tubulin antibody, which stains centrosomes (green). (E) GPF expression in the eye of a zebrafish embryo following injection of a DNA construct containing the GFP gene under the control of a heat-shock promoter. The transgene is expressed in only a small subpopulation of cells. (F) A confocal z-series through the eye of a living transgenic zebrafish, carrying a GFP transgene under the control of a rod opsin promoter (Fadool, 2003). Bright expression is present in rod photoreceptor cells (ca. 3 dpf). Scale bar, 50 μm. pe, pigmented epithelium; le, lens. Panel E reprinted from Malicki *et al.* (2002) with permission from Elsevier.

of ganglion cell dendrites (Burrill and Easter, 1995; Malicki and Driever, 1999; Mangrum *et al.*, 2002). Since DiI and DiO have different emission spectra, they can be used simultaneously to label different cell populations (Baier *et al.*, 1996).

4. Fluorescent Proteins

Fluorescent proteins (hereafter FPs), frequently fused to other polypeptides, offer a very rich source of markers to visualize tissues, cells, and even subcellular structures.

These can be expressed in embryos either transiently or from stably integrated transgenes. Numerous derivatives of two FPs—green fluorescent protein (GFP, from jellyfish, *Aequorea victoria*) and red fluorescent protein (RFP, from coral species)—are currently available (reviewed in Shaner *et al.*, 2007) and differ in brightness as well as emission spectra. Many of them have been applied in zebrafish. The uses of FPs can be grouped in at least three categories:

1. **Visualization of gene activity**. The purpose of these experiments is to determine where and when a gene of interest is transcribed. Although the same goal can be accomplished using *in situ* hybridization, the use of FP fusions may result in higher sensitivity of detection (see for example a sonic hedgehog study by Neumann and Nuesslein-Volhard, 2000), and, importantly, allow one to create time-lapse images tracking spatial-temporal changes in gene expression. The biggest challenge in this type of study is the need to include all regulatory elements in the transgene to faithfully recapitulate the expression of the endogenous transcript. The best way to accomplish that is to insert an FP coding sequence into the open reading frame of a gene derived from a phage or bacterial artificial chromosome (PAC or BAC). For example, to study the expression of zebrafish green opsin genes, a modified PAC clone of ca. 85 kb was used to generate transgenic lines. To visualize expression, the first exon after the initiation codon was replaced with a GFP sequence in each of these genes (Tsujimura *et al.*, 2007). The use of artificial chromosomes is frequently necessary as distant regulatory elements are likely to affect the expression of a given gene. One has to note, however, that even using an artificial chromosome does not assure that all relevant regulatory elements will be included in the transgene.

   In some experiments, when temporal characteristics of expression need to be faithfully reproduced, excessive stability of FP may pose a problem. FPs tend to be stable in the cell's cytoplasm and may persist for much longer than the transcript of the gene being studied, making it difficult to determine when the gene of interest is turned off. To circumvent this difficulty, FPs characterized by reduced stability, such as dRFP (destabilized RFP) or short half-life GFP, are available (Yeo *et al.*, 2007; Yu *et al.*, 2007). dRFP was used, for example, to study Notch pathway activity in the zebrafish retinal neuroepithelium (Del Bene *et al.*, 2008).

2. **Visualization of the subcellular localization of proteins**. In this type of experiment, it is not necessary to recapitulate the tissue distribution of the protein being studied and thus expression can be driven ubiquitously. Consequently, transient expression methods based on mRNA or DNA injection are preferred. Although they usually do not allow for the targeting of expression to particular tissues, they are much less time consuming, compared to using stable transgenic lines. The expression of FP fusions is especially valuable when antibodies are difficult to generate, as has been the case for the Elipsa protein, for example (Omori *et al.*, 2008). This procedure is not without drawbacks, however. First, adding GFP polypeptide to a protein may change its binding properties, and thus cause aberrant localization in the cell. Second, as FP fusions are frequently expressed at a higher level compared to their wild-type

counterparts, they may display nonspecific binding. Finally, fusion proteins may be toxic to cells. These problems can be largely, although not entirely, eliminated by placing FP tags in multiple locations and testing whether the resulting fusion proteins can rescue mutant/morphant phenotypes.

3. **Monitoring of fate, differentiation, and cell physiology**. In these studies, FP fusions are used solely to mark cells and/or subcellular structures. In the simplest case, this approach can be used to monitor the gross morphology of the cell and its survival. In more sophisticated variants of this technique, one monitors cell division patterns, migration trajectories, or specific aspects of cell morphology, such as the shape of dendritic processes, subcellular distribution of organelles, or intracellular transport. Zebrafish FP transgenic lines have been generated to monitor the differentiation of fine morphological features of various retinal cell classes, including bipolar interneurons (Schroeter *et al.*, 2006), horizontal interneurons (Shields *et al.*, 2007), amacrine interneurons (Godinho *et al.*, 2005; Kay *et al.*, 2004), ganglion cells (Xiao *et al.*, 2005), and Müller glia (Bernardos and Raymond, 2006) (Table II). These transgenic lines allow one to continuously observe fine features of cells, and even follow the entire trajectory of the retinotectal projection, or the phylopodia of differentiating bipolar cell axon terminals. In most studies conducted so far, FP fusions were expressed from stably integrated transgenes, although in some cases the GAL4–VP16-based system (Koster and Fraser, 2001, see below) is used to drive transient expression in retinal interneurons (Mumm *et al.*, 2006; Shields *et al.*, 2007). While generating stable transgenic lines, it is necessary to compare expression patterns from at least two different transgenic lines since the integration of same construct can produce very different expression patterns in different lines, due to position-specific effects. For example, depending on the integration site, a hexamer of the DF4 regulatory element of the Pax6 gene can drive expression either throughout the retina or in subsets of amacrine cells (Godinho *et al.*, 2005; Kay *et al.*, 2004).

In some experimental contexts, FPs can also be used to monitor the behavior of cellular organelles. This is accomplished by generating FPs fused to subcellular localization signals or to entire proteins that display a desirable subcellular localization. The H2A-GFP transgene, for example, allows one not only to visualize cell nuclei but also to distinguish when cells undergo mitosis, and even to determine the orientation of mitotic spindles in the retinal neuroepithelium (Cui *et al.*, 2007; Pauls *et al.*, 2001). Similarly, GFP-centrin can be used to monitor the position of the centrosome in differentiating ganglion cells (Zolessi *et al.*, 2006), and GFP fused to a mitochondrial localization sequence can be applied to observe the distribution of mitochondria (Kim *et al.*, 2008). GFP fused to the 44 C-terminal amino acids of rod opsin is targeted to the photoreceptor outer segment and can be used as a specific marker of this structure (Perkins *et al.*, 2002). FPs can also be applied to mark specific cell membrane domains: PAR-3/EGFP fusion, for example, labels the apical surface of retinal neuroepithelial cells (Zolessi *et al.*, 2006).

Photoconvertible FPs are yet another class of markers that can be used to visualize cell morphology. Kaede and Dronpa have been used most frequently in the zebrafish so

far (Aramaki and Hatta, 2006; Hatta *et al.*, 2006; Sato *et al.*, 2006). Kaede is irreversibly converted from green to red fluorescence using UV irradiation, whereas Dronpa green fluorescence can be reversibly activated and deactivated multiple times by irradiating it with blue and UV light, respectively. The advantage of these FPs is that they can be used to reveal morphology of single neurons by selective photoconversion in the cell soma (anterograde labeling) or in cell processes (retrograde labeling). This is particularly useful when appropriate regulatory elements are not available to drive FP expression in specific cell populations. Moreover, one can potentially use photoactivatable FPs to trace the journey of tagged proteins within cells. Although as yet this approach has not been applied in the zebrafish retina, it is potentially useful to analyze protein trafficking in photoreceptor cells.

The number of different FPs and the variety of their applications in zebrafish have been growing at a breathtaking pace. Given the multitude of available promoter sequences, the diversity of spectral variants, and the variety of methods for protein expression in the zebrafish embryo, one is frequently confronted with the task of generating multiple combinations of regulatory elements and FP tags. This is made easier by recombination cloning approaches (Kwan *et al.*, 2007; Villefranc *et al.*, 2007; see the description of the Gateway cloning system on page 241). The use of FPs to monitor the divisions, movements, and differentiation of cells and their organelles has been one of the fastest growing approaches in the studies of zebrafish embryogenesis.

C. Analysis of Cell Movements and Lineage Relationships

The best-established and the most versatile approach to cell labeling in living zebrafish embryos is iontophoresis. This technique was applied in numerous zebrafish cell fate studies (Collazo *et al.*, 1994; Devoto *et al.*, 1996; Raible *et al.*, 1992). In the context of visual system development, iontophoretic cell labeling was used to determine the developmental origins of the optic primordium (Woo and Fraser, 1995) and later to study cell rearrangements that accompany optic cup morphogenesis (Li *et al.*, 2000b). A potentially very informative variant of cell fate analysis is to perform it in the retinae of mutant animals (Poggi *et al.*, 2005; Varga *et al.*, 1999). Iontophoretic cell labeling has been applied to study cell lineage relationships in the developing retina of *Xenopus laevis* (Holt *et al.*, 1988; Wetts and Fraser, 1988). Lineage analysis has been performed in the zebrafish retina to a very limited extent, perhaps because of the perception that it would be unlikely to add much to the results previously obtained in higher vertebrates (Holt *et al.*, 1988; Turner and Cepko, 1987; Turner *et al.*, 1990). An alternative to iontophoresis is the activation of caged fluorophores using a laser beam. Caged flourescein (Molecular Probes, Inc.) is particularly popular in this type of experiment, and was applied to study cell fate changes caused by a double knockdown of *vax1* and *vax2* gene function (Take-uchi *et al.*, 2003). One study of lineage relationships in the zebrafish eye also took advantage of a transgenic line that expresses GFP in retinal progenitor cells (Poggi *et al.*, 2005).

D. Analysis of Cell and Tissue Interactions

   Transplantation techniques are used to reveal cell or tissue interactions. The size of a transplant varies from a small group of cells, or even a single cell, to the entire organ. In the case of mutations that affect retinotectal projections, it is important to determine whether defects originate in the eye or in brain tissues. This can be accomplished by transplanting the entire optic lobe at 12 hpf, and allowing the animals to develop until desired stages (Fricke *et al.*, 2001). Smaller size fragments of tissue can be transplanted to document cell–cell signaling events within the optic cup. This approach was used to demonstrate inductive properties of the optic stalk tissue, and to test the presence of cell–cell interactions within the optic cup (Kay *et al.*, 2005; Masai *et al.*, 2000). Transplantation can also be used to study interactions between the lens and the retina. Lens transplantation is performed following a procedure similar to that developed for *Astyanax mexicanus* (Yamamoto and Jeffery, 2002) and recently applied to zebrafish (Zhang *et al.*, 2009)

   Mosaic analysis is a widely used approach that combines genetic and embryological manipulations (Ho and Kane, 1990). The goal of such experiments is to determine the site of the genetic defect responsible for a mutant phenotype. In simple terms, cell-autonomous phenotypes are caused by gene function defects within the affected cells, while cell-nonautonomous phenotypes are caused by defects in other (frequently neighboring) cells. In contrast to approaches used in *Drosophila*, genetic mosaics in zebrafish are generated via blastomere transplantation, essentially a surgical procedure performed on the early embryo (Ho and Kane, 1990; Westerfield, 2000). As this technique has been widely used in zebrafish, also in the context of eye development, we provide a more extensive description of how it is applied.

   In the first step, the donor embryos are labeled at the one- to eight-cell stage with a cell tracer. Dextrans conjugated with biotin or a fluorophore are the most commonly used tracers, and frequently a mix of both is used. The choice of the tracer depends on how it is going to be detected during later stages of the experiment, when the fate of donor-derived cells is analyzed. Within a few minutes after injection into the yolk, tracers diffuse throughout the embryo, labeling all blastomeres. Subsequently, starting at about 3 hpf, blastomeres are transplanted from tracer-labeled donor embryos to unlabeled host embryos using a glass needle. The number of transplanted blastomeres usually varies from a few to hundreds, depending on the experimental context. One donor embryo is frequently sufficient to supply blastomeres for several hosts. The transplanted blastomeres become incorporated into the host embryo and randomly contribute to various tissues, including those of experimental interest. To increase the frequency of donor-derived cells in the retina, blastomeres should be transplanted into the animal pole of a host embryo (Moens and Fritz, 1999). Cells in that region will later contribute to eye and brain structures (Woo and Fraser, 1995). Embryos that contain descendants of donor blastomeres in the eye are identified using UV illumination between 24 and 30 hpf, when the retina is only weakly pigmented and contains large radially oriented neuroepithelial cells (Fig. 4C and D). An elegant way to control cell autonomy tests is to transplant cells from two donor embryos—one wild type, one mutant—into a single host (Ho and Kane, 1990). In such a case, each of the donors has to be labeled with a different tracer.

Tracer purity and the quality of the transplantation needle are two important technical aspects of mosaic analysis. To increase the survival rate of donor embryos and transplanted cells, it is important to purify dextran by filtering it through a spin column several times (Microcon YM-3, Millipore Inc.). This procedure removes small molecular weight contaminants that are toxic for cells. The preparation of transplantation needle requires considerable manual dexterity, and is fairly time consuming. A good transplantation needle has several features: (1) a smooth opening with a diameter that is slightly larger than blastomeres at the "high" stage (Kimmel *et al.*, 1995); (2) a fairly constant width near the tip; (3) lumen free of glass debris, which frequently accumulate when the needle is beveled; and (4) a sharp glass spike at the very tip, to help in penetrating the embryo. Needle preparation requires two instruments: a beveler and a microforge, available from WPI and Narishige, respectively. Useful technical details of needle preparation and other aspects of blastomere transplantation protocol are provided in *The Zebrafish Book* (Westerfield, 2000).

Following successful transplantations, the analysis of donor-derived cells in mosaic embryos can proceed in several ways. In the simplest case, the donor-derived cells are labeled with a fluorescent tracer or a transgene and directly analyzed in whole embryos using conventional or confocal miscroscopy (Zolessi *et al.*, 2006). Such analysis is sufficient to provide information about the position and sometimes the morphology of donor-derived cells. When more detailed analysis is necessary, the donor-derived cells can be further analyzed on frozen or plastic sections (Avanesov *et al.*, 2005). In such cases, the donor blastomeres are usually labeled with both fluorophore- and biotin-conjugated dextrans. The fluorophore-conjugated tracer is used to distinguish which embryos contain donor-derived cells in the desired tissue as described above. The biotin-conjugated dextran, on the other hand, is used in detailed analysis at later developmental stages. The HRP-conjugated streptavidin version of the ABC kit (Vector Laboratories Inc.) or fluorophore-conjugated avidin (Jackson ImmunoResearch Inc., Molecular Probes, Inc.) can be used to detect biotinylated dextran (Fig. 4C and D, respectively). HRP detection can be performed on whole mounts and analyzed on plastic sections, as described above for histological analysis. In contrast to that, fluorophorc-conjugated avidin is preferably used after sectioning of the frozen tissue, owing to degradation of some flurophores during embedding of specimen in plastic. In these experiments, cryosections are prepared as described for antibody staining above. In some experiments, it is desirable to analyze the donor-derived cells for the expression of molecular markers (see Fig. 4D for an example). On frozen sections, avidin detection of donor-derived cells can be combined with antibody staining. Another way to visualize donor-derived cells and analyze expression at the same time is to combine HRP detection of donor-derived cells with *in situ* hybridization or antibody staining (Halpern *et al.*, 1993; Schier *et al.*, 1997). When HRP is used for the detection of donor-derived cells, the resulting reaction product inhibits the detection of the *in situ* probe with AP (Schier *et al.*, 1997). Because of this, the opposite sequence of enzymatic detection reactions is preferred: *in situ* probe detection first and HRP staining second.

When mosaic analysis is performed in the zebrafish retina at 3 dpf or later, the dilution of a donor-cell tracer can make the interpretation of the results difficult. This is because the descendants of a single transplanted blastomere divide a variable number of times. Thus in the donor-derived cells which undergo the highest number of divisions the label may be diluted so much that it is no longer detectable. In mosaic animals, such a situation can lead to the appearance of a mutant phenotype or to the rescue of a mutant phenotype in places seemingly not associated with the presence of donor cells and complicate the interpretation of experimental results. Increasing the concentration of the tracer or, in the case of whole-mount experiments, improving the penetration of staining reagents can sometimes alleviate this problem. Alternatively, collagenase treatment of fixed embryos improves reagent penetration during the detection of donor-derived cells (Doerre and Malicki, 2001). The amount of injected dextran should be increased carefully as excessively high concentrations are lethal for labeled cells.

If the dilution of tracer cannot be circumvented, an excellent alternative is the use of transgenes. An ideal transgene to mark donor cells in mosaic analysis would drive the expression of FP at a high level in all cells throughout development. In the context of the retina, the mCFP Q01 line largely meets these requirements, although its expression becomes somewhat dimmer as development advances (Godinho *et al.*, 2005). This line has been used, for example, to study photoreceptor and glia defects in *ale oko* mutant retinae (Jing and Malicki, 2009). An additional advantage of using transgenic FP tracers is that they eliminate the need for tracer injections into the donors, which decreases mechanical damage to embryos. Lastly, FP are relatively nontoxic, which increases the survival of donor-derived cells further. A disadvantage of transgene use in this context is that it takes one generation to in-cross an FP transgene into a mutant line. In summary, mosaic analysis is an important approach that has been widely used to study zebrafish retinal mutants (Avanesov *et al.*, 2005; Cerveny *et al.*, 2010; Doerre and Malicki, 2001; 2002; Goldsmith *et al.*, 2003; Jensen *et al.*, 2001; Jing and Malicki, 2009; Krock *et al.*, 2007; Link *et al.*, 2000; Malicki and Driever, 1999; Malicki *et al.*, 2003; Pujic and Malicki, 2001; Wei and Malicki, 2002; Yamaguchi *et al.*, 2010).

E. Analysis of Cell Proliferation

Several techniques are available to study cell proliferation in the retina. The amount of cell proliferation, the timing of cell cycle exit (birth date), and cell cycle length can be evaluated by H^3-thymidine labeling (Nawrocki, 1985) or via bromodeoxyuridine (BrdU) injections into the embryo (Hu and Easter, 1999). Such studies can be very informative in mutant animals (Kay *et al.*, 2001; Link *et al.*, 2000; Yamaguchi *et al.*, 2008). To identify the population of cells that exit the cell cycle in a particular window of time, BrdU labeling can be combined with iododeoxyuridine (IdU) (Del Bene *et al.*, 2008). Finally, another useful technique that can be used to test for cell cycle defects in mutant strains is fluorescence activated cell sorting (FACS) of dissociated retinal cells (Plaster *et al.*, 2006; Yamaguchi *et al.*, 2008).

F. Behavioral Studies

Several vision-dependent behavioral responses have been described in zebrafish larvae and adults: the optomotor response (Clark, 1981), the optokinetic response (Clark, 1981; Easter and Nicola, 1996), the startle response (Easter and Nicola, 1996), the phototaxis (Brockerhoff *et al.*, 1995), the escape response (Li and Dowling, 1997), and the dorsal light reflex (Nicolson *et al.*, 1998). Not surprisingly, larval feeding efficiency also depends on vision (Clark, 1981). While some of these behaviors are already present by 72 hpf, others have been described in adult fish only (for a review see Neuhauss, 2003). The vision-dependent behaviors of zebrafish proved to be very useful in genetic screening (see *Phenotype Detection Methods* on page 244). The optokinetic response appears to be the most robust and versatile. It is useful both in quick tests of vision and in quantitative estimates of visual acuity. In addition to genetic screens, behavioral tests have been used to study the function of the zebrafish optic tectum (Roeser and Baier, 2003).

G. Electrophysiological Analysis of Retinal Function

In addition to behavioral tests, measurements of electrical activity in the eye are another, more precise way to evaluate retinal function. Electrical responses of the zebrafish retina can be evaluated by electroretinography (ERG) already by 4 dpf (for example, Avanesov *et al.*, 2005). Similar to other vertebrates, the zebrafish ERG response contains two main waves: a small negative a-wave, originating from the photoreceptor cells, and a large positive b-wave, which reflects the function of the INL (Dowling, 1987; Makhankov *et al.*, 2004). The goal of an ERG study in zebrafish is no different from that of a similar procedure performed on the human eye. ERG can be used to evaluate the site of retinal defects in mutant animals. Ganglion cell defects do not affect the ERG response (Gnuegge *et al.*, 2001), whereas the absence of the a-wave or the b-wave suggests a defect in photoreceptors or in the INL, respectively. The a-wave appears small in ERG measurements because of an overlap with the b-wave. To measure the a-wave amplitude, the b-wave has to be blocked pharmacologically (Kainz *et al.*, 2003). An additional ERG wave, the d-wave, is produced when longer (ca. 1 s) flashes of light are used. Referred to as the OFF response, the d-wave is thought to reflect the activity of OFF-bipolar cells and photoreceptors (Kainz *et al.*, 2003; Makhankov *et al.*, 2004).

Retinal responses are usually elicited using a series of light stimuli that vary by several orders of magnitude in intensity (Allwardt *et al.*, 2001; Kainz *et al.*, 2003). This allows the evaluation of the visual response threshold, a parameter that is sometimes abnormal in mutant animals (Li and Dowling, 1997). Another important variable in ERG measurements is the level of background illumination. ERG measurements can be performed on light-adapted retinae using background illumination of a constant intensity, or on dark-adapted retinae, which are maintained in total darkness for at least 20 min prior to measurements (Kainz *et al.*, 2003). Most frequently recordings are performed on intact anesthetized animals (Makhankov *et al.*, 2004). Alternatively, eyes may be gently removed from larvae and bathed in

an oxygenated buffer solution. The latter ensures the oxygen supply to the retina in the absence of blood circulation (Kainz *et al.*, 2003). ERG recordings have become a standard assay when evaluating zebrafish eye mutants (Allwardt *et al.*, 2001; Avanesov *et al.*, 2005; Biehlmaier *et al.*, 2007; Brockerhoff *et al.*, 1998; Kainz *et al.*, 2003; Makhankov *et al.*, 2004; Morris *et al.*, 2005).

In addition to ERG, other more technically sophisticated electrophysiological measurements can be used to evaluate zebrafish (mutant) retinae. The ganglion cell function, for example, can be evaluated by recording action potentials from the optic nerve (Emran *et al.*, 2007). Such measurements revealed ganglion cell defects in the retinae of *nbb* and *mao* mutants (Gnuegge *et al.*, 2001; Li and Dowling, 2000). Similarly, photoreceptor function has been evaluated by measuring outer segment currents in isolated cells (Brockerhoff *et al.*, 2003).

H. Biochemical Approaches

Genetic experiments in animal models are frequently supplemented with studies of protein–protein interactions. Although this type of analysis has not been traditionally a strength of the zebrafish model, zebrafish embryos can be used to analyze binding interactions. In the context of the visual system, biochemical analysis has been largely applied to study the intraflagellar transport (IFT) in photoreceptor outer segment formation. As IFT occurs in many tissues, it can be studied via co-immunoprecipitation from embryonic or larval extracts (Krock and Perkins, 2008). Alternatively, extracts from the retinae of adult animals can be used in this type of experiment (Insinna *et al.*, 2008). A clear advantage of using larvae is that one can apply biochemical methods to analyze mutant phenotypes. As most zebrafish mutants are lethal at embryonic or larval stages, adult retinae are not suitable for this purpose. In addition to immunoprecipitation experiments, a more sophisticated but also more laborious and technically demanding approach is to identify binding partners by tandem affinity purification (TAP) (reviewed in Collins and Choudhary, 2008). The TAP tag procedure involves attaching a peptide tag to the protein of interest, and expressing it in zebrafish embryos. Following the preparation of embryonic extract, the peptide tag is used to purify the bait protein along with its binding partners using appropriate affinity columns. The identities of the binding partners are established using mass spectrometry. The TAP tag approach was applied in the zebrafish to identify the binding partners of Elipsa, a determinant of outer segment differentiation (Omori *et al.*, 2008). It is a relatively demanding technique, as it requires the expression of the bait protein in thousands of embryos. As more efficient affinity purification tags are engineered (Burckstummer *et al.*, 2006), TAP is likely to become easier to apply in the zebrafish.

I. Chemical Screens

Another approach that is gaining popularity in the zebrafish model is the screening of chemical libraries for compounds that affect developmental processes. The characteristics that render the zebrafish embryo suitable for genetic experiments—small

size, rapid development, and transparency—also make it exceptionally useful for small-molecule screening (Kokel *et al.*, 2010; Peterson *et al.*, 2000; Tran *et al.*, 2007; Zon and Peterson, 2005). In this type of experiment, hundreds or even thousands of small batches of embryos are each exposed to a different chemical compound, and analyzed for developmental or behavioral changes. Such an approach has been applied either to wild-type embryos or to carriers of genetic defects (Cao *et al.*, 2009; North *et al.*, 2007, 2009; Peterson *et al.*, 2004). In the latter case, compounds that rescue a mutant phenotype can be screened for. When mutations that resemble human abnormalities are used, this approach can be a powerful way to identify chemicals of potential therapeutic importance (Cao *et al.*, 2009; Hong *et al.*, 2006; Peterson *et al.*, 2004).

Chemical compound libraries ranging in size from hundreds to tens of thousands of molecules are commercially available. Phenotype detection approaches in a small-molecule screen are potentially as varied as in a genetic screen (Kaufman *et al.*, 2009; *Phenotype Detection Methods* on page 244). Gross evaluation of morphological features is the simplest option. Transgenic lines that express FPs in target tissues make it possible to detect subtle phenotypes. In a recent experiment, for example, an flk-GFP transgenic line was used to screen ca. 2000 small molecules for their effects on retinal vasculature (Kitambi *et al.*, 2009). Although little precedent exists at this time for small-molecule screens focusing on retinal development, this approach has been successful in the analysis of several zebrafish organs and behaviors (Hong *et al.*, 2006; Kokel *et al.*, 2010; North *et al.*, 2007; Sachidanandan *et al.*, 2008), and thus is also likely to find its way into the studies of the visual system.

IV. Analysis of Gene Function in the Zebrafish Retina

A. Reverse Genetic Approaches

A series of mutant alleles of varying severity is arguably the most informative tool of gene function analysis. Although a great variety of mutant lines have been identified in forward genetic sceens, for many loci chemically induced mutant alleles are not yet available. In these cases, other approaches must be applied to study gene function. In this section, we briefly discuss advantages and disadvantages of different loss-of-function and gain-of-function approaches in the context of the zebrafish visual system, and provide references to more comprehensive discussions of each.

1. Loss-of-Function Analysis

In the absence of loss-of-function mutations, antisense-based interference is by far the most common way to obtain information about gene function in the zebrafish embryo (Nasevicius and Ekker, 2000). The reasons for this popularity are low cost and low labor expense involved in their use. Although antisense morpholino-modified oligonucleotides have been shown to reproduce mutant phenotypes quite well, their use suffers from two main disadvantages. First, they become progressively less

effective as development proceeds, presumably because of degradation. Second, some morpholinos produce nonspecific toxicity, which must be distinguished from specific features of a morpholino-induced phenotype. Morpholino oligos can be used to interfere either with translation initiation or with splicing. Importantly, the efficiency of splice-site morpholinos can be monitored by reverse transcription polymerase chain reaction (RT-PCR) (Draper et al., 2001; Tsujikawa and Malicki, 2004b). In general, splice-site morpholinos reduce wild-type transcript expression below the level of RT-PCR detection throughout the first 36 h of development, although some have been reported to remain active until 3 or even 5 dpf (Tsujikawa and Malicki, 2004b). Most morpholinos are thus sufficient to interfere with genetic pathways involved in retinal neurogenesis but not to study later differentiation events or retinal function. Some help in designing morpholinos can be obtained from their manufacturer (Gene Tools LLC). Detailed protocols for the use of morpholinos, including their target site homology requirements, injection protocols, and methods to control for specificity, are available in literature (reviewed by Eisen and Smith, 2008; Malicki et al., 2002).

A powerful alternative to the use of morpholinos in loss-of-function studies is TILLING (targeted induced local lesions in genomes) (Colbert et al., 2001; McCallum et al., 2000; Wienholds et al., 2002). This approach combines chemical mutagenesis with a PCR-based protocol for detecting mutations in a locus of choice, and yields a series of mutant alleles that vary in strength. Its main disadvantage is the vast amount of preparation that needs to be done to initiate these experiments. One particularly labor-intensive step is the collection of thousands of sperm and DNA samples from F1 males. Because of this limitation, TILLING experiments are frequently performed by core facilities, which serve a group of laboratories, or the entire research community.

A recent addition to mutagenesis approaches in zebrafish is the use of zinc finger nucleases (ZFNs) to induce lesions in desired genes. ZFNs consist of a DNA recognition module, essentially a tandem array of two to four zinc finger-type DNA binding domains, and a catalytic module, which is usually derived from the Fok I restriction endonuclease (reviewed in Porteus and Carroll, 2005). ZFN binding to its target sequence induces double-stranded DNA breaks, which results in heritable defects because of improper repair. Needless to say, DNA binding specificity is critical for the application of ZFNs in animal models. The ability to manipulate binding is based on several findings: individual zinc fingers primarily interact with a single triplet of the DNA sequence; this interaction involves a significant degree of sequence specificity; and multiple zinc fingers can be assembled together to recognize longer target sequences (Porteus and Carroll, 2005). In zebrafish, pilot studies confirmed that ZFNs can be used to induce mutations in desired genes with good efficiency and specificity (Doyon et al., 2008; Meng et al., 2008). Nonetheless, the engineering of zinc finger binding domains of predetermined specificities remains laborious as it requires lengthy in vitro and/or in vivo selection procedures (Doyon et al., 2008; Meng et al., 2008). Detailed protocols for the selection of zinc finger combinations that will efficiently target predetermined DNA sequences and for the subsequent generation of mutant zebrafish have been described (Foley et al., 2009; Wright et al., 2006).

2. Approaches to Gene Overexpression

To obtain a comprehensive understanding of gene function, one often needs to supplement loss-of-function analysis with overexpression data. In the simplest scenario, this can be accomplished in zebrafish by RNA or DNA injections into the embryo. Several variants of this procedure exist, each with unique advantages and drawbacks (reviewed in Malicki *et al.*, 2002). The main disadvantage of injecting RNA into embryos is its limited stability. The injection of DNA constructs, on the other hand, produces expression for a much longer period of time but only in a small number of cells. The fraction of cells that express a gene of interest following the injection of a DNA construct into the embryo can be increased by placing the gene to be studied under the control of UAS (Upstream Activating Sequence, multiple copies are used in tandem) and driving its expression using GAL4–VP16 fusion protein expressed from either a ubiquitous or a tissue-specific promoter (Koster and Fraser, 2001). Alternatively, transgene integration efficiency (estimated as the fraction of cells that express DNA construct) can be greatly improved by using Tol2 transposon-based vectors (Kawakami, 2004). These are injected into one- to two-cell embryos (the earlier the better) along with transposase mRNA (Kawakami, 2004; Kwan *et al.*, 2007). The integration of these constructs into the genome relies on terminal transposon sequences, including the terminal inverted repeats (TIRs). The Tol2-derived sequences can be as short as 150–200 bp, but tend to be longer in older vectors, such as T2KXIG (Kawakami, 2004; Urasaki *et al.*, 2006). In addition to transposon terminal sequences, these vectors contain an FP marker that helps to follow the pattern of transgene inheritance in embryonic tissues. Genes of interest can also be placed in these vectors under the control of appropriate regulatory elements. The heat-shock promoter has been used, for example, to drive the expression of a *crumbs* gene from a Tol2-based vector in the zebrafish retinal neuroepithelium. This approach produced expression in nearly half of neuroepithelial cells (Omori *et al.*, 2008).

Overexpression phenotypes can also be studied in stable transgenic lines, provided that the resulting dominant phenotype is viable or can be conditionally induced. Several efficient methods for generating transgenic zebrafish are available. To develop a good understanding of its function, a gene under investigation may have to be expressed under the control of several regulatory elements and/or as a fusion with more than one tag (FP tags with different spectral characteristics and/or a myc tag, for example). As generating appropriate expression constructs using traditional cloning approaches is laborious, recombination cloning-based strategies have been specifically tailored for use in zebrafish (Kwan *et al.*, 2007; Villefranc *et al.*, 2007). These methods utilize a set of bacteriophage λ recombination enzymes to transfer DNA fragments from so-called entry vectors into so-called destination vectors, and are referred to as Gateway cloning (Hartley *et al.*, 2000). One of the most obvious advantages of the Gateway system is that it allows one to combine several different DNA elements relatively efficiently in a single enzymatic reaction. In one example of how this method can be applied, three entry clones were assembled in the correct configuration into a Tol2-based zebrafish destination vector

in a single step (Kwan *et al.*, 2007). The use of the Gateway system requires some preparatory work. Recombination sites need to be added to generate entry vectors, and, similarly, the destination vectors have to be prepared by inserting recombination sites and selection markers. These procedures are nonetheless straightforward, and most standard laboratory vectors can be fairly easily converted into destination vectors. To make this approach even more attractive, several destination vectors are already available for use in the zebrafish (Kwan *et al.*, 2007; Villefranc *et al.*, 2007).

A frequent limitation of overexpression studies is the pleiotropy of mutant phenotypes: for many loci, early embryonic phenotypes are so severe that they preclude the analysis of late developmental processes, such as retinal neurogenesis. Several experimental tools are available to overcome this problem, including the use of heat-shock promoters, the GAL4–UAS overexpression system, and caged nucleic acids. Similar to invertebrate model systems, the use of heat-shock-induced expression in zebrafish relies on the hsp70 promoter (Halloran *et al.*, 2000). An interesting variant of this protocol involves the activation of a heat-shock promoter-driven transgene in a small group of cells in a living embryo by heating them gently with a laser beam, which provides both temporal and spatial control of overexpression pattern (Halloran *et al.*, 2000).

GAL4–UAS system is another method to achieve spatial control of gene expression. Modeled after *Drosophila* (Brand and Perrimon, 1993), the GAL4–UAS overexpression approach takes advantage of two transgenic strains. The activator strain expresses the GAL4 transcriptional activator in a desired subset of tissues, while the effector strain carries the gene of interest under the control of a GAL4 responsive promoter (UAS, upstream activating sequence). The effector transgene is activated by crossing its carrier strain to a line that carries the activator transgene (Scheer *et al.*, 2002). One variant of this system involves a fusion of the Gal4 DNA binding domain to the viral VP16 activation domain and uses a multimer of 14 UAS sites in the reporter construct (Koster and Fraser, 2001; see also comments above). The GAL4–UAS system was initially used in the zebrafish eye to study *notch* function (Scheer *et al.*, 2001), and since then has gained popularity (Del Bene *et al.*, 2008; Godinho *et al.*, 2005; Mumm *et al.*, 2006; Yeo *et al.*, 2007). Importantly, enhancer trap screens have generated hundreds of transgenic strains that express the Gal4 activator in a variety of patterns and can be used to drive the expression of UAS effector transgenes in many organs, including the eye (Asakawa and Kawakami, 2008; Scott *et al.*, 2007). Finally, an interesting method to control gene overexpression patterns is the use of Bhc-caged nucleic acids (Ando *et al.*, 2001). In this approach, embryos are injected with an inactive form of an overexpression construct, which is then later activated in a selected tissue using UV illumination. Both RNA and DNA templates can be used to produce overexpression in this approach (Ando and Okamoto, 2003).

B. Forward Genetics

The use of zebrafish in genetic studies offers several obvious advantages. The most important of these is the possibility of performing efficient forward genetic screens. Genetic screening is feasible because adult zebrafish are highly fecund and are easily

maintained in large numbers in a fairly small laboratory space. Screens performed in on the zebrafish so far identified hundreds of visual system mutants (Baier *et al.*, 1996; Fadool *et al.*, 1997; Malicki *et al.*, 1996; Muto *et al.*, 2005; Neuhauss *et al.*, 1999). While designing a genetic screen, one has to consider three important variables: the type of mutagen to be used, the design of the breeding scheme, and mutant defect recognition criteria. Each of these is discussed below.

1. Mutagenesis Approaches

The majority of screens performed in zebrafish so far involved the use of *N*-ethyl-*N*-nitrosourea (ENU) (Mullins *et al.*, 1994; Solnica-Krezel *et al.*, 1994). This mutagenesis approach is very effective as evidenced by the fact that the vast majority of mutations isolated so far are ENU induced. A powerful alternative to chemical mutagenesis is insertional retroviral mutagenesis. Although the efficiency of this mutagenesis approach is still lower than that of chemical methods, an obvious advantage of a retroviral mutagen is that it provides means for very rapid identification of mutant genes (Amsterdam *et al.*, 1999; Golling *et al.*, 2002). Retroviral mutagenesis has also been applied on a large scale to identify hundreds of mutant strains (Golling *et al.*, 2002). The photoreceptor mutant *nrf* is an example of a retinal defect induced using this approach (Becker *et al.*, 1998). More recently, a rescreen of 250 retrovirus-induced mutants led to the identification of defects in several aspects of eye development (Gross *et al.*, 2005).

In addition to chemical mutagens and retroviral vectors, transposons provide another option for effective mutagenesis. Transposable elements of the *Tc-1/mariner* (*Sleeping beauty*) and *hAT* (*Tol2*) families integrate into the zebrafish genome in a transposase-dependent manner (Fadool *et al.*, 1998; Kawakami *et al.*, 2000; Raz *et al.*, 1998). Although initial efforts to induce mutations using transposon-based vectors were unsuccessful (Balciunas *et al.*, 2004; Kawakami *et al.*, 2004), recent experiments that rely on improved vector design generate mutants with high efficiency (Nagayoshi *et al.*, 2008; Sivasubbu *et al.*, 2006). Both *Tol2*- and *Sleeping beauty*-based constructs were used in these efforts. Transposon-based mutagenesis is an attractive alternative to retrovirus-mediated one because transposon-based vectors efficiently integrate into the zebrafish genome, and their mutagenicity (measured as the fraction of genome insertion events that lead to mutant phenotypes in homozygous animals) already exceeds that of retroviral mutagenesis (Nagayoshi *et al.*, 2008; Sivasubbu *et al.*, 2006). The use of transposons does not require technically difficult packaging of DNA into viral particles, and also appears to pose few safety concerns. An added bonus of using transposons is that they can be remobilized from preexisting lines to generate additional insertions (Kondrychyn *et al.*, 2009). One has to bear in mind, however, that just like in the case of viral insertions, transposon integration is not entirely random (Kondrychyn *et al.*, 2009). As the efficiency of transposable element-mediated mutagenesis is gradually improving, future genetic screens are likely to be performed with the help of transposons.

Transposon-mediated mutagenesis is usually performed using enhancer or gene trap vectors, which carry FP reporter genes (reviewed in Balciunas *et al.*, 2004; Nagayoshi

et al., 2008). Such a design is important for several reasons. First, it allows one to visually detect integration events that occur in the vicinity of genes because the nearby regulatory elements frequently drive FP reporter expression. Such integrations are much more likely to produce phenotypic defects, compared to insertions into non-transcribed regions of the genome. Second, as different integration events tend to produce different expression patterns, at least in some cases one can distinguish them from each other via simple inspection of living embryos. Consequently, potentially mutagenic insertions can be driven to homozygocity already in the F2 generation of a screen (Nagayoshi *et al.*, 2008). Moreover, as gene/enhancer trap expression patterns suggest the function for genes in which insertions have occurred, they may allow one to focus a genetic screen on a specific developmental or physiological process. Finally, trap-induced mutant alleles are easier to maintain as their presence can be selected for in heterozygotes based on expression pattern. Although retroviral mutagenesis vectors can also be engineered to function as traps (Ellingsen *et al.*, 2005), mutants generated using retroviral trap vectors have not been reported in zebrafish so far.

2. Breeding Schemes

The second important consideration is the type of breeding scheme that will carry genetic defects from mutagenized animals (G0) to the generation in which the screening for mutant phenotypes is performed. The most straightforward option, but also the most space- and time-consuming one, is screening for recessive defects in F3 generation embryos. This procedure was used in early large-scale genetic screens (Amsterdam *et al.*, 1999; Driever *et al.*, 1996; Haffter *et al.*, 1996). Its main disadvantage is that it requires a very large number of tanks to raise the F2 generation to adulthood. As the majority of laboratories do not have access to several thousands of fish tanks, more space-efficient procedures are frequently required. In this regard, the zebrafish offers some possibilities not available in other genetically studied vertebrates—haploid and early pressure screens (for a review see Malicki, 2000). The major asset of these screening strategies is that one generation of animals is omitted and consequently time and the amount of laboratory space required is dramatically reduced. Although there are obvious advantages, these two screening strategies also suffer from some limitations. The most significant disadvantage of haploids is that their development does not proceed in the same way as wild-type embryogenesis. Haploid embryos do not survive beyond 5 dpf, and even at earlier stages of development they display obvious defects. Although the eyes of haploid zebrafish appear fairly normal at least until 3 dpf, the architecture of their retinae tends to be disorganized. By 5 dpf, haploid embryos are markedly smaller than the wild type and display numerous abnormalities. In the context of the visual system, haploid screens appear useful to search for early patterning defects prior to the onset of neurogenesis.

Screening of embryos generated via the application of early pressure (Streisinger *et al.*, 1981) is another strategy that can be used to save both time and space. Similar to haploidization, this technique also allows one to screen for recessive defects in F2

generation embryos. The early pressure technique also involves some shortcomings. Embryos produced via this method display a high background of developmental abnormalities, which complicate the detection of mutant phenotypes, especially at early developmental stages. Another limitation of early pressure screens is that the fraction of homozygous mutant animals in a clutch of early pressure-generated embryos depends on the distance of a mutant locus from the centromere. For centromeric loci, the fraction of mutant embryos approaches 50%, whereas for telomeric genes it decreases below 10% (Streisinger et al., 1986). In other types of screens, mutant phenotypes can be distinguished from non-genetic developmental abnormalities based on their frequencies (25% in the case of screens on F3 embryos). Clearly, this criterion cannot be used in early pressure screens. Despite these limitations, early pressure screens are useful, especially in small-scale endeavors. The experimental techniques involved in haploid and early pressure screens have been previously reviewed in depth (Beattie et al., 1999; Walker, 1999).

While the approaches discussed above are used to identify recessive mutant phenotypes, an entirely different breeding scheme is used in searches for dominant defects. These can already be detected in embryos, larvae, or adults of the F1 generation. Although this category of screens requires just a single generation and consequently a very small amount of laboratory space, few experiments focusing on dominant defects have been performed in zebrafish so far (van Eeden et al., 1999). An example of a search for dominant defects of the visual system is provided by a small behavioral screen of adult animals for defects of visual perception, which identified a late-onset photoreceptor degeneration phenotype (Li and Dowling, 1997).

3. Phenotype Detection Methods

The third important consideration while designing a genetic screen is the mutant phenotype detection method. This aspect of screening allows for substantial creativity. Phenotype detection criteria range from very simple to very sophisticated. Ideally, the mutant phenotype recognition strategy should fulfill the following requirements: (1) involve minimal effort, (2) detect gross abnormalities as well as subtle changes, and (3) exclude phenotypes irrelevant to the targeted process. One class of irrelevant phenotypes are nonspecific defects. In large-scale mutagenesis screens performed so far, more than two-thirds of all phenotypes were classified as nonspecific (Driever et al., 1996; Golling et al., 2002; Haffter et al., 1996). The most frequent nonspecific phenotypes in zebrafish are early degeneration spreading across the entire embryo, and developmental retardation affecting brain, eyes, fins, and jaw. The latter class of mutants affects tissues that display robust proliferation between 3 and 5 dpf. Nonspecific phenotypes are not necessarily without value, but are usually considered uninteresting because they are likely to be produced by defects in a broad range of housekeeping mechanisms (such as metabolic pathways or DNA replication machinery; see for example Allende et al., 1996; Plaster et al., 2006). Another category of irrelevant phenotypes includes specific defects that are of no interest to investigators performing the screen. Such phenotypes are isolated when a screening procedure

detects mutations affecting multiple organs, only one of which is of interest. A good example of such a situation is provided by behavioral screens involving the optomotor response. Lack of the optomotor response may be due to defects of photoreceptor neurons or skeletal muscles. These two cell types are seldom interesting to the same group of investigators. It is one of the virtues of a well-designed screen that irrelevant phenotypes are efficiently selected against.

The simplest way to screen for mutant phenotypes is by visual inspection. The most significant disadvantage of this method is that it detects changes only in structures easily recognizable using a microscope (preferably a dissecting scope). Thus visual inspection screens are suitable to search for defects in trunk blood vessels (which are easy to see in larvae), but would not detect a loss of a small population of neurons hidden in the depths of the retina or the brain. Visual inspection criteria work well when the aim of a screen is to detect gross morphological changes. Within the eye, such changes may reflect specific defects in a single neuronal lamina. In many mutants, the changes of eye size are caused by a degeneration of photoreceptor cells (Doerre and Malicki, 2002; Jing and Malicki, 2009; Malicki *et al.*, 1996). In this case, the affected cell population is numerous enough to cause a major change of morphology. Most likely, a morphological screen would not detect abnormalities in a less numerous cell class.

Changes confined to small populations of cells cannot usually be identified in a visual inspection screen. To detect these changes, the target cell population must somehow be made accessible to inspection. Several options exist in this regard: analysis of histological sections, whole-mount antibody staining, *in situ* hybridization, retrograde or anterograde labeling of neurons, and cell class-specific FP transgenes. One technically simple but rather laborious approach is to embed zebrafish larvae in paraffin and prepare histological sections. This approach was used to screen more than 2000 individuals from ca. 50 clutches of F2 early pressure-generated mutagenized larvae and led to the identification of two photoreceptor mutants (Mohideen *et al.*, 2003). In addition to histological analysis, individual cell populations can be visualized in mutgenized animals using antibody staining or *in situ* hybridization. In one screening endeavor, staining of 700 early pressure-generated egg clutches with anti-tyrosine hydroxylase antibody led to the isolation of two retinal mutants (Guo *et al.*, 1999).

An excellent example of a genetic screen that involves labeling of a specific neuronal population has been performed to uncover defects of the retinotectal projection (Baier *et al.*, 1996; Karlstrom *et al.*, 1996; Trowe *et al.*, 1996). In this screen, two subpopulations of retinal ganglion cells were labeled with the carbocyanine tracers, DiI and DiO. Labeling procedures usually make screening much more laborious. To reduce the workload in this screen, DiI and DiO labeling were highly automated. For tracer injection, fish larvae were mounted in a standardized fashion in a temperature-controlled mounting apparatus. After filling the apparatus with liquid agarose and mounting the larvae, the temperature was lowered allowing the agarose to solidify. Subsequently, the blocks of agarose containing mounted larvae were transferred into the injection setup. Upon injection, the larvae were stored overnight at room temperature to allow for the diffusion of the injected tracer, and then transferred to a microscope

stage for phenotypic analysis. The authors of this experiment estimate that using this highly automated screening procedure allowed them to inspect over 2000 larvae per day and to reduce the time spent on the analysis of a single individual to less than 1 min (Baier et al., 1996). Other labeling procedures can also be scaled up to process many clutches of embryos in a single experiment. Antibody or in situ protocols, for example, involve multiple changes of staining and washing solutions. To perform these protocols on many embryos in parallel, one can use multiwell staining dishes with stainless steel mesh at the bottom. Such staining dishes can be quickly transferred from one solution to another. Since many labeling procedures are time consuming, it is essential that during a screen they are performed in parallel on many embryos.

Recent advances provide an additional way to label specific cell populations in a much less labor-intensive way by using FP transgenes, such as the ones described earlier in this chapter. Transgenic FP lines can be either directly mutagenized or crossed to mutagenized males. Then the resulting progeny is used to search for defects in fine features of retinal cell populations. In contrast to other cell labeling procedures, the use of FP transgenes requires very little additional effort, compared to simple morphological observations of the external phenotype.

Behavioral tests are yet another screening alternative. Several screens based on behavioral criteria have been performed in recent years, leading to the isolation of interesting developmental defects (Brockerhoff et al., 1997; 2003; Li and Dowling, 1997; Muto et al., 2005; Neuhauss et al., 1999). Behavioral screens allow one to detect subtle functional defects of the retina that might evade other search criteria. They can be used to search for both recessive and dominant defects in larvae as well as in adult fish (Li and Dowling, 1997). Similar to many labeling procedures, however, behavioral screens tend to be laborious. In one instance of a screen involving the optokinetic response, the authors estimate that screening of a single zebrafish larva took, on average, 1 min. (Brockerhoff et al., 1995). Since optomotor tests can be performed on populations of animals, they tend to be less time consuming, compared to optokinetic response tests. They do, however, produce more false-positive hits (Muto et al., 2005). In addition, since behavioral responses usually involve the cooperation of many cell classes, screens of this type tend to detect a wide range of defects. The optokinetic response screens, for example, may lead to the isolation of defects in the differentiation of lens cells, the specification of the retinal neurons or glia, the formation of synaptic connections, the mechanisms of neurotransmitter release, or the development of ocular muscles. Additional tests are necessary to assure that the isolated mutants belong to the desired category. To be useful for screening, the behavioral response should be robust and reproducible, and should involve the simplest possible neuronal circuitry. In light of these criteria, the optokinetic response appears to be superior to other behaviors; both optomotor and startle responses require functional optic tecta while the optokinetic response does not (Clark, 1981; Easter and Nicola, 1996). The optokinetic response also appears to be more robust than the optomotor response and phototaxis (Brockerhoff et al., 1995; Clark, 1981). The most extensive visual behavior-based screen conducted so far relied on two tests conducted in parallel: optokinetic and optomotor responses (Muto et al., 2005). Although the results of this experiment are quite informative, they

also illustrate inconsistencies associated with the use of behavioral tests as a screening tool. First, the initial round of screening was characterized by a very high false-positive rate ($> 90\%$ for the optomotor test). Second, surprisingly, the two behavioral tests used in this study uncovered largely non-overlapping sets of mutants. Following retests it turned out, however, that all mutants display both optomotor and optokinetic defects to varying degrees. Finally, as pointed out above, a broad range of phenotypic abnormalities in different cell classes were found in this experiment.

4. Positional and Candidate Cloning

Molecular characterization of defective loci is usually a crucial step that follows the isolation of mutant lines. The development of positional and candidate gene cloning strategies is one of the most significant advances in the field of zebrafish genetics within the last decade. These approaches are currently well established and have played a key role in many important contributions to the understanding of eye development and function. The positional cloning strategy involves a standard set of steps, such as mapping, chromosomal walking, transcript identification, and the delivery of a proof that the correct gene has been cloned. These steps are largely the same, regardless of the nature of a mutant phenotype. An example of a positional cloning strategy, laborious but eventually successful, is the cloning of the *nagie oko* locus (Wei and Malicki, 2002).

5. Mutant Strains Available

Large and small mutagenesis screens identified numerous genetic defects of retinal development in zebrafish. Mutant phenotypes affect a broad range of developmental stages, starting with the specification of the eye primordia, through optic lobe morphogenesis, the specification of neuronal identities, and include the final steps of differentiation, such as outer segment development in photoreceptor cells. Lists of mutant lines, excluding those that produce nonspecific degeneration of the entire retina, have been provided previously (Avanesov and Malicki, 2004; Malicki, 1999). Although these are still useful, many new mutants have been generated in recent years. The descriptions of these are available in the Zebrafish Model Organism Database (ZFIN, http://zfin.org).

V. Summary

Relative simplicity, rapid development, and accessibility to genetic analysis make the zebrafish retina an excellent model system for the studies of neurogenesis in the vertebrate CNS. Numerous genetic screens have led to isolation of many mutants affecting the retina and the retinotectal projection in zebrafish. Mutant phenotypes are being studied using a rich variety of markers: antibodies, RNA probes, retrograde and anterograde tracers, as well as transgenic lines. A particularly impressive progress has

been made in the characterization of the zebrafish genome. Consequently, positional and candidate cloning of mutant loci are now fairly easy in zebrafish. Many mutant genes have been cloned, and their analysis has provided insights into genetic circuitries that regulate retinal pattern formation, and the differentiation of retinal neurons and glia. Genetic screens for visual system defects will continue in the future, and progressively more sophisticated screening approaches will make it possible to detect an increasingly broad and varied assortment of mutant phenotypes. The remarkable evolutionary conservation of the vertebrate eye provides the basis for the use of the zebrafish retina as a model of human-inherited eye defects. As new techniques are being introduced and rapidly improved, the zebrafish will continue to be an important organism for the studies of the vertebrate visual system.

Acknowledgments

The authors are grateful to Brian Perkins, Ichiro Masai, and Brian Link for critical reading of earlier versions of this manuscript and helpful comments. The authors' research on the retina is supported by grants from the National Eye Institute to JM (R01 EY018176 and R01EY016859).

References

Allende, M. L., Amsterdam, A., Becker, T., Kawakami, K., Gaiano, N., and Hopkins, N. (1996). Insertional mutagenesis in zebrafish identifies two novel genes, pescadillo and dead eye, essential for embryonic development. *Genes Dev.* **10**, 3141–3155.

Allwardt, B. A., Lall, A. B., Brockerhoff, S. E., and Dowling, J. E. (2001). Synapse formation is arrested in retinal photoreceptors of the zebrafish nrc mutant. *J. Neurosci.* **21**, 2330–2342.

Altshuler, D., Turner, D., and Cepko, C. (1991). Specification of cell type in the vertebrate retina. *In* "Development of the Visual System" (D. Lam, and C. Shatz, eds.) pp. 37–58. The MIT Press, Cambridge, MA.

Alvarez, Y., Cederlund, M. L., Cottell, D. C., Bill, B. R., Ekker, S. C., Torres-Vazquez, J., Weinstein, B. M., Hyde, D. R., Vihtelic, T. S., and Kennedy, B. N. (2007). Genetic determinants of hyaloid and retinal vasculature in zebrafish. *BMC Dev. Biol.* **7**, 114.

Amsterdam, A., Burgess, S., Golling, G., Chen, W., Sun, Z., Townsend, K., Farrington, S., Haldi, M., and Hopkins, N. (1999). A large-scale insertional mutagenesis screen in zebrafish. *Genes Dev.* **13**, 2713–2724.

Ando, H., Furuta, T., Tsien, R. Y., and Okamoto, H. (2001). Photo-mediated gene activation using caged RNA/DNA in zebrafish embryos. *Nat. Genet.* **28**, 317–325.

Ando, H., and Okamoto, H. (2003). Practical procedures for ectopic induction of gene expression in zebrafish embryos using Bhc-diazo-caged mRNA. *Methods Cell Sci.* **25**, 25–31.

Aramaki, S., and Hatta, K. (2006). Visualizing neurons one-by-one *in vivo*: optical dissection and reconstruction of neural networks with reversible fluorescent proteins. *Dev. Dyn.* **235**, 2192–2199.

Asakawa, K., and Kawakami, K. (2008). Targeted gene expression by the Gal4–UAS system in zebrafish. *Dev. Growth. Differ.* **50**, 391–399.

Avanesov, A., Dahm, R., Sewell, W. F., and Malicki, J. J. (2005). Mutations that affect the survival of selected amacrine cell subpopulations define a new class of genetic defects in the vertebrate retina. *Dev. Biol.* **285**, 138–155.

Avanesov, A., and Malicki, J. (2004). Approaches to study neurogenesis in the zebrafish retina. *Methods Cell Biol.* **76**, 333–384.

Bahadori, R., Rinner, O., Schonthaler, H. B., Biehlmaier, O., Makhankov, Y. V., Rao, P., Jagadeeswaran, P., and Neuhauss, S. C. (2006). The zebrafish fade out mutant: a novel genetic model for Hermansky–Pudlak syndrome. *Invest. Ophthalmol. Vis. Sci.* **47**, 4523–4531.

Baier, H., Klostermann, S., Trowe, T., Karlstrom, R. O., Nusslein-Volhard, C., and Bonhoeffer, F. (1996). Genetic dissection of the retinotectal projection. *Development* **123**, 415–425.

Balciunas, D., Davidson, A. E., Sivasubbu, S., Hermanson, S. B., Welle, Z., and Ekker, S. C. (2004). Enhancer trapping in zebrafish using the sleeping beauty transposon. *BMC Genomics* **5**, 62.

Barthel, L. K., and Raymond, P. A. (1990). Improved method for obtaining 3-microns cryosections for immunocytochemistry. *J. Histochem. Cytochem.* **38**, 1383–1388.

Barthel, L. K., and Raymond, P. A. (1993). Subcellular localization of alpha-tubulin and opsin mRNA in the goldfish retina using digoxigenin-labeled cRNA probes detected by alkaline phosphatase and HRP histochemistry. *J. Neurosci. Methods* **50**, 145–152.

Baye, L. M., and Link, B. A. (2007). Interkinetic nuclear migration and the selection of neurogenic cell divisions during vertebrate retinogenesis. *J. Neurosci.* **27**, 10143–10152.

Beattie, C. E., Raible, D. W., Henion, P. D., and Eisen, J. S. (1999). Early pressure screens. *Methods Cell Biol.* **60**, 71–86.

Becker, T. S., Burgess, S. M., Amsterdam, A. H., Allende, M. L., and Hopkins, N. (1998). Not really finished is crucial for development of the zebrafish outer retina and encodes a transcription factor highly homologous to human nuclear respiratory factor-1 and avian initiation binding repressor. *Development* **125**, 4369–4378.

Bernardos, R. L., and Raymond, P. A. (2006). GFAP transgenic zebrafish. *Gene Expr. Patterns* **6**, 1007–1013.

Biehlmaier, O., Makhankov, Y., and Neuhauss, S. C. (2007). Impaired retinal differentiation and maintenance in zebrafish laminin mutants. *Invest. Ophthalmol. Vis. Sci.* **48**, 2887–2894.

Biehlmaier, O., Neuhauss, S. C., and Kohler, K. (2003). Synaptic plasticity and functionality at the cone terminal of the developing zebrafish retina. *J. Neurobiol.* **56**, 222–236.

Bodick, N., and Levinthal, C. (1980). Growing optic nerve fibers follow neighbors during embryogenesis. *Proc. Natl. Acad. Sci. U. S. A.* **77**, 4374–4378.

Branchek, T., and Bremiller, R. (1984). The development of photoreceptors in the zebrafish, *Brachydanio rerio*. I. Structure. *J. Comp. Neurol.* **224**, 107–115.

Brand, A. H., and Perrimon, N. (1993). Targeted gene expression as a means of altering cell fates and generating dominant phenotypes. *Development* **118**, 401–415.

Brennan, C., Monschau, B., Lindberg, R., Guthrie, B., Drescher, U., Bonhoeffer, F., and Holder, N. (1997). Two Eph receptor tyrosine kinase ligands control axon growth and may be involved in the creation of the retinotectal map in the zebrafish. *Development* **124**, 655–664.

Brockerhoff, S. E., Dowling, J. E., and Hurley, J. B. (1998). Zebrafish retinal mutants. *Vision Res* **38**, 1335–1339.

Brockerhoff, S. E., Hurley, J. B., Janssen-Bienhold, U., Neuhauss, S. C., Driever, W., and Dowling, J. E. (1995). A behavioral screen for isolating zebrafish mutants with visual system defects. *Proc. Natl. Acad. Sci. U. S. A.* **92**, 10545–10549.

Brockerhoff, S. E., Hurley, J. B., Niemi, G. A., and Dowling, J. E. (1997). A new form of inherited red-blindness identified in zebrafish. *J. Neurosci.* **17**, 4236–4242.

Brockerhoff, S. E., Rieke, F., Matthews, H. R., Taylor, M. R., Kennedy, B., Ankoudinova, I., Niemi, G. A., Tucker, C. L., Xiao, M., Cilluffo, M. C., Fain, G. L., and Hurley, J. B. (2003). Light stimulates a transducin-independent increase of cytoplasmic Ca^{2+} and suppression of current in cones from the zebrafish mutant nof. *J. Neurosci.* **23**, 470–480.

Burckstummer, T., Bennett, K. L., Preradovic, A., Schutze, G., Hantschel, O., Superti-Furga, G., and Bauch, A. (2006). An efficient tandem affinity purification procedure for interaction proteomics in mammalian cells. *Nat. Methods* **3**, 1013–1019.

Burrill, J. D., and Easter, S. S. Jr. (1994). Development of the retinofugal projections in the embryonic and larval zebrafish (*Brachydanio rerio*). *J. Comp. Neurol.* **346**, 583–600.

Burrill, J., and Easter, S. (1995). The first retinal axons and their microenvironment in zebrafish cryptic pioneers and the pretract. *J. Neurosci.* **15**, 2935–2947.

Cajal, S. R. (1893). La retine des vertebres. *Cellule.* **9**, 17–257.

Cameron, D. A., and Carney, L. H. (2000). Cell mosaic patterns in the native and regenerated inner retina of zebrafish: implications for retinal assembly. *J. Comp. Neurol.* **416**, 356–367.

Cao, Y., Semanchik, N., Lee, S. H., Somlo, S., Barbano, P. E., Coifman, R., and Sun, Z. (2009). Chemical modifier screen identifies HDAC inhibitors as suppressors of PKD models. *Proc. Natl. Acad. Sci. U. S. A.* **106**, 21819–21824.

Cayouette, M., and Raff, M. (2003). The orientation of cell division influences cell-fate choice in the developing mammalian retina. *Development* **130**, 2329–2339.

Cayouette, M., Whitmore, A. V., Jeffery, G., and Raff, M. (2001). Asymmetric segregation of Numb in retinal development and the influence of the pigmented epithelium. *J. Neurosci.* **21**, 5643–5651.

Cedrone, C., Culasso, F., Cesareo, M., Zapelloni, A., Cedrone, P., and Cerulli, L. (1997). Prevalence of glaucoma in Ponza, Italy: a comparison with other studies. *Ophthalmic. Epidemiol.* **4**, 59–72.

Cerveny, K. L., Cavodeassi, F., Turner, K. J., de Jong-Curtain, T. A., Heath, J. K., and Wilson, S. W. (2010). The zebrafish flotte lotte mutant reveals that the local retinal environment promotes the differentiation of proliferating precursors emerging from their stem cell niche. *Development* **137**, 2107–2115.

Chinen, A., Hamaoka, T., Yamada, Y., and Kawamura, S. (2003). Gene duplication and spectral diversification of cone visual pigments of zebrafish. *Genetics* **163**, 663–675.

Choi, J., Dong, L., Ahn, J., Dao, D., Hammerschmidt, M., and Chen, J. N. (2007). FoxH1 negatively modulates flk1 gene expression and vascular formation in zebrafish. *Dev. Biol.* **304**, 735–744.

Chuang, J. C., Mathers, P. H., and Raymond, P. A. (1999). Expression of three Rx homeobox genes in embryonic and adult zebrafish. *Mech. Dev.* **84**, 195–198.

Clark, T. (1981). Visual responses in developing zebrafish (Brachydanio rerio). Ph. D. dissertation, University of Oregon, Eugene, OR.

Colbert, T., Till, B. J., Tompa, R., Reynolds, S., Steine, M. N., Yeung, A. T., McCallum, C. M., Comai, L., and Henikoff, S. (2001). High-throughput screening for induced point mutations. *Plant Physiol.* **126**, 480–484.

Collazo, A., Fraser, S. E., and Mabee, P. M. (1994). A dual embryonic origin for vertebrate mechanoreceptors. *Science* **264**, 426–430.

Collins, M. O., and Choudhary, J. S. (2008). Mapping multiprotein complexes by affinity purification and mass spectrometry. *Curr. Opin. Biotechnol.* **19**, 324–330.

Connaughton, V. P., Behar, T. N., Liu, W. L., and Massey, S. C. (1999). Immunocytochemical localization of excitatory and inhibitory neurotransmitters in the zebrafish retina. *Vis. Neurosci.* **16**, 483–490.

Cui, S., Otten, C., Rohr, S., Abdelilah-Seyfried, S., and Link, B. A. (2007). Analysis of aPKCλ and aPKCζ reveals multiple and redundant functions during vertebrate retinogenesis. *Mol. Cell. Neurosci.* **34**, 431–444.

Dahm, R., Schonthaler, H. B., Soehn, A. S., van Marle, J., and Vrensen, G. F. (2007). Development and adult morphology of the eye lens in the zebrafish. *Exp. Eye Res.* **85**, 74–89.

Dapson, R. W. (2007). Glyoxal fixation: how it works and why it only occasionally needs antigen retrieval. *Biotech. Histochem.* **82**, 161–166.

Das, T., Payer, B., Cayouette, M., and Harris, W. A. (2003). *In vivo* time-lapse imaging of cell divisions during neurogenesis in the developing zebrafish retina. *Neuron* **37**, 597–609.

Del Bene, F., Wehman, A. M., Link, B. A., and Baier, H. (2008). Regulation of neurogenesis by interkinetic nuclear migration through an apical-basal notch gradient. *Cell* **134**, 1055–1065.

Devoto, S. H., Melancon, E., Eisen, J. S., and Westerfield, M. (1996). Identification of separate slow and fast muscle precursor cells *in vivo*, prior to somite formation. *Development* **122**, 3371–3380.

Doerre, G., and Malicki, J. (2001). A mutation of early photoreceptor development, *mikre oko*, reveals cell–cell interactions involved in the survival and differentiation of zebrafish photoreceptors. *J. Neurosci.* **21**, 6745–6757.

Doerre, G., and Malicki, J. (2002). Genetic analysis of photoreceptor cell development in the zebrafish retina. *Mech. Dev.* **110**, 125–138.

Dowling, J. (1987). "The Retina". Harvard University Press, Cambridge, MA.

Doyon, Y., McCammon, J. M., Miller, J. C., Faraji, F., Ngo, C., Katibah, G. E., Amora, R., Hocking, T. D., Zhang, L., Rebar, E. J., Gregory, P. D., and Urnov, F. D., *et al.* (2008). Heritable targeted gene disruption in zebrafish using designed zinc-finger nucleases. *Nat. Biotechnol.* **26** 702–708.

Draper, B. W., Morcos, P. A., and Kimmel, C. B. (2001). Inhibition of zebrafish fgf8 pre-mRNA splicing with morpholino oligos: a quantifiable method for gene knockdown. *Genesis* **30**, 154–156.

Drescher, U., Bonhoeffer, F., and Muller, B. K. (1997). The Eph family in retinal axon guidance. *Curr. Opin. Neurobiol.* **7**, 75–80.

Driever, W., Solnica-Krezel, L., Schier, A. F., Neuhauss, S. C., Malicki, J., Stemple, D. L., Stainier, D. Y., Zwartkruis, F., Abdelilah, S., Rangini, Z., Belak, J., and Boggs, C. (1996). A genetic screen for mutations affecting embryogenesis in zebrafish. *Development* **123**, 37–46.

Dryja, T., and Li, T. (1995). Molecular genetics of retinitis pigmentosa. *Hum. Mol. Genet.* **4**, 1739–1743.

Duldulao, N. A., Lee, S., and Sun, Z. (2009). Cilia localization is essential for *in vivo* functions of the Joubert syndrome protein Arl13b/Scorpion. *Development* **136**, 4033–4042.

Easter, S. S., Jr., and Malicki, J. J. (2002). The zebrafish eye: developmental and genetic analysis. *Result Probl. Cell Differ.* **40**, 346–370.

Easter, S., and Nicola, G. (1996). The development of vision in the zebrafish (*Danio rerio*). *Dev. Biol.* **180**, 646–663.

Eisen, J. S., and Smith, J. C. (2008). Controlling morpholino experiments: don't stop making antisense. *Development* **135**, 1735–1743.

Ellingsen, S., Laplante, M. A., Konig, M., Kikuta, H., Furmanek, T., Hoivik, E. A., and Becker, T. S. (2005). Large-scale enhancer detection in the zebrafish genome. *Development* **132**, 3799–3811.

Emran, F., Rihel, J., Adolph, A. R., and Dowling, J. E. (2010). Zebrafish larvae lose vision at night. *Proc. Natl. Acad. Sci. U. S. A.* **107**, 6034–6039.

Emran, F., Rihel, J., Adolph, A. R., Wong, K. Y., Kraves, S., and Dowling, J. E. (2007). OFF ganglion cells cannot drive the optokinetic reflex in zebrafish. *Proc. Natl. Acad. Sci. U. S. A.* **104**, 19126–19131.

Fadool, J. M. (2003). Development of a rod photoreceptor mosaic revealed in transgenic zebrafish. *Dev. Biol.* **258**, 277–290.

Fadool, J. M., Brockerhoff, S. E., Hyatt, G. A., and Dowling, J. E. (1997). Mutations affecting eye morphology in the developing zebrafish (*Danio rerio*). *Dev. Genet.* **20**, 288–295.

Fadool, J. M., Hartl, D. L., and Dowling, J. E. (1998). Transposition of the mariner element from *Drosophila mauritiana* in zebrafish. *Proc. Natl. Acad. Sci. U. S. A.* **95**, 5182–5186.

Fashena, D., and Westerfield, M. (1999). Secondary motoneuron axons localize DM-GRASP on their fasciculated segments. *J. Comp. Neurol.* **406**, 415–424.

Foley, J. E., Maeder, M. L., Pearlberg, J., Joung, J. K., Peterson, R. T., and Yeh, J. R. (2009). Targeted mutagenesis in zebrafish using customized zinc-finger nucleases. *Nat. Protoc.* **4**, 1855–1867.

Fraser, S. (1992). Patterning of retinotectal connections in the vertebrate visual system. *Curr. Oppin. Neurobiol.* **2**, 83–87.

Fricke, C., Lee, J. S., Geiger-Rudolph, S., Bonhoeffer, F., and Chien, C. B. (2001). Astray, a zebrafish roundabout homolog required for retinal axon guidance. *Science* **292**, 507–510.

Gnuegge, L., Schmid, S., and Neuhauss, S. C. (2001). Analysis of the activity-deprived zebrafish mutant macho reveals an essential requirement of neuronal activity for the development of a fine-grained visuotopic map. *J. Neurosci.* **21**, 3542–3548.

Godinho, L., Mumm, J. S., Williams, P. R., Schroeter, E. H., Koerber, A., Park, S. W., Leach, S. D., and Wong, R. O. (2005). Targeting of amacrine cell neurites to appropriate synaptic laminae in the developing zebrafish retina. *Development* **132**, 5069–5079.

Goldsmith, P., Baier, H., and Harris, W. A. (2003). Two zebrafish mutants, ebony and ivory, uncover benefits of neighborhood on photoreceptor survival. *J. Neurobiol.* **57**, 235–245.

Golling, G., Amsterdam, A., Sun, Z., Antonelli, M., Maldonado, E., Chen, W., Burgess, S., Haldi, M., Artzt, K., Farrington, S., Lin, S. -Y., Nissen R. M., *et al.* (2002). Insertional mutagenesis in zebrafish rapidly identifies genes essential for early vertebrate development. *Nat. Genet.* **31** 135–140.

Gray, M. P., Smith, R. S., Soules, K. A., John, S. W., and Link, B. A. (2009). The aqueous humor outflow pathway of zebrafish. *Invest. Ophthalmol. Vis. Sci.* **50**, 1515–21.

Gross, J. M., and Perkins, B. D. (2008). Zebrafish mutants as models for congenital ocular disorders in humans. *Mol. Reprod. Dev.* **75**, 547–555.

Gross, J. M., Perkins, B. D., Amsterdam, A., Egana, A., Darland, T., Matsui, J. I., Sciascia, S., Hopkins, N., and Dowling, J. E. (2005). Identification of zebrafish insertional mutants with defects in visual system development and function. *Genetics* **170**, 245–261.

Guo, S., Wilson, S. W., Cooke, S., Chitnis, A. B., Driever, W., and Rosenthal, A. (1999). Mutations in the zebrafish unmask shared regulatory pathways controlling the development of catecholaminergic neurons. *Dev. Biol.* **208**, 473–487.

Haffter, P., Granato, M., Brand, M., Mullins, M. C., Hammerschmidt, M., Kane, D. A., Odenthal, J., van Eeden, F. J., Jiang, Y. J., Heisenberg, C. P., Kelsh, R. N., Furutani-Seiki, M., *et al.* (1996). The identification of genes with unique and essential functions in the development of the zebrafish, *Danio rerio. Development* **123**, 1–36.

Halloran, M. C., Sato-Maeda, M., Warren, J. T., Su, F., Lele, Z., Krone, P. H., Kuwada, J. Y., and Shoji, W. (2000). Laser-induced gene expression in specific cells of transgenic zebrafish. *Development* **127**, 1953–1960.

Halpern, M., Ho, R., Walker, C., and Kimmel, C. (1993). Induction of muscle pioneers and floor plate is distinguished by the zebrafish no tail mutation. *Cell* **75**, 99–111.

Hanker, J. S. (1979). Osmiophilic reagents in electronmicroscopic histocytochemistry. *Prog. Histochem. Cytochem.* **12**, 1–85.

Hartley, J. L., Temple, G. F., and Brasch, M. A. (2000). DNA cloning using *in vitro* site-specific recombination. *Genome Res.* **10**, 1788–1795.

Hartong, D. T., Berson, E. L., and Dryja, T. P. (2006). Retinitis pigmentosa. *Lancet* **368**, 1795–1809.

Hatta, K., Tsujii, H., and Omura, T. (2006). Cell tracking using a photoconvertible fluorescent protein. *Nat. Protoc.* **1**, 960–967.

Hauptmann, G., and Gerster, T. (1994). Two-color whole-mount *in situ* hybridization to vertebrate and *Drosophila* embryos. *Trends Genet.* **10**, 266.

Hinds, J., and Hinds, P. (1974). Early ganglion cell differentiation in the mouse retina: an electron microscopic analysis utilizing serial sections. *Dev. Biol.* **37**, 381–416.

Hisatomi, O., Satoh, T., Barthel, L. K., Stenkamp, D. L., Raymond, P. A., and Tokunaga, F. (1996). Molecular cloning and characterization of the putative ultraviolet- sensitive visual pigment of goldfish. *Vision Res.* **36**, 933–939.

Hitchcock, P. F., Macdonald, R. E., VanDeRyt, J. T., and Wilson, S. W. (1996). Antibodies against Pax6 immunostain amacrine and ganglion cells and neuronal progenitors, but not rod precursors, in the normal and regenerating retina of the goldfish. *J. Neurobiol.* **29**, 399–413.

Ho, R. K., and Kane, D. A. (1990). Cell-autonomous action of zebrafish spt-1 mutation in specific mesodermal precursors. *Nature* **348**, 728–730.

Holt, C., Bertsch, T., Ellis, H., and Harris, W. (1988). Cellular determination in the *Xenopus* retina is independent of lineage and birth date. *Neuron* **1**, 15–26.

Hong, C. C., Peterson, Q. P., Hong, J. Y., and Peterson, R. T. (2006). Artery/vein specification is governed by opposing phosphatidylinositol-3 kinase and MAP kinase/ERK signaling. *Curr. Biol.* **16**, 1366–1372.

Honig, M. G., and Hume, R. I. (1986). Fluorescent carbocyanine dyes allow living neurons of identified origin to be studied in long-term cultures. *J. Cell Biol.* **103**, 171–187.

Honig, M. G., and Hume, R. I. (1989). DiI and diO: versatile fluorescent dyes for neuronal labelling and pathway tracing. *Trends Neurosci.* **12**(333–335), 340–341.

Hu, M., and Easter, S. S. (1999). Retinal neurogenesis: the formation of the initial central patch of postmitotic cells. *Dev. Biol.* **207**, 309 321.

Hudak, L. M., Lunt, S., Chang, C. H., Winkler, E., Flammer, H., Lindsey, M., and Perkins, B. D. (2010). The intraflagellar transport protein ift80 is essential for photoreceptor survival in a zebrafish model of jeune asphyxiating thoracic dystrophy. *Invest. Ophthalmol. Vis. Sci.* **51**, 3792–3799.

Hyatt, G. A., Schmitt, E. A., Marsh-Armstrong, N., McCaffery, P., Drager, U. C., and Dowling, J. E. (1996). Retinoic acid establishes ventral retinal characteristics. *Development* **122**, 195–204.

Humphrey, C., and Pittman, F. (1974). A simple methylene blue-azure II-basic fuchsin stain for epoxy-embedded tissue sections. *Stain Technol.* **49**, 9–14.

Insinna, C., Pathak, N., Perkins, B., Drummond, I., and Besharse, J. C. (2008). The homodimeric kinesin, Kif17, is essential for vertebrate photoreceptor sensory outer segment development. *Dev. Biol.* **316**, 160–170.

Jacobson, M. (1991). "Developmental Neurobiology". Plenum Press, New York.

Jensen, A. M., Walker, C., and Westerfield, M. (2001). Mosaic eyes: a zebrafish gene required in pigmented epithelium for apical localization of retinal cell division and lamination. *Development* **128**, 95–105.

Jing, X., and Malicki, J. (2009). Zebrafish ale oko, an essential determinant of sensory neuron survival and the polarity of retinal radial glia, encodes the p50 subunit of dynactin. *Development* **136**, 2955–2964.

Jowett, T. (2001). Double *in situ* hybridization techniques in zebrafish. *Methods* **23**, 345–358.

Jowett, T., and Lettice, L. (1994). Whole-mount *in situ* hybridizations on zebrafish embryos using a mixture of digoxigenin- and fluorescein-labelled probes. *Trends Genet.* **10**, 73–74.

Kainz, P. M., Adolph, A. R., Wong, K. Y., and Dowling, J. E. (2003). Lazy eyes zebrafish mutation affects Müller glial cells, compromising photoreceptor function and causing partial blindness. *J. Comp. Neurol.* **463**, 265–280.

Karlsson, J., von Hofsten, J., and Olsson, P. E. (2001). Generating transparent zebrafish: a refined method to improve detection of gene expression during embryonic development. *Mar. Biotechnol. (NY)* **3**, 522–527.

Karlstrom, R. O., Trowe, T., Klostermann, S., Baier, H., Brand, M., Crawford, A. D., Grunewald, B., Haffter, P., Hoffmann, H., Meyer, S. U. *et al.* (1996). Zebrafish mutations affecting retinotectal axon pathfinding. *Development* **123**, 427–438.

Kaufman, C. K., White, R. M., and Zon, L. (2009). Chemical genetic screening in the zebrafish embryo. *Nat. Protoc.* **4**, 1422–1432.

Kawahara, A., Chien, C. B., and Dawid, I. B. (2002). The homeobox gene mbx is involved in eye and tectum development. *Dev. Biol.* **248**, 107–117.

Kawakami, K. (2004). Transgenesis and gene trap methods in zebrafish by using the Tol2 transposable element. *Methods Cell Biol.* **77**, 201–222.

Kawakami, K., Shima, A., and Kawakami, N. (2000). Identification of a functional transposase of the Tol2 element, an Ac-like element from the Japanese medaka fish, and its transposition in the zebrafish germ lineage. *Proc. Natl. Acad. Sci. U. S. A.* **97**, 11403–11408.

Kawakami, K., Takeda, H., Kawakami, N., Kobayashi, M., Matsuda, N., and Mishina, M. (2004). A transposon-mediated gene trap approach identifies developmentally regulated genes in zebrafish. *Dev. Cell* **7**, 133–144.

Kay, J. N., Finger-Baier, K. C., Roeser, T., Staub, W., and Baier, H. (2001). Retinal ganglion cell genesis requires lakritz, a zebrafish atonal homolog. *Neuron* **30**, 725–736.

Kay, J. N., Link, B. A., and Baier, H. (2005). Staggered cell-intrinsic timing of ath5 expression underlies the wave of ganglion cell neurogenesis in the zebrafish retina. *Development* **132**, 2573–2585.

Kay, J. N., Roeser, T., Mumm, J. S., Godinho, L., Mrejeru, A., Wong, R. O., and Baier, H. (2004). Transient requirement for ganglion cells during assembly of retinal synaptic layers. *Development* **131**, 1331–1342.

Kikuchi, Y., Segawa, H., Tokumoto, M., Tsubokawa, T., Hotta, Y., Uyemura, K., and Okamoto, H. (1997). Ocular and cerebellar defects in zebrafish induced by overexpression of the LIM domains of the islet-3 LIM/homeodomain protein. *Neuron* **18**, 369–382.

Kim, M. J., Kang, K. H., Kim, C. H., and Choi, S. Y. (2008). Real-time imaging of mitochondria in transgenic zebrafish expressing mitochondrially targeted GFP. *Biotechniques* **45**, 331–334.

Kimmel, C. B., Ballard, W. W., Kimmel, S. R., Ullmann, B., and Schilling, T. F. (1995). Stages of embryonic development of the zebrafish. *Dev Dyn* **203**, 253–310.

Kimmel, C. B., Sessions, S. K., and Kimmel, R. J. (1981). Morphogenesis and synaptogenesis of the zebrafish Mauthner neuron. *J. Comp. Neurol.* **198**, 101–120.

Kitambi, S. S., McCulloch, K. J., Peterson, R. T., and Malicki, J. J. (2009). Small molecule screen for compounds that affect vascular development in the zebrafish retina. *Mech. Dev.* **126**, 464–477.

Kitambi, S., Peterson, R., and Malicki, J. (2008). Small molecule screen for compounds that affect vascular development in the zebrafish retina. *Mech. Dev.* **126**, 464–477.

Kokel, D., Bryan, J., Laggner, C., White, R., Cheung, C. Y., Mateus, R., Healey, D., Kim, S., Werdich, A. A., Haggarty, S. J., Macrae, C. A., Shoichet, B., *et al.* (2010). Rapid behavior-based identification of neuroactive small molecules in the zebrafish. *Nat. Chem. Biol.* **6**, 231–237.

Kondrychyn, I., Garcia-Lecea, M., Emelyanov, A., Parinov, S., and Korzh, V. (2009). Genome-wide analysis of Tol2 transposon reintegration in zebrafish. *BMC Genomics* **10**, 418.

Kosodo, Y., Toida, K., Dubreuil, V., Alexandre, P., Schenk, J., Kiyokage, E., Attardo, A., Mora-Bermudez, F., Arii, T., Clarke, J. D., and Huttner, W. B. (2008). Cytokinesis of neuroepithelial cells can divide their basal process before anaphase. *EMBO J.* **27**, 3151–3163.

Koster, R. W., and Fraser, S. E. (2001). Tracing transgene expression in living zebrafish embryos. *Dev. Biol.* **233**, 329–346.

Krock, B. L., Bilotta, J., and Perkins, B. D. (2007). Noncell-autonomous photoreceptor degeneration in a zebrafish model of choroideremia. *Proc. Natl. Acad. Sci. U. S. A.* **104**, 4600–4605.

Krock, B. L., and Perkins, B. D. (2008). The intraflagellar transport protein IFT57 is required for cilia maintenance and regulates IFT-particle-kinesin-II dissociation in vertebrate photoreceptors. *J. Cell Sci.* **121**, 1907–1915.

Kwan, K. M., Fujimoto, E., Grabher, C., Mangum, B. D., Hardy, M. E., Campbell, D. S., Parant, J. M., Yost, H. J., Kanki, J. P., and Chien, C. B. (2007). The Tol2kit: a multisite gateway-based construction kit for Tol2 transposon transgenesis constructs. *Dev. Dyn.* **236**, 3088–3099.

Laessing, U., Giordano, S., Stecher, B., Lottspeich, F., and Stuermer, C. A. (1994). Molecular characterization of fish neurolin: a growth-associated cell surface protein and member of the immunoglobulin superfamily in the fish retinotectal system with similarities to chick protein DM-GRASP/SC-1/BEN. *Differentiation* **56**, 21–29.

Laessing, U., and Stuermer, C. A. (1996). Spatiotemporal pattern of retinal ganglion cell differentiation revealed by the expression of neurolin in embryonic zebrafish. *J. Neurobiol.* **29**, 65–74.

Larison, K., and Bremiller, R. (1990). Early onset of phenotype and cell patterning in the embryonic zebrafish retina. *Development* **109**, 567–576.

Lawson, N. D., and Weinstein, B. M. (2002). *In vivo* imaging of embryonic vascular development using transgenic zebrafish. *Dev. Biol.* **248**, 567–576.

Li, L., and Dowling, J. E. (1997). A dominant form of inherited retinal degeneration caused by a non-photoreceptor cell-specific mutation. *Proc. Natl. Acad. Sci. U. S. A.* **94**, 11645–11650.

Li, L., and Dowling, J. E. (2000). Disruption of the olfactoretinal centrifugal pathway may relate to the visual system defect in night blindness b mutant zebrafish. *J. Neurosci.* **20**, 1883–1892.

Li, Z., Hu, M., Ochocinska, M. J., Joseph, N. M., and Easter, S. S. Jr. (2000a). Modulation of cell proliferation in the embryonic retina of zebrafish (*Danio rerio*). *Dev. Dyn.* **219**, 391–401.

Li, Z., Joseph, N. M., and Easter, S. S. Jr. (2000b). The morphogenesis of the zebrafish eye, including a fate map of the optic vesicle. *Dev. Dyn.* **218**, 175–188.

Link, B. A., Fadool, J. M., Malicki, J., and Dowling, J. E. (2000). The zebrafish young mutation acts non-cell-autonomously to uncouple differentiation from specification for all retinal cells. *Development* **127**, 2177–2188.

Liu, I. S., Chen, J. D., Ploder, L., Vidgen, D., van der Kooy, D., Kalnins, V. I., and McInnes, R. R. (1994). Developmental expression of a novel murine homeobox gene (Chx10): evidence for roles in determination of the neuroretina and inner nuclear layer. *Neuron* **13**, 377–393.

Lowery, L. A., and Sive, H. (2004). Strategies of vertebrate neurulation and a re-evaluation of teleost neural tube formation. *Mech. Dev.* **121**, 1189–1197.

Macdonald, R., Barth, K. A., Xu, Q., Holder, N., Mikkola, I., and Wilson, S. W. (1995). Midline signalling is required for Pax gene regulation and patterning of the eyes. *Development* **121**, 3267 3278.

Macdonald, R., Scholes, J., Strahle, U., Brennan, C., Holder, N., Brand, M., and Wilson, S. W. (1997). The Pax protein Noi is required for commissural axon pathway formation in the rostral forebrain. *Development* **124**, 2397–2408.

Macdonald, R., and Wilson, S. (1997). Distribution of Pax6 protein during eye development suggests discrete roles in proliferative and differentiated visual cells. *Dev. Genes Evol.* **206**, 363–369.

Mack, A. F., and Fernald, R. D. (1995). New rods move before differentiating in adult teleost retina. *Dev. Biol.* **170**, 136–141.

Makhankov, Y. V., Rinner, O., and Neuhauss, S. C. (2004). An inexpensive device for non-invasive electroretinography in small aquatic vertebrates. *J. Neurosci. Methods* **135**, 205–210.

Malicki, J. (1999). Development of the retina. *Methods Cell Biol.* **59**, 273–299.

Malicki, J. (2000). Harnessing the power of forward genetics—Analysis of neuronal diversity and patterning in the zebrafish retina. *Trends Neurosci.* **23**, 531–541.

Malicki, J., and Driever, W. (1999). *Oko meduzy* mutations affect neuronal patterning in the zebrafish retina and reveal cell–cell interactions of the retinal neuroepithelial sheet. *Development* **126**, 1235–1246.

Malicki, J., Jo, H., and Pujic, Z. (2003). Zebrafish N-cadherin, encoded by the *glass onion* locus, plays an essential role in retinal patterning. *Dev. Biol.* **259**, 95–108.

Malicki, J., Jo, H., Wei, X., Hsiung, M., and Pujic, Z. (2002). Analysis of gene function in the zebrafish retina. *Methods* **28**, 427–438.

Malicki, J., Neuhauss, S. C., Schier, A. F., Solnica-Krezel, L., Stemple, D. L., Stainier, D. Y., Abdelilah, S., Zwartkruis, F., Rangini, Z., and Driever, W. (1996). Mutations affecting development of the zebrafish retina. *Development* **123**, 263–273.

Mangrum, W. I., Dowling, J. E., and Cohen, E. D. (2002). A morphological classification of ganglion cells in the zebrafish retina. *Vis. Neurosci.* **19**, 767–779.

Marcus, R. C., Delaney, C. L., and Easter, S. S. Jr. (1999). Neurogenesis in the visual system of embryonic and adult zebrafish (*Danio rerio*). *Vis. Neurosci.* **16**, 417–424.

Martinez-Morales, J. R., Del Bene, F., Nica, G., Hammerschmidt, M., Bovolenta, P., and Wittbrodt, J. (2005). Differentiation of the vertebrate retina is coordinated by an FGF signaling center. *Dev. Cell* **8**, 565–574.

Masai, I., Lele, Z., Yamaguchi, M., Komori, A., Nakata, A., Nishiwaki, Y., Wada, H., Tanaka, H., Nojima, Y., Hammerschmidt, M., Wilson, S. W., and Okamoto, H. (2003). N-cadherin mediates retinal lamination, maintenance of forebrain compartments and patterning of retinal neurites. *Development* **130**, 2479–2494.

Masai, I., Stemple, D. L., Okamoto, H., and Wilson, S. W. (2000). Midline signals regulate retinal neurogenesis in zebrafish. *Neuron* **27**, 251–263.

Masai, I., Yamaguchi, M., Tonou-Fujimori, N., Komori, A., and Okamoto, H. (2005). The hedgehog–PKA pathway regulates two distinct steps of the differentiation of retinal ganglion cells: the cell-cycle exit of retinoblasts and their neuronal maturation. *Development* **132**, 1539–1553.

McCallum, C. M., Comai, L., Greene, E. A., and Henikoff, S. (2000). Targeted screening for induced mutations. *Nat. Biotechnol.* **18**, 455–457.

Meng, X., Noyes, M. B., Zhu, L. J., Lawson, N. D., and Wolfe, S. A. (2008). Targeted gene inactivation in zebrafish using engineered zinc-finger nucleases. *Nat. Biotechnol.* **26**, 695–701.

Metcalfe, W. K. (1985). Sensory neuron growth cones comigrate with posterior lateral line primordial cells in zebrafish. *J. Comp. Neurol.* **238**, 218–224.

Metcalfe, W. K., Myers, P., Trevarrow, B., Bass, M., and Kimmel, C. (1990). Primary neurons that express the L2/HNK-1 carbohydrate during early development in the zebrafish. *Development* **110**, 491–504.

Moens, C. B., and Fritz, A. (1999). Techniques in neural development. *Methods Cell Biol.* **59**, 253–272.

Mohideen, M. A., Beckwith, L. G., Tsao-Wu, G. S., Moore, J. L., Wong, A. C., Chinoy, M. R., and Cheng, K. C. (2003). Histology-based screen for zebrafish mutants with abnormal cell differentiation. *Dev. Dyn.* **228**, 414–423.

Morris, A. C., Schroeter, E. H., Bilotta, J., Wong, R. O., and Fadool, J. M. (2005). Cone survival despite rod degeneration in XOPS-mCFP transgenic zebrafish. *Invest. Ophthalmol. Vis. Sci.* **46**, 4762–4771.

Muller, H. (1857). Anatomisch-physiologische untersuchungen uber die Retina bei Menschen und Wirbelthieren. *Z. Wiss. Zool.* **8**, 1–122.

Mullins, M. C., Hammerschmidt, M., Haffter, P., and Nusslein-Volhard, C. (1994). Large-scale mutagenesis in the zebrafish: in search of genes controlling development in a vertebrate. *Curr. Biol.* **4**, 189–202.

Mumm, J. S., Williams, P. R., Godinho, L., Koerber, A., Pittman, A. J., Roeser, T., Chien, C. B., Baier, H., and Wong, R. O. (2006). *In vivo* imaging reveals dendritic targeting of laminated afferents by zebrafish retinal ganglion cells. *Neuron* **52**, 609–621.

Muto, A., Orger, M. B., Wehman, A. M., Smear, M. C., Kay, J. N., Page-McCaw, P. S., Gahtan, E., Xiao, T., Nevin, L. M., Gosse, N. J., Staub, W., Finger-Baier, K., *et al.* (2005). Forward genetic analysis of visual behavior in zebrafish. *PLoS Genet.* **1**, e66.

Nagayoshi, S., Hayashi, E., Abe, G., Osato, N., Asakawa, K., Urasaki, A., Horikawa, K., Ikeo, K., Takeda, H., and Kawakami, K. (2008). Insertional mutagenesis by the Tol2 transposon-mediated enhancer trap approach generated mutations in two developmental genes: Tcf7 and synembryn-like. *Development* **135**, 159–169.

Nasevicius, A., and Ekker, S. C. (2000). Effective targeted gene "knockdown" in zebrafish. *Nat. Genet.* **26**, 216–220.

Nawrocki, W. (1985). Development of the neural retina in the zebrafish (*Brachydanio rerio*). Ph. D. dissertation, University of Oregon, Eugine, OR.

Neuhauss, S. C. (2003). Behavioral genetic approaches to visual system development and function in zebrafish. *J. Neurobiol.* **54**, 148–160.

Neuhauss, S. C., Biehlmaier, O., Seeliger, M. W., Das, T., Kohler, K., Harris, W. A., and Baier, H. (1999). Genetic disorders of vision revealed by a behavioral screen of 400 essential loci in zebrafish. *J. Neurosci.* **19**, 8603–8615.

Neumann, C. J., and Nuesslein-Volhard, C. (2000). Patterning of the zebrafish retina by a wave of sonic hedgehog activity. *Science* **289**, 2137–2139.

Nicolson, T., Rusch, A., Friedrich, R. W., Granato, M., Ruppersberg, J. P., and Nusslein-Volhard, C. (1998). Genetic analysis of vertebrate sensory hair cell mechanosensation: the zebrafish circler mutants. *Neuron* **20**, 271–283.

Norden, C., Young, S., Link, B. A., and Harris, W. A. (2009). Actomyosin is the main driver of interkinetic nuclear migration in the retina. *Cell* **138**, 1195–1208.

Nornes, H. O., Dressler, G. R., Knapik, E. W., Deutsch, U., and Gruss, P. (1990). Spatially and temporally restricted expression of Pax2 during murine neurogenesis. *Development* **109**, 797–809.

North, T. E., Goessling, W., Peeters, M., Li, P., Ceol, C., Lord, A. M., Weber, G. J., Harris, J., Cutting, C. C., Huang, P. *et al.* (2009). Hematopoietic stem cell development is dependent on blood flow. *Cell* **137**, 736–748.

North, T. E., Goessling, W., Walkley, C. R., Lengerke, C., Kopani, K. R., Lord, A. M., Weber, G. J., Bowman, T. V., Jang, I. H., Grosser, T., Fitzgerald, G. A., Daley, G. Q., *et al.* (2007). Prostaglandin E2 regulates vertebrate haematopoietic stem cell homeostasis. *Nature* **447**, 1007–1011.

Novak, A. E., and Ribera, A. B. (2003). Immunocytochemistry as a tool for zebrafish developmental neurobiology. *Methods Cell Sci.* **25**, 79–83.

Omori, Y., Zhao, C., Saras, A., Mukhopadhyay, S., Kim, W., Furukawa, T., Sengupta, P., Veraksa, A., and Malicki, J. (2008). Elipsa is an early determinant of ciliogenesis that links the IFT particle to membrane-associated small GTPase Rab8. *Nature Cell Biol.* **10**, 437–444.

Oxtoby, E., and Jowett, T. (1993). Cloning of the zebrafish krox-20 gene (krx-20) and its expression during hindbrain development. *Nucleic Acids Res.* **21**, 1087–1095.

Passini, M. A., Levine, E. M., Canger, A. K., Raymond, P. A., and Schechter, N. (1997). Vsx-1 and Vsx-2: differential expression of two paired-like homeobox genes during zebrafish and goldfish retinogenesis. *J. Comp. Neurol.* **388**, 495–505.

Pathak, N., Obara, T., Mangos, S., Liu, Y., and Drummond, I. A. (2007). The zebrafish fleer gene encodes an essential regulator of cilia tubulin polyglutamylation. *Mol. Biol. Cell* **18**, 4353–4364.

Pauls, S., Geldmacher-Voss, B., and Campos-Ortega, J. A. (2001). A zebrafish histone variant H2A.F/Z and a transgenic H2A.F/Z: GFP fusion protein for *in vivo* studies of embryonic development. *Dev. Genes Evol.* **211**, 603–610.

Perkins, B. D., Kainz, P. M., O'Malley, D. M., and Dowling, J. E. (2002). Transgenic expression of a GFP-rhodopsin COOH-terminal fusion protein in zebrafish rod photoreceptors. *Vis. Neurosci.* **19**, 257–264.

Peterson, R. E., Fadool, J. M., McClintock, J., and Linser, P. J. (2001). Müller cell differentiation in the zebrafish neural retina: evidence of distinct early and late stages in cell maturation. *J. Comp. Neurol.* **429**, 530–540.

Peterson, R. T., Link, B. A., Dowling, J. E., and Schreiber, S. L. (2000). Small molecule developmental screens reveal the logic and timing of vertebrate development. *Proc. Natl. Acad. Sci. U. S. A.* **97**, 12965–12969.

Peterson, R. T., Shaw, S. Y., Peterson, T. A., Milan, D. J., Zhong, T. P., Schreiber, S. L., MacRae, C. A., and Fishman, M. C. (2004). Chemical suppression of a genetic mutation in a zebrafish model of aortic coarctation. *Nat. Biotechnol.* **22**, 595–599.

Peterson, R. E., Tu, C., and Linser, P. J. (1997). Isolation and characterization of a carbonic anhydrase homologue from the zebrafish (*Danio rerio*). *J. Mol. Evol.* **44**, 432–439.

Plaster, N., Sonntag, C., Busse, C. E., and Hammerschmidt, M. (2006). p53 deficiency rescues apoptosis and differentiation of multiple cell types in zebrafish flathead mutants deficient for zygotic DNA polymerase delta1. *Cell Death Differ.* **13**, 223–235.

Poggi, L., Vitorino, M., Masai, I., and Harris, W. A. (2005). Influences on neural lineage and mode of division in the zebrafish retina *in vivo*. *J. Cell Biol.* **171**, 991–999.

Porteus, M. H., and Carroll, D. (2005). Gene targeting using zinc finger nucleases. *Nat. Biotechnol.* **23**, 967–973.

Prince, V. E., Joly, L., Ekker, M., and Ho, R. K. (1998). Zebrafish hox genes: genomic organization and modified colinear expression patterns in the trunk. *Development* **125**, 407–420.

Pujic, Z., and Malicki, J. (2001). Mutation of the zebrafish *glass onion* locus causes early cell-nonautonomous loss of neuroepithelial integrity followed by severe neuronal patterning defects in the retina. *Dev. Biol.* **234**, 454–469.

Pujic, Z., and Malicki, J. (2004). Retinal pattern and the genetic basis of its formation in zebrafish. *Semin. Cell Dev. Biol.* **15**, 105–114.

Pujic, Z., Omori, Y., Tsujikawa, M., Thisse, B., Thisse, C., and Malicki, J. (2006). Reverse genetic analysis of neurogenesis in the zebrafish retina. *Dev. Biol.* **293**, 330–347.

Raible, D. W., Wood, A., Hodsdon, W., Henion, P. D., Weston, J. A., and Eisen, J. S. (1992). Segregation and early dispersal of neural crest cells in the embryonic zebrafish. *Dev. Dyn.* **195**, 29–42.

Raymond, P., Barthel, L., and Curran, G. (1995). Developmental patterning of rod and cone photoreceptors in embryonic zebrafish. *J. Comp. Neurol.* **359**, 537–550.

Raymond, P., Barthel, L., Rounsifer, M., Sullivan, S., and Knight, J. (1993). Expression of rod and cone visual pigments in godfish and zebrafish: a rhodopsin-like gene is expressed in cones. *Neuron* **10**, 1161–1174.

Raz, E., van Luenen, H. G., Schaerringer, B., Plasterk, R. H., and Driever, W. (1998). Transposition of the nematode *Caenorhabditis elegans* Tc3 element in the zebrafish *Danio rerio*. *Curr. Biol.* **8**, 82–88.

Rembold, M., Loosli, F., Adams, R. J., and Wittbrodt, J. (2006). Individual cell migration serves as the driving force for optic vesicle evagination. *Science* **313**, 1130–1134.

Ren, J. Q., McCarthy, W. R., Zhang, H., Adolph, A. R., and Li, L. (2002). Behavioral visual responses of wild-type and hypopigmented zebrafish. *Vision Res.* **42**, 293–299.

Riley, B. B., Chiang, M., Farmer, L., and Heck, R. (1999). The deltaA gene of zebrafish mediates lateral inhibition of hair cells in the inner ear and is regulated by pax2.1. *Development* **126**, 5669–78.

Robinson, J., Schmitt, E., and Dowling, J. (1995). Temporal and spatial patterns of opsin gene expression in zebrafish (*Danio rerio*). *Vis. Neurosci.* **12**, 895–906.

Rodieck, R. W. (1973). "The Vertebrate Retina: Principles of Structure and Function". W. H. Freeman & Co., San Francisco, CA.

Roeser, T., and Baier, H. (2003). Visuomotor behaviors in larval zebrafish after GFP-guided laser ablation of the optic tectum. *J. Neurosci.* **23**, 3726–3734.

Sachidanandan, C., Yeh, J., Peteerson, Q., and Peteerson, R. (2008). Identification of a novel retinoid by small molecule screening with zebrafish embryos. *PLoS One* **3**, 1–9.

Sandell, J., Martin, S., and Heinrich, G. (1994). The development of GABA immunoreactivity in the retina of the zebrafish. *J. Comp. Neurol.* **345**, 596–601.

Sanes, J. R. (1993). Topographic maps and molecular gradients. *Curr. Opin. Neurobiol.* **3**, 67–74.

Sato, T., Takahoko, M., and Okamoto, H. (2006). HuC:Kaede, a useful tool to label neural morphologies in networks *in vivo*. *Genesis* **44**, 136–142.

Scheer, N., and Campos-Ortega, J. A. (1999). Use of the Gal4–UAS technique for targeted gene expression in the zebrafish. *Mech. Dev.* **80**, 153–158.

Scheer, N., Groth, A., Hans, S., and Campos-Ortega, J. A. (2001). An instructive function for Notch in promoting gliogenesis in the zebrafish retina. *Development* **128**, 1099–1107.

Scheer, N., Riedl, I., Warren, J. T., Kuwada, J. Y., and Campos-Ortega, J. A. (2002). A quantitative analysis of the kinetics of Gal4 activator and effector gene expression in the zebrafish. *Mech. Dev.* **112**, 9–14.

Schier, A. F., Neuhauss, S. C., Helde, K. A., Talbot, W. S., and Driever, W. (1997). The one-eyed pinhead gene functions in mesoderm and endoderm formation in zebrafish and interacts with no tail. *Development* **124**, 327–342.

Schmitt, E., and Dowling, J. (1994). Early eye morphogenesis in the zebrafish, *Brachydanio rerio*. *J. Comp. Neurol.* **344**, 532–542.

Schmitt, E. A., and Dowling, J. E. (1996). Comparison of topographical patterns of ganglion and photo-receptor cell differentiation in the retina of the zebrafish, *Danio rerio*. *J. Comp. Neurol.* **371**, 222–234.

Schmitt, E. A., and Dowling, J. E. (1999). Early retinal development in the zebrafish, *Danio rerio*: light and electron microscopic analyses. *J. Comp. Neurol.* **404**, 515–536

Schmitz, B., Papan, C., and Campos-Ortega, J. (1993). Neurulation in the anterior trunk of the zebrafish *Brachydanio rerio. Roux's Arch. Dev. Biol.* **202**, 250–259.

Schroeter, E. H., Wong, R. O., and Gregg, R. G. (2006). *In vivo* development of retinal ON-bipolar cell axonal terminals visualized in nyx: MYFP transgenic zebrafish. *Vis. Neurosci.* **23**, 833–843.

Scott, E. K., Mason, L., Arrenberg, A. B., Ziv, L., Gosse, N. J., Xiao, T., Chi, N. C., Asakawa, K., Kawakami, K., and Baier, H. (2007). Targeting neural circuitry in zebrafish using GAL4 enhancer trapping. *Nat. Methods* **4**, 323–326.

Seddon, J. (1994). Age-relatred macular degeneration: epidemiology. *In* "Principles and Practice of Ophthalmology" (B. Albert, and F. Jakobiec, eds.) pp. 1266–1274. Philadelphia

Seo, H. C., Drivenes, O., Ellingsen, S., and Fjose, A. (1998). Expression of two zebrafish homologues of the murine Six3 gene demarcates the initial eye primordia. *Mech. Dev.* **73**, 45–57.

Shaner, N. C., Patterson, G. H., and Davidson, M. W. (2007). Advances in fluorescent protein technology. *J. Cell. Sci.* **120**, 4247–4260.

Shields, C. R., Klooster, J., Claassen, Y., Ul-Hussain, M., Zoidl, G., Dermietzel, R., and Kamermans, M. (2007). Retinal horizontal cell-specific promoter activity and protein expression of zebrafish connexin 52.6 and connexin 55.5. *J. Comp. Neurol.* **501**, 765–779.

Sivasubbu, S., Balciunas, D., Davidson, A. E., Pickart, M. A., Hermanson, S. B., Wangensteen, K. J., Wolbrink, D. C., and Ekker, S. C. (2006). Gene-breaking transposon mutagenesis reveals an essential role for histone H2afza in zebrafish larval development. *Mech. Dev.* **123**, 513–529.

Solnica-Krezel, L., Schier, A., and Driever, W. (1994). Efficient recovery of ENU-induced mutations from the zebrafish germline. *Genetics* **136**, 1–20.

Soules, K. A., and Link, B. A. (2005). Morphogenesis of the anterior segment in the zebrafish eye. *BMC Dev. Biol.* **5**, 12.

Strahle, U., Blader, P., Adam, J., and Ingham, P. W. (1994). A simple and efficient procedure for non-isotopic *in situ* hybridization to sectioned material. *Trends Genet.* **10**, 75–76.

Streisinger, G., Singer, F., Walker, C., Knauber, D., and Dower, N. (1986). Segregation analyses and gene–centromere distances in zebrafish. *Genetics* **112**, 311–319.

Streisinger, G., Walker, C., Dower, N., Knauber, D., and Singer, F. (1981). Production of clones of homozygous diploid zebra fish (*Brachydanio rerio*). *Nature* **291**, 293–296.

Stuermer, C. A. (1988). Retinotopic organization of the developing retinotectal projections in the zebrafish embryo. *J. Neurosci.* **12**, 4513–4530.

Take-uchi, M., Clarke, J. D., and Wilson, S. W. (2003). Hedgehog signalling maintains the optic stalk–retinal interface through the regulation of Vax gene activity. *Development* **130**, 955–968.

Takechi, M., Hamaoka, T., and Kawamura, S. (2003). Fluorescence visualization of ultraviolet-sensitive cone photoreceptor development in living zebrafish. *FEBS Lett.* **553**, 90–94.

Takechi, M., and Kawamura, S. (2005). Temporal and spatial changes in the expression pattern of multiple red and green subtype opsin genes during zebrafish development. *J. Exp. Biol.* **208**, 1337–1345.

Taylor, M. R., Hurley, J. B., Van Epps, H. A., and Brockerhoff, S. E. (2004). A zebrafish model for pyruvate dehydrogenase deficiency: rescue of neurological dysfunction and embryonic lethality using a ketogenic diet. *Proc. Natl. Acad. Sci. U. S. A.* **101**, 4584–4589.

Thisse, B., Heyer, V., Lux, A., Alunni, V., Degrave, A., Seiliez, I., Kirchner, J., Parkhill, J. P., and Thisse, C. (2004). Spatial and temporal expression of the zebrafish genome by large-scale *in situ* hybridization screening. *Methods Cell Biol.* **77**, 505–519.

Tran, T. C., Sneed, B., Haider, J., Blavo, D., White, A., Aiyejorun, T., Baranowski, T. C., Rubinstein, A. L., Doan, T. N., Dingledine, R., and Sandberg, E. M (2007). Automated, quantitative screening assay for antiangiogenic compounds using transgenic zebrafish. *Cancer Res.* **67**, 11386–11392.

Trowe, T., Klostermann, S., Baier, H., Granato, M., Crawford, A. D., Grunewald, B., Hoffmann, H., Karlstrom, R. O., Meyer, S. U., Muller, B., Richter, S., Nüsslein-Volhard, C., *et al.* (1996). Mutations disrupting the ordering and topographic mapping of axons in the retinotectal projection of the zebrafish, *Danio rerio. Development* **123**, 439–450.

Tsujimura, T., Chinen, A., and Kawamura, S. (2007). Identification of a locus control region for quadruplicated green-sensitive opsin genes in zebrafish. *Proc. Natl. Acad. Sci. U. S. A.* **104**, 12813–12818.

Tsujikawa, M., and Malicki, J. (2004a). Genetics of photoreceptor development and function in zebrafish. *Int. J. Dev. Biol.* **48**, 925–934.

Tsujikawa, M., and Malicki, J. (2004b). Intraflagellar transport genes are essential for differentiation and survival of vertebrate sensory neurons. *Neuron* **42**, 703–716.

Turner, D., and Cepko, C. (1987). A common progenitor for neurons and glia persists in rat retina late in development. *Nature* **328**, 131–136.

Turner, D., Snyder, E., and Cepko, C. (1990). Lineage-independent determination of cell type in the embryonic mouse retina. *Neuron* **4**, 833–845.

Urasaki, A., Morvan, G., and Kawakami, K. (2006). Functional dissection of the Tol2 transposable element identified the minimal *cis*-sequence and a highly repetitive sequence in the subterminal region essential for transposition. *Genetics* **174**, 639–649.

van Eeden, F. J., Granato, M., Odenthal, J., and Haffter, P. (1999). Developmental mutant screens in the zebrafish. *Methods Cell Biol.* **60**, 21–41.

Varga, Z. M., Wegner, J., and Westerfield, M. (1999). Anterior movement of ventral diencephalic precursors separates the primordial eye field in the neural plate and requires cyclops. *Development* **126**, 5533–5546.

Vihtelic, T. S., Doro, C. J., and Hyde, D. R. (1999). Cloning and characterization of six zebrafish photoreceptor opsin cDNAs and immunolocalization of their corresponding proteins. *Vis. Neurosci.* **16**, 571–585.

Vihtelic, T. S., and Hyde, D. R. (2000). Light-induced rod and cone cell death and regeneration in the adult albino zebrafish (*Danio rerio*) retina. *J. Neurobiol.* **44**, 289–307.

Villefranc, J. A., Amigo, J., and Lawson, N. D. (2007). Gateway compatible vectors for analysis of gene function in the zebrafish. *Dev. Dyn.* **236**, 3077–3087.

Walker, C. (1999). Haploid screens and gamma-ray mutagenesis. *Methods Cell Biol.* **60**, 43–70.

Watanabe, T., and Raff, M. (1988). Retinal astrocytes are immigrants from the optic nerve. *Nature* **332**, 834–837.

Wei, X., and Malicki, J. (2002). Nagie oko, encoding a MAGUK-family protein, is essential for cellular patterning of the retina. *Nat. Genet.* **31**, 150–157.

Westerfield, M. (2000). "The Zebrafish Book. A Guide for the Laboratory Use of Zebrafish (*Danio rerio*)". University of Oregon Press, Eugene.

Wetts, R., and Fraser, S. (1988). Multipotent precursors can give rise to all major cell types of the frog retina. *Science* **239**, 1142–1145.

Wienholds, E., Schulte-Merker, S., Walderich, B., and Plasterk, R. H. (2002). Target-selected inactivation of the zebrafish rag1 gene. *Science* **297**, 99–102.

Wise, G., Dollery, C., and Henkind, P. (1971). "The Retinal Circulation". Harper & Row, New York.

Woo, K., and Fraser, S. E. (1995). Order and coherence in the fate map of the zebrafish nervous system. *Development* **121**, 2595–2609.

Wright, D. A., Thibodeau-Beganny, S., Sander, J. D., Winfrey, R. J., Hirsh, A. S., Eichtinger, M., Fu, F., Porteus, M. H., Dobbs, D., Voytas, D. F., and Joung, J. K. (2006). Standardized reagents and protocols for engineering zinc finger nucleases by modular assembly. *Nat. Protoc.* **1**, 1637–1652.

Xiao, T., Roeser, T., Staub, W., and Baier, H. (2005). A GFP-based genetic screen reveals mutations that disrupt the architecture of the zebrafish retinotectal projection. *Development* **132**, 2955–2967.

Yamaguchi, M., Fujimori-Tonou, N., Yoshimura, Y., Kishi, T., Okamoto, H., and Masai, I. (2008). Mutation of DNA primase causes extensive apoptosis of retinal neurons through the activation of DNA damage checkpoint and tumor suppressor p53. *Development* **135**, 1247–1257.

Yamaguchi, M., Imai, F., Tonou-Fujimori, N., and Masai, I. (2010). Mutations in N-cadherin and a Stardust homolog, Nagie oko, affect cell-cycle exit in zebrafish retina. *Mech. Dev.* **127**, 247–264.

Yamamoto, Y., and Jeffery, W. R. (2002). Probing teleost eye development by lens transplantation. *Methods* **28**, 420–426.

Yazulla, S., and Studholme, K. M. (2001). Neurochemical anatomy of the zebrafish retina as determined by immunocytochemistry. *J. Neurocytol.* **30**, 551–592.

Yeo, S. Y., Kim, M., Kim, H. S., Huh, T. L., and Chitnis, A. B. (2007). Fluorescent protein expression driven by her4 regulatory elements reveals the spatiotemporal pattern of Notch signaling in the nervous system of zebrafish embryos. *Dev. Biol.* **301**, 555–567.

Yu, C. J., Gao, Y., Willis, C. L., Li, P., Tiano, J. P., Nakamura, P. A., Hyde, D. R., and Li, L. (2007). Mitogen-associated protein kinase- and protein kinase A-dependent regulation of rhodopsin promoter expression in zebrafish rod photoreceptor cells. *J. Neurosci. Res.* **85**, 488–496.

Zhang, Y., McCulloch, K., and Malicki, J. (2009). Lens transplantation in zebrafish and its application in the analysis of eye mutants. *J Vis. Exp.*

Zhao, X. C., Yee, R. W., Norcom, E., Burgess, H., Avanesov, A. S., Barrish, J. P., and Malicki, J. (2006). The zebrafish cornea: structure and development. *Invest. Ophthalmol. Vis. Sci.* **47**, 4341–4348.

Zolessi, F. R., Poggi, L., Wilkinson, C. J., Chien, C. B., and Harris, W. A. (2006). Polarization and orientation of retinal ganglion cells *in vivo*. *Neural Dev.* **1**, 2.

Zon, L. I., and Peterson, R. T. (2005). *In vivo* drug discovery in the zebrafish. *Nat. Rev. Drug Discov.* **4**, 35–44.

CHAPTER 7

Photoreceptor Structure and Development: Analyses using GFP Transgenes

Brian D. Perkins[*] *and* James M. Fadool[†]

[*]Department of Biology, Texas A&M University, College Station, Texas

[†]Department of Biological Science, Florida State University, Tallahassee, Florida.

Abstract

I. Introduction
  A. Photoreceptor Anatomy and Biochemistry
  B. Photoreceptor Development
  C. Improvements in Transgenic Technology
II. Transport Mechanisms
III. Regulation of Photoreceptor Size
IV. Photoreceptor Synapse Structure
V. Regeneration
VI. Future Directions
VII. Conclusions
References

Abstract

In recent years, studies of zebrafish rod and cone photoreceptors have yielded novel insights into the differentiation of distinct photoreceptor cell types and the mechanisms guiding photoreceptor regeneration following cell death, and they have provided models of human retinal degeneration. These studies were facilitated by the use of transgenic zebrafish expressing fluorescent reporter genes under the control of various cell-specific promoters. Improvements in transgenesis techniques (e.g., Tol2 transposition), the availability of numerous fluorescent reporter genes with different localization properties, and the ability to generate transgenes via recombineering (e.g., Gateway technology) have enabled researchers to quickly develop transgenic lines that improve our understanding of the causes of human blindness and ways to mitigate its effects.

I. Introduction

More is known about photoreceptor cells than any other cell in the vertebrate retina. From early studies in psychophysics (Hecht *et al.*, 1942) and on visual pigments (Wald, 1955), to the work on mechanisms of dark adaptation (Dowling, 1963), phototransduction (Yau, 1994), and inherited retinal disease (Dryja and Li, 1995), rod photoreceptors have been central to the understanding of retinal function and hereditary blindness disorders. Many of the mutations known to cause hereditary retinal degeneration affect rod-specific genes or otherwise interfere with rod function. For reasons still not fully understood, rod degeneration often leads to the secondary death of cone photoreceptors and the loss of color vision and eventually all vision. For decades, insights into the mechanisms of rod photoreceptor degeneration came from studies of naturally occurring mutations and gene knock-outs in nocturnal mammalian models such as rats and mice, whose retinas have ~95% rods. More recently, diseases primarily affecting cone photoreceptors, such as achromatopsia, have prompted investigations into the biochemical and physiological properties of cones. The zebrafish is a diurnal animal with a retina containing large numbers of diverse cone subtypes (Branchek, 1984; Branchek and Bremiller, 1984; Fadool, 2003; Raymond *et al.*, 1993, 1995), which make zebrafish an ideal model to study the pathological mechanisms underlying both rod and cone degeneration, as well as to examine the capacity of the vertebrate retina to regenerate rods and cones.

A. Photoreceptor Anatomy and Biochemistry

Both rod and cone photoreceptors are highly specialized cells with a unique morphology consisting of an elongated outer segment, connecting cilium, inner segment, cell body, and synaptic terminal (Fig. 1). The shape and morphology of the outer segments (OSs) usually distinguish the rods from the cones and provide the basis for their nomenclature. The OS consists of hundreds of tightly stacked membrane discs that contain the proteins necessary for phototransduction. Protein synthesis occurs in the inner segment, and molecules destined for the OS must be transported apically through the connecting cilium via a process known as intraflagellar transport (IFT) (see Insinna and Besharse, 2008 for a recent review). The inner segment also contains numerous mitochondria needed to provide the energy for the demands of protein synthesis, protein trafficking, and phototransduction. The synaptic terminals of rod photoreceptors, known as spherules, are typically smaller than the cone terminals, known as pedicles. Both spherules and pedicles are filled with synaptic vesicles, contain synaptic ribbons, and are presynaptic to the bipolar and horizontal cells.

Zebrafish possess one type of rod photoreceptor and four subtypes of cone photoreceptors. The zebrafish cone subtypes absorb light maximally in the red (570 nm), green (480 nm), blue (415 nm), and ultraviolet (362 nm) regions of the spectrum (Robinson *et al.*, 1993). These cone types are also distinguishable morphologically; short and long single cones contain the ultraviolet (UV)- and blue-absorbing visual pigments, respectively, whereas the red- and green-sensitive visual pigments are found

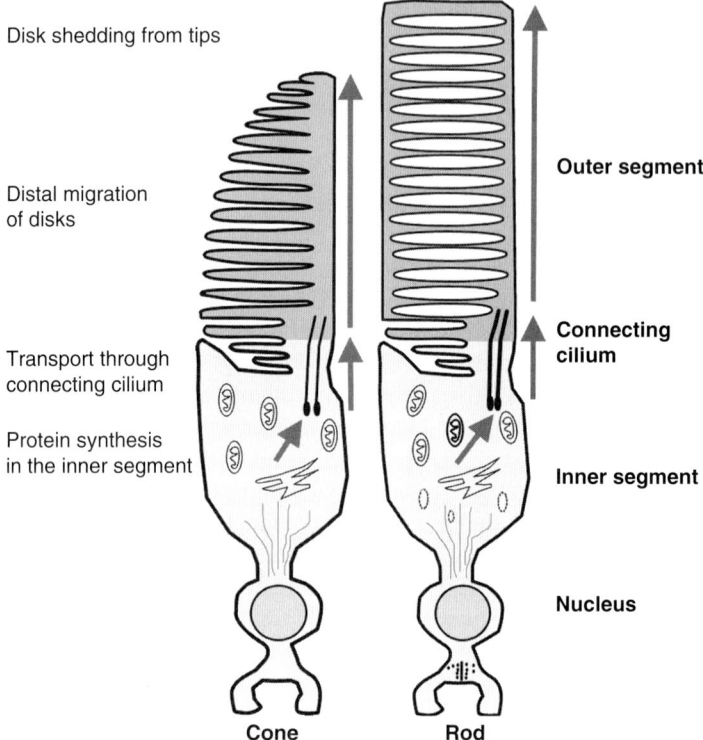

Disk shedding from tips

Distal migration
of disks

Transport through
connecting cilium

Protein synthesis
in the inner segment

Outer segment

**Connecting
cilium**

Inner segment

Nucleus

Cone   Rod

Fig. 1 Anatomy cartoon diagrams of vertebrate cone (left) and rod (right) photoreceptors. Arrows indicate the direction of protein trafficking for most outer segment proteins. The outer segments are more darkly shaded and cellular processes relevant for outer segment development are outlined on the left.

in the principal and accessory members of the double cones (Robinson *et al.*, 1993). The cone photoreceptors are arranged in a highly ordered crystalline mosaic, with rod inner segments projecting through the cone mosaic to surround the UV cones (Fadool, 2003). Additional details about the cone mosaic and cone physiology can be found in two recent review articles (Fadool and Dowling, 2008; Raymond and Barthel, 2004)

B. Photoreceptor Development

In vertebrates, neurogenesis of the retina begins when proliferating neuroblasts exit the cell cycle and begin the process of differentiation. In most species, including zebrafish, cone differentiation occurs before rod differentiation. The expression of the zebrafish rod opsin gene begins at approximately 50 h post-fertilization (hpf) (Schmitt and Dowling, 1996) in a region of precocious neurogenesis in the ventral nasal retina referred to as the ventral patch. The red and blue cone opsin genes are also expressed at about 50 hpf while expression of the UV opsin gene is observed about 5 h

later (Schmitt *et al.*, 1999). Similarly, small clusters of rods and the red/green double cones are labeled with the monoclonal antibodies ROS-1 and Zpr-1, respectively, in whole mount at 50 hpf (Raymond *et al.*, 1995). Given that the first cells in the outer nuclear layer (ONL) of the retina to become post-mitotic are seen between 43 and 48 hpf (Hu and Easter, 1999), it is likely that cell cycle exit precedes opsin expression by several hours. Cone differentiation occurs in a sweeping fashion from the ventro-nasal side of the choroid fissure to the dorsonasal retina and then to the dorsotemporal and ventrotemporal side of the choroid fissure (Raymond *et al.*, 1995; Schmitt and Dowling, 1996). On the other hand, rod differentiation initiates in the ventral retina on the nasal side of the choroid fissure but soon crosses directly to the temporal side of the fissure. This ventral patch of rods grows symmetrically across the choroid fissure and increases in density while slowly expanding. Between 50 and 72 hpf, rods fill in the dorsal retina in a seemingly random fashion that does not resemble the wave of differentiation exhibited by all other retinal cell types (Raymond *et al.*, 1995; Schmitt and Dowling, 1996). Despite similarities in timing of early rod and cone differentia-tion, zebrafish vision is predominantly cone driven for several days. Based upon behavioral studies and electroretinogram (ERG) recordings, rod function can first be demonstrated only between 14 and 21 days post-fertilization (dpf) (Bilotta *et al.*, 2001; Saszik *et al.*, 1999). Spectral sensitivity appears to be cone dominated through 15 dpf, and the rod contribution is not adult-like until close to 1 month of age (Bilotta *et al.*, 2001; Saszik *et al.*, 1999).

A number of intrinsic and extrinsic factors contribute to the specification, differ-entiation, and maturation of rods, but the story is far from complete. For example, the transcription factors neural retina leucine zipper (Nrl), cone–rod homeobox (*crx*) gene, *neuroD*, and *tbx2b* regulate distinct aspects of rod development (Alvarez-Delfin *et al.*, 2009; Furukawa *et al.*, 1997, 1999; Mears *et al.*, 2001). Extrinsic factors known to promote rod specification and differentiation include sonic hedgehog (Shh) (Shkumatava *et al.*, 2004; Stenkamp and Frey, 2003), taurine (Young and Cepko, 2004), fibroblast growth factor (Patel and McFarlane, 2000), and retinoic acid (Hyatt *et al.*, 1996; Perkins *et al.*, 2002). Given the potentially broad 12-h time window for cell cycle exit and differentiation of photoreceptors (~43–55 hpf), it is unclear if these extrinsic signals contribute equally to the specification and differentiation of rods and cones.

C. Improvements in Transgenic Technology

The techniques to generate transgenic zebrafish have significantly improved in the last 5 years. In the past, linearized plasmid DNA containing the transgene was injected into the yolks of 1–4-cell-stage zebrafish embryos. In a small percentage of cases (typically 5–10%), the DNA randomly integrates into the genome of the germ cells. These "founder fish" produced transgenic progeny. As described in a previous review (Perkins *et al.*, 2004), this approach was used to generate transgenic zebrafish lines that express GFP exclusively in rod photoreceptor cells (Fadool, 2003; Hamaoka *et al.*, 2002; Kennedy *et al.*, 2001; Perkins *et al.*, 2002), one line that expresses GFP in all

cone photoreceptors (Kennedy *et al.*, 2007), and one line that selectively marks UV cones (Takechi *et al.*, 2003). Today, transgenesis frequencies have been increased by creating plasmids that contain sequences recognized by a rare-cutting endonuclease, such as *I-SceI*, or by inserting the transgene into the Tol2 transposon (see Chapter 4). In both cases, the percentage of injected embryos transiently expressing the transgene is elevated severalfold, and the rate of germ line transmission has approached 70% (Balciunas *et al.*, 2006). Construction of Tol2 vectors for transgenesis is now facilitated by the use of site-specific recombination cloning (Gateway®) technology, as part of the "Tol2kit" (Kwan *et al.*, 2007). The Tol2kit consists of various plasmids containing Gateway recombination sequences. A functional Tol2 transposon may be rapidly assembled in a modular fashion by recombining a promoter-containing 5′ element, a 3′-tag, and a Tol2 transposon backbone. With these tools in hand, one can quickly construct Tol2 transposons and generate transgenic lines that express fluorescent markers behind cell-specific or inducible promoters and express proteins fused at the N- or C-terminus with fluorescent markers or protein purification tags, as well as other variations.

II. Transport Mechanisms

Photoreceptor OSs rapidly turn over throughout the lives of photoreceptor cells; thus, efficient protein transport mechanisms provide photoreceptors with a sufficient supply of protein to replenish that lost during OS turnover. It is estimated that rhodopsin constitutes 90% of the total protein in the rod OS and therefore plays important roles in both the physiology and structural integrity of the photoreceptor (Nathans, 1992). As such, investigations into this process have primarily focused on the transport of rhodopsin.

The photoreceptor OS can be considered a specialized sensory cilium that concentrates opsin for light detection. All cilia consist of a microtubule-based axoneme surrounded by the ciliary membrane. Beginning at the basal body, the photoreceptor axoneme extends into a transition zone, known as the connecting cilium, and continues distally into the OS where it terminates near the tip as singlet microtubules (Insinna and Besharse, 2008). Disc membrane assembly occurs at the base of the OS, just beyond the distal end of the transition zone. A process known as intraflagellar transport, or IFT, transports opsin molecules through the connecting cilium where they then incorporate into newly assembled disc membranes (Insinna *et al.*, 2008, 2009a; Marszalek *et al.*, 2000; Pazour *et al.*, 2002). IFT refers to the bidirectional movement of multisubunit IFT particles along the length of the ciliary axoneme. The IFT particle is composed of at least 18 distinct proteins and perhaps dozens of associated protein components and cargo molecules (Hao and Scholey, 2009). Although most OS proteins incorporate into disc membranes and are eventually shed from the distal tips of photoreceptors, IFT particles are believed to migrate along the length of the axoneme in both an antero-grade and a retrograde manner. Evidence for this comes from transient transgenesis experiments where constructs containing IFT proteins fused to GFP were expressed

using a zebrafish rhodopsin promoter (Insinna *et al.*, 2009b; Kennedy *et al.*, 2001; Luby-Phelps *et al.*, 2008). Rod-specific overexpression of IFT20-GFP, IFT52-GFP, IFT57-GFP, and IFT88-GFP found localization of IFT proteins along the entire length of the axoneme, as well as the basal body and the connecting cilium.

Cilia formation and anterograde IFT movement require two kinesin motor proteins. The heterotrimeric kinesin-II motor, which is composed of the Kif3a and Kif3b kinesin subunits and a kinesin-associated protein (KAP) subunit, was first shown to be required for rod photoreceptors (Marszalek *et al.*, 2000). Studies in *Caenorhabditis elegans*, however, revealed that a homodimeric kinesin-II motor, OSM-3, was also required for ciliogenesis in some sensory neurons. Insinna and colleagues (2008, 2009a) found that Kif17, the vertebrate homolog to OSM-3, was also essential for ciliogenesis and OS formation in zebrafish photoreceptors. The authors compared the roles of these different kinesin motors in photoreceptor function by creating new constructs that utilized the promoter for the zebrafish cone transducin alpha subunit (TαC) (Kennedy *et al.*, 2007) to drive dominant-negative forms of either the *Xenopus* Kif3b (Lin-Jones *et al.*, 2003) or the mouse Kif17 (Chu *et al.*, 2006). In both cases, the motor domain of the kinesin protein was replaced by GFP. The 3.2 kilobase fragment of the TαC promoter drives expression in all cone subtypes, although the first observed expression of a GFP transgene does not occur until ~70 hpf, which is several hours after endogenous opsin expression (Kennedy *et al.*, 2007). Overexpression of the dominant-negative Kif17 (DNKIF17) caused severe ablation of cone OSs, whereas the dominant-negative Kif3b (DNKIF3B) was less damaging to cones. In contrast, opsin mislocalization was observed in cells expressing DNKIF3B but not DNKIF17, suggesting that these two kinesins regulate trafficking of distinct cargoes (Insinna *et al.*, 2009b).

III. Regulation of Photoreceptor Size

The length of mature photoreceptor OSs remains almost constant through the daily balancing act of disc membrane renewal. Approximately 10% of the OS is shed from the apical tips and this material is replaced at the OS base on a daily basis. Almost nothing is known about the mechanisms that maintain OS at a constant size or the mechanisms that regulate the IFT process, thereby facilitating OS renewal. Recent work has suggested that the 4.1 protein, ezrin, radixin, moesin (FERM) protein Mosaic eyes may function to negatively regulate apical renewal of photoreceptors (Hsu *et al.*, 2006). The zebrafish *mosaic eyes* (*moe*) gene is essential for proper retinal lamination (Jensen *et al.*, 2001), thereby preventing analysis of photoreceptor morphology. The lamination defects could be rescued, however, through genetic mosaic analysis, which also revealed cell-autonomous functions for Moe. The rod-specific *(Tg)XOPS:GFP* transgene was placed on the *moe*[-/-] background to help visualize transplanted photo-receptors and to provide a means to measure cell volume. In the wild-type host retinas, GFP-expressing *moe*[-/-] rods adopted a normal morphology but the cell volume was nearly 50% greater than wild type at 6 dpf. Most of this increased volume appeared to

be due to an increase in OS volume. The Moe protein directly interacted with a number of Crumbs proteins and this interaction suggests a potential regulatory mechanism for OS length. In *Drosophila*, overexpression of *crumbs* expands the apical domains of ectodermal epithelia (Wodarz *et al.*, 1995) and lengthens the stalk region of photoreceptors (Pellikka *et al.*, 2002). These results suggest a conserved mechanism whereby Crumbs proteins function positively in OS growth and Moe negatively regulates Crumbs function (Hsu *et al.*, 2006).

IV. Photoreceptor Synapse Structure

Genetic screens utilizing the optokinetic response (OKR) assay have successfully identified mutants that affect synaptic transmission between cone photoreceptors and bipolar cells of the inner retina (Brockerhoff *et al.*, 1995; Muto *et al.*, 2005; Van Epps *et al.*, 2004). One such example is the *no optokinetic response c* (*nrc*) mutant, which disrupts the synaptojanin 1 protein. Synaptojanin 1 is a polyphosphoinositide phosphatase that regulates clathrin-mediated endocytosis at conventional synapses. In *nrc* mutant photoreceptor synapses, the ribbons formed normally but would "float" unanchored from the synaptic junction, fewer synaptic vesicles were present within the synapse, and the arciform density, which anchors the ribbon to the plasma membrane, was missing (Allwardt *et al.*, 2001; Van Epps *et al.*, 2004). Interestingly, changes in synaptic architecture could be observed using confocal microscopy of transgenic mutants (*Tg(TαC:GFP)nrc*). In wild-type cells expressing the *TαC:GFP* transgene, the cones form invaginating synapses that appear like a "donut" of fluorescence, with a dark center corresponding to the non-fluorescent bipolar cells. The mutant cone terminals were flattened and lacked the dark center, indicating that *nrc* mutant cones failed to form invaginating synapses with bipolar cells.

V. Regeneration

Persistent neurogenesis and regeneration in the visual system of adult teleost fish have been valuable models of neural development. Zebrafish, like many teleosts, continue to grow throughout their life, and the increase in body mass is matched by an increase in the size of the eye and the area of the retina (Fernald, 1990; Johns, 1982; Johns and Fernald, 1981; Marcus *et al.*, 1999; Otteson *et al.*, 2001). In the adult, neurogenesis occurs from two distinct stem cell populations (Hecht *et al.*, 1942): multipotent neural progenitors at the retinal margin, which can give rise to all retinal cell types except rods, and (Wald, 1955) slowly dividing stem cells within the ONL that serve as rod precursors. At the retinal margin, mitotic cells possess properties of stem cells, maintaining a balance between the self-renewal and the generation of multipotent neuroblasts that differentiate into all classes of neurons and glia. The spatial expression of proneural genes, cell-to-cell signaling molecules, and cellular differentiation markers recapitulates the temporal sequence observed during embryonic

development. The identification of neurogenesis at the margin of the hatchling chick retina and the isolation of proliferative cells from the ciliary margin of rodents suggested an evolutionarily conserved system, although further experimental evidence is warranted (Ahmad, 2001; Cicero *et al.*, 2009; Fischer and Reh, 2000; Haruta *et al.*, 2001; Tropepe *et al.*, 2000).

During post-embryonic growth in teleosts, new rods are generated in the central retina from a population of mitotic cells referred to as the rod progenitor lineage. As the animal grows, the retina is gradually stretched within the expanding optic cup, and cell density decreases for all cells except rod photoreceptors. Visual acuity is maintained by increasing the size of the retinal image proportional to that of the increase in eye size. Visual sensitivity, however, is maintained by the generation of new rods in the central retina from a population of mitotic cells referred to as rod progenitor cells (Johns, 1982; Johns and Fernald, 1981). The addition of new rods ensures a constant density across the ONL, thereby preserving scotopic sensitivity (Fernald, 1990; Johns, 1982). Rod progenitors were initially identified as proliferating cells that were distributed across the ONL and served as a source of newly generated rods (Johns, 1982; Johns and Fernald, 1981). These rod precursors proliferate at a low rate and subsequently differentiate into rod photoreceptors. Subsequent studies using multiple injections of tritiated thymidine or exposure to thymidine analogues revealed groups of mitotically active cells arranged in radial arrays spanning the inner nuclear layer (INL) and ONL. In histological sections, these "neurogenic clusters" appeared to migrate along Müller glia. The author proposed that the clusters of proliferating cells were rod progenitors that migrated to the outer retina and became the source of the rod precursors (Johns, 1982).

Perhaps more interesting is that in teleosts, including zebrafish, cell death resulting from mechanical damage, chemical toxicity, phototoxicity, or genetic lesions stimulates an increase in cellular proliferation in the INL and ONL, followed by regeneration of lost cell types, including damaged rod photoreceptors (Braisted *et al.*, 1994; Fausett and Goldman, 2006; Vihtelic and Hyde, 2000; Wu *et al.*, 2001), and recovery of vision (Mensinger and Powers, 1999). Until recently, the cellular and molecular processes underlying regeneration remained largely unknown; however, transgenic analysis of adult neurogenesis in zebrafish has contributed significantly to the understanding of the origin of the cells that maintain the regenerative potential and the genes that regulate regeneration of photoreceptor cells.

Several lines of evidence have identified a subpopulation of Müller glia as the origin of the "neurogenic clusters" that give rise to both the rod progenitor lineage and the multipotent stem cells observed in regenerating fish retinas (Bernardos *et al.*, 2007; Fausett *et al.*, 2008; Fimbel *et al.*, 2007; Morris *et al.*, 2008; Thummel *et al.*, 2008; Yurco and Cameron, 2005). In both the intact juvenile retina and the following acute photoreceptor cell damage, double-labeling experiments demonstrated that these proliferating cells co-label for definitive markers of Müller cells, including glial fibrillary acid protein (GFAP), carbonic anhydrase, and glutamine synthetase. The most convincing evidence of the single origin for both of these processes was provided by Barnados *et al.* (Bernardos and Raymond, 2006; Bernardos *et al.*, 2007) who used *Tg*

(gfap:GFP)^{mi2002} transgenic zebrafish, in which regulatory elements of the zebrafish GFAP gene drove expression of cytoplasmic or nuclear-targeted GFP specifically in Müller glia. The authors used the persistence of GFP fluorescence as a lineage tracer to track the fates of cells derived from the Müller glia. Following bromodeoxyuridine (BrdU) labeling to identify mitotically active cells in the uninjured retina, the authors observed small numbers of BrdU+ cells in the INL and ONL, some of which were also GFP+. Following light damage, the number of mitotically active (i.e., BrdU+) GFP+ cells increased considerably and could be traced as clusters extending from the INL to the ONL where some co-labeled for the transcription factor Crx and other markers of differentiating photoreceptors. The authors concluded that the co-labeled cells were de-differentiated Müller glia and their progeny, which migrate from the INL into the ONL to form rod precursors and regenerated photoreceptors, respectively. It has also been demonstrated that constant intense light treatment of dark-adapted albino zebrafish selectively kills rod and cone photoreceptors in the central retina (Qin *et al.*, 2009; Vihtelic and Hyde, 2000) and induces approximately 50% of the Müller glia to co-label for mitotic markers such as proliferating cell nuclear antigen (PCNA) (Thummel *et al.*, 2008). Injection and electroporation of antisense morpholinos complementary to PCNA prior to retinal lesion led to a significant increase in the number of dying cells in the INL and reduced both the number of proliferating cells and the number of Müller glia in the region of the light damage (Thummel *et al.*, 2008). These data suggest that following retinal lesion, asymmetric cell division of Müller glia generates a mitotic progenitor that gives rise to the neurogenic cluster while maintaining a constant population of Müller glia.

As regeneration progresses, the Müller-derived stem cells expressed numerous genes consistent with a photoreceptor developmental program. Following retinal damage, the Müller glial-derived mitotic cells initiate expression of retinal stem/ progenitor cell markers, including *ascl1a, pax6, rx1, neurogenin1*, and *chx10* (Fausett and Goldman, 2006; Raymond *et al.*, 2006; Thummel *et al.*, 2010). As the retinal progenitors migrate from the INL to the ONL, Pax6 expression is lost and other transcription factors are expressed, most notably the basic helix-loop-helix gene *neurod*, and the cone–rod homeobox gene *crx*. Similarly, cell signaling molecules of the Notch–Delta pathway and its downstream effectors are also expressed (Raymond *et al.*, 2006). These data suggest that progenitor cells produced in the INL in response to acute injury pass through a series of intervening, less-committed states prior to adopting a photoreceptor cell fate.

To uncover genes specifically expressed by the Müller glial-derived mitotic progenitors, the gene expression profiles were compared with those of GFP+ cells isolated from intact and light-lesioned *Tg(gfap:GFP)^{mi2002}* zebrafish retinas (Qin *et al.*, 2009). Using fluorescence-activated cell sorting, the GFP+ cells were isolated from control retinas and retinas up to 36 h post light damage. Similar to other studies of changes in gene expression following acute damage, gene networks associated with increased proliferation, stress response, and neurogenesis were activated. The regeneration of the retina shares features common to other regenerating tissues including the heart and fin. Two genes required for fin and heart regeneration in zebrafish—*hspd1*,

which encodes heat shock protein 60 and *mps1*, a protein kinase involved in mitotic checkpoint regulation—are upregulated in injury-activated Müller glia (Qin *et al.*, 2009). Co-labeling confirmed that both genes were expressed in the mitotically activated Müller glia and their progeny. Quite fortuitously, mutant alleles exist in zebrafish for both of these genes. The mutant *no blastema* (*nbl*) is a temperature-sensitive null allele of *hspd1* that disrupts chaperone activity (Makino *et al.*, 2005). *nightcap* (*ncp*) has a missense substitution in the conserved kinase domain of *mps1* and also exhibits a temperature-sensitive phenotype (Poss *et al.*, 2002). By rearing fish at the non-permissive temperature, the authors demonstrated that both were required at different stages of cone photoreceptor regeneration following light damage.

Whereas current work in numerous species has focused on the Müller cells and the INL stem cells, understanding the regulation of rod progenitor cell activity may offer an alternative avenue to investigate the fundamental processes of photoreceptor replacement. Two previous limitations of such studies have been that rod precursors are relatively few in number and no definitive marker specific for the rod progenitors is available. The *XOPS-mCFP* transgenic line experiences selective degeneration of the rod photoreceptor cells due to the toxic effects of high-level expression of a rod-targeted fluorescent reporter gene (Morris and Fadool, 2005). This rod degeneration resulted in a loss of rod-mediated electrophysiological responses, but did not cause any secondary cone pathology. It was, however, accompanied by a significant increase in rod progenitor mitotic activity (Morris and Fadool, 2005). Similar to cone regeneration, the transcription factors NeuroD and Crx were upregulated in cells in the ONL. But, in contrast, expression of *nr2e3*—a rod determination gene found exclusively in post-mitotic differentiating rod photoreceptors—was also upregulated, whereas expression of *pax6* was not observed (Morris and Fadool, 2005). Immunolabeling of retinal cryosections with anti-BrdU showed no significant increase in INL BrdU+ cells or PCNA expression. The expression of genes with demonstrated roles in photoreceptor development, such as *neurod, crx*, and *nr2e3*, combined with the lack of *pax6* expression, suggests an early commitment of the mitotic rod precursors to the rod cell fate (Morris and Fadool, 2005). Supporting this conclusion, knock-down of both Pax6a and Pax6b resulted in increased numbers of rod precursors in the ONL without affecting rod regeneration (Thummel *et al.*, 2010). Therefore, it appears that the rod progenitors in the ONL maintain the capacity to respond to rod photoreceptor degeneration without relying on increased activity of progenitor cells in the INL or the Müller glia. By contrast, cone-specific cell death resulting from mutation of the *pde6c* gene stimulates Müller glial proliferation in the absence of rod cell death (Morris *et al.*, 2008) which we inferred that the retina responds to rod and cone cell death differently.

Others have demonstrated mitosis of Müller glia following a specific form of rod death. Regeneration following acute and specific loss of all mature rod photoreceptors was tested using the bacterial nitroreductase (NTR)/metronidazole cell ablation system (Montgomery *et al.*, 2010). The authors describe two independent transgenic lines of zebrafish that express an NTR-EGFP fusion protein from the zebrafish rod opsin promoter. Similar lines have also been generated to induce the specific loss of bipolar cells in the retina (Montgomery *et al.*, 2010). In the former studies, uniform expression

of NTR-EGFP in all rod photoreceptors was observed in the $Tg(zop:nfsB-EGFP)^{nt19}$ line, whereas a subset of rods express NTR-EGFP in $Tg(zop:nfsB-EGFP)^{nt20}$ line. Treatment with metronidazole resulted in the death of all rods expressing the NTR-EGFP fusion protein. In contrast to what we observed in the $Tg(xops:mCFP)$ transgenic line, metronidazole treatment of the $Tg(zop:nfsB-EGFP)^{nt19}$ fish increased Müller glial cell proliferation. However, the authors reported that only increased rod precursor cell proliferation was observed following treatment of the $Tg(zop:nfs-B-EGFP)^{nt20}$ fish. Thus, acute loss of the majority of rod photoreceptors was sufficient to induce a Müller glial regeneration response, suggesting that the level of cell death is an important factor regulating increased proliferation of Muller glial versus rod progenitors.

These studies highlight the significant gains made possible by combining advances in many areas of zebrafish genetics such as the development of novel transgenic lines, application of cell sorting, gene profiling arrays, and cell ablation technologies. The ever-increasing utility of transgenic tools to investigate photoreceptor biology in zebrafish has mirrored many of the advances in other organ systems. The continuation of these studies should allow a more detailed characterization of the intrinsic properties of retinal stem cells, identification of the environmental signals that direct regeneration of the photoreceptor cells, and development of methods to manipulate cells to direct photoreceptor cell replacement.

VI. Future Directions

Transgenic zebrafish are powerful tools to analyze photoreceptor morphology, cell biology, and regeneration. To date, studies have utilized transgenic technology to express intrinsic proteins behind very strong promoters (e.g., *Xenopus* opsin or cone transducin promoters). While deleterious effects of overexpressing some proteins have not yet been reported, it is likely that overexpression of some proteins cannot be tolerated by rods or cones. Future efforts to identify and characterize weakly and moderately expressing promoters will enhance the toolkit available to zebrafish researchers and reduce the chances of overexpression toxicity.

The role of extrinsic factors on photoreceptor development and differentiation has largely relied on loss of function strategies (e.g., mutagenesis and morpholino approaches) and overexpression following mRNA injection. In both cases, analysis on photoreceptors is often limited because many signaling molecules have essential roles much earlier in development and the embryos may not survive to later time points, or photoreceptor phenotypes may be reflect non-specific effects from general developmental abnormalities. Temporally regulated expression of transgenes is one potential method to bypass these limitations. The promoter of the heat shock protein, hsp70, allows transgene expression to be limited to those times when the animals are exposed to a brief heat shock. This temporal control of transgene expression can be used to drive expression of signaling molecules or dominant-negative forms of growth factor receptors during windows critical for photoreceptor development but well after

critical developmental events, such as gastrulation or optic cup formation. These approaches will significantly enhance our understanding of how some molecules contribute to both photoreceptor differentiation and regeneration.

VII. Conclusions

The study of photoreceptor cell structure and development has benefited from the use of transgenic zebrafish expressing cell-specific GFP reporter genes. At the moment there are several rod- and cone-specific GFP lines, and one cone subtype-specific line, namely that for UV opsin-expressing cones. Transgenic technology will greatly facilitate future studies of neuronal structure and function in zebrafish. Dozens of different transgenic zebrafish lines currently exist (Udvadia and Linney, 2003) and more specialized lines will no doubt appear in the future. The classic studies of Ramón y Cajal used Golgi staining to provide tremendous insights into the structure of retinal neurons in fixed tissue (Ramon and Cajal, 1911). With the increasing power of fluorescent imaging technology, questions about the development and function of neurons *in vivo* and in real time are being addressed. With the appropriate promoters, it should be possible to generate individual transgenic lines that express reporter genes within all subsets of the major classes of retinal cells. These lines could be used for rapid identification of specific cell types prior to electrophysiological recordings. Additionally, time-lapse imaging of individual cells or whole layers of cells could be used to study the timing and control of synaptogenesis *in vivo*, as seen from the report by Kay *et al.* (2004). Certainly the combination of these techniques with the wide assortment of zebrafish mutants will facilitate our understanding of neural development and function.

References

Ahmad, I. (2001). *Invest. Ophthalmol. Vis. Sci.* **42**, 2743–2748.

Allwardt, B. A., Lall, A. B., Brockerhoff, S. E., and Dowling, J. E. (2001). *J. Neurosci.* **21**, 2330–2342.

Alvarez-Delfin, K., Morris, A. C., Snelson, C. D., Gamse, J. T., Gupta, T., Marlow, F. L., Mullins, M. C., Burgess, H. A., Granato, M., and Fadool, J. M. (2009). *Proc. Natl. Acad. Sci. U.S.A* **106**, 2023–2028.

Balciunas, D., Wangensteen, K. J., Wilber, A., Bell, J., Geurts, A., Sivasubbu, S., Wang, X., Hackett, P. B., Largaespada, D. A., McIvor, R. S., and Ekker, S. C. (2006). *PLoS Genet.* **2**, e169.

Bernardos, R. L., Barthel, L. K., Meyers, J. R., and Raymond, P. A. (2007). *J. Neurosci.* **27**, 7028–7040.

Bernardos, R. L., and Raymond, P. A. (2006). *Gene Expr. Patterns* **6**, 1007–1013.

Bilotta, J., Saszik, S., and Sutherland, S. E. (2001). *Dev. Dyn.* **222**, 564–570.

Braisted, J. E., Essman, T. F., and Raymond, P. A. (1994). *Development* **120**, 2409–2419.

Branchek, T. (1984). *J. Comp. Neurol.* **224**, 116–122.

Branchek, T., and Bremiller, R. (1984). *J. Comp. Neurol.* **224**, 107–115.

Brockerhoff, S. E., Hurley, J. B., Janssen-Bienhold, U., Neuhauss, S. C., Driever, W., and Dowling, J. E. (1995). *Proc. Natl. Acad. Sci. U.S.A* **92**, 10545–10549.

Chu, P. J., Rivera, J. F., and Arnold, D. B. (2006). *J. Biol. Chem.* **281**, 365–373.

Cicero, S. A., Johnson, D., Reyntjens, S., Frase, S., Connell, S., Chow, L. M., Baker, S. J., Sorrentino, B. P., and Dyer, M. A. (2009). *Proc. Natl. Acad. Sci. U.S.A* **106**, 6685–6690.

Dowling, J. E. (1963). *J. Gen. Physiol.* **46**, 1287–1301.

Dryja, T. P., and Li, T. (1995). *Hum. Mol. Genet.* **4**, 1739–1743.

Fadool, J. M. (2003). *Dev. Biol.* **258**, 277–290.

Fadool, J. M., and Dowling, J. E. (2008). *Prog. Retin. Eye Res.* **27**, 89–110.

Fausett, B. V., and Goldman, D. (2006). *J. Neurosci.* **26**, 6303–6313.

Fausett, B. V., Gumerson, J. D., and Goldman, D. (2008). *J. Neurosci.* **28**, 1109–1117.

Fernald, R. D. (1990). *J. Exp. Zool. Suppl.* **5**, 167–180.

Fimbel, S. M., Montgomery, J. E., Burket, C. T., and Hyde, D. R. (2007). *J. Neurosci.* **27**, 1712–1724.

Fischer, A. J., and Reh, T. A. (2000). *Dev. Biol.* **220**, 197–210.

Furukawa, T., Morrow, E. M., and Cepko, C. L. (1997). *Cell* **91**, 531–541.

Furukawa, T., Morrow, E. M., Li, T., Davis, F. C., and Cepko, C. L. (1999). *Nat. Genet.* **23**, 466–470.

Hamaoka, T., Takechi, M., Chinen, A., Nishiwaki, Y., and Kawamura, S. (2002). *Genesis* **34**, 215–220.

Hao, L., and Scholey, J. M. (2009). *J. Cell. Sci.* **122**, 889–892.

Haruta, M., Kosaka, M., Kanegae, Y., Saito, I., Inoue, T., Kageyama, R., Nishida, A., Honda, Y., and Takahashi, M. (2001). *Nat. Neurosci.* **4**, 1163–1164.

Hecht, S., Schlaer, S., and Pirenne, M. H. (1942). *J. Optic. Soc. Amer.* **38**, 196–208.

Hsu, Y. C., Willoughby, J. J., Christensen, A. K., and Jensen, A. M. (2006). *Development* **133**, 4849–4859.

Hu, M., and Easter, S. S. (1999). *Dev. Biol.* **207**, 309–321.

Hyatt, G. A., Schmitt, E. A., Fadool, J. M., and Dowling, J. E. (1996). *Proc. Natl. Acad. Sci. U.S.A* **93**, 13298–13303.

Insinna, C., and Besharse, J. C. (2008). *Dev. Dyn.* **237**, 1982–1992.

Insinna, C., Humby, M., Sedmak, T., Wolfrum, U., and Besharse, J. C. (2009a). *Dev. Dyn.* **238**, 2211–2222.

Insinna, C., Luby-Phelps, K., Link, B. A., and Besharse, J. C. (2009b). *Methods Cell Biol.* **93**, 219–234.

Insinna, C., Pathak, N., Perkins, B., Drummond, I., and Besharse, J. C. (2008). *Dev. Biol.* **316**, 160–170.

Jensen, A. M., Walker, C., and Westerfield, M. (2001). *Development* **128**, 95–105.

Johns, P. R. (1982). *J. Neurosci.* **2**, 178–198.

Johns, P. R., and Fernald, R. D. (1981). *Nature* **293**, 141–142.

Kay, J. N., Roeser, T., Mumm, J. S., Godinho, L., Mrejeru, A., Wong, R. O., and Baier, H. (2004). *Development* **131**, 1331–1342.

Kennedy, B. N., Alvarez, Y., Brockerhoff, S. E., Stearns, G. W., Sapetto-Rebow, B., Taylor, M. R., and Hurley, J. B. (2007). *Invest. Ophthalmol. Vis. Sci.* **48**, 522–529.

Kennedy, B. N., Vihtelic, T. S., Checkley, L., Vaughan, K. T., and Hyde, D. R. (2001). *J. Biol. Chem.* **276**, 14037–14043.

Kwan, K. M., Fujimoto, E., Grabher, C., Mangum, B. D., Hardy, M. E., Campbell, D. S., Parant, J. M., Yost, H. J., Kanki, J. P., and Chien, C. B. (2007). *Dev. Dyn.* **236**, 3088–3099.

Lin-Jones, J., Parker, E., Wu, M., Knox, B. E., and Burnside, B. (2003). *Invest. Ophthalmol. Vis. Sci.* **44**, 3614–3621.

Luby-Phelps, K., Fogerty, J., Baker, S. A., Pazour, G. J., and Besharse, J. C. (2008). *Vision Res.* **48**, 413–423.

Makino, S., Whitehead, G. G., Lien, C. L., Kim, S., Jhawar, P., Kono, A., Kawata, Y., and Keating, M. T. (2005). *Proc. Natl. Acad. Sci. U.S.A* **102**, 14599–14604.

Marcus, R. C., Delaney, C. L., and Easter, S. S., Jr. (1999). *Vis. Neurosci.* **16**, 417–424.

Marszalek, J. R., Liu, X., Roberts, E. A., Chui, D., Marth, J. D., Williams, D. S., and Goldstein, L. S. (2000). *Cell* **102**, 175–187.

Mears, A. J., Kondo, M., Swain, P. K., Takada, Y., Bush, R. A., Saunders, T. L., Sieving, P. A., and Swaroop, A. (2001). *Nat. Genet.* **29**, 447–452.

Mensinger, A. F., and Powers, M. K. (1999). *Vis. Neurosci.* **16**, 241–251.

Montgomery, J. E., Parsons, M. J., and Hyde, D. R. (2010). *J. Comp. Neurol.* **518**, 800–814.

Morris, A. C., and Fadool, J. M. (2005). *Physiol. Behav.* **86**, 306–313.

Morris, A. C., Scholz, T. L., Brockerhoff, S. E., and Fadool, J. M. (2008). *Dev. Neurobiol.* **68**, 605–619.

Muto, A., Orger, M. B., Wehman, A. M., Smear, M. C., Kay, J. N., Page-McCaw, P. S., Gahtan, E., Xiao, T., Nevin, L. M., Gosse, N. J., Staub, W., Finger-Baier, K., *et al.* (2005). *PLoS Genet.* **1**, e66.

Nathans, J. (1992). *Biochemistry* **31**, 4923–4931.

Otteson, D. C., D'Costa, A. R., and Hitchcock, P. F. (2001). *Dev. Biol.* **232**, 62–76.

Patel, A., and McFarlane, S. (2000). *Dev. Biol.* **222**, 170–180.

Pazour, G. J., Baker, S. A., Deane, J. A., Cole, D. G., Dickert, B. L., Rosenbaum, J. L., Witman, G. B., and Besharse, J. C. (2002). *J. Cell Biol.* **157**, 103–113.

Pellikka, M., Tanentzapf, G., Pinto, M., Smith, C., McGlade, C. J., Ready, D. F., and Tepass, U. (2002). *Nature* **416**, 143–149.

Perkins, B. D., Fadool, J. M., and Dowling, J. E. (2004). *Methods Cell Biol.* **76**, 315–331.

Perkins, B. D., Kainz, P. M., O'Malley, D. M., and Dowling, J. E. (2002). *Vis. Neurosci.* **19**, 257–264.

Poss, K. D., Nechiporuk, A., Hillam, A. M., Johnson, S. L., and Keating, M. T. (2002). *Development* **129**, 5141–5149.

Qin, Z., Barthel, L. K., and Raymond, P. A. (2009). *Proc. Natl. Acad. Sci. U.S.A* **106**, 9310–9315.

Ramon y Cajal, S. (1911). "Histologie du Systeme Nerveax de l'Homme et des Vertebres", Maloine, Paris.

Raymond, P. A., and Barthel, L. K. (2004). *Int. J. Dev. Biol.* **48**, 935–945.

Raymond, P. A., Barthel, L. K., Bernardos, R. L., and Perkowski, J. J. (2006). *BMC Dev. Biol.* **6**, 36.

Raymond, P. A., Barthel, L. K., and Curran, G. A. (1995). *J. Comp. Neurol.* **359**, 537–550.

Raymond, P. A., Barthel, L. K., Rounsifer, M. E., Sullivan, S. A., and Knight, J. K. (1993). *Neuron* **10**, 1161–1174.

Robinson, J., Schmitt, E. A., Harosi, F. I., Reece, R. J., and Dowling, J. E. (1993). *Proc. Natl. Acad. Sci. U.S. A* **90**, 6009–6012.

Saszik, S., Bilotta, J., and Givin, C. M. (1999). *Vis. Neurosci.* **16**, 881–888.

Schmitt, E. A., and Dowling, J. E. (1996). *J. Comp. Neurol.* **371**, 222–234.

Schmitt, E. A., Hyatt, G. A., and Dowling, J. E. (1999). *Vis. Neurosci.* **16**, 601–605.

Shkumatava, A., Fischer, S., Muller, F., Strahle, U., and Neumann, C. J. (2004). *Development.* **131**, 3849–3858

Stenkamp, D. L., and Frey, R. A. (2003). *Dev. Biol.* **258**, 349–363.

Takechi, M., Hamaoka, T., and Kawamura, S. (2003). *FEBS Lett.* **553**, 90–94.

Thummel, R., Enright, J. M., Kassen, S. C., Montgomery, J. E., Bailey, T. J., and Hyde, D. R. (2010). *Exp. Eye. Res.* **90**, 572–582.

Thummel, R., Kassen, S. C., Enright, J. M., Nelson, C. M., Montgomery, J. E., and Hyde, D. R. (2008). *Exp. Eye. Res.* **87**, 433–444.

Tropepe, V., Coles, B. L., Chiasson, B. J., Horsford, D. J., Elia, A. J., McInnes, R. R., and van der Kooy, D. (2000). *Science* **287**, 2032–2036.

Udvadia, A. J., and Linney, E. (2003). *Dev. Biol.* **256**, 1–17.

Van Epps, H. A., Hayashi, M., Lucast, L., Stearns, G. W., Hurley, J. B., De Camilli, P., and Brockerhoff, S. E. (2004). *J. Neurosci.* **24**, 8641–8650.

Vihtelic, T. S., and Hyde, D. R. (2000). *J. Neurobiol.* **44**, 289–307.

Wald, G. (1955). *Am. J. Ophthalmol.* **40**, 18–41.

Wodarz, A., Hinz, U., Engelbert, M., and Knust, E. (1995). *Cell* **82**, 67–76.

Wu, D. M., Schneiderman, T., Burgett, J., Gokhale, P., Barthel, L., and Raymond, P. A. (2001). *Invest Ophthalmol. Vis. Sci.* **42**, 2115–2124.

Yau, K. W. (1994). *Invest. Ophthalmol. Vis. Sci.* **35**, 9–32.

Young, T. L., and Cepko, C. L. (2004). *Neuron* **41**, 867–879.

Yurco, P., and Cameron, D. A. (2005). *Vision Res.* **45**, 991–1002.

CHAPTER 8

Physiological Recordings from Zebrafish Lateral-Line Hair Cells and Afferent Neurons

Josef G. Trapani *and* Teresa Nicolson

Howard Hughes Medical Institute, Oregon Hearing Research Center and Vollum Institute, Oregon Health and Science University, Portland, Oregon

Abstract

I. Introduction

II. Zebrafish Mounting and Immobilizing

    A. Anesthesia and Mounting of Larvae

    B. Immobilizing Larvae with α-Bungarotoxin

III. Microphonics

    A. Stimulation of Neuromast Hair Cells

    B. Microphonic Recordings from Neuromast Hair Cells

    C. Signal Collection and Analysis

    D. Confirming a Biologically Relevant Signal

IV. Action Currents

    A. Electrode and Recording Details

    B. Analysis of Action Currents

V. Summary

VI. Discussion

Acknowledgments

References

Abstract

Sensory signal transduction, the process by which the features of external stimuli are encoded into action potentials, is a complex process that is not fully understood. In fish and amphibia, the lateral-line organ detects water movement and vibration and is critical for schooling behavior and the detection of predators and prey. The

978-0-12-384892-5

DOI: 10.1016/B978-0-12-384892-5.00008-6

lateral-line system in zebrafish serves as an ideal platform to examine encoding of stimuli by sensory hair cells. Here, we describe methods for recording hair-cell microphonics and activity of afferent neurons using intact zebrafish larvae. The recordings are performed by immobilizing and mounting larvae for optimal stimulation of lateral-line hair cells. Hair cells are stimulated with a pressure-controlled water jet and a recording electrode is positioned next to the site of mechanotransduction in order to record microphonics—extracellular voltage changes due to currents through hair-cell mechanotransduction channels. Another readout of the hair-cell activity is obtained by recording action currents from single afferent neurons in response to water-jet stimulation of innervated hair cells. When combined, these techniques make it possible to probe the function of the lateral-line sensory system in an intact zebrafish using controlled, repeatable, physiological stimuli.

I. Introduction

The zebrafish has many features that make it a versatile model system for physiological studies. Larvae are optically transparent and genetically tractable, enabling one to express transgenic fluorescent proteins in target cells and subsequently visualize them in the intact organism. Furthermore, they are prolific egg layers, lines are simple to maintain, and large-scale mutagenesis screens are feasible. Finally, zebrafish larvae are well suited for electrophysiology, which provides a key method for answering fundamental questions in neurobiology (Ono *et al.*, 2001; Ribera and Nüsslein-Volhard, 1998; Smear *et al.*, 2007; Westerfield *et al.*, 1990). Researchers have used these advantages to advance fields ranging from development to cell biology and neuroscience (Feldman *et al.*, 1998; Patton and Zon, 2001; Söllner *et al.*, 2004b; Sumbre *et al.*, 2008; Whitfield *et al.*, 2002).

An important sensory feature of the larval zebrafish is its lateral-line system. Unlike mammals, in addition to auditory and vestibular organs, all fish and amphibia possess a lateral-line organ that is composed of sensory hair cells (Freeman, 1928; Ryder, 1884). Similar to auditory and vestibular sensory detection, the lateral line detects and encodes water motion through hair-cell mechanotransduction (Hoagland, 1933; Suckling and Suckling, 1950). Lateral-line hair cells are arranged together with support cells into rosette-like structures termed neuromast organs. These neuromasts are located along the surface of the animal (superficial neuromasts) and in fluid-filled subepidermal canals in adults (canal neuromasts). The accessibility of zebrafish neuromasts makes them an ideal platform for studying the molecular, cellular, and physiological features of sensory hair cells.

Here we describe in detail our method for recording activity from the lateral line of larval zebrafish. To date, we have recorded extracellular potentials from individual neuromasts and extracellular action currents from single afferent neurons. This preparation is highly suited for performing physiological studies because it utilizes intact animals and biologically relevant stimuli.

II. Zebrafish Mounting and Immobilizing

The lateral-line system of larvae at 120 hours post fertilization (hpf) is composed of superficial neuromasts arranged around the head to form the anterior lateral line and along the trunk to form the posterior lateral line (Dambly-Chaudière *et al.*, 2003). At this stage, all primary neuromasts are present along with a few immature secondary neuromasts still forming (Grant *et al.*, 2005; Nuñez *et al.*, 2009). Studies with primary neuromasts of the posterior lateral line are ideal because their planar polarity results in hair-cell activation with simple anterior–posterior deflections (López-Schier *et al.*, 2004; Nicolson *et al.*, 1998).

In order for routine, stereotyped access to neuromasts and afferent cell bodies of the posterior lateral line, larvae are mounted to a recording chamber with a circular opening (PC-R; Siskiyou, Inc. Grants Pass, OR, USA). This opening is covered by a square cover glass and coated with a 2 mm layer of silicone elastomer, Sylgard (#184; Dow Corning, Midland, MI, USA). Coating the chamber with Sylgard is ideal because of its elastic properties and relative optical clarity, which can be maintained with frequent recoating. The recording chamber fits within an adapter plate (PC-A; Siskiyou, Inc.), which allows the mounted larvae to be rotated 360°. Free rotation alleviates the need to mount the larva at a precise position relative to the recording chamber and pipette holders. Free rotation of the recording chamber also facilitates the proper alignment of all electrodes (Fig. 1A).

Prior to establishing a recording, an individual larva must be immobilized and mounted to the recording chamber. This is accomplished in two steps: first, the larva

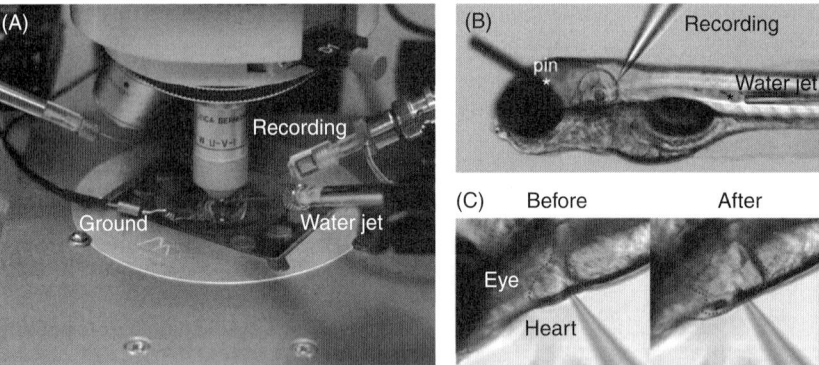

Fig. 1 Setup for mounting and immobilizing zebrafish larva. (A) Photograph illustrating the position of the recording electrode and water-jet pipettes. A third pipette holder is shown on the left side of the image and can be used for dual recordings or as a stimulating electrode. Note the bath ground attaches to the recording chamber, which fits into the circular adapter plate allowing for 360° rotation of the preparation. (B) A 120 hpf larva is pinned to the Sylgard-lined chamber. In this image (10× objective), the recording electrode is positioned for action-current recordings from the posterior lateral-line ganglion. The water jet is positioned just posterior to a neuromast (black asterisk). Note the insertion point of the anterior pin (white asterisk). (C) Before and after α-bungarotoxin injection, which expanded the heart cavity (red dashed outline).

is anesthetized with tricaine and pinned to the Sylgard-lined chamber. Second, the larva is injected with a paralytic (α-bungarotoxin) into its heart and the anesthetic is removed before recording.

A. Anesthesia and Mounting of Larvae

An individual larva is anesthetized in embryo media containing 0.02% tricaine methanesulfonate (MS-222; Sigma-Aldrich, St. Louis, MO, USA) for 30 s. Then, using a transfer pipette, the larva and a dropper full of solution are expelled onto the Sylgard-lined chamber. Next, with a dissecting microscope and two watchmaker's forceps (No. 5 Dumont), the larva is positioned on its right side in the center of the chamber and pinned with two tungsten pins. One pin is inserted just posterior to the eye and anterior to the ear (Fig. 1B). The other pin is inserted into the notochord near the end of the tail of the larva.

Pins are fabricated from tungsten rod (0.002 × 3 in., A-M Systems, Sequim, WA, USA). To aid in puncturing the skin, we first electrolytically sharpen one end of the rod using a 1 N NaOH solution and a 9 V battery. Then, a 90° bend is made 1 mm up from the tip end and the rod is cut to form a small, sharpened "L."

Using these techniques, the larva remains immobilized for the length of a given recording (2–30 min), although careful monitoring and readjustments of electrode position may be necessary for long-duration recordings. Conveniently, since larvae do not have gills at this stage, a perfusion setup is not necessary. During recordings, the heartbeat should remain regular, as hair cells are very sensitive to oxygen depletion.

B. Immobilizing Larvae with α-Bungarotoxin

Electrophysiological recordings are precluded by the presence of the anesthetic tricaine, which blocks both neuronal activity and mechanotransduction channels. Therefore, to record lateral-line activity, larvae are immobilized with a paralytic so that the tricaine can be washed away. The paralytic, α-bungarotoxin, blocks the acetylcholine receptor at the zebrafish neuromuscular junction (Westerfield *et al.*, 1990). This toxin is a suitable paralytic for lateral-line recordings since immunohistochemistry with α-bungarotoxin antibodies confirmed labeling of muscle and did not label hair cells of the ear or neuromasts (unpublished observations).

The anesthetized larva is injected with α-bungarotoxin (125 μM) into the heart using a patch pipette with a tip diameter of 1–3 μm. The larva is positioned, and the heart injection is visualized, using the wide-field upright microscope used for our recordings. The α-bungarotoxin pipette is then mounted to the pipette holder that is used for water-jet stimulation; the output tubing of the pipette holder is temporarily switched to a pressure injector (Pressure System IIe, Toohey Company, Fairfield, NJ, USA). Once mounted, the α-bungarotoxin pipette is aligned perpendicular to the heart under a 10× microscope objective. Next, the pipette is advanced toward the heart, pressed against the skin, and then advanced further until the skin is penetrated. After the pipette is inside the heart cavity, a bolus of α-bungarotoxin is injected. Successful injection will result in an obvious expansion of the heart cavity (Fig. 1C). If necessary, inclusion of

phenol red (10%) can increase visualization of the expelled α-bungarotoxin. Following injection, the pipette is retracted and the pinned larva is rinsed several times with standard extracellular solution [in mM: 130 NaCl, 2 KCl, 2 CaCl$_2$, 1 MgCl$_2$, and 10 4-(2-hydroxyethyl)-1-piperazineethanesulfonic acid (HEPES); 290 mOsm; pH 7.8]. The larva is left in 0.5–1 mL of solution for several minutes prior to recording to allow for effective tricaine removal.

III. Microphonics

The term "microphonics" was first used (Adrian, 1931) to describe the cochlear AC potentials that Wever and Bray (1930) observed in recordings from cats. Microphonic potentials have since been found to represent the changes in extracellular potential that result from the inward flow of cations during gating of mechanotransduction channels located at the tips of hair-cell stereocilia (Beurg et al., 2009; Corey and Hudspeth, 1983; Jaramillo and Hudspeth, 1991). As a measure of the activity of mechanotransduction channels, microphonic recordings are useful for determining the amplitude and frequency of the transduction response in hair cells. The following methods can be used for recording and analyzing microphonics from individual neuromasts of the larval lateral line.

A. Stimulation of Neuromast Hair Cells

At 120 hpf, larval neuromasts contain 9–14 hair cells. At the apical end of each hair cell is the stereociliary hair bundle. Each hair bundle is connected to a single kinocilium that extends roughly 20 μm into the aqueous medium. A gelatinous cupula covers and encompasses all of the kinocilia. Deflection of the cupula results in the concerted deflection of all kinocilia, which in turn deflects attached stereocilia (Netten and Kroese, 1987). Shearing of hair bundles ultimately opens mechanotransduction channels located at their tips (Beurg et al., 2009; Jaramillo and Hudspeth, 1991).

Stimulation of neuromast hair cells is performed with a pipette filled with extracellular solution that is driven by a pressure clamp (HSPC-1, ALA Scientific). This pipette, termed a water jet, should have a circular opening with a diameter of approximately 30 μm. To achieve this pipette shape, thick-walled (1.5 mm OD and 0.86 mm ID) borosilicate glass is fabricated from a single, hard pull (micropipette puller P-97; Sutter Instruments) that results in two pipettes with long, tapering ends. These two long ends are rubbed against each other until one end scores and cleanly breaks the other. Note that this process requires non-filament glass in order to have a clean tip break and laminar fluid flow.

The water-jet pipette is positioned approximately 100 μm from a given neuromast cupula. The height of the water jet should be slightly above the cupula and deflection by the water jet should be equal in forward and reverse directions.

B. Microphonic Recordings from Neuromast Hair Cells

Electrodes for microphonic recordings are pulled to resemble standard patch electrodes using filament glass that is otherwise the same as the water-jet pipettes. Electrode outer tip diameters are of 1–3 μm and have resistances of approximately 3 MΩ when filled with extracellular solution. The recording electrode is mounted at a 45° angle from the water-jet pipette and positioned (using a 40× objective) at a height even with the stereocilia (Fig. 2A). Since mechanotransduction channels are located at the tips of stereocilia, the position of the recording electrode is critical for obtaining the largest possible microphonic signal. Proper position is important because lateral-line stereocilia are very short and difficult to visualize even with a 40× objective, where they appear as a dark spot at the apical surface of each hair cell (white-dashed circle in Fig. 2A).

On day 5, in addition to the primary posterior lateral-line neuromasts, secondary neuromasts are also forming (Grant *et al.*, 2005). Care should be taken to record exclusively from primary neuromasts, as secondary neuromasts activate with deflections of different polarities, which may give confounding results. Also, note that all solutions (water jet, electrode, and bath) should be identical to avoid junction potential errors.

C. Signal Collection and Analysis

Microphonic potentials are recorded in current-clamp mode, sampled at 10 kHz, and filtered at 1 kHz. The voltage signal is then further amplified (10,000× final) and

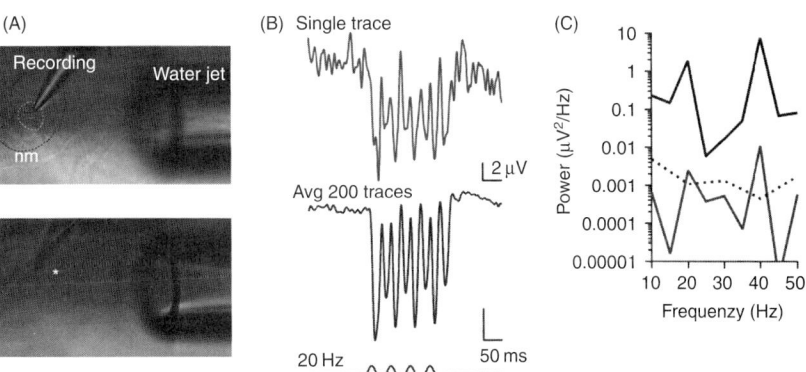

Fig. 2 Microphonic recordings from posterior lateral-line neuromasts. (A) Images (40× objective) of the location of the recording electrode (top panel) and water-jet pipette (bottom panel) relative to a neuromast (nm; black dashed circle). The recording electrode is at the level of the hair-cell stereocilia, which are dark spots under DIC differential interference contrast (white dashed circle). The water-jet pipette is positioned approximately 100 μm from the neuromast. Note that the water-jet height is even with the top of the neuromast cupula (white asterisk). (B) An individual trace (red trace) from a microphonic recording, compared to an average of 200 consecutive traces (black trace), which illustrates the increase in signal-to-noise ratio with averaging. (C) Power spectrum from the individual trace (red line) and the average trace (black line) during the 200 ms sine wave stimulation. Note that the individual-trace power is barely distinguished from the power spectrum of the noise portion of the average trace (black dashed line).

filtered (50–100 Hz; eight-pole Bessel) by an additional amplifier (Model 440, Brown-lee Precision). In order to drive the pressure clamp and record the microphonic potential simultaneously, a sinusoidal voltage command is delivered to the pressure clamp via an analog output from the recording amplifier. The pressure of the water jet can also be monitored by a feedback sensor on the pressure-clamp headstage and recorded simultaneously with acquisition software.

As microphonic signals are typically 10 μV, one should take care to effectively remove all electrical noise from the recording setup. Under ideal conditions, recording noise can be held to under 5 μV peak to peak, and microphonic signals can be seen without averaging traces (Fig. 2B). To overcome the variability of neuromast structure and the position of the recording electrode, one can quantify microphonic signals by determining the power of the total signal per unit frequency. The stimulus portion of the signal is analyzed with a power spectral density function (Fig. 2C). Power spectra will contain primary peaks at the stimulus frequency (1f) and at twice the frequency (2f); for a 20 Hz stimulus, this would result in peaks at 20 and 40 Hz, respectively (see Fig. 2C). These frequency components result from the presence of two groups of hair cells of different orientation that respond to stimuli of opposite polarity (Flock and Wersall, 1962; Ghysen and Dambly-Chaudière, 2007).

D. Confirming a Biologically Relevant Signal

There are several qualifications and tests that help to determine whether a recorded microphonic signal is biologically relevant. If bidirectional stimuli (e.g., sine waves) are used, then the signal should contain the 2f frequency component. In addition, a biologically relevant signal should not be detected with stimulation perpendicular to the excitation axis of a neuromast. Furthermore, the microphonic signal should be blocked by mechanotransduction channel antagonists such as dihydrostreptomycin and amiloride (Farris et $al.$, 2004). Microphonics should also be blocked by disruption of stereocilia tip links—components necessary for transduction. For example, chelating extracellular calcium with 1,2-bis(o-aminophenoxy)ethane-N,N,N',N'-tetraacetic acid (BAPTA) or ethylene glycol tetraacetic acid (EGTA), both of which break tip links, abolishes microphonics (Assad et $al.$, 1991). In addition, microphonics should be absent in larvae lacking mechanotransduction such as $cdh23$ or $pcdh15$ mutants (Nicolson et $al.$, 1998; Seiler et $al.$, 2005; Söllner et $al.$, 2004a).

IV. Action Currents

The neurotransmitter output of hair cells, which encodes the features of sensory stimuli, results from a unique process of graded transduction. Briefly, gating of mechanotransduction channels results in graded changes in hair-cell membrane potential, which drives graded synaptic vesicle fusion at specialized ribbon synapses. These graded signals are ultimately encoded as sequences of all-or-none action potentials (spikes) in afferent neurons.

When recorded in an extracellular loose-patch configuration, action potentials are measured as action currents. Typically, action-current recordings are less invasive and easier to establish than whole-cell recordings. Furthermore, lateral-line afferent neurons are ideal for this type of recording, as their dendrites are myelinated and spiking is absent without hair-cell neurotransmission (Obholzer *et al.*, 2008). These qualities imply that afferent spiking is evoked from spontaneous and stimulation-dependent release of neurotransmitter from hair cells. Thus, action-current recordings are sufficient for examining the temporal sequence of spikes, which provides a direct readout of the activity of innervated hair cells.

A. Electrode and Recording Details

Extracellular recording electrodes are pulled in a similar fashion as microphonic electrodes except with longer shafts and outer tip diameters slightly smaller than 1 μm. Recording electrode resistances should range between 5 and 10 MΩ in extracellular solution. The recording electrode is positioned at an approximately 45° angle to the dorsal side of the pinned larva (Fig. 1B). The electrode should be positioned in the same optical plane as the middle of the posterior lateral line ganglion (PLLg). In a manner similar to α-bungarotoxin injection, the electrode is moved ventrally with increasing pressure against the fish skin until it advances through it. Often following the skin puncture, the pipette will move rapidly past the ganglion. If this occurs, then the electrode should be slowly retracted until it is within the membranous sack that encompasses the cell bodies. The electrode's position within the PLLg is confirmed by inflation of this sack with positive pipette pressure delivered via the electrode holder. Note that if necessary, the PLLg can be visualized using transgenic animals with fluorescently labeled lateral-line neurons (see both Nagiel *et al.*, 2008 and Obholzer *et al.*, 2008 for constructs).

To record from an individual lateral-line neuron, the electrode is moved against a desired cell and negative pipette pressure is applied (Fig. 3A).Recordings are made with the amplifier in voltage-clamp mode (command voltage = 0) with a 10 kHz sample rate and filtered at 1 kHz. For typical action-current recordings, the series resistance lies within a range of 30–80 MΩ. Confirmation of an established recording is seen first by the appearance of spontaneous spiking, after which the water jet is used to scan for the innervated neuromast, resulting in the conversion of spontaneous to phase-locked spiking.

Occasional application of positive pipette pressure is useful for both increasing the clarity of the ganglion and freeing any skin, cell, or tissue debris that may be obstructing the pipette tip. Also, on occasion, electrodes will rapidly form gigaohm seals, which predicts that whole-cell recordings of afferent neurons will be easily obtained.

B. Analysis of Action Currents

In response to bidirectional stimulation of neuromast hair cells (e.g., sine waves), action-current recordings reveal phase-locked spiking to one direction of stimulation

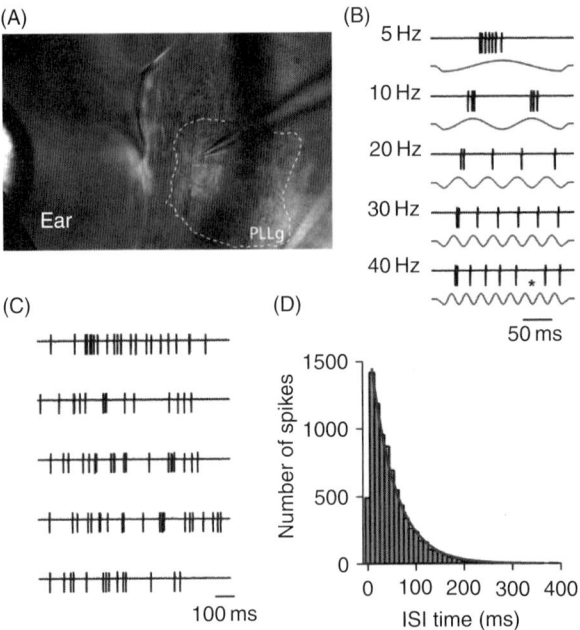

Fig. 3 Action-current recordings from individual posterior lateral-line neurons. (A) Under DIC differential interference contrast optics with a 40× objective, the posterior lateral-line ganglion (PLLg; dashed-with line) is visible just posterior to the ear. The recording electrode is positioned in the anterior-dorsal portion of the PLLg, where the majority of cell bodies for the primary neuromasts are located. Note the otolith in the anterior side of the image. (B) Single traces of action currents in response to hair-cell stimulation at various frequencies. Note the spikes are phase-locked to just one direction of deflection. (C) Five consecutive 1 s traces of spontaneous spiking. The mean interspike interval (ISI) for 400 s of spiking was 50 ms. (D) Histogram representing the distribution of ISIs (10 ms bins) from 400 s of spiking by the cell shown in C. The histogram is best fit by a single-phase exponential decay equation (red line) with a tau of 50 ms.

(Obholzer *et al.*, 2008). This occurs because an individual afferent neuron innervates only one of the two populations of hair cells within a neuromast that all display the same hair-bundle polarity (Faucherre *et al.*, 2009; Nagiel *et al.*, 2008; Obholzer *et al.*, 2008). Spiking remains phase-locked with rates of stimulation approaching 100 Hz. At lower frequencies, spiking occurs with each deflection, while at higher frequencies, spikes do not occur with every deflection. In both cases, there is no loss of phase locking to the water-jet stimulus (Fig. 3B). Spike trains in response to continuous stimulation can be visualized with period histograms where individual spike times are binned and plotted relative to the unit period for a given stimulus frequency. The resulting series of unit spike times can be further transformed into unit vectors with corresponding, specific phase angles. Thus, the collection of spikes, now represented by a series of vectors, can be quantified for the degree of synchrony between the stimulus and response by calculating the vector strength (r) of the series (Goldberg and Brown, 1969). We recently utilized this method of analysis to quantify phase-locked activity in larvae with disrupted hair-cell endocytosis (Trapani *et al.*, 2009).

In the absence of sensory stimuli, spontaneous spiking is observed in afferent neurons of the auditory, vestibular, and lateral-line systems across species. This spontaneous spiking apparently results from neurotransmitter release from innervated hair cells (Annoni *et al.*, 1984; Flock and Russell, 1976; Furukawa and Ishii, 1967; Siegel and Dallos, 1986; Starr and Sewell, 1991; Zimmerman, 1979). The most meaningful feature of spontaneous spiking is the time interval between two consecutive spikes, which is termed the interspike interval (ISI). The temporal pattern of the spontaneous activity can be analyzed using ISI histograms, where features of the histogram describe the nature of the activity. For example, if spikes are generated by a Poisson process, then the ISI histogram will be best fit by a single-phase exponential decay equation with the time constant (tau) corresponding to the recording's mean ISI time. Recording spontaneous action currents from a PLLg neuron results in a diverse range of ISI times (Fig. 3C). The ISI histogram of spontaneous spiking in Fig. 3D is best fit by a single exponential decay equation.

V. Summary

In this chapter, we have described our methods for recording activity from the posterior lateral-line system of the larval zebrafish. While some of the details are specific for this particular preparation, a majority of the techniques are translatable to physiology in other areas of the fish. The mounting and immobilization of larvae, as well as the basic principles of the recording, will benefit any investigator looking to establish physiological recordings in live, intact larvae.

VI. Discussion

The zebrafish is a genetically tractable vertebrate and a powerful platform for electro-physiological studies. The above preparation is extremely valuable in assessing sensory hair-cell function in various auditory/vestibular mutants. The ability to record both microphonics from hair cells and action currents from lateral-line neurons can help to pinpoint the nature of the defect in mutants and reveal the biological role of the protein of interest. Using these techniques provides important information about the impact of mutations affecting genes that are conserved in function from fish to humans. In addition, the preparation allows one to mechanically stimulate hair cells in an undissected animal with intact circuitry that receives sensory input for higher order processing. Such recordings are currently not possible with mammalian model systems.

Future experiments using the lateral-line system may include whole-cell patch clamp recordings from the afferent neuron and potentially from individual hair cells, which will further increase the capability of this preparation. Furthermore, the afferent neuron forms an unmyelinated basket beneath a given, innervated neuromast, and this presents the potential for postsynaptic recordings similar to those achieved in mice (Glowatzki and Fuchs, 2002). Experiments with a double patch-clamp amplifier, where one could

simultaneously record activity at both the neuromast and the afferent neuron, would be extremely powerful for examining sensory signal encoding.

Over the past 10 years, the zebrafish community has developed a large number of mutant and transgenic zebrafish lines. Using either endogenous regulatory sequences or Gal4/UAS targeting allows for spatially and temporally restricted expression of proteins of interest and has created a large set of optogenetic tools: photoconvertible fluorescent proteins to visualize cells, light-gated ion channels to activate cells, genetically encoded voltage and calcium sensors to examine cellular activity, and genetically encoded proteins that poison synaptic transmission to deactivate cells. The combination of transgenic lines with hair cell and afferent-specific expression of these reporters together with the electrophysiology techniques described in this chapter presents the potential for experiments aimed at understanding of the fundamental aspects of hair-cell processing and sensory signaling.

Acknowledgments

Many thanks go to P. Brehm and H. Wen for initial instruction on mounting and paralyzing larvae, as well as technical advice. Thanks also to L. Trussell and K. Bender for comments and advice on recordings and data analysis.

References

Adrian, E. D. (1931). The microphonic action of the cochlea: an interpretation of Wever and Bray's experiments. *J. Physiol.* **71**, 28–29.

Annoni, J. M., Cochran, S. L., and Precht, W. (1984). Pharmacology of the vestibular hair cell-afferent fiber synapse in the frog. *J. Neurosci.* **4**, 2106–2116.

Assad, J. A., Shepherd, G. M., and Corey, D. P. (1991). Tip-link integrity and mechanical transduction in vertebrate hair cells. *Neuron* **7**, 985–994.

Beurg, M., Fettiplace, R., Nam, J.-H., and Ricci, A. J. (2009). Localization of inner hair cell mechanotransducer channels using high-speed calcium imaging. *Nat. Neurosci.* **12**, 553–558.

Corey, D. P., and Hudspeth, A. J. (1983). Analysis of the microphonic potential of the bullfrog's sacculus. *J. Neurosci.* **3**, 942–961.

Dambly-Chaudière, C., Sapède, D., Soubiran, F., Decorde, K., Gompel, N., and Ghysen, A. (2003). The lateral line of zebrafish: a model system for the analysis of morphogenesis and neural development in vertebrates. *Biol. Cell* **95**, 579–587.

Farris, H. E., LeBlanc, C. L., Goswami, J., and Ricci, A. J. (2004). Probing the pore of the auditory hair cell mechanotransducer channel in turtle. *J. Physiol.* **558**, 769–792.

Faucherre, A., Pujol-Martí, J., Kawakami, K., and López-Schier, H. (2009). Afferent neurons of the zebrafish lateral line are strict selectors of hair-cell orientation. *PLoS One* **4**, e4477.

Feldman, B., Gates, M. A., Egan, E. S., Dougan, S. T., Rennebeck, G., Sirotkin, H. I., Schier, A. F., and Talbot, W. S. (1998). Zebrafish organizer development and germ-layer formation require nodal-related signals. *Nature* **395**, 181–185.

Flock, A. and Russell, I. (1976). Inhibition by efferent nerve fibres: action on hair cells and afferent synaptic transmission in the lateral line canal organ of the burbot *Lota lota. J. Physiol.* **257**, 45–62.

Flock, A., and Wersall, J. (1962). A study of the orientation of the sensory hairs of the receptor cells in the lateral line organ of fish, with special reference to the function of the receptors. *J. Cell Biol.* **15**, 19–27.

Freeman, W. (1928). The function of the lateral line organs. *Science* **68**, 205.

Furukawa, T., and Ishii, Y. (1967). Neurophysiological studies on hearing in goldfish. *J. Neurophysiol.* **30**, 1377–1403.

Ghysen, A., and Dambly-Chaudière, C. (2007). The lateral line microcosmos. *Genes Dev.* **21**, 2118–2130.

Glowatzki, E., and Fuchs, P. A. (2002). Transmitter release at the hair cell ribbon synapse. *Nat. Neurosci.* **5**, 147–154.

Goldberg, J. M. and Brown, P. B. (1969). Response of binaural neurons of dog superior olivary complex to dichotic tonal stimuli: some physiological mechanisms of sound localization. *J. Neurophysiol.* **32**, 613–36.

Grant, K. A., Raible, D. W., and Piotrowski, T. (2005). Regulation of latent sensory hair cell precursors by glia in the zebrafish lateral line. *Neuron* **45**, 69–80.

Hoagland, H. (1933). Electrical responses from the lateral-line nerves of catfish. I. *J. Gen Physiol.* **16**, 695–714.

Jaramillo, F., and Hudspeth, A. J. (1991). Localization of the hair cell's transduction channels at the hair bundle's top by iontophoretic application of a channel blocker. *Neuron* **7**, 409–420.

López-Schier, H., Starr, C. J., Kappler, J. A., Kollmar, R., and Hudspeth, A. J. (2004). Directional cell migration establishes the axes of planar polarity in the posterior lateral-line organ of the zebrafish. *Dev. Cell* **7**, 401–412.

Nagiel, A., Andor-Ardó, D., and Hudspeth, A. J. (2008). Specificity of afferent synapses onto plane-polarized hair cells in the posterior lateral line of the zebrafish. *J. Neurosci.* **28**, 8442–8453.

Nicolson, T., Rüsch, A., Friedrich, R. W., Granato, M., Ruppersberg, J. P., and Nüsslein-Volhard, C. (1998). Genetic analysis of vertebrate sensory hair cell mechanosensation: the zebrafish circler mutants. *Neuron* **20**, 271–283.

Nuñez, V. A., Sarrazin, A. F., Cubedo, N., Allende, M. L., Dambly-Chaudière, C., and Ghysen, A. (2009). Postembryonic development of the posterior lateral line in the zebrafish. *Evol. Dev.* **11**, 391–404.

Obholzer, N., Wolfson, S., Trapani, J. G., Mo, W., Nechiporuk, A., Busch-Nentwich, E., Seiler, C., Sidi, S., Söllner, C., Duncan, R. N., Boehland, A., and Nicolson, T. (2008). Vesicular glutamate transporter 3 is required for synaptic transmission in zebrafish hair cells. *J. Neurosci.* **28**, 2110–2118.

Ono, F., Higashijima, S., Shcherbatko, A., Fetcho, J. R., and Brehm, P. (2001). Paralytic zebrafish lacking acetylcholine receptors fail to localize rapsyn clusters to the synapse. *J Neurosci* **21**, 5439–5448.

Patton, E. E., and Zon, L. I. (2001). The art and design of genetic screens: zebrafish. *Nat. Rev. Genet.* **2**, 956–966.

Ribera, A. B., and Nüsslein-Volhard, C. (1998). Zebrafish touch-insensitive mutants reveal an essential role for the developmental regulation of sodium current. *J. Neurosci.* **18**, 9181–9191.

Ryder, J. A. (1884). The pedunculated lateral-line organs of *Gastrostomus*. *Science* **3**, 5.

Seiler, C., Finger-Baier, K. C., Rinner, O., Makhankov, Y. V., Schwarz, H., Neuhauss, S. C. F., and Nicolson, T. (2005). Duplicated genes with split functions: independent roles of protocadherin 15 orthologues in zebrafish hearing and vision. *Development* **132**, 615–623.

Siegel, J. H. and Dallos, P. (1986). Spike activity recorded from the organ of Corti. *Hear. Res.* **22**, 245–248.

Smear, M. C., Tao, H. W., Staub, W., Orger, M. B., Gosse, N. J., Liu, Y., Takahashi, K., Poo, M.-M., and Baier, H. (2007). Vesicular glutamate transport at a central synapse limits the acuity of visual perception in zebrafish. *Neuron* **53**, 65–77.

Starr, P. A., and Sewell, W. F. (1991). Neurotransmitter release from hair cells and its blockade by glutamate-receptor antagonists. *Hear. Res.* **52**, 23–41.

Suckling, E. E., and Suckling, J. A. (1950). The electrical response of the lateral line system of fish to tone and other stimuli. *J. Gen. Physiol.* **34**, 1–8.

Sumbre, G., Muto, A., Baier, H., and Poo, M.-M. (2008). Entrained rhythmic activities of neuronal ensembles as perceptual memory of time interval. *Nature* **456**, 102–106.

Söllner, C., Rauch, G.-J., Siemens, J., Geisler, R., Schuster, S. C., Müller, U., Nicolson, T., and Tübingen 2000 Screen Consortium (2004a). Mutations in cadherin 23 affect tip links in zebrafish sensory hair cells. *Nature* **428**, 955–959.

Söllner, C., Schwarz, H., Geisler, R., and Nicolson, T. (2004b). Mutated otopetrin 1 affects the genesis of otoliths and the localization of Starmaker in zebrafish. *Dev. Genes Evol.* **214**, 582–590.

Trapani, J. G., Obholzer, N., Mo, W., Brockerhoff, S. E., and Nicolson, T. (2009). Synaptojanin 1 is required for temporal fidelity of synaptic transmission in hair cells. *PLoS Genet.* **5**, e1000480.

Westerfield, M., Liu, D. W., Kimmel, C. B., and Walker, C. (1990). Pathfinding and synapse formation in a zebrafish mutant lacking functional acetylcholine receptors. *Neuron* **4**, 867–874.

Wever, E. G. and Bray, C. W. (1930). Action currents in the auditory nerve in response to acoustical stimulation. *Proc. Natl. Acad. Sci. U. S. A.* **16**, 344–350.

Whitfield, T. T., Riley, B. B., Chiang, M.-Y., and Phillips, B. (2002). Development of the zebrafish inner ear. *Dev. Dyn.* **223**, 427–458.

Zimmerman, D. M. (1979). Onset of neural function in the lateral line. *Nature* **282**, 82–84.

van Netten, S. M., and Kroese, A. B. (1987). Laser interferometric measurements on the dynamic behaviour of the cupula in the fish lateral line. *Hear. Res.* **29**, 55–61.

CHAPTER 9

Zebrafish Kidney Development

Iain A. Drummond[*] *and* Alan J. Davidson[†]

[*]Departments of Medicine and Genetics, Harvard Medical School and Nephrology Division, Massachusetts General Hospital, Charlestown, Massachusetts

[†]Center for Regenerative Medicine, Massachusetts General Hospital, Boston, Massachusetts

Abstract
I. Introduction
II. Structure of the Zebrafish Pronephros
III. Formation of the Pronephros
    A. Origin of the Nephrogenic Mesoderm
    B. Early Subdomains within the Nephrogenic Mesoderm
    C. Role of Retinoic Acid Signaling
    D. Role of Pax2a
    E. Role of the Endoderm
    F. Differentiation of the Tubular Epithelium
    G. Formation of the Glomerulus
    H. Formation of the Cloaca
IV. Methods to Study Pronephros Function
    A. Embryo Dissociation
    B. Isolation of Fluorescently Labeled Cells by Fluorescence Activated Cell Sorting (FACS)
    C. A Simple Assay for Glomerular Filtration
    D. Gentamicin Induced Kidney Tubule Injury in Embryos and Adults
    E. Adult Kidney Isolation
    F. Non-Lethal Surgical Access to the Adult Kidney
    G. Detecting and Imaging Zebrafish Cilia
    H. Histological Sectioning of Whole mount Stained Embryos
    I. Electron Microscopy Methods for Zebrafish
V. Conclusions
Acknowledgments
References

Abstract

The zebrafish pronephric kidney provides a useful and relevant model of kidney development and function. It is composed of cell types common to all vertebrate kidneys and pronephric organogenesis is regulated by transcription factors that have highly conserved functions in mammalian kidney development. Pronephric nephrons are a good model of tubule segmentation and differentiation of epithelial cell types. The pronephric glomerulus provides a simple model to assay gene function in regulating cell structure and cell interactions that form the blood filtration apparatus. The relative simplicity of the pronephric kidney combined with the ease of genetic manipulation in zebrafish makes it well suited for mutation analysis and gene discovery, in vivo imaging, functional screens of candidate genes from other species, and cell isolation by FACS . In addition, the larval and adult zebrafish kidneys have emerged as systems to study kidney regeneration after injury. This chapter provides a review of pronephric structure and development as well as current methods to study the pronephros.

I. Introduction

The kidney has two principal functions: to remove waste from the blood and to balance ion and metabolite concentrations in the blood within physiological ranges that support proper functioning of all other cells (Vize *et al.*, 2002). Kidney function is achieved largely by first filtering the blood and then recovering useful ions and small molecules by directed epithelial transport. This work is performed by nephrons, the functional units of the kidney (Fig. 1). The nephron is comprised of a blood filter, called the glomerulus, attached to a tubular epithelium (Fig. 1C and D). The glomerulus contains specialized epithelial cells called podocytes that form a basket-like extension of cellular processes around a capillary tuft. The basement membrane between podocytes and capillary endothelial cells together with the specialized junctions between the podocyte cell processes (slit diaphragms) function as a blood filtration barrier, allowing passage of small molecules, ions, and blood fluid into the urinary space, while retaining high molecular weight proteins in the vascular system (Fig. 1; see also Fig. 12D). The blood filtrate travels down the lumen of the kidney tubule, encountering distinct proximal and distal tubule segments that modify the composition of the urine via specific solute transport activities. The urine is drained by the collecting ducts, which further modify its salt and water composition, until eventually being voided outside the body (Fig. 1; Vize *et al.*, 2002).

In the course of vertebrate evolution, three distinct forms of kidneys of increasing complexity have been generated: the pronephros, mesonephros, and metanephros (Saxén, 1987). The pronephros is the first kidney to form during embryogenesis. In vertebrates with free-swimming larvae, including amphibians and teleost fish, the pronephros is the functional kidney of early larval life (Howland, 1921; Tytler, 1988; Tytler *et al.*, 1996; Vize *et al.*, 1997) and is required for proper osmoregulation (Howland, 1921). Later, in juvenile stages of fish and frog development, a mesonephros forms around and along the length of the pronephros and later serves as the final adult kidney. The metanephric kidney forms

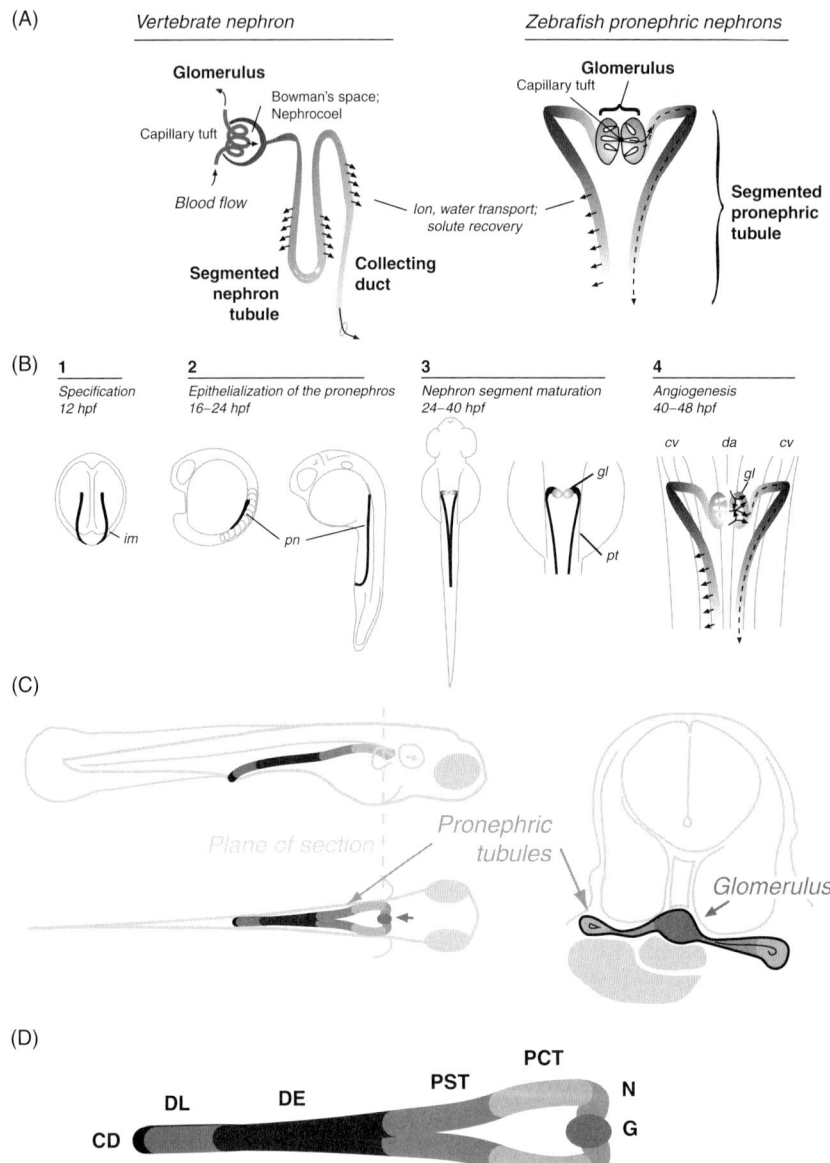

Fig. 1 The zebrafish pronephros. (A) Functional features of the vertebrate nephron and the zebrafish pronephric nephrons. See text for details. (B) Stages in zebrafish pronephric kidney development. (1) Specification of mesoderm to a nephric fate: expression patterns of pax2.1 and lim-1 define a posterior region of the intermediate mesoderm (im) and suggest that a nephrogenic field is established in early development. (2) Epithelialization of the pronephros (pn) follows somitogenesis and is complete by 24 hpf. (3) Patterning of the nephron gives rise to the pronephric glomerulus (gl) and pronephric tubules (pt). (4) Angiogenic sprouts from the dorsal aorta (da) invade the glomerulus and form the capillary loop. The cardinal vein (cv) is apposed to the tubules and receives recovered solutes. (C) Diagram of the mature zebrafish pronephric kidney in 3-day larva. A midline compound glomerulus connects to the segmented pronephric tubules that run laterally. The nephrons are joined at the cloaca where they communicate with the exterior. (D) Patterning of the pronephric nephron generates discrete segments: neck (N), proximal convoluted tubule (PCT), proximal straight tubule (PST), distal early (DE), late distal (DL), and collecting duct (CD).

exclusively in the amniotes (mammals, birds, and reptiles) and, in the case of mammals, is adapted for water retention and producing concentrated urine. Despite some differences in organ morphology between the various kidney forms, many common elements exist at the cellular and molecular level that can be exploited to further our general understanding of renal development and biology. In particular, the zebrafish pronephros has provided a useful model of nephrogenic mesoderm differentiation, kidney cell type differentiation, nephron patterning, kidney, vasculature interactions, glomerular function, and diseases affecting glomerular filtration and tubule lumen size, i.e., cystic kidney disease. While much remains to be done, the basic features of zebrafish pronephric development and patterning have emerged from studies using simple histology, cell lineage tracing, gene expression patterns, and analysis of zebrafish mutants affecting this process.

II. Structure of the Zebrafish Pronephros

The zebrafish pronephros consists of only two nephrons with glomeruli fused at the embryo midline just ventral to the dorsal aorta (Fig. 1C) (Agarwal and John, 1988; Armstrong, 1932; Balfour, 1880; Drummond, 2000; Drummond et al., 1998; Goodrich, 1930; Hentschel and Elger, 1996; Marshall and Smith, 1930; Newstead and Ford, 1960; Tytler, 1988; Tytler et al., 1996). Historically, much of the tubular epithelium extending from the glomerulus to the cloaca has been referred to as pronephric duct. This nomenclature was based on the similar anatomical location of the pronephric or Wolffian ducts in amphibians, chickens, and mammals (Vize et al., 2002). However, based on new molecular data there is now a consensus that the tubular epithelium of the zebrafish pronephros is actually subdivided into two proximal tubule segments (proximal convoluted tubule (PCT) and proximal straight tubule (PST)) and two distal tubule segments (distal early (DE) and distal late (DL)) that are homologous in many ways to the segments of the mammalian nephron (Wingert and Davidson, 2008). What was previously considered "tubule" is now believed to represent a "neck" segment, such as that described in the adult kidneys of other teleosts (Kamunde and Kisia, 1994).

The PCT segment is structurally similar to the proximal tubules of the mammalian kidney, displaying a well-developed brush border and high columnar epithelial cells (Seldin and Giebisch, 1992). The proximal tubule in other vertebrates plays a major role in reabsorbing the bulk of the salts, sugars, and small proteins that pass through the glomerular filtration barrier (Vize et al., 2002). The zebrafish PCT expresses the endocytic receptors megalin and cubalin and takes up small fluorescent dextrans that pass through the glomerulus, consistent with a conserved absorptive function (Anzenberger et al., 2006). The PCT also expresses the chloride/bicarbonate anion exchanger AE2, the sodium/bicarbonate cotransporter NBC1, and the sodium/hydrogen exchanger NHE (Fig. 2A) (Nichane et al., 2006; Shmukler et al., 2005; Wingert et al., 2007), indicating a role in acid/base homeostasis which is also shared with proximal tubules in mammals.

The function of the PST segment is less clear. Markers of this segment include an aspartoacylase homolog (Fig. 2B), the zebrafish *starmaker* gene (Fig. 2C) (Sollner et al., 2003; Sprague et al., 2008), *slc13a1* (a sodium/sulfate symporter), and *trpm7*

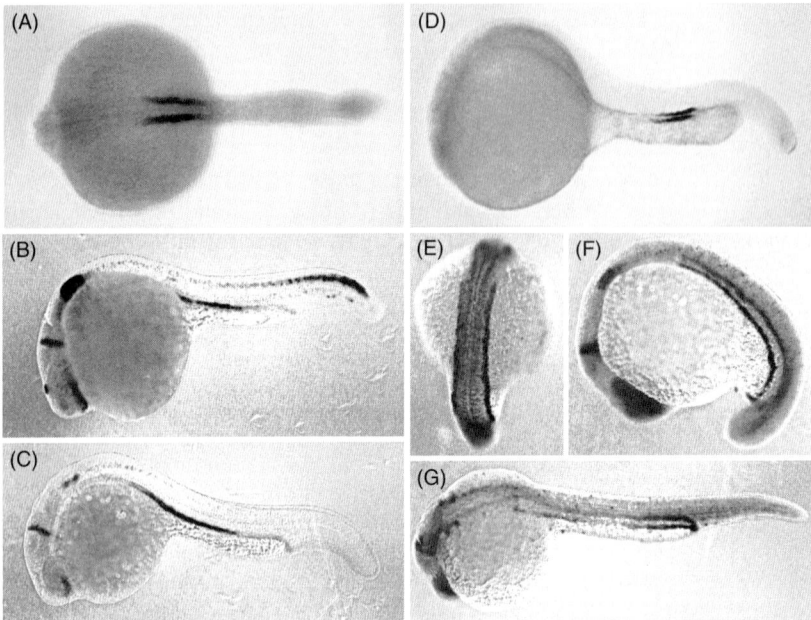

Fig. 2 Ion transporter mRNA expression defines pronephric nephron segments. (A) The chloride–bicarbonate anion exchanger (AE2) is expressed in the proximal convoluted tubule. The proximal straight segment specifically expresses the zebrafish starmaker gene (B) and an *aspartoacylase* homolog (C). Early distal segments express the Na–K–Cl symporter *slc12a1* (D). Expression of a putative ABC transporter (ibd2207) is observed initially throughout most of the forming pronephric tubules at the 15-somite stage (E and F) but becomes restricted primarily to the late distal segment by 24 hpf (G). Embryos in parts B, C, E, F, and G are counterstained with *pax2a* probe in red for reference. (A and D, courtesy of Alan Davidson *et al.*; B, C, E, F, and G, courtesy of Neil Hukreide.)

(a monovalent cation channel). The nutria mutant lacks a functional trpm7 gene and exhibits kidney stone formation and skeletal defects (Elizondo *et al.*, 2005) suggesting a role for TrpM7 in calcium uptake from the tubular fluid. The observation that the PST specifically expresses a sulfate (slc13a3) and calcium (*trpm7*) transporter suggests that this segment may specialize in the uptake of particular ions.

The DE segment expresses *slc12a1* (aka the NKCC2 (Na–K–Cl) symporter) (Fig. 2D), which in mammals is exclusively expressed in the thick ascending limb (TAL) portion of the distal tubule (Igarashi *et al.*, 1995). This segment is also known as the "diluting segment" as it reduces the osmolarity of the urinary filtrate (Guggino *et al.*, 1988). Diluting segments in freshwater fish and terrestrial vertebrates play an important role in NaCl conservation (and in the case of birds and mammals, urine concentration; Dantzler, 2003). The activity of Slc12a1 is dependent upon the recycling of K^+ ions back to the tubular fluid via the apical Romk2 potassium channel and the transport of Cl^- ions out of the cell via the Clckb chloride channel (Simon and Lifton, 1998). Consistent with the zebrafish DE segment functioning as a diluting segment, homologues of *Romk2* and *Clck* are also expressed by the DE (Wingert *et al.*, 2007).

The DL segment expresses *slc12a3*, encoding a NaCl cotransporter. In the mammalian nephron, this cotransporter is expressed in the distal convoluted tubule segment, which follows the TAL, and fine-tunes sodium and chloride absorption under hormonal regulation (Mastroianni *et al.*, 1996; Simon *et al.*, 1996). It is likely that the DL segment of the zebrafish pronephros has an analogous function, although this has yet to be confirmed. Some overlap exists between expression of what would be collecting duct markers in mammals (*c-ret, gata3*) and distal tubule markers (*slc12a3, clck*; Wingert *et al.*, 2007) in the DL segment of the pronephros. Since freshwater vertebrates do not have concentrating kidneys, the lack of a significant segment expressing markers consistent with collecting duct identity is not unusual.

III. Formation of the Pronephros

A. Origin of the Nephrogenic Mesoderm

Cell labeling and lineage tracing in zebrafish gastrula stage embryos have demonstrated that cells destined to form the pronephros arise from the ventral mesoderm, in a region partially overlapping with cells fated to form blood (Fig. 3A) (Kimmel *et al.*, 1990). These cells emerge shortly after the completion of epiboly as a band of tissue, the intermediate mesoderm (IM), at the posterior lateral edge of the paraxial mesoderm (Fig. 3B and C). In zebrafish, unlike other non-teleost vertebrates, the IM gives rise to both kidney and blood cells. The size and positioning of the IM are significantly influenced by dorsoventral and anterior–posterior axis patterning molecules, such as the ventralizing factors bone morphogenetic proteins (BMPs) and their inhibitors, and the Cdx family of homeobox genes (see Table I for a summary of zebrafish mutants with pronephric defects).

B. Early Subdomains within the Nephrogenic Mesoderm

By the early stages of somitogenesis, the nephrogenic mesoderm component of the IM is clearly defined by the expression of renal markers such as the transcription factors *pax2a, pax8*, and *lhx1a*, which extend from the level of somite 3 to the cloaca (Carroll *et al.*, 1999; Drummond, 2000; Heller and Brandli, 1999; Krauss *et al.*, 1991; Majumdar

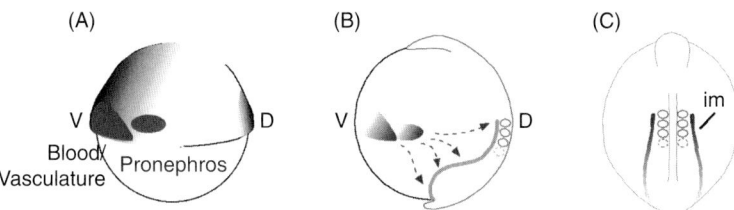

Fig. 3 Origins of the intermediate mesoderm. (A) Approximate positions of cells in a shield stage embryo destined to contribute to the blood/vasculature and pronephric lineages in the ventral (V) germ ring. (D; dorsal shield). (B) Migration of cells during gastrulation to populate the intermediate mesoderm (im) (C).

TABLE I
Zebrafish Mutants with Defects in Early Pronephros Formation

Mutant/gene	Gene product	Kidney phenotype	Reference
swirl/bmp2b	BMP ligand	Absent or reduced	Hild *et al.* (1999)
snailhouse/bmp7a	BMP ligand	Reduced	Kishimoto *et al.* (1997)
somitabun/smad5	BMP signal transducer	Reduced	Nguyen *et al.* (1998)
lost-a-fin/alk8	BMP receptor	Reduced	Mullins *et al.* (1996)
chordino/chordin	BMP antagonist	Expanded	Hammerschmidt *et al.* (1996b)
kugelig/cdx4 and cdx1a	Homeobox transcription factors	Posteriorly shifted	Davidson *et al.* (2003); Wingert *et al.* (2007)
ntla and spadetail/tbx16 double mutants	Mesoderm inducing T-box transcription factors	Absent	Amacher *et al.* (2002)

et al., 2000; Mauch *et al.*, 2000; Pfeffer *et al.*, 1998; Puschel *et al.*, 1992). Based on the overlapping but distinct expression patterns of the transcription factor genes *wt1a*, *pax2a*, and *sim1a* in the IM, together with fate mapping analyses, it was initially shown that podocytes, neck (previously "tubule"), and proximal tubule cells (previously "duct") could be defined as sequential anterior to posterior subdomains of the IM (Fig. 4) (Serluca and Fishman, 2001). Further refinement of this observation suggests that podocytes and neck cells arise from the IM adjacent to somites 3–4 whereas the first proximal tubule segment descends from the IM level with somites 5–8 (Bollig *et al.*, 2009; Wingert *et al.*, 2007). It is likely that the other tubule segments are derived sequentially from more posterior subdomains of the IM, although this has yet to be confirmed by lineage labeling. The notion that IM at all axial levels contributes to the pronephros in zebrafish is in contrast to other non-teleost vertebrates where only the anterior portion of the IM adopts a renal fate and the duct elongates to the cloaca.

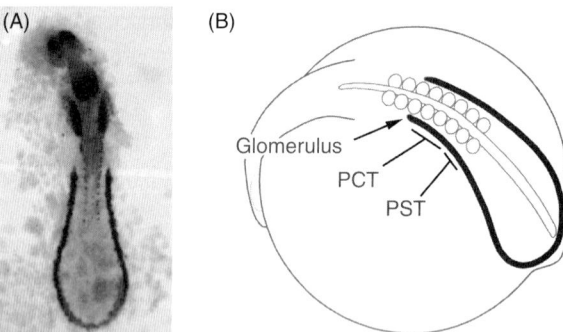

Fig. 4 Derivation of the pronephros from the intermediate mesoderm. (A) The pax2.1 expression domain in early somitogenesis stage embryos defines a stripe of intermediate mesoderm fated to become the pronephric epithelia. (B) Fate map of the nephrogenic intermediate mesoderm derived from fluorescent dye uncaging lineage experiments. Proximal fates previously referred to as "tubule" are now more accurately defined as proximal convoluted tubule (PCT) and more distal fates previously referred to as "duct" are now more accurately termed proximal straight tubule (PST).

C. Role of Retinoic Acid Signaling

Recently it was shown that retinoic acid (RA) signaling plays a major role in establishing the proximo-distal segmentation pattern of the pronephros (Wingert *et al.*, 2007). Exposure of embryos to high RA doses induces the formation of expanded proximal tubule segments at the expense of the distal segments. Conversely, inhibition of RA synthesis with diethylaminobenzaldehyde (DEAB), a competitive inhibitor of the aldehyde dehydrogenase enzymes, favors distal nephron cell fates. DEAB could elicit these patterning effects when added at the end of gastrulation through to the beginning of somitogenesis, consistent with a requirement for RA during early IM patterning. In support of this, DEAB-treated embryos show a loss of proximally restricted genes such as *jagged-2a* and *delta-c* and a concomitant expansion in the distal marker *meccom* at the 8-somite stage (Wingert *et al.*, 2007). RA may also function later in pronephric development as transgenic RA reporter embryos show significant activity in the pronephric tubules at the 18-somite stage (Perz-Edwards *et al.*, 2001). The source of the RA is presumed to be the paraxial mesoderm, which expresses high levels of *aldh1a2* (aka *retinaldehyde dehydrogenase-2*).

In addition to tubule patterning defects, DEAB-treated embryos are also characterized by a loss of podocytes. Expression of *wt1a*, implicated in podocyte differentiation (Perner *et al.*, 2007), is absent in DEAB-treated embryos from the subdomain of the IM fated to give rise to the glomerulus. Analysis of the *wt1a* promoter identified an RA-responsive element consistent with *wt1a* being a direct target of RA signaling (Bollig *et al.*, 2009). Additional RA target genes that specify podocytes or maintain podocyte function (Vaughan *et al.*, 2005) remain to be explored in zebrafish.

D. Role of Pax2a

The paired domain transcription factor *pax2a* is now known to play a key role in establishing the boundary between podocytes and the neck segment. After initially being expressed throughout the nephrogenic mesoderm, *pax2a* becomes highly expressed in the neck segment (Krauss *et al.*, 1991; Majumdar *et al.*, 2000). Fish with mutations in *pax2a* (*no isthmus, noi*) show an expansion in the expression of the podocyte markers *wt1a* and *vegf* into the neck and possibly proximal tubule domain (Fig. 5). Although these proximal tubule cells maintain an epithelial character (no transdifferentiation to a podocyte morphology is observed), they fail to express normal markers of this segment (NaK ATPase and 3G8, a brush border marker) (Majumdar *et al.*, 2000). The data suggest that *pax2a* plays an important role in defining the podocyte/neck/proximal tubule boundaries, possibly by repressing podocyte-specific genes in the neck and proximal tubule segments.

E. Role of the Endoderm

Although pronephric development does not require proper development of the endoderm, overdevelopment of the endoderm can alter pronephric development (Mudumana *et al.*, 2008). Knockdown of odd-skipped related1 gene (*osr1*), encoding

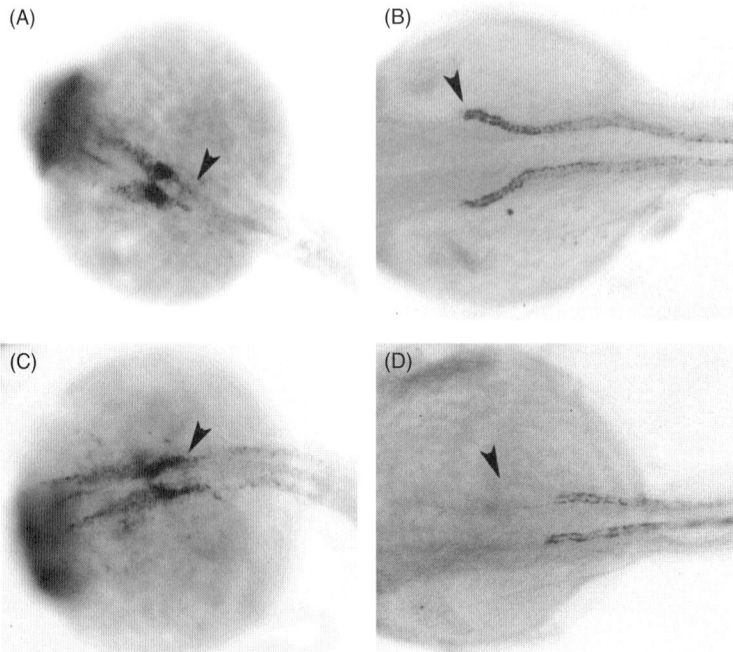

Fig. 5 Formation of the glomerulus–tubule boundary is disrupted in *no isthmus* (*noi; pax2a*) mutants. Wholemount *in situ* hybridization with *wt1a* marks the presumptive podocytes in wildtype embryos (A). *wt1a* expression is caudally expanded into the anterior pronephric tubules in noi[tb21] mutant embryos (C) at 24 hpf (arrow in C). Wholemount antibody staining of wild-type (B) and noi[tb21] (D) embryos with mAb alpha6F which recognizes a Na^+/K^+ ATPase alpha1 subunit. alpha6F marks the pronephric epithelia in wild-type (arrows in B) and proximal tubule Na^+/K^+ ATPase expression is missing in noi[-/tb21] mutant embryos at 2.5 dpf (D).

a zinc finger transcription factor, causes expansion of endoderm that subsequently inhibits epithelialization of the proximal portion of the pronephros.

F. Differentiation of the Tubular Epithelium

Tubule formation is mediated by a mesenchyme to epithelial transition, a process central to kidney formation in all vertebrates (Saxén, 1987). By the end of this transition, which is complete by 24 h post fertilization, the epithelial cells of the pronephros are polarized with apical and basolateral domains containing ion transport proteins (Fig. 6A) (Drummond *et al.*, 1998). In addition, individual multiciliated cells, induced by notch signaling during mid-somitogenesis, are interspersed with transporting epithelial cells along the pronephros (Liu *et al.*, 2007). Thus, tubule formation occurs simultaneously with patterning events that define functionally distinct epithelial cell types.

Establishment of cell–cell junctions is an essential step in separating apical and basolateral membrane domains and giving an epithelium its vectorial property. Cadherins are the major proteins of the adherens junction that maintain the integrity

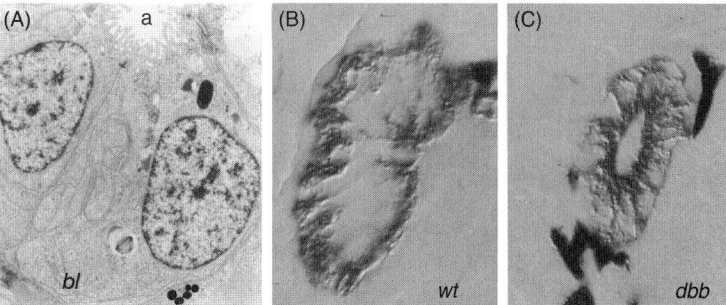

Fig. 6 Epithelial cell polarity in the pronephric tubules. (A) Electron micrograph of 2.5-day pronephric tubule epithelial cells showing apical (A) brush border and basolateral cell surfaces and infoldings. Polarized distribution of the NaK ATPase in 2.5-day pronephric tubule epithelial cells visualized by the alpha6F monoclonal antibody is shown. The apical cell surface is devoid of staining while staining is strong on the basolateral cell surface and membrane infoldings (B). Double bubble mutant embryos (C) aberrantly express the NaK ATPase on the apical cytoplasm and cell membrane.

of epithelial sheets and separate apical and basolateral membrane domains. *cadherin17* is specifically expressed in the zebrafish nephrogenic mesoderm and persists in the pronephric epithelium (Horsfield *et al.*, 2002). Knockdown of *cadherin17* causes a loss in renal epithelial cell-to-cell adhesion, failure of the ducts to fuse at the cloaca, and gaps between epithelial cells (Horsfield *et al.*, 2002), thus demonstrating an essential role for *cadherin17* in tubule and duct morphogenesis.

Cell polarity and proper targeting of membrane transporters are essential for proper kidney ion transport and function. Several zebrafish mutants have been found to mistarget the NaK ATPase in the tubules from its normal basolateral membrane location to the apical membrane (Fig. 6) (Drummond *et al.*, 1998). The activity of the NaK ATPase provides the motive force for many other coupled transport systems (Seldin and Giebisch, 1992); its mislocalization suggests that severe problems in osmoregulation exist in these mutants. In fact, these mutants later develop cysts in the pronephric tubule and the embryos eventually die of edema (Drummond *et al.*, 1998).

G. Formation of the Glomerulus

A major feature of the glomerular blood filter is the podocyte slit-diaphragm, a specialized adherens junction that forms between the finger-like projections of podocytes (podocyte foot processes) (Reiser *et al.*, 2000). Failure of the slit-diaphragm to form results in leakage of high molecular weight proteins into the filtrate, a condition called proteinuria in human patients. Several disease genes with known function in the slit-diaphragm have been cloned. Nephrin is a transmembrane protein present in the slit diaphragm itself and is thought to contribute to the zipper-like extracellular structure between foot processes (Ruotsalainen *et al.*, 1999). Podocin is a podocyte junction-associated protein (Roselli *et al.*, 2002) that resembles stomatin proteins which play a role in regulating mechanosensitive ion channels (Tavernarakis and Driscoll, 1997).

Electron microscopy of the zebrafish pronephric glomerulus reveals that, like mammalian podocytes, zebrafish podocytes form slit diaphragms between their foot processes (Fig. 7D). Zebrafish homologs of *podocin* and *nephrin* are specifically expressed in podocytes as early as 24 hpf (Fig. 7E and F) and have been shown to be required for proper slit-diaphragm formation in pronephric podocytes (Kramer-Zucker *et al.*, 2005b).

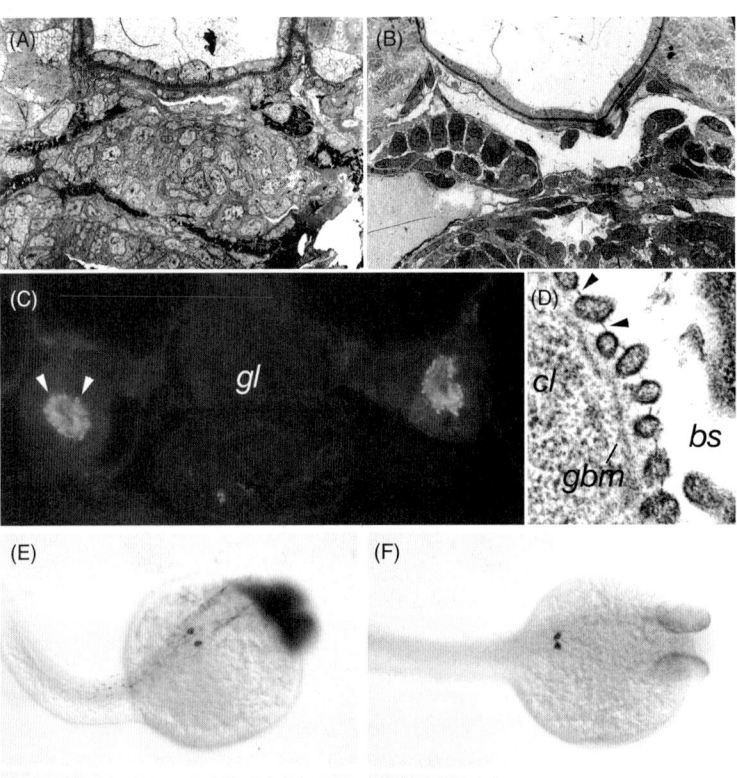

Fig. 7 The glomerular capillary tuft and podocyte slit diaphragms. (A) An electron micrograph of the forming glomerulus at 2.5 dpf with invading endothelial cells from the dorsal aorta shaded in red and podocytes shaded in blue (image false-colored in Adobe Photoshop). (B) A similar stage glomerulus in the mutant island beat which lacks blood flow due to a mutation in an L-type cardiac-specific calcium channel. The endothelial cells and podocytes are present but the aorta has a dilated lumen surrounding the podocytes with no sign of glomerular remodeling and morphogenesis. (C) Rhodamine dextran filtration and uptake by pronephric epithelial cells. Lysine-fixable rhodamine dextran (10 kDa) injected into the general circulation can be seen as red fluorescence in glomerular capillaries (gl), and filtered dye is seen in apical endosomes of pronephric tubule cells (arrowheads). Counterstain: FITC wheat germ agglutinin. (D) Electron micrograph of the glomerular basement membrane region in the glomerulus. Individual profiles of podocyte foot processes resting on the glomerular basement membrane (gbm) are connected by slit-diaphragms (arrowheads at top). cl, capillary lumen; bs, Bowman's space. Wholemount *in situ* hybridization shows expression of zebrafish podocin (E) and nephrin (F) specifically in the forming podocytes. (See Plate no. 9 in the Color Plate Section.)

Vascularization of the glomerulus occurs relatively late in development, after pronephric tubule development is complete (Armstrong, 1932; Drummond *et al.*, 1998; Tytler, 1988). The bilateral glomerular primordia coalesce at 36–40 hpf ventral to the notochord, bringing the presumptive podocytes into contact with endothelial cells of the overlying dorsal aorta (Fig. 8) (Drummond, 2000; Drummond *et al.*, 1998; Majumdar and Drummond, 1999). Podocytes express two known mediators of angiogenesis: *vascular endothelial growth factor* (*vegf*) and *angiopoietin2* (Carmeliet *et al.*, 1996; Ferrara *et al.*, 1996; Majumdar and Drummond, 1999, 2000; Pham *et al.*, 2001; Shalaby *et al.*, 1995). In a complementary manner, capillary-forming endothelial cells express *kdrl* (aka *flk1*), a VEGF receptor (Majumdar and Drummond, 1999). Between 40 and 48 hpf, kdrl-positive endothelial cells invade the glomerular epithelium and become surrounded by podocytes (Fig. 8) (Drummond *et al.*, 1998). Vascular shear force is required to drive capillary formation as mutants lacking cardiac function, such as silent heart/cardiac troponin T and island beat/L-type cardiac calcium channel, fail to form a proper glomerular capillary tuft (Fig. 7A and B) (Rottbauer *et al.*, 2001; Sehnert *et al.*, 2002). Glomerular filtration begins around 48 hpf but is leaky at this time, allowing large molecular weight dextrans to pass into the tubules. Full maturation and size-selectivity occur at 4 days post-fertilization (dpf), concomitant with well-developed podocyte foot processes and endothelial cell fenestrations (Kramer-Zucker *et al.*, 2005b).

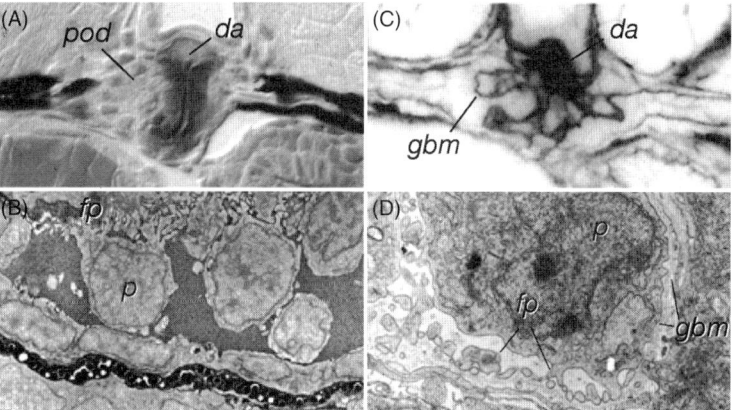

Fig. 8 Interaction of pronephric podocytes with the vasculature. (A) Apposition of nephron primordia at the embryo midline in a 40 hpf zebrafish embryo. Aortic endothelial cells in the cleft separating the nephron primordia are visualized by endogenous alkaline phosphatase activity. Podocytes, pod; dorsal aorta, da. (B) Ultrastructure of the forming zebrafish glomerulus at 40 hpf. A longitudinal section shows podocytes (p) extending foot processes (fp) in a dorsal direction and in close contact with overlying capillary endothelial cells. (C) Rhodamine dextran (10,000 Dalton MW) injected embryos show dye in the dorsal aorta (da) and in the glomerular basement membrane (gbm) shown here graphically inverted from the original fluorescent image. (D) Podocyte foot process formation does not require signals from endothelial cells as evidenced by the appearance of foot processes (fp) in cloche mutant embryos which lack all vascular structures. Glomerular basement membrane, gbm; podocyte cell body, p.

H. Formation of the Cloaca

Formation of the caudal pronephric opening or cloaca requires BMP signaling (Pyati *et al.*, 2006). The ligand involved here appears to be BMP4 since mutants lacking functional BMP4 exhibit failure to complete the fusion of the pronephric ducts with the epidermis and an absence of a pronephric opening (Stickney *et al.*, 2007), similar to the phenotype seen when a dominant negative BMP receptor is ectopically expressed late in development (Pyati *et al.*, 2006). These studies revealed that cloaca formation is likely to involve developmentally programmed cell death accompanied by cellular rearrangements within the terminus of the pronephric duct and the epidermis. Although zebrafish do not have a urinary bladder, the terminus of the pronephros may be homologous to the end of the common nephric duct in mammals which inserts into the bladder by a mechanism involving programmed cell death (Batourina *et al.*, 2005).

IV. Methods to Study Pronephros Function

A. Embryo Dissociation

Historically, the functional aspects of kidney epithelial ion transport have been studied using isolated single epithelial tubules in primary culture. This has not yet been achieved for zebrafish pronephric tubules. However, a useful first step in considering such an approach is larval tissue fractionation and tubule isolation (Fig. 9). Two to three day old zebrafish larvae show a remarkable resistance to collagenase digestion. However, a 1 h preincubation in dithiothreitol (DTT) or *N*-acetyl-cysteine, which degrade the protective mucous layer around the embryo, allows subsequent incubation in collagenase to be effective.

1. Anesthetize 2–3 day old larvae with 0.2% Tricaine.
2. Incubate larvae in 10 mM DTT or *N*-acetyl-cysteine in egg water for 1 h at room temperature.

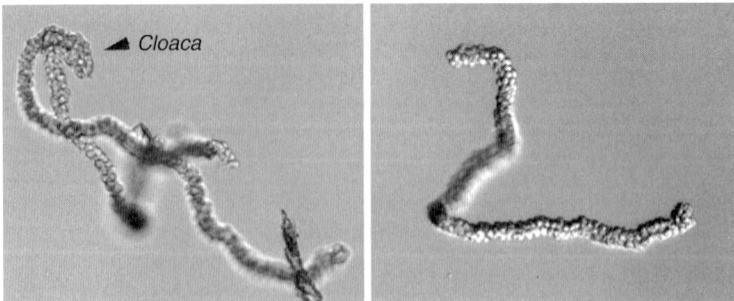

Fig. 9   Isolated pronephric tubules. Two day old larvae were treated with 10 mM DTT followed by incubation in 5 mg/ml collagenase for 3 h at 28°C. Individual pronephric nephrons can be dissected away from trunk tissue, often with the cloaca intact (arrowhead), joining the bilateral tubules at the distal segment.

3. Wash the larvae 3–4 times with egg water to remove the DTT.
4. Incubate larvae at 28.5°C in 5 mg/ml collagenase in tissue culture medium or Hanks saline with calcium (Worthington) for 3–6 h.
5. Triturate the larvae gently 5 times with a "blue tip" 1000 µl pipet tip. The larvae should disaggregate into chunks of tissue.
6. Disperse the cell/tissue suspension into a 10 cm petri dish containing 10 ml of tissue culture medium or Hanks buffer.
7. Collect pronephric tubules by visual identification under a dissecting microscope.

B. Isolation of Fluorescently Labeled Cells by Fluorescence Activated Cell Sorting (FACS)

The isolation of specific cells from transgenic embryos on the basis of fluorescent marker expression has numerous applications, including purification of cells for transplantation, preparation of cDNA libraries, and quantification of cell types in mutant or morpholino "knockdown" embryos.

1. Collect and dechorinate (if necessary) both transgenic embryos and non-transgenic controls and anesthetize with 0.2% Tricaine.
2. Transfer embryos to 1.5 ml microfuge tubes (no more than a 500 µl packed volume of embryos per tube) and wash 3 times with FACS buffer ($0.9\times$ phosphate buffered saline (PBS) + 5% fetal calf serum).
3. Homogenize the embryos with a microfuge pestle.
4. Spin down the cells for 3 min at 1500 g in a microcentrifuge.
5. Resuspend the cells in 500 ml FACS buffer.
6. Repeat the spinning and resuspending steps an additional 3 times.
7. Filter the cells by pipetting them through a 40 mm nylon cell strainer (Falcon 2340) into a 5 ml round-bottom tube (Falcon 2054) on ice.
8. Add propidium iodide (PI) to a final concentration of 1 µg/ml (to stain dead cells).
9. FACS the control non-transgenic cells on the basis of PI exclusion and reporter gene fluorescence (e.g., green fluorescence protein) in order to be able to set the gate for the transgenic sample. Repeat for the transgenic cells.
10. Collect positive cells in 500 ml of FACS buffer.

Note: A single sort usually gives only 70–80% purity. For ~95% purity it is recommended that the cells be double sorted. If RNA is to be collected from the cells then pellet the cells at 12,000 *g* in a microcentrifuge for 15 min and resuspend in a small volume of extraction buffer (e.g., Trizol).

C. A Simple Assay for Glomerular Filtration

Filtration of blood by the glomerulus can be detected by injections of fluorescent compounds (10–70 kDa rhodamine dextran) into the general circulation and then monitoring the appearance of fluorescent endosomes in the apical cytoplasm of pronephric tubule cells (Fig. 7C) (Drummond *et al.*, 1998; Majumdar and Drummond, 2000). The ability to detect filtered dextrans in endocytic vesicles of the tubules has

also been adapted to create an assay for disruption of the filtration barrier (i.e., proteinuria or nephrotic syndrome). Kramer-Zucker *et al.* demonstrated that large dextrans (500 kDa) do not significantly pass a normal glomerular filter while in embryos where orthologs of human nephrotic syndrome genes have been knocked down, passage of large dextrans could be observed as accumulation in tubule endocytic vesicles (Kramer-Zucker *et al.*, 2005b). Filtered fluorescent dextrans can also be observed directly as they exit the pronephros at the cloaca in live larvae and used as a qualitative assay of the rate of pronephric fluid output (Kramer-Zucker *et al.*, 2005a).

1. Embryos:

1. Anesthetize 84 hpf embryos in 0.2% Tricaine and position them ventral side up in a 1% agarose injection mold. The age of the embryo is critical for this assay since younger embryos have leaky glomeruli (Kramer-Zucker, *et al.*, 2005b).
2. Using a microinjection device (e.g., Nanoject II from Drummond Scientific, Broomall, PA) fitted with a pulled glass capillary needle inject 5 nl of fluorescently labeled dextran (40 mg/ml in 150 mM NaCl; Invitrogen) into the cardiac venous sinus or alternatively into the cardinal vein in the posterior trunk (Fig. 10).
3. Return embryo to egg water to recover.
4. Uptake of the 40 kDa dextran by the PCT occurs within 24 hpf and can be imaged in live embryos. To detect the 500 kDa dextran it is necessary to fix and section the embryos (see Section IV-H) at the level of the PCT. If the glomerulus is leaky, fluorescent endosomes can be visualized in endocytic vesicles of PCT epithelial cells (similar to Fig. 7C).

2. Adults:

1. Anesthetize an adult fish in 0.2% Tricaine.
2. Lay the fish on its back on a wet tissue or paper towel.

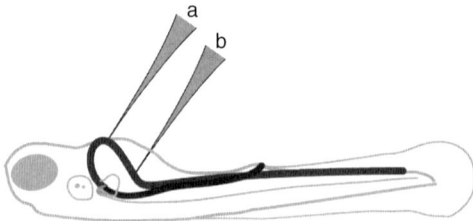

Fig. 10   Assaying glomerular filtration by injection of labeled dextrans. The diagram shows a zebrafish larva positioned dorsal side down, exposing the sinus venosus/inflow tract of the heart circulation. Vasculature is depicted in red; the pronephros is in blue. (A) Sinus venosus injections are feasible in 2–3 dpf embryos. (B) By 3.5 dpf the inflow tract is shifted forward, making it necessary to inject dye in the descending cardinal vein. (See Plate no. 10 in the Color Plate Section.)

3. Using a 0.5 cc insulin syringe, inject 20 μl of 40 kDa fluorescein-labeled dextran (40 mg/ml; Invitrogen) into the intraperitoneal space by carefully inserting the needle into the abdomen at the ventral midline. Insert the needle at a shallow angle to avoid injecting the gut. Once through the skin, pull the needle slightly away from the fish to "tent" the skin and create a void space to inject into. The entire abdomen of the fish should swell as the dextran is injected.

4. Return fish to the tank to recover. Uptake of the dextran by the proximal tubule occurs within a few hours and persists for several days.

D. Gentamicin Induced Kidney Tubule Injury in Embryos and Adults

Zebrafish and other teleosts have a remarkable capacity for kidney regeneration. Gentamicin is an antibiotic that in high enough doses induces necrosis of proximal tubule epithelial cells. Kidney damage in this model is repaired by the formation of new nephrons.

1. Embryos:

1. Dechorinate 50–55 hpf embryos and anesthetize in 0.2% Tricaine and positioned on their back in a 1% agarose injection mold.

2. Using a microinjection device (e.g. Nanoject II from Drummond Scientific, Broomall, PA) fitted with a pulled glass capillary needle inject 5 nl of gentamicin (5 mg/ml in 150 mM NaCl; Sigma) into the cardiac venous sinus or alternatively into the cardinal vein in the posterior trunk (Fig. 10).

3. Return embryo to egg water to recover. Marked pericardial and intracranial edema caused by a failure to osmoregulate will be apparent between 72 and 96 hpf. Histologically the PCT segment shows lysosomal phospholipidosis, flattening of the brush border, an accumulation of cellular debris in the lumen, and distention (Hentschel *et al.*, 2005).

Note: If necessary, phenol red (0.25%) can be added to the gentamicin to positively identify successfully injected embryos.

2. Adults:

1. Anesthetize an adult fish in 0.2% Tricaine.

2. Lay the fish on its back on a wet tissue or paper towel.

3. Using a 0.5 cc insulin syringe, inject 20 μl of gentamicin (5 mg/ml) into the intraperitoneal space by carefully inserting the needle into the abdomen at the ventral midline. Insert the needle at a shallow angle to avoid injecting the gut. Once through the skin, pull the needle slightly away from the fish to "tent" the skin and create a void space to inject into. The entire abdomen of the fish should swell as the gentamicin is injected.

4. Return fish to the tank to recover. Necrosis of the proximal tubule occurs within 24 h.

Note: Expect 5–10% mortality. Do not feed the fish for 24 h to increase the survival rate.

E. Adult Kidney Isolation

The adult (mesonephric or opisthonephric) kidney is located along the dorsal wall of the body cavity and can be divided into head, trunk (or saddle), and tail portions going from anterior to posterior (Fig. 11A). Two major collecting ducts run the length of the kidney and drain hundreds of nephrons. As in the pronephros, the mesonephric nephron is made up of a glomerulus, a neck segment, and two proximal and two distal tubule segments. The nephrons are branched predominantly at the DL segment, with occasional branching occurring at the DE segment.

1. Euthanize an adult fish by Tricaine overdose.
2. Using scissors, remove the head by cutting at the level of the gills (be careful to not cut beyond the posterior limit of the gill flap as the head kidney is located in this region).
3. Using scissors, make an excision along the ventral midline of the fish.
4. Carefully remove the gut and associated organs (liver, pancreas, etc.), gonads (testes or ovaries), and swim bladder using forceps. The kidney appears as a thin reddish organ with black melanocytes and is located along the dorsal side of the body wall.
5. Using forceps, score alongside each side of the kidney to sever the blood vessels entering the kidney and also to rupture a thin translucent membrane covering the kidney.
6. Starting at the head kidney, gently prize the tissue off the body wall. Although the kidney tissue has a gelatinous consistency it is possible to remove it intact. At the caudal end, use forceps to sever the relatively tough collecting ducts (these fuse at a urinary sinus near the cloaca).

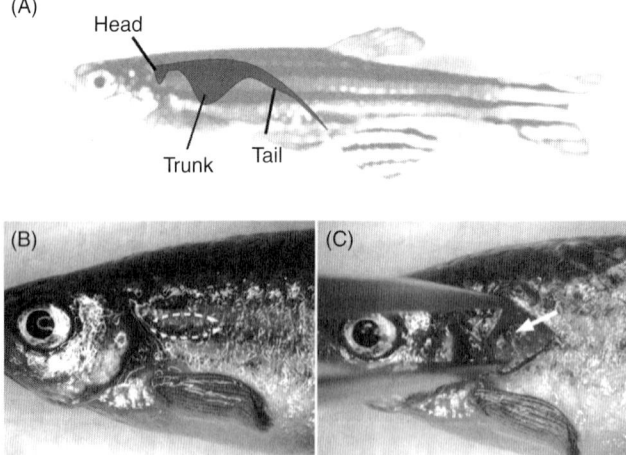

Fig. 11 The adult (mesonephric or opisthonephric) kidney. (A) The adult kidney is located along the dorsal wall of the body cavity and can be divided into head, trunk (or saddle), and tail portions going from anterior to posterior. (B) The head kidney can be accessed surgically by making an incision in the upper part of the middle blue stripe near the gills (dotted region) and appears as a red mass (C; arrow).

F. Non-Lethal Surgical Access to the Adult Kidney

Despite being located deep in the body cavity, the kidney can be non-lethally accessed. The head kidney, due to its lobular shape and anatomical position, can be exposed by a small lateral excision in the side of the fish.

1. Anesthetize an adult fish in 0.2% Tricaine.
2. Using forceps, remove the scales in the region just posterior to the gill flap.
3. Using a scalpel, make a 1–2mm cut level with the top of the middle blue stripe (see Fig. 11B). Initially make a shallow cut.
4. Use forceps to open the incision and cut progressively deeper until the dark red head kidney and underlying silver pigment (lining the anterior body cavity) are visible. Cells or dyes can be injected into the kidney using a fine gauge Hamilton syringe or glass needle.
5. Provided the excision is small, a suture is not needed to keep the wound closed and the fish can be returned to the tank to recover.

G. Detecting and Imaging Zebrafish Cilia

Zebrafish are ideally suited for analysis of genes required for ciliogenesis and cilia function. Pronephric cilia have a "9 + 2" microtubule doublet organization and dynein arms characteristic of motile cilia and flagella and consistent with their proposed function in propelling fluid down the tubules (Kramer-Zucker et al., 2005a; Liu et al., 2007; Pathak et al., 2007). Cilia can be visualized using the anti-acetylated tubulin monoclonal antibody 6-11-B1 (Sigma T6793) as well as anti-gamma tubulin (GTU-88) and anti-polyglutamyl tubulin (B3).

Anti-acetylated tubulin (clone 6-11b-1)	Sigma	T6793
Anti-polyglutamylated tubulin (clone B3)	Sigma	T9982
Anti-gamma-tubulin (clone GTU-88)	Sigma	T6657

The protocol that follows is designed for use of these antibodies but can be adapted for other antibodies and other fixation methods.

1. Materials

PBS
DMSO
Tween 20
Normal goat serum (Sigma G9023)
Bovine serum albumin (BSA; Sigma A8022) or gelatin from cold water fish skin (Sigma G7765)
Methanol
Benzyl alcohol

Glycerol
N-propyl gallate
Formaldehyde
Hydrogen peroxide
SDS

2. Solutions

PBST:	PBS + 0.5% tween20
Blocking solution:	PBS with
	1% DMSO
	0.5% Tween20
	1% BSA or 0.3% gelatin from cold water fish skin
	10% normal goat serum
Incubation solution:	PBS with
	1% DMSO
	0.5% Tween20
	2% normal goat serum
High-salt wash:	PBS with
	1% DMSO
	0.5% Tween20
	2% normal goat serum
	0.18% NaCl (final NaCl = 0.27%)
Dent's fixative:	80% methanol
	20% DMSO

Formaldehyde (BT) fix 4% formaldehyde (from paraformaldehyde)
0.1 M phosphate buffer, pH 7.2
3% sucrose
0.12 mM $CaCl_2$

Rehydration solutions:
75:25 MeOH/PBST
50:50 MeOH/PBST
25:75 MeO/PBST

Antigen retrieval solution:	1% SDS in PBS
Mounting medium:	53% benzyl alcohol (by weight)
	45% glycerol (by weight)
	2% N-propyl gallate

3. Methods

One of the main advantages of using zebrafish immunofluorescence is the transparency of their embryos. To achieve maximum embryo transparency, development of

pigmentation can be blocked by raising embryos in 0.003% phenylthiourea (PTU) egg water. Alternatively embryos can be bleached with hydrogen peroxide after fixation (see below).

4. Fixation

1. Fix embryos in Dent's fixative for 3 h to overnight at room temperature. If necessary, remove pigment by bleaching fixed embryos overnight in 10% H_2O_2. After fixation embryos can be stored in 100% methanol at $-20°C$.
2. Rehydrate Dent's fixed embryos with graded changes of methanol/PBT:

   75:25 MeOH/PBT 15 min
   50:50 MeOH/PBT 15 min
   25:75 MeOH/PBT 15 min
   PBT

   Dent's fixative works well for antibodies that preferentially recognize denatured epitopes. If formaldehyde fixation is preferable, antigen retrieval can be performed by denaturing the embryos after formaldehyde fixation. For formaldehyde fixation fix 2 h to overnight at 4°C in BT fix and then wash twice in PBS. Permeabilize the embryos in cold acetone for 20 min at $-20°C$. Warm to room temperature and wash with PBS. For antigen retrieval (denaturation), incubate fixed embryos in 1% SDS/PBS for 15 min at room temperature followed by washing with PBS 4×5 min each.

5. Antibody Staining

3. Block non-specific binding by incubating fixed embryos for 2 h to overnight in Blocking solution. Incubations are done in eppendorf tubes at 4°C on a nutator rocking platform.
4. Incubate with primary antibody (6-11-B1 at 1:1000) in incubation solution overnight at 4°C. Monoclonal supernatants can be used at 1:50 to 1:25 dilution. Be aware that any primary antibody raised in rabbits should be affinity purified on antigen to avoid high background staining of larval fish skin. If necessary to economize on primary antibodies the incubation can be done in 50–100 µl without agitation.
5. After incubation with primary antibody, wash at least 4×30 min with incubation solution. Normal goat serum (2%) is included in all steps to reduce background. If background staining becomes a problem, the first wash after incubation with primary antibody can employ the "high-salt" wash. Subsequent washes are in the standard incubation solution.
6. Incubate with secondary antibody in incubation solution overnight. Alexa anti-mouse secondary antibodies (Invitrogen) work well at 1:1000 dilution. Following incubation, wash at least 4×30 min with incubation solution on a rocking platform. Although background staining with secondary antibodies is not common, a high-salt wash can be used to minimize non-specific staining.

6. Mounting the Sample for Confocal Microscopy

To minimize optical distortion caused by mismatch in refractive index of the sample, coverslips, and immersion oil, use a mounting medium that has the same refractive index (1.513) as the immersion oil. We make a mounting medium developed by Gustafsson *et al.* (1999) that is a mixture of glycerol and benzyl alcohol and contains an antifade compound (*N*-propyl gallate) that is essential for preventing signal bleaching, especially when using 488 nM fluorophores (fluroescein (FITC), Alexa 488) on large Z image stacks.

7. Embryos can be placed in mounting medium directly after washing. The difference in the density of the mounting medium and PBS is significant and causes turbulence but does not damage the sample. It is best to transfer embryos to mounting medium in a depression slide or directly on a microscope slide since the embryos become essentially invisible and can be hard to find in an eppendorf tube. Change to fresh mounting medium and transfer embryos to a standard microscope slide. Using small balls of modeling clay, support the edges of a coverslip to provide space for the embryo and coverslip the sample. Alternatively, make a coverslip bridge with additional coverslips as spacers. The orientation of the embryo can often be shifted by moving the coverslip. This sample configuration is suited for viewing with an upright microscope using a 63× oil immersion objective.

7. Alternate Protocols

a. Visualization of Cilia with HRP/DAB   If visualization of cilia by light microscopy is desired, a horseradish peroxidase (HRP)-coupled secondary antibody can be used. To visualize HRP after washing out the secondary antibody:

1. Wash once in PBSBT (PBS/0.1% tween/0.2% BSA) for 5 min.
2. Incubate with 0.3 mg/ml Diamion Benzidine (DAB) and 0.5% $NiCl_2$ in PBSBT for 20 min.
3. Add H_2O_2 to 0.03% (1:1000 dilution of 30% stock) and monitor color development (check at 10 min).
4. Stop the reaction by rinsing in PBST and then PBS. Post fix in 2% formaldehyde/ 0.1% glutaraldehyde.

H. Histological Sectioning of Whole mount Stained Embryos

Fluorescent signal from secondary antibodies is preserved in glycol methacrylate (JB4) embedded samples. This allows for histological sectioning of wholemount stained embryos and viewing 3–10 μm sections by standard wide-field epifluorescence. Longer wavelength excitation secondary antibodies, for instance Alexa 546 or rhodamine, should be used instead of FITC since fluorescein excitation wavelengths result in significant autofluorescence from the glycol methacrylate/JB4.

1. After the final wash following secondary antibody incubation, dehydrate embryos in sequential changes of 30, 50, 75, 85, and 95% ethanol.

2. Prepare catalyzed JB4 (Polysciences or Electron Microscopy Sciences) following the manufacturer's instructions. Dissolve 0.625 g powdered catalyst to 50 ml monomer solution A in a 50 ml tube. Wrap the tube in foil to protect it from light and store it at 4°C. The catalyzed JB4 making can be used for 1–2 weeks.

3. Remove final ethanol and add catalyzed JB4 solution to the dehydrated wholemount stained embryos. The embryos will float at the top of the solution and gradually sink as they become infiltrated. Leave at 4°C overnight.

4. Draw off the first solution of JB4 making sure to remove traces of ethanol that can inhibit hardening. Add the hardener (solution B) at 40 μl/ml to the embryos. Pour embryos in JB4 into a plastic mold, orient the embryos, and allow the JB4 to harden. Humidity and oxygen will inhibit hardening, so this step is best done in a sealed, desiccated chamber.

5. To embed embryos to cut them in cross section, first half fill a mold by pouring a bed of JB4 (0.7 ml) in the mold (15 mm × 15 mm) and let it harden. Then add stained embryos in final catalyzed JB4 plus hardener and orient them to be headed to the side of the mold. The "half blocks" of JB4 adhere well even if polymerized separately and this allows for rotating and mounting the final block to have embryos facing the knife for sectioning.

6. JB4 blocks require sectioning with glass knives. Use a Leica RM2255 or equivalent style microtome. JB-4 embedding kit and embedding molds can be purchased from Polysciences or Electron Microscopy Sciences (EMS).

I. Electron Microscopy Methods for Zebrafish

Electron microscopy is a standard method required for assessing cellular structural defects. Electron microscopy on zebrafish is performed essentially the same as for other vertebrate tissues but with some alterations that accommodate the lower osmolarity of extracellular fluids in fish. Other modifications include use of tannic acid in the fixative to enhance contrast and use of partially reduced osmium that produces a less grainy final image.

1. Materials

(All available from Electron Microscopy Sciences)

Glutaraldehyde
Paraformaldehyde
Phosphate or cacodylate buffer
Sucrose
Tannic acid

2. Methods

1. Fix overnight at 4°C in

| 1.5% glutaraldehyde | 7.5 ml 10% stock |
| 1% paraformaldehyde | 3.125 ml 16% stock |

70 mM NaPO$_4$, pH 7.2	3.5 ml 1 M stock
3% sucrose	5 ml 30% stock
0.1% tannic acid	0.05 g
	Water to 50 ml

EMS (Electron Microscopy Sciences) premade stocks can be used for the aldehydes; these come in sealed glass vials. Addition of tannic acid enhances the final contrast of the image, particularly of microtubules and other filamentous structures. Add the tannic acid just before fixation; it does not keep well. The buffer has a slightly lower osmolarity than a fixative for mammalian tissue.

2. Wash 3×5 min in 0.1 M cacodylate or phosphate buffer, pH 7.4.

   Avoid the use of mechanical rocking or mixing during the fixation and washes; the overall concern is to be as gentle as possible with the tissue to preserve structure.

3. Post fix in 1% OsO$_4$ plus potassium ferrocyanide for 1 h on ice:

   1 ml 4% OsO$_4$
   3 ml H$_2$O
   0.06 g potassium ferrocyanide

   This is partially reduced osmium; it produces less grainy images. If desired the potassium ferrocyanide may be omitted.

4. Wash 2×10 min in 0.1 M cacodylate or phosphate buffer, pH 7.4.

5. En bloc stain fixed embryos in 1% uranyl acetate in 0.1 M cacodylate pH 7.4 for 1 h at room temperature.

6. Wash 3×10 min in 0.1 M cacodylate or phosphate buffer, pH 7.4.

7. Dehydrate in 25, 50, 70, 80, 95, and 100% EtOH for 20–30 min each.

8. Dehydrate with 2×100% EtOH 20–30 min each.

9. Infiltrate with propylene oxide for 15–20 min (in a fume hood).

10. Mix the Epon 812 (or substitute epoxy embedding medium) according to the manufacturer's instructions.

11. Infiltrate with

25%	30 min (25% Epon in propylene oxide)
50%	40 min
75%	Overnight
100%	4 h
100%	1 h

12. Embed embryos in fresh Epon in flat molds. Position embryo swimming toward the tip of the block for acquiring cross sections.

13. Harden at 60°C for 16 h.

The blocks are now ready for sectioning and staining. We use formvar-coated slot grids (Electron Microscopy Sciences; catalog # FF2010-Cu) to maximize visibility of the tissue in the sections.

V. Conclusions

The zebrafish pronephric kidney represents one of the many vertebrate kidney forms that have evolved to solve the problem of blood fluid and electrolyte homeostasis in an osmotically challenging environment. Despite differences in organ morphology between the mammalian and teleost kidneys, many parallels exist at the cellular and molecular levels that can be exploited to further our understanding of kidney cell specification, epithelial tubule formation, and the tissue interactions that drive nephrogenesis. The same genes (for instance *pax2*) and cell types (for instance podocytes, endothelial cells, and tubular epithelial cells) are employed in the development and function of fish, frog, chicken, and mammalian kidneys. Genes mutated in human disease are also essential for the formation and function of the zebrafish pronephros. The zebrafish thus presents a useful and relevant model of vertebrate kidney development: its principal strengths lie in the ease with which it can be genetically manipulated and phenotyped so as to rapidly determine the function of genes and cell–cell interactions that underlie the development of all kidney forms.

Acknowledgments

IAD was supported by NIH grants DK53093, DK071041, and DK070263 and by grants from the PKD foundation. AJD was supported by the NIH grant DK077186 and grants from the Harvard Stem Cell Institute, American Society of Nephrology, and the Cystinosis Research Foundation.

References

Agarwal, S., and John, P. A. (1988). Studies on the development of the kidney of the guppy, Lebistes reticulatus. Part 1. The development of the pronephros. *J. Anim. Morphol. Physiol.* **35**, 17–24.

Amacher, S. L., Draper, B. W., Summers, B. R., Kimmel, C. B., (2002). The zebrafish T-box genes no tail and spadetail are required for development of trunk and tail mesoderm and medial floor plate. *Development.* **129**, 3311–3323.

Anzenberger, U., Bit-Avragim, N., Rohr, S., Rudolph, F., Dehmel, B., Willnow, T. E., and Abdelilah-Seyfried, S. (2006). Elucidation of megalin/LRP2-dependent endocytic transport processes in the larval zebrafish pronephros. *J. Cell Sci.* **119**, 2127–2137.

Armstrong, P. B. (1932). The embryonic origin of function in the pronephros through differentiation and parenchyma–vascular association. *Am. J. Anat.* **51**, 157–188.

Balfour, F. M. (1880). "A Treatise on Comparative Embryology". Macmillan and Co., London.

Batourina, E., Tsai, S., Lambert, S., Sprenkle, P., Viana, R., Dutta, S., Hensle, T., Wang, F., Niederreither, K., McMahon, A. P., Carroll, T. J., and Mendelsohn, C. L. (2005). Apoptosis induced by vitamin A signaling is crucial for connecting the ureters to the bladder. *Nat. Genet.* **37**, 1082–1089.

Bollig, F., Perner, B., Besenbeck, B., Kothe, S., Ebert, C., Taudien, S., and Englert, C. (2009). A highly conserved retinoic acid responsive element controls wt1a expression in the zebrafish pronephros. *Development* **136**, 2883–2892.

Carmeliet, P., Ferreira, V., Breier, G., Pollefeyt, S., Kieckens, L., Gertsenstein, M., Fahrig, M., Vandenhoeck, A., Harpal, K., Eberhardt, C., Declercq, C., Pawling, J., *et al.* (1996). Abnormal blood vessel development and lethality in embryos lacking a single VEGF allele. *Nature* **380**, 435–439.

Carroll, T. J., Wallingford, J. B., and Vize, P. D. (1999). Dynamic patterns of gene expression in the developing pronephros of *Xenopus laevis. Dev. Genet.* **24**, 199–207.

Dantzler, W. H. (2003). Regulation of renal proximal and distal tubule transport: sodium, chloride and organic anions. *Comp. Biochem. Physiol. A Mol. Integr. Physiol.* **136**, 453–478.

Davidson, A. J., Ernst, P., Wang, Y., Dekens, M. P., Kingsley, P. D., Palis, J., Korsmeyer, S. J., Daley, G. Q., Zon, L. I., (2003). cdx4 mutants fail to specify blood progenitors and can be rescued by multiple hox genes. *Nature.* **425**, 300–306.

Drummond, I. A. (2000). The zebrafish pronephros: a genetic system for studies of kidney development. *Pediatr. Nephrol.* **14**, 428–435.

Drummond, I. A., Majumdar, A., Hentschel, H., Elger, M., Solnica-Krezel, L., Schier, A. F., Neuhauss, S. C., Stemple, D. L., Zwartkruis, F., Rangini, Z., Driever, W., and Fishman, M. C. (1998). Early development of the zebrafish pronephros and analysis of mutations affecting pronephric function. *Development* **125**, 4655–4667.

Elizondo, M. R., Arduini, B. L., Paulsen, J., MacDonald, E. L., Sabel, J. L., Henion, P. D., Cornell, R. A., and Parichy, D. M. (2005). Defective skeletogenesis with kidney stone formation in dwarf zebrafish mutant for trpm7. *Curr. Biol.* **15**, 667–671.

Ferrara, N., Carver-Moore, K., Chen, H., Dowd, M., Lu, L., O'Shea, K. S., Powell-Braxton, L., Hillan, K. J., and Moore, M. W. (1996). Heterozygous embryonic lethality induced by targeted inactivation of the VEGF gene. *Nature* **380**, 439–442.

Goodrich, E. S. (1930). "Studies on the Structure and Development of Vertebrates". Macmillan, London.

Guggino, W. B., Oberleithner, H., and Giebisch, G. (1988). The amphibian diluting segment. *Am. J. Physiol.* **254**, F615–F627.

Gustafsson, M. G., Agard, D. A., and Sedat, J. W. (1999). I5M: 3D widefield light microscopy with better than 100 nm axial resolution. *J. Microsc.* **195**, 10–16.

Hammerschmidt, M., Pelegri, F., Mullins, M. C., Kane, D. A., van Eeden, F. J., Granato, M., Brand, M., Furutani-Seiki, M., Haffter, P., Heisenberg, C. P., Jiang, Y. J., Kelsh, R. N., et al., (1996b). dino and mercedes, two genes regulating dorsal development in the zebrafish embryo. *Development.* **123**, 95–102.

Heller, N., and Brandli, A. W. (1999). Xenopus Pax-2/5/8 orthologues: novel insights into Pax gene evolution and identification of Pax-8 as the earliest marker for otic and pronephric cell lineages. *Dev. Genet.* **24**, 208–19.

Hentschel, H., and Elger, M. (1996). Functional morphology of the developing pronephric kidney of zebrafish. *J. Am. Soc. Nephrol.* **7**, 1598.

Hentschel, D. M., Park, K. M., Cilenti, L., Zervos, A. S., Drummond, I., and Bonventre, J. V. (2005). Acute renal failure in zebrafish: a novel system to study a complex disease. *Am. J. Physiol. Renal Physiol.* **288**, F923–F929.

Hild, M., Dick, A., Rauch, G. J., Meier, A., Bouwmeester, T., Haffter, P., Hammerschmidt, M., (1999). The smad5 mutation somitabun blocks Bmp2b signaling during early dorsoventral patterning of the zebrafish embryo. *Development.* **126**, 2149–2159.

Horsfield, J., Ramachandran, A., Reuter, K., LaVallie, E., Collins-Racie, L., Crosier, K., and Crosier, P. (2002). Cadherin-17 is required to maintain pronephric duct integrity during zebrafish development. *Mech. Dev.* **115**, 15–26.

Howland, R. B. (1921). Experiments on the effect of the removal of the pronephros of Ambystoma punctatum. *J. Exp. Zool.* **32**, 355–384.

Igarashi, P., Vanden Heuvel, G. B., Payne, J. A., and Forbush, B. III. (1995). Cloning, embryonic expression, and alternative splicing of a murine kidney-specific Na–K–Cl cotransporter. *Am. J. Physiol.* **269**, F405–F418.

Kamunde, C. N., and Kisia, S. M. (1994). Fine structure of the nephron in the euryhaline teleost, *Oreochromis niloticus. Acta Biol. Hung.* **45**, 111–121.

Kimmel, C. B., Warga, R. M., and Schilling, T. F. (1990). Origin and organization of the zebrafish fate map. *Development* **108**, 581–594.

Kishimoto, Y., Lee, K. H., Zon, L., Hammerschmidt, M., Schulte-Merker, S., (1997). The molecular nature of zebrafish swirl: BMP2 function is essential during early dorsoventral patterning. *Development.* **124**, 4457–4466.

Kramer-Zucker, A. G., Olale, F., Haycraft, C. J., Yoder, B. K., Schier, A. F., and Drummond, I. A. (2005a). Cilia-driven fluid flow in the zebrafish pronephros, brain and Kupffer's vesicle is required for normal organogenesis. *Development* **132**, 1907–1921.

Kramer-Zucker, A. G., Wiessner, S., Jensen, A. M., and Drummond, I. A. (2005b). Organization of the pronephric filtration apparatus in zebrafish requires Nephrin, Podocin and the FERM domain protein Mosaic eyes. *Dev. Biol.* **285**, 316–329.

Krauss, S., Johansen, T., Korzh, V., and Fjose, A. (1991). Expression of the zebrafish paired box gene pax[zf-b] during early neurogenesis. *Development* **113**, 1193–1206.

Liu, Y., Pathak, N., Kramer-Zucker, A., and Drummond, I. A. (2007). Notch signaling controls the differentiation of transporting epithelia and multiciliated cells in the zebrafish pronephros. *Development* **134**, 1111–1122.

Majumdar, A., and Drummond, I. A. (1999). Podocyte differentiation in the absence of endothelial cells as revealed in the zebrafish avascular mutant, cloche [In Process Citation]. *Dev. Genet.* **24**, 220–229.

Majumdar, A., and Drummond, I. A. (2000). The zebrafish floating head mutant demonstrates podocytes play an important role in directing glomerular differentiation. *Dev. Biol.* **222**, 147–157.

Majumdar, A., Lun, K., Brand, M., and Drummond, I. A. (2000). Zebrafish no isthmus reveals a role for pax2.1 in tubule differentiation and patterning events in the pronephric primordia. *Development* **127**, 2089–2098.

Marshall, E. K., and Smith, H. W. (1930). The glomerular development of the vertebrate kidney in relation to habitat. *Biol. Bull.* **59**, 135–153.

Mastroianni, N., De Fusco, M., Zollo, M., Arrigo, G., Zuffardi, O., Bettinelli, A., Ballabio, A., and Casari, G. (1996). Molecular cloning, expression pattern, and chromosomal localization of the human Na–Cl thiazide-sensitive cotransporter (SLC12A3). *Genomics* **35**, 486–493.

Mauch, T. J., Yang, G., Wright, M., Smith, D., and Schoenwolf, G. C. (2000). Signals from trunk paraxial mesoderm induce pronephros formation in chick intermediate mesoderm. *Dev. Biol.* **220**, 62–75.

Mudumana, S. P., Hentschel, D., Liu, Y., Vasilyev, A., and Drummond, I. A. (2008). odd skipped related1 reveals a novel role for endoderm in regulating kidney versus vascular cell fate. *Development* **135**, 3355–3367.

Mullins, M. C., Hammerschmidt, M., Kane, D. A., Odenthal, J., Brand, M., van Eeden, F. J., Furutani-Seiki, M., Granato, M., Haffter, P., Heisenberg, C. P., Jiang, Y. J., Kelsh, R. N., et al., (1996). Genes establishing dorsoventral pattern formation in the zebrafish embryo: the ventral specifying genes. *Development.* **123**, 81–93.

Newstead, J. D., and Ford, P. (1960). Studies on the development of the kidney of the Pacific Salmon, *Oncorhynchus forbuscha* (Walbaum). 1. The development of the pronephros. *Can. J. Zool.* **36**, 15–21.

Nguyen, V. H., Schmid, B., Trout, J., Connors, S. A., Ekker, M., Mullins, M. C. (1998). Ventral and lateral regions of the zebrafish gastrula, including the neural crest progenitors, are established by a bmp2b/swirl pathway of genes. *Dev. Biol.* **199**, 93–110.

Nichane, M., Van Campenhout, C., Pendeville, H., Voz, M. L., and Bellefroid, E. J. (2006). The Na$^+$/PO^4 cotransporter SLC20A1 gene labels distinct restricted subdomains of the developing pronephros in *Xenopus* and zebrafish embryos. *Gene Expr. Patterns* **6**, 667–672.

Pathak, N., Obara, T., Mangos, S., Liu, Y., and Drummond, I. A. (2007). The zebrafish fleer gene encodes an essential regulator of cilia tubulin polyglutamylation. *Mol. Biol. Cell* **18**, 4353–4364.

Perner, B., Englert, C., and Bollig, F. (2007). The Wilms tumor genes wt1a and wt1b control different steps during formation of the zebrafish pronephros. *Dev. Biol.* **309**, 87–96.

Perz-Edwards, A., Hardison, N. L., and Linney, E. (2001). Retinoic acid-mediated gene expression in transgenic reporter zebrafish. *Dev. Biol.* **229**, 89–101.

Pfeffer, P. L., Gerster, T., Lun, K., Brand, M., and Busslinger, M. (1998). Characterization of three novel members of the zebrafish Pax2/5/8 family: dependency of Pax5 and Pax8 expression on the Pax2.1 (noi) function. *Development* **125**, 3063–3074.

Pham, V. N., Roman, B. L., and Weinstein, B. M. (2001). Isolation and expression analysis of three zebrafish angiopoietin genes. *Dev. Dyn.* **221**, 470–474.

Puschel, A. W., Westerfield, M., and Dressler, G. R. (1992). Comparative analysis of Pax-2 protein distributions during neurulation in mice and zebrafish. *Mech. Dev.* **38**, 197–208.

Pyati, U. J., Cooper, M. S., Davidson, A. J., Nechiporuk, A., and Kimelman, D. (2006). Sustained Bmp signaling is essential for cloaca development in zebrafish. *Development* **133**, 2275–2284.

Reiser, J., Kriz, W., Kretzler, M., and Mundel, P. (2000). The glomerular slit diaphragm is a modified adherens junction. *J. Am. Soc. Nephrol.* **11**, 1–8.

Roselli, S., Gribouval, O., Boute, N., Sich, M., Benessy, F., Attie, T., Gubler, M. C., and Antignac, C. (2002). Podocin localizes in the kidney to the slit diaphragm area. *Am. J. Pathol.* **160**, 131–139.

Rottbauer, W., Baker, K., Wo, Z. G., Mohideen, M. A., Cantiello, H. F., and Fishman, M. C. (2001). Growth and function of the embryonic heart depend upon the cardiac-specific L-type calcium channel alpha1 subunit. *Dev. Cell* **1**, 265–275.

Ruotsalainen, V., Ljungberg, P., Wartiovaara, J., Lenkkeri, U., Kestila, M., Jalanko, H., Holmberg, C., and Tryggvason, K. (1999). Nephrin is specifically located at the slit diaphragm of glomerular podocytes. *Proc. Natl. Acad. Sci. U. S. A.* **96**, 7962–7967.

Saxén, L. (1987). "Organogenesis of the Kidney". Cambridge University Press, New York.

Sehnert, A. J., Huq, A., Weinstein, B. M., Walker, C., Fishman, M., and Stainier, D. Y. (2002). Cardiac troponin T is essential in sarcomere assembly and cardiac contractility. *Nat. Genet.* **31**, 106–110.

Seldin, D. W., and Giebisch, G. H. (1992). "The Kidney: Physiology and Pathophysiology". Raven Press, New York.

Serluca, F. C., and Fishman, M. C. (2001). Pre-pattern in the pronephric kidney field of zebrafish. *Development* **128**, 2233–2241.

Shalaby, F., Rossant, J., Yamaguchi, T. P., Gertsenstein, M., Wu, X. F., Breitman, M. L., and Schuh, A. C. (1995). Failure of blood-island formation and vasculogenesis in Flk-1-deficient mice. *Nature* **376**, 62–66.

Shmukler, B. E., Kurschat, C. E., Ackermann, G. E., Jiang, L., Zhou, Y., Barut, B., Stuart-Tilley, A. K., Zhao, J., Zon, L. I., Drummond, I. A., Vandorpe, D. H., Paw, B. H., *et al.* (2005). Zebrafish slc4a2/ae2 anion exchanger: cDNA cloning, mapping, functional characterization, and localization. *Am. J. Physiol. Renal Physiol.* **289**, F835–F849.

Simon, D. B., and Lifton, R. P. (1998). Ion transporter mutations in Gitelman's and Bartter's syndromes. *Curr. Opin. Nephrol. Hypertens.* **7**, 43–47.

Simon, D. B., Nelson-Williams, C., Bia, M. J., Ellison, D., Karet, F. E., Molina, A. M., Vaara, I., Iwata, F., Cushner, H. M., Koolen, M., Gainza, F. J., Gitleman, H. J., *et al.* (1996). Gitelman's variant of Bartter's syndrome, inherited hypokalaemic alkalosis, is caused by mutations in the thiazide-sensitive Na–Cl cotransporter. *Nat. Genet.* **12**, 24–30.

Sollner, C., Burghammer, M., Busch-Nentwich, E., Berger, J., Schwarz, H., Riekel, C., and Nicolson, T. (2003). Control of crystal size and lattice formation by starmaker in otolith biomineralization. *Science* **302**, 282–286.

Sprague, J., Bayraktaroglu, L., Bradford, Y., Conlin, T., Dunn, N., Fashena, D., Frazer, K., Haendel, M., Howe, D. G., Knight, J., Mani, P., Moxon, S. A., *et al.* (2008). The Zebrafish Information Network: the zebrafish model organism database provides expanded support for genotypes and phenotypes. *Nucleic Acids Res.* **36**, D768–D772.

Stickney, H. L., Imai, Y., Draper, B., Moens, C., and Talbot, W. S. (2007). Zebrafish bmp4 functions during late gastrulation to specify ventroposterior cell fates. *Dev. Biol.* **310**, 71–84.

Tavernarakis, N., and Driscoll, M. (1997). Molecular modeling of mechanotransduction in the nematode *Caenorhabditis* elegans. *Annu. Rev. Physiol.* **59**, 659–689.

Tytler, P. (1988). Morphology of the pronephros of the juvenile brown trout, *Salmo trutta*. *J. Morphol.* **195**, 189–204.

Tytler, P., Ireland, J., and Fitches, E. (1996). A study of the structure and function of the pronephros in the larvae of the turbot (*Scophthalmus maximus*) and the herring (*Clupea harengus*). *Mar. Fresh. Behav. Physiol.* **28**, 3–18.

Vaughan, M. R., Pippin, J. W., Griffin, S. V., Krofft, R., Fleet, M., Haseley, L., and Shankland, S. J. (2005). ATRA induces podocyte differentiation and alters nephrin and podocin expression *in vitro* and *in vivo*. *Kidney Int.* **68**, 133–144.

Vize, P. D., Seufert, D. W., Carroll, T. J., and Wallingford, J. B. (1997). Model systems for the study of kidney development: use of the pronephros in the analysis of organ induction and patterning. *Dev. Biol.* **188**, 189–204.

Vize, P. D., Woolf, A. S., and Bard, J. B. L. (2002). "The Kidney: From Normal Development to Congenital Diseases". Academic Press, Amsterdam and Boston.

Wingert, R. A., and Davidson, A. J. (2008). The zebrafish pronephros: a model to study nephron segmentation. *Kidney Int.* **73**, 1120–1127.

Wingert, R. A., Selleck, R., Yu, J., Song, H. D., Chen, Z., Song, A., Zhou, Y., Thisse, B., Thisse, C., McMahon, A. P., and Davidson, A. J. (2007). The cdx genes and retinoic acid control the positioning and segmentation of the zebrafish pronephros. *PLoS Genet.* **3**, 1922–1938.

CHAPTER 10

Molecular Regulation of Pancreas Development in Zebrafish

Robin A. Kimmel *and* **Dirk Meyer**

Institute of Molecular Biology, University of Innsbruck, A-6020 Innsbruck, Austria

Abstract

I. Introduction
II. Pancreas Development
    A. Morphogenesis
    B. Pancreas Specification
    C. Pancreas Cell Segregation into DB and VB
    D. Formation of the Dorsal Bud
    E. Beta-Cell Proliferation
III. Analysis of Beta-Cell Migration and Proliferation
    A. Time-Lapse Imaging of Beta-Cell Migration
    B. Pancreatic Beta-Cell Proliferation
IV. Future Directions
    Acknowledgments
    References

Abstract

The pancreas is a vertebrate-specific organ of endodermal origin which is responsible for production of digestive enzymes and hormones involved in regulating glucose homeostasis, in particular insulin, deficiency of which results in diabetes. Basic research on the genetic and molecular pathways regulating pancreas formation and function has gained major importance for the development of regenerative medical approaches aimed at improving diabetes treatment. Among the different model organisms that are currently used to elucidate the basic pathways of pancreas development and regeneration, the zebrafish is distinguished by its unique opportunities to combine genetic and pharmacological approaches with sophisticated live-imaging methodology,

978-0-12-384892-5
DOI: 10.1016/B978-0-12-384892-5.00010-4

and by its ability to regenerate the pancreas within a short time. Here we review current perspectives and present methods for studying two important processes contributing to pancreas development and regeneration, namely cell migration via time-lapse micropscopy and cell proliferation via incorporation of nucleotide analog EdU, with a focus on the insulin-producing beta cells of the islet.

I. Introduction

Pancreas from zebrafish and mammals show striking similarities in the molecular control of development, share cellular and subcellular architecture, and have a conserved physiological function. All of these criteria validate zebrafish as a relevant model for studying basic mechanisms of pancreas development and function. As in mammals, the zebrafish pancreas mostly consists of exocrine tissues, the acinar glands, that secrete inactive precursors of digestive enzymes (zymogens) into a branched intrapancreatic duct system. The intrapancreatic duct system transports these zymogens into the gut where they are activated by proteolytic cleavage. The endocrine hormones, including insulin, are produced in the pancreatic islets, which in mammals are termed islets of Langerhans. In fish and most mammals, these islets have a characteristic cellular architecture with a core of insulin-producing beta cells surrounded by a smaller number of up to four different cell types producing glucagon (alpha cells), somatostatin (delta cells), ghrelin (epsilon cells), and pancreatic polypeptide (PP cells). Islets are embedded in exocrine tissue and tightly connected to the vasculature through which they release endocrine hormones directly into the bloodstream.

Many adult tissues have the capacity to regenerate during a lifetime or after injury by proliferation of remaining cells, differentiation from multipotent precursor populations, or conversion from another mature cell type. From studies in rodents, there is evidence for all of these mechanisms contributing to homeostasis and response to injury in the pancreas, depending on the experimental manipulation and method of analysis. However, direct observation of transdifferentiation and migration to assemble into new islets is hampered by the inaccessibility of mouse embryos to direct observation during development. Studies in zebrafish have provided evidence for latent endocrine potential in cells of the pancreatic ductal system (Chung *et al.*, 2010; Dong *et al.*, 2007; Moro *et al.*, 2009; Parsons *et al.*, 2009) and emerging microscopic methods will enable close examination of regenerative processes.

II. Pancreas Development

A. Morphogenesis

In 2003, Field *et al.* (2003) showed that the zebrafish pancreas originates from two morphologically distinct endodermal structures that were initially termed posterior and anterior buds. In analogy to the dorsal and ventral pancreatic buds that build the pancreas

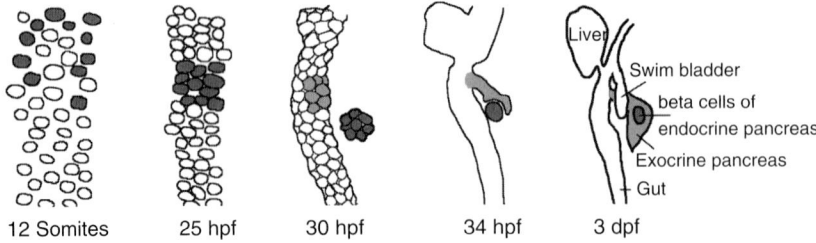

Fig. 1 Zebrafish pancreas development. Dorsal bud endocrine precursors (dark gray) emerge within the endoderm during early somite stages. These cells cluster to form the islet at around 24 hpf, while the ventral bud (light gray) forms after 34 hpf, surrounds the islet, and expands to form the exocrine pancreas.

in amniotes, these terms later changed to dorsal bud (DB) and ventral bud (VB), even though the buds from zebrafish and amniotes have different cell morphological and differentiation properties. In particular, the earlier forming zebrafish DB is distinguished from its amniotic counterpart in that cells exclusively differentiate into endocrine cells. The DB becomes morphologically apparent at 24 hours post-fertilization (hpf) on the dorsal side of the gut where a first set of early forming endocrine cells assemble into the primary islet (Fig. 1). After 24 hpf the process of gut looping leads to displacement of the gut to the left side and of the DB islet to the right side of the embryo. The VB forms after 34 hpf from a population of gut cells positioned anterior to the islet. Between 34 and 48 hpf, VB cells migrate from the gut toward and later around the DB and thereby build a connection between the gut and the islet. Similar to the fate of both pancreatic buds in amniotes, the VB cells give rise to exocrine, duct, and late-forming endocrine cells (Field *et al.*, 2003). Following cell-type specification, the exocrine (acinar) tissue expands posteriorly along the intestine and in parallel the intrapancreatic ductal system is established (Wan *et al.*, 2006). After only 5 days, the zebrafish pancreas has a cellular composition and histology very similar to that of the adult mammalian pancreas (Pack *et al.*, 1996; Parsons *et al.*, 2009; Pauls *et al.*, 2007; Yee *et al.*, 2005). At this stage, the pancreas consists mainly of exocrine acinar tissue connected to the gut through a branching ductal system and a single islet with about 60 endocrine cells including about 35 insulin-producing beta cells. During postembryonic growth, the primary islet increases in size and additional smaller secondary islets form along the exocrine duct system (Parsons *et al.*, 2009). In the adult animal, this expansion leads to the formation of several thousand beta cells distributed in a huge primary islet (diameter $> 240\,\mu m$) and several dozen secondary islets with diameters ranging from below $30\,\mu m$ to more than $100\,\mu m$ (Chen *et al.*, 2007).

B. Pancreas Specification

Over the past several years, enormous progress has been made in defining extrinsic and intrinsic factors that regulate pancreas formation starting with endodermal regionalization, through tissue specification, cell differentiation, and organ morphogenesis. At the onset of gastrulation, TGF-beta/Nodal signaling induces a subset of marginal

cells to adopt an endodermal fate after which they initiate expression of endodermal markers such as *sox17, cas/sox32*, and *foxA2* (reviewed in Zorn and Wells, 2007). Endodermal cells internalize together with mesodermal cells, but remain in contact with the yolk surface where they adopt a flat morphology. Recently, it was shown that during early gastrulation, endoderm cells spread out by "random walk" which leads to the establishment of a non-continuous monolayer between the yolk and the mesoderm (Pezeron *et al.*, 2008). After midgastrulation, endodermal cells change their behavior from random to directed migration which follows overall convergence-extension movement within the embryo. During these early stages, cells are exposed to various patterning signals that regionalize endodermal cells along the different body axes. Among them Hedgehog (Hh) signaling factors expressed in axial mesoderm and retinoic acid produced in presomitic mesoderm are both required during gastrulation to establish a region with competence for pancreatic differentiation. Hh-signaling mutants are characterized by a lack of DB with maintained but bilaterally duplicated VB and embryos deficient in retinoic acid (RA) synthesis or signaling are missing the entire pancreas (diIorio *et al.*, 2002; Roy *et al.*, 2001; Stafford and Prince, 2002).

The requirement for Hh signaling appears unique for zebrafish since in amniotes pancreas induction requires absence of Hh signaling. While the evolutionary causes for this difference remain to be investigated, experimental data suggest a role of early HH signals not for inducing DB fates but rather for preventing ventral signaling that blocks DB fates. In particular, in was shown that *smo/smoothened* mutant endoderm cells that are defective in intracellular transmission of Hh signaling are able to adopt DB fates when transplanted into an environment with wild-type endoderm (Chung and Stainier, 2008). Furthermore, cell autonomy studies with transgenic cells expressing a heat shock-inducible dominant-negative version of the bone morphogenetic protein (BMP) receptor Alk8 revealed that the inhibition of intracellular BMP signaling is sufficient to induce insulin expression in anterior endoderm, even in hh-deficient embryos (Chung *et al.*, 2010). The data suggest that in fish, Hh signaling is required to block BMP signals on the dorsal side, thus restricting BMP signals to the ventral side and thereby preventing cell-autonomous block of pancreatic differentiation by BMP signaling.

Retinoic acid is required cell-autonomously during late gastrulation to refine the competence field for pancreas to a small region of anterior endoderm (Stafford and Prince, 2002; Stafford *et al.*, 2006). A complex interplay between spatially and temporally regulated expression of RA-signaling agonists and antagonists regulates positioning of pancreatic precursors in the endoderm underlying the most anterior somites of the early-somite-stage embryo. Involved in this interplay are RA-producing enzyme (RALDH), RA-metabolizing enzymes (Cyp27a, b, c), different RA receptors, and Cdx transcription factors that inhibit RA functions on the transcriptional level (Kinkel and Prince, 2009; Kinkel *et al.*, 2008).

C. Pancreas Cell Segregation into DB and VB

As revealed by lineage tracing of individual endoderm cells from 12 hpf (6–8-somite stage) to 50 hpf, the precursors of both pancreatic buds, DB and VB, originate from the

endodermal cells that underlie the first three somites of the early-somite-stage embryo (Ward *et al.*, 2007). Cell-tracking studies further showed that only medial cells positioned directly adjacent to the notochord build the DB, while VB cells arise from the more lateral cells (Chung and Stainier, 2008; Ward *et al.*, 2007). Notably, most of the medial cells contributed only to the DB while all lateral cells produced a larger progeny that contributed to the VB, intestine, and in the case of the most lateral cells also to liver. This implies that DB cells are specified before or during early somitogenesis while VB fates are determined later, after a few more rounds of proliferation. Furthermore, a nucleosome-targeted H2B-RFP fusion protein is retained and labels nondividing cells in the DB at 52 hpf following injection of the coding mRNA at the one-cell stage, but is diluted and undetectable in VB cells, evidence supporting a difference in proliferative activity between these precursor populations (Hesselson *et al.*, 2009).

Consistent with an early onset of endocrine differentiation in medial cells, they initiate expression of various transcription factors with conserved function in endocrine differentiation already during early somite stages (Argenton *et al.*, 1999). Although not formally demonstrated, the notochord is implicated in promoting the separation of DB and VB fates and in inducing endocrine differentiation in flanking cells. Consistent with the latter, loss of the notochord in *flh* mutants causes a loss of DB as suggested by a lack or strong reduction of beta cells at 24 hpf (Biemar *et al.*, 2001). Endocrine differentiation of DB cells is associated with the temporally and spatially restricted expression of various transcriptional regulators of endocrine fates. The recent reviews of Kinkel and Prince (2009) and Tiso *et al.* (2009) together with two recent reports on pancreatic *pax6* and *nkx6* functions (Binot *et al.*, 2010; Verbruggen *et al.*, 2010) provide an excellent overview of our current understanding of the functions of these genes and their interplay during early endocrine differentiation in zebrafish.

D. Formation of the Dorsal Bud

Between 17 and 24 hpf, the endocrine cell progenitors rearrange from a linear assembly along the notochord into a single cluster, the DB, directly dorsal to the intestine and ventral to the notochord (Fig. 2). Time-lapse analysis of this process in embryos expressing GFP under the control of the *insulin* promoter (*Tg(ins:GFP)*) revealed that beta cells actively migrate to the level of the third somite where they aggregate to form the core of the islet (Huang *et al.*, 2001; Kim *et al.*, 2005). The migration path is consistent with the previously observed y-shaped expression of early endocrine markers, indicating that cells stay in close contact to the notochord during migration (Fig. 2).

The molecular mechanisms regulating migration and clustering of the DB cells are still unknown. Genetic screens, including two unpublished screens in our lab, have yielded only a few mutants with specific defects in islet morphogenesis (reviewed by Kinkel and Prince, 2009). These included mutants for *glypican4/knypek* and *tbx16/spadetail*, which both display a bilaterally split islet phenotype (Biemar *et al.*, 2001)

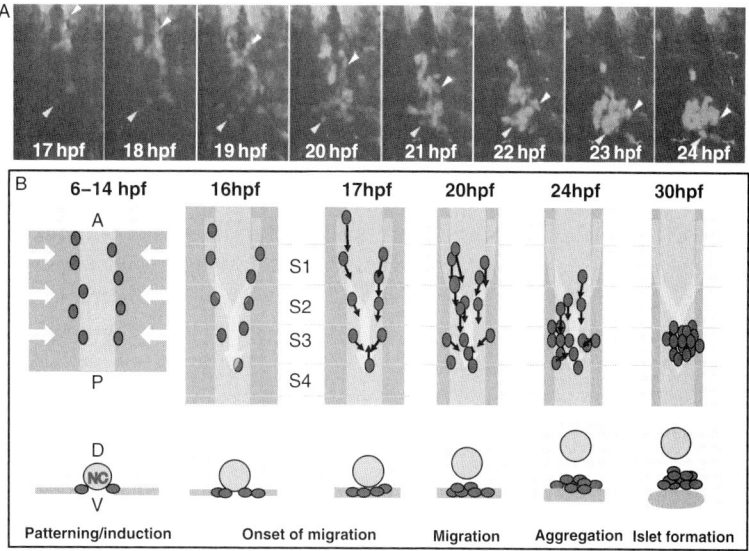

Fig. 2   Beta-cell migration during islet formation. (A) Time-lapse series showing beta-cell migration using a *Tg(mnx1:GFP)* transgenic embryo. Arrowheads trace movement of individual cells. (B) Beta cells form in two domains located lateral to the notochord. Cells migrate posteriorly and converge between 17 and 24 hpf to the level of the third somite (S3) where they form a cluster ventral to the notochord (NC) and dorsal to the developing intestine. A, anterior; D, dorsal; P, posterior; V, ventral; S, somite. (See Plate no. 11 in the Color Plate Section.)

and affect cell polarity in the context of non-canonical Wnt signaling (*knypek*) and formation of mesoderm structures, respectively. Currently, only one pancreas-specific screen has been reported. Among other mutants, this screen yielded five novel mutants of currently unknown molecular nature in which endocrine cells are formed in normal numbers but aggregate in multiple scattered clusters (Kim *et al.*, 2006). While these scattered clusters are smaller in comparison to the wild-type DB, they still display the cellular architecture of the wild-type islet, indicating that posterior migration and clustering are regulated by separate pathways.

More recently, chemokine signaling by the Cxcr4a/Crxl12(Sdf1) receptor/ligand pair was shown to be required for correct pancreas morphogenesis (Mizoguchi *et al.*, 2008; Siekmann *et al.*, 2009). Similar to the phenotype of the type III mutants described by Kim *et al.* (2006), loss of chemokine function either by morpholino knock-down or in *cxcr4a* mutants causes scattered insulin expression and duplication or disrupted assembly of the later forming VB (Mizoguchi *et al.*, 2008). Currently, it is not clear if Cxcr4a has a direct role in migrating endocrine progenitor cells as expression of this receptor has not been reported for the time window of migration. However, *cxcr4* is expressed in the gastrula endoderm and Cxcr4a-Crxl12 signaling was shown to coordinate the movement between mesoderm and endoderm during gastrulation (Mizoguchi *et al.*, 2008; Nair and Schilling, 2008). As endoderm cells in

cxcr4a-depleted embryos fail to align close to the notochord, observed pancreas phenotypes in *cxcr4a* mutants might result from the inability of endodermal cells to receive inductive signals from the notochord. Alternatively, Cxcr4a function in meso-derm adjacent to the islet could be required to establish the necessary environment for proper migration.

A similar mode of function has been suggested for non-canonical Wnt signaling. Interference with this pathway caused failure of anterior gut tube fusion with sub-sequent bilateral duplication of endocrine and exocrine pancreatic primordia (Matsui *et al.*, 2005). In addition, morpholino knock-down of either the *frizzled2* receptor or its potential ligand *wnt5b* leads to scattered endocrine cells (Kim *et al.*, 2005). As *frizzled2* is not expressed in the migrating cells but in neighboring cells, the data imply that *wnt5b/frizzled2* helps to establish the correct environment for directed migration.

E. Beta–Cell Proliferation

Beta-cell number increases significantly during the transition from late embryonic to postnatal and juvenile stages in zebrafish as well as in mammals (Chen *et al.*, 2007; Hanley *et al.*, 2008; Moro *et al.*, 2009; Parsons *et al.*, 2009). Although recovery from beta-cell loss can occur during development, in patients with diabetes, the regenerative response is not sufficient to reestablish glucose homeostasis. In experimental models, re-expansion of the beta-cell population occurs following manipulations such as pancreatectomy or drug-induced beta-cell ablation (Curado *et al.*, 2007; Levine and Itkin-Ansari, 2008; Moss *et al.*, 2009; Pisharath *et al.*, 2007). The existence of a proliferative compartment of beta cells or progenitors has been proposed to account for beta-cell expansion and regeneration, but has not been convincingly demonstrated. Studies in mouse have documented beta-cell self-renewal (Dor *et al.*, 2004) and an absence of specialized progenitors (Teta *et al.*, 2007), while other studies have found evidence for endogenous progenitor cells that can give rise to new beta cells (Inada *et al.*, 2008; Xu *et al.*, 2008). Beta cells in adult mice have extremely slow rates of replication (Teta *et al.*, 2005), and no activation of beta-cell replication is seen in human diabetes patients (Butler *et al.*, 2003, 2007), but an increase of duct-associated beta cells is associated with obesity (Butler *et al.*, 2003).

The molecular control of beta-cell proliferation is not clearly defined, but, as in all cell types, involves a complex interplay of growth factors and external signals impact-ing on positive and negative regulators of the cell cycle. Specific D-type cyclin and cyclin-dependent kinases (Cdks), which interact to promote entry into the cell cycle, are expressed in the islet and their manipulation impacts beta-cell growth and prolif-eration (reviewed in Levine and Itkin-Ansari, 2008). For example, mice deficient in CDK4 or cyclin D2 showed impaired beta-cell proliferation and signs of insulin-dependent diabetes (Georgia and Bhushan, 2004; Rane *et al.*, 1999; Tsutsui *et al.*, 1999). A recent study demonstrated that chromatin modification regulates transcription of negative cell-cycle regulators from the *Ink4a/Arf* locus, causing changes in beta-cell proliferation during regeneration and aging (Dhawan *et al.*, 2009).

Table I
Studies of beta-cell proliferation in zebrafish

Treatment/detection	Stage	Finding	Reference
Pdx1-morpholino,[a] BrdU	48–60 hpf	BrdU+ in pancreas, none in islet	Yee *et al.* (2001)
MTZ-ablation,[b] BrdU	120 hpf	BrdU+/Ins+ in islet	Pisharath *et al.* (2007)
Time-lapse *Tg(Ins:GFP)*	20–25 hpf	No beta-cell mitoses	Moro *et al.* (2009)
Time-lapse *Tg(Ins:GFP)*	48–60 hpf	Beta-cell mitoses detected	Moro *et al.* (2009)
PCNA antibody	48 hpf, 7 dpf	Rare PCNA+/Ins+ in islet	Moro *et al.* (2009)
Thymidine-kinase (TK)-ablation *Tg(Ins:TKGFP)*	20 hpf–7 dpf	Reduced number of beta cells after treatment with ganciclovir (GCG)	Moro *et al.* (2009)
Adult islet homeostasis, BrdU	Adult (5, 9 months)	Rare proliferation of beta cells	Moro *et al.* (2009)
STZ-ablation,[c] MTZ-ablation, PCNA antibody	Adult	PCNA+ cells (peri-islet) increased, PCNA+/Ins+ unchanged	Moro *et al.* (2009)
Pancreatectomy, PCNA antibody	Adult	increased PCNA+/Ins+ cells	Moro *et al.* (2009)
EdU (injection)	25–60 hpf, 60–120 hpf	EdU+ in pancreas, no EdU+ beta cells	Hesselson *et al.*, 2009

[a] Pdx1 morpholino treatment results in disruption of pancreas development at 2 dpf, with restoration by 5 dpf.
[b] MTZ treatment in zebrafish expressing nitroreductase (Ntr) from an insulin promoter causes beta cell apoptosis.
[c] STZ is a chemical toxin that specifically ablates beta cells.
dpf, days post-fertilization; hpf, hours post-fertilization; MTZ, metronidazole; PCNA, proliferating cell nuclear antigen; STZ, streptozotocin.

Studies in zebrafish, examining normal development and regeneration, have variably found proliferative beta cells at low frequencies, with findings depending on stage and experimental protocol used (Table I). Recent studies quantitatively assessed beta-cell proliferation dynamics through embryonic development and into adulthood (Moro *et al.*, 2009), and following drug treatment and surgical manipulations widely used in mouse regeneration studies (Moss *et al.*, 2009), expanding the applicability of the zebrafish as a model organism for investigating beta-cell homeostasis and regeneration. In zebrafish, dividing cells can be detected in the three-dimensional context of a developing embryo and even within the living organism (Moro *et al.*, 2009) to help dissect both intrinsic and environmental regulators.

III. Analysis of Beta-Cell Migration and Proliferation

A. Time-Lapse Imaging of Beta-Cell Migration

Due to their small size and transparency, zebrafish embryos are a favored model for microscopic analyses of embryonic beta-cell formation and regeneration. Cells of interest tagged by expression of fluorescent proteins can be visualized in both fixed specimens and living embryos. The spectrum of fluorescent proteins (FPs) available for use in live imaging experiments have been steadily expanding to include not only a broad range of colors but also offering variants of other relevant properties such as brightness, photostability, phototoxicity, and maturation speed (for an extensive review, see Davidson and Campbell, 2009). The subcellular localization of the expressed FP(s) can reveal different

aspects of biological processes and facilitate later analysis. Cytoplasmically expressed FPs render a cell's full morphological complexity but can cause difficulties in cell tracking and quantitative studies as particle detection algorithms may be unable to recognize individual cells within a closely spaced cluster. FPs fused to a nuclear localization signal (NLS) mark distinct entities which helps cell identification and tracking (Nowotschin *et al.*, 2009). However, the cell may be lost to tracking during cell division because after nuclear envelope breakdown the fluorescent signal becomes weak and diffusely distributed throughout the cell. This problem is avoided by using an FP–histone fusion, which becomes incorporated into the nucleosome and produces a robust signal throughout all stages of the cell cycle (Kanda *et al.*, 1998). FPs can also be localized to the plasma membrane or secretory apparatus through addition of sequences that direct lipid modification (Rhee *et al.*, 2006). This enables visualization of cell membrane dynamics during migration, including extension and retraction of membrane protrusions and intercellular bridge formation. The rapidly expanding research on pancreas development and regeneration has resulted in the generation of numerous transgenic fish lines that express FPs in beta cells or endocrine progenitors, or in associated cell types (Table II). Simultaneous demarcation of plasma membrane and cytoplasm or nucleus is possible through combinations of spectrally distinct FPs (Fig. 3); labeling of related cell populations by different fluorophores helps to orient the migration process within the embryo and provides a point of reference during cell tracking (Fig. 4).

1. Method: Analysis of DB Morphogenesis by Time-Lapse Laser-Scanning Microscopy

a. Equipment   *Microscope:* Upright microscope (Zeiss LSM5 Exciter) with water dipping lens (40×).

Software: Images collected using Zen 2008 software and analyzed with LSM Image Browser (Zeiss) or Imaris (Bitplane).

b. Materials   *Observation chamber*—Prepare an observation chamber by drilling a 1 cm diameter hole into the bottom of a 3.5 cm petri dish. Apply silicon glue around the edge of the circle on the bottom of the plate. Seal the opening by pressing a coverslip onto the glue.

Egg water—Dissolve 7.5 g Coral Pro Salt (Red Sea) in 25 l reverse osmosis H_2O.

Low-melt agarose (Biozym, Vienna, Austria)—Prepare 1.2% low-melt agarose in egg water and place in a prewarmed 37°C waterbath.

Tricaine (Sigma-Aldrich, Vienna, Austria)—Prepare 25× stock solution by dissolving 400 mg in ~97.9 ml H_2O, plus 2.1 ml Tris-HCl (pH 9.0). Adjust pH to 7.0. Store aliquots at –20°C.

c. Imaging

1. *Temperature control:* Imaging room should be maintained at a constant temperature. Adjust air conditioner/heater to give room temperature of approximately 28°C. Allow the room and equipment temperature to equilibrate before starting to image.

Table II
Transgenic lines that label pancreatic cells

Gene	Name of transgenic line	Location	Cell type	References
ClaudinB	*Tg(-8.0cldnb:lynEGFP)zf106*	Membrane	Epithelia	Haas and Gilmour, (2006)
Elastase	*Tg(ela3l:EGFP)gz2*	Cytoplasm	Exocrine	Wan *et al.* (2006)
Endoderm	*Tg(XlEef1a1:GFP)s854*	Cytoplasm	Endoderm	Field *et al.* (2003)
Glucagon	*Tg(gcga:GFP)ia1*	Cytoplasm	Alpha cells	Zecchin *et al.* (2007)
Hb9/mnx1	*Tg(mnx1:GFP)ml2*	Cytoplasm	Beta cells	Flanagan-Steet *et al.* (2005)
	Tg(mnx1:mGFP)ml3	Membrane	Beta cells	Flanagan-Steet *et al.* (2005)
	Tg(mnx1:nucRFP)ml4	Nuclear	Beta cells	Arkhipova and Meyer (personal communication)
Insulin	*Tg(-4.0ins:GFP)zf5*	Cytoplasm	Beta cells	Huang *et al.* (2001)
	Tg(ins:CFP-NTR)s892	Cytoplasm	Beta cells	Curado *et al.* (2007)
	Tg(-1.0ins:eGFP)sc1	Cytoplasm	Beta cells	Moro *et al.* (2009)
	Tg(ins:dsRed)m1018	Cytoplasm	Beta cells	Hesselson *et al.*, 2009
	Tg(ins:eGFP)jh3/jh3	Cytoplasm	Beta cells	Pisharath *et al.* (2007)
	Tg(ins:Kaede)jh6/jh6	Cytoplasm	Beta cells	Pisharath *et al.* (2007)
	Tg(ins:mCherry)jh2	Cytoplasm	Beta cells	Pisharath *et al.* (2007)
	Tg(ins:nfsB-mCherry)jh4	Cytoplasm	Beta cells	Pisharath *et al.* (2007)
	Tg(ins:nfsB-mCherry)jh5/jh5	Cytoplasm	Beta cells	Pisharath *et al.* (2007)
	Tg(T2Kins:EGFP-mCherry)jh8	Cytoplasm	Beta cells	Provost *et al.* (2007)
	Tg(-1.2ins:TKGFP)	Nuclear	Beta cells	Moro *et al.* (2009)
	Tg(T2Kins:Hmgb1-eGFP)jh10	Nuclear	Beta cells	Parsons *et al.* (2009)
	Tg(ins:Gal4)m1080	Nuclear	Beta cells	Zecchin *et al.* (2007)
NeuroD	*TgBAC(neurod:EGFP)nl1*	Cytoplasm	Endocrine	Obholzer *et al.* (2008)
Nkx2.2a	*Tg(-3.5nkx2.2a:GFP)ia3*	Cytoplasm	Duct	Pauls *et al.* (2007)
	Tg(-8.5nkx2.2a:GFP)ia2	Cytoplasm	Duct, endocrine	Pauls *et al.* (2007)
	Tg(nkx2.2a:mEGFP)vu16	Membrane	Endocrine	Ng *et al.* (2005)
Pax6b	*Tg(P0-pax6b:DsRed)ulg302*	Cytoplasm	Endocrine	Delporte *et al.* (2008)
	Tg(P0-pax6b:GFP)ulg515	Cytoplasm	Endocrine	Delporte *et al.* (2008)
Pdx1	*Tg(-6.5pdx1:GFP)zf6*	Cytoplasm	Endocrine/beta cells	Huang *et al.* (2001)
Ptf1/P48	*Tg(ptf1a:eGFP)jh1*	Cytoplasm	Exocrine	Godinho *et al.* (2005)
	Tg(ptf1a:eGFP)jh7	Cytoplasm	Exocrine	Park *et al.* (2008)
Somatostatin	*Tg(sst2:gfp)*	Cytoplasm	Delta cells	Li *et al.* (2009)
	Tg(sst2:rfp)	Cytoplasm	Delta cells	Li *et al.* (2009)
Sox17	*Tg(-5.0sox17:EGFP)zf99*	Cytoplasm	Endoderm	Wilkins *et al.* (2008)
	Tg(sox17:DsRed)s903	Cytoplasm	Endoderm	Chung and Stainier (2008)
Tp1[a]	*Tg(Tp1bglob:eGFP)um14*	Cytoplasm	Notch-responsive cells	Parsons *et al.* (2009)
	Tg(T2KTp1bglob:hmgb1-mCherry)ih11	Nuclear	Notch-responsive cells	Parsons *et al.* (2009)

[a] Tp1 is an artificial Notch-responsive element with 12 RBP-Jk binding sites
For further details, see also ZFIN (http://zfin.org).

2. ***Staging:*** Migration occurs between 16 hpf (14-somite stage) and 24 hpf.

   a. If the aquarium is coordinated with daylight, this is normally between 12 midnight and 8 AM

   b. To speed up development: If a 32 °C incubator is used, time-lapse can be started at 10 PM

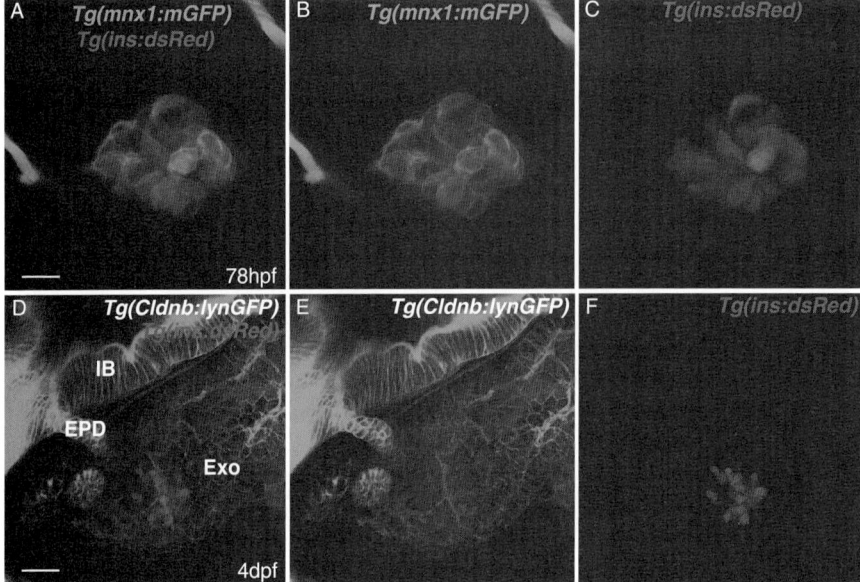

Fig. 3 Visualizing beta cells using transgenic zebrafish lines. (A–C) z-stack projection obtained by imaging a living double transgenic *Tg(Hb9:mGFP; ins:dsRed)* embryo, lateral view at 78 hpf, showing both membrane-targeted GFP and cytoplasmic dsRed (A) or single fluorophores (B, C). (D–F) z-stack projection, ventral view of fixed embryo, double transgenic *Tg(Cldnb:lynEGFP; ins:dsRed)*. Membrane-targeted GFP expression from a *ClaudinB* promoter fragment (Haas and Gilmour, 2006) outlines epithelia of the developing gastrointestinal system, delineating the intestinal bulb (IB), extrapancreatic duct (EPD), and exocrine pancreas (Exo), while *Tg(ins:dsRed)* cells are seen in the islet. Scale bar in A = 10 μM and in D = 20 μM.

   c. To slow down development:
- Collect embryos after 11 AM.
- Grow embryo overnight at 24°C.
- Embryos reach 12–14-somite stage at 8 AM.

3. Manually remove chorion from embryo.
4. Fill opening in the observation dish with 0.2 ml low-melt agarose solution.
5. Immediately after, pick up an embryo in a minimal volume of egg water with a fire polished Pasteur pipette and let embryo sink to the outer surface.
6. Dip the tip of the pipette into the liquid agarose solution and the embryo will sink into the agarose.
7. For a dorsal view, orient the embryo under a stereomicroscope using a blunt probe such that the first somite is closest to the objective and let agarose harden.
8. Fill observation chamber with a maximal volume of 1X Tricaine-containing egg water.

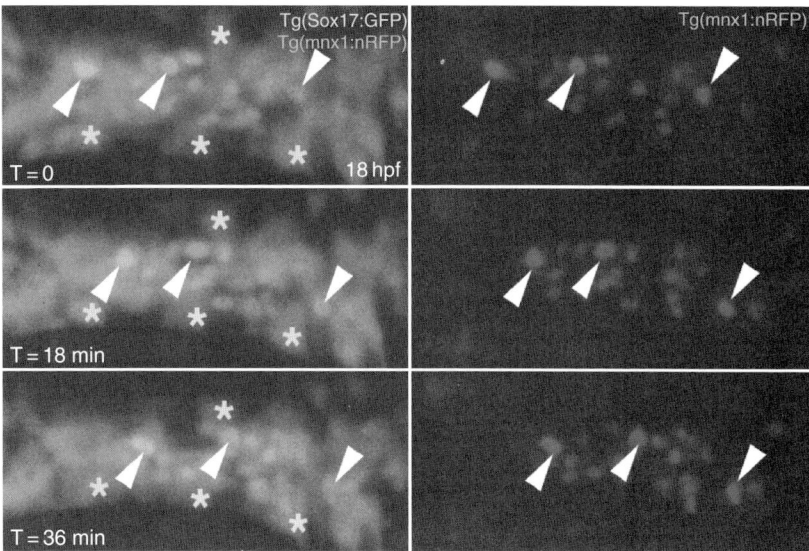

Fig. 4   Time-lapse analysis of beta-cell migration. Three-dimensional projections from a time-lapse series imaging a double transgenic *Tg(sox17:GFP; mnx1:nucRFP)* embryo starting at 18 hpf. Images shown are spaced 18 min apart. Individual nucRFP-positive nuclei of beta cells (arrowheads) migrate relative to Sox17: GFP-labeled endoderm cells (asterisks).

9. Locate embryo and region of interest using bright-field and fluorescent illumination. Start imaging.

10. Due to the morphological changes during somitogenesis, the region of intertest may shift. Therefore, it is often necessary to frequently check on the scan, and if required to adjust microscope settings and stage.

d. Imaging Parameters   Adjustable settings for time-lapse imaging include pinhole size, laser power, thickness of z-slice, number of slices, and time interval. Optimal settings must be determined empirically for each transgene and particular imaging experiment. Illumination is set depending on strength of signal to achieve an adequate signal-to-noise ratio while limiting photobleaching and phototoxicity. The necessary time interval depends on the speed of the process under observation. (For a discussion of optimizing parameters for image acquisition during time-lapse experiments, see Koster and Fraser, 2004.)

For an imaging experiment as shown in Figure 4, we typically collect 15–20 z-sections of 2–5 µM, with a total z-stack of 50–60 µM. The z-range is set to include several slices above and below the region of interest, which helps to ensure the full z-stack contains the area of interest even after some drift of the sample. For following individual cells during beta-cell migration, a time interval of 3–5 min is used.

B. Pancreatic Beta–Cell Proliferation

Techniques to label and localize proliferative cells are well established in mice and have been adapted and utilized in zebrafish studies of pancreas development. Incorporation of the nucleotide analog BrdU into DNA establishes that a cell has passed through the cell cycle, which is detected following tissue fixation by antibody immunohistochemistry. The advantage of BrdU labeling is that the labeled cell and its progeny retain the label in their DNA, and incorporated BrdU remains detectable for up to six or seven cell divisions (Mandyam *et al.*, 2007). Therefore, proliferative progenitors, once labeled, can be identified at later stages of differentiation and followed during migration.

However, there are several confounding factors that can complicate the interpretation of BrdU-labeling experiments. Bioavailability of BrdU, that is, the time window after administration in which incorporation into DNA can occur, has not been systematically studied in zebrafish embryos. In adult fish, it has been reported to be 4 h after injection (Reimer *et al.*, 2008), while values ranging from 15 min up to 2 h have been estimated for labeling adult and perinatal mouse brain following systemic BrdU injection (Mandyam *et al.*, 2007; Taupin, 2007). The duration of BrdU labeling after injection depends on route of delivery, permeability of tissue, as well as metabolism by the organism. Overestimating the labeling duration can lead to proliferation events going undetected.

Alternative markers for proliferative cells include antibodies for PCNA and Ki67. PCNA is a cofactor of DNA polymerase expressed during S-phase, but also during DNA repair, and has a long half-life. Therefore, PCNA may remain detectable long after cell-cycle exit, while Ki-67, a nuclear protein expressed throughout the cell cycle, has a very short half-life (Kee *et al.*, 2002). Thus, Ki-67 is a more reliable marker of recent proliferation compared to PCNA, but as expression is soon lost, the newly generated cells can no longer be identified.

EdU is a recently introduced thymidine analog that, unlike BrdU, does not require DNA denaturation or antibody staining for its detection. The harsh denaturing treatment required to expose the BrdU epitope for antibody labeling damages the tissue and decreases reproducibility of the procedure, and may alter antibody epitopes which limit the ability to identify cell types based on co-expression of other markers. By contrast, EdU can be detected within intact double-stranded DNA with reagents that easily penetrate tissue (Salic and Mitchison, 2008), making this method highly suitable for labeling in whole mount. EdU can be visualized in combination with cells expressing transgenic reporters, or with proteins detected by antibody staining. Labeling of the developing pancreas by injection of EdU into zebrafish embryo yolk or pericardial space has been described (Anderson *et al.*, 2009; Hesselson *et al.*, 2009). With extended incubation in EdU, we have detected rare proliferation events in the slowly dividing beta-cell population (Fig. 5).

1. Method: Detection of Proliferation in Pancreas Using EdU

a. Materials

Click-It EdU Imaging Kit (product C10085, Invitrogen, Lofer, Austria)

EdU—Dissolve in 2 ml $1\times$ PBS (final 10 mM). Store in aliquots at $-20°C$ protected from light.

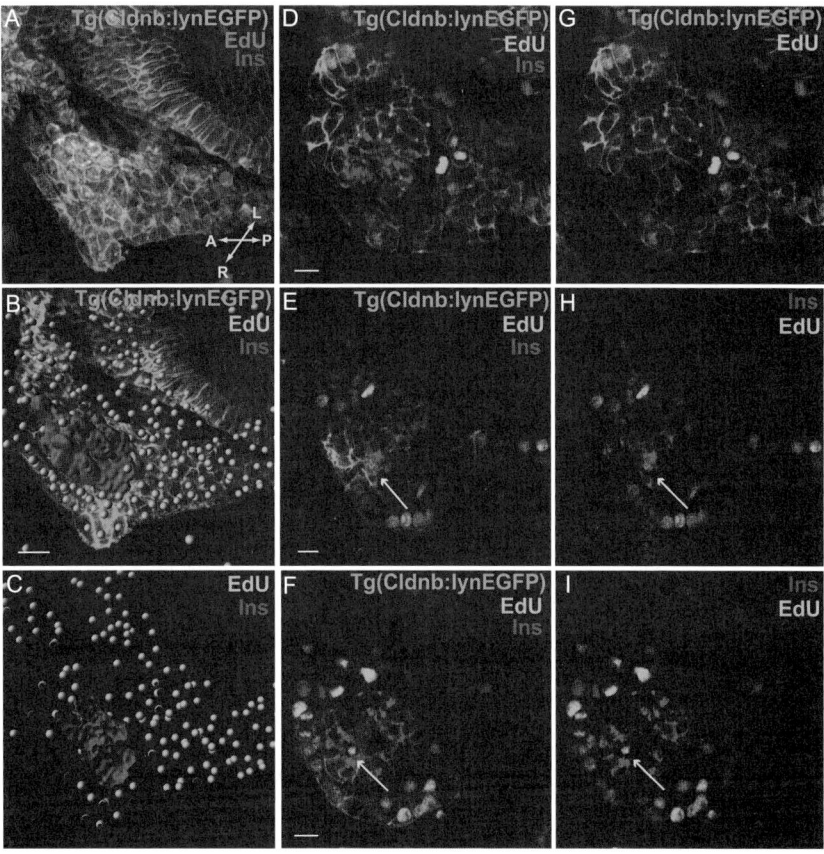

Fig. 5 Identification of proliferating cells in developing pancreas. (A) Confocal projection of *Tg(Cldnb:lynEGFP)* larva at 72 hpf following 48 h of EdU labeling. Most exocrine cells are EdU+. (B) Same data set as in A, with a surface rendering of the insulin domain and EdU+ nuclei represented by a spot-detection algorithm. (C) Cutting plane through the insulin surface shown in B to demonstrate the absence of EdU+ nuclei within the insulin domain. (D–I) Single plane views of embryos treated as in A, showing an example of no proliferating cells in the islet (D, G), an EdU+/insulin+ cell (E, H; arrow), and an EdU+/insulin- islet cell (F, I; arrow). Scale bar in B = 15 µM and in D–F = 10 µM. A, anterior; P, posterior; L, left; R, right. (See Plate no. 12 in the Color Plate Section.)

AlexaFluor Azide—Dissolve in 70 µl DMSO. Store in aliquots at −20°C protected from light.

Reaction Buffer D (10×) stock solution—Store at 4°C.

Additive F—Add 2 ml ddH₂O. Mix well and store in aliquots at −20°C

CuSO4—Store at 4°C

Other Reagents

BSA

DMSO
MeOH
4% PFA (prepared fresh)
PBS
Proteinase K—10 mg/ml stock solution, aliquot and store at –20°C
Triton X-100
Tween-20
Normal Goat Serum
PBST—PBS + 0.1% Tween-20
PTU—For 10× stock solution, dissolve 304 mg powder in 1 l ddH$_2$O
Blocking solution—1% DMSO/1% Triton X-100/2% NGS/1% BSA in PBS

1. Collect embryos. From 22 hpf onward, include 1X PTU in egg water to inhibit pigmentation. Incubate to desired stage. (This protocol has been tested on embryos 48 hpf and older.)
2. Transfer embryos to EdU solution (final concentration 0.5 mM in 0.4% DMSO/1X PTU/egg water). Incubate at 28°C.
3. After desired incubation period, wash embryos in egg water, transfer to fresh egg water (with PTU), or continue to fixation.
4. Fix in 4% PFA (made fresh) at room temperature for 3 h, or 4°C overnight on rocker.
5. Wash 3× in PBST for 5 min each.
6. Wash in 100% MeOH for 5 min. Change to fresh MeOH. Incubate at –20°C for 2 h (longer storage at –20°C is possible).
7. Rehydrate through MeOH/PBST series: 75% MeOH/PBST, 50% MeOH/PBST, 25% MeOH/PBST. Incubate for 5 min each.
8. Wash 3× in PBST.
9. Manually deyolk embryos.
10. Treat with Proteinase K diluted to 10 µg/ml in PBST at room temperature for 45 min.
11. Refix for 15 min in 4%PFA/PBST at room temperature. Wash 3× in PBST.
12. Incubate embryos in 1% DMSO/0.5% Triton in PBS × 20 min at RT.
13. Wash with PBST. Remove as much PBST as possible.
14. Add *staining solution*: (for 500 µl/100 µl per tube of embryos)
    44 µl   1× Click-iT Reaction Buffer
    10 µl   CuSO$_4$
    2.5 µl   Fluorescent dye azide
    50 µl   Reaction Buffer Additive
    394 µl   ddH$_2$O
15. Incubate for 2 h at room temperature in the dark.
16. Rinse 3× with PBST.
17. Proceed to antibody staining, starting with a minimum of 1 h incubation in blocking solution. We typically incubate with primary antibody diluted in blocking solution overnight at 4°C, followed by washes in 1% BSA/0.3% Triton

in PBS, overnight incubation with secondary antibody at 4°C followed by washes in PBST. (Keep embryos protected from light throughout.)
18. Imaging. For an experiment as shown in Figure 5, we use Alexa-Azide 647 (Invitrogen), and the following antibodies: rabbit anti-GFP (Torrey Pines Biolabs, East Orange, NJ, USA), guinea-pig anti-insulin (Dako, Vienna, Austria), anti-rabbit Alexa 488 (Invitrogen), anti-guinea pig Alexa-555 (Invitrogen).

IV. Future Directions

Cell proliferation and migration are developmental events critical for assembly of the mature pancreatic islet. Recent advances in labeling methods and microscopy, combined with the visualization benefits of zebrafish, enable high-resolution detection of these processes during normal development and following genetic manipulations. Fluorescent probes fused to proteins subject to ubiquitin-mediated degradation at particular parts of the cell cycle provide a real-time read-out of the cell-cycle phase in living cells (Sakaue-Sawano *et al.*, 2008). By using transgenic lines expressing these proteins, it becomes possible to visualize cell-cycle dynamics in real-time within living organisms, in conjunction with other biological processes such as migration. A better understanding of how beta-cell proliferation and migration are coordinately regulated will provide important insights for diabetes therapies for which not only production of sufficient numbers of replacement beta cells is required, but also the assembly of an appropriate three-dimensional structure is a necessary prerequisite for fully functional regulation of glucose homeostasis.

Acknowledgments

The authors would like to thank Mayra Eduardoff and Valeriya Arkhipova for help with protocols, technical assistance, and useful discussions; Pia Aanstad for insightful comments; Jochen Holzschuh and lab members for sharing protocols; and Darren Gilmour for the *Tg(cldnb:lynGFP)* line. This work is supported in part by FWF grants to D.M. and R.A.K.

References

Anderson, R. M., Bosch, J. A., Goll, M. G., Hesselson, D., Dong, P. D., Shin, D., Chi, N. C., Shin, C. H., Schlegel, A., Halpern, M., *et al.* (2009). Loss of Dnmt1 catalytic activity reveals multiple roles for DNA methylation during pancreas development and regeneration. *Dev. Biol.* **334**, 213–223.

Argenton, F., Zecchin, E., and Bortolussi, M. (1999). Early appearance of pancreatic hormone-expressing cells in the zebrafish embryo. *Mech. Dev.* **87**, 217–221.

Biemar, F., Argenton, F., Schmidtke, R., Epperlein, S., Peers, B., and Driever, W. (2001). Pancreas development in zebrafish: Early dispersed appearance of endocrine hormone expressing cells and their convergence to form the definitive islet. *Dev. Biol.* **230**, 189–203.

Binot, A. C., Manfroid, I., Flasse, L., Winandy, M., Motte, P., Martial, J. A., Peers, B., and Voz, M. L. (2010). Nkx6.1 and nkx6.2 regulate alpha- and beta-cell formation in zebrafish by acting on pancreatic endocrine progenitor cells. *Dev. Biol.* **340**, 397–407.

Butler, A. E., Galasso, R., Meier, J. J., Basu, R., Rizza, R. A., and Butler, P. C. (2007). Modestly increased beta cell apoptosis but no increased beta cell replication in recent-onset type 1 diabetic patients who died of diabetic ketoacidosis. *Diabetologia* **50**, 2323–2331.

Butler, A. E., Janson, J., Bonner-Weir, S., Ritzel, R., Rizza, R. A., and Butler, P. C. (2003). Beta-cell deficit and increased beta-cell apoptosis in humans with type 2 diabetes. *Diabetes* **52**, 102–110.

Chen, S., Li, C., Yuan, G., and Xie, F. (2007). Anatomical and histological observation on the pancreas in adult zebrafish. *Pancreas* **34**, 120–125.

Chung, W. S., Andersson, O., Row, R., Kimelman, D., and Stainier, D. Y. (2010). Suppression of Alk8-mediated Bmp signaling cell-autonomously induces pancreatic beta-cells in zebrafish. *Proc. Natl. Acad. Sci. U.S.A* **107**, 1142–1147.

Chung, W. S., and Stainier, D. Y. (2008). Intra-endodermal interactions are required for pancreatic beta cell induction. *Dev. Cell* **14**, 582–593.

Curado, S., Anderson, R. M., Jungblut, B., Mumm, J., Schroeter, E., and Stainier, D. Y. (2007). Conditional targeted cell ablation in zebrafish: A new tool for regeneration studies. *Dev. Dyn.* **236**, 1025–1035.

Davidson, M. W., and Campbell, R. E. (2009). Engineered fluorescent proteins: Innovations and applications. *Nat. Methods* **6**, 713–717.

Delporte, F. M., Pasque, V., Devos, N., Manfroid, I., Voz, M. L., Motte, P., Biemar, F., Martial, J. A., and Peers, B. (2008). Expression of zebrafish pax6b in pancreas is regulated by two enhancers containing highly conserved cis-elements bound by PDX1, PBX and PREP factors. *BMC Dev. Biol.* **8**, 53.

Dhawan, S., Tschen, S. I., and Bhushan, A. (2009). Bmi-1 regulates the Ink4a/Arf locus to control pancreatic beta-cell proliferation. *Genes Dev.* **23**, 906–911.

diIorio, P. J., Moss, J. B., Sbrogna, J. L., Karlstrom, R. O., and Moss, L. G. (2002). Sonic hedgehog is required early in pancreatic islet develo. *Dev. Biol.* **244**, 75–84.

Dong, P. D., Munson, C. A., Norton, W., Crosnier, C., Pan, X., Gong, Z., Neumann, C. J., and Stainier, D. Y. (2007). Fgf10 regulates hepatopancreatic ductal system patterning and differentiation. *Nat. Genet.* **39**, 397–402.

Dor, Y., Brown, J., Martinez, O. I., and Melton, D. A. (2004). Adult pancreatic beta-cells are formed by self-duplication rather than stem-cell differentiation. *Nature* **429**, 41–46.

Field, H. A., Dong, P. D., Beis, D., and Stainier, D. Y. (2003). Formation of the digestive system in zebrafish. II. Pancreas morphogenesis. *Dev. Biol.* **261**, 197–208.

Flanagan-Steet, H., Fox, M. A., Meyer, D., and Sanes, J. R. (2005). Neuromuscular synapses can form in vivo by incorporation of initially aneural postsynaptic specializations. *Development* **132**, 4471–4481.

Georgia, S., and Bhushan, A. (2004). Beta cell replication is the primary mechanism for maintaining postnatal beta cell mass. *J. Clin. Invest.* **114**, 963–968.

Godinho, L., Mumm, J. S., Williams, P. R., Schroeter, E. H., Koerber, A., Park, S. W., Leach, S. D., and Wong, R. O. (2005). Targeting of amacrine cell neurites to appropriate synaptic laminae in the developing zebrafish retina. *Development* **132**, 5069–5079.

Haas, P., and Gilmour, D. (2006). Chemokine signaling mediates self-organizing tissue migration in the zebrafish lateral line. *Dev. Cell* **10**, 673–680.

Hanley, N. A., Hanley, K. P., Miettinen, P. J., and Otonkoski, T. (2008). Weighing up beta-cell mass in mice and humans: Self-renewal, progenitors or stem cells? *Mol. Cell. Endocrinol.* **288**, 79–85.

Hesselson, D., Anderson, R. M., Beinat, M., and Stainier, D. Y. (2009). Distinct populations of quiescent and proliferative pancreatic beta-cells identified by HOTcre mediated labeling. *Proc. Natl. Acad. Sci. U.S.A* **106**, 14896–14901.

Huang, H., Vogel, S. S., Liu, N., Melton, D. A., and Lin, S. (2001). Analysis of pancreatic development in living transgenic zebrafish embryos. *Mol. Cell. Endocrinol.* **177**, 117–124.

Inada, A., Nienaber, C., Katsuta, H., Fujitani, Y., Levine, J., Morita, R., Sharma, A., and Bonner-Weir, S. (2008). Carbonic anhydrase II-positive pancreatic cells are progenitors for both endocrine and exocrine pancreas after birth. *Proc. Natl. Acad. Sci. U.S.A* **105**, 19915–19919.

Kanda, T., Sullivan, K. F., and Wahl, G. M. (1998). Histone-GFP fusion protein enables sensitive analysis of chromosome dynamics in living mammalian cells. *Curr. Biol.* **8**, 377–385.

Kee, N., Sivalingam, S., Boonstra, R., and Wojtowicz, J. M. (2002). The utility of Ki-67 and BrdU as proliferative markers of adult neurogenesis. *J. Neurosci. Meth.* **115**, 97–105.

Kim, H. J., Schleiffarth, J. R., Jessurun, J., Sumanas, S., Petryk, A., Lin, S., and Ekker, S. C. (2005). Wnt5 signaling in vertebrate pancreas development. *BMC Biol.* **3**, 23.

Kim, H. J., Sumanas, S., Palencia-Desai, S., Dong, Y., Chen, J. N., and Lin, S. (2006). Genetic analysis of early endocrine pancreas formation in zebrafish. *Mol. Endocrinol.* **20**, 194–203.

Kinkel, M. D., Eames, S. C., Alonzo, M. R., and Prince, V. E. (2008). Cdx4 is required in the endoderm to localize the pancreas and limit beta-cell number. *Development* **135**, 919–929.

Kinkel, M. D., and Prince, V. E. (2009). On the diabetic menu: Zebrafish as a model for pancreas development and function. *Bioessays* **31**, 139–152.

Koster, R. W., and Fraser, S. E. (2004). Time-lapse microscopy of brain development. *Methods Cell Biol.* **76**, 207–235.

Levine, F., and Itkin-Ansari, P. (2008). Beta-cell regeneration: Neogenesis, replication or both? *J. Mol. Med.* **86**, 247–258.

Li, Z., Wen, C., Peng, J., Korzh, V., and Gong, Z. (2009). Generation of living color transgenic zebrafish to trace somatostatin-expressing cells and endocrine pancreas organization. *Differentiation* **77**, 128–134.

Mandyam, C. D., Harburg, G. C., and Eisch, A. J. (2007). Determination of key aspects of precursor cell proliferation, cell cycle length and kinetics in the adult mouse subgranular zone. *Neuroscience* **146**, 108–122.

Matsui, T., Raya, A., Kawakami, Y., Callol-Massot, C., Capdevila, J., Rodriguez-Esteban, C., and Izpisua Belmonte, J. C. (2005). Noncanonical Wnt signaling regulates midline convergence of organ primordia during zebrafish development. *Genes Dev.* **19**, 164–175.

Mizoguchi, T., Verkade, H., Heath, J. K., Kuroiwa, A., and Kikuchi, Y. (2008). Sdf1/Cxcr4 signaling controls the dorsal migration of endodermal cells during zebrafish gastrulation. *Development* **135**, 2521–2529.

Moro, E., Gnugge, L., Braghetta, P., Bortolussi, M., and Argenton, F. (2009). Analysis of beta cell proliferation dynamics in zebrafish. *Dev. Biol.* **332**, 299–308.

Moss, J. B., Koustubhan, P., Greenman, M., Parsons, M. J., Walter, I., and Moss, L. G. (2009). Regeneration of the pancreas in adult zebrafish. *Diabetes* **58**, 1844–1851.

Nair, S., and Schilling, T. F. (2008). Chemokine signaling controls endodermal migration during zebrafish gastrulation. *Science* **322**, 89–92.

Ng, A. N., de Jong-Curtain, T. A., Mawdsley, D. J., White, S. J., Shin, J., Appel, B., Dong, P. D., Stainier, D. Y., and Heath, J. K. (2005). Formation of the digestive system in zebrafish: III. Intestinal epithelium morphogenesis. *Dev. Biol.* **286**, 114–135.

Nowotschin, S., Eakin, G. S., and Hadjantonakis, A. K. (2009). Live-imaging fluorescent proteins in mouse embryos: Multi-dimensional, multi-spectral perspectives. *Trends Biotechnol.* **27**, 266–276.

Obholzer, N., Wolfson, S., Trapani, J. G., Mo, W., Nechiporuk, A., Busch-Nentwich, E., Seiler, C., Sidi, S., Sollner, C., Duncan, R. N., *et al.* (2008). Vesicular glutamate transporter 3 is required for synaptic transmission in zebrafish hair cells. *J. Neurosci.* **28**, 2110–2118.

Pack, M., Solnica-Krezel, L., Malicki, J., Neuhauss, S. C., Schier, A. F., Stemple, D. L., Driever, W., and Fishman, M. C. (1996). Mutations affecting development of zebrafish digestive organs. *Development* **123**, 321–328.

Park, S. W., Davison, J. M., Rhee, J., Hruban, R. H., Maitra, A., and Leach, S. D. (2008). Oncogenic KRAS induces progenitor cell expansion and malignant transformation in zebrafish exocrine pancreas. *Gastroenterology* **134**, 2080–2090.

Parsons, M. J., Pisharath, H., Yusuff, S., Moore, J. C., Siekmann, A. F., Lawson, N., and Leach, S. D. (2009). Notch-responsive cells initiate the secondary transition in larval zebrafish pancreas. *Mech. Dev.* **126**, 898–912.

Pauls, S., Zecchin, E., Tiso, N., Bortolussi, M., and Argenton, F. (2007). Function and regulation of zebrafish nkx2.2a during development of pancreatic islet and ducts. *Dev. Biol.* **304**, 875–890.

Pezeron, G., Mourrain, P., Courty, S., Ghislain, J., Becker, T. S., Rosa, F. M., and David, N. B. (2008). Live analysis of endodermal layer formation identifies random walk as a novel gastrulation movement. *Curr. Biol.* **18**, 276–281.

Pisharath, H., Rhee, J. M., Swanson, M. A., Leach, S. D., and Parsons, M. J. (2007). Targeted ablation of beta cells in the embryonic zebrafish pancreas using E. coli nitroreductase. *Mech. Dev.* **124**, 218–229.

Provost, E., Rhee, J., and Leach, S. D. (2007). Viral 2A peptides allow expression of multiple proteins from a single ORF in transgenic zebrafish embryos. *Genesis* **45**, 625–629.

Rane, S. G., Dubus, P., Mettus, R. V., Galbreath, E. J., Boden, G., Reddy, E. P., and Barbacid, M. (1999). Loss of Cdk4 expression causes insulin-deficient diabetes and Cdk4 activation results in beta-islet cell hyperplasia. *Nat. Genet.* **22**, 44–52.

Reimer, M. M., Sorensen, I., Kuscha, V., Frank, R. E., Liu, C., Becker, C. G., and Becker, T. (2008). Motor neuron regeneration in adult zebrafish. *J. Neurosci.* **28**, 8510–8516.

Rhee, J. M., Pirity, M. K., Lackan, C. S., Long, J. Z., Kondoh, G., Takeda, J., and Hadjantonakis, A. K. (2006). In vivo imaging and differential localization of lipid-modified GFP-variant fusions in embryonic stem cells and mice. *Genesis* **44**, 202–218.

Roy, S., Qiao, T., Wolff, C., and Ingham, P. W. (2001). Hedgehog signaling pathway is essential for pancreas specification in the zebrafish embryo. *Curr. Biol.* **11**, 1358–1363.

Sakaue-Sawano, A., Kurokawa, H., Morimura, T., Hanyu, A., Hama, H., Osawa, H., Kashiwagi, S., Fukami, K., Miyata, T., Miyoshi, H., *et al.* (2008). Visualizing spatiotemporal dynamics of multicellular cell-cycle progression. *Cell* **132**, 487–498.

Salic, A., and Mitchison, T. J. (2008). A chemical method for fast and sensitive detection of DNA synthesis in vivo. *Proc. Natl. Acad. Sci. U.S.A* **105**, 2415–2420.

Siekmann, A. F., Standley, C., Fogarty, K. E., Wolfe, S. A., and Lawson, N. D. (2009). Chemokine signaling guides regional patterning of the first embryonic artery. *Genes Dev.* **23**, 2272–2277.

Stafford, D., and Prince, V. E. (2002). Retinoic acid signaling is required for a critical early step in zebrafish pancreatic development. *Curr. Biol.* **12**, 1215–1220.

Stafford, D., White, R. J., Kinkel, M. D., Linville, A., Schilling, T. F., and Prince, V. E. (2006). Retinoids signal directly to zebrafish endoderm to specify insulin-expressing beta-cells. *Development* **133**, 949–956.

Taupin, P. (2007). BrdU immunohistochemistry for studying adult neurogenesis: Paradigms, pitfalls, limitations, and validation. *Brain Res. Rev.* **53**, 198–214.

Teta, M., Long, S. Y., Wartschow, L. M., Rankin, M. M., and Kushner, J. A. (2005). Very slow turnover of beta-cells in aged adult mice. *Diabetes* **54**, 2557–2567.

Teta, M., Rankin, M. M., Long, S. Y., Stein, G. M., and Kushner, J. A. (2007). Growth and regeneration of adult beta cells does not involve specialized progenitors. *Dev. Cell* **12**, 817–826.

Tiso, N., Moro, E., and Argenton, F. (2009). Zebrafish pancreas development. *Mol. Cell. Endocrinol.* **312**, 24–30.

Tsutsui, T., Hesabi, B., Moons, D. S., Pandolfi, P. P., Hansel, K. S., Koff, A., and Kiyokawa, H. (1999). Targeted disruption of CDK4 delays cell cycle entry with enhanced p27(Kip1) activity. *Mol. Cell. Biol.* **19**, 7011–7019.

Verbruggen, V., Ek, O., Georlette, D., Delporte, F., Von Berg, V., Detry, N., Biemar, F., Coutinho, P., Martial, J. A., Voz, M. L., *et al.* (2010). The Pax6b homeodomain is dispensable for pancreatic endocrine cell differentiation in zebrafish. *J. Biol. Chem.* **285**, 13863–13873.

Wan, H., Korzh, S., Li, Z., Mudumana, S. P., Korzh, V., Jiang, Y. J., Lin, S., and Gong, Z. (2006). Analyses of pancreas development by generation of gfp transgenic zebrafish using an exocrine pancreas-specific elastaseA gene promoter. *Exp. Cell Res.* **312**, 1526–1539.

Ward, A. B., Warga, R. M., and Prince, V. E. (2007). Origin of the zebrafish endocrine and exocrine pancreas. *Dev. Dyn.* **236**, 1558–1569.

Wilkins, S. J., Yoong, S., Verkade, H., Mizoguchi, T., Plowman, S. J., Hancock, J. F., Kikuchi, Y., Heath, J. K., and Perkins, A. C. (2008). Mtx2 directs zebrafish morphogenetic movements during epiboly by regulating microfilament formation. *Dev. Biol.* **314**, 12–22.

Xu, X., D'Hoker, J., Stange, G., Bonne, S., De Leu, N., Xiao, X., Van de Casteele, M., Mellitzer, G., Ling, Z., Pipeleers, D., *et al.* (2008). Beta cells can be generated from endogenous progenitors in injured adult mouse pancreas. *Cell* **132**, 197–207.

Yee, N. S., Lorent, K., and Pack, M. (2005). Exocrine pancreas development in zebrafish. *Dev. Biol.* **284**, 84–101.

Yee, N. S., Yusuff, S., and Pack, M. (2001). Zebrafish pdx1 morphant displays defects in pancreas development and digestive organ chirality, and potentially identifies a multipotent pancreas progenitor cell. *Genesis* **30**, 137–140.

Zecchin, E., Filippi, A., Biemar, F., Tiso, N., Pauls, S., Ellertsdottir, E., Gnugge, L., Bortolussi, M., Driever, W., and Argenton, F. (2007). Distinct delta and jagged genes control sequential segregation of pancreatic cell types from precursor pools in zebrafish. *Dev. Biol.* **301**, 192–204.

Zorn, A. M., and Wells, J. M. (2007). Molecular basis of vertebrate endoderm development. *Int. Rev. Cytol.* **259**, 49–111.

CHAPTER 11

Monitoring Sleep and Arousal in Zebrafish

Jason Rihel[*], **David A. Prober**[†], *and* **Alexander F. Schier**[*,‡,§,¶]

[*]Department of Molecular and Cellular Biology, Harvard University, Cambridge, Massachusetts

[†]Division of Biology, California Institute of Technology, Pasadena, California

[‡]Division of Sleep Medicine, Harvard University, Cambridge, Massachusetts

[§]Center for Brain Science, Harvard University, Cambridge, Massachusetts

[¶]Harvard Stem Cell Institute, Harvard University, Cambridge, Massachusetts

I. Introduction
II. Behavior, Genetics, and Pharmacology of Zebrafish Sleep
   A. Behavior
   B. Genetics and Pharmacology
III. Methods for Monitoring Sleep/Wake Behavior in Zebrafish
   A. Methodological Considerations—from *Drosophila* to *Danio*
   B. Experimental Design and Setup
   C. Monitoring and Analysis of Sleep/Wake Behaviors
IV. Conclusion
   References

Abstract

Zebrafish has emerged in the past 5 years as a model for the study of sleep and wake behaviors. Experimental evidence has shown that periods of behavioral quiescence in zebrafish larvae and adults are sleep-like states, as these rest bouts are regulated by the circadian cycle, are associated with decreases in arousal, and are increased following rest deprivation. Furthermore, zebrafish share with mammals a hypocretin/orexin system that promotes wakefulness, and drugs that alter mammalian sleep have similar effects on zebrafish rest. In this chapter, we review the zebrafish sleep literature and describe a long-term, high-throughput monitoring system for observing sleep and wake behaviors in larval zebrafish.

I. Introduction

Sleep is essential, time consuming, and conserved across the animal kingdom, yet it remains one of the major mysteries of biology. What is the function of sleep, and how is it regulated by genes and neurons? Since the discovery of characteristic electroencephalographic (EEG) signatures for states of sleep and waking in the late 1930s (Davis *et al.*, 1937), mammalian model systems have dominated sleep research. However, behavioral observations over the past decade have demonstrated that non-mammalian systems, including *Drosophila* (Hendricks *et al.*, 2000; Shaw *et al.*, 2000), *Caenorhabditis elegans* (Raizen *et al.*, 2008; Van Buskirk and Sternberg, 2007), and zebrafish (Prober *et al.*, 2006; Yokogawa *et al.*, 2007; Zhdanova *et al.*, 2001), have sleep-like states. These "simple" model systems allow researchers to bring large-scale genetics and *in vivo* imaging to bear on fundamental questions of sleep biology. Zebrafish is an attractive model because it combines the facile genetics of invertebrates with brains that are morphologically and molecularly analogous to mammals. In this chapter, we review the progress of sleep studies in zebrafish and discuss the high-throughput methods (Fig. 1) we have developed to study larval zebrafish sleep/wake behaviors.

II. Behavior, Genetics, and Pharmacology of Zebrafish Sleep

A. Behavior

Sleep is a period of reversible, inattentive behavioral quiescence that can be distinguished from quiet wakefulness using several behavioral criteria (Borbely and

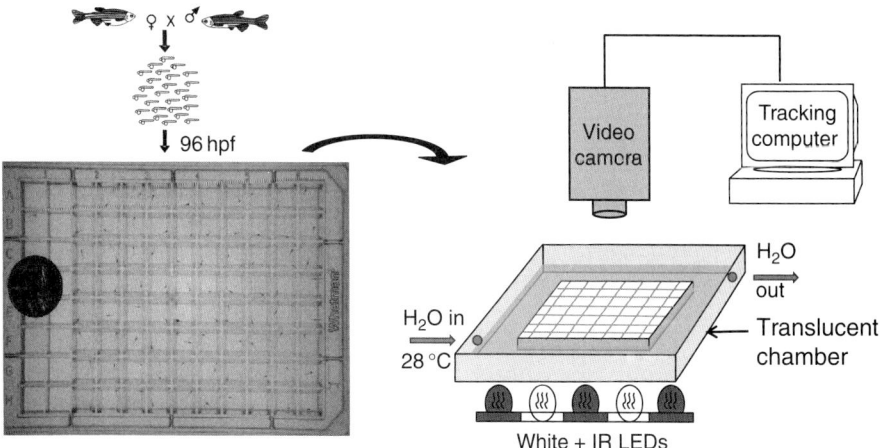

Fig. 1   High-throughput tracking of zebrafish locomotor behavior. Zebrafish larvae older than 96 hpf are placed into a 96-well plate filled with blue water. The plate is then placed into a translucent chamber that is illuminated from the bottom continuously by infrared (IR) light emitting diodes (LEDs) and from 9:00 AM to 11:00 PM with white LEDs. Water heated to 28°C is pumped continuously through (water goes in and out of the chamber) the plate chamber to maintain temperature and humidity. The plate is observed by a video camera connected to a computer with video tracking software. The chamber, lights, and camera are housed inside a box (not shown) to prevent extraneous light from interfering with the experiment. Modified from Fig. 4 in Prober *et al.*, 2006.

Tobler, 1996; Campbell and Tobler, 1984; Cirelli, 2009). Sleep-like rest periods are typically associated with a species-specific posture and location, and are regulated in a circadian (about 24 h) manner (Borbely and Tobler, 1996). Importantly, sleep is associated with an increased arousal threshold in response to stimuli, although strong stimuli can still wake the animal (thereby distinguishing sleep from coma or stupor). Finally, sleep is thought to be under homeostatic control, such that depriving an animal of sleep results in a subsequent increase in the duration or intensity of sleep, known as sleep rebound (Borbely and Tobler, 1996). Using automated tracking systems to observe individual animals for several days (see Section III), a variety of non-mammalian model systems, including fruit flies (Hendricks et al., 2000; Shaw et al., 2000), C. elegans (Raizen et al., 2008; Van Buskirk and Sternberg, 2007), and both larval (Prober et al., 2006; Zhdanova et al., 2001) and adult zebrafish (Yokogawa et al., 2007), have been shown to exhibit sleep-like states that meet these behavioral criteria.

1. Criterion 1—Quiescent Behavior Regulated by an Endogenous Circadian Rhythm

Zebrafish larvae first exhibit active spontaneous locomotor activity around 96 h post fertilization (hpf), shortly after inflation of the swim bladder (Hurd and Cahill, 2002; Prober et al., 2006). When maintained on a 14 h:10 h light:dark cycle, these swim bouts occur maximally during the lights-on phase and are tightly synchronized with the light stimulus (Fig. 2A). If raised on a light:dark cycle and then transferred to constant dark conditions, the spontaneous locomotor activity of >96 hpf larvae and adults continues to cycle with a circadian rhythm of ~25–25.5 h, with the phase set by the prior entraining light:dark cycle (Hurd and Cahill, 2002; Hurd et al., 1998). These observations demonstrate that zebrafish have an endogenously controlled circadian rhythm behavior that can be entrained by light. Consistent with the behavioral observations, rhythmic components of the molecular machinery that controls circadian rhythms in Drosophila and mammals also exhibit rhythmic expression with a light-entrainable circadian period in zebrafish. Such genes include the zebrafish orthologs of Period1 (Dekens and Whitmore, 2008), Period3 (Kaneko and Cahill, 2005; Pando et al., 2001), Clock (Whitmore et al., 1998), Cryptochrome (Kobayashi et al., 2000), and the clock-binding partner Bmal (Cermakian et al., 2000).

2. Criterion 2—Increased Arousal Threshold/Decreased Responsiveness during Quiescence

Throughout the 24 h light:dark cycle, zebrafish larvae exhibit bouts of quiescence that can last for several minutes or longer (Fig. 2B and C) and that occur maximally at night. During these quiescent periods, both zebrafish larvae and adults have an accompanying increase in arousal threshold, as measured by a decreased responsiveness of larvae to taps at night versus the day (Zhdanova et al., 2001), an increase in response time to large changes in light intensity (Prober et al., 2006), or a decreased responsiveness to electrical stimuli in adults (Yokogawa et al., 2007). By correlating the length of a quiescent bout with the concomitant change in arousal state, sleep is defined as a quiescent bout lasting at least 1 min in larvae and at least 6 s in adults. This definition of a sleep-like state has served well in the analysis of molecules that regulate sleep and has been effective in uncovering conserved pathways (see Section II-B). The definition can now be refined by testing changes in arousal across multiple sensory modalities,

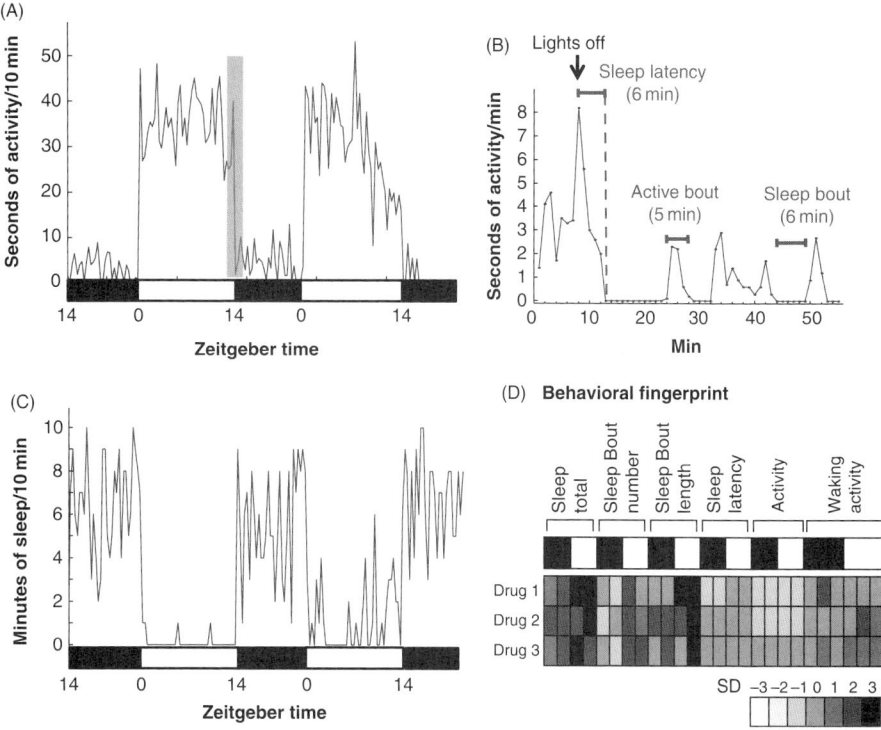

Fig. 2   Zebrafish sleep/wake data. (A) The average activity of a single wild type larva is plotted per 10 min for two light:dark cycles starting at ~110 hpf. Activity occurs maximally during each day period. The gray area is expanded in (B). Zeitgeber time 0 = lights on; 14 = lights off. (B) The average activity of the same fish in (A) (gray area), expanded and replotted per 1 min. Examples of sleep latency, sleep bout, sleep bout length, active bout, and active bout length are indicated. (C) The behavior of the same fish shown in (A) and (B), plotted as minutes of sleep per 10 min. Sleep occurs maximally during each night period. (D) By normalizing each behavioral parameter to wild type controls, the data can be transformed into a behavioral fingerprint. Each square of the fingerprint represents the average relative value in standard deviations (black, higher than controls; white, lower than controls) for a single behavioral measurement. The black and white bars across the top represent night and day measurements, respectively. In this example, fingerprints are shown for three different drugs that increase sleep bout lengths, leading to increased total sleep.

experimental conditions, and behavioral situations, including a detailed analysis of changes in arousal states during different times of the day and night. For example, a recent report notes that larval respiration rate is reduced and arousal threshold elevated in nighttime rest compared to daytime rest (Zhdanova, 2006), indicating that these two quiescent states may not be equivalent. We also observe changes in average sleep bout length and in waking activity (see section III.C for details) during the day and night that may reflect underlying differences between these states (Table I).

3. Criterion 3—Sleep Rebound Following Sleep Deprivation (Homeostatic Regulation of Sleep)

Two studies have demonstrated an increase in total sleep amount in larval and adult zebrafish following nighttime sleep deprivation. Zhdanova and colleagues used a

TABLE I

Typical Sleep/Wake Measures Obtained from Wild Type Larvae (TLAB × TLAB Cross) in a 14 h:10 h Light:Dark Cycle

	Mean	(± SD)	Median
TOTAL SLEEP (minutes)			
Day 6	189	138	135
Day 7	110	127	60
Night 6	252	125	240
Night 7	219	118	202
# SLEEP BOUTS			
Day 6	29	21	24
Day 7	37	34	26
Night 6	58	23	57
Night 7	64	20	63
SLEEP LENGTH (minutes)			
Day 6	7.7	5.4	6.6
Day 7	2.8	1.9	2.3
Night 6	4.5	2.8	3.9
Night 7	3.5	2.4	3.1
SLEEP LATENCY (minutes)			
Day 6	24.8	70.6	4
Day 7	59.2	147.0	7
Night 6	21.2	28.0	13
Night 7	19.6	27.0	14
AVERAGE ACTIVITY (seconds/minute)			
Day 6	4.04	1.88	3.90
Day 7	4.18	1.95	4.00
Night 6	0.85	0.46	0.78
Night 7	0.85	0.42	0.79
WAKING ACTIVITY (seconds/minute)			
Day 6	5.01	1.67	4.80
Day 7	4.58	1.88	4.30
Night 6	1.29	0.50	1.21
Night 7	1.21	0.41	1.19

$n = 321$; 14 h Day, 10 h Night

Each parameter (±SD) was calculated from the behavioral data of 321 wild type larvae. Day 6 starts at 120 hpf, and night 6 starts at 134 hpf. Day 7 starts at 144 hpf, and night 7 starts at 158 hpf. Sleep latency at night is calculated as the time from lights out to the first sleep bout; sleep latency during the day is calculated as the time from lights on to the first sleep bout.

mechanical shaker to deprive larvae of rest during the day and the night and found that night time deprivation decreased subsequent larval locomotor activity more than day time deprivation (Zhdanova *et al.*, 2001). Furthermore, following nighttime sleep deprivation, the larvae had an increased arousal threshold response to a tap stimulus. Together, these data suggest that nighttime deprivation causes a sleep rebound in both the amount and the depth of sleep. However, the decrease in general locomotion may indicate that the mechanical deprivation had caused some harm or stress. A long-term behavioral assessment demonstrating a return to baseline activity levels would help

clarify this point. Furthermore, it is unclear whether the arousal threshold changes occurred specifically during times of sleep-like states (see Section II-A, criterion 2). Large changes in arousal states during waking activity may also be an indicator of poor health. In another study, adult zebrafish were deprived of rest for 6 h with either electric shocks or light (Yokogawa *et al.*, 2007). Shock-induced deprivation at night resulted only in a modest rebound in total sleep relative to yoked controls; changes in arousal threshold following deprivation were not examined. Curiously, light-induced deprivation resulted in no observed sleep rebound, which may reflect a strong masking effect of light or the effect of light on the circadian rhythm in zebrafish.

Although these results are encouraging, much more work needs to be done to firmly establish that rest is under homeostatic control in zebrafish. In particular, it remains to be demonstrated that either the depth or duration of sleep rebound increases with increasing amounts of sleep deprivation. Also underexplored are the behavioral consequences of short- and long-term sleep deprivation, including the time course of the return to normal sleep patterns and altered performance in behavioral tasks (see Zhdanova, 2006). Finally, rebound sleep following short-term exposure to arousing drugs or genetic manipulations that dramatically reduce total sleep (see Section II-B) should also be investigated, as these treatments may represent less invasive and more reproducible methods for sleep deprivation.

B. Genetics and Pharmacology

Genetic and pharmacological experiments in zebrafish indicate that mechanisms that regulate mammalian sleep are conserved in zebrafish. Of these, the most studied is the hypocretin/orexin (Hcrt) system, which has been shown to increase wakefulness in mammals (Sakurai, 2007). Hcrt peptide is produced by a population of hypothalamic neurons that project throughout the brain, particularly to other known wake-promoting centers (Peyron *et al.*, 1998). Deficiency in Hcrt signaling can be caused by loss-of-function mutations in the peptide (Chemelli *et al.*, 1999) or its receptors (Lin *et al.*, 1999; Willie *et al.*, 2003) or by a selective loss of Hcrt-producing neurons (Hara *et al.*, 2001; Peyron *et al.*, 2000; Thannickal *et al.*, 2000). In mammals, loss of Hcrt signaling leads to narcolepsy, a disease characterized by excessive daytime sleepiness, unstable sleep/wake states, and sudden loss of muscle tone during waking. Larval and adult zebrafish express their single *hcrt* ortholog in a small number of hypothalamic neurons that project to putative wake-promoting centers of the brain and down the spinal cord, regions that also express the single *hcrt* receptor (Appelbaum *et al.*, 2009; Faraco *et al.*, 2006; Kaslin *et al.*, 2004; Prober *et al.*, 2006). As expected for a wake-promoting peptide, overexpression of *hcrt* in larval zebrafish leads to increased wakefulness at the expense of rest (Prober *et al.*, 2006). Furthermore, *in vivo* observation of Hcrt neural activity using the bioluminescent reporter GFP-Aequorin reveals that they are maximally active during episodes of spontaneous locomotor activity and inactive during rest (Naumann *et al.*, 2010), consistent with results obtained in mammals (Lee *et al.*, 2005; Mileykovskiy *et al.*, 2005). Finally, adult zebrafish with mutations in the Hcrt receptor exhibit sleep fragmentation (Yokogawa *et al.*, 2007), another hallmark of

narcolepsy (Mochizuki *et al.*, 2004; Overeem *et al.*, 2001). Taken together, these data confirm that at least some aspects of Hcrt's regulatory role in sleep/wake behavior are preserved in zebrafish.

There are some discrepancies between the Hcrt data obtained in larvae and adults that should be noted. Paradoxically, Hcrt receptor mutant adults were reported to have a small decrease in sleep at night and to be slightly hyperactive at night, although the latter effect was only significant compared to unrelated non-mutagenized wild type animals (Yokogawa *et al.*, 2007). These effects were not seen in Hcrt receptor mutant larvae under constant dim light conditions (Appelbaum *et al.*, 2009). In addition, some behavioral effects (e.g., hypoactivity and increased sleep in constant dark) appear only in Hcrt receptor heterozygotes (Yokogawa *et al.*, 2007). The reasons for these discrepancies are unclear but might include background mutations or the small number of adults that were analyzed. In addition, injection of a very high dose (280–2800 pmol/g body weight) of Hcrt peptide into adult zebrafish brains led to a slight decrease in activity (Yokogawa *et al.*, 2007). This observation contrasts with the long-term and strong increase in wakefulness following Hcrt overexpression in zebrafish larvae (Prober *et al.*, 2006) and the arousing effects of moderate levels (2.8–28 pmol/g body weight) of Hcrt peptide injected into adult goldfish (Nakamachi *et al.*, 2006). Additional genetic studies and more extensive behavioral analyses will be needed to resolve these discrepancies. Currently, the preponderance of evidence indicates that Hcrt has arousing effects in teleosts, and that loss of Hcrt receptor in adults leads to fragmented sleep/wake states, as in mammals.

Other evidence for the conservation of sleep-regulatory mechanisms in zebrafish comes from pharmacological studies. Tests of known sedative and hypnotic compounds, including the hormone melatonin (Zhdanova *et al.*, 2001), the GABA receptor agonists baclofen, phenobarbital, and diazepam (Renier *et al.*, 2007; Zhdanova *et al.*, 2001), and the alpha-2 adrenergic receptor agonist clonidine (Ruuskanen *et al.*, 2005), demonstrated that these compounds dose-dependently decrease locomotor activity and increase rest in zebrafish larvae 2–3 h following drug treatment. We have confirmed and extended these results through an unbiased screen of the long-term (3 days) effects of nearly 4000 small molecules on larval zebrafish sleep/wake behaviors. We found that many modulators of neurotransmitter systems, including the noradrenaline, serotonin, dopamine, GABA, glutamate, histamine, adenosine, and melatonin systems, induce similar sleep/wake phenotypes in zebrafish as observed in mammals (Rihel *et al.*, 2010).

Overall, the behavioral, genetic, and pharmacological evidence indicates that a sleep-like state exists in both larval and adult zebrafish and that this state is regulated by mechanisms that are conserved among vertebrates. Now that the conceptual groundwork for studying sleep in zebrafish has been established, future screens and experiments can begin to dissect novel mechanisms of sleep/wake regulation in earnest. Indeed, the pharmacological screen has already identified several pathways, including the ether-a-go-go related gene (ERG) potassium channel, verapamil-sensitive L-type calcium channels, and the immunomodulatory nuclear factor of activated T cells (NFAT) and nuclear factor kappa B (NF-κB) pathways, as targets for future zebrafish sleep/wake studies (Rihel *et al.*, 2010).

III. Methods for Monitoring Sleep/Wake Behavior in Zebrafish

A. Methodological Considerations—from *Drosophila* to *Danio*

Two major methods have been used to track locomotor behavior of animals, either by counting the number of times an animal breaks an infrared beam or by direct analysis of movements captured by video. Pioneering work on circadian rhythms (Konopka and Benzer, 1971) and sleep (Hendricks *et al.*, 2000; Shaw *et al.*, 2000) in *Drosophila* predominantly used the infrared beam break method, which measures when the fly crosses an infrared beam in the center of a tube. Some work on zebrafish (Hurd *et al.*, 1998) and goldfish (Azpeleta *et al.*, 2010; Iigo and Tabata, 1996; Vera *et al.*, 2007) also used this method to assess locomotion. Although useful for low time-resolution analysis of circadian rhythms, the beam break method suffers from blind spots, where an animal may move without crossing the beam. Indeed, a direct comparison of results from infrared beam breaks to direct video recording suggests that the beam break method can overestimate total sleep in flies by 10–90% (Zimmerman *et al.*, 2008). More fine-scale measures of sleep structure, including sleep latency, sleep bout number, and sleep bout length (see Section III-C for descriptions of these measures), can be even more dramatically overestimated (Zimmerman *et al.*, 2008). In addition, to obtain an accurate assessment of these important sleep parameters, individual animals must be tracked unambiguously throughout the experiment. Because sleep bouts occur non-simultaneously among individuals, methods that average activity across a population of animals lack details of sleep architecture. While methods that allow for the simultaneous and unambiguous tracking of animals within the same arena are under development (Branson *et al.*, 2009; Grover *et al.*, 2009; Kato *et al.*, 2004; Straw *et al.*, 2010), most available methods require animals to be individually housed.

There are a growing number of methods for automated detection of zebrafish locomotion using video-based analysis. These include commercially available zebrafish tracking systems from Noldus Information Technology (http://www.noldus.com) and Viewpoint Life Sciences, Inc. (http://www.vplsi.com) as well as custom algorithms designed for the analysis of short-term responses to stimuli captured by a high-speed camera (e.g., Burgess and Granato, 2007; Fontaine *et al.*, 2008). To simultaneously observe hundreds of animals over several days with minimal user input, we use a relatively simple frame-by-frame background subtraction method within an analysis suite from Viewpoint Life Sciences. We find that counting pixel changes per frame at a low frame rate (15 frames per second) gives a reliable readout of the timing and duration of each larva's locomotor activity and rest for days or weeks of continuous behavioral recording.

B. Experimental Design and Setup

Because sleep behavior can be strongly influenced by prior environmental conditions and experiences (Ganguly-Fitzgerald *et al.*, 2006 and personal observation), the conditions for raising larvae prior to behavioral testing must be rigorously maintained. Following fertilization, embryos and larvae are raised from a single cell on a 14 h:10 h

light:dark (LD) cycle at 28°C in petri dishes with conventional embryo water at a density of no more than 50 larvae per 100 mm dish. If the effect of a mutation or a transgene (e.g., heat shock driving Hcrt overexpression; Prober *et al.*, 2006) on behavior is being tested, all comparisons are ideally done within the same clutch or batch, raised in the same petri dish, and not pre-sorted by genotype. Each day, the dishes are cleared of any sick larvae, water levels are readjusted as necessary, and the chorions are removed post hatching.

Between 96 and 110 hpf, single larvae are placed into each well of a flat-bottom, square-well 96-well plate (650 μl well volume, Whatman) filled with embryo blue water. For the best optical properties, each well is filled so that the meniscus is flat and nearly flush with the top of the well. To test the long-term behavioral effects of small molecules, drugs may be added directly to the wells at this time by pipetting compound dissolved in DMSO. Usually the desired final concentration of drug in each well is between 100 nM and 1 mM, and the final DMSO concentration should not exceed 1% (above this level, DMSO can have behavioral consequences).

The 96-well plate is then placed into the zebrafish tracking setup (custom modified from Viewpoint Life Sciences; see Fig. 1). Inside a box, the plate chamber is illuminated continuously with an infrared LED panel and from 9:00 AM to 11:00 PM with white LEDs. The plate is monitored by a video camera (Dinion one-third inch Monochrome Bosch camera) fitted with a fixed-angle megapixel lens (50 mm focal length; Computar) and a filter that transmits infrared light. To slightly humidify the box and maintain a constant temperature, distilled water heated to 28°C is continuously pumped through the plate chamber. Embryo water is added daily to each well to maintain high water levels. Although the larvae need not be fed until the 7th day of development, paramecia can also be added to each well daily (they will not be detected by the software). By adding paramecia to the wells, we have monitored larvae in the same 96-well plate continuously from day 4 to day 14 of development without a noticeable decline in health (JR, unpublished data). Older animals can be monitored in plates with larger well sizes (e.g., 6-, 12-, and 24-well plates).

C. Monitoring and Analysis of Sleep/Wake Behaviors

To collect the movement data from each larva in the 96-well plate, we use Viewpoint Life Sciences Videotrack software running in the quantization mode. In this software package, a detection threshold is set to distinguish the dark fish from the white background (in our setup, the threshold is 40, although this value depends on the infrared lighting used). For each camera frame, any pixels darker than this threshold that change are detected as a movement and stored in a raw data file as pixels changed per frame for each larva. These data are further processed by setting a threshold value for the number of pixels that must change to constitute larval movement instead of random pixel noise (the "freeze" threshold; for our experimental setups, a cutoff of 4 pixels). The data are then converted into total seconds spent moving per minute for each larva by summing the total time of pixel changes that exceed the threshold. A sample 56 h activity trace from a single fish is plotted in Fig. 2A.

From the activity data of each larva, we use custom-designed Matlab code to extract multiple additional parameters that measure the amount and structure of sleep for each day and night (Fig. 2B and Table I).

1. Total Sleep

We define sleep in larval zebrafish as a continuous period of inactivity that lasts at least 1 min, because 1 min of inactivity at night is associated with increased response latencies to changes in light intensity (Prober *et al.*, 2006). We measure the total sleep for each day and night period and plot the average sleep per 10 min to generate a sleep time course (Fig. 2C). Based on recent wild type data from our lab ($n = 321$ animals, TLAB × TLAB cross), zebrafish larvae sleep at night on average 235 min each 10 h night (23.5 min/h), but individual larvae can vary considerably from this value (Table I).

2. Sleep Bout Number and Sleep Bout Length

A sleep bout is defined as a continuous period of inactivity lasting 1 min or longer (see Fig. 2B). Because each larva has numerous sleep bouts that are interrupted by brief awakenings, plotting the sleep bout length distribution for each larva can also be useful. At night, larvae have on average 61 sleep bouts, with each bout lasting an average of 4 min (Table I).

3. Sleep Latency

Sleep latency is defined as the amount of time from the start of each day and night period until the first sleep bout (see Fig. 2B). Following lights out, wild type (TLAB × TLAB cross) larvae have sleep latencies at night averaging about 20 min (Table I).

4. Average Activity per Waking Minute

Changes in a larva's locomotor behavior could be due to perturbations in muscle control and coordination, altered stress or arousal state, or other general health deficits. For example, an unhealthy larva with increased sleep may also move considerably less during active swimming. By measuring the average activity only during bouts of waking activity, we can assess whether the overall health and swimming ability of the fish have been compromised. Average activity per waking minute is calculated for each day and night period by summing the total activity and dividing by the number of active minutes (total active minutes = total time - total sleep time). This measure can also be used to determine whether an experimental perturbation causes a larva to be hyperactive when awake. In our experimental apparatus, wild type larvae have an average waking activity of 4.8 s per minute during the day and 1.3 s per minute at night (Table I).

These measures of sleep structure are biologically important, providing information about the initiation, maintenance, and timing of sleep. Given that these parameters can

be selectively perturbed by pharmacological agents (Rihel *et al.*, 2010), they are likely controlled by at least partially independent regulatory mechanisms. These parameters can be presented as a multi-dimensional "behavioral fingerprint" to facilitate comparisons among multiple genotypes, experimental manipulations, or small molecules. Each measurement is normalized to matched controls and then combined to create a vector that accounts for all of the parameters (see Fig. 2D; Rihel *et al.*, 2010). Clustering algorithms and principal component analyses can be used to organize large datasets by phenotype and to uncover small molecules or genotypes that have similar effects across multiple zebrafish sleep/wake behavioral parameters.

IV. Conclusion

Only a decade old, the study of sleep in non-mammalian systems is still in its infancy. While early zebrafish sleep studies have focused on establishing the existence of behavioral sleep regulated by conserved mechanisms, the challenge ahead is to use the zebrafish sleep model to uncover heretofore unsuspected aspects of the neuronal and genetic control of sleep/wake regulation. Recent studies that potentially link Hcrt signaling to pineal gland regulation (Appelbaum *et al.*, 2009) and that uncover novel small molecule regulators of sleep/wake states (Rihel *et al.*, 2010) may represent two such discoveries.

A major advantage of the zebrafish system is the ability to efficiently perform genetic and pharmacological screens in a cost-, space-, and labor-effective manner. With this in mind, we have described an automated high-throughput method for observing long-term sleep/wake behavior in larval zebrafish. This methodology is highly flexible and can easily be adapted to other behaviors, for example to observe larval responses to temperature, vibration, noxious chemicals, and changes in light intensity (Emran *et al.*, 2007, 2008, 2010; Prober *et al.*, 2008). In principle, the behavioral space of future screens could be expanded by testing these and other behavioral modalities in conjunction with long-term sleep/wake behavioral monitoring, incorporating all the data into a single multi-dimensional behavioral fingerprint. Such screens could not only uncover novel mutants that affect specific behaviors but also identify correlated sets of behaviors that are regulated by similar underlying mechanisms. With new techniques to directly observe neural activity in behaving zebrafish, including neuroluminescence, whole-brain calcium imaging, and optogenetic techniques to manipulate the activity of neurons with light, future studies will also begin to directly elucidate the activities and functions of neural circuits that underlie sleep/wake behaviors.

References

Appelbaum, L., Wang, G. X., Maro, G. S., Mori, R., Tovin, A., Marin, W., Yokogawa, T., Kawakami, K., Smith, S. J., Gothilf, Y., Mignot, E., and Mourrain, P. (2009). Sleep–wake regulation and hypocretin–melatonin interaction in zebrafish. *Proc. Natl. Acad. Sci. U. S. A.* **106**, 21942–21947.

Azpeleta, C., Martinez-Alvarez, R. M., Delgado, M. J., Isorna, E., and De Pedro, N. (2010). Melatonin reduces locomotor activity and circulating cortisol in goldfish. *Horm. Behav.* **57**, 323–329.

Borbely, A. A., and Tobler, I. (1996). Sleep regulation: Relation to photoperiod, sleep duration, waking activity, and torpor. *Prog. Brain Res.* **111**, 343–348.

Branson, K., Robie, A. A., Bender, J., Perona, P., and Dickinson, M. H. (2009). High-throughput ethomics in large groups of *Drosophila. Nat. Methods* **6**, 451–457.

Burgess, H. A., and Granato, M. (2007). Modulation of locomotor activity in larval zebrafish during light adaptation. *J. Exp. Biol.* **210**, 2526–2539.

Campbell, S. S., and Tobler, I. (1984). Animal sleep: A review of sleep duration across phylogeny. *Neurosci. Biobehav. Rev.* **8**, 269–300.

Cermakian, N., Whitmore, D., Foulkes, N. S., and Sassone-Corsi, P. (2000). Asynchronous oscillations of two zebrafish CLOCK partners reveal differential clock control and function. *Proc. Natl. Acad. Sci. U. S. A.* **97**, 4339–4344.

Chemelli, R. M., Willie, J. T., Sinton, C. M., Elmquist, J. K., Scammell, T., Lee, C., Richardson, J. A., Williams, S. C., Xiong, Y., Kisanuki, Y., et al. (1999). Narcolepsy in orexin knockout mice: Molecular genetics of sleep regulation. *Cell* **98**, 437–451.

Cirelli, C. (2009). The genetic and molecular regulation of sleep: From fruit flies to humans. *Nat. Rev. Neurosci.* **10**, 549–560.

Davis, H., Davis, P. A., Loomis, A. L., Harvey, E. N., and Hobart, G. (1937). Changes in human brain potentials during the onset of sleep. *Science* **86**, 448–450.

Dekens, M. P., and Whitmore, D. (2008). Autonomous onset of the circadian clock in the zebrafish embryo. *EMBO J.* **27**, 2757–2765.

Emran, F., Rihel, J., Adolph, A. R., and Dowling, J. E. (2010). Zebrafish larvae lose vision at night. *Proc. Natl. Acad. Sci. U. S. A.* **107**, 6034–6039.

Emran, F., Rihel, J., Adolph, A. R., Wong, K. Y., Kraves, S., and Dowling, J. E. (2007). OFF ganglion cells cannot drive the optokinetic reflex in zebrafish. *Proc. Natl. Acad. Sci. U. S. A.* **104**, 19126–19131.

Emran, F., Rihel, J., and Dowling, J. E. (2008). A behavioral assay to measure responsiveness of zebrafish to changes in light intensities. *J. Vis. Exp* **20**, http://www.jove.com/index/details.stp?id=923, doi: 10.3791/923.

Faraco, J. H., Appelbaum, L., Marin, W., Gaus, S. E., Mourrain, P., and Mignot, E. (2006). Regulation of hypocretin (orexin) expression in embryonic zebrafish. *J. Biol. Chem.* **281**, 29753–29761.

Fontaine, E., Lentink, D., Kranenbarg, S., Muller, U. K., van Leeuwen, J. L., Barr, A. H., and Burdick, J. W. (2008). Automated visual tracking for studying the ontogeny of zebrafish swimming. *J. Exp. Biol.* **211**, 1305–1316.

Ganguly-Fitzgerald, I., Donlea, J., and Shaw, P. J. (2006). Waking experience affects sleep need in *Drosophila. Science* **313**, 1775–1781.

Grover, D., Yang, J., Ford, D., Tavare, S., and Tower, J. (2009). Simultaneous tracking of movement and gene expression in multiple *Drosophila melanogaster* flies using GFP and DsRED fluorescent reporter transgenes. *BMC Res. Notes* **2**, 58.

Hara, J., Beuckmann, C. T., Nambu, T., Willie, J. T., Chemelli, R. M., Sinton, C. M., Sugiyama, F., Yagami, K., Goto, K., Yanagisawa, M., and Sakurai, T. (2001). Genetic ablation of orexin neurons in mice results in narcolepsy, hypophagia, and obesity. *Neuron* **30**, 345–354.

Hendricks, J. C., Finn, S. M., Panckeri, K. A., Chavkin, J., Williams, J. A., Sehgal, A., and Pack, A. I. (2000). Rest in *Drosophila* is a sleep-like state. *Neuron* **25**, 129–138.

Hurd, M. W., and Cahill, G. M. (2002). Entraining signals initiate behavioral circadian rhythmicity in larval zebrafish. *J. Biol. Rhythms* **17**, 307–314.

Hurd, M. W., Debruyne, J., Straume, M., and Cahill, G. M. (1998). Circadian rhythms of locomotor activity in zebrafish. *Physiol. Behav.* **65**, 465–472.

Iigo, M., and Tabata, M. (1996). Circadian rhythms of locomotor activity in the goldfish *Carassius auratus. Physiol. Behav.* **60**, 775–781.

Kaneko, M., and Cahill, G. M. (2005). Light-dependent development of circadian gene expression in transgenic zebrafish. *PLoS Biol.* **3**, e34.

Kaslin, J., Nystedt, J. M., Ostergard, M., Peitsaro, N., and Panula, P. (2004). The orexin/hypocretin system in zebrafish is connected to the aminergic and cholinergic systems. *J. Neurosci.* **24**, 2678–2689.

Kato, S., Nakagawa, T., Ohkawa, M., Muramoto, K., Oyama, O., Watanabe, A., Nakashima, H., Nemoto, T., and Sugitani, K. (2004). A computer image processing system for quantification of zebrafish behavior. *J. Neurosci. Methods* **134**, 1–7.

Kobayashi, Y., Ishikawa, T., Hirayama, J., Daiyasu, H., Kanai, S., Toh, H., Fukuda, I., Tsujimura, T., Terada, N., Kamei, Y., Yuba, S., Iwai, S., et al. (2000). Molecular analysis of zebrafish photolyase/cryptochrome family: Two types of cryptochromes present in zebrafish. *Genes Cells* **5**, 725–738.

Konopka, R. J., and Benzer, S. (1971). Clock mutants of *Drosophila melanogaster. Proc. Natl. Acad. Sci. U. S. A.* **68**, 2112–2116.

Lee, M. G., Hassani, O. K., and Jones, B. E. (2005). Discharge of identified orexin/hypocretin neurons across the sleep–waking cycle. *J. Neurosci.* **25**, 6716–6720.

Lin, L., Faraco, J., Li, R., Kadotani, H., Rogers, W., Lin, X., Qiu, X., de Jong, P. J., Nishino, S., and Mignot, E. (1999). The sleep disorder canine narcolepsy is caused by a mutation in the hypocretin (orexin) receptor 2 gene. *Cell* **98**, 365–376.

Mileykovskiy, B. Y., Kiyashchenko, L. I., and Siegel, J. M. (2005). Behavioral correlates of activity in identified hypocretin/orexin neurons. *Neuron* **46**, 787–798.

Mochizuki, T., Crocker, A., McCormack, S., Yanagisawa, M., Sakurai, T., and Scammell, T. E. (2004). Behavioral state instability in orexin knock-out mice. *J. Neurosci.* **24**, 6291–6300.

Nakamachi, T., Matsuda, K., Maruyama, K., Miura, T., Uchiyama, M., Funahashi, H., Sakurai, T., and Shioda, S. (2006). Regulation by orexin of feeding behaviour and locomotor activity in the goldfish. *J. Neuroendocrinol.* **18**, 290–297.

Naumann, E. A., Kampff, A. R., Prober, D. A., Schier, A. F., and Engert, F. (2010). Monitoring neural activity with bioluminescence during natural behavior. *Nat. Neurosci.* **13**, 513–520.

Overeem, S., Mignot, E., van Dijk, J. G., and Lammers, G. J. (2001). Narcolepsy: Clinical features, new pathophysiologic insights, and future perspectives. *J. Clin. Neurophysiol.* **18**, 78–105.

Pando, M. P., Pinchak, A. B., Cermakian, N., and Sassone-Corsi, P. (2001). A cell-based system that recapitulates the dynamic light-dependent regulation of the vertebrate clock. *Proc. Natl. Acad. Sci. U. S. A.* **98**, 10178–10183.

Peyron, C., Faraco, J., Rogers, W., Ripley, B., Overeem, S., Charnay, Y., Nevsimalova, S., Aldrich, M., Reynolds, D., Albin, R., Li, R., Hungs, M., et al. (2000). A mutation in a case of early onset narcolepsy and a generalized absence of hypocretin peptides in human narcoleptic brains. *Nat. Med.* **6**, 991–997.

Peyron, C., Tighe, D. K., van den Pol, A. N., de Lecea, L., Heller, H. C., Sutcliffe, J. G., and Kilduff, T. S. (1998). Neurons containing hypocretin (orexin) project to multiple neuronal systems. *J. Neurosci.* **18**, 9996–10015.

Prober, D. A., Rihel, J., Onah, A. A., Sung, R. J., and Schier, A. F. (2006). Hypocretin/orexin overexpression induces an insomnia-like phenotype in zebrafish. *J. Neurosci.* **26**, 13400–13410.

Prober, D. A., Zimmerman, S., Myers, B. R., McDermott, B. M. Jr., Kim, S. H., Caron, S., Rihel, J., Solnica-Krezel, L., Julius, D., Hudspeth, A. J., and Schier. A. F. (2008). Zebrafish TRPA1 channels are required for chemosensation but not for thermosensation or mechanosensory hair cell function. *J. Neurosci.* **28**, 10102–10110.

Raizen, D. M., Zimmerman, J. E., Maycock, M. H., Ta, U. D., You, Y. J., Sundaram, M. V., and Pack, A. I. (2008). Lethargus is a *Caenorhabditis elegans* sleep-like state. *Nature* **451**, 569–572.

Renier, C., Faraco, J. H., Bourgin, P., Motley, T., Bonaventure, P., Rosa, F., and Mignot, E. (2007). Genomic and functional conservation of sedative-hypnotic targets in the zebrafish. *Pharmacogenet. Genomics* **17**, 237–253.

Rihel, J., Prober, D. A., Arvanites, A., Lam, K., Zimmerman, S., Jang, S., Haggarty, S. J., Kokel, D., Rubin, L. L., Peterson, R. T., and Schier., A. F. (2010). Zebrafish behavioral profiling links drugs to biological targets and rest/wake regulation. *Science* **327**, 348–351.

Ruuskanen, J. O., Peitsaro, N., Kaslin, J. V., Panula, P., and Scheinin, M. (2005). Expression and function of alpha-adrenoceptors in zebrafish: Drug effects, mRNA and receptor distributions. *J. Neurochem.* **94**, 1559–1569.

Sakurai, T. (2007). The neural circuit of orexin (hypocretin): Maintaining sleep and wakefulness. *Nat. Rev. Neurosci.* **8**, 171–181.

Shaw, P. J., Cirelli, C., Greenspan, R. J., and Tononi, G. (2000). Correlates of sleep and waking in *Drosophila melanogaster. Science* **287**, 1834–1837.

Straw, A. D., Branson, K., Neumann, T. R., and Dickinson, M. H. (2010). Multi-camera real-time three-dimensional tracking of multiple flying animals. *J. R. Soc. Interface*, doi:10.1098/rsif.2010.0230.

Thannickal, T. C., Moore, R. Y., Nienhuis, R., Ramanathan, L., Gulyani, S., Aldrich, M., Cornford, M., and Siegel, J. M. (2000). Reduced number of hypocretin neurons in human narcolepsy. *Neuron* **27**, 469–474.

Van Buskirk, C., and Sternberg, P. W. (2007). Epidermal growth factor signaling induces behavioral quiescence in *Caenorhabditis elegans*. *Nat. Neurosci.* **10**, 1300–1307.

Vera, L. M., De Pedro, N., Gomez-Milan, E., Delgado, M. J., Sanchez-Muros, M. J., Madrid, J. A., and Sanchez-Vazquez, F. J. (2007). Feeding entrainment of locomotor activity rhythms, digestive enzymes and neuroendocrine factors in goldfish. *Physiol. Behav.* **90**, 518–524.

Whitmore, D., Foulkes, N. S., Strahle, U., and Sassone-Corsi, P. (1998). Zebrafish Clock rhythmic expression reveals independent peripheral circadian oscillators. *Nat. Neurosci.* **1**, 701–707.

Willie, J. T., Chemelli, R. M., Sinton, C. M., Tokita, S., Williams, S. C., Kisanuki, Y. Y., Marcus, J. N., Lee, C., Elmquist, J. K., Kohlmeier, K. A., Leonard, C. S., Richardson, J. A., et al. (2003). Distinct narcolepsy syndromes in Orexin receptor-2 and Orexin null mice: Molecular genetic dissection of non-REM and REM sleep regulatory processes. *Neuron* **38**, 715–730.

Yokogawa, T., Marin, W., Faraco, J., Pezeron, G., Appelbaum, L., Zhang, J., Rosa, F., Mourrain, P., and Mignot, E. (2007). Characterization of sleep in zebrafish and insomnia in hypocretin receptor mutants. *PLoS Biol.* **5**, e277.

Zhdanova, I. V. (2006). Sleep in zebrafish. *Zebrafish* **3**, 215–226.

Zhdanova, I. V., Wang, S. Y., Leclair, O. U., and Danilova, N. P. (2001). Melatonin promotes sleep-like state in zebrafish. *Brain Res.* **903**, 263–268.

Zimmerman, J. E., Raizen, D. M., Maycock, M. H., Maislin, G., and Pack, A. I. (2008). A video method to study *Drosophila* sleep. *Sleep* **31**, 1587–1598.

CHAPTER 12

Use of Flatbed Transparency Scanners in Zebrafish Research: Versatile and Economical Adjuncts to Traditional Imaging Tools for the *Danio rerio* Laboratory

Charles A. Lessman[*], **Michael R. Taylor**[†], **Wilda Orisme**[†], *and* **Ethan A. Carver**[‡]

[*]Department of Biological Sciences, The University of Memphis, Memphis, Tennessee

[†]Department of Chemical Biology and Therapeutics, St. Jude Children's Research Hospital, Memphis, Tennessee

[‡]Department of Biological and Environmental Sciences, The University of Tennessee-Chattanooga, Chattanooga, Tennessee

Abstract
I. Introduction
II. Scanner Basics
   A. Resolution and Bit Depth
   B. Temperature and Fluidics
   C. Image Acquisition with Scanners
   D. Macroscheduler: A Mouse–Keyboard Macro for Automation of Scans
   E. ImageJ: Image Analysis Program for Animation, Manipulation, and Analysis of Image Stacks
III. Motility Analysis
   A. Computer-aided Screening: Survey of Development after Microinjection or Other Treatment
   B. Motility of Wild-type Zebrafish Embryos and Larvae
   C. Computer-aided Larval Motility Screen (CALMS)
   D. Determining Embryo Movement over Time (*x–y* Coordinates)
   E. Drug Screens: Effect on Motility

978-0-12-384892-5
DOI: 10.1016/B978-0-12-384892-5.00012-8

F.   Vision Mutants: Mutant Phenotype versus Wild Type
G.   Colorimetric Potential of Scanners: pH Indicators
IV.   Oocyte and Egg Assays
A.   Computer–aided Meiotic Maturation Assay: Quantification of Oocyte Maturation
B.   Osmoregulation during Oocyte Maturation
C.   Egg Activation Assay (EggsAct)
V.   Using Scanners to Count and Measure
A.   Oocyte, Egg, and Embryo Counting by Scanners
B.   Scanner Imaging of Juveniles and Adults
VI.   Other Potential Applications of Scanners
VII.   Summary: Inexpensive Adjunct to Microscopy
Acknowledgments
References

Abstract

Flatbed transparency scanners are typically relegated to routine office tasks, yet they do offer a variety of potentially useful imaging tools for the zebrafish laboratory. These include motility screens, oocyte maturation and egg activation assays as well as counting and measuring tasks. When coupled with Macroscheduler (http://www.mjtnet.com) and ImageJ (http://rsbweb.nih.gov/ij), the scanner becomes a stable platform for imaging large arrays of zebrafish oocytes, embryos, larvae, and adults. Such large arrays are a prerequisite to the development of high-throughput screens for small molecules as potential therapeutic drugs in the treatment of many diseases including cancer and epilepsy. Thus the scanner may have a role in adapting zebrafish to future drug and mutagenesis screening. In this chapter, some of the uses of scanners are outlined to bring attention to the potentials of this simple-to-use, flexible, inexpensive device for the zebrafish research community.

I. Introduction

One of the remarkable features of the zebrafish embryo is its transparent optical properties, which provides exceptional opportunities for imaging scientists. An under-utilized and perhaps overlooked imaging tool for the zebrafish community is the common office scanner used routinely to image documents, photos, and films. In this chapter, some of the imaging possibilities will be reviewed in order to bring attention to the range of uses such a commonly available device can have for the zebrafish researcher. The goal here is to demonstrate that the transparency scanner can provide additional imaging capabilities to the laboratory that complement the stereoscope, fluorescent, and confocal microscopes used routinely in zebrafish work. A particularly important feature of these scanners is their large imaging surface upon which may be scanned one or more multiwell plates or Petri dishes or large arrays of

immobilized adults. While digital cameras fitted with macro lenses may also be used for this type of imaging, the scanner, by contrast, provides a stable platform complete with both transmitted and reflective lighting and constant 1× magnification with a focal plane that will not change over time.

We have used scanners to automate screens of zebrafish motility mutants (Lessman, 2002, 2004) to complement the earlier procedures involving manual observation methods of screening (Granato *et al.*, 1996). In addition, we have adapted scanners for analysis of zebrafish oocyte maturation (Lessman *et al.*, 2007). In this chapter we expand on these methods and introduce new methods to image zebrafish with flatbed transparency scanners.

II. Scanner Basics

Flatbed transparency scanners consist of a glass sheet that makes up the imaging surface below which is driven a linear charge-coupled device (CCD) by a binary stepper motor with associated belt drive. In the reflected mode, light comes from a bulb below the glass to reflect the image onto the CCD as it passes under the object. In the transmitted mode, a light source (transparency adapter) produces light from above the object and passes through to the CCD moving below. For most zebrafish work, especially embryos and oocytes, the transmitted mode provides the most informative imaging. However, for adults and in some cases larvae, the reflective mode is more appropriate. Depending on the scanner, the bed can be the size of a standard sheet of paper or larger in reflected mode while the transmitted mode generally provides a smaller, yet useful scanned area such as one or more 96- or 384-well plates. Scanners are particularly well suited for following large arrays of oocytes, embryos, or larvae over time (i.e., minutes to days); however, they are not very useful for capturing rapid events (seconds or less). Scans take a few seconds to a few minutes depending upon the resolution, bit depth, speed of data pipeline to the computer as well as the CPU speed. An 8-bit, 1200-dpi scan of a 96-well plate should be possible in about 20 s with most scanners and computer systems available today.

We have successfully used a variety of scanners over the last 10 years including a high-resolution (for its day) Relisys scanner, HP scanjets, Umax Astra 6700, and most recently the Epson photo V500. The scanner software bundled with the scanner was compatible with Macroscheduler to allow for automated scanning. Most of the software included scanner-specific drivers, although even MS Windows has some limited capability to drive scanners. We have used ArtScan (Relisys), HP Director (HP scanjet), SilverFast SE (Umax), and Epson Scan (Epson photo V500) scanner software.

A. Resolution and Bit Depth

Scanner resolution along the *x*-axis depends on the number of individual elements in the linear CCD, whereas in the *y*-axis it is the stepper motor and drive system that

determines resolution. Optical resolution is the limit imposed by the hardware and in most modern scanners is 1200×1200 dpi or better. It should be noted that scanners may be advertised with significantly higher resolution but this is usually interpolated resolution or empty magnification that replaces a single pixel with multiple smaller, but identical pixels. Bit depth refers to the binary array of intensity bins producing an image. An 8-bit grayscale image has 255 shades of gray, from white to black. A 1-bit image is either black or white with no shades of gray; this is the type of image used by image analysis programs for measuring areas or other quantitative data in an image. A red, blue, green (RGB) image is a 24-bit image with 8 bits for each color, i.e., 255 shades of intensity for each of the three primary colors. Higher bit depth such as 12- or 16-bit grayscale or 48-bit color is advertised in the scanner literature, but, like interpolation, does not add significantly to the image quality in terms of suitability for image analysis and quantification. Higher resolution and bit depth quickly produces very large image files that may overwhelm even rather substantial modern workstations. So in this case, bigger is not necessarily better.

B. Temperature and Fluidics

Scanners produce some heat when powered up, which is useful in zebrafish work because the optimal temperature is 28°C for these animals, especially the embryos. Use of a simple thermistor probe on the operating scanner bed gives a convenient temperature reading that will average about 2°C above ambient. In air-conditioned laboratories, a black plastic sheet or an opaque box fitted over the scanner bed may be used to further elevate the bed temperature and eliminate stray ambient light. We have used small thermostatically controlled ceramic space heaters or small incandescent desk lamps to provide additional heat when needed. Another issue is fluid media temperature. Cold media, for example fish salines stored in refrigerators until use for oocyte work, should be 28°C prior to addition to plates or dishes. Otherwise gas bubbles will form that obscure imaging and, in the case of oocytes and embryos, the attached bubbles can actually cause the cells to float around in the wells, which produces disconcerting, spurious movement between scans. For most imaging tasks, an uncovered plate provides the best images; however, long-term scanning (i.e., overnight) requires covered plates or the addition of media via peristaltic pump. Condensation in the case of covered wells can obscure images if temperature differentials are great. Media reservoirs, in the latter case, should also be warmed to 28°C before pumping onto plates. Obviously, a siphon or overflow port should be provided to vent the excess liquid (Fig. 1).

The edges of the glass scanner bed should be sealed to the scanner chassis with a thin film of clear silicone glue; this will prevent spills from entering the underlying compartment containing the electronic components. Various vessels may be used to hold oocytes, embryos, larvae, and adults during scanning. Nunc rectangular (cat # 176600) plates fitted with plastic screen (plastic canvas or needlepoint screen #10 mesh with 10 square perforations per inch, available at most hobby stores) provide a suitable substrate for perfusing media over oocytes or chorionated embryos in the same

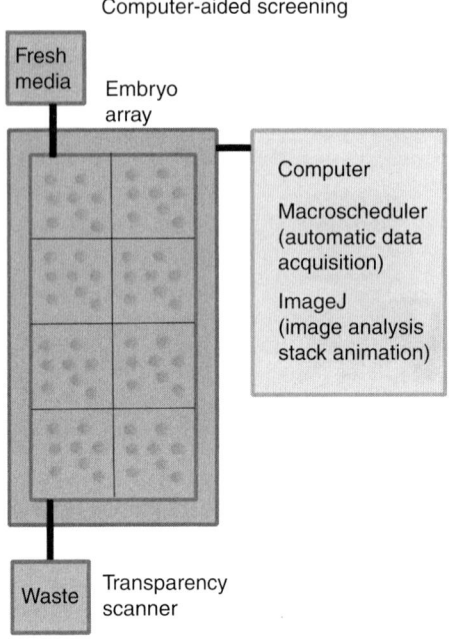

Fig. 1 Diagram of a flatbed transparency scanner and computer system for zebrafish imaging. Embryos are arrayed in a multiwell plate on the scanner bed (transparency adapter has been removed); shown is the optional media input and outflow via peristaltic pumps. The scanner is controlled by a macroprogram (Macroscheduler: http://www.mjtnet.com) to automatically acquire images at user-defined intervals. ImageJ (http://rsbweb.nih.gov/ij) is subsequently used to import the image sequence into a stack for animation and image analysis.

liquid compartment, yet maintaining relative positions on the plate. Gluing the ~1-mm mesh screen to the plate bottom provides shallow wells to keep embryos separated and in place (Lessman, 2004). Such plates can carry 100 oocytes or embryos per square inch, so that a Nunc plate with an area of 15 square inches could hold 1500 embryos in individual wells from which particular embryos could be retrieved after scanning. Multiwell plates with round-bottom or U-shaped wells help center the oocyte or embryo in the well and reduce wall effects; however, most common multiwell plates can be used for scanning (Fig. 2). Bubbles are a particularly troublesome problem when using 384-well plates; however, brief centrifugation of the media-filled plate prior to oocyte or embryo loading will remove most of the bubbles.

C. Image Acquisition with Scanners

For most scanner software, avoid using the full-auto scanning features. Instead select manual mode to set up the scanner for research scanning. There are three main settings to choose prior to image acquisition. (1) The initial decision is to choose

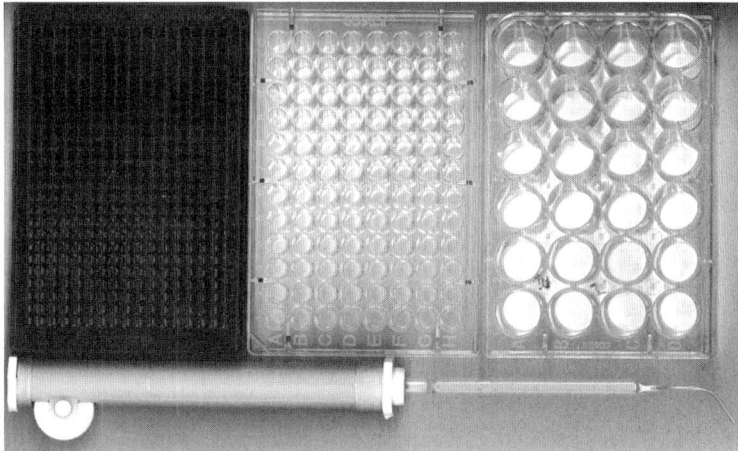

Fig. 2 Multiwell plates and pipette pump for use in imaging arrays of zebrafish oocytes, embryos, and larvae with transparency scanners. A black-walled 384-well plate with clear flat bottom (left) is useful for high-density embryo imaging; 96-well round-bottom plate (middle) works especially well with oocytes, embryos, and larvae; the 24-well plate (right) may be used to image multiple oocytes per well and allows greater free movement for single larva. The pipette pump (bottom) has a Pasteur pipette with tip bent in flame for reaching into individual wells. This figure was produced by imaging with an Umax Astra 6700 reflective mode scanner at 1200 dpi and 8-bit grayscale.

reflective or transmitted mode for the scan; this decision will be dictated by the zebrafish stage being imaged. The opaque adults and older juveniles are best imaged with reflective lighting, while oocytes, embryos, and larvae are visualized in more detail by transmitted light. Most scanners that have transparency adapters will have this option; if not, the machine is likely a reflected mode-only scanner and may require the purchase of an optional transparency adapter. Many of the newer scanners have transparency adapters built into the lids; these will work provided the lid can accommodate the thickness of the Petri plate or multiwall plate holding the fish. (2) The second option to select is the resolution (i.e., 1200 dpi or higher). There is little to be gained by setting the resolution higher than the optical resolution of the scanner. Also, file size and scan time increase with higher resolution settings. (3) Third is a choice of 8-bit grayscale or 24-bit color; use color only if there is a need, as color scans will be three times larger and take longer to scan. The other options available, such as sharpening or automatic brightness–contrast adjustment, are not recommended; these can always be applied later. If incorporated in the scan, these options are applied after the image is already in memory but before it has been saved to disk, thus adding to scan time. Run a prescan to bring the scanned image onto the computer screen. Using the rectangular marquee tool, select the area of the image you want to scan. Select the area carefully since this saves time in initial scanning and reduces the need to crop unwanted areas later. Finally, click on the scan button; a dialog box will appear to request a filename or a default sequentially numbered filename will be used.

D. Macroscheduler: A Mouse–Keyboard Macro for Automation of Scans

For sequential time-lapse scan series, a macro program such as Macroscheduler (http://www.mjtnet.com) is needed for unattended scanner operation. After installing the program, Macroscheduler will reside in memory; click on the gear icon found in the resident program toolbar to bring a dialog box on screen. Select record, give the macro a filename, and then begin recording by clicking on scan (the initial scan must be done manually to set up the conditions and area to be scanned as well as the directory and default filename series). A dialog box will appear with the default sequential filename automatically assigned; click on OK or save and the scanner will scan the area and save the file. After the scanner is finished, stop recording the macro by pressing ctrl-alt-s. Double click on the newly created macro that should appear on a list of available macros and select run when from the menu. Select the day of week, time to begin scanning, and scan interval. If multiple days are selected, the scanner will run through-out that time frame. Minimize the macroscheduler window and monitor the computer to ensure that the automated scanning will proceed. The gear icon will flash when the macro is running. Before the macro has been initiated, it is recommended to disconnect the computer from the Internet and shut off antivirus or other programs that might produce unwanted popups or other dialog-status boxes that can interfere with the macro. Since the macro will carry out mouse moves and keystrokes, moving the scanner windows after the initial macroprogramming or using the computer while the macro is running will prematurely stop the scan series.

E. ImageJ: Image Analysis Program for Animation, Manipulation, and Analysis of Image Stacks

ImageJ is freely available from the NIH Web site (http://rsb.info.nih.gov/ij) bundled with Java (Abramoff *et al.*, 2004). Most PC users will need the 32-bit version for Windows machines; additional versions are available for Macs and Unix machines as is a 64-bit version for Windows 7. ImageJ is being constantly updated and we have used versions 1.40 through 1.43 for most of the work presented here. Image stacks are sequential images in memory that ImageJ can animate, manipulate, and analyze; the stack will be the sum of all the image data, so if a single image is 5 MB then an image stack of 100 images will be 0.5 GB. In order to address enough RAM to put a large stack in memory, the memory option will need to be specified under the edit menu. Up to 75% of RAM may be specified, with the upper limit for the 32-bit version being around 1.6 GB. Nevertheless, we have found 1.3 GB to be a useful maximum, and, depending upon the particular computer, instability often occurs above this limit. However, with both 64-bit Windows and 64-bit Java, ImageJ may address over 1.7 GB according to the website. A number of plugins and macros are available for ImageJ and are listed at the Web site. Some that are needed for this chapter include bij plugin (http://webscreen.ophth.uiowa.edu/bij) for registering stacks (aligns the stack on an inanimate region of interest (ROI)), and stack combiner (puts two stacks together in parallel) and stack concatenator (puts two stacks together in series). The stepper motors in scanners are binary driven and some uncertainty is inherent in the exact

positioning of the scan head; thus on animation the image stack may chatter one or more pixels reflecting the hardware limitations. The bji plugin, specifically the **register ROI** plugin, will remove this chatter by aligning the image stack on an inanimate portion of the image such as a portion of the plate between wells. Newer versions of ImageJ come bundled with many useful plugins and macros; plugins are usually initiated from the plugins menu when an image or stack is open and some selection made. Macros are usually compiled by the user in the plugin menu prior to their first use when ImageJ is opened. The documentation for ImageJ is found on the Web site, and an active user-group is available for help with specific issues pertaining to ImageJ.

III. Motility Analysis

A. Computer-aided Screening: Survey of Development after Microinjection or Other Treatment

This procedure has been described previously (Lessman, 2002, 2004) and allows detection of motility mutants during the first 72 h of zebrafish development. Here we will describe using computer-aided screening (CAS) to screen embryo development after injection. Following injection of early embryos with various substances (e.g., morpholinos, transgene constructs, antibodies, or drugs), arraying them in 96-well, round-bottom plates and scanning in transparency mode at intervals from 10 to 60 min will provide the researcher with a developmental record of their experiment. Since the yolk cell undergoes cytolysis and turns opaque at death (Fig. 3), yolky zebrafish embryos are well suited for toxicity assays.

B. Motility of Wild-type Zebrafish Embryos and Larvae

Use of transparency scanners for continual or intermittent imaging of zebrafish embryos and larvae permits time-lapse visualization of development and of the movement of the developing fish over time. Image stacks compressed by *z*-projection of minimum density (i.e., dark pixels) in ImageJ produce single images depicting a summary of movement at different embryonic stages (Fig. 4). The *z*-projections are useful when a single image is needed to portray the extent of movement over time. The stack of images is particularly useful since it may be animated and the researcher can see the motility directly, and the center of mass (density) function of ImageJ may be used to generate quantitative data on the movement of individuals over time.

C. Computer-aided Larval Motility Screen (CALMS)

The protocol for detecting motility mutants during the larval stage (i.e., after 72 h of development) has been detailed earlier (Lessman, 2004). Using computer-aided larval motility screen (CALMS), movement of individual larva is followed over time, and the center of mass (image density) is computed. The standard deviation of the *x*- and *y*-coordinates gives an indication of degree of movement within the well. These

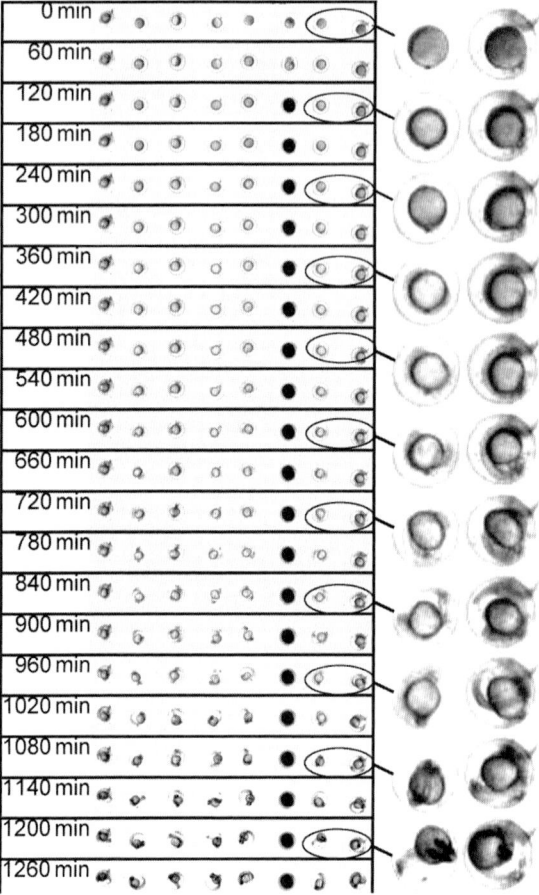

Fig. 3 Use of CAS to follow microinjected embryos through development. In this example, eight early cleavage stage embryos were injected into the yolk cell with 1 nl of distilled water and allowed to recover before distributing into individual wells of a 96-well round-bottom plate. Hourly scanning (1200 dpi, 8-bit grayscale) was started at the late blastula or early gastrula stage. This montage was created from the image stack by selecting **stack** from the **image** menu of ImageJ using the **montage** option. Cytolysis occurred to the third embryo from the right. Major developmental events such as gastrulation, neurulation, somitogenesis, pigment cell migration, and motility are readily apparent in the other embryos. Insets on the right are enlargements of the circled embryos.

are plotted, and clustering of points is indicative of similar phenotypes in mutagenesis screens or, for pharmacological studies, similar drug-treatment effects.

D. Determining Embryo Movement over Time (*x–y* Coordinates)

Once an image stack is loaded in ImageJ, the position of each embryo can be determined using the center of mass function in the particle analyzer menu (Protocol 1).

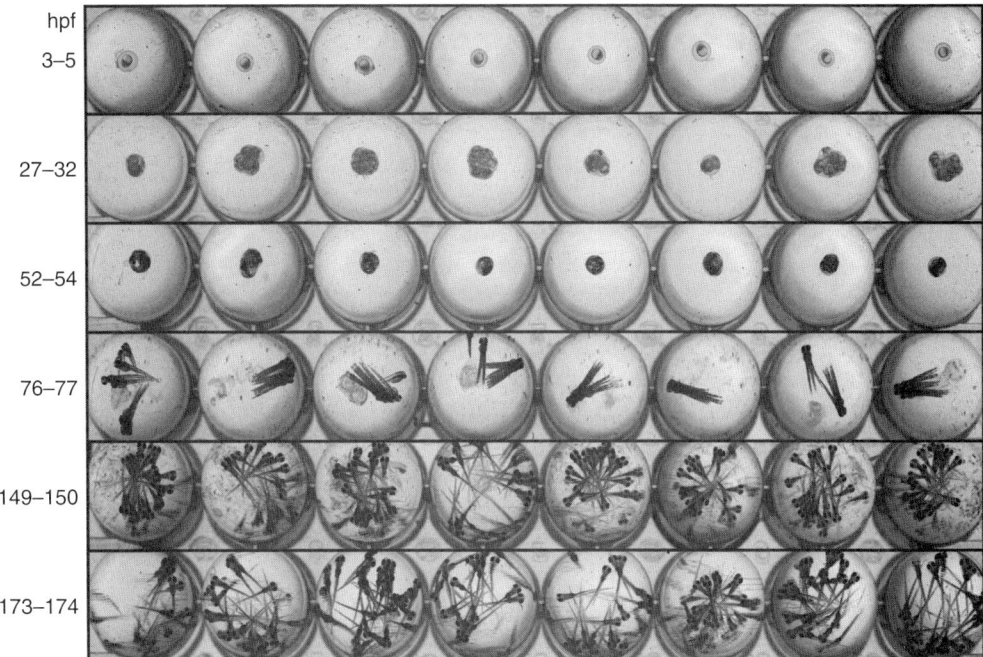

Fig. 4 *z*-Projection of scanner images of eight wild-type zebrafish embryos at different hours post fertilization (hpf). The apparent multiple exposures are due to the single embryo or larva moving between scans. Scans at 1200-dpi, 8-bit grayscale were made every 10 min during the times indicated on the left, and a minimum density *z*-projection (darkest pixels) was made to summarize the stage-specific movements (Protocol 1, step 16).

The program calculates the coordinates of the greatest optical density for each ROI and all images in the stack. In order to specify what wells are to be analyzed, a thresholded 1-bit binary image of the wells should be made and the resulting mask sent to the ROI manager. This set of ROIs can be saved and reused with other similar plates. A simple way to do this is to fill an empty plate with a dark dye such as neutral red, trypan blue, or methylene blue and scan it, carefully selecting the area of the plate from edge to edge and top to bottom. This is important since it will be used as a template for all other similar plates subsequently scanned; therefore, the procedure is reiterated here and in the outline protocol below. Open the image in ImageJ and **adjust threshold** under the **image** menu such that the dyed wells are highlighted. After pressing OK the image will become a 1-bit black and white binary image. Open the **analyze particles** option under the analyze menu and check the **add to manager** box. After pressing OK the ROI manager window should appear with the ROIs for the plate. **Show all** will highlight the ROIs in numbered circles on the original image. **Save** the ROIs as **96well ROI.zip** or other identifying filename for subsequent use. If extraneous portions of the plate appear highlighted the remedy is to go back to analyze particles and specify 0.8–1.0

circularity to specify the circular wells. The area size range of the wells can also be specified to further eliminate unwanted ROIs.

Protocol 1. Determination of relative movement in time-lapse scans of zebrafish embryos and larvae.

1. Load image stack in ImageJ by selecting **import** from **file** menu; if necessary, register the stack (bij plugin; http://webscreen.ophth.uiowa.edu/bij) by specifying an ROI around a portion of the plate or dish (inanimate object) using the rectangle tool. Use **register ROI** to align all images in the stack from **plugin** menu under **bji plugins**.

2. Create a mask of the wells to be analyzed as follows: Threshold an image of the type of plate being used (e.g., 96-well) under **image** menu **adjust** > **threshold** as described above in Section D. Adjust sliders until the wells are highlighted—it may be necessary to fill wells with a dark dye such as methylene blue or trypan blue. Once wells are selected, press **apply** and the image should revert to a binary black and white image. In **analyze** menu, press **set measurements** and check box for area. Under the **analyze** menu, select **analyze particles** and select **outlines** in show dialog box; press **OK** to get results table and drawing showing numbered outlines of particles measured. Note the area size of the wells; extraneous portions will also be measured but they will be eliminated in the next step.

3. Again go to **analyze particles** in **analyze** menu and insert a range of area values that approximate the well areas from the results table. Delete the results table and the drawing showing outlines from the preliminary run. Also under **circularity** set the range to be 0.8–1.0 (high circularity) and check box for **add to manager**, then press OK. The ROI manager window should open along with a new results table and drawing showing the measured outlines of particles, or in this case the wells.

4. In the ROI manager select **save** to put the ROIs of wells into an ROI.zip file.

5. Open **ROI manager** (**analyze** menu under **tools**).

6. In **analyze** menu select **set measurements**; select **center of mass**.

7. In **ROI manager**, open ROI.zip file from thresholded 1-bit plate image.

8. Use **show all** in ROI manager to determine whether the ROIs match the image stack.

9. Use **multi-measure** in ROI manager to obtain measurements specified in set measurements.

10. Data will be arrayed in the results window; click **select all** under **edit** menu and **copy** to spreadsheet for further manipulation.

11. In **MS Excel**, use standard deviation (**stdev**) function to determine the variation in x- and y-coordinates for each embryo.

12. Copy and paste special (check values and transpose options) to give a column of data on new sheet.

13. Create a macro to reformat the single column into x, y data pairs as follows (under **tools** in excel 2003 or **developer** menu in 2007 version). Use shortcut **ctrl-z**, be sure **relative cell reference** option is checked and cut the y value (second number in the column) and paste into the empty column next to the x value; go to the second y value in the first column and press **ctrl-z**. Stop recording the macro.

14. Go to the third y value in the original column and press **ctrl-z**, the macro should loop and move all the remaining y values next to their respective x values. The spaces will be ignored when plotted.

15. Create a scatterplot indicating the x and y columns in the plot window.

16. Create a summary z-projection of image stack; under **image** menu select **stacks** and **z-projection.** A window will appear with options for the type of projection desired; generally we project the darkest pixels of each slice in the stack. For bright objects (e.g., birefringence) the maximum density projection results in the most satisfactory image.

E. Drug Screens: Effect on Motility

Scanners may be used to analyze the effects of drugs and small molecules on the motility of larval zebrafish. Variations in responses to a pharmacological agent may be visualized and quantified. Subsequent mating and generational breeding could lead to lines of fish with different responses to specific drugs and the genes that contribute to the overall response. Researchers could use this technology to screen for genes that are involved in complex behavioral traits.

Here we test pentylene tetrazole (ptz) for its effect on day 3 larval motility (Figs. 5 and 6). This convulsant agent has been shown to elicit seizures and promote increased motility in zebrafish (Baraban *et al.*, 2005, 2007). Furthermore, zebrafish provide a

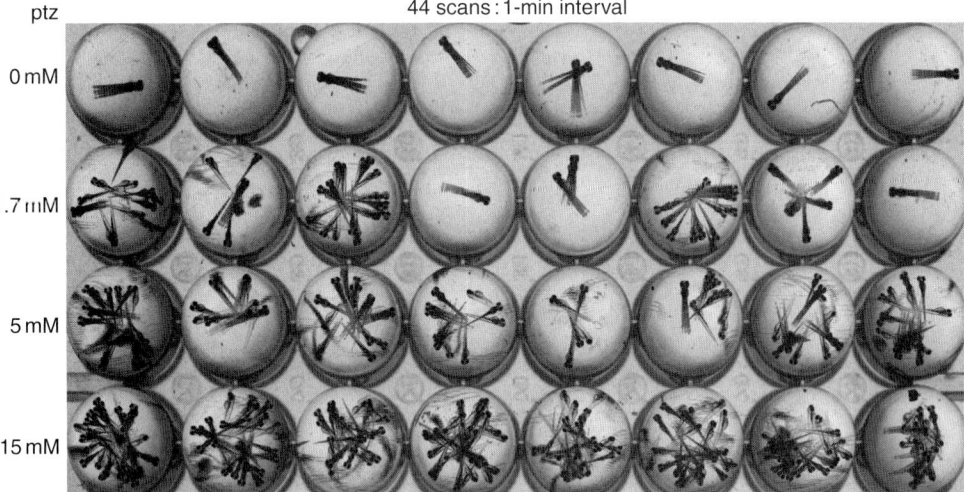

Fig. 5 CALMS z-projection of 3-day larvae treated with different doses of the convulsant agent pentylene tetrazole (ptz). Scans acquired every minute for 44 min at 1200-dpi, 8-bit grayscale. A single larva is placed in each well. The resulting image stack may be animated (Supplement Movie 1; http://www.elsevierdirect.com/companions/9780123848925) or in this case a z-axis projection using the minimum density function (i.e., darkest pixels) was made (Protocol 1, step 16).

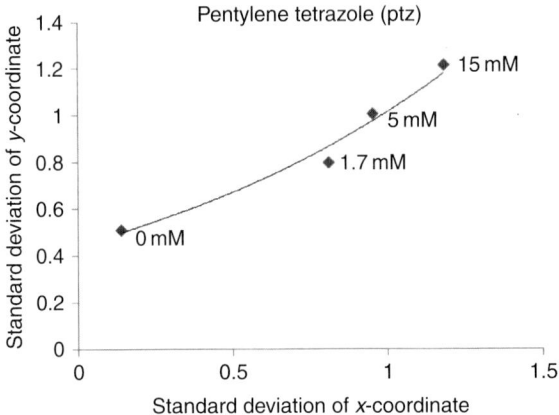

Fig. 6 Mean of standard deviation of *x*- and *y*-coordinates of center of mass (density) with different doses of pentylene tetrazole. *N* equals eight larvae per dose. Scanned every minute for 44 min at 1200-dpi, 8-bit grayscale (Protocol 1).

tractable vertebrate system to screen for antiepileptic drugs and genes involved in seizure disorders (Baraban *et al.*, 2005, 2007; Berghmans *et al.*, 2007).

The dose–response curve (Fig. 6) indicates that ptz induces a dose-dependent increase in motility (hypermotility) in day 3 larvae (Protocol 1).

F. Vision Mutants: Mutant Phenotype versus Wild Type

Motility defects in mutant zebrafish larvae may be found in screens using CALMS (Lessman, 2004). Here we apply CALMS as an additional screen for vision-defective mutants, since we hypothesize that visually impaired or blind larvae may have altered motility (Protocol 1). To test this hypothesis, the vision mutant *laj*[s304] (Muto *et al.*, 2005) was screened at day 7 for the optokinetic response (OKR) (Brockerhoff *et al.*, 1995) and then subsequently assayed using CALMS (Fig. 7). The difference between the motilities of mutant and wild-type embryos was statistically significant, but there was some overlap between the mutant and wild-type sibs. This result suggests that additional cues, such as lateral line input, compensate or provide primary directives for movement within the confines of a well. This also agrees with the normal (i.e., 1.0) score this mutant obtained in a spontaneous swimming assay (Muto *et al.*, 2005).

G. Colorimetric Potential of Scanners: pH Indicators

In theory, color scanners should be able to detect color changes while monitoring movement or development simultaneously. To test this, phenol red pH indicator dye was used in the bathing media of the larvae in 96-well plates. Preliminary trials with concentrations between 0.001% and 0.25% indicated that a level of 0.01% phenol red

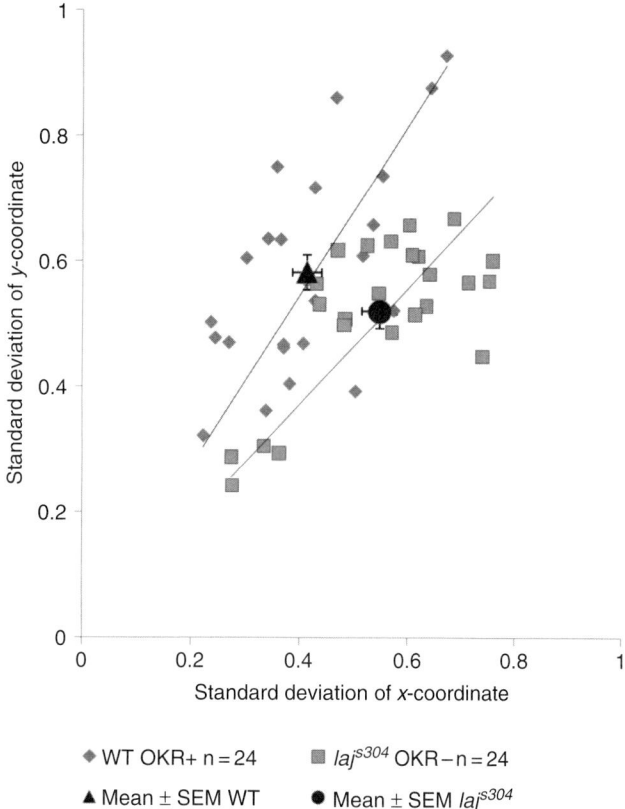

Fig. 7 Motility differences in *laj*[s304] mutant (Muto *et al.*, 2005) and wild-type (wt) sibs using CALMS. A single spawn from a *laj*[s304] heterozygote cross was screened using the optokinetic response (OKR) assay (Brockerhoff *et al.*, 1995) and the mutants (24) and wt sibs (24) were arrayed into a 96-well round-bottom plate and scanned every minute for 33 min at 1200 dpi and 8-bit grayscale. The *x–y* coordinates were computed from the stack by ImageJ and the standard deviations were computed in MS Excel and the result plotted. The data are shown as standard deviation of *x-* and *y*-coordinates with the mean, SEM, and trendlines shown (Protocol 1).

provided sufficient detectable color without affecting the viability of embryos. Since 24-bit images are not compatible for thresholding and image density measurements, the stack was split into separate RGB channels (in ImageJ: **image** menu, select **color** then **split channels**). The green channel showed the greatest intensity change as expected for a red-to-yellow color change with acid production by larvae. The green channel stack was used to determine *x–y* coordinate changes (Protocol 1) and green intensity change for each corresponding well (Fig. 8). Plumes of yellow, denoting acid associated with larvae especially near the gills, could be seen upon stack animation (Fig. 9; Supplement Movie 2).

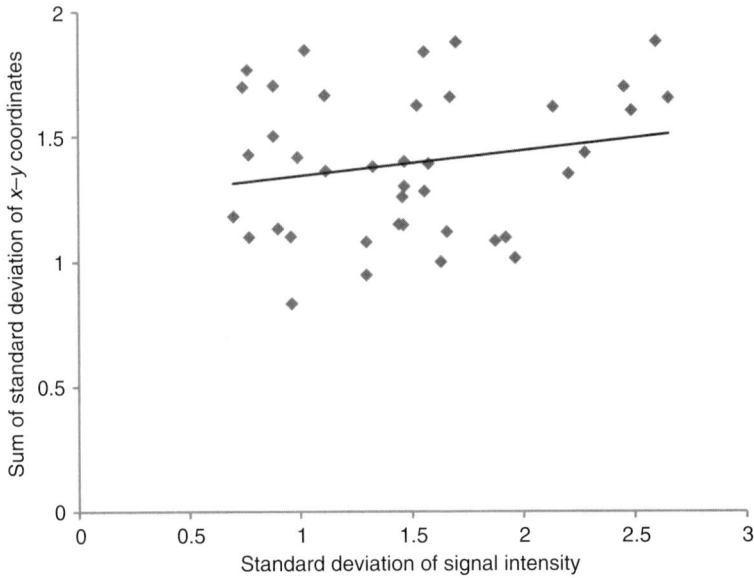

Fig. 8 General relationship between degree of larval movement (*y*-axis) (Protocol 1) and change in pH as a function of green intensity (*x*-axis). Some correlation between the extent of movement and pH indicator color change is apparent, but additional factors, such as basal metabolism, likely account for much of the variation.

IV. Oocyte and Egg Assays

A. Computer-aided Meiotic Maturation Assay: Quantification of Oocyte Maturation

This technique was described previously (Lessman *et al.*, 2007). The basis for this assay is the clearing of oocytes as they undergo meiotic maturation (Lessman *et al.*, 2007; Selman *et al.*, 1993, 1994). Fully grown oocytes (~0.5–0.7 mm) are harvested from the ovary of a killed gravid female and placed into multiwell plates in groups (i.e., 24-well plate) or singly (i.e., 96-well round-bottom plate) with an appropriate media such as Cortland's fish saline containing (g/l) NaCl 7.25, $CaCl_2$ 0.23, $MgSO_4$ 0.23, KCl 0.38, HEPES acid 1.9, HEPES salt 3.1, penicillin 30 mg/l, and streptomycin 50 mg/l, pH 7.8 (Wolf and Quimby, 1969) or Leibovitz L-15 cell culture medium. The normal inducer of maturation is 17α-20β-dihydroxyprogesterone (DHP) which elicits oocyte clearing in a dose-dependent manner (Fig. 10). Previously, oocyte maturation was scored as a quantal endpoint assay—that is, cleared oocytes were counted along with those that remained opaque at the end of a predetermined incubation period (Selman *et al.*, 1993, 1994). Computer-aided meiotic maturation assay (CAMMA) follows each oocyte over time. Thus, the investigator can determine the clearing of individual oocytes and quantify the temporal changes (Fig. 11); this provides more useful information about treatment effects. CAMMA for multiple

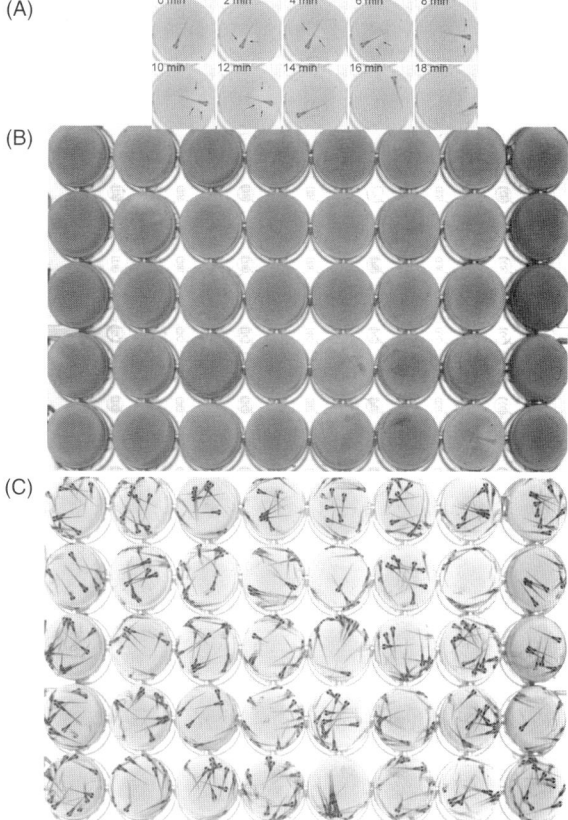

Fig. 9 Use of colorimetric capability of scanners to relate movement and pH change. Wild-type AB 7-day larvae were loaded individually into the wells of a 96-well round-bottom plate containing 0.01% phenol red, 0.015 M NaCl, 0.1 mM KCl, 0.1 mM $CaCl_2$, 0.1 mM $MgSO_4$, and 1 mM Tris (pH 7.2), and the plate was scanned at 2-min intervals (1200 dpi, 24-bit color). (A) Eight-bit green channel showing a single well with larva surrounded by plumes of lighter (yellow in color image) area (arrows) (Supplement Movie 2; http://www.elsevierdirect.com/companions/9780123848925). (B) Eight-bit z-projection of green channel maximum intensity (brightest pixels) showing areas of change, especially around larva. (C) Eight-bit z-projection of minimum intensity showing larval position changes over 24 min of observation period.

oocytes per well is given in Protocol 2. CAMMA for single oocytes in 96-well round-bottom plates is similar to the protocol for CAS or CALMS except that average density is the main measurement desired rather than center of mass density (Fig. 12).

Protocol 2. Semi-automated CAMMA for 24-well plate or other cultureware containing more than one cell per well.

1. **Import** images in **image** menu into stack and animate (backslash or **image** menu **stacks** then **animation options**) to visualize results.
2. Register stack if necessary (register ROI plugin).

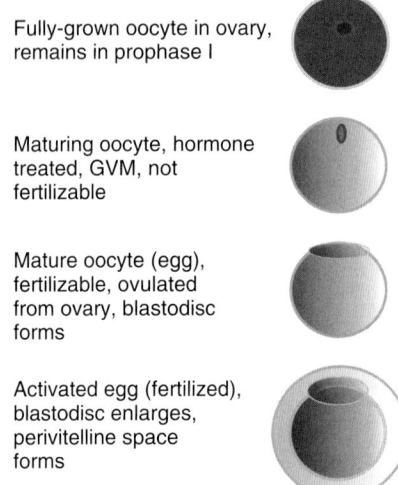

Fully-grown oocyte in ovary, remains in prophase I

Maturing oocyte, hormone treated, GVM, not fertilizable

Mature oocyte (egg), fertilizable, ovulated from ovary, blastodisc forms

Activated egg (fertilized), blastodisc enlarges, perivitelline space forms

Fig. 10 Diagram of changes during zebrafish oocyte maturation and egg activation. During oocyte maturation, the prophase I oocyte (dark oocyte nucleus or germinal vesicle, GV) is converted to a metaphase II egg (GV breaks down after migrating to the animal pole) and the ooplasm clears. During egg activation the blastodisc at the animal pole enlarges and the elevation of the chorion produces the perivitelline space.

3. Use oval selection tool and produce round selection by pressing **shift and left mouse button** to encompass all oocytes in a well.
4. Under **image** menu, select **duplicate**, check box for entire stack.
5. Repeat for each well providing a filename for the well (e.g., well A1).
6. To make an ROI set for each well, activate the stack and in **image** menu **duplicate** the first image (do not check the box for entire stack).

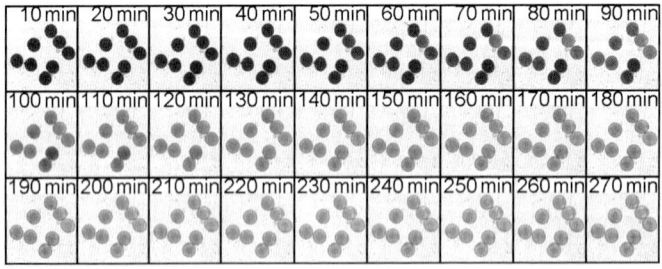

Fig. 11 Maturation of fully grown, prophase I zebrafish ovarian oocytes treated with 10 ng/ml 17α-20β-dihydroxyprogesterone. Oocytes from a single well of a 24-well plate (28°C) were scanned every 10 min at 1200 dpi and 8-bit grayscale in transparency mode (Supplement Movie 3; http://www.elsevierdirect.com/companions/9780123848925). Oocytes clear and develop a blastodisc at the animal pole. This figure shows a group of eight oocytes from the well that were subjected to signal intensity analysis (Protocol 2); the results are plotted in Fig. 12.

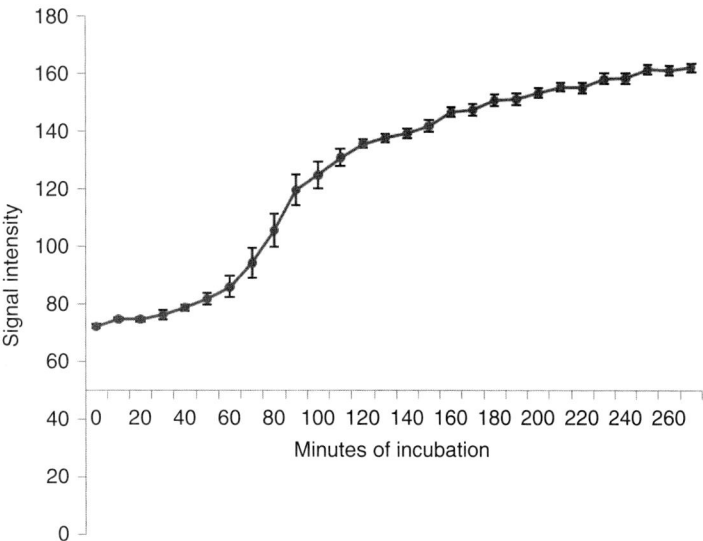

Fig. 12 CAMMA analysis of the maturation (clearing) of the eight oocytes depicted in Fig. 11. Signal intensity data are given as mean±SEM in 8-bit grayscale (i.e., 0 equals black, 255 equals white).

7. Under **edit** menu, select **clear outside.**
8. Use **adjust threshold** under **image** menu; move sliders to highlight oocytes and **apply.**
9. Under **process** menu, select **binary** and **fill holes** to uniformly highlight oocytes.
10. Oocytes that are touching need to be separated using the **watershed** function in the **binary** selection of the **process** menu.
11. Produce ROI set by using **particle analyzer** under **analyze** menu and selecting **direct to manager**; the ROI set may be saved.
12. **Set measurements** of average density, area, and options such as feret's diameter under **analyze** menu.
13. In **ROI manager**, select **more** then **multiple measure**. Selecting **show all** will indicate the numbered areas to be measured.
14. Results will appear in a new window. The data are then transferred to a spreadsheet for analysis.

The temporal change in the signal intensity of individual oocytes may also be followed by CAMMA as shown in Fig. 13.

B. Osmoregulation during Oocyte Maturation

As oocytes mature and are ovulated, they leave the ionic environment of the ovary and are deposited in freshwater during oviposition. Immature oocytes, taken directly from the ovary by dissection, cytolyse when placed in freshwater (Lessman,

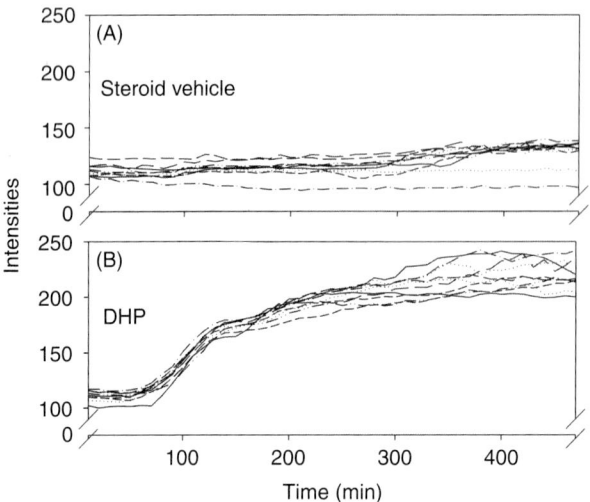

Fig. 13 Signal intensity plots of individual zebrafish oocytes ($n = 10$) versus time of incubation with either vehicle control (A) or 10 ng/ml 17α-20β-dihydroxyprogesterone (DHP) (B). Each line represents a single oocyte analyzed by CAMMA.

unpublished). The osmoregulation assay determines the time course of acquisition by maturing oocytes of the capacity to osmoregulate in freshwater. In this example, fully grown oocytes were incubated with 10 ng/ml DHP in 100% Cortland's and, at 1-h intervals, eight oocytes were placed individually into the wells of successive rows (96-well round-bottom plate) containing 200 μl of 5% Cortland's saline. Scans were made every 10 min at 1200 dpi, 8-bit. A montage of the resulting images is shown in Fig. 14 and a count of cytolysed eggs is given in Fig. 15.

C. Egg Activation Assay (EggsAct)

Zebrafish eggs obtained from anesthetized gravid females or from progestogen-matured oocytes may be assayed for their ability to undergo a cortical response and formation of the perivitelline space (Fig. 16) by following Protocol 3 (Fig. 17). This can be combined with *in vitro* fertilization if sperm suspensions are added to the freshwater used to initiate activation.

Protocol 3. Egg activation assay (EggsAct).

1. Place individual eggs into wells containing 50 μl 100% Cortland's in a 96-well round-bottom plate.
2. Position the plate on the scanner bed.
3. Replace media in wells with 200 μl 5% Cortland's using a multichannel pipette.
4. Begin a scan series at 1200 dpi, 8-bit, every 1–5 min for at least 30 min total.
5. Stack the resulting scans in ImageJ.

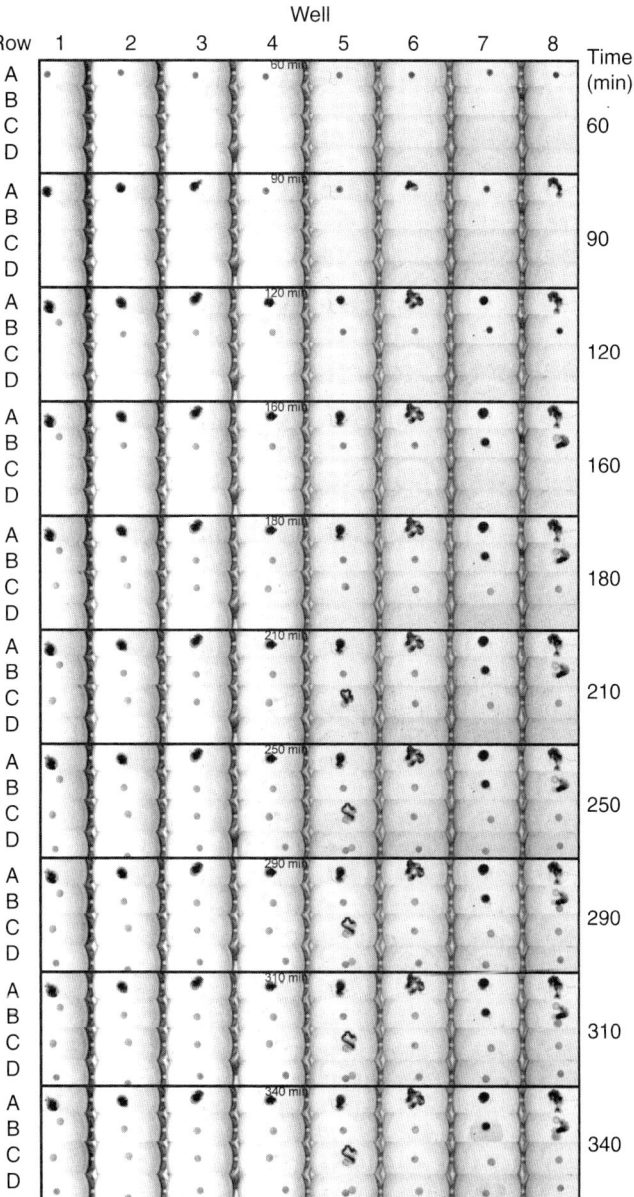

Fig. 14 Development of osmoregulation in maturing zebrafish oocytes. Fully grown ovarian oocytes were treated with 10 ng/ml 17α-20β-dihydroxyprogesterone in 100% Cortland's solution at time 0, and at hourly intervals, eight oocytes were transferred individually into the wells of successive rows of a 96-well plate. Each well contained 5% Cortland's solution (Wolf and Quimby, 1969). Oocytes that have not developed osmoregulatory capacity undergo cytolysis and turn opaque (i.e., all at 60 min); increasing numbers of oocytes remain viable after 120-min incubation with DHP, coincident with oocyte maturation (Supplement Movie 4; http://www.elsevierdirect.com/companions/9780123848925).

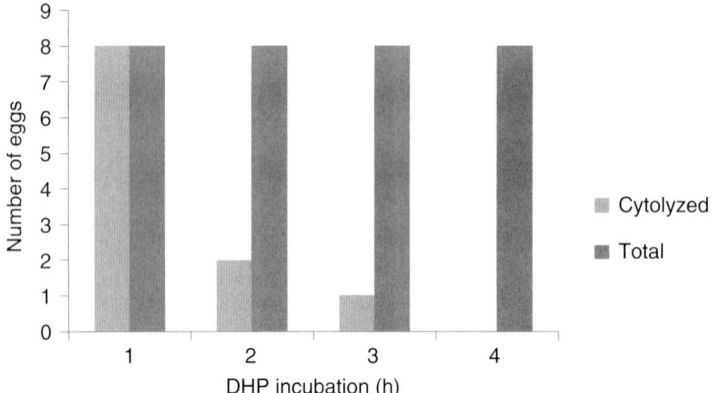

Fig. 15 Cytolysis of zebrafish oocytes subjected to osmoregulatory shock as a function of maturation. The number of cytolysed eggs is plotted as a function of time of incubation with 17α-20β-dihydroxyprogesterone (DHP) prior to osmotic shock. Data taken from last image of Fig. 14 (340 min).

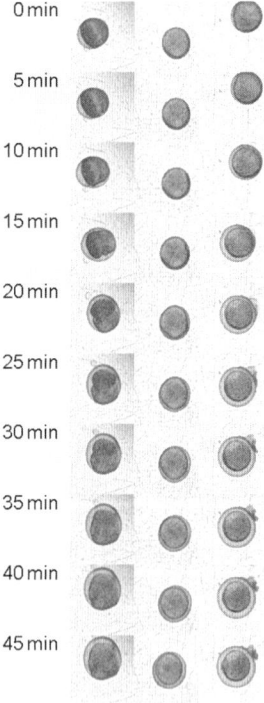

Fig.16 Montage of 17α-20β-dihydroxyprogesterone-matured zebrafish oocytes (eggs) undergoing activation in low ionic strength media (5% Cortland's balanced fish saline). Note the formation of the perivitelline space and elevation of the chorion (Supplement Movie 5; http://www.elsevierdirect.com/companions/9780123848925).

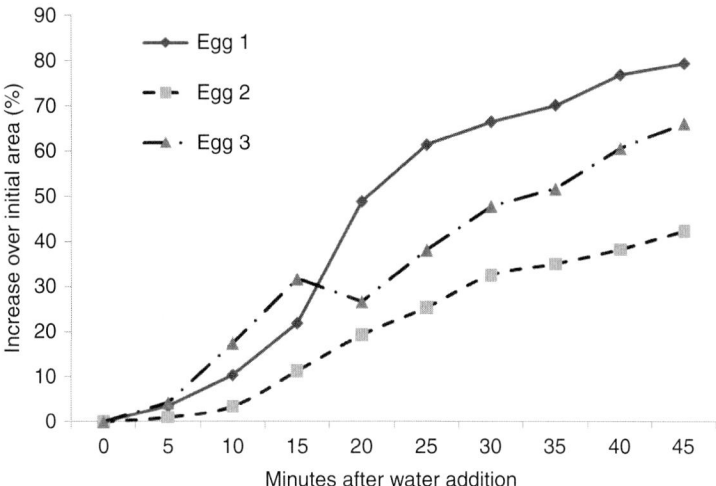

Fig. 17   Results of the egg activation assay (EggsAct) for the three eggs imaged in Fig. 16 (Protocol 3).

6. For each egg to be analyzed, select and duplicate the well containing the egg.
7. Make a **montage** under the **stacks** selection of **image** menu using 1 as **scale factor**.
8. **Adjust** the **threshold** and apply to make 1-bit binary image; if necessary use **fill holes** option from **process** menu.
9. **Set measurements** in **analyze** menu to include area; use the **analyze particles** selection under **analyze** menu to return areas of eggs in montage.
10. Copy the results into MS Excel and calculate the percent area change by dividing the area at $t = 0$ into the area obtained at each time point multiplied by 100.
11. Repeat steps 6–10 for each egg and plot results (Fig. 17).

V. Using Scanners to Count and Measure

A. Oocyte, Egg, and Embryo Counting by Scanners

Scanned images of dishes containing oocytes, eggs, or embryos can be easily counted using Protocol 4 (Fig. 18). It is important to remove, as much as possible, extraneous material such as scales, fish waste, uneaten food, and debris that may produce errors in embryo counts. In the case of oocytes, complete dissociation of the ovarian fragments produces the most accurate counts.

Protocol 4. Semi-automated zebrafish embryo counting with ImageJ.

1. Open ImageJ program. In **file** menu, **open** the image to be analyzed. Use the **magnify tool** to increase or decrease magnification so that you can see the embryos.

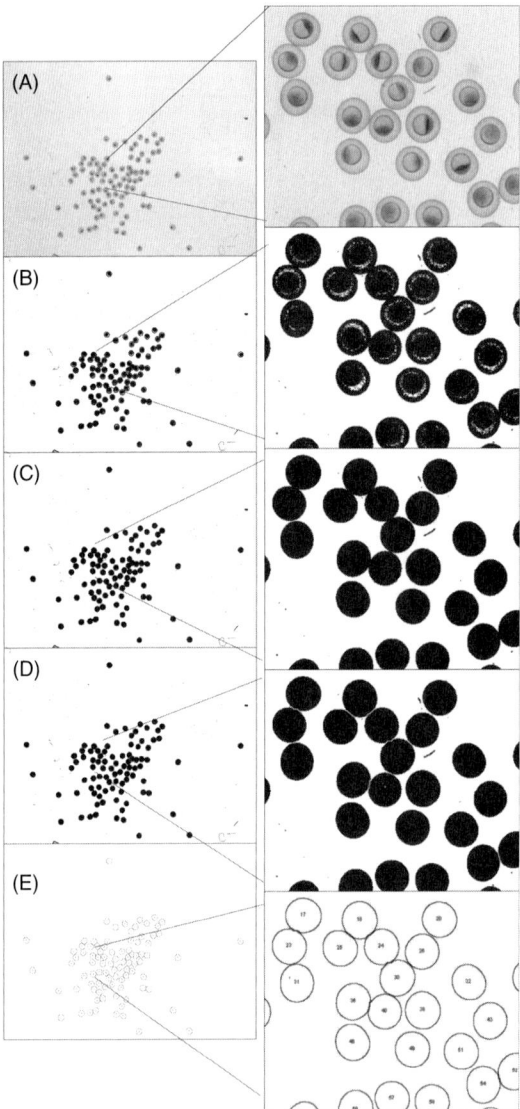

Fig. 18 Counting zebrafish embryos. Scanned image (A) of blastula-stage embryos was thresholded to produce a 1-bit binary image (B). Under process menu, selection of the fill holes function of the binary option transforms embryos to filled circles (C). Next the watershed function of the binary option is used to separate touching embryos (D). Finally the analyze particles' function under the analyze menu is used to count embryos (E). A total of 90 embryos are present and accurately counted (Protocol 4).

2. **Adjust threshold** (under **image** menu select **adjust**), set sliders so that only embryos are highlighted, then select **apply**. It may be impossible to exclude all debris, but try to minimize selecting debris while maximizing embryo selection.

3. Crop image using rectangle tool to remove extra areas if possible (under **image** menu: **crop**).

4. Under **process** menu select **binary** then **fill holes.**

5. Under **process** menu select **binary** then **watershed** to separate those embryos that are still touching.

6. Run a preliminary analysis: under **analyze** menu select **analyze particles** (under **show**, select **outlines** and check **show results** box); known embryos should give areas of about 2000 pixels (at 1200-dpi scan) on the **results** window and **circularity** should be set for 0.6–1.0. This will give particles both larger and smaller than embryos, but look at the numbered outlines and pick some that are known embryos and check their area on the results list. Delete the preliminary results and the window showing particle outlines.

7. Under **analyze** menu select **analyze particles** and set minimum and maximum size to include all embryos (try 1600–2600 to start). Compare the original image with the drawing to see if all embryos have in fact been counted.

8. In **results** window, **select all** under **edit** menu and paste to excel spreadsheet.

B. Scanner Imaging of Juveniles and Adults

1. Morphometrics and Archival Images

Protocol 5 provides a rapid method to obtain archival images of anesthetized larvae, juveniles, and adults (Fig. 19). Different images collected at a single resolution (i.e., 1200 dpi) may be compared directly, and morphometrics may be obtained from the images as outlined in this protocol.

Protocol 5. Imaging adult zebrafish using the reflected mode at 1200 dpi and using ImageJ to measure standard length (snout to caudal peduncle).

1. Obtain an image of anesthetized adults (0.04% tricaine methane sulfonate).
2. **Open** the image in ImageJ and under **analyze** menu, select **calibrate**.
3. In window type in 1200 for pixel number and 25.4 in actual distance.
4. Use mm as units.
5. Using the line tool, draw a line from the snout to caudal peduncle and under the **analyze** menu, select **measurement**.
6. A results window will appear with the measurement in mm.
7. Repeat for other fish in image or make other morphometric determinations.

2. Tumors or Abnormal Morphologies

Scanners may also provide a convenient imaging tool for routine measurements of tumors or other abnormalities (Fig. 20). An example used here is the nerve sheath and abdominal tumor prone tp53[-/-] line (Berghmans *et al.*, 2005). Anesthetized adults may be

Fig. 19 Morphometric measurements and archival imaging of zebrafish. Reflected mode scanning of adult zebrafish (1200 dpi, 24-bit color). Two spawning pairs of adults were anesthetized in 0.04% tricaine methane sulfonate (females on top). The lines above or below each fish represent the standard length from snout to caudal peduncle determined with ImageJ from the scanned image (Protocol 5).

scanned in reflective mode at weekly intervals, for example, to follow tumor development and to allow measurement of tumor extent or size change. A scanner in the fish room could be devoted to this task. Efforts in the zebrafish research community are under way to develop high-throughput zebrafish bioassays for human cancers (Snaar-Jagalska and ZF-Cancer Consortium, 2009) as well as tumor assays after cancer cell transplantation (Taylor and Zon, 2009). Scanners may prove useful in these future endeavors.

VI. Other Potential Applications of Scanners

Theoretically, scanners should be able to detect fluorescent molecules in zebrafish. Light-emitting diodes (LEDs) of the appropriate wavelength could be used along with

Fig. 20   Documentation of disease and abnormal morphology in adult zebrafish. Adult tp53[-/-] zebrafish with spontaneous eye tumor (top) and abdominal tumor (bottom) were scanned at 1200-dpi, 24-bit color in reflected mode. Some of the affected areas are circled.

an appropriate barrier filter placed between the plate bearing the fish and the CCD to allow detection. Diode lasers may also be suitable for illuminating samples via fiber optics or prisms on the scanner bed. Developing a suitable scanner system for detection of fluorescent molecules in zebrafish embryos, larvae, and adults in large arrays is a worthwhile future endeavor. The ability to quickly identify embryos expressing fluorescent molecules from large arrays would be useful in high-throughput assays, for example, in the nile red obesity assay (Jones *et al.*, 2008).

Myofibrils in zebrafish larval muscle illuminated by plane-polarized light produce anisotropy or birefringence (Granato *et al.*, 1996). Transparency scanners can easily produce birefringence by placing multiwell plates between crossed polarizing film on the scanner bed. Two sheets of the polarizing film, when turned 90° from each other, prevent light from passing through unless the light is refracted by something between the polarizers.

Sheets of colored filter (e.g., Wratten filters) may also be used to illuminate embryos or larvae with different wavelengths by placing the sheet between the light source and the animals. This could be useful in behavioral studies. With the appropriate combination of filters above and below the fish, various fluorescent or colored molecules may be visualized.

Scanners may also be used for imaging developing embryos of other fish species, especially if the embryos have some degree of transparency. We have successfully imaged channel catfish embryos (*Ictalurus punctatus*) with transparency scanners. These embryos are relatively large (~3 mm) and semi-transparent, and require about 7 days at 25°C to develop. In order to scan during the entire course of development, peristaltic pumps were used to produce fresh laminar flow of media over the embryos. Other important food species such as rainbow trout or salmon could also be imaged;

however, these cold water species would require scanning in a cold room or low-temperature incubator. We have tested scanners in environments as low as 2°C and found them to be functional and reliable over the course of days.

VII. Summary: Inexpensive Adjunct to Microscopy

The flatbed transparency scanner is inexpensive, simple to use, and provides a stable imaging platform for large arrays of zebrafish oocytes, embryos, larvae, and adults. Coupled with Macroscheduler and ImageJ, the scanner can produce quantitative data for numerous research endeavors. We have outlined a few of the possible uses of this commonly available device and look forward to new uses that can arise from the vibrant, creative zebrafish research community.

Acknowledgments

This chapter was written while CAL was on sabbatical from the University of Memphis and hosted by St. Jude Children's Research Hospital and the University of Tennessee-Chattanooga. We thank the members of the Taylor and Carver laboratories for their suggestions and assistance. We thank the personnel of the Stoneville, MS USDA Catfish Facility, for providing channel catfish embryos. This work was supported in part by a Hartwell grant to MRT and an NIH grant to EAC.

References

Abramoff, M. D., Magelhaes, P. J., and Ram, S. J. (2004). Image processing with ImageJ. *Biophotonics Int.* **11**, 36–42.

Baraban, S. C., Dinday, M. T., Castro, P. A., Chege, S., Guyenet, S., and Taylor, M. R. (2007). A large-scale mutagenesis screen to identify seizure-resistant zebrafish. *Epilepsia* **48**, 1151–1157.

Baraban, S. C., Taylor, M. R., Castro, P. A., and Baier, H. (2005). Pentylenetetrazole induced changes in zebrafish behavior, neural activity and c-fos expression. *Neuroscience* **131**, 759–768.

Berghmans, S., Hunt, J., Roach, A., and Goldsmith, P. (2007). Zebrafish offer the potential for a primary screen to identify a wide variety of potential anticonvulsants. *Epilepsy Res.* **75**, 18–28.

Berghmans, S., Murphey, R. D., Wienholds, E., Neuberg, D., Kutok, J. L., Fletcher, C. D. M., Morris, J. P., Liu, T. X., Schulte-Merker, S., Kanki, J. P., Plasterk, R., Zon, L. I., *et al.* (2005). tp53 mutant zebrafish develop malignant peripheral nerve sheath tumors. *Proc. Natl. Acad. Sci. U. S. A.* **102**, 407–412.

Brockerhoff, S. E., Hurley, J. B., Janssen-Bienhold, U., Neuhauss, S. C. F., Driever, W., and Dowling, J. E. (1995). A behavioral screen for isolating zebrafish mutants with visual system defects. *Proc. Natl. Acad. Sci. U. S. A.* **92**, 10545–10549.

Granato, M., van Eeden, F. J., Schach, U., Trowe, T., Brand, M., Furutani-Seiki, M., Haffter, P., Hammerschmidt, M., Heisenberg, C. P., Jiang, Y. P., Kane, D. A., Kelsh, R. N., *et al.* (1996). Genes controlling and mediating locomotion behavior of the zebrafish embryo and larva. *Development* **123**, 399–413.

Jones, K. S., Alimov, A. P., Rilo, H. L., Jandacek, R. J., Woollett, L. A., and Penberthy, W. T. (2008). A high throughput live transparent animal bioassay to identify non-toxic small molecules or genes that regulate vertebrate fat metabolism for obesity drug development. *Nutr. Metab.* **5**, 23–33.

Lessman, C. A. (2002). Use of computer-aided-screening (CAS) for detection of motility mutants in zebrafish embryos. *Real-Time Imaging* **8**, 189–201.

Lessman, C. A. (2004). Computer-aided screening for zebrafish embryonic motility mutants. *Methods Cell Biol.* **76**, 285–313.

Lessman, C. A., Nathani, R., Uddin, R., Walker, J., and Liu, J. (2007). Computer-aided meiotic maturation assay (CAMMA) of zebrafish (*Danio rerio*) oocytes *in vitro*. *Mol. Reprod. Dev.* . **74**, 97–107.

Muto, A., Orger, M. B., Wehman, A. M., Smear, M. C., Kay, J. N., Page-McCaw, P. S., Gahtan, E., Xiao, T., Nevin, L. M., Gosse, N. J., Staub, W., Finger-Baier, K., *et al.* (2005). Forward genetic analysis of visual behavior in zebrafish. *PLoS Genet.* **1**, 575–588.

Rasband, W. S. (1997–2009). ImageJ. U. S. National Institutes of Health, Bethesda, Maryland, USA, http://rsb.info.nih.gov/ij/.

Selman, K., Petrino, T. R., and Wallace, R. A. (1994). Experimental conditions for oocyte maturation in the zebrafish, *Brachydanio rerio*. *J. Exp. Zool.* **269**, 538–550.

Selman, K., Wallace, R. A., Sarka, A., and Qi, X. (1993). Stages of oocyte development in the zebrafish, *Brachydanio rerio*. *J. Morphol.* **218**, 203–224.

Snaar-Jagalska, E. B. and ZF-CANCER Consortium (2009). ZF-CANCER: developing high-throughput bioassays for human cancers in zebrafish. *Zebrafish* **6**, 441–443.

Taylor, A. M., and Zon, L. I. (2009). Zebrafish tumor assays: the state of transplantation. *Zebrafish* **6**, 339–346.

Wolf, K., and Quimby, M. C. (1969). Fish cell and tissue culture. *In* "Fish Physiology" (Hoar, W. S., and Randall, D. J., eds.) **Vol. 3**, pp. 293–312, Academic Press, New York.

SUBJECT INDEX

Note: The letters 'f' and 't' following the locators refer to figures and tables respectively.

A

Achromatopsia, 155, 206
Action currents
    analysis of, 226–228
        hair-cell endocytosis, 227
        interspike interval (ISI), 228
        neuromast hair cells (sine waves), 226–227
        spiking, 227
    electrode and recording details, 226
        from individual posterior lateral-line neurons, 226f
        posterior lateral line ganglion (PLLg), 226
        voltage-clamp mode recordings, 226
Adult kidney isolation, 249
Albinism, 60
Alkaline phosphatase (AP) staining for 3 dpf
    embryos, 28, 38–39
    materials, 38–39
        fixation buffer, 38
        rinse buffer, 38
        staining buffer, 39
        staining solution, 39
    protocol, 39
Anaplastic lymphoma kinase *(ALK)* gene, 145
Angiopoietin2, 244
Anterior neural plate (ANP), 76f, 77, 167
Antibodies, 166–172
    antibody staining, 166, 171
    anti-GABA staining, 171
    antigen retrieval, 171
    confocal microscopy, 170
    conventional epifluorescence microscopy, 171
    GABA staining protocol, 172
    glyoxal-based fixatives, 170
    labeling with, 9
        Alcam-a (zn-5/zn-8, monoclonal antibodies), 9
        anti-acetylated tubulin (Sigma), 9
        anti-DsRed (mCherry, Clontech), 9
        anti-GFP (Invitrogen), 9
        anti-tagRFP (Invitrogen) antibodies, 9
        standard whole-mount antibody staining techniques, 9
        zn-5/8 staining, 9

Anti-gamma aminobutyric acid (GABA) staining, 170
Anti-gamma tubulin (GTU-88), 171, 250
Antigen retrieval, pretreatments, 104–105
    citrate retrieval, 104
    hydrochloric acid pretreatment, 104
Anti-polyglutamyl tubulin (B3), 278
Artificial cerebrospinal fluid (ACSF), 107
Auerbach plexus (myenteric plexus), 130
Autonomic nervous system (ANS), 128–130

B

BAPTISM (birthdating analysis by photoconverted
    fluorescent protein tracing in vivo combined
    with subpopulation markers), 96
Behavior/genetics/pharmacology
    criterion 1—quiescent behavior regulation, 283
        endogenous circadian rhythm, 283
        genes in zebrafish orthologs, 283
        light:dark cycle, 283
    criterion 2—quiescence, 283–284
        decreased responsiveness, 283
        increased arousal threshold, 283
        sleep-like state, definition, 283
        sleep/wake measures from wild type larvae, 285t
        zebrafish sleep/wake data, 284f
    criterion 3—sleep deprivation, 284–286
        homeostatic regulation of sleep, 284–286
        light deprivation, 286
        nighttime sleep deprivation, 284–285
        rebound sleep, 286
        shock deprivation at night, 286
    hypocretin/orexin (Hcrt) system, 286
    larval/adult zebrafish, sleep-like state in, 287
    neurotransmitter systems, 287
    non-mammalian model systems, 283
    pharmacological screen, pathways, 287
    sleep-like rest periods, 283
Berlin blue dye, 31, 36
Beta cell migration and proliferation, 268–276
    studies of, 268t
    time-lapse imaging of, 268–273
    visualization, 271f

Blastomere transplantation, 93, 100, 177–178
Blood vessels in zebrafish, imaging, 27–52
  imaging vascular gene expression, 29–30
  non-vital blood vessel imaging
    AP staining for 3 dpf embryos, 38–39
    micro-dye/-resin injection, 31–38
  vital imaging of blood vessels
    microangiography, 41–45
    in transgenic zebrafish, 45–51
Bone morphogenetic proteins (BMP) signaling, 130, 238, 264
  PHOX2B, 132
  proneural gene, Mash-1, 132
BRAD1 (BRCA1-associated domain 1 located at 2q35), 145
Breeding schemes, 187–188
  early pressure technique, 188
  in F1/F3 generation embryos, 187
  mutagenized animals (G0), 187
  screening strategies, 187–188
Bromodeoxyuridine (BrdU), 95
  injections, 179
  intraperitoneal injections of adult fish, 105–106
  labeling, 179
  tracing and marker analysis
    quiescent radial glial cells (state I progenitors), 80
    radial glial cells (state II progenitors), 80

C

Cadherins
  cadherin17, 242
  N-cadherin, 15t, 98t, 132
  VE-cadherin, 29
Caged fluorescent dyes, 87
CAMP and MAPK signaling, 133
Carbocyanine dyes
  DiI, DiO, DiA, or DiD (Invitrogen), 9
Cell ablation methods, 102
  diphtheria toxin, use of, 102
  ganciclovir, 103
  KillerRed, 103
  laser-mediated cell ablation, 102
  prodrug (metronidazole), conversion of, 103
  thymidine kinase–ganciclovir, 103
Cell-autonomy of gene function, 17–19
  early transplants at blastula stage, 17
  entire eye primordia, transplanting, 17
  transplants at later stage, 17–19
Cell culture derivation, 59–60
  blastomere cells, 59

medaka karyotype, 60
nucleolar cycle, 60
pure haploid ES cells, 60, 61f
uniparental haploidy
  Albinism, 60
  haploid metaphases, 60
  haploid syndrome, 60
Cell cycle events, 95–96
  BAPTISM, 96
  Fucci, 96
  S-phase, BrdU incorporation, 96
Central nervous system (CNS), 74–75, 80, 89, 94, 96, 100, 130, 139–140, 142, 143f, 154–155
Charge-coupled device (CCD), 297, 320
Chemokine signaling, 266
Chlorodeoxyuridine (CldU), 96
Chromaffin cells, 129
  neural crest (NC), 129
  sympathoadrenal (SA) cell, 129
Cilia, zebrafish, 250
  alternate protocols, 253
  confocal microscopy, 253
  fixation, 252
  materials, 250–251
  methods, 251–252
  solutions, 251
Cloaca formation, 245
Cloning
  animal, 63
  Gateway
    destination vectors, 185
    entry vectors, 185
    Tol2-based zebrafish destination vector, 184
  positional and candidate, 191
    of *nagie oko locus*, 191
    standard steps, 191
Computer-aided larval motility screen (CALMS), 302–303, 306f, 307, 308f, 310
Computer-aided meiotic maturation assay (CAMMA)
  quantal endpoint assay, 309
  quantification of oocyte maturation, 309–312
  semi-automated CAMMA, 310–312
Computer-aided screening (CAS)
  survey of development after microinjection or other treatment, 302
  use of, 303f
Cone–rod homeobox *(crx)* gene, 208, 213–214
Confocal microscopy, 5f, 6, 6f–7f, 13, 19f, 95, 170–171, 211, 253, 296
Cyclin-dependent kinases (Cdks), 267
Cystic kidney disease, 236

D

DAB (3,3'-diaminobenzidine) staining
  antibody staining, 166
  photoconversion, 166
  retrograde labeling, 166
Danio rerio, 134
  See also Flatbed transparency scanners in
    zebrafish research, use of
dHAND, 131*f*, 133, 136*t*, 137*f*
Diethylaminobenzaldehyde (DEAB), 240
Digital scanned laser light sheet fluorescence
  microscopy (DSLM), 95
Diluting segment, 237
Dopamine-β-hydroxylase (dbh), 131*f*, 133, 139
Dorsal bud (DB), 263
  endocrine differentiation, 265
  formation, 265–267
Drosophila to *Danio,* methodological
    considerations, 288
  direct analysis of movements by video, 288
  fine-scale measures of sleep structure
    sleep bout length, 288
    sleep bout number, 288
    sleep latency, 288
  infrared beam break method, 288
  video-based analysis
    Noldus Information Technology, 288
    Viewpoint Life Sciences, 288
Drug screens, effect on motility, 306–307
  CALMS z-projection of 3-day larvae, 306*f*
  dose–response curve, 307
  mean of standard deviation of x- and y-coordinates,
    307*f*
  pentylene tetrazole (ptz), testing of, 306
Duct markers, 238
Dye injection method
  of embryos and early larvae, 36–38
  of juvenile and adult zebrafish, 38
  materials, 36
  and mounting of developing zebrafish, 37*f*

E

Egg activation assay (EggsAct), 313–316
  17α-20β-dihydroxyprogesterone-matured
    zebrafish oocytes, 315*f*
  results of, 316*f*
eHAND, 133
Electroencephalographic (EEG) signatures, 282
Electron microscopy, 254–255
  materials, 254
  methods, 254–255

Electroretinography (ERG), 180
  d-wave, 180
  large positive b-wave, 180
  small negative a-wave, 180
Embryo dissociation, 245–246
Embryogenesis, 62, 95, 134–135, 154, 161–162,
    166, 176, 187, 234
Embryo movement over time (x–y coordinates),
    303–306
  particle analyzer menu, 303–304
  in time-lapse scans, determination of, 305–306
Embryonic stem (ES) cells, 55–68
Enhanced green fluorescent protein (EGFP), 5*f*, 7,
    18*f*–19*f*, 20, 39, 40–41, 40*t*, 45–46, 50, 90*t*–
    91*t*, 101*t*, 102*t*, 175, 214–215, 270*t*, 271*f*, 274*f*
Enhancer trap (ET) screens, 21, 92*t*, 185
Enteric nervous systems (ENS), 128–129, 137, 143
Epson photo V500, 297
  Epson Scan scanner software, 297
Ethynyldeoxyuridine (EdU), 96, 262, 268*t*, 273,
    274*f*, 275

F

Fate-mapping studies
  late gastrulation, 155
  neurulation, 155
    neural keel, 155
    optic lobes, 155
Fibroblast growth factors (FGFs), 130
  Fgf signaling, 79, 92*t*, 130
Fight-or-flight response, 129
Flatbed transparency scanners in zebrafish research,
    295–321, 299*f*
  motility analysis
    CALMS, 302–303
    CAS, 302
    drug screen effects on, 306–307
    embryo movement over time (x–y coordinates),
      303–306
    pH indicators, 306–307
    vision mutants, mutant phenotype *vs.* wild
      type, 306
    of wild-type zebrafish embryos and larvae, 302
  oocyte and egg assays
    CAMMA, 309–312
    egg activation assay (EggsAct), 313–316
    osmoregulation during oocyte maturation,
      312–313
  potential applications of scanners, 319–321
  scanner basics
    image acquisition with scanners, 299–300

Flatbed transparency scanners in zebrafish research
(cont.)
imageJ, 301–302
mouse–keyboard macro for automation of
scans, 301
resolution and bit depth, 297–298
temperature and fluidics, 298–299
using scanners to count and measure
oocyte/egg/embryo counting by scanners,
316–318
scanner imaging of juveniles and adults,
318–319
Fluorescence activated cell sorting (FACS), 179,
213, 246
Fluorescent proteins (FPs), 7, 173–176, 268–269
advantages, 176
Dronpa green fluorescence, 176
green fluorescent protein (GFP), 174
photoconvertible FPs, 175
recombination cloning approaches, 176
red fluorescent protein (RFP), 174
uses of
monitoring of fate/differentiation/cell
physiology, 175
visualization of gene activity, 174
visualization of subcellular localization of
proteins, 174–175
Forward genetics, 185–191
breeding schemes, 187–188
mutagenesis approaches, 186–187
mutant strains, 191
phenotype detection methods, 188–191
positional and candidate cloning, 191
Fucci (fluorescent ubiquitination-based cell cycle
indicator), 99

G

Gal4-specific Upstream Activating Sequence (UAS)
promoter (UAS:GFP), 90
Gateway cloning
destination vectors, 185
entry vectors, 185
Tol2-based zebrafish destination vector, 184
Gene expression, manipulating, 101–103
gain-of-function approaches, 97
loss-of-function studies, 97
mutagenesis screens, 97
spatial control of genetic manipulations, 100–101
blastomere transplantation, 100
cell-type-specific manipulations, 101, 101*t*
Cre-loxP system, 101

Gal4-UAS bipartite system, 100
genetic perturbations, 100
temporally controlled genetic manipulations,
101–103
chemical compound, administration of, 101
Cre-LoxP system, 101
drug tamoxifen, administration of, 101
heat shock promoter (hsp70), 101
steroid Mifepristone, administration of, 101
Tet system in mammals, 101
time controlled cell-type-specific manipulation,
102*t*
TILLING libraries, 97
Gene function, 65–67
germline chimera formation, 65
in haploid ES cells, 66
in theory, 66–67
genomic PCR, 67
p53, 66–67
RT-PCR, 67
Gene function in zebrafish retina
forward genetics, 185–191
breeding schemes, 187–188
mutagenesis approaches, 186–187
mutant strains, 191
phenotype detection methods, 188–191
positional and candidate cloning, 191
reverse genetic approaches, 182–185
approaches to gene overexpression, 184–185
loss-of-function analysis, 182–183
Gene overexpression, 184–185
GAL4–UAS overexpression system, 185
GAL4–VP16 fusion protein, 184
Gateway cloning, 184
recombination cloning-based strategies, 184
RNA or DNA injections, 184
terminal inverted repeats (TIRs), 184
T2KXIG, 184
Tol2 transposon-based vectors, 184
Gentamicin induced kidney tubule injury, 248
adults, 248
embryos, 248
Ghrelin, 262
Glial cells generation, 74–75
astrocytic function in brain, 80
of oligodendrocytes, 80
radial glial cells, molecular markers, 80
Glial fibrillary acid protein (GFAP), 78*t*, 82*t*, 90*t*,
89–90, 88*f*, 169*t*, 212–213
Gliogenesis, 74
Glomerular filtration, 244
assay for, 246–247

  adults, 247–248
  embryos, 247
  blood filter, 242
  capillary tuft, 243f
  formation of glomerulus, 242–244
  glomerulus–tubule boundary, 241f
Glucagon, 262, 270
Green fluorescent protein (GFP), 8t, 9, 16, 19f, 40t, 45, 63, 65f, 88f, 90t–92t, 94–95, 101–102, 101t, 167–169f, 173f, 174–176, 182, 190, 205–216, 265, 266f, 268f, 270t, 271f, 272f, 276, 286

H

Hank's buffered salt solution (HBSS), 106
Haploid embryos, production of, 59, 59f
  androgenesis, 59
  gynogenesis, 59
  albino medaka strains (i1 and i3), 59
  haploid syndrome, 59
Haploid ES cell
  characterization of, 62
  culture, 57–58
  lines, generation of stable, 61–62
  clonal growth, 61
  flow cytometry analyses, 62
  medaka gynogenetic ES cells, 61
Haploid genetic screens, 67
  chemical mutagenesis, 67
  gene targeting, 67
  gene transfer technology, 67
  mutagenesis screening, 67
Haploid syndrome, 59–60
Heat-shocks to induce misexpression, 16–17
  focal heat shock method, 17
  using optical fiber, 17
  using sharpened soldering iron, 17
  global heat shocks, 17
hsp70l promoter, 16–17
Hedgehog (Hh) signaling, 264
Helix-loop-helix gene neurod, 213
Holly, 65
Horseradish peroxidase (HRP) staining, 173f, 178
HP scanjets, 297
HP Director, 297
H³-thymidine labeling, 179

I

ImageJ, 301–302
  bij plugin, 301
  register ROI plugin, 302
  32-bit version for Windows machines, 301
  dark pixels, 302
  image stacks, 301
  stack combiner, 301
  stepper motors, 301–302
Imaging blood vessels in transgenic zebrafish, 45–51
  confocal microangiography, 45
  germline transgenic zebrafish
  flia:EGFP transgenic lines, 45–46
  murine Tie2-GFP, 45
  tissue-specific expression of fluorescent proteins, 45
  long-term mounting for time-lapse imaging
  materials, 46–47
  method, 47–49
  mounting animals in imaging chambers, 49–50
  multiphoton time-lapse imaging, 50–51
Immunohistochemistry
  fixation of adult brain, 103
  on vibratome sections, 103
Immunostaining, 62, 81, 104
Induced pluripotent stem (iPS) cells, 56
In situ hybridization
  fixation of adult brain, 103
  on gelatine–albumin sections, 104
  on whole mount adult brains, 104
Insulin, 105, 248, 262–267, 270t, 274f, 276
Interkinetic nuclear migration (INM), 75, 96
Intermediate mesoderm (IM), 238
  pronephros, derivation of, 239f
  sequential anterior, 239
Intraflagellar transport (IFT), 181, 206, 209–210
Intrapancreatic duct system, 262
In vivo imaging, 93–95
  confocal microscopy, 95
  DSLM, 95
  fate tracking, 94
  light microscopy, 95
melanin formation
  iridophores (roy orbison embryos), 93
  melanophores (nacre embryos), 93
  multicolor imaging, 94
  multiphoton microscopy, 95
  pigmentation mutants, 94
  spinning disk confocals, 95
  two-photon excitation, 94
  UV light irradiation, 94
  in vivo time-lapse imaging, 94
In vivo single cell electroporation, 10
Iododeoxyuridine (IdU), 96, 179
Islets of Langerhans, 262

J

June, Joubert, and Hermansky–Pudlak syndromes, 155
Juvenile/adult stages, neurogenesis at, 80
  BrdU tracing and marker analysis
    quiescent radial glial cells (state I progenitors),
      80
    radial glial cells (state II progenitors), 79
  generation of neurons, 80
  neural progenitors in adult brain, 88f
  proneural genes, 80
  radial glial/early differentiation markers, 80
  zones of proliferation, 80

K

Kidney
  epithelial ion transport, 245
  functions, 234
  mesonephros, 234
  metanephros, 234
  nephrons, 234
  pronephros, 234

L

Labeling methods, protocols for, 11–13
  precise labeling with intra-retinal injection of
    lipophilic dyes
    protocol, 11–12
    solutions needed, 11
  single cell in vivo electroporation
    protocol, 12–13
    solutions needed, 12
Lens transplantation, 177
Light-emitting diodes (LEDs), 319
Light-sensing photoreceptors, 155
Lipofections/electroporations of adult brain in vivo,
  96f, 106
Lipophilic dyes, labeling with, 9–10
  advantages, 10
  DiI (red) and DiO (green)
    dye-coated microneedle, 10
    vibrating-needle injection apparatus, 9
    "whole eye fills" technique, 9
  lipophilic carbocyanine dyes
    DiI, DiO, DiA, or DiD (Invitrogen), 9
Lipophilic tracers, 228–229
  carbocyanine dyes (DiI and DiO), 228
Live cells labeling methods, 81–88
  blastomere transplantation, 93
  caged fluorescent dyes, 87
  electroporation and lipofections, 93

lineage analysis, 81
mosaic labeling, 88
photolysis, 87
stable transgenic lines, 89–90
  expression of KalTA4, 90
  Gal4-specific UAS promoter (UAS:GFP), 90
  green fluorescent protein (GFP), 89
transient expression of RNA and DNA, 88–89
  targeting neural plate by injection, 89f
ubiquitous labeling, 88
in vivo imaging, See In vivo imaging
Long-term, high-throughput sleep/wake monitoring,
  291–292
  DMSO concentration, 289
  infrared LED panel, 289
  single cell on 14/10 h light/dark (LD) cycle,
    288–289
Long-term mounting for time-lapse imaging
  materials, 46–47
  method, 47–49
    mounting animals in imaging chambers, 49–50
    preparation of imaging chambers, 47–49
    mounting zebrafish embryos and larvae for,
      48f–49f
Loss-of-function analysis, 182–183
  antisense-based interference, 182
  morpholino-modified oligonucleotides
    disadvantages, 183
    Gene Tools LLC (designing morpholinos), 183
    RTPCR (splice-site morpholinos), 183
  mutagenesis approaches in zebrafish
    DNA sequences, 183
    ZFNs, 183
    TILLING, 183

M

Macroscheduler, 301
Mammalian achaete-scute homolog (Mash-1), 131f,
  132–133, 137
MAPK pathway, 145
Meccom, 240
Medaka haploid embryonic stem cells, 55–68
  applications of
    analysis of gene function, 65–67
    haploid genetic screens, 67
    semicloning, 63–65
  haploid ES cell culture, 57–58
    in Drosophila, 57
    first mouse ES cells, 58
    in frog, 57
    in human, 57

haploidy in evolution, 57
   asexual reproduction, 57
   honeybee (invertebrates), 57
   meiosis, 57
   plantlets, 57
   yeast (single-celled eukaryotic organisms), 57
  methods
   cell culture derivation, 59–60
   characterization of haploid ES cells, 62
   culture condition, 60
   generation of stable haploid ES cell lines, 61–62
   production of haploid embryos, 59
   stable growth and genetic stability, 62–63
  rationale, 58
Meissners plexus (submucosal plexus), 130
Mesonephros, 234
Metanephric kidney, 234
Metanephros, 234
Microangiography, 41–45
  of developing zebrafish embryos and larvae, 43*f*
  experimental procedure, 43–45
  materials, 41–42
   PEG-coated non-targeted QDs, 42
   Quantum dots (QD), 42
  preparation of apparatus, 42–43
   glass microinjection needles, 42
   for holding embryos, 42–43
   holding pipettes, 42
   for microinjection, 42
Micro-dye and micro-resin injection, 31–38
  classical dye or resin injection methods, 31
  corrosive resin casting method, 31
  dye injection method
   materials, 36
   protocol, 36–38
  Florence Sabin (vascular embryologists), 31
  resin injection method
   materials, 31
   protocol, 31
Microphonics
  biologically relevant signal, 225
   bidirectional stimuli (sine waves), 225
   dihydrostreptomycin and amiloride, 225
  recordings from neuromast hair cells, 224
   mechanotransduction channels, 224
   posterior lateral-line neuromasts, 224*f*
   secondary neuromasts, 224
   standard patch electrodes, 224
  signal collection and analysis, 224–225
  stimulation of neuromast hair cells, 223
   stereociliary hair bundle, 223
   water jet pipette, 223

Microsurgery tool (Fine Science Tools 10055-12),
  106
Midgastrulation, 264
Molecular markers, 81
  antibodies and antisense RNA probes, 81
   distinct cellular states, 82*t*–87*t*
   neuronal identity markers, 84*t*–87*t*
  immunostaining and *in situ* hybridization, 81
  phosphohistone H3, 81
  proliferating cell nuclear antigen (PCNA), 81
  use of, 166–176
   antibodies, 166–172
   cell class-specific markers, 166
   endogenous transcripts and proteins, 166
   fluorescent proteins (FPs), 173–176
   lipophilic tracers, 172–173
   mRNA Probes, 172
   to study zebrafish retina, 167*t*–169*t*
   transverse sections through center of zebrafish
    eye, 170*f*–171*f*
Monitoring sleep and arousal in zebrafish, 281–291
  behavior
   criterion 1—quiescent behavior regulation,
    283
   criterion 2—quiescence, 283–284
   criterion 3—sleep deprivation, 284–286
  Drosophila to *Danio,* methodological
   considerations, 288
  genetics and pharmacology, 286–287
  high-throughput tracking of zebrafish locomotor
   behavior, 282*f*
  long-term, high-throughput sleep/wake
   monitoring, 289–290
  of sleep/wake behaviors
   average activity per waking minute, 290–291
   sleep bout number/length, 290
   sleep latency, 290
   total sleep, 290
Morphogenesis, 28, 52, 75, 76*f*, 88, 135, 158–159,
  161, 176, 191, 243*f*, 262–263, 265–266, 269
Morphogenetic events, early, 155–157, 156*f*
  choroid fissure forms, 157
  fate-mapping studies, 155
  invagination, 157
  lens rudiment, 156*f*, 157
  optic cup, sheets of cells in
   cuboidal pigmented epithelium (pe), 157
   pseudostratified columnar neuroepithelium (rne),
    157
  optic stalk, 157
  retinal pigmented epithelium (RPE), 157
Morpholino knockdown embryos, 266–267

Morpholino oligonucleotides (MOs) injections,
    13, 16
  hsp70l heat shock promoter, 16
  morpholino oligonucleotides (MOs), 16
  protein translation, 16
  splicing of pre-mRNAs, 16
  transcription factor *atoh7,* 16
Mosaic analysis
  genetic approaches in *Drosophila,* 177
  in mosaic animals, 179
  quality of transplantation needle, 178
    features, 178
    preparation (beveler and microforge), 178
  tracer purity, 178
  UV illumination, 177
  The Zebrafish Book, 178
  in zebrafish retina, 179
Mounting and immobilizing, zebrafish, 221*f*
  anesthesia and mounting of larvae, 222
  coating chamber with Sylgard, 221
  immobilizing larvae with α-Bungarotoxin, 222–223
    α-bungarotoxin (125 μM) injection, 222
    electrophysiological recordings, 222
    2-hydroxyethyl-1-piperazineethanesulfonic
      acid (HEPES), 223
    inclusion of phenol red, 222–223
  larva anesthetized with tricaine, 221
  lateral-line system of larvae, 221
  paralytic (α-bungarotoxin), 222
  recording chamber with circular opening (PC-R),
    221
  studies with primary neuromasts, 221
mRNA Probes, 172
  alkaline phosphatase (AP), 172
  in situ hybridization, 172
  in situ reagents, 172
Multiphoton time-lapse imaging, 50–51
  4-D imaging, 50
  multiphoton transgenic blood vessel imaging, 51
  PTU, 50
  three dimensionality, 50
  tricaine (MS-222), use of, 50
Multiple endocrine neoplasia syndromes type IIA
    and type IIB (MEN IIA/B), 129
Mutagenesis approaches, 186–187
  N-ethyl-*N*-nitrosourea (ENU), use of, 186
  insertional mutagenesis strategies, 186
  retroviral mutagenesis, 186
  transposon-based mutagenesis
    enhancer or gene trap vectors, 186
    hAT (Tol2) families, 186
    Tc-1/mariner (Sleeping beauty), 186

N

Nephrin, 242, 243*f*
Nephrogenic mesoderm
  origin of, 238
  subdomains, early, 238–239
Nephrons, 234, 235*f*, 236, 245*f*, 248–249
Neural crest (NC) cells
  chromaffin cells, 129
  development and migration, 135–136
    fate decisions, 135–136
    migration and cell fate specification, 135
  formation and fate specification of, 130
  neuregulin-1 pathway, 132
  premigratory stage, 130
  progenitors, 130
    bone morphogenetic proteins (BMPs), 130
    fibroblast growth factors (FGFs), 130
    retinoic acid (RA), 130
    Snail1/2/Tfap2α/Foxd3, genes, 130
    Wnt signaling, 130
  restricting NC cell migration, 132
  ventrolateral migration, 132
Neural plate
  fate map/neurogenesis/early proneural clusters, 76*f*
  molecular determinants/patterning of, 77
    anterior neural plate (ANP), 77
    BMP signaling, 77
    neural crest and Rohon-Beard neurons, 77
    neural progenitors, 77
    SoxB transcription factors, 77
  and tube formation, 75
    brain regions and markers, 75
    fate mapping study, 75
    interkinetic nuclear migration (INM), 75
    Par3, localization of, 75
    6–10 somite stages, 75
    structural proteins, localization of, 75
    two-dimensional neural plate (gastrulation), 75
    ZO1 and N cadherin, localization of, 75
Neural retina leucine zipper (Nrl), 208
Neuroblastoma (NB)
  aberrant MYCN expression, 145
  allelic losses, 145
  anaplastic lymphoma kinase *(ALK)* gene, 145
  cytogenetic analyses, 145
  embryonic tumor of PSNS, 144
  PHOX2B gene, 145
  signaling pathways
    MAPK pathway, 145
    PI3K/Akt pathway, 145
    STAT3 pathway, 145

single nucleotide polymorphisms, 145
  BRAD1 (BRCA1-associated domain 1 located
    at 2q35), 145
  putative gene FLJ22536 at 6p22, 145
studies in zebrafish, 145–146
  morpholinos and shRNA techniques, 146
  murine transgenic mice, 145–146
  new zebrafish model of NB, 146
  TILLING, 146
  in vivo animal model of NB, 145
  Zinc-Finger Nuclease strategies, 146
tumor suppressor genes (KIF1B and miR34a), 145
Neurogenesis, 73–107, 157–159
  cell cycle, changes in, 158
    lakritz gene, 158
  cell fate determination, 159
  classes of neurons, 159
    ganglion cell differentiation, 159
  in development and adult brain
    formation of neural plate/tube, 75
    generation of glial cells, 79–80
    molecular determinants/patterning of neural
      plate, 77
    neurogenesis at juvenile/adult stages, 80
    proneural domains/lateral inhibition/
      neurogenic cascade, 77–84
    signaling centers and downstream neuronal
      specification, 79
    zones of delayed differentiation, 78
  methods in developing and adult brain
    expression of genes, manipulation, 97–99
    to follow cell cycle events, 95–96
    methods to label live cells, 81–88
    molecular markers, 981
    for targeted cell ablations, 103
    in vivo imaging, 93–95
  mitotic divisions, 158
  photoreceptor morphogenesis, 159
    photoreceptor mosaic, 160
    types of photoreceptor cells, 159–160
  postmitotic neurons, 159
  protocols to study adult neurogenesis
    fixation of adult brain, 103
    immunohistochemistry on vibratome sections,
      103
    intraperitoneal injections of BrdU, 105–106
    lipofections/electroporations of adult brain
      in vivo, 105
    pretreatments, antigen retrieval, 104
    in situ hybridization, 104
    slice culture, 107
    sterile filter, 107

retinal neuroepithelium, 157*f*, 158
  two-photon imaging studies, 158
  in zebrafish retina, 173*f*
Neurotransmitters, 129
Non-neuronal tissues, 161
  choroidal and retinal vasculatures, 161
  other non-neuronal ocular tissues, 161
  transgenic lines, 161
Non-vital blood vessel imaging
  alkaline phosphatase staining for 3 dpf embryos,
    38–39
  micro-dye and micro-resin injection, 31–38
Noradrenergic neurons, differentiation of, 139–141
  expression of Hu proteins, 139
  expression of noradrenalin and genes
    dopamine-β-hydroxylase (dbh), 139
    tyrosine hydroxylase (th), 138–140
    phenylethanolamine-*N*-methyltransferase (PNMT),
      140
    mRNA *in situ* hybridization assays, 140
  sympathoadrenal derivatives in embryonic and
    juvenile zebrafish, 141*f*
  tyrosine hydroxylase gene *(th2),* 139
Normal goat serum (NGS), 104–105, 250–252, 275
Nuclear localization signal (NLS), 269
Nucleolar silver staining, 60

O

Oocyte and egg assays
  CAMMA, 309–310, 312, 312*f*, 313*f*
    changes during oocyte maturation and egg
      activation, 311*f*
    prophase I zebrafish ovarian oocytes, 311*f*
  egg activation assay (EggsAct), 313–316
    individual zebrafish oocytes *vs.* time of
      incubation, 313*f*
    osmoregulation during oocyte maturation, 312–313,
      314*f*
    cytolysis of zebrafish oocytes, 315*f*
Oocyte/egg/embryo counting by scanners, 316–318
  counting zebrafish embryos, 317*f*
  semi-automated zebrafish embryo counting with
    ImageJ, 316–318
Optokinetic response (OKR), 164*t*, 180, 190, 211,
  307, 308*f*
Organ morphogenesis, 263
Oryzias latipes (fish medaka), 57
Osmoregulation during oocyte maturation, 312–313
  fully grown oocytes, 313
  immature oocytes, 312–313
  oviposition, 312

P

Pancreas development, 263*f*
  beta-cell proliferation, 267–268, 268*t*, 273–276
  cell segregation into DB and VB, 264–265
  dorsal bud, formation, 265–267
  morphogenesis, 262–263
  specification, 261–262
Paraformaldehyde (PFA), 11, 38–39, 103, 107, 170,
    251, 254, 275
Pedicles, 206
Peripheral autonomic nervous system, 128–129
Peripheral sympathetic nervous system (PSNS)
    development, 127–146
  molecular pathways of, 130–134
    bHLH transcription factor HAND2 (dHAND),
      133
    bilateral migration of SA cells, 130
    dose-dependent BMP signaling, 132, 134
    HAND1 (eHAND), 133
    homeobox proteins PHOX2A, 133
    NC cells, *See* Neural crest (NC) cells
    neuronal and noradrenergic differentiation of SA
      progenitors, 130
    neurotrophic factors (NGF and NT-3), 133
    PHOX2A and PHOX2B, 133
    PSNS neurons, 130
    zebrafish *(Danio rerio)*, 134
    zinc-finger proteins GATA-2/3, 133
  mutations affecting, 142–144
    isolation of PSNS mutants in zebrafish, 143*f*
    lockjaw or mount blanc mutant (tfap2a), 143
    NC fate specification, 143
    nosedive mutant, 143
    retinoic acid (RA) signaling pathways, 144
    SCG formation, 142
    sympathetic mutant 1 (sym1), 142–143
    th mRNA whole-mount in situ hybridization,
      142
    zebrafish colorless (cls) mutant, 142–143
    zebrafish hands off mutant, 144
  peripheral autonomic nervous system, 128–129
  PSNS development in zebrafish, 134–144, 137*f*
    differentiation of noradrenergic neurons,
      139–141
    gene expression in migrating SA progenitors,
      136–138
    modeling of sympathetic ganglia, 141–142
    murine gene knockout models, 135
    neural crest development and migration,
      135–136
    neuronal differentiation and coalescence, 138

new gene identification for, 135
    phenotype-based genetic screens, 134
Phenotype detection methods, 188–191
  behavioral tests, 190
  DiI and DiO labeling, 189
  genetic screen, 189
  large-scale mutagenesis screens, 188
  metabolic pathways or DNA replication
      machinery, 188
  mutant phenotype recognition strategy, 188
  optokinetic response, 190
  optomotor response, lack of, 189
  transgenic GFP lines, 190
  visual inspection screens, 189
Phenylethanolamine-*N*-methyltransferase (PNMT),
    136*t*, 139, 142
1-Phenyl-2-thiourea (PTU), 12, 20–21, 43, 46–47,
    49–50, 94, 162, 252, 275
Pheochromocytomas, 129
pH indicators, colorimetric potential of scanners,
    307–309
  24-bit images, 308
  color scanners, 307
  green channel stack, 308
  phenol red pH indicator dye, 307
  relationship between degree of larval movement,
    309*f*
  use of colorimetric capability of scanners, 310*f*
Phosphate-buffered saline (PBS), 11–12, 104–105,
    107, 170, 246, 250–253, 273
Photoreceptor cells
  cones
    double cones, 160
    green and red opsin genes, 160
    long and short single cones, 160
  rods, 159
Photoreceptor structure and development: analyses
    using GFP transgenes, 205–216
  anatomy and biochemistry, 206–207
    cone photoreceptors, 206
    cone subtypes, 206
    intraflagellar transport (IFT), 206
    outer segments (OSs), 206
    spherules and pedicles, 206
    vertebrate cone (left) and rod (right)
      photoreceptors, 206, 207*f*
  development, 207–208
    cone differentiation, 207–208
    electroretinogram (ERG) recordings, 208
    intrinsic and extrinsic factors, 208
    rod differentiation, 208
  regeneration, 211–215

bromodeoxyuridine (BrdU) labeling, 213
cell death, 212
co-labeling, 214
damaged rod photoreceptors, 212
markers of Müller cells, 212
neurogenic clusters, 212
Notch–Delta pathway, 213
ONL and INL, 213–214
post-embryonic growth in teleosts, 212
proliferating cell nuclear antigen (PCNA), 213
recovery of vision, 212
XOPS-mCFP transgenic line, 214
size, regulation of, 210–211
crumbs in *Drosophila,* overexpression of, 211
FERM protein Mosaic eyes, 210
rod-specific *(Tg)XOPS:GFP* transgene, 210
zebrafish mosaic eyes (moe) gene, 210
synapse structure, 211
confocal microscopy of transgenic mutants
(Tg(TαC:GFP)nrc), 211
no optokinetic response c (nrc) mutant, 211
optokinetic response (OKR) assay, 211
synaptojanin 1, 211
transgenic technology, improvements in, 208–209
"founder fish", 208
I-SceI, rare-cutting endonuclease, 209
site-specific recombination cloning
(Gateway®) technology, 209
transport mechanisms, 209–210
connecting cilium, 209
heterotrimeric kinesin-II motor, 210
intraflagellar transport (IFT), 209–210
Kif3b (DNKIF3B), 210
photoreceptor OSs, 209
studies in *Caenorhabditis elegans,* 210
studies in *Xenopus* Kif3b or mouse Kif17, 210
PI3K/Akt pathway, 145
Podocin, 242–243
Podocytes, 234, 239–244, 256
Proliferating cell nuclear antigen (PCNA), 81, 96,
213–214, 268, 273
Proliferation in pancreas using EdU, 273–276
Pronephric cilia, 250
Pronephric duct, 236, 245
Pronephric podocytes, 243, 244*f*
Pronephros, 234, 235*f*, 247
division
distal early (DE), 236
distal late (DL), 236
proximal convoluted tubule (PCT), 236
proximal straight tubule (PST), 236
structure of, 236–237

Pronephros formation
cloaca formation, 245
endoderm, role of, 240
glomerulus formation, 242–244
nephrogenic mesoderm
origin of, 238
subdomains, early, 238–239
pax2a, role of, 240
retinoic acid signaling, role of, 240
tubular epithelium, differentiation, 241–242
zebrafish mutants with defects, 239*t*
Pronephros function, methods to study
adult kidney isolation, 249
electron microscopy methods for zebrafish
materials, 254
methods, 254–255
embryo dissociation, 245–246
fluorescently labeled cells by FACS, 246
gentamicin induced kidney tubule injury, 248
adults, 248
embryos, 248
glomerular filtration, assay for, 246–247
adults, 247–248
embryos, 247
histological sectioning of wholemount stained
embryos, 253–254
non-lethal surgical access to adult kidney, 250
zebrafish cilia, 250
alternate protocols, 253
antibody staining, 252
confocal microscopy, mounting sample for,
253
fixation, 252
materials, 250–251
methods, 251–252
solutions, 251
Proneural domains/lateral inhibition/neurogenic
cascade, 77–84
antibodies and antisense probes, 82*t*
distinct cellular states, 82*t*–87*t*
neuronal identity markers, 84*t*–85*t*
Cdkn1c and histone deacetylase, 78
2 days post-fertilization (dpf), 77
family of Hairy/Enhancer of Split genes, 78
18 h post-fertilization (hpf), 77
Notch ligands and receptors, 78
proneural/neurogenic genes, 78
"salt and pepper" pattern, 78
Proteinuria or nephrotic syndrome, 242, 247
Proximal convoluted tubule (PCT), 235–237, 239,
247–248
Proximal straight tubule (PST), 236–237, 267

R

RA-producing enzyme (RALDH), 264
Recordings from lateral-line hair cells/afferent
        neurons, physiological, 219–229
    action currents
        analysis of action currents, 226–228
        electrode and recording details, 226
    microphonics
        biologically relevant signal, 225
        recordings from neuromast hair cells, 224
        signal collection and analysis, 224–225
        stimulation of neuromast hair cells, 223, 226
    zebrafish mounting and immobilizing, 221–223
        anesthesia and mounting of larvae, 222
        immobilizing larvae with α–Bungarotoxin,
            222–223
Regeneration, 211–215
    expression of *nr2e3,* 214
    fin and heart regeneration
        hspd1, 213–214
        mps1, 214
    GFP fluorescence, 213
    identification of neurogenesis, 212
    inner nuclear layer (INL), 212
    metronidazole treatment, 215
    Müller glial-derived mitotic cells, 213
    neurogenic clusters, 212
    no blastema (nbl), 214
    NTR-EGFP fusion protein, 214
    pax6 expression, 213–214
    post-embryonic growth in teleosts, 212
    proliferating cell nuclear antigen (PCNA), 213
    rod progenitor cells, 212, 214
    subpopulation of Müller glia, 212
    uncovering genes, 213
    visual sensitivity, 212
    XOPS-mCFP transgenic line, 214
Region of interest (ROI), 96, 272, 301–302, 304–305,
        310–312
Relisys scanner
    ArtScan (Relisys), 297
Resin injection method, 31*f*, 31–32, 38, 40
    experimental procedure, 34–36
        digestion of tissue, 36
        fixation with 2% glutaraldehyde, 35
        mixing resin, 35
        saline buffer, 34–35
        scanning electron microscopy, 36
    materials, 31
    preparation of apparatus, 31–33
        glass needles, 32–33, 32*f*

    for injecting fixative, 33, 33*f*
    for injecting physiological saline buffer, 33, 33*f*
    for injecting resin, 33
    paraffin bed, 31–32, 32*f*
    resin for injection, preparation of, 33–34, 34*f*
Retina in zebrafish model, 153–192
    analysis of gene function
        forward genetics, 185–191
        reverse genetic approaches, 182–185
    analysis of wild-type and mutant visual system
        behavioral studies, 180
        biochemical approaches, 181
        cell and tissue interactions, 177–179
        cell movements and lineage relationships, 176
        cell proliferation, 179
        chemical screens, 181–182
        electrophysiological analysis of retinal
            function, 180–181
        histological analysis, 162–166
        molecular markers, use of, 166–176
    development of
        early morphogenetic events, 155–157
        neurogenesis, 158–160
        non-neuronal tissues, 161
        retinotectal projections, 160–161, 172, 177
    histology of, 157*f*
Retinal axon guidance in zebrafish, analyzing, 3–22
    perturbing retinotectal system
        injecting DNA or MOs, 16
        protocols for transplants, 19–21
        retinotectal mutants, 13–16
        transplanting to test cell-autonomy of function,
            17–19
        using heat-shocks to induce misexpression, 16–
            17
        zebrafish retinotectal projection, 5*f*, 18*f*–19*f*
    visualizing retinal axons
        labeling with antibodies, 9
        labeling with lipophilic dyes, 9–10
        methods for, 6*f*–7*f*
        protocols for labeling methods, 11–13
        time-lapse imaging, 11
        transgenic lines, 7–9
        transiently expressing DNA constructs, 10
        in vivo single cell electroporation, 10
Retinal ganglion cells (RGCs), 4–10, 12–13, 17, 19–
        21, 84, 86, 161, 189
Retinoic acid (RA), 130, 264
    signaling, 144, 240, 264
Retinotectal mutants, 13–16, 14*t*–15*t*
    ace/fgf8, 16
    astray/robo2 genes, 16

gdf6a and collagenIVa5, 16

pou4f3: mGFP transgenic line, 16

projection, 5f–7f, 18f–19f

regulators of axon guidance or brain patterning

  adhesion molecule *N-cadherin,* 13

  receptor *robo2 or patched1,* 13

  transcription factor *lhx2,* 13

Retinotectal projections, 160–161

  ganglion cell axons, 160–161

  optic nerve head, 160

  optic tectum, 160–161

  optic tract, 161

Retinotectal system, perturbation

  cell-autonomy, 17–19

  heat-shocks to induce misexpression, 16–17

  injecting DNA or MOs, 16

  protocols for transplants, 19–21

  retinotectal mutants, 13–16, 14t–15t

  zebrafish retinotectal projection, 5f–7f

Reverse genetic approaches, 182–185

  approaches to gene overexpression, 184–185

  loss-of-function analysis, 182–183

Reverse transcription polymerase chain reaction
  (RTPCR), 183

S

SA progenitors, gene expression in migrating,
  136–138

  cyclops mutant, 137

  expression of bmp, 137f, 138

  expression of conserved PSNS genes, 136t

  gene knockdown techniques, 138

  noradrenergic differentiation, 138

  no tail mutants, 137

  phox2a/phox2b/gata-2/3/hand2 genes, 137

  soulless, 138

  zash1a gene (homolog of Mash-1), 137

Scanners

  to count and measure

    oocyte/egg/embryo counting by scanners,
      316–318

    scanner imaging of juveniles and adults, 318–319

  image acquisition with, 299–300

    8-bit grayscale or 24-bit color, 300

    rectangular marquee tool, 300

    reflective or transmitted mode, 300

    resolution, selection of, 300

  imageJ, 301–302

  imaging of juveniles and adults, 318–319

    documentation of disease and abnormal
      morphology, 320f

    morphometrics and archival images, 319f,
      320f

    tumors or abnormal morphologies, 318–319

  Macroscheduler, 301

  multiwell plates/pipette pump in imaging arrays,
    300f

  pH indicators, 307–309

  potential applications of, 319–321

    catfish embryos *(Ictalurus punctatus),* 320

    diode lasers, 320

    light-emitting diodes (LEDs), 319–320

    myofibrils in zebrafish larval muscle, 320

    sheets of colored filter (Wratten filters), 320

    transparency scanners, 320

  resolution and bit depth, 297–298

    8-bit grayscale image, 298

    optical resolution, 298

    12- or 16-bit grayscale or 48-bit color (higher
      bit depth), 298

    red, blue, green (RGB) image (24-bit image),
      298

  temperature and fluidics, 298–299

    cold media, 298

    long-term scanning, 298

    multiwell plates, 299, 300f

    Nunc rectangular plates, 298

    simple thermistor probe, 298

Semicloning, 63–65

  animal cloning, 63

  first semicloned medaka (Holly), 65

  and germline transmission, 64f–65f

  human ES cell derivation, 63

  i1 and i3 males, test-crosses to, 65

  intracytoplasmic sperm injection, 63

  NT1, 63

  nuclear reprogramming, 63

Signaling centers and downstream neuronal
  specification, 79

  FGF signaling activity, 79

  manipulation of whole embryo, 97t–98t

  mutants affecting neuronal specification, 99t

  transgenic lines to visualize distinct states and
    fates, 90t–91t

  Uas effector lines, 92t

  V3 and KA''/KA' interneurons, 79–80

  Wnt signaling activity, 79

Sleep, zebrafish

  behavior, 282–286

  genetics and pharmacology, 286–287

Sleep bout, 284–285, 288, 290

Sleep latency, 284–285, 288, 290

Sleep rebound, 283–286

Sleep/wake behaviors
  average activity per waking minute, 290–291
  sleep bout number/length, 290
  sleep latency, 290
  total sleep, 290
Slice culture, 74, 81, 107
Smo/smoothened mutant endoderm cells, 264
Somatostatin, 82, 168*t*, 262, 270
Somitogenesis, 235, 238–239, 241, 265, 272, 303
Sonic hedgehog (Shh), 15, 79, 132, 174, 208
Spherules, 206
Stable growth and genetic stability
  diploidization and haploidization, 62–63
  ES cell lines, 62–63
  HX cell lines, 62–63
Standard whole-mount antibody staining, 9
Starmaker gene, 236, 237f
STAT3 pathway, 145
Sterile filter
  brain embedding, cutting, and culture, 107
  fixation, 107
Sympathetic ganglia
  modeling of, 141–142
    neurotrophic factors (NGF and NT-3), 141
    pan-neuronal marker, 16A11, expression of, 141
    Rohon-Beard sensory neurons, 141
    TrkA expression, 141
  neuronal differentiation and coalescence into, 138
    expression of BMPs, 138
    pan-neuronal antibody 16A11, 138
    superior cervical ganglion (SCG) complex, 138
Sympathetic nervous system
  adenosine triphosphate, 129
  long adrenergic postganglia, 129
  preganglionic neurons, 129
  smooth muscle cells (cardiac muscle cells), 129

T

Teleost retinae, 154
Terminal inverted repeats (TIRs), 184
TGF-beta/Nodal signaling, 263
Thick ascending limb (TAL), 237
TILLING (Targeting induced local lesions in
    genomes), 97, 99*t*, 146, 163, 183
Time-lapse imaging, 11, 268–272, 300*f*
  confocal or two-photon microscopy, 11
Tissue specification, 263
Tol2kit, 209
Transgenesis, 209
Transgenic lines, 7–9
  advantages, 7

enhanced green FP (EGFP), 7
fluorescent proteins (FPs), 7
Gal4-VP16, 7
GAP-43, N-terminal palmitoylation sequence, 7
isl2b and *atoh7* genes, 7
labeling RGCs, 8*t*
and pancreatic cells, 270*t*
pou4f3 genes, 7
Transplants, protocols for, 19–21
  blastula transplants, 19–20
  late topographic transplants
    protocol, 20–21
  solutions needed, 20
Transport mechanisms, 209–210
Tubule formation, 241
Tyrosine hydroxylase (th), 81, 84*t*, 131, 133, 139,
    145, 168*t*, 171*f*, 189

U

Umax Astra 6700, 297, 300
  SilverFast SE (Umax), 297

V

Vascular endothelial growth factor (vegf), 244
Vascular gene expression, imaging, 29–30
  Disabled-2 (Dab2), cytosolic adaptor, 30
  ephb4, venous marker, 30
  immunohistochemistry, 29
  marker genes used in zebrafish vasculature
    research, 29*t*
  Prox-1 and Lyve-1, lymphatic endothelial
    markers, 30
  in situ hybridization, 29
    fli1a and *scl* genes, 29
    kdrl and *flt4* genes, 29–30
    other genes *(efnb2/grl/Dll4/Tbx20/notch5)*, 30
    tie2 and *cdh5* genes, 29
Ventral bud (VB), 263–266
Vision mutants, mutant phenotype *vs.* wild type, 307
  motility differences in laj[s304] mutant and
    wild-type (wt) sibs, 308f
  optokinetic response (OKR), 307
Visualizing retinal axons
  labeling with antibodies, 9
  labeling with lipophilic dyes, 9–10
  protocols for labeling methods, 11–13
  time-lapse imaging, 11
  transgenic lines, 7–9
  transiently expressing DNA constructs, 10
  in vivo single cell electroporation, 10

Vital dyes, 135
Vital imaging of blood vessels
    confocal microangiography, 39–41, 45
    confocal or multiphoton microscope, 44
    high-resolution imaging, 40
    imaging blood vessels in transgenic zebrafish,
        45–51
    microangiography, 41–45
    multiphoton time-lapse imaging, 41, 50–51
    transgenic lines expressing nuclear-targeted EGFP
        artery/venous sprouts *Tg(flt1:YFP; kdrl:*
            mCherryRed), 41
        Tg(fli1a:EGFP–cdc42wt)^{y48}, 41
        Tg(fli1a:EGFP; gata1:DsRed), 41
        Tg(fli1a:EGFP; kdrl:ras-cherry), 41
        Tg(fli1a:nEGFP)^{y7}, 41
        Tg(gata1:DsRed)^{sd2}, 41
    zebrafish *fli1a:EGFP/kdrl:EGFP*, 40
    zebrafish transgenic lines for time-lapse vascular
        imaging, 40*t*

W

Wild-type and mutant visual system, analysis of
    behavioral studies, 180
        escape response, 180
        optokinetic response, 180
        optomotor response, 180
        phototaxis, 180
        startle response, 180
    biochemical approaches, 181
        co-immunoprecipitation, 181
        tandem affinity purification (TAP) tag, 181
    cell and tissue interactions, 177–179
        advantages and disadvantages, 179
        conventional or confocal miscroscopy, 178
        enzymatic detection reactions, 178
        fluorophore-conjugated tracer, 178
        HRP-conjugated streptavidin version, 178
        mCFP Q01 line, 179
        mosaic analysis, 177–179
        transplantation techniques, 177
    cell movements and lineage relationships, 176
        Caged flourescein, 176
        iontophoresis, 176

Xenopus laevis, iontophoretic cell labeling
        in, 176
    zebrafish retina, lineage analysis in, 176
cell proliferation, 179
    bromodeoxyuridine (BrdU) injections, 179
    fluorescence activated cell sorting (FACS), 179
    H^3-thymidine labeling, 179
chemical screens, 181–182
    chemical compound libraries, 182
    flk-GFP transgenic line, 182
    phenotype detection approaches, 182
    small-molecule screening, 182
electrophysiological analysis of retinal function,
        180–181
    electroretinography (ERG), 180–181
    ganglion cell function, 181
    OFF response, 180
    retinal responses, 180
histological analysis, 162–166
    DAB-labeled, 166
    electron microscopy, 162, 166
    epoxy resins, 162
    light microscopy, 162
molecular markers, use of, 166–176
    antibodies, 166–172
    cell class-specific markers, 166
    endogenous transcripts and proteins, 166
    fluorescent proteins, 173–176
    lipophilic tracers, 172–173
    mRNA Probes, 172
    studies of zebrafish retina and their sources/
        examples of use, 163*t*–165*t*
Wild-type zebrafish embryos and larvae, motility
    of, 302
    z-Projection of scanner images, 304*f*
Wnt signaling, 79, 130, 265, 267
Wolffian ducts, 236

Z

The Zebrafish Book, 178
Zebrafish Model Organism Database (ZFIN), 8, 191
Zinc finger nucleases (ZFNs), 97, 183
Zn-5/8 staining, 9
Zymogens, 262

VOLUMES IN SERIES

Founding Series Editor
DAVID M. PRESCOTT

Volume 1 (1964)
Methods in Cell Physiology
Edited by David M. Prescott

Volume 2 (1966)
Methods in Cell Physiology
Edited by David M. Prescott

Volume 3 (1968)
Methods in Cell Physiology
Edited by David M. Prescott

Volume 4 (1970)
Methods in Cell Physiology
Edited by David M. Prescott

Volume 5 (1972)
Methods in Cell Physiology
Edited by David M. Prescott

Volume 6 (1973)
Methods in Cell Physiology
Edited by David M. Prescott

Volume 7 (1973)
Methods in Cell Biology
Edited by David M. Prescott

Volume 8 (1974)
Methods in Cell Biology
Edited by David M. Prescott

Volume 9 (1975)
Methods in Cell Biology
Edited by David M. Prescott

Volume 10 (1975)
Methods in Cell Biology
Edited by David M. Prescott

Volume 11 (1975)
Yeast Cells
Edited by David M. Prescott

Volume 12 (1975)
Yeast Cells
Edited by David M. Prescott

Volume 13 (1976)
Methods in Cell Biology
Edited by David M. Prescott

Volume 14 (1976)
Methods in Cell Biology
Edited by David M. Prescott

Volume 15 (1977)
Methods in Cell Biology
Edited by David M. Prescott

Volume 16 (1977)
Chromatin and Chromosomal Protein Research I
Edited by Gary Stein, Janet Stein, and Lewis J. Kleinsmith

Volume 17 (1978)
Chromatin and Chromosomal Protein Research II
Edited by Gary Stein, Janet Stein, and Lewis J. Kleinsmith

Volume 18 (1978)
Chromatin and Chromosomal Protein Research III
Edited by Gary Stein, Janet Stein, and Lewis J. Kleinsmith

Volume 19 (1978)
Chromatin and Chromosomal Protein Research IV
Edited by Gary Stein, Janet Stein, and Lewis J. Kleinsmith

Volume 20 (1978)
Methods in Cell Biology
Edited by David M. Prescott

Advisory Board Chairman
KEITH R. PORTER

Volume 21A (1980)
Normal Human Tissue and Cell Culture, Part A: Respiratory, Cardiovascular, and Integumentary Systems
Edited by Curtis C. Harris, Benjamin F. Trump, and Gary D. Stoner

Volume 21B (1980)
Normal Human Tissue and Cell Culture, Part B: Endocrine, Urogenital, and Gastrointestinal Systems
Edited by Curtis C. Harris, Benjamin F. Trump, and Gray D. Stoner

Volume 22 (1981)
Three-Dimensional Ultrastructure in Biology
Edited by James N. Turner

Volume 23 (1981)
Basic Mechanisms of Cellular Secretion
Edited by Arthur R. Hand and Constance Oliver

Volume 24 (1982)
The Cytoskeleton, Part A: Cytoskeletal Proteins, Isolation and Characterization
Edited by Leslie Wilson

Volume 25 (1982)
The Cytoskeleton, Part B: Biological Systems and *In Vitro* Models
Edited by Leslie Wilson

Volume 26 (1982)
Prenatal Diagnosis: Cell Biological Approaches
Edited by Samuel A. Latt and Gretchen J. Darlington

Series Editor

LESLIE WILSON

Volume 27 (1986)
Echinoderm Gametes and Embryos
Edited by Thomas E. Schroeder

Volume 28 (1987)
***Dictyostelium discoideum:* Molecular Approaches to Cell Biology**
Edited by James A. Spudich

Volume 29 (1989)
Fluorescence Microscopy of Living Cells in Culture, Part A: Fluorescent Analogs, Labeling Cells, and Basic Microscopy
Edited by Yu-Li Wang and D. Lansing Taylor

Volume 30 (1989)
Fluorescence Microscopy of Living Cells in Culture, Part B: Quantitative Fluorescence Microscopy—Imaging and Spectroscopy
Edited by D. Lansing Taylor and Yu-Li Wang

Volume 31 (1989)
Vesicular Transport, Part A
Edited by Alan M. Tartakoff

Volume 32 (1989)
Vesicular Transport, Part B
Edited by Alan M. Tartakoff

Volume 33 (1990)
Flow Cytometry
Edited by Zbigniew Darzynkiewicz and Harry A. Crissman

Volume 34 (1991)
Vectorial Transport of Proteins into and across Membranes
Edited by Alan M. Tartakoff

Selected from Volumes 31, 32, and 34 (1991)
Laboratory Methods for Vesicular and Vectorial Transport
Edited by Alan M. Tartakoff

Volume 35 (1991)
Functional Organization of the Nucleus: A Laboratory Guide
Edited by Barbara A. Hamkalo and Sarah C. R. Elgin

Volume 36 (1991)
Xenopus laevis: **Practical Uses in Cell and Molecular Biology**
Edited by Brian K. Kay and H. Benjamin Peng

Series Editors
LESLIE WILSON AND PAUL MATSUDAIRA

Volume 37 (1993)
Antibodies in Cell Biology
Edited by David J. Asai

Volume 38 (1993)
Cell Biological Applications of Confocal Microscopy
Edited by Brian Matsumoto

Volume 39 (1993)
Motility Assays for Motor Proteins
Edited by Jonathan M. Scholey

Volume 40 (1994)
A Practical Guide to the Study of Calcium in Living Cells
Edited by Richard Nuccitelli

Volume 41 (1994)
Flow Cytometry, Second Edition, Part A
Edited by Zbigniew Darzynkiewicz, J. Paul Robinson, and Harry A. Crissman

Volume 42 (1994)
Flow Cytometry, Second Edition, Part B
Edited by Zbigniew Darzynkiewicz, J. Paul Robinson, and Harry A. Crissman

Volume 43 (1994)
Protein Expression in Animal Cells
Edited by Michael G. Roth

Volume 44 (1994)
Drosophila melanogaster: **Practical Uses in Cell and Molecular Biology**
Edited by Lawrence S. B. Goldstein and Eric A. Fyrberg

Volume 45 (1994)
Microbes as Tools for Cell Biology
Edited by David G. Russell

Volume 46 (1995)
Cell Death
Edited by Lawrence M. Schwartz and Barbara A. Osborne

Volume 47 (1995)
Cilia and Flagella
Edited by William Dentler and George Witman

Volume 48 (1995)
Caenorhabditis elegans: **Modern Biological Analysis of an Organism**
Edited by Henry F. Epstein and Diane C. Shakes

Volume 49 (1995)
Methods in Plant Cell Biology, Part A
Edited by David W. Galbraith, Hans J. Bohnert, and Don P. Bourque

Volume 50 (1995)
Methods in Plant Cell Biology, Part B
Edited by David W. Galbraith, Don P. Bourque, and Hans J. Bohnert

Volume 51 (1996)
Methods in Avian Embryology
Edited by Marianne Bronner-Fraser

Volume 52 (1997)
Methods in Muscle Biology
Edited by Charles P. Emerson, Jr. and H. Lee Sweeney

Volume 53 (1997)
Nuclear Structure and Function
Edited by Miguel Berrios

Volume 54 (1997)
Cumulative Index

Volume 55 (1997)
Laser Tweezers in Cell Biology
Edited by Michael P. Sheetz

Volume 56 (1998)
Video Microscopy
Edited by Greenfield Sluder and David E. Wolf

Volume 57 (1998)
Animal Cell Culture Methods
Edited by Jennie P. Mather and David Barnes

Volume 58 (1998)
Green Fluorescent Protein
Edited by Kevin F. Sullivan and Steve A. Kay

Volume 59 (1998)
The Zebrafish: Biology
Edited by H. William Detrich III, Monte Westerfield, and Leonard I. Zon

Volume 60 (1998)
The Zebrafish: Genetics and Genomics
Edited by H. William Detrich III, Monte Westerfield, and Leonard I. Zon

Volume 61 (1998)
Mitosis and Meiosis
Edited by Conly L. Rieder

Volume 62 (1999)
Tetrahymena thermophila
Edited by David J. Asai and James D. Forney

Volume 63 (2000)
Cytometry, Third Edition, Part A
Edited by Zbigniew Darzynkiewicz, J. Paul Robinson, and Harry Crissman

Volume 64 (2000)
Cytometry, Third Edition, Part B
Edited by Zbigniew Darzynkiewicz, J. Paul Robinson, and Harry Crissman

Volume 65 (2001)
Mitochondria
Edited by Liza A. Pon and Eric A. Schon

Volume 66 (2001)
Apoptosis
Edited by Lawrence M. Schwartz and Jonathan D. Ashwell

Volume 67 (2001)
Centrosomes and Spindle Pole Bodies
Edited by Robert E. Palazzo and Trisha N. Davis

Volume 68 (2002)
Atomic Force Microscopy in Cell Biology
Edited by Bhanu P. Jena and J. K. Heinrich Hörber

Volume 69 (2002)
Methods in Cell-Matrix Adhesion
Edited by Josephine C. Adams

Volume 70 (2002)
Cell Biological Applications of Confocal Microscopy
Edited by Brian Matsumoto

Volume 71 (2003)
Neurons: Methods and Applications for Cell Biologist
Edited by Peter J. Hollenbeck and James R. Bamburg

Volume 72 (2003)
Digital Microscopy: A Second Edition of Video Microscopy
Edited by Greenfield Sluder and David E. Wolf

Volume 73 (2003)
Cumulative Index

Volume 74 (2004)
Development of Sea Urchins, Ascidians, and Other Invertebrate Deuterostomes: Experimental Approaches
Edited by Charles A. Ettensohn, Gary M. Wessel, and Gregory A. Wray

Volume 75 (2004)
Cytometry, 4th Edition: New Developments
Edited by Zbigniew Darzynkiewicz, Mario Roederer, and Hans Tanke

Volume 76 (2004)
The Zebrafish: Cellular and Developmental Biology
Edited by H. William Detrich, III, Monte Westerfield, and Leonard I. Zon

Volume 77 (2004)
The Zebrafish: Genetics, Genomics, and Informatics
Edited by William H. Detrich, III, Monte Westerfield, and Leonard I. Zon

Volume 78 (2004)
Intermediate Filament Cytoskeleton
Edited by M. Bishr Omary and Pierre A. Coulombe

Volume 79 (2007)
Cellular Electron Microscopy
Edited by J. Richard McIntosh

Volume 80 (2007)
Mitochondria, 2nd Edition
Edited by Liza A. Pon and Eric A. Schon

Volume 81 (2007)
Digital Microscopy, 3rd Edition
Edited by Greenfield Sluder and David E. Wolf

Volume 82 (2007)
Laser Manipulation of Cells and Tissues
Edited by Michael W. Berns and Karl Otto Greulich

Volume 83 (2007)
Cell Mechanics
Edited by Yu-Li Wang and Dennis E. Discher

Volume 84 (2007)
Biophysical Tools for Biologists, Volume One: *In Vitro* Techniques
Edited by John J. Correia and H. William Detrich, III

Volume 85 (2008)
Fluorescent Proteins
Edited by Kevin F. Sullivan

Volume 86 (2008)
Stem Cell Culture
Edited by Dr. Jennie P. Mather

Volume 87 (2008)
Avian Embryology, 2nd Edition
Edited by Dr. Marianne Bronner-Fraser

Volume 88 (2008)
Introduction to Electron Microscopy for Biologists
Edited by Prof. Terence D. Allen

Volume 89 (2008)
Biophysical Tools for Biologists, Volume Two: *In Vivo* Techniques
Edited by Dr. John J. Correia and Dr. H. William Detrich, III

Volume 90 (2008)
Methods in Nano Cell Biology
Edited by Bhanu P. Jena

Volume 91 (2009)
Cilia: Structure and Motility
Edited by Stephen M. King and Gregory J. Pazour

Volume 92 (2009)
Cilia: Motors and Regulation
Edited by Stephen M. King and Gregory J. Pazour

Volume 93 (2009)
Cilia: Model Organisms and Intraflagellar Transport
Edited by Stephen M. King and Gregory J. Pazour

Volume 94 (2009)
Primary Cilia
Edited by Roger D. Sloboda

Volume 95 (2010)
Microtubules, in vitro
Edited by Leslie Wilson and John J. Correia

Volume 96 (2010)
Electron Microscopy of Model Systems
Edited by Thomas Müeller-Reichert

Volume 97 (2010)
Microtubules: In Vivo
Edited by Lynne Cassimeris and Phong Tran

Volume 98 (2010)
Nuclear Mechanics & Genome Regulation
Edited by G. V. Shivashankar

Volume 99 (2010)
Calcium in Living Cells
Edited by Michael Whitaker

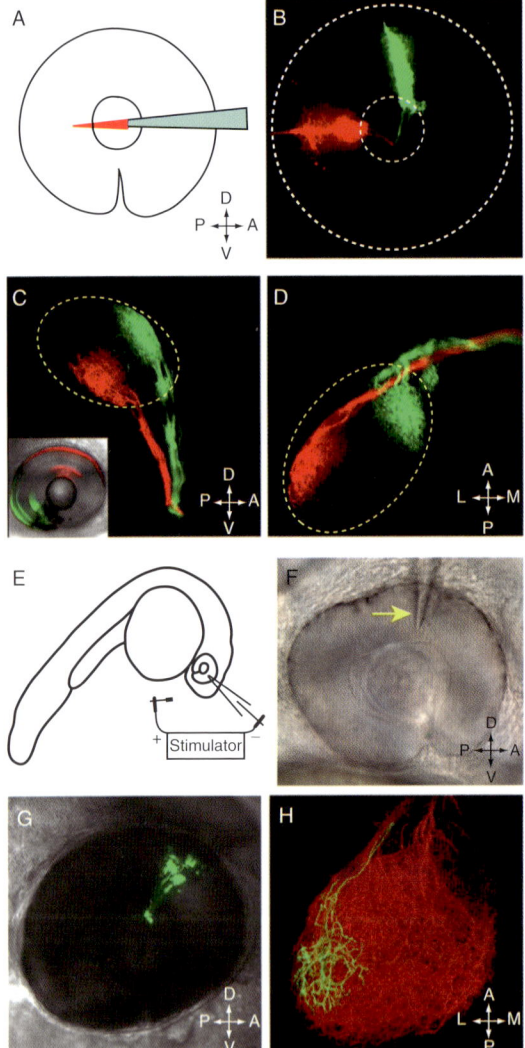

Plate 1 (Figure 2 on page 6 of this volume)

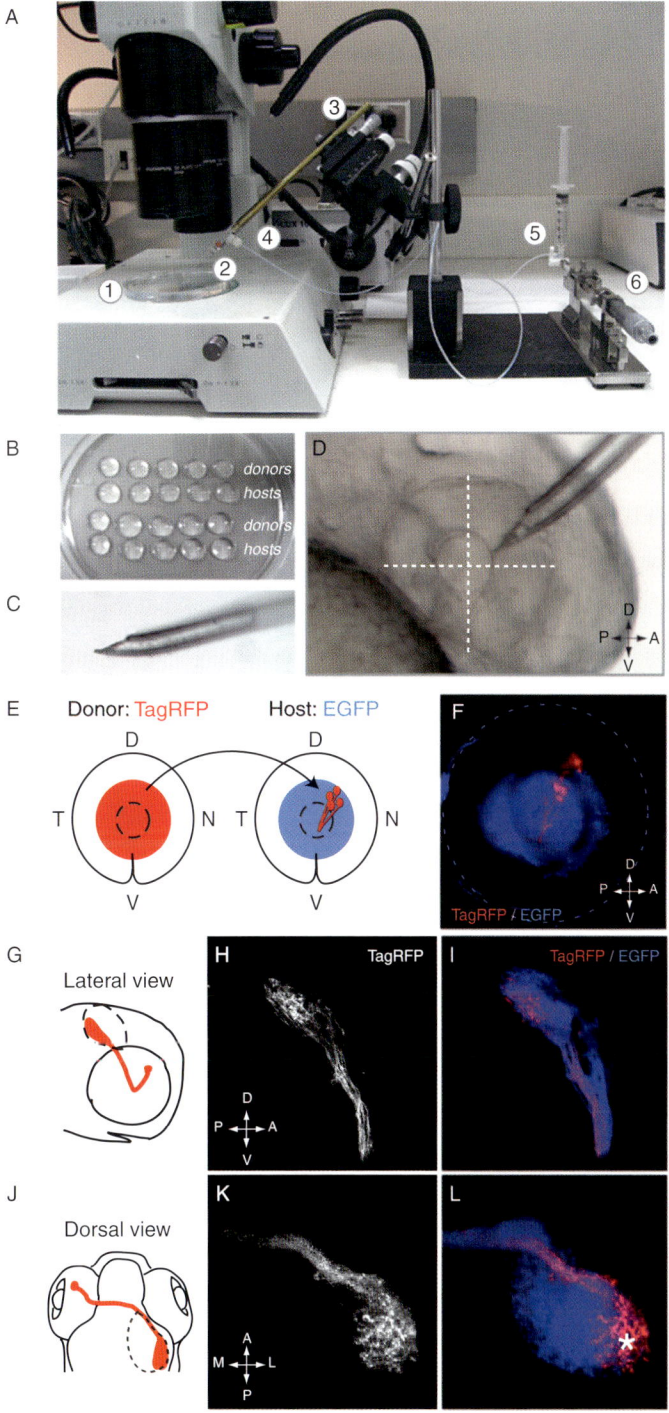

Plate 2 (Figure 3 on page 18 of this volume)

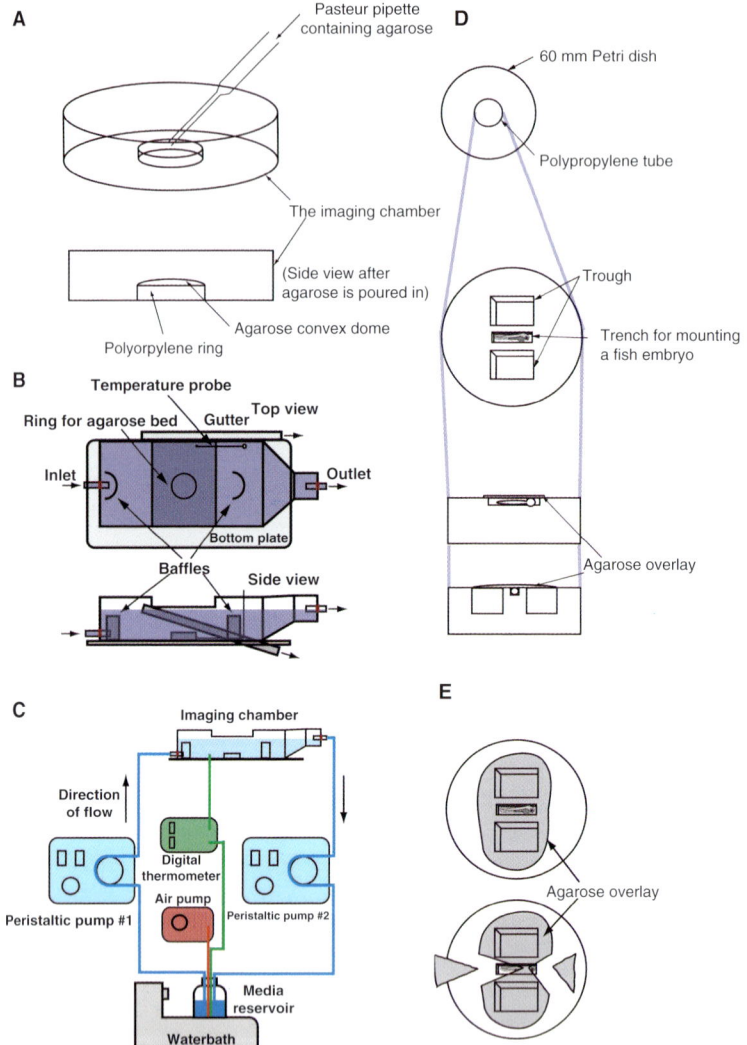

Plate 3 (Figure 8 on page 48 of this volume)

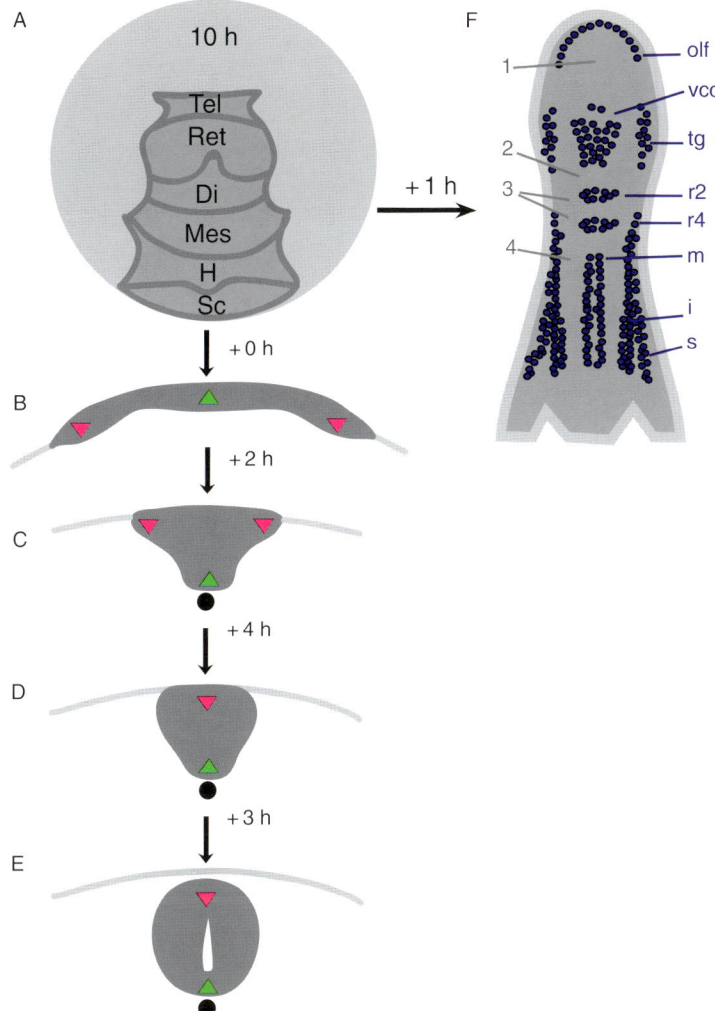

Plate 4 (Figure 1 on page 76 of this volume)

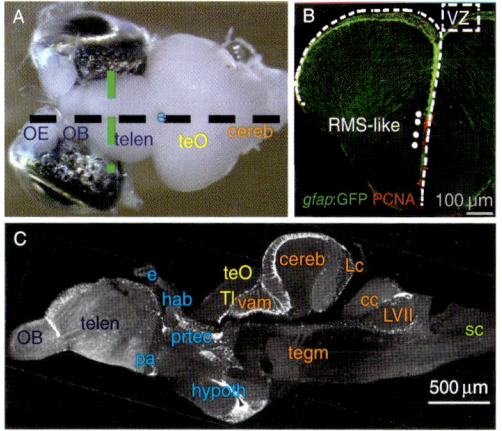

Plate 5 (Figure 2 on page 88 of this volume)

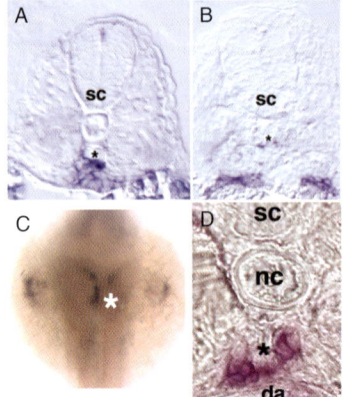

Plate 6   (Figure 2 on page 137 of this volume)

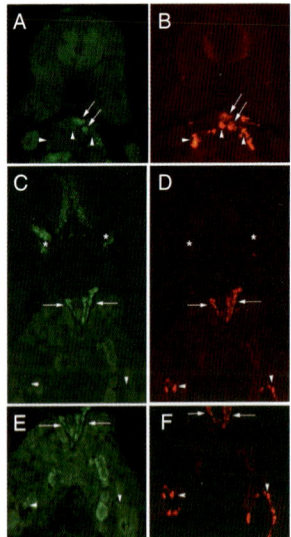

Plate 7   (Figure 4 on page 141 of this volume)

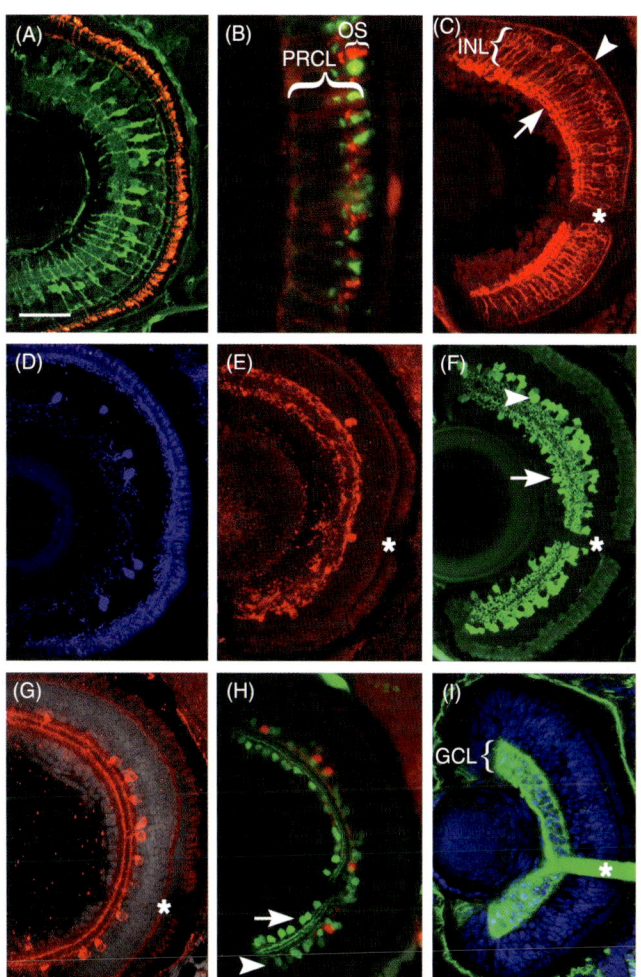

Plate 8 (Figure 3 on page 170 of this volume)

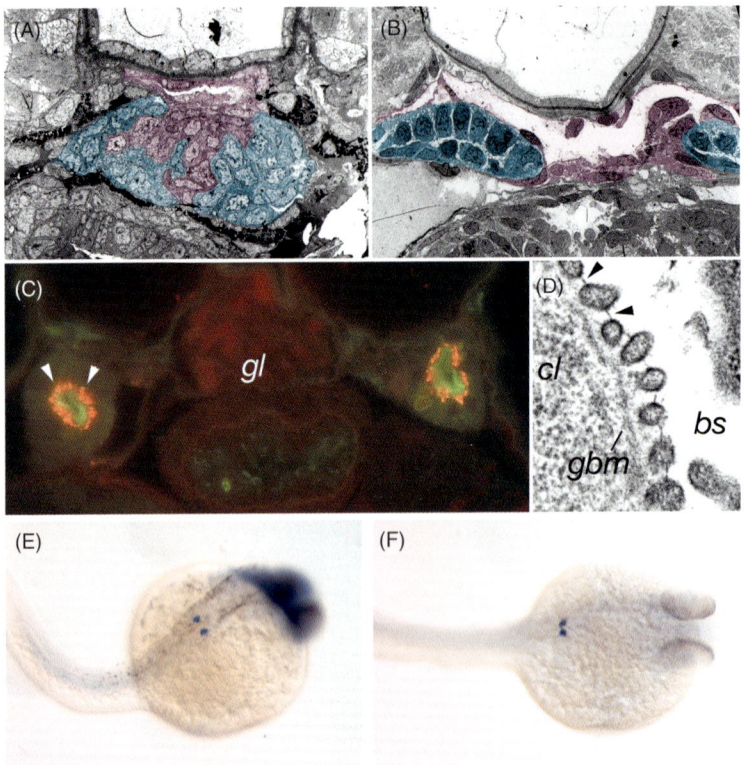

Plate 9   (Figure 7 on page 243 of this volume)

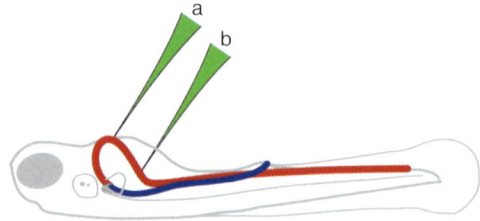

Plate 10   (Figure 10 on page 247 of this volume)

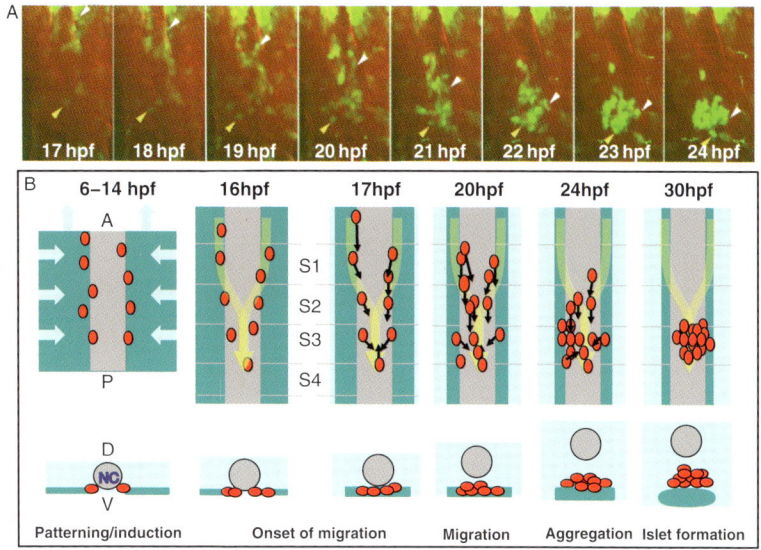

Plate 11   (Figure 2 on page 266 of this volume)

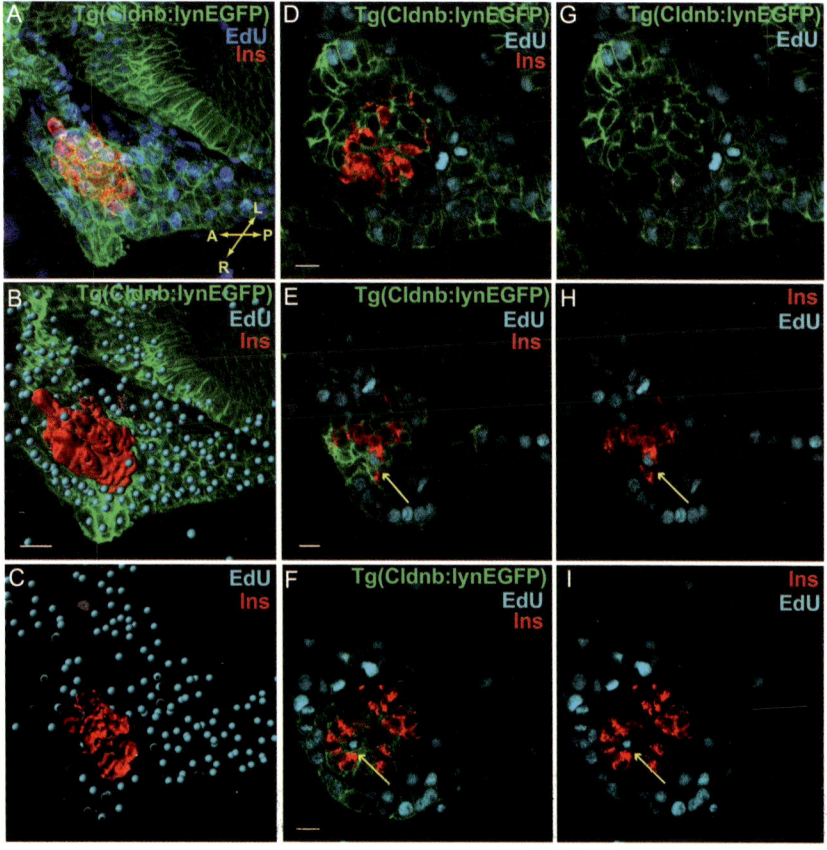

Plate 12   (Figure 5 on page 274 of this volume)